# International Conference on Science, Technology, Engineering and the Industrial Revolution (ICSTEIR 2020)



IOP Conference Series: Materials Science and Engineering
Volume 1115

Online
14 - 15 November 2020

## Editors:

**Sundarapandian Vaidyanathan**
**Irene Moroz**
**Chang-Hua Lien**
**Mustafa Mamat**
**Esmeralda C. Djamal**
**Sukono**
**Aceng Sambas**

ISBN: 978-1-7138-4604-8
ISSN: 1757-8981

**Printed from e-media with permission by:**

Curran Associates, Inc.
57 Morehouse Lane
Red Hook, NY 12571

proceedings.com

**Some format issues inherent in the e-media version may also appear in this print version.**



Institute of Physics
Dirac House, Temple Back
Bristol BS1 6BE UK

Phone: 44 1 17 929 7481
Fax: 44 1 17 920 0979

techtracking@iop.org

**Additional copies of this publication are available from:**

Curran Associates, Inc.
57 Morehouse Lane
Red Hook, NY 12571 USA
Phone: 845-758-0400
Fax: 845-758-2633
Email: curran@proceedings.com
Web: www.proceedings.com

# TABLE OF CONTENTS

IOP Conf. Series: Materials Science and Engineering 1115 (2021) 011001 doi:10.1088/1757-899X/1115/1/011001

# Preface

We are proud to present the **first International Conference on Science, Technology, Engineering and the Industrial Revolution (ICSTEIR 2020), 14-15 November 2020**, West Java, Indonesia, conducted entirely online. Initially the conference was planned to take place in Bandung, Indonesia. However, due to the fact that COVID-19 has significantly affected the travels between countries and in many countries total lockdowns have prohibited the organizing of events such as this, it was decided to go on-line. Also, the ICSTEIR conference was planned to be held in November 2020 with a face-to-face presentation. However, due to the uncertain circumstances of the COVID-19, the conference was postponed to February 23-24, 2021, with the video-conferencing presentation.

The conference has managed to draw the attention of many researchers from different country, i.e., Malaysia, Nigeria, India, Algeria, China, Ukraine, Czech Republic and Taiwan. In the conference, 103 papers were selected by the Scientific Committee to be presented, after double-blind review from at least two reviewers. All papers were presented on-line on zoom platform in about 10 minutes plus 5 minutes for Q&A, managed by the University of Informatics and Business Indonesia (UNIBI) and Ma'soem University. Furthermore, 3 opening speeches and 6 on plenary speeches interesting topics on Science and engineering. Keynote presentation of research results about 30 minutes and 15 minutes for Q&A. During the plenary session, a streaming process was also carried out via youtube (https://www.youtube.com/watch?v=79eY3kdyW5Q) so that the Speaker at ICSTEIR 2020 presented all information can be seen by the wider community, especially the Science, Technology, Engineering and the Industrial Revolution. Meanwhile, the Contribution Session was divided into 10 parallel rooms using the Zoom breakout feature, which was attended by 103 presenters.

The keynote speakers' presentations are provided, mainly to show the contribution of Science, Technology, Engineering and the Industrial Revolution towards research and knowledge sharing. We have four keynote speakers coming from the Czestochowa University of Technology, Poland, Prof. Sebastian Kot, PhD Eng.; Vel Tech University, India, Prof. Dr. Sundarapandian Vaidyanathan; Universiti Sultan Zainal Abidin (UnisZa) Malaysia, Assoc. Prof. Dr. Abdul Aziz Abdullah; Universiti Malaysia Terengganu, Malaysia, Assoc. Professor Dr. Suriyani Muhamad.

The conference brings up issues that inspire thought about the most relevant subjects in science, technology and engineering. With the theme "Post-COVID Community Development: A Holistic Approach in Collaboration in Science, Engineering and Technology," Participants from several countries have joined us to add to the conditions for open learning at the Conference. We have received high quality paper submissions and non-presentation participant registrations. Join us; we can't wait to meet you at the event!

The 1st ICSTEIR 2020 is held in collaboration between the University of Informatics and Business Indonesia (UNIBI), Ma'soem University, and the Science, Technology, and Engineering International Community (STEIC). We have also invited more universities to collaborate with us and co-host this event.

The following are figures that show the ICSTEIR 2020 process online.

Figure 1. Opening Ceremony ICSTEIR 2020

Figure 2. Prof. Dr. Sundarapandian Vaidyanathan from Vel Tech University, India, delivering his keynote speech

Figure 3. Assoc. Professor Dr. Suriyani Muhamad.from Universiti Malaysia Terengganu, Malaysia, delivering his keynote speech

Figure 4. Prof. Sebastian Kot, PhD Eng from Czestochowa University of Technology, Poland, delivering his keynote speech

Figure 5. One of the participants delivering his speech in parallel session

Figure 6. Q & A session with the Keynote Speaker

We thank you for the support and participation of all parties, so that the Conference was held successfully.

IOP Conf. Series: Materials Science and Engineering 1115 (2021) 011001 doi:10.1088/1757-899X/1115/1/011001

# Organizing Committee

**Conference Chair:**
Drs. H. Muhamad Deni Johansyah, M.M, Universitas Informatika dan Bisnis Indonesia (UNIBI), Indonesia

**Co-Conference Chair**
Asep Sudjana, Drs., M.M, Ma'soem University, Indonesia

**Secretariat**
Indarta Priyana, S.T, M. M, Universitas Informatika dan Bisnis Indonesia (UNIBI), Indonesia
Ir. H. Tonton Taufik Rachman, M.BA, Ma'soem University, Indonesia
Yudhy, Drs., M.Ag, Ma'soem University, Indonesia
Drs. Elan Rusnendar, M.M, Universitas Informatika dan Bisnis Indonesia (UNIBI), Indonesia

**Editorial Board:**

Prof. Sundarapandian Vaidyanathan
Research and Development Centre, Vel Tech University, India

Prof. Irene Moroz
Mathematical Institute, University of Oxford, United Kingdom

Prof. Chang-Hua Lien
National Kaohsiung University of Science and Technology, Taiwan

Prof. Mustafa Mamat
Faculty of Informatics and Computing, Universiti Sultan Zainal Abidin, Malaysia

Assoc. Prof. Esmeralda C. Djamal
Department of Informatics, Universitas Jenderal Achmad Yani, Indonesia

Assoc. Prof. Sukono
Department of Mathematics, Universitas Jenderal Achmad Yani, Indonesia

Aceng Sambas, Ph.D
Department of Mechanical Engineering, Universitas Muhammadiyah Tasikmalaya, Indonesia

# Peer review declaration

All papers published in this volume of *IOP Conference Series: Materials Science and Engineering* have been peer reviewed through processes administered by the Editors. Reviews were conducted by expert referees to the professional and scientific standards expected of a proceedings journal published by IOP Publishing.

**Type of peer review:** All the conference papers used a double-blind review process.

**Conference submission management system:** Conference submission management system: All the submitted materials, timely received in http://www.steic.org/icsteir2021/home/engineering, and by mail (icsteir2021@steic.org).

**Number of submissions received:** The number of applications received is 230, of which 103 were accepted for work.

**Number of submissions sent for review:** The number of applications sent for review process is 204

**Number of submissions accepted:** 103 were accepted for the publication in IOP Conference series

**Acceptance Rate (Number of Submissions Accepted / Number of Submissions Received X 100): 44.78**

**Average number of reviews per paper:** The article has submitted to two independent peer reviewers with a scientific degree

**Total number of reviewers involved:** The total number of reviewers involved is 50 experts in this particular subject area.

**Any additional info on review process:** We received a total of 230 paper submissions for the ICSTEIR 2020. All the papers were first screened for related topics in mechanical engineering, electrical engineering, industrial engineering, information engineering, Mathematics and Physics. Plagiarism checking was also carried for all the papers by using Turnitin software with recommended similarity index of no more than 25%. The papers were then sent to 90 reviewers with expertise in the science and engineering. From the first review, only 103 papers were accepted and 127 papers were rejected due to poor quality and out of scope. In order to maintain the quality of the published paper, the review process is made into two rounds. In the first round, the articles were evaluated in terms of the novelty and depth of material of the authors. The second review process was to make sure that the authors have made all the necessary corrections as recommended by the reviewers.

**Contact person for queries:** Aceng Sambas Ph.D, Department of Mechanical Engineering, Universitas Muhammadiyah Tasikmalaya

# Integral Sliding Mode Controller Design for the Global Chaos Synchronization of a New Finance Chaotic System with Three Balance Points and Multi-Stability

**S Vaidyanathan[1,*], H M Hutapea[2], B Feng[3], A Sambas[4], M. Mamat[5], Subiyanto[6], and F Renaldi[7]**

[1]Research and Development Centre, Vel Tech University, Avadi, Chennai, India
[2]Department of Military Electrical Engineering, Universitas Pertahanan, Indonesia
[3]Department of Economic Mathematics, Southwestern University of Finance and Economics, Chengdu 611130, Sichuan, China
[4]Department of Mechanical Engineering, Universitas Muhammadiyah Tasikmalaya, Indonesia
[5]Faculty of Informatics and Computing, Universiti Sultan Zainal Abidin, Kuala Terengganu, Malaysia
[6]Department of Marine Science, Faculty of Fishery and Marine Science, Universitas Padjadjaran, Indonesia
[7]Department of of Informatics, Universitas Jenderal Achmad Yani, Indonesia

*Corresponding author: sundarvtu@gmail.com

**Abstract.** This paper conveys new results of a new finance chaotic system with three unstable balance points of which one balance point is a hyperbolic saddle while the other balance points are saddle-foci. The new finance chaotic system involves three nonlinearities of which one is quadratic, the other is quartic and the third is absolute function nonlinearity. As an application of the new finance chaotic system, integral sliding mode control is invoked to achieve synchronization of master-slave finance chaotic systems. MATLAB simulations are carried out to explain the main results of this research report.

Keywords: Finance system, chaos, chaotic system, sliding mode control, balance points, etc.

## 1. Introduction

Chaos applications of dynamical models arise in various engineering domains such as nonlinear oscillatory systems [1-6], biological models [7-8], circuit devices [9-12], ANN models [13-14], chemical models [15-16], fuzzy models [17-18], robotics [19-20], etc.

Dynamical systems with $\tau = 1$ are called chaotic dynamical systems, where $\tau$ is the number of positive Lyapunov index numbers. Chaotic systems have high sensitivity to changes in the initial states.

Chaotic systems possess engineering applications such as secure devices [21-23], crypto-communication devices [24-25], memristor devices [26-28], neural networks [29-30], etc.

Chaos theory has also many applications in other areas such as finance systems [31-34], ecology [35], epidemics [36-37], biological systems [38-40], etc.

As an application of the new finance chaotic system, we obtain new results for the global chaos synchronization of master-slave finance systems using integral sliding mode control. Sliding mode control (SMC) is a useful control technique which has advantages like robustness [41-43]. SMC is a popular technique applied for synchronization of chaotic attractors [44-48]. MATLAB plots are shown to visualize the SMC results for the synchronization of the master-slave finance systems.

## 2. A New Finance Chaos System with Three Balance Points

A novel three-dimensional finance chaotic system with three nonlinear terms is proposed in this work.

$$\begin{cases} \dot{\xi} = \zeta + (\eta - a)\xi \\ \dot{\eta} = 1 - b\eta - |\xi| - d\xi^4 \\ \dot{\zeta} = -\xi - c\zeta \end{cases} \tag{1}$$

The three-dimensional vector $\mathrm{K} = (\xi, \eta, \zeta)$ designates the state of the finance system (1). It is remarked that there are three nonlinear terms in the dynamics (1). The first differential equation in (1) has a quadratic nonlinearity $\eta\xi$, and the second differential equation in (1) has an absolute function nonlinearity $|\xi|$ and a quartic nonlinearity $\xi^4$.

In the finance model (1), the phase $\xi$ represents the interest rate, the phase $\eta$ represents the investment demand and the phase $\zeta$ stands for the price exponent. The finance constant $a$ stands for the household savings and the finance constant $b$ stands for the investment cost. Also, the constant $c$ signifies the demand elasticity of commercial markets while the constant $d$ represents a real parametric uncertainty. We assume that all system constants $a, b, c, d$ are positive.

It will be established using Lyapunov Index (LI) values spectrum analysis in MATLAB that chaos exists in the finance system (1) when the parameters assume the values

$$a = 0.9, \quad b = 0.1, \quad c = 1.1, \quad d = 0.1 \tag{2}$$

In order for the time-series analysis of the state K of the system (1), we assume the initial state as

$$\xi(0) = 0.3, \quad \eta(0) = 0.2, \quad \zeta(0) = 0.3 \tag{3}$$

Using Wolf's procedure [49] for Lyapunov index numbers, the LI values of the 3-D finance model (1) are numerically found for $T = 1e4$ seconds (see Figure 1) as follows:

$$\mu_1 = 0.1407, \ \mu_2 = 0, \ \mu_3 = -0.6070 \tag{4}$$

The occurrence of a positive Lyapunov index value $\mu_1$ signifies that the model (1) has chaos behaviour. As the total of all LI values in the LI spectrum is seen to be negative, the model (1) has also dissipative motion of all its trajectories converging to the strange chaotic attractor.

The Kaplan-Yorke dimension of the finance chaotic system (1) is found as follows:

$$D_{KY} = 2 + \frac{\mu_1 + \mu_2}{|\mu_3|} = 2.2334 \tag{5}$$

Figures 1 and 2 show the LI values and signal plots of the model (1) simulated in MATLAB for the parameter set $(a, b, c, d) = (0.9, 0.1, 1.1, 0.1)$ and $\mathrm{K}_0 = (0.3, 0.2, 0.3)$.

MATLAB plots (Figures 1 and 2) show the high complexity of the finance chaos system (1).

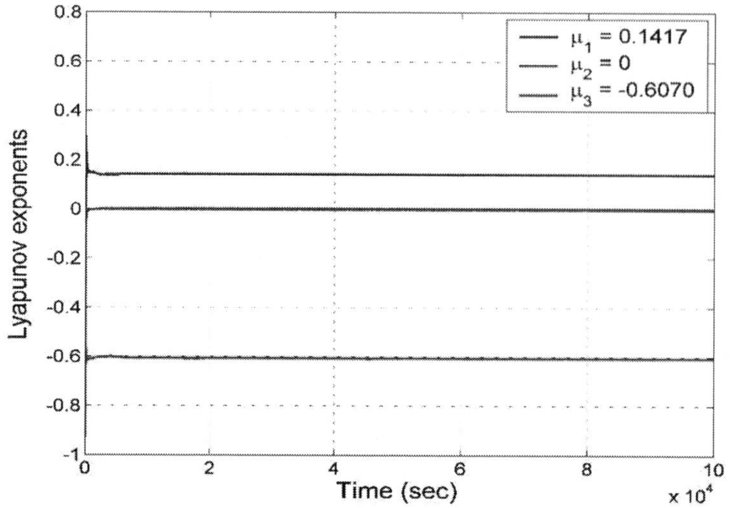


**Figure 1.** Lyapunov index values spectrum for the new finance chaos system (1)

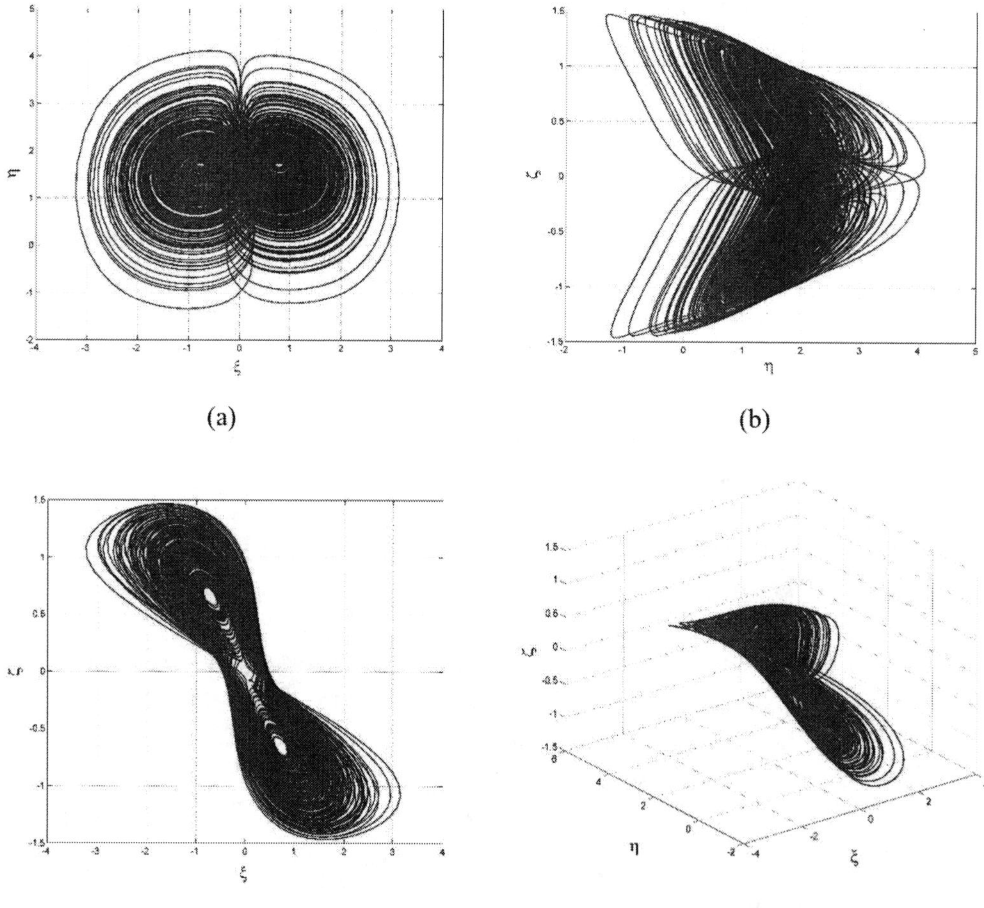

(a) (b)

(c) (d)

**Figure 2.** Signal plots of the new finance chaotic system (1) simulated in MATLAB

The balance points of the finance system (1) are got by seeking the roots of the following system:

$$\zeta + (\eta - a)\xi = 0 \tag{6a}$$

$$1 - b\eta - |\xi| - d\xi^4 = 0 \tag{6b}$$

$$-\xi - c\zeta = 0 \tag{6c}$$

The parameter values are taken as in the chaotic case, *viz.* $(a,b,c,d) = (0.9, 0.1, 1.1, 0.1)$.

Solving the system (6) numerically, three balance points are derived as given below:

$$\Omega_1 = \begin{bmatrix} 0 \\ 10 \\ 0 \end{bmatrix}, \ \Omega_2 = \begin{bmatrix} 0.7817 \\ 1.8091 \\ -0.7107 \end{bmatrix}, \ \Omega_3 = \begin{bmatrix} -0.7817 \\ 1.8091 \\ 0.7107 \end{bmatrix} \tag{7}$$

The Jacobian matrix of the finance system (1) at $\Omega_1$ has the following spectral values:

$$\alpha_1 = -0.1, \quad \alpha_2 = -1, \quad \alpha_3 = 9.001 \tag{8}$$

Thus, it is concluded that the balance point $\Omega_1$ is a saddle-point and unstable.

The Jacobian matrix of the finance system (1) at $\Omega_2$ has the following spectral values:

$$\alpha_1 = -0.7737, \quad \alpha_{2,3} = 0.2414 \pm 1.1249i \tag{9}$$

The Jacobian matrix of the finance system (1) at $\Omega_3$ has the following spectral values:

$$\alpha_1 = -0.7737, \quad \alpha_{2,3} = 0.2414 \pm 1.1249i \tag{10}$$

From (9) and (10), it is concluded that the balance points $\Omega_2$ and $\Omega_3$ are saddle-foci and unstable points. Since all the three balance points of the finance system (1) are established as unstable points, the chaotic attractor exhibited by the finance system (1) for the chaotic case $(a,b,c,d) = (0.9, 0.1, 1.1, 0.1)$ is a self-excited chaotic attractor.

The finance chaotic system (1) remains the same under the transformation of coordinates

$$(\xi, \eta, \zeta) \mapsto (-\xi, \eta, -\zeta) \tag{11}$$

This shows that the finance chaotic system (1) possesses rotation symmetry about the $\eta$ – axis.

As a consequence of the rotation symmetry about the $\eta$ – axis, every non-trivial trajectory of the finance chaotic system (1) must have a twin trajectory. The rotation symmetry is also observed in the balance points $\Omega_2$ and $\Omega_3$.

Multistability means two or more attractors coexist together with different initial conditions and has been found in many nonlinear systems [11].

Let the parameters be fixed as $a = 0.9$, $b = 0.1$, $c = 1.1$ and $d = 0.1$.

We suppose that the initial states of the finance chaotic system (1) are picked as $K_0 = (0.3, 0.2, 0.3)$ and $\Lambda_0 = (-0.6, -0.6, 0.6)$.

Figure 3 shows multistability of the finance chaotic system (1) with two coexisting chaotic attractors emanating from $K_0$ (blue color) and $\Lambda_0$ (red color), respectively.

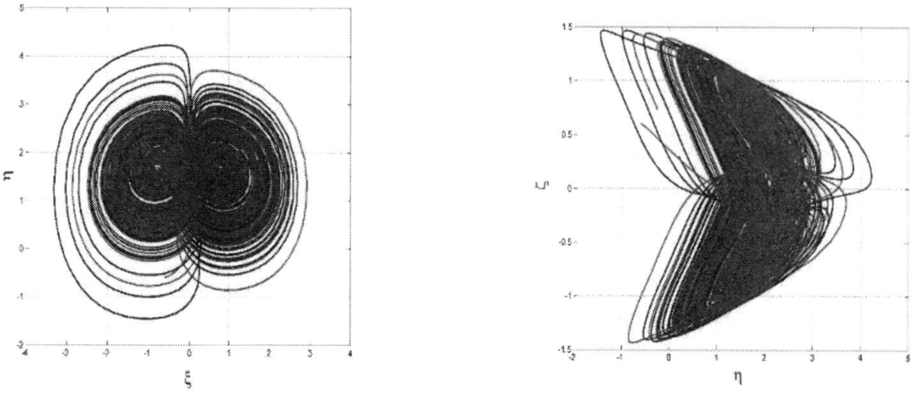

(a) (b)

**Figure 3.** Phase portraits of the coexisting chaos attractors of the finance system (1) for the initial states $K_0 = (0.3, 0.2, 0.3)$ (blue color) and $\Lambda_0 = (-0.6, -0.6, 0.6)$ (red color).

## 3. Global Synchronization of Master-Slave Finance Chaotic Systems

This section applies Integral Sliding Mode Control (ISMC) to achieve global synchronization of master-slave finance chaotic systems and we use Lyapunov stability theory [50] to prove the main control result.

As the master system, we consider the finance chaotic system

$$\begin{cases} \dot{\xi}_1 = \zeta_1 + (\eta_1 - a)\xi_1 \\ \dot{\eta}_1 = 1 - b\eta_1 - |\xi_1| - d\xi_1^4 \\ \dot{\zeta}_1 = -\xi_1 - c\zeta_1 \end{cases} \tag{12}$$

As the slave system, we consider the finance chaotic system equipped with three controls

$$\begin{cases} \dot{\xi}_2 = \zeta_2 + (\eta_2 - a)\xi_2 + u_x \\ \dot{\eta}_2 = 1 - b\eta_2 - |\xi_2| - d\xi_2^4 + u_y \\ \dot{\zeta}_2 = -\xi_2 - c\zeta_2 + u_z \end{cases} \tag{13}$$

In the system (12), $K = (\xi_1, \eta_1, \zeta_1)$ is the state and $a, b, c, d$ are constant, positive, parameters. In the system (13), $\Omega = (\xi_2, \eta_2, \zeta_2)$ is the state and $u_x, u_y, u_z$ are controls.

The synchronization error between the finance systems (12) and (13) is quantified as

$$\varepsilon = \Omega - \mathrm{K} \tag{14}$$

Mathematically, the error can be represented as follows:

$$\begin{cases} \varepsilon_x = \xi_2 - \xi_1 \\ \varepsilon_y = \eta_2 - \eta_1 \\ \varepsilon_z = \zeta_2 - \zeta_1 \end{cases} \tag{15}$$

The synchronization error dynamics is obtained with a simple calculation as follows:

$$\begin{cases} \dot{\varepsilon}_x = \varepsilon_z - a\varepsilon_x + \eta_2\xi_2 - \eta_1\xi_1 + u_x \\ \dot{\varepsilon}_y = -b\varepsilon_y - |\xi_2| + |\xi_1| - d(\xi_2^4 - \xi_1^4) + u_y \\ \dot{\varepsilon}_z = -\varepsilon_y - c\varepsilon_z + u_z \end{cases} \tag{16}$$

An integral sliding surface connected with each error variable can be defined as follows:

$$\begin{cases} \alpha_x = \varepsilon_x + \beta_x \int_0^t \varepsilon_x(\tau)d\tau \\ \alpha_y = \varepsilon_y + \beta_y \int_0^t \varepsilon_y(\tau)d\tau \\ \alpha_z = \varepsilon_z + \beta_z \int_0^t \varepsilon_z(\tau)d\tau \end{cases} \tag{17}$$

Using differentiation, we obtain at the following set of differential equations:

$$\begin{cases} \dot{\alpha}_x = \dot{\varepsilon}_x + \beta_x \varepsilon_x \\ \dot{\alpha}_y = \dot{\varepsilon}_y + \beta_y \varepsilon_y \\ \dot{\alpha}_z = \dot{\varepsilon}_z + \beta_z \varepsilon_z \end{cases} \tag{18}$$

Combining (16) and (18) leads to the following dynamics:

$$\begin{cases} \dot{\alpha}_x = \varepsilon_z - a\varepsilon_x + \eta_2\xi_2 - \eta_1\xi_1 + \beta_x\varepsilon_x + u_x \\ \dot{\alpha}_y = -b\varepsilon_y - |\xi_2| + |\xi_1| - d(\xi_2^4 - \xi_1^4) + \beta_y\varepsilon_y + u_y \\ \dot{\alpha}_z = -\varepsilon_y - c\varepsilon_z + \beta_z\varepsilon_z + u_z \end{cases} \tag{19}$$

Thus, we derive the sliding mode controls as follows:

$$\begin{cases} u_x = -\varepsilon_z + a\varepsilon_x - \eta_2\xi_2 + \eta_1\xi_1 - \beta_x\varepsilon_x - \kappa_x \operatorname{sgn}(\alpha_x) - \theta_x\alpha_x \\ u_y = b\varepsilon_y + |\xi_2| - |\xi_1| + d(\xi_2^4 - \xi_1^4) - \beta_y\varepsilon_y - \kappa_y \operatorname{sgn}(\alpha_y) - \theta_y\alpha_y \\ u_z = \varepsilon_y + c\varepsilon_z - \beta_z\varepsilon_z - \kappa_z \operatorname{sgn}(\alpha_z) - \theta_z\alpha_z \end{cases} \tag{20}$$

In (20), we take the constants $\beta_x$, $\beta_y$, $\beta_z$, $\kappa_x$, $\kappa_y$, $\kappa_z$, $\theta_x$, $\theta_y$, $\theta_z$ to be all positive.

By plugging in (20) into (19), we get the closed-loop (CL) dynamics

$$\begin{cases} \dot{\alpha}_x = -\kappa_x \operatorname{sgn}(\alpha_x) - \theta_x\alpha_x \\ \dot{\alpha}_y = -\kappa_y \operatorname{sgn}(\alpha_y) - \theta_y\alpha_y \\ \dot{\alpha}_z = -\kappa_z \operatorname{sgn}(\alpha_z) - \theta_z\alpha_z \end{cases} \tag{21}$$

Next, we establish the following global synchronization result for the new finance chaotic systems (12) and (13).

**Theorem 1.** The integral SMC law (20) globally synchronizes the states of the new finance chaotic systems (12) and (13), where we take the control constants $\beta_x$, $\beta_y$, $\beta_z$, $\kappa_x$, $\kappa_y$, $\kappa_z$, $\theta_x$, $\theta_y$, $\theta_z$ to be all positive.

**Proof.** We use Lyapunov stability theory to establish the proposed result. We consider the closed-loop system (21), which is obtained by plugging in the integral SMC law (20) into the sliding dynamics (19).

We use the Lyapunov function candidate given by

$$V(\alpha_x, \alpha_y, \alpha_z) = \frac{1}{2}\alpha_x^2 + \frac{1}{2}\alpha_y^2 + \frac{1}{2}\alpha_z^2 \tag{22}$$

By calculating the time-derivative of (22) along the CL dynamics (21), we get

$$\dot{V} = -\kappa_x|\alpha_x| - \kappa_y|\alpha_y| - \kappa_z|\alpha_z| - \theta_x\alpha_x^2 - \theta_y\alpha_y^2 - \theta_z\alpha_z^2 \tag{23}$$

Since $\dot{V}$ is a negative definite function on $R^3$, we conclude from Lyapunov stability theory that the CL dynamics (21) is globally asymptotically stable. Hence, it follows that $\varepsilon_x(t) \to 0$, $\varepsilon_y(t) \to 0$, $\varepsilon_z(t) \to 0$ as $t \to \infty$ for all values of $\varepsilon_x(0)$, $\varepsilon_y(0)$, and $\varepsilon_z(0)$. This ends the proof. ■

**Numerical Simulations:**

We consider the system parameter values as $a = 0.9$, $b = 0.1$, $c = 1.1$, $d = 0.1$.

We take $\beta_x = \beta_y = \beta_z = 10$. We also take $\alpha_1 = \alpha_2 = \alpha_3 = 0.2$ and $\theta_1 = \theta_2 = \theta_3 = 8$.

We regard the initial values of the master and slave finance chaotic systems represented by (12) and (13) as follows.

$$\xi_1(0) = 2.4, \quad \eta_1(0) = 3.9, \quad \zeta_1(0) = 6.5$$

$$\xi_2(0) = 1.9, \quad \eta_2(0) = 7.2, \quad \zeta_2(0) = 3.1$$

Figure 4 shows the global synchronization between the master and slave finance chaotic systems represented by the equations (12) and (13). The integral sliding mode controller is very efficient as we notice that the synchronization between the finance chaotic systems is achieved in less than 5 seconds.

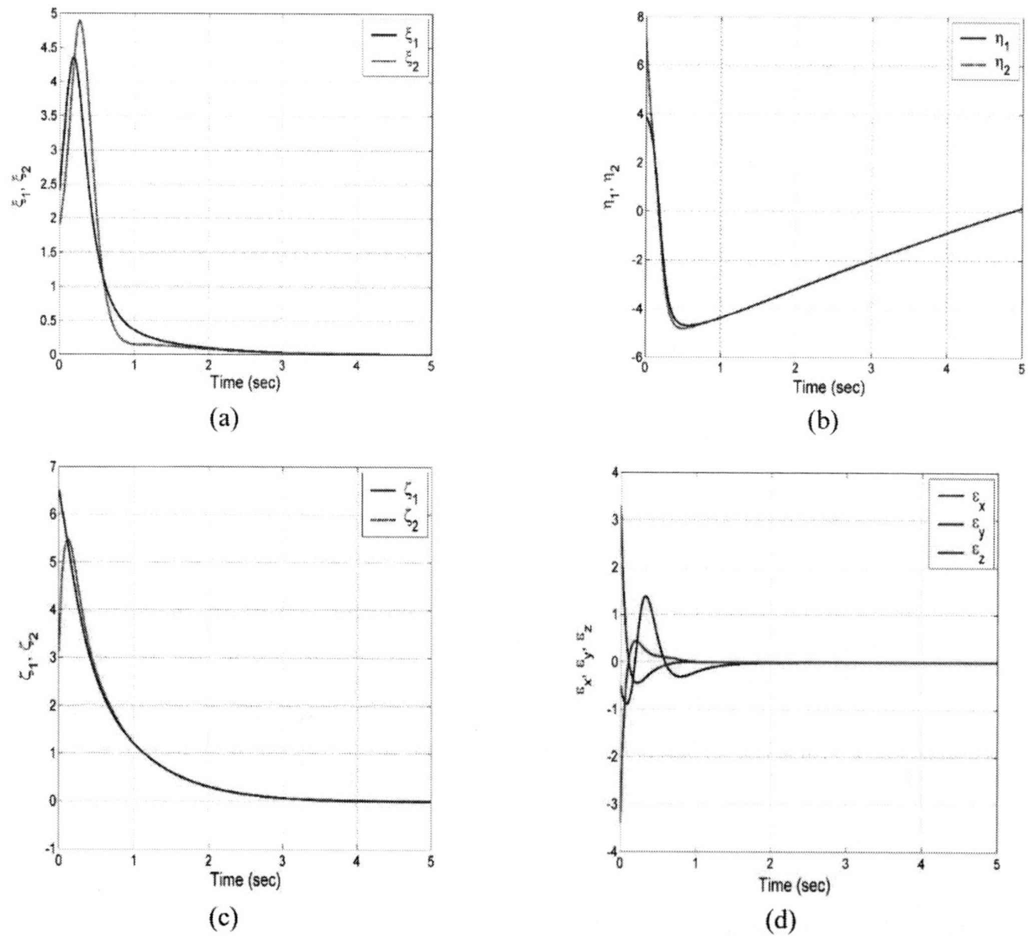


**Figure 4.** Global synchronization between the master and slave finance chaotic systems (12) and (13)

## 4. Conclusions

In this research paper, new results of a new finance chaotic system with three unstable balance points are reported. Multistability of the new finance chaotic system was proved by exhibiting two coexisting chaotic attractors for the same parameter set but different initial states. As an application of the new finance chaotic system, integral sliding mode control was applied to achieve global chaos synchronization of the master-slave finance chaotic systems. Lyapunov stability theory was invoked to ascertain the main control result of this research work. MATLAB simulations were shown to explain the main results of this research report.

## References

[1] Vaidyanathan S 2015 *Kyungpook Mathematical Journal* **55** 563-586
[2] Sundarapandian V 2013 *Lecture Notes in Electrical Engineering* **131** 319-327
[3] Vaidyanathan S 2011 *Communications in Computer and Information Science* **133** 98-107
[4] Vaidyanathan S 2011 *Communications in Computer and Information Science* **198** 1-9
[5] Vaidyanathan S 2015 *International Journal of PharmTech Research* **8** 156-166
[6] Vaidyanathan S 2015 *International Journal of PharmTech Research* **8** 956-963
[7] Awal N M and Epstein I R 2020 *Physical Review E* **101** 042222
[8] Vaidyanathan S 2015 *International Journal of PharmTech Research* **8** 117-127
[9] Sambas A, Vaidyanathan S, Tlelo-Cuautle E, Zhang S, Guillen-Fernandez O, Sukono, Hidayat Y and Gundara G 2019 *Electronics* **8** 1211
[10] Sambas A, Vaidyanathan S, Tlelo-Cuautle E, Abd-El-Atty B, El-Latif, A A A, Guillen-Fernandez O, Sukono, Hidayat Y and Gundara G 2020 *IEEE Access* **8** 137116-137132
[11] Sambas A, Vaidyanathan S, Zhang S, Zeng Y, Mohamed M A and Mamat M 2019 *IEEE Access* **7** 115454-115462
[12] Volos C, Maaita J O, Vaidyanathan S, Pham V T, Stouboulos I and Kyprianidis I 2017 *IEEE Transactions on Circuits and Systems II: Express* Briefs **64** 339-343.
[13] Xiu C, Zhou R and Liu Y 2020 *Chaos Solitons and Fractals* **141** 110316
[14] Vaidyanathan S 2015 *International Journal of PharmTech Research* **8** 61-73
[15] Vaidyanathan S 2015 *International Journal of ChemTech Research* **8** 209-221
[16] Vaidyanathan S and Azar A T 2016 *International Journal of Intelligent Engineering Informatics* **4** 135-150
[17] Sukono, Sambas A, He S, Liu H, Vaidyanathan S, Hidayat Y and Saputra J 2020 *Advances in Difference Equations* **2020** 674
[18] Wali W A 2021 *International Journal of Electrical and Computer Engineering* **11** 328-335
[19] Bazzi S and Sternad D 2020 *Advanced Robotics* **34** 1137-1155
[20] Moysis L, Petazratzis E, Marwan M, Volos C, Nistazakis H and Ahmad S 2020 *Complexity* **2020** 2826850
[21] Faragallah O S, El-sayed H S, Afifi A and El-Shafai W 2021 *Optics and Lasers in Engineering* **137** 106333
[22] Zhong C and Pan M S 2020 *Chinese Journal of Liquid Crystals and Displays* **35** 91-97
[23] Li Z, Jiang A and Mu Y 2020 *Lecture Notes in Electrical Engineering* **571** 796-810
[24] Ding L and Ding Q 2020 *Electronics* **9** 1-20
[25] Shakiba A 2020 *Multimedia Tools and Applications* **79** 32575-32605
[26] Vaidyanathan S, Azar A T, Akgul A, Lien C H, Kacar S and Cavusoglu U 2019 *International Journal of Automation and Control* **13** 644-667
[27] Raj B and Vaidyanathan S 2017 *Studies in Computational Intelligence* **701** 449-476
[28] Vaidyanathan S, Pham V T and Volos C 2017 *Studies in Computational Intelligence* **701** 101-130
[29] Swathy P S and Thamilmaran K 2014 *Nonlinear Dynamics* **78** 2639-2650
[30] Lin H, Wang C and Tan Y 2020 *Nonlinear Dynamics* **99** 2369-2386
[31] Gao Q and Ma J 2009 *Nonlinear Dynamics* **58** 209-216

[32] Vaidyanathan S, Volos C K, Tacha O I, Kyprianidis I M, Stouboulos I N and Pham V T 2016 *Studies in Computational Intelligence* **636** 495-512
[33] Zhao X S, Li Z B and Li S A 2011 *Applied mathematics and Computation* **217** 6031-6039
[34] Pereira-Pinto F H I and Savi M A 2020 *Nonlinear Dynamics* **102** 1151-1171
[35] Mahmoud E E, Trikha P, Jahanzaib L S and Almaghrabi O A 2020 *Chaos, Solitons and Fractals* **141** 110348
[36] Wang X, Wang Z, Lu J and Meng B 2021 *Mathematics and Computers in Simulation* **182** 182-194
[37] Postavaru O, Anton S R and Toma A 2021 *Mathematics and Computers in Simulation* **181** 138-149
[38] Khan A Q and Khalique T 2020 *International Journal of Biomathematics* **13** 2050022
[39] Yamaguchi H Q, Ode K L and Ueda H R 2021 *iScience* **24** 101946
[40] Vandermeer J 2020 *Theoretical Ecology* **13** 177-182
[41] Teng Q, Xu D, Yang W, Li J and Shi P 2021 *Energy Reports* **7** 1-9
[42] Liu Y-A, Tang S, Liu Y, Kong Q and Wang J 2021 *Applied Mathematics and Computation* **396** 125901
[43] Fujimoto K, Sakata N, Maruta I and Ferguson J 2021 *IEEE Control Systems Letters* **5** 839-844
[44] Takhi H, Kemih K, Moyis L and Volos C 2021 *Mathematics and Computers in Simulation* **181** 150-169
[45] Wu X, Bao H and Cao J 2021 *Journal of the Franklin Institute* **358** 1002-1020
[46] Almatroud A O 2020 *Advances in Difference Equations* **2020** 78
[47] Jia H, Guo C, Zhao L and Xu Z 2020 *Algorithms* **13** 346
[48] Li W, Bai G and Imani Marrani H 2020 *Journal of Control, Automation and Electrical Systems* **31** 1375-1385
[49] Wolf A, Swift J B, Swinney H L and Vastano J A 1985 *Physica D* **16** 285-317
[50] Khalil H K 2002 *Nonlinear Systems* Pearson Education (New York: USA)

IOP Conf. Series: Materials Science and Engineering 1115 (2021) 012002 doi:10.1088/1757-899X/1115/1/012002

# Derivative Free Levenberg-Marquardt Method for Solving Fuzzy Nonlinear Equation

**A U Omesa[1], I M Sulaiman[2,*], M Mamat[2], M Y Waziri[3], A Shadi[2], M A Zaini[4], Kalfin[5], and I Sumiati[6]**

[1]Department of Mathematics, College of Agriculture Zuru, Kebbi, Nigeria.
[2]Faculty of Informatics and Computing, Universiti Sultan Zainal Abidin, Campus Kuala Nerus, 22200, Terengganu, Malaysia.
[3]Department of Mathematics, Bayero University, Kano, Nigeria.
[4]Institute of Engineering Mathematics, Universiti Malaysia Perlis, Kangar, Malaysia
[5]Doctor Program of Mathematics, Faculty of Mathematics and Natural Sciences, Universitas Padjadjaran, Indonesia
[6]Master Program in Mathematics, Faculty of Mathematics and Natural Science, Universitas Padjadjaran, Indonesia

Corresponding author: sulaimanib@unisza.edu.my

**Abstract.** In this paper, we present a derivative-free computational method for solving fuzzy nonlinear equation. Derivative-free technique avoids computing the derivative by generating an estimate to the derivative. This is made possible by inserting the estimate of $F'(x_k)$ in Levenberg-Marquardt's method. Numerical experiments are carried out which shows that, the method is efficient.

## 1. Introduction

In recent time, most real-life problems are modelled in form of nonlinear system equations as

$$F(x) = 0 \tag{1}$$

$F: R^n \rightarrow R^n$ and that it is required to find $x^* \in R^n$ such that $F(x^*) = 0$ . When the coefficients of (1) are written in crisp number which can be conveniently represented by fuzzy numbers. Zadeh [13] introduced and investigated the idea of numbers having fuzzy variables with their arithmetic operations. Among the widely used application of arithmetic of number with fuzzy variables is systems of nonlinear equations with a parametric fuzzy number [4,5,8]. The numerical solution to fuzzy nonlinear equation with fuzzy coefficient involving fuzzy variable is on when the Jacobian is non-singular near exact root( $x^*$), in particular [1] solved the parameterized fuzzy equations via Newton's method. [2] obtained the solution of nonlinear least squares approximation via a derivative free scheme of Levenberg-Marquardt and Gauss algorithms. [9] employ Levenberg-Marquardt method for the solution of parameterized fuzzy equations and [10] used a modification of Shamanskii's steps to obtained the solutions of Dual fuzzy nonlinear equations. [11] employ a Quasi Newton algorithm to obtain the solution of parameterized fuzzy equation. [12] employ Chord approach for the same problems. Most of the above discussed methods are variants of Newton's method that require computing for the Jacobian or approximate Jacobian at every iteration. To overcome these shortcomings, numerous studies suggested evaluating the Jacobian matrix once throughout the iterations or once after every few iterations [3].

In this paper, we consider forward difference and central difference approach applied to Levenberg-Marquardt that does not compute the Jacobian for solving systems of fuzzy nonlinear equations. The Levenberg-Marquardt method is an iterative method whose iterative function is generated either by computing the Jacobian or by a derivative estimation. we introduce the finite difference approach to the method that avoids computing the derivative of the function $f$. This is made possible by inserting the estimate g(x) of $F'(x)$ in Levenberg-Marquardt method. The expectation is to reduce the burden of computing for the Jacobian matrix at every iteration. Section 2 presents a brief overview and some fundamental of fuzzy nonlinear equations. The Levenberg-Marquardt method is presented in Section 3 followed by the proposed method in section 4. And finally, we report our numerical results and conclusion in sections 5 and 6 respectively.

## 2. Preliminaries

Some basic definition numbers with fuzzy variables are presented in this section.

**Definition 1.** A fuzzy number can be defined as a set like $u: R \to I = [0,1]$ satisfying the criteria below [13,15],

(1) $u$ is upper semicontinuous,
(2) outside some interval say $[c, d]$, then, $u(x) = 0$
(3) $\exists\ a, b$ such that $c \leq a \leq b \leq d$, where $a, b, c, d$ are real numbers and
  (3.1) on the interval $[c, a]$, we say $u(x)$ is monotonic increasing
  (3.2) on the interval $[b, d]$, we say $u(x)$ is monotonic decreasing
  (3.3) for $a \leq x \leq b$, $u(x) = 1$.

The set of all number with fuzzy variables are represented by $E$ with its parameterized form defined in [8] as follows.

**Definition 2.** The parameterized fuzzy number $u$ is a pair $(\underline{u}, \overline{u})$ of function $\underline{u}(r), \overline{u}(r), 0 \leq r \leq 1$, satisfying the criteria below [14,15]:

(1) The function $\underline{u}(r)$ is bounded and monotonic increasing left continuous,
(2) The function $\overline{u}(r)$ is bounded and monotonic decreasing left continuous,
(3) For $0 \leq r \leq 1$, $\underline{u}(r) < \overline{u}(r)$.

The following $\underline{u}(r) = \overline{u}(r) = \alpha, 0 \leq r \leq 1$ is used to represent the crisp number $\alpha$. One of the well-known number with fuzzy variables is the trapezoidal number with fuzzy variable given as $u = (x_0, y_0, \alpha, \beta)$ with $[x_0, y_0]$ as the interval defuzzifier with $\alpha$ and $\beta$ as left and right fuzziness respectively, where

$$u(x) = \begin{cases} \frac{1}{\alpha}(x - x_0 + \alpha), & x_0 - \alpha \leq x \leq x_0, \\ 1 & x \in [x_0, y_0] \\ \frac{1}{\beta}(y_0 - x + \beta), & y_0 \leq x \leq y_0 + \beta, \\ 0 & otherwise. \end{cases}$$

is the membership function whose parameterized form is

$$\underline{u}(r) = x_0 - \alpha + \alpha r, \quad \overline{u}(r) = y_0 + \beta - \beta r.$$

Let the set of all triangular numbers with fuzzy variables be denoted as Let $TF(R)$. Then, arithmetic operations of addition and scalar multiplication of numbers with fuzzy variables are defined by the extension principle which are denoted as [14].

For $u = (\underline{u}, \overline{u}), v = (\underline{v}, \overline{v})$ and $k > 0$ as arbitrary, then, the addition $u + v$ and multiplication by real number $k > 0$ is given as

$$\underline{(u+v)}(r) = \underline{u}(r) + \underline{v}(r), \quad \overline{(u+v)}(r) = \overline{u}(r) + \overline{v}(r),$$

$$(\underline{ku})(r) = k\underline{u}(r), \qquad (\overline{ku})(r) = k\overline{u}(r).$$

For more reference of fuzzy nonlinear systems, researchers should refer to (see [14,15,16,17])

## 3. Levenberg-Marquadt method

The Levenberg-Marquadt method [6,7] is an iterative scheme that depends on a damping parameter $\mu_k$ and generates a sequence of approximation to the minimum

$$x_{n+1} = x_n - [J(x)^T J(x) + \mu_k I]^{-1} J(x)^T F(x) \tag{2}$$

where $\mu_k$ is a scalar that controls both the magnitude and direction $d_k$. The direction which is given by

$$d_k = -(J_k^T J_k + \mu_k I)^{-1} g_k \tag{3}$$

with $g_k = J_k^T F_k$. The Jacobian matrix $J(x)$ will be approximated using $J(x^k, Fx^k)$ and is given as

$$d_k = -(D_k^T D_k + \mu_k I)^{-1} D_k^T F_k \tag{4}$$

where the matrix $D_k$ is component wise computed by two possible choices of forward difference or central difference and is defined as

$$D_k^{ij} = \frac{F_i(x_k + h_k^j e_j) - F(x_k)}{h_k^j} \tag{5}$$

and

$$D_k^{ij} = \frac{F_i(x_k + h_k^j e_j) - F_i(x_k - h_k^j e_j)}{h_k^j} \tag{6}$$

with $e_j$ the $j-th$ unit column vector and $\mu_k$ and $h_k^j$ given by

$$\mu_k = c\frac{\|J_k\|}{\|J_k\|^2}$$

with

$$c = \begin{cases} 10, & if\ 10 \le \frac{\|J_k\|}{\|J_k\|^2} \\ 1, if\ 1 < \frac{\|J_k\|}{\|J_k\|^2} < 10 \\ 0.01, & if\ \frac{\|J_k\|}{\|J_k\|^2} \le 1 \end{cases}$$

and

$$h_k^j = min\left\{\frac{\|J_k\|}{\|J_k\|^2}, \delta_k^j\right\}$$

Notice that $h_k$ and $\delta_k$ are $n$ dimensional vectors, the index $j$ denotes the $j-th$ component of these vectors, and

$$\delta_k = \begin{cases} 10^{-9}\ |x_k^j| < 10^{-6} \\ 0.001\ |x_k^j|, otherwise \end{cases}$$

In this paper we consider the Derivative free approach to solving Levenberg-Marquardt method which we present via the following Algorithm.

**Algorithm 1:** Derivative-free Levenberg-Marquadt method Given $x_0$
*Step 1*: Solve $J(x_k, F(x_k))$ and $F(x_k)$
*Step 2:* Compute

$$x_{k+1} = x_k - [J(x_k, h_k)^T J(x_k, h_k) + \mu_k I]^{-1} J(x_k, h_k)^T F(x), k = 0,1,2, \dots \quad (7)$$

## 4. Iterative Approach for solving fuzzy nonlinear equations

This section presents the proposed scheme for solving nonlinear equation

$$F(x) = 0$$

whose parameterized form is:

$$\underline{F}(\underline{x}, \overline{x}; r) = 0$$
$$\overline{F}(\underline{x}, \overline{x}; r) = 0 \quad . \quad \forall r \in [0,1] . \quad (8)$$

Assume $\alpha = (\underline{\alpha}, \overline{\alpha})$ represents the solution to (5), that is

$$\underline{F}(\underline{\alpha}, \overline{\alpha}; r) = 0,$$
$$\overline{F}(\underline{\alpha}, \overline{\alpha}; r) = 0, \quad \forall r \in [0,1]$$

Now, suppose $x_0 = (\underline{x}_0, \overline{x}_0)$ is the approximate solution of the above nonlinear system, at that point, $\forall r \in [0,1]$, there are $h(r), k(r)$ with

$$\underline{\alpha}(r) = \underline{x}_0(r) + h(r),$$
$$\overline{\alpha}(r) = \overline{x}_0(r) + k(r) \quad .$$

Applying the Taylor series of $\underline{F}, \overline{F}$ of $(\underline{x}_0, \overline{x}_0)$, at that point $\forall r \in [0,1]$,

$$\underline{F}(\underline{\alpha}, \overline{\alpha}; r) = \underline{F}(\underline{x}_0, \overline{x}_0, r) + h\, \underline{F}_{\underline{x}}(\underline{x}_0, \overline{x}_0, r) + \ g\, \underline{F}_{\overline{x}}(\underline{x}_0, \overline{x}_0, r) + 0(h^2 + hk + h^2) = 0$$
$$\overline{F}(\underline{\alpha}, \overline{\alpha}; r) = \overline{F}(\underline{x}_0, \overline{x}_0, r) + h\, \overline{F}_{\underline{x}}(\underline{x}_0, \overline{x}_0, r) + g\overline{F}_{\overline{x}}(\underline{x}_0, \overline{x}_0, r) + 0(h^2 + hk + h^2) = 0$$

Also, suppose $\underline{x}_0$ and $\overline{x}_0$ are defined close to $\underline{\alpha}$ *and* $\overline{\alpha}$, at that point, $h(r)$ *and* $k(r)$ are sufficiently small. Let suppose all partial derivatives that are needed exist and bounded. Then, for sufficiently small $h(r)$ *and* $k(r)$, $\forall r \in [0,1]$, we get,

$$\underline{F}(\underline{x}_0, \overline{x}_0, r) + h\, \underline{F}_{\underline{x}}(\underline{x}_0, \overline{x}_0, r) + g\, \underline{F}_{\overline{x}}(\underline{x}_0, \overline{x}_0; r) = 0$$
$$\overline{F}(\underline{x}_0, \overline{x}_0, r) + h\, \overline{F}_{\underline{x}}(\underline{x}_0, \overline{x}_0, r) + g\overline{F}_{\overline{x}}(\underline{x}_0, \overline{x}_0; r) = 0$$

and hence the unknown quantities $h(r)$ *and* $k(r)$ are obtained using the equation that follows, $\forall r \in [0,1]$,

$$J(\underline{x}_0, \overline{x}_0, r) \begin{pmatrix} h(r) \\ g(r) \end{pmatrix} = \begin{pmatrix} -\underline{F}(\underline{x}_0, \overline{x}_0, r) \\ -\overline{F}(\underline{x}_0, \overline{x}_0, r) \end{pmatrix} \quad (9)$$

where

$$J(\underline{x}_0, \overline{x}_0, r) = \begin{bmatrix} \underline{F}_{\underline{x}}(\underline{x}_0, \overline{x}_0, r) & \underline{F}_{\overline{x}}(\underline{x}_0, \overline{x}_0, r) \\ \overline{F}_{\underline{x}}(\underline{x}_0, \overline{x}_0, r) & \overline{F}_{\overline{x}}(\underline{x}_0, \overline{x}_0, r) \end{bmatrix}$$

is the Jacobian of $F = (\underline{F}, \overline{F})$ computed at $x_0 = (\underline{x}_0, \overline{x}_0)$. However, $J(\underline{x}_0, \overline{x}_0, r)$ in (6) is derived by a derivative estimation $J(\underline{x}_k, \overline{x}_k, F(x_k), r)$ for $k = 0,1,2 \dots$ and for all $r \in [0,1]$

Thus, the subsequent approximations for $\underline{x}(r)$*and* $\overline{x}(r)$ are

$$\underline{x}_1(r) = \underline{x}_0(r) + h(r),$$
$$\overline{x}_1(r) = \overline{x}_0(r) + k(r),$$

for all $r \in [0,1]$.

Now, using the iterative scheme, the approximated solution is obtained as follows where $r \in [0,1]$,

$$\underline{x}_{n+1}(r) = \underline{x}_n(r) + h_n(r),$$
$$\overline{x}_{n+1}(r) = \overline{x}_n(r) + k_n(r), \quad (10)$$

when $n = 1,2, \ldots$ corresponding to (5)

$$J(\underline{x}_n, \overline{x}_n, r)\begin{pmatrix} h(r) \\ g(r) \end{pmatrix} = \begin{pmatrix} -\underline{F}(\underline{x}_0, \overline{x}_0, r) \\ -\overline{F}(\underline{x}_0, \overline{x}_0, r) \end{pmatrix}$$

Now, if $J(\underline{x}_n, \overline{x}_n, r)$ is non-singular, at this point, applying (6), we can obtain the iterative scheme of Newton's method as,

$$\begin{bmatrix} \underline{x}_{n+1}(r) \\ \overline{x}_{n+1}(r) \end{bmatrix} = \begin{bmatrix} \underline{x}_n(r) \\ \overline{x}_n(r) \end{bmatrix} - J(\underline{x}_n, \overline{x}_n, r)^{-1}\begin{bmatrix} \underline{F}(\underline{x}_n, \overline{x}_n, r) \\ \overline{F}(\underline{x}_n, \overline{x}_n, r) \end{bmatrix}$$

which is the proposed method whose algorithm is defined as follows

**Algorithm 2**: Derivative free method
*Step 1.* Initialization: Given a parameterized nonlinear equation with fuzzy variables
*Step 2.* Compute for $x_0$ by solving the parameterized equations for $r = 0,1$ and $k = 0,1,2 \ldots$
*Step 3.* Evaluate $J(x_k, F(x_k))^{-1}$ *and* $F(x_k)$.
*Step 4.* Compute $[J(x_k, h_k)^T J(x_k, h_k) + \mu_k I]^{-1}$ via (4) or (5) above
*Step 5.* Compute $x_{k+1} = x_k - [J(x_k, h_k)^T J(x_k, h_k) + \mu_k I]^{-1} J(x_k, h_k)^T F(x)$
*Step 6.* Continue the process with step 3 to 5 using the next $k$ until $\epsilon \le 10^{-4}$ are satisfied.

## 5. Numerical Results

This section considered some nonlinear equations with fuzzy variables from [1] and presented the solutions to the parameterized equations illustrate the performances of forward difference and central difference methods applied to Levenberg-Marquardt iterative method. The experiments are performed on MATLAB (R2015a) programming software using double precision computer.

**Example 1 [1]**: Given a nonlinear equation with fuzzy variables as

$$(3,3,4,5)x^2 + (1,2,3)x = (1,1,2,3)$$

Let $x$ be positive, then, without any loss of generality, we parameterized the equation as follows:

$$(3+r)\underline{x}^2(r) + (1+r)\underline{x}(r) = (1+r)$$
$$(5-r)\overline{x}^2(r) + (3-r)\overline{x}(r) = (3-r)$$

The initial guess is obtained by letting $r = 0$ and $r = 1$ in the parameterized system,

$$r = 1$$
$$4\underline{x}^2(1) + 2\underline{x}(1) = 2$$
$$4\overline{x}^2(1) + 2\overline{x}(1) = 2$$

$$r = 0$$
$$3\underline{x}^2(0) + \underline{x}(0) = 1$$
$$5\overline{x}^2(0) + 3\overline{x}(0) = 3$$

when $r = 0$, then, $\underline{x}(0) = 0.4343, \overline{x}(0) = 0.5307$ and when $r = 1$, we have $\underline{x}(1) = \overline{x}(1) = 0.5000$. We consider $x_0 = (0.4,0.4,0.5,0.6)$, as our initial guess. Via Algorithm 2 with $x_0 = (0.4,0.4,0.5,0.6)$ and approximate Jacobian $J(x^k, Fx^k)$ the number of iteration and execution time for forward difference and central difference are 2(0.6443) and 6(0.6676) respectively with an error $> 10^{-5}$. Please, refer to figure 1 for details of the solution for problem 1 $\forall r \in [0,1]$.

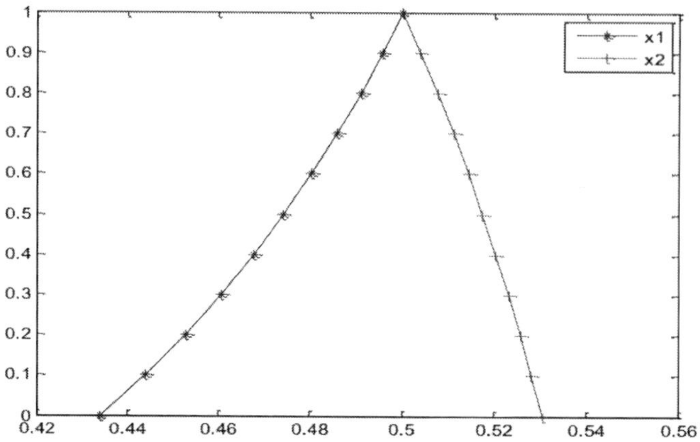


**Figure 1.** Iterative solution of example 1

**Example 2 [12]:** Given a nonlinear equation with fuzzy variables as

$$(2,2,1,1)x^3 + (3,3,1,1)x^2 + (4,1,1) = (8,8,3,5)$$

Let $x$ be positive, then, without any loss of generality, we parameterized the equation as follows:

$$(1+r)\underline{x}^3(r) + (2+r)\underline{x}^2(r) + (3+r) = (5+3r)$$
$$(3-r)\overline{x}^3(r) + (4-r)\overline{x}^2(r) + (5-r) = (13-5r)$$

or equality

$$(1+r)\underline{x}^3(r) + (2+r)\underline{x}^2(r) = (2+2r)$$
$$(3-r)\overline{x}^3(r) + (4-r)\overline{x}^2\,(r) = (8-4r)$$

let $r = 0$ and $r = 1$. Then, the initial guess is obtained as follows

$$\underline{x}^3(0) + 2\underline{x}^2(0) = 2$$
$$3\overline{x}^3(0) + 4\overline{x}^2\,(0) = 8$$

and

$$2\underline{x}^3(1) + 3\underline{x}(1) = 4$$
$$2\overline{x}^3(1) + 3\overline{x}\,(1) = 4$$

The initial guess $x_0 = (0.8,0.8,0.9,0.9)$ is considered. Applying Algorithm 2, the number of iteration and execution time for forward difference and central difference are 2(0.0636) and 7(0.0604) respectively with an error $> 10^{-5}$. Please, refer to figure 2 for details of the solution for problem 2 $\forall r \in [0,1]$.

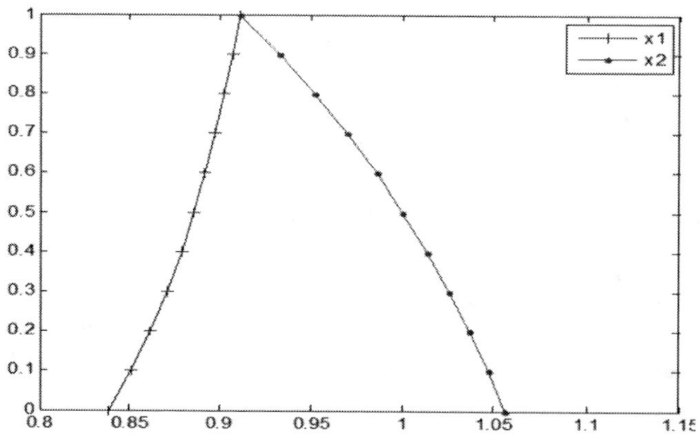


**Figure 2.** Iterative solution of example 2

## 6. Conclusion

This paper studied the forward difference and central difference methods applied to the Levenberg-Mardquardt iterative method for solution of nonlinear equation having fuzzy varaibles. We were mainly interested in reducing the experiment cost of the Jacobian by computing the approximation to the Jacobian matrix throughout the iteration process. This was achieved by parameterizing the nonlinear equation with fuzzy variables and then solved via Levenberg-Mardquardt method. The numerical result presented in section five illustrates that forward difference approach is very promising in all the tested problems used.

## References

[1] Abbasbandy S and Asady B 2004 Newton's method for solving fuzzy nonlinear equations *Applied Mathematics and Computation* **159** (2) 349-356

[2] Brown K M and Dennis J E 1971 Derivative free analogues of the Levenberg-Marquardt and Gauss algorithms for nonlinear least squares approximation *Numerische Mathematik* **18** (4) 289-297

[3] Broyden C G 1965 A class of methods for solving nonlinear simultaneous equations *Mathematics of computation* **19** (92) 577-593

[4] Buckley J J and Qu 1990 Y Solving linear and quadratic fuzzy equations *Fuzzy sets and systems* **38** (1) 43-59

[5] Fang J X 2002 On nonlinear equations for fuzzy mappings in probabilistic normed spaces *Fuzzy Sets and Systems* **131** (3) 357-364

[6] Levenberg K A 1944 method for the solution of certain non-linear problems in least squares *Quarterly of applied mathematics* **2** (2) 164-168

[7] Marquardt D W 1963 An algorithm for least-squares estimation of nonlinear parameters *Journal of the society for Industrial and Applied Mathematics* **11** (2) 431-441

[8] Peeva K 1992 Fuzzy linear systems *Fuzzy Sets and Systems* **49** (3) 339-355

[9] Sulaiman I M, Mamat M, Waziri M Y, Mohamed M A and Mohamad F S 2018 Solving Fuzzy Nonlinear Equation via Levenberg-Marquardt Method *Far East Journal of Math Sci* **103** (10) 1547-1558

[10] Umar A O, Waziri M Y and Sulaiman I M 2018 Solving Dual Fuzzy Nonlinear Equations via a Modification of Shamanskii Steps *Malaysian Journal of Computing and Applied Mathematics* **1** (2) 1-9

[11] Omesa U A, Mamat M, Sulaiman I M and Sukono 2020 On Quasi Newton Method for Solving Fuzzy Nonlinear Equations *International Journal of Quantitative Research and Modelling* 1 (1) 1-10

[12] Waziri M Y and Moyi A U 2016 An alternative approach for solving dual fuzzy nonlinear equations *International Journal of Fuzzy Systems* **18** (1) 103-107

[13] Zadeh L A 1965 Fuzzy sets *Information and control* **8** (3) 338-353

[14] Mamat M, Sulaiman I M and Ghazali P L 2020 An accelerated scheme for solving parameterized fuzzy nonlinear equations *International Journal of Advanced Science and Technology* **29** (5) 248-255

[15] Sulaiman I M and Mamat M 2021 Shamanskii Method for Solving Parameterized Fuzzy Nonlinear Equations *An International Journal of Optimization and Control: Theories & Applications*. In press.

[16] Sulaiman I M, Mamat M, Zamri N and Ghazali P L 2018 Solving dual fuzzy nonlinear equations via Shamanskii method *International Journal of Engineering and Technology* **7** (28) 89-91

[17] Sulaiman I M, Mamat M, Waziri M Y, Fadhilah and Kamfa K 2016 Regula Falsi method for solving fuzzy nonlinear equation *Far East Journal of Mathematical Sciences* 100 (6) 873-88

# Adaptive Control and Multi-variables Projective Synchronization of Hyperchaotic Finance System

**O S Onma[1,*], Heryanto[2], B Foster[3] and Subiyanto[4]**

[1] Department of Physics, Federal University of Agriculture Abeokuta, Nigeria
[2]Department of Military Electrical Engineering, Universitas Pertahanan, Indonesia
[3]Universitas Informatika dan Bisnis Indonesia, Indonesia
[4]Department of Marine Science, Faculty of Fishery and Marine Science, Universitas Padjadjaran, Indonesia

[*]Corresponding author: onmaokpabisunday@yahoo.com

**Abstract.** This paper introduces a new hyperchaotic finance system and show that it is a self-excited hyperchaotic attractor with the coexistence of double attractor with difference initial conditions for the same set of parameter values. Nonlinear feedback control function $u_i(t)$, $(i = 1,2,3,4)$ via adaptive control laws are design for the control and synchronization of the 4D-hyperchaotic finance system with the multiple values of the economics scaling factor $\alpha_i$ , $(i = 1,2,3,4)$, a special case of projective synchronization. The proposed schemes are able to stabilize as well as globally synchronize the two identical finance systems evolving from different initial conditions with uncertain parameter for the different values of $\alpha_i$ , $(i = 1,2,3,4)$. The analytical results were verified through numerical simulation.

## 1. Introduction

Chaos phenomenon is an interesting subject of discuss among the researchers of different fields today due to it wide applications in biological, physical, chemical, and financial systems as well as electronic engineering, secure communication, signal processing, fibre optical communication scheme [1-5].

For economics systems, the chaotic behavior was found in 1985 [6], since then researches on the complicated economics system via nonlinear method have been successful [7-8].

Lyapunov exponent is a critical prerequisite use to distinguish the 3D chaotic system from a hyperchaotic system. The 3D chaotic system has only one positive Lyapunov exponent; thus it is difficult to meet some certain requirement in practical application such as in secure communication [9-10]. But hyperchaotic system has at least two positive Lyapunov exponents thus has more prominent advantage over the 3D chaotic system due to its higher dimensions and more unpredictable behaviors [11-12].

In order to achieve financial control and synchronization of both developed and developing countries and different areas, we need to solve more problems of high order finance chaotic systems. Different approaches have been reported in the literature for chaos control and synchronization such as; active control [13], adaptive control [14], backstepping technique [15]-[16], sliding mode control [17-18]. The methods can be used for systems with either known or unknown parameters.

Multistability is a system property which refers to systems that neither stable nor totally instable, but that alternate between two or more mutually exclusive states (attractors) over time [19]. Multistability, namely, the coexistence of different attractors is a feature of many nonlinear dynamical models, ranging from the geophysical fluid dynamics to model of finance in economics, together with engineering applications. Multistability depends on the choice of initial conditions, as well as small changes in parameter, so that a sudden transition can occur to a different attractor [20].

Since the finance chaotic and hyperchaotic systems were proposed, many works has been done on them (see the references therein). However, there are some limitations in the existing results. For instance, in the stabilization problem of such systems, the designed control functions $u_i(t)$ are too complex to use in practical application. On the other hand, the coexistence (multistability) attractors with different initial conditions for the same set of parameter values have not been investigated.

Motivating by the above limitations, the present work address the problems by investigated the multistability of the system, reducing the controller complexity using the scaling factor $\alpha_i$ which could be varies in order to achieve a stable economic growth. The numerical simulation results are presented to validate effectiveness of the designed controller.

## 2. Mathematical Formulation

In constructing the hyperchaotic finance system model, four factors are involves the: money, production, labor force and stock , and are expressed by four-order nonlinear ordinary differential equations shown in system (1) .

The mathematical model considered in this paper is 4D hyperchaotic finance system as described in equation (1).

The system describes the time variations of the state variables: the interest rate $x$ , the investment demand $y$ , the price exponent $z$ and the average profit margin $w$ .

$$\begin{aligned} \dot{x} &= -a(x+y)+w \\ \dot{y} &= -y-axz \\ \dot{z} &= b+axy \\ \dot{w} &= -cxz-dw \end{aligned} \tag{1}$$

Where $a$ is the saving, $b$ is the per investment cost, $c$ is the elasticity of demands of commercials and $d$ is the governing parameter of the system.

System (1) exhibit complex hyperchaotic behavior with the real constant parameter values $a=3.0$, $b=15.0$, $c=0.5$ and $d=0.12$ with the portrait of the phase attractor displayed in Figure 1 .

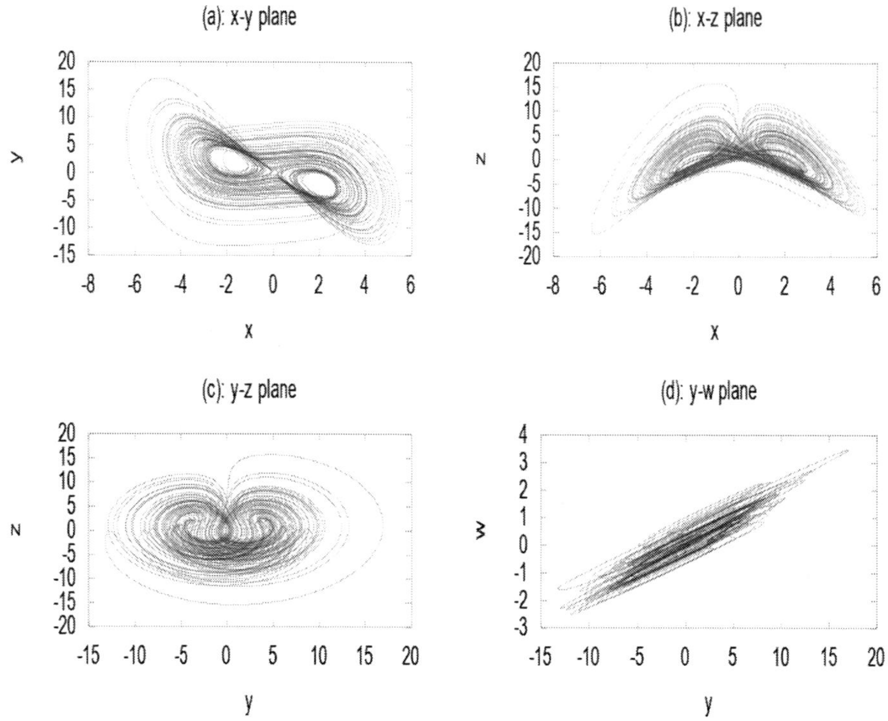


**Figure 1.** Phase portrait of the hyperchaotic financial system (1) attractor with the parameter values $a = 3.0, b = 15.0$, $c = 0.5$ and $d = 0.12$.

## 3. Adaptive Control Technique

### 3.1 Design of the Adaptive control function $u_i(t)$

$$\begin{aligned} \dot{x} &= -a(x+y)+w+u_1 \\ \dot{y} &= -y-axz+u_2 \\ \dot{z} &= b+axy+u_3 \\ \dot{w} &= -cxz-dw+u_4 \end{aligned} \tag{2}$$

Where $u_i(t)$, $(i = 1,2,3,4)$ is the control input to design.

The result of the system (2) is validated using Lyapunov stability theory (LST) by construct a positive Lyapunov function $V(x, y, z, w)$ as follows;

$$V = 0.5\left(x^2 + y^2 + z^2 + w^2 + \tilde{a}^2 + \tilde{b}^2 + \tilde{c}^2 + \tilde{d}^2\right) \tag{3}$$

Where; $\tilde{a} = a - \bar{a}$, $\tilde{b} = b - \bar{b}$, $\tilde{c} = c - \bar{c}$ and $\tilde{d} = d - \bar{d}$ are the estimated values of the unknown parameter respectively.

The time derivative of equation (3) along the trajectories of the system (1) is:

$$\dot{V} = x\dot{x} + y\dot{y} + z\dot{z} + w\dot{w} + \tilde{a}\dot{\tilde{a}} + \tilde{b}\dot{\tilde{b}} + \tilde{c}\dot{\tilde{c}} + \tilde{d}\dot{\tilde{d}} \tag{4}$$

To ensure the convergences of the controller $u_i(t)$ $(i = 1,2,3,4)$ to the origin asymptotically, the control function $u_i(t)$ is choose as follows;

$$
\begin{aligned}
u_1 &= a(x+y) - w - x \\
u_2 &= y + axz - y \\
u_3 &= -b - axy - z \\
u_4 &= cxz + dw - w
\end{aligned} \tag{5}
$$

The substitution of equation (2) into the equation (3) yields the equation below.

$$
\begin{aligned}
\dot{V} &= x[-a(x+y) + w + u_1] + y[-y - axz + u_2] + z[b + axy + u_3] + \\
& w[-cxz - dw + u_4] + \tilde{a}[-\dot{\hat{a}} - x^2 - xy - xyz] + \tilde{b}[-\dot{\hat{b}} + z] + \tilde{c}[-\dot{\hat{c}} - xzw] + \\
& \tilde{d}[-\dot{\hat{d}} - w^2]
\end{aligned} \tag{6}
$$

From equation (6), the parameter update law is estimated in equation (7).

$$
\begin{aligned}
\dot{\hat{a}} &= -(x^2 + xy + xyz) + xyz \\
\dot{\hat{b}} &= z \\
\dot{\hat{c}} &= -xzw \\
\dot{\hat{d}} &= -w^2
\end{aligned} \tag{7}
$$

The substitution of equations (5) and (7) respectively into equation (4) result to equation (8)

$$\dot{V} = -x^2 - y^2 - z^2 - w^2 - \tilde{a}^2 - \tilde{b}^2 - \tilde{c}^2 - \tilde{d}^2 < 0 \tag{8}$$

Remarkably, $V$ is a quadratic positive definite Lyapunov function and it time derivative $\dot{V}$ is a quadratic negative definite function.

According to the Lyapunov stability theory (LST), the system (2) can converge to the unstable equilibrium $E_0$ with the control function $u_i(t)$, $(i = 1,2,3,3)$ in equation (5) and the parameter update law in equation (7).

### 3.2 Numerical Simulation Results

In this section, fourth-order Runge-kutta algorithm is adopted with the following initial conditions of the state variables $x(0) = 0.8$, $y(0) = 0.6$ $z(0) = -0.9$ and $w(0) = -0.8$ with time grid of $0.001$. Fixing the parameters values of $a$, $b$, $c$ and $d$ of the system as in figure 1 above, the system (2) is solved with the control input $u_i(t)$, $(i = 1,2,3,4)$ as defined in equation (5).

The results display in figure 2 shows that the state variables move hyperchaotically when the control function $u_i(t)$ are switched off and when the activated at $t = 50$, the state variables are converges asymptotically at the unstable equilibrium point $E_0(0,0,0,0)$ according to the Lyapunov stability theory (LST).

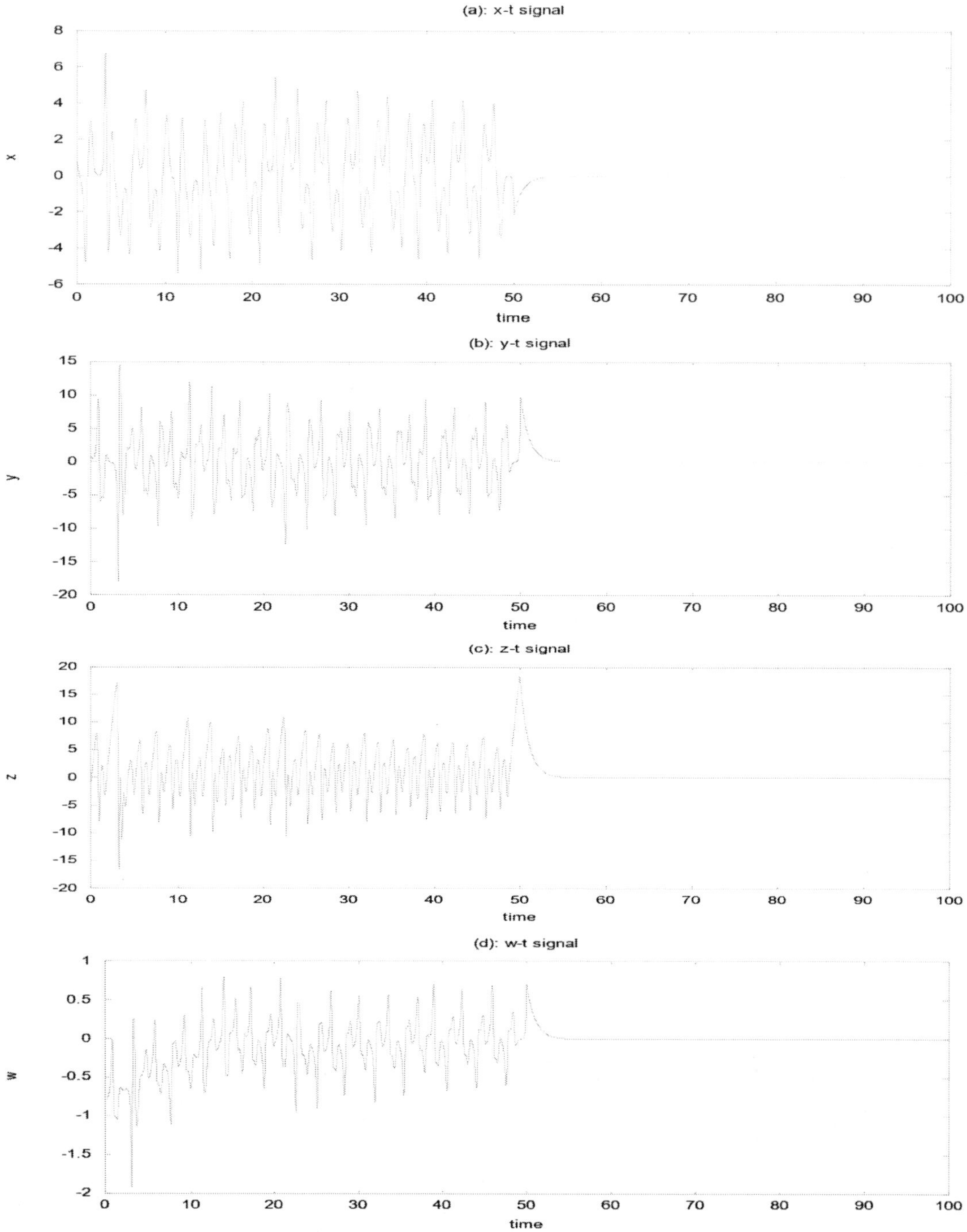


**Figure 2.** Time responses of the state variables $(x, y.z, w)$ for hyperchaotic finance system when the controller is activated at $t = 50$ to stabilize at the origin.

IOP Conf. Series: Materials Science and Engineering 1115 (2021) 012003 doi:10.1088/1757-899X/1115/1/012003

## 4. Multi-variable Projective Synchronization

### 4.1 Adaptive Synchronization with multiple values of $\alpha_i$

From system (1), we let; $x = x_1, y = x_2, z = x_3$ and $w = x_4$.

Hence;

$$\begin{aligned}\dot{x}_1 &= -a(x_1 + x_2) + x_4 \\ \dot{x}_2 &= -x_2 - ax_1x_3 \\ \dot{x}_3 &= b + ax_1x_2 \\ \dot{x}_4 &= -cx_1x_3 - dx_4\end{aligned} \tag{9}$$

Equation (9) is designated as the drive or master which will transmit the signal to the system (10) the response or the slave which act as the signal receiver during the transmission.

$$\begin{aligned}\dot{y}_1 &= -a(y_1 + y_2) + y_4 + \alpha_1 u_1 \\ \dot{y}_2 &= -y_2 - ay_1y_3 + \alpha_2 u_1 \\ \dot{y}_3 &= b + ay_1y_2 + \alpha_3 u_3 \\ \dot{y}_4 &= -cy_1y_3 - dy_4 + \alpha_4 u_4\end{aligned} \tag{10}$$

Where $\alpha_i$ $(i = 1,2,3,4)$ is the economic scaling factor to be estimated appropriately and $u_i(t)$ $(i = 1,2,3,4)$ is the control function to be derives.

The error vector between the drive (master) in equation (9) and the response (slave) in equation (10) is defined as;

$$e_i = y_i - x_i \tag{11}$$

By subtracting equation (9) from equation (10) using the definition of the error vector in equation (10) the error dynamical variables is obtained in equation (12).

$$\begin{aligned}\dot{e}_1 &= -a(e_1 + e_2) + e_4 + \alpha_1 u_1 \\ \dot{e}_2 &= -e_2 - a(x_1e_3 + x_3e_1 + e_1e_3) + \alpha_2 u_2 \\ \dot{e}_3 &= a(x_1e_2 + x_2e_1 + e_1e_2) + \alpha_3 u_3 \\ \dot{e}_4 &= -c(x_1e_3 + x_3e_1 + e_1e_3) - de_4 + \alpha_4 u_4\end{aligned} \tag{12}$$

We select the positive Lyapunov function $V(t)$ as;

$$V = 0.5\left(e_1^2 + e_2^2 + e_3^2 + e_4^2 + \tilde{a}^2 + \tilde{b}^2 + \tilde{c}^2 + \tilde{d}^2\right) \tag{13}$$

Where $\tilde{a} = a - \bar{a}$, $\tilde{b} = b - \bar{b}$, $\tilde{c} = c - \bar{c}$ and $\tilde{d} = d - \bar{d}$ are the estimate values of the unknown parameter $a$, $b$, $c$ and $d$ respectively.

The time derivative of equation (13) along the trajectories of equation (12) is given in equation (14).

$$\dot{V} = e_1\dot{e}_1 + e_2\dot{e}_2 + e_3\dot{e}_3 + e_4\dot{e}_4 + a\dot{\tilde{a}} + b\dot{\tilde{b}} + c\dot{\tilde{c}} + d\dot{\tilde{d}} \tag{14}$$

The substitution of equation (12) into equation (14) is shown in equation (15).

$$\begin{aligned}\dot{V} &= e_1[-a(e_1 + e_2) + e_4 + \alpha_1 u_1] + e_2[-e_2 - a(x_1e_3 + x_3e_1 + e_1e_3) + \alpha_2 u_2] \\ &+ e_3[a(x_1e_2 + x_2e_1 + e_1e_2) + \alpha_3 u_3] + e_4[-c(x_1e_3 + x_3e_1 + e_1e_3) - de_4 + \alpha_4 u_4] \\ &+ \tilde{a}[\dot{\bar{a}} - e_1^2 - e_1e_2 - e_2(x_1e_3 + x_3e_1 + e_1e_3) + e_3(x_1e_2 + x_2e_1 + e_1e_2)] \\ &+ \tilde{b}[\dot{\bar{b}}] + \tilde{c}[\dot{\bar{c}} - e_4(x_1e_3 + x_3e_1 + e_1e_3)] + \tilde{d}[\dot{\bar{d}} - e_4^2]\end{aligned} \tag{15}$$

The control input $u_i(t)$ together with the scaling factor and the parameter update law are respectively estimated from equation (15) as shown in equations (16) and (17) in that order.

$$\begin{aligned} \alpha_1 u_1 &= a(e_1+e_2)-e_4-k_1e_1 \\ \alpha_2 u_2 &= e_2 + a(x_1e_3+x_3e_1+e_1e_3)-k_2e_2 \\ \alpha_3 u_3 &= -a(x_1e_2+x_2e_1+e_1e_2)-k_3e_3 \\ \alpha_4 u_4 &= c(x_1e_3+x_3e_1+e_1e_3)+de_4-k_4e_4 \end{aligned} \qquad (16)$$

Where $k_i$ $(i=1,2,3,4)$ is the arbitrary feedback control gain to be determine

$$\begin{aligned} \dot{\tilde{a}} &= -e_1^2 - e_1e_2 - e_2(x_1e_3+x_3e_1+e_1e_3)+e_3(x_1e_2+x_2e_1+e_1e_2)-a \\ \dot{\tilde{b}} &= -b \\ \dot{\tilde{c}} &= -e_4(x_1e_3+x_3e_1+e_1e_3)-c \\ \dot{\tilde{d}} &= -e_4^2 - d \end{aligned} \qquad (17)$$

For the error dynamics system (12) to converge to the equilibrium point asymptotically, the condition $\dot{V} = -\sum k_i e_i^2 < 0$ must be satisfied.

**Theorem**: For a given nonzero economic scaling factor $\alpha_i$ the multi variable projective synchronization between two systems (9) and (10) will occur by the adaptive control law (19) and the parameter update law [17].

By substitute equations (16) and (17) respectively into equation (14) gives the result in equation (18) which satisfied the above condition.

$$\dot{V} = -k_1e_1^2 - k_2e_2^2 - k_3e_3^2 - k_4e_4^2 - a^2 - b^2 - c^2 - d^2 < 0 \qquad (18)$$

The Lyapunov function $V$ is a positive definite and its derivative $\dot{V}$ is negative definite in the neighborhood of the zero solution for system (12). Based on the Lyapunov stability theory (LST), the error dynamical system (12) can converge to the origin asymptotically. Thus, by Barbalat's lemma in Lyapunov stability theory [21], one can conclude that $e_i(t)$ $(i=1,2,3,4)$ (equation (12)) is globally and exponentially stable. Consequently, the drive (master) and response (slave) in equations (9) and (10) respectively are globally and exponentially synchronized for all initial conditions $x_i(0)$, $y_i(0)$ $\in R^4$ and update law (17).

In order to vary the value of the scaling factor $\alpha_i$ $(i=1,2,3,4)$, the control function $u_i(t)$ in equation (16) is reduce to equation (19) as illustrated below.

$$\begin{aligned} u_1 &= \frac{1}{\alpha_1}\left[a(e_1+e_2)-e_4-k_1e_1\right] \\ u_2 &= \frac{1}{\alpha_2}\left[e_2+a(x_1e_3+x_3e_1+e_1e_3)-k_2e_2\right] \\ u_3 &= \frac{1}{\alpha_3}\left[-a(x_1e_2+x_2e_1+e_1e_2)-k_3e_3\right] \\ u_4 &= \frac{1}{\alpha_4}\left[c(x_1e_3+x_3e_1+e_1e_3)+de_4-k_4e_4\right] \end{aligned} \qquad (19)$$

### 4.2 Numerical Simulation Results

In order to demonstrate the effective of the proposed techniques, the classical fourth-order Runge-Kutta routine is employed to solve the control law (19) and update law (17). The initial conditions of the drive (master) equation (9) are; $x_1(0) = -0.5$, $x_2(0) = 0.6$, $x_3(0) = 0.6$ and $x_4(0) = 0.8$ while that of the response (slave) equation (10) are; $y_1(0) = -10.0$, $y_2(0) = 1.0$, $y_3(0) = 0.2$ and $y_4(0) = 10.0$. A time step of $0.001$ is used with the values of the constant parameters $a$, $b$, $c$ and $d$ as in Figure 1 and the control feedback gain $k_i = 1$ to ensure complex hyperchaotic dynamics of the state variables in systems (9) and (10).

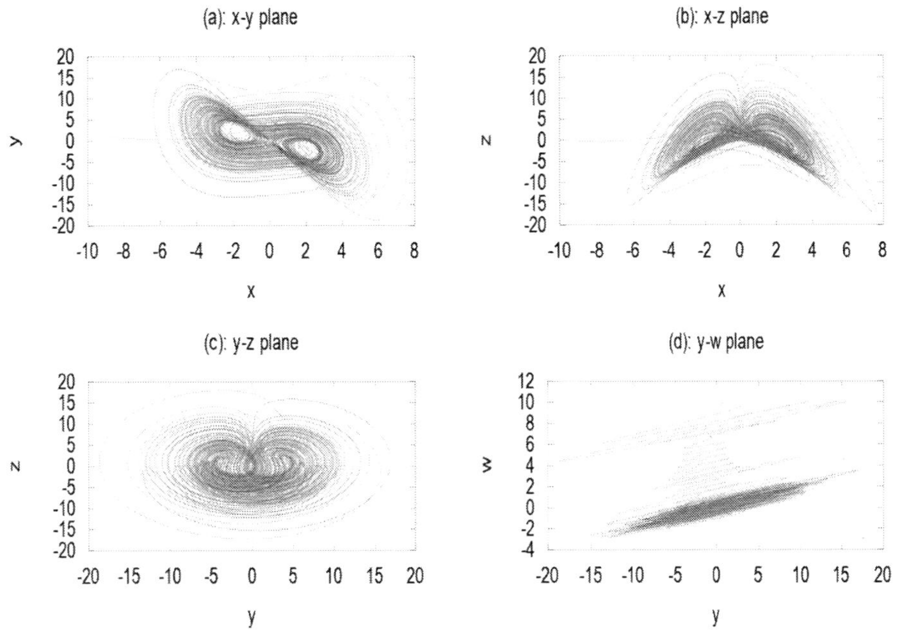


**Figure 3.** Multi-stability with double coexistence of the hyperchaotic attractors for the different initial conditions.

Figure 3 showed the multi-stability of the system (1) with double coexistence of the hyperchaotic attractors for the different initial conditions of the drive (in red) $(x_1, x_2, x_3, x_4) = (-0.5, 0.6, 0.6, 0.8)$ and the response (in green) $((y_1, y_2, y_3, y_4) = (-10.0, 1.0, 0.2, 10.0)$ for the same parameter values $(a,b,c,d) = (3.0, 15.0, 0.5, 0.12)$. This is a special property of a system which refers to neither stable nor totally instable, but alternate between two or more mutually exclusive states (attractors) over time. However, for complete synchronization $\alpha_i = 1$.

The results of the simulation displayed the states trajectories of the drive (master) system (9) and the response (slave) system (10) in figure 4 ((a), (b) , (c) and (d)). The states of the response (slave) (10) track the dynamics of the drive (master) (9) when the controller is activated at $t = 50$ .The error vectors $e_i(t)$ $(i = 1,2,3,4)$ (equation (12)) converges to the origin at $t = 50$ as shown in figure 5 and the synchronization norm $e = \sqrt{e_1^2 + e_2^2 + e_3^2 + e_4^2}$ is depicted in in Figure 6.

IOP Conf. Series: Materials Science and Engineering 1115 (2021) 012003 doi:10.1088/1757-899X/1115/1/012003

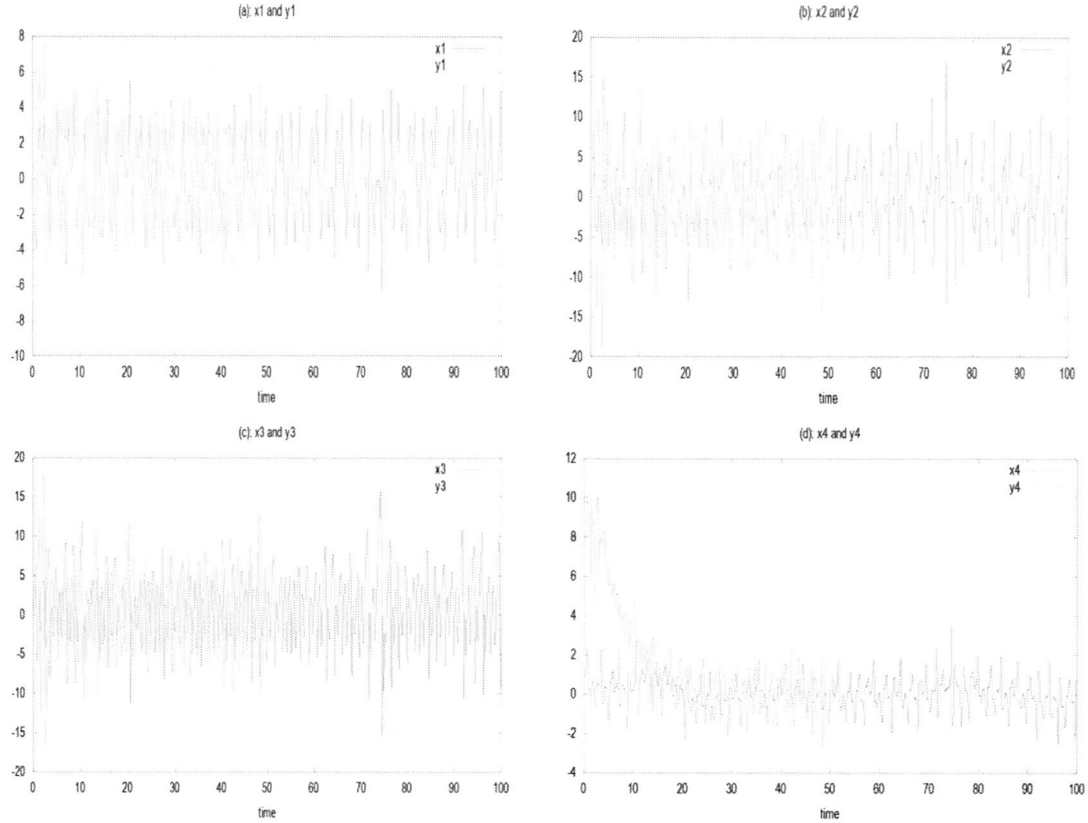


**Figure 4.** The time response of the state variables drive (master) and response (slave) for complete synchronization ( $\alpha_i = 1$ )

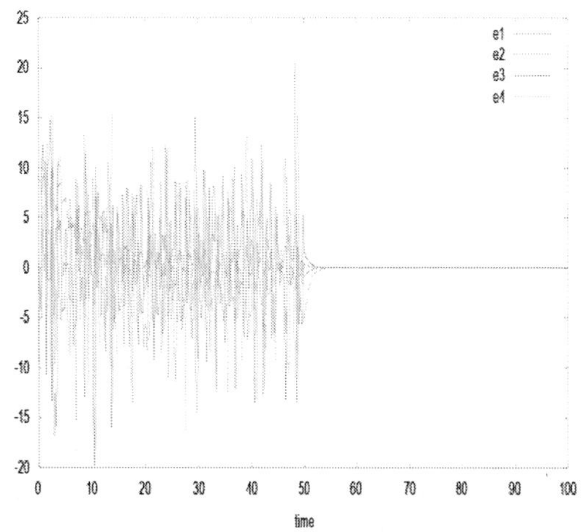


**Figure 5.** The error vectors between the drive (master) and the response (slave) for complete synchronization ( $\alpha = 1$ )

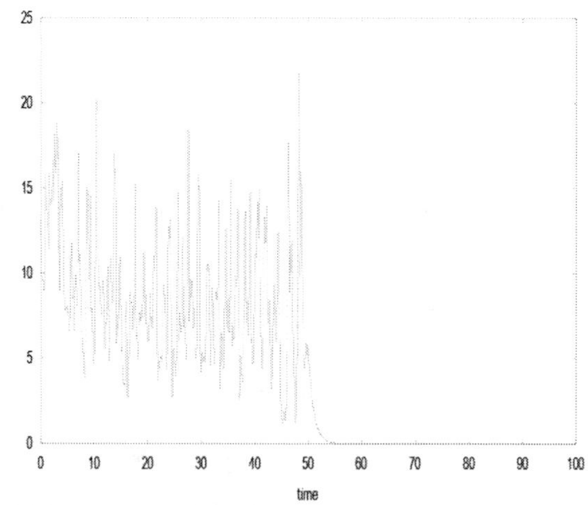


**Figure 6.** Synchronization norm for complete synchronization ( $\alpha = 1$ )

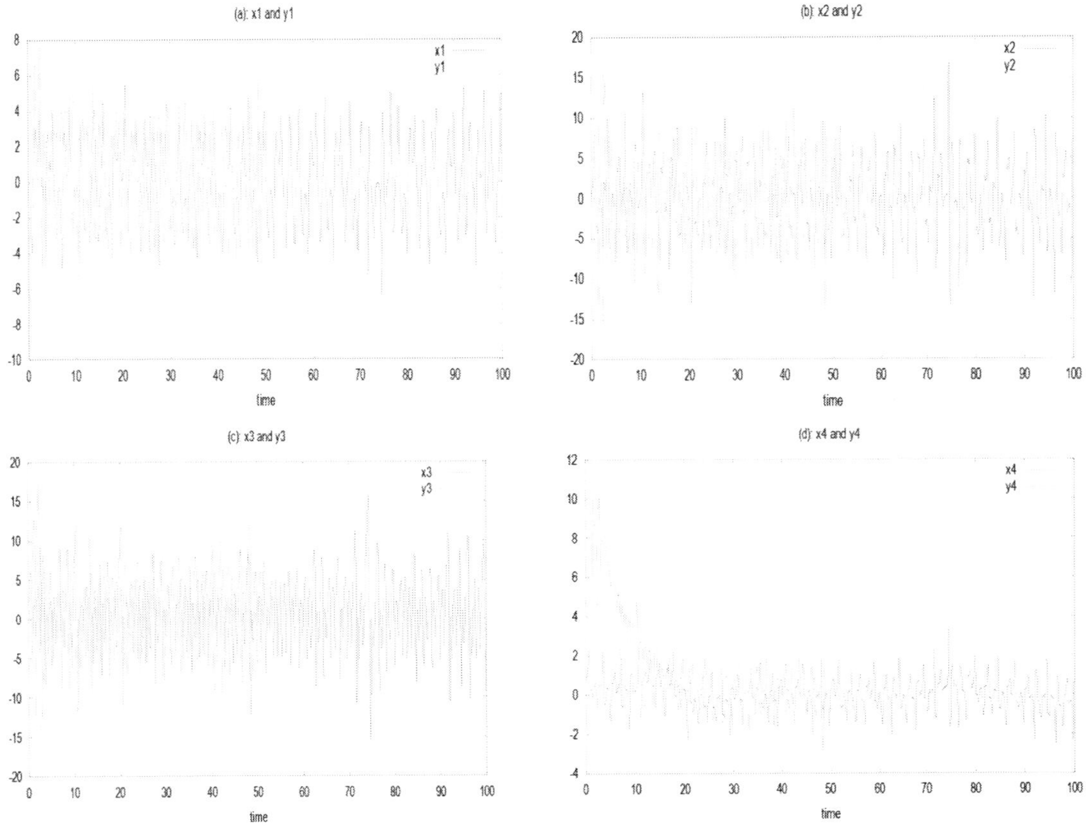


**Figure 7.** The time response of the state variables drive (master) and response (slave) for multi-variables projective synchronization $(\alpha_1 = 1, \alpha_2 = 2, \alpha_3 = 3, \alpha = -4)$

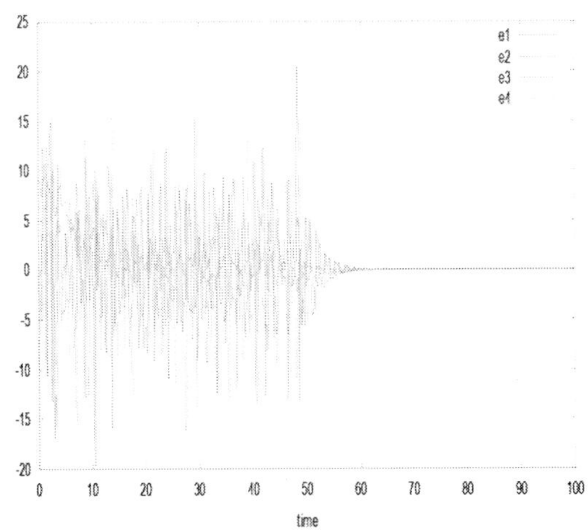


**Figure 8.** The error vectors between the drive (master) and the response (slave) for multi-variables projective synchronization $(\alpha_1 = 1, \alpha_2 = 2, \alpha_3 = 3, \alpha = -4)$.

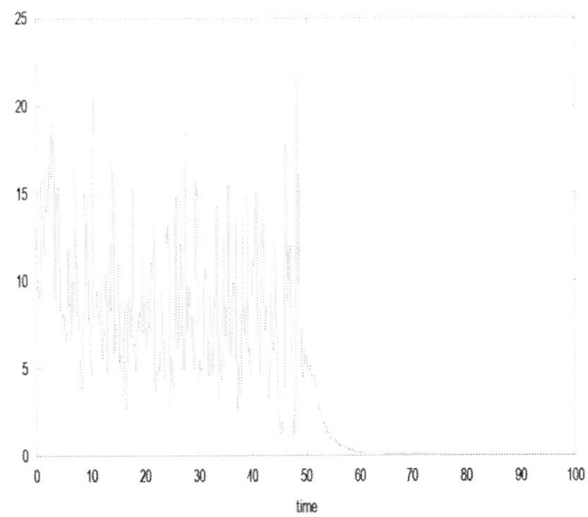


**Figure 9.** Synchronization norm for anti-synchronization for multi-variables projective synchronization $(\alpha_1 = 1, \alpha_2 = 2, \alpha_3 = 3, \alpha = -4)$.

For multi-variable projective synchronization, $\alpha_1 = 1$, $\alpha_2 = 2$, $\alpha_3 = 3$ and $\alpha_4 = -4$. The states of the drive (master) and the response (slave) are depicted in Figure 7 ((a), (b), (c) and (d)). The states of the response (slave) (10) track the dynamics of the drive (master) (9) when the controller is switched on at $t = 50$. The error vectors $e_i(t)$ ($i = 1,2,3,4$) (equation (12)) converges to the origin at $t = 50$ as shown in Figure 8 and the synchronization norm $e = \sqrt{e_1^2 + e_2^2 + e_3^2 + e_4^2}$ is depicted in in Figure 9.

Figure 10 present the dynamics of the parameter estimation errors. The parameter $a, b, c$ and $d$ converges from $(3.0,15.0,0.5,0.12)$ to the origin $(0,0,0,0)$. However, figure 11 display the time response of the parameter update law (17) updated from the initial values $a_1(0) = 3.0$, $b_1(0) = 15.0$, $c_1(0) = 0.5$ and $d_1 = 0.12$ to the estimated values $a = 5.0$, $b = 10.0$, $c = 2.5$ and $d = 1.5$ respectively.

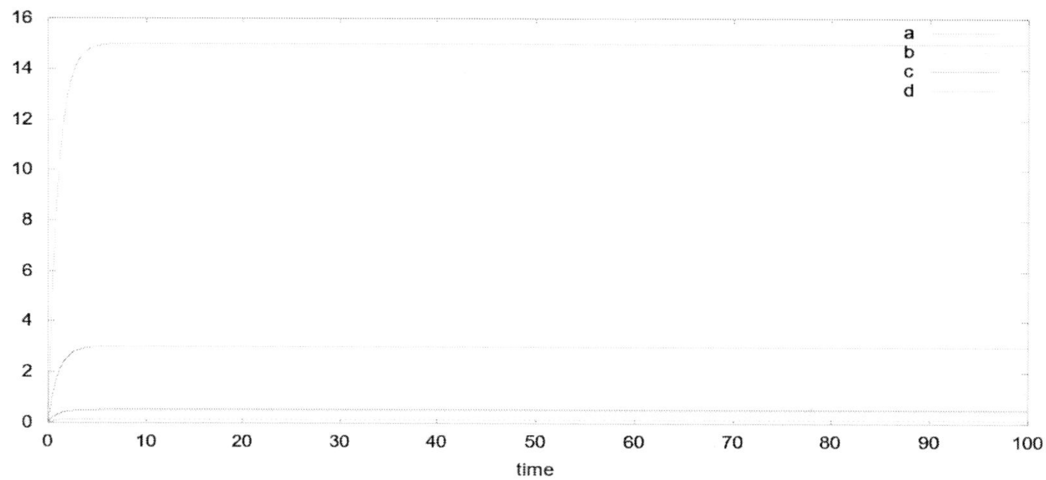


**Figure 10.** Time response of the parameter estimation errors.

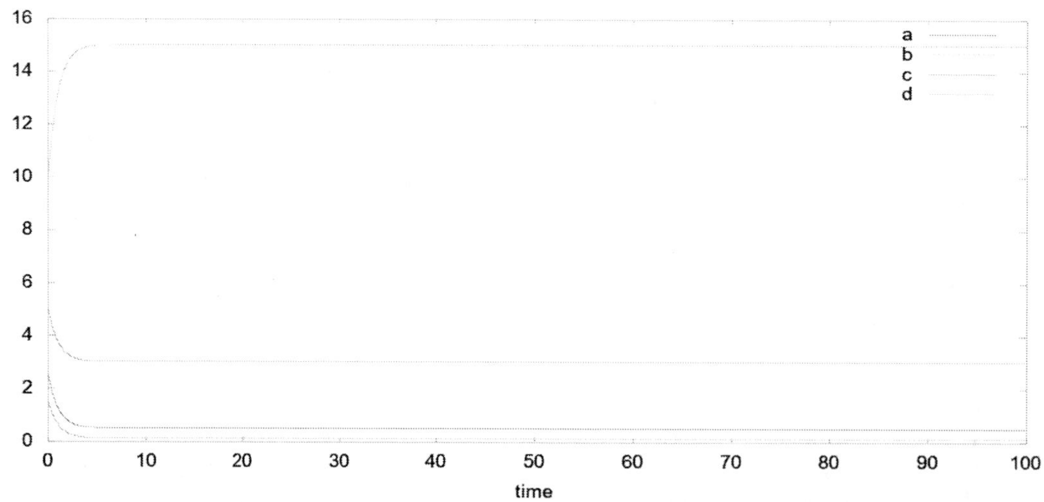


**Figure 11.** Time response of the parameter update law.

## 5. Conclusions

This work, studied the 4D hyperchaotic finance system. The multistability property of the system was discussed. The adaptive control method with parameter update laws has been applied to control the finance hyperchaotic system while the multi-variables projective technique with the update is used to synchronization between the two identical financial systems.

The novel adaptive controller $u_i(t)$ stabilizes as well as globally synchronized the two identical systems at different initial values of $x_i(t)$ and $y_i(t)$. The orbits of the error vector (12) converges to the origin asymptotically when the controller is activated at $t = 50$ as $t \to \infty$, that is the stabilization of the error dynamic (12) is achieved by the controller (19).

The stabilization of the error vector (12) indicates an effective way to regulate and control the interest rate ( $x$ variable) when the inflation occurs and chaos arises in nonlinear finance system. When the saving (parameter $a$ ) is small, the inflation will increase hence chaos will appear. Therefore, there is a need for parameter updating in finance system.

In various economic systems, efforts are made towards reducing the investment cost ( $b$ ), thus practical implementation of the update laws shall be very useful.

We established that the proposed method is an effective measure to; regulate, revive as well as track the economy of the nations in particular during the global crisis of COVID-19 pandemic that paralyzes the world economy.

## References

[1] Sambas A, Vaidyanathan S, Tlelo-Cuautle E, Abd-El-Atty B, Abd El-Latif A A, Guillén-Fernández O, Hidayat Y and Gundara G 2020 A 3-D multi-stable system with a peanut-shaped equilibrium curve: Circuit design, FPGA realization, and an application to image encryption *IEEE Access* **8** 137116-137132.

[2] Fu C, Meng W H, Zhan Y F, Zhu Z L, Lau F C, Chi K T and Ma H F 2013 An efficient and secure medical image protection scheme based on chaotic maps *Computers in biology and medicine* **43** (8) 1000-1010

[3] Zang X, Iqbal S, Zhu Y, Liu X and Zhao J 2016 Applications of chaotic dynamics in robotics *International Journal of Advanced Robotic Systems* **13** (2) 60

[4] Sambas A, Vaidyanathan S, Tlelo-Cuautle E, Zhang S, Guillen-Fernandez O, Hidayat Y and Gundara G 2019 A novel chaotic system with two circles of equilibrium points: multistability, electronic circuit and FPGA realization *Electronics* **8** (11) 1211

[5] Sambas A, Vaidyanathan S, Zhang S, Zeng Y, Mohamed M A and Mamat M 2019 A new double-wing chaotic system with coexisting attractors and line equilibrium: bifurcation analysis and electronic circuit simulation *IEEE Access* **7** 115454-115462

[6] Grandmont, J.M. (1985) On Endogenous Competitive Business Cycles. Econometrica 53: 994-1045.

[7] Chen W C 2008 Nonlinear dynamics and chaos in a fractional-order financial system *Chaos Solitons & Fractals* **36** (5) 1305-1314

[8] Sukono, Sambas A, He S, Liu H, Vaidyanathan S, Hidayat Y and Saputra J 2020 Dynamical analysis and adaptive fuzzy control for the fractional-order financial risk chaotic system *Advances in Difference Equations* 674 (1) 1-12

[9] Okpabi S O, Oke A A, Akinlami J O and Opeifa S T 2017 Backstepping Control and Synchronization of hyperchaotic Lorenz-Stenflo system with application to secure communication *Far East Journal of Dynamical Systems* **29** (1) 1-23.

[10] Mobayen S, Vaidyanathan S, Sambas A, Kacar S and Çavuşoğlu Ü 2019 A novel chaotic system with boomerang-shaped equilibrium, its circuit implementation and application to sound encryption *Iranian Journal of Science and Technology, Transactions of Electrical*

*Engineering* **43** (1) 1-12
[11] Muthuswamy B 2010 Implementing memristor based chaotic circuits *International Journal of Bifurcation and Chaos* **20** (05) 1335-1350
[12] Rusyn V and Khrapko S 2018 Memristor: Modeling and research of information properties. *Chaotic Modeling and Simulation International Conference* 229-238
[13] Bai E W and Lonngren K E 1997 Synchronization of two Lorenz systems using active control *Chaos, Solitons & Fractals* **8** (1) 51-58
[14] Idowu B A, Olusola O I, Onma O S, Vaidyanathan S, Ogabi C O and Adejo O A 2019 Chaotic financial system with uncertain parameters-its control and synchronisation *International Journal of Nonlinear Dynamics and Control* **1** (3) 271-286
[15] Sambas A, Vaidyanathan S, Zhang S, Mohamed M A, Zeng Y and Azar A T 2021 A new 4-D chaotic hyperjerk system with coexisting attractors, its active backstepping control, and circuit realization *Backstepping Control of Nonlinear Dynamical Systems* 73-94
[16] Onma O S, Olusola O I and Njah A N 2014 Control and synchronization of chaotic and hyperchaotic Lorenz systems via extended backstepping techniques *Journal of nonlinear dynamics* **2014** 21-32
[17] Medhaffar H, Feki M and Derbel N 2019 Stabilizing periodic orbits of chua's system using adaptive fuzzy sliding mode controller *International Journal of Intelligent Computing and Cybernetics* **12** (1) 102-126
[18] Pan Y, Yang C, Pan L and Yu H 2017 Integral sliding mode control: performance, modification, and improvement *IEEE Transactions on Industrial Informatics* **14** (7) 3087-3096
[19] Sambas A, Vaidyanathan S, Bonny T, Zhang S, Hidayat Y, Gundara G and Mamat M 2021 Mathematical Model and FPGA Realization of a Multi-Stable Chaotic Dynamical System with a Closed Butterfly-Like Curve of Equilibrium Points *Applied Sciences* **11** (2) 788
[20] Vaidyanathan S, Tlelo-Cuautle E, Anand P G, Sambas A and Guillen-Fernandez O 2021 A new conservative chaotic dynamical system with lemniscate equilibrium, its circuit model and FPGA implementation *Int. J. Automation and Control* **15** (2) 128-148
[21] Hhalil H K 2002 *Nonlinear Systems* New Jersey: Prentice Hall of Indian

# Performance Analysis of a Modified Conjugate Gradient Algorithm for Optimization Models

**S E Olowo[1], I M Sulaiman[2,*], M Mamat[2], A E Owoyemi[3], M A Zaini[4], Kalfin[5], and S H Yuningsih[6]**

[1]Department of Mathematics, Kano University of science and Technology, Kano Nigeria.
[2]Faculty of Informatics and Computing, Universiti Sultan Zainal Abidin, Campus Kuala Nerus, 22200, Terengganu, Malaysia.
[3]Department of General Studies. Federal College of Agricultural Produce Technology, Kano, Nigeria
[4]Institute of Engineering Mathematics, Universiti Malaysia Perlis, Kangar, Malaysia.
[5]Doctor Program of Mathematics, Faculty of Mathematics and Natural Sciences, Universitas Padjadjaran, Indonesia
[6]Master Program of Mathematics, Faculty of Mathematics and Natural Sciences, Universitas Padjadjaran, Indonesia

*Corresponding author: kademi1985@gmail.com

**Abstract.** The Conjugate gradient (CG) algorithms is very important and widely used in solving optimization models. This is due to its simplicity as well as global convergence properties. Various line search procedures as usually employ in the analysis of the CG methods. Recently, many studies have been done aimed at improving the CG method. In this paper, an alternative formula for conjugate gradient coefficient has been proposed which possesses the global convergence properties under exact minimization condition. The result of the numerical computation has shown that this new coefficient performs better than the existing CG methods.

## 1. Introduction

The CG algorithms play an essential role when obtaining the solution of optimization models, particularly, when the problems involve large-dimension. This is as a result of its low memory requirement. The general formula of this method is given as

$$\min f(x), \quad x \in R^n, \tag{1}$$

with $f: R^n \to R$ been smooth & $g \equiv \nabla f(x)$ denotes the gradient. The CG approaches generates an iterative sequence $\{x_k\}$via a method given as

$$x_k = x_{k-1} + \alpha_{k-1} d_{k-1}, \forall k \geq 1 \tag{2}$$

where $\alpha_{k-1} \geq 1$ is obtained with any line search method. One of the commonly used line search scheme is defined as

$$f(x_k + \alpha_k d_k) = \min f(x_k + \alpha d_k), \tag{3}$$

and referred to the exact minimization condition which is computed along the direction of search $d_k$ with formula

$$d_k = \begin{cases} -g_k, & \text{if} \quad k = 0, \\ -g_k + \beta_k d_{k-1}, & \text{if} \quad k \geq 1, \end{cases} \tag{4}$$

The scalar $\beta_k \in R$ is known as the coefficient of the CG algorithm with different parameter characterizing the CG algorithms. Some of the classical formulas for CG coefficients $\beta_k$ are defined below

$$\beta_k^{FR} = \frac{\| g_k \|^2}{\| g_{k-1} \|^2}, \ \beta_k^{PRP} = \frac{g_k^T (g_k - g_{k-1})}{\| g_{k-1} \|^2}, \ \beta_k^{HS} = \frac{g_k^T (g_k - g_{k-1})}{d_{k-1}^T (g_k - g_{k-1})},$$

$$\beta_k^{LS} = \frac{g_k^T (g_k - g_{k-1})}{-d_{k-1}^T g_{k-1}}, \ \beta_k^{DY} = \frac{\| g_k \|^2}{d_{k-1}^T (g_k - g_{k-1})}, \ \beta_k^{CD} = -\frac{\| g_k \|^2}{d_{k-1}^T g_{k-1}}.$$

$g_k$ is known as the gradients with the Euclidean norm denoted by $\|.\|$. The algorithms define above are Liu-Storey (LS) [21], Conjugate Descent (CD) [15], Hestenes-Steifel (HS) [9], Polak-Ribiere-Polyak (PRP) [1, 4], Dai-Yuan (DY) [20], and Fletcher-Reeves (FR) [16]. These methods have been shown to be equivalent if the quadratic functions $f(x)$ is strongly convex. However, their behaviours differ for non-quadratic functions [22].

Several studies of conjugate gradient methods have been focusing on its convergence behaviour under several line search procedures. These include Zoutendijk [5], who prove the global converges of FR using exact minimization condition. However, Powell [8], illustrates through a counter example that FR performs poorly in experimentations. Further studies on the convergence of methods linked to FR algorithm were presented by Touati-Ahmed and Storey [2], Al-Baali [7] via the strong Wolfe conditions. For more references on current findings of the CG algorithms, we refer researchers [11,13,14,17,18,19].

In this study, the authors present an efficient type of the CG parameter followed by its algorithm in the next section. The converge analysis of the proposed method is discussed under exact minimization condition in section 3. Numerical results and discussion of our new formula with other well-known formulas followed in section 4 and concluding in section 5.

## 2. New modification with its algorithm

Lately, Rivaie et al. [10], developed an alternative coefficient of the CG method by replacing the denominator of PRP [1, 4], and HS [9] while retaining the numerator as follows

$$\beta_k^{RMIL} = \frac{g_k^T (g_k - g_{k-1})}{d_{k-1}^T (d_{k-1}^T - g_k)} \tag{5}$$

Motivated by this idea, we construct a simple variant of RMIL coefficient by introducing a new numerator of Rivaie et al. [10] as follows.

$$\beta_k^{SMO} = \frac{\|g_k\|^2 - \frac{\|g_k\|}{\|g_{k-1}\|} |g_k^T d_{k-1}|}{d_{k-1}^T (d_{k-1}^T - g_k)} \tag{6}$$

and present its algorithms below

**Algorithm 1. Variant of RMIL**

Step 1. With $x_0 \in R^n$, $d_k = -g_k, k = 0$, check if $\| g_k \| \leq \varepsilon$, terminate.

Step 2. Calculate $\beta_k^{SMO}$ using (6).

Step 3. Obtain $d_k$ via (4). If $\| g_k \| \leq \varepsilon$, terminate

Step 4. Calculate $\alpha_k$ via (3).
Step 5. Update $x_k$ via (2).
Step 6. Check if $\| g_k \| \le \varepsilon$ , terminate, else, set $k = k+1$ and start from the beginning.

## 3. Global convergence analysis

For any method to be considered effective and robust, it must satisfy the descent condition and also the convergence criteria. Here, the convergence of the proposed $\beta_k^{SMO}$ scheme will be studied.

### 3.1 Sufficient descent condition

This condtion is true if

$$g_k^T d_k \le -C \| g_k \|^2 \text{ for } k \ge 0 \text{ and } C > 0 \tag{7}$$

The Deduction that follows would be used to show the proposed method possess (7) under exact minimization criterion.

**Theorem 1**
Given a CG algorithm 1, then the sufficient descent property (7) is true $\forall k \ge 0$.

**Proof:** For $k = 0$, it's obvious $g_0^T d_0 = -C \| g_0 \|^2$. Thus, (7) holds. Next is to prove (7) also holds for $k \ge 1$. From search direction in equation (4), it implies that

$$d_{k+1} = -g_{k+1} + \beta_{k+1}^{SMO} d_k$$

Multiplying through (4) by $g_{k+1}$ gives

$$g_{k+1}^T d_{k+1} = g_{k+1}^T(-g_{k+1} + \beta_k^{SMO} d_K) = -\| g_{k+1} \|^2 + \beta_{k+1}^{SMO} g_{k+1}^T d_k \tag{8}$$

But $g_{k+1}^T d_k = 0$ [10]. Therefore, (8) becomes

$$g_{k+1}^T d_{k+1} = -\| g_{k+1} \|^2 .$$

and thus, (7) holds true. This completes the prove. ∎

### 3.2 Global convergence properties

We need the simplification of our coefficient $\beta_k^{ASO}$ to make the global convergence proof easier

$$\beta_k^{SMO} = \frac{\|g_k\|^2 - \frac{\|g_k\|}{\|g_{k-1}\|}\left|g_k^T d_{k-1}\right|}{d_{k-1}^T(d_{k-1}^T - g_k)} = \frac{\|g_k\|^2 - \frac{\|g_k\|}{\|g_{k-1}\|}\left|g_k^T d_{k-1}\right|}{\|d_{k-1}\|^2 - d_{k-1}^T g_k} \tag{9}$$

Since $g_k^T d_{k-1} = 0$, then (9) reduces to

$$0 \le \beta_k^{SMO} \le \frac{\|g_k\|^2}{\|d_{k-1}\|^2} \tag{10}$$

Now, we need the assumptions below for the convergence analysis of these types of algorithms.

**Assumption A.** On the level set $\Omega = \{x \in R^n / f(x) \le f(x_0)\}$, the function $f(x)$ is bounded from below.

**Assumption B.** The function $f$ is smooth and its gradient $g(x)$ is Lipchitz continuous in some neighbourhood $N$ of $\Omega$, such that $\exists L > 0$ (constant) which satisfy;

$$\|g(x) - g(y)\| \leq L\|x - y\| \ \forall x, y \in N.$$

The following important lemma follows from Assumptions A and B, whose proof can be found in Zoutendijk [5].

**Lemma 1**

Assume Assumption A and B holds, for a given algorithm defined by algorithm 1. Then, the Zoutendijk assertion [5] that follow is true

$$\sum_{k=0}^{\infty} \frac{(g_k^T d_k)^2}{\|d_k\|^2} < \infty \tag{11}$$

Based on the above lemma and Assumption B, we have the convergence theorem as follows

**Theorem 2**

For any given CG proccess defined using algorithm 1, let Assumptions A and B holds, at this point

$$\lim_{k\to\infty} \| g_k \| = 0 \ \text{or} \ \sum_{k=0}^{\infty} \frac{(g_k^T d_k)^2}{\|d_k\|^2} < \infty \tag{12}$$

**Proof.**

Rewriting search direction equation in (4), gives

$$d_{k+1} + g_{k+1} = \beta_{k+1}^{SMO} d_k,$$

If we square through the above equation, we have

$$\|d_{k+1}\|^2 = \left(\beta_{k+1}^{SMO}\right)^2 \|d_k\|^2 - 2g_{k+1}^T d_{k+1} - \|g_{k+1}\|^2 \tag{13}$$

Substitute (10) in (13), will give

$$\|d_{k+1}\|^2 = \left(\frac{\|g_{k+1}\|^2}{\|d_k\|^2}\right)^2 \|d_k\|^2 - 2g_{k+1}^T d_{k+1} - \|g_{k+1}\|^2$$

$$\|d_{k+1}\|^2 = \frac{\|g_{k+1}\|^4}{\|d_k\|^2} - 2g_{k+1}^T d_{k+1} - \|g_{k+1}\|^2 \tag{14}$$

Considering (7), it follows

$$g_{k+1}^T d_{k+1} = -C\|g_{k+1}\|^2$$

and thus, (14) becomes

$$\|d_{k+1}\|^2 = \frac{\|g_{k+1}\|^4}{\|d_k\|^2} + 2c\|g_{k+1}\|^2 - \|g_{k+1}\|^2$$

$$\|d_{k+1}\|^2 = \frac{\|g_{k+1}\|^4}{\|d_k\|^2} - \|g_{k+1}\|^2 (1 - 2c) \tag{15}$$

Multiplying through (15) by $\frac{\|g_{k+1}\|^2}{\|d_{k+1}\|^2}$ gives

$$\|d_{k+1}\|^2 \frac{\|g_{k+1}\|^2}{\|d_{k+1}\|^2} = \frac{\|g_{k+1}\|^2}{\|d_{k+1}\|^2}\left(\frac{\|g_{k+1}\|^4}{\|d_k\|^2} - \|g_{k+1}\|^2(1-2c)\right)$$

$$\frac{\|d_{k+1}\|^2\|g_{k+1}\|^2}{\|d_{k+1}\|^2} = \frac{\|g_{k+1}\|^4}{\|d_{k+1}\|^2}\left((2c-1) + \frac{\|g_{k+1}\|^2}{\|d_k\|^2}\right)$$

$$\frac{\|d_{k+1}\|^2\|g_{k+1}\|^2}{\|d_{k+1}\|^2} \le \frac{\|g_{k+1}\|^4}{\|d_{k+1}\|^2} \tag{16}$$

which implies $\lim_{k\to\infty} \frac{\left(g_{k+1}^T d_{k+1}\right)^2}{\|d_{k+1}\|^2} < 0$ from Lemma 1.

If Theorem 2 is not true, then, it will imply that $\lim_{k\to\infty} \frac{\left(g_{k+1}^T d_{k+1}\right)^2}{\|d_{k+1}\|^2} = \infty$, using (16), which gives

$\infty \le \frac{\|g_{k+1}\|^4}{\|d_{k+1}\|^2}$. Therefore, for sufficiently large $k$, Theorem 2 holds true. ■

## 4. Numerical results

This subdivision is the outcome of the computational experiment from iteration number and CPU time. Most of the benchmark problems used in the analysis are from Andrei [12], as shown in Table 1. The terminating condition is set as $\| g_k \| < 10^{-6}$. Four initial guesses are considered for each problem starting with guess close to the solution point to points far away. All problems are solved on MATLAB (R 2015b) programming. The results are analysed by Dolan and More [3] performance profile as presented in Figure 1 and Figure 2. In the performance profile, they evaluated the set of solvers $S$ on performance $P$ which is the test problems.

**Table 1**: List of Benchmark problems

| No | Functions | DIM | Initial guess |
|---|---|---|---|
| 1 | Booth | 2 | (50, 50), (100, 100), (10, 10), (25, 25) |
| 2 | Three-Hump Camel | 2 | (21, 21), (23, 23), (13, 13), (17, 17) |
| 3 | Six Hump Camel | 2 | (10, 10), (-10, -10), (8, 8), (-8, -8) |
| 4 | Treccani | 2 | (50, 50), (100, 100), (5, 5), (7, 7). |
| 5 | Ext Freud and Roth | 2, 4 | (20, -20, -20), (33, -33, 33), (0, 0), (10,10) |
| 6 | Dixon and Price | 2, 4 | (5, 5), (13, 13), (30, 30), (33, 33) |
| 7 | Generalized Trig 1 | 2, 4, 100 | (33, 33), (135,135), (15, 15), (23,23). |
| 8 | Ext Maratos | 2, 4, 10, 100 | (25,25), (55,55), (15, 15), (16, 16 ,...16). |
| 9 | Hager | 2, 4, 10, 100 | (22, 22, ..., 22), (-23, 23, ...,23), (-1, -1, ..., -1), (21, 21, ..., 21). |
| 10 | Ext Penalty | 2, 4, 10, 100 | (135,135,...,135),(200,200,...,200), (100, 100,...,100),(105,105,...,105). |
| 11 | Quadratic Penalty QP2 | 2, 4, 10, 100 | (-15, 15, ..., 15), (-25, 25, ..., 25), (-11, 11, ..., 11), (13, 13, ..., 13) |

| | | | |
|---|---|---|---|
| 12 | Ext Beale | 2, 4, 10, 100 | (5, 5, …,5), (100, 100, …, 100), (2, 2, …, 2), (3, 3, …, 3). |
| 13 | Diagonal 4 | 2, 4, 10, 100,500, 1000 | (10, 10, …, 10), (15, 15, …, 15), (2, 2, …, 2), (5, 5, …,5). |
| 14 | Ext Tridiagonal 1 | 2, 4, 10, 100,500, 1000 | (20, 20, …, 20),(30, 30, …, 30), (12, 12, …, 12), (17, 17, …,17). |
| 15 | Generalized Quartic | 2, 4, 10, 100,500, 1000, 10,000 | (3, 3, …, 3), (5, 5, …, 5), (0.5, 0.5, …, 0.5), (1, 1, …, 1). |
| 16 | Rosenbrock | 2, 4, 10, 100,500, 1000, 10,000 | (13, 13, …, 13), (24, 24, …,24), (31, 31, …, 31), (35, 35 , …, 35) |

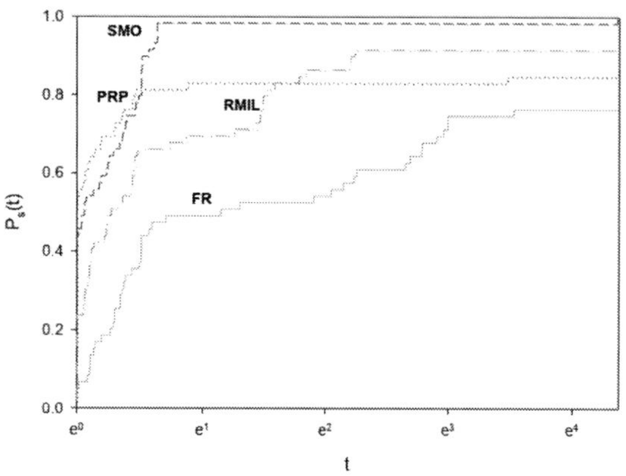


**Figure 1**: Performance analysis for iteration number

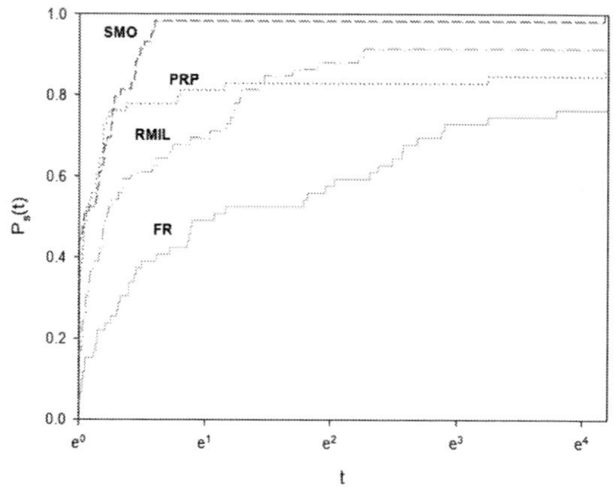


**Figure 2**: Performance analysis for CPU time

From the above Fig 1 and Fig 2, it has been clearly revealed the new SMO algorithm performed efficiently both in terms of CPU time and iteration number by solving the whole test functions successful with 100% success. On the other hand, FR, PRP, and RMIL methods respectively solved 76%, 86%, and 91% of the functions. Thus, deduce that the proposed SMO algorithm is more efficient that the existing methods used in the analysis.

## 5. Conclusion

Recently, several researches have modified the CG algorithms. This paper presented an improved modification of the CG method base on modification of RMIL coefficient. The obtained computational results illustrate that the new algorithm is more reliable and competitive compare to methods of FR, PRP, and AMRI. The exact minimization condition was applied for the study the global convergence of SMO. In future, we intend to prove that the method also converge globally under other line searches.

## References

[1] Polyak B T 1969 The conjugate gradient method in extreme problems, USSR *Comp. Math. Phys.***9** 94-112
[2] Touati-Ahmed D and Storey C 1990 Efficient hybrid conjugate gradient techniques *J. Optim. Theory Appl* **64** 379–397

[3] Dolan E and More J J 2002 Benchmarking optimization software with performance profile, *Math. Prog.* **91**, 201–213,.
[4] Polak E and Ribière G 1969 Note sur la convergence de directions conjuguée, *Rev. Francaise Informat Recherche Operationelle* **16** 35-43
[5] Zoutendijk G 1970 Nonlinear programming, computational methods, in Integer and Nonlinear Programming *J. Abadie, ed., North- Holland, Amsterdam* 37-86
[6] Gilbert J C and Nocedal J 1992 Global convergence properties of conjugate gradient methods for optimization *SIAM. J. Optim.* **2** 21-42
[7] Al-Baali M 1985 Descent property and global convergence of the Fletcher-Reeves method with inexact line search *IMA J. Numer. Anal.* **5** 121-124
[8] Powell M J D 1986 Convergence properties of algorithm for nonlinear optimization *SIAM Rev.* **28** 487-500
[9] Hestenes M R and Stiefel E L 1952 Methods of conjugate gradients for solving linear systems *J. Research Nat. Bur. Standard* **49** 409-436
[10] Rivaie M, Mamat M, Leong W J and Ismail M 2012 A new Conjugate gradient coefficientfor large scale Nonlinear unconstrained optimization *Int.Journal of Math.Analysis* **6** (23) 1131-1146
[11] Rivaie M, Abashar A, Mamat M and Mohd I 2014 The convergence properties of a new type of conjugate gradient methods *Appl. Math. Sci.* **8** 33-44
[12] Andrei N 2008 An unconstrained optimization test functions collection *Advanced Modelling and optimization* **10** (1) 147-161
[13] Andrei N 2009 Accelerated conjugate gradient algorithm with finite difference Hessian / vector product approximation for unconstrained optimization *J. comput. Appl. Math* **30** 361–369
[14] Sulaiman I M, Mamat M, Abashar A, Rivaie M, Mohamad F S and Saleh Z 2015 The global convergent properties of an improved conjugate gradient method *Applied Mathematical Science* **9** (38) 1857-1868
[15] Fletcher R 1987 *Practical Method of Optimization Unconstrained Optimization* New York: John Wiley & Sons
[16] Fletcher R and Reeves C 1964 Function minimization by conjugate gradients *Comput. J.* **7** 149-154
[17] Olowo S E, Abudulhadi A A and Sulaiman I M 2016 An improved version of conjugate gradient Algorithms with global convergence properties for unconstrained optimization.*International journal of Engineering and Research Technology* 75-86
[18] Sulaiman I M and Mamat M 2020 A New Conjugate Gradient Methods with Descent Properties and its Application to Regression analysis *Journal of Numerical Analysis, Industrial and Applied Mathematics* **14** (3-4) 1-15
[19] Hager W W and Zhang H C 2005 A new conjugate gradient method with guaranteed descent and efficient line search *SIAM J. Optim.* **16** 170–192
[20] Dai Y H and Yuan Y 1999 A nonlinear conjugate gradient with a strong global convergence properties *SIAM J. Optim.* **10** 177-182
[21] Liu Y and Storey C 1992 Efficient generalized conjugate gradient algorithms part 1: Theory *J. Comput. Appl. Math.* **69** 129-137
[22] Yuan Y and Sun W 1999 *Theory and methods of optimization* Beijing: Science Press

# Fertilizer Transportation Problem Using Vogel Approximation Method

**R H Kankarofi[1], U AYakubu[1,*], I M Sulaiman[2], M Mamat[2], Sukono[3], and M P A Saputra[4]**

[1]Department of Mathematics, Yusuf Maitama Sule University, Kano, Nigeria
[2]Faculty of Informatics and Computing, Universiti Sultan Zainal Abidin, Campus Kuala Nerus, 22200, Terengganu, Malaysia.
[3]Department of Mathematics, Faculty of Mathematics and Natural Science, Universitas Padjadjaran, Indonesia
[4]Master Program in Mathematics, Faculty of Mathematics and Natural Science, Universitas Padjadjaran, Indonesia

Corresponding author: usman.abbas84@yahoo.com

**Abstract.** This research work was designed to optimize the processes of transporting fertilizer from Kano State Agricultural Supply Company, KASCO, Maiduguri Road, Kano, also implement the distribution route that lead to the reduction of operating cost of transportation problems that usually occurs due to the nature of bad roads to some selected local government in Kano state Nigeria. To minimize the cost of shipping fertilizer from the company to the destinations a mathematical optimization model (Vogel Approximation method) has been used to ease the cost and the distribution of fertilizer to the various locations.

## 1. Introduction

The transportation problem was first formulated and developed by [6], and solved optimally for complex business problem. The transportation problem usually appears in a transportation tableau and often described by means of Linear Mathematical Programming model. By utilizing the classical methods such as Least-Cost method, Simplex method or North-West corner method, researchers are optimistic that the initial basic feasible solution of any transportation problem can be obtained [8]. Finally, in this research the optimality of the given transportation problem is verified by Vogel Approximation method.

A transportation problem includes $m$ sources, in which each of the source contain available $S_i$ where $i = 1, 2, 3, 4, \ldots, m$ unit of a homogeneous products with $n$ dimensions, each of which requires $d_j$ where $j = 1, 2, 3, 4, \ldots, n$ units for some positive integers $S_i$ and $d_j$. The cost $C_{ij}$ used to transport a unit of product to the $j^{th}$ destination from the source $i^{th}$ is given for every $i$ and $j$. The main aim is developing an integral transportation plan that would satisfy all demands at minimum total shipping cost which is considered from the current inventory. It is supposed that total demand and total supply are equal, i.e.

$$\sum_{j=1}^{n} d_j = \sum_{i=1}^{m} S_i \quad (1)$$

Equation (1) is certain by forming either a false destination with a demand equivalent to the supply, if the overall demand is less than total supply, or a fictions source with a supply equivalent to the shortage, or with the whole demand exceeding the total supply. The problem of transportation often arises when planning for the distribution of goods and services to different destinations which is originating from different supply locations.

## 2. Method for transportation problems

The mathematical models have been discovered to minimize the cost of transportation. This can be seen as a special case of the minimum cost flow problem. [2] said that the application of the transportation problem tends to involve a very large number of variable and constraints, thus, a straight forward simplex method computer application may need an excessive computation effort, for instance, George B. Dantzing (1951) solved the transportation model using the idea of Linear Programming [1,6]. Beamo [1] present the simplex method standard for solving linear programming formulation of transportation problem. Since then the problem has the classical common literature in most text. This demonstration is regards as among the first significant contribution for solving the transportation problem. Hakim [9], studied independently, which was not related to Koopmans (1947), and referred to it as "the transportation system optimum utilization". These contributions aided in developing the methods of transportation methods involving a number of number destinations and a number of shipping source. Each unit transportation cost function is a linear in nature. Frank [12] constructed an algorithm and use it to reach the transportation problem optimal solution for the convex case. Fortunately, a key characteristic is most of the $X_{ij}$ coefficient in the constraints are zeros and relatively few none-zero coefficients give the idea in a distinctive pattern [2, 3, 4, 5, 7, 10, 11, 13].

The general model for the transportation problem from source to the destinations is to minimize the total cost, i.e. sources $i = 1,2,3, \dots, m$ has demanded for $d_j$ being received from the source $i$ to destination $j$ this process is directly proportional to the number distributed, with $C_{ij}$ denoting the per unit distribution cost, and letter $Z$ is the cost of total distribution and $X_{ij}$ for $i = 1,2,3,4,\dots,m,$ $and\, j = 1,2,3,4,\dots,n$, represent the units number to be distributed to destination $j$ from source $i$, the problem of linear programming is as follows:

$$minimize = \sum_{j=1}^{n}\sum_{i=1}^{m} C_{ij}X_{ij} \tag{2}$$

$$subject\ to \sum_{j=1}^{n} X_{ij} = S_i \qquad for\, j = 1,2,3,\dots,m$$

$$and \sum_{i=1}^{m} X_{ij} = d_j \qquad for\, i = 1,2,3,\dots,n$$

$for\ all\ X_{ij}$ non – negative, for all $i$ and $j$.

### 2.1. Northwest Corner Method

Commencing by the transportation tableau upper left corner and the set $X_{11}$ being as large as possible, then, obviously, $X_{11}$ can no more be larger than the smaller of $S_i$ and $d_j$.

- If $X_{11} = S_1$, then, crossing out the tableau first row is required. More, there is need for changing $d_1$ to $d_1 - S_1$.
- If $X_{11} = d_1$, then, crossing out the tableau first column is required. Also, there is need for changing $S_1$ to $S_1 - d_1$.
- If $X_{11} = S_1 = d_1$, then, crossing out either column 1 or row 1 but not both of them is needed.
    - By crossing out row, then, change $d_1$ to 0.
    - By crossing out the column, then, you change $S_1$ to 0.

This procedure is continuously applied to the most tableau Northwest cells not laying in a crossed-out column or row. Ultimately, a continuous process of this procedure will arrive at certain point with only one cell which can be assigned a value. Assigning the last cell, a value equivalent to its column or row demand, and crossing-out these cell's column or row, produce a basic formulating solution.

## 2.2. Simplex Method

This method was summarized by the following steps:

- If the have an unbalance problem, then, balance the problem first.
- Employ any of the existing methods to find a basic formulating solution.
- For all basic variables, apply the fact that $U_1 = 0$ and $U_i + V_j = C_{ij}$ to obtain the $U^{,S}$ and $V^{,S}$ for current basic formulation solution.
- Suppsoe $U_i + V_j - C_{ij} \leq 0 \,\forall$ non-basic variables, then the current basic formulation solution is optimal. If this is not the case, then enter the variable with the most positive $U_i + V_j - C_{ij}$ into the basic using the pivoting procedure. This yield new basic formulating solution, then return to step three for a maximization problem. Continue as specified, but replacing the fourth step by the step that follows.
- If $U_i + V_j - C_{ij} \geq 0, \forall$ nonbasic variables, then the present basic formulating solution is optimal, else, input the variable with negative $U_i + V_j - C_{ij}$ into the base sign of the pivoting procedure. This yield a new basic formulating solution. Then return to step three.

## 2.3. Vogel Approximation Method

Calculating each column and row, a penalty equivalent to difference between the two least costs in column and row. Next is finding the column or row having the major penalty, select as the first basic variable, i.e. the variable in this column or row that has the least costs. For the method of Northwest corner, there is need to make this variable as large as possibly, and also crossing out column or row, change the demand or supply associated with the basic variable. Now re-compute new penalties and repeating the procedure continuously until only one uncrossed cell remains. Set this variable equal to the supply or do demand associates with the variable and cross out the variable's row and column [10].

# 3. Data Analysis

The data was collected from the marketing department of Kano State Agricultural and Supply Company (KASCO) Maiduguri road Kano. The request on the fertilizer to various destinations are always very high nowadays due to closure of Nigerian land border. This research will focus on six (6) local government area in Kano state which include Kano Municipal, Kura, Tudun Wada, Madobi, Gaya and Wudil). The information obtained are summarized and presented in table 2 and table 3 below. The general Transportation problem is detailed using the information that follows:

- A set of $m$ supply points from which a good is shipped. Supply point $i$ can supply at most $S_i$ units.
- A set of $n$ demand points to which the good is shipped. Demand point $j$ must receive at least $d_j$ units of the shipped good.
- Each unit product at supply point $i$ and shipped to demand point $j$ gives available cost of $C_{ij}$.

The relevant data can be formulated in a transportation in Table 1.

If total supply equals to the total demand, then the problem is said to be a balanced transportation problem. Let $X_{ij}$ be a number of units shipped from supply point $i$ to demand point $j$ then the general programming representation of a transportation problem is

$$minimize = \sum_{j=1}^{n} \sum_{i=1}^{m} C_{ij} X_{ij}$$

$$such\ that \sum_{j} X_{ij} \leq S_i \ (i = 1,2,3, \dots, m) \rightarrow supply\ constraints$$

$$and \sum_{i} X_{ij} \geq d_j \ (j = 1,2,3, \dots, n) \ \rightarrow demand\ constraints$$

where $X_{ij} \geq 0$

**Table 1**: Transportation Tableau

| | Demand 1 | Demand 2 | ... | ... | Demand $n$ | Supply |
|---|---|---|---|---|---|---|
| Supply 1 | $C_{11}$ | $C_{12}$ | ... | ... | $C_{1n}$ | $S_1$ |
| Supply 2 | $C_{21}$ | $C_{22}$ | ... | ... | $C_{2n}$ | $S_2$ |
| Supply 3 | ... | ... | ... | ... | ... | $S_3$ |
| . | ... | ... | ... | ... | ... | . |
| . | | | | | | . |
| . | $C_{m1}$ | $C_{m2}$ | ... | ... | $C_{mn}$ | . |
| $m$ | | | | | | $S_m$ |
| Demand | $d_1$ | $d_2$ | ... | ... | $d_n$ | |

## 3.1. Data presentation

Total bags of fertilizer supplied from the KASCO company and total bags demanded for the respective local government.

**Table 2** Fertilizer demand by local governments

| Unit | Municipal | Kura | Wudil | Madobi | Tudun Wada | Gaya | Total Supply |
|---|---|---|---|---|---|---|---|
| 1 | 20,100 | 16,700 | 15,000 | 11,700 | 14,800 | 11,000 | 89,300 |
| 2 | 19,300 | 16,000 | 14,000 | 11,200 | 13,800 | 10,100 | 84,400 |
| 3 | 19,300 | 16,100 | 13,600 | 11,400 | 13,700 | 10,300 | 84,400 |
| 4 | 19,000 | 16,100 | 14,100 | 11,900 | 13,900 | 10,400 | 85,400 |
| 5 | 19,700 | 16,000 | 14,600 | 11,000 | 14,600 | 10,500 | 86,400 |
| **Total Demand** | 97,400 | 80,900 | 71,300 | 57,200 | 70,800 | 52,300 | |

**Table 3.** Cost of transportation per bag in naira (₦)

| Unit | Municipal | Kura | Wudil | Madobi | Tudun Wada | Gaya |
|---|---|---|---|---|---|---|
| 1 | 20 | 30 | 39 | 31 | 50 | 51 |
| 2 | 25 | 36 | 45 | 35 | 55 | 56 |
| 3 | 26 | 35 | 47 | 36 | 54 | 54 |
| 4 | 21 | 28 | 39 | 29 | 51 | 50 |
| 5 | 18 | 31 | 40 | 29 | 50 | 49 |

The Method of generating optimal solution considered in this research work is Vogel approximation method. Considering Table 4

**Table 4.** Destination Table 1

| | 1 | 2 | 3 | 4 | ... | n | Supply | $U_i$ |
|---|---|---|---|---|---|---|---|---|
| 1 | $C_{11}$ $X_{11}$ | $C_{12}$ $X_{12}$ | $C_{13}$ $X_{13}$ | $C_{14}$ $X_{14}$ | ... ... | $C_{1n}$ $X_{1n}$ | $S_1$ | $U_1$ |
| 2 | $C_{21}$ $X_{21}$ | $C_{22}$ $X_{22}$ | $C_{23}$ $X_{23}$ | $C_{24}$ $X_{24}$ | ... ... | $C_{2n}$ $X_{2n}$ | $S_2$ | $U_2$ |
| 3 | $C_{31}$ $X_{31}$ | $C_{32}$ $X_{32}$ | $C_{33}$ $X_{33}$ | $C_{34}$ $X_{34}$ | ... ... | $C_{3n}$ $X_{3n}$ | $S_3$ | $U_3$ |
| ⋮ | ⋮ ... | ⋮ ... | ⋮ ... | ⋮ ... | ... ... | ⋮ ... | ⋮ | ⋮ |
| $m$ | $C_{m1}$ $X_{m1}$ | $C_{m2}$ $X_{m2}$ | $C_{m3}$ $X_{m3}$ | $C_{m4}$ $X_{m4}$ | ... ... | $C_{mn}$ $X_{mn}$ | $S_m$ | $U_m$ |
| $V_j$ | $V_1$ | $V_2$ | $V_3$ | $V_4$ | ... | $V_n$ | | |

where,

- $C_{ij}$ is the cost of transporting a unit of the products from $i^{th}$origin to $j^{th}$destination.
- $X_{ij}$ represents the unknown number of units to be shipped from source $i$ to destination $j$.
- $U_i$ is the quantity of the commodity available at source $i$.
- $V_j$ is the quantity of the commodity needed at destination $j$.
- Basic variables can be found by using $C_{ij} = U_i + V_j$ and non-basic variable by $X_{ij} = C_{ij} - (U_i + V_j)$

Therefore, the optimal solution $Z$ can be calculated using

$$minimize = \sum_{j=1}^{n}\sum_{i=1}^{m} C_{ij} X_{ij}$$

where, $X_{11} = 11{,}000$, $X_{13} = 71{,}300$, $X_{15} = 7{,}000$, $X_{24} = 48{,}200$, $X_{25} = 36{,}200$, $X_{35} = 27{,}600$, $X_{36} = 52{,}300$, $X_{42} = 80{,}900$, $X_{44} = 4{,}500$, $X_{51} = 86{,}400$

$$Z = C_{11}X_{11} + C_{13}X_{13} + C_{15}X_{15} + C_{24}X_{24} + C_{25}X_{25} + C_{35}X_{35} + C_{36}X_{36} + C_{42}X_{42} + C_{44}X_{44} + C_{51}X_{51}$$

$$= 20(11{,}000) + 39(71{,}300) + 50(7{,}000) + 35(52{,}700) + 55(31{,}700) + 54(32{,}100) + 54(52{,}300) + 28(80{,}900) + 29(4{,}500) + 18(86{,}400)$$

$$= 220{,}000 + 2{,}780{,}700 + 350{,}000 + 1{,}844{,}500 + 1{,}743{,}500 + 1{,}733{,}400 + 2{,}824{,}200 + 2{,}265{,}200 + 2{,}265{,}200 + 130{,}500$$

$$Z = 15{,}447{,}200 \quad (3)$$

updating the result in naira we have ₦15,447,200.

**Table 5.** Destination Table 2

| | 1 | 2 | 3 | 4 | 5 | 6 | Supply | $U_i$ |
|---|---|---|---|---|---|---|---|---|
| 1 | 20 | 30 | 39 | 31 | 50 | 51 | 89,300 | $U_1$ |
| 2 | 25 | 36 | 45 | 35 | 55 | 56 | 84, 400 | $U_2$ |
| 3 | 26 | 35 | 47 | 36 | 54 | 54 | 84, 400 | $U_3$ |
| 4 | 21 | 28 | 39 | 29 | 51 | 50 | 85, 400 | $U_4$ |
| 5 | 18 | 31 | 40 | 29 | 50 | 49 | 86, 400 | $U_5$ |
| Demand | 97, 400 | 80, 900 | 71, 300 | 57, 200 | 70, 800 | 52, 300 | D=S429,900 | |
| $V_j$ | $V_1$ | $V_2$ | $V_3$ | $V_4$ | $V_5$ | $V_6$ | | |

where $1 \leq i \leq 5$ is the number of row and $1 \leq j \leq 6$ the number of column. Therefore, the total demand is equal to the total supply i.e. D = S = 429,900.

**Table 6.** Row Penalty

| | PEN 1 | PEN 2 | PEN 3 | PEN 4 | PEN 5 | PEN 6 | PEN 7 | PEN 8 | PEN 9 | PEN 10 |
|---|---|---|---|---|---|---|---|---|---|---|
| ROW 1 | 10 | (10) | 1 | 8 | (11) | (1) | - | - | - | - |
| ROW 2 | 10 | 10 | 1 | 10 | 10 | 1 | 1 | 1 | 1 | 55 |
| ROW 3 | 9 | 9 | 1 | (11) | 7 | 0 | 0 | 0 | - | - |
| ROW 4 | 7 | 7 | 1 | 10 | 11 | 1 | 1 | - | - | - |
| ROW 5 | (11) | - | - | - | - | - | - | - | - | - |

**Table 7.** Column Penalty

| | COL 1 | COL 2 | COL 3 | COL 4 | COL 5 | COL 6 |
|---|---|---|---|---|---|---|
| PEN 1 | 2 | 2 | 0 | 0 | 1 | 1 |
| PEN 2 | 1 | 2 | 0 | 2 | 1 | 1 |
| PEN 3 | - | (2) | 0 | 2 | 1 | 1 |
| PEN 4 | - | - | 0 | 2 | 1 | 1 |
| PEN 5 | - | - | 0 | - | 1 | 1 |
| PEN 6 | - | - | - | - | 1 | 1 |
| PEN 7 | - | - | - | - | 3 | (4) |
| PEN 8 | - | - | - | - | 1 | (2) |
| PEN 9 | - | - | - | - | 55 | (56) |
| PEN10 | - | - | - | - | 55 | - |

where, PEN=penalty and COL=column. From table 5 above we have

- Row 5 has the highest penalty and 18 is the smallest cost, therefore we assign 86,400 to it and cancel row 5.
- Row 1 has the highest penalty also 20 is the smallest cost, so, we assign 11,000 and cancel column 1, since it is balanced.
- Column 2 has the highest penalty and 28 is the smallest cost, therefore, we assign 80,900 and cancel column 2.
- Row 3 has the highest penalty and 36 is the smallest cost, therefore, we assign 57,200 and cancel column 4.
- Row 1 has the highest penalty and 39 is the smallest cost, therefore, we assign 71,300 and cancel column 3.
- Row 1 has the highest penalty and 50 is the smallest cost, again we assign 7,000 and cancel row 1.
- Column 6 has the highest penalty and 50 is the smallest cost, now we have to assign 4,500 and cancel row 4.
- Column 6 has the highest penalty and 54 is the smallest cost. Therefore, put 27,200 and cancel row 3.
- Column 6 has the highest penalty and assigns 20,600 to 56 cells since it is the only remaining cell in that column, then cancel column 6.
- The only remaining cell is 55, so, we assign 63,800 and cancel both row 2 and column 5.

## 4. Result and Discussion

In this section, we clearly discuss the cost of transportation of fertilizer from the KASCO company to the required destinations. Considering the table 4 Destination table 1 after assigned the values, it indicated that the fertilizer distribution profile is as follows:

- 11,000 bags of fertilizer from unit (1) should be sent to Municipal, 71,300 bags should be sent to Wudil and 7,000 bags should be sent to Tudun Wada while no any bag of fertilizer should be sent to Kura, Madobi, or Gaya local government.
- From unit (2) 52,700 bags should be sent to Madobi, 31,700 bags to Tudun Wada, and no any bag should be sent to Municipal, Kura, Wudil and Gaya local government.
- In the unit (3) 32,100 bags of fertilizer should be sent to Tudun Wada, 52, 300 to Gaya, but no any bag to Municipal, Kura, Wudil and Madobi local government.
- From unit (4) 80,900 bags should be sent to Kura, 4,500 bags should be sent to Madobi, and no any bag to Municipal, Gaya, Tudun Wada and Wudil local government.
- Unit (5) should send 86,400 bags of fertilizer to Municipal but no any bag to the remaining local governments.

Therefore, from table 2 and table 3. Initially, the cost of transportation was as follows:

$$\begin{aligned} Initial\ Cost = {} & (20{,}100 \times 20) + (16{,}700 \times 30) + (15{,}000 \times 39) + (11{,}700 \times 31) + (14{,}800 \\ & \times 50) + (11{,}000 \times 51) + (19{,}300 \times 25) + (16{,}000 \times 36) + (14{,}000 \times 45) \\ & + (11{,}200 \times 35) + (13{,}800 \times 55) + (10{,}100 \times 56) + (19{,}300 \times 26) + (16{,}100 \\ & \times 35) + (13{,}600 \times 47) + (11{,}400 \times 36) + (13{,}700 \times 54) + (10{,}300 \times 54) \\ & + (19{,}000 \times 21) + (16{,}100 \times 28) + (14{,}100 \times 39) + (11{,}900 \times 29) + (13900 \\ & \times 51) + (10{,}400 \times 50) + (19{,}700 \times 18) + (16{,}000 \times 31) + (14{,}600 \times 40) \\ & + (11{,}000 \times 29) + (14{,}600 \times 50) + (10{,}500 \times 49 \\ & = 15{,}939{,}500 \end{aligned} \tag{4}$$

The cost of transportation was amount to ₦15,939,500. Nevertheless, the Vogel approximation method generally gives a better initial solution. Now, by comparing the total cost of transportation from the optimal solution equation (3) and the initial cost of transportation equation (4) we observe that there is discount of **₦492,300**. Table 8 shows the number of bags of fertilizer demanded from selected local government in Kano state and the number of bags that should be sent from the KASCO Company after using Vogel approximation method of transportation.

**Table 8.** Demand before and after

| Local Government | Number of Bags Demanded | Number of Bags Supply | Total |
|---|---|---|---|
| Municipal | 20,100 from unit (1), 19,300 from unit (2), 19,300 from unit (3), 19,000 from unit (4) and 19,700 from unit (5). | 11,000 from unit (1) and 86,400 from unit (5) | 97,400 |
| Kura | 16,700 from unit (1), 16,000 from unit (2), 16,100 from unit (3), 16,100 from unit (4) and 16,000 from unit (5). | 80,900 from unit (4) | 80,900 |
| Wudil | 15,000 from unit (1), 14,000 from unit (2), 13,600 from unit (3), 14,100 from unit (4) and 14,600 from unit (5). | 71,300 from unit (1) | 71,300 |
| Madobi | 11,700 from unit (1), 11,200 from unit (2), 11,400 from unit (3), 11,900 from unit (4) and 11,000 from unit (5). | 52,700 from unit (2) and 4,500 from unit (4) | 57,200 |
| Tudun Wada | 14,800 from unit (1), 13,800 from unit (2), 13,700 from unit (3), 13,900 from unit (4) and 614,600 from unit (5). | 7,000 from unit (1), 31,700 from unit (2) and 32,100 from unit (3) | 70,800 |
| Gaya | 11,000 from unit (1), 10,100 from unit (2), 10,300 from unit (3), 10,400 from unit (4) and 10,500 from unit (5). | 52,300 from unit (3) | 52,300 |

Nonetheless each and every local government get the same number of bags demanded from the company, but only units have been changed for some local government in order to reduce the cost of transportation. Municipal demand 97,400 bags of fertilizer which will cost the company ₦2,139,900 but after using this method their demand change to 11,000 bags from unit (1) and 86,400 from unit (5) which cost ₦1,775,200. Kura demand 80,900 bags which will cost ₦2,587,300 if their demand has been change to 80,900 bags form unit (4) then the price will have reduced to ₦2,265,200.

Wudil demand 71,300 which will cost ₦2,988,100, now their demand changes to 71,300 from unit (1) which cost ₦2,780,700. Madobi demand 57,200 bags which will cost ₦1,829,200 but after using Vogel approximation method their demand changed to 52,700 bags from unit (2) and 4,500 bags from unit (4) which cost ₦1,975,000. Tudun Wada demanded 70,800 bags which will cost ₦3,677,700 but if their demand has been change to 7,000 bags from unit (1), 31,700 bags from unit (2) and 2,100 bags from unit (3) the transportation cost was increased to ₦3,826,900.

Lastly, Gaya local government demanded 52,300 bags which will cost ₦2,717,600 but if their demand change to 52,300 bags from unit 3 then the cost of transportation becomes ₦2,824,200.

The Table 9 shows the cost of transportation of fertilizer demanded before and after using Vogel's method.

Now we observed that in order to minimize the cost of transportation and distribution of fertilizer by using Vogel approximation method, we have to change some request and increase others without reducing the demand of each and every local government.

**Table 9:** Cost before and after

| Local Government | Transportation Cost Before | Transportation Cost After |
|---|---|---|
| Municipal | ₦2,139,900 | ₦1,775,200 |
| Kura | ₦2,587,300 | ₦2,265,200 |
| Wudil | ₦2,988,100 | ₦2,780,700 |
| Madobi | ₦1,829,200 | ₦1,975,000 |
| Tudun Wada | ₦3,677,700 | ₦3,826,900 |
| Gaya | ₦2,717,300 | ₦2,824,200 |
| **Total** | ₦15,939,500 | ₦15,447,200 |

## 5. Conclusion

This research was carried out in order to obtain the minimum cost of transporting fertilizer from source to the destinations. The KASCO Company was able to know how much they have to spend to supply their customers based on the cost of transportation to minimize the total cost of transport the Vogel approximation method was utilized and found to be efficient and it is determined towards profitability of the company.

## References

[1] Dantzing G B 1951 Application of the simple method to a transformation *Problem chapter* **22** 359 - 373

[2] Beamo BM and Chem V C K 2001Performance analysis of conjoined supply chain *international journal of production research* **39** 3195-3218.

[3] Gupta P K and Man Mohan 1988 *Linear Programming and Theory of Games* New Delhi: Sultan Chand & Sons

[4] Usman A Y, Abdul I, Abba V M and Salisu M 2019 Scale Parameter of a Spectral PRP Conjugate Gradient Method for Unconstrained Optimization *Malaysian Journal of Computing & Applied Mathematics* **2** (1) 34-41

[5] Koopmans T C 1947 Efficient allocation resources *Proceedings of the international statistical conferences* 34-41

[6] Kantorovich L V 1942 Pereshcheni mass *De L'Academic des sciences* **37** 199-201

[7] Usman AY, Mamat M, Mohamad A M, Sukono and Rivaie M 2018 Modification on spectral conjugate gradient method for unconstrained optimization *International Journal of Engineering & Technology* **7** (3.28) 307-311

[8] Hillier F S, Lieberman G J, Bodhibrata N and Preetam B 2014 *Introduction to operations research (Mathematics in science and engineering)* New Delhi: Sultan Chand & Sons

[9] Hakim M A 2012 An Alternative Method to Find Initial Basic Feasible Solution of a Transportation Problem *Annals of Pure and Applied Mathematics* **2** (1) 203-209

[10] Goyal S K 1984 Improving VAM for unbalanced transportation problems *Journal of Operational Research Society* **35** (12) 1113-1114

[11] Sulaiman I M, Usman A Y and Mamat M 2020 Application of Spectral Conjugate Gradient Methods for Solving Unconstrained Optimization Problems *International Journal of Optimization and Control: Theories & Applications* **10** (2) 198-205

[12] Frank S J 1970 Decomposition algorithm for mul-tifacility production transportation problem with nonlinear production costs *Economist* **38** (3) 112-122
[13] Usman A Y and Uğur Y 2015 Necessary and Sufficient Conditions for First Order Differential Operators to be Associated with a Disturbed Dirac Operator in Quaternionic Analysis *Adv. Appl. Clifford Alg.* **25** 1-12

# An Accelerated Shamanskii-like Scheme for Some Conservative problems

**I M Sulaiman[1,*], M Mamat[1], M Y Waziri[2], M A Alomari[1], A O Umar[3], M Malik[1], Sukono[4], and S H Yuningsih[5]**

[1]Faculty of Informatics and Computing, Universiti Sultan Zainal Abidin, Terengganu, Malaysia.
[2]Department of Mathematics, Faculty of Physical Sciences, Bayero University, Kano, Nigeria.
[3]Department of Mathematics, College of Agriculture Zuru, Kebbi State, Nigeria
[4]Department of Mathematics, Faculty of Mathematics and Natural Science, Universitas Padjadjaran, Indonesia
[5]Master Program in Mathematics, Faculty of Mathematics and Natural Science, Universitas Padjadjaran, Indonesia

*Corresponding author: kademi4u@yahoo.com

**Abstract.** In this paper, we studied some conservative problems, and constructed a low memory approach to its solution. By means of component-wise approximation approach, a diagonal update to the Jacobian matrix was derived. Numerical examples are given to illustrate the proposed scheme.

## 1. Introduction

Some practical and real-life examples of nonlinear problems which include; distribution of water in the pipeline, equation of radiation transfer, and power flow are traced to field of engineering and sciences [1]. These problems can be modelled in the form of a nonlinear equation

$$F(x) = 0, \quad (1)$$

where $F$ a sufficiently smooth map defined on a convex subset $D$ of a Banach space $X$ with values in another Banach space $Y$ and $x^* \in X$ is a solution to

$$F(x^*) = 0. \quad (2)$$

Problems of the form (1) are often solved by variations of numerical methods [2], and the most famous method is known as Newton's method. This is due to its computational efficiency [3]. Starting from an approximation $x_0$, this method generates a sequence of points $\{x_k\}$ that approximate the solution of (1) via

$$x_{k+1} = x_k - F'(x_k)^{-1}F(x_k), \quad (3)$$

which converge quadratically to $x^*$ if the starting approximation $x_0$ is in the neighborhood of solution $x^*$ and $F(x^*)$ is nonsingular, [4, 5]. The major concern about Newton's method is high cost evaluation of the first derivative of $F(x)$ and solving the linear equation for Newton's step $s_k = -F'(x_k)^{-1}F(x_k)$ at every iteration [1, 3].

Over time, a quiet number of alternative approaches that almost entirely avoids the computational cost of Newton's method including; Chord method, Shamanskii method, quasi-Newton, have been proposed. These techniques likewise require the evaluation and also store the approximate Jacobian during the iteration process [6-11]. A diagonal updating scheme, is often applied to overcome the inconvenient of these methods. However, this may reduce the speed of convergence.

In this paper, we proposed a diagonal updating scheme to the Jacobian of Shamanskii method. The component-wise approximation approach was considered from the geometrical interpretation of previous schemes. The attractive feature of this approach is that the method does not require the evaluation of Jacobian matrix and thus, reduce computational cost and storage requirement. In Section 2, the conservative problems are presented, and section 3 present the Shamanskii method. In Section 4, we present the proposed scheme using component-wise approximation approach and in Section 5, we illustrate the numerical performance of proposed method. Finally, conclusion is given in section 6.

## 2. Conservative problems

The second order differential equation of the form

$$\ddot{x} = f(x, \dot{x}, t) \tag{4}$$

with condition $x(t_0)$ and $\dot{x}(t_0)$ is an example of a dynamic system which often arise in the study of motion [12]. Generally, these systems are initial-value problems whose energy are often dissipated through some form of friction. Though, in most circumstances, this dissipation is so negligible that it can be ignored over relatively short periods of time. However, in such cases, the law of conservation of energy is employed. This system can be applied to problems that possess conserved quantities which is often referred to as conservative systems [3,13,14].

Let assume $\rho$ and $\omega$ are arbitrary functions of $\rho(0) = 0$ and $\omega(0) = 0$ respectively. The equation

$$m\frac{d^2x(t)}{dt^2} = -\propto \frac{dt}{dt^2} - \rho(x) \tag{5}$$

can be construed as a nonlinear equation of motion for a mechanical system of unit mass $m$ under the action of applied force $-\rho(x)$ and a damping force $-\propto \frac{dt}{dt}$. Equation (5) which is rewritten as

$$m\frac{d^2x(t)}{dt^2} + \propto \frac{dt}{dt^2} + \rho(x) = 0 \tag{6}$$

is referred to the basic equation of nonlinear mechanics. The main idea of the proposed scheme is to improve the convergence of the existing methods using the diagonal updating technique and to limit the overall complexity which would be applied to a special case of a nonlinear conservative system defined by,

$$m\frac{d^2x(t)}{dt^2} + \rho(x(t)) = 0 \tag{7}$$

with zero damping force and having a boundary condition

$$x(0) = x(1) = 0 \tag{8}$$

and thus, no dissipation of energy. Further discussions on (6), with applications to various problems, please refer to [5,15,16]. Next, we need to solve for $t \in [0, 1]$ in (7) and (8). Transforming equation (7) and (8) into (6), and applying (8), we have

$$\frac{d^2x(t)}{dt^2} + x(t)^{1+p} + Q = 0, p \in [0,1], Q\epsilon R \tag{9}$$

Using the finite-difference approximations, we replace the derivate of (7) and thus, transforming the differential equations into systems of nonlinear equations which is described as follows.

$$x''(t_i) = \frac{x_{i+1} - 2x_i + x_{i-1}}{h^2} \quad i = 1,2, \dots, n-1 \tag{10}$$

where $n$ is the dimension and $h = \frac{1}{n}$. Substitute (10) into (9), it becomes,

$$2x_1 - x_2 - h^2x_1^{p+1} - Qh^2 = 0$$
$$2x_i - x_{i+1} - h^2x_i^{p+1} - x_{i-1} - Qh^2 = 0 \quad i = 1,2, \dots, n-2$$
$$2x_{n-1} - x_n - h^2x_{n-1}^{p+1} - x_{n-2} - Qh^2 = 0$$

Applying (8) we have $x_n = 0$, and substituting in the above systems of equations, we obtain the discretized nonlinear systems described as follows,

$$2x_1 - x_2 - h^2x_1^{p+1} - Qh^2 = 0$$
$$2x_i - x_{i+1} - h^2x_i^{p+1} - x_{i-1} - Qh^2 = 0 \quad i = 1,2, \dots, n-2$$
$$2x_{n-1} - h^2x_{n-1}^{p+1} - x_{n-2} - Qh^2 = 0.$$

Recently, numerous researchers have applied various iterative techniques to solve the conservative problems. This include Hernandez et al. [3] whom applied the secant like method under the restrictions that $n = 10$. However, researchers suggested the extension of this approach to solve large-scale problem for $10 \leq n \leq 5000$. This includes study by Waziri [17]. In this paper, a new diagonal updating Shamanskii algorithm is employed to approximate the solution of the corresponding equations to this problem. This scheme would also be used on large scale problems with the values $p = 0.5$, $Q = 0.25$, $Q = 0.125$, $x_0 = 0$, $x_0 = 0.2$ and $x_0 = 0.3$. The scheme would be able to limit the overall complexity for solving the discretized nonlinear equations which often arise as dimension increases.

## 3. Shamanskii Method

As discussed earlier, the Newton method (3) converge quadratically when the starting approximation is sufficiently near to the root [4,7,23,24]. The Newton's scheme is defined as

$$x_{k+1} = x_k - F'(x_0)^{-1}F(x_k). \tag{11}$$

Its main drawback is computing and storing the Jacobian at each iteration. A method that almost entirely overcome these drawbacks is the Chord scheme. The Chord process computes the Jacobian only once during iteration process [4]. Motivated by this, a multiple pseudo-Newton method due originally to Shamanskii, Shamasnkii [18], was developed, and analyzed in detail by [4,8,19,20,21] proceeds as follows

$$x_{k+\frac{1}{2}} = x_k - F'(x_k)^{-1}F(x_k) \tag{12}$$

$$x_{k+1} = x_{k+\frac{1}{2}} - F'(x_k)^{-1}F(x_{k+\frac{1}{2}}) \tag{13}$$

which can be rewritten as

$$x_{k+1} = x_k - F'(x_k)^{-1}\left[F(x_k) + F\left(x_k - F'(x_k)^{-1}F(x_k)\right)\right] \tag{14}$$

This method is very efficient and reduce the limit of complexity by applying just one Jacobian approximation for more than one pseudo-Newton iterations [21]. The convergence theorem of this method is demonstrated below.

**Theorem** 1. Let $F: D \subset R^n \to R^n$ obey hypotheses H1(2), H2, and H3. Then, we can say that the solution point $x^*$ is an attraction point of the Shamanskii iteration i.e. equation (10), and possesses at least cubic convergence order.

**Proof**. Please refer to Ortega and Rheinboldt [22] for the proof of this theorem.

## 4. Method Formulation

For the Jacobian matrix $F'(x_k)$, we proposed a diagonal approximation say $D_k$ which would be updated at each iteration. i.e.;

$$F'(x_k) \approx D_k \tag{15}$$

Let $U_k = (x_{k+1}) - (x_k)$ and $V_k = (\alpha_{k+1}) - (\alpha_k)$, then by mean value theorem (15) becomes

$$D_{k+1}U_k \approx V_k \tag{16}$$

Since the diagonal matrix $D_{k+1}$ is denoted by $D = diag(d_1, d_2, \dots, d_n)$, then its diagonal element is given as the component of the vector $\frac{V_k}{U_k}$, and (16) becomes

$$d_k^i = \frac{V_k^i}{U_k^i} \tag{17}$$

and

$$D_{k+1} = diag\left(d_k^{(i)}\right) \tag{18}$$

which $i = 1,2,3, \dots, n$, and $k = 0,1,2, \dots, n$.

However, suppose $U_k^i$in (17) equal to zero, then $D_{k+1}$ would be undefined. Therefore, to prevent this, we impose a condition as follows

$$D_{k+1} = \begin{cases} diag\left(\frac{V_k^{(i)}}{U_k^{(i)}}\right), V_k^{(i)} \neq 0\ U_k^{(i)} \neq 0 \text{ for all } i \\ D_k; \text{ otherwise} \end{cases} \tag{19}$$

Therefore, the proposed diagonal Shamanskii method is described as follows.

$$x_{k+1} = x_k - D_k^{-1}\left[F(x_k) + F\left(x_k - D_k^{-1}F(x_k)\right)\right]. \tag{20}$$

## 5. Numerical Results

The solution of discretized nonlinear systems of some conservative problems will be presented here to illustrate the efficiency and robustness of the developed (DSM) algorithm. This is based on number of iterations as well as CPU time with; $\|U_k\| + \|F(x_k)\| \leq 10^{-4}$ as the truncation criteria. The algorithms are coded on MATLAB 7.0 (R2015a) subroutine programming via double precision computer. When the method fails to solve a problem due to insufficient memory or unable to satisfy the truncations criteria, we denote that point with "- " to indicate failure. The computational results are presented in the Table 1.

**Table 1**: Results of Conservative Problem based on Number of iterations with Q=0.125.

| Dimension (n) | Initial guess ($x_0$) | DSM NOI | CPU Time |
|---|---|---|---|
| 10 | (0, 0, …,0) | 3 | 0.0026 |
| 20 | (0, 0, …,0) | 3 | 0.0342 |
| 50 | (0, 0, …,0) | 4 | 0.0396 |
| 100 | (0.2, …,0.2) | 4 | 0.0821 |
| 500 | (0.2, …,0.2) | 4 | 0.4139 |
| 1000 | (0.2, …,0.2) | 4 | 0.6826 |
| 5000 | (0.3, …,0.3) | 5 | 2.3577 |
| 10000 | (0.3, …,0.3) | 5 | 198.611 |

**Table 2:** Results of Conservative Problem based on Number of iterations with Q=0.25.

| Dimensio n (n) | Initial guess ($x_0$) | DSM NOI | CPU time |
|---|---|---|---|
| 10 | (0, 0, …,0) | 3 | 0.0029 |
| 20 | (0, 0, …,0) | 3 | 0.0372 |
| 50 | (0, 0, …,0) | 3 | 0.0412 |
| 100 | (0.2, … ,0.2) | 4 | 0.0883 |
| 500 | (0.2, …,0.2) | 4 | 0.4321 |
| 1000 | (0.2, …,0.2) | 4 | 0.7153 |
| 5000 | (0.3, …,0.3) | 5 | 2.6235 |
| 10000 | (0.3, …,0.3) | 5 | 215.420 |

Shamanskii's method is among the efficient variants of Newton's method [25]. In this paper, we defined a diagonal updating formula to replace the Jacobian of the Shamanskii method. This proposed Shamanskii-like approach is analyzed based on number of iterations, CPU time, storage capacity. Results presented in Tables 1 and Table 2, has shown the performance of the method on the discretized nonlinear problem. While the dimension is large, the proposed DSM method was able to obtain the solution for $10 \leq n \leq 5000$. This is due to the diagonal updating scheme employed which saves a lot of computational cost. Moreover, $x = 0$, $x = 0.2$, and $x = 0.3$ are selected as the initial points due to their closeness to the solution point. Also, the performance results indicate that the proposed method increased sequentially when solving the given problems.

## 6. Conclusion

In this paper, we defined an accelerated shamanskii scheme for solution of conservative problems. This approach employed a diagonal updating scheme that replaces the diagonal in every iteration. The proposed method reduced the computational burden encountered in other numerical methods with low memory storage requirement. The proposed SDM have shown great efficiency in solving large-scale discretized nonlinear problem.

## References

[1] Dennis Jr JE and Schnabel R B 1996 *Numerical method for unconstrained optimization and nonlinear equations* London: SIAM

[2] Chong E K P and Zak S H 2013 *An introduction to optimization*, Wiley series in discrete mathematics and optimization New York: John Wiley
[3] Hernandez M A, Rubio M J and Ezquerro J 2005 A Solving a special case of conservative problems by Secant-like methods *Applied Mathematics and Computation* **169** 926-942
[4] Kelley C T 1986 A Shamanskii-Like Acceleration Scheme for Nonlinear Equations at Singular Roots *Mathematics of Computation* **47** 609-623
[5] Jordan D W and Smith P 2007 *Nonlinear Ordinary Differential Equations: An introduction for Scientists and Engineers* London: Oxford University Press Inc
[6] Broyden C G 1967 Quasi-Newton Methods and their Application to Function Minimization *Mathematics of Computation* **21** (99) 368-38
[7] Gutierrez J M and Hernandez M 1997 A New recurrence relations for Chebyshev method *Appl. Math. Lett,* **10** 63-65
[8] Sulaiman I M, Mamat M, Nurnadiah Z and Puspa L G 2018 Solving Dual Fuzzy Nonlinear Equations via Shamanskii method *International Journal of Engineering Technology* **7** (3) 89-91
[9] Sulaiman I M, Mamat M and Puspa L G 2020 Shamanskii method for solving parameterized fuzzy nonlinear equations *An International Journal of Optimization and Control: Theories & Applications* (Accepted).
[10] Sulaiman I M, Mamat M, Puspa L G, Foziah H M and Kamfa K 2020 An Improved Shamanskii Method for solving Nonlinear Equation. *Journal of Advanced Research in Dynamical & Control Systems* **12** (2) 591-594
[11] Mamat M, Sulaiman I M and Puspa L G 2020 An accelerated scheme for solving parameterized fuzzy nonlinear equations *International Journal of Advanced Science and Technology* **29** (5) 248-255
[12] Stoker J J 1950 *Nonlinear Vibrations* New York: Interscience-Wiley
[13] Xin C *A* 2001 Conservative Difference Scheme for Conservative Differential Equation with Periodic Boundary *Applied Mathematics and Mechanics* **22** (10) 1210-1215
[14] Xin C and Peng-cheng L 1992 Conservative difference scheme of conservative equation with a small parameter *Journal of Huaqiao University (Natural Science)* **13** (2) 170-176
[15] Wan A T S, Bihlo A and Nave J 2016 The Multiplier Method to Construct Conservative Finite Difference Schemes for Ordinary and Partial Differential Equations *SIAM J. NUMER. ANAL,* **54** (1) 86–119
[16] Ezquerro J A 2000 Hernandez M A and Salanova M A A discretization scheme for some conservative problems *Journal of Computational & Applied Mathematics* **115** 181–192
[17] Waziri M Y 2018 On Solving Differential Equation of Conservative Problems-type in Real Dynamical Problems via Low Memory approach *Malaysian Journal of Computing and Applied Mathematics* **1**(1) 20-28
[18] Shamanskii V E 1967 A modification of Newton's method *Ukrain. Mat. Zh.* **19** 133-138
[19] Brent R P 1973 Some efficient algorithms for solving systems of nonlinear equations *J. Nucleic Acids* **10** (2) 327–344
[20] Traub J F 1964 *Iterative Methods for the Solution of Equations* London: Prentice-Hall
[21] Kchouk B and Dussault J 2013 The Chebyshev–Shamanskii Method for Solving Systems of Nonlinear Equations *J Optim Theory Appl* **157** 148–167
[22] Ortega J M and Rheinboldt W C 1970 *Iterative Solution of Nonlinear Equations in Several Variables* New York: Academic Press
[23] Kamfa K, Waziri M Y, Sulaiman I M, Ibrahim M A H and Mamat M 2020 A quasi-newton like method via modified rational approximation model for solving system of nonlinear equation. *Journal of Advanced Research in Dynamical and Control Systems* **12** (2) 1019-1026
[24] Aini N, Rivaie M, Mamat M and Sulaiman I M 2019 A Hybrid of Quasi-Newton Method with CG Method for Unconstrained Optimization *J. Phys.: Conf. Ser.* **1366** 012079
[25] Sulaiman I M, Mamat M, Zamri N and Ghazali P L 2018 Solving dual fuzzy nonlinear equations via Shamanskii method *International Journal of Engineering and Technology (UAE)* **7** (3.28) 89-91

IOP Conf. Series: Materials Science and Engineering 1115 (2021) 012007 doi:10.1088/1757-899X/1115/1/012007

# Monitoring System for Early Detection of Fire in Wetlands based Internet of Things (IoT) using Fuzzy Methods

**A F Zulkarnain, Y Sari* and R Rakhmadani**

Department of Information Technology, Faculty of Engineering, Universitas Lambung Mangkurat, Banjarmasin, Indonesia

*Email: yuzlena@ulm.ac.id

**Abstract.** In Indonesia, especially the province of South Kalimantan is a province with a high frequency of forest and wetland fires. These fires seem to be a routine annual natural disaster. BNPB (National Disaster Management Agency) and BPBD (Regional Disaster Management Agency) Central Kalimantan recorded that during 2014 to early 2019, the largest area of forest and wetlands fires in Central Kalimantan occurred in 2015 with a total burned area of 583,833.44 ha. The fire early detection monitoring system that will be made based on Arduino Uno uses the MQ-2, DHT-22 sensor, and flame detector sensor and is integrated with the Thingspeak IoT, which functions to receive data sent from the Arduino Uno microcontroller using the ESP8266 module, then Thingspeak will send the data to an Android smartphone in the form of a real-time display and also a pop-up notification when a potential fire is detected. Finally, the fuzzy method is also added so that later the reading data from the MQ-2, DHT-22, and flame detector sensors will be more accurate before being sent to Thingspeak. The level of accuracy obtained from the readings of the three sensors against the tests that have been carried out is 90%.

Keywords: Fire Monitoring System, Arduino Uno, Fuzzy Method, Monitoring System

## 1. Introduction
The impacts of forest and wetland fires range from decreased environmental conditions, material loss to human health, which can lead to respiratory infections, eye pain, and coughing from fire smoke. Based on data from BNPB (National Disaster Management Agency) and BPBD (Regional Disaster Management Agency) Central Kalimantan, from 2014 to early 2019, the largest area of forest and wetlands fires in Central Kalimantan occurred in 2015 with a total burned area of 583,833,44 ha. Then from 12 districts/cities in Central Kalimantan, from early January to August 2018, Palangkaraya City was the city with the most incidents of forest and wetlands fires with 76 incidents, followed by East Kotawaringin with 53 incidents.

Because the impact of the fires caused is so significant, there must be severe handling of this forest and wetlands fire disaster. The solution that will be made is to implement a monitoring system for forest and wetlands fires so that it can provide quick action before and when fires occur and minimize the impact of losses caused by fires [1]-[2].

Based on these conditions, a similar monitoring system will be built by adding Thingspeak. Thingspeak is a web-based on open API IoT (source platform) information which is reliable in storing sensor data from various IoT applications. Thingspeak functions to retrieve, store, analyze, observe, and work on data sent from sensors connected to a microcontroller. The resulting output data is in the form of graphs and conclusions and can be accessed via the web desktop and Android-based mobile applications [3]-[6].

Research on fire monitoring systems has been carried out in ref [7]. In his research entitled "Building Fire Management Systems Using Fuzzy Methods". The study used an MQ-2 smoke sensor, a DHT-22 temperature sensor and a flame detector sensor. This study applies the Fuzzy Logic method to the Uno Atmega328P microcontroller.

Another research was also carried out in ref [8], entitled "Prototype of Land Fire Fire Spot Detection Based on Arduino Uno R3 with Early Warning through the Website". This research concluded that the MQ-2, DHT-11 sensors and flame detector sensors were able to detect fire light very quickly and make it easier for users to access information, especially in forest prone to fires.

Similar research was also carried out in ref [9], entitled "Monitoring and Early Warning System for Forest and Peatland Fires Based on Arduino with Website Interface and Short Message Service (SMS)". This research was designed to monitor forest and peatland fires by measuring the level of temperature, humidity, and smoke. The tool used to measure the high temperature and humidity of the air is the DHT-11 sensor, while to measure the level of smoke density using the MQ-7 sensor. Tool testing is done by measuring temperature, humidity, and smoke, the results of which will be displayed on the website interface and SMS.

In addition, a fuzzy method is also added, which will later function to process input data into output data so that the data sent is more accurate. Referring to the above discussion, a "Monitoring System for Early Detection of Fires on Wetlands. Based on the Arduino Uno Microcontroller and the Internet of Things using the Fuzzy method" was created.

## 2. Research Methods

The research procedure used in this study can be seen in Figure 1.

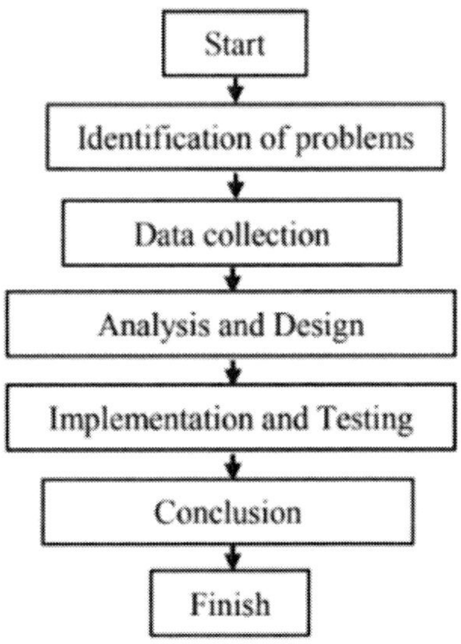


**Figure 1.** Research Process Flow

### 2.1 System planning

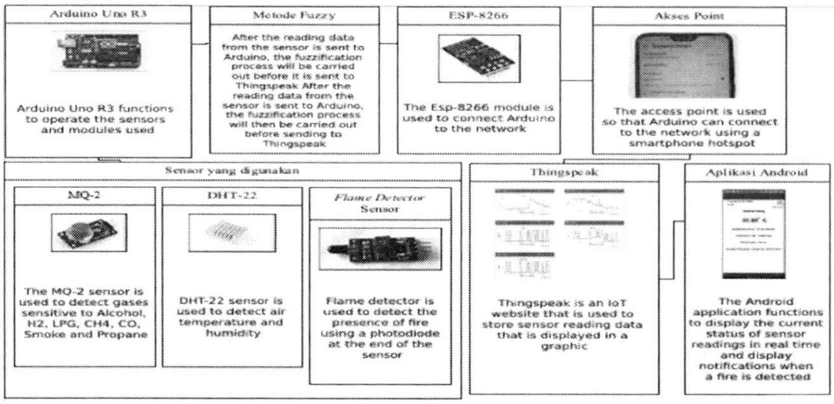


**Figure 2.** General Description of the System

The flow of the system in this study can be seen in Figure 2:

1. The Arduino is turned on, and then the MQ-2, DHT-22 sensors, and flame detector sensors will detect the state of temperature, smoke, and fire around the sensor.
2. Data that has been seen by the sensor will be forwarded to Arduino for processing
3. Arduino processes the data that has been received by the sensor using the fuzzy method
4. After the fuzzification process is complete, Arduino then sends the data to the Thingspeak website via ESP8266
5. Thingspeak receives, after the data, and this data will be sent and also displayed in an Android-based application that has a pop-up notification as a monitoring medium.

### 2.2 Trial Scenarios

At this stage, tests will be carried out on the system that has been created. Prototype testing aims to ensure that the sensors used are functioning correctly and can be connected to the system so that monitoring can be done via the Thingspeak website and the Android application. The test was carried out in a place that resembled the original position, namely forest and wetlands but on a smaller scale.

### 2.3 Implementation

At this stage, fuzzy logic will be incorporated into the system that has been built. There are two types of input data entered, namely, data on smoke levels and temperature. The data is sent from a sensor that is connected to the Arduino IDE software, and then it will be stored and then processed using the fuzzy method to find out which value of the sensor readings is more likely to be a member. After going through the fuzzy process, it will produce output in the form of data that is ready to send the sensor to the user. The input smoke level uses variable part per million (ppm) value data [10]-[12]. So, the membership list for the smoke variable is [0 10000]. Smoke uses 2 fuzzy sets, namely smokeless and smoky. Fires that occur are usually identified by the increase in smoke from smoke in normal air, which changes the conditions to the air above normal, which is greater than the value of 4500 ppm. With smokeless [0 4000] and smoky [4500 10000] domains [13]. List of smoke levels membership can be seen in Figure 3.

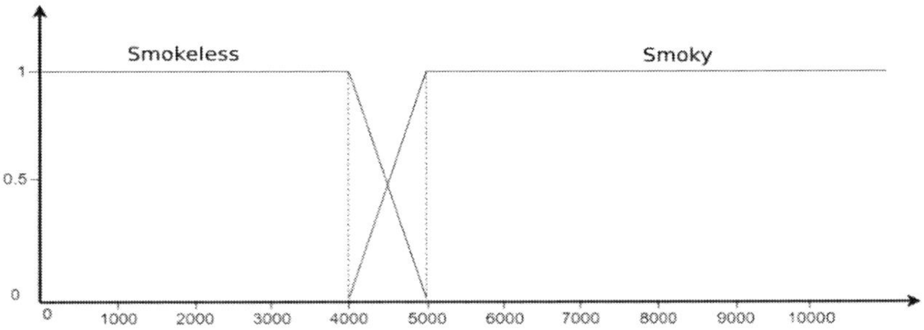


**Figure 3.** Membership List of Smoke Density

As for the temperature using three fuzzy sets, namely normal, warm, and hot, fires occur usually known by an increase in temperature from normal air which rises above normal temperatures, which is greater than 30 ° C. Thus, in this study, the membership list for temperature variables was [20 40]. With normal [20 30], warm [25 35] and hot [30 40]. The list of temperature membership can be seen in Figure 4.

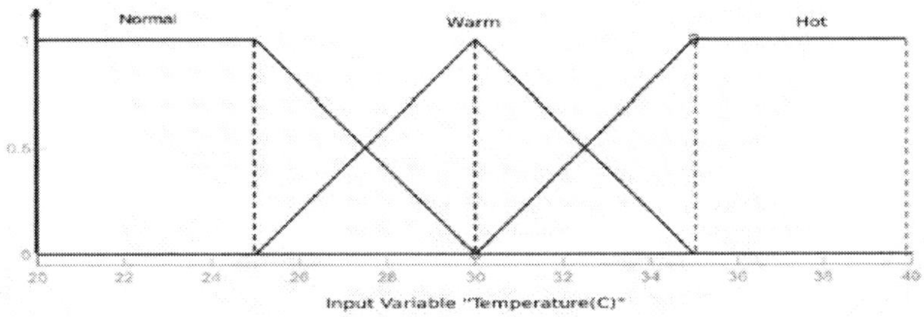


**Figure 4.** List of Temperature Membership

Fuzzy logic has several stages starting with the process of fuzzification (formation of fuzzy sets), rules, the composition of rules, and defuzzification (affirmation) . The steps of the fuzzy logic process can be seen in the Figure 5.

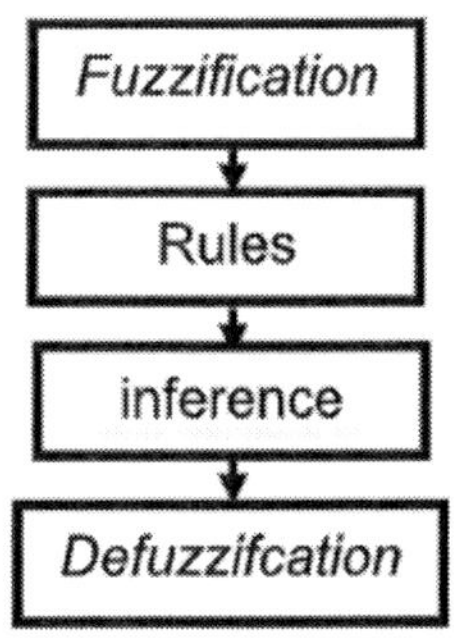


**Figure 5.** Fuzzy Logic Process Structure

Figure 5 is the structure of the Fuzzy logic process. Where according to the flow of the image, the process starts from fuzzification, inference, rules, and defuzzification as the final result [14]. From the above process flow, it can be described as follows:

IOP Conf. Series: Materials Science and Engineering 1115 (2021) 012007 doi:10.1088/1757-899X/1115/1/012007

1. Fuzzification
   In the first process, where the fuzzification process is the process of changing the input/output system through fuzzy. A range of input variable values are grouped into fuzzy sets; each fuzzy has a membership list that shows the mapping of data input points into its membership value in the form of a membership list and has an interval between 0 to 1. The fuzzy membership list is calculated based on the membership formula fuzzification triangles that match the parameter results. The form of fuzzification determines the membership list in an input/output range [15].
2. Rules
   If the system consists of several rules, then the inference is obtained from the collection and correlation between the rules. There are three input variables and one output variable that have been determined by the formation of rules by analyzing the boundaries in a fuzzy set. The output consists of 12 fuzzy rules, which can be seen
3. Inference
   The inference stage part is looking for the MIN function first.
4. Defuzzification
   The input of the defuzzification process is a fuzzy set obtained from the fuzzy rules' composition so that the resulting output is in the form of a number in the fuzzy set. If given a fuzzy set within a certain range, a certain (crisp) value must be taken as output.

## 3. Results and Discussion

At this stage of processing or processing the data, it becomes a conclusion in the form of information conveyed by the system to its users (Users). At this stage, data processing is carried out using a fuzzy inference system in the form of a computational framework and is based on IF-THEN fuzzy rules, fuzzy reasoning, and fuzzy set theory [16].

In the system that has been created, there are two interfaces, namely the system interface on the web/desktop and the system interface on an Android smartphone. Both have different functions. Display on the web/desktop using the Thingspeak platform. The data presented is in the form of graphical data, the results of data processing sent from each sensor.

1. Temperature Graph Display

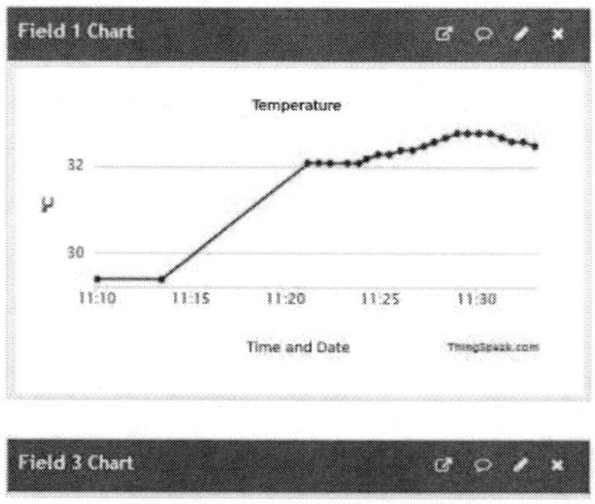


**Figure 6.** Temperature Graph

Figure 6 show the displays real-time temperature monitoring with a time range that is every 5 minutes.

2. Smoke Graph Display

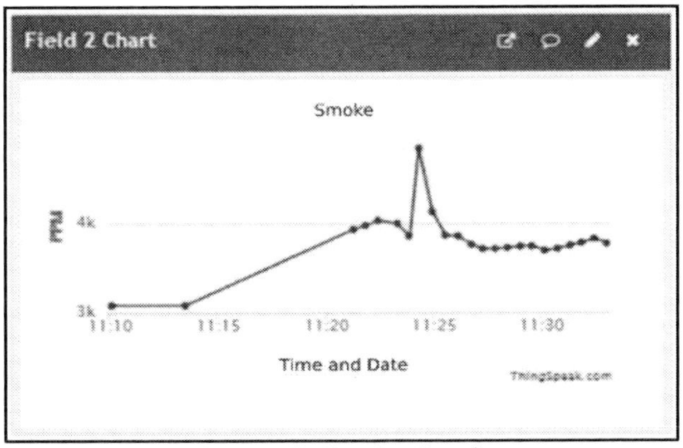


**Figure 7.** Smoke Graph

Figure 7 displays the smoke monitoring in real-time with a time range every 5 minutes.

3. Fire Graph Display

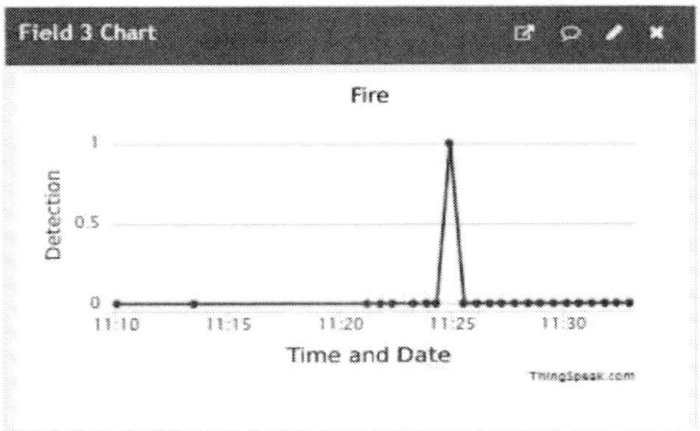


**Figure 8.** Fire Graph

Figure 8 show the real-time fire monitoring with a time range, which is every 5 minutes.

4. Status Graph Display

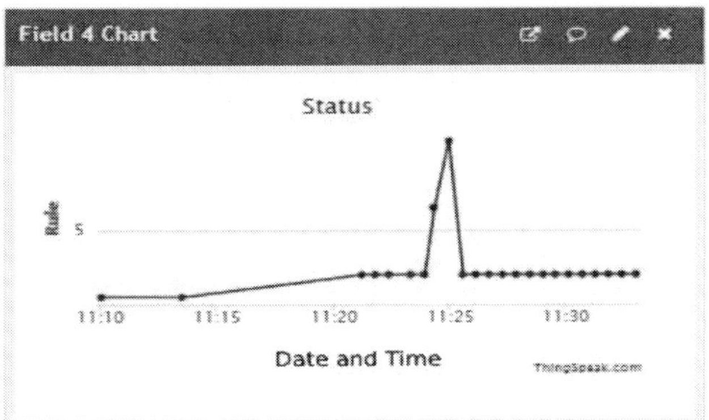


**Figure 9.** Status Graph Display

IOP Conf. Series: Materials Science and Engineering 1115 (2021) 012007 doi:10.1088/1757-899X/1115/1/012007

Figure 9 show the real-time monitoring of fuzzy conclusions with a time range that is every 5 minutes. The data displayed is the result of processing data sent by each sensor in the form of a status. The status is divided into 4 conclusions, namely normal status, alert status 1, alert status 2 and danger status.

Meanwhile, the interface on an Android smartphone is in the form of today's time and date, temperature conditions, smoke and fire, current status, last update, and notification page which contains unread data and read data.

5. Android Interface Page

Figure 10 is an implementation of an Android mobile device to display the details of the read data. Pieces of the data include time, gas value, temperature value, fire, and the condition's status.

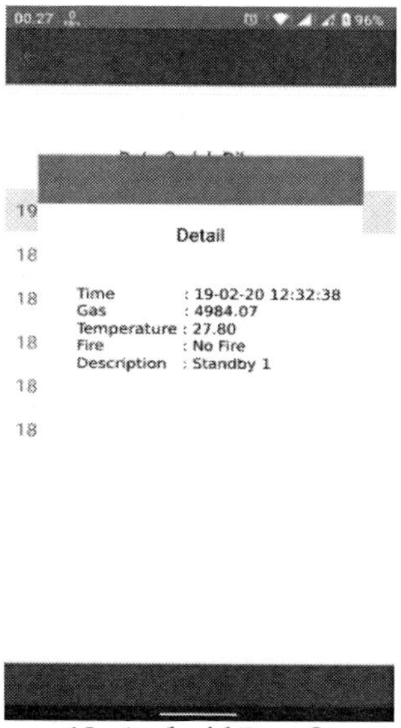


**Figure 10.** Android Interface Page

6. Notification Interface Page

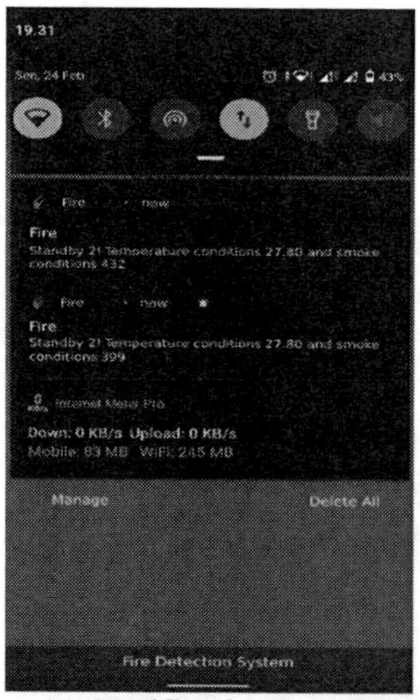


**Figure 11.** Notification Interface Page

Figure 11 is a display of the user (user) getting information in the form of the latest status of a state or condition detected by the system and stored in Thingspeak.

System testing is carried out aimed at ensuring the system that has been created is in accordance with the results of analysis and design and produces conclusions as expected. Before the system is implemented, what must be done is to ensure the program is free from errors that may occur when the system is created.

The following Figure 12 is a view of the sensor that has been assembled on a breadboard and connected to a laptop.

**Figure 12.** Sensor Circuit Above Breadboard

The first stage is to install the MQ-2 sensor, DHT-22 sensor, and flame detector sensor on top of the breadboard. Next is to connect each sensor to the Arduino Uno Atmega328P using a jumper cable. Then attach the LED light to the breadboard, connect each LED light to the Arduino Uno Atmega 328P so that the program can give the LED light command to turn on based on the setup that has been done in the program. The last step is to connect the Arduino Uno Atmega 328P to the laptop using a USB cable.

At the sensor testing stage, the assembled sensors will be tested with the materials provided, namely dry grassroots, dry twigs, used paper, and leftover cigarette butts. The test is carried out in a place that resembles the original place on a smaller scale. Testing time is carried out from noon (11.00 WITA) to (12.00 WITA). Each test has a time interval of 15 minutes. It aims to condition changes in environmental temperature as the original.

The following is a photo when testing the sensor:

1. Testing prototypes of dry grass

   In the Figure 13, the prototype can detect the presence of fire but has not detected the smoke due to the direction of the smoke away from the prototype. The blue LED for standby status two lights up when the prototype detects a fire from dry grass.

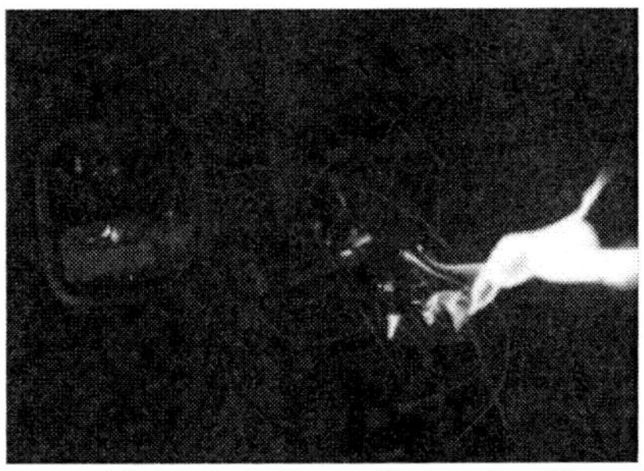

**Figure 13.** Dried Grass Detection Test

2. Testing prototypes of dry branches
   In the Figure 14, the prototype can detect the presence of fire and also detect smoke due to the direction of the smoke that is approaching the sensor installed on the prototype. The red LED light for hazard status lights up when the prototype detects fire and smoke coming from dry branches.

**Figure 14.** Dry Twig Detection Test

3. Prototype testing of residual cigarette butts is smoke and fire
   In the Figure 15, the prototype can detect smoke and embers from leftover cigarette butts because the coals from cigarettes can be detected by the sensor installed on the prototype. The red LED light for hazard status lights up when the prototype detects smoke and flames from leftover cigarette butts.

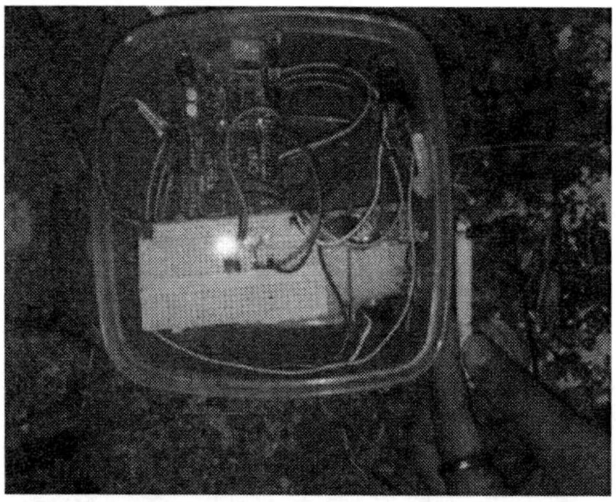

**Figure 15.** Detection Testing of Remaining Cigarette Butts, There is Smoke and Fire

After the system testing is complete, the next stage is the evaluation stage. The evaluation aims to get the level of accuracy of the system that has been made. Evaluation of the system is done by comparing the suitability of the sensor output with the results displayed by the system. The results of the accuracy of Prototype testing that has been carried out on each test material can be seen in the Table 1.

**Table 1.** Test Accuracy Results

| Test Time (WITA) | Precondition | MQ-2 Smoke Sensor | Temperature Sensor DHT-22 | *Flame Detector Sensor* | LED Indicator Light | Evaluation | Conclusion |
|---|---|---|---|---|---|---|---|
| 11.00 | • Burn the dry grass<br>• Burn dry twigs<br>• Burn waste paper<br>• Turn on leftover cigarette butts | 33.80 °C<br>47.90 °C<br>50.90 °C<br>42.20 °C | 3685.04 PPM<br>4794.05 PPM<br>4083.28 PPM<br>6814.08 PPM | Flame detected<br>Flame detected<br>Flame detected<br>Turns on but not detected | Standby 2<br>Danger<br>Standby 2<br>Standby 1 | Each sensor can detect well | Corresponding<br>Corresponding<br>Corresponding<br>Corresponding |
| 11.15 | • Burn the dry grass<br>• Burn dry twigs<br>• Burning waste paper<br>• Lit up leftover cigarette butts | 36.90 °C<br>50.00 °C<br>37.20 °C<br>43.50 °C | 5628.84 PPM<br>5107.22 PPM<br>4064.32 PPM<br>4453.08 PPM | Flame detected<br>Flame detected<br>Flame detected<br>No flame detected | Standby 2<br>Danger<br>Standby 2<br>Secure | The fire sensor experienced a one-time detection read error | Corresponding<br>Corresponding<br>Corresponding<br>It is not in accordance with |
| 11.30 | • Burn the dry grass<br>• Burn dry twigs<br>• Burning waste paper | 37.20 °C<br>51.60 °C<br>49.40 °C | 4699.61 PPM<br>5503.21 PPM<br>3041.06 PPM | Flame detected<br>Flame detected<br>Flame detected | Standby 2<br>Danger<br>Standby 2 | The smoke sensor made a mistake | Corresponding<br>Corresponding<br>It is not in accordance with |
| Test Time (WITA) | Precondition | MQ-2 Smoke Sensor | Temperature Sensor DHT-22 | *Flame Detector Sensor* | LED Indicator Light | Evaluation | Conclusion |
| 11.45 | • Burn the dry grass<br>• Burn dry twigs<br>• Burning waste paper<br>• Lit up leftover | 39.60 °C<br>50.40 °C<br>48.70 °C | 4614.27 PPM<br>5806.74 PPM<br>3997.95 PPM<br>5249.56 PPM | Flame detected<br>Flame detected<br>Flame detected<br>Flame detected | Standby 2<br>Danger<br>Standby 2 | Each sensor can detect well | Corresponding<br>Corresponding<br>Corresponding |

| | | | | | | | |
|---|---|---|---|---|---|---|---|
| | cigarette butts | 32.60 $^{o}C$ | | | Danger | | Correspond ing |
| 12.00 | • Burn the dry grass | 47.90 $^{o}C$ | 4632.50 PPM | Flame detected | Standby 2 | Each sensor can detect well | Correspond ing |
| | • Burn dry twigs | 49.80 $^{o}C$ | 6023.14 PPM | Flame detected | Danger | | Correspond ing |
| | • Burning waste paper | 48.50 $^{o}C$ | 4248.13 PPM | Flame detected | Danger | | Correspond ing |
| | • Lit up leftover cigarette butts | 31.40 $^{o}C$ | 6482.21 PPM | No flame detected | Standby 1 | | Correspond ing |

In the Table 1, it can be seen that the comparison of the results of all system tests to measure the level of accuracy of the system can be calculated with the following equation:

$$Classification\ Accuracy = \frac{TP}{TP+FN} \text{ x } 100\ \% \quad (1)$$
$$= \frac{18}{20} \text{ x } 100\ \%$$
$$= 90$$

Information:
• TP (true positive) = indicates a measurement that is classified as correct
• FN (false negative) = indicates unclassified measurement.

So, from the above calculations, it is known that the level of accuracy of the reading of the system is 90%. This result was caused by an error in sensor readings that occurred twice when the test was carried out. The errors that occur are in the MQ-2 sensor and also the flame detector sensor. The error on the MQ-2 sensor occurs due to the wind blowing the smoke from the combustion away from the sensor so that the sensor cannot read the smoke changes that occur, while for the flame detector sensor the error occurs because the coals of residual cigarette butts are too small so that the sensor cannot read the existing flame. So from the above calculation, it is known that the accuracy of the system reading is 90%. This result was caused by an error in sensor readings that occurred twice when the test was carried out. The errors that occur are in the MQ-2 sensor and also the flame detector sensor. The error on the MQ-2 sensor occurs because the wind blows the smoke from the combustion away from the sensor so that the sensor cannot read the smoke changes that occur, while for the flame detector sensor, the error occurs. Because the coals of residual cigarette butts are too small so that the sensor cannot read the existing flame.

## 4. Conclusion

Prototypes for early detection of forest and wetlands fires have been successfully created using the MQ-2 Smoke sensor, DHT-22 Temperature Sensor, and Flame Detector Sensor based on Arduino Uno Atmega328P. The prototype successfully applies the fuzzy method in its calculations. The prototype created is able to send sensor readings to the Thingspeak website using the ESP8266 module and is able to display data from Thingspeak to an Android smartphone and is able to display pop-up notifications when 2 alert status and danger is detected. The level of accuracy obtained from the readings of the three sensors against the tests that have been carried out is 90%.

## References

[1] Narasimha Rao G, Jagadeeswara Rao, Duvvuru R S, Bendalam, and Gemechu R 2018 Fire detection in Kambalakonda Reserved Forest *Mater Today Proc* **5** (1) 1162–1168
[2] Mahbub M, Hossain M M, and Gazi M S A 2021 Cloud-Enabled IoT-based embedded system and software for intelligent indoor lighting, ventilation, early stage fire detection and prevention *Comput Networks* **184** 107673
[3] Sharma A, Singh P K, and Kumar Y 2020 An integrated fire detection system using IoT and image processing technique for smart cities *Sustain. Cities Soc* **61** 102332
[4] Yu L, Nazir B and Wang Y 2020 Intelligent power monitoring of building equipment based on Internet of Things technology *Comput Commun* **157** 76–84
[5] Mujiarto, Djohar A, Komaro M, Mohamed M A, Rahayu D S, Sanjaya W S M, Mamat M, Sambas A and Subiyanto 2019 Colored object detection using 5 dof robot arm based adaptive neuro-fuzzy method *Indonesian Journal of Electrical Engineering and Computer Science* **13** (1) 293-299
[6] Trisnawan, Sanjaya W S M, Sambas A, Jannah M, Rahayu D S, Mamat M and Mohamed M A 2019 Design of 4 Dof Robot ARM Based on Adaptive Neuro-Fuzzy (ANFIS) using Vision in Detecting Color Objects *International Journal of Recent Technology and Engineering* **8** (2S7) 224-227
[7] Munadhif I, Adianto A, and Mustofa A A 2018 Sistem Pengendalian Penanganan Kebakaran Gedung Menggunakan Metode Fuzzy *Rekayasa* **11** (2) 171–183
[8] Kurniawan S, Marindani E D, and Priyatman H 2018 Prototipe Pendeteksi Titik Api Kebakaran Lahan Berbasis Arduino Uno R3 Dengan Peringatan Dini Melalui Website *Jurnal Teknik Elektro Universitas Tanjungpura* **1** (1) 1-10
[9] Leven T S, Rismawan T, Nirmala L 2017 Sistem Monitoring dan Peringatan Dini Kebakaran Hutan dan Lahan Gambut Berbasis Arduino Dengan Antarmuka Website dan Short Message Service (SMS) *Coding Junal Komputer dan Aplikasi* **5** (3) 72-79
[10] Kumar Sai K B, Mukherjee S, and Parveen Sultana H 2020 Low Cost IoT Based Air Quality Monitoring Setup Using Arduino and MQ Series Sensors with Dataset Analysis *Procedia Comput Sci* **165** (X) 322–327
[11] Hakim L and Halim J 2018 Peringatan Kebakaran Hutan Menggunakan Sensor Api, Suhu Dan Asap *Seminar Nasional Teknologi Informasi dan Komunikasi (SEMNASTIK)* **1** (1) 26–38
[12] Kuncoro A H, Mellyanawaty M, Sambas A, Maulana D, Subiyanto and Mamat M 2020 Air Quality Monitoring System in the City of Tasikmalaya based on the Internet of Things (IoT) *Jour of Adv Research in Dynamical & Control Systems* **12** (2) 2473-2479
[13] Lekić M and Gardašević G 2018 IoT sensor integration to Node-RED platform, *international symposium infoteh-jahorina (infoteh)* 1-5
[14] Launda A P, Mamahit D J and Allo E K 2017 Prototipe System Pengering Biji Pala Berbasis Mikrokontroler Arduino Uno *Jurnal Teknik Elektro dan Komputer* **6** (3) 141-147
[15] Sanubary I 2019 Perbandingan Kinerja Sensor TGS2610, MQ2, dan MQ6 Pada Alat Pendeteksi Kebocoran Tabung Liquified Petroleum Gas (LPG) Menggunakan ATMega2560, *PRISMA FISIKA* **7** (1) 14-19
[16] Sukono, Sambas A, He S, Liu H, Vaidyanathan S, Hidayat Y and Saputra J 2020 Dynamical analysis and adaptive fuzzy control for the fractional-order financial risk chaotic system *Advances in Difference Equations* **1** 1-12

# Android-Based Intangible Cultural Heritage Expert System In Banten Province Using Backward Chaining Method

**I Purwanto, H Y Fauziah*, M Subali, M Andriansyah, D Saptono, A I Sukowati and H Hanifah**

Informatics engineering, Sekolah Tinggi Teknik Multimedia Cendekia Abditama (STTMCA), Tangerang - Indonesia

*Email : helmi@cendekia.ac.id

**Abstract.** Expert system is a system used to assist decision making that works by using the knowledge and analysis methods that have been defined by the expert. Intangible cultural heritage is a living cultural heritage that is recognized by some communities, groups or individuals. Banten Province has an intangible cultural heritage, but many people do not know there is many kinds of intangible cultural heritage that exist in Banten Province. The purpose of this paper is to make an Android-Based Intangible Cultural Heritage Expert System in Banten Province Using the Backward Chaining which is useful as a medium for the introduction of intangible cultural heritage that can be accessed easily. The software used is the Android Studio IDE which is installed with the Android SDK. The process of making an intangible cultural heritage expert system uses several stages, namely analysis, design, coding, implementation, and testing. The output of this research is that the application can be installed on an Android smartphone and data can be accessed offline.

## 1. Introduction

Intangible cultural heritage is scattered throughout Indonesia, which has become the identity of the area, one of which is in Banten Province. Banten is rich in intangible cultural heritage which contains ancestral values, for example, such as traditional musical instruments that have distinctive sound compositions and those instruments used for certain ritual activities. However, in the modern era, intangible cultural heritage has begun to be shifted.

The public interest in recognizing intangible cultural heritage (intangible) is also reduced due to social changes triggered by technological sophistication. Along with the times, various kinds of systems have emerged and were created to help identify a problem. The system not only provides information about an object, but can also handle object recognition problems based on characteristics and symptoms. This system is called the Expert System.

Expert system is a computer based system using knowledge, facts, and reasoning techniques in solving problems that usually can only be solved by an insider certain fields. Expert system add value to technology to assist in handle the increasingly information age sophisticated [1]. Expert System is a system that is intended to assist decision making that works using the knowledge and analysis methods that have been defined by the expert. This system has an absolute component that must exist, namely a knowledge base that comes from experts, a database that contains all the facts, an inference engine that contains a system of thinking and reasoning mechanisms used by experts, and a user interface that connects the expert system with the user with questions. "Yes / no" or in the form of an option. In the inference engine, there are two patterns of reasoning, one of which is Backward Chaining.

Backward chaining is a reasoning backward, where work from the goal or goal which is in the section THEN, then match the facts in the section IF whether it is in accordance with the objectives or not to get a conclusion. If it matches it will be processed, otherwise it will be repeated [2]. The process ends when the destination is found. In backward chaining, reasoning starts from consequent to antecedent. Trace back to work backward to get facts that support the hypothesis [3]. In this research, the Depth First Search method is used which will carry out a deep search of the rules from the root node moving down to the inner level sequentially [4]. Backward Chaining is a search strategy in the opposite direction of forward chaining. The research process starts from the goal, namely the conclusion that becomes the solution to the problems at hand [5].

The research conducted by Cantika determined backward chaining as a method used to determine the nutritional composition system for cattle feed [6]. In a study conducted by Maradesa, it was explained that an expert system can help cataract patient to be diagnose the impact and to know the solutions and problems [7].

Based on this background, the author plans to make an Android-based Intangible Cultural Heritage Expert System in Banten Province Using the Backward Chaining Method which can be used as a medium for the introduction of intangible cultural heritage in Banten Province more easily. Expert system created using the android studio application as an IDE (Integrated Development Environment), and the Java Programming Language.

## 2. Methodology

### 2.1. Expert System

Some definitions of expert systems are:

- According to Durkin, the expert system is a computer program designed to model the problem-solving skills by an expert.
- According Ignizio, an expert system is a model and procedures relating, in a particular domain, where the level of expertise that can be compared with the expertise of a specialist.
- According Giarrantano and Riley, an expert system is a computer system that could match or mimic the ability of an expert.
- According to Turban, an expert system is a software package of decision making or problem solving that can achieve performance levels comparable or even better with a human expert in some specialized fields, and usually narrow problem area [8].

Expert systems or knowledge-based systems are part of artificial intelligence that allows computers to process and derive conclusions from a set of rules [9]. This system works by using knowledge and analysis methods that have been defined in advance by experts in accordance with their field of expertise. This system is called an expert system because it has the same function and role as an expert who must have knowledge and experience in solving a problem. The system usually functions as an important key that will help a decision support system or executive support system [10].

### 2.2. Intangible Cultural Heritage

Intangible cultural heritage is a practice, representation, expression, and knowledge and skills (including instruments, objects, artifacts, cultural space), that communities, groups and in some cases individuals recognize as part of their cultural heritage. Individuals who claim to be a cultural heritage are called living cultural heritage. UNESCO classifies intangible cultural heritage into several categories as follows; oral traditions and expressions, performing arts, practices (social, rituals and festivals), knowledge and practice about nature and the universe, traditional skills [11].

### 2.3. Backward Chaining

Backward Chaining is An inference technique. Goal-driven technique starts from the goal to reach the data and proves if it supports the hypothesis [12]. In this approach, tracking starts from the goal, then searches for rules that have that goal for the conclusion. Next process tracking using the premise for the rule as a new goal and looking for another rule with a new goal as the conclusion. The process continues until all possibilities are found. Backward Chaining is inference that works backward towards the initial conditions. The process begins with Goal (which is in the THEN section of the IF-THEN rule), then a search is started to match whether the facts match the premises in the IF section. If it fits, the rule is executed, then the hypothesis in the THEN section is placed in the database as a new fact. If they don't match, save the premise in the IF section to the stack as a sub Goal. Process ends if Goal is found or there are no rules that can prove the truth of the subGoal or Goal [13].

### 2.4. Android Studio

Android Studio is Environment The official Integrated Development Environment (IDE) for Android application development, which is based on IntelliJ IDEA. Apart from being a powerful code editor and IntelliJ developer tools, Android Studio offers many features that increase your productivity in building Android applications [14].

### 2.5. System Development Life Cycle (SDLC)

SDLC is a work stage that aims to produce a high quality system in accordance with the wishes of the customer or the purpose of the system. SDLC is a framework that contains the steps that must be taken to process the development of a software. This method contain complete plan to develop, maintain, and replace specific software [15].

## 3. Research Methodology

### 3.1. Research Object

Legacy culture not object (intangible) is a living cultural heritage that is recognized by some communities, groups and individuals. UNESCO classifies intangible cultural heritage into several categories, namely traditions and oral expressions, performing arts, practices (social, rituals and festivals), knowledge and practice about nature and the universe, traditional skills.

### 3.2. Research Methods

The method that researchers used in this research is the System Development Life Cycle (SDLC) with the Waterfall model. The method used consists of 6 parts, namely:

- Needs Analysis
- Design
- Encoding
- Implementation
- Testing

### 3.3. Stage of Making the Applications

The following are the stages of making an application which can be seen in the Figure 1.

- Data and Information Collection
  The stage begins with collecting data and information related to making Android applications.
- Android Design
  At this stage the authors designed the workflow and work system to be used in the APK.
- Interface Design
  The Android Interface design stage is to make the application comfortable to use by the user.
- Creating an Android Interface
  After designing, then the writer made the display design and logo.
- Preparing Data
  The amount of data prepared was 30 intangible cultural heritages in Banten Province.
- Analyze Data
  The next step is to classify intangible cultural heritage by creating a database and a decision tree.
- Data Results
  At this stage, the data obtained from the analysis will be imported into the Android Studio application.
- Import Android UI design and Data to Android Studio
  After the data, logos and images are obtained, they are imported into Android Studio.
- Android-Based Application Development
  The next step is to create an Android-based application by entering program listings and setting the UI appearance in Android Studio.
- Implementation of Coding
  After the design and program listing is complete, the next stage is coding with Java and XML to arrange the application according to the wishes both in terms of appearance and program listing.
- Build into APK
  The build stage is the application project stage that Android Studio builds can run on an Android device.
- Trial APK on Android
  The last stage is the implementation and testing of finished applications on Android smartphones.

## 4. Results and Discussions

### 4.1. Need Analysis

The data needed in designing the Expert System application is intangible cultural heritage in Banten Province. The hardware and software requirements include a laptop with a minimum of 4GB RAM, the Android Studio application.

### 4.2. Design

#### *4.2.1. Flowchart*

The flowchart is the flow of the Android application that will be created as shown in Figure 2. The Flowchart model is an illustration of the APK file workflow that is installed on an Android device. There is a main page with 5 buttons that represent their respective functions, there is a start button to start an expert system, a guide button, information and about which has a flow for reading how to use, reading information and about the application being made. And finally the exit button is to end the use of the application (See Figure 1).

**Figure 1**. Application Making Stages

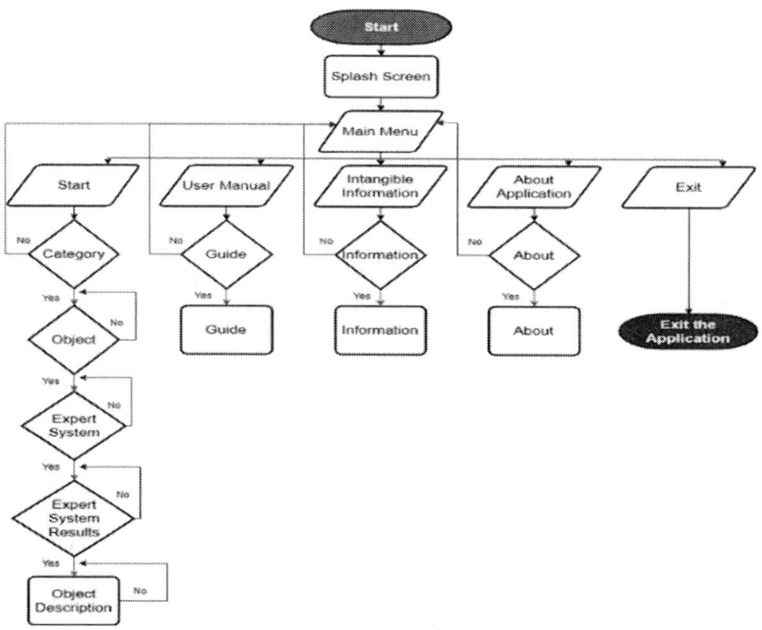


**Figure 2**. Flowchart Model

### *4.2.2. Data Table*

The data in the table is a representation of knowledge that comes from facts obtained from an expert in row and columns. From Table 1, we can see that the musical instruments and their characteristics are related, for example CAM1 (bamboo) is one of the characteristics of AM1 (Angkulung Buhun) and AM5 (Pantun Bamboo).

**Table 1.** Musical Instrument Data Table

| Musical Instrument Characteristics | Musical Instrument | | | | |
|---|---|---|---|---|---|
| | AM1 | AM2 | AM3 | AM4 | AM5 |
| CAM1 | x | | | | x |
| CAM2 | | x | | x | |
| CAM3 | | | x | | |
| CAM4 | x | | | | |
| CAM5 | | x | | x | |
| CAM6 | | | x | | |
| CAM7 | | | | | x |
| CAM8 | x | | | | |
| CAM9 | | x | | | |
| CAM10 | | | x | | |
| CAM11 | | | | | x |
| CAM12 | x | | | | |
| CAM13 | | x | | | |
| CAM14 | | | x | | |
| CAM15 | | | | x | |
| CAM16 | | | | | x |

**Table 1**. Description:
Musical Instrument
AM1 = Angklung Buhun
AM2 = Bedug
AM3 = Bendrong Lesung
AM4 = Dogdog Lojor
AM5 = Pantun Bamboo
Musical Instrument Characteristics
CAM1 = Bamboo
CAM2 = Wood and Skins
CAM3 = Wood
CAM4 = Shaken
CAM5 = Beaten
CAM6 = Pounded with Pestle
CAM7 = Beaten with a Stick
CAM8 = Once a Year
CAM9 = Month of Ramadhan

CAM10 = Rice Processing
CAM11 = After Working in Rice Fields
CAM12 = Ornamental Rice Trunks
CAM13 = Professional Person
CAM14 = Like a Boat
CAM15 = Sounds “dog.. dog..”
CAM16 = Sounds Similar to Patingtung Accompaniment

#### *4.2.3. Rule Table*

Rule table is information of new facts built from existing facts. Rule uses an IF-THEN rule base to present expert knowledge on a particular problem [16]. For detail can be seen in Table 2.

**Table 2**. Table of Musical Instrument Rules

| No. | Rule |
|---|---|
| 1. | **IF** Bamboo **and** Shaken **and** Once a Year **and** Ornamental Rice Trunks **THEN** Angklung Buhun |
| 2. | **IF** Wood&Skins **and** Beaten **and** Month of Ramadhan **and** Professional Person **THEN** Bedug |
| 3. | **IF** Wood **and** Pounded with Pestle **and** Rice Processing **and** Like a Boat **THEN** Bendrong Lesung |
| 4. | **IF** Wood&Skins **and** Beaten **and** Sounds “dog..dog..” **THEN** Dogdog Lojor |
| 5. | **IF** Bamboo **and** Beaten with a Stick **and** Sounds Similar to a Patingtung Accompaniment **and** After Working in Rice Fields **THEN** Pantun Bamboo |

#### *4.2.4. Decision Tree*

Decision Tree is similar to the human decision-making process and so that it is easy to understand. It can solve in both situations whether one has discrete or continuous data as input [17]. A tree is a hierarchical structure consisting of nodes that store information and branches that connect the nodes. The decision tree makes it easy to make decisions. For detail be seen in Figure 3.

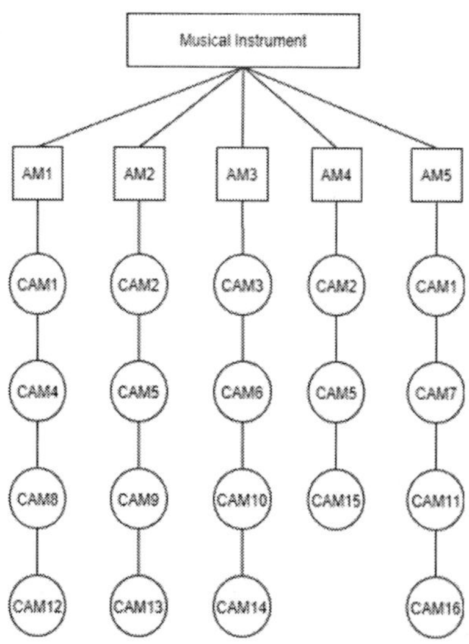


**Figure 3**. Musical Instrument Decision Tree [18]

### 4.3. Encoding

Coding, namely entering programming language script coding into the application project that is being created, the author uses the Java programming language and Extensible Markup Language (XML) as a programming language to create Android applications through Android Studio.

### 4.4. Implementation

Implementation is carried out after the design is complete done and next will implemented in the programming language that will be used. The aim of implementation is to confirm the design module is running as it should. Can be seen in Figure 4 until Figure 12.

| *Splash Screen Interface* | *Main Page Interface* | *Category Page Interface* |
|---|---|---|
|   |   | 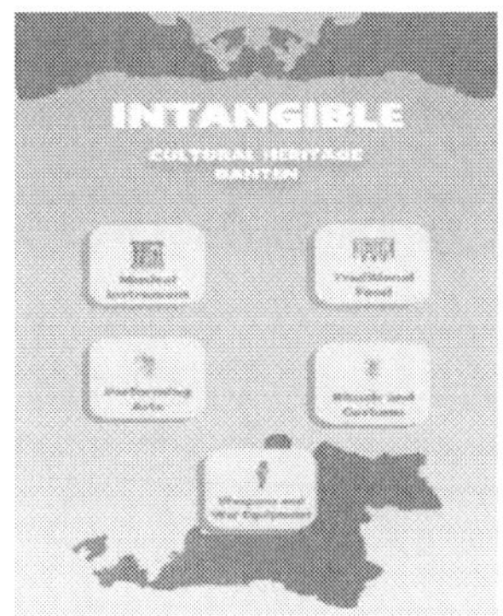  |
| **Figure 4**. Splash Screen Page | **Figure 5**. Main Page | **Figure 6**. Category Page |

| *Object Page Interface* | *Expert System Page Interface* | *Object Description Page Interface* |
|---|---|---|
| 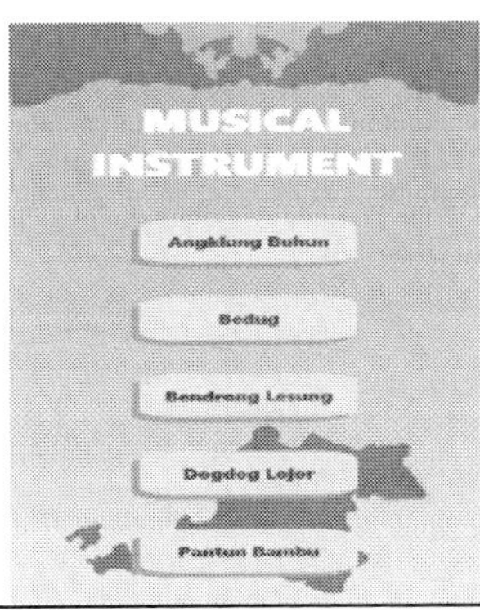  | 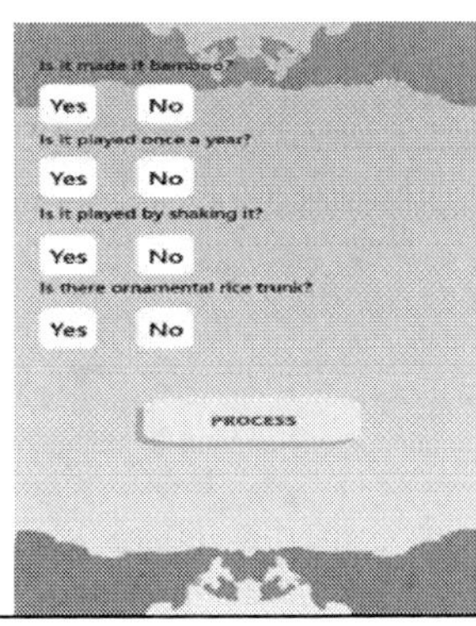  | 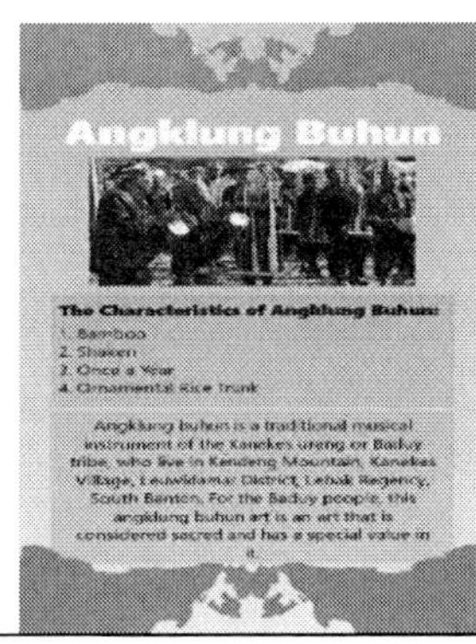  |
| **Figure 7**. Object Page | **Figure 8.** Expert System Page | **Figure 9**. Object Description Page |

*Manual Page Interface*

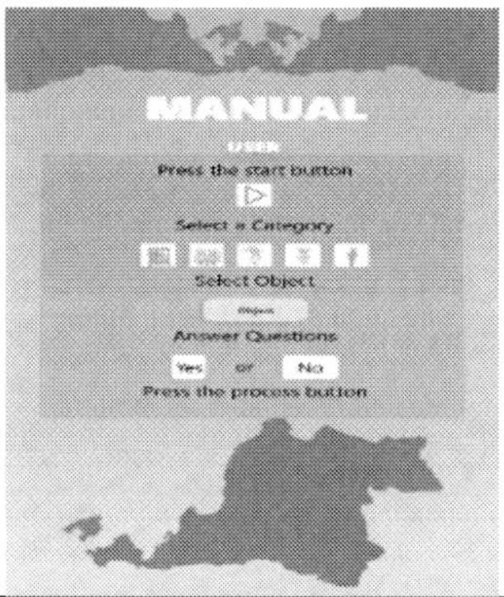


**Figure 10.** Manual Page

*Information Page Interface*

**Figure 11.** Information Page

*The About Page Interface*

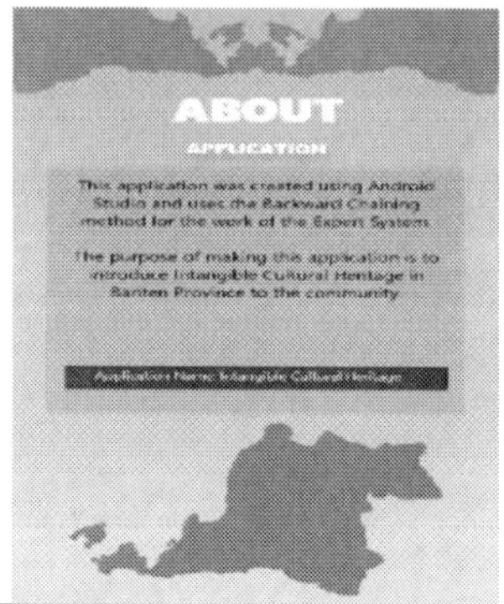


**Figure 12.** The About Page

## 4.5. Application Testing

At this stage the writer will test the application that has been completed. Application testing is a critical element of application quality assurance and provides key reviews of specification, design and coding. Testing will be done by testing the application in alpha and beta.

### *4.5.1. Alpha Testing Results*

Based on the test plan that has been prepared, the following test results can be seen in Table 3.

**Table 3**. Blackbox Test Results

| No. | Test Item | Test Details | Test Results |
|---|---|---|---|
| 1. | Splash Screen | Displays the title of the application | Successful |
| 2. | Main Menu | Displays the main page of the application | Successful |
| 3. | Category | Displays the category menu | Successful |
| 4. | Object | Displays the object menu | Successful |
| 5. | Expert System | Displays questions with "yes/no"<br>Process inquiries | Successful<br>Successful |
| 6. | Object Description | Displays the object description | Successful |
| 7. | Guide | Displays a guide on how to use the application | Successful |
| 8. | Information | Display intangible information | Successful |
| 9. | About | Displays information about the application | Successful |
| 10. | Exit | Exit the application | Successful |

Based on the results testing with blackbox that has been done, the authors conclude that the application can work and produce the expected output, no errors or errors are seen.

### 4.5.2. Beta Testing Results

Based on a questionnaire that has been made with fast variables, namely an assessment of the display, user convenience, system performance, and content, the results can be seen in Table 4.

**Table 4.** The Results of The Questionnaire [19]

| | | Answer Frequency | | | | | | |
|---|---|---|---|---|---|---|---|---|
| No. | Display | SS | S | N | TS | STS | Total | Index % |
| 1. | Color composition | 8 | 11 | 1 | | | 84 | 84 |
| 2. | Clarity of existing text | 6 | 14 | | | | 86 | 86 |
| 3. | Display / image variations | 5 | 11 | 4 | | | 81 | 81 |
| 4. | Display quality | 5 | 11 | 3 | 1 | | 80 | 80 |
| 5. | Clarity of the questions given | 6 | 12 | 2 | | | 84 | 84 |
| | Index Average Display | | | | | | | 83 |
| | | Answer Frequency | | | | | | |
| No. | User Convenience | SS | S | N | TS | STS | Total | Index % |
| 1. | Easy of installing applications | 7 | 9 | 3 | 1 | | 80 | 80 |
| 2. | Easy of operating the applications | 6 | 13 | 1 | | | 85 | 85 |
| 3. | Easy of understanding the results | 4 | 15 | 1 | | | 83 | 83 |
| | Index Average User Convenience | | | | | | | 82.67 |
| | | Answer Frequency | | | | | | |
| No. | System Performance | SS | S | N | TS | STS | Total | Index % |
| 1. | Feedback given after giving an answer | 6 | 13 | 1 | | | 82 | 82 |
| 2. | Speed to display the information | 8 | 10 | 2 | | | 86 | 86 |
| | Index Average System Performance | | | | | | | 84 |
| | | Answer Frequency | | | | | | |
| No. | Content | SS | S | N | TS | STS | Total | Index % |
| 1. | System Objectives | 3 | 15 | 1 | 1 | | 80 | 80 |
| 2. | The information provided is in accordance with the needs and objectives | 6 | 12 | 2 | | | 84 | 84 |
| | Index Average Content | | | | | | | 82 |

From Table 4, it can be concluded that the average index value of display variables is 83%, user convenience variables are 82.67%, system performance variables are 84%, and content variables are 82%. The four variables fall into the "Strongly Agree" category so that the application is feasible to use. Beta testing is done using a Likert scale. The original Likert scale is a set of statements (items) offered for a real or hypothetical situation under study. Participants are asked to show their level of agreement (from strongly disagree to strongly agree) with the given statement (items) on a metric scale. Here all the statements in combination reveal the specific dimension of the attitude towards the issue, hence, necessarily inter-linked with each other [20]. The percentage field of the four variables is shown in Figure 13.

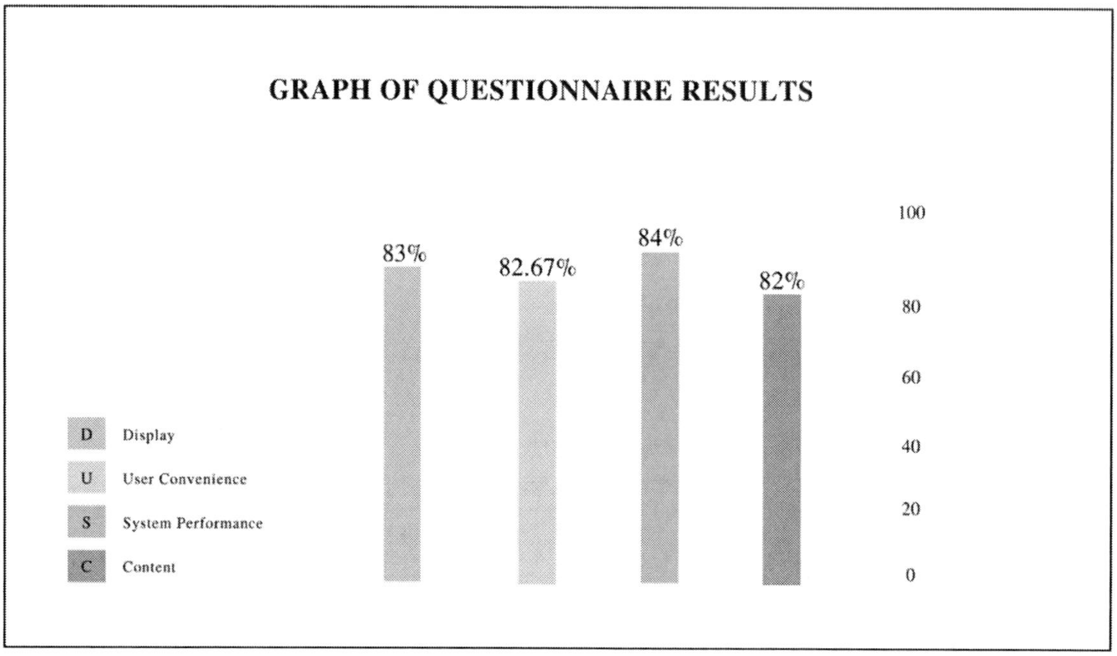


**Figure 13.** Percentage of Four Variable Questionnaire Results

## 5. Conclusion

From this research, several conclusions can be drawn, namely:

- Expert System for Intangible Cultural Heritage in Banten Province Based on Android Using the Backward Chaining Method, successfully created with Android developer tools Studio, and with the Java and XML programming languages.
- Based on alpha testing using the blackbox method, this application has been able to work and produce the expected output. Based on the beta test using the Likert scale, this application falls into the very agreeable category in terms of display 83%, convenience 82.67% users, 84% system performance, and 82% content so that the application is suitable for use as a medium for the introduction of intangible cultural heritage in Banten Province.

## References

[1] Tarigan A F 2014 Sistem Pakar untuk Mendiagnosa Penyakit Ginjal dengan Metode Backward Chaining *TIMES Journal* **3** (2) 25- 29

[2] Bahl N and Kapoor N 2016 Comparative Study of Forward and Backward Chaining in Artificial Intelligence. International Journal Of Engineering And Computer Science *International Journal of Engineering and Computer Science* **5** (4) 16239-16242

[3] Mukhtar N and Samsudin 2015 Sistem Pakar Diagnosa Dampak Penggunaan Softlens Menggunakan Metode Backward Chaining *Buana Informatics Journal* **6** (2) 21-30

[4] Iriani S 2015 Penerapan Metode Backward Chaining pada Sistem Pakar Diagnosa Penyakit Tulang Manusia *IJNS – Indonesian Journal on Networking and Security* **4** (1) 51-55

[5] Al-Ajlan A 2015 The Comparison between Forward and Backward Chaining *International Journal of Machine Learning and Computing* **5** (2) 106-113

[6] Cantika Y E, Witanti W and Renaldi F 2016. *Proceedings of the 7th SNST 2016.*

[7] Maradesa E 2012 *Application of Backward Chaining Method for Disease Diagnosis Cataract* Gorontalo: Universitas Negeri Gorontalo

[8] Sasmito G W 2011 Application Expert System of Forward Chaining and The Rule Based Reasoning For Simulation Diagnose Pest and Disease Red Onion and Chili Plant.*Proceedings of The 1st International Conference on Information Systems For Business Competitiveness* 392-398

[9] Prasetyadi G C 2017 Web-Based Expert System Application To Recommend Computer Specifications For Gaming Using Backward Chaining Inference Method *Journal of Information Systems* **13** (2) 110-117

[10] Salman F M and Abu-Naser S S 2019 Expert System for Castor Diseases and Diagnosis. International Journal of Engineering and Information Systems *International Journal of Engineering and Information Systems* **3** (3) 1-10

[11] del Barrio M J, Devesa M and Herrero L C 2012 Evaluating intangible cultural heritage: the case of cultural festivals *City, Culture and Society* **3** (4) 235-244

[12] Alhammadi D A A 2010 *Developing Expert System for Diabetes Mellitus Patients* Riyadh: King Saud University

[13] Maiyendra N A 2018 Perancangan Sistem Pakar Mendiagnosa Penyakit Kulit Pada Anak Dengan Menggunakan Metode Backward Chaining *Jursima* **6** (2) 6–17

[14] Gestwicki P and Ahmad K 2011 App inventor for Android with studio-based learning *Journal of Computing Sciences in Colleges* **27** (1) 55-63

[15] Guntamukkala V, Wen H J and Tarn J M 2006 An Empirical study of selecting software development life cycle models. *Human Systems Management* **25** (4) 265-278

[16] Wahyudi R and Prasetyo W D 2018 Implementing Forward, Backward Chaining and Certainty Factor in Responsive Web-Based Expert System of Cow Disease *International Journal of Informatics and Computer Science* **2** (1) 7-19

[17] Patel H H and Prajapati P 2018 Study and Analysis of Decision Tree Based Classification Algorithms *International Journal Of Computer Sciences And Engineering* **6** (10) 74-78

[18] Erdani Y 2012 Improving the Knowledge Performance using Ternary Grid Knowledge Acquisition and Model, WSEAS Transactions (Journals) on Information Science and Application *International Journal of Advanced Computer Science and Applications* **3** (2) 1-12

[19] Mohd C K and Shahbodin F 2015. Personalized Learning Environment: Alpha Testing, Beta Testing & User Acceptance Test.*Procedia-Social and Behavioral Sciences* **195** 837-843

[20] Joshi A, Kale S, Chandel S and Pal D K 2015 Likert Scale: Explored and Explained *Current Journal of Applied Science and Technology* **7** (4) 396-403

IOP Conf. Series: Materials Science and Engineering 1115 (2021) 012009 doi:10.1088/1757-899X/1115/1/012009

# Deep learning pre-trained model as feature extraction in facial recognition for identification of electronic identity cards by considering age progressing

**M Usgan*, R Ferdiana and I Ardiyanto**

Department of Electrical and Information Engineering, Gadjah Mada University, Yogyakarta, Indonesia

*Email: muhammadusgan@mail.ugm.ac.id

**Abstract.** The usage of the face recognition system is various, one of which is to identify missing people. Cases of missing persons can usually happen to people who do not carry an identity card, such as people with mental disorders. Identity cards have some information, namely name, address, date of birth, and a photo ID of a face. This information has stored at the civil registry office, which is an Indonesian government agency. Therefore to solve this problem, we use a photo ID as a dataset to identify someone's identity. Photo ID as a dataset has age progressing factors in facial recognition systems. Several previous studies have discussed age progressing factors. Previous research used MORPH, CACD, and FGNET datasets. The dataset has several photos of faces consisting of various age levels for the same subject, but it is different from the photo ID that we use. For the training data process, we only use one data, namely a photo ID used to recognize faces at this point. Therefore we use the pre-trained VGGFace2 model, and then, fine-tuning the data during the training process, we use AM-Softmax loss as a loss function. Then the classification is done using SVM. The method we use than compared with the Moustafa method, which also uses a pre-trained model. The results we obtained have a better accuracy performance of 0.9351 compared to the Moustafa method of 0.733

## 1. Introduction

The facial recognition system is a method used to recognize a person's identity, and it is a topic that has been studied in an unconstrained environment which has made rapid progress in recent studies. However, facial recognition caused by large age variation is still an intriguing subject for study. Face aging variation is one of the facial recognition system problems due to the large amount of intra-personal variation caused by age progression, which changes the face's shape and texture [1].

The identity card is indeed owned by every citizen, especially in the Republic of Indonesia's territory. The identity cards have some information, namely name, address, date of birth, and photo ID, which is stored in the civil registry. The information on the identity card is related to family data. Many missing person cases occurred where they did not carry the identity card, causing them to be stranded on the road or accommodated by the foundation. These cases of missing persons usually last for years. His family must have tried to find his relatives, but it is not easy to find them with the vast territory of Indonesia. These missing people are usually people who have mental disorders or people who have lost their memories. This study uses ID photos from identity cards as training data to identify someone's identity to overcome this problem. Facial recognition using identity cards is an example of facial recognition caused by age progressing. This is because the photo ID on the identity card has a different age from the current face.

In general, there are two main parts of the facial recognition system, which are feature extraction and classification. The feature extraction method used in the facial recognition system can use a pre-trained model was done by [1, 2, 3]. The pre-trained model [2] used is VGGFace [4] with the VGG-Very-Deep-16 CNN architecture model. The pre-trained model is then fine-tuned to solve the face recognition case caused by age progressing. In their research [2] used the feature fusion to perform a combination of features obtained from several layers of the model VGGFace, namely flatten, fc6, and fc7, with each vector length, is 25088, 4096, 4096. The feature fusion used is Multimodal Discriminant Correlation Analysis (MCDA) [5]. The output of the feature fusion is a feature that is used for classification.

Another method that uses a pre-trained model is [1]. The pre-trained model architecture used is AlexNet. The feature vectors generated from the model are then encode based on their proposed algorithm. However, their method is not suitable for use in the dataset proposed in this study because the method we are proposing uses only one dataset for each subject in the training data process. In contrast, the method they are proposing uses the fusion feature on the same identity for different images, requiring several data for the same identity.

Xiangyu Zhu et al. [3] proposed research using photo ID as a dataset known as the ID vs Spot (IvS) dataset. The research they are doing uses a pre-trained model where they build themselves using the LFW dataset. The method procedure they propose consists of three stages, namely classification (pre-learning), verification (transfer learning), and classification (fine-grained learning). However, what is different from what we are going to do is that our research uses only one data for each subj ect during the training process, in contrast to [3], which used two-sample data for each subject in the second stage of training data. Based on this description, it is known that the pre-trained model can be used to build facial recognition caused by age progressing.

VGGFace [6] is a pre-trained model used in facial recognition systems to identify electronic identity cards caused by age development. VGGFace2 uses the ResNet-50 [7] architecture. In addition to the pre-trained model used for feature extraction, a loss function when training data affects the results to be obtained. Therefore, this study uses the Addictive Margin Softmax (AM-Softmax) loss [8]. This loss function minimizes intra-class variation and maximizes inter-class features. This is so important that it produces a discriminating feature. This loss function has been compared with several other loss function approaches, including Softmax [9], Softmax + 75% dropout, Center Loss [10], NormFace [11], and A-Softmax [12]. Therefore, the softmax loss used by the pre-trained VGGFace2 model is replaced by AM-Softmax loss when fine-tuning. Furthermore, the classification method used is the Support Vector Machine (SVM) [1, 5, 7].

Based on the description above, this study used a pre-trained model from VGGFace2 and then performed fine-tuning. Next, we conduct training data using AM-Softmax loss as a loss function then classify it using SVM. This study's results were compared with [2], who also discussed facial recognition systems due to increasing age. The method they proposed was applied to the dataset we used

## 2. Related work

### 2.1. VGGFACE2

VGGFace2 [6] is also termed a new large-scale face dataset. VGGFace2 is a pre-training dataset trained with a collection of images obtained from Google Image Search, which contains 3.31 million images of celebrities from 9131 subjects, covering various ethnicities such as Chinese and Indians have more number than VGGFace. The gender ratio in the VGGFace2 dataset is 59.3% male and 48.7% female. Furthermore, each subject consists of 90 to 843 images, with an average of 362 images.

The pictures are composed of various poses, age, lighting, ethnicity, and professions. There are three main objectives to be achieved from collecting this dataset are

- It has a large number of both identities and many pictures of each identity
- It has a wide range of poses, ethnicity, and age
- To minimize label noise

The dataset from VGGFace2 is then trained using the ResNet-50 and SE-ResNet-50(SENET) architectures. The output of this network architecture is a face descriptor, which is then classified using the SVM (Support Vector Machine). The Cosine Similarity is used to represent the similarity between two vectors.

The first test of VGGFace2 is facing probing across poses and face probing across age. This first test uses the VGGFace and MS1M dataset as a comparison. Face probing across poses testing uses 3 conditions, namely front, three-quarter, and profile. Testing for probing across age uses 2 conditions, namely young and mature. Cosine similarity is used to obtain the similarity score. From the test results, it was found that VGGFace2 has better performance than the other two datasets.

The second test is the performance evaluation on the IJB-A, IJB-B, IJB-C dataset. The first thing to do is create a pre-trained model from VGGFace, MS1M, and VGGFace2. The architecture used is ResNet-50. Then evaluate the performance of the pre-trained model using IJB-A, IJB-B, IJB-C. From the results of the trial conducted, VGGFace has better performance than other pre-trained models. This is because VGGFace2 consists of various poses and ages.

### 2.2. AM-softmax (Additive Margin Softmax)

AM-Softmax loss function development is used for deep face verification, which can be seen as a metric learning problem. AM-Softmax is a learning large-margin face features that minimize intra-class and maximize inter-class features. This is important to do to get good performance. AM-Softmax is not the first approach that optimizes the margin face features. Previously, some approaches modified the softmax loss function, namely Angular softmax (A-Softmax) [12] and Large Margin Softmax(L-Softmax) [14]. The large margin property $\psi(\theta)$ of AM-Softmax [8] is more straightforward than that of A-Softmax and L-Softmax.

$$\psi(\theta) = \cos\theta - m \quad (1)$$

The value of m is a number that ranges from 0.35 to 0.4. AM-Softmax normalizes the weight and input features parameters on the last fully connected layer( $f_i$ ) as done by A-Softmax. Besides, AM-Softmax also introduced a new hyperparameter s, which scales the cosine value. In general, the equation of AM-Softmax is as follows

$$\mathcal{L}_{AMS} = -\frac{1}{N}\sum_{i=1}^{n}\log\frac{e^{s(\cos\theta_{yi} - m)}}{e^{s(\cos\theta_{yi}-m)} + \sum_{j=1;\, j\neq yi}^{c} e^{s\cos\theta_j}}$$

$$= -\frac{1}{N}\sum_{i=1}^{n}\log\frac{e^{s\left(W_{yi}^{T} f_i - m\right)}}{e^{s\left(W_{yi}^{T} f_i - m\right)} + \sum_{j=1;\, j\neq yi}^{c} e^{s W_j^T f_j}} \quad (2)$$

Where the value of s is the scaling factor, the scaling factor set is 30. $W_{yi}^T f_i$ is called the logit target. AM-Softmax was then tested using the LFW and MegaFace datasets. Subsequently, an appeal was made against several other types of loss functions, namely Softmax, Softmax + 75% dropout, Center Loss, NormFace, and A-Softmax. The results of this test show that AM-Softmax obtained good performance compared to other loss functions.

### 2.3. SVM (Support Vector Machine)

SVM is a technique introduced by Vapnik in 1992 as an efficient classification technique for non-linear problems. SVM works by trying to find hyperplanes by maximizing the distance between classes. Hongling Chen[13], in their research, discusses the effect of dimensional reduction on face recognition using the VGG-16 model as feature extraction and SVM is used as a classification. The dimension reduction used is PCA (Principal Component Analysis). The results of feature extraction from the VGG-16 model were then reduced in PCA dimensions and then classified using SVM. The value of the dimensional reduction used was 400. The proposed system was then tested on the CelebA dataset. Their research shows that with the reduction dimension of the vector features on the VGG-16 Net with SVM as a classification, the accuracy increases. Compared to some of the proposed methods such as DeepFace, TL Joint Bayesian, DFD, and several other methods, the proposed method has a better accuracy level.

Moustafa [2] discusses age invariant face recognition. The proposed method uses the VGG-Face model as feature extraction then performs feature fusion using the DCA method. The resulting fusion features are then classified using two methods: the SVM and KNN classifier. This proposed method was then tested on the FGNET and MORPH datasets. The trials conducted showed that the accuracy of SVM was superior to that of the KNN on the FGNET dataset, while on the MORPH dataset, the KNN classifier had better accuracy.

## 3. Proposed method

This section proposes a method to solve the facial recognition case for electronic ID cards, taking into account age progressing. The proposed method consists of 3 stages, namely image preprocessing, feature extraction, classification. The results of the proposed method were then compared with the method [2]. In general, the proposed method is shown in Figure 1.

### 3.1. Image preprocessing

The first step is image preprocessing, which is to crop the face area for training data and testing data. The dimensions of the cropped image adopted from VGGFace2 are 224 x 224 pixels where Multitask Cascaded Convolutional Networks (MTCNN) [15] is used to crop the face area. After cropping the face, then changing the image whose shape is RGB is converted to grayscale form. Then perform data augmentation [16] on the training data by rotating the image. The value of the image rotation is 4 ° and -4 °. An example of data augmentation is shown in Figure 2

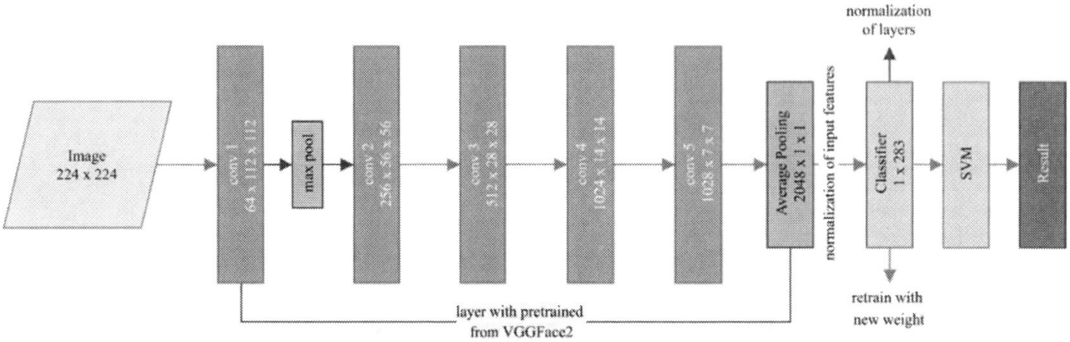


**Figure 1.** ResNet 50 Model where the input feature classifier layer and its layers are normalized

**Figure 2.** Data augmentation performed on training data is by rotating -4 ° and 4 ° in the image

### 3.2 Fine-tuning and feature extraction

After preparing the data to be used, the next step is to prepare the model and do fine-tuning. The model used is VGGFace2, which uses ResNet-50 architecture. This model has been provided publicly by [6]. Furthermore, the model architecture is modified in the classifier layer, changing the number of output classes to 1 x 283. The output class's value is adjusted to the number of data classes we have, which is 283 identities sourced from primary data and tertiary data.

The next step is to add a normalization function to the weight layer classification parameter and normalize the feature input on the classification layer. Next, change the required gradient parameter for convolution layer one to convolution layer five, namely weight and bias to false, while in the required gradient layer classifier is true. This is done because the model uses model parameters that have been previously trained, while the pre-trained layer parameters are updated during the training process. An overview of the model architecture is shown in Figure 1. The classifier uses a convolution layer based on the VGGFace2 model architecture, while the AM-Softmax loss function uses a fully connected layer. Therefore, the tests are carried out using the original model from VGGFace2 and then comparing it with the modified VGGFace2 architecture in the classifier layer using fully connected layers.

The next procedure is to perform training data, but beforehand it is necessary to prepare some optimizer parameters and loss function. The optimizer used is Stochastic Gradient Descent (SGD). The properties applied to the SGD optimizer are a learning rate of 0.001 and momentum of 0.9. Furthermore, the SGD optimizer parameters are the parameters of the weight and bias of the classifier layer. Next is the loss function used is AM-Softmax[8] with the parameter scale factor (s) 30 and margin (m) 0.4. Then the data training was carried out for 500 epochs.

### 3.3 Classification

The third step is to perform classification using a Support Vector Machine (SVM). Before using SVM, it is necessary to prepare vector features to fit the SVM. This vector feature results from feature extraction of the layer classifier features from the model that has been training 500 epochs. The SVM classifier property uses a linear kernel function [1, 7]

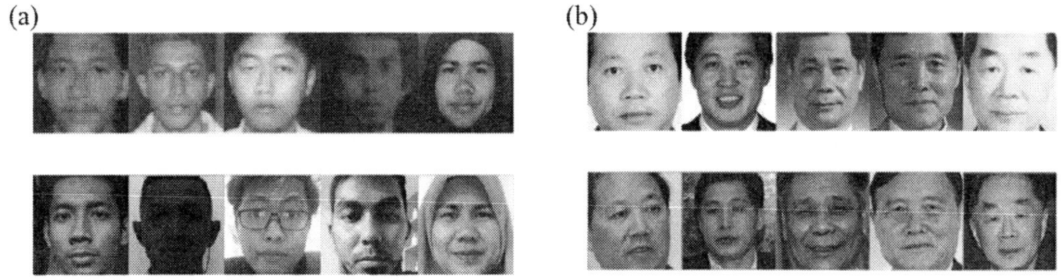

**Figure 3.** The dataset used during the training and testing process.
(a) is a primary dataset and (b) is a tertiary dataset

## 4. Experiment

This section explains the data used and the results of the experiments that have been carried out

### 4.1. Datasets

We used two types of datasets to test the proposed method, namely the primary dataset for electronic identity cards and the tertiary dataset IvS [3]. The primary dataset is data that has been collected manually through questionnaires, while the tertiary dataset is made available publicly. Both the primary and tertiary datasets that have been collected are then filtered against the usable data. A description of the number of objects between primary data and tertiary data is shown in Table 1. After filtering the data, the next process is image preprocessing, as described in section 3.1. Examples of primary and tertiary datasets are shown in Figure 3

**Table 1**. Description of the dataset

| Dataset | Total | | |
|---|---|---|---|
| | Subject | Training | Testing |
| Primary | 43 | 43 | 66 |
| Tertiary | 240 | 240 | 381 |
| Total | 283 | 283 | 447 |

### 4.2. Experiment setup

This section explains preparation when conducting experiments. Experiments were carried out using primary and tertiary datasets. The primary dataset is obtained either directly or through questionnaires from participants, and the tertiary dataset is the public IvS dataset. The two datasets are then carried out, preprocessing and separating the training data and test data. The dataset structure used is shown in Figure 4. The amount of training data used is one sample per subject. The training data used is a photo with a straight face facing the front with a plain background. An example of training data is a photo of an electronic identity card. So the scenario is to predict the current photo based on the photo ID card that has been taken in the past.

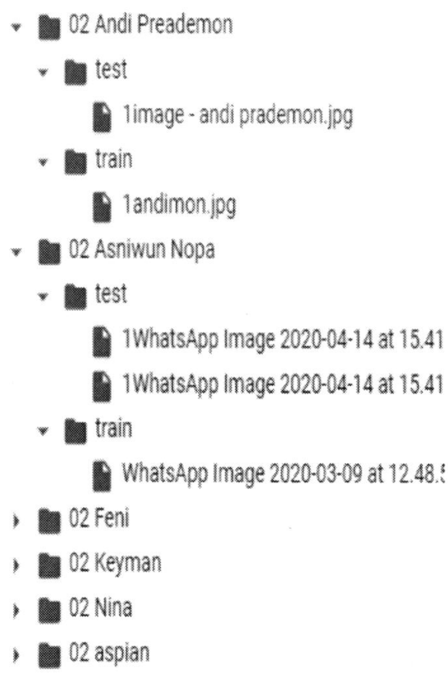


**Figure 4.** The dataset structure is used where the label is the name of the person. An example of a label is 02 Andi Prademon. Furthermore, in the label there is a train and test folder where the forlder contains the photos used

Besides, there also be compared the pre-trained VGGFace2 model on the use of AM-Softmax loss, namely, comparing the original VGGFace2 model with the modified VGGFace2 model in the classifier layer section by replacing it with a fully connected layer. In addition to seeing the accuracy obtained, but also seeing the similarity score obtained between the test data vector feature and predictive data, which is a training data vector feature. To perform this task, we use the public dataset LFW [17] and AFAD [18] as an evaluation in which the dataset is not part of the training model. Cosine similarity is used as a method to measure the similarity score between two vectors. The experiment uses Google Colabs tools with python 3 specifications and takes advantage of the CUDA runtime available on Google Colabs to speed up computation. The framework used is Pytorch.

### 4.3. Experimental results

The experimental results are obtained after carrying out the steps discussed in section 3. In this initial section, we discussed the comparison between the original VGGFace2 and VGGFace2 architecture, which in the classifier section uses the fully connected layer, or it can be called the comparison of the original VGGFace2 and modified VGGFace2. Both models use AM-Softmax loss as a loss function, which is different only in the classifier part. The model is then carried out by training data on the primary and tertiary datasets. Then evaluated by looking at the accuracy obtained and measuring the similarity score of vector features using the cosine similarity between the test data and the prediction results, which are training data. The evaluation data used is the public dataset from LFW [17] and AFAD [18]. This is done to justify a prediction based on the specified threshold score. A comparison of the VGGFace2 model between the original and modified is shown in Figure 5.

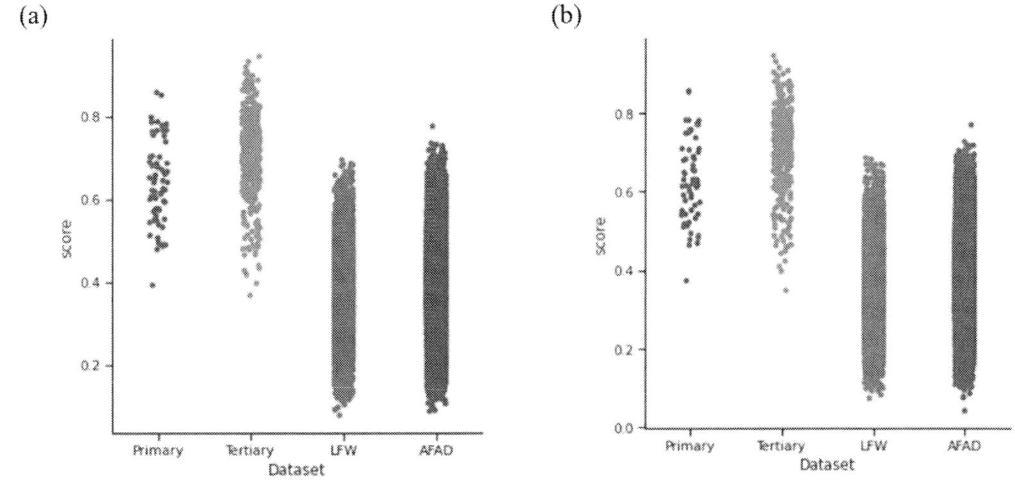


**Figure 5.** The similarity score is obtained from the vector feature of the test data on the prediction results (vector features from the training data). (a) is original VGGFace2 and (b) is Modified VGGFace2. Based on the good picture, the original VGGFace2 and Modified VGGFace2 tend to get a similarity score which tends to be greater than 0.5

Sometimes we need to justify an object based on its similarity and disimilarity values. Figure 5 shows a comparison between original VGGFace2 and modified VGGFace2 using primary, tertiary, LFW, and AFAD datasets. The primary and tertiary datasets represent similarity scores. The LFW and AFAD datasets represent dissimilarity scores. The threshold score to be determined in this experiment is 0.5. These results are shown in Table 2 and Table 3

**Table 2.** The similarity score of test data from primary and tertiary datasets with threshold values above 0.5

| Model | Classifier | Threshold | Primary | Tertiary |
|---|---|---|---|---|
| Original VGGFace2 | Convolutional layer | 0.5 | 0.8906 | 0.9518 |
| Modified VGGFace2 | Fully connected layers | 0.5 | 0.8906 | 0.9519 |

**Table 3.** The dissimilarity score of the test data from the LFW and AFAD datasets with a threshold value below 0.5

| Model | Classifier | Threshold | LFW | AFAD |
|---|---|---|---|---|
| Original VGGFace2 | Convolutional layer | 0.5 | 0.920 | 0.875 |
| Modified VGGFace2 | Fully connected layers | 0.5 | 0.9240 | 0.8988 |

Table 2 and Table 3 show the similarity score above 0.5 and dissimilarity score below 0.5. Table 2 shows that modified VGGFace2 is 0.9519 has better value than original VGGFace2 is 0.9518 in the tertiary dataset, and Table 3 shows modified VGGFace2 also has a better value than original VGGFace2, both the LFW and AFAD datasets. Furthermore, the accuracy obtained is shown in Table 4

**Table 4.** The accuracy obtained by Original VGGFace2 and Modified VGGFace2

| Metode | Dataset | | |
|---|---|---|---|
| | Primary | Tertiary | Primary + Tertiary |
| Original VGGFace2 | 0.9696 | 0.9265 | 0.9328 |
| Modified VGGFace2 | 0.9696 | 0.9291 | 0.9351 |

A comparison of the accuracy obtained between original and Modified VGGFace2 in Table 4 shows that Modified VGGFace2 is 0.9291 has better accuracy on the tertiary dataset than original VGGFace2 is 0.9265. Based on Table 4, it can be concluded that modifying the classifier into fully connected layers on the classifier layer increases the accuracy obtained and the similarity score.

After determining the model to be used, the next step is to compare the proposed method with the other method [2] to determine the suitable method for the facial recognition system based on the dataset we use. Both our proposed method and [2] method also use a pre-trained model. The experimental results are shown in Table 5.

**Table 5.** Experimental results

| Metode | Dataset | | |
|---|---|---|---|
| | Primary | Tertiary | Primary + Tertiary |
| ResNet + AM-Softmax Loss (proposed method) | 0.939 | 0.918 | 0.9217 |
| ResNet + AM-Softmax Loss + SVM (proposed method) | 0.9696 | 0.9291 | 0.9351 |
| Moustafa [2] | 0.65 | 0.748 | 0.733 |

The experimental results in Table 5 show that our proposed method, namely ResNet + AM-Softmax Loss + SVM, shows better accuracy performance than the method [2]. Besides, the fine-tuning result model as feature extraction using AM-Softmax as a loss function and SVM as a classifier obtained better accuracy performance compared to without using SVM as a classifier. Thus our proposed method is suitable for the electronic identity card photo dataset in facial recognition systems.

## 5. Conclusion

Based on the experimental results that have been stated in section 4, it can be concluded that using the VGGFace2 pre-trained model then adjusts the classifier section, namely the input of the classifier feature, and the classifier layer is normalized. Next, do training data for 500 epochs and AM-Softmax as a loss function. Then the model is used as feature extraction and uses SVM as a classifier. It should be noted that this experiment conducted a comparison between the original VGGFace2 (layer classifier using convolution layer) and modified VGGFace2 (layer classifier using fully connected layer). The comparison of the two shows that modified VGGFace2 has a better accuracy of 0.9351 compared to the original VGGFace2 0.9328. The accuracy performance obtained from modified VGGFace2 and then compared with method [2] obtained an accuracy of 0.733, which also uses a pre-trained model. based on that explanation, this shows that modified VGGFace2 has better accuracy performance than the method [2] so that it can be used as a facial recognition system method for the electronic photo identity card.

## References

[1] Shakeel M S and Lam K M 2019 Deep-Feature Encoding-Based Discriminative Model for Age-Invariant Face Recognition *Pattern Recognit.* **93** (16) 442–457

[2] Moustafa A A, Elnakib A and Areed N F F 2020 Age-Invariant Face Recognition Based on Deep Features Analysis *Signal, Image Video Process.* **14** (2) 1027–1034

[3] Zhu X, Liu H, Lei Z, Shi H, Yang F, Yi D, Qi G and Li S Z 2019 Large-Scale Bisample Learning on ID Versus Spot Face Recognition *International Journal of Computer Vision* **127** (4) 684–700

[4] Parkhi O M, Vedaldi A and Zisserman A 2015 Deep Face Recognition *Proceedings of the British Machine Vision Conference (BMVC)* **41** 1-12

[5] Haghighat M, Abdel-Mottaleb M and Alhalabi W 2016 Discriminant Correlation Analysis: Real-Time Feature Level Fusion for Multimodal Biometric Recognition *IEEE Trans. Inf. Forensics Secure* **11** (9) 1984–1996

[6] Cao Q, Shen L, Xie W, Parkhi O M and Zisserman A 2018 VGGFace2: A dataset for recognizing faces across pose and age in *Proceedings - 13th IEEE International Conference on Automatic Face and Gesture Recognition* 67-74

[7] He K, Zhang X, Ren S and Sun J 2016 Deep residual learning for image recognition in *Proceedings of the IEEE Computer Society Conference on Computer Vision and Pattern Recognition* 770–778

[8] Wang F, Cheng J, Liu W and Liu H 2018 Additive Margin Softmax for Face Verification *IEEE Signal Processing Letter* **25** (7) 926–930

[9] Sun Y, Wang X and Tang X 2014 Deep learning face representation from predicting 10,000 classes in *Proceedings of the IEEE Computer Society Conference on Computer Vision and Pattern Recognition* 1891–1898

[10] Wen Y, Zhang K, Li Z and Qiao Y 2016 A discriminative feature learning approach for deep face recognition in *Computer Vision -- ECCV 2016* 499–515

[11] Wang F, Xiang X, Cheng J and Yuille A L 2017 NormFace: L2 hypersphere embedding for face verification in *MM 2017 - Proceedings of the 2017 ACM Multimedia Conference* 1041–1049

[12] Liu W, Wen Y, Yu Z, Li M, Raj B and Song L 2017 SphereFace: Deep hypersphere embedding for face recognition in *Proceedings - 30th IEEE Conference on Computer Vision and Pattern Recognition, CVPR 2017* 6738–6746

[13] Chen H and Haoyu C 2019 Face Recognition Algorithm Based on VGG Network Model and SVM *Journal Physics Conference Series* **1229** (1) 012015

[14] Liu W, Wen Y, Yu Z and Yang M 2016 Large-Margin Softmax Loss for Convolutional Neural Networks in *International Conference on Machine Learning* 507–516

[15] Zhang K, Zhang Z, Li Z and Qiao Y 2016 Joint Face Detection and Alignment Using Multitask Cascaded Convolutional Networks *IEEE Signal Process. Lett.* **23** (10) 1499–1503

[16] Wang X, Wang K and Lian S 2020 A survey on face data augmentation for the training of deep neural networks *Neural Comput. Appl.* **32** (19) 15503–15531

[17] Huang G B, Ramesh M, Berg T and Learned-Miller E 2008 Labeled Faces in the Wild: A Database for Studying Face Recognition in Unconstrained Environments *Workshop on Faces*

*in 'Real-Life' Images: Detection, Alignment, and Recognition* 1-14

[18] Niu Z, Zhou M, Wang L, Gao X and Hua G 2016 Ordinal regression with multiple output CNN for age estimation in *Proceedings of the IEEE Computer Society Conference on Computer Vision and Pattern Recognition* 4920-4928

IOP Conf. Series: Materials Science and Engineering 1115 (2021) 012010 doi:10.1088/1757-899X/1115/1/012010

# Development of Electromyography Signal Function to Control Robot Arm Working Adaptively in the Reverse Direction of the Human Arm Movement

**B Siregar*, O S Sitompul, M Zarlis and Fahmi**

Computer Science Postgraduate Program, Faculty of Computer Science and Information Technology, Universitas Sumatera Utara, Indonesia

*Corresponding author: baihaqi@usu.ac.id

**Abstract**. Malfunctioning conditions in one arm can be experienced by humans due to congenital defects, accidents, or side effects of experienced health problems. This has an impact on reducing the effectiveness of work because in normal conditions humans often need a pair of arms that are able to function actively. This situation was tried to be overcome by utilizing a mechanical robotic arm controlled by an Arduino microcontroller and Myo Armband as a sensor for orientation and an electromyography signal from one of the user's normal arms. The results of sensing from the user's arm that have been successfully read are then processed using an algorithm that is built to provide a decision in the form of what action the robot arm should take. The test results obtained show that the robot arm responds real-time to input from the orientation of the user's arm, although it is also found that sometimes the robot arm does not respond to input from the electromyography signal.

## 1. Introduction

Humans in normal conditions have a pair of arms that function systematically, both of which are moved to help with daily human activities. However, there are conditions when one of these arms becomes malfunctioning or is no longer able to move, which among others is caused by congenital defects, accidents experienced, or due to health problems that interfere with the nervous system such as stroke or Parkinson's. Humans in this condition certainly have difficulty carrying out daily activities without help from other people. One of the things that can be done to help humans in conditions like this is to implement robotics technology as a mechanical means. The development of robotics technology is a necessity that must be accepted by the world community today. This technology can really help human life if it is functioned wisely and efficiently. The robot arm is a product of robotics technology. This product is a robot that is shaped like a human arm and is also capable of moving like an arm in humans. There are several types of robot arms, but what is generally used in the industrial and educational world and is available in the market is a robot arm with a Degree of Freedom (DoF) of 3 to 7. DoF is the number of independent parameters that define the condition of the mechanical system in the robot. The more DoF the robot arm has, the more possible movements the robot can do.

Some of the previous research related to this research topic has resulted in various interesting topics. Among them are what Gordleva et al. did in (2020), namely recording and processing multimodal signals that were collected using foot motor images and electromyography (EMG) signals from the feet [1]. In the case of lower limb representation, there is still a problem in differentiating the intentions of the leg movements and differentiating them in the built system. Research conducted is based on directions, based on, and classification of multimodal data. The developed system can analyze up to 15 signals occurring simultaneously and real-time during the movement. Ouyang et al. (2013) collect electroencephalography (EEG) data to then classify it into one of four control signals: raise, lower, turn clockwise and rotate counter clockwise [2]. In addition, the system observes the collected data for EMG signals indicating facial muscle movements. A personal set of EEG data patterns was trained for each individual, with each control signal taking only a few minutes to train initially. EMG signal detection is measured based on a common threshold for all users. After the user trains his personal data into the system, each positive detection triggers a signal to the linked robotic arm to perform a separate action accordingly. Sarker et al. (2017) developed hardware capable of acquiring and transmitting electrocardiography (ECG) and EMG data wirelessly in real-time while keeping overall energy consumption low [3]. The device signal acquisition performance has been evaluated for ECG and EMG at 8 channels 24-bit resolution / channel with a configuration of 500 samples / second. Energy consumption measurements have been taken and it was found that the device can continue to work for up to 13.6 hours with a 3.7V 1700 mAh battery. Bird et al. (2020) measuring signals from the brain and forearm muscles [4]. EMG data was collected from a 4-class movement classification experiment via Myo Armband, and a 3-grade mental state EEG dataset was obtained through Muse EEG Headband. The hyper heuristic multi-objective evolutionary search method is used to find the best network hyperparameter. They then used this optimized deep neural network topology to classify the EMG and EEG signals, which obtained an accuracy of 84.76% and 62.37%, respectively. Borgul et al. (2012) conducted research using an intuitive control system of the upper limb exoskeleton and other mechatronic devices with EMG and EEG for people with different degrees of musculoskeletal system damage [5]. The technology built allows controlling the equipment by thinking about it. Benalcázar et al. (2017) conducted research for new models of real-time hand gesture recognition [6]. The input of this model is surface electromyography as measured by Myo Armband positioned on the forearm. The output is a requirement label which is executed by the user at any time. Model based on k-Nearest Neighbor algorithm and dynamic time warping. The evaluation results show that this model gets an accuracy value of 86% while the Myo system is accurate at 83%. Silva et al. (2016) describes the use of sensor data fusion techniques to combine data from the Myo Armband and Leap Motion sensors to create real-time 3D simulations of arm movements, including forearms, hands and fingers [7]. The experimental results show that the monitoring algorithm is able to combine data from the two sensors resulting in considerable precision and arm tracing.

## 2. Material and Method

Myo Armband is used as a device for sensing EMG signals from the human arm muscles as a user of the system built on this research. Myo Armband is mounted on the right arm and left arm of the user in turn to test the adaptive nature of the algorithm developed in analyzing arm position and determining the directional movement of the robot arm. Myo Armband is a gesture controller consisting of 8 medical grade EMG muscle sensors and an Inertial Measurement Unit (IMU) [8]. When the Myo Armband is attached to a human arm, this device senses signals from the muscles in the arm and reads the orientation of the arm movement using a combination of sensing results from the gyroscope, accelerometer, and magnetometer sensors. The Myo Armband device can be seen in Figure 1.

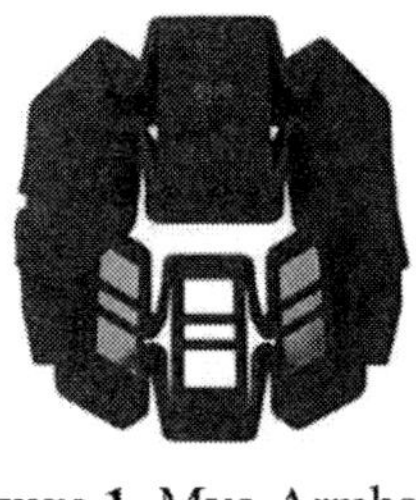

**Figure 1.** Myo Armband.

The robot arm used in this research is coded 5AL, has 5 DoF with each servo motor functioning for waist rotation (A), shoulder rotation (B), elbow rotation (C), wrist rotation (D), and gripper (E) as shown in Figure 2.

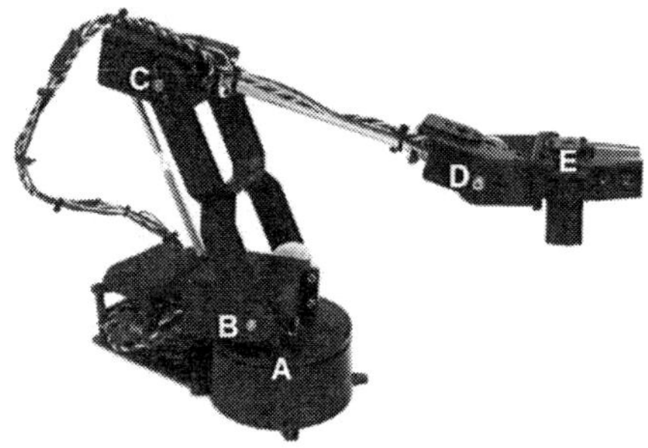


**Figure 2.** Robot Arm.

The general architecture of the system built in this research can be seen in Figure 3. It can be seen in Figure 3 that the system built consists of three stages, namely the sensing stage, the computing stage, and the actuating stage.

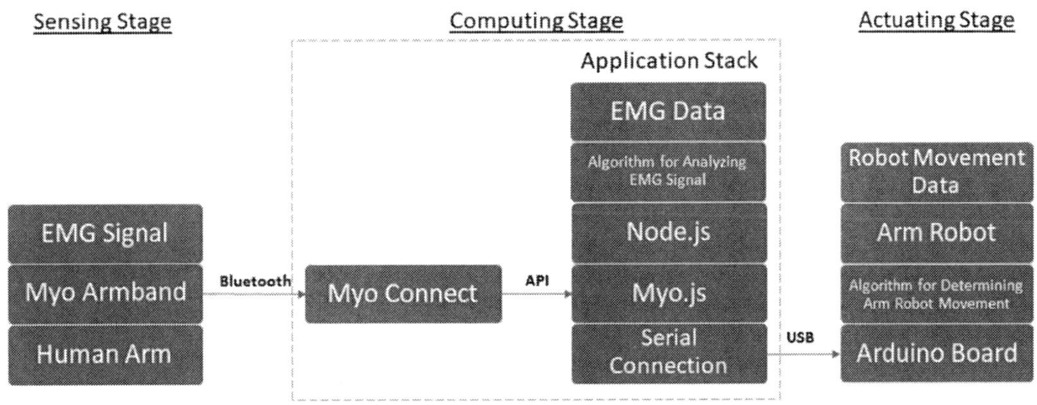


**Figure 3.** General Architecture.

### 2.1. Sensing Stage

At this stage the Myo Armband is attached to the user's arm so that the movement of the muscles from the user's arm can be read as an EMG signal to be transmitted to the computing stage via the Bluetooth communication line for further processing.

### 2.2. Computing Stage

At this stage, the EMG signal on the sensing stage is processed by the Myo Connect application to then recognize the pattern of the gesture from the palm and the orientation of the user's arm. The palm gestures used in this research are the default gestures from Myo Armband, namely pinch, fist, open, wave in, and wave out. The EMG pattern and signal are then read through the Application Programming Interface (API) by an application built using Node.js using the Myo.js library which is then processed by an algorithm built by the researchers. The algorithm is useful for analyzing and determining decisions that must be made by the next robot arm on the actuating stage [9-14]. The application built also produces raw data recording of the EMG signal for further, more complex analysis purposes. An example of a successfully read EMG signal can be seen in Table 1.

**Table 1.** EMG Signal Reading Sample Value.

| Time | Muscle Sensor | | | | | | | |
|---|---|---|---|---|---|---|---|---|
| (hh:mm:ss:ms) | #1 | #2 | #3 | #4 | #5 | #6 | #7 | #8 |
| 18:31:44:4 | 0 | 0 | -9 | -10 | -1 | -1 | -2 | -1 |
| 18:31:44:23 | 1 | -3 | -2 | 30 | 9 | 5 | 1 | 1 |
| 18:31:44:24 | 1 | 0 | -34 | -31 | -3 | -1 | -1 | 1 |
| ... | ... | ... | ... | ... | ... | ... | ... | ... |
| 18:31:56:515 | 3 | -4 | -17 | -29 | -7 | -3 | -15 | 1 |

Some of the code used at this stage to analyze the orientation of the user's arm appears in the following line of code:

```
}
function calc(x, in_min, in_max, out_min, out_max){
  return (x - in_min) * (out_max - out_min) / (in_max - in_min) + out_min;
}
var dataSend = "0:0:0";
var lastDataSend = "0:0:0";
var lastTime = 0;
Myo.on('imu', function(data) {
        var time = (new Date()).getTime();
        if (lastTime < time - 50) {
                var orz = data.orientation.z.toFixed(2);
                var orzt = calc(orz, -0.6, 0.6, 0, 180);
                var orx = data.orientation.y.toFixed(2);
                var orxt = calc(orx, -0.6, 0.6, 0, 180);
                dataSend = "M:"+orzt.toFixed()+":"+orxt.toFixed()+"\n";
                send_serial(dataSend);
                lastTime = time;
        }
```

### 2.3. Actuating Stage

At this stage, a robotic arm is prepared which is connected to the Arduino microcontroller. The schematic diagram of the robot arm and the microcontroller connected and its power supply can be seen in Figure 4.

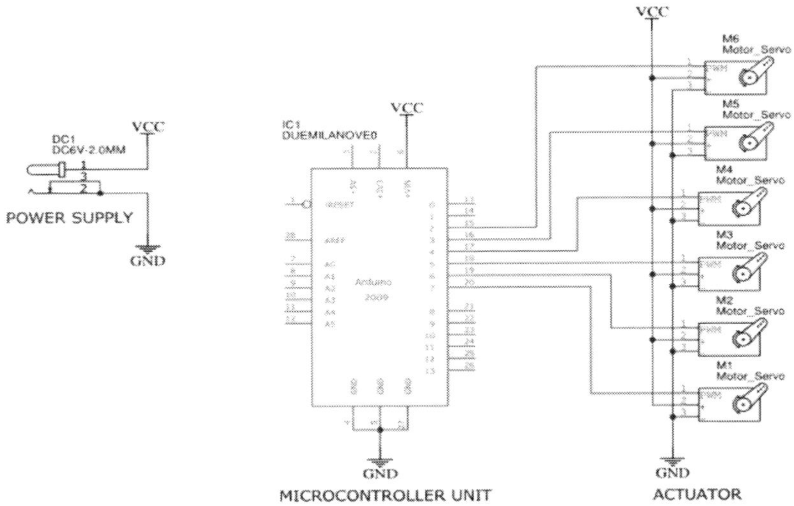


**Figure 4.** Schematic Diagram of Robot Arm.

The algorithm developed by the researchers is also prepared so that it can be run on a microcontroller. The function of this algorithm is to give commands from the microcontroller to the robot arm to make movements according to the user's arm movement, with the exception of the user's right arm (which means that the Myo Armband is attached to the user's right arm) moves to the left of the user's body, the robot arm will move to the right direction of the robot. Likewise, if the user's left arm (which means Myo Armband is attached to the user's left arm) moves to the right of the user's body, the robot's arm will move to the left of the robot. This concept is useful so that if the user's left arm malfunctions, it can be replaced by a robot arm and vice versa if the user's right arm malfunctions it can be replaced by a robotic arm. So that users are helped in activities using a pair of hands. The detection of the location of the Myo Armband attached to the right or left arm is automatically detected so that adaptive movement of the robot arm also applies in accordance with the objectives of this research. Some of the code that is executed at this stage appears in the following line of code:

```
event = getValue(inputString, ':', 0);
data0 = getValue(inputString, ':', 1);
data1 = getValue(inputString, ':', 2);
if (event == "P") {
 if (data0 == "G") {
  gripper.write(160);
  delay(100);
 }
 else if (data0 == "R") {
  gripper.write(0);
  delay(100);
 }
 else if (data0 == "T") {
  int pos = wrist.read();
  if(pos<=120 && pos>=60){
   wrist.write(pos-30);
  }
  else{
   wrist.write(60);
  }
  delay(100);
 }
```

```
      else if (data0 == "N") {
        int pos = wrist.read();
        if(pos<=120 && pos>=60){
          wrist.write(pos+30);
        }
        else{
          wrist.write(120);
        }
      }
    }
    else if (event == "M") {
      int z = data0.toInt();
      int x = data1.toInt();
      x = abs(x - 180);
      shoulder.write(x);
      //elbow.write(x);
      //z = abs(z - 180);
      waist.write(z);
    }
  }
```

## 3. Result and Discussion

Tests are carried out to determine whether the objectives of this research have been achieved or not. Before testing, it is necessary to ensure that all the services required at the computing stage are running, Myo Armband has been connected via the Bluetooth interface to the main computer, and the robot arm is on with sufficient power supply. Next is the calibration process to determine the center point of the user's arm against the robot arm, as shown in Figure 5. Calibration process by positioning the user's arm and pinch the palm of the hand.

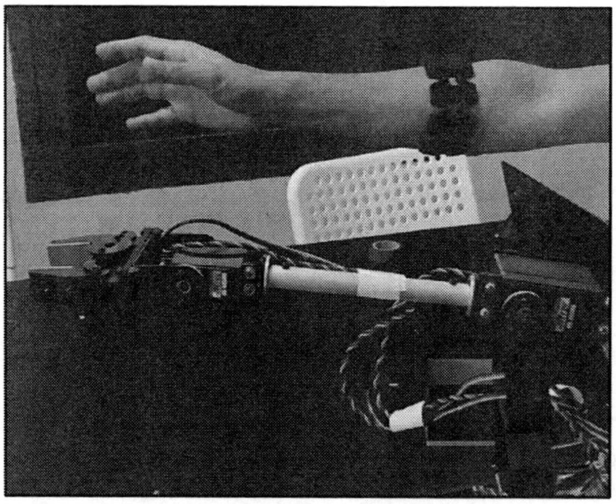

**Figure 5.** Calibration Process.

After the calibration process has been successfully carried out, the next step is to make several movements on the user's arms and palms, namely moving forward and backward on the X-axis (surge), moving left and right on the Y-axis (sway), moving up and down on the Z-axis (heave), tilting forward and backward on the Y-axis (pitch), and turning left and right on the Z-axis (yaw). The results of the test can be seen in Table 2.

**Table 2.** Testing Results.

| No. | User Hand Movement and Palm Gestures | Movement of the Robot Arm | Robot Arm Response |
|---|---|---|---|
| 1 | Surge | Surge | Real-time |
| 2 | Sway | Sway | Real-time |
| 3 | Pitch | Pitch | Real-time |
| 4 | Heave | Heave | Real-time |
| 5 | Yaw to the Left (Myo Armband on the Right Arm) | Yaw to the Right | Real-time |
| 6 | Yaw to the Right (Myo Armband on the Left Arm) | Yaw to the Left | Real-time |
| 7 | Pinch | Move to the Starting Point | Sometimes it does not Respond |
| 8 | Open | Gripper Opens | Sometimes it does not Respond |
| 9 | Fist | Gripper Clamping | Sometimes it does not Respond |
| 10 | Wave In | Wrist Move Down | Sometimes it does not Respond |
| 11 | Wave Out | Wrist Move Up | Sometimes it does not Respond |

The testing process carried out to determine the adaptive behavior of the robot arm that functions as the left arm can be seen in Figure 6.

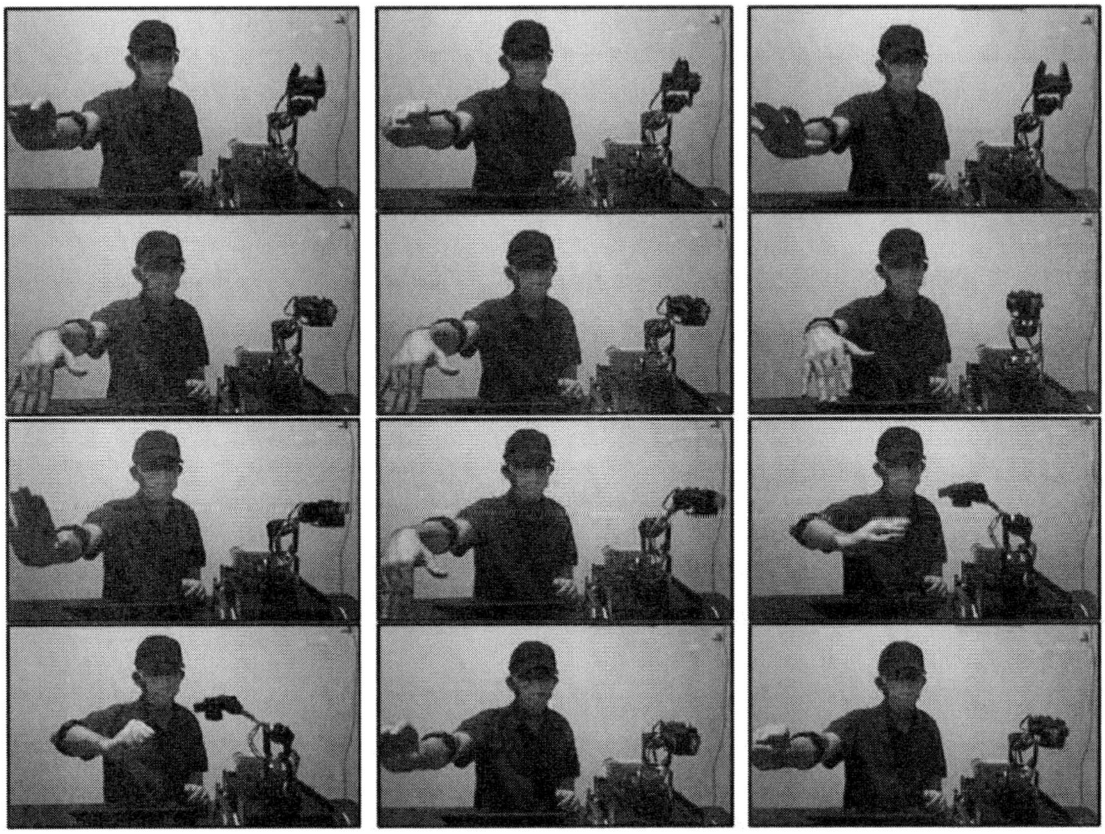

**Figure 6.** Robot Arm Serves as Left Arm.

While the testing process that determines the robot arm to function as the right arm of the user can be seen in Figure 7.

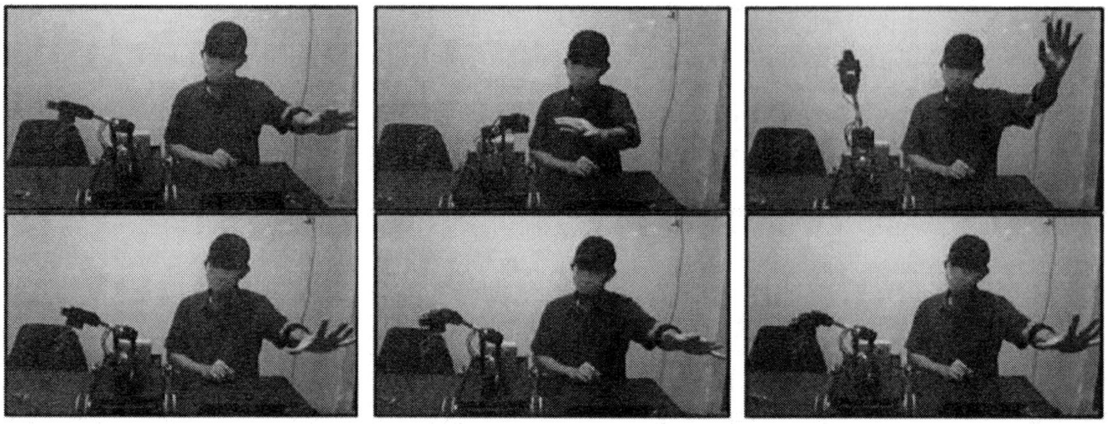

**Figure 7.** Robot Arm Serves as Right Arm.

In the testing process with the robot arm serves as the user's right arm, it is more often found that the robot arm fails to respond to the user's left palm gesture. The failure conditions that often occur are when the user's palms are fisted which should be responded to with the gripper clamping and when the user's palms wide open, which should be responded to by the robot arm with the gripper open. After evaluation, this occurs because the stimulus from the left arm of the user is weaker than the stimulus from the right arm. So that the results of sensing the EMG signal from muscle gesture fail to be translated by the system.

## 4. Conclusion

The conclusions that can be drawn from the results of this research are as follows:

- The robot arm is able to respond real-time to input in the form of sensing the motion of the user's arm, namely for the movement of surge, sway, pitch, heave, and yaw. This response is the result of sensing from IMU on Myo Armband;
- The adaptive response of the robot arm which functions as the left or right arm also takes place in real-time. In this case, the system that was built successfully identified the location of Myo Armband and then processed it using an algorithm developed by the researcher by giving an opposite response to the arm moving to the left or right of the user's body;
- The robot arm also succeeds in responding to input in the form of an EMG signal generated from sensing the user's arm muscle, but sometimes the robot arm also doesn't respond. This is due to noise in the EMG signal or the tired condition of the user's arm when the test is carried out several times, which affects the processing of the read EMG signal and results in the system failing to make decisions based on the range of input values in the algorithm used. It can also be due to the incorrect position and movement of the arm and palm gesture of the user during the calibration process.

## References

[1] Gordleeva S Y, Lobov S A, Grigorev N A, Savosenkov A O, Shamshin M O, Lukoyanov M V, Khoruzhko M A and Kazantsev V B 2020 Real-Time Eeg-Emg Human-Machine Interface-Based Control System for A Lower-Limb Exoskeleton *IEEE Access* **8** 84070–84081

[2] Ouyang W, Cashion K and Asari V K 2013 Electroencephelograph Based Brain Machine Interface for Controlling a Robotic Arm *IEEE Applied Imagery Pattern Recognition Workshop: Sensing for Control and Augmentation (AIPR 2013)*

[3] Sarker V K, Jiang M, Gia T N, Anzanpour A, Rahmani A M and Liljeberg P 2017 Portable Multipurpose Bio-Signal Acquisition and Wireless Streaming Device for Wearables *IEEE Sensors Application Symposium Procceding* 3–8

[4] Bird J J, Kobylarz J, Faria D R, Ekart A and Ribeiro E P 2020 Cross-Domain MLP and CNN Transfer Learning for Biological Signal Processing: EEG and EMG *IEEE Access* **8** 54789–54801

[5] Borgul A, Margun A, Zimenko K, Kremlev A and Krasnov A 2012 Intuitive Control for Robotic Rehabilitation Devices by Human-Machine Interface with EMG and EEG Signals *17th International Conference on Methods and Models in Automation and Robotics MMAR* 308-311

[6] Benalcázar M E, Jaramillo A G, Zea A and Páez A 2017 Hand Gesture Recognition Using Machine Learning and the Myo Armband *European Signal Processing Conference (EUSIPCO)* **25** 1075–1079

[7] Silva E C P, Clua E W G and Montenegro A A 2016 Sensor Data Fusion for full Arm Tracking using Myo Armband and Leap Motion *14th Brazilian Symposium on Computer Games and Digital Entertainment (SBGames)* 128-134

[8] Uzair M, Khadija A S, Waqas M and Shanshan T 2020 Applications of Myo Armband using EMG and IMU Signals *IEEE 3rd International Conference on Mechatronics, Robotics and Automation (ICMRA 2020)*

[9] Chen J, Bi S, Zhang G and Cao G 2020 High-Density Surface EMG-Based Gesture Recognition Using a 3D Convolutional Neural Network *Sensors* **20** (4) 1201

[10] Ding Z, Yang C, Tian Z, Yi C, Fu Y and Jiang F 2018 sEMG-Based Gesture Recognition with Convolution Neural Networks *Sustainability* **10** (6) 1865

[11] Trisnawan, Sanjaya W S M, Sambas A, Jannah M, Rahayu D S, Mamat M and Mohamed M A 2019 Design of 4 Dof Robot ARM Based on Adaptive Neuro-Fuzzy (ANFIS) using Vision in Detecting Color Objects *International Journal of Recent Technology and Engineering* **8** (2S7) 224-227

[12] Mujiarto, Djohar A, Komaro M, Mohamed M A, Rahayu D S, Sanjaya W S M, Mamat M, Sambas A and Subiyanto 2019 Colored object detection using 5 dof robot arm based adaptive neuro-fuzzy method *Indonesian Journal of Electrical Engineering and Computer Science* **13** (1) 293-299

[13] Morone G, Cocchi I, Paolucci S and Iosa M 2020 Robot-assisted therapy for arm recovery for stroke patients: state of the art and clinical implication *Expert review of medical devices* **17** (3) 223-233

[14] Jiang B, Karimi H R, Yang S, Gao C and Kao Y 2020 Observer-Based Adaptive Sliding Mode Control for Nonlinear Stochastic Markov Jump Systems via T–S Fuzzy Modeling: Applications to Robot Arm Model *IEEE Transactions on Industrial Electronics* **68** (1) 466-477

IOP Conf. Series: Materials Science and Engineering 1115 (2021) 012011 doi:10.1088/1757-899X/1115/1/012011

# Investigation of landslides and floods utilizing a geographic information system (A case study: Trenggalek Regency, East Java)

**A R Putri***, **A Widodo and M S Purwanto**

Department of Geophysical Engineering, Faculty of Planning and Geo-Engineering, Sepuluh Nopember Institute of Technology, Surabaya

*Corresponding Author Email: ameliaadict@gmail.com

**Abstract.** Trenggalek Regency is an area that is affected by floods and landslides every year. This study aimed to map areas that are prone to floods and landslides in each district using Geographic Information System (GIS) and Remote Sensing. In its processing, this disaster hazard map used secondary data sourced from the Shapefile Ina-Geoportal, soil type map, DEM data, BMKG data, Google Earth imagery, and BPS data for Trenggalek Regency which are then processed using ArcGis 10.7 software. These produced parameters for making flood and landslide hazard maps in the form of rainfall, slope, soil type, land use, and exposed population in Trenggalek Regency. After classifying, scoring, and overlaying the existing parameters, flood and landslide hazard maps were generated. The flood hazard map covered seven districts of the Trenggalek Regency, including Tugu, Karangan, Trenggalek, Pogalan, Gandusari, and a small part of Kampak are very prone to floods with an average of the districts in Trenggalek Regency have a result of fairly prone to floods. Meanwhile, the landslide hazard map shows that several districts such as Trenggalek, Watulimo, and Panggul have a high landslide hazard level, while the dominance of other areas is landslides with a moderate hazard level. With this processing, it can be seen that areas with high populations and residential areas are prone to floods or landslides.

## 1. Introduction

Nobody can predict exactly when a natural disaster may occur. Natural disasters in Indonesia have become commonplace as seen in terms of geology, geomorphology, or climatology. Even though Indonesia is a very large country, Indonesian people do not understand the significance of disaster mitigation before a disaster occurs. It is known that during the rainy season, the rainfall is very high in various areas, causing floods and landslides. However, during the dry season, there is drought in most areas. It is because Indonesia is also included in the tropical zone which has two seasons of the rainy season and the dry season [1].

Floods and landslides are disasters that are frequently experienced in almost all parts of the country every year, although these disasters cause losses for affected residents [2]. Based on research conducted by CRED (Center for Research on the Epidemiology of Disasters) in 1907-2007, floods had the highest frequency and was in the first position as the largest natural disaster that frequently occurred, while landslides were in the fourth position [3]. In an area, floods and landslides commonly occur simultaneously, for example in Trenggalek Regency.

IOP Conf. Series: Materials Science and Engineering 1115 (2021) 012011 doi:10.1088/1757-899X/1115/1/012011

According ref [4], most of the areas in the Trenggalek Regency are prone to both medium and high levels of floods and landslides. There are two reasons why Trenggalek Regency is prone to floods and landslides, including due to natural and human factors. With advances in geographic information system technology and the ease of obtaining data, comprehensive observations can be made about areas prone to floods and landslides in Trenggalek Regency. Therefore, this paper aimed to map the areas prone to floods and landslides in Trenggalek Regency.

## 2. Theoretical Basis

### 2.1. Regional Geology

Van Bemmelen (1949) stated that Trenggalek Regency was located in the south of East Java which was still included in the southern mountainous zone. Trenggalek Regency is part of East Java Province, that according to BNPB in 2019 amounted to 689.034 people. If seen from the provincial capital of Surabaya, Trenggalek is 180 km away and an area of approximately 1.205,22 square km [6]. If seen from the morphology, the structure of the Trenggalek area tends to be in the form of faults, folds, or intrusions. According reference [7], Trenggalek Regency is divided into three original forms, including structural, fluvial, and volcanic origin.

### 2.2. Floods

Flood is a natural disaster that occurs when it rains in an area for a long period with high intensity but the surrounding riverbanks are unable to accommodate the falling water. Besides, another cause of floods is that the embankment or dam no longer accommodates the flow of water that falls when there is very heavy rain. This flood disaster usually raises a lot of environmental damage, changes in soil shape, and a decrease in the volume of rainwater storage [8]. Floods are known to come rapidly, even in less than 5 hours. If heavy rains begin to fall, floods can occur. It is defined as a flash flood. Because floods can arrive at any time, before a flood occurs, residents should be aware of the areas that are prone to floods [9].

### 2.3. Landslides

Landslides are common in upland areas and the rainy season. They are the process of soil movement accompanied by rocks and gravel or their mixture because of the instability of the soil around the slope. Landslides suddenly cause casualties, and the process occurs when water sinks into the ground. Soil weight will automatically increase. If the infiltration has reached the slip point, which is the soil is not waterproof, then the soil will be slippery and will make it easier to weathering it. At this time, the soil condition will be prone to landslides [1]. Wang (2017) argued that landslides would be related to factors, such as geology, vegetation, fracture, and precipitation [10].

### 2.4. Rainfall

Rainfall is the height of rainwater when it falls in an area, assuming the water does not infiltrate, flow, or evaporate [11]. it is included in the parameter affecting both floods and landslides. Commonly, we can see rainfall data from an area on the website of the Meteorological, Climatological, and Geophysical Agency (BMKG). Rainfall is known to have units of millimeters and has criteria and scores according to the Procedures for the Preparation of the Technical Plan for Forest and Watershed Rehabilitation (RTkRLH-DAS) in 2009 can be seen in Table 1.

**Table 1.** Table of Rainfall

| Rainfall | Notes | Score |
|---|---|---|
| 0.5 – 20 mm/day | Light rain | 1 |
| 20 – 50 mm/day | Moderate rain | 2 |
| 50 – 100 mm/day | Heavy rain | 3 |
| 100 – 150 mm/day | Very heavy rain | 4 |
| >150 mm/day | Extremely rain | 5 |

### 2.5. Slopes

A slope is a crucial parameter in a disaster hazard map because it can describe the surface runoff or erosion in an area. The slope is commonly expressed either in percentage or in degrees and has a definition of the angle formed by the difference between the height of land and the plane of land [12]. There are two forms of slopes proposed on ref [12], including concave and convex. If the slope in an area is tilted more, it is known that it will accelerate erosion because of the large number of grains of soil, sand, rock, or a mixture of the three that falls to the plains and is commonly carried by rainwater.

The slope is expressed in percentage and degrees with the equation (1) [13]:

$$a = \frac{b}{c}\ x\ 100\% \tag{1}$$

In which a is the slope, b is the height difference, and c is the distance. The following is a Table 2 of slopes in ref [14], which is based on the 1986 Land Rehabilitation and Soil Conservation Compilation.

**Table 2.** Slope Table

| Criteria | Notes | Score |
|---|---|---|
| (>40%) | Very Steep | 5 |
| (25-40%) | Steep | 4 |
| (15-25%) | Slightly Steep | 3 |
| (8-15%) | Sloping | 2 |
| (0-8%) | Even | 1 |

## 3. Methodology

### 3.1. Research Site

In this paper, the study was conducted using secondary data of 2020 in Trenggalek Regency.

### 3.2. Tools and Materials

In making the flood and landslide hazard maps, the following tools and materials were needed:

1) ArcGis 10.7 Software
2) The secondary data were :
    - Shapefile Ina-Geoportal
    - Soil Type Map
    - DEM Earth Explorer USGS
    - BMKG Data
    - 2020 Google Earth Imagery
    - BPS of Trenggalek Regency in 2018

IOP Conf. Series: Materials Science and Engineering 1115 (2021) 012011 doi:10.1088/1757-899X/1115/1/012011

### 3.3. Research Methods

In the research method, there were three stages carried out to create the flood and landslide hazard maps as follows (See Figure 1 and Figure 2):

a. Data Collection
To create a flood hazard map, the authors used four parameters so that secondary data was needed in the form of Ina-Geoportal Shapefile, Soil Type Map, DEM Earth Explorer USGS, and the latest Google Earth Image in 2020. When making a landslide hazard map, the authors used three parameters so that DEM Earth Explorer, Population Data from BPS of Trenggalek Regency, and BMKG data related to Rainfall were needed.

b. Classification and Scoring
When determining areas that are prone to floods or landslides, it took stages called classification and scoring. This classification and scoring aimed to analyze the results of all the factors assessed for each land unit. The classification of flood hazard areas is divided into five classes based on the references mentioned. Meanwhile, the classification of landslide hazard areas is divided into three classes. To determine the level of hazard to floods and landslides, the following equation was used:

$$K = \frac{JT - JR}{JK} \quad (2)$$

In which K is the Class Interval, JT is the highest number obtained from (Weight x Highest Score for each parameter), and JK is the number of classes.

c. Overlay
In the spatial analysis, there is a final step to combine several layers that have been generated in the ArcGis software, called the overlay stage. The overlay can only be performed for at least two layers created in ArcGis software. These two or more layers are combined from previously known parameters [15]. Therefore, we can create a disaster hazard map based on what has been combined from several overlays.

**Figure 1.** Methodology of the Flood Hazard Map

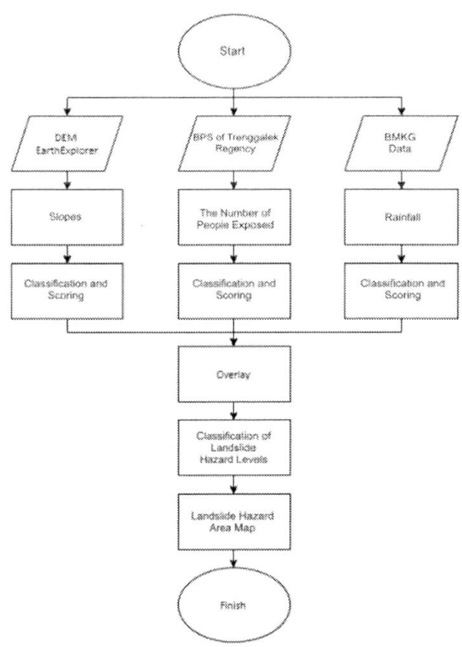


**Figure 2.** Methodology of the Landslide Hazard Map

## 4. Results and Discussion

### 4.1. Flood Hazard Map

The first discussion is the results of the flood hazard map that had been processed in ArcGis software, where each parameter was described through a map, including Figure 3 to Figure 6. Figure 3 is an image of the Rainfall Map of Trenggalek Regency whose data was obtained from BMKG. Subsequently, the process followed by classification and scoring that were performed in accordance with the reference for 2009 the Procedures for the Preparation of the Technical Plan for Forest and Watershed Rehabilitation (RTkRLH-DAS) so that a flood hazard map was obtained. It can be seen in the depiction of this rainfall parameter map that 14 sub-districts in Trenggalek Regency have rainfall less than 2500 mm per year. Rainfall less than 2500 mm per year is classified as 'very low' so that the rainfall map in Trenggalek Regency was assigned a weight of 1.

Figure 4 depicts the parameter, which is the slope. This slope data were obtained from the DEM Earth Explorer USGS data from Trenggalek Regency. The DEM data used must be combined from the surrounding DEM data (Ponorogo and Tulungagung) considering that Bendungan District (which borders Ponorogo) and Durenan District (which borders Tulungagung ) are not defined in the DEM data for Trenggalek Regency. Based on the preparation of Land Rehabilitation and Soil Conservation in 1986, the slope map-making is classified into five types, including <8% of even, 8-15% of sloping, 15-25% of slightly steep, 25-45% of steep, and >45% of very steep. Therefore, on this slope map, the classification of six districts in Trenggalek Regency tends to be even, including in the Tugu, Karangan, Gandusari, Trenggalek, Pogalan, and Durenan Districts. Meanwhile, the other ten districts have a classification that tends to vary from sloping to steep.

Figure 5 depicts a picture of the land use map in Trenggalek Regency. This land cover map is divided into five classifications referring to 2009 the Procedures for the Preparation of the Technical Plan for Forest and Watershed Rehabilitation (RTkRLH-DAS) from densely forested areas to settlements and paddy fields. This land cover map is essential to map whether a place is densely populated or not. The forest area here is given the smallest score of 1 because the impact of floods in the forested areas is known to be very small. Meanwhile, settlements and rice fields are given the highest score because settlement areas are highly influential on floods. Settlements and paddy fields are given the highest score according to the order of 5. Based on the results of processing, it is known that the dominant Trenggalek Regency is an area with dense forest. Furthermore, if observed using Google Earth imagery, it is known that the dominant residential area includes the Tugu, Karangan, Trenggalek, Pogalan, Durenan, and Gandusari Districts.

Figure 6 is a map of the soil types in Trenggalek Regency. The soil type map is a useful parameter to validate the areas that are prone to floods because the soil types will affect the infiltration process. If the soil type is smooth, then it will have a high infiltration power and vice versa. If the infiltration power is high, the surface flow in an area will be higher. When processing is conducted, classification and scoring were performed concerning the 2009 RTkRLH-DAS. The results show that Trenggalek Regency is dominant with litosol soil with a large infiltration, followed by reddish-brown latosol soil, which when classified, the latosol soil infiltration is rather small. In a densely populated area, it is known that the soil type is alluvial so that the infiltration in this area is known to be in a small category.

Figure 7 is a map of flood hazards in the Trenggalek Regency. The Flood Hazard Map of Trenggalek Regency is obtained from the overlay of the four parameters in Figure 1 to Figure 4. These four parameters include Rainfall, Slope, Land Use, and Soil Types of Trenggalek Regency. According to the available citations, it is known that the Trenggalek Regency area is an area prone to floods and landslides. It has been proven from the results of existing processing that in the seven districts of Trenggalek Regency of Tugu, Karangan, Trenggalek, Pogalan, Gandusari, and a small part of Kampak are areas that are very prone to floods. The districts in Trenggalek Regency are mostly quite prone to floods. It can be seen that the land use parameter greatly influences the flood hazard map because if the area is residential, it will be more prone to floods.

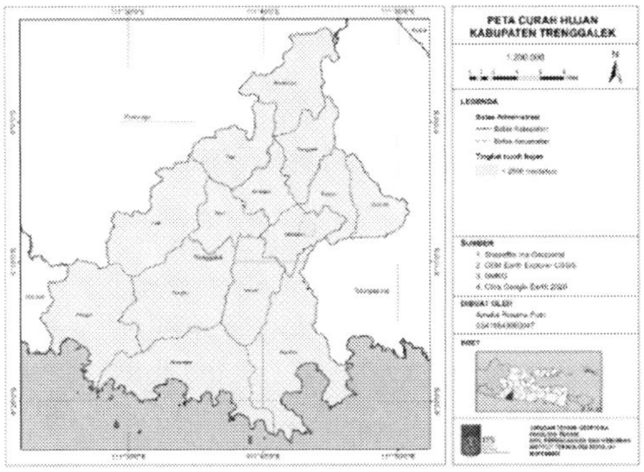


**Figure 3.** Rainfall Map of Trenggalek Regency

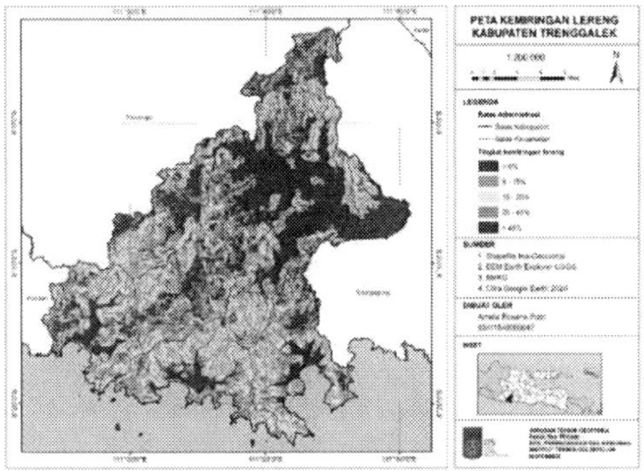


**Figure 4.** Slope Map of Trenggalek Regency

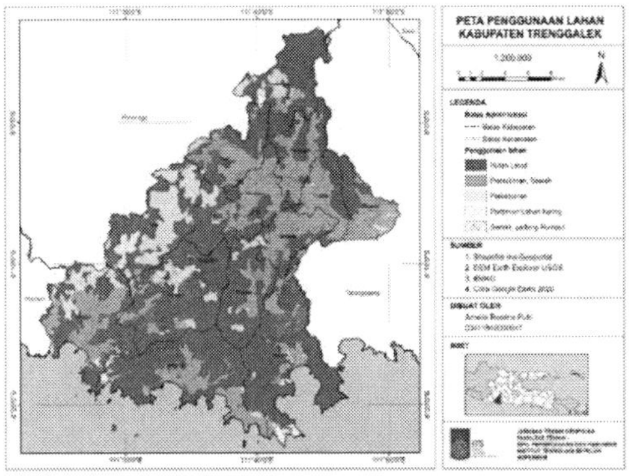


**Figure 5.** Land Use Map of Trenggalek Regency

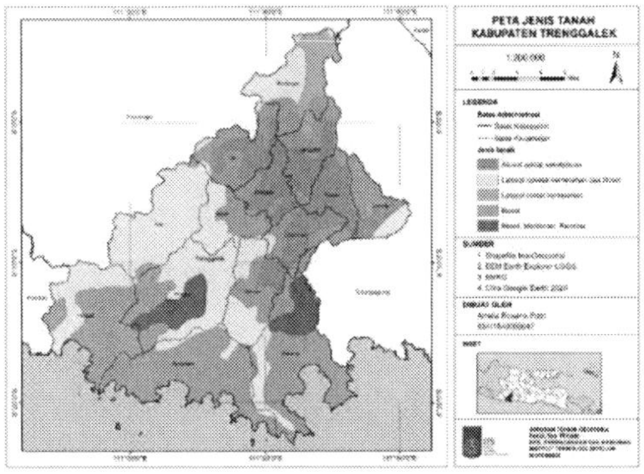


**Figure 6.** Soil Type Map of Trenggalek Regency

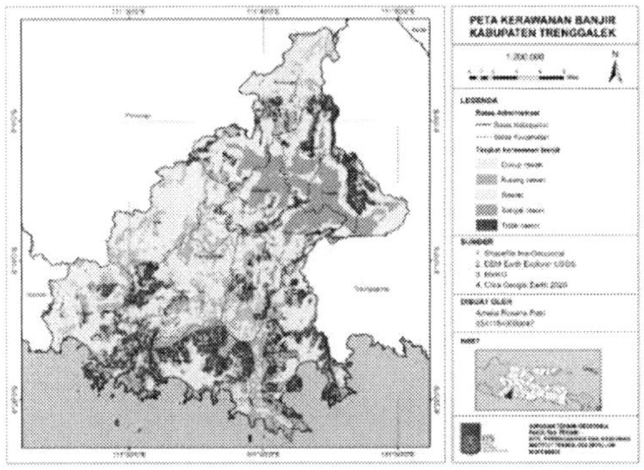


**Figure 7.** Flood Hazard Map of Trenggalek Regency

### 4.2. Landslide Hazard Map

The second discussion is the result of processing the landslide hazard map. When generating the landslide hazard map, it is known that three parameters were described one by one in Figure 8 to Figure 10, including the parameters of rainfall, slope, and the number of exposed populations in Trenggalek Regency. The first parameter is rainfall. The parameter of rainfall is highly influential in landslides with units of millimeters per year. In making a rainfall map, areas with low rainfall (<50 mm) were given a low score of 1, while moderate rainfall (50-150 mm) was given a score of 2, and finally, areas with high rainfall (>150 mm) were given a score of 3. The rainfall map for this landslide disaster was sourced on the DAS II of Rainfall Distribution Map in December 2020.

In Figure 8, the rainfall levels are classified into three classes, including <50 mm, 50-150 mm, and >150 mm. Based on the classification results, the dominant Trenggalek Regency has an average rainfall value of 50-150 mm. Meanwhile, the Districts of Bendungan and Trenggalek are known as areas with rainfall of more than 150 mm. In a small part of the Gandusari area, it is known that the rainfall is less than 50 mm.

Figure 9 shows a slope map in Trenggalek Regency. The slope is also included in the parameter of making the landslide hazard map. Discussion on the slope map in Figure 11 is the same as that in Figure 4 where this slope is a parameter for making a disaster hazard map to floods or landslides. The last parameter for making a landslide hazard map is the number of exposed populations shown in Figure 10. According to the BNPB in 2019, the population in Trenggalek Regency was 689.034 people. Therefore, when classifying the number of exposed populations, the Regency is divided into three classifications, including low with < 30.000 people, medium with the population ranges from 30.000 to 60.000 and high with > 60.000. If the population is classified as low in the processing process, then it is given a score of 1. If it has a medium population, the score is 2, and if it has a high population, then the score is 3. Referring to BPS data for Trenggalek Regency in 2018, in the map of the number of people exposed to landslides, 3 districts have > 60.000 people, including Watulimo, Panggul, and Trenggalek Districts. Meanwhile, the population of < 30.000 people are in the Bendungan and Suruh Districts. The rest, areas with a population ranging from 30.000 to 60.000 people are exposed to landslides.

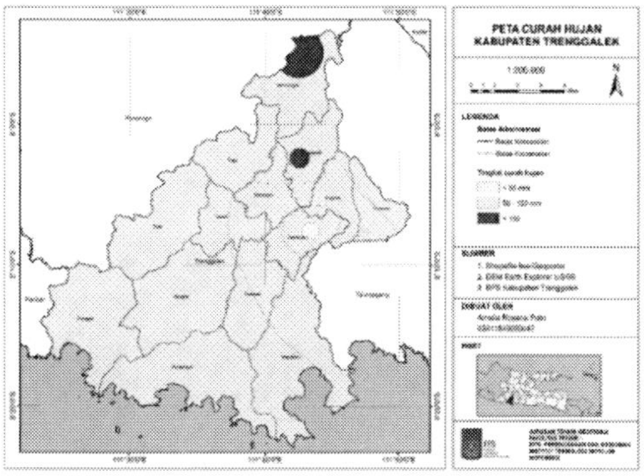


**Figure 8.** Rainfall Map of Trenggalek Regency

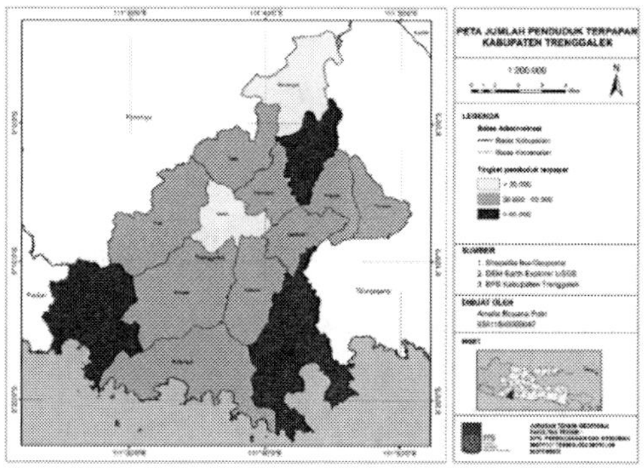


**Figure 9.** A number of People Exposed Map in Trenggalek Regency

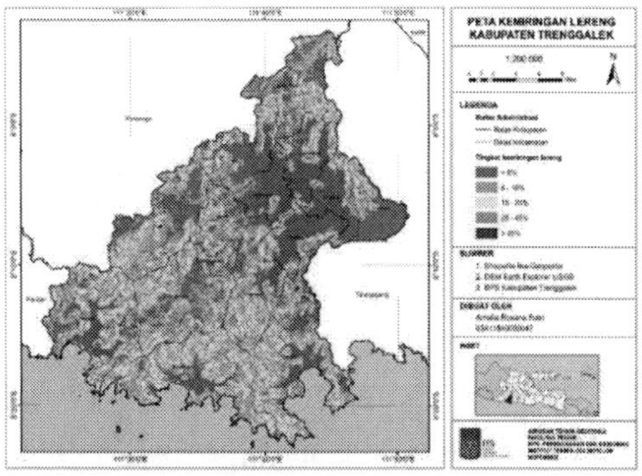


**Figure 10.** Slope Map of Trenggalek Regency

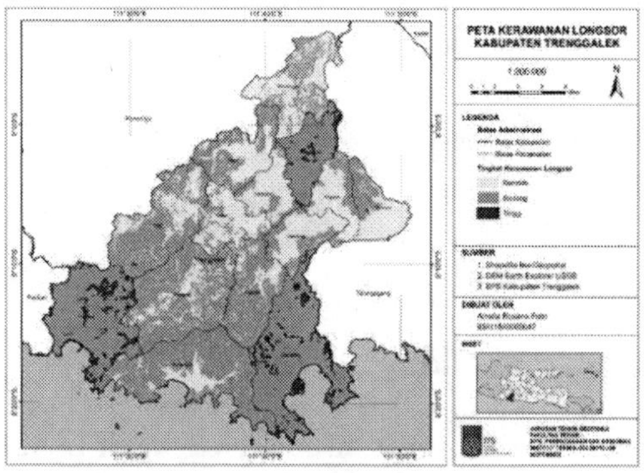


**Figure 11.** Landslide Hazard Map of Trenggalek Regency

Figure 11 is an image of the landslide hazard map in Trenggalek Regency. The Landslide Hazard Map in Trenggalek Regency is obtained from the overlay between the parameters of rainfall, slope, and the number of exposed populations. Based on the existing references, Trenggalek Regency is dominant with areas that are prone to medium to high landslides. The map of the processing results shows that Trenggalek Regency has a moderate level of landslide hazard. The results of the map overlay can also be concluded if areas with >60.000 people, then these are prone to moderate landslides, for instance, Trenggalek, Watulimo, and Panggul Districts. Meanwhile, areas with a low population such as Suruh and Bendungan Districts are known to have a low level of landslide hazard. Several points in Districts such as Trenggalek, Watulimo, and Panggul have a high level of landslide hazard. Landslide hazard map shows to be very proportional to the number of people. The greater the population in an area, the higher the hazard level for landslides.

## 5. Conclusion

Based on the processing using ArcGis 10.7 software, two maps of disaster hazard were obtained, including the Flood and Landslides which have been disasters with the highest frequency in Trenggalek Regency. Based on the parameters of the floods, including Rainfall, Slope, Land Use, and Soil Type, the results show that areas dominated by settlements have a very high level of flood hazards, such as in the seven districts of Trenggalek Regency of Tugu, Karangan, Trenggalek, Pogalan, Gandusari, and a small portion of the Kampak. Regarding the landslide, it shows that districts with a population of >60.000 people at some points also have a high level of landslide hazards such as in Trenggalek, Watulimo, and Panggul Districts. Therefore, the greater the population in an area, the higher the hazard level for landslides.

After processing based on the references that have been mentioned, results are obtained in the form of flood and landslide hazard maps in Trenggalek Regency. Hopefully, this paper can be a reference source for the general public to determine the disaster hazard points in Trenggalek Regency. It is expected that in the future, there will be a more and more detailed reference of soil type maps in each Indonesian Regency because the soil types in an area are also included in the required parameters. This paper was made with the hope of reducing these two disasters because both always hit people every year. If the areas have the potential hazard of floods or landslides, it is hoped that the community and local government will be active and responsive in protecting the surrounding environment and ready to mitigate the disaster to minimize casualties because the population will be directly proportional to the potential disaster hazard areas in Trenggalek Regency.

## References

[1] Liao Z, Hong Y, Wang J, Fukuoka H, Sassa K, Karnawati D and Fathani F 2010 Prototyping an experimental early warning system for rainfall-induced landslides in Indonesia using satellite remote sensing and geospatial datasets. *Landslides* 7 (3) 317–324
[2] Petley D 2012 Global patterns of loss of life from landslides *Geology* **40** (10) 927–930
[3] Murdiyanto and Gutomo T 2015 Bencana alam banjir dan tanah longsor *Journal PKS* **14** (4) 437-452
[4] Faturahman B M 2018 strategy of emergency response of flood and landslide disaster in pacitan regency *Jurnal Ilmu Administrasi (JIA)* **15** (2) 133-147
[5] Van Bemmelen R W 1949 *The geology of Indonesia IA* General Geology: The Hague
[6] Yumna 2017 Anut Grubyuk in the Voting Process: The Neglected Explanation of Javanese Voters (Preliminary Findings) *Southeast Asian Studies Center Kyoto University* **5** (2) 239–268
[7] Nahrowi T, Suratman Y, Namida S and Hidayat S 1978 *Geologi Pegunungan Selatan Jawa Timur* Eksplorasi PPTKGD Lemigas Cepu : East Java
[8] Iskandar D and Sugandi D 2015 Flood Mitigation Efforts in the Capital Region of Jakarta *International Journal of Conservation Science* **6** (4) 685-696

[9] Klijn F, Kreibich H, de Moel H and Penning-Rowsell E 2015 Adaptive flood risk management planning based on a comprehensive flood risk conceptualization *Mitigation and Adaptation Strategies for Global Change* **20** (6) 845-864

[10] Wang F, Xu P, Wang C, Wang N and Jiang N 2017 Application of a GIS-based slope unit method for landslide susceptibility mapping along the Longzi River, Southeastern Tibetan Plateau, China *ISPRS International Journal of Geo-Information* **6** (6)172 1-20

[11] Roth B E, Slatton K C and Cohen M J 2007 On the Potential for High-Resolution Lidar to Improve Rainfall Interception Estimates in Forest Ecosystems *Frontiers in Ecology and the Environment* **5** (8) 421–428

[12] Cotton C A 1952 The Erosional Grading of Convex and Concave Slopes *The Geographical Journal* **118** (2) 197-204

[13] Walton R S, Volker R E, Bristow K L and Smettem K R J 2000 Solute transport by surface runoff from low-angle slopes: theory and application *Hydrological Processes* **14** (6) 1139-1158

[14] Bojo J and Cassells D 1995 Land Degredation and Rehabilitation in Ethiopia: a Reassessment *AFTES Working Paper* **17** 1-48

[15] Manjula K R, Jyothi S and Varma S A K 2010 Digitizing the Forest Resource Map Using ArcGIS *International Journal of Computer Science Issues* **7** (6) 300-306

# The Effect of Cr (VI) Metal On Aerobic Stability of (WWTP) Process CED UGM Laboratories

**S Darmayani[1], S Alaydrus[2] and T Yuniarty[1]**

[1] Department of Medical Laboratory Technology, Health Polytechnic of Kendari, Southeast Sulawesi, Kendari 93232, Indonesia
[2] Department of Pharmacy, STIFA Pelita Mas Palu, Central Sulawesi, Palu 94112, Indonesia

*Corresponding author: satya.darmayani@gmail.com

**Abstract.** Chemical Engineering Department, Gadjah Mada University, has a wastewater treatment plant to process wastewater produced by its laboratories. The WWTP consists of equalization, anaerobic, aerobic and phytoremediation ponds. WWTP operates with the adding of starch as substrate and NPK as nutrient. However this WWTP could not treat wastewater which contained a variety of chemical compounds, so that it did not operate properly. Therefore WWTP revitalization should be done in order to make it operate optimally. The purposes of the research are to determine the effect of Cr on the WWTP stability conditions, and the performance of aerobic activated sludge. The experiment was done by controlling substrates flow as much as 3 litters per day to the aerobic pond in a laboratory scale pond made from plastic with the capacity of 10 litters. The pond was run continuously, the samples of wastewater was taken to measure its BOD, COD, MLSS and MLVSS along with the waste feeding. The results showed that the higher Cr concentration the longer time required by aerobic bacteria to degrade it into Cr. Aerobic bacteria are able to reduce the BOD - COD of waste-water that contains Cr up to 50% on various concentrations.

## 1. Introduction

Laboratory is an important part of a university especially engineering faculty, in more specifically in chemical engineering. In this laboratory, activities produce waste which can lead to environmental pollution if not managed properly.

In the laboratory waste treatment efforts Chemical Engineering Department (CED), University of Gadjah Mada (UGM), has a wastewater treatment system to treat waste from all laboratories at CED UGM. However, the WWTP is not able to process waste containing various types of chemical compounds, so it is no longer in operation. It is necessary for the revitalization of the WWTP, in order to make operate optimally again. Based on the survey data, there are known that the most common pollutants found in wastewater laboratory is Cr. Therefore, the presence of Cr becomes a major concern in the wastewater treatment process. Some research [1-3] showed that Cr(VI) in low concentration can be reduced in activated sludge

Activated sludge is a biological treatment process that has been applied widely for treating wastewater. The process becomes an option because it has the effectiveness and efficiency of the process as well as has low operating costs [4].

The research premeditated the core of Cr (VI) on the aerophilic bacteria stableness in the wastewater treatment process using activated sludge at laboratory scale WWTP.

## 2. Literature Review

### 2.1. Laboratory wastewater Characteristic

The operational activities of Laboratory generate waste consists of solid, liquid and gas. Liquid waste containing big metals much as Hg, Ag and Cr can harm the environment because they are toxic to animal and human [5]. According ref [6], laboratorial waste generally contains more complex pollutants comparing with industrial waste.

### 2.2. Wastewater Processing Technology

Wastewater treatment technologies are very widely, depending on the water-waste characteristics, the desired quality of processed product and operational costs. According to [7], the techniques of water-waste treatment are generally divided into three methods that are:

a. Physical Wastewater Treatment
b. Chemical Wastewater Treatment
c. Biological Wastewater Treatment

### 2.3. Laboratorial Wastewater Treatment with Aerobic Suspended Culture

Based on the technique of microorganisms controlling in growth media, biological wastewater treatment can be divided into [8]:

a. Attached Growth Processes
b. Suspended Growth Processes

### 2.4. Activated Sludge

Activated sludge method is one of biological treatment types where the microorganisms are in suspended growth. It is aerobic process that requires oxygen for aerobic biological reactions. In the biological system, the microorganisms live and grow in colonies that form easily sedimented clumps [9]

### 2.5 Aerobic Bacteria Stability on Waste Treatment Using Activated Sludge Method

Microorganism's stability is strongly influenced by the availability of nutrients and environmental conditions, especially degree of acidity (pH) and dissolved oxygen (DO). According to [10], the degree of acidity values for aerobic bacteria in activated sludge systems ranged from 6.5 to 8.5. To maintain aerobic conditions in the floc core of the dissolved oxygen concentration, it takes approximately 4-6 mg / L [11].

## 2. 6. The Effect of Chromium (Cr) Addition on Aerobic Bacteria Stability in Activated Sludge Wastewater Treatment

The effect of chromium (Cr) addition on aerobic bacteria stability in water-waste treatment described as: Cr (VI) in the cell membrane is rock-bottom to Cr (III) and reacts with intracellular bimolecular. Cr6 + concentration on activated sludge that has concentration above 25 mg / L can inhibit the growth of activated sludge, and the lethal dose of the activated goo has concentrations supra 150 mg / l [1].

## 2.7. Theoretical Basis

The main principles of wastewater treatment by aerobic process is the supply of oxygen in the wastewater treatment ponds, the appropriate ratio of substrate composition and substrate feeding period to keep the bacteria remaining in a stable condition. Stability of the bacteria will be disrupted if there is a foreign substance whether deliberately or not inserted in aerobic pond. The research will be added $K_2Cr_2O_7$ as Cr (VI) substance that will act as an inhibitor for aerobic bacteria.

In order to Cr (VI) can be degraded by aerobic bacteria optimally, it is necessary to make Cr (VI) acclimatization. Acclimatization process intended to aerobic bacteria, so aerobic bacteria can adapt to the presence of Cr (VI).

The Effects of Cr (VI) (as chromate, $[CrO_4]^{2-}$ include efficient absorption by the cells through the anion (such as sulfuric line), and subsequent responses with intracellular reduction attorney (such as glutathione, ascorbate, or NADPH), which lead to intermediate phase that can damage DNA and organic radicals [12].

# 3. Experimental Method

## 3.1. Equipment

Equipment's used are: rectangular plastic tubs which has 10-liter capacity, pump U-9900 220-240V. The apparatus set is shown in Figure 1.

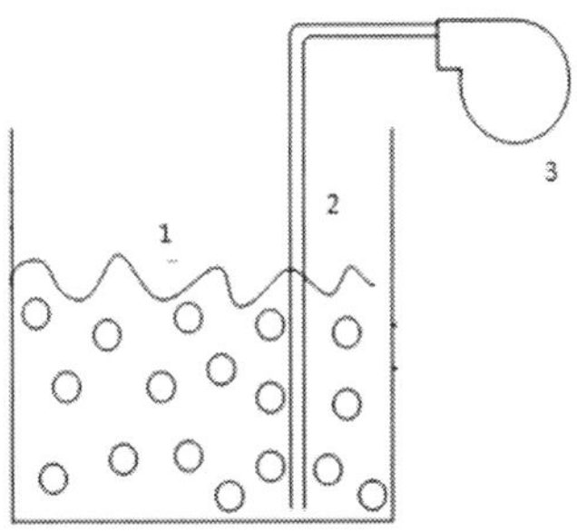


**Figure 1.** Aerobe bacteria growth medium

Description:
1. Aeration tank
2. Aerator pipe
3. Aerator

Analysis tools used are: UV-Vis spectrophotometer GBC Scientific Equipment with serial No. V4131, Lutron pH meter pH-201, Lutron DO meter 5501, oven (memmert 854 schwabach type) 220V that has Tmax 220°C, furnace mod L51/s that has Tmax 1100°C 220V, analytical scale ohaus type PA214C max cap: 210 gram.

### 3.2. Material Procedure

The materials used are: Waste milk from the Waste Water Treatment Plant (WWTP) PT. Sari Husada Yogyakarta, Aerobic bacterial culture inoculums of Waste Water Treatment Plant (WWTP) PT. Sari Husada, Yogyakarta, urea as a source of natural phosphorous and nitrogen source $K_2Cr_2O_7$ as Cr (VI) source.

### 3.3. Research Procedure

The steps performed are:

a. Aerobic bacterial cultures.
b. Dairy wastewater as substrate included in the aeration basin as much as 3 liters / day.
c. In the process of aeration, wastewater is put in aeration pond, further the transfer of air from the aerator is given. Phosphate and Nitrate was added as microorganism's nutrient. Operating conditions are pH between 6.5 to 8.5, and temperature (25°C). besides, the aeratin process had to be controlled to avoid the sediment on the base of the ponds.
d. The volume f the wastewater is kept constant by taking and adding with the same amount of 3L along with the nutrient adding with the ratio of BOD : N : P = 100 : 5 : 1 once a day. The bacteria in aeration pond were aclimated for ± 20 days in order to make the bacteria abe to adapt the new environment, indicatd from constant COD reduction eficiency value.
e. The addition of $K_2Cr_2O_7$ as an inhibitor as 0.05 grams was added. The feeding is done intermittently. The process is done until the value of BOD, COD and chromium levels obtained are constant. After Cr concentration was constant, it was then followed by the addition of $K_2Cr_2O_7$ 0,025 gram with the same analytical procedure.

### 3.4. Samples Analysis

Determination of Cr (VI) concentration using UV-Vis spectrophotometer with difenilkarbazid method, BOD analysis used BOD5 method, COD analysis used open reflux method, mixed liquor suspended solids (MLSS) is done by evaporation and drying of biomass at temperatures of 103-105°C, Mixed liquor volatile suspended solid (MLVSS) performed by heating biomass at a temperature of 600°C.

### 3.5. Product Analysis

Sampling of waste water in the reactor was conducted to determine the concentration of COD and BOD in wastewater. The data are used to determine the effect of Cr (VI) addition on aerobic bacteria stability in degrading organic substances. Analysis of the MLSS and MLVSS also conducted to determine the decreasing of aerobic bacteria after addition of Cr (VI).

## 4. Result and Discussion

### 4.1. The Effect Of Cr(VI) Adding Toward Activated Sludge

This reserch was devided into to steps, those are aerobe bacteria acimatization and $K_2Cr_2O_7$ adding. In aclimatizatin process the mmicroorganisms in the reactor were grown for ± 20 days. After the breeding of microorganisms in the reactor run well, $K_2Cr_2O_7$ as an inhibitor as 0.025 grams was added. After Cr concentration was constant, it was then followed by varying the addition of Cr (VI) to 0.05 gram. After Cr concentation was constant subsequently the $K_2Cr_2O_7$ adding was variated as 0.025 gram and so on.

Total suspended solid concentration which contains in the waste is measured from MLSS level. Th analysis result of the solid during the waste treatment is shown on Figure 2. The wastewater treatment which contains Cr(VI) using aerobic bacteria can reduce MLSS level. This phenomenon occured because it is dificut for aerobe bacteria to reduce the waste which contains Cr(VI).

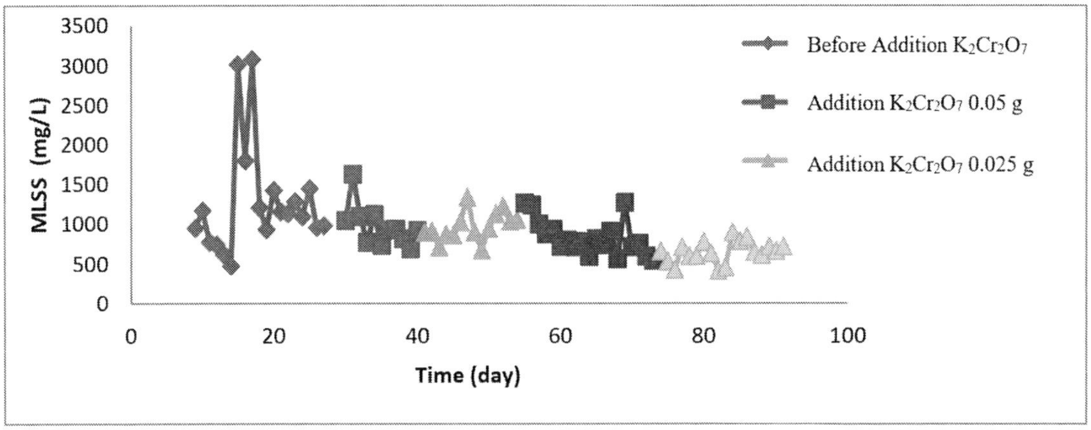


**Figure 2.** The effect of $K_2Cr_2O_7$ addition as the function of time to MLSS level

MLVSS is the suspended solid level which only consist of bacteria component. MLVSS measurements by a further heating for the filtered solids MLSS, with high heat the components of microorganisms will break down and evaporate. The difference in weight before and after a heated indicates the number of microorganisms [13]. Basically the value of MLVSS is proportional to the value of MLSS. Figure 3 shows the $K_2Cr_2O_7$ additional to produced MLVSS level. MLVSS level in reactor without Cr addition is higher compare to the addition of Cr.

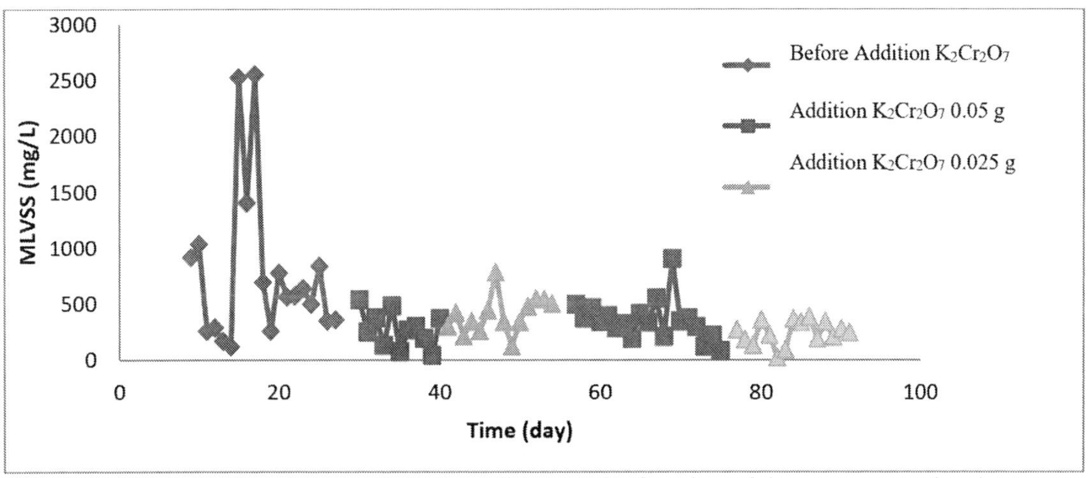


**Figure 3**. The effect of $K_2Cr_2O_7$ addition as the function of time to MLVSS level

### 4.2 The Efffect Of $K_2Cr_2O_7$ Adding Toward Cr(VI) Concentration

This is due to high levels of Cr (VI) were feed as influent to be treated by aerobic bacteria are not comparable with the levels of aerobic bacteria that will process the metal, the measured data obtained for the addition of Cr (VI) 0.025 grams with the lowest measurement data which is equal to 4.614 mg/L and a decrease in the highest concentration of 2.064 mg/L.

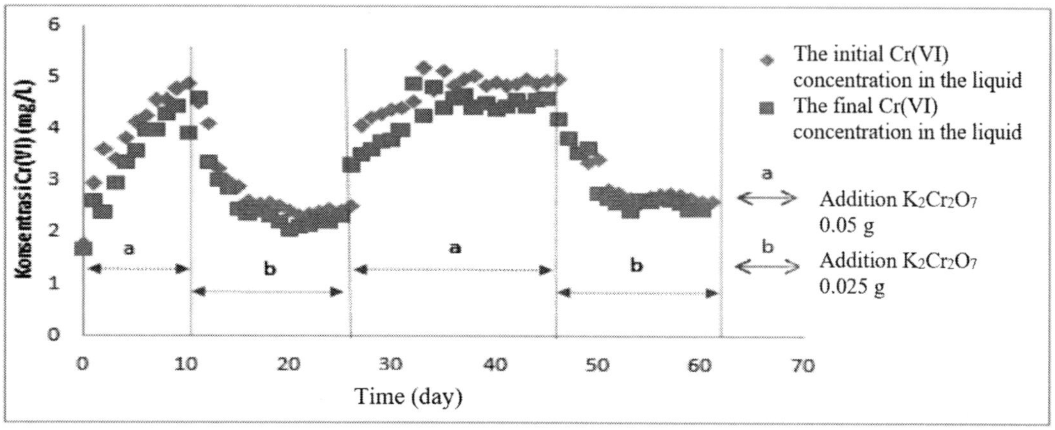


**Figure 4.** Calculated Cr(VI) denseness abating and data of Cr (VI) denseness as a office of age

Figure 4 shows that the calculated initial concentration of Cr (VI) in liquid was greater than the concentration of Cr (VI) final measurement results it is because in the bacterial cell the Cr has been reduced into Cr (III) through sulphate stream [14].

$Cr_2O_7^{2-} + H_2O \rightleftharpoons 2\,CrO_4^{2-} + 2H$ (outside of cell)

$CrO_4^{2-} + 4\,H_2O + 3e^- \longrightarrow Cr(OH)_3 + 5\,OH^-$ (inside cell)

$Cr(OH)_3 \longrightarrow Cr^{3+} + 3\,OH^-$ (inside cell)

### 4.3. The Effect Of Cr(VI) Toward The Ability Of Aerob Bacteria To Reduce COD Of Wastewater

COD is an analysis to measure the oxygen demand of the organic substances which are difficult to be destroyed by oxidation, so that it needs support from a strong oxidizing agent in acid [15-16]. In this research, the treatment for wastewater by using aerobe bacterica showed the COD concentration decreasing with the initial concentration as 56.23 mg/L became 25.16 mg/L. Figure 5 shows the comparison of COD concenration before and after $K_2Cr_2O_7$ adding as 0.05 g and 0.025 g.

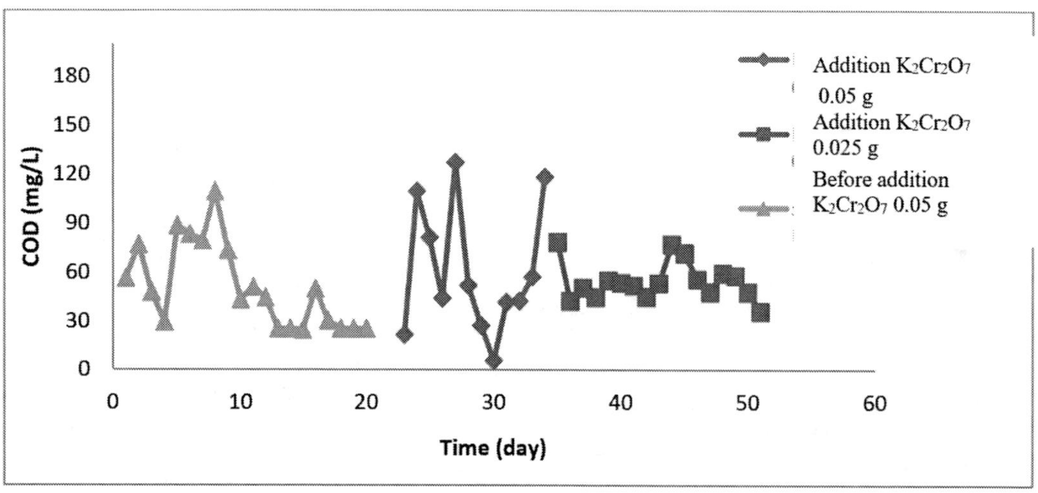


**Figure 5.** Wastewater COD value declining as a function of time before and after $K_2Cr_2O_7$ adding

Decreased levels of COD is caused by the degradation of organic matter due to a simle component of bacterial [15]. The chart shows that constant COD levels was obtain on the day of 18, 19 and 20 as 25.16 mg/L. So it can be followed by adding $K_2Cr_2O_7$ into the pond.In the $K_2Cr_2O_7$ adding as 0.05 g showed the fluctuate value of COD, with the highest value as 127.18 mg/L and the lowest value as 5.54 mg/L. Meanwhile at the $K_2Cr_2O_7$ adding as 0.025 g, the value of COD obtained was tend to be constant with the highest value of COD as 78.31 mg/L and the lowest as 35.77. This constant condition is because the adding of $K_2Cr_2O_7$ as 0.025 g dis not give significant effect toward bacterial activity.

### 4.4 The Effect Of Cr(VI) Toward The Ability Of Aerobe Bacteria To Reduce BOD Of Wastewater

The ability of aerobic bacteria in wastewater treatment was considered good, because it fufill the requirement of wastewater as environmental ministry No. 03 year 2010 the maximum BOD of wastewater as 50 mg/L. This is proofed by the value of BOD has been reduced from 84 mg/L into 4.5 mg/L.

The BOD decreasing due to the oxygen used by bacteria to degradate organic compound as its nutrient.the aerobe bacteria can reduce wastewater BOD which contains up to 94%. Figure 6 show a comparison of wastewater BOD levels before and after the addition of $K_2Cr_2O_7$.

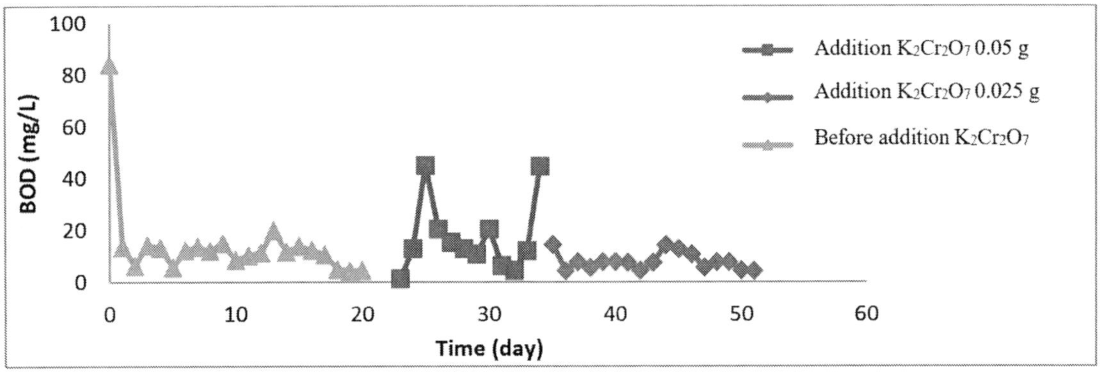


**Figure 6**. Wastewater BOD value declining as a function of time before and after $K_2Cr_2O_7$ adding

### 4.5. Acidity (pH)

From the pH analysis in sludge in the reactor with the various resident times was shown in Figure 7.

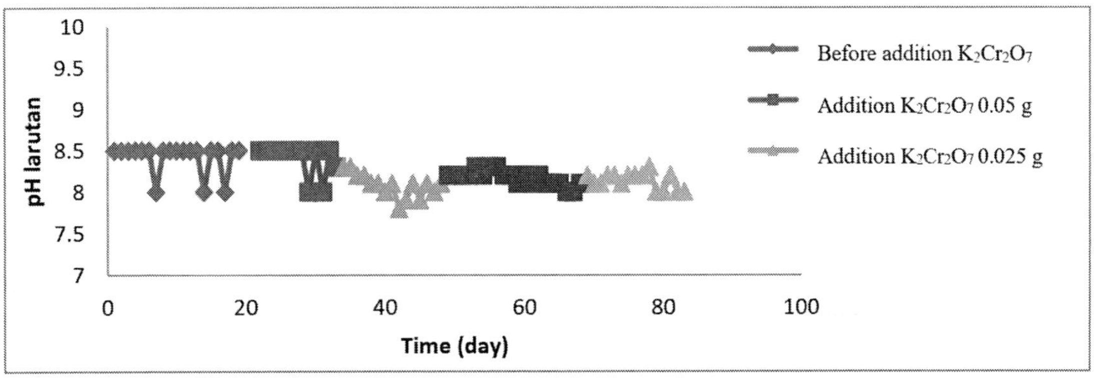


**Figure 7**. The effect of $K_2Cr_2O_7$ as function of time toward the pH of solution before and after $K_2Cr_2O_7$ adding

From Figure 7, there was an increased pH in the aeration to 55, 56, and 78 day and a decrease in pH on day 43. A decrease and an increase in the pH value is not significant, which still remains in the range of pH for aerobic bacteria in activated sludge systems, which are in the range of 6.5 to 8.5 [16].

## 5. Conclusion

On the addition of $K_2Cr_2O_7$ 0.05 g it shows that it inhibit the growth of microorganisms indicated from MLSS and MLVSS analysis results which shows the microorganism was reduced for about 35%.
2. The time required to obtain a new stable state with the addition of $K_2Cr_2O_7$ 0.05 g of $K_2Cr_2O_7$ and 0.025 g were 14 days and 12 days to reach a constant point.

There is no information on the repeated use of a maximum of sludge that requires further research on the characteristics of the sludge on a few times it is used. It is necessary to know the trend of change in sludge characteristics associated with the ability of aerobic bacteria in reducing waste especially waste metal.

Further optimization of the operating unit can be done, for example by using a variation of the concentration of other metals, the contact time optimization for wastewater produced continuously in large quantities and need to be processed continuously, the continuous assessment needs to be done for the optimization process.

## References

[1] Vaiopoulou A E and Petros G B 2012 Effects of Chromium on Activated Sludge and on The Performance of Wastewater Treatment Plants *Water Research* **46** (3) 549-570
[2] Stasinakis A S, Mamais D, Thomaidis N S and Lekkas T D 2002 Effect of chromium (VI) on bacterial kinetics of heterotrophic biomass of activated sludge *Water Research* **36** (13) 3341-3349
[3] Moore W A, Mc Dermott G N, Post M A and Ettinger M B 1961 Effects of chromium on the activated sludge process *Journal of Water Pollution Control Federation* **33** 54 - 72.
[4] Wulandari Y M, Rasmito A and Caroline J 2010 MSG Waste Biomass Concentration On Membrane Bioreactor Submerged *Jurnal Teknik Kimia* **5** (1) 352 - 362
[5] Hussein I, Abdel-Shafya, Mona S M M 2018 Solid waste issue: Sources, composition, disposal, recycling, and valorization 27 (4) 1275-1290
[6] Rohaeti E, Trie N F and Irmanida B 2009 Pengolahan Limbah Cair dari Kegiatan Praktikum Analisis Spot Test Dengan Koagulasi Menggunakan Polialuminium Klorida *Prosiding Seminar Nasional Teknologi Pengelolaan Limbah IX* **9** 141 - 148
[7] Tchobanoglous N 1997 *Wastewater Treatment Pocket Handbook* New York: Pudvan Publishing
[8] Metcalf and Eddy 2003 *Wastewater Engineering : Treatment* Disposal and Reuse 4th edition New York: McGraw Hill Book Co
[9] Najafpour G and Atieh E 2016 Biological Treatment Processes: Suspended Growth vs. Attached Growth *Iranica Journal of Energy & Environment* **7** (2) 114-123
[10] Eckenfelder W W 2000 *Industrial Water Pollution Control* 3rd edition Singapore: McGraw HillBook Co
[11] Klopping, Paul H, Mashall Jr, Richard H, Richard and Michael G 1995 *Activated Sludge Operations for Pulp & Papermills* Oregon: Callan and Brooks Publishing Company
[12] Tumolo M, Ancona V, Paola D D, Losacco D, Claudia C, Carmine M and Uricchio V F Chromium Pollution in European Water, Sources, Health Risk, and Remediation Strategies: An Overview International *Journal Environmental Research and Public Health* **17** (15) 1-24
[13] Alaerts G, Sri S and Santika 1987 *Metoda Penelitian Air* Surabaya: Penerbit Usaha Nasional
[14] Sastrohamidjojo H 2005 *Kimia Dasar* 2nd Edition Yogyakarta: Gadjah Mada University Press
[15] Mes de T Z D, Stams A J M, Reith J H and Zeeman G 2003 *Methane Production by Anaerobic Digestion of Wastewater and Solid Wastes* Netherland: Dutch Biological Hydrogen

Foundation
[16] Pisutpaisal N and Sirisukpoca U 2014 Development of Rapid Chemical Oxygen Demand Analysis Using Ozone as Oxidizing *Agent Energy Procedia* **50** 711-718

# Design of Management Information System for Employee Performance Appraisal Using Service Oriented Architecture (SOA) Method

**R Purnamafajari, Budiman and Z Niqotaini***

Department Information System, Faculty of Technology and Informatics, Universitas Informatika dan Bisnis Indonesia, Bandung, Indonesia.

*Email: zatinniqotaini@gmail.com

**Abstract.** Information Technology and Information Systems are able to assist the performance appraisal process and assist decision-making by the management of an organization. Bank BTN Bogor Branch Office itself has carried out the performance appraisal process for employees every quarter, semester and annually every year. The performance appraisal process uses the key performance indicator method using several indicators, where the assessment is carried out by superiors to subordinates after a performance target has been set by the Central Bank BTN Office. The performance appraisal process takes quite a long time because the accountant has to collect every performance appraisal result from each work unit. In addition, the ongoing system has another weakness, namely that it cannot be monitored via the dashboard in real time by the Management division which aims to assist in decision making in finding top talent. This study designed a Management Information System for employee performance appraisal at Bank BTN Bogor Branch Office which is designed for 2 levels of management, namely middle level management and bottom level management. The design of this SIM includes 3 stages, namely the data collection method, the Rapid Application Development (RAD) system development method and the Service Oriented Architecture (SOA) method. The application of SOA is used so that communication between systems can run smoothly without being hindered by different platforms. The results of the SIM design in the form of a mockup that functions as an assessment system and monitoring media for employee performance with several features in it both mobile and website. Then there are several additional features in designing a mobile-based information system, namely employee performance which aims to see top talent candidate employees.

## 1. Introduction

In the current era of the digital 4.0 industry, Information Technology and Information Systems are able to change the human perspective in completing a job. One of the conveniences with the presence of technology is that it is able to assist the performance appraisal process and assist decision making by the management of an organization or company.

PT. Bank Tabungan Negara (Persero) Bogor Branch Office is one of the state-owned corporation engaged in the banking sector through housing finance and industrial activities by distributing mortgages and loans. Bank BTN Bogor Branch Office itself has carried out the performance appraisal process for employees every quarter, semester and annually every year. The performance appraisal process uses the key performance indicator method by using several indicators as a two-way measurement benchmark, where the assessment is carried out by superiors to subordinates after the performance targets have been set by the Central Bank BTN Office. On-going employee performance appraisal uses a semi-manual program, namely the Performance Management System (individual) and the Annual Performance Target Planning (PSKT) for Branch Office performance. The performance appraisal process takes quite a long time because the accountant has to collect every performance appraisal result from each work unit. Futhermore, the ongoing system has another weakness, namely that it cannot be monitored via the dashboard in real time by the Management division which aims to assist in decision making in finding top talent among employees.

There are many studies related to information system management, one of which is research on surveillance in the campus environment. This study aims to produce a website-based information system that supports integrated management of supervision in higher education[1]. Subsequent research discusses the design of a management information system for project management which aims to make it easier for consultants to assist their clients and design the application process for intensive programs [2]. Subsequent research explains that the data management information system provides accurate and real-time information to assist the decision-making process for managers and planning effectively and control the operations of an organization [3]. The research on the dashboard to display KPI achievements. This study produces a management information system that is able to present important information through a media dashboard, such as employee operational information, a dashboard for tactical and strategy for managers [4]. The next research discusses the Service Oriented Architecture. This research resulted in a service modeling paradigm by adopting the Basic Model Interface and developing a set of web API services to demonstrate this functionality [5]. Furthermore, research that discusses the integration of services in electric companies. This research produces a service modeling that can integrate several services at once [6].

From the problems mentioned above, the management requires an information system capable of monitoring employee performance efficiently and effectively, one of which is a management information system that will be designed based on web and mobile. However, SIM requires communication with other information systems in the Bank BTN KC Bogor ERP for data collecting data. The integration will be designed using Service Oriented Architecture (SOA) technology and the NuSOAP library. This SOA design aims to distribute data from ERP without being hindered by differences in platforms, differences in databases, differences in programming languages and differences in server locations. Based on the background above, there are 3 objectives in this research including the following :

a. To design a Management Information System that is able to assist the Management of Bank BTN Bogor Branch Offices in employee performance appraisal and monitoring of KPI achievement in real time.
b. To design a Management Information System so that the Management of Bank BTN Bogor Branch Office can make decisions in finding top talent among employees.
c. To design a Management Information system that is integrated with the ERP Bank BTN Bogor Branch Office using the Service Oriented Architecture (SOA) method.

## 2. Background Theory

### 2.1. Management Information System

Management Information Systems are information systems that provide support for organizations represented by managers in a management level or in a particular field [7].

Management Information System is an information system that is poured through reports and visualization of information to be distributed to managers or superiors. Examples are sales analysis, employee performance and financing reporting systems [8]. Therefore, the type of management information system has a management level in the following Figure 1 [9] :

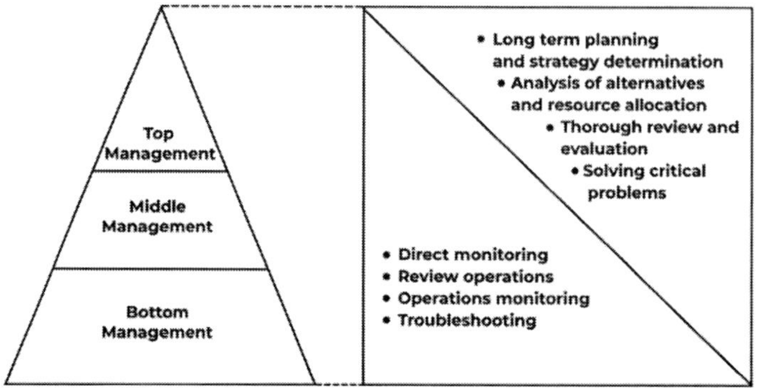


**Figure 1.** Type of Information and Management Activities

In the Figure 1, there is an explanation of the level of management, including the following:

a. Top level management is the executive who is in charge of strategic planning, policy and decision making.
b. Middle level management is the management who is in charge of preparing tactical planning.
c. Lower level management is the management in charge of planning and monitoring business process operations.
d. Operators are tasked with processing all forms of transactions and answering all forms of requests.

### 2.2. Performance Appraisal

Performance appraisal is an evaluation of the performance of employees at a company or organization by comparing the results of the achievement with the targets to be achieved through the employee performance appraisal instrument and displaying the quality of work in the organization [10].

### 2.3. Key Performance Indicator

Key Performance Indicator is one type of performance measurement used by management in an organization to create reports, measure and manage performance. In other words, KPI is an indicator that presents the performance results in a process that has been done [11].

### 2.4. Dashboard

Dashboard is a form of visualization of information presented in the form of tables, charts, reports, visual indicators that aim to monitor organization performance on a single screen. Visual displays must present the best possible information, so that the human eye can see quickly and the human brain can understand it correctly [12].

## 2.5 Service Oriented Architecture (SOA)

Service Oriented Architecture is a collection of several services that communicate with each other and coordinate several activities involving data through one or more services that have been converted into business service[13]. By using this service business, every function of each connected service can be used by anyone without changing the components of a system if a service is changed. SOA also works by converting a software into a reusable service so that the service can be used without being interrupted by changes in business processes[14]. In SOA there are also several components can be seen in Figure 2.

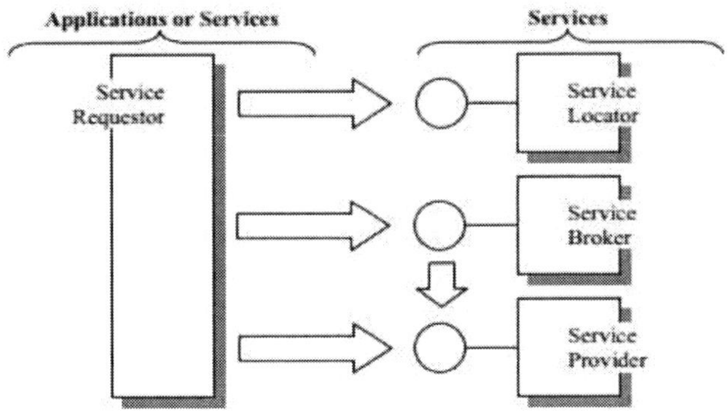


**Figure 2.** SOA components

In Figure, there are several explanations as follows:

a. Service: services in SOA must be loose-coupling, meaning that the service does not depend on differences in platforms, programming languages or server locations.
b. Loose-coupling: serves to maintain communication between services in case of changes in the system when there is a change in service.
c. Service provider: entity in software whose job is to run software specifications.
d. Clients: entities in the software that make requests to service.
e. Service locator: Registry on the type of service provider and functions in finding the service location and service provider interface.
f. Service broker: A type of service provider that is able to answer requests from service to one or more service providers.

## 2.6 Rapid Application Development (RAD)

Rapid Application Development (RAD) is one type of object-oriented system development method that includes development methods and software [16]. RAD aims to shorten the time in a traditional system development life cycle between planning and implementing an information system (See Figure 3).

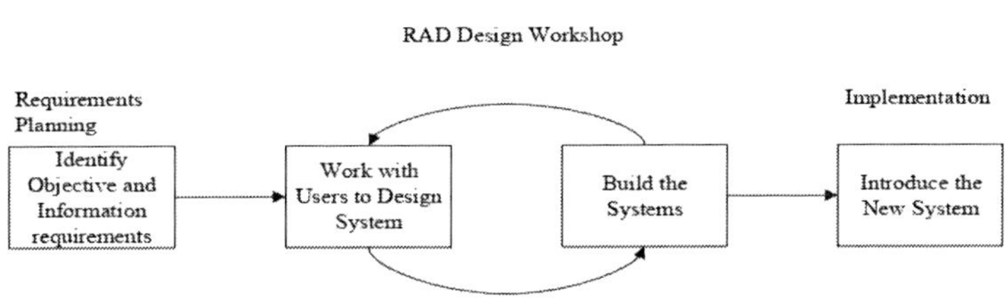


**Figure 3.** Rapid Application Development Model

In RAD there are 3 phases or stages, namely [10]:

a. Requirements planning
Stages of data collection where users and analysts discuss the objectives of the information system to be built by identifying information needs. This stage focuses on meeting the two parties, especially the involvement of users from various levels of the organization in order to meet user needs properly.
b. Design workshop
At this stage, it focuses on designing the system to be built and making some improvements if needed (there are design deficiencies) between the user and the system analyst. Users play an important role because they actively participate in providing input to each design process.
c. Implementation
The last stage where after a discussion between the user and the system analyst and the design of the information system to be built. So, next is the process of building information systems by programmers.

### 2.7 Unified Modeling Language (UML)

Unified Modeling Language or commonly abbreviated as UML is a form of notation used in the design and development of information systems or software covering analysis and design to architecture in object-oriented programming [17].

UML which will be notated in the form of a diagram to represent the design of a system, namely:

a. Usecase diagram
Usecase diagram is a modeling of the properties or functions of the information system to be designed. Usecase describes the interaction between actors and the system to be designed.
b. Activity diagram
Activity diagram is an activity modeling that represents the work flow of a system or menu in an information system.
c. Class diagram
Class diagram is a system structure from the translation of classes contained in an information system design. In a class, there are attributes and methods or operations.
- Attributes are variables owned by a class.
- Operations or methods are functions that belong to a class.

## 3. Methodology

The research methodology is depicted in a flow chart, the researcher uses the Rapid Application Development (RAD) system development method. The stages of this research are described as shown in Figure 4.

The principle of developing a research methodology is based on the development of information systems by Kendall (2008) which has been modified by the addition of modeling and service from the Service Oriented Architecture (SOA) method. For an explanation of the research methodology, it will be discussed in general below:

a. Identification of Needs
The initial step taken was a process of observation and interviews to dig up any technical information from employee performance appraisals and identification of problems related to employee performance appraisals at Bank BTN, Bogor Branch Office.

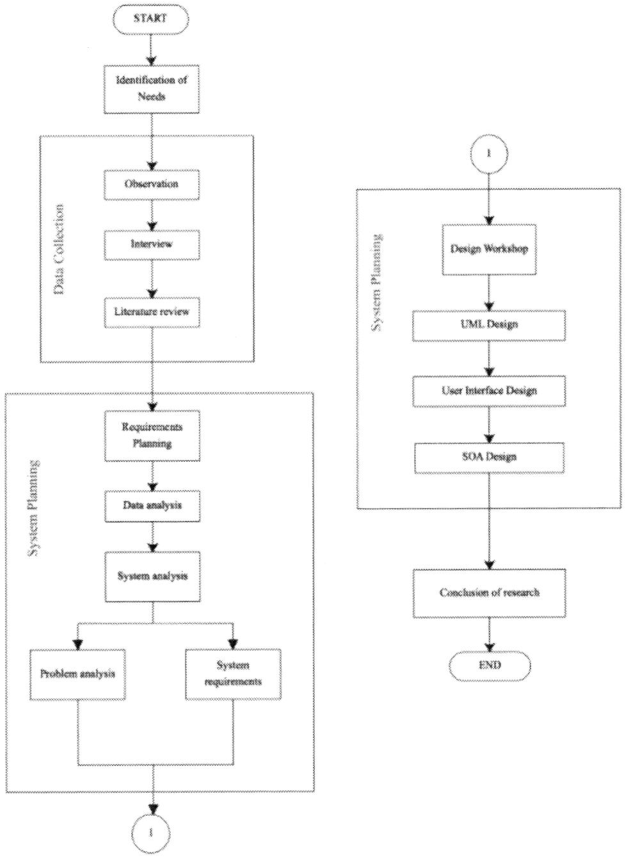


**Figure 4.** Research methodology

b. Data collection
Data collection aims to collect data made in designing a system. Researchers do the following 3 ways:
- Observation: researchers conducted observations for 2 weeks at Bank BTN Bogor Branch Office. The data obtained is a semi-manual system using Microsoft Excel for the employee performance appraisal process and making observations on existing business processes in the company.
- Interview: researchers conducted interviews with the Deputy Branch Manager in order to obtain more relevant information about the issues discussed in this study.
- Literature study: the last step to find supporting literature for this research.

c. System planning
The next stage is designing a system consisting of 2 subsystems, namely Requirements planning and Design Workshop. For general explanation as follows:
- Requirements planning: to describe the objectives of the information system which will be designed to identify information requirements that arise from the objectives that have been previously made.
- Design Workshop: this stage consists of 3 designs, namely UML design, then user interface design, and the third is Service Oriented Architecture (SOA) design.

d. Conclusion of Research Results
The final stage in this research is a mockup of an employee performance appraisal management information system and is expected to be a reference for Bank BTN to be able to implement this research.

# 4. Results and Discussions

## 4.1 Requirements Planning

Requirements Planning consists of 2 stages, namely data analysis and system analysis. For data analysis, researchers will discuss the user needs in system design in order to obtain the required data. While the system analysis is divided into 2 stages, namely problem analysis which will discuss the analysis of problems that exist in the system that is currently running at BTN KC Bogor while the system requirements focus on management level functionality.

a. Data analysis

Data analysis was carried out in order to get an overview of the user's needs in designing this management information system. User requirements in this process are as follows:

- Determine the object of the system to be built.
- Learning the KPI of Bank BTN Bogor Branch Office.
- Studying the organizational structure of Bank BTN Bogor Branch Office.
- Analysis of pre-existing outputs.
- Learning the running system.

b. System Analysis

Systems analysis is used to solve a problem by decomposing the system into its constituent components to find out how these components work and interact with each other.

- Problem Analysis

  The problems that exist in the system that is currently running at Bank BTN Bogor Branch Office will be described in the form of an activity diagram in Figure 5.

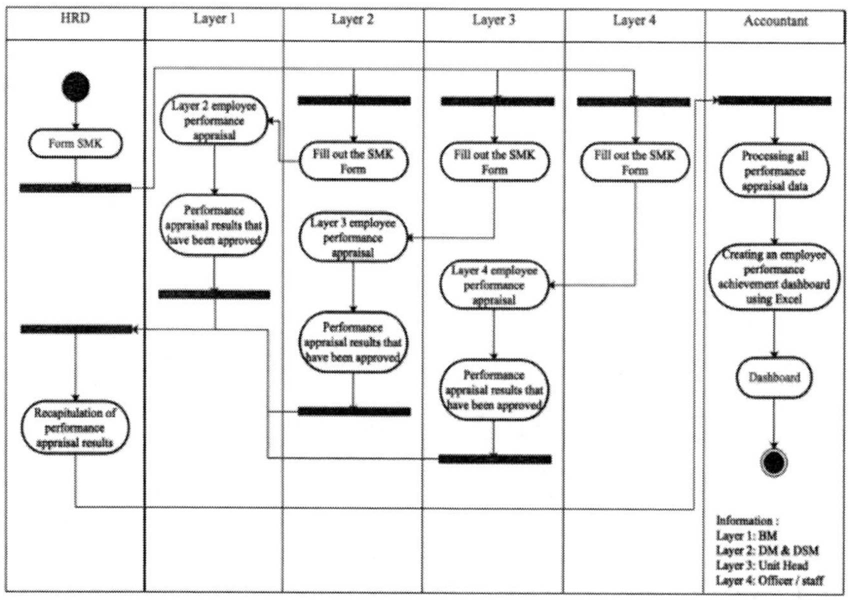


**Figure 5.** Activity diagram of the employee performance appraisal process

- System Requirements

  The system requirements are the result of a system analysis that is currently running at Bank BTN Bogor Branch Office which then determines the functional requirements of the system to be built into the SIM functionality Table 1.

**Table 1.** Activity diagram of the employee performance appraisal process

| No | Users | SIM Functionality | |
|---|---|---|---|
| | | **Main Functions** | **Additional Functions** |
| | | ***Middle Level Management*** | |
| 1 | *Branch Manager* | Monitoring of KPI achievements and subordinate approval. | Making decisions in determining top talent |
| 2 | *Deputy Branch Manager* | Monitoring of KPI achievements and subordinate approval. | Making decisions in determining top talent |
| 3 | *Deputy Service Manager* | Monitoring of KPI achievements and subordinate approval. | Making decisions in determining top talent |
| | | ***Bottom Level Management*** | |
| 4 | *Priority Banking* | Monitoring of KPI achievements and employee PSKT approval in work units. | - |
| 5 | *BCLU Subsidi* | Monitoring of KPI achievements and employee PSKT approval in work units. | - |
| 6 | *BCLU Non Subsidi* | Monitoring of KPI achievements and employee PSKT approval in work units. | - |
| 7 | *BCSU* | Monitoring of KPI achievements and employee PSKT approval in work units. | - |
| 8 | *BCFU* | Monitoring of KPI achievements and employee PSKT approval in work units. | - |
| 9 | *Operation* | Monitoring of KPI achievements and employee PSKT approval in work units. | - |
| 10 | *Accounting & Reporting* | Monitoring of KPI achievements and employee PSKT approval in work units. | - |
| 11 | *CCRU* | Monitoring of KPI achievements and employee PSKT approval in work units. | - |
| 12 | *Teller Service* | Monitoring of KPI achievements and employee PSKT approval in work units. | - |
| 13 | *Customer Service* | Monitoring of KPI achievements and employee PSKT approval in work units. | - |
| 14 | *Loan Admin* | Monitoring of KPI achievements and employee PSKT approval in work units. | - |
| 15 | *General Administration* | Monitoring of KPI achievements and employee PSKT approval in work units. | - |
| 16 | *TP* | Monitoring of KPI achievements and employee PSKT approval in work units.. | - |
| 17 | *Sub Branch Head* | Monitoring of KPI achievements and employee PSKT approval in work units. | Making decisions in determining top talent |
| 18 | *Cash Office Head* | Monitoring of KPI achievements and employee PSKT approval in work units. | Making decisions in determining top talent |
| 19 | *Officer* | Fill in KPI achievements on the PSKT form | |

## 4.2 Design Workshop

The Design Workshop consists of 3 stages, namely the UML design consisting of 2 diagrams (usecase diagrams and class diagrams), the second is the design of the user interface (mock up), the third is the design of Service Oriented Architecture which consists of SOA modeling and service modeling.

a. Usecase diagram

Usecase diagrams are used to find out what functions are in the management information system design and who has access to use these functions (See Figure 6).

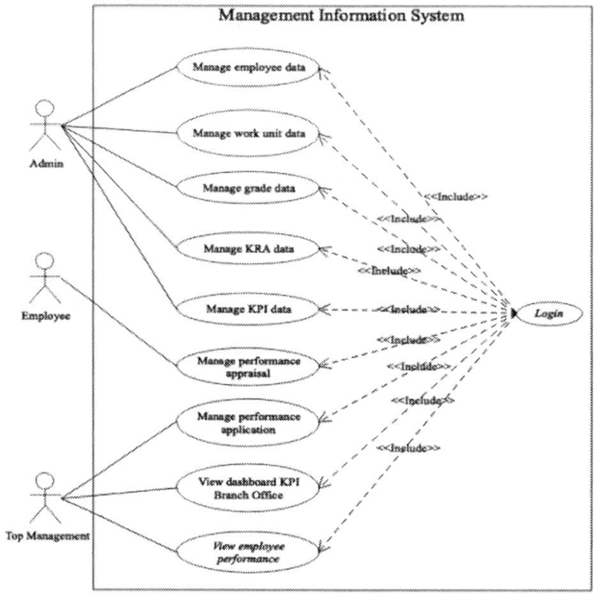


**Figure 6.** Usecase diagram

b. Class diagram

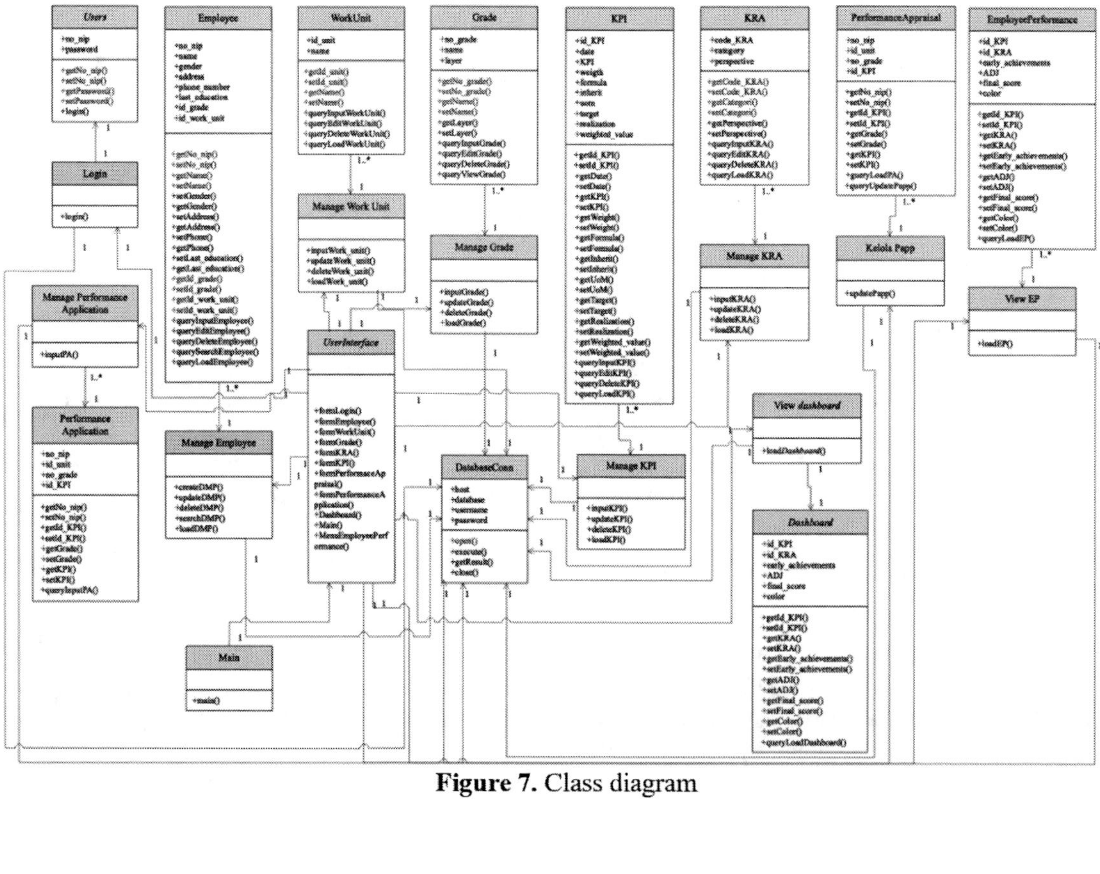


**Figure 7.** Class diagram

c. User Interface Design
User Interface design is built based on website and mobile (Android) using Adobe XD application. This UI design is based on the functional system that has been designed before. The following is the result of designing a user interface management information system for employee performance appraisal that researchers have created:

- Website design can be accessed by all users, namely Admin, employees and top management of Bank BTN Bogor Branch Office by writing the website address in the url in the browser. After the login page appears, the user can enter the ID number and password that was provided previously (See Figure 8).

**Figure 8.** Login page

- After the system validates the login, the system will display the dashboard page as below. The information management menu is provided on the sidebar which will appear when the user presses the word dashboard which is on the left of the page (See Figure 9).

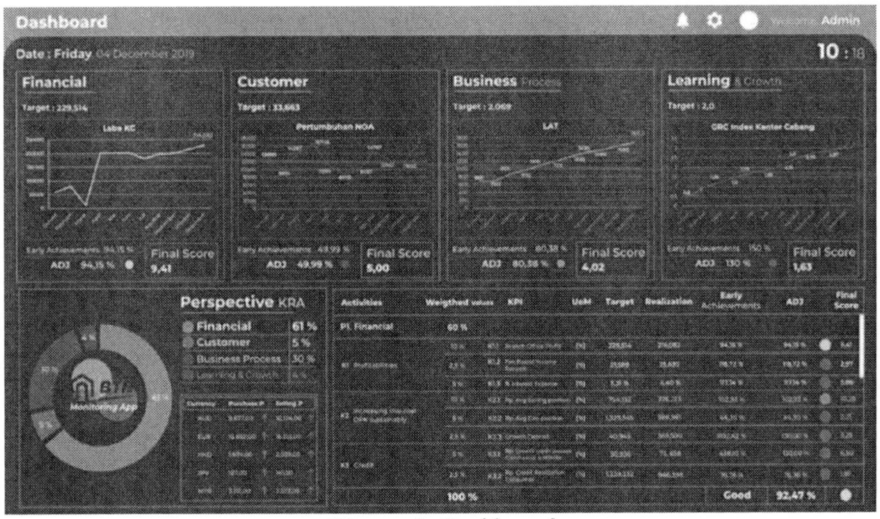


**Figure 9.** Dashboard

- For admin, there are 5 main menus, namely managing employee data, managing work unit data, managing grade data, managing Key Result Area (KRA) data and the last menu managing Key Performance Indicator (KPI) data. For UI design, the researcher included one example, namely managing employee data as seen in Figure 10.

IOP Conf. Series: Materials Science and Engineering 1115 (2021) 012013 doi:10.1088/1757-899X/1115/1/012013

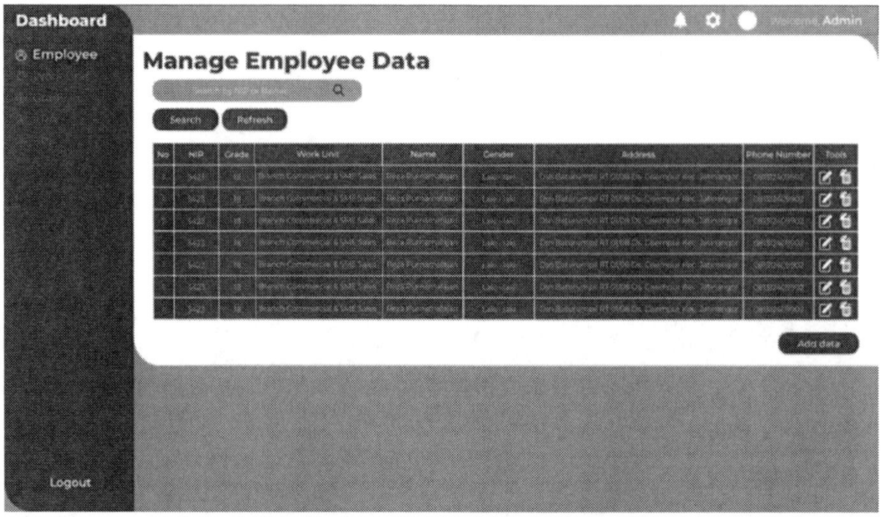


**Figure 10.** Manage Employee Data Page

- For employees, only the performance appraisal menu is available which aims at the employee performance appraisal process. In this menu, there is a table that must be filled in by each employee which will then be assessed by their respective superiors according to the work unit (See Figure 11).

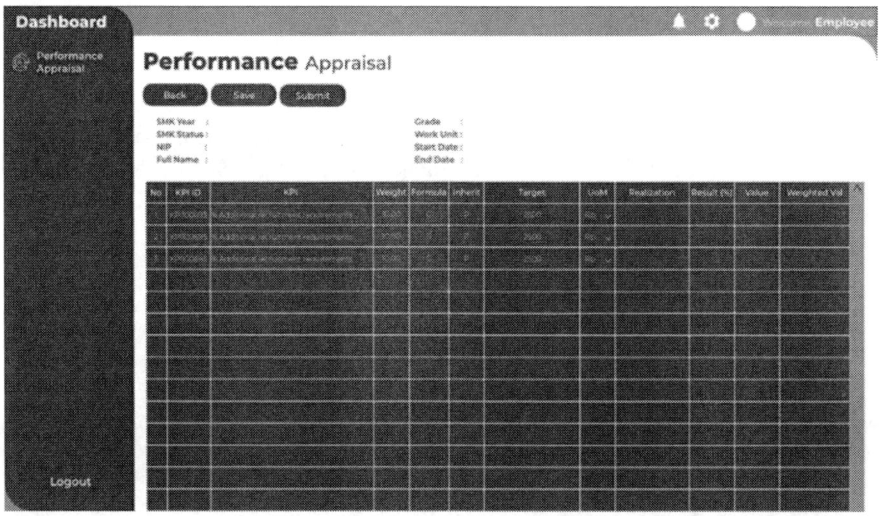


**Figure 11.** Performance Appraisal page

- Top management who logs in through the website will provide a performance application menu that aims to approve the results of employee performance achievements. If there is an improvement, the top management as the superior can make changes to the data for further approval process (See Figure 12).

IOP Conf. Series: Materials Science and Engineering 1115 (2021) 012013 doi:10.1088/1757-899X/1115/1/012013

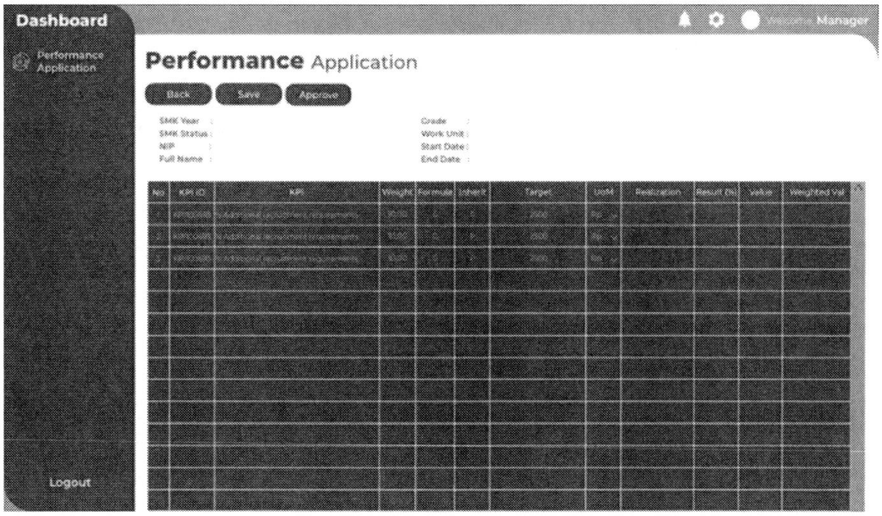


**Figure 12.** Performance Application page

- For managerial levels, namely Branch Managers, Deputy Branch Managers and Deputy Service Managers, they can monitor the achievement of employee performance through the mobile SIM application by inputting the provided ID number and password, after which the system will validate and direct to the main menu or dashboard (See Figure 13).

**Figure 13.** Login menu and dashboard menu

- The manager can also see details of performance achievement through the menu provided on the sidebar or by pressing the KRA percentage or KC percentage icon, this facility uses a drill down (See Figure 14).

**Figure 14.** Detailed menu of employee performance achievement

- Similar to the detailed performance achievement facility, the manager can monitor information based on each KRA, such as financial, customer, business process, and learning and growth. In the picture on the right, an employee performance facility is provided to see the achievements of each employee in order to make it easier to make decisions in choosing top talents for promotional activities and employee education (See Figure 15).

**Figure 15.** KRA (financial) Menu and Employee Performance Menu

d. Designing a Service Oriented Architecture (SOA

Designing a Service Oriented Architecture (SOA) consists of 2 stages, namely SOA modeling and service modeling.

- SOA modeling aims to represent a system architecture using existing components. SOA uses the Enterprise Service Bus (ESB) for communication media between data. In this study, SOA is used as a bridge between the ERP Bank BTN Bogor Branch Office with the design of this employee performance appraisal management information system (See Figure 16).

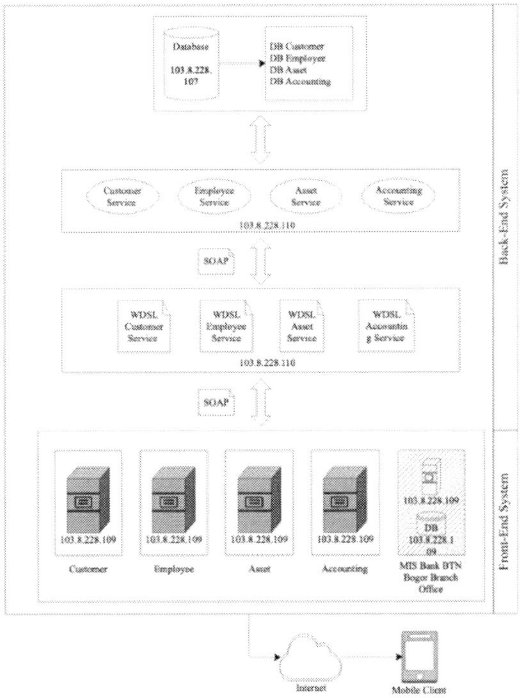


**Figure 16.** SOA Modeling

- Service modeling is used to model a service in a management information system so that it can communicate with other systems. Service modeling also serves to define the data to be sent on the requester. Most of the data returns in this function are arrays and strings (See Figure 17).

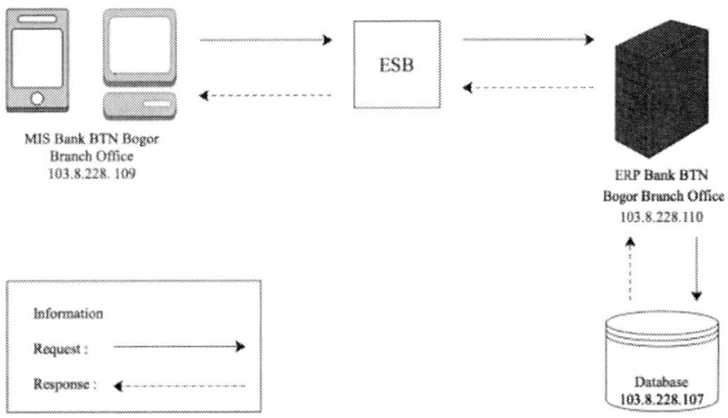


**Figure 17.** Modeling Service

## 5. Conclusions

Based on the results of the research that has been done, the researcher can present the following conclusions:

a. The management of Bank BTN Bogor Branch Office can conduct employee performance appraisals and monitor KPI achievement using the design of the Management Information System for employee performance appraisals.

b. The management of Bank BTN Bogor Branch Office can make decisions in finding top talent among employees using the design of the Management Information System for employee performance appraisals.
c. Produce the form of Service Oriented Architecture (SOA) modeling and service modeling so that the Management Information system can be integrated with the ERP Bank BTN Bogor Branch Office.

## References

[1] Rupilele F G J, Soulisa I, Palilu A, Hasibuan A, Winesty O F, Goraph F A and Tondo S 2018 Management Information System for Monitoring and Inspection of the Implementation of Universities *International Journal of Engineering & Technology* **7** (2) 451-456.
[2] Teixeira L, Xambre A R, Figueiredo J and Alvelos H 2016 Analysis and Design of a Project Management Information System: practical case in a consulting company *Procedia Computer Science* **100** 171-178
[3] Reddy G S, Srinivasu R, Rikkula S R and Rao V S 2009 Management information system to help managers for providing decision making in an organization *International Journal of Reviews in Computing* **5** (1) 1-6
[4] Tokola H, Gröger C, Järvenpää E and Niemi E 2016 Designing manufacturing dashboards on the basis of a Key Performance Indicator survey *Procedia CIRP* **57** 619-624
[5] Jiang P, Elag M, Kumar P, Peckham S D, Marini L and Rui L 2017 A service-oriented architecture for coupling web service models using the Basic Model Interface (BMI) *Environmental Modelling & Software* **92** 107-118
[6] Li D, Chen Z, Deng Z, Huang W, Tang J, Di F and Jiang X 2018 A wide area service oriented architecture design for plug and play of power grid equipment *Procedia Computer Science* **129** 353-357
[7] Laudon K C and Laudon J P 2014 *Management Information Systems Managing The digital Firm* New York: Person
[8] James A O 2003 *Introduction to Information System : Essentials for the E-Business Enterprise* New York: McGraw Hill Inc
[9] Jogiyanto H M 2005 *Analisis Dan Desain Sistem Informasi Pendekatan Terstruktur Teori dan Praktek Bisnis,* Yogyakarta: Andi
[10] Kurtz and Boone 2007 *Pengantar Bisnis Kontemporer* Jakarta: Salemba Empat
[11] Parmenter D 2010 *Key Performance Indicators : Developing, Implementing and Using Winning KPIs* Canada: John Wiley & Sons, Inc
[12] Few S 2006 *Information Dashboard Design* CA: O'Reilly Media
[13] Thomas E 2004 Erl Service Oriented Architecture: A Field Guide to Integrating XML and Web Services United States: Prentice Hall
[14] Safuwan R S and Akbar R J 2010 Integrasi Perangkat Lunak Enterprise Resource Planning (ERP) Dengan Menggunakan Metode Service Oriented Architecture (SOA) *Seminar Institute Teknologi Sepuluh Nopember*
[15] Sen H 2010 *Service Oriented Architecture (SOA) and Their Application and Usage in Healthcare* Connecticut: Computer Science and Engineering Department
[16] Kendal 2008 *Rapid Application Development (RAD)* Klaten: PT Indeks
[17] Rosa A S and Shalahuddin M 2013 *Rekayasa Perangkat Lunak* Bandung: Informatika

# Survival Analysis Competing Risk using Fine-Gray Subdistribution Model with the Maximum Partial Likelihood Estimation Method

**A Ainurrochmah*, S W Purnami and Purhadi**

Department of Statistics, Faculty of Science and Data Analytics,
Institut Teknologi Sepuluh Nopember (ITS), Surabaya, Indonesia

* Email: alifta.aaa@gmail.com

**Abstract.** Survival analysis is a statistical method which the variable of concern is the time until the event occurs. In survival analysis, there is a situation where an individual can experience more than one type of event and the occurrence of these events prevents the occurrence of other types of events. This situation is called competing risk. The situation causes the Kaplan Meier method, which is a method of estimating the survival function, cannot be used. The Cumulative Incidence Function method is proposed as a solution to solve competing risk events uses the probability of each type of event. The Cox regression model is also modified to allow for competing risk is called the Fine-Gray subdistribution model using the Maximum Partial Likelihood Estimation. This study examines the estimation of parameter Fine-Gray subdistribution model and applies it to melanoma case. Melanoma is a type skin cancer that can spread to other organs in the body. In the case of death, melanoma patients who experience death with other causes are called a competing risks and death with melanoma are considered as the event of interest. Study showed that factors that influence melanoma mortality are age, sex, tumor thickness, and cultured epidermal autografts. Patients death with melanoma was 1.400 times higher risk than death with other causes.

## 1. Introduction

Survival analysis is a statistical procedure, which is related to time and is often referred to as time to event. Survival analysis can be used to analyse the time of an individual survival time starting from the initial observation until an event occurs [1]. The response variable in the survival analysis is the initial time of observation until a certain event occurs, while the predictor variables are factors that might influence an event to occur.

In the survival analysis, there are several situations where the observation is not suitable for survival method which is commonly used time to event analysis. One such situation is when an individual can experience more than one type of event and these events prevent other events from occurring. In general, this situation is called a competing risk [2]. The Cumulative Incidence Function (CIF) method is proposed as a solution to solve more than one event. The CIF method uses the probability of each event that occurs, and then is partitioned into probabilities for each type of event [3].

Survival analysis can also estimate the effect predictor variables on survival time. The analysis that is commonly used to determine the effect of predictor variables to the response variable is regression analysis. Hazard regression analysis can be used to estimate the effect of predictor variables on survival time. The hazard regression model that is often used is the Cox regression model. Fine-Gray (1999) modified the Cox proportional hazard regression model to allow for competing risk events [4]. Fine-Gray modelling is performed based on estimating the hazard subdistribution. The Fine-Gray subdistribution model can estimate the effect of predictor variables on survival time based on the cumulative incidence function at each event with competing risk, but in interpreting the Fine-Gray subdistribution model and cumulative incidence, more attention must be paid to determine the effect of the predictor variables [5]. If they do not fulfill the proportional hazard assumptions, we have to use basic regression model with stratified Cox [6] or extended Cox regression [7].

Some of the events which are competing risk for example in clinical oncology, cancer-related mortality may be of primary interest, but other causes of death can prevent its occurrence and deaths caused by reasons other than cancer are typical examples of competing risk. Melanoma is one of the most invasive and metastatic human cancers and accounts for the majority of skin cancer deaths despite comprising less than 5% of all cutaneous malignancies [8].

Based on the description above, this study will describe a different approach to survival analysis in the presence of competing risks with Fine-Gray subdistribution model and the application using Melanoma data.

## 2. Methodology

### 2.1 Survival Analysis

Survival analysis is a statistical method with the variable of concern is the time until an event occurs or it is called the survival time. Time used in the survival analysis can be in the form of days, weeks, months or even years obtained from the initial observation until the event occurs, while events can be positive events such as healing and negative events such as contracting a disease, recurrence of a disease and death [1].

In the survival analysis, there are two main functions, namely the survival function and the hazard function. The survival function is used to determine the probability of individual survival time from the initial observation time to time $t$. In the survival function equation, if the survival time is denoted by $T$ which is a non-negative random variable and has a chance density function, then the probability density function can be written with the following equation:

$$f(t)=\lim_{\Delta t\to 0}\frac{P(t\le T<t+\Delta t)}{\Delta t} \tag{1}$$

where the cumulative distribution function can be written as follows:

$$F(t)=P(T\le t)=\int_0^t f(t)dt \tag{2}$$

The survival function is expressed in terms of the cumulative distribution function as follows:

$$S(t)=P(T>t)=1-P(T\le t)=1-F(t) \tag{3}$$

The hazard function $h(t)$ describes the failure rate of an individual that occurs in the time interval $t$ to $t+\Delta t$ the individual still survives until time $t$. The hazard function can be written as follows:

$$\begin{aligned} h(t) &= \lim_{\Delta t\to 0}\frac{P(t\le T<t+\Delta t\,|\,T\ge t)}{\Delta t} \\ &= \frac{f(t)}{S(t)} \end{aligned} \tag{4}$$

## 2.2 Cumulative Incidence Function and Gray's Method

Cumulative Incidence Function (CIF) describes the estimated marginal probability of survival of an individual at each event. The CIF method can separately estimate the failure rate for each event. The CIF value for the type *c* event at time *t* can be written as follows:

$$CIF_c(t_{(f)}) = \sum_{f=1}^{f} I_c(t_{(f)}) = \sum_{f=1}^{f} S(t_{(f-1)}) \times \mathrm{h}_c(t_{(f)}) \tag{5}$$

Gray's test is an extension of the Log Rank test which is used to evaluate the CIF similarity between two or more groups. The concept of Gray's test is to compare hazard subdistribution functions. Gray's test can be written as follows:

$$\chi^2 = \frac{U^2}{\widehat{Var(U)}} \tag{6}$$

with,

$$U = \sum_{all\ t_{(f)}} R_{1(f)} \left( \frac{d_{1(f)}}{R_{1(f)}} - \frac{d_{1(f)} + d_{2(f)}}{R_{1(f)} + R_{2(f)}} \right) \tag{7}$$

and

$$\widehat{\boldsymbol{Var(U)}} = \sum_{i=1}^{f} \frac{d_1(t_{(f)}) + d_2(t_{(f)})}{m_1(t_{(f)}) + m_2(t_{(f)})} \tag{8}$$

where:

$R_{1(f)} = n_1\left(t_{(f)}\right) \frac{1 - F_1(t_{(f-1)})}{S_1(t_{(f-1)})}$

$m_c(t_{(f)}) = \frac{n_c(t_{(f)})}{S_c(t_{(f-1)})}$

$d_c(t_f)$ = number of events of interest in the type *c* event group at time *t*

The test criterion is that $H_0$ is rejected if $\chi^2 > \chi^2_{\alpha,g-1}$ where *g* is many groups of event types or if *p-value* $< \alpha$ [9].

## 2.3 Fine-Gray Subdistribution Model

Fine and Gray (1999) modified the Cox model to allow for competing risk by using a hazard subdistribution regression model by modeling the cumulative incidence function (CIF) curve with covariates. The CIF curve comes from the subdistribution function. This model is basically a development of the Cox model so that there is a proportional hazard assumption that must be tested [9]. Based on equation (1), which is the opportunity density function, it can be seen that the opportunity density function for event type *c* can be written as follows:

$$f_c(t) = \lim_{\Delta t \to 0} \frac{P(t \le T < t + \Delta t, C = c)}{\Delta t} \tag{9}$$

Fine-Gray subdistribution hazard function for event type *c* can be written as follows:

$$h_c(t) = \lim_{\Delta t \to 0} \frac{P(t \le T < t + \Delta t, C = c \mid T \ge t)}{\Delta t} \tag{10}$$

Then the CIF based Fine-Gray subdistribution model can be defined as follows:

$$\hat{h}_c(t, \mathbf{X}_i) = \hat{h}_{0c}(t) \exp\left( \sum_{j=1}^{p} \hat{\beta}_{jc} X_{ij} \right) \tag{11}$$

where:

$\hat{h}_c(t,\mathbf{X}_i)$ : estimate hazard subdistribution function event type $c$

$\hat{h}_{0c}(t)$ : estimate baseline hazard function event type $c$

$\hat{\beta}_1\ \hat{\beta}_2\ \ldots\ \hat{\beta}_p$ : regression parameter estimation event type $c$

$X_{i1}\ X_{i2}\ \ldots\ X_{ip}$ : predictor variable object $i$ event type $c$

The partial likelihood model for the Fine-Gray subdistribution model is as follows:

$$L(\boldsymbol{\beta})=\prod_{i=1}^{n}\frac{\exp\left(\boldsymbol{\beta}^T\mathbf{X}_i\right)}{\sum_{l\in R(t_i)} w_{il}\exp\left(\boldsymbol{\beta}^T\mathbf{X}_l\right)} \tag{12}$$

A collection of risks $R(t_i)$ defined by adding weights, which is based on non-censored objects. It can be concluded that collection of risks is formed based on the event of interest object with the following definition:

$R(t_i)=\{l;T_l\geq t$ or $(T_l\leq t)$ and an individual in a competing risk$\}$

$w_{il}$ = the weight of the object $i$ in the event of interest, the weight can be defined as follows:

$$w_{il}=\frac{\hat{G}(t_i)}{\hat{G}(\min(t_i,t_l))} \tag{13}$$

where:

$\hat{G}(t_i)$ = Estimated of survival function from survival time $i$ object

$\hat{G}(\min(t_i,t_l))$ = Estimation of the survival function from the minimum value between the survival time of the object $i$ and the object on the event $l$ or event of interest

The weight will be 1 if the object has a survival time $t_i\leq t_l$ and has a value $\leq 1$ if $t_l>t_i$ [10].

### 2.4 Proportional Hazard Assumption

The proportional hazard assumption is used to measure whether the relationship between cumulative hazards is constant over time [1]. A subdistribution Fine-Gray regression based on the Cox regression cox requires the assumption of a proportional hazard function over time because the Cox regression does not accommodate variables that varying over time [11].

Testing the proportional hazard assumption on competing risk case can be tested with goodness of fit. Goodness of fit testing used Schoenfeld residuals with modifications to the weight [12].

$$T(D)=n^{-1}\mathbf{U}_{1\theta}^T\left(\boldsymbol{\theta}_0,\hat{\boldsymbol{\beta}}\right)\mathbf{A}^{-1}\mathbf{U}_{1\theta}\left(\boldsymbol{\theta}_0,\hat{\boldsymbol{\beta}}\right) \tag{14}$$

with,

$$\mathbf{A}=n^{-\frac{1}{2}}\mathbf{U}_{1\theta}\left(\boldsymbol{\theta}_0,\hat{\boldsymbol{\beta}}\right) \tag{15}$$

where,

$\mathbf{U}_{1\theta}\left(\boldsymbol{\theta}_0,\hat{\boldsymbol{\beta}}\right)$ = parameter estimation vectors $\hat{\boldsymbol{\beta}}$ and time covariate effect $\boldsymbol{\theta}$

The test criterion is that $H_0$ is rejected if $T(D)\geq\chi^2_{\alpha;m}$ where $m$ is covariates effect or if *p-value* $<\alpha$, it means that the proportional hazard assumption is not fulfilled [13].

### 2.5 Parameter Significance Test

The parameter significance test was conducted to determine whether there is an effect of a predictor variable on the response variable. There are two tests in testing the significance of parameters, namely the test simultaneously using the likelihood ratio test and partially test using the Wald test.

1. Likelihood Ratio Test

   The likelihood ratio (LR) test denoted by $G^2$ is used to test the effect of the predictor variables in the regression model simultaneously.

$$G^2 = -2\left(\frac{\max\limits_{\omega} \ell(\omega)}{\max\limits_{\Omega} \ell(\Omega)}\right) = -2\left(\frac{\ell(\hat{\omega})}{\ell(\hat{\Omega})}\right) \tag{16}$$

   The $G^2$ test follows the chi-square distribution with degrees of freedom $p$ so that the testing criterion, namely $H_0$ is rejected if the value is $G^2 \geq \chi^2_{\alpha,p}$, or if *p-value* $< \alpha$, meaning that at least one predictor variable has a significant effect on survival time.

2. Wald Test

   Wald test is used to test the effect of parameter $\beta_j$ individually

$$W^2 = \left(\frac{\hat{\beta}_j}{SE(\hat{\beta}_j)}\right)^2 \tag{17}$$

   The Wald test statistic follows the chi-square distribution with degrees of freedom 1. The test criterion is that $H_0$ is rejected if the value is $W^2 \geq \chi^2_{(\alpha,1)}$ or if *p-value* $< \alpha$, meaning the *j* predictor variable has a significant effect on survival time.

## 3. Result and Discussion

### 3.1 Estimated Fine-Gray Subdistribution Model Parameters

Estimated subdistribution Fine-Gray model parameters can be done using the maximum partial likelihood estimation. The sensor status in the survival data is defined as censored data and uncensored data so that the survival data can have a certain distribution such as the Logistic distribution with the opportunity density function as follows [3].

$$f(t) = \frac{\lambda\theta t^{\theta-1}}{\left(1+\lambda t^{\theta}\right)^2} \quad \text{where } \lambda, \theta > 0 \text{ and } t \geq 0 \tag{18}$$

Then the likelihood function is obtained from multiplying the probability density function for the type $c$ event based on equations (9) and (10) as follows.

$$\begin{aligned} L(\boldsymbol{\beta}) &= \prod_{i=1}^{n} f_c(t_i, X_i) \\ &= \prod_{l \in R(t_i)}^{n} \left[h_c(t_i, X_l)\right] \prod_{i=1}^{n} \left[S(t_i, X_i)\right] \\ &= \prod_{l \in R(t_i)}^{n} \left[h_c(t_i, X_l)\right] \exp \prod_{i=1}^{n} \left[-\int_0^t h(t, X_i) dt\right] \\ &= \prod_{l \in R(t_i)}^{n} \left[h_{0c}(t_i) \exp\left(\boldsymbol{\beta}^T \mathbf{X}_l\right)\right] \exp \prod_{i=1}^{n} \left[-H_0(t_i) \exp\left(\boldsymbol{\beta}^T \mathbf{X}_i\right)\right] \end{aligned} \tag{19}$$

Based on equation (12), the partial likelihood function of the subdtribution Fine-Gray model can be known. Then form that, the individual opportunities for censored and uncensored data can be defined as follows.

$$P(\text{cencored}) = \frac{\exp\left(\boldsymbol{\beta}^T \mathbf{X}_i\right)}{\sum_{l \in R(t_i)} w_{il} \exp\left(\boldsymbol{\beta}^T \mathbf{X}_l\right)} \qquad P(\text{uncencored}) = \frac{1}{\sum_{l \in R(t_i)} w_{il} \exp\left(\boldsymbol{\beta}^T \mathbf{X}_l\right)} \tag{20}$$

Therefore, the log-partial likelihood function as follows.

$$\begin{aligned} \ell(\boldsymbol{\beta}) &= \log\left\{\prod_{i=1}^{n} \frac{\exp\left(\boldsymbol{\beta}^T \mathbf{X}_i\right)}{\sum_{l \in R(t_i)} w_{il} \exp\left(\boldsymbol{\beta}^T \mathbf{X}_l\right)}\right\} \\ &= \sum_{i=1}^{n}\left\{\log\left(\exp\left(\boldsymbol{\beta}^T \mathbf{X}_i\right)\right) - \log\left(\sum_{l \in R(t_i)} w_{il} \exp\left(\boldsymbol{\beta}^T \mathbf{X}_l\right)\right)\right\} \\ &= \sum_{i=1}^{n}\left[\sum_{j=1}^{p}\left(\beta_j X_{ji}\right) - \log\left(\sum_{l \in R(t_i)} w_{il} \exp\sum_{j=1}^{p}\left(\beta_j X_{jl}\right)\right)\right] \end{aligned} \tag{21}$$

After obtaining the log partial likelihood function in equation (21), here are the steps for estimating parameters $\boldsymbol{\beta}$ and the hypothesis test statistics on the Fine-Gray subdistribution model. To estimate parameters $\boldsymbol{\beta}$, the following steps are required.

1. Obtain the first partial derivative of the log-partial likelihood function based on equation (21) as follows.

$$\begin{aligned} \frac{\partial \ell(\boldsymbol{\beta})}{\partial \beta_p} &= \frac{\partial\left(\sum_{i=1}^{n}\left[\sum_{j=1}^{p}\left(\beta_j X_{ji}\right) - \log\left(\sum_{l \in R(t_i)} w_{il} \exp\sum_{j=1}^{p}\left(\beta_j X_{jl}\right)\right)\right]\right)}{\partial \beta_p} \\ &= \sum_{i=1}^{n}\left[X_{pi} - \frac{\sum_{l \in R(t_i)} w_{il} X_{pl} \exp\sum_{j=1}^{p}\left(\beta_j X_{jl}\right)}{\sum_{l \in R(t_i)} w_{il} \exp\sum_{j=1}^{p}\left(\beta_j X_{jl}\right)}\right] \end{aligned} \tag{22}$$

So that it forms a vector $g(\boldsymbol{\beta})$ which contains the first derivative of the log-partial likelihood function of size $p \times 1$.

2. Obtain the second partial derivative of the log-partial likelihood function against $\boldsymbol{\beta}$ based on equation (22).

$$\begin{aligned} \frac{\partial \ell(\boldsymbol{\beta})}{\partial \beta_k \partial \beta_{k*}} = &-\sum_{i=1}^{n}\left(\frac{\left(\sum_{l \in R(t_i)} w_{il} X_{kl} X_{k*l} \exp\sum_{j=1}^{p}\left(\beta_j X_{jl}\right)\right)}{\left(\sum_{l \in R(t_i)} w_{il} \exp\sum_{j=1}^{p}\left(\beta_j X_{jl}\right)\right)}\right) \\ &+\sum_{i=1}^{n}\left(\frac{\left(\sum_{l \in R(t_i)} w_{il} X_{k*l} \exp\sum_{j=1}^{p}\left(\beta_j X_{jl}\right)\right)\left(\sum_{l \in R(t_i)} w_{il} X_{kl} \exp\sum_{j=1}^{p}\left(\beta_j X_{jl}\right)\right)}{\left(\sum_{l \in R(t_i)} w_{il} \exp\sum_{j=1}^{p}\left(\beta_j X_{jl}\right)\right)^2}\right) \end{aligned} \tag{23}$$

where $k, k^* = 1, 2, ..., p$

So that it can form a Hessian matrix $H(\boldsymbol{\beta})$ which contains the second derivative of the log-partial likelihood function of size $p \times p$.

3. Then perform the Newton-Raphson iteration using the following equation.

$$\hat{\boldsymbol{\beta}}^{(z+1)} = \hat{\boldsymbol{\beta}}^{(z)} - \mathbf{H}^{-1}\left(\hat{\boldsymbol{\beta}}^{(z)}\right)g\left(\hat{\boldsymbol{\beta}}^{(z)}\right) \tag{24}$$

where $\mathbf{H}^{-1}\left(\hat{\boldsymbol{\beta}}^{(z)}\right)$ is the inverse of the Hessian matrix $\mathbf{H}\left(\hat{\boldsymbol{\beta}}^{(z)}\right)$.

4. Newton Raphson's iteration will stop if the value of $\left\|\hat{\boldsymbol{\beta}}^{(z+1)} - \hat{\boldsymbol{\beta}}^{(z)}\right\| \leq \varepsilon$, where $\varepsilon$ is a very small positive number.

After getting the parameter estimate $\hat{\boldsymbol{\beta}}$, then a Fine-Gray subdistribution model can be formed based on equation (11). Hypothesis testing and statistical tests simultaneously on the parameters of the Fine-Gray subdistribution model are carried out using the maximum likelihood ratio test (MLRT) method. This method involves two log-likelihood functions, namely $\ell(\Omega)$ which is the log-likelihood function of the parameter set under the model population by involving the predictor variable and $\ell(\omega)$ which is the log-likelihood function of the parameter set under $H_0$ model without involving the existing predictor variables, with steps as follows.

1. Obtain the log-likelihood function from the set of parameters under population $\Omega = \{\boldsymbol{\beta}, \lambda, \theta\}$ model by involving predictor variables based on the likelihood function in equation (12).

$$\begin{aligned}
\ell(\hat{\Omega}) &= \log\left\{\prod_{i=1}^{n} \frac{\exp\left(\hat{\boldsymbol{\beta}}^T \mathbf{X}_i\right)}{\sum_{l \in R(t_i)} w_{il} \exp\left(\hat{\boldsymbol{\beta}}^T \mathbf{X}_l\right)}\right\} \\
&= \sum_{i=1}^{n}\left\{\log\left(\exp\left(\hat{\boldsymbol{\beta}}^T \mathbf{X}_i\right)\right) - \log\left(\sum_{l \in R(t_i)} w_{il} \exp\left(\hat{\boldsymbol{\beta}}^T \mathbf{X}_l\right)\right)\right\} \\
&= \sum_{i=1}^{n}\left[\sum_{j=1}^{p}\left(\hat{\beta}_j X_{ji}\right) - \log\left(\sum_{l \in R(t_i)} w_{il} \exp \sum_{j=1}^{p}\left(\hat{\beta}_j X_{jl}\right)\right)\right]
\end{aligned} \tag{25}$$

2. Obtain the log-likelihood function from the set of parameters under the $H_0$, $\omega = \{h_{0c}(t)\}$ model without involving the existing predictor variables based on the likelihood function in equation (12)

$$\begin{aligned}
\ell(\hat{\omega}) &= \log\left\{\prod_{i=1}^{n} \frac{1}{\sum_{l \in R(t_i)} w_{il}}\right\} \\
&= \sum_{i=1}^{n}\left\{\log(1) - \log\left(\sum_{l \in R(t_i)} w_{il}\right)\right\} \\
&= -\sum_{i=1}^{n}\left(\log\left(\sum_{l \in R(t_i)} w_{il}\right)\right)
\end{aligned} \tag{26}$$

3. Obtain the likelihood ratio test statistics $G^2$ according to equation (16) as follows.

$$
\begin{aligned}
G^2 &= -2\left(\frac{\ell(\hat{\omega})}{\ell(\hat{\Omega})}\right) \\
&= 2\left[\left(\sum_{i=1}^{n}\left[\sum_{j=1}^{p}\left(\hat{\beta}_j X_{ji}\right) - \log\left(\sum_{l\in R(t_i)} w_{il}\exp\sum_{j=1}^{p}\left(\hat{\beta}_j X_{jl}\right)\right)\right]\right) - \left(-\sum_{i=1}^{n}\left(\log\left(\sum_{l\in R(t_i)} w_{il}\right)\right)\right)\right] \qquad (27) \\
&= 2\left[\left(\sum_{i=1}^{n}\left[\sum_{j=1}^{p}\left(\hat{\beta}_j X_{ji}\right) - \log\left(\sum_{l\in R(t_i)} w_{il}\exp\sum_{j=1}^{p}\left(\hat{\beta}_j X_{jl}\right)\right) + \log\left(\sum_{l\in R(t_i)} w_{il}\right)\right]\right)\right]
\end{aligned}
$$

The $G^2$ test follows the chi-square distribution with degrees of freedom $p$ so that the testing criterion, namely $H_0$ is rejected if the value is $G^2 \geq \chi^2_{\alpha,p}$ or if *p-value* $< \alpha$. If there are predictor variables that have a significant effect on survival time together, then it can be done using partial hypothesis testing. Hypothesis testing of parameter $\hat{\beta}_j$ is partially carried out to determine the significance of the partial test parameters, with steps as follows:

1. Obtain the parameter estimate of the *j* predictor variable $\hat{\beta}_j$ using the Newton-Raphson iteration.
2. Obtain the standard error estimation of the parameter of the *j* predictor variable. The standard error estimation of the parameter of the j predictor variable is obtained from the second partial derivative of the log-partial likelihood function which is the diagonal element of the Hessian matrix $H(\boldsymbol{\beta})$.
3. Get the Wald test statistics according to equation (17).

The Wald test statistic follows the chi-square distribution with degrees of freedom 1. The test criterion is that $H_0$ is rejected if the value is $W^2 \geq \chi^2_{(\alpha,1)}$ or if *p-value* $< \alpha$.

### 3.2 Application on Melanoma Data

Data were used is data on patients with Melanoma (cancer of skin) had a radical surgery performed at the University Hospital of Odense, Denmark. All patients were followed which time 134 were still alive while 71 had died (of out whom 57 had died from cancer and 14 from other causes). The dataset can be seen in software *R*. The research variables are shown in Table 1.

**Table 1.** Research Variables

| Variables | Describes |
|---|---|
| Status | 0= Cencored Data<br>1= Death with Melanoma (event of interest)<br>2= Death with other causes (competing risk) |
| Age | Age at operation (years) |
| Sex | 0= Female ; 1= Male |
| Tumor Thickness | Tumor thickness (1/100 mm) |
| Cultured Epidermal Autografts | 0= Not present Epicel ; 1= Present Epicel |

Therefore, modeling was carried out using a Fine-Gray subdistribution to determine death with melanoma of competing risks case, namely death with other causes and estimate the CIF of death with melanoma and death with other causes. The cumulative curves of the categorical predictor variables such as sex and cultured epidermal autografts (epicel) can be seen in Figure 1.

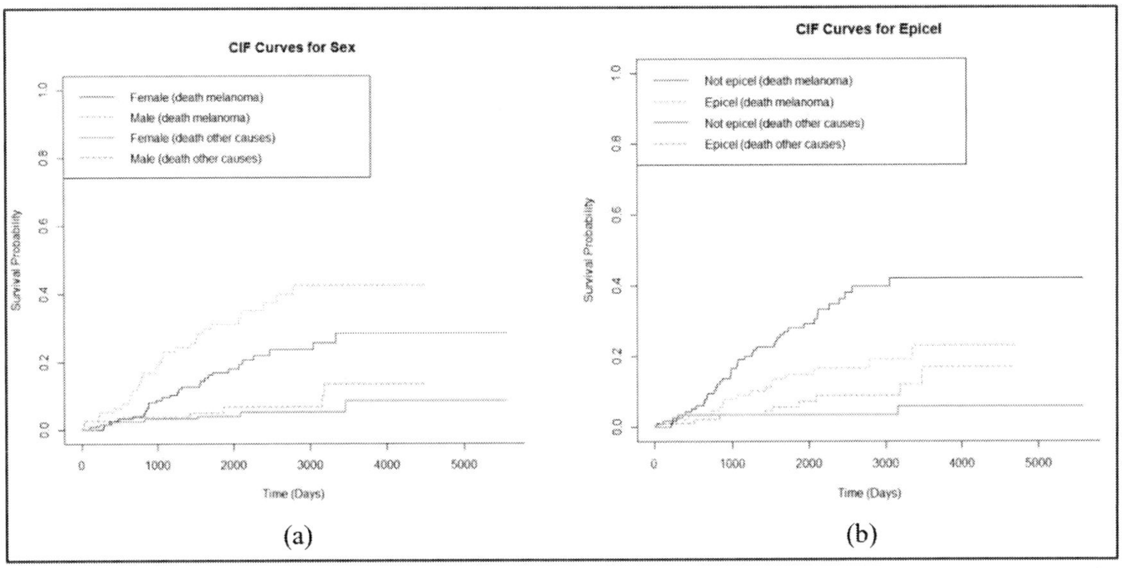


**Figure 1.** Cumulative Incidence Curve for Sex and Epicel

In Figure 1(a), the risk of death based on sex can be seen, namely male patients have higher risk of death with melanoma and other causes than female. Figure 1(b) shown that patients who did not have cultured epidermal autografts have higher risk of death with melanoma than death with other causes and patients who had cultured epidermal autografts have higher risk of death with melanoma than death with other causes. The difference between the cumulative incidence curve can be determined with certainty using the Gray's test statistic based on equation (6) are shown in Table 2.

**Table 2.** Gray's Test

| Predictor Variables | | $\chi^2$ | *p-value* |
|---|---|---|---|
| Sex | Male | 5.814 | 0.015 |
| | Female | 0.854 | 0.355 |
| Cultured Epidermal Autografts | Not Present Epicel | 7.524 | 0.016 |
| | Present Epicel | 2.629 | 0.355 |

Based on Table 2, it can be seen that both patient, male patients and patients who did not cultured epidermal autografts have a *p-value*$<\alpha$, therefore there is a significant difference between the cumulative incidence curve of patients who experienced death with melanoma and death with other causes evaluated by male sex and did not have cultured epidermal autografts. Then modeling using the Fine-Gray subdistribution model and the following results are obtained.

Fine-Gray subdistribution model for patients who experienced death with melanoma:

$$\hat{h}_M(t,\mathbf{X}_i)=\hat{h}_{0c}(t)\exp\left(0.669\text{male}+0.013\text{age}+0.126\text{thick}-0.947\text{epicel}\right) \quad (28)$$

Fine-Gray subdistribution model for patients who experienced death with other causes:

$$\hat{h}_{OC}(t,\mathbf{X}_i)=\hat{h}_{0c}(t)\exp\left(0.121\text{male}+0.054\text{age}+0.029\text{thick}-0.683\text{epicel}\right) \quad (29)$$

Based on Fine-Gray subdistribution model for patients death with melanoma and death with other causes, the likelihood ratio test is used to test the effect of the predictor variables in the regression model simultaneously based on equation (16) are shown in Table 3.

**Table 3.** Likelihood Ratio Test

| Models | $G^2$ |
|---|---|
| Death with Melanoma | 31.600 |
| Death with Other Causes | 13.000 |

In Table 3, it can be seen that Fine-Gray subdistribution model for patients death with melanoma and death with other causes have a likelihood ratio test value $\geq \chi^2_{0,05;4} = 0.711$, therefore there is a significant influence of predictor variables on survival time. Then a partial test is using Wald test based on equation (17) are shown in Table 4.

**Table 4.** Wald Test

| Models | Predictor Variables | *p-value* |
|---|---|---|
| Death with Melanoma | Sex (male) | $1.5\times10^{-2}$ * |
| | Age | $1.9\times10^{-1}$ * |
| | Tumor thickness | $3.6\times10^{-5}$ * |
| | Cultured epidermal autografts (did epicel) | $2.7\times10^{-3}$ * |
| Death with Other Causes | Sex (male) | 0.830 |
| | Age | $1\times10^{-4}$ * |
| | Tumor thickness | 0.720 |
| | Cultured epidermal autografts (did epicel) | 0.300 |

Based on Table 4, it can be seen that the predictor variables that affect the risk of death with melanoma patients are sex, age, tumor thickness, and cultured epidermal autografts, while the predictor variable that affects the risk of death with other causes patients is age. There is a basic assumption that a subdistribution has a covariate value is constant, it can be seen based on schoenfeld residual plots as follows (See Figure 2):

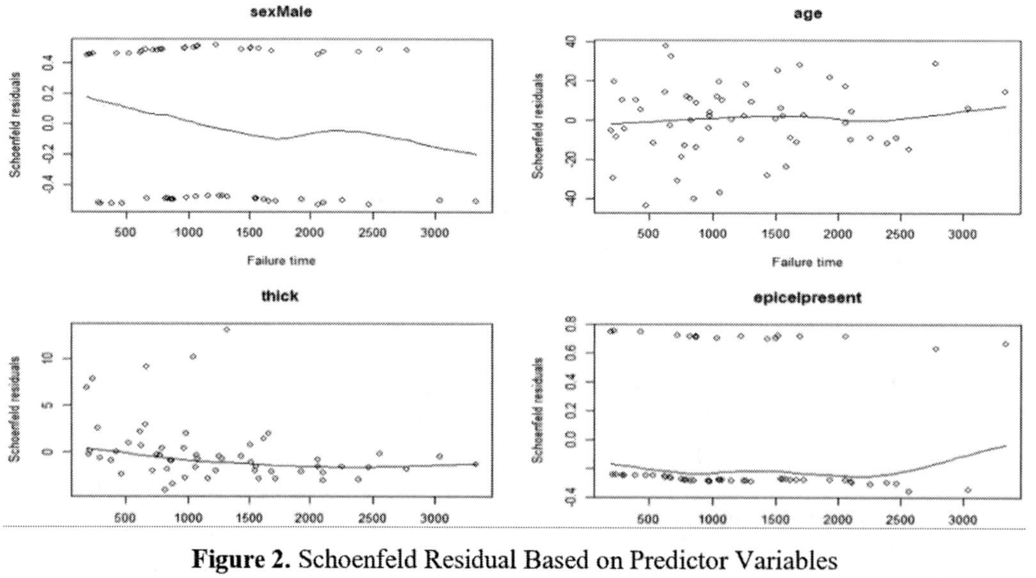


**Figure 2.** Schoenfeld Residual Based on Predictor Variables

Schoenfeld residuals against failure time for each covariate, if the proportional assumption is true, the residual should have a constant mean across time. In Figure 2, it appears that the schoenfeld residuals based predictor variabels constant residuals across time, indicating potential proportional assumptions is fulfilled and based on hazard ratio in models. Then tested the proportional hazard assumption with goodness of fit can be seen in Table 5.

**Table 5.** Goodness of fit for proportional hazard assumption

| Predictor Variables | *p-value* |
|---|---|
| Tumor thickness x $t$ | 0.330 |
| Tumor thickness x $t^2$ | 0.800 |
| Age x $t$ | 0.880 |
| Age x $t^2$ | 0.900 |

Based on Table 5, it can be seen that the predictor variables for tumor thickness and age (continuous predictor variables) did not have a significant effect on the time function, so the proportional hazard assumption has been fulfilled. Then, based on Fine-Gray subdistribution model for patients who experienced death with melanoma and other causes, the hazard ratio of patient mortality can be found as follows.

$$\begin{aligned} HR &= \frac{\hat{h}_M(t,\mathbf{X}_i)}{\hat{h}_{OC}(t,\mathbf{X}_i)} \\ &= \frac{\hat{h}_{0c}(t)\exp\left(0.669\text{male}+0.013\text{age}+0.126\text{thick}-0.947\text{epicel}\right)}{\hat{h}_{0c}(t)\exp\left(0.121\text{male}+0.054\text{age}+0.029\text{thick}-0.683\text{epicel}\right)} \\ &= \frac{\hat{h}_{0c}(t)\exp\left(0.669+0.013+0.126-0.947\right)}{\hat{h}_{0c}(t)\exp\left(0.121+0.054+0.029-0.683\right)} \\ &= 1.400 \end{aligned} \quad (30)$$

Based on equation (30), it can be seen that the patients death with melanoma was 1.400 times higher risk than death with other causes.

## 4. Conclusion

Based on the analysis described above, then it can be concluded that, the estimation the Fine-Gray subdistribution model can be written as follows.

$$\hat{h}_c(t,\mathbf{X}_i) = \hat{h}_{0c}(t)\exp\left(\sum_{j=1}^{p}\hat{\beta}_{jc}X_{ij}\right)$$

Fine-Gray subdistribution model for patients who experienced death with melanoma as follows.

$$\hat{h}_M(t,\mathbf{X}_i) = \hat{h}_{0c}(t)\exp\left(0.669\text{male}+0.013\text{age}+0.126\text{thick}-0.947\text{epicel}\right)$$

Fine-Gray subdistribution model for patients who experienced death with other causes as follows.

$$\hat{h}_{OC}(t,\mathbf{X}_i) = \hat{h}_{0c}(t)\exp\left(0.121\text{male}+0.054\text{age}+0.029\text{thick}-0.683\text{epicel}\right)$$

Predictor variables that affect the risk of patients who experienced death with melanoma are sex, age, tumor thickness, and cultured epidermal autografts, while the predictor variable that affects the risk of patients who experienced death with other causes is age. The patients death with melanoma was 1.400 times higher risk than death with other causes.

## References

[1] Kleinbaum D G and Klein M 2012 *Survival Analysis A Self Learing Text Third Edition* New York: Springer

[2] Gichangi A and Vach W 2005 The Analysis of Competing Risks Data: A Guided Tour *Stat Med* **132** (4) 1–41

[3] Pintilie M 2011 An Introduction to Competing Risks Analysis *Revista Espanola de Cardiología (English Edition)* **64 (7)** 599–605

[4] Fine J P and Gray R J 1999 A Proportional Hazards Model for the Subdistribution of a Competing Risk *Journal of the American Statistical Association* **94** (446) 496–509

[5] Austin P, Latouche A and Fine J 2019 A Review of the use Time-Varying Covariate in the Fine–Gray Subdistribution Hazard Competing Risk Regression Model *Statistics in Medicine* **39** (8) 103-113

[6] Purnami S W, Inayati K D, Sari N W W, Chosuvivatwong V and Sriplung H 2016 Survival Analysis of Cervical Cancer using Stratified Cox Regression *AIP Conference Proceedings* **1723** (1) 030018

[7] Edina A I, Purnami S W, Purhadi and Sukur E 2020 Estimation of Baseline Hazard Function on Cox Model using Maximum Penalized Likelihood *International Journal of Advanced Science and Technology* **29** (7) 3253-3263

[8] Ward W H and Farma J M 2017 *Cutaneous Melanoma: Etilogy and Therapy* Brisbane: Codon Publications

[9] Zhang Z, Geskus R B and Kattan M 2017 Nomogram Survival Analysis in the Presence of Competing Risks *Annals of Translational Medicine* **5** (20) 1–9

[10] Pintilie M 2006 *Competing Risks A Practical Perspective* John London: John Wiley and Sons

[11] Collett D 2003 *Modelling Survival Data in Medical Research* (*Second Edition*) London: Chapman and Hall

[12] Zhou B, Fine J and Laird G 2013 Goodness of Fit Test for Proportional Subdistribution Hazards Model *Statistics in Medicine* **32** (22) 3804–3811.

[13] Husain H, Thamrin S A, Tahir S, Mukhlisin A and Apriani M M 2018 The Application of Extended Cox Proportional Hazard Method forEstimating Survival Time of Breast Cancer *Journal of Physics: Conference Series* **979** 012087 1-8

# Bernoulli Fractional Differential Equation Solution Using Adomian Decomposition Method

**M D Johansyah[1,*], A K Supriatna[2], E Rusyaman[3] and J Saputra[4]**

[1,2,3]Department of Mathematics, Universitas Padjadjaran, Indonesia
[4]School of Social and Economic Development, Universiti Malaysia Terengganu, Malaysia

*Email: muhamad.deni@unpad.ac.id

**Abstract.** Fractional calculus relates with derivatives, integrals, and differential equations of order not integers. The Bernoulli Differential Equation is a form of the first-order ordinary differential equation. This paper aims to solve the Bernoulli Differential Equation with $\alpha$ fractional-order using the Adomian Decomposition Method, where $0 < \alpha \leq 1$. The fractional derivative used in this paper is the fractional derivative of Caputo. Based on several numerical examples presented in this paper, the results show that the Adomian Decomposition Method is easy and very effective to use for solving Bernoulli Differential Equations with fractional order $\alpha$.

## 1. Introduction

Fractional calculus is the area of calculus that deals with derivatives, integrals, and differential equations of fractional order. In calculus, the derivatives are usually of the order of a non-negative integer denoted by $D_x^n y(x)$ or $\frac{d^n y}{dx^n}$ , where $n = 1, 2, 3, \ldots$. This notation can be generalized to the derivative of the fractional-order $\alpha$ denoted by $D_x^\alpha y(x)$ or $\frac{d^\alpha y}{dx^\alpha}$, where $\alpha \in \mathbb{R}$. Most people assume that fractional calculus is an abstract area of mathematics that is of little use and almost no application. Nowadays, various researches on the application of fractional calculus, particularly fractional differential equations, are starting to emerge in various fields, such as physics, engineering, chemistry, biology, environment, economics, and finance [1]-[4].

The Bernoulli Differential Equation is a form of the first-order ordinary differential equation [5]. This differential equation has a general form, viz

$$\frac{dy}{dx} + P(x)y = Q(x)y^m, \tag{1}$$

where $P$ and $Q$ are functions of $x$ or a constant, and $m$ is a real number. Bernoulli Differential Equation is a first-order differential equation which can be linear or nonlinear. For $m = 0$ or $m = 1$, equation (1) becomes the Linear Bernoulli Differential Equation. For $m \geq 2$, equation (1) becomes the Nonlinear Bernoulli Differential Equation. Equation (1) can be generalized to Bernoulli Differential Equation with a fractional-order which is written as follows

$$D_x^\alpha y(x) + P(x)y = Q(x)y^m, \tag{2}$$

where $\alpha \in \mathbb{R}$, with $0 < \alpha \leq 1$.

This paper aims to solve the Bernoulli Differential Equation with $\alpha$ fractional-order using the Adomian Decomposition Method, where the fractional derivative used is the fractional derivative of Caputo. Previous research has been conducted by Agom and Ogunfiditimi (2014) [6] who used the Adomian Decomposition Method to solve the classical Bernoulli Differential Equation, in other words, it has no fractional order.

George Adomian (1980) [7] first introduced the Adomian Decomposition Method for solving systems of stochastic equations. This Adomian Decomposition Method can be an effective and easy procedure for solving differential or integral equations without linearization, perturbation, discretization, or transformation [8]. This method can solve differential equations with natural or fractional order, ordinary or partial, with initial or boundary value problems, with constant or variable coefficients, linear or nonlinear, homogeneous or non-homogeneous [9]-[11].

Other research on Fractional Bernoulli Differential Equation can be seen in [12]-[16].

## 2. Fractional Calculus

This section presents the basic theories and properties that are related to and support this research, such as derivatives and fractional integrals.

**Definition 1.** (Podlubny, 1999) [17] Caputo fractional derivative of the function $f$ to $t$ of order $\alpha$, where $\alpha > 0$ and $n-1 < \alpha \leq n$, is defined as

$$ {}_a^C D_t^\alpha f(t) = \frac{1}{\Gamma(n-\alpha)} \int_a^t (t-u)^{n-\alpha-1} f^{(n)}(u) du. $$

**Theorem 2.** (Mathai and Haubold, 2017) [18] Caputo fractional derivative of order $\alpha > 0$, with $n - 1 < \alpha \leq n$, from the function $f(t) = t^\beta$, where $\beta \geq 0$, is

$$ {}^C D_t^\alpha t^\beta = \begin{cases} \dfrac{\Gamma(\beta+1)}{\Gamma(\beta-\alpha+1)} t^{\beta-\alpha} & ;\beta > n-1, \\ 0 & ;\beta \leq n-1. \end{cases} $$

Based on Theorem 2, Caputo fractional integral of order $\alpha > 0$, with $n-1 < \alpha \leq n$, from the function $f(t) = t^\beta$, where $\beta \geq 0$, is

$$ I^\alpha t^\beta = {}^C D_t^{-\alpha} t^\beta = \begin{cases} \dfrac{\Gamma(\beta+1)}{\Gamma(\beta+\alpha+1)} t^{\beta+\alpha} & ;\beta > n-1, \\ 0 & ;\beta \leq n-1. \end{cases} $$

**Theorem 3.** (Mathai dan Haubold, 2017) [18] Let $f$ be continuous in $[a,b]$ and $n-1 < \alpha \leq n$, then

$$ I^\alpha D_t^\alpha f(t) = D_t^{-\alpha} D_t^\alpha f(t) = f(t) - \sum_{k=0}^{n-1} \frac{t^k}{k!} f^{(k)}(0). $$

## 3. Adomian Decomposition Method

The Adomian Decomposition Method assumes that the solution is decomposed into an infinite series, the nonlinear form (if any) is decomposed into an Adomian polynomial, and an iterative algorithm is constructed to recursively determine the solution. This method can solve differential equations with natural or fractional order, ordinary or partial, with initial or boundary value problems, with constant or variable coefficients, linear or nonlinear, homogeneous or non-homogeneous. The solution of the Fractional Ordinary Differential Equation using the Adomian Decomposition Method can refer to [19]-[22].

## 4. Numeric Examples

This section presents two examples of numerical solutions to Bernoulli Fractional Differential Equation using the Adomian Decomposition Method.

**Example 1.** Given the Linear Bernoulli Fractional Differential Equation as follows

$$ D_x^\alpha y - 2y = 0, \tag{3} $$

with $0 < \alpha \leq 1$ and initial condition $y(0) = 1$. The exact solution of equation (3) is

$$y = \exp(2x). \tag{4}$$

Meanwhile, the approximate solution of equation (3) using the Adomian Decomposition Method based on the recursive formula in equation (2) is

$$y_0 = 1,$$
$$y_{n+1} = I^{\alpha}[2y_n], n = 0,1,2,.... \tag{5}$$

Figure 1 depicts a numerical simulation for the comparison of the approximate solutions of the Linear Bernoulli Fractional Differential Equation using the Adomian Decomposition Method (5), where $\alpha = 1$ and $n = 5$, with the exact solution (4). The results show that in the initial iteration the approach solution for $\alpha = 1$ is close to the exact solution.

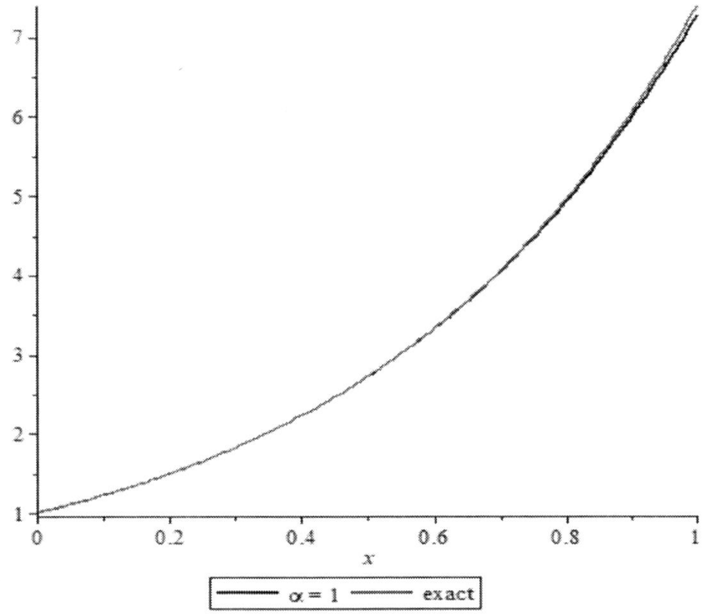


**Figure 1.** Comparison of the approximate solutions of the Linear Bernoulli Fractional Differential Equation using the Adomian Decomposition Method with its exact solution

Furthermore, Figure 2 illustrates a numerical simulation for the approximate solution of the Linear Bernoulli Fractional Differential Equation using the Adomian Decomposition Method (5), where $n = 5$, with different $\alpha$ values. The results show that if the $\alpha$ value increases, then the $y$ graph decreases.

**Example 2.** Given the Nonlinear Bernoulli Fractional Differential Equation as follows

$$6D_x^{\alpha}y - 2y = xy^4, \tag{6}$$

with $0 < \alpha \leq 1$ and initial condition $y(0) = -2$. The exact solution of equation (6) is

$$y = -\frac{2}{(4x - 4 + 5\exp(-x))^{\frac{1}{3}}}. \tag{7}$$

Meanwhile, the approximate solution of equation (6) using the Adomian Decomposition Method based on the recursive formula in equation (2) is

$$y_0 = -2,$$
$$y_{n+1} = I^{\alpha}\left[\frac{1}{6}xA_n\right] - I^{\alpha}\left[-\frac{1}{3}y_n\right], n = 0,1,2,..., \tag{8}$$

where $A_n$ is an Adomian polynomial, with the nonlinear operator $Ny = y^4$.

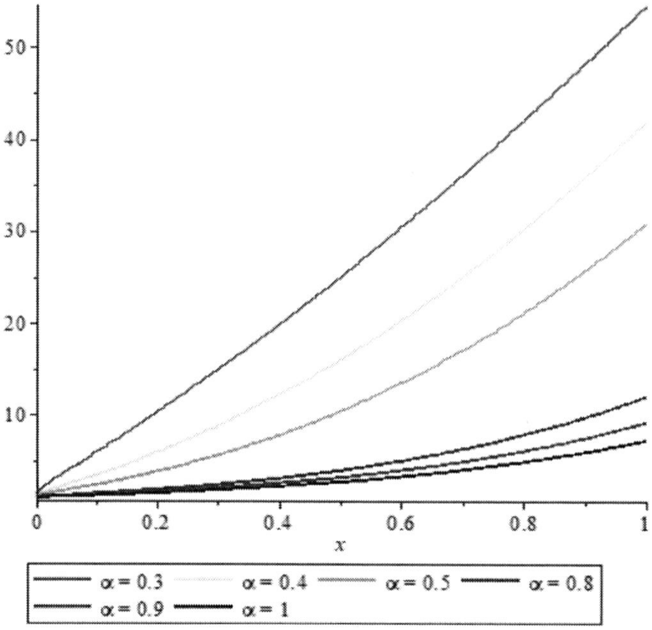


**Figure 2.** The solution to the approach of the Linear Bernoulli Fractional Differential Equation using the Adomian Decomposition Method with different $\alpha$ values

Figure 3 illustrates the numerical simulation for comparison of the approximate solutions of the Nonlinear Bernoulli Fractional Differential Equation using the Adomian Decomposition Method (5), where $\alpha = 1$ and $n = 5$, with the exact solution (7). The results show that in the initial iteration the approach solution for $\alpha = 1$ is close to the exact solution.

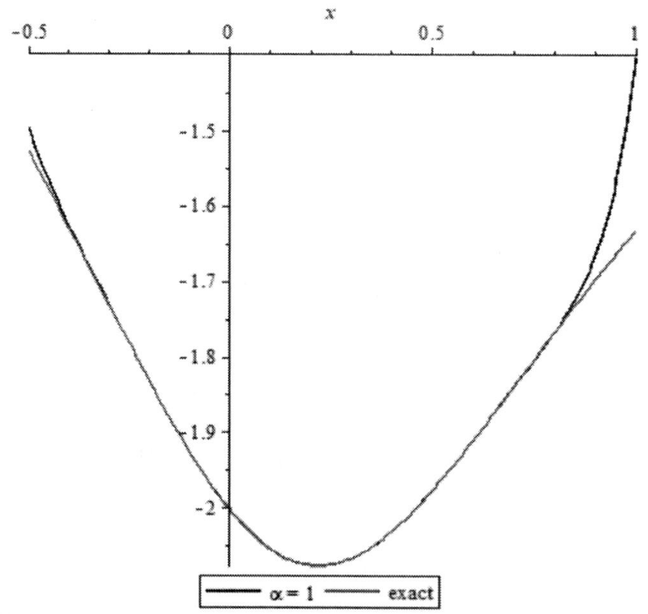


**Figure 3.** Comparison of the approximate solution of the Nonlinear Bernoulli Fractional Differential Equation using the Adomian Decomposition Method with its exact solution

Next, Figure 4 depicts a numerical simulation for the approximate solution of the Nonlinear Bernoulli Fractional Differential Equation using the Adomian Decomposition Method (6), where $n = 5$, with different $\alpha$ values. The results show that if the $\alpha$ value increases, then the $y$ graph decreases.

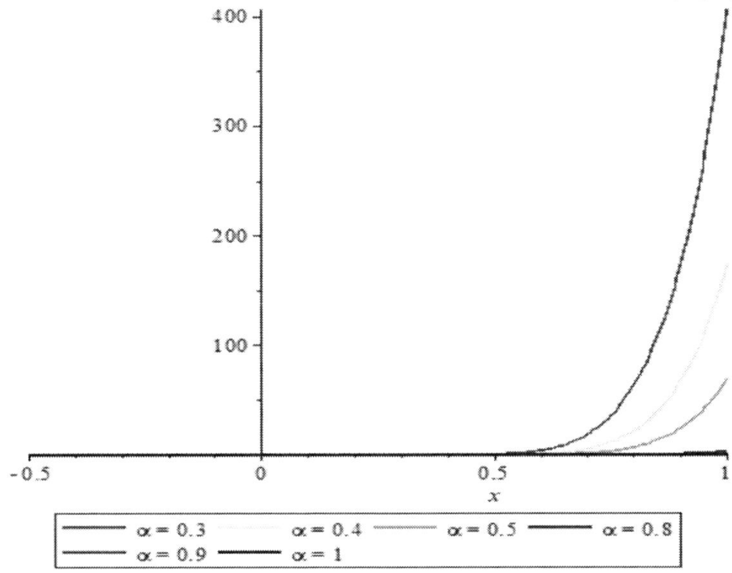


**Figure 4.** The approach solution of the Nonlinear Bernoulli Fractional Differential Equation using the Adomian Decomposition Method with different $\alpha$ values

## 5. Conclusion

The Bernoulli Differential Equation is a form of the first-order ordinary differential equation. This differential equation can be of integer or fractional order, it can also be linear or nonlinear. The Adomian Decomposition Method is an approach method that can solve ordinary and partial differential equations, linear or nonlinear, and can be of integer or fractional order. Two numerical examples presented in this paper show that the Adomian Decomposition Method is reliable and easy to use to solve the Linear and Nonlinear Fractional Bernoulli Differential Equations.

## References

[1] David S A, Linares J L and Pallone E M J A 2011 *Revista Brasileira de Ensino de Física* **33** 4302(1-7)

[2] Katsikadelis J T 2015 *Archive of Applied Mechanics* **85** 1307-1320

[3] Giusti A and Colombaro I 2018 *Communications in Nonlinear Science and Numerical Simulation* **56** 138-143

[4] Sun H, Zhang Y, Baleanu D, Chen W and Chen Y 2018 *Communications in Nonlinear Science and Numerical Simulation* **64** 213-231

[5] Azarian M K 2012 *Applied Mathematical Sciences* **6** 6799-6803

[6] Agom E U and Ogunfiditimi F O 2013 *International Journal of Science and Research* **4** 1581-1584

[7] Adomian G 1980 *Applied Stochastic Processes* 1-17

[8] Adomian G 1988 *Journal of Mathematical Analysis and Applications* **135** 501-544

[9] Cheng J and Chu Y 2011 *Mathematical Problems in Engineering* 587068

[10] Duan J, Rach R, Baleanu D and Wazwaz A 2012 *Communications in Fractional Calculus* **3** 73-99

[11] El-Borai M M, El-Sayed W G and Jawad A M 2015 *International Research Journal of Engineering and Technology* **2** 296-306

[12] Bulut H, Yel G and Baskonus H M 2016 *Turkish Journal of Mathematics and Computer Science* **5** 1-7
[13] Ilie M, Biazar J and Ayati Z 2017 *International Journal of Applied Mathematical Research* **6** 49-51
[14] Ordokhani Y, Rahimkhani P and Babolian E 2017 *International Journal of Nonlinear Analysis and Applications* **8** 277-292
[15] Belgacem R, Bokhari A and Amir A 2018 *General Letters in Mathematics* **5** 32-46
[16] Dababneh A, Albarmawi B, Hammad M A, Zraiqat A and Hamadneh T 2019 *IEEE Jordan International Joint Conference on Electrical Engineering and Information Technology* 421-424
[17] Podlubny, I. 1999. *Fractional Differential Equations*. California: Academis Press.
[18] Mathai A M and Haubold H J 2017 *An Introduction to Fractional Calculus*. New York: Nova Science Publishers
[19] Wazwa A M 2001 *Computers & Mathematics with Applications* **41** 1237-1244
[20] Zeid S S, Yousefi M and Yousefi M 2016 *Journal of Applied Mathematics, Statistics and Informatics* **12** 83-94
[21] Sumiati I, Johansyah M D, Rusyaman E and Supriatna A K 2020 *AIP Conference Proceedings* **2264** 040004(1-9)
[22] Sukono, Sambas A, He S, Liu H, Vaidyanathan S, Hidayat Y and Saputra J 2020 *Advances in Difference Equations* **674** 1-12

# Comparison Analysis of the SAW Method and TOPSIS Method in the Decision Support System for Determining Permanent Teachers in SMK Pasundan 2 Banjaran

**S D Jayanti[1], Budiman[2] and T P Yoga[3*]**

[1]Student of Faculty Technology and Informatics, Universitas Informatika dan Bisnis Indonesia, Jalan Soekarno-Hatta 643, Bandung 40285, Indonesia.
[2,3]Faculty of Technology and Informatics, Universitas Informatika dan Bisnis Indonesia, Jalan Soekarno-Hatta 643, Bandung 40285, Indonesia.

*Email: titanparama@unibi.ac.id

**Abstract.** The decision-making process is quite burdensome because it is necessary to issue the right decisions so many people do it carefully. Nowadays with the development of decision making system can help in the decision making process because it is done with measurable stages and with methods that have a high level of accuracy. By using the Decision Making System, the process of assigning teachers remains in SMK Pasundan 2 Banjaran can be done quickly, precisely and accurately so that it is expected that the results obtained can be accounted for. But here will be tried to do the decision-making process by using SAW method and TOPSIS method by doing comparison or comparison with the two methods with the aim of getting the right results who can be submitted as a permanent teacher of SMK Pasundan 2 Banjaran but also get advantages and disadvantages and which method is appropriately used for the determination of permanent teachers in SMK Pasundan 2 Banjaran from each method by conducting process trials. Each method certainly has stages in the process of determining results to assist in decision making and this can be a reference in looking at the accountable calculation process. Based on the results of comparison or comparison of SAW method and TOPSIS method which then carried out sensitivity test resulted in a method that is similar to this case study namely SAW method. With saw change value of 10.03168% and TOPSIS by 0.62590%.

## 1. Introduction

Not all teachers of SMK Pasundan 2 Banjaran have been appointed as Civil Servants (Civil Servants) and Permanent Teachers Foundation (GTY), but there are still teachers who are permanent teachers or honorary teachers. To become a Permanent Teacher of Yayasan (GTY), it must be done assessment and evaluation activities in advance. This assessment and evaluation activity of honor teachers is useful to know the results of teaching honor teachers to their students. Assessment, evaluation, and awarding can be done aimed at spurring the performance of teachers in the teaching learning process. Assessments are carried out to obtain honorary teachers who will then be appointed as Permanent Teachers Foundation (GTY) [1]. Teachers recruited to become permanent teachers by foundations or schools must demonstrate good performance.

SMK Pasundan 2 Banjaran in the appointment of The Permanent Teacher Foundation (GTY) still has constraints that are done manually and the appointment must be in accordance with the approval of the principal. To facilitate the appointment of permanent teachers, a system known as the decision support system is needed (DSS). DSS aims to assist decision makers in producing a decision on the appointment of permanent teachers at SMK Pasundan 2 Banjaran.

The use of decision support systems has many methods that can be used. For the appointment of Yayasan Permanent Teacher at SMK Pasundan 2 Banjaran will use Simple Additive Weighting (SAW) Method which will be compensated with Technique For Order Preference by Similarity to Ideal Solution (TOPSIS) method in the hope that it will be presented which method is appropriate in decision making.

The comparison of the two methods mentioned above has also been used but in different cases such as the selection of exemplary teachers [2], the selection of residential destinations [3], the selection of tourist destinations in West Java [4], the selection of tourist destinations in Indonesia [5], then based on the reference is expected to be the basis of this research reference although certainly the criteria will be different.

## 2. Background Theory

### 2.1. Decision Support System

The concept of a decision support system was first introduced by Michael S. Scott Morton in the early 1970s as the Management Decision System. It is a computer-based system aimed at assisting decision-making by utilizing certain data and models to solve unstructured problems.

System is an interconnected set of elements responsible for processing inputs resulting in output. Decision is the activity of choosing a strategy or action in solving the problem [6]. The criteria of the decision are many options/alternatives, there are constraints/conditions, following a pattern/model of behavior, many inputs/variables, there are risk factors, it takes speed and accuracy.

Decision Support System (DSS) is an interactive information system that provides information, modeling, and manipulation of data [7]. The system is used to assist decision-making in semi-structured situations and unstructured situations, where no one knows exactly how the decision should be.

DSS serves as an add-on or advocate for decision makers, can expand knowledge and possibilities, but does not replace judgment. The system is intended for decisions that require assessment and decisions that can be processed with algorithms or technically.

### 2.2 Simple Additive Weighting (SAW) Method

The SAW method is the simplest and most widely used MADM method [8]. This method is also the easiest method to apply, because it has an algorithm that is not very complicated. SAW methods are often also known as weighted summation methods. The basic concept of the SAW method is to look for weighted summation of performance ratings on each alternate on all attributes. The SAW method requires the process of normalizing the decision matrix (X) to a scale that can be compared to all existing alternative rating.

This SAW method requires decision makers to determine the weight for each attribute. The total score for an alternative is obtained by summing up the entire multiplication between the rating (which can be compared across attributes) and the weight of each attribute. The rating of each attribute must be dimension-free in the sense that it has gone through the normalization process of the previous matrix.

### 2.3 Technique for Order Preference by Similarity to Ideal Solution (TOPSIS) Method

The TOPSIS completion method is based on a concept in which the best selected alternative not only has the shortest distance from the positive ideal solution, but also has the longest distance from the negative ideal solution [9]-[10].

## 3. Methodology

Research methodology is essentially a scientific way to obtain information with specific purposes and uses. Research methods are a way used by researchers in collecting their research data. In the preparation of this research will go through several stages that form a systematic flow in Figure 1.

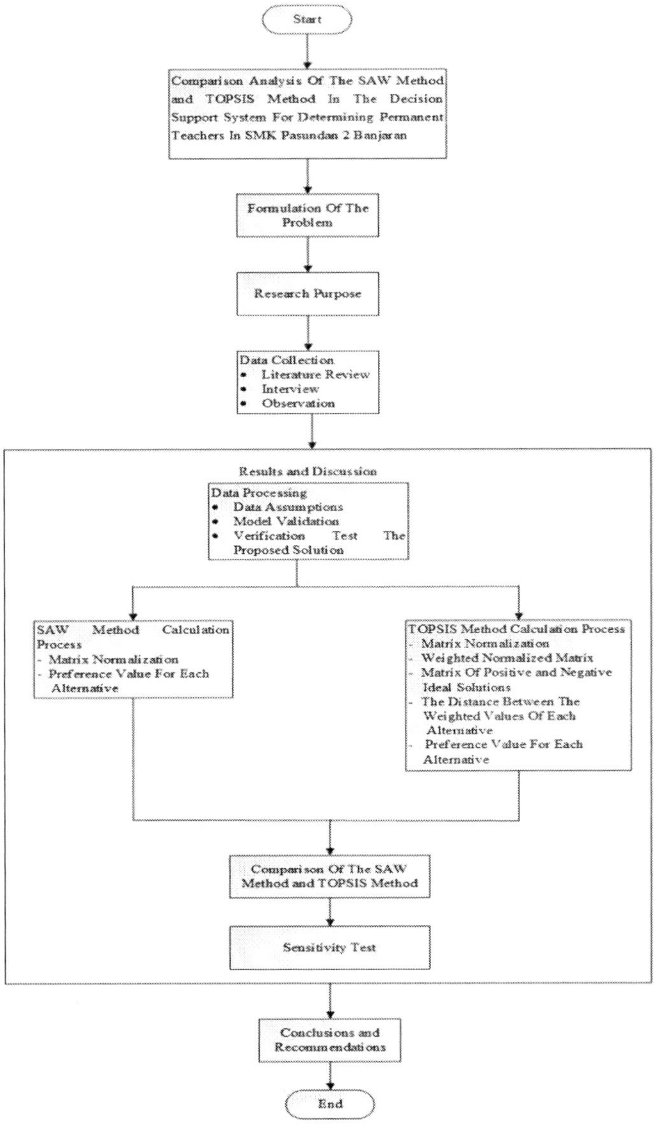


**Figure 1.** Research Methods

The stage of problem identification aims to know the problem that is being experienced in the research. At this stage spelled out what problems were raised for research. The issues raised will then be reviewed and obtained information related to the issue.

Furthermore, the authors conducted a literature study to look for references to existing problems and support the research conducted. By looking for some good materials from books, scientific journals, and research results related to the application of fuzzy logic using mamdani method [11]-[14].

The Third Stage is to perform the data collection stage is the author's activity of collecting the data needed for application development. The data collection technique performed by the author is an observation on the website of pikobar. The purpose of this observation is to obtain data used for application development, namely in the form of data on the number of ODP, the number of PDP, and the positive amount of Covid-19.

The Fourth phase is the analysis and planning phase is the beginning of the implementation of the system development method rapid application development (RAD) in this study. Activities to plan needs include system planning and analysis. At this stage the author analyzes all the needs that have been collected before to understand the nature of the program to be built. Then the author defines the needs of the application to be built in the form of user characteristics, data needs, and functional and non-functional needs. After the needs of the system have been defined, the authors created a fuzzy logic model of mamdani method on predicting an increase in the number of positive cases of Covid-19 14 days later. Based on prior studies on prediction using the fuzzy logic of the mamdani method and the limitations of the author's ability, the study used linear representation for membership functions, the MAX (Maximum) method for the method of rule composition, and the centroid (Composite Moment) method for the defuzzification method. The design architecture used to design the applications used in this study was Object Oriented Analysis and Design (OOAD), so the authors used Unified Modelling Language (UML) to create a software design object model. The UML diagrams to be created are use case diagrams, class diagrams, sequence charts, and state diagrams. The author only created the four diagrams because they are the main diagrams that are quite used for applications that are not very large. At this stage also the author creates a prototype in the form of an application mockup to be built. After the fuzzy logic was created, the authors then tested the results of the mamdani method in predicting covid-19 cases with field sample data.

The last stage of this research is the implementation and testing phase is the advanced stage of rad implementation, namely construction and cut over. Construction is the author's activity of implementing the user model into an application. Applications are built using java programming languages with Netbeans and MySQL tools for database management. If during the application development stage there is a discrepancy with the needs of the application that has been defined in the stage of planning the needs then a model and prototype improvement is carried out at the user model stage. If the application built at the construction stage is in accordance with the needs of the application that has been defined at the stage of planning the need, then the next step is to test the application or cut over. The testing method that the authors did on this study was the black-box method. At this stage, a comparison of the calculation of prediction results using the application built on this study with data on the number of covid-19 positive cases that have occurred to look for the error rate.

## 4. Results and Discussions

### 4.1 Data Assumptions

The decision-making process in the appointment of Yayasan Permanent Teacher at SMK Pasundan 2 Banjaran still has constraints that are done manually and the appointment must be in accordance with the approval of the principal. To facilitate the appointment of permanent teachers, a system known as

IOP Conf. Series: Materials Science and Engineering 1115 (2021) 012016 doi:10.1088/1757-899X/1115/1/012016

the decision support system (DSS) is required. SPK aims to assist decision makers in producing a decision on the appointment of permanent teachers at SMK Pasundan 2 Banjaran.

### 4.2 Model Validation

One of the steps in the creation of a decision support system is to validate the model which means entering the necessary facts. In the form of criteria, weight values, and honorary teacher data that will be used as an alternative.

### 4.3 Determining the Criteria and Weight of Permanent Teacher Appointment

In the process of appointment of permanent teachers carried out using saw method and TOPSIS method, criteria are required, weight of interest of each criterion and alternative match rating to the criteria to perform the calculation so that the best alternative will be obtained (See Table 1). There are 7 benefit criteria used in conducting assessment, namely as follows:

1. C1 = The length of work, in the appointment of the teacher still takes 4 years of work.
2. C2 = The level of education, higher education includes bachelor's or master's. If his education is vocational school then appointed as a permanent teacher to guide the extracurricular.
3. C3 = Attendance, teacher attendance in teaching is at least 75%, so it is seen the teacher's dedication to the world of education.
4. C4 = Evaluation of learning results, evaluation of learning results is done to know the quality of the teacher's performance in conducting teaching and learning activities.
5. C5 = Competency, a teacher must be able to master the science to the teacher and be able to apply it in learning strategies.
6. C6 = Discipline, a teacher must be disciplined, so as to be an example for his protégé.
7. C7 = Responsible, responsible in devising interesting learning strategies that are loved by students.

**Table 1.** Criteria

| No | Criteria | Description | Type |
|---|---|---|---|
| 1. | C1 | Length of Work | Benefit |
| 2. | C2 | Education Level | Benefit |
| 3. | C3 | Presence | Benefit |
| 4. | C4 | Evaluation of Learning Outcomes | Benefit |
| 5. | C5 | Competence | Benefit |
| 6. | C6 | Discipline | Benefit |
| 7. | C7 | Responsible | Benefit |

### 4.4 Rank Order Centroid (ROC) Method

The ROC algorithm is based on the level of importance or priority of the criteria [15]. The ROC techniques give weight to each criterion according to the rankings assessed based on priority level. It is usually formed with the statement "Criterion 1 is more important than criteria 2, which is more important than criteria 3" and so on until the nth criterion.

Based on the above criteria, it will then be given to the weight value by applying the Rank Order Centroid (ROC) method, using the equation (1) as follows:

$$W_k = \frac{1}{k}\sum_{i=1}^{k}\left(\frac{1}{i}\right) \quad (1)$$

Description:
Wj = j attribute weighting value
K = number of attributes
i= attribute priority sequence value

After going through the process of calculating the weight value by applying the Rank Order Centroid (ROC) method obtained the results displayed in the form of a table of weight values for each criterion on the decision support system for the SAW and TOPSIS methods can be seen in Table 2.

**Table 2.** Criteria Weight Values

| No | Criteria | Description | Type | Weights |
|---|---|---|---|---|
| 1. | C1 | Length of Work | Benefit | 0.37 |
| 2. | C2 | Education Level | Benefit | 0.23 |
| 3. | C3 | Presence | Benefit | 0.16 |
| 4. | C4 | Evaluation of Learning Outcomes | Benefit | 0.11 |
| 5. | C5 | Competence | Benefit | 0.07 |
| 6. | C6 | Discipline | Benefit | 0.04 |
| 7. | C7 | Responsible | Benefit | 0.02 |

## 4.5 Teacher Honorer Data That Becomes An Alternative

The teacher honourer data that will be used as an alternative to be assigned as a teacher can still be seen in Table 3.

**Table 3.** Honourer Teacher Data as an Alternative

| No | Alternative | Teacher's Name |
|---|---|---|
| 1. | A1 | Sunaeni Haryati, S.Si |
| 2. | A2 | David Ramadhan Pratama |
| 3. | A3 | Drs. Undang Hidayat |
| 4. | A4 | Heri Kiswanto, S.Sn |
| 5. | A5 | Muhamad Ridwan, S.Pd |
| 6. | A6 | Sena Dwi Septoaji Suherman, S.Pd |
| 7. | A7 | Dewi Kartika, S.Pd |
| 8. | A8 | Mohamad Gelar Cipta, S.Pd |
| 9. | A9 | Ilham Halimi Akbar, S.Pd |
| 10. | A10 | Nurdin, S.Pd |
| 11. | A11 | Riki Agus Setiawan |

## 4.6 Determining Each Alternate Match Rating

From the specified criteria, a level of criteria interest is created based on the alternative stipulated based on the match rating of each alternate on each criterion converted to the crips number: VL = 0; L = 0.25; E = 0.5; H = 0.75; VH = 1.

a. Working Length Criteria
The old working variable is converted with a fuzzy number like Table 4.

**Table 4.** Crips Value Of Old Working Criteria

| Criteria (C1) Length of Work | Crips Value |
|---|---|
| 6 Months | 0 |
| 1 Years | 0.25 |
| 2 Years | 0.5 |
| 3 Years | 0.75 |
| 4 Years | 1 |

b. Education Level Criteria
Educational level variables are converted with fuzzy numbers such as Table 5.

**Table 5.** Crips Value Criteria For Education Level

| Criteria (C2) Education Level | Crips Value |
|---|---|
| Elementary School | 0 |
| Junior High School | 0.25 |
| Vocational High School | 0.5 |
| Strata 1 | 0.75 |
| Strata 2 | 1 |

c. Attendance Criteria
Presence variables are converted with fuzzy numbers such as Table 6.

**Table 6.** Crips Attendance Criteria Values

| Criteria (C3) Presence | Crips Value |
|---|---|
| Presence < 25% | 0 |
| Presence > 25% | 0.25 |
| Presence = 50% | 0.5 |
| Presence = 75% | 0.75 |
| Presence = 100% | 1 |

d. Study Results Evaluation Criteria
The evaluation variables of the study results are converted with fuzzy numbers such as Table 7.

**Table 7.** Crips Value Evaluation Criteria For Study Results

| Criteria (C4) Evaluation of Learning Outcomes | Crips Value |
|---|---|
| Very Low | 0 |
| Low | 0.25 |
| Enough | 0.5 |
| High | 0.75 |
| Very High | 1 |

e. Competency Criteria

Competency variables are converted with fuzzy numbers such as Table 8.

**Table 8.** Competency Crips Values

| Criteria (C5) | Crips Value |
|---|---|
| Competence | |
| Very Low | 0 |
| Low | 0.25 |
| Enough | 0.5 |
| High | 0.75 |
| Very High | 1 |

f. Disciplinary Criteria

Discipline variables are converted with fuzzy numbers such as Table 9.

**Table 9.** Crips Disciplinary Criteria Values

| Criteria (C6) | Crips Value |
|---|---|
| Discipline | |
| Very Low | 0 |
| Low | 0.25 |
| Enough | 0.5 |
| High | 0.75 |
| Very High | 1 |

g. Responsible Criteria

The variables responsible are converted with fuzzy numbers such as Table 10.

**Table 10.** Crips Values Of Responsible Criteria

| Criteria (C7) | Crips Value |
|---|---|
| Responsible | |
| Very Low | 0 |
| Low | 0.25 |
| Enough | 0.5 |
| High | 0.75 |
| Very High | 1 |

Based on the criteria and rating of each alternate match on each specified criteria, then the alternate description of each criterion has been converted with a crisp value. Based on the data of the teacher name above can be formed matrix of decision X that has been converted with crisp value, such as Table 11.

**Table 11.** Match Ratings of Each Alternate on Each Criteria

| Alternative | Criteria | | | | | | |
|---|---|---|---|---|---|---|---|
| Name | C1 | C2 | C3 | C4 | C5 | C6 | C7 |
| A1 | 1 | 0.75 | 0.75 | 0.75 | 0.75 | 0.5 | 0.75 |
| A2 | 1 | 0.5 | 0.5 | 0.75 | 0.5 | 0.5 | 0.5 |
| A3 | 0.25 | 0.75 | 1 | 0.75 | 0.75 | 0.75 | 0.75 |
| A4 | 1 | 0.75 | 0.75 | 0.5 | 0.75 | 0.75 | 0.75 |
| A5 | 0.25 | 0.75 | 0.75 | 1 | 0.75 | 0.5 | 0.5 |
| A6 | 0.5 | 0.75 | 0.5 | 0.75 | 0.5 | 0.5 | 0.5 |
| A7 | 1 | 0.75 | 1 | 0.75 | 0.75 | 0.75 | 0.5 |
| A8 | 0.25 | 0.75 | 0.5 | 0.5 | 0.5 | 0.5 | 0.75 |
| A9 | 0.5 | 0.75 | 0.5 | 0.75 | 0.5 | 0.5 | 0.75 |
| A10 | 1 | 0.75 | 1 | 0.75 | 0.75 | 0.75 | 0.5 |
| A11 | 0.75 | 0.5 | 0.75 | 0.5 | 0.5 | 0.5 | 0.5 |

## 4.7 Verification Testing Proposed Solutions

Verification testing the proposed solution is the implementation of the decision solution to get the best alternative decision. Where in this case to get the best alternative decision is done by comparatively simple additive weighting (SAW) method and Technique for Order Preference by Similarity to Ideal Solution (TOPSIS) method.

## 4.8 The Calculation Process of Simple Additive Weighting (SAW) Method

### *4.8.1 Matrix Normalization*

Normalize the matrix by calculating the normalized performance rating value (rij) of the Ai alternative to the Cj attribute based on an equation adjusted to the attribute type (profit/benefit attribute = MAXIMUM or cost attribute = MINIMUM).

The normalize the matrix given the following equation (2):

$$r_{ij} \begin{cases} \dfrac{xij}{Max^{xij}} & \text{If j is an advantage attribute (benefit)} \\ \dfrac{Min^{xij}}{xij} & \text{If j is a cost attribute (cost) xij} \end{cases} \quad (2)$$

where with rij is the nominee performance rating of the Ai alternative on the $C_j$ : i = 1,2,......,m and j = 1,2,..,n.

Description:
$M_{ax}\ x_{ij}$ = The greatest value of any criteria i
$M_{in}\ x_{ij}$ = Smallest value of each criteria i
$x_{ij}$ = Attribute values held by each criteria
Benefit = If the greatest value is best
Cost = If the smallest value is best

After going through the calculation process, the normalized performance matrix is obtained as follows:

$$\text{Matrix R} = \begin{bmatrix} 1 & 1 & 0.75 & 0.75 & 1 & 0.67 & 1 \\ 1 & 0.67 & 0.5 & 0.75 & 0.67 & 0.67 & 0.67 \\ 0.25 & 1 & 1 & 0.75 & 1 & 1 & 1 \\ 1 & 1 & 0.75 & 0.5 & 1 & 1 & 1 \\ 0.25 & 1 & 0.75 & 1 & 1 & 0.67 & 0.67 \\ 0.5 & 1 & 0.5 & 0.75 & 0.67 & 0.67 & 0.67 \\ 1 & 1 & 1 & 0.75 & 1 & 1 & 0.67 \\ 0.25 & 1 & 0.5 & 0.5 & 0.67 & 0.67 & 1 \\ 0.5 & 1 & 0.5 & 0.75 & 0.67 & 0.67 & 1 \\ 1 & 1 & 1 & 0.75 & 1 & 1 & 0.67 \\ 0.75 & 0.67 & 0.75 & 0.5 & 0.67 & 0.67 & 0.67 \end{bmatrix}$$

*4.8.2 Preference Values for Each Alternative (Vi)*

Specifies the reverence value for each alternate (Vi) by summing the result times between the normalized matrix (R) and the weight value (W). So it gets the greatest value chosen as the best alternative (Ai) as a solution.

According to Alonso et al. (2008) states that to calculate the preference weight value for each alternative (Vi) given the following equation [16]:

$$V_i = \sum_{j=1}^{n} w\, j\, rij \qquad (3)$$

Description:
$V_i$ = Rank for each alternative.
$w_j$ = Rank weight value (of each criteria).
$r_{ij}$ = Normalized performance rating value.

Based on the calculation of equation (3), the result can be seen in Table 12.

**Table 12.** Saw Method Warking

| Alternative Name | Preference Weight Value (Vi) | Ranking |
|---|---|---|
| A7 | 0.96583 | 1 |
| A10 | 0.96583 | 1 |
| A1 | 0.91917 | 2 |
| A4 | 0.90500 | 3 |
| A2 | 0.77250 | 4 |
| A3 | 0.69500 | 5 |
| A11 | 0.69250 | 6 |
| A9 | 0.67083 | 7 |
| A6 | 0.66417 | 8 |
| A5 | 0.66250 | 9 |
| A8 | 0.55083 | 10 |

Based on the process of SAW method, it can be concluded that the largest value is alternative code A7 and A10 with a value of 0.96583 namely Dewi Kartika, S.Pd and Nurdin, S.Pd is appointed as the alternative chosen honorary teacher who will be appointed as a permanent teacher in SMK Pasundan 2 Banjaran.

## 4.9 The Calculation Process of Technique For Order Preference by Similarity to Ideal Solution (TOPSIS) Method

### *4.9.1 Matrix Normalization*

According to Tong et al. (2005) states that to determine the matrix of normalized decisions (R) TOPSIS of each ai alternative and cj criteria given the following equation (4) [17]:

$$rij = \frac{x_{ij}}{\sqrt{\sum_{i=1}^{m} x_{ij}}} \quad (4)$$

with i = 1,2,...,m; dan j = 1,2,...,n.

Description:
$r_{ij}$ = a value that must be normalized to a comparable scale.
$x_{ij}$ = comparison form pairs each Ai alternative in each Cj criteria

Based on the equation (4), the calculation results of each criterion can be seen in Table 13.

**Table 13.** Normalized Values R

| Alternative | Criteria | | | | | | |
|---|---|---|---|---|---|---|---|
| | C1 | C2 | C3 | C4 | C5 | C6 | C7 |
| A1 | 0.43644 | 0.318 | 0.3 | 0.31449 | 0.34874 | 0.25 | 0.36116 |
| A2 | 0.43644 | 0.212 | 0.2 | 0.31449 | 0.23250 | 0.25 | 0.24077 |
| A3 | 0.10911 | 0.318 | 0.4 | 0.31449 | 0.34874 | 0.375 | 0.36116 |
| A4 | 0.43644 | 0.318 | 0.3 | 0.20966 | 0.34874 | 0.375 | 0.36116 |
| A5 | 0.10911 | 0.318 | 0.3 | 0.41931 | 0.34874 | 0.25 | 0.24077 |
| A6 | 0.21822 | 0.318 | 0.2 | 0.31449 | 0.23250 | 0.25 | 0.24077 |
| A7 | 0.43644 | 0.318 | 0.4 | 0.31449 | 0.34874 | 0.375 | 0.24077 |
| A8 | 0.10911 | 0.318 | 0.2 | 0.20966 | 0.23250 | 0.25 | 0.36116 |
| A9 | 0.21822 | 0.318 | 0.2 | 0.31449 | 0.23250 | 0.25 | 0.36116 |
| A10 | 0.43644 | 0.318 | 0.4 | 0.31449 | 0.34874 | 0.375 | 0.24077 |
| A11 | 0.32733 | 0.212 | 0.3 | 0.20966 | 0.23250 | 0.25 | 0.24077 |

### *4.9.2 Weighted Normalized Matrix*

According to Tong et al. (2005) states that to determine the matrix of weighted normalized decisions or matrix Y is a multiplication between the matrix of normalized decisions (R) with weight (W) given the following equation (5) [17]:

$$y_{ij} = w_{ij}\, r_{ij}\,; \quad (5)$$

with i = 1,2.... m; j = 1,2,..., n
Based on the equation (5), the calculation result can be seen in Table 14

**Table 14.** Normalized Values Y

| Alternative | Criteria | | | | | | |
|---|---|---|---|---|---|---|---|
| | Y1 | Y2 | Y3 | Y4 | Y5 | Y6 | Y7 |
| A1 | 0.16148 | 0.07314 | 0.048 | 0.03459 | 0.02441 | 0.01 | 0.00722 |
| A2 | 0.16148 | 0.04876 | 0.032 | 0.03459 | 0.01627 | 0.01 | 0.00482 |
| A3 | 0.04037 | 0.07314 | 0.064 | 0.03459 | 0.02441 | 0.015 | 0.00722 |
| A4 | 0.16148 | 0.07314 | 0.048 | 0.02306 | 0.02441 | 0.015 | 0.00722 |
| A5 | 0.04037 | 0.07314 | 0.048 | 0.04612 | 0.02441 | 0.01 | 0.00482 |
| A6 | 0.08074 | 0.07314 | 0.032 | 0.03459 | 0.01627 | 0.01 | 0.00482 |
| A7 | 0.16148 | 0.07314 | 0.064 | 0.03459 | 0.02441 | 0.015 | 0.00482 |
| A8 | 0.04037 | 0.07314 | 0.032 | 0.02306 | 0.01627 | 0.01 | 0.00722 |
| A9 | 0.08074 | 0.07314 | 0.032 | 0.03459 | 0.01627 | 0.01 | 0.00722 |
| A10 | 0.16148 | 0.07314 | 0.064 | 0.03459 | 0.02441 | 0.015 | 0.00482 |
| A11 | 0.12111 | 0.04876 | 0.048 | 0.02306 | 0.01627 | 0.01 | 0.00482 |

*4.9.3 Positive Ideal Solution Matrix ($A^+$) and Matrix of Negative Ideal Solutions ($A^-$)*

According to Tong et al. (2005) stated that to find a positive ideal solution ($A^+$) given the following equation (6) [17]:

$$A^+ = (y1^+ + y2^+ +, \dots, yn^+); \quad (6)$$

with

$$y_j^+ = \begin{cases} \max yij; & \text{if j is an advantage attribute} \\ \min yij; & \text{if j is a cost attribute} \end{cases}$$

with value j = 1,2,...,n.

Based on equation (6), then the result of the calculation **$A^-$ = {0.04037; 0.04876; 0.032; 0.02306; 0.01627; 0.01; 0.00482}.**

*4.9.4 Distance Between Weighted Values of Each Alternative*

Determining the distance between the value of each alternative and the matrix of positive ideal solutions and the matrix of negative ideal solutions requires a weighted normalized decision matrix value ($y_{ij}$) and profit attributes ($y_j^+$), and cost attributes ($y_j^-$) in the calculation process.

a. Positive Ideal Solution

According to Tong et al. (2005) states that to calculate the distance between the alternative and the matrix of positive ideal solutions given the following equation (7) [17]:

$$D_i^+ = \sqrt{\sum_{j=1}^{n} (y_i^+ - y_{ij})^2} \quad (7)$$

with i = 1, 2,..., m.

Based on equation (7), then the calculation results can be seen in Table 15.

**Table 15.** Positive Ideal Solutions

| Alternative | Positive Ideal Solutions | On Behalf of | Value |
|---|---|---|---|
| 1 | $D_1^+$ | Sunaeni Haryati, S.Si | 0.02035 |
| 2 | $D_2^+$ | David Ramadhan Pratama | 0.04299 |
| 3 | $D_3^+$ | Drs. Undang Hidayat | 0.12166 |
| 4 | $D_4^+$ | Heri Kiswanto, S.Sn | 0.02807 |
| 5 | $D_5^+$ | Muhamad Ridwan, S.Pd | 0.12229 |
| 6 | $D_6^+$ | Sena Dwi Septoaji Suherman, S.Pd | 0.08816 |
| 7 | $D_7^+$ | Dewi Kartika, S.Pd | 0.01178 |
| 8 | $D_8^+$ | Mohamad Gelar Cipta, S.Pd | 0.12773 |
| 9 | $D_9^+$ | Ilham Halimi Akbar, S.Pd | 0.08813 |
| 10 | $D_{10}^+$ | Nurdin, S.Pd | 0.01178 |
| 11 | $D_{11}^+$ | Riki Agus Setiawan | 005576 |

b. Negative Ideal Solution

According to Tong et al. (2005) states that to calculate the distance between the alternative and the matrix of negative ideal solutions given the following equation (8) [17]:

$$D_i^- = \sqrt{\sum_{j=1}^{n}(y_{ij} - y_i^-)^2} \tag{8}$$

with : i = 1,2,...,m.

Based on equation (8), then the calculation results can be seen in Table 16.

**Table 16** Negative Ideal Solutions

| Alternative | Positive Ideal Solutions | On Behalf of | Value |
|---|---|---|---|
| 1 | $D_1^-$ | Sunaeni Haryati, S.Si | 0.12539 |
| 2 | $D_2^-$ | David Ramadhan Pratama | 0.12166 |
| 3 | $D_3^-$ | Drs. Undang Hidayat | 0.04299 |
| 4 | $D_4^-$ | Heri Kiswanto, S.Sn | 0.12496 |
| 5 | $D_5^-$ | Muhamad Ridwan, S.Pd | 0.03806 |
| 6 | $D_6^-$ | Sena Dwi Septoaji Suherman, S.Pd | 0.18257 |
| 7 | $D_7^-$ | Dewi Kartika, S.Pd | 0.17716 |
| 8 | $D_8^-$ | Mohamad Gelar Cipta, S.Pd | 0.02450 |
| 9 | $D_9^-$ | Ilham Halimi Akbar, S.Pd | 0.04861 |
| 10 | $D_{10}^-$ | Nurdin, S.Pd | 0.12849 |
| 11 | $D_{11}^-$ | Riki Agus Setiawan | 0.08231 |

*4.9.5 Preference Values for Each Alternative (Vi)*

According to Tong et al. (2005) states that to determine the preference value for each alternative given the following equation (9) [17]:

$$V_i = \frac{D_i^-}{D_i^- + D_i^+} \quad (9)$$

with : i = 1,2,...,m.

Based on equation (9), then the calculation results can be seen in Table 17.

**Table 17** Topsis Method Warkingan

| Alternative Name | Preference Weight Value (Vi) | Ranking |
|---|---|---|
| A7 | 0.93765 | 1 |
| A10 | 0.91602 | 2 |
| A1 | 0.86039 | 3 |
| A4 | 0.81658 | 4 |
| A2 | 0.73889 | 5 |
| A6 | 0.67435 | 6 |
| A11 | 0.59615 | 7 |
| A9 | 0.35549 | 8 |
| A3 | 0.26111 | 9 |
| A5 | 0.23735 | 10 |
| A8 | 0.16093 | 11 |

Based on the process of TOPSIS method, it can be concluded that the largest value is the alternative code A7 with a value of 0.93765 namely Dewi Kartika, S.Pd is designated as an alternative chosen honorary teacher who will be appointed as a permanent teacher in SMK Pasundan 2 Banjaran.

## 4.10 Comparison of SAW Method and TOPSIS Method

Based on the SAW method and TOPSIS method, the results of the comparison of the two methods are obtained in Table 18.

**Table 18.** Results of Comparison of SAW Methods and TOPSIS Methods

| SAW Method | | | TOPSIS Method | | |
|---|---|---|---|---|---|
| Alternative Name | Preference Weight Value (Vi) | Ranking | Alternative Name | Preference Weight Value (Vi) | Ranking |
| A7 | 0.96583 | 1 | A7 | 0.93765 | 1 |
| A10 | 0.96583 | 1 | A10 | 0.91602 | 2 |
| A1 | 0.91917 | 2 | A1 | 0.86039 | 3 |
| A4 | 0.90500 | 3 | A4 | 0.81658 | 4 |
| A2 | 0.77250 | 4 | A2 | 0.73889 | 5 |
| A3 | 0.69500 | 5 | A6 | 0.67435 | 6 |
| A11 | 0.69250 | 6 | A11 | 0.59615 | 7 |
| A9 | 0.67083 | 7 | A9 | 0.35549 | 8 |
| A6 | 0.66417 | 8 | A3 | 0.26111 | 9 |
| A5 | 0.66250 | 9 | A5 | 0.23735 | 10 |

| SAW Method | | | TOPSIS Method | | |
|---|---|---|---|---|---|
| Alternative Name | Preference Weight Value (Vi) | Ranking | Alternative Name | Preference Weight Value (Vi) | Ranking |
| A8 | 0.55083 | 10 | A8 | 0.16093 | 11 |

In the process of SAW method can be concluded that that that has the largest value there are two alternative codes A7 and A10 with a value of 0.96583 on behalf of Dewi Kartika, S.Pd and Nurdin, S.Pd. While in the process of topsis method can be concluded that that has the largest value there is only one alternative code A7 with a value of 0.93765 in the name of Dewi Kartika, S.Pd. The final result obtained from the calculation of the two methods can be concluded that there are differences in results, but basically the two methods used in this study play a role in establishing honorary teachers who will be appointed permanent teachers in SMK Pasundan 2 Banjaran.

The result of the comparison of SAW method and TOPSIS method obtained that the name Dewi Kartika, S.Pd emerged as a more chosen alternative as an honorary teacher who will be appointed as a permanent teacher in SMK Pasundan 2 Banjaran.

## 5. Conclusions

In the process of SAW method can be concluded that that that has the largest value there are two alternative codes A7 and A10 with a value of 0.96583 on behalf of Dewi Kartika, S.Pd and Nurdin, S.Pd. Based on the results of comparison or comparison of SAW method and TOPSIS method named Dewi Kartika, S.Pd emerged as a more chosen alternative as honorary teacher to be appointed as a permanent teacher in SMK Pasundan 2 Banjaran. After further comparison or comparison, sensitivity test was conducted which resulted in that saw method is the most appropriate method in the determination of permanent teachers in SMK Pasundan 2 Banjaran because it has the highest percentage of change scores. This is seen in the SAW method sensitivity test result value with a change value of 10.03168%. While in the TOPSIS method the percentage value changes by 0.65290%.

## References

[1] Juhri A M, Tri Y H and Agus S 2017 The Effectiveness of Teacher Work between Permanent and Non-Permanent Teachers on the Implementation of School-Based Management in Muhammadiyah Metro Elementary School *Educational Research and Reviews* **12** (20) 980-987

[2] Suyanti S and Roestam R 2018 Analisis Perbandingan Metode Simple Additive Weighting (SAW) dan TOPSIS dalam Pemilihan Guru Teladan pada SMA Negeri 4 Sarolangun *Jurnal Manajemen Sistem Informasi* **3** (3) 1208-1225

[3] Ameri A A, Pourghasemi H R and Cerda A 2018 Erodibility prioritization of sub-watersheds using morphometric parameters analysis and its mapping: A comparison among TOPSIS, VIKOR, SAW, and CF multi-criteria decision making models *Science of the Total Environment* **613** 1385-1400

[4] Sarraf A Z, Mohaghar A and Bazargani H 2013 Developing TOPSIS method using statistical normalization for selecting Knowledge management strategies *Journal of industrial engineering and management* **6** (4) 860-875

[5] Sundari J, Anggraeni S and Siahaan F B 2018 Comparison Topsis And Saw Method In The Selection Of Tourism Destination In Indonesia. In *2018 Third International Conference on Informatics and Computing*

[6] Sugiyarti E, Jasmi K A, Basiron B, Huda M, Shankar K and Maseleno A 2018 Decision support system of scholarship grantee selection using data mining *International Journal of Pure and Applied Mathematics* **119** (15) 2239-2249

[7] Hassan A R and Subasi A 2017 A decision support system for automated identification of sleep stages from single-channel EEG signals *Knowledge-Based Systems* **128** 115-124
[8] Puspa M 2019 Decision Support System For Supplementary Food Recipients (PMT) By Using The Simple Additive Weighting (SAW) Method *Jurnal Teknik Informatika CIT Medicom* **11** (2) 37-44
[9] Garg H and Kumar K 2019 An extended technique for order preference by similarity to ideal solution group decision-making method with linguistic interval-valued intuitionistic fuzzy information *Journal of Multi-Criteria Decision Analysis* **26** (1-2) 16-26
[10] Kurniawan D E 2018 Pemilihan wisata menggunakan technique for order preference by similarity to ideal solution (topsis) dengan visualisasi lokasi objek *Klik-Kumpulan Jurnal Ilmu Komputer* **5** (1) 75-86
[11] Mujiarto, Djohar A, Komaro M, Mohamed M A, Rahayu D S, Sanjaya W S M, Mamat M, Sambas A and Subiyanto 2019 Colored object detection using 5 dof robot arm based adaptive neuro-fuzzy method *Indonesian Journal of Electrical Engineering and Computer Science* **13** (1) 293-299
[12] Trisnawan, Sanjaya W S M, Sambas A, Jannah M, Rahayu D S, Mamat M and Mohamed M A 2019 Design of 4 Dof Robot ARM Based on Adaptive Neuro-Fuzzy (ANFIS) using Vision in Detecting Color Objects International Journal of Recent Technology and Engineering 8 (2S7) 224-227
[13] Sri Kusumadewi H P 2010 *Aplikasi Logika Fuzzy Pendukung Keputusan* Yogyakarta: Graha Ilmu
[14] Wong B K and Lai V S 2011 A survey of the application of fuzzy set theory in production and operations management: 1998–2009 *International Journal of Production Economics* **129** (1) 157-168
[15] Ahn B S 2011 Compatible weighting method with rank order centroid: Maximum entropy ordered weighted averaging approach *European Journal of Operational Research* **212** (3) 552-559
[16] Alonso S, Chiclana F, Herrera F, Herrera-Viedma E, Alcalá-Fdez J and Porcel C 2008 A consistency-based procedure to estimate missing pairwise preference values *International journal of intelligent systems* **23** (2) 155-175
[17] Tong L I, Wang C H and Chen H C 2005 Optimization of multiple responses using principal component analysis and technique for order preference by similarity to ideal solution *The International Journal of Advanced Manufacturing Technology* **27** (3-4) 407-414

# Designing Souvenir Products Berastagi Clothes with Kansei Engineering Method

**N Sembiring*, B Febrilliandika, H Oktaviani, L S Siregar and N N Azmi**

Department of Industrial Engineering, Faculty of Engineering, Universitas Sumatera Utara, Medan, Indonesia

*Email: nurhayatipandia68@usu.ac.id

**Abstract.** Consumer desire is something that is very vital to be criticized in product design. Clothing products are also known as the time changing industry, which is an industry that must continue to innovate to meet consumer desires. Berastagi clothing products are one of the typical souvenirs of the Berastagi tourism area. Based on a preliminary survey conducted on 84 tourists, 70.24 percent stated that the dress design was not attractive. As many as 70 percent of the tourists state that the design of these clothes is outdated. Therefore it is necessary to design new clothes designs. The method used in this study is kansei engineering to identify the kansei, items, and product categories that are consumer's preferences. In this study, four questionnaires were used. The first questionnaire is an open questionnaire that is used to analyze the product word kansei. The results are obtained in the form of nineteen kansei words which will be used as input to the second questionnaire. Each kansei word was assessed by the respondent using a semantic differential scale. The results of the data adequacy test for the second questionnaire obtained N 'value of 91 while the number of samples was 93 so that the data was sufficient. The results of the validity test state that all kansei words are valid. The reliability test yields a value of 0.907 or greater than 0.60 so that the data is considered reliable. MSA and KMO test results obtained all valid data. Questionnaire III was used to determine product stimulus. The results of the second questionnaire were analyzed for conjoint analysis in order to obtain 8 product stimuli. Questionnaire IV is used to find out the models used to fulfill each customer's kansei words. In addition, an analysis of the main preferences of customers was also carried out. The final result is that the model with the highest utility which is the customer's preference is clothing with cotton materials, ikat motif, cardigan models, white color, and long sleeves.

## 1. Introduction

Consumer desire is very important in designing a product in order to be widely accepted by the market [1]. Product design is done to create quality products and can meet the wishes of consumers so that products can be accepted and compete with competitors [2]. A good product should meet two basic characteristics. The first characteristic is physical factors consisting of basic performance needs, quality, capacity, and appearance. The second characteristic is the mental factor which is an aesthetic display that can attract consumers [3]. Product designs that have unique values or characteristics are more able to compete in the world of industry fashion [4].

The Fashion industry is an industry that never stops innovating at all times or is also known as the time-changing industry. Innovation in product appearance is necessary to maintain customer interest in the fashion product. Berastagi City is one of the iconic tourist attractions in Karo Regency, North Sumatra. In 2017, tourist visits to Berastagi reached 837,476 visits [5]. Berastagi has a temperature between 16.40C-23.90C so that the air feels cool [6]. In addition to nature tourism, Berastagi is also known for its cultural tourism. Various souvenirs with Batak Karo Culture pattern can be found here. One of the typical souvenirs of berastagi tourism area is a dress with a typical batak pattern. However, based on a preliminary survey conducted on 84 tourists, 70.24 percent stated that the design of the clothes is less attractive. A total of 70 percent stated that the design of the shirt is very old-fashioned and does not correspond to the present era. The design of clothes available today in the market can not meet the expectations and desires of consumers so it is necessary to design new clothing products that are by the wishes of consumers.

One method that can be used to design products is kansei engineering which is a system of assessment based on feelings to produce a real component [7]. This method is used to translate consumer needs based on psychological feelings costumer into product parameters [8]. Kansei was introduced by Mitsuo Nagamachi (Dean of Hiroshima International University) as a method of engineering that involves sensitivity and feeling integrated by senses of vision, hearing, smell, taste, and skin sensation [9]. The application of Kansei Engineering Approach in product design will produce products that are effective and have high aesthetic value [10]. Product design is one of the main factors that are considered by consumers in choosing batik because the design of a product will affect the appearance, as stated by Kotler that design is a totality of privileges that affect the appearance and function of a product in terms of consumer needs [11]. Kansei engineering is used to determine the look and design of clothes that are in following the wishes of consumers.

## 2. Methodology

This research was conducted using questionnaires. There are 4 questionnaires used and there are several stages of processing of each quiz. The first is an open questionnaire that is used to collect customer perceptions that are transformed into kansei words from souvenir products typical of berastagi clothes. A total of three souvenir designs berastagi clothes in the market as an example to be responded by consumers. Kansei words from emotionally related consumers are used as a second questionnaire input.

In the second questionnaire, kansei's words compared to his opponent's words were then assessed by respondents using semantic differential 1-5. There are 4 steps of data processing carried out in this second questionnaire. The first step is to test the adequacy of the data to determine whether the collected data already represents the population. If the value $N' < N$, the data is considered sufficient [12]. The second step is to test the validity of the data with a significance of 5% to determine the invalid kansei word. If invalid, the kansei word is selected and a follow-up iteration validity test is performed until all kansei word is valid. The third step is a reliability test to see the consistency of a measuring instrument to measure the same symptoms. A measuring instrument is considered consistent if the Alpha Cronbach coefficient is above 0.6 [13]. The last step is the analysis of factors carried out to reduce variables that are related to each other [14]. Before factor analysis is performed, Kaiser Mayer Olkin (KMO) testing needs to be done as a condition. Factor analysis can be done if the KMO value is between 0.5 to 1.0 and vice versa if the KMO value is less than 0.5, the factor analysis cannot be done [15]. After the KMO analysis is fulfilled, a Measure of Sampling Adequacy (MSA) analysis is carried out [16]. The end result is valid kansei words and can be used at a later stage.

The third questionnaire is the preliminary questionnaire of the fourth questionnaire. The third questionnaire contains items and categories that have been made based on the preferences of existing Baju Berastagi souvenir products. In this questionnaire, each respondent chooses five categories and two items from each category that are considered by consumers in buying souvenir products Berastagi clothes. Five items with each of the next two categories are carried out conjoin analysis processing to determine the product stimuli. Conjoint analysis is done with a full profile approach.

In the fourth questionnaire, stimuli products are judged based on kansei words that have been formed on a semantic differential scale of 1-5. Number 1 is chosen if the design is considered very unattractive, number 2 is chosen if the design is considered unattractive, number 3 is chosen if the design is considered ordinary, number 4 is chosen if the design is considered attractive, and number 5 is chosen if the design is considered very attractive Respondents were asked to assess the suitability of the product model in each product stimuli against the words kansei. The processing of the fourth questionnaire is done by looking for utility values. The category that most supports kansei word is marked with the highest utility value. The result is a category of models of souvenir clothes berastagi from each kansei word.

## 3. Result and Discussion

### 3.1. Data Collection Questionnaire I

The questionnaire I is an open questionnaire used to collect kansei words based on consumer perception. A total of 40 respondents were asked to give an assessment and specification of the clothes they wanted. Words related to a user's feelings are taken as kansei words. The final hasi obtained as many as 19 kansei words shown in Table 1.

**Table 1.** Kansei Words

| No | Kansei Words | No | Kansei Words | No | Kansei Words |
|---|---|---|---|---|---|
| 1 | Closed | 8 | Cheap | 15 | Flexible |
| 2 | Varied | 9 | Colorful | 16 | Natural |
| 3 | Elegant | 10 | Bright | 17 | Not Slippery |
| 4 | Smooth | 11 | Calm | 18 | Not Sexy |
| 5 | Absorb | 12 | Simple | 19 | Long |
| 6 | Big | 13 | Modern | | |
| 7 | Thick | 14 | Causal | | |

### 3.2. Data Collection Questionnaire II

Questioner II is used to assess the kansei word that is the preference and desire of consumers using the semantic differential scale. A questionnaire was distributed to 93 people who had visited the tourist area Berastagi and had seen souvenirs directly Berastagai clothes. Respondents were asked to rate each kansei word with 5 semantic differential scales consisting of 1 (very disagree), 2 (disagree), 3 (neither), 4 (agree), or 5 (very agree). Furthermore, data is carried out data adequacy test, validity test, and relatability test.

- Data Adequacy Test Results

The data adequacy test is carried out to see if the data can represent the population. The data adequacy test was conducted with a confidence level of 95%. The final result is an N' value of 91 while the number of samples is 93. It can be concluded that N > N' so that the data is said to be sufficient

- Validity Test Results

A validity test is performed to determine a valid kansei word. The validity test is carried out with a significant rate of 5%. Based on the validity test, all kansei word is declared valid because the value of R calculates greater than the value of R table. The R-value of each kansei word can be seen in Table 2.

**Table 2.** Validity Test

| Kansei Words | R | $R_{table}$ | Description | Kansei Words | R | $R_{table}$ | Description |
|---|---|---|---|---|---|---|---|
| Closed | 0.771 | 0.202 | Valid | Calm | 0.531 | 0.202 | Valid |
| Varied | 0.693 | 0.202 | Valid | Simple | 0.759 | 0.202 | Valid |
| Elegant | 0.299 | 0.202 | Valid | Modern | 0.760 | 0.202 | Valid |
| Smooth | 0.908 | 0.202 | Valid | Causal | 0.757 | 0.202 | Valid |
| Absorb | 0.869 | 0.202 | Valid | Flexible | 0.772 | 0.202 | Valid |
| Big | 0.362 | 0.202 | Valid | Natural | 0.302 | 0.202 | Valid |
| Thick | 0.460 | 0.202 | Valid | Not Slippery | 0.509 | 0.202 | Valid |
| Cheap | 0.743 | 0.202 | Valid | Not Sexy | 0.845 | 0.202 | Valid |
| Colorful | 0.268 | 0.202 | Valid | Long | 0.682 | 0.202 | Valid |
| Bright | 0.207 | 0.202 | Valid | | | | |

- Reliability Test Results

Reliability tests are performed to see the consistency of questionnaires. Based on reliability test, the alpha cronbach value is 0.907. It can be concluded that the value is greater than 0.6 so that the data is declared reliable.

- Factor Analysis

Factor analysis is carried out to reduce variables that are related to each other. The first step in performing factor analysis is the KMO test. Based on the results of data processing, the kmo value was obtained by 0.877. The KMO value is greater than 0.5 so it is feasible to proceed to the next stage. MSA tests are performed to reduce invalid word kansei. The MSA test results are shown in Table 3.

**Table 3.** MSA Test

| Kansei Words | $MSA_{count}$ | MSA | Description | Kansei Words | $MSA_{count}$ | MSA | Description |
|---|---|---|---|---|---|---|---|
| Closed | 0.895 | 0.5 | Valid | Calm | 0.797 | 0.5 | Valid |
| Varied | 0.844 | 0.5 | Valid | Simple | 0.903 | 0.5 | Valid |
| Elegant | 0.700 | 0.5 | Valid | Modern | 0.941 | 0.5 | Valid |
| Smooth | 0.926 | 0.5 | Valid | Causal | 0.952 | 0.5 | Valid |
| Absorb | 0.940 | 0.5 | Valid | Flexible | 0.958 | 0.5 | Valid |
| Big | 0.637 | 0.5 | Valid | Natural | 0.584 | 0.5 | Valid |
| Thick | 0.710 | 0.5 | Valid | Not Slippery | 0.837 | 0.5 | Valid |
| Cheap | 0.910 | 0.5 | Valid | Not Sexy | 0.932 | 0.5 | Valid |
| Colorful | 0.571 | 0.5 | Valid | Long | 0.883 | 0.5 | Valid |
| Bright | 0.670 | 0.5 | Valid | | | | |

### 3.3. Data Collection Questionnaire III

Questionnaire III is used to determine stimuli in questionnaire IV. Some of the steps taken in collecting and processing data questionnaire III are as follows:

- Product Item and Category Determination

The first step is to determine the items and categories of products that will be the consumer's choice. The determination is based on the products in the market. Items and categories of souvenir products Baju Berastagi shown in Table 4.

**Table 4.** Item and Category Determination

| Item | Category | Item | Category | Item | Category | Item | Category |
|---|---|---|---|---|---|---|---|
| Materials | Polyester | Color | Red | Motif | Mandala | Arm | Arm ¾ |
| | Spandex | | Chocolate | | Geometric | | Long |
| | Rayon | | Sky Blue | | Batik | | Short |
| | Cotton | | Purple | | Tie Dye | | |
| | Hyget | | Gray | | Floral | | |
| | Woll | | Green Grass | | Connective | | |
| Ornament | Embroidery | Model | T-shirt | Collar | O Neck | | |
| | Screen printing | | Shirt | | U Neck | | |
| | Write | | Polo | | V Neck | | |
| | Printing | | Cardigan | | Y Neck | | |

Each respondent was asked to choose five items with each of the two categories that become the preference in buying souvenirs Berastagi clothes from all items and categories. Items and categories with the most preferences are shown in Table 5.

**Table 5.** Selected Items and Categories

| Item | Category | Item | Category | Item | Category |
|---|---|---|---|---|---|
| Materials | Rayon | Color | Black | Motif | Connective |
| | Cotton | | White | | Mandala |
| Arm | Long | Model | Cardigan | | |
| | Short | | T-shirt | | |

- Conjoint Analysis

Conjoint analysis is carried out to determine the stimuli of the product. Conjoint analysis is done with a full profile approach. The results of conjoint analysis of rice products can be seen in Table 6.

**Table 6.** Conjoint Analysis Results

| Stimulus | Material | Motif | Model | Colour | Arm |
|---|---|---|---|---|---|
| 1 | Cotton | Mandala | Cardigan | Black | Short |
| 2 | Rayon | Mandala | T-shirt | Black | Short |
| 3 | Rayon | Connective | Cardigan | Black | Long |
| 4 | Rayon | Connective | T-shirt | White | Short |
| 5 | Rayon | Mandala | Cardigan | White | Long |
| 6 | Cotton | Connective | T-shirt | Black | Long |
| 7 | Cotton | Mandala | T-shirt | White | Long |
| 8 | Cotton | Connective | Cardigan | White | Short |

### 3.4. Data Collection Questionnaire IV

In questionnaire IV, respondents were asked to rate each stimulus based on kansei words using the semantic differential scale. The result is a utility assessment to determine the categories that can meet the words of the customer kansei. The category with the highest utility is shown in Table 7.

**Table 7.** Utility

| Kansei Words | Material (Utility) | Motif (Utility) | Model (Utility) | Colour (Utility) | Arm (Utility) |
|---|---|---|---|---|---|
| Closed | Cotton (0.221) | Mandala (0.007) | T-shirt (0.186) | Black (0.057) | Long (0.032) |
| Varied | Rayon (0.082) | Mandala (0.025) | Cardigan (0.082) | White (0.096) | Long (0.032) |
| Elegant | Rayon (0.018) | Mandala (0.060) | Cardigan (0.246) | Black (0.067) | Long (0.032) |
| Smooth | Cotton (0.086) | Connective (0.007) | Cardigan (0.043) | Black (0.021) | Short (0.028) |
| Absorb | Rayon (0.075) | Connective (0.032) | T-shirt (0.075) | White (0.003) | Short (0.003) |
| Big | Rayon (0.028) | Mandala (0.043) | T-shirt (0.014) | White (0.071) | Long (0.057) |
| Thick | Cotton (0.057) | Mandala (0.079) | T-shirt (0.057) | White (0.021) | Long (0.100) |
| Cheap | Rayon (0.100) | Mandala (0.021) | Cardigan (0.014) | White (0.092) | Short (0.043) |
| Colorful | Rayon (0.046) | Mandala (0.003) | T-shirt (0.046) | White (0.154) | Short (0.060) |
| Bright | Rayon (0.003) | Connective (0.046) | Cardigan (0.017) | White (0.553) | Long (0.017) |

| | | | | | |
|---|---|---|---|---|---|
| Calm | Rayon (0.082) | Connective (0.046) | T-shirt (0.025) | Black (0.039) | Long (0.075) |
| Simple | Rayon (0.064) | Mandala (0.071) | T-shirt (0.028) | Black (0.050) | Short (0.035) |
| Modern | Rayon (0.092) | Connective (0.014) | Cardigan (0.014) | White (0.050) | Long (0.043) |
| Causal | Rayon (0.017) | Connective (0.017) | T-shirt (0.132) | White (0.032) | Short (0.017) |
| Flexible | Cotton (0.011) | Mandala (0.018) | Cardigan (0.0035) | Black (0.039) | Short (0.053) |
| Natural | Rayon (0.021) | Connective (0.021) | T-shirt (0.028) | Black (0.028) | Short (0.021) |
| Not Slippery | Rayon (0.003) | Connective (0.003) | T-shirt (0.118) | White (0.060) | Short (0.0530 |
| Not Sexy | Cotton (0.028) | Mandala (0.028) | T-shirt (0.085) | Black (0.071) | Short (0.050) |
| Long | Cotton (0.164) | Connective (0.014) | Cardigan (0.286) | White (0.036) | Long (0.250) |

In addition to calculating the category utilization for each kansei words, utility calculations are also performed to determine the overall design that is the consumer's preference. The results of the overall utility calculation can be seen in Table 8. Based on utility calculations, categories that are in accordance with consumer desires are clothes with cotton materials, ikat motifs, cardigan models, white color, and long sleeves

**Table 8.** Overall Utility

| Materials (Utility) | Motif (Utility) | Model (Utility) | Colour (Utility) | Arm (Utility) |
|---|---|---|---|---|
| Cotton (-0.367) | Mandala (-0.709) | T-shirt (-0.615) | Black (0.025) | Long (0.727) |
| Rayon (0.291) | Connective (0.419) | Cardigan (0.368) | White (0.479) | Short (-0.572) |

## 4. Conclusion

Based on the research obtained 19 kansei word that can meet the needs of consumers consisting of closed, varied, elegant, smooth, absorbing, large, thick, cheap, colorful, bright, calm, simple, modern, casual, flexible, natural, not slippery, not sexy, and long. There are 5 items with each consisting of two categories to meet the wishes of consumers. Utility calculations obtained that the design of clothes that can meet the needs of consumers are clothing with ikat motifs, cardigan models, white colors, and long sleeves. Research is recommended to continue by comparing the proposed design motif with the motif of the shirt that is currently on the market whether there is a difference in consumer preferences in buying both products.

## References

[1] Janari D, Rakmawati A 2016 Developing Baby Bag Design by Using Kansei Engineering Method *IOP Conf. Series: Materials Science and Engineering* **105** (1) 012031

[2] Hasibuan C F 2020 Design Sling Bag Using Kansei Engineering Method *IOP Conf. Series: Materials Science and Engineering* **1003** (1) 012003

[3] Nagamachi M 1999 Kansei Engineering the Implication and Applications to Product Development *IEEE SMC '99 Conference Proceedings* **6** 273-278.

[4] Tama I P, Azlia W and Hardiningtyas D 2015 Development of Customer Oriented Product Design Using Kansei Engineering and Kano Model: Case Study of Ceramic Souvenir *Procedia Manufacturing* **4** 328-335.

[5] Putri D D B, Yoza D and Budiani E S 2017 Nilai Ekonomi Wisata Gunung Sibayak Berdasarkan Metode Biaya Perjalanan di Berastagi Sumatera Utara *JOM Faperta UR* **4** (2) 1-6

[6] Sembiring S A B 2018 Agricultural Practice and Social Change in Berastagi Area *IOP Conf. Series: Earth and Environmental Science* **126** (1) 1-8

[7] Pambudi A T, Suryoputro M R, Sari A D and Kurnia R D 2016 Design of Lesehan Chair by Using Kansei Engineering Method and Anthropometry Approach *IOP Conference Series: Material Science and Engineering* **105** (1) 012036.

[8] Schutee Simon T W, Eklund J, Axelsson J R C and Nagamachi M 2007 Concepts, Methods and Tools in Kansei Engineering *Theoretical Issues in Ergonomics Science* **5** (3) 214-231

[9] Prasetyo L R and Ulfa S K 2017 Determination of Standard Time in Packaging Processing Using Stopwatch Time Study o Find Output Standard *Journal Of Engineering And Managemenet Industrial System* **5** (2) 87-94

[10] Kjetil G, Gangwar M, Baggerud B and Narain R 2010 Kansei Engineering Based Emotional Product Design: An Illustration by a Battery Drill Case-Study *Kansei Engineering International Journal* **9** (2) 67-79

[11] Naeni H S and Heidaripour Maryam 2011 Kansei Engineering and Ergonomic Design of Products *International Journal of Occupational* **3** (2) 81-84

[12] Mitsuo N 2002 Kansei Engineering in Consumer Product Design *Ergonomic in Design The Quarterly of Human Factors Applications* **10** (2) 5-9

[13] Mat D K A, Khidir N Z, Ismail A R and Abdullah A F A 2018 Validity and Reliability of Instrument to Measure Social Media Skill Among Small And Medium Entrepreneurs at Pangkalan Datu River *International Journal of Development and Sustainability* **7** (3) 1026-1037

[14] Almagro L M and Llabre M X T 2011 Statistical Methods in Kansei Engineering : A Case of Statisctical Engineering Barcelona Tech *Departement of Statistical and Operation Research Universitas Politecnica de Catalunya* ENBIS **11** 1-6

[15] Okka A, Jatmiko H A and Erni 2019 Development of Food Packaging Design With Kansei Engineering Approach *International Journal of Scientific & Technology* **8** (12) 1778-1780

[16] Baroroh D K, Mya A and Lestari N P 2019 Kansei Engineering Approach for Developing Electric Motorcycle *Journal of Communications in Science and Technology* **4** (2) 50-56

IOP Conf. Series: Materials Science and Engineering 1115 (2021) 012018 doi:10.1088/1757-899X/1115/1/012018

# Breast Cancer Lesion Detection and Classification in mammograms using Deep Neural

**A R J Silalahi***

Department of Medical Physics, Matana University, Tangerang, Indonesia

*corresponding author: alexander.silalahi@matanauniversity.ac.id

**Abstract.** A method to automatically detect and classify a lesion into either malignant or non-malignant is presented in this work. The dataset used is obtained from INbreast database and are in format of full-field digital mammography (FFDM). Some of the key challenges in detecting cancerous lesion in mammography are the low contrast between cancerous lesion and its surrounding tissues, apparent contrast similarities between lesion and pectoral muscle, presence of calcifications that may disrupt the detection process, and some level of morphological similarities between the lesion and some normal tissues. The work here consists of two main parts. The first part is the image processing section that aims to sample the lesion with intended lesion-to-surrounding ratio (0.4-0.6) and to avoid sampling from unintended regions such as pectoral muscle. Another key challenge is that the database is relatively small while machine learning requires a relatively large dataset. To improve size of samples, eighty fixed-size images (250 pixels x 250 pixels) are randomly cropped out of each of the previously processed image. The second part is to build the machine learning application based on deep neural network framework to classify samples into two classes, malignant and non-malignant. In present work we apply two different frameworks known as Plain and Residual Net (ResNet). Our calculations show that both models can detect a single lesion with more than 90% accuracy and area under ROC curve >0.94.

## 1. Introduction

Breast cancer is the most prevalent cancer among women and among the top of all population. According to USA statistics one in every 8 women will develop some type of breast cancer in their life [1]. Mammography is the gold standard used in medical field to detect and diagnose breast cancer in various stages and conditions. In developed countries, early mammography is conducted at age of 40 and is continued with forward diagnose time interval adjusted to the findings. Mammography has shown to reduce the breast cancer mortality in the range of 25-30% [2]-[3]. However, the trial findings have met some concerns regarding the validity of the results [4]-[7]. In the typical mammography examination, the X-ray images are captured from two different angles of breast. The images are then further analyzed by two radiologists and raised malignant suspicion is followed up with further diagnostic evaluation.

The currently implemented screening evaluation of mammogram images require radiologists to read and analyze the entire image area, which is a tiring and monotonous job, and thus are prone to errors. Computer-aided detection (CAD) programs were developed to identify suspicious locations within the mammograms, which are then continued with further readings by the radiologists. These CAD programs [8] have shown to be useful and helpful for radiologists, in which the use of CAD has similar performance to double reading. However, it also is still controversial because its usage has not shown a performance improvement of radiologists in the last decade in USA. It also is rather expensive with current cost around USD 400 million every year [9]. The current state-of-art CAD program uses hand-crafted features of an X-ray images which are then sent to machine learning for identification and classification.

The manual process in selecting the most relevant features can result in wrongly chosen features and results in wrong representation of the mammogram. The exciting and rapid development in the field of computer vision brings the deep neural network, a technique based on artificial neural network, capable of detecting, classifying and localizing objects in images with very high accuracy, without hand crafting the relevant features. The algorithm simply learns the "correct" features from the mammogram samples. There have been studies using machine learning classifiers [10]-[11] and deep neural networks [12]-[15] on mammography images. The present work is an attempt build a deep neural network model and to improve its predicting power over its predecessors.

The paper is organized into four sections. The required dataset and methods are presented in Section 2 this includes the formulation of deep neural net framework used in the present work. Section 3 contains the results of simulations which mostly includes the performance of deep neural net model in terms of accuracy, convergency, and the attempt to improve the performance by using resnet. Finally, Section 4 discusses the future works to further improve the performance of such CNN method.

## 2. Materials and Methods

The use of machine learning (ML) to detect and classify malignant breast cancer is not new subject of research and has been applied with various types ranging from rather traditional type of artificial neural network to convolutional neural network. The scope of detection also ranges from patch detection to whole slide image of mammogram. In the current work we use two CNN-based machine learning frameworks with two different schemes: VGGnet and Resnet. These two methods have very good predicting power with relatively simple networks.

### 2.1. *Whole slide image processing*

The mammograms were obtained from INbreast database [16] which provides FFDM of 116 patients. According to American College of Radiology [17], breast cancer can be classified using BI-RADS into seven categories (see Table 1).

**Table 1.** Bi-Rads Classification

| BI-RADS index | Interpretation |
|---|---|
| 0 | Insufficient study |
| 1 | Negative |
| 2 | Benign findings |
| 3 | Likely benign findings |
| 4 | Suspicious findings |
| 5 | Highly likely of malignancy |
| 6 | Biopsy proven malignancy |

In the present work we only try to classify the patches of mammograms to either fully non-malignant or to partially contain malignant lesion. One of the key challenges with using electronic medical record (EMR) is the unavailability of large public database. In the present work, we try to improve the number of samples by randomly choosing quite large number of patches within a mammogram image. The randomness is imposed to make sure that there is no correlation between patches of the same mammogram slide. The algorithm is also tuned to avoid region of pectoral muscle, because previous researches have shown degradation of detection accuracy if it is included. We also know from previous findings that the contrast between tumor region and its surrounding tissue is one of the most important features in defining a tumor. Because of this reason, random patches that include tumor region must contain both tumor area and enough surrounding area. In this work we define ratio (R) as tumor area divided by surrounding area.

$$R = 0.4 - 0.6 \quad (1)$$

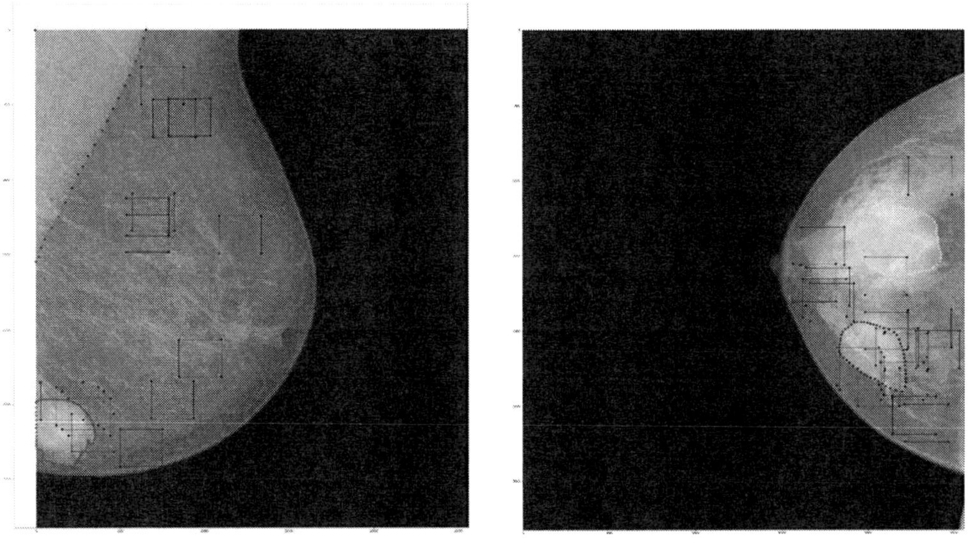

**Figure 1.** Cropping lesions from malignant and non-malignant areas (excluding pectorial muscle region).

### 2.2. *ConvNet Parameterization*

In the present work we use two deep neural net framework: plain and resnet. The two schemes utilize ConvNet blocks that include modules such as: nonlinear activation function, max pooling, batch normalization, dropout, and classification layer. The resnet general structure is similar to VGGNet but is modified to include a shortcut link that connects two previously identified modules. One of the main reasons for this modification is because in general, deep neural network suffers from vanishing gradient issue. The shortcut link ensure that the deeper neural network always performs better than its shallower version.

The activation functions used is the rectified linear unit (ReLU) with form:

$$(x) = \max(0, x) \quad (2)$$

Each of the convolutional layer is then processed with a batch normalization layer. A batch normalization does similar thing to hidden layer weights to what normalization does to input data. This process reduces how much hidden values shift around and helps each layer to learn by itself a little bit more independently. Dropout layer is used to avoid overfitting by probabilistically removing numbers of input to a layer. In present work 0.25 fraction of inputs is dropped randomly. Max pooling layer is used to reduce the dimensionality of the inputs and to help the representation becomes slightly invariant to small translation of the input data [18].

### 2.3. *ConvNet Architectures*

Here we will employ two frameworks, plain and resnet [19]. The resnet building block is built similarly to VGGNet [20] in which we have serially connected modules. Building blocks are defined to represent convolutional neural network and other necessary modules. The shortcut link connection is applied every two of these building blocks, starting at the input of building block 1 and ending at right before the ReLu module of building block 2.

IOP Conf. Series: Materials Science and Engineering 1115 (2021) 012018 doi:10.1088/1757-899X/1115/1/012018

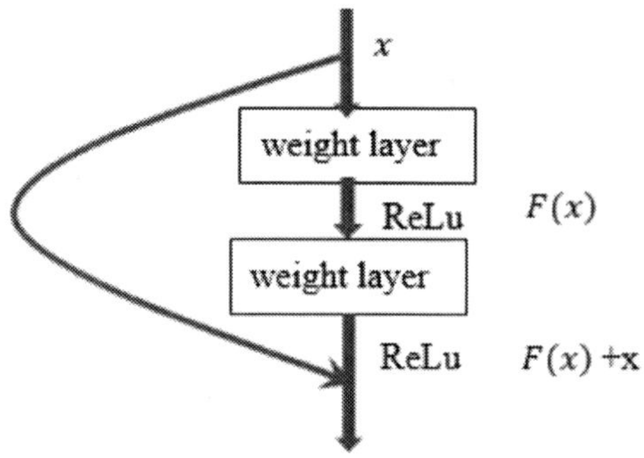


**Figure 2.** Shortcut link in resnet.

Shortcut link contains addition of $x$ and $(x)$, thus their dimensions must be equal. If they are not, we need to do matrix linear projection to equalize the two.

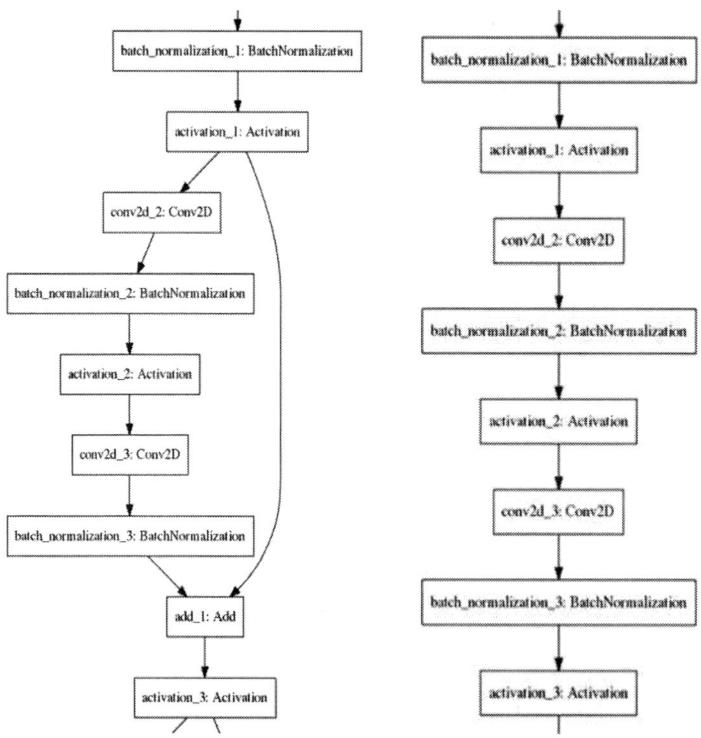


**Figure 3.** Plain and resnet framework.

## 3. Result and Discussion

### *3.1. Accuracy and Precision*

The first calculations show that plain and vgg-16 model seems to show a better predicting power than resnet, as shown in Figure 4. The plain and resnet models were used with depth of layer equals to 6n+2.

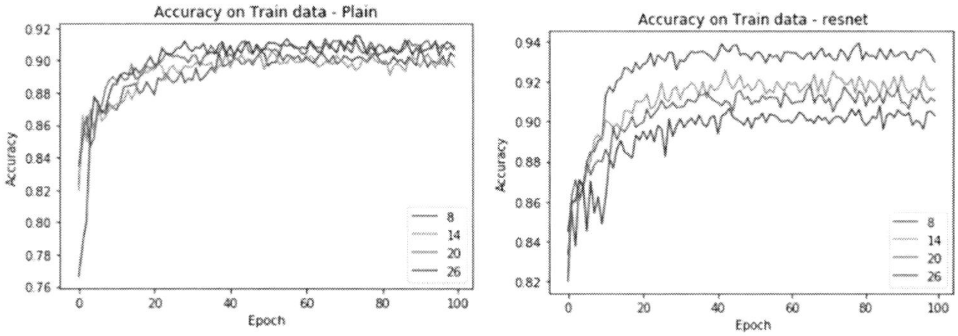


**Figure 4.** Accuracy and Precision on Train data.

### *3.2. Overfitting test*

Here we want to see how both models suffer from overfitting issue in which the predicting accuracy over train data is better than test data.

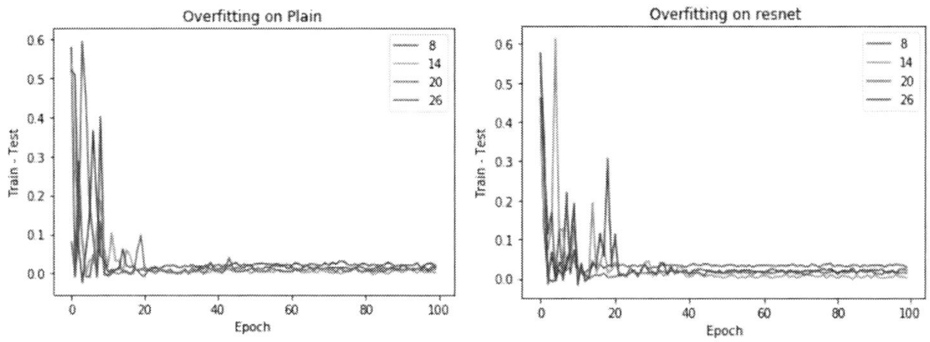


**Figure 5.** Overfitting plot on plain and resnet.

The calculation indicated that both models suffer from overfitting as it gets worse as depth of layer increases as shown in figure 5. It is also clear that overfitting issue is worsened in resnet model. This conclusion is in agreement with the He [19].

### *3.3. Augmented Data*

Machine learning requires a relatively large dataset to train the model to have high predicting power. One of the issues in predicting breast cancer is the limited availability of mammogram dataset. Although we have used independent lesion sampling to improve our size of input data, it is still possible that the model doesn't completely capture all the relevant features as predicting basis. One of the possible remedies to this problem is to produce additional data from the current training set. This new data is generated by modifying the images through horizontal and vertical flipping and through images shifting. This method is called augmented data and is applied only on training dataset.

In Figure 6 we show that the predicting accuracy of resnet are the same for n=1 and 4 which converge to around 90%. The calculation also shows that predicting accuracies are roughly the same for test and training data. The robust convergence of resnet is in agreement with previous finding [19].

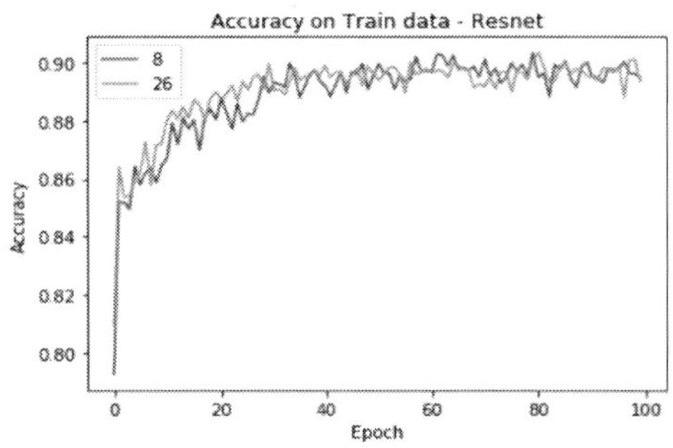


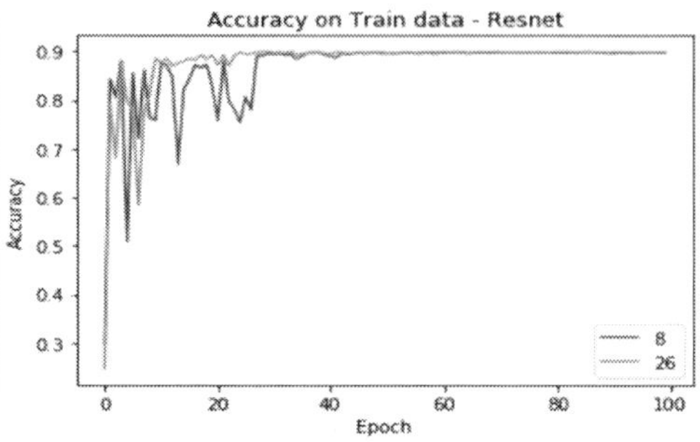


**Figure 6.** Accuracies on Train data: Left (plain) & Right (resnet).

### *3.4. Area Under Curve (ROC)*

ROC curve [21] is a plot of true positive rate vs false positive rate; thus each point of data represents a pair of sensitivity/specificity. A perfect predictor would have an ROC that passes thought the most left upper point and area under curve is equal to one. Area under the curve represents how well the model can distinguish lesion with malignancy and non-malignancy. Our calculations indicate that the resnet with deeper layer performs better than plain and shallow resnet as shown in Figure 7.

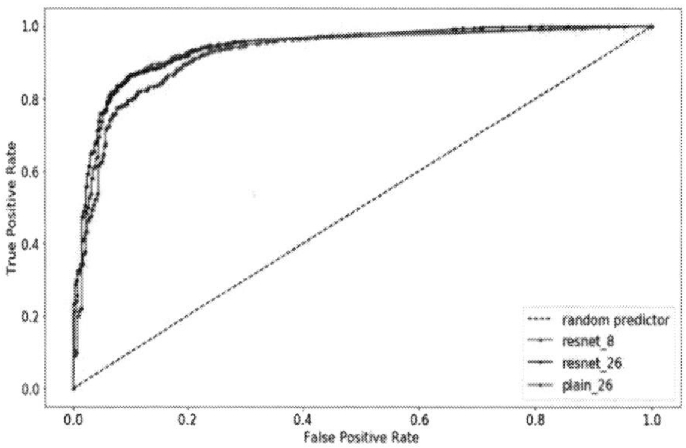


**Figure 7.** ROC Curve for resnet and plain with depths of 8 and 26.
Random Predictor is shown as straight line.

### *3.5. False Predictions*

In this section we want to understand why and how our model fails to detect malignancy or normal lesions. The two possible miss classifications are the false positive (normal identified as malignant) and false negative (malignant identified as normal). We sort the mammograms based on how large the predictions deviate from the true annotations and how many lesions are misclassified in a single mammogram. Here we show the WSI with highest number of wrongly identified lesions, ones that are belong to the false positive and false negative. The false positive predictions tend to occur in whiteish mammogram image that is usually associated with dense breast tissue. On the other hand, the false negative predictions tend to occur in greyish mammogram image that is associated with less dense breast tissue. It is interesting to note that the high dense vs low dense breast tissue has been identified in previous study as one of prominent feature differences between Asian and Caucasian women [22]-[23].

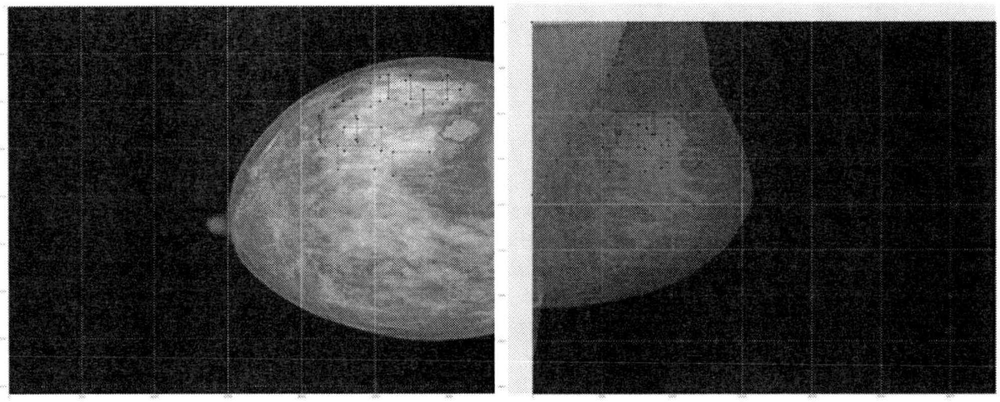

**Figure 8.** False Positve (left) and False Negative (right)

## 4. Conclusions

In the present work we have been able to automatically detect a lesion in mammogram and classify it into either malignant or non-malignant. This automated method is based on a class of machine learning known as deep neural network, specifically plain and resnet. The two frameworks were applied with varying depths and were demonstrated to reach accuracies more than 90%. The convergence of resnet prediction is ensured with applying augmented data which includes vertical and horizontal flipping, and image shifting. We have also shown that the resnet has the improved ROC curve over the plain's. The areas under the roc curve are 0.943 and 0.923 for resnet and plain with depths of 26. The present work, which is lesion detection, is part one of our project and should be continued with part two, in which we incorporate this resnet model into faster r-cnn and mask r-cnn method [24]-[25] to automatically detect, locate, and segmentize malignant patches in a mammogram WSI.

## References

[1] Siegel R L, Miller K D and Jemal A 2019 Cancer statistics, 2019 *CA. Cancer J. Clin.* **69** 7–34
[2] Bleyer A, Baines C and Miller A B 2016 Impact of screening mammography on breast cancer mortality *Int. J. Cancer* **138** 2003–12
[3] Mayor S 2017 Mammography screening reduces breast cancer deaths in Ireland, study finds *BMJ* **359**
[4] Kopans D B 2003 Opinion. The most recent breast cancer screening controversy about whether mammographic screening benefits women at any age: Nonsense and nonscience *Am. J. Roentgenol.* **180** 21–6
[5] Feig S A 2014 Screening Mammography Benefit Controversies. Sorting the Evidence *Radiol. Clin. North Am.* **52** 455–80
[6] Lee C H 2002 Screening mammography: proven benefit, continued controversy *Radiol. Clin. North Am.* **40** 395–407
[7] Green B B and Taplin S H 2003 Breast cancer screening controversies *J. Am. Board Fam. Pract.* **16** 233–41
[8] Anon Computer-Aided Detection (CAD) in Screening Mammography: Sensitivity of Commercial CAD Systems for Detecting Architectural Distortion : American Journal of Roentgenology : Vol. 181, No. 4 (AJR)
[9] Lehman C D, Wellman R D, Buist D S M, Kerlikowske K, Tosteson A N A and Miglioretti D L

2015 Diagnostic accuracy of digital screening mammography with and without computer-aided detection *JAMA Intern. Med.* **175** 1828–37

[10] Wei L, Yang Y, Nishikawa R M and Jiang Y 2005 A study on several machine-learning methods for classification of malignant and benign clustered microcalcifications *IEEE Trans. Med. Imaging* **24** 371–80

[11] Ramos-Pollán R, Guevara-López M A, Suárez-Ortega C, Díaz-Herrero G, Franco-Valiente J M, Rubio-Del-Solar M, González-De-Posada N, Vaz M A P, Loureiro J and Ramos I 2012 Discovering mammography-based machine learning classifiers for breast cancer diagnosis *J. Med. Syst.* **36** 2259–69

[12] Kooi T, Litjens G, van Ginneken B, Gubern-Mérida A, Sánchez C I, Mann R, den Heeten A and Karssemeijer N 2017 Large scale deep learning for computer aided detection of mammographic lesions *Med. Image Anal.* **35** 303–12

[13] Dhungel N, Carneiro G and Bradley A P 2015 Automated Mass Detection in Mammograms Using Cascaded Deep Learning and Random Forests *2015 International Conference on Digital Image Computing: Techniques and Applications, DICTA 2015* (Institute of Electrical and Electronics Engineers Inc.)

[14] Wang J, Yang X, Cai H, Tan W, Jin C and Li L 2016 Discrimination of Breast Cancer with Microcalcifications on Mammography by Deep Learning *Sci. Rep.* **6** 1–9

[15] Becker A S, Marcon M, Ghafoor S, Wurnig M C, Frauenfelder T and Boss A 2017 Deep Learning in Mammography *Invest. Radiol.* **52** 434–40

[16] Moreira I C, Amaral I, Domingues I, Cardoso A, Cardoso M J and Cardoso J S 2012 INbreast: Toward a Full-field Digital Mammographic Database. *Acad. Radiol.* **19** 236–48

[17] D'orsi C, Bassett L, Berg W, Feig S, … V J-A C of and 2003 undefined Breast imaging reporting and data system: ACR BI-RADS-mammography

[18] Aloysius N and Geetha M 2018 A review on deep convolutional neural networks *Proceedings of the 2017 IEEE International Conference on Communication and Signal Processing, ICCSP 2017* vol 2018-January (Institute of Electrical and Electronics Engineers Inc.) pp 588–9

[19] He K, Zhang X, Ren S and Sun J 2016 Deep residual learning for image recognition *Proceedings of the IEEE Computer Society Conference on Computer Vision and Pattern Recognition* vol 2016-Decem pp 770–8

[20] Chung C, Patel S, Lee R, Fu L, Reilly S, Ho T, Lionetti J, George M D and Taylor P 2018 VGGNET *Am. J. Heal. Pharm.* **75** 398–406

[21] Bradley A P 1997 The use of the area under the ROC curve in the evaluation of machine learning algorithms *Pattern Recognit.* **30** 1145–59

[22] Bae J M and Kim E H 2016 Breast density and risk of breast cancer in Asian women: A meta-analysis of observational studies *J. Prev. Med. Public Heal.* **49** 367–75

[23] Mariapun S, Li J, Yip C H, Taib N A M and Teo S H 2015 Ethnic differences in mammographic densities: An Asian cross-sectional study *PLoS One* **10**

[24] Ren S, He K, Girshick R and Sun J 2017 Faster R-CNN: Towards Real-Time Object Detection with Region Proposal Networks *IEEE Trans. Pattern Anal. Mach. Intell.* **39** 1137–49

[25] He K, Gkioxari G, Dollár P and Girshick R 2020 Mask R-CNN *IEEE Trans. Pattern Anal. Mach. Intell.* **42** 386–97

# New Era Chain Supply on Retail Food Industry in Indonesia

**Bob Foster**

Universitas Informatika dan Bisnis Indonesia, Bandung, Indonesia

*Email: bobriset@unibi.ac.id

**Abstract.** The purpose of this study is to find out a new era in the supply chain of the food retail business in Indonesia. The development of the number of retail food actors is a phenomenon that occurs due to the support for the ease of opening retail food. Factors that support the growth of retail food at this time is the high level of mobility of people who are busy demanding speed and practicality, in any case, is no exception to the fulfilment of basic needs such as food. In the era of disruption, business change is very visible, the way of doing business which used to emphasize owning (ownership) becomes sharing (sharing roles, collaboration resources). The existence of phenomena such as go food, go send, go shop, goat, grab food, and other application-based media are developing in the latest era. This phenomenon changes the view of the supply chain, one of which is affected by the retail food business. The discussion in this research is about a new era in which there are supply chains, food retail business, the role of technology, marketing, distribution channels, and descriptive effectiveness. The study was conducted using qualitative descriptive methods, namely using literature or literature studies, using journals, books, magazines, and the web that are relevant to research. The conclusion in this research is the development in this era in distribution, warehousing, transportation, and services in food retailing supported by internet application-based providers including go food, grab food, etc. making it easier for retailers and consumers to transact. The development of this new era experienced many improvements in terms of delivery, cost, and time so that the more days the more effective the supply chain in the food retail industry.

## 1. Introduction

The business sector that is currently developing is the trade sector. The trade sector is the non-agricultural sector which provides the smallest contribution in total income [1]. The trade sector, including the retail industry, is an important industry sector and should be taken into account, especially in its contribution to the Indonesian economy. The growth of modern retail every year records a range of 10% to 30%. This is indicated by the expansion of modern retail to rural areas and into people's residential areas [2].

Trading activities in the culinary business are currently busy with the community. The development of the number of retail food actors is a phenomenon that occurs due to the support for the ease of opening retail food. Food retailing is the choice of many people because of the type of business that is easier to do than other businesses. Besides that, retail food is much in demand by the community because it does not require too much capital costs and does not require that someone who runs it must be highly educated. The era of economic globalization accompanied by the rapid development of technology, the impact of very tight competition, and rapid changes in the business environment [3].

Factors that support the growth of retail food at this time is the high level of mobility of people who are busy demanding speed and practicality, in any case, is no exception to the fulfillment of basic needs such as food. This requires good supply chain management, in supply chain management, the most important ideal strategy is to emphasize the essence and ability to manage accuracy to respond to consumer demand [4]. This strategy is realized by the application of company policy in handling the six factors driving supply chain performance, namely facilities, inventory, transportation, information, resources, and prices [5].

At this time the company has entered the era of disruption. Many companies try to make adjustments to the era. Because the ability to adjust it directs the company to victory. The era of disruption is an era that will change not only the way of its business but up to its business fundamentals, starting from the cost structure to its culture, even to the industrial ideology [6]. In the era of disruption, business change is very visible, the way of doing business which used to emphasize owning (ownership) becomes sharing (sharing roles, collaboration resources). One of the units in the company that is very affected in disruption is marketing. The marketing sector spearheads the company that is most quickly affected. One form of the influence of disruption in marketing is the presence of an Online marketing system. The marketing system is called online, the arena uses the internet as a means of marketing goods or service products to consumers.

Explanation of the era of disruption by the above explanation has occurred in almost all industries, both service industries, and manufacturing industries. And it also happens in almost all professions, such as government, medicine, administration, front office, cashier, banking, marketing, lecturers, teachers, and other professions. Almost all professions are required to learn a lot about environmental change. Because the learning process will lead us to understand renewal in change. There are five important things in the era of disruption, namely, First, disruption results in saving a lot of costs through business processes that become simpler. Second, it makes whatever quality it produces better than before. Third, disruption has the potential to create new markets, or make those who have been excluded so far have been excluded. Fourth, the products/services resulting from this disruption must be more accessible or accessible to its users. Fifth, disruption makes everything smart. Smarter, more time saving, and more accurate [6].

In the past, all businesses need to be owned by themselves, controlled by them, if now they share roles. Thus the main strength of the company in the era of disruption is, lies in how much the company's ability to share roles among stakeholders. Therefore, the mindset or mindset that develops is not just a phenomenon today but must be able to think tomorrow or, brought by the reformer to the present [6]. This mindset results in every business actor must have the ability to learn so that we realize that every time there is a change and every time a renewal is needed.

The existence of phenomena such as go food, go send, go shop, go mart, grab food, and other application-based media are developing in the latest era. This phenomenon changes the view of the supply chain, one of which is affected by the retail food business. Technology facilitates the alignment and integration of all key supply chain actors, increasing the level of visibility and transparency so that resource estimates will be more certain. This is possible because the development of the volume and movement of goods can be known at any time. Resource planning management procedures, in accordance with the adoption of Industry 4.0 paradigm, will increase productivity, flexibility, and overall agility towards changes that might occur in the supply chain. Proper alignment and integration between key supply chain actors and increased levels of visibility and transparency will ensure more precise estimates of resources (people, goods, and equipment). Thus, there will be an optimization of resources, procurement time, and marketing, and increasing equipment usability.

## 2. Research Methodology

The study was conducted using qualitative descriptive methods, namely using literature or literature studies, using journals, books, magazines, and the web that are relevant to research.

## 3. Result and Discussion

Several previous studies have been analyzed about the effectiveness of online marketing, such studies include the Effectiveness of Advertising on Social Networks as One of the Marketing Strategies of Small and Medium Enterprises (SMEs) Using the method, compiled by Prisma Lisawati, the research was conducted to test the effectiveness of advertising through social networks as one of the marketing strategies of SME products using the EPIC (Empathy, Persuasion, Impact, Communication) method. The results of the study delivered that the marketing strategy of product advertising through social networks is very effective. Furthermore, the researcher suggests that producers actively marketing their products through social networks need to update or update advertisements and product information so that it will continue to stimulate consumers' memories of products that have been seen and continue to do sales promotions because promotional activities have a very effective impact in increasing sales.

Utilization of Social Media as Media Product Marketing and Indonesian Potential in Efforts to Support the ASEAN Community 2015 (Social Media Marketing Study on Twitter Kemenparekraf RI and Facebook Disparbud West Java Province), compiled by Ita Suryani, the research was conducted to provide a complete and in-depth description of the use of social media as marketing media to support the ASEAN Community 2015. The results of the study revealed that a number of smart companies, not only use social media as marketing media but have become supporters of business activities, facilitating and strengthening the function of communication to the public.

The research on the supply chain in one company entitled, analysis of the supply chain of sausage food industries from producers to consumers in the city of Bandung, was compiled by Desra Isma Diana. The study describes the effectiveness of the supply chain in one company using the scorecard method. The results of the study that the planned supply chain went well.

Research on a literature review of the effectiveness of product marketing using online marketing systems in the era of disruption compiled by Teguh Suripto in 2018. The research explains the strength of online marketing in the era of disruption has a positive impact on companies, because companies have good product marketing strategies to achieve goals.

From some of the previous studies, we tried to conduct further research by analyzing a new era in the food industry supply chain in Indonesia. Our research will complement the results of previous studies with the main studies we will undertake including supply chains in food retailing during the era of disruption.

### 3.1 Supply Chain

The definition of the supply chain (supply chain) is: "*A supply chain encompasses all activities associated with the flow and transformation of goods and services from the raw material stage to the end-user (customer), as well as the associated information flows. A supply chain is also an integrated group of processes to source, make, and deliver products* " [7].

The supply chain is all integrated activities including information flow relating to three aspects, namely: (1) sources; (2) the production process, and (3) the product delivery process.

There are three components in the supply chain, namely:

1) Upstream supply chain,
   Includes various company activities with suppliers, including procurement of raw materials and companion materials.
2) Internal supply chain,
   Includes all processes of importing goods into warehouses that are used up to the production process. Its main activities include production and inventory control.
3) Downstream supply chain,

Covers all activities that involve shipping products to customers. The main focus of its activities is distribution, warehousing, transportation, and services.

The simulation of the flow of goods and information in the supply chain can be described in Figure 1 as follows:

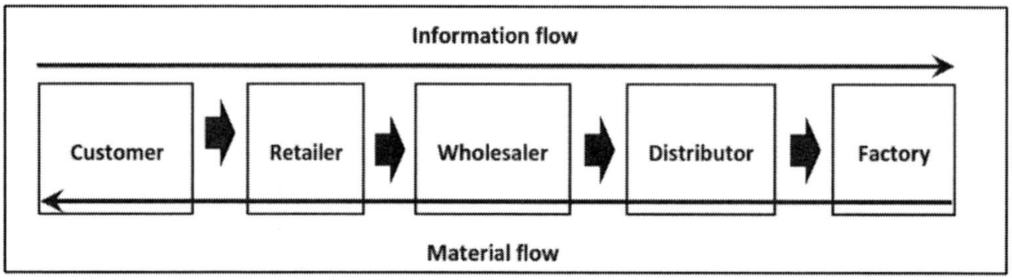


**Figure 1.** The flow of Goods and Information in the Supply Chain [7]

Supply chain management is one of the efforts to meet customer needs because basically supply chain management pays attention to how the process of goods to the hands of consumers by paying attention to the quality of goods, such as durability, the integrity of goods, response time ordering. Besides supply chain management also emphasizes the quality of services that are facilitated to customers such as response time and labor efficiency to meet customer needs.

The function of supply chain management is to coordinate the flow of materials, information, and money between all related companies such as supply companies and other companies related to material supply, manufacturing companies that process materials supplied, distributors and retailers [8].

From the above concept retail business is included in the downstream supply chain which focuses on distribution, warehousing, transportation, and services. This new era in downstream supply presents new breakthroughs based on internet applications that enable companies to achieve the functions of supply chain management.

### 3.2 Food Retail

The word retail comes from the French, "ritellier", which means to cut or break something. In Indonesian, the word retail can also be interpreted as "retail". Related to the activities carried out, retails describe activities to break down goods or products that are produced and distributed in large quantities and in bulk to be consumed by end consumers in small quantities according to their needs [9]. The retail marketing mix is a set of marketing tools that companies use to pursue their marketing goals [10]. The retail mix includes all actions a company can take to influence the demand for its own products and all actions that a company might take can be summed up as a group of variables including products, location, price, and promotion.

Meanwhile, retailing is a business venture that seeks to market goods and services to end consumers who use it for personal and household purposes. Products sold in the retailing business are goods, services, or a combination of both [11]. From some of the definitions above, it can be concluded that retail is all business ventures that directly direct their marketing capabilities to satisfy the final consumer.

According to Berman and Evans (2001) [11], there are three characteristics of retailing, namely:

1. *Small* Average Sale
   The level of retailing sales in these stores is relatively small because the target is the final consumer who buys in small quantities.
2. *Impulse Purchase*

Purchases that occur in retailing are mostly unplanned. This must be observed by retailers, namely how to find the right strategy to maximize purchases to optimize revenue.

3. *Popularity Of Stores*
The success of retailing depends very much on the popularity and image of the store or company. The more famous the store or company, the higher the level of visits, which in turn has an impact on revenue.

Online Retailing Online retailing, commonly called e-retailing or e-tailing, is a retail activity that is carried out online through the internet. Retail electronics is a retail format in which retailers and consumers communicate with each other through interactive electronic networks [12]. Meanwhile, retail electronics is retail that is done through the internet [13].

In its evolution, modern retailers increase market share and concentration, thereby increasing purchasing power which causes an imbalance of bargaining power in the supply chain, reducing supplier competition, influencing consumers through reduced innovation, long-term losses, reduced choices, and higher prices. These effects are often neglected, thus affecting the continuity of consumption and production [14].

Modern retailers reduce the strength of their suppliers' bargaining power, first increase their market power, by controlling market share in the downstream market. Market domination and dominant position can be used as a means of negotiating requirements on the supplier market (upstream market), by applying conditions that provide more benefits, including price cuts. The two-sided market example above is in the credit card market, where banks will have two sides of the market faced, namely the credit card holder's customer market and the merchant market [15].

The minimum degree of cooperation needed for a relationship and cooperation more closely reflects the level of trust and mutual assistance. Interdependence is influenced by competitive institutions, market structures and behaviours that determine the availability of choices and the strength of prospective partners' bargaining power resources in the market [16].

The basic concept above directs that food retailing in the new era has the overall characteristics of the retail business. Food retailing in the age of internet applications such as this demands products, locations, prices, and promotions that are increasingly competitive and growing due to the fast pace of food retailing companies compared to consumers.

## 3.3 The Role of Technology

The implications of Industry 4.0 on logistics are very large and important [17]. Demand for individual products and services will continue to increase. Thus, logistics within and outside the industrial system must adapt to a changing environment. The logistics system will be increasingly complex so that it cannot be handled with usual planning and control practices.

One application to handle supply chain performance is through the use of the internet and computer-based information technology. This phenomenon was first mentioned in 2011 in a proposal for the development of a new concept of German economic policy based on a high-tech strategy [18]. The industry as a sector of human activity that is mostly related to the movement, processing, and marketing of materials (materials), as well as transactions and exchange of information (data), experienced many changes that will continue to occur, then enter a new stage of the industrial-technological revolution called Industry 4.0 (smart industry) ) or Industry 4.0 (smart industry).

Zezulka, et al. (2016) uses Industry terminology 4.0 in three interrelated factors, namely [19]: (i) digitalization and simple technical integration - economic relations with complex technical - complex economic networks, (ii) digitization of product and service offerings, and (iii) new market model. All of these human activities are interconnected with many communication systems today. The most widely used technologies are the Internet of Things (IoT), the Internet of Service (IoS), and the Internet of People (IoP) which rely on Physical-Cyber-Systems (Cyber-Physical-Systems). This technology enables communication entities (in an Industry 4.0 environment) to link to each other and utilize data from producers during the system life cycle without being limited by the company and country barriers. All related parties can obtain relevant information and data at any time so that they can more clearly know the developments that occur in supply, processing, and transportation as the basis for business planning, management, and evaluation. The pattern of communication also changes not only between humans (Customer to Customer / C2C) but also between humans and machines (Costumer to Machine / C2M) and between machine to machine (M2M) [20]. M2M and smart products are not treated as independent parts but are supporters of IoT. Smart products are sub-components of the Physical-Cyber System [21].

This transformation will dramatically affect the management of the organization according to incentives, environmental configuration, and new contexts. This revolution caused profound changes, not only in industry but also in society, in rhythm and economic outlook, in the way work was planned and operationalized, as well as ways of working that had to be oriented to human-machine interaction, etc.

The internet transformation of the digital industry is still on-going, but artificial intelligence, big data, and connectivity show certainty of a new round of digital revolution [22]-[23]. Industry 4.0 is on its way and will have an important influence on the complete transformation of the industry because it represents progress on three points, namely: (i) Digitizing production - information systems for production management and planning; (ii) Automation - a system for data acquisition from production lines and using machines; and (iii) Linking manufacturing sites in a comprehensive supply chain.

The company corrects its weaknesses using partnership relationship management to maximize the performance of its supply chain. It is evident from the results of research that by sharing information resources and using e-processes, companies can improve their close relationships with partners to improve competitiveness and can maximize the performance of their supply chains [24].

This era makes the production process of the company to be low cost, this era can create abundance which is characterized by free economy and sharing economy. Free economy including sending free letters via e-mail, free long-distance calls via WA, free books via pdf drive, free movies, and music via youtube, all this will be massive happening in all fields [25]. The digital economy is fully supported by the birth of the internet while changing people's shopping patterns, namely by the existence of e-commerce [26]. E-commerce as one of the revolutions in the field of trade has become a medium of buying and selling from books to heavy equipment; from food to clothing accessories

The role of technology in this era is changing the way humans do many jobs ranging from communicating, transacting, producing, innovating, and traveling. Many of these activities can be accomplished through digital technology using computers and smartphones to speed up the process and save resources. The new era also gave rise to a technology system based on artificial intelligence that made the supply chain more effective.

### 3.4 Marketing and Services

Every product produced by the company will then be distributed to consumers, the process of distributing these goods through a process that is often referred to as marketing. Marketing is defined as a process of identification and meeting between people and their needs, the process through the management of functions and creative processes, communication, and delivery of value of the product to consumers, to produce a mutually beneficial relationship [27]. Meanwhile, Marketing Management is defined as an art and science of choosing marketing targets and taking, capturing, and developing consumers through creativity in communicating with consumers.

With this understanding, it can be understood that the key to success in marketing depends on two things, namely the consumer market and the product market. The consumer market will receive many goods and services with various brands. The strength of the brand or brand will depend on the quality, appearance (packing), availability, and support it with attractive communication and reliable service. While the product market must be able to provide education and information in a good and professional manner, with the hope that producers will be able to make and sell products to consumers by making a profit. In this process, there is one important thing, namely communication. Communication between producers and consumers can be done one of them through advertising. With these advertisements, consumers understand the product and strive to buy it.

While marketing communication can be interpreted "All of the marketing mixes that involve communication between organizations/companies and the target audience elements in all forms that are intended for performance marketing" [28]. From the basic concept of marketing, we can understand the relationship between producers and consumers very important in the company's success in marketing its products in the market.

One of the units in the company that is very affected in disruption is marketing. The marketing sector spearheads the company that is most quickly affected. One form of the influence of disruption in marketing is the presence of an online marketing system. The marketing system is called online because it uses the internet as a means of marketing goods or service products to consumers. With the ease of accessing the internet, everyone can see, choose, and finally decide to buy, just by looking at the various types of products displayed at various prices. And in a matter of seconds, anyone can decide whether or not to buy the goods he has seen.

The increasing number of internet users in Indonesia has had a big impact on the company. Many companies make changes to their business processes, so the products they sell can be easily absorbed in the market. One of the changes made by many companies is a product marketing medium or media. The company seeks to bring its product-market quickly to consumers, they use the internet as a means to market and sell their products. Digital marketing is a marketing activity including branding that uses various web-based media such as blogs, websites, e-mails, AdWords, or social networking [29]. Of course, digital marketing isn't just talking about internet marketing. while the internet network is a wide and large worldwide computer network, connecting computer users from one place to another around the world. The internet which stands for inter and networking is a collection of computer networks and various types of types that communicate with one another using a standard of communication [30].

The market which was conventional in nature has changed into a digital form of market, without form, but has a wider range without limits. Of course, this is a form of new innovation in the industry. However, in the era of disruption of new innovations that have been formed, it is not enough to just stop there, online marketing innovations must continue to be developed, as changes accelerate. It is not enough for a company to only innovate by changing its shape, size, and design but more importantly is the ability to read the future [31]

This marketing system becomes very simple and simple but is able to provide complete, clear, and interesting information for consumers. Consumers are given a wider shopping space without limits of space and time, wherever whenever consumers can see and choose the items to be purchased. This is very different from conventional marketing, conventional marketing is very dependent on space and time, because the market cannot be moved anywhere, and time to access is very limited.

Marketing and food retail services in this modern era can use online media including gofood, grabfood, twiter, instagram, facebook, and several other media, and based on buying and selling applications such as Tokopedia, Bukalapak, and others. By using this online media, a product can be marketed and communicated to consumers quickly without knowing the time limit. The excellence of social media is the main key to marketing success for goods and services. This makes it easier for marketing and food retail services in this modern era.

Using online media services such as gofood & grabfood has several advantages that make it easy for consumers, among others, as follows:

1. Service Performance
   The service performance factor consists of easy-to-understand online order attributes, fast order confirmation messages, fast loading time, error-free order processing, fast order delivery, and order return policies.

After the customer completes an order, the next gofood / grabfood system looks for a driver for the customer. An order confirmation will be made by the driver either by telephone or through the chatroom on the platform page. Some of these steps are carried out to suppress the processing of order errors made by the driver so that order returns can be avoided.

2. Ease Of Access
A variety of partner product offerings, promotional product offers, other unique product recommendations, and support in various ways of payment. This service category is very helpful for customers because it contains the latest merchant or partner information around customers. Therefore, customers can not only enjoy food but get the latest culinary information updates and promotional product offers.
Support of various payment methods is done through gofood / grabfood user service categories, including cash and using virtual money with the names go-pay, ovo, funds. This virtual or electronic money payment method is more practical, many attractive promos are offered to motivate consumers to use the payment system.

3. Interaction
Availability of online chatrooms, availability of product images, and good communication interactions. Communication interaction is very important in go-food platform services. The availability of online chatrooms that makes it easy for customers and drivers to communicate. Questions about customer food orders have become easier and more practical.
The availability of product images in retail online is a very important thing in influencing consumers to buy. Based on internal go-jek surveys, transactions can increase by 22.85%, in addition, 70.4% of the menu added photos have higher-order numbers. Restaurants / go-jek partners who put photos for their menus on the go-food application, get a very good response related to the sale of their products.

4. Information
Information on product availability and information provided is accurate. The importance of menu information and partner or restaurant information is a major consideration for consumers using online application services. The assessment information can be a reference for consumers to buy quality food in terms of taste and service through the platform. On the other hand, this assessment system can help the development of partners who are just starting a business. It is intended that new sellers are able to learn so that they can be trusted by customers.

5. Securities
With each use of an online account, security is a key factor in maintaining consumer confidence. With the existence of virtual money payments on the platform gofood / grabfood, which is known as a go-pay account, ovo, funds. The ownership of go-pay, ovo, and fund accounts is a form of consumer trust in service providers in the form of money and customer information.

### 3.5 Warehouse, Distribution and Transportation

American marketing association defines distribution channels as the structure of inter-company organizational units and extra-company, wholesale, and retail agency agents and dealers, through the name of a commodity, product or service marketed services [32]. From the expert understanding above, it can be concluded that the distribution channel is an activity of moving goods from producers to consumers effectively and efficiently where, the process of channeling the goods can pass through various collaborations between producers, agents, large traders, and small traders and even producers can also cut off supply chain paths to sell directly to consumers.

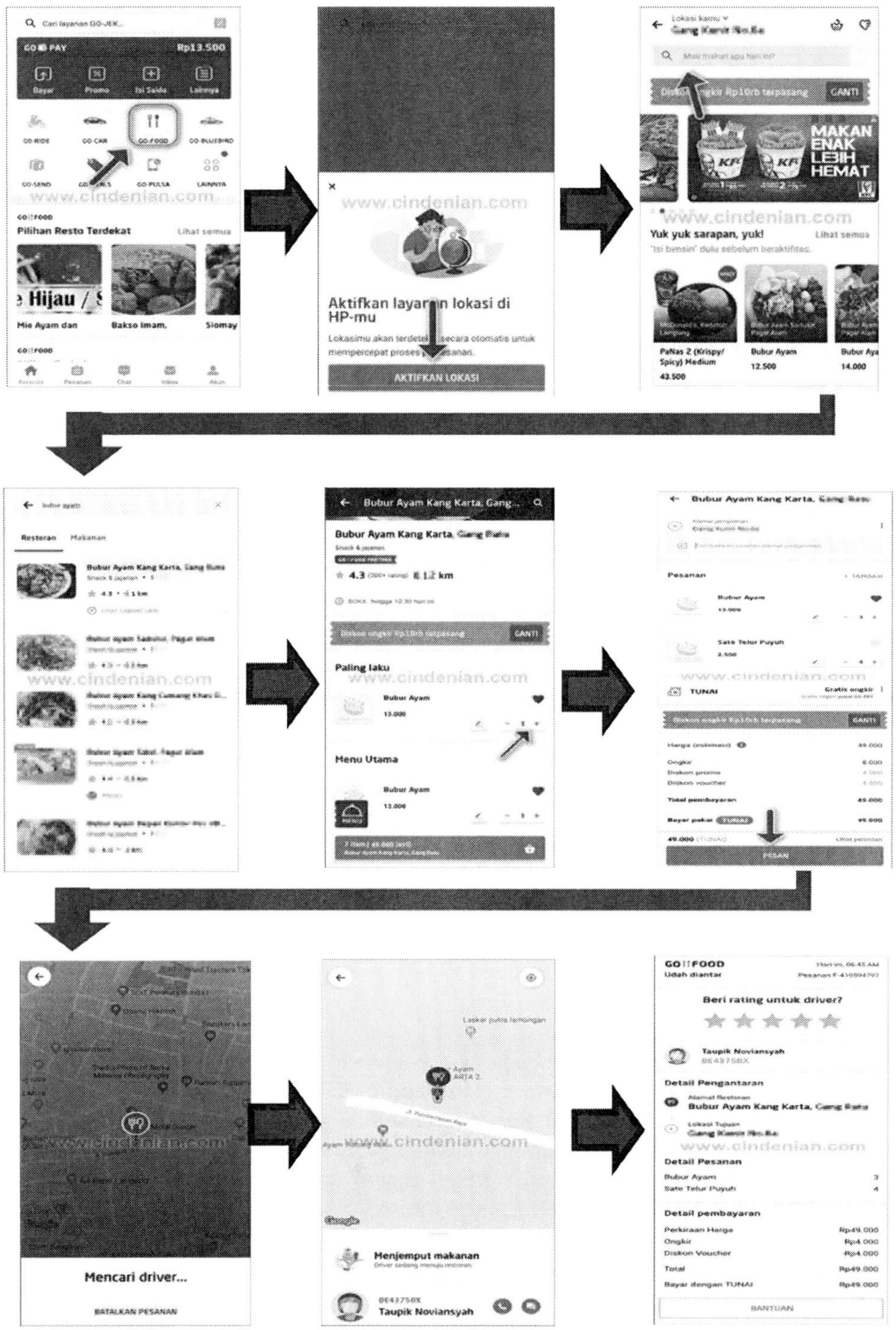


Figure 2. New Era of Supply Chain Flow in Food Retail Industry in Indonesia

The logistics chain starts from and ends at the warehouse which was before the factory and after it to the consumer. Therefore, warehouses have always been an important centre of the flow of goods in and determine the performance of the supply chain. Industry 4.0 requires that warehouses serve as the main source of competitive advantage for logistics providers. The adoption of the Industry 4.0 paradigm will introduce extraordinary changes in the way warehouse works today. The introduction of 'smart' management during the adoption and implementation of an appropriate Warehouse Management System that will transform warehouse activities into future logistical requirements in accordance with Industry 4.0 paradigm. Actors and stakeholders in the supply chain need to be integrated to ensure overall coordination and harmony between all phases of the value chain. For example, transportation will be able to communicate their position and predict arrival times to an intelligent warehouse management system, so that they can select and prepare docking slots, optimize the delivery of just-in-time and just-in-sequence.

The development of information and communication technology is growing rapidly, including online-based transportation and distribution services. Easily sending goods in this case food with the time, distance, and costs that have been determined to make the company and consumers trust each other. The process of distribution of goods is in collaboration with application-based providers such as go-food. Grab-food. Go-sent, go-mart, and other applications that make it easy for retailers and consumers.

### 3.6 The Effectiveness of the New Era of Supply Chains on Retail Food

Effectiveness is a measure that states how far targets (quantity, quality and time) have been achieved by management, which targets have been determined in advance [33].

The Main Purpose of Supply Chain Management, among others:

1. Delivery or delivery of products on time to satisfy consumers Main Purpose of Supply Chain Management, among others.
2. Reducing costs
3. Improve all results of the entire supply chain (not just one company).
4. Reducing time.
5. Concentrate on planning and distribution activities

From the objectives of supply chain management and the definition of effectiveness above descriptively, the development of this new era has experienced many improvements in terms of delivery, cost, and time so that the more effective the supply chain in this food retail industry.

## 4. Conclusion

From the results of this study, it can be concluded that the new era of the supply chain in the retail food industry in Indonesia is included in the downstream supply chain which focuses on distribution, warehousing, transportation, and service. The role of technology in this era is changing the way humans do many jobs ranging from communicating, transacting, producing, innovating, and traveling. Many of these activities can be accomplished through digital technology using computers and smartphones to speed up the process and save resources.

Marketing and food retail services in this modern era can use online media including gofood, grabfood, twiter, instagram, facebook, and several other media, and based on buying and selling applications such as Tokopedia, Bukalapak, and others. By using this online media, a product can be marketed and communicated to consumers quickly without knowing the time limit.

An appropriate Warehouse Management System that will transform warehouse activities into future requirements of logistics by Industry 4.0 paradigm. The process of distribution of goods is in collaboration with application-based providers such as go-food, Grab-food, Go-sent, go-mart, and other applications that make it easier for retailers and consumers to transact. The development of this new era has experienced many improvements in terms of shipping, cost, time so that the more days the more effective the supply chain in the food retail industry.

## References

[1] Lanjouw P, Quizon J and Sparrow R 2001 Non-agricultural earnings in peri-urban areas of Tanzania: evidence from household survey data *Food policy* **26** (4) 385-403
[2] Pandin M L 2009 Potret Bisnis Ritel Di Indonesia: Pasar Modern *Economic Review* **215** 1-12
[3] Arifin Z and Kuncoro M 2003 Spatial Concentration and Dynamics of The Growth of Manufacturing Industries In East Java *Sosiohumanika* **16** (1) 119-137
[4] Chopra S and Meindl P 2007 Supply chain management. Strategy, planning & operation *Das summa summarum des management* 265-275
[5] Trisilawaty C, Marimin M and Achsani N A 2011 Analisis Optimasi Rantai Pasok Beras dan Penggunaan Gudang Di Perum BULOG Divre DKI Jakarta *Jurnal Pangan* **20** (2) 177-198
[6] Loebbecke C and Picot A 2015 Reflections on societal and business model transformation arising from digitization and big data analytics: A research agenda *The Journal of Strategic Information Systems* **24** (3) 149-157
[7] Russell R S and Taylor B W 2019 *Operations and supply chain management* New York: John Wiley & Sons
[8] Sinulingga S 2009 *Perencanaan dan Pengendalian Produksi* Yogyakarta: Graha Ilmu
[9] Wrigley N, Warm D and Margetts B 2003 Deprivation, diet, and food-retail access: Findings from the Leeds 'food deserts' study *Environment and Planning A* **35** (1) 151-188
[10] Kotler P and Lane K K 2006 *Manajemen Pemasaran* Jakarta: Ghalia
[11] Berman Barry and R Evans Joel 2001 *Retail management, a strategic approach* New York: Pearson Education Limited.
[12] Levy M and Weitz B A 2001 *Retailing Management* Berlin: Irwin Inc.
[13] Turban E 2006 *Decision support and expert system: management support system* New Jersey: Prentice-Hall international inc
[14] Špička J 2016 Market concentration and profitability of the grocery retailers in Central Europe *Central European Business Review* **5** (3) 5-24
[15] La Cour L F and Møllgaard H P 2002 Meaningful and measurable market domination *Journal of Law and Economics* **14** (2) 99-127
[16] Scott W R 2005 Institutional theory: Contributing to a theoretical research program *Great minds in management: The process of theory development* **37** (2) 460-484
[17] Hofmann E and Rüsch M 2017 Industry 4.0 and the current status as well as future prospects on logistics *Computers in industry* **89** 23-34
[18] Qrunfleh S and Tarafdar M 2014 Supply chain information systems strategy: Impacts on supply chain performance and firm performance *International Journal of Production Economics* **147** 340-350
[19] Zezulka F, Marcon P, Vesely I and Sajdl O 2016 Industry 4.0–An Introduction in the phenomenon *IFAC-PapersOnLine* **49** (25) 8-12
[20] Roblek V, Meško M and Krapež A 2016 A complex view of industry 4.0 *Sage Open* **6** (2) 2159.
[21] Alam K M and El Saddik A 2017 C2PS: A digital twin architecture reference model for the cloud-based cyber-physical systems *IEEE access* **5** 2050-2062
[22] Schwertner K 2017 Digital transformation of business *Trakia Journal of Sciences* **15** (1) 388-393
[23] Frank A G, Mendes G H, Ayala N F and Ghezzi A 2019 Servitization and Industry 4.0 convergence in the digital transformation of product firms: A business model innovation perspective *Technological Forecasting and Social Change* **141** 341-351
[24] Robinson C J and Malhotra M K 2005 Defining the concept of supply chain quality management and its relevance to academic and industrial practice *International journal of*

*production economics* **96** (3) 315-337
[25] Etgar M 2008 A descriptive model of the consumer co-production process *Journal of the academy of marketing science* **36** (1) 97-108
[26] Mesenbourg T L 2001 Measuring the digital economy. *US Bureau of the Census* **1** 1-19
[27] Calantone R J and Mazanec J A 1991 Marketing management and tourism *Annals of Tourism Research* **18** (1) 101-119
[28] Prisgunanto I 2006 *Komunikasi Pemasaran: strategi dan taktik* Bogor: Ghalia Indonesia
[29] Royle J and Laing A 2014 The digital marketing skills gap: Developing a Digital Marketer Model for the communication industries *International Journal of Information Management* **34** (2) 65-73
[30] Litan R E and Rivlin A M 2001 Projecting the economic impact of the Internet *American Economic Review* **91** (2) 313-317
[31] Wang Y, Hong A, Li X and Gao J 2020 Marketing innovations during a global crisis: A study of China firms' response to COVID-19 *Journal of Business Research* **116** 214-220
[32] Sainathuni B, Parikh P J, Zhang X and Kong N 2014 The warehouse-inventory-transportation problem for supply chains *European Journal of Operational Research* **237** (2) 690-700
[33] Swaminathan J M and Tayur S R 2003 Models for supply chains in e-business *Management Science* **49** (10) 1387-1406

IOP Conf. Series: Materials Science and Engineering 1115 (2021) 012020 doi:10.1088/1757-899X/1115/1/012020

# Optimal Knots Point and Bandwidth Selection in Modeling Mixed Estimator Nonparametric Regression

**A T R Dani, V Ratnasari*, and I N Budiantara**

Department of Statistics, Faculty of Science and Data Analytics, Institut Teknologi Sepuluh Nopember, Surabaya, East Java, Indonesia

*Corresponding author: vita_ratna@statistika.its.ac.id

**Abstract.** The nonparametric regression model that is currently developing is limited to only one estimator form that is used. This happens, due to the assumption from the researcher that each of these predictors has the same data pattern. However, when we model the predictor variable with the response variable, what happens is that each predictor may have a different pattern, so it is necessary to develop a mixed estimator. The mixed estimator used in this study is the truncated spline and the Gaussian Kernel. Furthermore, it becomes a separate problem, when combining the truncated spline with the Gaussian Kernel, so it must determine the optimal knot point and bandwidth. Determine the optimal knot point and bandwidth is very important. The methods used are Cross-Validation (CV), Generalized Cross-Validation (GCV), and Unbiased Risk (UBR). In this study, the nonparametric regression model was applied to the data on the percentage of poor people in Districts/Cities on Borneo Island in 2019. Based on analysis, the modeling of the nonparametric regression approach with a mixed estimator of the truncated spline and the Gaussian Kernel uses the GCV method which gives the best results. This is supported by the coefficient of determination obtained by 90.52% and the MSE value of 1.10. This means that the nonparametric regression model built can explain the effect of the predictor variable on the response variable, namely the percentage of poor people by 90.52%.

## 1. Introduction

Regression analysis is a statistical method used to determine the pattern of relationships between one or more variables. The variables used in the regression analysis consist of response and predictor variables [1]. The purpose of regression analysis is to find the form of parameter estimation that matches the shape of the regression curve [2]. There are several approaches developed by researchers in statistical modeling using regression analysis, namely parametric regression, nonparametric regression, and semiparametric regression [3]. Recently, the approach using nonparametric regression has received a lot of attention from researchers [4]. The nonparametric regression approach does not depend on the assumption of a certain regression curve shape, so it will provide high flexibility, where it is expected that the data will adjust itself to the estimated form of the regression curve without being influenced by the researcher's subjectivity [5].

In nonparametric regression modeling, there are several estimators developed by researchers including Kernel [6, 7], Spline [8, 9], and Fourier series [10, 11]. Among the several methods, nonparametric regression with the spline and Kernel approach is an estimator method that is usually used in research. One of the advantages of the spline is that it is flexible, meaning that this function tends to find its own form of data estimation wherever the data pattern moves [12]. Apart from spline, the nonparametric regression approach with Kernel functions is also of interest. The kernel has a more flexible form, very good at modeling data that does not have a certain pattern [13].

The nonparametric regression model that is currently developing is limited to only one form of an estimator for all predictor variables used. But in fact, when we model predictor variables with response variables, what happens is that each predictor may have a different pattern [14]. Then a mixed estimator was developed by [14, 15, 16].

In nonparametric regression truncated spline, determine the optimal knot point is very important and crucial [17]. The same thing applies to Kernel regression, where the thing that must be considered is the choice of bandwidth [13]. It becomes a separate problem when using a mixed estimator model, especially combining the spline with the Kernel, so it must determine the optimal knot point and bandwidth. Determination of the optimal point of knots and bandwidth will greatly affect the regression curve that will be formed. To choose the optimal knot point and bandwidth parameters, you can use various methods that have been developed by researchers, for example Cross-Validation (CV) by [18], Generalized Cross-Validation (GCV) by [8], and Unbiased Risk (UBR) by [19].

In this study, a nonparametric regression model was applied to analyze problems related to poverty. Based on BPS data (2020), in March 2019 the percentage of poor people in Indonesia was 9.41% with the number of poor people in March 2019 of 25.14 million people. Poverty in brief is often understood as a low level of welfare [20]. The problem of poverty is a very serious problem faced by countries in the world, especially developing countries including Indonesia [21]. Poverty eradication is the main goal in the Sustainable Development Goals (SDG's) for 2016 to 2030.

The purpose of this research is to examine the estimator form of a mixed nonparametric regression truncated spline and Gaussian Kernel, which is then applied to the data on the percentage of poor people in Districts/Cities on Borneo Island in 2019. The method of selecting the optimal knot point and bandwidth uses CV, GCV, and UBR. The nonparametric regression approach, especially the estimator of the mixed of the truncated spline and the Gaussian Kernel, using the correct predictor variables, is expected to be a statistical model capable of analyzing poverty problems.

## 2. Methodology

### 2.1. Truncated Spline Nonparametric Regression

A truncated spline nonparametric regression approach is used if the regression curve between the response and the predictor variable does not form a pattern [22]. The data characteristic of the truncated spline estimator is a changing pattern at certain sub-intervals [2]. The study of nonparametric regression truncated multivariable spline on paired data $(x_{1i}, x_{2i}, \dots, x_{qi}, y_i)$ which assumed the relationship between the predictor variable and the response variable followed a nonparametric regression model.

$$y_i = f(x_{1i}, x_{2i}, \dots, x_{qi}) + \varepsilon_i, i = 1, 2, \dots, n \tag{1}$$

Furthermore, the regression curve is assumed to be additive:

$$f(x_{1i}, x_{2i}, \dots, x_{qi}) = f_1(x_{1i}) + f_2(x_{2i}) + \dots + f_q(x_{qi}) = \sum_{p=1}^{q} f_p(x_{pi}), i = 1, 2, \dots, n \tag{2}$$

The component of degree 1 (linear) truncated spline nonparametric regression for each predictor variable can be written in general:

$$f_p(x_p) = \delta_{0p} + \sum_{j=1}^{m}\sum_{p=1}^{q} \delta_{jp} x_p^j + \sum_{k=1}^{r}\sum_{p=1}^{q} \phi_{(m+k)p}(x_p - K_{kp})_+^m \tag{3}$$

Equation (3) can be presented in the form of a matrix:

$$\sum_{p=1}^{q} \vec{f}_p(x_{pi}) = \mathbf{X_0}\vec{\delta} + \mathbf{X_1}(\mathbf{K_1})\vec{\phi}_1 + \mathbf{X_2}(\mathbf{K_2})\vec{\phi}_2 + \ldots + \mathbf{X_q}(\mathbf{K_q})\vec{\phi}_q \quad (4)$$

Based on Equation (4), so $\sum_{p=1}^{q} \vec{f}_p(x_{pi}) = \mathbf{X}(\phi)\vec{\beta}$.

### 2.2. Gaussian Kernel Nonparametric Regression

The Kernel estimator is a linear estimator that is similar to other estimators, but the difference is that there is bandwidth in the Kernel [9]. Suppose given paired data $t_i$ and $y_i$, where $i = 1, 2, \ldots, n$ follows a nonparametric regression model. Furthermore, the relationship $t_i$ and $y_i$ can be modeled functionally into:

$$y_i = h(t_i) + \varepsilon_i ; i = 1, 2, \ldots, n \quad (5)$$

The regression curve of $h(t_i)$ is an unknown shape. It to be approximated by the Kernel function, with the regression curve estimation is presented in Equation (6):

$$\hat{h}_\alpha(t) = n^{-1}\left[\sum_{i=1}^{n} W_{\alpha i}(t)\right] y_i \quad (6)$$

where $W_{\alpha i}(t) = \dfrac{K_\alpha(t - t_i)}{n^{-1}\sum_{i=1}^{n} K_\alpha(t - t_i)}$ and $K_\alpha(t - t_i) = \dfrac{1}{\alpha} K\left(\dfrac{t - t_i}{\alpha}\right)$.

The Kernel function used is the Gaussian Kernel: $K(z) = \dfrac{1}{\sqrt{2\pi}} \exp\left(\dfrac{1}{2}\left(-z^2\right)\right); I_{[-\infty,\infty]}(z)$.

Based on the Kernel function that applies to each $t = t_1, t = t_2, \ldots, t = t_n$, then Equation (6) can be written in form of a matrix to be $\mathbf{D}(\alpha)\vec{Y}$.

## 3. Result and Discussion

### 3.1. Mixed Estimator Nonparametric Regression Truncated Spline and Gaussian Kernel

Given a pair of data with a predictor variable of $q$ for the truncated spline component and only 1 for the Kernel component such that it can be written $(x_{1i}, x_{2i}, \ldots, x_{qi}, t_i, y_i)$, with $i = 1, 2, \ldots, n$. A nonparametric regression model of truncated spline and Gaussian Kernel can be written:

$$y_i = \sum_{p=1}^{q} f_p(x_{pi}) + h(t_i) + \varepsilon_i \quad (7)$$

Based on Equation (7), it can be written briefly $y_i = \mu(x_i, t_i) + \varepsilon_i$. Furthermore, it can be presented in the form of a matrix to be:

$$\vec{Y} = \mathbf{X}(\phi)\vec{\beta} + \mathbf{D}(\alpha)\vec{Y} + \vec{\varepsilon} \quad (8)$$

The method of parameter estimation used in this study is Ordinary Least Squares (OLS). The sum squared of error can be found with:

$$
\begin{aligned}
Q(\vec{\beta} \mid \phi, \alpha) &= \sum_{i=1}^{n} \vec{\varepsilon}_i^{\,2} = \vec{\varepsilon}^{\,T} \vec{\varepsilon} \\
&= \left\| [\mathbf{I} - \mathbf{D}(\alpha)] \vec{Y} \right\|^2 - 2\vec{\beta}^T \mathbf{X}(\phi)^T [\mathbf{I} - \mathbf{D}(\alpha)] \vec{Y} + \vec{\beta}^T \mathbf{X}(\phi)^T \mathbf{X}(\phi) \vec{\beta}
\end{aligned} \tag{9}
$$

To get an estimator of $\vec{\beta}$, you can do a partial derivative of Equation (9) on $\vec{\beta}$ as follows:

$$
\begin{aligned}
\frac{\partial Q(\vec{\beta} \mid \phi, \alpha)}{\partial \vec{\beta}} &= \frac{\partial \left( \left\| [\mathbf{I} - \mathbf{D}(\alpha)] \vec{Y} \right\|^2 - 2\vec{\beta}^T \mathbf{X}(\phi)^T [\mathbf{I} - \mathbf{D}(\alpha)] \vec{Y} + \vec{\beta}^T \mathbf{X}(\phi)^T \mathbf{X}(\phi) \vec{\beta} \right)}{\partial \vec{\beta}} \\
&= -2\mathbf{X}(\phi)^T [\mathbf{I} - \mathbf{D}(\alpha)] \vec{Y} + 2\mathbf{X}(\phi)^T \mathbf{X}(\phi) \vec{\beta}
\end{aligned} \tag{10}
$$

Equation (10) is equal to zero:

$$
\frac{\partial Q(\vec{\beta} \mid \phi, \alpha)}{\partial \vec{\beta}} = \vec{0} \tag{11}
$$

Then it will be obtained:

$$
\mathbf{X}(\phi)^T \mathbf{X}(\phi) \hat{\vec{\beta}} - \mathbf{X}(\phi)^T [\mathbf{I} - \mathbf{D}(\alpha)] \vec{Y} = \vec{0} \tag{12}
$$

So that

$$
\mathbf{X}(\phi)^T \mathbf{X}(\phi) \hat{\vec{\beta}} = \mathbf{X}(\phi)^T [\mathbf{I} - \mathbf{D}(\alpha)] \vec{Y} \tag{13}
$$

Then we will get an estimator for the following:

$$
\hat{\vec{\beta}} = \left[ \mathbf{X}(\phi)^T \mathbf{X}(\phi) \right]^{-1} \mathbf{X}(\phi)^T [\mathbf{I} - \mathbf{D}(\alpha)] \vec{Y} \tag{14}
$$

Equation (14) can be written as $\hat{\vec{\beta}} = \mathbf{C}(\phi, \alpha) \vec{Y}$, with $\mathbf{C} = \left[ \mathbf{X}(\phi)^T \mathbf{X}(\phi) \right]^{-1} \mathbf{X}(\phi)^T [\mathbf{I} - \mathbf{D}(\alpha)]$.
Truncated spline curve estimator form is:

$$
\sum_{p=1}^{q} \vec{f}_p(x_p) = \mathbf{X}(\phi) \vec{\beta} \tag{15}
$$

And we can write it:

$$
\sum_{p=1}^{q} \vec{f}_p(x_p) = \mathbf{X}(\phi) \vec{\beta} \rightarrow \mathbf{X}(\phi) \left[ \mathbf{X}(\phi)^T \mathbf{X}(\phi) \right]^{-1} \mathbf{X}(\phi)^T [\mathbf{I} - \mathbf{D}(\alpha)] \vec{Y} \tag{16}
$$

Equation (16) can be witten as $\sum_{p=1}^{q} \vec{f}_p(x_{pi}) = \mathbf{A}(\phi, \alpha) \vec{Y}$. Then we can write the Gaussian Kernel estimator form $\hat{\vec{h}}_\alpha(t) = \mathbf{D}(\alpha) \vec{Y}$.

Based on Equation (14) and the estimator form of each component, we will get a mixed estimator of the nonparametric regression truncated spline and Gaussian Kernel as follows:

$$
\begin{aligned}
\hat{\vec{\mu}}_{\phi,\alpha}(x,t) &= \sum_{p=1}^{q} \hat{\vec{f}}_p(x_{pi}) + \hat{\vec{h}}_\alpha(t) \\
&= [\mathbf{A}(\phi, \alpha) + \mathbf{D}(\alpha)] \vec{Y} \\
&= \mathbf{B}(\phi, \alpha) \vec{Y}
\end{aligned} \tag{17}
$$

Matrix $\mathbf{B}(\phi,\alpha)$ is highly dependent on $\mathbf{A}(\phi,\alpha)$ which is a truncated spline component with knots $\phi=(K_1,K_2,...,K_r)^T$ and $\mathbf{D}(\alpha)$ which is a component of the Gaussian Kernel with bandwidth.

### 3.2. Methods for Selecting Knot Points and Bandwidth

The method used in the process of selecting optimal knots and bandwidth points is Cross-Validation (CV), Generalized Cross-Validation (GCV), and Unbiased Risk (UBR). The CV, GCV, and UBR equations have been modified based on the nonparametric regression model used, namely the mixed estimator.

1. *Cross-Validation (CV)*

$$CV(\phi_{opt},\alpha_{opt})=\frac{1}{n}\sum_{i=1}^{n}\left[\frac{(y_i-\hat{y}_i)}{\left(1-\left[\mathbf{A}_{ii}(\phi,\alpha)+\mathbf{D}_{ii}(\alpha)\right]\right)}\right]^2 \quad (18)$$

2. *Generalized Cross-Validation (GCV)*

$$GCV(\phi_{opt},\alpha_{opt})=\left[\frac{MSE(\phi,\alpha)}{\left(n^{-1}trace\left\{\mathbf{I}-\left[\mathbf{A}(\phi,\alpha)+\mathbf{D}(\alpha)\right]\right\}\right)^2}\right] \quad (19)$$

3. *Unbiased Risk (UBR)*

$$UBR(\phi_{opt},\alpha_{opt})=n^{-1}\left\{\begin{array}{l}\left\|\left(\mathbf{I}-\left[\mathbf{A}(\phi,\alpha)+\mathbf{D}(\alpha)\right]\right)\vec{Y}\right\|^2+\frac{\hat{\sigma}^2}{n}trace\left[\mathbf{I}-\left[\mathbf{A}(\phi,\alpha)+\mathbf{D}(\alpha)\right]\right]^2+\\ \frac{\hat{\sigma}^2}{n}trace\left[\left[\mathbf{A}(\phi,\alpha)+\mathbf{D}(\alpha)\right]^2\right]^2\end{array}\right\} \quad (20)$$

with $\hat{\sigma}^2=\dfrac{\left\|\left(\mathbf{I}-\left[\mathbf{A}(\phi,\alpha)+\mathbf{D}(\alpha)\right]\right)\vec{Y}\right\|^2}{trace\left[\left(\mathbf{I}-\left[\mathbf{A}(\phi,\alpha)+\mathbf{D}(\alpha)\right]\right)\vec{Y}\right]}$

### 3.3. Data Application

The data used in this study are secondary data for 2019 published by the Badan Pusat Statistik (BPS). The observation unit is the Districts/Cities on the Borneo Island, which currently of 5 Provinces. The variables used in this study are shown in Table 1.

The initial observations in this study were 56 Districts/Cities on the Borneo Island. Furthermore, the cities of Bontang, Banjarmasin, and Pontianak were not used in research observations because they were outlier data. The observations in this study were left 53 Districts/Cities.

The pattern of relationship between the response variable and each predictor variable can be seen from the scatter plot in Figure 1.

Based on Figure 1, a nonparametric regression model will be applied using a mixed estimator of the truncated spline and Gaussian Kernel. This is because there is a pattern of relationship between predictor and response variables that tend to fluctuate in certain sub-intervals $(x_1-x_6)$ and there are also undefined pattern $(t_1)$.

**Table 1** Research variable

| Variable | Notation | Description |
|---|---|---|
| Response | $y$ | Percentage of Poor Population |
| Predictor | $x_1$ | Life Expectancy |
| | $x_2$ | Open Unemployment Rate |
| | $x_3$ | Average Length of Schooling |
| | $x_4$ | Economic Growth Rate |
| | $x_5$ | Percentage of Households with Adequate Access to Sanitation |
| | $x_6$ | Percentage of Households with Access to Adequate Drinking Water |
| | $t_1$ | Population Density |

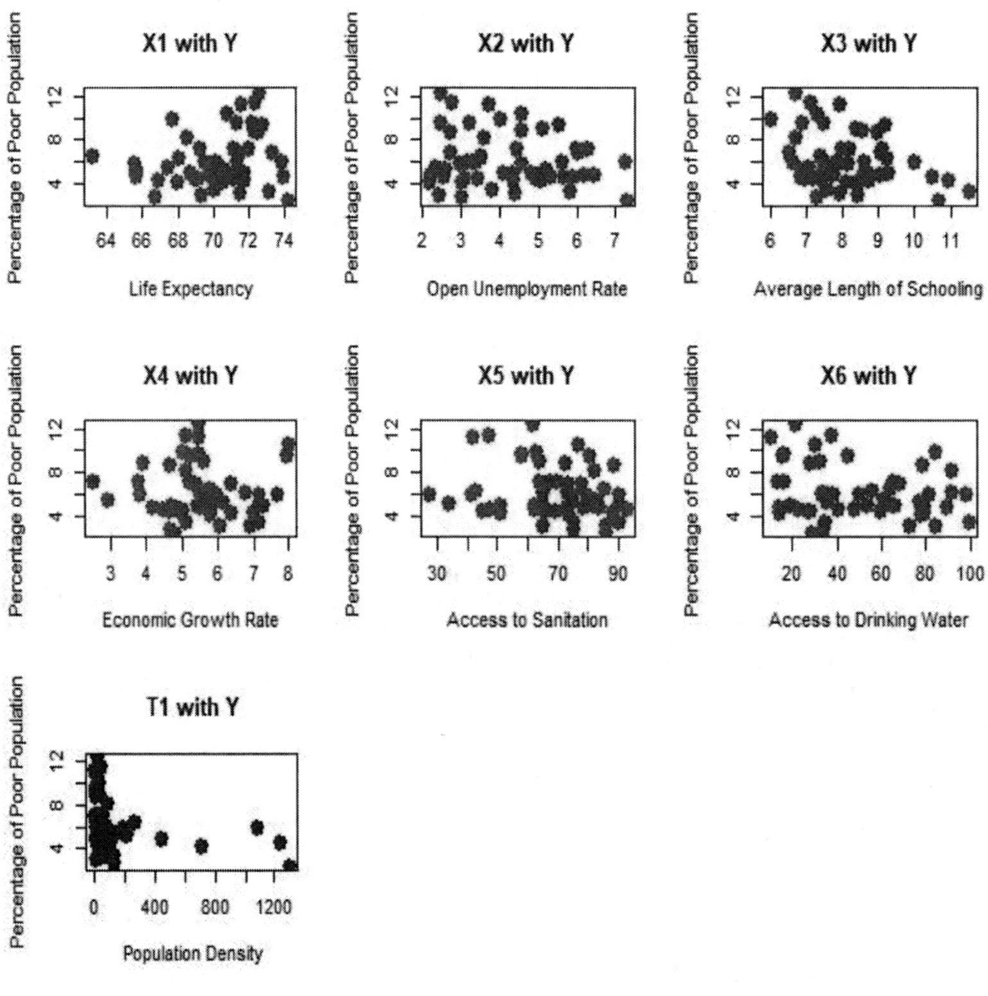


**Figure 1.** Scatter plot between predictor and response variables

The best estimator model for the mixed of truncated spline and Gaussian Kernel, will be selected by looking at the smallest CV, GCV and UBR values from several models based on the number of knot points and bandwidth. This study uses the same number of points of knots for each predictor variable used, namely 1 to 3 points of knots.

**Table 2** Cross-Validation (CV) method

| | Knot Point Locations | | | | | | Bandwidth | CV | MSE | $R^2$ |
|---|---|---|---|---|---|---|---|---|---|---|
| | $x_1$ | $x_2$ | $x_3$ | $x_4$ | $x_5$ | $x_6$ | $t_1$ | | | |
| 1 knot | 69.65 | 5.16 | 9.23 | 5.73 | 65.71 | 62.46 | 2.55 | 4.27 | 2.53 | 66.00% |
| 2 knots | 67.01 | 3.92 | 7.90 | 4.40 | 49.85 | 40.97 | 2.18 | 4.27 | 2.31 | 73.35% |
| | 73.04 | 6.93 | 11.13 | 7.61 | 88.36 | 93.17 | | | | |
| 3 knots | 65.12 | 3.03 | 6.95 | 3.46 | 38.53 | 25.61 | 1.41 | 4.18 | 1.82 | 83.80% |
| | 66.63 | 3.74 | 7.71 | 4.22 | 47.59 | 37.89 | | | | |
| | 72.67 | 6.58 | 10.75 | 7.23 | 83.83 | 87.03 | | | | |

**Table 3** Generalized Cross-Validation (GCV) method

| | Knot Point Locations | | | | | | Bandwidth | GCV | MSE | $R^2$ |
|---|---|---|---|---|---|---|---|---|---|---|
| | $x_1$ | $x_2$ | $x_3$ | $x_4$ | $x_5$ | $x_6$ | $t_1$ | | | |
| 1 knot | 69.65 | 5.16 | 9.23 | 5.72 | 65.71 | 62.47 | 2.55 | 5.88 | 2.52 | 66.00% |
| 2 knots | 71.16 | 5.87 | 9.99 | 6.48 | 74.77 | 74.75 | 1.82 | 3.09 | 1.29 | 85.10% |
| | 72.29 | 6.40 | 10.56 | 7.05 | 81.56 | 83.96 | | | | |
| 3 knots | 64.37 | 2.68 | 6.57 | 3.08 | 34.00 | 19.47 | 1.09 | 1.51 | 1.10 | 90.52% |
| | 71.54 | 6.05 | 10.18 | 6.66 | 77.03 | 77.82 | | | | |
| | 72.30 | 6.40 | 10.56 | 7.05 | 81.56 | 83.96 | | | | |

**Table 4** Unbiased Risk (UBR) method

| | Knot Point Locations | | | | | | Bandwidth | UBR | MSE | $R^2$ |
|---|---|---|---|---|---|---|---|---|---|---|
| | $x_1$ | $x_2$ | $x_3$ | $x_4$ | $x_5$ | $x_6$ | $t_1$ | | | |
| 1 knot | 73.80 | 7.11 | 11.32 | 7.80 | 90.62 | 96.24 | 3.09 | $2.68\times10^{-5}$ | 2.79 | 60.13% |
| 2 knots | 71.92 | 6.22 | 10.37 | 6.86 | 79.30 | 80.89 | 2.58 | $1.33\times10^{-5}$ | 2.48 | 70.12% |
| | 73.80 | 7.11 | 11.32 | 7.80 | 90.62 | 96.24 | | | | |
| 3 knots | 64.37 | 2.68 | 6.57 | 3.08 | 34.00 | 19.47 | 1.92 | $1.79\times10^{-6}$ | 2.18 | 79.23% |
| | 71.92 | 6.22 | 10.37 | 6.86 | 79.30 | 80.89 | | | | |
| | 73.80 | 7.11 | 11.32 | 7.80 | 90.62 | 96.24 | | | | |

Based on Tables 2, 3, and 4, it is known that the CV, GCV, and UBR methods produce high $R^2$ values and low MSE values when using 3 knots.

Based on Table 3, it can be seen that by using the GCV method in the process of selecting optimal knots and bandwidth points in the mixed estimator nonparametric regression model, better results are obtained than the CV and UBR methods. This is because, with the GCV method, high performance and accuracy are obtained, the $R^2$ value is 90.52% and the MSE value is 1.10. Visualization of the value with the best mixed estimator nonparametric regression model is shown in Figure 2.

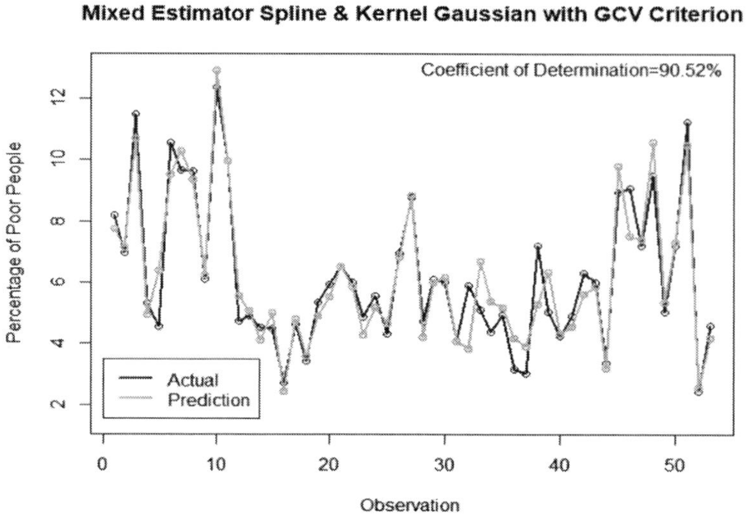


**Figure 2.** Visualization the $y$ and $\hat{y}$ of the best model

Based on Figure 2, it can be seen that the value obtained from nonparametric regression modeling used a mixed estimator of the truncated spline and the Gaussian Kernel with the GCV method tends to follow the actual data pattern. Furthermore, the nonparametric regression model using mixed estimator with three knot points obtained from the GCV method is written in Equation (21).

$$\begin{aligned}\hat{y} &= \hat{\delta}_0^* + \hat{\delta}_{11}x_{1i} + \hat{\delta}_{12}x_{2i} + \hat{\delta}_{13}x_{3i} + \hat{\delta}_{14}x_{4i} + \hat{\delta}_{15}x_{5i} + \hat{\delta}_{16}x_{6i} + \hat{\phi}_{21}(x_{1i} - 64,37)_+ + \hat{\phi}_{31}(x_{1i} - 71,54)_+ + \\ &\hat{\phi}_{41}(x_{1i} - 72,30)_+ + \hat{\phi}_{22}(x_{2i} - 2,68)_+ + \hat{\phi}_{32}(x_{2i} - 6,05)_+ + \hat{\phi}_{42}(x_{2i} - 6,40)_+ + \hat{\phi}_{23}(x_{3i} - 6,57)_+ + \\ &\hat{\phi}_{33}(x_{3i} - 10,18)_+ + \hat{\phi}_{43}(x_{3i} - 10,56)_+ + \hat{\phi}_{24}(x_{4i} - 3,08)_+ + \hat{\phi}_{34}(x_{4i} - 6,66)_+ + \hat{\phi}_{44}(x_{4i} - 7,05)_+ + \\ &\hat{\phi}_{25}(x_{5i} - 34,00)_+ + \hat{\phi}_{35}(x_{5i} - 77,03)_+ + \hat{\phi}_{45}(x_{5i} - 81,56)_+ + \hat{\phi}_{26}(x_{6i} - 19,47)_+ + \hat{\phi}_{36}(x_{6i} - 77,82)_+ + \\ &\hat{\phi}_{46}(x_{6i} - 83,96)_+ + \frac{1}{53}\sum_{i=1}^{53}\left[\frac{\frac{1}{1,09}K\left(\frac{t-t_i}{1,09}\right)}{\frac{1}{53}\sum_{i=1}^{n}\frac{1}{1,09}K\left(\frac{t-t_i}{1,09}\right)}\right]y_i\end{aligned} \quad (21)$$

This means that the mixed estimator nonparametric regression model built is able to explain the effect of the predictor variable on the response variable, namely the percentage of poor people based on the $R^2$ value of 90.52%.

## 4. Conclusion

Based on the analysis, we get a nonparametric regression model using mixed estimator truncated spline and Gaussian Kernel as follows:

$\vec{Y} = \mathbf{X}(\phi)\vec{\beta} + \mathbf{D}(\alpha)\vec{Y} + \vec{\varepsilon}$

where estimate results from $\hat{\vec{\beta}}$ is $\left[\mathbf{X}(\phi)^T\mathbf{X}(\phi)\right]^{-1}\mathbf{X}(\phi)^T\left[\mathbf{I} - \mathbf{D}(\alpha)\right]\vec{Y}$.

In general, the nonparametric regression model using mixed estimator can be written as:

$$\hat{\mu}_{\phi,\alpha}(x,t) = \sum_{p=1}^{q} \hat{f}_p(x_{pi}) + \hat{h}_\alpha(t)$$
$$= \left[\mathbf{A}(\phi,\alpha) + \mathbf{D}(\alpha)\right]\vec{Y}$$

The results of the nonparametric regression model using mixed estimator truncated spline and Gaussian Kernel using 3 knot points with the GCV method, obtained a $R^2$ value of 90.52% and MSE value of 1.10. This means that the nonparametric regression model built is able to explain the effect of the predictor variable on the response variable, namely the percentage of poor people by 90.52%.

## Acknowledgments

Thanks to the Education and Culture Office of East Borneo Province for providing "Kaltim Tuntas Scholarship" assistance to complete Postgraduate Education at the Sepuluh Nopember Institute of Technology Surabaya.

## References

[1] Draper N R and Smith H 1992 *Applied Regression Analysis: Third Edition* USA: John Wiley and Sons, Inc
[2] Budiantara I N 2019 *Regresi Nonparametrik Spline Truncated* Surabaya: ITS Press
[3] Dani A T R, Adrianingsih N Y and Ainurrochmah A 2020 Pengujian Hipotesis Simultan Model Regresi Nonparametrik Spline Truncated dalam Pemodelan Kasus Ekonomi *JAMBURA Journal of Probability and Statistics* **1** 98-106
[4] Prahutama A, Suparti and Utami T W 2018 Modelling Fourier Regression for Time Series Data with Case Study Modelling Inflation in Foods Sector in Indonesia *IOP Conf. Series: Journal of Physics* **974** 1-9
[5] Devi A R, Budiantara I N and Ratnasari V 2018 Unbiased Risk and Cross-Validation Method for Selecting Optimal Knots in Multivariable Nonparametric Regression Spline Truncated AIP Conf. Proc **2021** 060003
[6] Nadaraya E A 1989 *Nonparametric Estimation of Probability Densities and Regression Curves* London: Kluwcer Academic Publisher
[7] Hardle W 1994 *Applied Nonparametric Regression* Berlin: Humboldt-Universitat zu Berlin
[8] Wahba G 1990 *Spline Models for Observational Data* Pennsylvania: SIAM
[9] Eubank R L 1999 *Nonparametric Regression and Spline Smoothing* New York: Marcel Dekker
[10] Bilodeau M 1992 Fourier Smoother and Additive Models *Canadian Journal of Statistics* **20** 257-269
[11] Morton J and Silverberg L 1999 Fourier Series of Half-range Functions by Smooth Extension *Applied Mathematical Modelling* **33** 812-821
[12] Saputro D R S, Demu K R and Widyaningsih P 2018 Nonparametric Truncated Spline Regression Model on the Data of Human Development Index (HDI) in Indonesia *IOP Conf. Series: Journal of Physics* **1028** 1-4
[13] Aydin D, Guneri O I and Fit A 2016 Choice of Bandwidth for Nonparametric Regression Models using Kernel Smoothing: A Simulation Study *International Journal of Sciences Basic and Applied Research* **26** 47-61
[14] Budiantara I N, Ratnasari V, Madu R and Zain I 2015 The Combination of Spline and Kernel Estimator for Nonparametric Regression and Its Properties *Applied Mathematical Sciences* **9** 6083-6094
[15] Ratnasari V, Budiantara I N, Ratna M and Zain I 2016 Estimation of Nonparametric Regression Curve Using Mixed Estimator of Multivariable Truncated Spline and Multivariable Kernel *Global Journal of Pure and Applied Mathematics* **12** 5047-5057

[16] Hidayat R, Budiantara I N, Otok B W and Ratnasari V 2019 Kernel-Spline Estimation of Additive Nonparametric Regression Model *IOP Conf. Series: Materials Science and Engineering* **546** 1-9
[17] Setiawan R N S, Budiantara I N and Ratnasari V 2017 Application of Confidence Intervals for Parameters of Nonparametric Spline Truncated Regression on Index Development Gender in East Java *IPTEK Journal of Sciences* **2** 49-55
[18] Craven P and Wahba G 1979 Smoothing Noisy Data with Spline Functions *Numerische Mathematics* **31** 377-403
[19] Wang Y 1997 GRKPACK: Fitting Smoothing Spline Analysis of Variances Models to Data from Exponential Families *Communication in Statistics: Simulation and Computation* **26** 756-782
[20] Badan Pusat Statistik 2020 *Provinsi Kalimantan Timur dalam Angka 2020* Kaltim: BPS
[21] Ferezagia D V 2018 Analisis Tingkat Kemiskinan di Indonesia *Jurnal Sosial Humaniora* **1** 1-6
[22] Montoya E L, Ulloa N and Miler V 2014 A Simulation Study Comparing Knot Selection Methods with Equality Spaced Knots in a Penalized Regression Spline *International Journal of Statistics and Probability* **3** 96-110

# Determination of Aquifer Layer Using Geoelectric Resistivity Method in Lappae, Bone Regency, South Sulawesi, Indonesia

**M A Massinai[1], M F I Massinai[1*], E Syamsuddin[1], AH Baso[1] and H Cottang[2]**

[1]Department of Geophysics, Faculty of Mathematics and Natural Sciences, Hasanuddin University, Makassar, Indonesia.
[2]Teacher of Junior High School 2 Awangpone, Bone, Indonesia.

*Corresponding author: altin@science.unhas.ac.id

**Abstract**. This research was carried out in Lappae, Bone Regency, an area that was supposedly having difficulty with water sources, remarkably in the dry season. This research aims to determine the water source location and the type of aquifer layer using the Geoelectric resistivity method with the Wenner-Schlumberger configuration. The results showed that the aquifer zone, which was thought to be a sandstone layer , was located at various depths ranging from 25-60 meters with a resistivity value of 25-150 Ωm. Meanwhile, bedrock layers are limestone with a resistivity value of 300 - 1200 Ω m. Based on the type of aquifer layer in the area, namely sandstone and limestone, the aquifer can be classified as a free aquifer because it only has a waterproof layer underneath

## 1. Introduction

Water is essential in the process of life because all types of creatures depend on water. With increasing population growth, the need for water is also increasing. An area with water-restricted difficulty to meet the needs of the population is high, primarily when the summer drought occurs. Water the soil is one of the sources will need water for living creatures on the face of the earth [1].

Water is the soil stored in a container (aquifers), namely the formation geology of saturated water that can store and release water in quantity sufficient and economical [2]. One method that can be used in the search for groundwater is the resistivity method [3]. This method adheres to the principle that the layering of rock has a value that detainee types (resistivity) are different. The resistivity value of each layer of rock is determined by factors the type of material constituent, the content of water in the rock, the nature of chemical water, and the porosity of the rock [4].

Rock resistivity values have been known to be used to study rock material types, subsurface layers, and subsurface water distribution in certain areas [5]. Furthermore, the use of resistivity geoelectric methods and describing subsurface conditions well, can also be done in a time-efficient procedure and not too expensive in all costs [6]. The geoelectric survey of resistivity mapping and sounding methods produces information on resistivity value variations in both the lateral and vertical directions [7][8].

The research location is in Bone Regency, one of the strength districts in the field of agriculture. Water demand in this region is very high for drinking and irrigation purposes. Therefore, proper management of groundwater utilization and the search for new sources of water remain carried out to ensure the continuity of human life and increased production in agriculture; therefore, it is essential to do a geoelectric survey to assist in the management of groundwater [8].

This study is limited to the identification of aquifer layers using the Wenner Schlumberger configuration of resistivity geoelectric method in Lappae, Bone Regency. This study aims to determine the subsurface layer based on resistivity and aquifer distribution values in the study area.

## 2. Research Methodology

This research was conducted in the Lappae Village, Bone Regency of South Sulawesi in the coordinates position of 4 ° 21'52 "S - 4 ° 24'10" S and 120 ° 11'56 "- 120 ° 13'52" E. Data have collected at six the trajectory. The length of each stretch is 300 meters, with the distance of each electrode is 5 meters. There are three measurement lines in the East-West direction, one trajectory in the East-West direction, one trajectory in the Northwest-Southeast direction, one trajectory in the Northeast-Southwest direction, and one trajectory in the North-East-East direction. The research location can be seen in Figure 1.

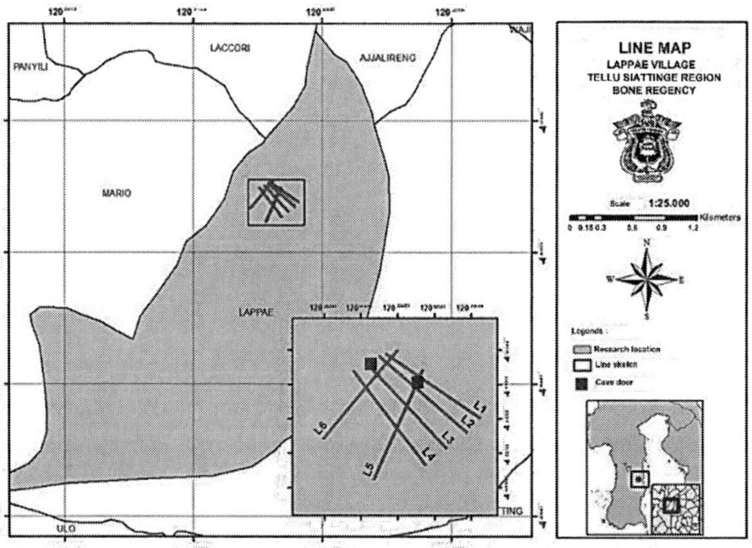


**Figure 1**. Research Location Map (modification from [9]).

Resistivity measurement is generally done by injecting current into the ground through two current electrodes (C1 and C2), and measuring the potential difference caused by two potential electrodes (P1 and P2).

The data obtained the amount of electric current (I), potential difference (V), elevation, and local geology information. Electric current data and potential differences are processed to obtain apparent resistivity, and then the actual resistivity values represent subsurface sections. This cross-section contains the distribution of resistivity values represented by different colour images [7][10]. Thus the qualitative data analysis was carried out on a 2D cross-sectional resistivity map in order to obtain a cross-section of the surface layers carrying groundwater.

## 3. Result and Discussion

Through reading electric current and the electric potential difference of electrode configuration each line, the apparent resistivity value will be obtained. The apparent value than processed using RES2DINV software. The inversion results with the software will produce a 2D cross-section showing the relationship between depth and the actual resistivity of the soil layers. The Table 1 is the result of measurement inversion.

**Table 1.** Results of inversion geoelectric resistivity measurement.

| Line | Track Length (m) | Space (m) | Resistivity Interval (Ωm) | Elevation (m) | Depth (m) |
|---|---|---|---|---|---|
| 1 | 300 | 5 | 25.58 - 2340.50 | 20 – 90 | 60 |
| 2 | 300 | 5 | 18.42 - 12066.90 | 20 – 90 | 60 |
| 3 | 300 | 5 | 35.74 - 9041.60 | 30 – 110 | 60 |
| 4 | 300 | 5 | 28.27 - 8992.60 | 40 – 130 | 60 |
| 5 | 300 | 5 | 47.87 - 8128.70 | 40 – 120 | 60 |
| 6 | 300 | 5 | 0.23 - 68575.70 | 10 – 80 | 60 |

The nature of the layer, distribution, and aquifer can be determined by analyzing the resistivity cross-section based on the resistivity value, regional geological data, and boundary layers. The following will discuss the types of layers and the distribution of aquifers in each line.

### 3.1. Line 1

Based on the results of measurements on track 1, each has a length of track 300 meters with the number of electrodes as many as 60 spaces; each electrode is 5 meters range. The coordinate electrode was 4°22'39" S - 120°13'26" E with the direction of the extent leading from the west sea to the east. Resistivity section obtained in Figure 2 with a range of values of resistivity 25.58 Ωm -2340.50 Ωm and reach a depth of 60 meters. Line 1 intersects the line 5 and 6.

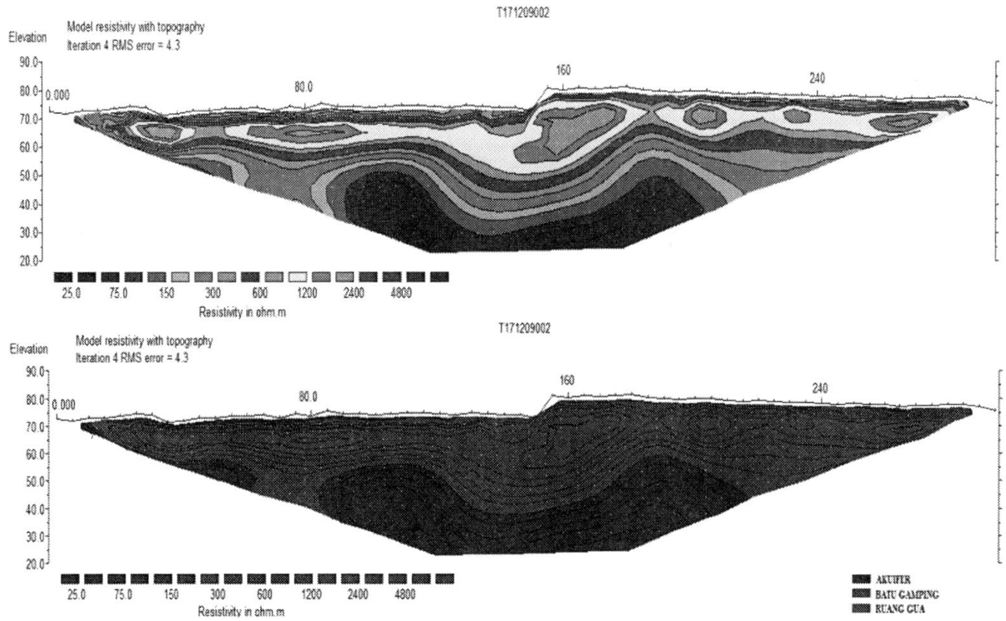


**Figure 2.** Cross-section resistivity of line 1.

Figure 2 shows the value of resistivity of 25 Ωm - 150 Ωm interpreted as aquifer contained in sandstones were imaged with the colour blue which is contained in the electrode 9 to 11 (point measuring 40 meters - 55 meters) and the electrodes 19 to 42 (point measuring 95 meters - 210 meters) at a depth of ± 25 meters. The value of resistivity of 300 Ωm - 1200 Ωm interpreted as layers of rock limestone imaged as green. The values of resistivity 1800 Ωm - 6400 Ωm interpreted as Lappae caves space were imaged as red. This section is composed of two layers of sandstone and limestone.

### 3.2. Line 2

Line 2 intersects with lines 5 and 6. The line leads from north-west-east with the coordinates of the initial 4°22'39' 'S - 120°13'25" E, in the measurement of the resistivity of the track 2, spaced smallest inter-electrode 5 meters and a length of track 300 meters with a depth of 60 m. Based on the measurement results obtained, a resistivity cross-section (Figure 3). The resistivity cross-section 2 shows the lowest resistivity value of 18.42 Ωm and the highest resistivity value of 12066.90 Ωm.

Based on the image cross-section of the resistivity of line 2 (Figure 3) are interpreted there are two layers. Layer 1 with a value of resistivity of 25 Ωm - 150 Ωm (colour blue) is interpreted as sandstone indicated serves as an aquifer. Layer 2 with a value of resistivity of 300 Ωm - 1200 Ωm (colour green) are interpreted as limestone based on the value of resistivity in the cross-section of the resistivity, resistivity table, regional geological data and geology survey. As for the distribution of the aquifer on a cross- resistivity Tracks 2 contained in the electrodes 21 to 40 (point measuring 105 meters - 200 meters) at a depth of ± 40 meters.

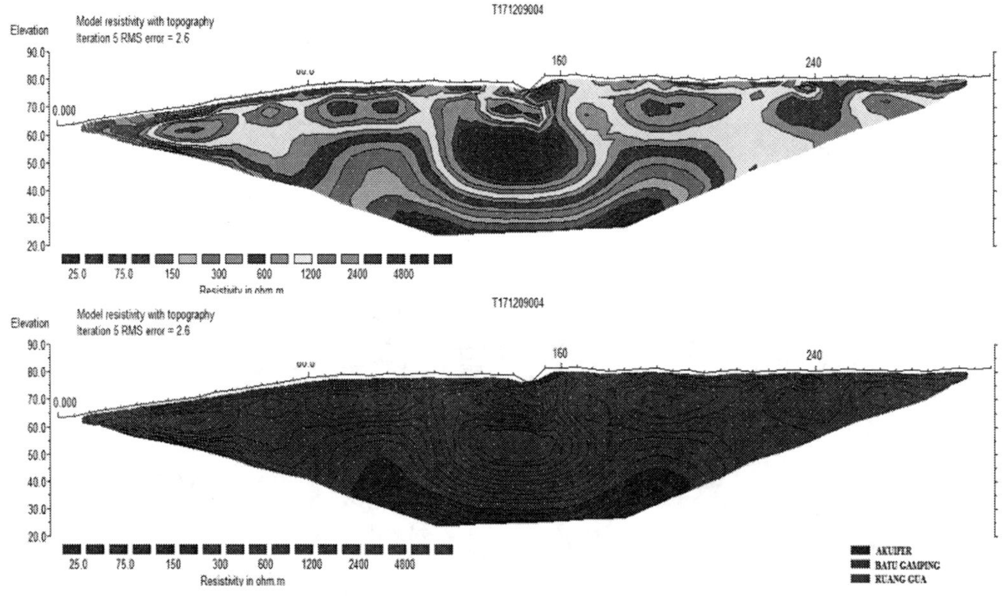


**Figure 3**. Cross-section resistivity of track 2 (top) and interpretation (bottom).

### 3.3. Line 3

The results of measurements on line 3, which have a length of 300 meters with the number of electrodes as many as 60 spaces, each electrode 5 meters. Coordinate electrode first 4°22'40" S - 120°13'24" E with the direction of the stretch leading from the west sea to the east. Resistivity section obtained (Figure 4) with a range of values of resistivity 35.74 Ωm - 9041.60 Ωm and reach a depth of 60 meters. Tracks 1 intersect the trajectory 5 and 6.

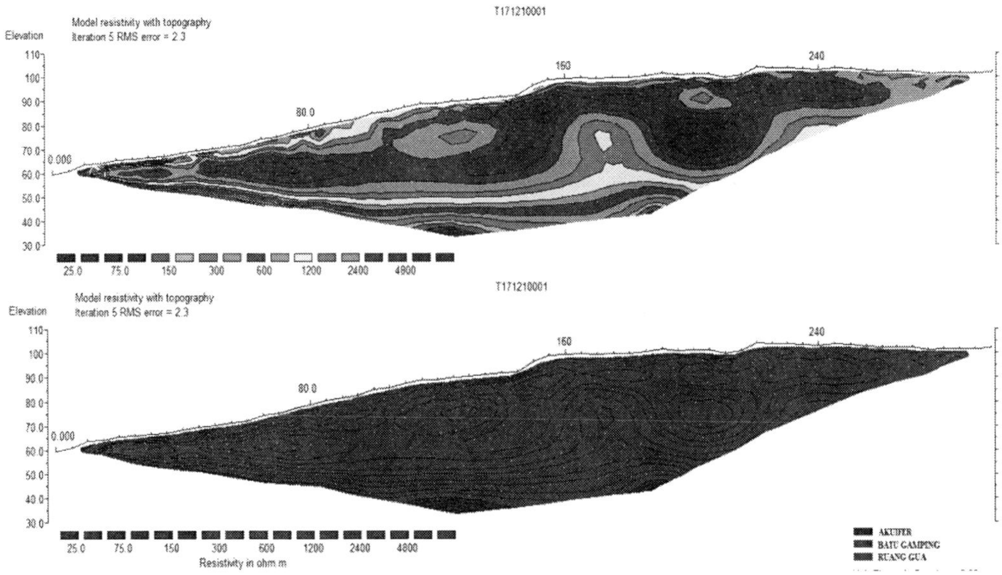


**Figure 4**. Cross-section resistivity of track 3 (top) and interpretation (bottom).

Figure 4 shows the value of resistivity of 25 Ωm - 150 Ωm interpreted as an aquifer in a layer of clay that is imaged with the colour blue elderly who are on the electrodes 23 to 30 (point measuring 115 meters - 150 meters) and 37 to 39 (point measuring 185 meters - 195 meters) at a depth of ± 60 meters. The value of resistivity of 300 Ωm - 1200 Ωm interpreted as layers of rock limestone were imaged with colour green dark. Values of resistivity 1800 Ωm - 6400 Ωm interpreted as Lappae caves space were imaged with the colour red. This section is composed of two layers of the clay and limestone.

### 3.4. Line 4

Line 4 intersects with the trajectory 5 and 6. Tracks 4 leads from the direction west-northwest to the southeast with the coordinates of the initial 4°22'41" S - 120°13'23" E, in the measurement of the resistivity of the track 4, spaced smallest inter-electrode 5 meters and a length of track 300 meters with a depth of 60 m. Based on the measurement results obtained, a resistivity cross-section (Figure 5). The resistivity profile (Figure 5) shows the lowest resistivity value of 28.27 Ωm and the highest resistivity value of 8992.60 Ωm.

Based on the image cross-section of the resistivity of line 4 (Figure 5) is interpreted, there are two layers. Layer 1 with a value of resistivity of 25 Ωm - 150 Ωm (colour blue) is interpreted as stone sand indicated serves as an aquifer. Layer 2 with a value of resistivity of 300 Ωm - 1200 Ωm (colour green) are interpreted as rock limestone based on the value of resistivity in the cross-section of the resistivity, resistivity table, regional geological data and geology survey. As for the distribution of the aquifer on a cross- resistivity Tracks 4 contained in the electrodes 23 to 30 (point measuring 115 meters - 150 meters) and 37 to 39 (point measuring 185 meters - 195 meters) at a depth of ± 60 meters.

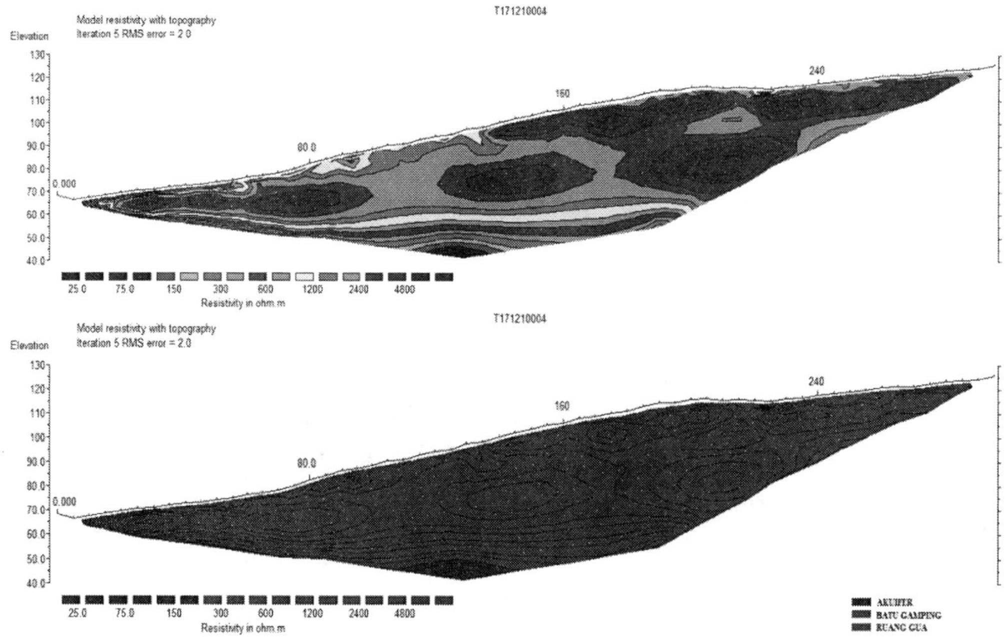


**Figure 5**. Cross-section resistivity of track 4 (top) and interpretation (bottom).

### 3.5. Line 5

Measurement on line 5, which has a length of 300 meters, requires 60 electrodes with a distance of 5 meters each. Electrode coordinates 4 ° 22'40.83 " S - 120 ° 13'30 " E, in the direction of the stretch from northeast to southwest. The resistivity values obtained were in the range of 47.87 Ωm to 8128.70 Ωm and reaching a depth of 60 meters (Figure 6). Line 5 intersects with lines 1, 2, 3, and 4.

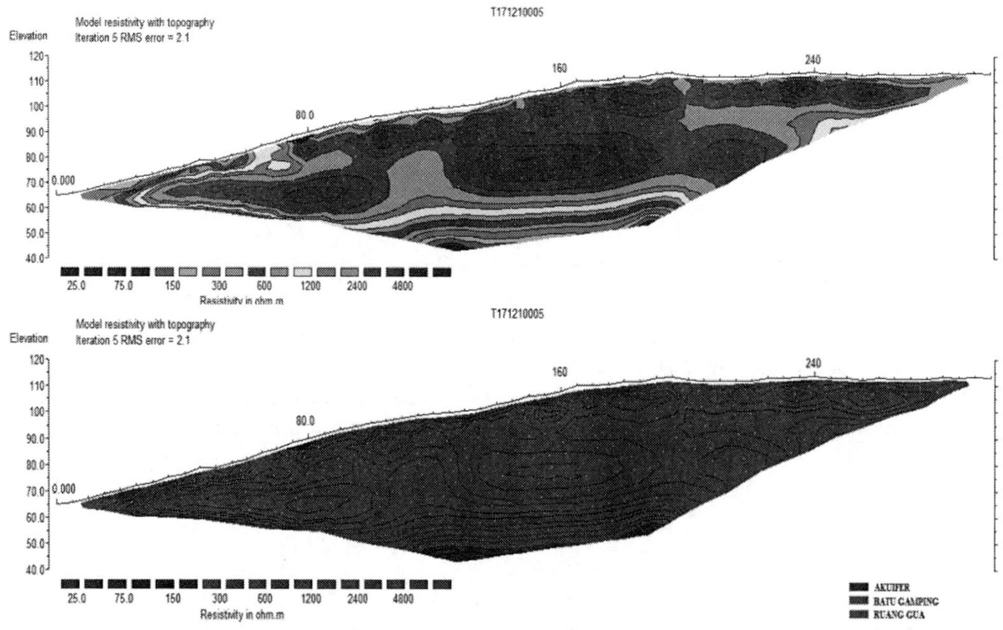


**Figure 6**. Cross-section resistivity of track 5 (top) and interpretation (bottom)

Figure 6 shows the resistivity values of 25 Ωm to 150 Ωm interpreted as sandstone aquifers, imaged in dark blue. The aquifer is at electrodes position number 24 to 39 or the measurement point of 120 to 195 meters at a depth of ± 55 meters. Resistivity values in the range of 300 Ωm to 1200 Ωm are interpreted as limestone layers, imaged in dark green. Whereas the high resistivity value of 1800 Ωm-6400 Ωm imaged in red is interpreted as Lappae Cave.

### 3.6. Line 6

Line 6 intersects with lines 1, 2, 3, and 4. The line leads from northwest to east in coordinates of 4°22'38.78" S - 120°13'27.20" E, in the measurement of the resistivity of the line, spaced the smallest inter-electrode 5 meters and a length of track 300 meters with a depth of 60 m. Based on the measurement results obtained, resistivity cross-section shown in Figure 7. The segment of resistivity confirmed that the value of resistivity ranging from the lowest 0.23 Ωm to the highest 68,575.70 Ωm.

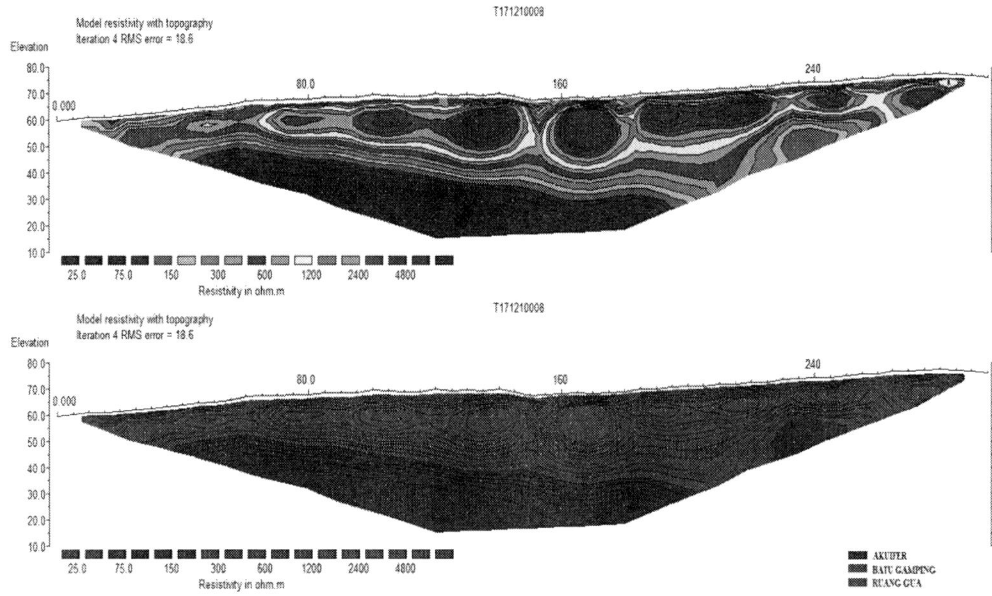


**Figure 7**. Cross-section resistivity of line 6 (top) and interpretation (bottom)

Based on the image cross-section of the resistivity of line 6, shown in Figure 7, are interpreted as two layers. Layer 1, with a value of resistivity of 25 Ωm - 150 Ωm, imaged in blue, is interpreted as sandstone indicated as an aquifer. Layer 2, with a value of resistivity of 300 Ωm - 1200 Ωm, imaged as green, are interpreted as limestone based on resistivity in the section of the resistivity, resistivity table, and regional geological data, and geology survey. Then, the distribution of the aquifer for line 4 contained in the number of the electrode of 14 to 30 (point measuring 70 meters - 150 meters) at a depth of ± 30 meters.

## 4. Conclusions

The layer below the surface in the area of research consists of sandstones and limestones. It has identified several aquifers divided at several points on each line. On line 1 there are two points of the aquifer, the position of the aquifer that located on the electrode number 9 to 11, measurements point 40 meters - 55 meters) and the electrodes 19 to 42 (point measuring 95 meters - 210 meters) at a depth of ± 25 meters. Line 2 has one aquifer section; its position is on electrodes number 21 to 40 (measuring point 105 meters - 200 meters) at a depth of ± 40 meters. Line 3 identified two points of the aquifer, the position electrodes number 23 to 30 (point measuring 115 meters - 150 meters) and 37 to 39 (point measuring 185 meters - 195 meters) at a depth of ± 60 meters. At line 4, 1 aquifer point identified, its position is at electrodes 23 to 29 (115 meters - 145 meters measuring point) at ± 60 meters depth. At line 5, one identified aquifer located on the electrodes of number 24 to 39 (point measuring 120 meters - 195 meters) at a depth of ± 55 meters. Line 6 identified one point of the aquifer, the position on the electrodes at number 14 to 30 (point measuring 70 meters - 150 meters) at a depth of ± 30 meters.

## References

[1] George N J, Ibuot J C and Obiora D N 2015 Geoelectrohydraulic parameters of shallow sandy aquifer in Itu, Akwa Ibom State (Nigeria) using geoelectric and hydrogeological measurements *J. African Earth Sci.* **110** 52–63

[2] Massinai M A 2015 *Geomorfologi Tektonik* Yogyakarta: Pustaka Ilmu

[3] Falco P, Negro F, Szalai S and Milnes E 2013 Fracture characterisation using geoelectric null-arrays *J. Appl. Geophys.* **93** 33–42

[4] Todd D K 2006 *Groundwater Hydrology* New York: Wiley

[5] Loke M H 2004 *Tutorial: 2D and 3D electrical imaging surveys*

[6] Syamsuddin E, Haris N A, Mustafa N S, Swastika D A and Nurdin N H 2019 Application of Geoelectrical Resistivity Method for Groundwater Exploration Using Wenner-Schlumberger Configuration in Sampakang, Maros Regency, Indonesia *Journal of Physics: Conference Series* **1341** 082046

[7] Telford W M, Geldart L P and Sheriff R E 1990 *Applied Geophysics* Cambridge: Cambridge University Press

[8] Azhar and Handayani G 2004 Penerapan Metode Geolistrik Konfigurasi Schlumberger untuk Penentuan Tahanan Jenis Batubara *J. Natur Indones.* **6** 122–126

[9] Virman, Massinai M A and Massinai M F I 2019 Groundwater Identification at East Sentani Auwena Formation (Tema) Based on Geoelectric *IOP Conf. Ser. Earth Environ. Sci.* **279** 1–6

[10] Sukamto R 1982 Geological map of the Pangkajene and Western Part of Watampone Quadrangles, Sulawesi Bandung: Geol. *Res. Dev. Centre,*

[11] Massinai M A, Lantu, Latuconsina H and Massinai M F I 2018 Distribution Analysis of Sulphide Mineral (Pyrite) Using Induced Polarization Method in Libureng, Bone, South Sulawesi *IOP Conf. Ser. Earth Environ. Sci.* **132** 1–7

IOP Conf. Series: Materials Science and Engineering 1115 (2021) 012022 doi:10.1088/1757-899X/1115/1/012022

# Effect of Connected Non-Linear Household Appliances on the Disturbance at Frequency Range of 9-150 kHz in the On- Grid Photovoltaic System

**Faizal*** **and B Sudiarto**

Department of Electrical Engineering, Universitas Indonesia, Indonesia

*Corresponding author: faizal81@ui.ac.id

**Abstract.** Nowadays, the usage of power electronic circuit which is called inverter is increasing especially in household appliances. This inverter technology is the technology that can convert alternating current (AC) to direct current (DC), and this technology may produce disturbance at frequency range 9-150 kHz. In the residential area, a house can be built one of on-grid photovoltaic (PV) system, and the usage of household appliances inside the house system can be connected together at same time and operated simultaneously. This research is focused on investigating the effect of connected household appliances, such as induction cooker. The result shows that the connected two types of induction cooker in same system may either increase or decrease the disturbance value in the system. The connected appliances affect the disturbance through by the power consumption.

## 1. Introduction

Power electronic is an electronic circuit that can control a power converter with aim to produce alternating current (AC) or direct current (DC) which can be designed for the differences load. And because of the main target power electronic system is to transfer power efficiently and the size of the device should be smaller, power electronic circuit that called by inverter technology is using inside the photovoltaic (PV) system, such as PV inverter [1]. In this inverter technology, it has some switching devices that switching in the frequency range 9-150 kHz, it means this PV inverter produced distortion in that frequency range 9-150 kHz [2]-[3]. Furthermore, renewable energy is widely used in PV system, not only in industrial building, also installed at residential areas, such as rooftop PV system and this system usually using grid-connected inverter. Nowadays, caused of power electronic circuit is using inside all electronic equipment, especially household appliances. It may influence the PV system including other equipment of household appliances that connected or used inside the PV system which installed in the house [4]-[5].

Some of researcher reported that the power electronic devices generated distortion in waveform of voltage and current in some frequencies. In international standardization distortion that generated below 2 kHz called harmonics, and the distortion that generated above 2 kHz until 150 kHz is called disturbance. However, the standardization of this frequency range 9-150 kHz is still not covered, so the research to find the standard of disturbance and to limit this disturbance in order to improve power quality [6]-[7].

In residential area or PV system that installed in the house, there must be some of loads connected in the same time inside the system, these connected loads are not only generates disturbance, also influenced the PV system, and may be influenced the other appliances [8]. The household appliances also have some of power electronic circuit inside the equipment, because the appliances operated in differences voltage level and power consumption. Some of researcher also recognized and reported that the non-linear household appliances, especially induction cooker, were generated harmonic [9]-[10]. This study will observe the effect of connected household appliances in the on-grid PV system, in this case is two types of induction cooker.

## 2. Methodology

This measurement is aimed to find the influence of disturbances at frequency 9-150 kHz while non-linear household appliances loads, such as two types of induction cooker are connected in the same system. The system that used is on-grid PV system which the system is system generated by PV and also connected to utility grid.

The first part of the measurement was observed characteristic of each induction cooker which is using two types of induction cooker, brand C and brand P. After that, for the next part, these two brands of induction cooker connected to the system at the same time and find the changes and the differences disturbances between the first part of measurement and the second measurement.

The data taken using PicoScope which used for measured the disturbance in frequency 9-150kHz, and the data processed using Matlab. So that, the data of disturbance may be known.

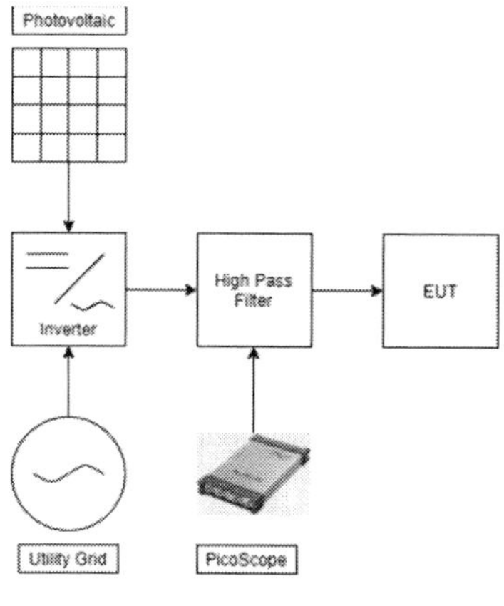


**Figure 1.** Scheme of Measurement

Based on Figure 1, it can be seen output of load measurement point was connected with high pass filter and connected to PicoScope.

## 3. Result and Discussion

### *3.1. Disturbance Characteristics of each Induction Cooker*

This measurement is using two types of induction cooker. For each induction cooker had disturbance characteristics and this study aimed to look at the disturbance characteristics of each induction cooker before connecting these two types of induction cooker to the system and measuring its, so the differences and changes of disturbance can be compared between of each induction cooker and two types of induction cooker that connected together in same system. The disturbance characteristics of each induction cooker shown in figures below

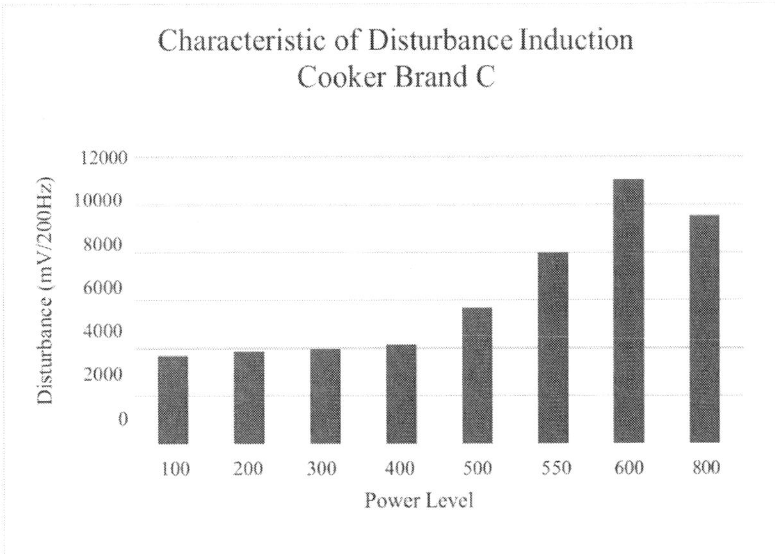


**Figure 2.** Disturbance of Induction Cooker Brand C

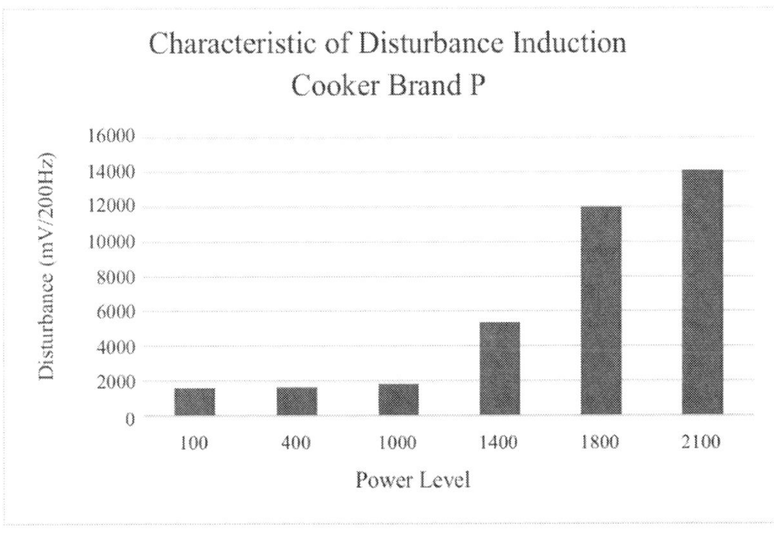


**Figure 3.** Disturbance of Induction Cooker Brand P

Figure 2 and 3 shows each of disturbance characteristics of two types induction cooker, which brand C disturbance at power 100 W is 3699.09 mV; power 200 W is 3868.42 mV; power 300 W is 3965.14 mV; power 400 W is 4158.87 mV; power 500 W is 5691.53; power 550 W is 7991.61 mV; power 600 W is 11037.38 mV; and 800 W is 9536.96 mV. And for brand P disturbance at power 100 W is 1608.58 mV; power 400 W is 1644.56 mV; power 1000 W is 1845.10 mV; power 1400 W is 5361.03 mV; power 1800 W is 12009.20 mV; and power 2100 W is 14122.20 mV.

### *3.2. The Effect of Connected Appliance to the Disturbance*

The effect of connected two types of induction cooker to the disturbance appeared in this on-grid PV system. In this measurement, used two methods of measurement, the first one is giving randomly changes of power level the two induction cookers and the next method is raising the power level step by step from the lowest power level to the highest power level. The disturbance for this measurement can be obtained as shown in Table 1. and Table 2.

**Table 1.** The Disturbance measured in the system using variation power

| Power (W) | Disturbance Brand P (mV) | Power (W) | Disturbance Brand C (mV) | Power (W) | Brand P and C (mV) |
|---|---|---|---|---|---|
| 100 | 1495.00 | 100 | 3495.99 | 100 100 | 2477.80 |
| 400 | 1595.42 | 200 | 3253.72 | 400 200 | 2418.01 |
| 400 | 1595.42 | 400 | 3877.24 | 400 400 | 2928.68 |
| 1000 | 1633.76 | 800 | 8594.55 | 1000 800 | 8471.73 |
| 2100 | 13385.08 | 400 | 3877.24 | 2100 400 | 5900.99 |
| 1000 | 1633.764 | 400 | 3877.24 | 1000 400 | 2809.05 |
| 2100 | 13385.07 | 800 | 8594.55 | 2100 800 | 8664.99 |

**Table 2.** The Disturbance measured in the system by rising power step by step

| Power (W) | Disturbance Brand P (mV) | Power (W) | Disturbance Brand C (mV) | Power (W) | Brand P and C (mV) |
|---|---|---|---|---|---|
| 100 | 1495.00 | 100 | 3495.99 | 100 100 | 2615.04 |
| 400 | 1595.42 | 200 | 3253.72 | 400 200 | 2497.16 |
| 1000 | 1633.76 | 400 | 3877.24 | 1000 400 | 3107.79 |
| 2100 | 13385.08 | 800 | 8594.55 | 2100 800 | 9333.62 |

According to the Table 1 and 2, there are some changes and influence of connected the two types of induction cooker. Even the measurement used random variation of power usage or rising the power level step by step.

The Table 3 has relation with the two table before, table 1 and 2. In this table 3 can be seen that there are some changes in disturbance. In 2nd, 4th, and 6th column shown the disturbance change of each power level by percentage. 3rd and 5th column shown the disturbance change of connected two induction cooker and before connected the induction cooker by percentage. Also, there is the difference definition of negative mark on the table 3. In 2nd, 4th, and 6th column, the negative sign means the changes of the disturbance caused by power consumption is decreasing the disturbance value, and in 3rd and 5th column, the negative sign means the changes that caused by connected appliances is increasing the value. For the graphic of the table before can be seen in Figure 4.

Figure 5 and 6 which are showing the graph of disturbance changes by two methods and the figures has a relation with the table above. It can be seen there are some changes on disturbances value even the measurement used any method. And even the other load absorbs the disturbance, the disturbance value should be got the attention.

**Table 3.** Percentage changes of disturbance value

| Power (W) | Disturbance Change (%) | Disturbance Change (%) | Disturbance Change (%) | Disturbance Change (%) | Disturbance Change (%) |
|---|---|---|---|---|---|
| 100/100 | | -133.85% | | 25.20% | |
| 400/200 | 6.29% | -103.94% | -7.45% | 23.25% | -4.72% |
| 1000/400 | 2.35% | -137.32% | 16.08% | 19.85% | 19.65% |
| 2100/800 | 87.79% | 35.79% | 54.89% | -8.60% | 66.70% |

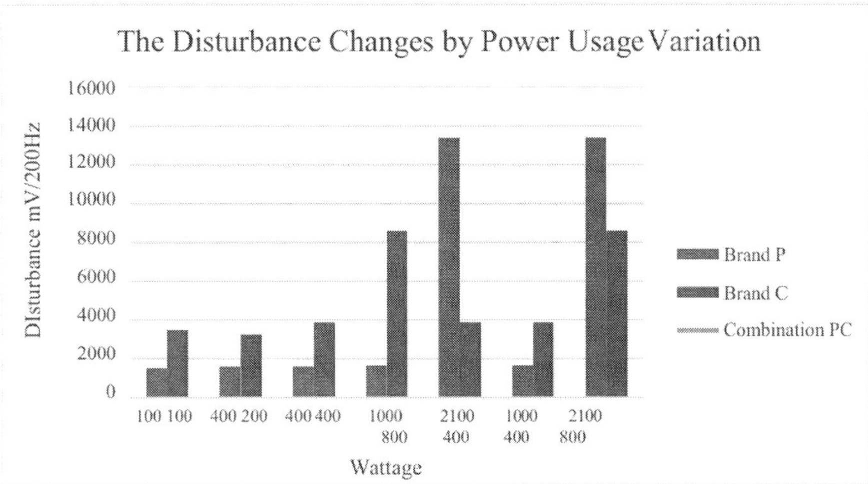


**Figure 4.** Disturbance Changes by Power Usage Variation

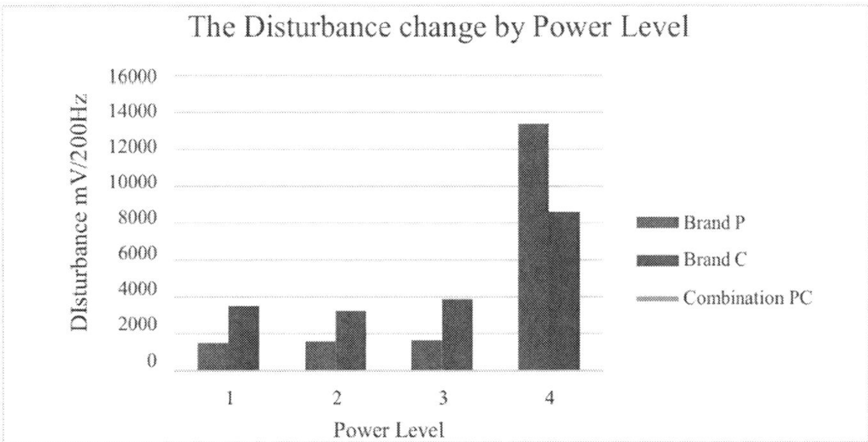


**Figure 5.** Disturbance Changes by Power Level

## 4. Conclusion

The connected household appliances, in this case connected two types of induction cooker affect the disturbance through power consumption. The power consumption of connected appliances may cause changes the disturbance generated by appliances in the system. The disturbance generated by the appliances could be higher or lower depended on the disturbance behaviour generated by the appliance responding to the system.

In this study, connected appliance caused the disturbance in the system either decrease or increase. For the method of raising power consumption, each induction cooker has different changes. Disturbance changes brand P from 100 W to 400 W is 6.29%; from 400 W to 1000 W is 2.35%; from 1000 W to 2100 W is 87.79%. Disturbance changes brand C from 100 W to 200 W is -7.45%; from 200 W to 400 W is 16.08%; from 400 W to 800 W is 54.89%. The effect of connected two types of induction cooker is only for equipment used in this measurement. Because every appliance has unique disturbance behaviour and component, the effect of connected other appliances may get the difference results.

## Acknowledgement

This research was supported by Universitas Indonesia (UI) through PUTI grant 2020 (International Indexed Publication) launched by DRPM UI.

## References

[1] Flicker J, Gonzalez S 2015 Performance and Reliability of PV Inverter Component and Systems due to Advanced Inverter Functionality *2015 IEEE 42nd Photovoltaic Specialist Conference* 1-5
[2] Roy J, Mather B 2019 Study of Voltage-Dependent Harmonic Characteristics of Residential Appliances *2019 IEEE Texas Power and Energy Conference* 1-6
[3] Wu W, Zhang C 2015 The Design of New High Efficiency Photovoltaic Grid and Independent Power Supply Inverter *2015 International Conference on Computational Intelligence and Communication Networks* 1579-1582
[4] Ortenzi G, Pomilio J A 2009 Switch Mode Power Supply Applied to Very Low Cost Electronic Board of Home Appliances *2009 Brazilian Power Electronics Conference* 291-297
[5] Zahra A, Jufri FH, Ardirta IM, Hudaya C 2019 Characteristics of Disturbance in Frequency 9-150 kHz of Photovoltaic System under Fluctuated Solar Irradiance *2019 IEEE 2nd International Conference on Power and Energy Applications* 217-221
[6] Sudiarto B, Widyanto AN, Hirsch H 2017 The Effect of Mains Voltage Level Variations on The Disturbances Produced by Household Appliances in The Frequency Range of 9-150 kHz *2017 International Symposium on Electromagnetic Compatibility* 1-5
[7] Sudiarto B, Widyanto AN, Hirsch H 2017 The Mutual Influence of Appliances on The Disturbance in The Frequency Range of 9-150 kHz Produced by HouseholdAppliances in The Low Voltage Network *2017 International Symposium on Electromagnetic Compatibility* 1-6
[8] A. Gil-de-Castro, S. K. Rönnberg and M. H. J. Bollen 2014 A study about harmonic interaction between devices *2014 16th International Conference on Harmonics and Quality of Power* 728-732
[9] Dario R, Javier MG, Gonzalo G, Augusto R, Ricardo A 2019 Impact of Induction Stoves Penetration over Power Quality in Ecuadorian Households *Revista Espacios* **40** (13) 1-8
[10] Sanna C, Gawronska M, Musio M, Damiano A 2014 Performance Analysis of Non-Linear Load Effects on A Micro-Grid Supplied by Photovoltaic Power Plants *29th European Photovoltaic Solar Energy Conference* 1-8

IOP Conf. Series: Materials Science and Engineering 1115 (2021) 012023 doi:10.1088/1757-899X/1115/1/012023

# Automatic License Plate Recognition Using Optical Character Recognition

**U Salimah, V Maharani and R Nursyanti**

Faculty of Technology and Informatics, Indonesian University of Informatics and Business, Soekarno-Hatta Street no. 643, Bandung, Jawa Barat, Indonesia 40285

*Email: u.salimah@mhs.unibi.ac.id

**Abstract.** Automatic plate recognition is useful for everyday life, for example for law enforcement, traffic control, access to restricted areas, E-Tolls, or for checking in parking areas. The police number plate or TNKB is the main identity of a motorized vehicle that is roadworthy or operating on the highway. Thus, we can use the license plate to identify the vehicle owner. This study aims to recognize vehicle number plates using an Android smartphone. Using a camera on a handphone to scan the plate image, and it is processed to find out the vehicle owner's information from the numbered plate being scanned. This study also uses Optical Character Recognition in processing images scanned by the camera where the steps taken are image acquisition, pre-processing, segmentation, normalization, feature extraction and recognition. The results show that the camera on an Android smartphone can be used to read and display information on number plate owners and also show that vehicle plate recognition using Optical Character Recognition based on the tests conducted shows the percentage of successful character recognition on motorized vehicle plates by 75% while the characters are equal to 97.36%.

## 1. Introduction

Automatic plate reading is useful for everyday life. There are many reasons why it is important for both individuals and organizations to identify vehicles or their owners, for example for law enforcement, traffic control, access to restricted areas, E-Tolls, or for checking in parking areas. License Plate Recognition (LPR) has the advantage of recording a license plate number to fight crime, if an additional camera is given which is focused on the driver's face so that it can simultaneously recognize the identity of the car driver.

ALPR is a new technology in the transportation system automation world. This technology utilizes image processing to identify vehicles from the license plate image [1], however variations on the license plate and the surrounding environment such as font size, font type, font color, number plate location, and differences in intensity due to lights or environment can cause problems during recognition. license plate [2]. In Indonesia, ALPR is still being developed, especially for the detection of number plate objects. Although in other countries ALPR has been developed a lot, this ALPR system cannot be used directly in Indonesia due to differences in the format or type, color, and structure of number plates.

Research on Automatic License Plate Recognition (ALPR) or automatic number plate recognition in various parts of the world has been carried out, but those who conduct research on number plates in Indonesia are still developing today. In the research conducted by Michael, the method used in image conversion is Optical Character Recognition. Tests carried out resulted in an accuracy of 78.57% if the condition of the number plate was in good condition and for the condition of the number plate in a bad condition it resulted in an accuracy of 57.41% [3]. Referring to further research conducted by Manshur, in 2018, stated that the position of the camera will determine the accuracy of the final result [4]. Budianto, in his research he compared the level of accuracy of various methods on automatic license plate recognition and various factors that influence it, and the results of his research stated that the Neural Network Backpropagation algorithm produced an accuracy rate of 96 percent and there were many factors that influenced this accuracy, one of which was the camera quality and also the distance between the camera and the vehicle [5].

The purpose of the study expected by the authors was to produce a transportation automation system capable of recognizing motor vehicle license plates using Optical Character Recognition and recognizing motor vehicle license plate owners in real time using mobile devices as well as knowing the accuracy of the systems tested and creating applications capable of identifying vehicle owners.

## 2. Material and Methods

### 2.1 Automatic License Plate Recognition System

Automatic License Plate Recognition or ALPR for short is a method used as vehicle surveillance that uses optical character recognition in images, this method works by reading vehicle license plates. ALPR has an important role in its application in daily life, such as for law enforcement, traffic control, access to restricted areas, E-Tolls, or for checking in parking areas. Vehicle license plates recognition from the ALPR system is taken from an image or image taken through a camera, whether in color or not [6]. Make use of many techniques such as object detection, image processing, and pattern recognition. ALPR is also known as vehicle identification (AVI), car plate recognition (CPR), automatic number plate recognition (ANPR), and optical character recognition (OCR) for vehicles. ALPR system that extracts a license plate number from a given image can be composed of four stages [7].

The first step is to acquire the image using the camera. The second stage is to extract the license plate from the image based on several features such as borders, colors, or the presence of characters. The third stage is to segment number plates and extract characters by projecting their color information, by labeling them, or by matching their position to a template. And the last stage is to recognize extracted characters by matching the template or by using classifiers such as neural networks and fuzzy classifiers [6].

### 2.2 Optical Character Recognition

OCR stands for Optical Character Recognition. This technology allows the machine to recognize characters automatically via optics. Humans themselves have optics, namely eyes. The image seen by the eye is input to the brain. The ability to recognize this input varies from person to person due to many factors [8]. OCR is a technology that functions much like human reading ability (see Figure 1).

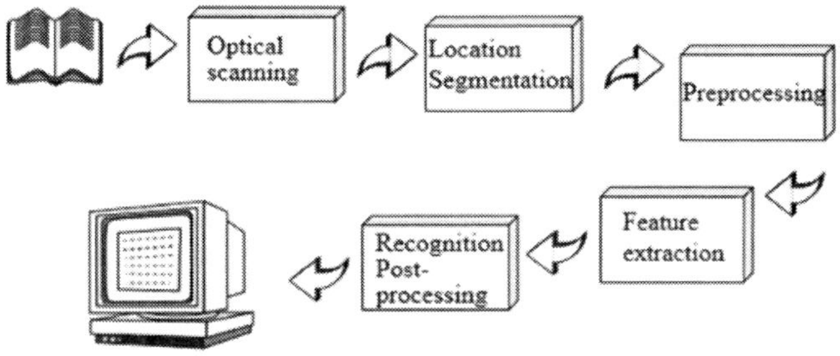


**Figure 1.** OCR

In OCR the pattern is letters, numbers and some special symbols such as commas, question marks, etc., whereas different classes correspond to different characters. The typical OCR system consists of several components [9]-[10]. In Figure 1, the general configuration is illustrated in OCR. The first step in this process is to digitize analogue documents using an optical scanner. When a region containing text is found, each symbol is extracted through a segmentation process. Extracted symbols can then be processed first, eliminating noise, to facilitate the extraction of features in the next step.

## 3. Purposed System

Image acquisition is an early stage in the system for acquiring digital imagery. License plate image scan via Vivo 1610 smartphone with camera resolution of 13 megapixels (mp) with sampling distance of 5 – 10 cm and camera height of 150 – 170 cm. For system design can be seen in Figure 2.

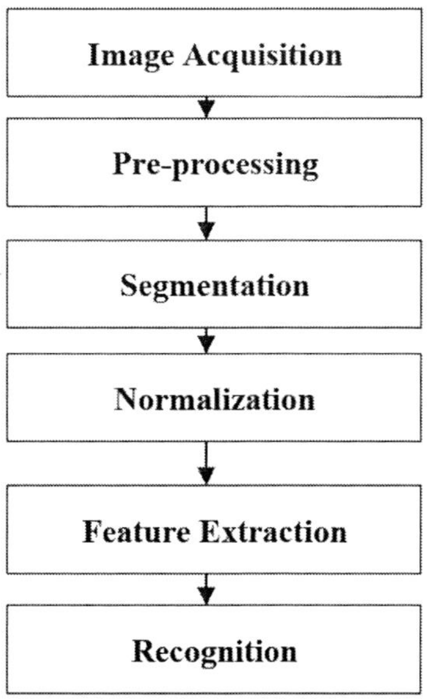


**Figure 2.** System design

In the pre-processing stage, the system will adjust the image that has been taken to suit the next process. Pre-processing is done in the form of cropping or cutting the plate number area with the background area. This cutting is done by thresholding by setting the threshold value between the

IOP Conf. Series: Materials Science and Engineering 1115 (2021) 012023 doi:10.1088/1757-899X/1115/1/012023

background and the number plate area value. The preprocessing process consists of grayscaling and binaryization processes. Both of these processes are carried out to change the pixel intensity. This process produces an image with binary colors, namely black and white (see Figure 3).

**Figure 3.** Original Image (a), Binnarizing (b), Grayscaling (c)

At the segmentation stage, the position of each number and character will be determined. In this section, if there are related candidate digits, it is likely that they will be extracted simultaneously. That way, to detect each row, each row and each column will be checked. If the number of consecutive black pixels is less than or more than the specified threshold limit, the pixels will be converted to white. With this process, it can be done by detecting a column containing white and all matrices equal to 0 so it is determined as the initial cutoff limit. This detection is carried out continuously until a black color is found with all matrices equal to 1. If a black color is found, the process will go back one column again and the last white color will be displayed with label number one. the next detection process is carried out the same thing until the last labeling process has been completed. And pixels containing black will be deleted and not saved. That way the characters that have been labeled will be carried out the next process (see Figure 4).

**Figure 4.** Segmentation

The process that will be done at the normalization stage is scaling against the image so that the image has a fixed resolution. As we know, the camera on the phone has different resolutions so it must be determined the optimal resolution so that the resulting image is not too large, as it will affect the memory and reading process of the image. However, it should not be too small, as it will affect the text reading result because the resolution of the image is too small resulting in the characters in the text being split. In this study, the resolution taken on the camera will be converted into dimensions of 1024 x 960 pixels. After obtaining the edge detection image, each character will be matched to the imagery in the training image database. The process of matching by adjusting the character image size to the character template that's in the library (see Figure 5).

**Figure 5.** Expresses the number 9 character

After obtaining the image from the edge detection results, each character will be matched with the image in the training image database. The matching process is by adjusting the size of the character image with the character template in the library. The pattern shown in Figure 6 expresses the number 9 character. The black box is worth 1 while the white box is 0. The same is done for the other characters.

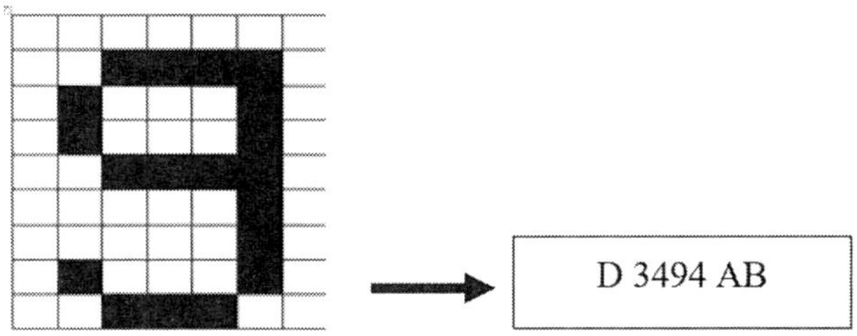


**Figure 6.** Feature Extraction

On Stage Process Introduction, Collection Fitur Results Extraction Citra On the training process Have Stored Will Used To Process Classification With data input. In This Process, dataset gambar Character To Training (training) Especially First Prepared By Character yang Will Determined.

## 4. Implementation and Testing

From the research conducted, the implementation for this system uses ML-Kit as an additional library to extract number plate text using an android camera. Which is then connected to the database to access the detected number plate owner data. This implementation is implemented and tested on android devices. The interface for this application can be seen in the image below:

The scanned plate will display the introductory resulting text on the next page, also displaying the vehicle owner's information, if the scanned plate results in a perfect introduction and the license plate is registered on the database.

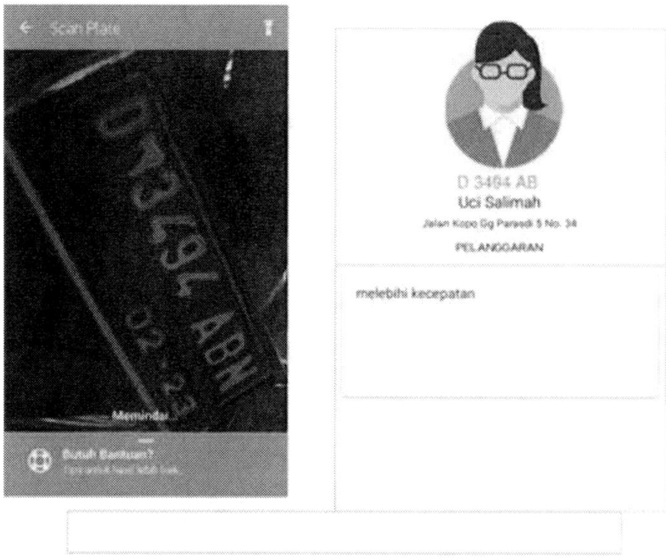


**Figure 7.** Interface implementation

Testing has been conducted to test the accuracy of this system. Test images are taken during the day under various conditions as well as shooting angles. As a result of the proposed implementation, it can be seen in Table 1.

**Table 1**. Test Results in this table are displayed 10 test results from 100 tests.

| No. | Citra Flat | No. Plat | j. experiment | Introduction Results |
|---|---|---|---|---|
| |  | E 3996 R | 3 | E 3996 R |
| |  | E 5142 TW | 3 | 142 TU |
| |  | E 4993 TI | 3 | E 4993 H |
| |  | E 3115 SN | 3 | E 3115 SN |
| |  | E 4564 SZ | 3 | E 4564 S |
| |  | E 1245 RG | 3 | E 1245 RG |
| |  | E 6295 QL | 3 | E 6295 OL |
| |  | E 5361 TI | 3 | E 5361 TI |
| |  | E 2956 RP | 3 | E 2956 RP |
| |  | E 2393 TT | 3 | E 2393 TT |

After testing carried out on 100 number plate samples, it can be concluded with the percentage of success in the recognition process from 100 number plates, 75 number plates can be recognized, namely 75%. Character recognition accuracy with 722 total characters and 703 number of successful character recognition, amounting to 97.36%.

## 5. Conclusion

In this study, the program was able to recognize license plates using Optical Character Recognition (OCR) techniques. The application of the OCR method on the android operating system results in good accuracy in number plate readings, where from the test results are obtained 75% accuracy, where the accuracy for the reading of each character is 97.36 %. The system is able to display vehicle owner data that is successfully recognise correctly. Recognition failures are due to the resemblance of characters to each other, such as B and 8, 2 and Z, 1and I, D and 0, A and 4. Imperfections in character segmentation can cause characters to be truncated that can change the perception of character recognition.

## References

[1] Singh R V and Randhawa N 2014 Automobile Number Plate Recognition AndExtraction Using Optical Character Recognition *International Journal of Scientific & Technology Research* **3** 37-39

[2] Ramachandran R 2015 Multi Depot Vehicle Routing Problem using ODV EV Based Genetic Algorithm *International Journal Of Engineering And Computer Science* **4** 10779-10786

[3] Michael A 2016 Pengenalan Plat Kendaraan Berbasis Android menggunakan Viola Jones dan Kohonen Neural Network *Ilkom Scientific Journal* **8** 95-102

[4] Manshur Y 2018 *Optical Character Recognition For Detection of Car Plates And Motor Vehicles On Gowa Engineering Campus* Bachelor Degree 54 – 55

[5] Budianto A 2018 Automatic License Plate Recognition: A Review with Indonesian Case Study *Scientific Journal of Informatics* **5** 258-270

[6] Du S, Ibrahim M, Shehata M and Badawy W 2012 Automatic license plate recognition (ALPR): A state-of-the-art review *IEEE Transactions on circuits and systems for video technology* **23** 311-325

[7] Sarfraz M, Ahmed M J and Ghazi S A 2003 Saudi Arabian license plate recognition system *International Conference on Geometric Modeling and Graphics, 2003. Proceedings* 36-41 IEEE.

[8] Chiou Y C, Lan L W, Tseng C M and Fan C C 2011 Optimal locations of license plate recognition to enhance the origin-destination matrix estimation *Proceedings of the Eastern Asia Society for Transportation Studies* **8** 297-297

[9] Sobh T and Elleithy K 2010 *Innovations in Computing Sciences and Software Engineering* Berlin: Springer Science & Business Media

[10] Shreya S, Upadhyay Y, Manchanda M, Vohra R and Singh G D 2019 Optical character recognition using convolutional neural network *2019 6th International Conference on Computing for Sustainable Global Development* 55-59

# Modeling with Mixed Kernel, Spline Truncated and Fourier Series on Human Development Index in East Java

**N Y Adrianingsih, I N Budiantara* and J D T Purnomo**

Department of Statistics, Institut Teknologi Sepuluh November, Surabaya 60111, Indonesia

*Corresponding author: nyomanbudiantara65@gmail.com

**Abstract**. The pattern of the relationship between the response variable and the unknown predictor can be determined using a nonparametric regression approach. Nonparametric regression allows it to be used for response variables following different curves between one variable and another. In this regression, there are several types of approaches including the kernel, spline, and Fourier series. In its use, there is not only one type of approach, but can be in the form of a mixture, such as a mixture of a spline and Fourier series, a kernel and a Fourier series, and so on. In this study, in modeling the HDI cases in East Java Province, nonparametric regression mixed kernels, Spline Truncated, and Fourier series were used. The results of research that have been applied to HDI in East Java with the predictor variables for APM SMA, Morbidity, and GRDP per Capita are using the mixed kernel, Spline Truncated, and Fourier series nonparametric regression approach with three-knot points and three oscillations. Good because the coefficient of determination of the estimator is 75.1041%.

## 1. Introduction

One of the successes in regional development can be seen from the resolution or failure of basic problems in the community, for example, poverty, illiteracy, unemployment, and so on. Recently, the Indonesian government has been very interested in achieving the success of human development. The United Nations (United Nations) has established a measure for the achievement of human development standards, namely the Human Development Index (IPM). HDI is a summary measure of the average achievement in the main dimensions of human development. This human development can be done by increasing the aspects which are very important for life, namely the age of life, education, and a decent standard of living. There are 4 indicators formed in this index, namely life expectancy, literacy rate, the average length of schooling, and purchasing power. In human development, there are important factors that a region needs to have, including social and economic factors. From BPS data in East Java, HDI reached a value of 70.27 in 2017. This value is influenced by several factors, including the open unemployment rate, education level, and per capita GRDP.

In modeling HDI with the factors that influence it, there are several ways, one of which is by using nonparametric regression. Nonparametric regression is a regression analysis used to determine the relationship between the response variable and the predictor variable, for which no information is available from before about the relationship between these two types of variables. This regression is very good to use for unknown data patterns because it is flexible [1]. There are many types of estimators in this regression, namely kernel, wavelet, Fourier series, and Spline Truncated. There are many nonparametric regression analyzes used in a study can be seen in ref [2]-[7].

The kernel estimator is quite in demand by researchers because it can handle data that has a good ability to model data that does not have a certain pattern [8], its mathematical form is easy, and is more flexible, and can achieve relatively fast convergence [9], and from a computational point of view, it is easier to do. The spline truncated estimator can handle data characteristics that have variable behavior at certain sub-intervals [10]. The Fourier series estimator has the flexibility to adapt effectively to local data properties [11], and there is a recurring trend [12]. Based on the advantages possessed by the three estimators, the researchers are interested in modeling the human development index in East Java by using a nonparametric regression additive mixed model between the spline, kernel, and Fourier series, because the pattern between the response variables and each predictor, in this case, has The pattern is unknown, follows a repeating pattern at certain intervals, and has a changing pattern at certain sub-intervals.

## 2. Methodology

### 2.1. Kernel Nonparametric Regression

Initially, the most widely used nonparametric regression research is kernel estimator. The first researchers were Watson [13] and Nadaraya [14]. The regression curve is approximated by the kernel function, the regression curve estimate can be written in equation (1).

$$g_\alpha(t) = n^{-1}\sum_{i=1}^{n}\left[\frac{K_\alpha(t-t_i)}{n^{-1}\sum_{i=1}^{n}K_\alpha(t-t_i)}\right]y_i = n^{-1}\sum_{i=1}^{n}W_{\alpha i}(t)y_i \tag{1}$$

If the form of the addition in the equation is described completely, it can be written in equation (2).

$$g_\alpha(t) = n^{-1}W_{\alpha 1}(t)y_1 + n^{-1}W_{\alpha 2}(t)y_2 + \ldots + n^{-1}W_{\alpha n}(t)y_n \tag{2}$$

If notated in the form of a matrix, it will be like equation (3).

$$\mathbf{g}_\alpha(t) = \mathbf{V}(\alpha)\mathbf{y} \tag{3}$$

### 2.2. Spline Truncated Nonparametric Regression

Spline Truncated is a very popular estimator that is often used, because it has a good visual interpretation, can handle smooth functions, and is flexible [1] [15]. In general, the Spline Truncated function with degrees and knots is a function that can be written in equation (4).

$$h(z_i) = \sum_{j=0}^{M}\alpha_j z_i^j + \sum_{k=1}^{r}\beta_k(z_i-\xi_k)_+^M \tag{4}$$

with truncated function

$$(z_i-\xi_k)_+^M = \begin{cases}(z_i-\xi_k)^M & ,z_i \geq \xi_k\\ 0 & ,z_i \leq \xi_k\end{cases}$$

In general, the Spline Truncated model can be written in equation (5).

$$\begin{aligned} y_i &= \sum_{j=0}^{M}\alpha_j z_i^j + \sum_{k=0}^{r}\beta_k(z-\xi_k)_+^M + \varepsilon_i \\ &= \alpha_0 + \alpha_1 z_i + \ldots + \alpha_M z_i^M + \beta_1(z_i-\xi_1)_+^M + \ldots + \beta_r(z_i-\xi_r)_+^M + \varepsilon_i \end{aligned} \tag{5}$$

Equation (5) can be denoted in the form of a matrix written in equation (6).

$$\mathbf{h}(z_i) = \mathbf{G}(\xi)\boldsymbol{\theta} \tag{6}$$

### 2.3. Fourier Series

Fourier series is an estimator whose regression curve shows sine and cosine waves and has high flexibility. If the data being investigated has an unknown pattern and there is a seasonal trend, the Fourier series estimator can be used [12] [16]. Recently, many alternatives have been developed and studied by researchers. The first researcher of the Fourier series was De Jong [17]. The Fourier series equation is in equation (7).

$$f(x_i) = bx_i + \frac{1}{2}\alpha_0 + \sum_{k=1}^{K}\alpha_k \cos kx_i \tag{7}$$

where $b, a_0, a_k, k = 1,2,....,K$ are the model parameters.

Equation (7) can be written as equation (8).

$$f(x_i) = bx_i + \frac{1}{2}\alpha_0 + \alpha_1 \cos x_i + \alpha_2 \cos 2x_i + ... + \alpha_K \cos Kx_i \tag{8}$$

If equation (8) is written in matrix form in equation (9).

$$\mathbf{f}(x) = \mathbf{N}(K)\boldsymbol{\rho} \tag{9}$$

### 2.4. Kernel Mix Estimator, Spline Truncated and and Fourier Series

In this mixed estimator, there are three predictor components, the first predictor component, the regression curve is approached using the Spline Truncated, the second uses the kernel, the third uses the Fourier series. Paired data $(z_{1i}, z_{2i}, ..., z_{pi}, t_{1i}, t_{2i}, ..., t_{qi}, x_{1i}, x_{2i}, ..., x_{ri}, y_i)$ with $i = 1,2,...,n$ which has a relationship following a nonparametric regression model. The variable $z_{1i}, z_{2i}, ..., z_{pi}, t_{1i}, t_{2i}, ..., t_{qi}, x_{1i}, x_{2i}, ..., x_{ri}$ is the predictor variable and $y_i$ is the response variable. In general, the nonparametric regression model for the mixture of the kernel, spline, and Fourier series is in equation (10).

$$\begin{aligned} y_i &= \mu(z_{1i}, z_{2i}, ..., z_{pi}, t_{1i}, t_{2i}, ..., t_{qi}, x_{1i}, x_{2i}, ..., x_{ri}) + \varepsilon_i \\ &= \mu(z_i, t_i, x_i) + \varepsilon_i \end{aligned} \tag{10}$$

where $z_i = (z_{1i}, z_{2i}, ..., z_{pi})'$, $t_i = (t_{1i}, t_{2i}, ..., t_{qi})'$, and $x_i = (x_{1i}, x_{2i}, ..., x_{ri})'$. Random error $\varepsilon_i$ is normally distributed with $E(\varepsilon_i) = 0$ and $Var(\varepsilon_i) = \sigma^2$. The regression curve of $\mu(z_i, t_i, x_i)$ is assumed to be additive, so it can be written as in equation (11).

$$\mu(z_i,t_i,x_i) = h(z_{1i}) + h(z_{2i}) + ... + h(z_{pi}) + g(t_{1i}) + g(t_{2i}) + ... g(t_{qi}) + f(x_{1i}) + f(x_{2i}) + ... + f(x_{ri}) \tag{11}$$

The regression curve $\mu(z_i, t_i, x_i)$ is called a nonparametric mixed regression curve which is grouped into three components of the regression curve, namely the Spline Truncated, kernel, and Fourier series regression curve components. The equation can be written like equation (12).

$$\mu(z_i,t_i,x_i) = \sum_{s=1}^{p} h(z_{si}) + \sum_{u=1}^{q} g(t_{ui}) + \sum_{v=1}^{r} f(x_{vi}) \tag{12}$$

## 3. Result and Discussion

This section discusses the results of the modeling process using a mixed nonparametric regression of the kernel, Spline Truncated, and Fourier series which will be applied to the human development index case in East Java Province in 2017.

Table 1 shows that each Regency / City in East Java Province has different characteristics for all variables, namely Human Development Index, morbidity, APM SMA, GRDP Per Capita.

**Table 1.** Descriptive Statistics of Response Variables and Predictor Variables

| Variable | Number of Observations | Min | Max | Range | Mean | Standard Deviation |
|---|---|---|---|---|---|---|
| (1) | (2) | (3) | (4) | (5) | (6) | (7) |
| Y | 38 | 59.90 | 81.07 | 21.17 | 70.35 | 5.31 |
| X1 | 38 | 8.48 | 27.04 | 18.56 | 14.33 | 3.72 |
| X2 | 38 | 34.22 | 84.33 | 50.11 | 63.75 | 11.89 |
| X3 | 38 | 16.96 | 408.66 | 391.70 | 51.84 | 66.33 |

Identification using a scatter plot will determine the pattern of the relationship between the response variable and each predictor variable in Figure 1.

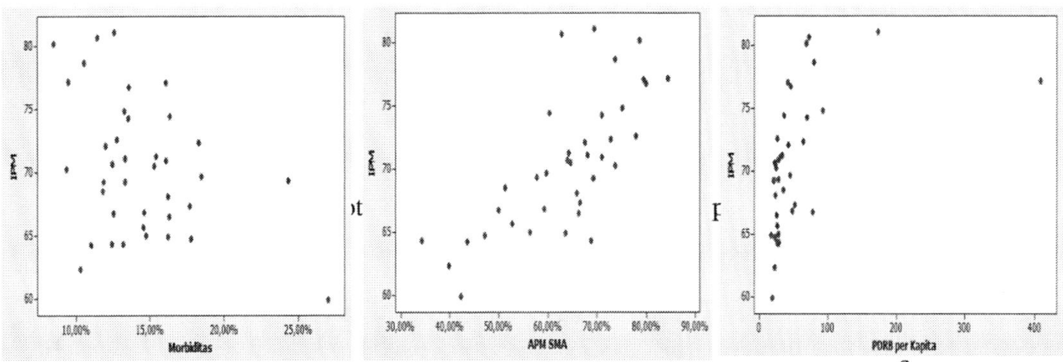


**Figure 1.** Scatter plot for each response variable with predictor variables

In Figure 1a, it can be seen that the relationship pattern of $Y$ and $X_1$ shows a relationship pattern that does not follow a certain pattern so that it can use the Fourier series nonparametric regression approach. Figure 1b shows that the relationship pattern of $Y$ and $X_2$ shows a relationship pattern that does not follow a certain pattern so that it can use the nonparametric spline regression approach. Figure 1c shows the relationship pattern of $Y$ and $X_3$ shows a relationship pattern that does not follow a certain pattern so that it can use the nonparametric kernel regression approach.

Judging from the scatterplot above, the mixed estimation model of the Kernel, Spline Truncated, and Fourier series is an appropriate method, because the pattern of the relationship between predictor variables and response variables is unknown, changes in certain sub-intervals, and there is a loop with a downward trend. The mixed estimation model of the Kernel, Spline, and Fourier series are selected by looking at the smallest GCV. This study uses knot points 1 to 3 and oscillations 1 to 3 so that the smallest GCV and MSE values, and maximum $R^2$, optimal knot points, bandwidth, and oscillations, as well as the parameter values of the best models.

The results obtained from the HDI modeling process on the factors that are thought to influence it, namely morbidity, APM SMA, and GRDP per capita using a mixed estimator of the kernel, spline, and Fourier series with one oscillation and one-knot point are shown in Table 2. Table 2 shows that the smallest GCV and MSE values are 6.6273 and 28.3273, respectively. In Table 3, the GCV and MSE values are generated using two knots and two oscillations. Table 3 shows that the smallest GCV and MSE values are 14.9144 and 7.1923, respectively.

**Table 2.** GCV and MSE values with one-knot point and one oscillation

| No | $\xi$ | $\varphi$ | k | MSE | GCV |
|---|---|---|---|---|---|
| 1 | 35.5049 | 7.9939 | 1 | 7.1773 | 16.2666 |
| 2 | 36.7897 | 7.9939 | 1 | 7.1773 | 17.6161 |
| 3 | 38.0746 | 7.9939 | 1 | 7.1773 | 18.4572 |
| 4 | 35.5049 | 7.9939 | 1 | 7.1773 | 19.2121 |
| 5 | 39.3595 | 7.9939 | 1 | 7.1773 | 19.8570 |

**Table 3.** GCV and MSE values with two knots and two oscillations

| No | $\xi_1$ | $\xi_2$ | $\varphi$ | k | MSE | GCV |
|---|---|---|---|---|---|---|
| 1 | 44.4990 | 48.3536 | 7.9939 | 2 | 7.1923 | 14.9144 |
| 2 | 50.9233 | 59.9174 | 7.9939 | 2 | 6.9389 | 14.9214 |
| 3 | 53.4931 | 61.2023 | 7.9939 | 2 | 7.0015 | 14.9541 |
| 4 | 48.3536 | 63.7721 | 7.9939 | 2 | 7.0298 | 14.9702 |
| 5 | 40.6444 | 44.499 | 7.9939 | 2 | 7.0561 | 14.9786 |

**Table 4.** GCV and MSE values with three knots and three oscillations

| No | $\xi_1$ | $\xi_2$ | $\xi_3$ | $\varphi$ | k | MSE | GCV |
|---|---|---|---|---|---|---|---|
| 1 | 53.4931 | 58.6326 | 66.3418 | 7.9939 | 3 | 6.3569 | 10.2322 |
| 2 | 47.0687 | 59.9174 | 65.0569 | 7.9939 | 3 | 6.3407 | 10.2618 |
| 3 | 41.9292 | 63.7720 | 70.1964 | 7.9939 | 3 | 6.2501 | 10.2800 |
| 4 | 47.0687 | 53.4931 | 70.1964 | 7.9939 | 3 | 6.5296 | 10.2803 |
| 5 | 53.4931 | 66.3418 | 70.1964 | 7.9939 | 3 | 6.8384 | 10.2947 |

In Table 4, the GCV and MSE values are generated using three knots and three oscillations. Table 4 shows that the smallest GCV and MSE values are 10.2322 and 6.3569, respectively. The comparison of the minimum MSE and GCV values obtained from one-knot point and one oscillation, two knot points and two oscillations, and three knot points and three oscillations are shown in Table 5.

IOP Conf. Series: Materials Science and Engineering 1115 (2021) 012024 doi:10.1088/1757-899X/1115/1/012024

**Table 5.** GCV, MSE, and R2 values for each model

| Model | MSE | GCV | $R^2$ |
|---|---|---|---|
| 1 Knot Point and 1 Oscillation | 6.6273 | 28.3273 | 75.8877 |
| 2 Knot Point and 2 Oscillation | 7.1923 | 14.9144 | 73.8321 |
| 3 Knot Point and 3 Oscillation | 6.3570 | 10.2328 | 75.1041 |

Based on Table 5, it can be seen that the smallest MSE and GCV values, as well as the largest $R^2$, are the three-point knot and three oscillation models with an MSE value of 6.3570, the GCV value is 10.2328, and R2 is 75.1041 so it shows that the best model used in the case of HDI in East Java Province is with using a mixed regression model of the spline, kernel, and Fourier series with three knots and three oscillations. The parameter estimates are in Table 6.

The model is formed in equation (13).

$$\hat{y} = \hat{\alpha}_0^* + \hat{\alpha}_{11} z_{1i} + \hat{\beta}_{11}(z_{1i} - \xi_{11})_+ + \hat{\beta}_{21}(z_{1i} - \xi_{21})_+ + \hat{\beta}_{31}(z_{1i} - \xi_{31})_+ + n^{-1}\sum_{i=1}^{n}\left[\frac{K_\phi(t-t_i)}{n^{-1}\sum_{j=1}^{n} K_\phi(t-t_j)}\right] y_i + \hat{b}x_{1i} + \frac{1}{2}\hat{a}_0 + \sum_{k=1}^{3}\hat{a}_1 \cos kx_{1i} \quad (13)$$

**Table 6.** Estimated parameters

| Parameter | Estimate | Parameter | Estimate |
|---|---|---|---|
| $\alpha_0$ | -5.280 | $a_0$ | -0.143 |
| $\alpha_1$ | 0.070 | $a_1$ | -0.327 |
| $\beta_{11}$ | 0.855 | $\Phi$ | 7.994 |
| $\beta_{21}$ | -1.035 | $\xi_{11}$ | 53.493 |
| $\beta_{31}$ | 0.295 | $\xi_{21}$ | 58.633 |
| b | 0.000 | $\xi_{31}$ | 66.342 |

Thus, the mixed nonparametric regression estimator model of Spline Truncated, Kernel, and Fourier series in equation (14).

$$\hat{y} = -5.280 + 0.070 z_{1i} + 0.855(z_{1i} - 53.493)_+ - 1.035(z_{1i} - 58.633)_+ + 0.295(z_{1i} - 66.342)_+ + \sum_{i=1}^{38}\left[\frac{\frac{1}{7.994}K(\frac{t-t_i}{7.994})}{\sum_{j=1}^{38}\frac{1}{7.994}K(t-t_j)}\right] y_i - 0.072 + \sum_{k=1}^{3}(-0.327)\cos kx_{1i} \quad (14)$$

where K is the Gaussian kernel function. With $R^2$ is 75.1041. Visualization of $y$ and $\hat{y}$ values for each possible knot point and the tested oscillations are as follows:

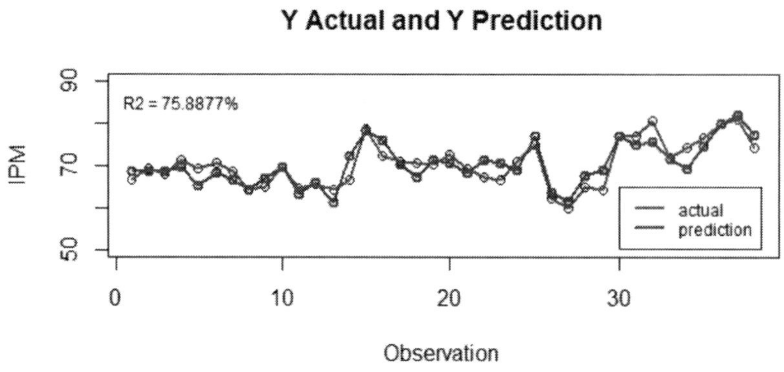


**Figure 2**. Visualization of values from 1 knots and 1 oscillations

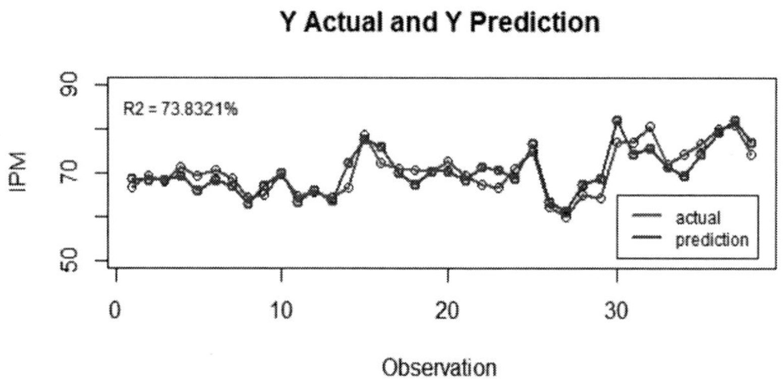


**Figure 3.** Visualization of values from 2 knots and 2 oscillations

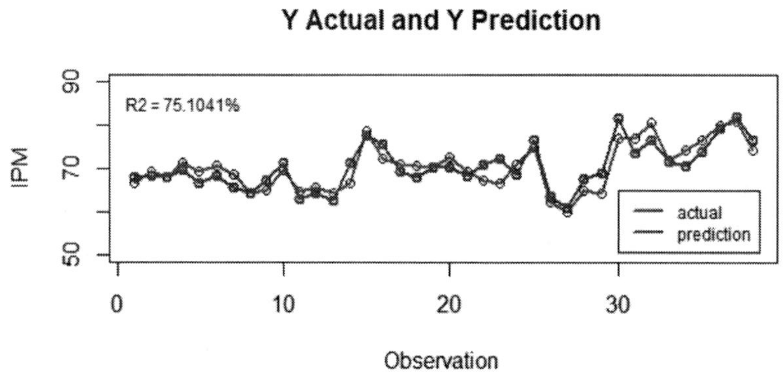


**Figure 4.** Visualization of values from 3 knots and 3 oscillations

It can be seen that Figure 4, Y the prediction is more similar to the actual Y, compared to Figure 2 and Figure 3.

## 4. Conclusions

Based on the results and discussion, it can be concluded that the case modeling of District / City Human Development Index in East Java Province in 2017 using a nonparametric regression approach of a mixture of the kernel, Spline Truncated, and Fourier series with three knots and three oscillations gave quite good results because the coefficient of determination of the estimator is 75.1041%, with the estimator model is

$$\hat{y} = -5.280 + 0.070 z_{1i} + 0.855(z_{1i} - 53.493)_+ - 1.035(z_{1i} - 58.633)_+ + 0.295(z_{1i} - 66.342)_+$$
$$+ \sum_{i=1}^{38} \left[ \frac{\frac{1}{7.994} K(\frac{t - t_i}{7.994})}{\sum_{j=1}^{38} \frac{1}{7.994} K(t - t_j)} \right] y_i - 0.072 + \sum_{k=1}^{3} (-0.327) \cos k x_{1i}$$

## References

[1] Eubank R L 1999 *Nonparametric Regression and Spline Smoothing Second Edition* New York: Marcel Dekker
[2] Saputro D R S, Sukmayanti A, Widyaningsih P 2019 The nonparametric regression model using Fourier series approximation and penalized least squares (PLS) (case on data proverty in East Java) *Journal of Physics: Conf. Series*, **1188** 012019
[3] Saputro D R S, Demu K R, and Widyaningsih, P 2018 Nonparametric Spline Truncated Regression on the Data of Human Development Index (HDI) in Indonesia *Journal of Physics: Conference Series* **1028** (1) 012219
[4] Budiantara I N, Mariati N P A M, Ratnasari V, Zain I, Ratna M, Sudiarsa I W, Mardianto M F F and Hendayanti N P N 2015 Comparison Spline Truncated and Fourier Series in Multivariable Nonparametric Regression Models (Application: Data of Poverty in Papua, Indonesia) *International Journal of Basic and Applied Sciences* **15** 9-12
[5] Mariati N P A M, Budiantara I N and Ratnasari V 2019 Modeling Poverty Percentages in the Papua Islands using Fourier Series in Nonparametric Regression Multivariable *Journal of Physics: Conference Series* **1397** (1) 012071
[6] Sudiarsa I W, Budiantara I N and Purnami S W 2015 Combined estimator Fourier series and spline truncated in multivariable nonparametric regression *Applied Mathematical Sciences* **9** (100) 4997-5010
[7] Aydin D 2007 A comparison of the nonparametric regression models using smoothing spline and kernel regression *World Academy of Science, Engineering and Technology* **36** 253-257
[8] Hardle W 1994 *Applied Nonparametric Regression* Berlin: Springer-Verlag
[9] Amini M and Roozbeh M 2015 Optimal partial ridge estimation in restricted semiparametric regression models *Journal of Multivariate Analysis* **136** 26-40
[10] Budiantara I N 2009 *Spline in Nonparametric and Semiparametric Regression: A Present and Future Model of Statistics*, Inagural Speech fot Professor's Position at the Department of Statistics Surabaya: ITS.
[11] Asrini L J and Budiantara I N 2014 Fourier Series Semiparametric Regression Models (Case Study: The Production of Law Land Rice Irrigation in Central Java). *ARPN Journal of Engineering and Applied Sciences* **9** 1501-1506
[12] Bilodeau M 1992 Fourier Smoother and Additive Models *The Canadian Journal of Statistics* **3** 257-259
[13] Watson G S 1964 Smooth regression analysis *Sankhya: The Indian Journal of Statistic, Series A* **26** (4) 359-372

[14] Nadaraya E A 1964 On Estimating regression. *Theory of Probability and its Applications* **9** (1) 1412
[15] Budiantara I N 2007 Inferensi Statistik untuk Model Spline *Jurnal Mat-Stat* **7** 1-14
[16] Tripena A and Budiantara I N 2006 Fourier Estimator in Nonparametric Regression *International Conference On Natural and Applied Natural Scienes. Ahmad Dahlan University*
[17] De Jong P 1977 The Fast Fourier Transform Spectral Estimator *Journal of Statistical Soc.* **39** (3) 327-330

IOP Conf. Series: Materials Science and Engineering 1115 (2021) 012025 doi:10.1088/1757-899X/1115/1/012025

# Analysis of temporal change in land use and impact of water resource in Delta's Jeneberang Makassar, South Sulawesi

**M F I Massinai, K Khatimah*, T Razin, N S Arifuddin, M N A Rahman, M A F Dzaky**

Department of Geophysics, Faculty of Mathematics and Natural Sciences, Hasanuddin University, Makassar, Indonesia.

*Coressponding author: text.khusnul@gmail.com

**Abstract**. Twelve groundwater samples were collected in the delta Jeneberang River to study the possible impact of municipal solid waste (MSW) cause by land use cover change. The groundwater samples were collected for assessment in terms of physicochemical parameters such as pH and turbidity. A statistical analysis has been done with the calculation of the correlation coefficient and the water quality index (WQI) was used to analyze the groundwater quality of the study area. The study has revealed that the correlation between these parameters is about -0.476155325. Further, the WQI at the first location has a rating scale of C known as a poor water quality with an index value of 61.5 and the worst is at the 12th location has an index value of 90.58, or with a D rating as very poor water quality.

## 1. Introduction

Makassar is the 5th largest industrial and metropolitan city with a population of 1.6 million and the enhancement of industrial activity resulted in a sharp increase in urban land demand along with an increase in urban population. Therefore, the amount of municipal solid waste (MSW) generation has increased with the population. Municipal solid waste (MSW) disposal in the surrounding environment has increased a large amount due to rapid urbanization and industrialization. Disposal of solid waste and sewage, urban runoff, agricultural activities and polluted surface water are major contributors to deteriorate urban groundwater resources [1].

Since the era of industrialization and rapid population growth, land-use change phenomena have strongly accelerated in many regions which directly impacting the hydrology of the catchment area [2]. Groundwater supplies are replenished, or recharged, by rain and snowmelt that seeps down through the rocks from the land surface. The quality of groundwater is a result of all processes and reactions that have acted on the water from the moment it condensed in the atmosphere until it is discharged by a well [3]. Overpopulation has resulted in an increase in demand which eventually has lead to an accelerated rate of urbanization which results in a reduction in infiltration, and thus affecting the groundwater recharge and storage [4] as water resources.

The determination of physical and chemical parameters of water samples that also dictate various other life processes should be taken as an environmentally viable study. Statistical studies have been done on physicochemical parameters such as pH and turbidity to find out the correlation coefficient of samples. The aim of this study is to analyze land use cover change implicates the quantity and quality of groundwater as a water resource in the delta Jeneberang River.

## 2. Materials and Methods

### 2.1. Study Area

The delta Jeneberang River location was selected for this study. It is 3 km far from the downtown and directly adjacent to Gowa district. Groundwaters samples were collected from tube-wells from 12 locations around the delta (See Figure 1).

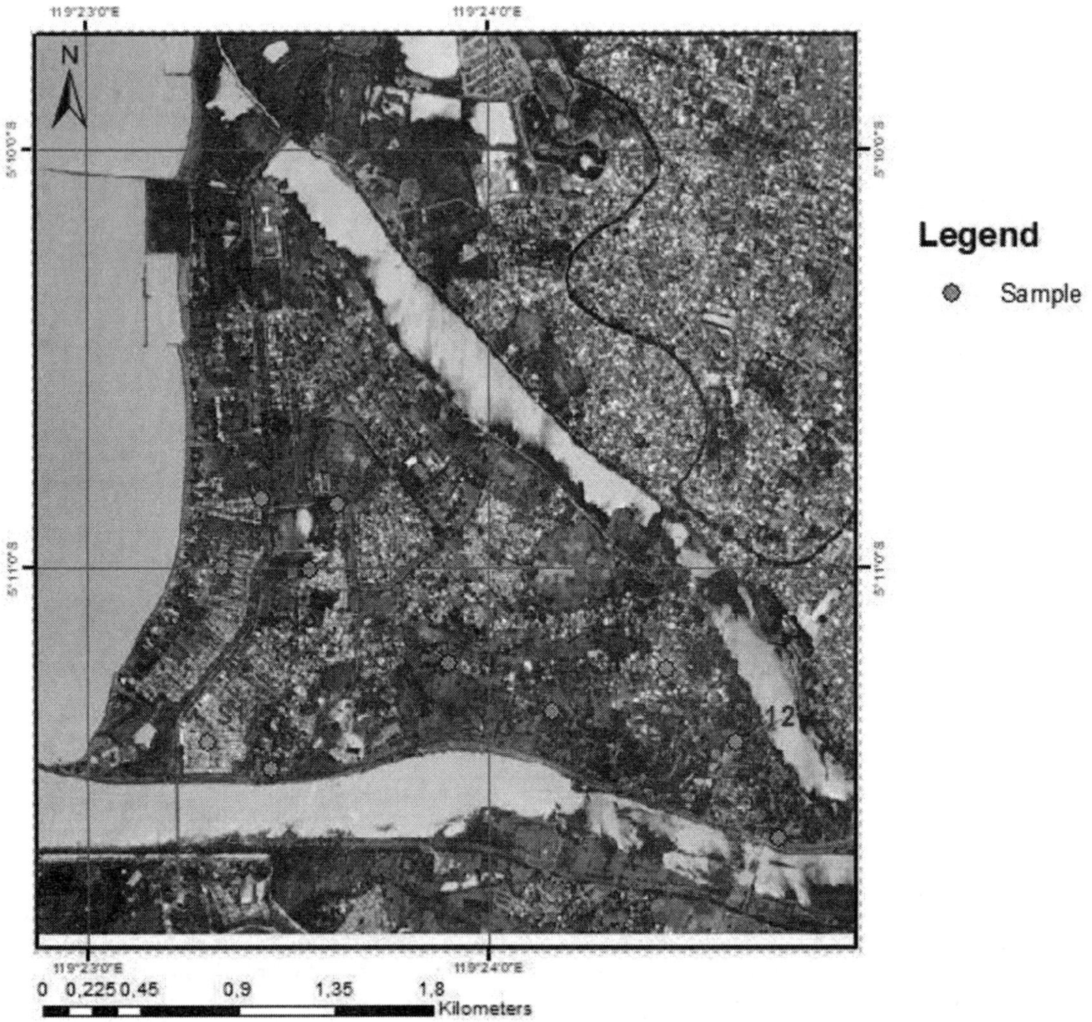


**Figure 1.** Selected study area at Jeneberang Delta.

### 2.2. Landsat Data Analysis

The addition of Landsat-8 data from the USGS website and processing the data in ArcGIS (Hasanuddin University licensed) for land use cover change analysis by comparing two or more imageries data in the same area but with different time taking [5].

### 2.3. Analytical Method

The collected groundwater samples then measured in a laboratory. In case of physicochemical parameters includes pH was measured by PHT-026 Multi-function water quality meter and the turbidity parameter was measured by Turbidity Meter Lutron TU-2016.

### 2.4. Statistical Anaysis

The correlation between the calculated parameters is pH and turbidity water samples were analyzed statically through Microsoft Excel by equation (1).

$$r = \frac{n(\sum xy)-\sum x*\sum y}{\sqrt{[n\sum x^2-(\sum x)^2]*[n\sum y^2-(\sum y)^2]}} \tag{1}$$

with,

r = Coefficient of correlation
x = Individual reading of 1[st] parameter
y = Individual reading of 2[nd] parameter
n = Number of values of single parameter

The correlation among the different parameters will be true when the value of correlation coefficient (r) is high and approaching one [6].

### 2.5. Drinking Water Quality Index (DWQI)

Water quality assessment is carried out using WQI, which is widely used to evaluate the quality of drinking water. WQI was initially discovered by [7] and then modified by [8]. According to the report of the World Health Organization (WHO) in 2004, using WQI could help to clarify the combinatorial effects of each parameter and also the qualitative parameters of drinking water quality [9]. Each parameter has a relative role in determining the quality of drinking water. The relative weight was calculated using the following equation (2) :

$$Wi = \sum \frac{wi}{\sum_{i=1}^{n} wi} \tag{2}$$

with,

Wi = Relative weight of each parameter
wi = Weight of each parameters
n = Number of parameter

For each parameter, the quality rating scale was calculated by dividing its concentration in each water sample to its respective standards (released by World Health Organization 2011) and then multiplied the results by 100 through equation (3):

$$qi = \left(\frac{Ci}{Si}\right) \times 100 \tag{3}$$

with,

qi = The quality rating
Ci = Concentration of each parameter (mg/L)
Si = Standard limit (mg/L) according to WHO released in 2011

Water quality is categorized into 5 types which are shown in Table 1. To calculate the sub-index of each parameter, the weight (Wi) of each parameter can be seen in Table 2.

**Table 1.** Rating of water quality index [10].

| WQI Value | Rating of water quality | Grading |
|---|---|---|
| 0 – 25 | Excellent water quality | A |
| 26 – 50 | Good water quality | B |
| 51 – 75 | Poor water quality | C |
| 76 – 100 | Very poor water quality | D |
| Above 100 | Unsuitable for drinking purposes | E |

**Table 2.** The weight (wi) and WHO standard values for drinking water.

| Parameters | Concentration (mg/L) (Ci) | WHO (mg/L) (Si) | Weight (wi) |
|---|---|---|---|
| pH | 7.64 | 6.5 – 8.5 | 4 |
| Turbidity | 3.01 | 4 | 3 |
| | | Sum = | 7 |

The final part of the WQI calculation is to computing the Sli for each parameter and the sum of Sli values is the final value of WQI, indicated by equations (4) and (5):

$$Sli = Wi \times qi \quad (4)$$

with, Sli is the sub-index of each parameter

$$WQI = \sum_{i=1}^{n} Sli \quad (5)$$

with, WQI is the water quality index.

## 3. Results and Discussion

Analysis results of the physicochemical parameters namely pH and turbidity of the groundwater samples are shown in Table 3. Further, some statistical analysis has done with the results shown in Table 4.

Results from the laboratory test, the values of pH and turbidity parameters in the 12 groundwater samples in the study area obtained several parameters for statistical analysis such as the average value, standard deviation, variance, *n* (total sample), and correlation. The average value of the pH parameter is 6.605 in accordance with WHO standard whereas for turbidity parameter obtained 2.094 NTU. For variance, the measurement differences are relatively the same in both parameters. However, the correlation between pH and turbidity parameters did not have a significant value obtained by -0.476.

**Table 3.** Values of Jeneberang Delta's data sample with coordinate.

| Sample | Lat. | Long. | Turbidity Meter (NTU) | pH | | | | |
|---|---|---|---|---|---|---|---|---|
| | | | | 1 | 2 | 3 | Average | Stdev |
| 1 | -5.1941600 | 119.41188 | 2.31 | 6.54 | 6.52 | 6.53 | 6.53 | 0.0100000 |
| 2 | -5.1891178 | 119.40258 | 1.02 | 6.64 | 6.63 | 6.64 | 6.64 | 0.0057735 |
| 3 | -5.1874341 | 119.40724 | 2.08 | 6.81 | 6.82 | 6.64 | 6.76 | 0.1011599 |
| 4 | -5.1871768 | 119.39834 | 2.19 | 6.44 | 6.43 | 6.42 | 6.43 | 0.0100000 |
| 5 | -5.1807760 | 119.39376 | 2.31 | 6.83 | 6.83 | 6.82 | 6.83 | 0.0057735 |
| 6 | -5.1913695 | 119.39091 | 1.89 | 6.85 | 6.86 | 6.85 | 6.85 | 0.0057735 |
| 7 | -5.1833331 | 119.38896 | 1.79 | 6.95 | 6.94 | 6.93 | 6.94 | 0.0100000 |
| 8 | -5.1833331 | 119.38896 | 1.73 | 6.86 | 6.87 | 6.88 | 6.87 | 0.0100000 |
| 9 | -5.1833933 | 119.39265 | 1.77 | 6.92 | 6.93 | 6.90 | 6.92 | 0.0152753 |
| 10 | -5.1805609 | 119.39053 | 1.94 | 7.02 | 7.00 | 7.01 | 7.01 | 0.0100000 |
| 11 | -5.1902899 | 119.38834 | 1.82 | 5.63 | 5.62 | 5.61 | 5.62 | 0.0100000 |
| 12 | -5.1903389 | 119.41013 | 4.82 | 5.88 | 5.87 | 5.86 | 5.87 | 0.0100000 |

After the statistical analysis has done, the final stage of the computation is calculating the WQI (Water Quality Index) using equations (2), (3), (4), and (5). The result is shown in Table 5 and Table 6.

The value of WQI in the first location till location 11 was found variation values in the ranges of 51 to 75, so could be said that the water quality is poor and grading as C (Table. 1). Except for the 5th location, reveals WQI value 76.76269841 and it indicates in that location had very poor water quality and grading as D. Lastly, for location 12 same as the 5th location the water quality in this location is very poor and not suitable for drinking purpose because the QWI value shown 90.58095238 which is between the range of 76 to 100 and grating as D (Table. 1).

**Table 4.** Values of physicochemical parameters.

| Location | pH | Turbidity |
|---|---|---|
| 1 | 6.53 | 2.31 |
| 2 | 6.64 | 1.02 |
| 3 | 6.76 | 2.08 |
| 4 | 6.43 | 2.19 |
| 5 | 6.83 | 2.31 |
| 6 | 6.85 | 1.89 |
| 7 | 6.94 | 1.79 |
| 8 | 6.87 | 1.73 |
| 9 | 6.92 | 1.77 |
| 11 | 7.01 | 1.94 |
| 10 | 5.62 | 1.82 |
| 12 | 5.87 | 4.82 |
| Sum | 79.26 | 25.13 |
| Average | 6.605 | 2.0941667 |
| Standard Deviation (STDEV) | 0.4395372 | 0.7690071 |
| Variance | 0.1931929 | 0.591372 |
| n | 12 | 12 |
| Correlation | -0.476155325 | |
| WHO Standard | 6.5 – 8.5 | 4 |

**Table 5.** Results of WQI for Location 1.

| Parameters | Concentration (mg/L) (Ci) | WHO (mg/L) (Si) | Weight (wi) | Relative Weight (Wi) | qi (Ci/Si)*100 | SIi (Wi*qi) |
|---|---|---|---|---|---|---|
| pH | 6.53 | 6.5 – 8.5 | 4 | 0.57142857 | 87.0666667 | 49.75238 |
| Turbidity | 2.31 | 4 | 3 | 0.42857143 | 57.75 | 24.75 |
| | | Sum = | 7 | | WQI = | 74.50238 |

**Table 6.** Results of WQI for all location sample.

| Sample Location | Result WQI |
|---|---|
| 1 | 74.50238095 |
| 2 | 61.49365079 |
| 3 | 73.76507937 |
| 4 | 72.45476190 |
| 5 | 76.76269841 |
| 6 | 72.46587302 |
| 7 | 72.05476190 |
| 8 | 70.87857143 |
| 9 | 71.66269841 |
| 10 | 74.19523810 |
| 11 | 62.31904762 |
| 12 | 90.58095238 |

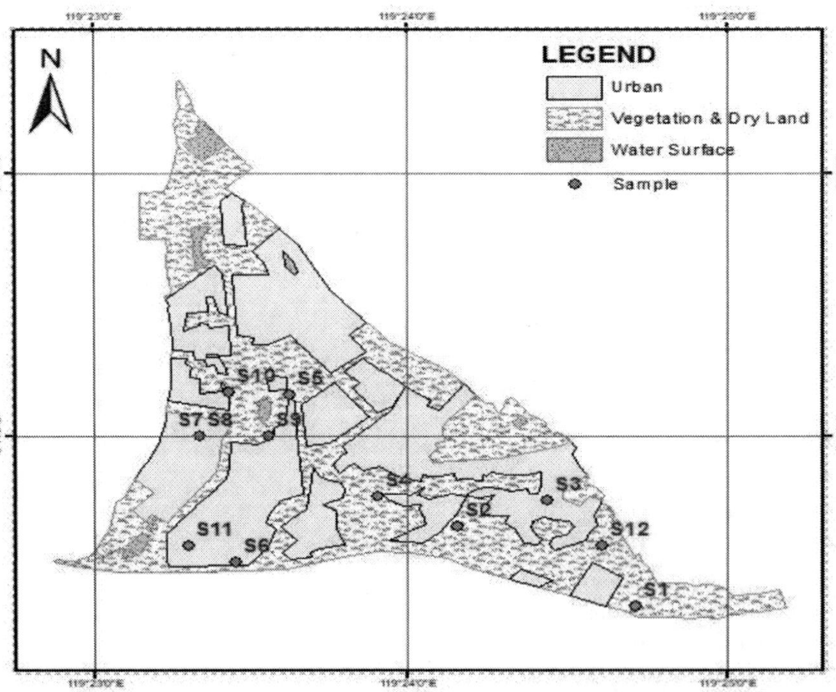


**Figure 2.** Spatial digitations of land use Jeneberang Delta's.

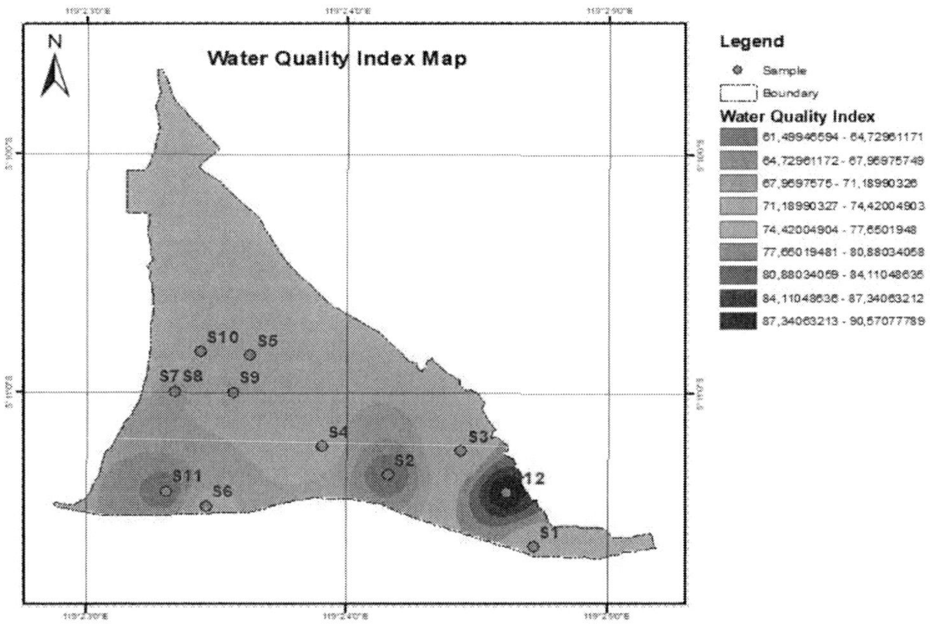


**Figure 3.** Spatial variation of Water Quality Index using IDW.

The varied results obtained are clearly caused by several factors like municipal solid waste generation increased since there is a land-use change from agricultural land purpose to urban land and impact the groundwater catchment area and also the study area directly adjacent to the Jeneberang river flow.

Digitization was carried out on ArcGIS which produced a map of land use cover change in the study area shown in Figure 2 and also has done an Inverse Distance Weighted (IDW) interpolation for all WQI values, the graph in Figure 3 represents the spatial variation of WQI around the delta Jeneberang River.

## 4. Conclusion

From the analysis of temporal change in land use and how it impacts water resources in the delta Jeneberang River, it is found that the correlation coefficient (a value between -1 and 1) could use to display how strongly two variables are related to each other. Statistical analysis used for indicates the value of physicochemical parameters is a good measure before detailed observation about groundwater pollution because its time saving also cost-effective. WQI was used to analyze the underground water quality of the study area. Hence we can conclude that an accelerated rate of urbanization could implicate the quantity and quality of groundwater as water resources. The result showed that overall the groundwater quality in the study area needs to be treated before use and for more detailed research, parameters of water quality should be a variation to find out how much environmental damage has occurred the same as the correlation results can vary between parameters.

## Acknowledgment

This research was supported by Lembaga Penelitian dan Pengabdian Masyarakat (LPPM) Hasanuddin University through research grant "Hibah Penelitian Internal 2020: Penelitian Dosen Penasehat Akademik Scheme" and Geophysics Department, Faculty of Mathematics and Natural Science Hasanuddin University.

## References

[1] Jain C K, Bhatia K K S and Vijay T 1995 *Ground water quality monitoring and evaluation in and around Kakinada Andhra Pradesh, Technical Report, CS (AR)172* (Roorkee)

[2] Dams J, Woldeamlak S T and Batelaan O 2008 Predicting Land-use Change and Its Impact on the Groundwater System of the Kleine Nete Catchment, Belgium *Hydrol. Earth Syst. Sci.* **12** 1369–1385

[3] Saxena S, Ramakrishnan S, Soni D, Meena K K and Arora S 2020 Water quality monitoring of residential area affected by solid waste landfill site in Jaipur city *Test Eng. Manag.* **82** 14609–22

[4] Sharma P, Gupta G, Prabhakar P, Tiwari S, Kathait P, Pathak Y, Mishra N and Kumar S 2017 Land Use Land Cover Change Impact on Water Resources-a Review *Int. J. Adv. Eng. Sci. Res.* **4** 7–14

[5] Zhu Z 2017 Change detection using landsat time series: A review of frequencies, preprocessing, algorithms, and applications *ISPRS J. Photogramm. Remote Sens.* **130** 370–84

[6] Joshi D M, Bhandari N S, Kumar A and Agrawal N 2009 Statistical Analysis of Physicochemical Parameters of Water of River Ganga in Haridwar District *Rasayan J. Chem.* **2** 579–87

[7] Brown R M, McClelland N I, Deininger R A and Tozer R G 1970 A Water Quality Index-Do We Dare? *Water Sew. Work.* **117** 339–43

[8] Backman B, Bodiš D, Lahermo P, Rapant S and Tarvainen T 1998 Application of a groundwater contamination index in Finland and Slovakia *Environ. Geol.* **36** 55–64

[9] Abbasnia A, Yousefi N, Mahvi A H, Nabizadeh R, Radfard M, Yousefi M and Alimohammadi M 2019 Evaluation of groundwater quality using water quality index and its suitability for assessing water for drinking and irrigation purposes: Case study of Sistan and Baluchistan province (Iran) *Hum. Ecol. Risk Assess. An Int. J.* **25** 988–1005

[10] Tyagi S, Sharma B, Singh P and Dobhal R 2013 Water Quality Assessment in Terms of Water Quality Index *Am. J. Water Resour.* **1** 34–38

IOP Conf. Series: Materials Science and Engineering 1115 (2021) 012026 doi:10.1088/1757-899X/1115/1/012026

# Professional Engineering and World Class Perspectives of Industrial Engineering and Digital Transformation

**K E N Soebandrija***

Industrial Engineering Department, Faculty of Engineering, Bina Nusantara University, Jakarta, Indonesia 11480

*Corresponding author: Knugroho@Binus.edu

**Abstract.** This paper conveys the theoretical discourses of Industrial Engineering and Digital Transformation. It proposes perspectives of Professional Engineering and World Class' Industrial implementation by combining Indonesian Local Wisdom and Global Benchmark Settings. The objective of this paper is to intertwine both theoretical and industrial implementation among Academician, Business and Government (ABG). Precisely, this harmonized industrial implementation is deemed indispensable by considering several pillars. First pillar constitutes the prior trigger of Industry 4.0 (IR 4.0), as one of Global Benchmark Settings. Subsequently, second pillar refers to the Indonesian Local Wisdom, in term of Making Indonesia 4.0 (MI 4.0). Ultimately, third pillars are pertaining global benchmark settings, in term of Industry X.0, Society 5.0, Hallyu 2.0. The research methodology in this paper refers to the quantitative method that enables the achievement of the objective of this paper. The discussion of this paper elaborates the methods and methodology to achieve the objective of this paper. Subsequently, the mentioned discussion proceeds to the necessity to have a holistic industrial implementation among ABG. The result of this paper is expected to equip current professional engineers and future generations. They are expected to be able to cope with new norm of new paradigm in the future.

## 1. Introduction

This paper conveys the theoretical discourses of Industrial Engineering and Digital Transformation. It proposes perspectives of Professional Engineering and World Class' Industrial implementation by combining Indonesian Local Wisdom and Global Benchmark Settings.

Theoretical discourses of Industrial Engineering refer to the most cited scholar works within Industrial Engineering, in particular within Industry 4.0 study, as illustrated in Table 1.

Those most cited scholar works are originated mostly from the sources of the following but not limited to the Manufacturing Letters; Procedia CIRP; Business and Information Systems Engineering; Proceedings of 2014 IEEE Automation, Quality and Testing, Robotics, AQTR 2014; and eventually from IEEE International Conference on Industrial Engineering and Engineering Management.

**Table 1.** Most Cited Scholar Works within Industrial Engineering – Industry 4.0

| *No* | *Document title* | *Authors* | *Year* | *Sources* | *Number of Citation* | *Annual Citation* |
|---|---|---|---|---|---|---|
| 1 | A Cyber-Physical Systems architecture for Industry 4.0-based manufacturing systems | Lee, J., Bagheri, B., Kao, H.-A. [1] | 2015 | Manufacturing Letters | 583 | 194 |
| 2 | Service innovation and smart analytics for Industry 4.0 and big data environment | Lee, J., Kao, H.-A., Yang, S. [2] | 2014 | Procedia CIRP | 272 | 68 |
| 3 | Industry 4.0 | Lasi, H., Fettke, P., Kemper, H.-G., Feld, T., Hoffmann, M. [3] | 2014 | Business and Information Systems Engineering | 245 | 61 |
| 4 | Cyber physical systems in the context of Industry 4.0 | Jazdi, N. [4] | 2014 | Proceedings of 2014 IEEE Automation, Quality and Testing, Robotics, AQTR 2014 | 109 | 27 |
| 5 | Smart factories in Industry 4.0: A review of the concept and of energy management approached in production based on the Internet of Things paradigm | Shrouf, F., Ordieres, J., Miragliotta, G. [5] | 2014 | IEEE International Conference on Industrial Engineering and Engineering Management | 109 | 27 |

Furthermore, the subsequent theoretical discourses of Digital Transformation refer to the most cited scholar works within Digital Transformation, through the lens of its Digital Transformation's Systematic Literature Review, as illustrated in Table 2 [6].

**Table 2.** Most Cited Scholar Works within Digital Transformation's Systematic Literature Review

| *No* | *Document sources* | *Authors* | *Year* | *Number of Citation* | *Annual Citation* |
|---|---|---|---|---|---|
| 1 | Business and information Systems Engineering | C.Matt et al. [7] | 2015 | 114 | 28.50 |
| 2 | Government Information Quarterly | T.Janowski et al.[8] | 2015 | 96 | 24.00 |
| 3 | MIS Quarterly Executive | T.Hess et al.[9] | 2016 | 71 | 23.67 |
| 4 | MIS Quarterly: Management Information Systems | A.Majchrzak et al [10] | 2016 | 53 | 17.67 |
| 5 | MIS Quarterly Executive | R.Hansen et al [11] | 2015 | 43 | 10.75 |

The perspectives of Professional Engineering refer to the "primary project", known as capstone project. As part of global citizen, professional engineers are required to perform several engineers' quality of work. The aforementioned quality of work refers to the ability to solve a problem in which there are uncertain methods to implement, nor the appropriate solutions [12].

This paper is entitled Professional Engineering and World Class Perspectives of Industrial Engineering and Digital Transformation. The prior paragraphs elaborate Professional Engineering and its Perspectives of Industrial Engineering and Digital Transformation. Subsequently, the discourses of World Class Perspectives are as indispensable as the one of Professional Engineering's Perspectives.

To some extent, the World Class Perspectives are intertwined with its orchestrated title with World Class Manufacturing, as the result of Bibliometric Analysis from 2000 until 2016. The aforementioned intertwined aspects refer to three categories. First, it refers to merely the Title of either Knowledge Management or World Class Manufacturing. Second, it refers to combination of Title, keyword or abstract of World Class Manufacturing. Third, it refers to the merely title of Knowledge Management and World Class Manufacturing. Fourth, it refers to the combination of Knowledge Management and World Class Manufacturing.

The Bibliometric Analysis from 2000 until 2016 is elaborated and discussed in its illustration within Table 3 pertaining the academic papers within scholar works. Its triangulation database and terms are further elaborated by discourses of World Class in term of sustainability [13] and [14].

**Table 3.** Academic Papers on Combination of Knowledge Management and World Class

| DATABASE SOURCES | Knowledge Management (Merely Title) | World Class Manufacturing ( Merely Title) | World Class Manufacturing (Combination of Title, keyword or abstract) | Knowledge Management + World Class Manufacturing (Merely Title) | Knowledge Management + World Class Manufacturing (Combination of Title, keyword or abstract) |
|---|---|---|---|---|---|
| Emerald | 1375 | 11 | 110 | 0 | 5 |
| SciELO | 376 | 0 | 5 | 0 | 0 |
| Spell | 170 | 0 | 0 | 0 | 0 |
| Ebsco | 1880 | 7 | 24 | 0 | 0 |
| Web of Science | 3783 | 14 | 261 | 1 | 2 |
| Scopus | 6913 | 41 | 508 | 1 | 3 |
| **TOTAL** | **14497** | **73** | **908** | **2** | **10** |

## 2. Methodology

The research methodology in this paper refers to the quantitative method that enables the achievement of the objective of this paper. Subsequently, the quantitative method is augmented its quality through the bibliometric analysis to generate the timeline perspectives on the theoretical discourses and its empirical implementation [15], [16], [17], [18].

Both theoretical discourses and empirical implementations provide the cutting edge discussion in this paper toward the Professional Engineering and Digital Transformation Perspectives of Industrial Engineering and Digital Transformation.

The discussion of this paper elaborates the methods and methodology to achieve the objective of this paper. Subsequently, the mentioned discussion proceeds to the necessity to have a holistic industrial implementation among Academician, Business and Government.

The result of this paper is expected to equip current professional engineers and future generations. They are expected to be able to cope with new norm of new paradigm in the future.

## 3. Results and Discussion

The discussion of this paper elaborates the methods and methodology to achieve the objective of this paper. Subsequently, the mentioned discussion proceeds to the necessity to have a holistic industrial implementation among ABG. The result of this paper is expected to equip current professional engineers and future generations. They are expected to be able to cope with new norm of new paradigm in the future.

This paper is entitled Professional Engineering and World Class Perspectives of Industrial Engineering and Digital Transformation. The prior paragraphs elaborate Professional Engineering and its Perspectives of Industrial Engineering and Digital Transformation. Subsequently, the discourses of World Class Perspectives are as indispensable as the one of Professional Engineering's Perspectives.

Theoretical discourses of Industrial Engineering refer to the most cited scholar works within Industrial Engineering, in particular within Industry 4.0 study.

Those most cited scholar works are originated mostly from the sources of the following but not limited to the Manufacturing Letters; Procedia CIRP; Business and Information Systems Engineering; Proceedings of 2014 IEEE Automation, Quality and Testing, Robotics, AQTR 2014; and eventually from IEEE International Conference on Industrial Engineering and Engineering Management.

To some extent, the World Class Perspectives are intertwined with its orchestrated title with World Class Manufacturing, as the result of Bibliometric Analysis from 2000 until 2016.

The discussion in prior paragraphs further more orchestrates the three pillars that were mentioned before. First pillar constitutes the prior trigger of Industry 4.0 (IR 4.0), as one of Global Benchmark Settings. Subsequently, second pillar refers to the Indonesian Local Wisdom, in term of Making Indonesia 4.0 (MI 4.0). Ultimately, third pillars are pertaining global benchmark settings, in term of Industry X.0, Society 5.0, Hallyu 2.0.

## 4. Conclusion

This paper conveys the theoretical discourses of Industrial Engineering and Digital Transformation. It proposes perspectives of Professional Engineering and World Class' Industrial implementation by combining Indonesian Local Wisdom and Global Benchmark Settings.

The objective of this paper is to intertwine both theoretical and industrial implementation among Academician, Business and Government (ABG). Precisely, this harmonized industrial implementation is deemed indispensable by considering several pillars. First pillar constitutes the prior trigger of Industry 4.0 (IR 4.0), as one of Global Benchmark Settings. Subsequently, second pillar refers to the Indonesian Local Wisdom, in term of Making Indonesia 4.0 (MI 4.0). Ultimately, third pillars are pertaining global benchmark settings, in term of Industry X.0, Society 5.0, Hallyu 2.0.

Theoretical discourses of Industrial Engineering refer to the most cited scholar works within Industrial Engineering, in particular within Industry 4.0 study.

Those most cited scholar works are originated mostly from the sources of the following but not limited to the Manufacturing Letters; Procedia CIRP; Business and Information Systems Engineering; Proceedings of 2014 IEEE Automation, Quality and Testing, Robotics, AQTR 2014; and eventually from IEEE International Conference on Industrial Engineering and Engineering Management.

To some extent, the World Class Perspectives are intertwined with its orchestrated title with World Class Manufacturing, as the result of Bibliometric Analysis from 2000 until 2016.

The discussion of this paper elaborates the methods and methodology to achieve the objective of this paper. Subsequently, the mentioned discussion proceeds to the necessity to have a holistic industrial implementation among Academician, Business and Government.

The result of this paper is expected to equip current professional engineers and future generations. They are expected to be able to cope with new norm of new paradigm in the future.

## References

[1] Lee, J, Kao H A and Yang S 2014 Service innovation and smart analytics for Industry 4.0 and big data environment *6th CIRP Conference on Industrial Product Service Systems* 4-8
[2] Cohen Y, Faccio M, Galizia F G, Mora C and Pilati F 2017 Assembly system configuration through Industry 4.0 principles: the expected change in the actual paradigms *IFAC-PapersOnLine* **50** (1) 14958–14963
[3] Jazdi N 2014 Cyber physical systems in the context of Industry 4.0 *International Conference on Automation, Quality and Testing, Robotics* ) 1–4
[4] Shrouf F, Ordieres J and Miragliotta G 2014 Smart factories in Industry 4.0: A review of the concept and of energy management approached in production based on the Internet of Things paradigm *International Conference on Industrial Engineering and Engineering Management* 697–701
[5] Gorecky D, Schmitt M, Loskyll M and Zühlke D 2014 Human-machine-interaction in the industry 4.0 era *12th IEEE International Conference on Industrial Informatics* 289–294
[6] Mahraz I, Benabbou L and Berrado A 2019 A Systematic literature review of Digital Transformation *Proceedings of the International Conference on Industrial Engineering and Operations Management Toronto* 1–4
[7] Matt C 2015 Digital Transformation Strategies *Business and Information Systems Engineering* **57** (5) 339-343
[8] Jedlowski A 2015 Across media: Mobility and transformation of cultural materials in the digital age *Journal of African Media Studies* **7** (1) 3-9
[9] Hess T 2016 Options for formulating a digital transformation strategy *MIS Quarterly Executive* **15** (2) 123-139
[10] Majchrzak A 2016 Designing for digital transformation: Lessons for information systems research from the study of ICT and societal challenges *MIS Quarterly: Management Information Systems* **40** (2) 267-277
[11] Hansen R 2015 Hummel's digital transformation toward omnichannel retailing: Key lessons learned *MIS Quarterly Executive* **14** (2) 51-66
[12] Gratton P M and Gratton G B 2020 *Achieveing Success with The Engineering Dissertation* Switzerland: Springer Nature
[13] World Class Manufacturing: an initial approach based on a literature review *Perspectivas em Ciência da Informação* **22** 244-263
[14] Fu R, Tang Y and Chen G 2020 Chief sustainability officers and corporate social (Ir) responsibility *Strat Mgmt J* **41** 656–680
[15] Basias N and Pollais Y 2018 Quantitative and Qualitative Research in Business & Technology: Justifying a Suitable Research Methodology *Review of Integrative Business and Economics Research* **7** 91-105
[16] Goertz G and Mahoney J 2012 *A Tale of Two Cultures: Qualitative and Quantitative Research in the Social Sciences* New York: Princeton University Press.
[17] Martin W and Bridgmon K 2012 *Quantitative and Statistical Research Methods. From Hypothesis to Results* New York: John Wiley & Sons
[18] Yin, R 2011 *Applications of Case Study Research* New Yoek: SAGE Publication

# User Acceptance of Virtual Hotel Operator Applications in Indonesia

**I N Ramadhiana[1], G Prakarsa[2,*] and V M Nasution[3]**
[1]Students of Department Information System, Universitas Informatika dan Bisnis Indonesia, Bandung, Indonesia.
[2,3]Faculty of Technology and Informatics, Universitas Informatika dan Bisnis Indonesia, Bandung, Indonesia

*Email: gprakarsa@gmail.com

**Abstract.** This research is based on the main problem, namely the lack of measurement of the integration of mobile factors in the VHO application. This research is categorized as quantitative research using descriptive methods, and using the technology acceptance model / TAM. The technique used is multiple linear regression, starting with the validity and reliability tests, as well as the heteroscedasticity test. The purpose of this research is to determine how much acceptance of VHO application users based on the Technology Acceptance Model (TAM). This research approach uses a quantitative approach with survey methods. The data collection technique is using a questionnaire that distributed randomly to users of the VHO application. The data analysis method used is descriptive analysis, using multiple linear regression techniques. The results of this study indicate that the variables of mobile perceived compatibility, mobile perceived financial resources, and mobile perceived risk have a simultaneous effect on the mobile variable perceived usefulness, the mobile variable perceived ease of use and the intention to use variable.

## 1. Introduction

The industrial revolution 4.0 is a phase that changes the way people think and act. The main principle is the combination of machines, workflows and systems by implementing an intelligent network in the production process chain. This technological development encourages changes in human behavior and creates business opportunities and creates new jobs. With this technological innovation, competition in the business world is also inevitable, including competition in hotel services with applications, both *device* applications and websites. Supporting this, now there is an online room reservation service provider business, commonly known as VHO (Virtual Hotel Operator).

To get many users on a hotel booking application, the application must consider what convenience factors will be adapted in use. The application will not be used continuously if the user finds it difficult to use it or make transactions. As affordable as the price offered in the application, if users find it difficult it will be in vain. If users find a technology easy to use, then they will adapt the technology [1]. Regarding the facts of previous research, we are interested in conducting a study of the VHO application with the aim of knowing whether or not a technology is accepted based on what the user feels. The method used to measure the usefulness of these technologies uses TAM (Technology

Acceptance Model). TAM was first introduced by Fred Davis in 1986 which is a method designed to analyze what factors influence the acceptance of the use of a technology [2].

There have been many studies conducted using TAM. TAM is used to empirically test a research model that combines the antecedents of mobile shopping loyalty and the context of hotel room reservations [3]. Furthermore, TAM is used to identify external factors used in the context of e-learning adoption [4]. Furthermore, regarding TAM, there is a study that aims to analyze the factors that influence student interest in mobile banking applications [5]. Subsequent research was carried out with the object of PDAM Surya Sembada Surabaya which measures the reliability of a mobile application [6]. Two years ago, research using TAM was carried out with the Bukalapak e-marketplace object, this research focuses on measuring Bukalapak user acceptance with a quantitative descriptive method [7]. TAM is also used to build a model used to measure technology acceptance in the e-marketplace Lazada [8] and Shopee applications [9].

VHO is a startup in the technology sector, especially the hotel operator sector that collaborates with hotels (generally budget hotels, but has penetrated into villas and star hotels), property owners are offered to register their properties for accommodation [10]. One of the studies regarding VHO is research that focuses on the influence of service quality on customer satisfaction and loyalty of VHO [11]. Then there is a study that aims to analyze the effect of service quality through tangible variables, reliability, responsiveness, assurance, and empathy for consumers staying at hotels in Surabaya [12] . Furthermore, there is VHO research that analyzes the effect of digital marketing on brand advocacy with perceived quality and customer engagement as intervening variables in VHO Airy [13]. The VHO trend this year is increasingly developing in Yogyakarta in line with the tourism sector with a minimal budget [14].

Furthermore, research on risks with the object of mobile payment focuses on examining the factors that determine the acceptance of mobile payments for consumers [15]. There is also research that analyzes online purchasing behavior in the hotel industry using an integrative framework with variables that are rarely used in previous studies [16]. In the same year, a study was conducted with the aim of proposing the acceptance of AR (Augmented Reality) in the context of urban heritage tourism [17].

## 2. Hypothesis Development

This study uses three main variables adopted from the study [17] three external variables. The main variables used are *mobile perceived usefulness, mobile perceived ease of use* and *consumer intention to use*. Meanwhile, the three external variables are *mobile perceived compatibility, mobile perceived financial resources* and *mobile perceived risk*.

*Consumer intention to use is* defined as an individual's interest or desire to perform certain behaviors [18], it can also be interpreted as how long the user will use this technology [19], [20]. *Mobile perceived usefulness* refers to the extent to which *users* expect a technology to increase work effectiveness [15]. In addition, *mobile perceived usefulness* also leads to the perception of increased usability given to *users* when adopting technology [1].

**H1**: *mobile perceived usefulness* has a significant relationship to *consumer intention to use*

*Mobile perceived ease of use* refers to how easy this technology is to use, when consumers feel that a technology is easy to use and learn, they will adopt this new innovation, especially if it is supported by a *friendly* interface , and how to use it easily [1]. The less *users* do activities related to social media, the more likely they will find social media sites useful. Results of research on *mobile perceived ease of use* previously has shown a significant relationship between the variables *of mobile perceived ease of use* to the consumer wishes to use (*consumer intention to use*) [21].

**H2**: *mobile perceived ease of use* has a significant positive relationship to *consumer intention to use*

*Mobile perceived usefulness is* defined as the extent to which a person can believe that using a system will improve his job performance. *Mobile perceived ease of use* has become an important concept in many technology acceptance models. This variable is defined as the extent to which a person can believe that using a system will be free of effort [18]. In addition, consumers may not have a strong motivation to adopt *mobile* technology unless the service creates new innovations where mobility is very important and manages to positively affect people's lives [22].

**H3**: *mobile perceived usefulness* and *mobile perceived ease of use* have a simultaneous relationship to *consumer intention to use*

*Mobile perceived compatibility* refers to the extent to which mobile phone innovation can be adapted to potential needs and *user* behavior patterns [1]. In his research, it was stated that consumers have adopted *smartphones* to buy products and services, access television, banking, exchange information simply by bringing their cellphones closer together, and playing *online games* together. Thus, past experience using smart phones can be considered an indicator of compatibility. *Perceived compatibility* has a positive effect on *perceived usefulness* [23]. Based on the research above, the hypothesis was found:

**H4**: *mobile perceived compatibility* has a significant relationship to *mobile perceived usefulness*

*Mobile perceived compatibility* leads to the extent to which innovation is considered consistent with existing values, previous experiences, and the needs of potential *users*. In his research, it was stated that to increase the use of *mobile* technology , the system must create the impression that the technology is easy to use and compatible with the *user's* lifestyle [3]. Consumers are more accepting if they have previous experience using technology, consumers may think that new technology will have difficulty using it [1]. Based on the research above, the hypothesis was found:

**H5**: *mobile perceived compatibility* has a significant relationship with *mobile perceived ease of use.*

*Mobile perceived financial resources* refer to financial resources which usually hinder the application of new innovations to consumers. Devices that adapt to these new innovations are usually more expensive, this is due to high production costs as they require redesign of existing ones. Given financial considerations, these innovations may be out of reach of consumers due to limited financial resources [1]. *Mobile perceived financial resources are* used to measure *perceived financial resources* and assess the level of financial difficulties felt by technology developers [24]. Based on the research above, the hypothesis was found:

**H6**: *mobile perceived financial resources* have a significant relationship with *mobile perceived ease of use*

**H7**: *mobile perceived financial* has a significant relationship to *mobile perceived usefulness*

*Mobile perceived compatibility* refers to the extent to which mobile phone innovation can be adapted to potential needs and *user* behavior patterns [1]. *Mobile perceived financial resources are* used to measure *perceived financial resources* and assess the level of financial difficulties felt by technology developers [24]. This *mobile* variable *perceived risk* refers to most applications on *smartphones* endangering the privacy and security of their users [17].

**H8**: *mobile perceived compatibility, mobile perceived financial resources,* and *mobile perceived risk* have a simultaneous relationship to *mobile perceived usefulness.*

This *mobile* variable *perceived risk* refers to most applications on *smartphones* endangering the privacy and security of their users [17] . If an application asks for too many personal details, most *users are* afraid that the data will be shared with other parties. The public is concerned about the risk of who will have access to their data information, because nowadays anyone can access any information on the internet [25].

**H9**: *mobile perceived risk* has a significant relationship to *mobile perceived usefulness*
**H10**: *mobile perceived risk* has a significant relationship to *mobile perceived ease of use*

*Mobile perceived compatibility* leads to the extent to which innovation is considered consistent with existing values, previous experiences, and the needs of potential *users* [3]. *Mobile perceived financial resources* refer to financial resources which usually hinder the application of new innovations to consumers. Given financial considerations, these innovations may be out of reach of consumers due to limited financial resources [1]. *Mobile perceived risk* refers to when new technology is adapted, in addition to the technology being easy to use as well as *user* data requests which are often unsettling for the public, they are worried about who will access their data on these technologies [25].

**H11**: *mobile perceived compatibility, mobile perceived financial resources,* and *mobile perceived risk* have a simultaneous relationship to *mobile perceived ease of use.*

Based on the above hypothesis, a research paradigm has been built as shown in Figure 1.

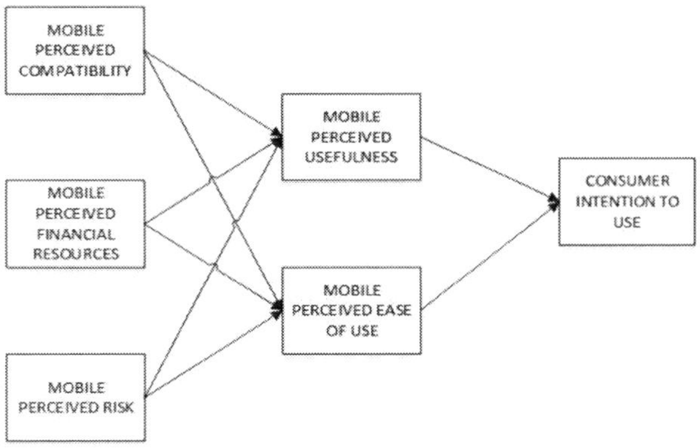


**Figure 1.** Research Paradigm

The Figure 1 contains 3 external variables; mobile perceived compatibility, mobile perceived financial resources, mobile perceived risk. The main variables are mobile perceived usefulness, mobile perceived ease of use and consumer intention to use. The following indicators can be formulated in Table 1.

**Table 1.** Operational Research Variables

| Constructs | Number Of Items | Sources |
|---|---|---|
| Mobile Perceived Compatibility | 4 | [15], [26], [1], [3], |
| Mobile Perceived Financial | 3 | [9], [27], [3], [8], |
| Resources | 4 | [16], [19], [28], (7), |
| Mobile Perceived Risk | 5 | [20], [18], [17], [29], |
| Mobile Perceived Usefulness | 4 | [25], [23], [21], [24], |
| Mobile Perceived Ease Of Use | 5 | [28], [1], [8] |
| Consumer Intention To Use | 3 | [9], [1] |

Each variable has at least 3 indicators that become the reference for making questionnaire questions. The written indicators refer to previous research journals.

## 3. Methodology

### 3.1. Participants and instuments

The object of this research is the users of the VHO application that are widespread in Indonesia. Researchers want to find out how compatible the VHO application has been downloaded and used, how much risk occurs when transacting on the VHO application, how easy the VHO application is to use , and also the researcher wants to find out whether the VHO application can ease the user's work or not. Based on previous research data, the researchers are interested in conducting research "Integration of Mobile Perceived Compatibility, Mobile Perceived Financial Resources, and Mobile Perceived Risk with TAM in Virtual Hotel Operator Applications in Indonesia". This study aims to develop and expand TAM with external compatibility, financial resources, and risk variables which will be tested to determine how much influence the user's intention to use hotel room reservations online.

This study uses a survey method, while the survey method itself is a method that studies data based on a sample to measure the relationship between variables in a population [30]. This study took a number of samples from the population through questionnaires to collect basic data included in the survey research category. If described, this study adapts a framework that has been modified by [1], [9], [17] by using 3 main variables and 3 external variables. The main variables used are; *mobile perceived usefulness, mobile perceived ease of use,* and *consumer intention to use*. The external variables are; *mobile perceived compatibility, mobile perceived financial resources*, and *mobile perceived risk*.

### 3.2. Data Analysis

The statistical software SPSS 25 was used to calculate these collected data. For the research questions, validity and reliability test, correlation analyses and multiple regressions were used to answer the research question.

## 4. Results and Discussions

The data obtained has shown a validity level that is valid and consistent reliability, where the resulting r value has met the requirements so that it is significant and reliable as shown in Table 2. Hypotheses testing is done through F test for simultaneous and the T test for partial.

Based on the results of partial testing in Table 2, it is shown that all hypotheses are shown a significant value, and also they are proven valid. This means there are no unrelated partial relationships.

Based on the results of partial testing in Table 3, it is shown that all simultaneous hypotheses have been proven, where each hypothesis has shown a calculated F value that is greater than the F table, and a significant number less than 0.05. The coefficient of determination in this study for the effect of mobile perceived usefulness and mobile perceived ease of use on consumer intention to use is 0.636 or 63.6%. For the effect of mobile perceived compatibility, mobile perceived financial resources and mobile perceived risk to mobile perceived usefulness of 0.502 or 50.2%. Meanwhile, the effect of mobile perceived compatibility, mobile perceived financial resources and mobile perceived risk on mobile perceived ease of use is 0.610 or 61%. Thus, based on the results of research to increase consumer intention to use, it is necessary to pay attention to the factors built in this study.

**Table 2.** T Test result

| Independent Variable | Dependant Variable | T Value | T Table | Significant Value |
|---|---|---|---|---|
| Mobile Perceived Usefulness | Consumer Intention To Use | 18,351 | 0.726 | .000 |
| Mobile Perceived Ease Of Use | Consumer Intention To Use | 23,183 | 0.726 | .000 |
| Mobile Perceived Compatibility | Mobile Perceived Usefulness | 17,199 | 0.726 | .000 |
| Mobile Perceived Financial Resources | Mobile Perceived Ease Of Use | 16,406 | 0.726 | .000 |
| Mobile Perceived Compatibility | Mobile Perceived Ease Of Use | 21,107 | 0.726 | .000 |
| Mobile Perceived Financial Resources | Mobile Perceived Usefulness | 14,645 | 0.726 | .000 |
| Mobile Perceived Risk | Mobile Perceived Usefulness | 11,762 | 0.726 | .000 |
| Mobile Perceived Risk | Mobile Perceived Ease Of Use | 14,437 | 0.726 | .000 |

**Table 3.** F Test Result

| Independent Variable | Dependent Variable | F Value | F Table | Significant Value |
|---|---|---|---|---|
| Mobile Perceived Usefulness, Mobile Perceived Ease Of Use | Consumer Intention To Use | 324,453 | 2,238 | .000[b] |
| Mobile Perceived Compatibility, Mobile Perceived Financial Resources, Mobile Perceived Risk | Mobile Perceived Usefulness | 125,369 | 2,238 | .000[b] |
| Mobile Perceived Compatibility, Mobile Perceived Financial Resources, Mobile Perceived Risk | Mobile Perceived Ease Of Use | 194,184 | 2,238 | .000[b] |

## 5. Conclusions

Based on the research results, it shows that the developed model has been able to answer the problem regarding the absence of a model that can be used specifically to measure user acceptance of the Virtual Hotel Operator Application in Indonesia. In addition to supporting the model produced in Davis 1989 research, this study also provides updates by adding several variables through research support [1] and [17]. The practical benefit obtained from this research is the need to pay attention to risk factors of a VHO application. With good system security, users will feel safe when making hotel booking transactions in the VHO application.

## References

[1] Ontiveros-Robles E, Melin P and Castillo O 2018 Comparative analysis of noise robustness of type 2 fuzzy logic controllers *Kybernetika* **54** (1) 175-201
[2] Ooi K B and Tan G W H 2016 Mobile Technology Acceptance Model: An Investigation using Mobile Users to Explore Smartphone Credit Card *Expert Systems with Applications* **59** 33-46
[3] Apriyanto 2015 Kajian Penerimaan Sistem Informasi Kepesertaan dengan Pendekatan Technology Acceptance Model *Jurnal Pilar Nusa Mandiri* **11** (1) 30–42
[4] Ozturk A B, Bilgihan A, Nusair K and Okumus F 2016 What keeps the mobile hotel booking users loyal? Investigating the roles of self-efficacy, compatibility, perceived ease of use, and perceived convenience. *International Journal of Information Management* **36** (6) 1350–1359
[5] Abdullah F and Ward R 2016 Developing a General Extended Technology Acceptance Model for E-Learning (GETAMEL) by analysing commonly used external factors. *Computers in Human Behavior* **56** 238–256
[6] Suyanto and Kurniawan T A 2019 Faktor yang Mempengaruhi Tingkat Kepercayaan Penggunaan FinTech pada UMKM Dengan Menggunakan Technology Acceptance Model ( TAM ) **16** (1) 175-186
[7] Utomo Y and Walujo D A 2018 Penerapan Konstruk Technology Acceptance Model ( TAM ) Pada Layanan Mobile Application Di PDAM Surya Sembada Kota Surabaya *WAKTU : Jurnal Teknik UNIPA* **16** (1) 39–48
[8] Prakarsa G 2018 Analisis Technology Acceptance Model terhadap Pengguna E-Marketplace Bukalapak *In Search* **17** (02) 121-130
[9] Prakarsa G 2019 Analisis Faktor-faktor Penerimaan Konsumen Pada Aplikasi E-Marketplace Lazada Menggunakan TAM. *Sainteks: Jurnal Sains dan Teknik* 1(2) 104-116.
[10] Prakarsa G 2019Analisis Faktor-Faktor yang Mempengaruhi Penggunaan E-Marketplace Shopee *Journal SisInfo* **1** (1) 1-11
[11] Wiastuti R D and Erna S M 2020 Virtual Hotel Operator ; Is It Disruption for Hotel Industry? *Jurnal Hospitality dan Pariwisata* **2** (2) 201-215.
[12] Dara W C M 2019 Pengaruh Kualitas Layanan terhadap Kepuasan dan Loyalitas Pelanggan Virtual Hotel Operator *S1 Thesis Universitas Negeri Jakarta*.
[13] Gunawan E, Sebastian G O and Harianto A 2019 Analisa Pengaruh Kualitas Layanan Terhadap Kepuasan Pelanggan Menginap di Empat Virtual Hotel Operator Di Surabaya. *Journal of Indonesia Tour Hospitality and Recreation* **2** (2) 145–53
[14] Dharmayanti D and Prasojo S 2016 Analisa Pengaruh Digital Marketing terhadap Brand Advocacy dengan Perceived Service Quality dan Customer Engagement Sebagai Variabel Intervening Jurnal Strategi Pemasaran **7** (1)
[15] Fitri K 2020 Tren Virtual Hotel Operator (Vho) Di Yogyakarta (Studi Kasus Hotel Oyo). *Jurnal AMPTA* **18** (1) 1–20
[16] Chen L D 2008 A model of consumer acceptance of mobile payment *International Journal of Mobile Communications*. **6** (1) 32–52
[17] Indiani N L P, Rahyuda I K, Kerti Y N N and Sukaatmadja I P G 2015 Perceived Risk and Trust as Major Determinants of Actual Purchase, Transcending The Influence of Intention. *ASEAN Marketing Journal* **7** (1) 1–13.
[18] Dieck M C T and Jung T 2018 A theoretical model of mobile augmented reality acceptance in urban heritage tourism. *Current Issues in Tourism* **21** (2) 154–74
[19] Zarmpou T, Saprikis V, Markos A and Vlachopoulou M 2012 Modeling users' acceptance of mobile services *Electron Commer Res* **12** (2) 225–48.
[20] Boonsiritomachai W and Pitchayadejanant K 2019 Determinants affecting mobile banking adoption by generation Y based on the unified theory of acceptance and use of technology model modified by the technology acceptance model concept. *Kasetsart Journal of Social Sciences* **40** (2):349–58

[21] Tan G W H, Ooi K B, Chong S C and Hew T S 2013 NFC mobile credit card: The next frontier of mobile payment? *Telematics Informatics* **31** (2) 292–307
[22] Dumpit D Z and Fernandez C J 2017 Analysis of the use of social media in Higher Education Institutions (HEIs) using the Technology Acceptance Model. International Journal Educational Technology in Higher Education **14** (1) 1-16
[23] Virgianne Y, Ariani N M and Suarka F M The Influence of E-Service Quality towards Domestic Tourists Satisfaction in AiryRooms Kuta Bali District *Jurnal Kepariwisataan dan Hospitalitas* **3** (1) 108-125
[24] Ismail H A 2016 Intention to Use Smartphone Through Perceived Compatibility, Perceived Usefulness, and Perceived Ease of Use *Jurnal Dinamika Manajemen* **7** (1) 1-10
[25] Hong M and Harrington D 2016 The Effects of Caregiving Resources on Perceived Health among Caregivers *Health and Social Work* **41** (3) 155–163
[26] Schnall R, Higgins T, Brown W, Carballo-Dieguez A and Bakken S 2015 Trust, Perceived Risk, Perceived Ease of Use and Perceived Usefulness as Factors Related to mHealth Technology Use. *Studies in Health Technology and Informatics* **216** 467–471
[27] Chung H-H, Chen S-C and Kuo M-H 2015 A Study of EFL College Students' Acceptance of Mobile Learning *Procedia - Social and Behavioral Sciences* **176** 333–339
[28] Leong L Y, Hew T S, Tan G W H and Ooi K B 2013 Predicting the determinants of the NFC-enabled mobile credit card acceptance: A neural networks approach. *Expert Systems with Applications* **40** (14) 5604–5620
[29] Davis F D 1989 Perceived Usefulness, Perceived Ease of Use, and User Acceptance of Information Technology *MIS Quarterly* **13** (3) 319–340
[30] Chen H, Rong W, Ma X, Qu Y and Xiong Z 2017 An Extended Technology Acceptance Model for Mobile Social Gaming Service Popularity Analysis. Hindawi Mobile Information System **2017** (3) 1-12
[31] Sugiyono 2016 Metode Penelitian Kuantitatif, Kualitatif dan R&D Bandung : PT. Alfabet

# Preparation of Polypropylene/PVDF Composite Membrane by Dip-Coating Method

**P T P Aryanti*, G Trilaksono, A Hotmaida, M A Afifah, F P Pratiwi, N Udini and F A Nugroho**

Chemical Eng. Dept., Universitas Jenderal Achmad Yani, Jl. Ters. Jend. Sudirman, Cimahi

*Email: teta.aryanti@yahoo.com

**Abstract.** In this research, a composite membrane was prepared by dipping a hollow fiber polypropylene (PP) membrane in a polyvinylidene fluoride (PVDF) solution. A small concentration of volatile solvent (acetone and tetrahydrofuran/THF) was added to the PVDF solution to form a tight structure in the membrane surface. The influence of PVDF concentration, type of volatile solvents (acetone and THF), and the dip-coating cycle on membrane characteristics, including porosity, swelling degree, and contact angle, were investigated. The flux and rejection of the membranes were also investigated during the filtration of hexane/oil solution. The experimental results showed that low membrane porosity (3%), low swelling degree (0.1%), and high water contact angle (140°) were obtained when 8%wt PVDF was blended with 10%wt acetone and then coated on PP hollow fiber membrane surface with 3 (three) dip-coating cycles. The oil rejection was 68% after 45 minutes of filtration time with a permeate flux of 20 $L.m^{-2}.h^{-1}$. Higher membrane porosity and permeate flux can be obtained by reducing the dip-coating cycle; however, a lower oil rejection was achieved.

## 1. Introduction

Polypropylene (PP) has increasingly been used as membrane material since it shows excellent chemical resistance and thermal stability, and low cost [1-3]. Most commercial PP membranes are hollow fiber and have been widely used in several separation processes, including water treatment, gas separation, water-oil separation, vegetable oil membrane filtration, and other applications [4-7]. The PP membranes are generally fabricated by melt-spinning and stretching method, which forms microporous structures under controlled operating temperature [8]. Since the PP membranes are microporous, characterized by their pore size from 0.1 to 10 μm, the PP membranes have low selectivity to the soluble compounds. Therefore, modifications on the PP membrane surface have increasingly been carried out. This condition improves the selectivity and hydrophobicity or hydrophilicity, subject to the membrane application [9, 10].

There are two standard surface modification methods to improve membrane selectivity, namely interfacial polymerization (or crosslinking) [11] and physical coating methods [12]. The polymerization reaction occurs at two immiscible solvents' interface, which contains reactive monomers and initiators/catalysts [13, 14]. Heat treatment is often used in this technique to complete the polymerization or crosslinking of the monomers on the porous membrane support [15]. This method is generally used to manufacture reverse osmosis (RO) and nanofiltration (NF) membrane. Compared to the chemically modified membrane, the physical coating is relatively more comfortable to be applied in large scale production with no complicated operating condition [16]. Dip-coating is one of the physical coating methods by immersing the dried PP membrane in a homogenous polymer solution for a few seconds and then lifting it vertically at a constant speed. A thin film membrane was formed on the substrate surface due to solvent evaporation from the polymer solution to the atmosphere [17].

Although it is considered a simple method for modifying PP membranes, few publications focus on the dip-coating method. Himma et al. [18] used the dip-coating method to alter the hollow fiber PP membrane surface. The dried hollow fiber PP was first immersed in a non-solvent solution (water), and then followed by immersing the membrane in the PP solution. A high hydrophobic PP membrane was obtained with a contact angle of 149°. When it was used for oil/water separation with a ratio of 1:9 v/v, the rejection of water and volatile matter were 64.35 and 67.15%, respectively. The membrane hydrophobicity was improved to 151.3° by using methyl ethyl ketone (MEK) as a non-solvent instead of water [16]. Wenten et al. [12] coated polysulfone(PSf)/PEG400/ZnO layer on the hollow fiber PP membrane surface and then used it for peat water treatment. ZnO's presence enhanced the membrane's hydrophilicity, which reduced organic compounds' fouling tendency during the peat water treatment. The modified membrane removed 70% of humic substances from peat water.

In this research, the hollow fiber PP membranes were also modified by the dip-coating method. The dried hollow fiber membrane was immersed in PVDF solution and then lifted vertically at a constant speed. The influence of PVDF concentration, type of volatile solvent, and dip-coating cycle on the composite PP/PVDF membrane characteristics was investigated. The flux and rejection of the composite membrane were determined during the hexane/oil separation process.

## 2. Material and Methods

### 2.1. Materials

Polypropylene hollow fiber was supplied by GDP Filter, Indonesia, which had an inside diameter (ID) of 0.2 mm and an outside diameter (OD) of 0.4 mm. The PVDF was supplied by a local supplier (PT. KGC Saintifik) with a purity of 99.5% and a molecular weight of 600,000 Da. Meanwhile, the NN-dimethylacetamide (DMAc) was supplied by Shanghai Jingsan Jingwei Chemical CO., Ltd, with a purity of 99.0%. The tetrahydrofuran (THF, ≥99.9%) was provided by Sigma Aldrich, while a local supplier supplied the acetone. The palm oil and hexane were provided by local suppliers and then mixed with a ratio of 1:9 (v/v).

### 2.2. Dip-coating of hollow fiber PP membrane

The PVDF solution was prepared by blending polymer in DMAC at room temperature until the solution was homogenous. The PVDF concentration was varied from 0 – 8% by weight. A dip coater was used to immerse the PP hollow fiber membrane in PVDF solution vertically. The PP membrane was immersed for 10 seconds and then lifted at a constant speed of 15 $mm.s^{-1}$. The coated PP membrane was dried at room temperature for 24 hours until all solvent was evaporated. The schematic dip-coating process of the hollow fiber PP membrane is shown in **Figure 1**.

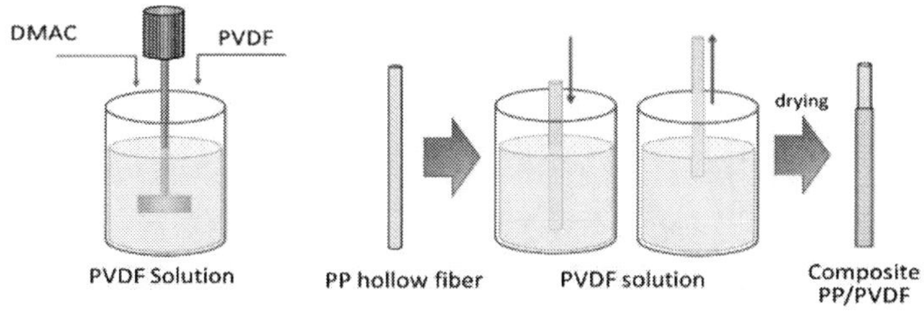


**Figure 1**. Dip-coating process of hollow fiber PP membrane in PVDF solution.

### 2.3. Determination of porosity and swelling degree of the composite PP/PVDF membrane.

The porosity of the composite PP/PVDF membrane was determined by weighing dry and wet membrane. Distilled water was flowed through the dried membrane pore using a centrifugal pump for a few minutes until all membrane pore was filled with water. The wetted membrane was drained off until there is no water on the membrane surface and then weighed. The composite membrane porosity (ε) was determined by the following equation:

$$Porosity = {(W_w - W_d)}/{(V\ x\ \rho_w)} \quad (1)$$

where $W_d$ is the weight of dry membrane (g), $W_w$ is the weight of wet membrane (g), $V$ is the volume of composite hollow fiber membrane (cm$^3$), and $\rho_w$ is the density of water (0.99708 g/cm$^3$).

The membrane's swelling degree was determined by immersing the membrane in a solvent (hexane) for 48 hours. The wet membrane was wiped to remove water on the membrane surface and then weighed. The effect of hexane on the membrane swelling degree was calculated using the following equation.

$$Swelling\ degree\ (\%) = \left[{(W_w - W_d)}/{W_w}\right] x\ 100\% \quad (2)$$

### 2.4. Determination of contact angle of the composite membrane

The water contact angle on the composite membrane surface was measured using a digital microscope (Genesys Logic Microscope 1600xUSB). A micro-syringe was used to drop and place the water droplets on the membrane surface. The contact angles were measured from a digital image of the water droplet using a graphic software (Corel Draw) program. The contact angle measurement was conducted 3 (three) times at different locations to minimize experimental errors.

### 2.5. Flux and oil rejection of the composite PP/PVDF membrane

The experimental set-up for the filtration of hexane/oil solution is presented in **Figure 2**. The hollow fiber membrane was placed in a U-shape hollow fiber membrane module and operated under dead-end mode (inside-out flow) at different operating pressure (2 and 4 bars). The permeate flux was determined by measuring the volume of permeate produced per time and membrane area, as described in the following equation.

$$J = {V}/{A\ x\ t} \quad (3)$$

Where $J$ is the permeate flux (L.m$^{-2}$.h$^{-1}$), $V$ is the permeate volume (L), $A$ is the surface membrane area (m$^2$), and $t$ is filtration time (h).

Meanwhile, oil rejection (R, %) was calculated by the following equation:

$$R = \left[1 - {C_p}/{C_f}\right] \text{x } 100 \quad (4)$$

$C_p$ is the concentration of oil in the feed solution, and $C_f$ is the concentration of oil in permeate solution.

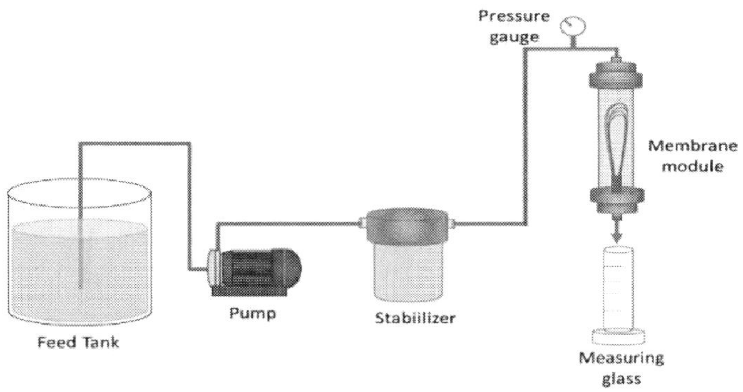


**Figure 2**. Experimental set-up for hexane/oil filtration

## 3. Result and Discussion

### 3.1. The influence of operating parameters on the composite membrane characteristics

Figure 3 shows the influence of PVDF concentrations, solvent types, and dip-coating cycles on the composite membrane porosities. The experimental results showed that the increase of PVDF concentration reduced membrane porosity. It has been known that the concentration of polymer is related to the intrinsic viscosity of the polymer solution. The higher density of the polymer solution inhibited the membrane pore's growth, and consequently, it reduced the pore size formed in the membrane structure (**Figure 3a**). Besides, the increase of polymer concentration reduced the void fraction in the membrane structure [19]. The composite membrane's porosity was significantly decreased when 8%wt PVDF was used in the coating solution with three dip-coating cycles. Figure 3b shows that the addition of volatile solvent as an additive improves the composite membrane's porosity (**Figure 3b**). The presence of a volatile solvent reduced the polymer and solvent's miscibility. It contributed to phase separation phenomena in the membrane solution and led to the pore nuclei formation in the membrane structure [20]. Besides dip-coating cycles and polymer concentration, the membrane porosity was also influenced by the volatile solvent type (**Figure 3c**). It has been reported that the acetone was more volatile than THF. The delayed phase separation in THF/DMAc/PVDF solution allowed the pore nuclei to form and grow in the membrane structure. Therefore, the membrane porosity that used THF as an additive was higher than acetone. The highest membrane porosity (25%) was achieved when 4%wt PVDF was blended with 10%wt THF with 2 (two) times of dip-coating process. Meanwhile the lowest membrane porosity (3%) was resulted when 8%wt PVDF was blended with 10%wt acetone.

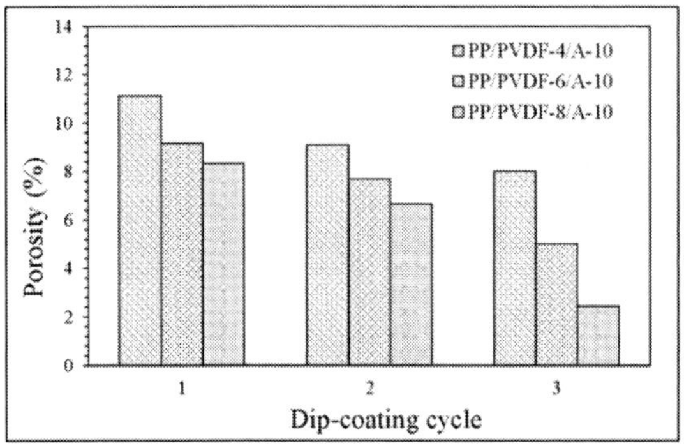


(a)

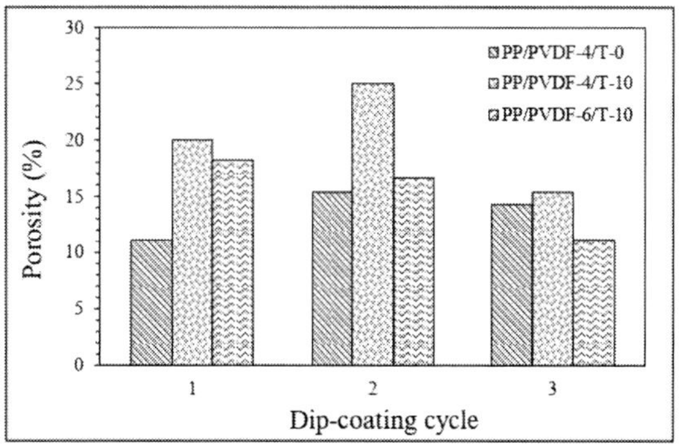


(b)

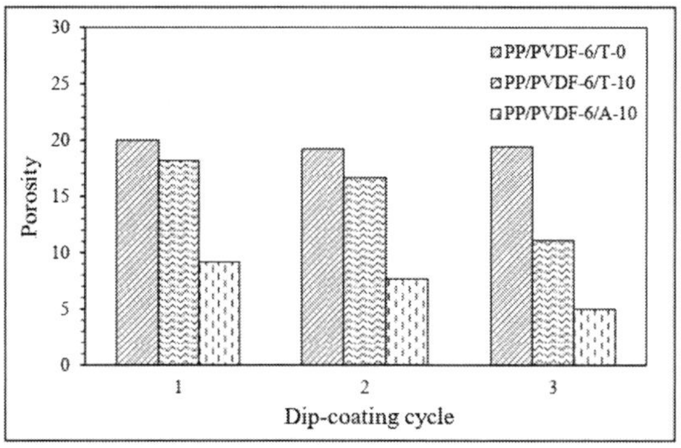


(c)

**Figure 3.** The influence of dip-coating cycles on composite membrane porosity at different concentrations of PVDF: (a) with acetone, (b) at various concentrations of PVDF in THF, and (c) at different types of solvents.

The influence of operating parameters on the swelling degree of the composite PP/PVDF membrane is shown in **Figure 4**. Swelling of the membrane structure is attributed to the solvent's penetration into the polymer network and leads to the polymer volume change [21]. A high concentration of solvent in membrane structure allowed the solvent to penetrate rapidly into the polymer chain. Therefore, the composite membrane with high porosity was susceptible to swelling by hexane. Otherwise, the increase of polymer concentration and dip-coating cycle reduced the swelling degree of the composite membrane. The low concentration of solvent in the membrane structure minimized the solvent's possibility to penetrate the polymer chain. Due to the low porosity of the resulting membrane, the composite membranes that used acetone as an additive resulted in a lower swelling degree than THF.

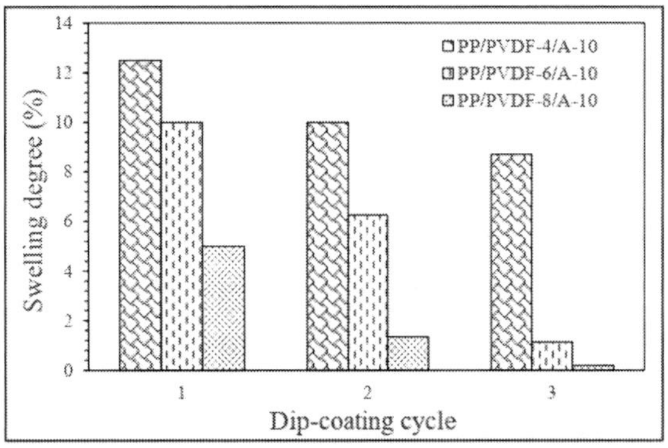


(a)

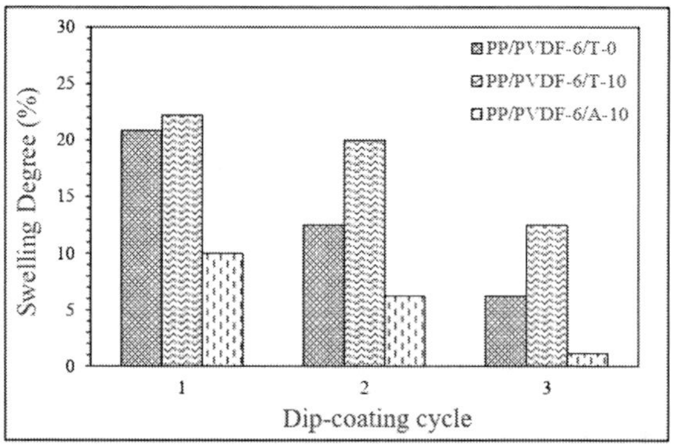


(b)

**Figure 4**. The influence of dip-coating cycle on the swelling degree of the composite membrane in hexane: (a) at various PVDF concentration and (b) at the different solvent type

## 3.2. The influence of PVDF concentration and solvent type the composite membrane hydrophobicity

**Figure 5** presents the influence of the dip-coating cycle and polymer concentration on water contact angle. The water contact angle indicates the hydrophilicity or hydrophobicity of the membrane [12]. It shows that the composite PVDF/PP membranes had a higher water contact angle than the unmodified PP membrane. A high-water contact angle between 109 and 140° was achieved by blending 4%wt PVDF and 10%wt THF in the coating solution. Meanwhile, the coating solution that used acetone as an additive produced a contact angle 121°. When the coating solution was prepared without volatile solvent as an additive, the resulted membrane had a contact angle between 102 and 106°. It was suggested that the higher contact angle was attributed to the increased surface roughness of the membrane [16].

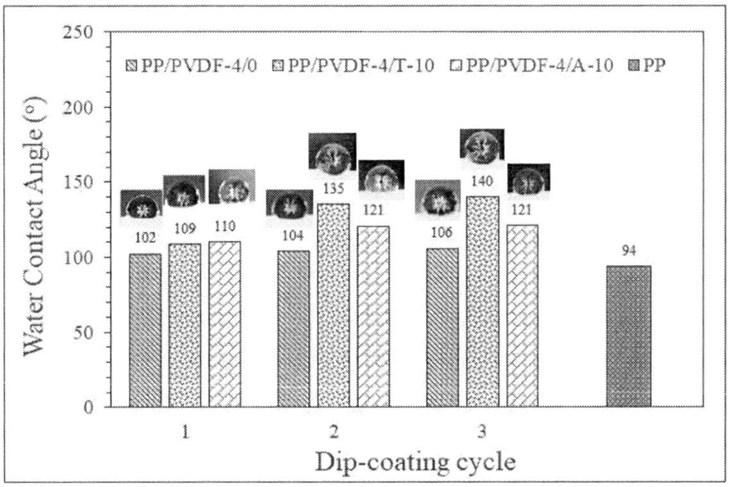


**Figure 5.** The influence of dip-coating cycle and solvent type on the water contact angle of the membrane

## 3.3. The influence of dip-coating cycle on permeate flux and oil rejection during the filtration of hexane/oil solution

The permeate flux the resulted composite membrane were presented in **Figure 6** and **7**. The experiments were conducted at two operating pressures, i.e., 2 and 4 bars. The coating polymer was prepared by blending 8%wt PVDF and 10%wt acetone. When the composite membrane was prepared by three times dip-coating cycle, the permeate flux was 20 $L.m^{-2}.h^{-1}$. Higher permeate flux was obtained when the dip-coating process was reduced. It has been explained that the low concentration of polymer resulted in higher membrane porosity, and as a result, the volume of permeate that passes through the membrane became larger (**Figure 6a**). The increase of volume fraction in the membrane structure reduced the intrinsic resistance of the membrane, and thus, the solution can more easily pass through the membrane to the permeate side. Also, the increase of operating pressure enhanced the permeate flux (**Figure 6b**).

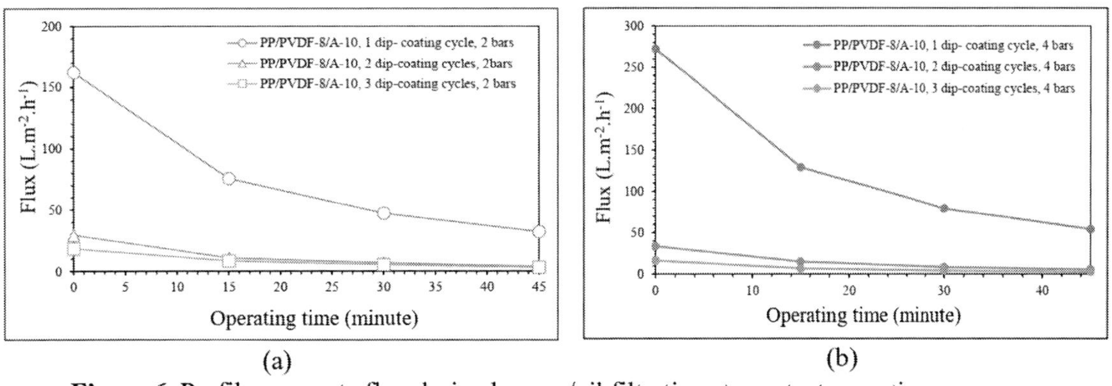


**Figure 6.** Profile permeate flux during hexane/oil filtration at constant operating pressure: (a) 2 bars and (b) 4 bars

The oil rejection of the resulted composite membranes is shown in **Figure 7**. The figure showed high rejection of oil when the PP membrane was prepared by three times coating process. A high concentration of polymer resulted in lower membrane porosity. Therefore, the rejection of solute (oil) was improved. Higher rejection of oil was obtained when the operating pressure was raised to 4 bars. Around 50% of oil rejection was achieved when 8% PVDF and 10% acetone were blended and coated on the PP membrane with three times of the dip-coating process. The increase in operating time enhanced the oil rejection. It was attributed to the fouling formation of the membrane surface, which reduced the membrane surface layer's sufficient pore size.

Consequently, it improved the oil rejection. Around 68% oil rejection has resulted after 45 minutes of hexane/oil filtration at an operating pressure of 4 bars. Compact fouling on the membrane surface due to a high concentration of solutes at high working pressure increased the oil rejection.

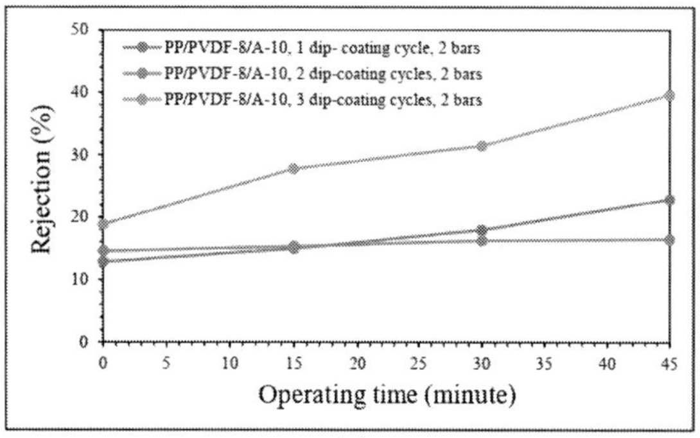


(a)

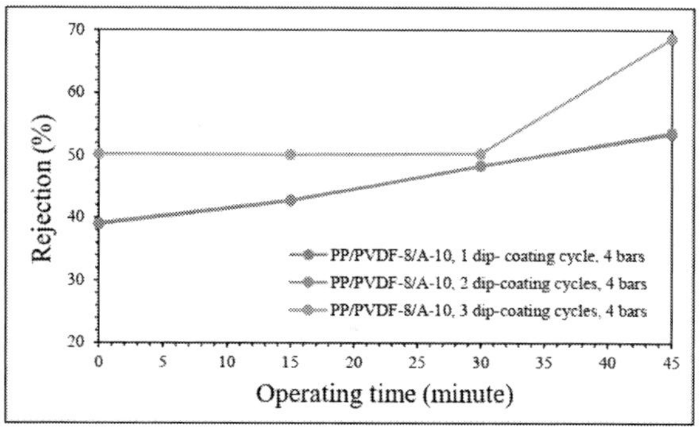


(b)

**Figure 7.** Oil rejection during hexane/oil filtration at constant operating pressure: (a) 2 bars and (b) 4 bars

## 4. Conclusion

A composite membrane has been prepared by dipping a hollow fiber polypropylene (PP) membrane in a polyvinylidene fluoride (PVDF) solution. Two types of volatile solvents (acetone and tetrahydrofuran/THF) are added to the PVDF solution to form a tight structure in the membrane surface. The influence of PVDF concentration, type of volatile solvents (acetone and THF), and the dip-coating cycle on membrane characteristics, including porosity, swelling degree, and contact angle, are investigated. The flux and rejection of the membranes are determined during the filtration of hexane/oil solution. The increase of PVDF concentration and dip-coating cycle reduces the composite PP/PVDF membrane's porosity and swelling degree. The lowest membrane porosity (3%) was achieved by blending 8%wt PVDF and 10%wt acetone in the coating solution and dipping the PP membrane 3 (three) times into the coating solution. The resulted membrane has a porosity of 3%, swelling degree of 0.1%, and permeate flux of 20 $L.m^{-2}.h^{-1}$. The oil rejection is 50% at the earlier of the filtration process and then increases to 68% after 45 minutes of hexane/oil filtration. Higher membrane porosity and permeate flux are obtained by reducing the dip-coating cycle. However, lower oil rejection has resulted.

### Acknowledgment

LPPM supported this work, Universitas Jenderal Achmad Yani, through "Internal Competitive Research Program 2020" (letter no. SKEP/129/UNJANI/V/2020).

## References

[1] Komaladewi A A I A S, Aryanti P T P, Lugito G, Surata I W dan Wenten I G 2018 Recent progress in microfiltration polypropylene membrane fabrication by stretching method *The 3rd International Tropical Renewable Energy Conference Sustainable Development of Tropical Renewable Energy* **67** 03018

[2] Yu Y, Liu M, Huang H, Zhao L, Lin P, Huang S, Xu J, Wang H and Wang L 2019 Low cost fabrication of polypropylene fiber composite membrane with excellent mechanical, super hydrophilic, antifouling and antibacterial properties for effective oil-in-water emulsion separation *Reactive and Functional Polymers* **142** 15-24

[3] Komaladewi A, Aryanti P, Surata I, Subagia I A and Wenten I 2019 Surface Modification of Microfiltration Polypropylene Membrane for Molecular Air Filtration *International Journal of Engineering and Emerging Technology* **4** (1) 74-80

[4] Wang Y, He G, Shao Y, Zhang D, Ruan X, Xiao W, Li X, Wu X and Jiang X 2019 Enhanced performance of superhydrophobic polypropylene membrane with modified antifouling surface for high salinity water treatment *Separation and Purification Technology* **214** 11-20

[5] Franco J A, Kentish S E, Perera J M and Stevens G W 2018 Fabrication of a superhydrophobic polypropylene membrane by deposition of a porous crystalline polypropylene coating *Journal of Membrane Science* **318** (1-2) 107-113

[6] Wenten I G, Victoria A V, Tanukusuma G, Khoiruddin K and Zunita M 2019 Simultaneous clarification and dehydration of crude palm oil using superhydrophobic polypropylene membrane *Journal of Food Engineering* **248** 23-27

[7] Ariono D, Wardani A, Widodo S, Aryanti P T and Wenten I 2018 Fouling mechanism in ultrafiltration of vegetable oil *Materials Research Express* **5** (3) 26-29

[8] Liu X D, Ni L, Zhang Y F, Liu Z, Feng X S and Ji. L 2012 Technology study of polypropylene hollow fiber membranes-like artificial lung made by the melt-spinning and cold-stretching method *Advanced Materials Research* **418** 26-29

[9] Li R, Liu J, Shi A, Luo X, Lin J, Zheng R, Fan H, Selasie S V and Lin H 2019 A facile method to modify polypropylene membrane by polydopamine coating via inkjet printing technique for superior performance *Journal of colloid and interface science* **552** 719-727.

[10] Changani Z, Razmjou A, Taheri-Kafrani A, Warkiani M E and Asadnia M 2020 Surface modification of polypropylene membrane for the removal of iodine using polydopamine chemistry *Chemosphere* **249** 126079

[11] Yang Y-F, Wan L-S, dan Xu Z-K 2009 Surface hydrophilization of microporous polypropylene membrane by the interfacial crosslinking of polyethylenimine *Journal of Membrane Science* **337** (1) 70-80

[12] Wenten I, Khoiruddin K, Wardani A, Aryanti P, Astuti D and Komaladewi 2020 A Preparation of antifouling polypropylene/ZnO composite hollow fiber membrane by dip-coating method for peat water treatment *Journal of Water Process Engineering* **34** 101158

[13] Shao H, Qi Y, Liang S, Qin S, and Yu J 2019 Polypropylene composite hollow fiber ultrafiltration membranes with an acrylic hydrogel surface by in situ ultrasonic wave-assisted polymerization for dye removal *Journal of Applied Polymer Science* **136** (8) 47099

[14] Abednejad A S, Amoabediny G and Ghaee A 2014 Surface modification of polypropylene membrane by polyethylene glycol graft polymerization *Materials Science and Engineering* **42** 443-450

[15] Lau W and Ismail A 2011 Progress in interfacial polymerization technique on composite membrane preparation *Proceedings of the 2nd International Conference on Environmental Engineering and Applications, Shanghai, China* **17** 173-177

[16] Himma N F, Wardani A K and Wenten I G 2017 The effects of non-solvent on surface morphology and hydrophobicity of dip-coated polypropylene membrane *Materials Research Express* **4** (5) 054001

[17] Asad A, Sameoto D and Sadrzadeh 2020 Overview of membrane technology *Nanocomposite*

*Membranes for Water and Gas Separation* **1** 1-28
[18] Himma N F, Wardani A K and Wenten I G 2017 Preparation of superhydrophobic polypropylene membrane using the dip-coating method: the effects of solution and process parameters *Polymer-Plastics Technology and Engineering* **56** (2) 184-194
[19] Ariono D, Aryanti P T P, Subagjo S and Wenten I G 2017 The effect of polymer concentration on flux stability of polysulfone membrane *AIP Conference Proceedings* **1788** (1) 030048
[20] Aryanti P T P, Ariono D, Hakim A N and Wenten I G 2018 Flory-Huggins Based Model to Determine Thermodynamic Property of Polymeric Membrane Solution *Journal of Physics: Conference Series* **1090** 012074
[21] Gugliuzza A 2016 *Solvent Swollen Polymer*, in *Encyclopedia of Membranes* Berlin Heidelberg: Springer

# Dominant Local Metric Dimension of Wheel Related Graphs

**R Umilasari[1], L Susilowati[2*] and Slamin[3]**

[1]Ph.D student of Mathematics and Natural Sciences, Faculty of Science and Technology, Universitas Airlangga, Jl. Mulyorejo Surabaya 60115, Indonesia
[2]Department of Mathematics, Faculty of Sciences and Technology, Universitas Airlangga, Jl. Mulyorejo Surabaya 60115, Indonesia
[3]Department of Informatics, Universitas Jember, Jl. Kalimantan 37 Jember 68121, Indonesia

*Email: liliek-s@fst.unair.ac.id

**Abstract.** Dominant local metric dimension is consist of two interesting topic in graph theory, they are dominating and metric dimension which be expanded as local metric dimension. A graph $G$ is said having dominant local metric dimension if $G$ be a connected graph and there is an ordered subset $W_l = \{w_1, w_2, \dots, w_n\} \subseteq V(G)$ where $W_l$ is a local resolving set as well as a dominating set of $G$. Minimum cardinality of dominant local resolving set of $G$ is called the dominant local basis of $G$. Then, cardinality of the dominant local basis of $G$ is called the dominant local metric dimension of $G$ which denoted by $Ddim_l(G)$. In this paper, we determine the dominant local metric dimension of wheel related graphs. That are Wheel, Jahangir, Friendship and Helm graph. Furthermore, we characterize the dominant local metric dimension of those graphs.

## 1. Introduction

Graph theory is a branch of discrete mathematics that can be used to solve a problem which can be represented as vertices and edges. Since it was first introduced in 1736 until now, graph theory has led to various new concepts that are always interesting to be researched, such as dominating set and metric dimension. The concept of dominating set developed since the beginning of 1850 [1], this theory studies the existence of a vertex set on a graph that causes every other vertex on the graph to be adjacent to at least one element in the vertex set which be studied and it called the dominating set. The cardinality of the dominating set is called the dominating number. The concept of basis and metric dimensions on a graph was first proposed in ref [2]. A basis on a graph is defined as a vertex set with minimum cardinality which results in each vertex on the graph having a different representation to the basis. Representations are presented using the concept of distance (metric) and the cardinality of basis is called the metric dimension of the graph. The development of the metric dimension concept includes the concept of the strong metric dimension [3], the local metric dimension [4], and dimensions local neighbourliness metric [5]. An application of the metric dimension theory in real life is determine the placement of fire sensors in a building which be written in ref [6].

Based on the explanation above, both the concepts of metric dimensions and dominating set on a graph have the same opportunity to be developed together. Some researchers who have developed this concept include Brigham et al. (2003) who introduced the term resolving domination number, which is a combination of the concept of metric dimensions and dominating set [7]. Next, Henning and

Oellarmann (2004) who developed the same definition but with different term, namely the metric location domination number [8]. A new research in this topic was introduced in ref [9], which determined the dominant metric dimension of some well-known graphs. Then, author combined the concept of the local metric dimension with the concept of dominating set which called the dominant local metric dimension. The concept of the dominant local metric dimension examines the vertex set on a graph which is a minimum local resolving set also as the dominating set of a graph. The results that have been obtained are some characteristics of dominant local metric dimension and its value in special graphs such as path, cycle, complete and bipartite graph [10]. Because the concept is still needed to developed, so this paper will determine the dominant local metric dimension of wheel related graphs. Specific graphs involved in this study include Wheel, Jahangir, Friendship and Helm graphs which described at the following definition below.

Definition 1. [11] Wheel graph $W_{n+1}$ is a graph that contains an *n-cycle* and one additional central vertex that is adjacent to all vertices of the cycle. Wheel graph $W_{n+1}$ has $(n+1)$-vertices and $2n$-*edges*.

Definition 2. [11] Jahangir graph $J_{3,n}$ is a graph with two vertex added between each pair adjacent graph vertices of the outer cycle. The Jahangir graph $J_{3,n}$ has $(3n+1)$-vertices and $4n$-*edges*.

Definition 3. [12] Friendship graph $F_n$ for $n \geq 2$ is a graph constructed by joining $n$ copies of the cycle graph $C_3$ with a common vertex. This graph has $(2n+1)$-vertices and $3n$-*edges*.

Definition 4. [13] Helm graph $H_n$ is the graph obtained from an *n-wheel graph* by adjoining a pendant edge at each vertex of the cycle. The helm graph $H_n$ has $(2n+1)$-vertices and $3n$-*edges*.

Dominating number of wheel related graphs are given in Table 1.

**Table 1.** Dominating Number of Wheel Related Graphs [15]-[18]

| Graph | Notation | Dominating Number |
|---|---|---|
| Wheel | $W_{n+1}$ | $\gamma(W_{n+1}) = 1, n \geq 3$ |
| Jahangir | $J_{3,n}$ | $\gamma(J_{3,n}) = n, n \geq 2$ |
| Friendship | $F_n$ | $\gamma(F_n) = n, n \geq 2$ |
| Helm | $H_n$ | $\gamma(H_n) = n, \geq 2$ |

The following definition and lemma are used to analyse the dominant local metric dimension of wheel related graphs. Some Lemma provides the property of local resolving set of a graph.

Definition 5. [10] Given a connected graph $G$. An ordered set $W_l = \{w_1, w_2, \dots, w_n\} \subseteq V(G)$ is called a dominant local resolving set if $W_l$ is local resolving set and dominating set of $G$. Dominant local resolving set with minimum cardinality is called dominant local basis. Number of vertex in dominant local basis of $G$ is called dominant local metric dimension and denoted by $Ddim_l(G)$.

Lemma 6. [10] Let $G$ be a connected graph and $S \subseteq V(G)$. Every set $S$ containing local resolving set is local resolving set.

Lemma 7. [10] Let $G$ be a connected graph and $W_l \subseteq V(G)$. For every $v_i, v_j \in W_l$ then $r(v_i|W_l) \neq r(v_j|W_l)$.

Lemma 8. [10] Let $G$ be a connected graph, then

$$max\{\gamma(G), dim_l(G)\} \leq Ddim_l(G) \leq min\{\gamma(G) + dim_l(G), |V(G)| - 1\}.$$

Theorem 9. [14] Let $F_n$ be a friendship graph, then for $n \geq 2$ $dim_l(F_n) = n$.

## 2. Methodology

This topic consist of some method or prosedure to get the dominant local metric dimension of some graphs related to wheel.

1. Study the definion of dominant local metric dimension.
2. Study some characteristic of dominant local metric dimension.
3. Determine vertex and edge set of the graphs (Wheel, Jahangir, Frienship and Helm graphs).
4. Observe the number of dominating set and local metric dimension of the graphs by this procedure:
   a. Determine the local resolving set of the graphs
   b. Calculating the distance of each vertex with a predetermined, such that every two different vertices adjacent to each other on the graphs has a different representation to the local resolving set.
   c. Determine the local metric dimension and the general pattern of local metric dimensions on the graph.
   d. Determine the dominating set and dominating number.
   e. Compare between the dominating number and local metric dimension
5. Determine the dominant local metric dimension and make proofs by compiling the terms and theorems based on the results obtained.

## 3. Result and Discussion

This section present the dominant local metric dimension of Wheel, Jahangir, Friendship and Helm graphs. We start by giving a property to show the dominant local metric dimension of wheel graph.

Lemma 10. Let $W_{n+1}$ be a wheel graph of order $n \geq 6$, the edge set is $E(W_{n+1}) = \{uv_i|1 \leq i \leq n\} \cup \{v_iv_{i+1}|1 \leq i \leq n-1\} \cup \{v_nv_1\}$, then $W_l = \{v_1, v_4\}$ is local resolving set of $W_{n+1}$.

Proof. Let $V(W_{n+1}) = \{u, v_i|1 \leq i \leq n\}$ and $E(W_{n+1}) = \{uv_i|1 \leq i \leq n\} \cup \{v_iv_{i+1}|1 \leq i \leq n-1\} \cup \{v_nv_1\}$ with $n \geq 6$. Let $W_l = \{v_1, v_4\}$, for every $v_i \in V(W_{n+1})$ with $i = 1, 2, 3, \ldots, n$, then:

$$d(u, v_1) = 1$$
$$d(u, v_4) = 1$$
$$d(v_i, v_1) = \begin{cases} 0 \ ; & i = 1 \\ 1 \ ; & i = 2, n \\ 2 \ ; & i = 3, \ldots, n-1 \end{cases}$$
$$d(v_i, v_4) = \begin{cases} 0 \ ; & i = 4 \\ 1 \ ; & i = 3,5 \\ 2 \ ; & i = 6,7, \ldots, n, 1,2 \end{cases}$$

Based on the description above, for any two adjacent vertices have different representation. Hence, $W_l$ is local resolving set of $W_{n+1}$ for $n \geq 6$. We can say that if $W_{n+1}$ is a wheel with the order $n \geq 6$ and $E(W_{n+1}) = \{uv_i|1 \leq i \leq n\} \cup \{v_iv_{i+1}|1 \leq i \leq n-1\} \cup \{v_nv_1\}$, then $W_l = \{v_1, v_4\}$ is a local resolving set of $W_{n+1}$.■

Theorem 11. Let $W_{n+1}$ $n \geq 6$, then $Ddim_l\ (W_{n+1}) = \left\lceil \frac{n}{3} \right\rceil$.

Proof. Let $W_{n+1}$ be a wheel graph with order $n \geq 6$, the vertex set of wheel consist of a center vertex and an outer cycle with order $n$, every vertex in cycle adjacent with a center vertex. Let the vertex set is $V(W_{n+1}) = \{u, v_i|1 \leq i \leq n\}$ and the edge set is $E(W_{n+1}) = \{uv_i|1 \leq i \leq n\} \cup \{v_iv_{i+1}|1 \leq i \leq n-1\} \cup \{v_nv_1\}$, then there exist three possibilities of $n$.

1. If $n \equiv 0\ (mod\ 3)$, choose $W_l = \{v_1, v_4, v_7, v_{10}, \dots, v_{3i-2}, \dots, v_{n-2}\}$ then $|W_l| = \left\lceil \frac{n}{3} \right\rceil$.
2. If $n \equiv 1\ (mod\ 3)$, choose $W_l = \{v_1, v_4, v_7, v_{10}, \dots, v_{3i-2}, \dots, v_n\}$ then $|W_l| = \left\lceil \frac{n}{3} \right\rceil$.
3. If $n \equiv 2\ (mod\ 3)$, choose $W_l = \{v_1, v_4, v_7, v_{10}, \dots, v_{3i-2}, \dots, v_{n-1}\}$ then $|W_l| = \left\lceil \frac{n}{3} \right\rceil$.

Because $\{v_1, v_4\} \subseteq W_l$, by Lemma 10, $W_l$ is local resolving set of $W_{n+1}$. Next, because $E(W_{n+1}) = \{uv_i | 1 \le i \le n\} \cup \{v_i v_{i+1} | 1 \le i \le n-1\} \cup \{v_n v_1\}$, we can see that $u$ adjacent to every vertex of $W_l$ and $v_{3i-1}$ adjacent to $v_{3i-2}$ and $v_{3i}$ adjacent to $v_{3(i+1)-2}$, then $W_l$ is dominating set of $W_{n+1}$. Hence, $W_l$ is dominant local resolving set of $W_{n+1}$. Next, we will show that $W_l$ is a dominant local resolving set with minimum cardinality. Choose any $S \subseteq V(W_{n+1})$ with $|S| = |W_l| - 1$, then there exist two possibilities:

a. $u \in S$
If $u \in S$, there exist $v_i, v_k$ with $d(v_i, v_k) > 3$ in cycle such that $v_i, v_{i+1}, v_{i+2}, \dots, v_{k-1}, v_k \notin S$. Therefore $r(v_{i+1}|S) = r(v_{i+2}|S) = (2,2, \dots, 2,1)$, where $v_{i+1}v_{i+2} \in E(W_{n+1})$. Hence, $S$ is not local resolving set of $W_{n+1}$ graph.

b. $u \notin S$
If $u \notin S$ there exist $v_i, v_k \in S$ such that $d(v_i, v_k) = 2$ in cycle. Therefore, $S$ is not dominating set of $W_{n+1}$ graph.

Based on the two possibilities in point a and b, this is contrary with the definition of dominant local metric dimension. Therefore, it is proven that $|W_l| = \left\lceil \frac{n}{3} \right\rceil$ is a dominant local resolving set with minimum cardinality. Thus, for $n \ge 6$ $Ddim_l(W_{n+1}) = \left\lceil \frac{n}{3} \right\rceil$. ■

We now ready to determine the dominant local metric dimension of Jahangir graph. At the first, we show the local resolving set of the graph which is presented in the following Lemma.

Lemma 12. Let $J_{3,n}$ be a Jahangir graph of order $3n+1, n \ge 2$. The edge set is $(J_{3,n}) = \{uv_i; 1 \le i \le n\} \cup \{v_i x_{i1}, x_{i1} x_{i2}; 1 \le i \le n\} \cup \{x_{i2} v_{i+1}; 1 \le i \le n-1\} \cup \{x_{n2} v_1\}$, then $W_l = \{v_1, v_2\}$ is local resolving set of $J_{3,n}$.

Proof. Let $V(J_{3,n}) = \{u, v_i, x_{i1}, x_{i2}; 1 \le i \le n\}$ and $E(J_{3,n}) = \{uv_i; 1 \le i \le n\} \cup \{v_i x_{i1}, x_{i1} x_{i2}; 1 \le i \le n\} \cup \{x_{i2} v_{i+1}; 1 \le i \le n-1\} \cup \{x_{n2} v_1\}$ with $n \ge 2$. Put $W_l = \{v_1, v_2\}$, for every $u, v_i, x_{i1}, x_{i2} \in V(J_{3,n})$ with $i = 1, 2, 3, \dots, n$, then:

$$d(u, v_1) = 1$$
$$d(u, v_2) = 1$$
$$d(v_i, v_1) = \begin{cases} 0 & ; \quad i = 1 \\ 2 & ; \quad 2 \le i \le n \end{cases}$$
$$d(v_i, v_2) = \begin{cases} 0 & ; \quad i = 2 \\ 2 & ; \quad i = 3,4, \dots, n-1, n, 1 \end{cases}$$
$$d(x_{i1}, v_1) = \begin{cases} 1 & ; \quad i = 1 \\ 2 & ; \quad i = n \\ 3 & ; \quad 2 \le i \le n-1 \end{cases}$$
$$d(x_{i1}, v_2) = \begin{cases} 1 & ; \quad i = 2 \\ 2 & ; \quad i = 1 \\ 3 & ; \quad 3 \le i \le n \end{cases}$$
$$d(x_{i2}, v_1) = \begin{cases} 1 & ; \quad i = n \\ 2 & ; \quad i = 1 \\ 3 & ; \quad 2 \le i \le n-1 \end{cases}$$
$$d(x_{i2}, v_2) = \begin{cases} 1 & ; \quad i = 1 \\ 2 & ; \quad i = 2 \\ 3 & ; \quad 3 \le i \le n \end{cases}$$

Based on the description above, for any two adjacent vertices have different representation. Hence, $W_l$ is local resolving set of $J_{3,n}$ for $n \geq 2$. We can say that if $J_{3,n}$ is a Jahangir with the order $3n+1$ for $n \geq 2$ and the edge set is $(J_{3,n}) = \{uv_i; 1 \leq i \leq n\} \cup \{v_i x_{i1}, x_{i1}x_{i2},; 1 \leq i \leq n\} \cup \{x_{i2}v_{i+1}; 1 \leq i \leq n-1\} \cup \{x_{n2}v_1\}$, then $W_l = \{v_1, v_2\}$ is a local resolving set of $J_{3,n}$.■

Theorem 13. Let $n \geq 2$, then $Ddim_l(J_{3,n}) = \gamma(J_{3,n})$.

Proof. Let $V(J_{3,n}) = \{u, v_i, x_{i1}, x_{i2}; 1 \leq i \leq n\}$ and $E(J_{3,n}) = \{uv_i; 1 \leq i \leq n\} \cup \{v_i x_{i1},\ x_{i1}x_{i2},; 1 \leq i \leq n\} \cup \{x_{i2}v_{i+1}; 1 \leq i \leq n-1\} \cup \{x_{n2}v_1\}$. Put $W_l = \{v_1, v_2, \dots, v_n\}$ then $|W_l| = n$. Because $\{v_1, v_2\} \subseteq W_l$, based on the Lemma 12, we get that $W_l$ is a local resolving set of $J_{3,n}$. Next, because $(J_{3,n}) = \{uv_i; 1 \leq i \leq n\} \cup \{v_i x_{i1}, x_{i1}x_{i2},; 1 \leq i \leq n\} \cup \{x_{i2}v_{i+1}; 1 \leq i \leq n-1\} \cup \{x_{n2}v_1\}$, we can see that $u$ is adjacent to $v_i$, $v_i$ adjacent to $x_{i1}$ for $1 \leq i \leq n$ and $v_{i+1}$ adjacent to $x_{i2}$ for $1 \leq i \leq n-1$. Hence, $W_l$ is a dominating set of $J_{3,n}$ and $W_l$ is a dominant local resolving set of $J_{3,n}$. Based on the Table 1, $\gamma(J_{3,n}) = n$, then it can be conclude that $W_l$ is a dominant local resolving set with minimum cardinality and $Ddim_l(J_{3,n}) = \gamma(J_{3,n}), n \geq 2$. ■

The following theorem presents the dominant local metric dimension of a Friendship graph of order $2n+1$ for $n \geq 2$

Lemma 14. Let $F_n$ be a Friendship graph of order $2n+1, n \geq 2$. The edge set is $E(F_n) = \{uv_i, ux_i, v_i x_i; 1 \leq i \leq n\}$, then $W_l = \{v_1, v_2\}$ is local resolving set of $F_n$.
Proof. Let $V(F_n) = \{u, v_i, x_i; 1 \leq i \leq n\}$ and $E(F_n) = \{uv_i, ux_i, v_i x_i; 1 \leq i \leq n\}$ with $n \geq 2$. Put $W_l = \{v_1, v_2\}$, then for every $u, v_i, x_i \in V(F_n)$ with $i = 1, 2, 3, \dots, n$, then:

$$d(u, v_1) = 1$$
$$d(u, v_2) = 1$$
$$d(v_i, v_1) = \begin{cases} 0 \ ; & i = 1 \\ 2 \ ; & 2 \leq i \leq n \end{cases}$$
$$d(v_i, v_2) = \begin{cases} 0 \ ; & i = 2 \\ 2 \ ; & i = 3,4, \dots, n-1, n, 1 \end{cases}$$
$$d(x_i, v_1) = \begin{cases} 1 \ ; & i = 1 \\ 2 \ ; & i = 2,3, \dots, n \end{cases}$$
$$d(x_i, v_2) = \begin{cases} 1 \ ; & i = 2 \\ 2 \ ; & i = 3,4, \dots, n-1, n, 1 \end{cases}$$

Based on the description above, for any two adjacent vertices have different representation. Hence, $W_l$ is local resolving set of $F_n$ for $n \geq 2$. We can say that if $F_n$ is a Friendship with the order $2n+1$ for $n \geq 2$ and $E(F_n) = \{uv_i, ux_i, v_i x_i; 1 \leq i \leq n\}$, then $W_l = \{v_1, v_2\}$ is a local resolving set of $F_n$.■

Theorem 15. Let $n \geq 2$, then $Ddim_l\ (F_n) = dim_l(F_n)$.

Proof. Let $V(F_n) = \{u, v_i, x_i; 1 \leq i \leq n\}$ and $E(F_n) = \{uv_i, ux_i, v_i x_i; 1 \leq i \leq n\}$. Put $W_l = \{v_1, v_2, \dots, v_n\}$ then $|W_l| = n$. Because $\{v_1, v_2\} \subseteq W_l$, based on Lemma 14, we get that $W_l$ is a local resolving set of $F_n$. Next, because $E(F_n) = \{uv_i, ux_i, v_i x_i; 1 \leq i \leq n\}$ we can see that $v_i$ is adjacent to $x_i$ and $u$. Hence, $W_l$ is a dominating set of $F_n$ and $W_l$ is a dominant local resolving set of $F_n$. Based on Theorem 9, $dim_l(F_n) = n$, then it can be conclude that $W_l$ is a dominant local resolving set with minimum cardinality and $Ddim_l(F_n) = dim_l(F_n)$, $n \geq 2$. ■

Theorem 16. Let $n \geq 3$, then $Ddim_l\ (H_n) = \gamma(H_n)$.
Proof. Let $V(H_n) = \{u, v_i, x_i; 1 \leq i \leq n\}$ and $E(H_n) = \{uv_i; 1 \leq i \leq n\} \cup \{v_i x_i; 1 \leq i \leq n\} \cup \{v_i v_{i+1}; 1 \leq i \leq n-1\} \cup \{v_n v_1\}$. Based on the Table 1, we have $\gamma(H_n) = n$. Let $|W_l| = n$,

choose $W_l = \{v_i; 1 \le i \le n\}$. We can see that $d(v_i, u) = d(v_i, x_i) = 1$, then $W_l$ is dominating set of $H_n$. Without loss of generality, representation of each vertex to $W_l$ can be written as follows.

a. Representation of vertex $u$ to $W_l$ is $r(u|W_l) = (1,1,1,\dots,1)$.
b. Representation of vertex $v_i$ to $W_l$: $r(v_1|W_l) = (0,1,1,\dots,1)$, $r(v_2|W_l) = (1,0,1,\dots,1),\dots,$ $r(v_n|W_l) = (1,1,1,\dots,0)$.
c. Representation of vertex $x_i$ to $W_l$: $r(u_1|W_l) = (1,2,3,3,\dots,3,2)$, $r(u_2|W_l) = (2,1,2,3,\dots,3,3)$, $r(u_3|W_l) = (3,2,1,2,3,\dots,3),\dots, r(u_n|W_l) = (2,3,\dots,3,2,1)$.

Representation of each vertex to $W_l$ is different and $W_l$ is dominating set, hence $W_l$ is dominant local resolving set of $H_n$. Because on the Table 1 we know that dominating number of Helm graph is $(H_n) = n$ , then $W_l = n$ is local resolving set with minimum cardinality. So, it is proven that $Ddim_l(H_n) = \gamma(H_n)$. ■

## 4. Conclusion

Because this topic still new and needed to expand, so we conclude this paper by giving an open problem: determine the dominant metric dimension of some product graphs such as join, corona, or comb product of graphs.

## References

[1] Chartrand G, Eroh L, Johnson M A and Oellermann O R 2000 Resolvability in Graphs and The Metric Dimension of A Graph. *Discrete Appl. Math.* **105** 99.
[2] Saputro S W, Mardiana N and Purwasih I A 2013 The metric dimension of comb product graphs. *Graph Theory Conference in honor of Egawa's 60th birthday September 10 to 14* **1-2**
[3] Oellarman and Peters-Fransen 2007 The strong metric dimension of graphs and digraphs *Discrete Applied Mathematics* **155** 356.
[4] Okamoto F,Crosse L,Phinezy B, Zhang P and Kalamazo 2010 The Local Metric Dimension of Graphs *Mathematica Bohemica* **135** 239.
[5] Rodriguez-Velazquez J A, Gomes C G and Ramirez G A B 2014 Computing the local metric dimension of a graph from the local metric dimension of primary subgraphs *International Journal of Computer Mathematics* **92** 1.
[6] Wahyudi S 2018 Aplikasi Dimensi Metrik untuk Meminimalkan Pemasangan Sensor Kebakaran Sebuah Gedung *Jurnal Mathematics and Its Application (Limits)* **15** 89.
[7] Brigham R C, Chartrand G, Dutton R D and Zhang P 2003 Resolving domination in graphs. *Mathematica Bohemica* **128** 25.
[8] Henning M A and Oellermann O R 2004 Metric-locating dominating sets in graphs *Ars Combinatoria -Waterloo then Winnipeg* **73** 1,
[9] Susilowati L, Sa'adah I, Fauziyyah R Z, Erfanian A and Slamin 2020 The Dominant Metric Dimension of Graph *Heliyon* **6** 1.
[10] Umilasari R, Susilowati L and Slamin 2020 the Dominant Local Metric Dimension of Graphs *CAUCHY* **4** 125.
[11] Aytaç A, Odabaş Z N 2011 Residual Closeness of Wheels and Related Networks. Int. J. Found. Comput. Sci. **22** 1229
[12] Gallian J A 2019 A Dynamic Survey of Graph Labelling *The Electronic Journal of Combinatoric* **DS6** 4.
[13] Javaid I, Shokat S 2008 On the Partition Dimension of Some Wheel Related Graphs *Journal of Prime Research in Mathematics* **4** 154.
[14] Cahyabudi A N and Kusmayadi T A 2017 On the local metric dimension of a lollipop graph, a web graph, and a friendship graph *Journal of Physics: Conference Series* **909** 012039
[15] Snyder K 2011 *c-Dominating Sets for Families of Graphs* Fredericksburg: University of Mary Washington.

[16] Mojdeh D A and Ghameshlou A N 2007 Domination in Jahangir Graph $J_{2,m}$. *Int. J. Contemp. Math. Sciences* **24** 193.
[17] Ayhan A K 2011 Determination and testing the Domination Numbers of Helm Graph, Web graph and Levi Graph Using MATLAB *Journal of Education and Science* **24** 103.
[18] Nagabhushana C S, Kavitha B N and Chudamani H M 2017 Split and Equitable Domination of Some Special Graph *International Journal of Science Technology & Engineering* **4** 50.
[19] Ariono D, Aryanti P T P, Subagjo S and Wenten I G 2017 The effect of polymer concentration on flux stability of polysulfone membrane *AIP Conference Proceedings* **1788** 030048.

# Disturbance Characteristics of Uninterruptible Power Supply System in the Frequency Range 9 – 150 kHz with Capacitive Load Level Variations

**R L Pratama, B Sudiarto* and R Setiabudy**

Departement of Electrical Engineering, Faculty of Engineering, University of Indonesia, Depok 16424, Indonesia

*Corresponding author : budi.sudiarto@ui.ac.id

**Abstract.** Nowadays the use of Uninterruptible Power Supply (UPS) is common because UPS is able to provide a backup electrical power when the main source is disconnected or interrupted. With the use of a UPS, the supply of electrical power to the system will not be interrupted. When the UPS operates, the voltage go through dual conversion process, from AC to DC by the rectifier and from DC to AC by the inverter. The DC voltage in the UPS is used to supply the battery charger and the AC voltage is used to supply the load system. The switching process of the inverter can produce disturbances at high frequency. In this study, a 1 KVA UPS is connected to a capacitive load 3.6 µF, 7.2 µF, 14.4 µF and 28.8 µF which generate the highest disturbances voltage of 309.53 mV at a frequency of 9.8 kHz at a load of 28.8 µF and the lowest of 1.33 mV at a frequency of 110.6 kHz at a load of 3.6 µF. The disturbances voltage occur at three dominant frequencies of 19-24 kHz, 35-44 kHz and 105- 114 kHz and increases as the increase of capacitive load level.

## 1. Introduction

The need for backup of electric power sources as a source that is able to supply electricity temporarily when electricity from the main source is interrupted or disrupted make the use of Uninterruptible Power Supply (UPS) devices increasing. Also in recent years the use of UPS increase as the development of power infrastructure needed to support IT [1]. Because the use of UPS devices ensures that all equipment connected to a UPS device will not interrupted for even a second [2]. The UPS is compatible to use for sensitive applications and it can adjust the input voltage so the output of the UPS suit with the load input ranges voltage [3]. In standard operation the UPS has three modes namely bypass mode, battery mode and normal mode [4]. Bypass mode is used when the UPS device is in the process of maintenance or under certain conditions it can automatically switch to bypass mode when the load carried by the UPS device exceeds its nominal capacity [5]. Battery mode or emergency mode occurs when the main power source is disconnected so the UPS device automatically switches to use a battery to supply voltage to the load without power cut [6]. In this mode, the direct current bus supplied by the battery [7]. Stored electric power will be distributed to electronic equipment when themain power source is lost. The process of supplying the battery will continue until the battery voltage drops. And the last is the normal mode, this mode is used when the UPS works in normal conditions. The alternating current from the main source enters the UPS device which is then converted into direct current by the rectifier.

Direct current is used to supply to the battery charger for charging the battery and the output from it is converted again into alternating current by the inverter to supply voltage to the load. High switching frequency of inverters can reduce the low order harmonic distortion, but it shifts emissions into higher frequencies which is more than 2 kHz [8]. With the use of an inverter, the UPS device becomes one of the devices that can cause disturbances on the output voltage that connected to the load [9]. The characteristic of an inverter is nonlinear, thus potentially causing disturbances voltage in the electrical system [10].

## 2. Research Methodology

This study uses a picoscope to record the data. Picoscope is a portable oscilloscope that is used by connecting to a computer so that it gets power through a USB connection from the computer. All measurement data files recorded by the oscilloscope are then stored in the computer. Data recorded by the oscilloscope is still in the time domain signal, so it needs to be converted into the frequency domain signal using fourier transforms. In order to facilitate the data processing, changing data from the time domain signal to the frequency domain signal using fourier transforms is done using the matlab application. To filter the measured frequency only in the range of 9 kHz to 150 kHz, this study uses a high pass filter that connects the output voltage of the UPS to the oscilloscope so that the recorded data is only the disturbances voltage that generated at frequencies from 9 kHz to 150 kHz. In result of this research can be seen the characteristics of the disturbances voltage produced when the UPS uses normal mode with capacitive loads in the frequency range of 9 kHz to 150 kHz. The research flowchart presented in Figure 1.

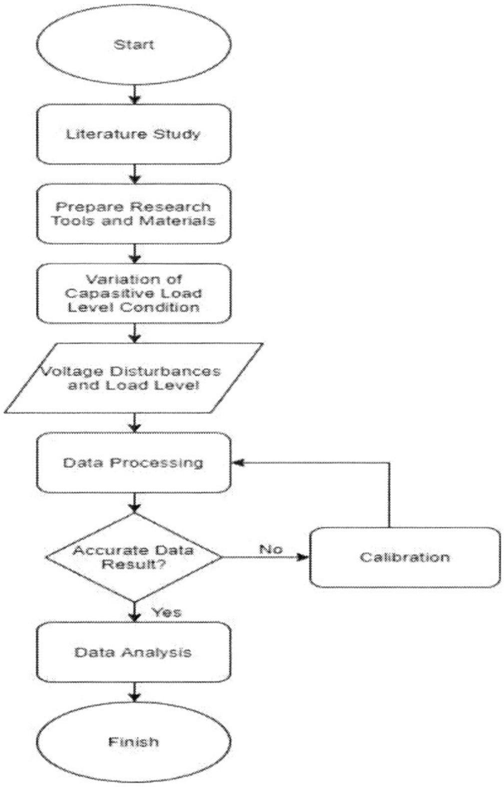


**Figure 1.** Research Flowchart

IOP Conf. Series: Materials Science and Engineering 1115 (2021) 012030 doi:10.1088/1757-899X/1115/1/012030

This research was conducted with a system configuration as Figure 2.

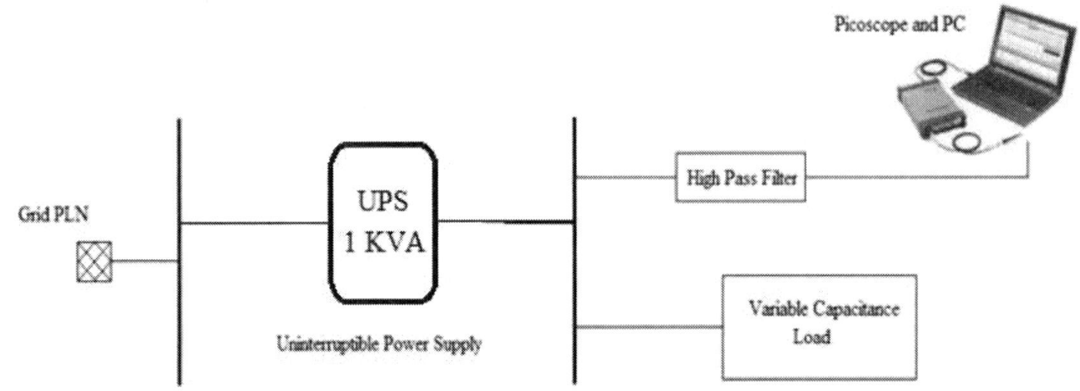


**Figure 2.** Measurement System Configuration

As in Figure 2, in this study the disturbances voltage generated from the output side of the UPS with a maximum capacity of 1 KVA are recorded continuously using an oscilloscope. Before carrying out data measurements, the UPS is first set to operate in normal mode so that the voltage supplied to the system comes from the output of an inverter. The dc voltage that comes from the rectifier is then converted to ac voltage using an inverter so that the voltage supplied to the system is in the form of an ac voltage with a voltage range of 200 volts to 240 volts. When the UPS operates in normal mode, alternately the capacitive load connected to the UPS is turned on so that the capacitive load connected to the UPS varies from 3.6 μF, 7.2 μF, 14.4 μF and 28.8 μF. Thus the data recorded by the oscilloscope has each characteristics according to the connected load level.

Frequencies that are below 9 kHz will be cut by a high pass filter so that the data recorded for analysis will only display disturbances voltage in the frequency range of 9 kHz to 150 kHz.

## 3. Result and Discussion

From the results of this study it was found that the disturbances voltage from 1 KVA UPS in the frequency range of 9 kHz to 150 kHz have different characteristics at each load level variation. The peak disturbances voltage generated in this study increase as the increase of capacitive load level and the result of disturbances voltage with capacitive load level variation accumulate with the disturbances voltage that generated from the switching frequency process from an inverter in UPS. The disturbances voltage that generated in this study are occur at three dominant frequencies of 15 kHz - 24 kHz, 35 kHz - 44 kHz and 105 kHz - 114 kHz as presented in Table 1.

**Tabel 1.** Disturbance Voltage at Three Dominant Frequencies

| Capasitive Load | Load Level | Description | Frequency Range (kHz) | | |
|---|---|---|---|---|---|
| | | | 15-24 | 35-44 | 105-114 |
| 3.6 μF | 10% | Frequency (kHz) | 19.6 | 39.4 | 110.6 |
| | | Voltage Peak (mV) | 195.07 | 71.07 | 1.33 |
| 7.2 μF | 20% | Frequency (kHz) | 19.6 | 39.4 | 110.4 |
| | | Voltage Peak (mV) | 243.85 | 214.20 | 61.47 |
| 14.4 μF | 40% | Frequency (kHz) | 19.8 | 39.6 | 110.4 |
| | | Voltage Peak (mV) | 288.53 | 274.59 | 66.91 |
| 28.8 μF | 80% | Frequency (kHz) | 19.8 | 39.6 | 110.4 |
| | | Voltage Peak (mV) | 309.53 | 291.94 | 83.60 |

As in Figure 3, it shown when the UPS is connected to a capacitive load of 3.6 μF. The highest disturbance voltage that generated is 195.07 mV at a frequency of 19.6 kHz then the next disturbance voltage peak is 71.07 mV at a frequency of 39.4 kHz and 1.33 mV at a frequency of 110.6 mV.

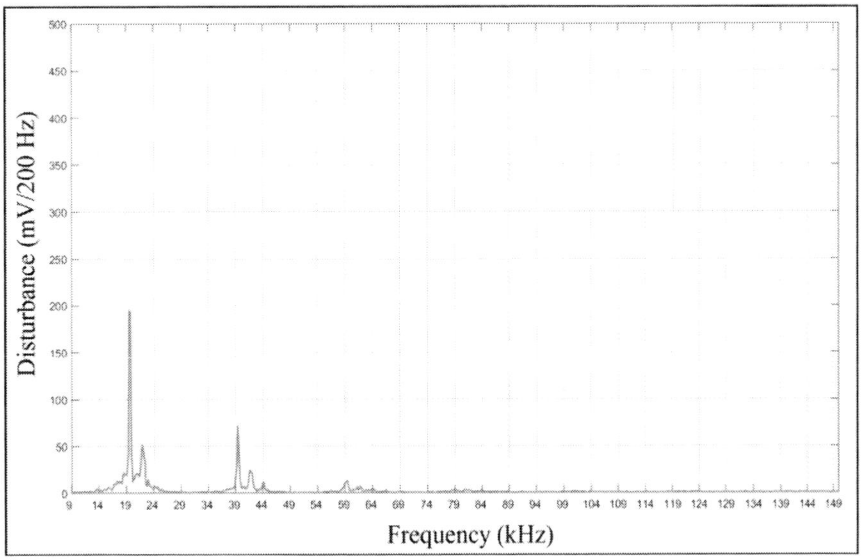


**Figure 3.** Disturbance Voltage at Load of 3.6 μF

As in Figure 4, it shown when the UPS is connected to a capacitive load of 7.2 μF. The highest disturbance voltage that generated is 243.85 mV at a frequency of 19.6 kHz then the next disturbance voltage peak is 214.20 mV at a frequency of 39.4 kHz and 61.47 mV at a frequency of 110.4 mV.

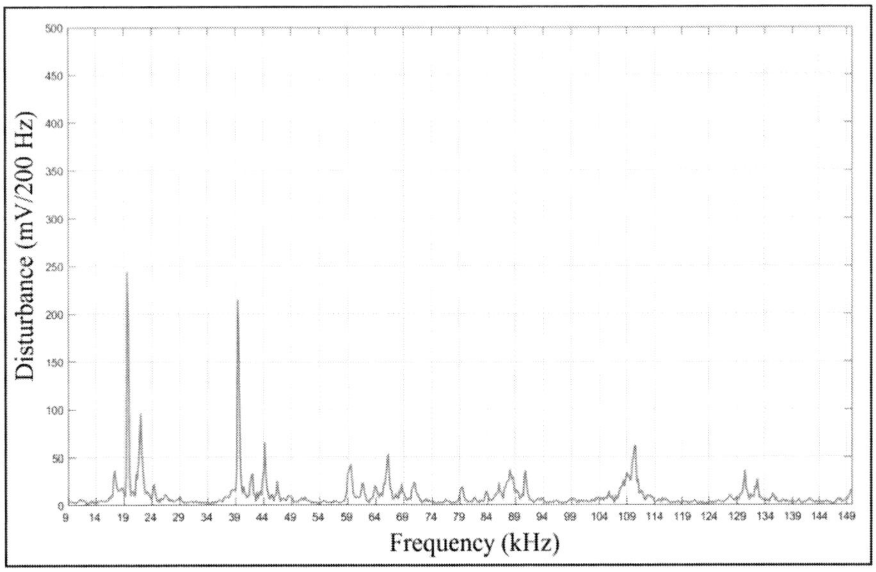


**Figure 4.** Disturbance Voltage at Load of 7.2 μF

As in Figure 5, it shown when the UPS is connected to a capacitive load of 14.4 μF. The highest disturbance voltage that generated is 288.53 mV at a frequency of 19.8 kHz then the next disturbance voltage peak is 274.59 mV at a frequency of 39.6 kHz and 66.91 mV at a frequency of 110.4 mV.

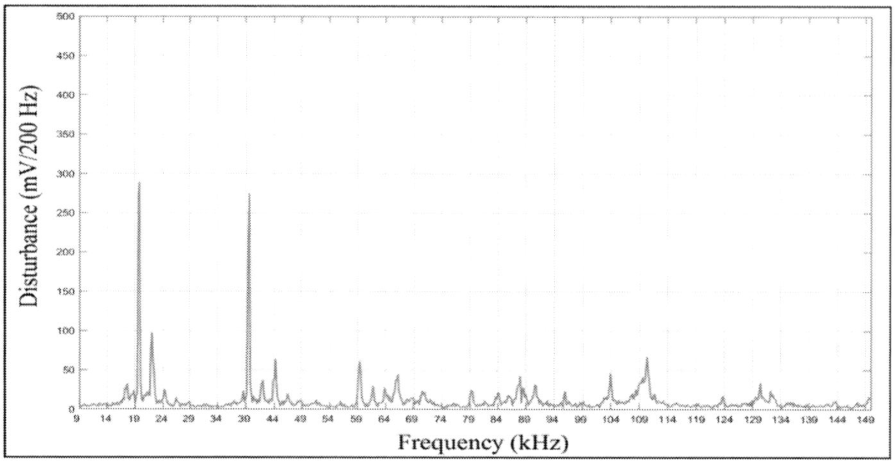


**Figure 5.** Disturbance Voltage at Load of 14.4 µF

As in Figure 6, it shown when the UPS is connected to a capacitive load of 28.8 µF. The highest disturbance voltage that generated is 309.53 mV at a frequency of 19.8 kHz then the next disturbance voltage peak is 291.94 mV at a frequency of 39.6 kHz and 83.6 mV at a frequency of 110.4 mV.

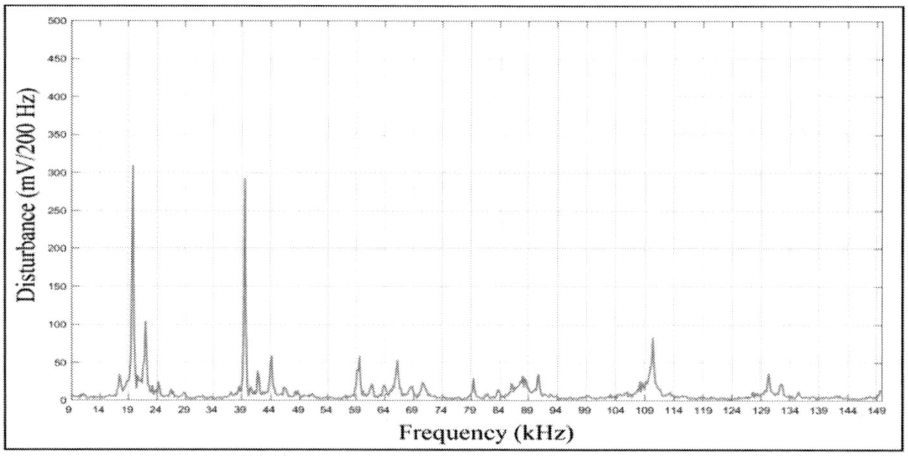


**Figure 6.** Disturbance Voltage at Load of 28.8 µF

## 4. Conclusion

Based on the results of measurements of disturbances voltage at the frequency of 9 kHz to 150 kHz in UPS system with capasitive load level variations, the following conclusions were obtained:

- UPS system has disturbances voltage peak which are dominated by frequency ranges of 15 – 24 kHz, 35 – 44 kHz and 105 – 114 kHz
- Disturbances voltage peak which are generated in UPS system that connected to capasitive load has the highest voltage peak at 19.8 kHz and 19.6 kHz.
- As the capasitive load level increases, the disturbances voltage also increase. When the load level is 10% or when load connected is 3,6 µF the disturbance voltage is 195.07 mV at frequency of 19.8 kHz and when the load level increase to 80% or when load connected is 28.8 µF the disturbance voltage increase to 309.53 mV at frequency of 19.8 kHz.

IOP Conf. Series: Materials Science and Engineering 1115 (2021) 012030 doi:10.1088/1757-899X/1115/1/012030

## Acknowledgment

This research was supported by Universitas Indonesia (UI) through PUTI grant 2020 (International Indexed Publication) launched by DRPM UI.

## References

[1] Hanaoka H, Matsuo H and Yanagisawa N 2015 The new concept for 3-phase 4-wire UPS system. In *2015 IEEE International Telecommunications Energy Conference* 1-5

[2] Singh S and Ghatakchoudhuri S 2016 Average Current Feed-Forward Control for Multi-Modular Single-Phase UPS Inverters System *2016 7th International Conference on Intelligent Systems, Modelling and Simulation* 244-249

[3] Giuntini L 2018 Ring Bus Architecture with Static UPS for Critical Power Applications. *2018 IEEE International Telecommunications Energy Conference* 1-6

[4] Liu X, Teng D, Wang D, Zhu Q and Liu Z 2017 Application of eco Mode UPS in data center. *2017 IEEE International Telecommunications Energy Conference* 30-34

[5] Giuntini L 2014 Impact of temperature on the efficiency of double-conversion UPS *7th IET International Conference on Power Electronics, Machines and Drives* 1–5

[6] Alqinsi P, Edward I J M, Ismail N and Darmalaksana W 2018 IoT-Based UPS Monitoring System Using MQTT Protocol *4th International Conference on Wireless and Telematics* 1–5

[7] Yu J, Ohn S, Burgos R, Sun B, Boroyevich D, Suryanarayana H and Belcastro C 2019 EMI Evaluation and Filter Design of a SiC-Based 3-Level UPS *IEEE Applied Power Electronics Conference and Exposition* 2390–2395

[8] Anne G, Jan M and Sarah R 2018 Comparison of Measurement Methods for the Frequency Range 2 – 150 kHz (Supraharmonics) *IEEE 9th International Workshop on Applied Measurements for Power Systems* 1–6

[9] Sudiarto B 2017 *The Properties and Behavior of Disturbances in the Frequency Range 9 kHz to 150 kHz Produced by Household Appliances in a Residential Network Environment* Doctoral Dissertation, Universität Duisburg-Essen, Duisburg,

[10] Purnamaputra R P and Sudiarto B 2019 Disturbance Frequency 9–150 kHz Characteristics towards Total Demand Distortion on On-Grid Solar Panel System in the Electrical System of Kuningan Gas Station *IOP Conference Series: Earth and Environmental Science* **353** (1) 012011

# Design and Development of Fuzzy Logic Application Mamdani Method in Predicting The Number of Covid-19 Positive Cases in West Java

**E F Yogachi, V M Nasution and G Prakarsa**

Faculty of Technology and Informatics, Universitas Informatika dan Bisnis Indonesia, Jalan Soekarno-Hatta 643, Bandung 40285, Indonesia.

*Email: egafaradhiayogachi@gmail.com

**Abstract.** The increase in covid-19 positive patients is unpredictable which results in being unprepared in dealing with Covid-19 cases. People in monitoring and patients under supervision are the category that is breastfeed to be covid-19 positive patients after passing the incubation period for 14 days. Artificial intelligence used to predict the increase in cases of covid-19 positive patients is fuzzy logic. The fuzzy logic method used in this study is the mamdani method. The enter variables used are the increase in ODP, PDP, and Covid-19 positive with 3 sets of low, medium, and high respectively. While the output variable is a positive prediction 14 days later. The curve used for fuzzification needs is a linear curve. The method of composition of the rules used is the MAX function and defuzzification using Centroid.

## 1. Introduction

In line with the development of information technology, the technology's ability to solve problems, among them the development of applications that can predict things. One example of an app that can predict something is an app that uses fuzzy logic. Fuzzy logic is a logic that can be used to analyze problems containing uncertainty, one example of the prediction process [1]. Fuzzy logic is a science that studies uncertainty. Fuzzy logic is considered capable of mapping an input into an output without ignoring existing factors. Fuzzy logic is believed to be very flexible and has a tolerance to existing data [2].

The mamdani method is one of the methods found in fuzzy logic. There have been several previous studies that used fuzzy logic with mamdani methods in terms of predicting things. In ref [3], system has an accuracy of 81.1% in predicting the procurement of hospital household appliances with the fuzzy logic of mamdani method. In ref [4], the predicts amount of rubber production using fuzzy logic mamdani method, has an accuracy rate of 87.82706% and error rate reaches 12.17294%. Pramana (2016) conducted a study predicting the number of dengue fever cases in Malang regency using fuzzy logic mamdani method with error rates for sub-districts in lowlands of 6%, medium plains by 12%, and plateaus of 14% [5]. There is also research comparing the fuzzy logic of Tsukamoto, Sugeno, and Mamdani methods in predicting the number of new student applicants of the Faculty of Science and Technology of Sunan Gunung Djati Bandung State Islamic University conducted irfan et al., (2018) with the conclusion of mamdani method has a smaller error rate of 19.76% compared to Tsukamoto method of 39.93% and Sugeno by 86.41% [6]. Based on the studies mentioned, it can be concluded the mamdani method has a small error rate because it is close to 0% and a large accuracy rate because it is close to 100% in predicting cases. The mamdani method also has the lowest error rate compared to the tsukamoto and sugeno methods on predicting.

Coronavirus pandemic 2019 (Covid-19) is an ongoing problem from 200 countries in the world. Covid-19 has been identified as the cause of the outbreak of infectious respiratory diseases in Wuhan, China. As of March 31, 2020, there were 719,758 confirmed cases worldwide. The number of Covid-19-related deaths also reached 33,673 worldwide. Indonesia confirms there are 1,528 cases of Covid-19 and 136 deaths related to the disease [7].

Covid-19 pandemic causes many hospitals around the world to have difficulty both management and infrastructure advice in providing services as patient numbers soar in a short period of time. Starting from the first case until now, there has been an increase and decrease in the number of patients infected with coronavirus. On Monday, March 9, 2020, the West Java Provincial Health Office acknowledged the lack of Covid-19 treatment facilities at 52 Regional General Hospitals (HOSPITALS) in West Java Province to closely monitor covid-19 positive patients cannot be predicted so that treatments such as providing services, facilities, and medical personnel cannot be predicted. Therefore, an application is needed that can predict the number of covid-19 positive cases 14 days later to help the Task Force team accelerate the handling of Covid-19 in West Java Province in order to provide appropriate policies for the handling of the provision of services, facilities, and medical personnel.

According to Abudi et al. (2020) there are 4 (four) mentions of covid-19 related people to help the government and health authorities analyze their patients, namely People in Supervision (ODP), Patients In Supervision (PDP), People Without Symptoms (OTG), and positive Covid-19 [8]. People who are categorized as ODP, PDP, and OTG have a positive chance of becoming Covid-19 because the 3 categories have the characteristics of people infected with coronavirus. Therefore, the government recommends that ODP, PDP, and OTG conduct self-quarantine for 14 days. Indicators that result in an increase in covid-19 positive patients are the increase in ODP, PDP, and covid-19 positive. These indicators can be divided into three groups: low, medium, and high. The three groups can make sets on ODP, PDP, positive case, and positive cases 14 days later for fuzzy logic needs. The argument is reinforced by research conducted in Mexico state examining the rise of Covid-19 using a combination of fractal dimension and fuzzy logic dividing the set on variable increases in covid-19 positive cases in the world into no group namely low, medium, and high [9]. Other research, namely the application of Fuzzy Tsukamoto in seeking the percentage of risk level hanger enter maintenance makes the research results better in terms of the process of calculation speed, accuracy of percentage results and accuracy in decision making [10].

Covid-19 case data in West Java Province such as data on the number of People in Monitoring (ODP), Patients Under Supervision (PDP), and the number of patients confirmed by the Covid-19 virus are recorded and informed on the Pikobar website. Pikobat is the information and coordination center of Covid-19 West Java Province. In addition to the covid-19 case information feature, pikobar also provides donation or social assistance, pretical info, and news about Covid-19 in West Java.

## 2. Background Theory

Fuzzy logic is an improvement from boolean logic that introduces the concept of partial truth. Where classical logic states that anything can be expressed in binary terms (0 or 1, black or white, yes or no), fuzzy logic replaces boolean truth with the level of truth. Therefore fuzzy logic can allow membership values between 0 and 1, black and white, and in linguistic form, uncertain concepts such as "little", "half" and "many". Fuzzy logic was first developed by Lotfi A. Zadeh an Iranian American scientist from the University of California at Berkeley [11].

The basis of fuzzy logic is the fuzzy set theory. In fuzzy set theory, the role of membership degrees as a determinant of the existence of elements in a set is very important. Membership value or membership function degree is the main feature of reasoning with the fuzzy logic [12]-[13].

### 2.1. Fuzzy Membership

Kusumadewi and Purnomo (2010) explains that on a crisp set, the membership value of an item x in an relationship, often written with μA[x], has the possibility, i.e. [14]:

- One (1), which means that an item is a member of a set, or Zero (0), which
- Means that an item is not a member of a set.

Sometimes the resemblance between fuzzy membership and probability raises confusion. Both have values at intervals of [0.1], but the interpretation of the values differs greatly between the two cases. Fuzzy membership provides a measure of an opinion or decision, while probability indicates a proportion to the likeness of an outcome of true value in the long run. The fuzzy set has 2 attributes, namely:

- Linguistics, i.e. naming a group that represents a particular state or condition using natural language, such as: YOUNG, PAROBAYA, and TUA.
- Numeris, which is a value (number) that indicates the size of a variable, such as: 40, 25, 50, etc.

There are several things to know in understanding the fuzzy system, namely [14]:

- A fuzzy variable, a fuzzy variable is a variable to be discussed in a fuzzy system. Examples: age, temperature, demand, etc.
- A fuzzy set, a fuzzy set is a group that represents a particular condition or state in a fuzzy variable. For example, the age variable, divided into 3 sets of fuzzy, namely: YOUNG, PAROBAYA, and TUA.
- The talk universe, the talk universe is the whole value that is allowed to be operated in a fuzzy variable. The talking universe is a set of real numbers that always rise monotonously from left to right. The value of the conversation universe can be either positive or negative. Sometimes the value of this universe of talks is not limited to its upper limit. For example, the universe talks to an age variable: $[0+\infty)$.
- Domains, fuzzy set domains are the entire allowed value in the talk universe and can be operated in a fuzzy set. As with the universe of talks, domains are a monotonous set of real numbers from left to right, domain values can be both positive and negative numbers. Examples of fuzzy set domains:
  YOUNG = [0 45]
  MIDDLE-AGED = [35 55]
  OLD = $[45 +\infty)$

### 2.2. Membership Functions

A membership function is a curve that shows the mapping of data input points into their membership values that have intervals between 0 and 1. One way that can be used to get membership value is through a function approach. There are several functions that can be used, namely as follows [14]:

#### *2.2.1. Linear Representation*

Mapping input to membership degrees on linear representation is described as a straight line. This form is simplest and a good choice to approach a less obvious concept. There are 2 linear fuzzy set states. First, the set incress starts at a domain value that has zero membership degrees (0) moving right towards a domain value that has a higher membership degree (See Figure 1).

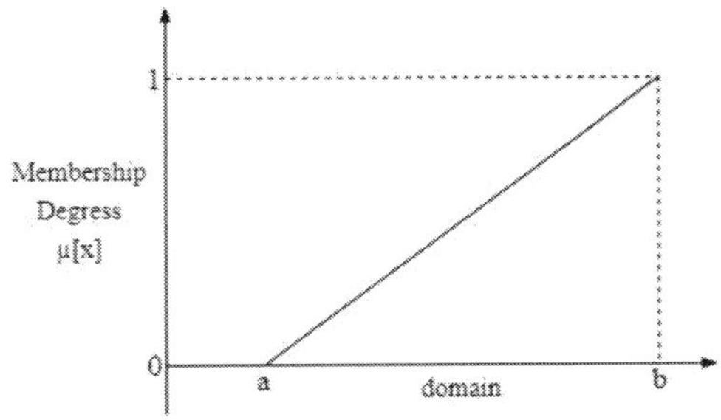


**Figure 1.** Linear Curve Rise

Membership functions:

$$\mu[x] = \begin{cases} 0; & x \leq a \\ (x-a)/(b-a); & a < x < b \\ 1; & x \geq b \end{cases} \tag{1}$$

Description:
x = the value of the domain to be searched for membership degree.
a = a domain value whose membership degree is equal to 0.
b = a domain value whose membership degree is equal to 1.

Second, it is the opposite of the first. A straight line starts from the domain value with the highest membership degree on the left side, then moves downhill to a domain value that has a lower membership degree (See Figure 2).

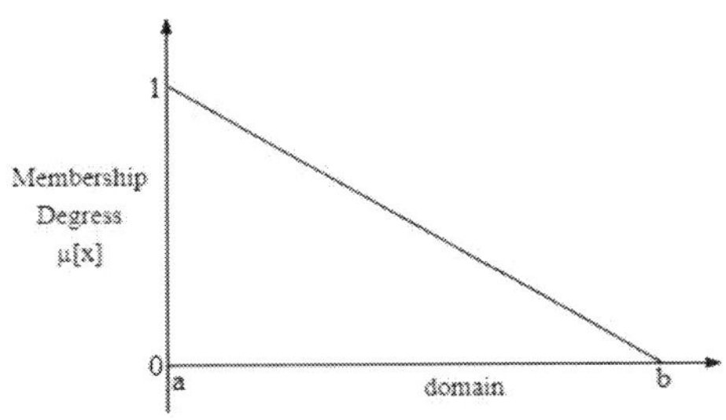


**Figure 2.** Linear Curve Down

Membership functions:

$$\mu[x] = \begin{cases} (b-x)/(b-a); & a < x < b \\ 0; & x \geq b \end{cases} \tag{2}$$

Description:
x = the value of the domain to be searched for membership degree.
a = a domain value whose membership degree is equal to 1.
b = a domain value whose membership degree is equal to 0.

### *2.2.2. Trapezoidal Curve Representation*

The representation of the trapezoidal curve is basically like a triangular shape, only there are some dots that have a membership value of 1. A representation of the trapezoidal curve can be seen in Figure 3.

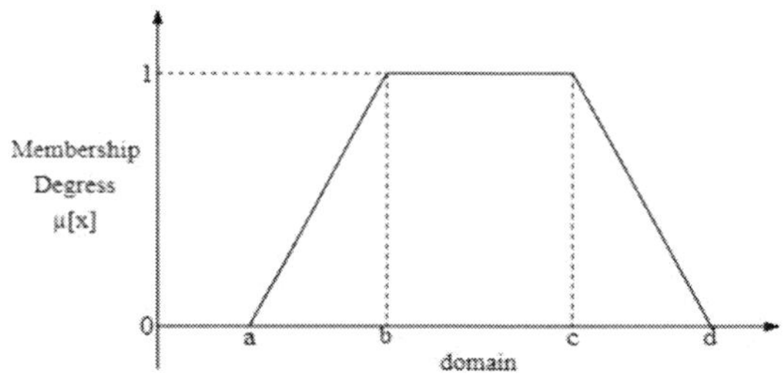


**Figure 3.** Phi Bell Curve

## 2.3. Fuzzy Set Base Operation

As with conventional sets, there are several specially defined operations for combining and modifying fuzzy sets. Membership value as a result of operation of 2 sets is often known as fire strength α predicate [15]-[16]. There are 3 basic operators created by Zadeh, namely:

- Operator AND, this operator relates to intersection operations on the set. α-predicate as a result of operation with the OPERATOR AND is obtained by taking the smallest membership value between elements on the relevant sets.
- Operator OR, operator ini berhubungan dengan operasi union pada himpunan α-predicate sebagai hasil operasi dengan operator OR diperoleh dengan mengambil nilai keanggotaan terbesar antar elemen pada himpunan-himpunan yang bersangkutan.
- Operator NOT, this operator relates to complementary operations on the set. α-predicate as a result of operations with the NOT operator is obtained by subtracting the membership value of the element on the corresponding set from 1.

## 2.4. Fuzzy Inference System (FIS) Mamdani Method

According to Kusumadewi & Purnomo (2010), mamdani method is often also known by the method name Max-Min. This method was introduced by Ebrahim Mamdani in 1975. To get output it takes 4 stages [14]:

- The formation of the Fuzzy Set, on the mamdani method, both input variables and output variables are divided into one or more fuzzy sets.
- Application Function Implications, on the mamdani method, the implication function used is Min.
- The composition of the Rules, unlike mononton reasoning, if the system consists of several rules, then inference is obtained from the set and correlation between rules. There are 3 methods used in performing fuzzy inference systems, namely: max, additive and probabilistic OR (probor).
- Deffuzification is a fuzzy set derived from the composition of fuzzy rules, while the resulting output is a number in the domain of the fuzzy set. So if given a fuzzy set in a certain range, it should be able to take a certain crisp value as output.

## 3. Methodology

The research method used by the authors in this study used a descriptive method, which is to explain and describe data related to the current state, the relationship between emerging variables, and the differences in existing facts and their impact on conditions. Descriptive methods also aim to compare what can be done to get the most suitable solution. The Figure 4 show the explanations of the research methods conducted in this study.

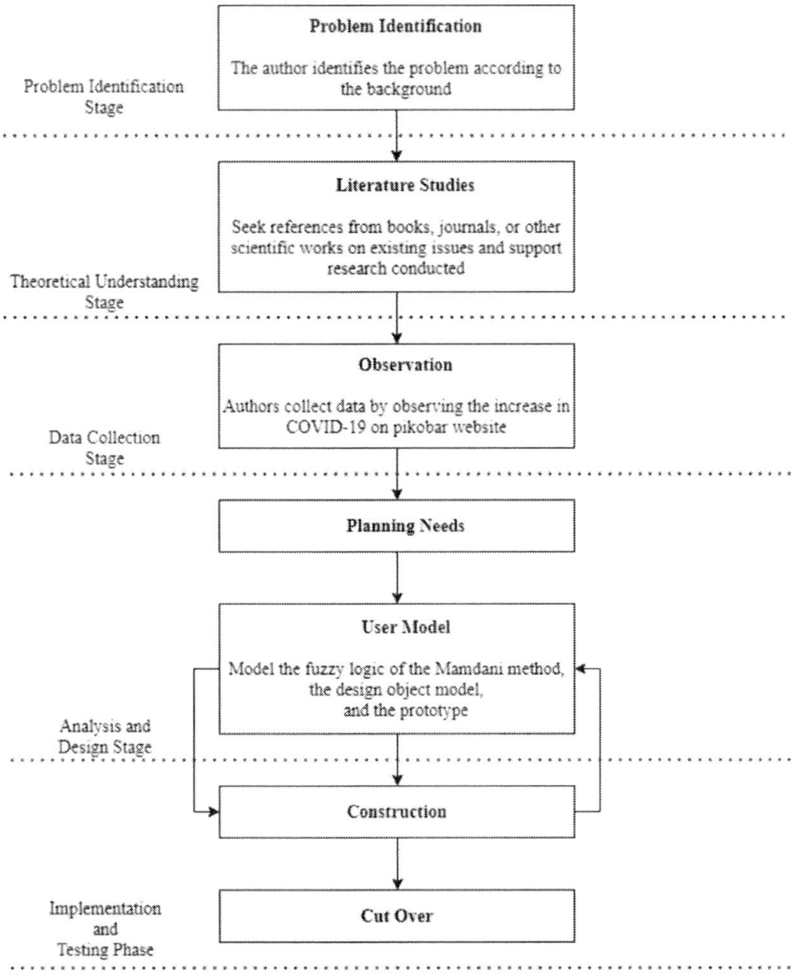


**Figure 4.** Research Methods

The stage of problem identification aims to know the problem that is being experienced in the research. At this stage spelled out what problems were raised for research. The issues raised will then be reviewed and obtained information related to the issue.

Furthermore, the authors conducted a literature study to look for references to existing problems and support the research conducted. By looking for some good materials from books, scientific journals, and research results related to the application of fuzzy logic using mamdani method.

The Third Stage is to perform the data collection stage is the author's activity of collecting the data needed for application development. The data collection technique performed by the author is an observation on the website of pikobar. The purpose of this observation is to obtain data used for application development, namely in the form of data on the number of ODP, the number of PDP, and the positive amount of Covid-19.

The Fourth phase is the analysis and planning phase is the beginning of the implementation of the system development method rapid application development (RAD) in this study. Activities to plan needs include system planning and analysis. At this stage the author analyzes all the needs that have been collected before to understand the nature of the program to be built. Then the author defines the needs of the application to be built in the form of user characteristics, data needs, and functional and non-functional needs. After the needs of the system have been defined, the authors created a fuzzy logic model of mamdani method on predicting an increase in the number of positive cases of Covid-19 14 days later. Based on prior studies on prediction using the fuzzy logic of the mamdani method and the limitations of the author's ability, the study used linear representation for membership functions, the MAX (Maximum) method for the method of rule composition, and the centroid (Composite Moment) method for the defuzzification method. The design architecture used to design the applications used in this study was Object Oriented Analysis and Design (OOAD), so the authors used Unified Modelling Language (UML) to create a software design object model. The UML diagrams to be created are use case diagrams, class diagrams, sequence charts, and state diagrams. The author only created the four diagrams because they are the main diagrams that are quite used for applications that are not very large. At this stage also the author creates a prototype in the form of an application mockup to be built. After the fuzzy logic was created, the authors then tested the results of the mamdani method in predicting covid-19 cases with field sample data.

The last stage of this research is the implementation and testing phase is the advanced stage of rad implementation, namely construction and cut over. Construction is the author's activity of implementing the user model into an application. Applications are built using java programming languages with Netbeans and MySQL tools for database management. If during the application development stage there is a discrepancy with the needs of the application that has been defined in the stage of planning the needs then a model and prototype improvement is carried out at the user model stage. If the application built at the construction stage is in accordance with the needs of the application that has been defined at the stage of planning the need, then the next step is to test the application or cut over. The testing method that the authors did on this study was the black-box method. At this stage, a comparison of the calculation of prediction results using the application built on this study with data on the number of covid-19 positive cases that have occurred to look for the error rate.

## 4. Results and Discussions

This study uses fuzzy logic model with mamdani method for the process of calculating the percentage prediction of the increase in the number of positive cases of Covid-19. The fuzzy logic model of the mamdani method that suda created, will then be applied into the application. Mamdani method consists of 4 stages namely the formation of fuzzy sets, application of implication functions, composition of rules, and affirmation (defuzzification). This activity will explain the stages of the implementation of fuzzy logic mamdani method to predict the positive increase of COVID-19 up to the example of the case.

The first thing to do in creating a fuzzy logic model is to make the determination of the speaker universe to be a fuzzy set. The variables needed to be used as crisp inputs are the number of percentage increases in ODP, PDP, and Positive Covid-19. While the crisp output is the amount of positive percentage increase. Data collection conducted by authors with vulnerable time from March 6, 2020 to May 6, 2020 obtained on the website pikobar. In search of the number of odp, PDP, and Positive covid-19 increases, the calculation of odp, PDP, and Positive each day is reduced by the data of the amount recovered and died on the date to be calculated. The data is processed to look for a percentage increase each day and look for the minimum and maximum values to be domaind for each variable that has been set, and is searched for the middle value of each domain (see Table 1 ).

**Table 1.** Speaker Universe

| Variable | Set | Domain |
|---|---|---|
| Increase in ODP | Low | [1,60; 20,35] |
| | Medium | [1,60; 20,35; 48,85; 67,60] |
| | High | [48,85; 67,60; ∞] |
| Increase in PDP | Low | [1,29; 16,58] |
| | Medium | [1,26; 16,58; 47,56; 62,86] |
| | High | [47,56; 62,86; ∞] |
| Increase in Positive Cases | Low | [5,19; 27,43] |
| | Medium | [5,19; 27,43; 33,32; 55,56] |
| | High | [33,32; 55,56; ∞] |
| Increase in Positive Cases of Covid-19 14 Days Later | Low | [5,19; 24,13] |
| | Medium | [5,19; 24,13; 29,75; 34,95] |
| | High | [29,75; 34,95; ∞] |

### 4.1 Formation of Fuzzy Associations

The first stage of the fuzzy logic of the mamdani method is the formation of a fuzzy set. Based on Table 1, a curved representation of each variable is made to know the membership functions of each set. The curve representation of each variable can be seen in Figures 5-8.

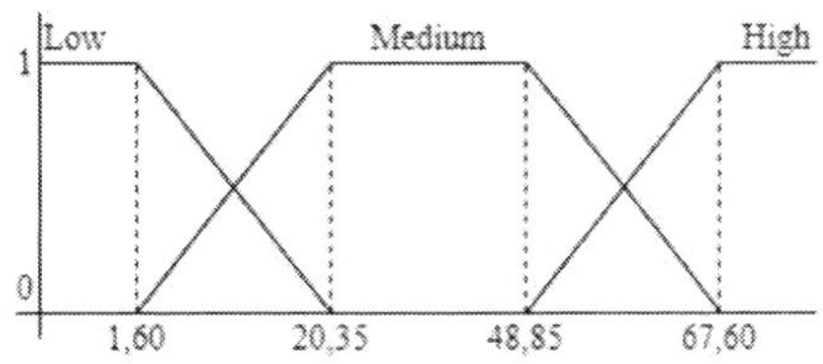


**Figure 5.** Increase in ODP set

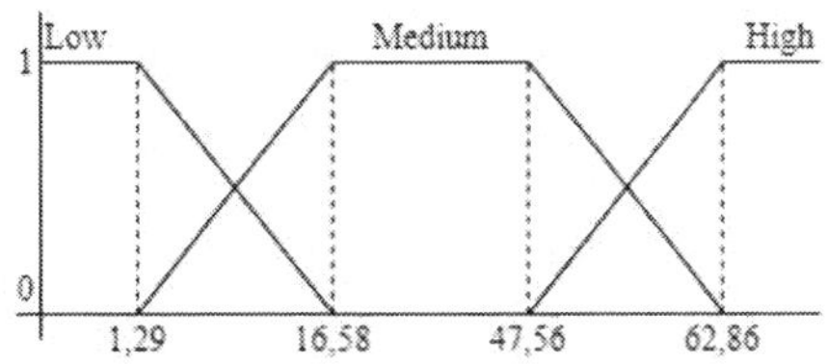


**Figure 6.** Increase in PDP set

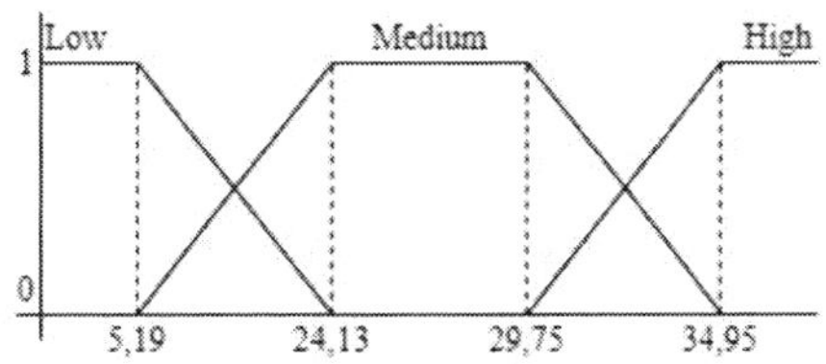


**Figure 7.** Increase in Positive cases set

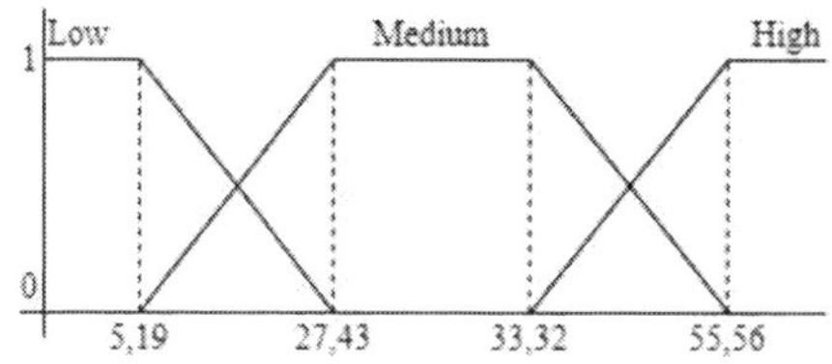


**Figure 8.** Increase in Positive cases Covid-19 14 Days Later set

### 4.2 Function Implications

The method used in the implication function in the mamdani method is the min method. The final rule of the fuzzy inference system in this study was established by combining existing sets, then the authors analyzed which rules had occurred based on the field data already collected. The final rules of the fuzzy inference system in this study can be seen in Table 2.

**Table 2.** Fuzzy Inference System Final Rules

| No. Rules | IF ODP | AND PDP | AND Positive | THEN Predictions |
|---|---|---|---|---|
| R1 | Low | Low | Low | Low |
| R2 | Low | Low | Low | Medium |
| R3 | Low | Low | Medium | Low |
| R4 | Low | Low | Medium | Medium |
| R5 | Low | Low | High | Medium |
| R6 | Low | Medium | Low | Low |
| R7 | Low | Medium | Low | Medium |
| R8 | Low | Medium | Medium | Medium |
| R9 | Low | Medium | Medium | High |
| R10 | Low | Medium | High | Low |
| R11 | Low | Medium | High | Medium |
| R12 | Low | High | Low | Low |
| R13 | Low | High | Low | Medium |
| R14 | Low | High | Medium | Low |
| R15 | Low | High | Medium | Medium |
| R16 | Low | High | High | Low |
| R17 | Low | High | High | Medium |
| R18 | Medium | Low | Low | Low |
| R19 | Medium | Low | Low | Medium |
| R20 | Medium | Low | Medium | Low |
| R21 | Medium | Low | Medium | Medium |
| R22 | Medium | Low | High | Low |
| R23 | Medium | Low | High | Medium |
| R24 | Medium | Medium | Low | Low |

IOP Conf. Series: Materials Science and Engineering 1115 (2021) 012031 doi:10.1088/1757-899X/1115/1/012031

| No. Rules | IF ODP | AND PDP | AND Positive | THEN Predictions |
|---|---|---|---|---|
| R25 | Medium | Medium | Medium | Low |
| R26 | Medium | Medium | Medium | Medium |
| R27 | Medium | Medium | High | Low |
| R28 | Medium | Medium | High | Medium |
| R29 | Medium | High | Low | Low |
| R30 | Medium | High | Low | Medium |
| R31 | Medium | High | Medium | Low |
| R32 | Medium | High | Medium | Medium |
| R33 | Medium | High | High | Low |
| R34 | Medium | High | High | Medium |
| R35 | High | Low | Low | Low |
| R36 | High | Low | Low | Medium |
| R37 | High | Low | Medium | Low |
| R38 | High | Low | Medium | Medium |
| R39 | High | Low | High | Low |
| R40 | High | Low | High | Medium |
| R41 | High | Medium | Low | Low |
| R42 | High | Medium | Low | Medium |
| R43 | High | Medium | Medium | Low |
| R44 | High | Medium | Medium | Medium |
| R45 | High | Medium | High | Medium |
| R46 | High | Medium | High | High |
| R47 | High | High | Low | Low |
| R48 | High | High | Low | Medium |
| R49 | High | High | Medium | Medium |
| R50 | High | High | Medium | High |
| R51 | High | High | High | Medium |
| R52 | High | High | High | High |

The results of the implications function are as follows:

[R3] IF ODP Low AND PDP Low AND POSITIVE Medium THEN PREDICTIONS Low

$$\begin{aligned} \alpha - predikat &= \mu_{ODPLow} \cap \mu_{PDPLow} \cap \mu_{PositiveMedium} \\ &= \min\left(\mu_{ODPLow}(10), \mu_{PDPLow}(3.45), \mu_{PositiveMedium}(50)\right) \\ &= \min(0.55; 0.86; 0.25) \\ &= 0.25 \end{aligned}$$

[R4] IF ODP Low AND PDP Low AND POSITIVE Medium THEN PREDICTIONS Medium

$$\begin{aligned} \alpha - predikat &= \mu_{ODPLow} \cap \mu_{PDPLow} \cap \mu_{PositiveMedium} \\ &= \min\left(\mu_{ODPLow}(10), \mu_{PDPLow}(3.45), \mu_{PositiveMedium}(50)\right) \\ &= \min(0.55; 0.86; 0.25) \\ &= 0.25 \end{aligned}$$

[R5] IF ODP Low AND PDP Low AND POSITIVE High THEN PREDICTIONS Medium

IOP Conf. Series: Materials Science and Engineering 1115 (2021) 012031 doi:10.1088/1757-899X/1115/1/012031

$\alpha - predikat = \mu_{ODPLow} \cap \mu_{PDPLow} \cap \mu_{PositiveHigh}$
$= \min\left(\mu_{ODPLow}(10), \mu_{PDPLow}(3.45), \mu_{PositiveHigh}(50)\right)$
$= \min(0.55; 0.86; 0.75)$
$= 0.55$

[R8] IF ODP Low AND PDP Medium AND POSITIVE Medium THEN PREDICTIONS Medium

$\alpha - predikat = \mu_{ODPLow} \cap \mu_{PDPMedium} \cap \mu_{PositiveMedium}$
$= \min(\mu_{ODPLow}(10), \mu_{PDPMedium}(3.45), \mu_{PositiveMedium}(50))$
$= \min(0.55; 0.14; 0.25)$
$= 0.14$

[R9] IF ODP Low AND PDP Medium AND POSITIVE Medium THEN PREDICTIONS High

$\alpha - predikat = \mu_{ODPLow} \cap \mu_{PDPMedium} \cap \mu_{PositiveMedium}$
$= \min(\mu_{ODPLow}(10), \mu_{PDPMedium}(3.45), \mu_{PositiveMedium}(50))$
$= \min(0.55; 0.14; 0.25)$
$= 0.14$

[R10] IF ODP Low AND PDP Medium AND POSITIVE High THEN PREDICTIONS Low

$\alpha - predikat = \mu_{ODPLow} \cap \mu_{PDPMedium} \cap \mu_{PositiveHigh}$
$= \min\left(\mu_{ODPLow}(10), \mu_{PDPMedium}(3.45), \mu_{PositiveHigh}(50)\right)$
$= \min(0.55; 0.14; 0.75)$
$= 0{,}14$

[R11] IF ODP Low AND PDP Medium AND POSITIVE High THEN PREDICTIONS Medium

$\alpha - predikat = \mu_{ODPLow} \cap \mu_{PDPMedium} \cap \mu_{PositiveHigh}$
$= \min\left(\mu_{ODPLow}(10), \mu_{PDPMedium}(3.45), \mu_{PositiveHigh}(50)\right)$
$= \min(0.55; 0.14; 0.75)$
$= 0.14$

[R20] IF ODP Medium AND PDP Low AND POSITIVE Medium THEN PREDICTIONS Low

$\alpha - predikat = \mu_{ODPMedium} \cap \mu_{PDPLow} \cap \mu_{PositiveMedium}$
$= \min(\mu_{ODPMedium}(10), \mu_{PDPLow}(3.45), \mu_{PositiveMedium}(50))$
$= \min(0.45; 0.86; 0.25)$
$= 0.25$

[R21] IF ODP Medium AND PDP Low AND POSITIVE Medium THEN PREDICTIONS Medium

$\alpha - predikat = \mu_{ODPMedium} \cap \mu_{PDPLow} \cap \mu_{PositiveMedium}$
$= \min(\mu_{ODPMedium}(10), \mu_{PDPLow}(3.45), \mu_{PositiveMedium}(50))$
$= \min(0.45; 0.86; 0.25)$
$= 0.25$

[R22] IF ODP Medium AND PDP Low AND POSITIVE High THEN PREDICTIONS Low

$\alpha - predikat = \mu_{ODPMedium} \cap \mu_{PDPLow} \cap \mu_{PositiveHigh}$
$= \min\left(\mu_{ODPMedium}(10), \mu_{PDPLow}(3.45), \mu_{PositiveHigh}(50)\right)$
$= \min(0.45; 0.86; 0.75)$
$= 0.45$

[R23] IF ODP Medium AND PDP Low AND POSITIVE High THEN PREDICTIONS Medium

$\alpha - predikat = \mu_{ODPMedium} \cap \mu_{PDPLow} \cap \mu_{PositiveHigh}$
$= \min\left(\mu_{ODPMedium}(10), \mu_{PDPLow}(3.45), \mu_{PositiveHigh}(50)\right)$

$= \min(0.45; 0.86; 0.75)$
$= 0.45$

[R26] IF ODP Medium AND PDP Medium AND POSITIVE Medium THEN PREDICTIONS Medium

$\alpha - predikat = \mu_{ODPMedium} \cap \mu_{PDPMedium} \cap \mu_{PositiveMedium}$
$= \min(\mu_{ODPMedium}(10), \mu_{PDPMedium}(3{,}45), \mu_{PositiveMedium}(50))$
$= \min(0.45; 0.14; 0.25)$
$= 0.14$

[R27] IF ODP Medium AND PDP Medium AND POSITIVE High THEN PREDICTIONS Low

$\alpha - predikat = \mu_{ODPMedium} \cap \mu_{PDPMedium} \cap \mu_{PositiveHigh}$
$= \min\left(\mu_{ODPMedium}(10), \mu_{PDPMedium}(3.45), \mu_{PositiveHigh}(50)\right)$
$= \min(0.45; 0.14; 0.75)$
$= 0.14$

[R28] IF ODP Medium AND PDP Medium AND POSITIVE High THEN PREDICTIONS Medium

$\alpha - predikat = \mu_{ODPMedium} \cap \mu_{PDPMedium} \cap \mu_{PositiveHigh}$
$= \min\left(\mu_{ODPMedium}(10), \mu_{PDPMedium}(3{,}45), \mu_{PositiveHigh}(50)\right)$
$= \min(0.45; 0.14; 0.75)$
$= 0.14$

### 4.3 Composition Rules

Based on the results of the rule implications function, the MAX method is used to perform the composition between all the following rules:

Variable Output Increase in Positive cases Covid-19 14 Days Later

- Low = MAX(R3; R10; R20; R22; R27)
  = MAX(0.25; 0.14; 0.25; 0.45; 0.14)
  = 0.45
- Medium = MAX(R4; R5; R8; R11; R21; R23; R26; R28)
  = MAX(0.25; 0.55; 0.14; 0.14; 0.25; 0.45; 0.14; 0.14)
  = 0.55
- High = MAX(R9)
  = MAX(0.14)
  = 0.14

Based on the calculation of the composition of the rules, the resulting area of the composition is formed as in Figure 9.

Based on Figure 9, it can be seen the resulting area is divided into 3 sections namely A1, A2, A3, A4, and A5. Then the author looks for the values a1, a2, a3, and a4 as follows:

a1 = 0.45 × (24.13 – 5.19) + 5.19 = 13.71
a2 = 0.55 × (24.13 – 5.19) + 5.19 = 15.61
a3 = 34.95 – (0.55 × (34.95 – 29.75) = 32.09
a4 = 34.95 – (0.14 × (34.95 – 29.75) = 34.22

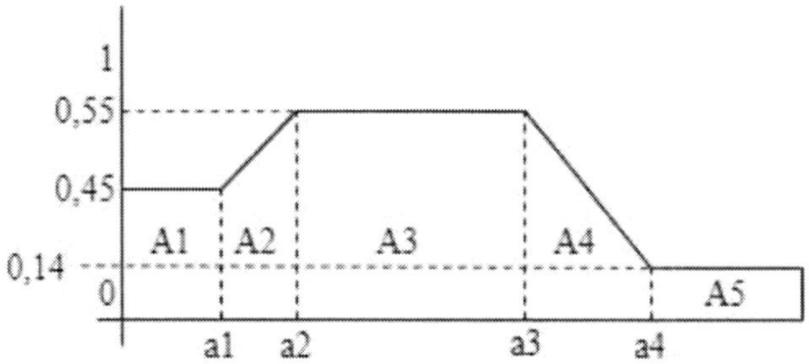


**Figure 9.** Composition Result Area

### 4.4 Defuzzification

The affirmation method used by the author is the centroid method. For that first the author calculates the moments for each area as follows:

$$M1 = \int_{0}^{13.71} (0.45)z\,dz = 42.29$$

$$M2 = \int_{13.71}^{15.61} \frac{z - 5.19}{24.13 - 5.19} z\,dz = 13.96$$

$$M3 = \int_{15.61}^{32.09} (0.55)z\,dz = 216.18$$

$$M4 = \int_{32.09}^{34.22} \frac{34.95 - z}{34.95 - 29.75} z\,dz = 24.22$$

$$M5 = \int_{34.22}^{34.95} (0.14)z\,dz = 3.53$$

Then calculated the area of each area:

A1 = 13.71 * 0.45 = 6.17
A2 = (½ * (15.61 – 13.71) * (0.55 – 0.45)) + ((15.61 – 13.71) * (0.45))
= 0.95
A3 = (32.09 – 15.61) * 0.55 = 1.17
A4 = (½ * (34.22 – 32.09) * (0.55 – 0.14)) + ((34.22 – 32.09) * (0.14))
= 0.73
A5 = (34.22 – 32.09) * 0.14 = 0.1

After the moment and the area of each area are finished, then the next step is to calculate the centre point as the final result of the calculation of the fuzzy logic of the mamdani method. This central point is used as the predicted value of Increase in Positive cases covid-19. The calculation of the central point is as follows:

$$z = \frac{42.29+13.96+216.18+24.22+3.53}{6.17+0.95+1.17+0.73+0.1} = \frac{300.17}{9.12} = 32.91$$

## 5. Conclusions

The results of this study to predict the percentage increase in covid-19 positive cases in West Java Province use fuzzy logic mamdani method with membership function in the form of Linear Curve and Trapezoid. The enter variables used are Increase in ODP, PDP, and Positive Covid-19 with 3 sets of low, medium, and high, respectively. Medium the output variable is a positive prediction 14 days later. The method of composition of the rules used is the MAX function and defuzzification using Centroid.

## References

[1] Ontiveros-Robles E, Melin P and Castillo O 2018 Comparative analysis of noise robustness of type 2 fuzzy logic controllers *Kybernetika* **54** (1) 175-201

[2] Dzitac I, Filip F G and Manolescu M J 2017 Fuzzy logic is not fuzzy: World-renowned computer scientist Lotfi A. Zadeh. *International Journal of Computers Communications & Control* **12** (6) 748-789

[3] Nurhayati S and Immanudin I 2019 Penerapan Logika Fuzzy Mamdani Untuk Prediksi Pengadaan Peralatan Rumah Tangga Rumah Sakit *Komputika: Jurnal Sistem Komputer*, **8** (2) 81-87.

[4] Rahakbauw D L, Rianekuay F J and Lesnussa Y A 2019 Penerapan metode fuzzy mamdani untuk memprediksi jumlah produksi karet (studi kasus: data persediaan dan permintaan produksi karet pada ptp nusantara xiv (persero) kebun awaya, teluk elpaputih, maluku-indonesia) *Jurnal ilmiah matematika dan terapan* **16** (1) 51-59

[5] Pramana I P A A 2016 Peramalan Jumlah Kasus Demam Berdarah di Kabupaten Malang Menggunakan Metode Fuzzy Inference System *Jurnal Teknik ITS* **5** (1) A1-A6

[6] Irfan M, Ayuningtias L P and Jumadi J 2017 Analisa Perbandingan Logic Fuzzy Metode Tsukamoto, Sugeno, Dan Mamdani (Studi Kasus: Prediksi Jumlah Pendaftar Mahasiswa Baru Fakultas Sains Dan Teknologi Uin Sunan Gunung Djati Bandung) *Jurnal Teknik Informatika* **10** (1) 9-16

[7] Setiati S and Azwar M K 2020 COVID-19 and Indonesia *Acta Medica Indonesiana* **52** (1) 84-89

[8] Abudi R, Mokodompis Y and Magulili A N 2020 Stigma Terhadap Orang Positif Covid-19 *Jambura Journal of Health Sciences and Research* **2** (2) 77-84

[9] Castillo O and Melin P 2020 Forecasting of COVID-19 time series for countries in the world based on a hybrid approach combining the fractal dimension and fuzzy logic *Chaos, Solitons & Fractals* **140** 110242

[10] Prakarsa G and Nasution V M 2019 Pengembangan Sistem Pendukung Keputusan Menggunakan Metode Tsukamoto *Jurnal RESTI (Rekayasa Sistem Dan Teknologi Informasi)* **3** (3) 414-421

[11] Zadeh L A 2008 Is there a need for fuzzy logic? *Information sciences* **178** (13) 2751-2779

[12] Mujiarto, Djohar A, Komaro M, Mohamed M A, Rahayu D S, Sanjaya W S M, Mamat M, Sambas A and Subiyanto 2019 Colored object detection using 5 dof robot arm based adaptive neuro-fuzzy method *Indonesian Journal of Electrical Engineering and Computer Science* **13** (1) 293-299

[13] Trisnawan, Sanjaya W S M, Sambas A, Jannah M, Rahayu D S, Mamat M and Mohamed M A 2019 Design of 4 Dof Robot ARM Based on Adaptive Neuro-Fuzzy (ANFIS) using Vision in Detecting Color Objects *International Journal of Recent Technology and Engineering* **8** (2S7) 224-227

[14] Sri Kusumadewi H P 2010 *Aplikasi Logika Fuzzy Pendukung Keputusan* Yogyakarta: Graha

Ilmu.
[15] Wong B K and Lai V S 2011 A survey of the application of fuzzy set theory in production and operations management: 1998–2009 *International Journal of Production Economics* **129** (1) 157-168
[16] Sukono, Sambas A, He S, Liu H, Vaidyanathan S, Hidayat Y and Saputra J 2020 Dynamical analysis and adaptive fuzzy control for the fractional-order financial risk chaotic system *Advances in Difference Equations* **674** (1) 1-12

# Disaster Tweets Classification in Disaster Response using Bidirectional Encoder Representations from Transformer (BERT)

**A K Ningsih*** **and A I Hadiana**

Department of Informatics, Universitas Jenderal Achmad Yani, Cimahi, Indonesia

*Corresponding author: Kania215@gmail.com

**Abstract.** The omnipresence of smartphones helps people to declare an emergency in real-time. In times of crisis, Twitter has become a big communication platform. As a result, more organizations are interested in tracking Twitter programmatically. Although governments and emergency management agencies work together through their respective national system for response to disasters, the sentiments of the people affected during and after the catastrophe decide the effectiveness of the disaster response and the recovery process. In recent years, sentiment analysis via Twitter-based machine learning has become a common subject. However, the detection of such tweets was often difficult due to the tweets' language structure's uncertainty. Thus, it is not always apparent if the words of a person announce a catastrophe. BERT (Bidirectional Encoder Representations by Transformers) is a profound learning model developed by Google. Since Google opened it, several scientists and companies have embraced it and have applied it to many text classification tasks. Therefore, we use BERT in this paper to some dataset disaster tweets. This research will help rescue and emergency responders establish effective knowledge management techniques for a rapidly evolving disaster environment.

## 1. Introduction

Various disasters in various parts of the world have raised awareness about internal awareness disaster reduction. The issue of disaster management is one of many countries and organizations. In this regard, research on disaster management is increasing. Social media have gained much coverage in recent years for their possible use of space and time events. Twitter is one that allows society to communicate in real-time an emergency. Twitter data processing for such necessary data is a challenge because of the vocabulary people use when tweeting. Analysis of feelings is a class of analytical techniques that automatically extracts and summarizes the thoughts of such an enormous data volume that the average human reader cannot process[1].

This ocean of opinion posts in social media is central to individuals' actions because they affect our behavior. Social media has played an increasingly important role. In the last few years, during emergencies and disasters, they have become primary alternative information to conventional media, making them the fourth most common information source in emergencies. In the same tragedies, people with different ethnic characteristics had different responses and actions. They can influence their actual harm during crises by other replies [2].

Sentiment analysis is a type of method used to characterize, isolate or define personal information, such as thoughts conveyed in a given text, based on computational and natural language processing techniques. The primary objective of sensation analysis is to describe the writer's emotion in positive, negative, or neutral categories for various subjects [1]. Research on sentiment analysis from social media has been widely carried out [3][4][5], but efforts to research this in disaster management have received less attention. Identifying these sentiments from social media will help respondents understand network dynamism, e.g., users' significant issues, panics, and the emotional effects of members' experiences [6].

Deep learning has recently been shown to be one of the best methods of natural language management. The supervised training process with a large volume of data offers a good result for deep learning. BERT (Bidirectional Encoder Representations from Transformers) is a Google-designed, deep learning model. Since Google opened it, several scientists and businesses have adopted it and applied it to several text classification tasks. BERT's release was one of the most recent milestones in developing natural language processing (NLP) and an event that marked the beginning of a new period in the NLP. By conditioning both the left and right contexts, BERT is built to establish a deep bi-directional representation of unmarked texts on every level. This condition allows the pre-trained BERT model to be finalized with one additional output layer for various tasks without significant task-specific changes, including questions answering or language inference [7].

## 2. Related Work

Social media has become an inseparable part of our lives. Especially with the extremely rapid development of internet technology. One of the critical roles of social media is felt when a disaster occurs. Where plays a role in the disclosure of information about disasters. This situation encourages research to conduct sentiment analysis on posts on social media, especially Twitter. Users are turned into a social sensor by collecting public opinion from social media. Social media data are thought to be essential for understanding the reactions and feelings of the public. Finding posts that indicate frustration, danger or concern, can be critical, particularly for disaster management. Therefore, such posts' processing is effectively assisted by a systematic classification and is categorized in feeling polarity and aspect-based category, which will aid various agencies, including non-governmental agencies or governments, to manage such crises.

Analysis of social media data in transforming disasters has been carried out in several studies [8] [9]. Current studies have used geotagged tweets to measure the degree of social media involvement in areas impacted by disaster to determine whether vulnerable groups remain silent on social media to explore relevant factors. The ref [10] suggests that social media consumers rely more on rescue and donation-based knowledge using a machine-learning tweet model. However, differences in subjects are consistent across various fields and differing levels of focus.

The study of sentiment in online review sites has been thoroughly explored to provide consumers with synthetic views on various product aspects. However, little work has been done to define the polarity of users' emotions during catastrophic incidents. Identification of such feelings on the websites may allow emergency respondents to understand the complexities of their network, such as users' key concerns, panics, and the emotional effects of interactions among participants [6]. On research conducted by ref [11], they trained sentiment classes in categorizing messages on a concentrated basis and then discussed the demographic variations in these classifiers. They conclude that social media research complements conventional approaches for understanding public awareness of an imminent catastrophe in real-time.

In ref [12], they propose that Twitter feeds data analytics to enhance emergency planning and rescue efforts and services rendered in unusual circumstances. They follow a different approach, which focuses on the thoughts, fears, and opinions of users shared in emergency tweets and evaluates these feelings and expectations within an incident group to ensure that emergency response staff and local authorities have sufficient input. They use analysis and detection methods to store, identify and infer users' spatiotemporal feelings.

BERT has become state of the art in NLP. Since its inception, BERT has been widely used and experienced [7]. BERT's main technological advancement is applying transformer bidirectional training, a standard language model, to language modeling. Compared to previous attempts, a text sequence was evaluated from left to right or combined from left to right and from right to left. To solve the fine-grained challenge of classification uses a promising profound learning model called BERT [13]. Experiments demonstrate that their model succeeds other standard models without a sophisticated architecture for this mission. It also explains how efficient transfer learning is in manipulating natural languages. Whereas in ref [14], BERT was applied to a series of tweets linked to the Jakarta Flood in 2020. In the Jakarta flood tragedy of early 2020, which became a trending issue on Twitter, they tore up the tweet details. From there, the goal is to find specific tweets that can provide valuable information on emergency response to disaster management. Experimental studies have shown positive results. However, according to them, the data collection's consistency substantially affects the efficiency of the system.

In ref [15], they explored how to boost performance on (T)ABSA by applying a context to self-attention models. We suggest two versions of CG-BERT, which are capable of spreading focus in various contexts. First, they adopt a context-conscious transformer to create a CG-BERT using context-controlled softmax. Then, they suggest an enhanced CG-BERT, almost attention compositional model, which enhances subtraction. Their work offers more evidence for the usefulness of incorporating context dependency for context-based natural language tasks to pre-trained self-attention language models. In ref [16], they learned to link the sentence to the aspect and the long-term dependencies using the pre-trained BERT model. They further detect a self-critical strengthening learning snippet of opinion. Experimental results show how effective this method is and demonstrate that our hard-selection approach exceeds the soft selection approach in dealing with multi-aspect sentences.

## 3. Methodology

BERT utilizes a transformer, a mechanism of attention that learns the link between words (or subwords) in a text. The encoder that reads text data and the decoder generates a job forecast in its vanilla form, and the converter comprises two different mechanisms. As BERT's purpose is to construct a language model, it just requires an encoder mechanism. Two steps are used in BERT: pre-training and fine-tuning. During pre-training, the model is conditioned on unlabelled data in separate pre-training activities. Pertained parameters launch BERT models to be fine targeted, and the named downstream data refines all parameters, whose procedure we can see in Figure 1.

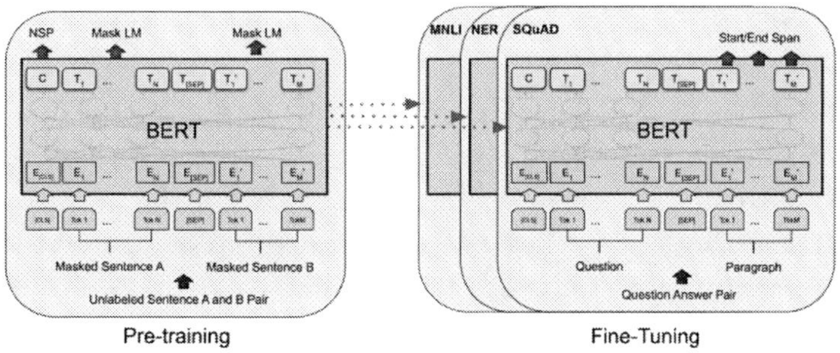


**Figure 1.** General pre-training and fine-tuning procedures for BERT[7].

IOP Conf. Series: Materials Science and Engineering 1115 (2021) 012032 doi:10.1088/1757-899X/1115/1/012032

### 3.1 Data Collection

In this study, we used the dataset from Kaggle. The dataset could be downloaded from https://www.kaggle.com/c/nlp-getting-started/data. We have 10873 comments, and we estimate whether or not a particular tweet is about a specific catastrophe. From 10,873 data, 57.03% were not real disasters. Figure 2 described the features and specifications.

- Id - tweet identifier
- text - text of the tweet
- location - location the tweet was sent from (maybe NaN)
- keyword - A relevant keyword in the tweet (maybe NaN)
- target - Output that tells if a tweet is a real disaster (1) or not (0)

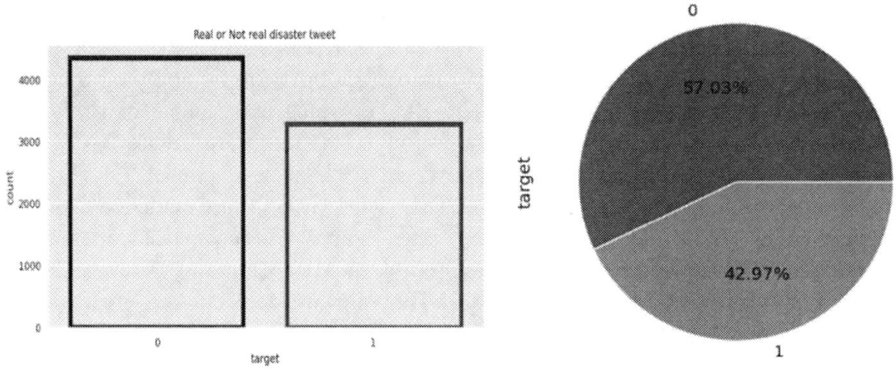


**Figure 2.** Real or not real disaster tweet from the dataset

### 3.2 Implementation

We will use Pandas to read data, preprocess_kgptalkie for data pre-processing. Seaborn for data visualization, all are running on Google Colab. By using pandas in Figure 3, we can see an example of data from the dataset.

| | id | keyword | location | text | target |
|---|---|---|---|---|---|
| **3040** | 4362 | earthquake | NaN | #USGS M 0.9 - Northern California: Time2015-08... | 1 |
| **3041** | 4363 | earthquake | TÌÁchira - Venezuela | #SCSeEstaPreparando Light mag. 4.4 earthquake ... | 1 |
| **3042** | 4365 | earthquake | California, USA | #USGS M 1.2 - 23km S of Twentynine Palms Calif... | 1 |
| **3043** | 4366 | earthquake | a box | @AGeekyFangirl14 's things she looks in a sign... | 0 |
| **3044** | 4368 | earthquake | Melbourne, Australia | Nepal earthquake 3 months on: Women fear abuse... | 1 |
| **3045** | 4372 | earthquake | Barcelona, Spain | ML 2.0 SICILY ITALY http://t.co/z6hxx6d2pm #eu... | 0 |
| **3046** | 4373 | earthquake | Hawaii, USA | USGS reports a M1.94 #earthquake 5km S of Volc... | 1 |
| **3047** | 4374 | earthquake | New Zealand | GNS sees unnecessary deaths resulting from ear... | 1 |
| **3048** | 4375 | earthquake | NaN | 'There was a small earthquake in LA but don't ... | 1 |

**Figure 3.** Sample dataset

We can get basic features of the dataset with the help of preprocess_kgptalkie, the results of which can be seen in Figure 4.

| text | target | char_counts | word_counts | avg_wordlength | stopwords_counts | hashtag_counts | mentions_counts | digits_counts | uppercase_counts |
|---|---|---|---|---|---|---|---|---|---|
| #USGS M 0.9 - Northern California: Time2015-08... | 1 | 122 | 15 | 8.133333 | 1 | 2 | 0 | 16 | 4 |
| #SCSeEstaPreparando Light mag. 4.4 earthquake ... | 1 | 118 | 17 | 6.941176 | 3 | 1 | 1 | 7 | 1 |
| #USGS M 1.2 - 23km S of Twentynine Palms Calif... | 1 | 119 | 17 | 7.000000 | 1 | 2 | 0 | 18 | 5 |
| @AGeekyFangirl14 's things she looks in a sign... | 0 | 115 | 21 | 6.476190 | 6 | 0 | 1 | 7 | 0 |
| Nepal earthquake 3 months on: Women fear abuse ... | 1 | 77 | 11 | 7.000000 | 1 | 0 | 1 | 3 | 0 |
| ML 2.0 SICILY ITALY http://t.co/z6hxx6d2pm #eu... | 0 | 48 | 6 | 8.000000 | 0 | 1 | 0 | 5 | 3 |
| USGS reports a M1.94 #earthquake 5km S of Volc... | 1 | 94 | 17 | 5.529412 | 3 | 2 | 1 | 11 | 4 |
| GNS sees unnecessary deaths resulting from ear... | 1 | 96 | 11 | 8.727273 | 1 | 0 | 0 | 2 | 1 |
| "There was a small earthquake in LA but don't ... | 1 | 83 | 17 | 4.882353 | 6 | 1 | 0 | 0 | 1 |
| USGS EQ: M 1.9 - 15km E of Anchorage Alaska. T... | 1 | 119 | 17 | 7.000000 | 1 | 1 | 0 | 19 | 5 |

**Figure 4.** Primary feature from the dataset

| | id | keyword | location | text | target | char_counts | word_counts |
|---|---|---|---|---|---|---|---|
| **3040** | 4362 | earthquake | NaN | usgs m 0 9 northern california time20150806 01... | 1 | 122 | 15 |
| **3041** | 4363 | earthquake | TÌÁchira - Venezuela | scseestapreparando light mag 4 4 earthquake 73... | 1 | 118 | 17 |
| **3042** | 4365 | earthquake | California, USA | usgs m 1 2 23km s of twentynine palms californ... | 1 | 119 | 17 |
| **3043** | 4366 | earthquake | a box | ageekyfangirl14 s things she looks in a signif... | 0 | 115 | 21 |
| **3044** | 4368 | earthquake | Melbourne, Australia | nepal earthquake 3 months on women fear abuse ... | 1 | 77 | 11 |
| **3045** | 4372 | earthquake | Barcelona, Spain | ml 2 0 sicily italy httpt coz6hxx6d2pm euroquake | 0 | 48 | 6 |
| **3046** | 4373 | earthquake | Hawaii, USA | usgs reports a m1 94 earthquake 5km s of volca... | 1 | 94 | 17 |
| **3047** | 4374 | earthquake | New Zealand | gns sees unnecessary deaths resulting from ear... | 1 | 96 | 11 |
| **3048** | 4375 | earthquake | NaN | there was a small earthquake in la but do not ... | 1 | 83 | 17 |
| **3049** | 4376 | earthquake | Alaska, USA | usgs eq m 1 9 15km e of anchorage alaska time2... | 1 | 119 | 17 |

**Figure 5.** Dataset after cleaning

The cleaning stage is done by removing email, URLs, HTML tags, special chars, and duplicates chars. Figure 5 shows the cleaning result data, especially in the text column, where there are already visible changes from the cleaning results.

## 4. Results and Discussion

We convert the raw document into the TF-IDF feature matrix. Next, by using LinearSVC, we are going to describe the classifier. The default settings of the class can be used. The parameters can be tailored to the data content of the classification. The results can be seen in the classification report in Figure 6.

```
Classification report
              precision    recall  f1-score   support

           0       0.79      0.89      0.84       869
           1       0.83      0.69      0.75       654

    accuracy                           0.81      1523
   macro avg       0.81      0.79      0.80      1523
weighted avg       0.81      0.81      0.80      1523
```

**Figure 6.** Classification Report

We can achieve a validation accuracy of 79% with an excellent F-1 Score for each of the predicted classes. The classification report includes a per-class representation of the significant grading metrics. This condition gives a more intimate understanding of global accuracy classification, which can mask functional deficiencies in a multiclass problem.

By using the TF-IDF Vectorizer, we transform the text into vectors. They are using a linear classification algorithm for a vector machine. The model in the binary classification classifier LinearSVC was fitted, and we predict the feeling that real data is either a real disaster or not a real disaster.

We added a dropout layer of 0.5 to the BERT layer's output to complete the pre-trained model, and then we said a Dense layer with relu activation. We use the relu and sigmoid activation functions. The architecture will generate opportunities for each class, real disaster (target) or non-real disaster (non-target). Figure 7 shows a summary of the model.

```
Model: "sequential"
_________________________________________________________________
Layer (type)                 Output Shape              Param #
=================================================================
embedding (Embedding)        (None, 40, 100)           2258200
_________________________________________________________________
conv1d (Conv1D)              (None, 39, 32)            6432
_________________________________________________________________
max_pooling1d (MaxPooling1D) (None, 19, 32)            0
_________________________________________________________________
dropout (Dropout)            (None, 19, 32)            0
_________________________________________________________________
dense (Dense)                (None, 19, 16)            528
_________________________________________________________________
global_max_pooling1d (Global (None, 16)                0
_________________________________________________________________
dense_1 (Dense)              (None, 1)                 17
=================================================================
Total params: 2,265,177
Trainable params: 2,265,177
Non-trainable params: 0
_________________________________________________________________
```

**Figure 7.** Model Summary

```
begin training using onecycle policy with max lr of 2e-05...
Epoch 1/8
108/108 [==============================] - 99s 747ms/step - loss: 0.6037 - accuracy: 0.6759 - val_loss: 0.4114 - val_accuracy: 0.8241
Epoch 2/8
108/108 [==============================] - 79s 730ms/step - loss: 0.4086 - accuracy: 0.8247 - val_loss: 0.3664 - val_accuracy: 0.8320
Epoch 3/8
108/108 [==============================] - 79s 729ms/step - loss: 0.3694 - accuracy: 0.8437 - val_loss: 0.3540 - val_accuracy: 0.8504
Epoch 4/8
108/108 [==============================] - 78s 727ms/step - loss: 0.2840 - accuracy: 0.8889 - val_loss: 0.3880 - val_accuracy: 0.8451
Epoch 5/8
108/108 [==============================] - 79s 729ms/step - loss: 0.2133 - accuracy: 0.9199 - val_loss: 0.4400 - val_accuracy: 0.8294
Epoch 6/8
108/108 [==============================] - 79s 728ms/step - loss: 0.1323 - accuracy: 0.9521 - val_loss: 0.5513 - val_accuracy: 0.8294
Epoch 7/8
108/108 [==============================] - 79s 728ms/step - loss: 0.0833 - accuracy: 0.9721 - val_loss: 0.6590 - val_accuracy: 0.8215
Epoch 8/8
108/108 [==============================] - 79s 728ms/step - loss: 0.0686 - accuracy: 0.9731 - val_loss: 0.6544 - val_accuracy: 0.8255
<tensorflow.python.keras.callbacks.History at 0x7fd596648be0>
```

**Figure 8.** Model training

```
data = ['Posted a new song: Earthquake']
predictor.predict(data)

['not_target']
```

```
data = ['There was a small earthquake in LA but dont worry Emily Rossum is fine']
predictor.predict(data)

['target']
```

**Figure 9.** Prediction

The predictions made are pretty good; the model created can predict accurately determine a real disaster (target) and which one is non-disaster (non-target), as shown in Figure 9.

## 5. Conclusion

BERT (Bidirectional Encoder Representations from Transformers) is a Google-designed, profound learning model. Since Google opened it, several scientists and companies have embraced it and have applied it to many text classification tasks. Therefore, we use BERT in this paper to some earthquake tweets. This research will help rescue and emergency responders establish effective knowledge management techniques for a rapidly evolving disaster environment. The predictions made are pretty good; the model created can predict accurately determine a real disaster (target) and which one is non-disaster (non-target).

## References

[1] Beigi G, Hu X, Maciejewski R and Liu H 2016 An overview of sentiment analysis in social media and its applications in disaster relief *Sentiment analysis and ontology engineering* 313-340

[2] Yuan F, Li M, Liu R, and Zhai W and Qi B 2021 Social media for enhanced understanding of disaster resilience during Hurricane Florence *International Journal of Information Management* **57** 102289

[3] Xiong S, Lv H, Zhao W and Ji D 2018 Towards Twitter sentiment classification by multi-level sentiment-enriched word embeddings *Neurocomputing* **275** 2459-2466

[4] Shelar A and Huang C Y 2018 Sentiment analysis of Twitter data *2018 International Conference on Computational Science and Computational Intelligence* 1301-1302

[5] Alharbi A S M and de Doncker E 2019 Twitter sentiment analysis with a deep neural network: An enhanced approach using user behavioral information *Cognitive Systems Research* **54** 50-61

[6] Neppalli V K, Caragea C, Squicciarini A, Tapia A and Stehle S 2017 Sentiment analysis during Hurricane Sandy in emergency response *International journal of disaster risk reduction* **21** 213-222

[7] Devlin J, Chang M. W, Lee K, and Toutanova K BERT: Pre-training of deep bidirectional transformers for language understanding *2019 Conference of the North American Chapter of the Association for Computational Linguistics: Human Language Technologies - Proceedings of the Conference*

[8] Zahera H M, Elgendy I A, Jalota R and Sherif M A 2019 Fine-tuned BERT Model for Multi-Label Tweets Classification *TREC* 1-7
[9] Frank M R, Mitchell L, Dodds P S and Danforth C M 2013 Happiness and the patterns of life: A study of geolocated tweets *Scientific reports* 3 (1) 1-9
[10] Fan C, Esparza M, Dargin J, Wu F, Oztekin B and Mostafavi A 2020 Spatial biases in crowdsourced data: Social media content attention concentrates on populous areas in disasters *Computers, Environment and Urban Systems* **83** 101514
[11] Mandel B, Culotta A, Boulahanis J, Stark D, Lewis B and Rodrigue J 2012 A demographic analysis of online sentiment during hurricane irene *Proceedings of the second workshop on language in social media* 27-36
[12] Singh N, Roy N and Gangopadhyay A 2018 Analyzing the sentiment of the crowd for improving the emergency response services *2018 IEEE International Conference on Smart Computing* 1-8
[13] Munikar M, Shakya and S and Shrestha A 2019 Fine-grained sentiment classification using BERT *2019 Artificial Intelligence for Transforming Business and Society* **1** 1-5
[14] Maharani W 2020 Sentiment Analysis during Jakarta Flood for Emergency Responses and Situational Awareness in Disaster Management using BERT *8th International Conference on Information and Communication Technology* 1-5
[15] Wu Z and Ong D C 2020 Context-Guided BERT for Targeted Aspect-Based Sentiment Analysis *Association for the Advancement of Artificial Intelligence* 1-9
[16] Hu M, Zhao S, Guo H, Cheng R and Su Z 2019 Learning to detect opinion snippet for aspect-based sentiment analysis *23rd Conference on Computational Natural Language Learning, Proceedings of the Conference* 1-10

IOP Conf. Series: Materials Science and Engineering 1115 (2021) 012033 doi:10.1088/1757-899X/1115/1/012033

# Auxiliary information based generally weighted moving coefficient of variation (AIB-GWMCV) control chart

**M A Nuriman*, M Mashuri, and M Ahsan**

Department of Statistics, Faculty of Science and Data Analytics, Institut Teknologi Sepuluh Nopember (ITS), Surabaya, Indonesia

*Email: alifianmuhammad13@gmail.com

**Abstract.** Statistical Process Control (SPC) is a statistical technique to accurately determine process performance. It has been widely used in manufacturing industries and services. The most powerful instrument used in SPC is the control chart. It is designed to observe and detect timely assignable causes of the process. In general, the control chart is employed to detect shifts in process location and process dispersion. Monitoring the coefficient of variation (CV) is efficient in SPC when the standard deviation process changes with the mean. Also, the mean itself is expected to fluctuate with time but considered to be in-control. Recently, some studies investigate using auxiliary information to enhance the sensitivity of the CV control chart. This study investigates auxiliary information based CV-GWMA (AIB-GWMCV) control chart by using log-normal transformation to detect the small to large shifts in process CV. It turns out the proposed control chart performs better than the CV-GWMA control chart using log-normal transformation without auxiliary information. Also, a real case is presented to display the application of the AIB-GWMCV control chart. Results show that the proposed control chart is faster to detect shifts in the process CV than the control chart without auxiliary information.

## 1. Introduction

Control charts become the most powerful instrument used in SPC. It is designed to observe the behavior of the process and to detect timely assignable causes of the process [1]. The principle of SPC is that the process cannot be declared as in-control until both the process dispersion and the process location is stable. When standard deviation changes with the mean and mean itself timely fluctuate but considered to be in-control then the CV based control charts can be efficiently used to keep the stability of the process. Shewhart's CV chart was first introduced by [2]. They conclude that it has an attractive performance than $\overline{X}$ and $S$ chart [3-4]. Adaptive Shewhart CV chart implementing VSS strategy was presented by [5]. Moreover, they implemented log-normal transformation to get the normal distribution of CV statistic. The proposed chart compared with Shewhart's CV chart [2], synthetic CV chart [6], and VSI chart [7] and showed that it is more efficient to detect shifts of CV than existing charts. Further, EWMA-type CV [8] and DEWMA-type CV [9] were investigated to increase the efficiency in detecting process CV. In a recent study, the log-normal transformation was applied in an EWMA-type CV chart named EWMCV chart [10]. They conclude that EWMA-type CV with RSS performs better than EWMA-type CV with SRS. The GWMA based CV control chart was evolved by [11] and they concluded that it has better performance than the CV-EWMA chart [8] and CV-DEWMA chart [9].

In these control charts, the monitoring process dispersion is constructed by using a single quality characteristic. Recently, researchers have been proposed a method to increase the detecting ability by adding an auxiliary variable that correlated with the study variable. Haq [12] proposed two charts for efficient monitoring of process dispersion by using the regression type auxiliary estimator of the population variance called AIB-EWMA-I and using the cumulative distribution transformation called AIB-EWMA-II. The result showed that the AIB-EWMA-I chart performs better than the AIB-EWMA-II chart when detecting process dispersion. Abbasi [13] presented some varieties of auxiliary information based ratio form, regression form, and hybrid estimator to enhance the shift detection ability of the CV charts. The result showed that it is significantly efficient than competing charts without auxiliary information for moderate to high correlation levels. Further, a study about utilizing auxiliary information in the EWMCV chart is investigated by [14]. They conclude that the AIB-EWMCV chart outperforms the EWMCV chart developed by [10] in terms of quickly detect out-of-control signals.

This study investigates new auxiliary information based CV-GWMA (AIB-GWMCV). We proposed the AIB-GWMCV control chart utilizing the log-normal transformation suggested by [5]. The capability of control charts in detecting shifts can be evaluated with the Average Run Length (ARL) value. In the current study, we discuss the capability of control charts with different sample sizes and different combination parameters of $q$ and $\omega$. Also, we display the demonstration of the proposed control chart to monitor process CV in the production process of NPK fertilizer.

## 2. Methodology

### 2.1. Control chart structure and log-normal transformation for CV statistic

This subsection describes the CV-GWMA control chart proposed by [11], one of the developed CV charts using the GWMA structure. Further, the CV statistic will be transformed with log-normal transformation suggested by [5].

*2.1.1. CV-GWMA control chart*

Let $\{Y_{i,1}, Y_{i,2}, \ldots, Y_{i,n}\}$ denote sample random variable $Y_{i,j}$ which normally distributed such that $Y_{i,j} \sim N(\mu, \sigma)$, where $\mu_0$ is in-control mean and $\sigma_0$ is in-control for $i$ time. Then, the CV of the population $(\gamma)$ is explained as $\gamma = \frac{\sigma}{\mu}$. Let the mean, SD, and CV of a random sample for $i$ time is explained as

$$\bar{Y}_i = \frac{1}{n}\sum_{j=1}^{n} Y_{i,j}, \quad S_i = \sqrt{\frac{1}{n-1}\sum_{j=1}^{n}\left(Y_{i,j} - \bar{Y}_i\right)^2}. \tag{1}$$

$$W_i = \frac{S_i}{\bar{Y}_i}, \quad -\infty < \gamma \leq \infty. \tag{2}$$

Let $M$ denotes count the number of samples until an event A first occurs since a previous of event A occurs, $P_i = P(M = i) = \bar{P}_{i-1} - \bar{P}_i$ denotes the probability that event A occurs in the first $i$ sample and $\bar{P}_i = P(M > i)$ denotes the probability that the event A does not occur in the first $i$ samples

$$\sum_{m=1}^{\infty} P(M = m) = P(M = 1) + P(M = 2) + \cdots + P(M = i) + P(M > i) = 1, \tag{3}$$

Where $P(M=1)$ denotes a weight of current observation, $P(M=2)$ denotes a weight of previous observation, $P(M=i)$ denotes a weight of the oldest observation, and $P(M>i)$ denotes a weight for target value. The GWMA-type CV statistic for $i$ time is explained as,

$$\begin{aligned} G_i &= P(M=1)W_i + P(M=2)W_{i-1} + \cdots + P(M=i)W_1 + P(M>i)G_0 \\ G_i &= \left(\overline{P}_0 - \overline{P}_1\right)W_i + \left(\overline{P}_1 - \overline{P}_2\right)W_{i-1} + \cdots + \left(\overline{P}_{i-1} - \overline{P}_i\right)W_1 + \overline{P}_i G_0. \end{aligned} \tag{4}$$

For easy computation, $q^{i^\omega}$ is modified into $\overline{P}_i$. Consequently, equation (2) can be written as

$$G_i = \left(q^{0^\omega} - q^{1^\omega}\right)W_i + \left(q^{1^\omega} - q^{2^\omega}\right)W_{i-1} + \cdots + \left(q^{(i-1)^\omega} - q^{i^\omega}\right)W_1 + q^{i^\omega}\mu_W \tag{5}$$

$q$ is defined as a design parameter in the interval $0<q<1$. The $\omega$ is a parameter determined by the practitioner and the target value is determined by $G_0 = \mu_W$. The expected value of $G_i$ can be computed as

$$\begin{aligned} E(G_i) &= E\left(\left(\overline{P}_0 - \overline{P}_1\right)W_i + \left(\overline{P}_1 - \overline{P}_2\right)W_{i-1} + \cdots + \left(\overline{P}_{i-1} - \overline{P}_i\right)W_1 + \overline{P}_i G_0\right) \\ E(G_i) &= E\left(\left(q^{0^\omega} - q^{1^\omega}\right)W_i + \left(q^{1^\omega} - q^{2^\omega}\right)W_{i-1} + \cdots + \left(q^{(i-1)^\omega} - q^{i^\omega}\right)W_1 + q^{i^\omega}\mu_W\right) \\ E(G_i) &= \mu_w, \end{aligned} \tag{6}$$

and variance of $G_i$,

$$\begin{aligned} Var(G_i) &= \left(\left(\overline{P}_0 - \overline{P}_1\right)^2 + \left(\overline{P}_1 - \overline{P}_2\right)^2 + \cdots + \left(\overline{P}_{i-1} - \overline{P}_i\right)^2\right)\sigma_W^2 \\ Var(G_i) &= \left(\left(q^{0^\omega} - q^{1^\omega}\right)^2 + \left(q^{1^\omega} - q^{2^\omega}\right)^2 + \cdots + \left(q^{(i-1)^\omega} - q^{i^\omega}\right)^2\right)\sigma_W^2 \\ Var(G_i) &= Q_i \sigma_W^2, \end{aligned} \tag{7}$$

where, $Q_i = \sum_{k=1}^{i}\left(q^{(k-1)^\omega} - q^{k^\omega}\right)^2, i \geq 1$. The lower control limit (LCL), central limit (CL), and upper control limit (UCL) of the CV-GWMA control chart is calculated as

$$LCL_i = \mu_W - L\sqrt{Q_i}\sigma_W, \tag{8}$$

$$CL_i = \mu_W, \tag{9}$$

$$UCL_i = \mu_W + L\sqrt{Q_i}\sigma_W, \tag{10}$$

where $L$ explains the width of control limits. The $\mu_W$ and $\sigma^2_W$ [15] calculated as

$$\mu_W \approx \gamma\left(1 - \frac{0.25 - \gamma^2}{n}\right), \tag{11}$$

$$\sigma^2_W \approx \frac{\gamma^2}{n}\left(0.5 + \frac{0.4375}{n} + \gamma^2\left(1 + \frac{1+9\gamma^2}{n}\right)\right) - \left(\mu_W - \gamma\right)^2. \tag{12}$$

### 2.1.2. Log-normal transformation for CV statistic

Log-normal transformation is used to transform the distribution of CV into a normal distribution. Castagliola et al [5] suggested this transformation form since the CV sample has a unimodal right-skewed distribution, so its distribution may become a preferred choice to transform CV statistic. The transformation form for $i$ time is explained as,

$$Q_i = a + b\ln\left(\gamma_i - c\right), \tag{13}$$

where, $\gamma_i = \frac{S_i}{\bar{Y}_i}$, $a,b > 0$ and $c$ can be obtained depend on $n$ and $(\gamma_0)$ of the process, such that $T_i \sim N(0,1)$.

$$b = \frac{F_N^{-1}(r)}{\ln\left(\frac{x_{0.5} - x_r}{x_{1-r} - x_{0.5}}\right)}, \tag{14}$$

$$a = -b\ln\left(\frac{x_{0.5} - x_r}{1 - \exp\left(\frac{F_N^{-1}(r)}{b}\right)}\right), \tag{15}$$

$$c = x_{0.5} - e^{-a/b}. \tag{16}$$

$F_N^{-1}(r)$ is the quantile function of $N(0,1)$. $x_r = F^{-1}(r \mid n_i, \gamma_0)$ is the quantile function noncentral-t distribution with the parameter of non-centrality $\gamma_0$, degrees of freedom $n_i$, and probability $r$. $x_{0.5} = F^{-1}(0.5 \mid n_i, \gamma_0)$ and $x_{1-r} = F^{-1}(1-r \mid n_i, \gamma_0)$ are the quantile order of 0.5 and $1-r$. The value of $r$ is determined by the practitioner in range $r$, $0.01 \le r \le 0.1$.

### 2.2. Proposed AIB-GWMCV control chart

Let $(Y_{ij}, X_{ij})$ denote random sample with $j = 1,2,\cdots,n$ as samples of size $n$ at time $i = 1,2,\cdots m$ bivariate normally distributed, i.e. $(Y,X) \sim N_2(\mu_Y, \mu_X, \sigma_Y, \sigma_X, \rho_{XY})$ where $\mu_Y, \sigma_Y$ are the process mean and process standard deviation variable of interest $Y$ and $\mu_X, \sigma_X$ are the process mean and process standard deviation auxiliary variable $X$. While $\rho_{XY}$ is the correlation coefficient between the study variable $Y$ and auxiliary variable $X$. A suitable selection of the auxiliary information is very important when working with the AIB control chart for the monitoring process. The auxiliary information is a variable that should be stable and does not change when a shift occurs in the study variable. If it occurs, that variable should be included in the list of study variables, and a multivariate control chart is more suitable for this condition [16]. Following Haq and Khoo [17], AIB difference estimators for CV is defined by

$$M_{Y,i} = Q_{Y,i} - \rho_{XY}^{*} Q_{X,i}, \tag{17}$$

where,

$$Q_{Y,i} = a + b\ln\left(\gamma_{Y,i}\right),\ Q_{X,i} = a + b\ln\left(\gamma_{X,i}\right), \tag{18}$$

$\rho^*_{XY}$ denotes the correlation of $Q_{Y,i}$ and $Q_{X,i}$, which is estimated by using Monte Carlo simulation. The expected value and variance $M_{Y,i}$ are defined as follows,

$$E\left(M_{Y,i}\right) = 0,\ V\left(M_{Y,i}\right) = \left(1 - \rho^{*2}_{XY}\right), \tag{19}$$

Now, $M_{Y,i}$ is transformed into standard normal transformation defined in equation (20),

$$M^*_{Y,i} = \frac{M_{Y,i} - E\left(M_{Y,i}\right)}{\sqrt{\left(1 - \rho^{*2}_{XY}\right)}}, \tag{20}$$

As a result, $M^*_{Y,i}$ is normally distributed, such that $M^*_{Y,i} \sim N(0,1)$. The statistic of AIB-GWMCV for $i$ time is defined as follows,

$$A_i = \left(q^{0^\omega} - q^{1^\omega}\right)M^*_{Y,i} + \left(q^{1^\omega} - q^{2^\omega}\right)M^*_{Y,i-1} + \cdots + \left(q^{(i-1)^\omega} - q^{i^\omega}\right)M^*_{Y,1} + q^{i^\omega}\mu_{M^*_Y}, \tag{21}$$

$q = 1 - \lambda$ is design parameter in the interval $0 < q < 1$. The $\omega$ is a parameter determined by the practitioner and the target value of the process is $A_0 = \mu_{M^*_Y} = 0$. The expected value and variance of $A_i$ can be computed as follows,

$$E(A_i) = E\left(\left(q^{0^\omega} - q^{1^\omega}\right)M^*_{Y,i} + \left(q^{1^\omega} - q^{2^\omega}\right)M^*_{Y,i-1} + \cdots + \left(q^{(i-1)^\omega} - q^{i^\omega}\right)M^*_{Y,1} + q^{i^\omega}\mu_{M^*_Y}\right)$$

$$E(A_i) = \mu_{M^*_Y}, \tag{22}$$

$$V(Q_i) = \left(\left(q^{0^\omega} - q^{1^\omega}\right)^2 + \left(q^{1^\omega} - q^{2^\omega}\right)^2 + \cdots + \left(q^{(i-1)^\omega} - q^{i^\omega}\right)^2\right)\sigma^2_{M^*_Y}$$

$$V(Q_i) = Z_i\sigma^2_{M^*_Y}, \tag{23}$$

supposing $Z_i = \sum_{k=1}^{i}\left(q^{(k-1)^\omega} - q^{k^\omega}\right)^2, i \geq 1$, control limits are defined as follows

$$LCL_i = \mu_{M^*_Y} - L\sigma_{M^*_Y}\sqrt{Z_i}, \tag{24}$$

$$CL_i = \mu_{M^*_Y}, \tag{25}$$

$$UCL_i = \mu_{M^*_Y} + L\sigma_{M^*_Y}\sqrt{Z_i}, \tag{26}$$

supposing $L$ is the width of control limits, a process is considered out-of-control when the $A_i$ value falls outside the range of control limits.

### 2.3. AIB-GWMCV control chart charting procedures

These steps are given below for executing the AIB-GWMCV control chart:

1. Consider the process parameters of CV $(\gamma_0)$ for the study variable $Y$ and auxiliary variable $X$. It can be also estimated from previous data sets or compute the sample of CV from the data set $(\gamma_0)$.
2. Compute $a,b,c$ based on $\gamma_0,n,r$ and width of control limits $L$. $L$ is determined such that $ARL_0 = 370$.
3. Compute $Q_{Y,i}$ and $Q_{X,i}$ by utilizing equation (18) for each time.
4. Compute $M_{Y,i}$ by utilizing equation (17) for each time.
5. Transform $M_{Y,i}$ into standard normal distribution by utilizing equation (20).
6. Compute $A_i$ by utilizing equation (21) for each time.
7. Compute the mean and variance of $A_i$ that used to determine control limits.
8. Compute LCL, CL, and UCL by utilizing equation (24), (25), (26), respectively.
9. Construct AIB-GWMCV chart by comparing statistic $A_i$ and control limits.
10. When $A_i$ values fall outside of the control limit value, then investigate the source of out-of-control signals. Thus, eliminate it to get the in-control process if it is detected as an assignable cause.

## 3. Result and Discussions

### 3.1 Simulation procedures

The performance of the proposed chart can be evaluated with ARL. ARL defines as an average number of points plotted until out-of-control signals are indicated. There are several numerical methods in the literature to measure ARL. We utilize Monte Carlo simulation to measure ARL that is categorized as $(ARL_0)$ and $(ARL_s)$. When the process under in-control, the $ARL_0$ value should be larger to keep away from false alarms. While $ARL_s$ values should be small indicating shifts are rapidly detected when out-of-control condition.

The algorithms to determine the ARL of the proposed control chart are given below:

1. Consider the shift of the process $\tau = \frac{\gamma_1}{\gamma_0}$. $\gamma_0$ denotes CV under the in-control condition and $\gamma_1$ denotes CV under out-of-control conditions. We use $\tau = 1$, 1.1,1.2, 1.3, 1.4, 1.5, 2.0, 3.0, 4.0. Note that $\tau = 1$ denotes no shift in the process.
2. Compute $a,b,c$ based on $\gamma_0,n,r$ and width of control limits $L$. $L$ is determined such that $ARL_0 = 370$.
3. Generate 1,000 subgroups of random samples that bivariate normally distributed from a shift process with samples of size $n$ and in-control parameters are $\mu_Y = \mu_X = \mu_0 = 10$, $\sigma_Y = \sigma_X = \sigma_0 = 1$ and $\gamma_Y = \gamma_X = \gamma_0 = 0.1$. The coefficient correlation between the study variable and the auxiliary variable is considered such that $\rho_{XY} = 0.3, 0.5, 0.7, 0.9$.
4. Compute mean, standard deviation, and CV for each subgroup/time of variable.
5. Compute statistic $Q_{Y,i}$ and $Q_{X,i}$ by utilizing equation (18) for each subgroup/time, and compute statistic $M_{Y,i}$ by utilizing equation (17).
6. Transform $M_{Y,i}$ into standard normal distribution by utilizing equation (20).

7. Compute $A_i$ by utilizing equation (21) for each subgroup/time.
8. Compute the expected value and the variance of $A_i$ that used for control limits.
9. Compute LCL, CL, and UCL defined in equation (24), (25), and (26).
10. Construct AIB-GWMCV chart by comparing statistic $A_i$ and control limits. When the process is considered out-of-control, record numbers of subgroups as run length.
11. Repeat the whole process 1,000 times to get $ARL_0$ and $ARL_s$.

### 3.2. Performance Evaluation of proposed control chart

We evaluate the performance of the control chart by measuring ARL. The smaller $ARL_s$ values, the better performance of the control chart. Here, we consider sample sizes $n=5,12$ by using two combination parameters $q$ and $\omega$, i.e. (0.75,0.8) and (0.9,0.9) at fixed $ARL_0=370$. Moreover, we compute the ARL value for the AIB-GWMCV chart at some different correlation levels. We used different correlation levels such that $\rho_{XY}=0.0,0.3,0.5,0.7,0.9$. Note that for correlation level $\rho_{XY}=0$ represents the CV-GWMA control chart using log-normal transformation without auxiliary variable. Table 1 displayed the parametric values $a,b,c$ for transformed statistics and the constant value of $L$ for the AIB-GWMCV control chart based on $\gamma_0=0.1$ $q,\omega$, and $n$.

Table 2 and Table 3 report the average run length (ARL) comparison of the AIB-GWMCV chart for different correlation levels and sample sizes. Table 2 presents the performance of the proposed control chart and CV-GWMA control chart using log-normal transformation without auxiliary variable by using combination parameters $q=0.75$ and $\omega=0.8$. According to Table 2, it is concluded that for sample sizes $n=5$, the proposed control chart is more efficient in detecting small to large shifts at low to high correlation levels than the CV-GWMA control chart using log-normal transformation without auxiliary variable. It can be seen that at the shift $\tau=1.1$, the CV-GWMA control chart using log-normal transformation without auxiliary variable has $ARL_s$ equal to 178.840. Meanwhile, the $ARL_s$ of proposed control charts with correlation levels $\rho_{XY}=0.3,0.5,0.7,0.9$ are 154.356, 150.154, 146.100, 76.472, respectively.

**Table 1.** Values of $L$ for different sample sizes and correlation levels

| $(q,\omega)$ | $n=5$ | | | | | $n=12$ | | | | |
|---|---|---|---|---|---|---|---|---|---|---|
| | $a=9.8864$, $b=6.8112$, $c=-0.1426$ | | | | | $a=15.412$, $b=9.9859$, $c=-0.1167$ | | | | |
| $\rho$ | 0.0 | 0.3 | 0.5 | 0.7 | 0.9 | 0.0 | 0.3 | 0.5 | 0.7 | 0.9 |
| (0.75,0.8) | 3.031 | 3.05 | 3.02 | 3.04 | 3.04 | 3.06 | 3.06 | 3.07 | 3.09 | 3.09 |
| (0.9,0.9) | 2.89 | 2.91 | 2.92 | 2.92 | 2.92 | 2.918 | 2.92 | 2.91 | 2.92 | 2.92 |

For sample sizes $n=12$, Also the proposed control chart demonstrates a better performance to detect small to large shifts at low to high correlation levels than the CV-GWMA control chart using log-normal transformation without auxiliary variable. For example, at the same shift, the $ARL_s$ for CV-GWMA control chart using transformation log-normal without auxiliary variable is 63.694. Meanwhile, the $ARL_s$ for the proposed control chart with correlation levels $\rho_{XY}=0.3,0.5,0.7,0.9$ are 62.070, 59.775, 51.164, 23.732, respectively. From the results, it indicates that an increase in sample sizes cause early detection of the proposed control chart at low to high correlation levels.

Table 3 provides the performance of the proposed control chart and CV-GWMA control chart using log-normal transformation without auxiliary variable with $n=5$ and $n=12$ for the same value of combination parameter $q=0.9$ and $\omega=0.9$ at correlation levels $\rho_{XY}=0.3,0.5,0.7,0.9$. From Table 3, It is implied that the increase in sample size will significantly increase the performance of the control chart in detecting shifts of process CV. For example, at shift 1.4 and sample of sizes $n=5$, the $ARL_s$ for CV-GWMA control chart using log-normal transformation without auxiliary variable is 13.680. Whereas, the $ARL_s$ for the proposed control chart with correlation levels $\rho_{XY}=0.3,0.5,0.7,0.9$ are 12.592, 11.750, 9.922, 5.920, respectively. Also, the proposed chart with sample sizes $n=12$ is more sensitive than the CV-GWMA control chart using log-normal transformation without auxiliary variable. It can be known at the same shift, the $ARL_s$ for CV-GWMA control chart using transformation log-normal without auxiliary variable is 5,046. Whereas, the $ARL_s$ for the proposed control chart with correlation levels $\rho_{XY}=0.3,0.5,0.7,0.9$ are 4.870, 4.764, 3.914, 2.276, respectively.

From these simulation studies, we can conclude that as $\rho_{XY}$ increases from low to high correlation levels, the performance of the proposed control chart gets better and better, as compared to the CV-GWMA control chart using log-normal transformation without auxiliary variable. Furthermore, larger $q$ enhances the sensitivity of the AIB-GWMCV control chart. A value of $\lambda$ in the interval $0.05 \le q \le 0.25$ is the practical approach [1]. It is equal to the value of $q$ in the interval $0.75 \le q \le 0.95$.

**Table 2.** ARL values of AIB-GWMCV for $q=0.75$ and $\omega=0.8$ at fixed $ARL_0=370$

| Shift $(\tau)$ | Sample Size $(n)$ | $\rho_{XY}$ | | | | |
|---|---|---|---|---|---|---|
| | | 0.0 | 0.3 | 0.5 | 0.7 | 0.9 |
| 1.0 | 5 | 375.668 | 382.996 | 376.568 | 369.450 | 372.588 |
| | 12 | 371.783 | 366.648 | 372.566 | 370.480 | 370.250 |
| 1.1 | 5 | 178.840 | 154.356 | 150.154 | 146.100 | 76.472 |
| | 12 | 63.694 | 62.070 | 59.774 | 51.164 | 23.732 |
| 1.2 | 5 | 51.960 | 51.416 | 48.592 | 40.456 | 19.822 |
| | 12 | 17.240 | 16.852 | 15.566 | 13.516 | 6.486 |
| 1.3 | 5 | 27.194 | 26.818 | 23.106 | 21.034 | 10.528 |
| | 12 | 8.466 | 8.462 | 8.450 | 6.886 | 3.766 |
| 1.4 | 5 | 18.098 | 15.906 | 14.564 | 12.242 | 6.810 |
| | 12 | 5.680 | 5.666 | 5.438 | 4.540 | 2.504 |
| 1.5 | 5 | 11.752 | 1.762 | 10.406 | 8.804 | 5.054 |
| | 12 | 3.992 | 3.954 | 3.926 | 3.356 | 1.902 |
| 2.0 | 5 | 5.118 | 4.784 | 4.474 | 3.872 | 2.598 |
| | 12 | 1.918 | 1.906 | 1.900 | 1.686 | 1.160 |
| 3.0 | 5 | 2.912 | 2.596 | 2.456 | 2.288 | 1.662 |
| | 12 | 1.228 | 1.218 | 1.210 | 1.128 | 1.006 |
| 4.0 | 5 | 2.210 | 2.132 | 2.032 | 1.894 | 1.512 |
| | 12 | 1.070 | 1.070 | 1.048 | 1.036 | 1.002 |

**Table 3.** ARL values of AIB-GWMCV for $q = 0.9$ and $\omega = 0.9$ at fixed $ARL_0 = 370$

| Shift $(\tau)$ | Sample Size $(n)$ | $\rho_{XY}$ | | | | |
|---|---|---|---|---|---|---|
| | | 0.0 | 0.3 | 0.5 | 0.7 | 0.9 |
| 1.0 | 5 | 367.260 | 369.038 | 374.432 | 368.504 | 372.624 |
| | 12 | 369.397 | 369.622 | 379.702 | 367.613 | 370.052 |
| 1.1 | 5 | 120.326 | 111.484 | 104.164 | 88.530 | 46.786 |
| | 12 | 45.874 | 39.022 | 38.004 | 31.676 | 14.968 |
| 1.2 | 5 | 35.698 | 34.222 | 32.800 | 27.862 | 15.224 |
| | 12 | 13.690 | 12.962 | 11.704 | 11.146 | 5.384 |
| 1.3 | 5 | 19.392 | 18.338 | 17.004 | 15.254 | 8.652 |
| | 12 | 7.298 | 7.060 | 6.610 | 6.034 | 3.152 |
| 1.4 | 5 | 13.680 | 12.592 | 11.750 | 9.922 | 5.920 |
| | 12 | 5.046 | 4.870 | 4.764 | 3.914 | 2.276 |
| 1.5 | 5 | 9.528 | 9.494 | 11.782 | 7.438 | 4.664 |
| | 12 | 3.976 | 3.522 | 3.432 | 3.014 | 1.792 |
| 2.0 | 5 | 4.410 | 4.156 | 4.101 | 3.656 | 2.418 |
| | 12 | 1.888 | 1.894 | 1.820 | 1.576 | 1.146 |
| 3.0 | 5 | 2.494 | 2.392 | 2.226 | 2.090 | 1.600 |
| | 12 | 1.208 | 1.200 | 1.152 | 1.094 | 1.006 |
| 4.0 | 5 | 2.040 | 1.968 | 1.906 | 1.794 | 1.412 |
| | 12 | 1.080 | 1.056 | 1.044 | 1.018 | 1.002 |

### 3.3 An illustrative example

This section presents an illustrative example using real data set to show the application of the AIB-GWMCV control chart for monitoring the NPK fertilizer production process. We consider nitrogen as the study variable $(Y)$ and phosphorus as an auxiliary variable $(X)$. These variables consist of 27 subgroups with samples of size $n = 5$, respectively. Assuming $\gamma_0 = 0.1$ for an in-control process, the coefficient $a, b$ and $c$ are calculated such that $a = 9.8864$, $b = 6.8112$, and $c = -0.1426$. The main quality characteristic $(Y)$ is normally distributed with parameters $\mu_Y = 14.157$, $\sigma_Y = 1.186$, $\gamma_Y = 0.08$ and rounded to $\gamma_0 = 0.1$ Also, the auxiliary variable $(X)$ is normally distributed with parameters $\mu_Y = 15.301$, $\sigma_Y = 1.754$, $\gamma_Y = 0.11$. It is known that $\rho_{XY} = 0.3$. We also consider combination parameters $q = 0.9$ and $\omega = 0.9$ for this application.

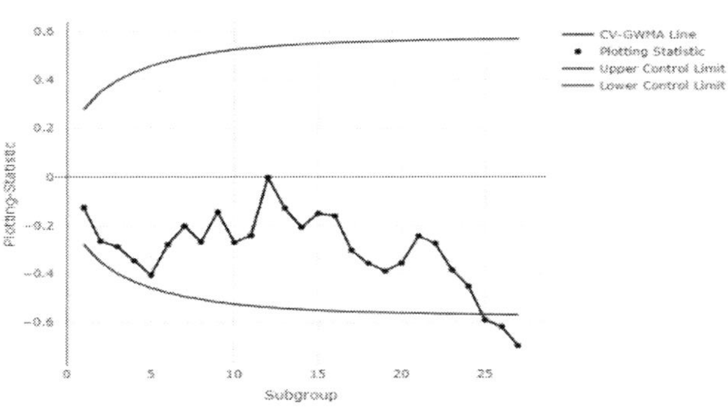


**Figure 1.** CV-GWMA control chart using log-normal transformation without auxiliary variable

According to Figure 1, it can be seen that the CV-GWMA control chart using log-normal transformation without auxiliary variable detects an out-of-control process at 25th observation. Meanwhile, based on figure 2, it is inferred that process is stated out-of-control at 7th observation when utilizing the auxiliary variable for monitoring process CV. According to these results, the proposed control chart can detect earlier out-of-control signals than the CV-GWMA control chart using log-normal transformation without auxiliary variable. Thus, based on Monte Carlo simulation and real application, the proposed control chart has better performance than the CV-GWMA control chart using log-normal transformation without auxiliary variable.

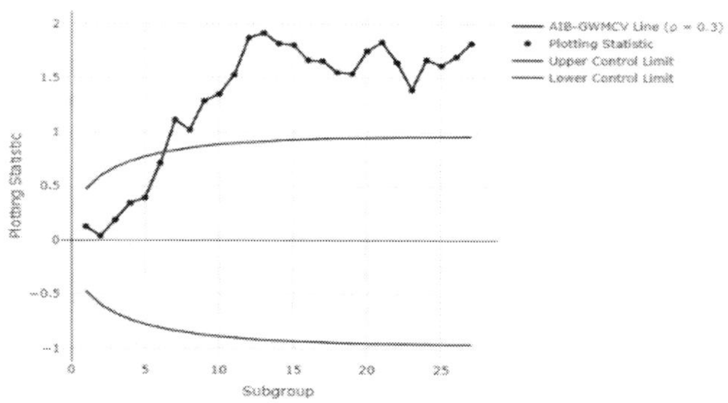


**Figure 2**. AIB-GWMCV control chart

## 4. Conclusion

We have proposed an AIB-GWMCV control chart to monitor the process CV. According to performance evaluation results, by utilizing the auxiliary variable, the detection ability of the control chart increases for each shift of process CV in all considered combination parameters $q$ and $\omega$. Furthermore, the detection ability of the proposed chart increases with an increase in levels of $\rho$. For small sample sizes, the proposed control chart quickly loses its ability to detect shifts, particularly at low correlation levels compared to large sample sizes. The application of the proposed control chart using a real data set supports the simulation results. Consequently, compared to the CV-GWMA control chart using log-normal transformation without auxiliary information, the AIB-GWMCV control chart has better performance in terms of faster detection of shifts process CV. Using more than one auxiliary variable in the proposed chart can be investigated in future research. Also, some auxiliary information estimators can be evaluated.

## References

[1] Montgomery D C 2009 *Introduction to Statistical Quality Control, Sixth Edition* Arizona: John Wiley and Sons

[2] Kang C W, Lee M S, Seong Y J and Hawkins D M 2007 A control chart for the coefficient of variation *Journal of Quaity Technology* **39** (2) 151–158

[3] Shewhart W A 1930 Economic Quality Control of Manufactured Product *Bell System Technical Journal* **9** (2) 364-389

[4] Nelson L S 1984 The Shewhart Control Chart—Tests for Special Causes *Journal of Quality Technology* **16** (4) 237-239

[5] Castagliola P, Achouri A, Taleb H, Celano G and Psarakis S 2015 Monitoring the coefficient of variation using a variable sample size control chart *The International Journal of Advanced Manufacturing Technology* **80** (3-4) 1561–1576

[6] Calzada M E and Scariano S M 2013 A synthetic control chart for the coefficient of variation

*Journal of Statistical Computation and Simulation* **83** (5) 853–867
[7] Castagliola P, Achouri A, Taleb H, Celano G and Psarakis S 2013 Monitoring the coefficient of variation using a variable sampling interval control chart *Quality and Reliability Engineering* **29** (8) 1135–1149
[8] Hong E P, Kang C W, Kang H W and Baik J W 2008 Development of CV Control Chart Using EWMA Technique *J. Soc. Korea Ind. Syst. Eng.* **31** (4) 114–120
[9] Hong E P, Kang H W and Kang C W 2011 DEWMA control chart for the coefficient of variation *In Proceedings of the Advanced Materials Research, Qinhuangdao, China,* 1682–1688
[10] Noor-ul-Amin M and Riaz A 2020 EWMA Control Chart for Coefficient of Variation Using Log-Normal Transformation Under Ranked Set Sampling *Iranian Journal of Science and Technology, Transaction A, Science* **44** (2) 155–165
[11] Hong E P, Kang H W, Kang C W and Baik J W 2011 CV control chart using GWMA technique *In Proceedings of the Advanced Materials Research, Qinhuangdao, China* 247-254
[12] Haq A 2017 New EWMA control charts for monitoring process dispersion using auxiliary information *Quality and Reliability Engineering.* **33** (3) 2597–614
[13] Abbasi S A 2020 Efficient control charts for monitoring process CV using auxiliary information *IEEE Access* **8** 46176–92
[14] Riaz A, Noor-ul-Amin M and Khan H 2020 Auxiliary information based exponentially weighted moving co-efficient of variation control chart: An application to monitor electric conductivity for water system *Communications in Statistics: Case Studies, Data Analysis and Applications* 1–20
[15] Bao Y 2009 Finite-sample moments of the coefficient of variation *Econom. Theory* **25** (1) 291-297
[16] Haq A and Khoo M B C 2018 Memory-type multivariate control charts with auxiliary information for process mean *Quality and Reliability Engineering* **35** (1) 1–12
[17] Haq A and Khoo M B C 2016 A new synthetic control chart for monitoring process mean using auxiliary information *Journal of Statistical Computation and Simulation* **86** (15) 3068–3092

IOP Conf. Series: Materials Science and Engineering 1115 (2021) 012034 doi:10.1088/1757-899X/1115/1/012034

# Industry X.0 and Sustainability through Lens of Industrial Engineering, Product Design Engineering and Business Engineering Perspectives

**K E N Soebandrija*, A K Samara, A W Wiratama, C D Sasabone, O Gunanjar and Z Toar**

Product Design Program, Industrial Engineering Department, BINUS ASO School of Engineering, Bina Nusantara University, Jakarta, Indonesia 11480

*Corresponding author: Knugroho@Binus.edu

**Abstract.** Industry X.0 orchestrates emerging, connected and smart technologies to digitally transform industry; as this term is coined by Accenture. Precisely, Industry X.0, according to Accenture, is defined as approach that uses advanced technologies to reinvent products and services from design and engineering to manufacturing and support, accelerating operational efficiency and enterprise-wide growth. The objective of this paper is to conduct balance research in term of theoretical and industrial perspectives within current and the most updated of real case implementation. The methodology in this paper orchestrates both research methodology and design methodology of quantitative approaches. The discussion lens of industrial engineering focuses on the transformation within manufacturing and production. Furthermore, the discussion lens of product design engineering refers to the transformation within design and product engineering. Subsequently, discussion lens of business engineering plays vital role on the transformation within innovation and growth; and support and services. In the arching synergy, the trilogy lens of industrial engineering, product design engineering and business engineering perspectives synergizes into sustainable transformation. Posterior to prior discussion of mentioned trilogy lens, this paper provides sustainable framework result to achieve the objective of this paper. Implicitly, this paper elaborates the significant discussion that relates Industry X.0 and Sustainability.

## 1. Introduction

Industry X.0 orchestrates emerging, connected and smart technologies to digitally transform industry; as this term is coined by Accenture. Precisely, Industry X.0, according to Accenture, is defined as approach that uses advanced technologies to reinvent products and services from design and engineering to manufacturing and support, accelerating operational efficiency and enterprise-wide growth.

Accenture indicates the important aspects of this Industry X.0. Precisely, these aspects convey the facts that merely 13 percent of businesses have been aware of the digital investment full impact. The aforementioned impacts are beneficial to conduct costs savings and to generate revenue growth. To some extent, the technologies involvement could generate corporate savings to the level of $16 billion.

Furthermore, nowadays corporate requires adjustment to the next new normal within business constellation. The current value chain constellates surroundings hyper-personalized experiences, products and services driven. The primary trigger for that constellation refers to innovative business models gear toward the new origins of revenue within blue ocean strategy.

## 2. Materials

Industry X.0 is identical with the escalating modern life that involves technologies within digital transformation. Modern intellectual structures define digital transformation as the orchestrated part of the modern worldwide economy within its rational resource management [1].

To some extent, the business constellations optimize and revamp its structural [2] and [3]. As its consequences, there are dynamic fluctuations in the complication of the technological processes. These processes accelerate its innovation speed through the supply chain added values [4] and [5].

Digital transformation in this digital era through Industry X.0 has conveyed it's the new normal within worldwide industries through its ecosystems [6] and [7]. Industry X.0 encompasses spectrum of not only product but also service sector. Similar to the product sector, Banking is one of service sectors and banking experiences unprecedented digitalization speed within its revolutionary information systems for financial services [8]. There are equilibrium changes in the banking mechanism in which it is possible to simultaneously proceed to cost reduction and augment financial services productivity. Simultaneously, tangible accumulation provide bubble phenomenon that triggers policy agitation toward regulation, control, supervision within banking financial sectors market.

The ubiquitous transformation of Industry X.0 is equivalent to the development of Industry 4.0 in each country. This development is illustrated in its geographical distribution of the aforementioned publication in Figure 1.

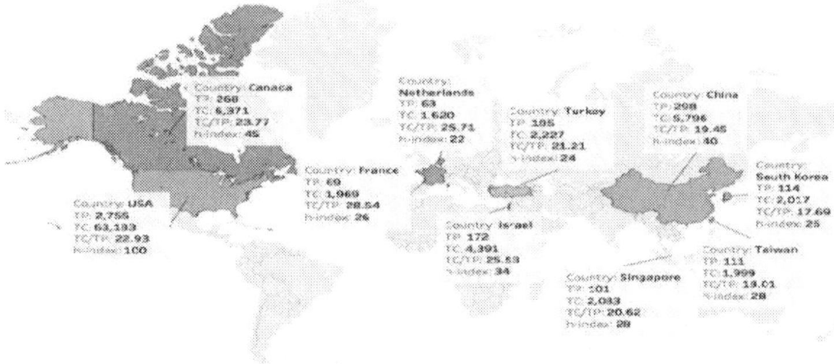


**Figure 1.** Geographical distribution in each country for Industry 4.0 publications

## 3. Methods

The objective of this paper is to conduct balance research in term of theoretical and industrial perspectives within current and the most updated of real case implementation. The methodology in this paper orchestrates both research methodology and design methodology of quantitative approaches. The research methodology in this paper refers to the quantitative method in order to achieve the objective of this paper in term of multidisciplinary lens. Those lens refer to Lens of Industrial Engineering, Product Design Engineering and Business Engineering Perspectives.

Subsequently, the quantitative method is intertwined with analysis to orchestrated theoretical implementation and its industrial implementation [9], [10], [11], [12]. Both theoretical discourses and empirical implementations provide the cutting edge discussion in this paper toward the Industry X.0, Sustainability and Digital Transformation Perspectives.

## 4. Results and Discussion

The discussion lens of industrial engineering focuses on the transformation within manufacturing and production. Furthermore, the discussion lens of product design engineering refers to the transformation within design and product engineering. Subsequently, discussion lens of business engineering plays vital role on the transformation within innovation and growth; and support and services.

In addition to the importance factors of digital transformation within Industry X.0, Corporate Top Management put high priority for sustainable competitive advantages as part of their decision making process [13]. The chronological aspect of micro foundation of dynamic capabilities for Industry X.0 can be seen in Table 1.

**Table 1.** Chronological aspect of Micro Foundation of Dynamic Capabilities for Industry X.0

| **Focal Micro-Foundation** | **Sensing** | **Seizing** | **Reconfiguring** |
|---|---|---|---|
| **Approach to History** | **Objective** | **Interpretive** | **Imaginative** |
| **Historical Competence** | **Empirical**<br>Ability to use objective micro-historical data to:<br>• Map the historical evolution of markets & technologies<br>• Trace the invisible thread of the grammar of technology | **Rhetorical**<br>Ability to use narrative history to:<br>• Mobilize commitment of internal and external stakeholders to a shared past and identification with a common mnemonic community | **Future Perfect Thinking**<br>Ability to use imaginary history to:<br>• Manage perceptions of change and continuity, and<br>• Create a vision of a shared utopian future |
| **Organizational Competence** | **Organizational Memory** | **Rhetorical History** | **Prospective History** |
| **Temporal Orientation** | **Past** | **Present** | **Future** |

The aforementioned sustainability within Industry X.0 is indispensable to provide corporate performance along with its benefit presence for society, including the concept of Society 5.0. This sustainability requires corporate's Micro Foundation of Dynamic Capabilities for Industry X.0 [14].

The Table 2 show the topic of scholar work in term of Industry 4.0 based upon its bibliometric since its since mapping in 2012, according to Ahmi and Elbardan. It is observed that the digital transformation plays vital role in the development of both Industry 4.0 and Industry X.0. The aforementioned table 3 is integral parts of prior Table 1 and 2 in term of its orchestrated publication between Industry 4.0 and Industry X.0

**Table 2.** Chronological aspect of Micro Foundation of Dynamic Capabilities for Industry X.0

| *No* | *Publication Titles* | *Reference* | *Year* | *Source of Publications* |
|---|---|---|---|---|
| 1 | Towards smart factory for industry 4.0: A self-organized multi-agent system with big data based feedback and coordination | [15] | 2016 | Automation, Quality and Testing, Computer Networks |
| 2 | Towards Industry 4.0 - Standardization as the crucial challenge for highly modular, multi-vendor production systems | [16] | 2015 | IFAC-PapersOnLine |
| 3 | Industry 4.0 - Potentials for creating smart products: Empirical research results | [17] | 2015 | Lecture Notes in Business Information Processing |
| 4 | A review of industrial wireless networks in the context of Industry 4.0 | [18] | 2017 | Wireless Networks |
| 5 | A Categorical Framework of Manufacturing for Industry 4.0 and beyond | [19] | 2016 | Procedia CIRP |
| 6 | Industry 4.0: Towards future industrial opportunities and challenges | [20] | 2016 | 2015 12th International Conference on Fuzzy Systems and Knowledge Discovery, FSKD 2015 |
| 7 | Change through digitization—value creation in the age of industry 4.0 (Book Chapter) | [21] | 2015 | Management of Permanent Change |
| 8 | Industry 4.0: A survey on technologies, applications and open research issues | [22] | 2017 | Journal of Industrial Information Integration |
| 9 | A dynamic model and an algorithm for short-term supply chain scheduling in the smart factory industry 4.0 | [23] | 2016 | International Journal of Production Research |
| 10 | Lean Automation enabled by Industry 4.0 Technologies | [24] | 2015 | IFAC-PapersOnLine |
| 11 | Understanding the implications of digitisation and automation in the context of Industry 4.0: A triangulation approach and elements of a research agenda for the construction industry | [25] | 2016 | Computers in Industry |

## 5. Conclusion

Industry X.0 orchestrates emerging, connected and smart technologies to digitally transform industry; as this term is coined by Accenture. Precisely, Industry X.0, according to Accenture, is defined as approach that uses advanced technologies to reinvent products and services from design and engineering to manufacturing and support, accelerating operational efficiency and enterprise-wide growth.

The objective of this paper is to conduct balance research in term of theoretical and industrial perspectives within current and the most updated of real case implementation. The methodology in this paper orchestrates both research methodology and design methodology of quantitative approaches.

Industry X.0 is identical with the escalating modern life that involves technologies within digital transformation. Modern intellectual structures define digital transformation as the orchestrated part of the modern worldwide economy within its rational resource management.

To some extent, the business constellations optimize and revamp its structural. As its consequences, there are dynamic fluctuations in the complication of the technological processes. These processes accelerate its innovation speed through the supply chain added values.

In addition to the importance factors of digital transformation within Industry X.0, Corporate Top Management put high priority for sustainable competitive advantages as part of their decision making process.

The aforementioned sustainability within Industry X.0 is indispensable to provide corporate performance along with its benefit presence for society, including the concept of Society 5.0. This sustainability requires corporate's Micro Foundation of Dynamic Capabilities for Industry X.0.

## References

[1] Antikainen M, Uusitalo T and Kivikytö-Reponen P 2018 Digitalisation as an enabler of circular economy *Procedia CIRP* **73** 45-49.

[2] Rachinger M, Rauter R, Müller C, Vorraber W and Schirgi E 2018 Digitalization and its influence on business model innovation *Journal of Manufacturing Technology Management* 30 (8) 1143-1160

[3] Heavin C and Power D J 2018 Challenges for digital transformation–towards a conceptual decision support guide for managers *Journal of Decision Systems* **27** 38-45.

[4] Latos B A, Harlacher M, Burgert F, Nitsch V, Przybysz P and Niewohner S M 2018 Complexity Driversin Digitalized Work Systems: Implications for Cooperative Formsof Work *Advances in Science, Technology and Engineering Systems Journal* 3 (5), 171-185.

[5] Srai J S and Lorentz H 2019 Developing design principles for the digitalisation of purchasing and supply management *Journal of Purchasing and Supply Management* **25** (1) 78-98

[6] Neubert M 2018 The impact of digitalization on the speed of internationalization of lean global startups *Technology Innovation Management Review* **8** (5) 44-54

[7] Alcácer V and Cruz-Machado V 2019 Scanning the industry 4.0: A literature review on technologies for manufacturing systems *Engineering Science and Technology, an International Journal* 22 (3) 889-919

[8] Carbó-Valverde S 2017 The impact on digitalization on banking and financial stability. *Journal of Financial Management, Markets and Institutions* **1** 133-140

[9] Basias N and Pollais Y 2018 Quantitative and Qualitative Research in Business & Technology: Justifying a Suitable Research Methodology *Review of Integrative Business and Economics Research* **7** 91-105
[10] Goertz G and Mahoney J 2012 *A Tale of Two Cultures: Qualitative and Quantitative Research in the Social Sciences* Berlin: Princeton University Press.
[11] Martin W and Bridgmon K 2012 *Quantitative and Statistical Research Methods: From Hypothesis to Results* New York: John Wiley & Sons
[12] Yin R 2011 *Applications of Case Study Research* New York: SAGE Publications
[13] Kagermann, H. 2014 *Chancen von Industrie 4.0* Deutschland: Springer Vieweg.
[14] Heavin C and Power D J 2018 Challenges for digital transformation–towards a conceptual decision support guide for managers *Journal of Decision Systems,* **27** (1) 38-45
[15] Basias, Wang S, Wan J, Zhang D, Li D and Zhang C 2018 Towards smart factory for industry 4.0: A self-organized multi-agent system with big data based feedback and coordination *Comput. Networks* **101** 158–168
[16] Weyer S, Schmitt M, Ohmer M and Gorecky D 2015 Towards Industry 4.0 - Standardization as the crucial challenge for highly modular, multi-vendor production systems *IFAC-Papers OnLine* **48** (3) 579–584
[17] Schmidt R, Möhring M, Härting R-C, Reichstein C, Neumaier P and Jozinović P 2015 Industry 4.0 - Potentials for creating smart products: Empirical research results *18th International Conference on Business Information Systems* 16–27 2015
[18] Li X, Li D, Wan J, Vasilakos A V, Lai C F, and Wang S 2017 A review of industrial wireless networks in the context of Industry 4.0 *Wirel. Networks* **23** (1) 23–41
[19] Qin J, Liu Y and Grosvenor R A 2016 Categorical Framework of Manufacturing for Industry 4.0 and beyond *6th International Conference on Changeable, Agile, Reconfigurable and Virtual Production* 173–178
[20] Zhou K, Liu T and Zhou L 2016 Industry 4.0: Towards future industrial opportunities and challenges *12th International Conference on Fuzzy Systems and Knowledge Discovery* **2** 2147–2152
[21] Kagermann H 2015 Change through digitization—value creation in the age of industry 4.0 *Management of Permanent Change* Germany: Springer
[22] Lu Y 2017 Industry 4.0: A survey on technologies, applications and open research issues *J.Ind. Inf. Integr.*, **6** 1–10
[23] Ivanov D, Dolgui A, Sokolov B, Werner F and Ivanova M 2016 A dynamic model and an algorithm for short-term supply chain scheduling in the smart factory industry 4.0 *Int. J. Prod. Res.* **54** 386–402
[24] Kolberg D and Zühlke D 2015 Lean Automation enabled by Industry 4.0 Technologies *IFAC-PapersOnLine* **48** 1870–1875
[25] Oesterreich T D and Teuteberg F 2016 Understanding the implications of digitisation and automation in the context of Industry 4.0: A triangulation approach and elements of a research agenda for the construction industry *Comput. Ind.* **83** 121–139

IOP Conf. Series: Materials Science and Engineering 1115 (2021) 012035 doi:10.1088/1757-899X/1115/1/012035

# HPLC Method for Simultaneous Determination of Dextromethorphan Hydrobromide, Chlorpheniramine Maleate and Potassium Sorbate in Cough Syrup

**T Yuliana[1,*], S S N Gustin[2], A Alamsyah[2], S Budiman[1], A Hardian[1], Y F Yun[1] and M Agma[1]**

[1]Department of Chemistry, Faculty of Sciences and Informatics, Universitas Jenderal Achmad Yani, Jl. Ters. Jend. Sudirman, Cimahi, Jawa Barat 40531, Indonesia
[2]PT. Combiphar, Jl. Ry. Caringin No. 383 Kertajaya Padalarang, KBB, Jawa Barat 40553, Indonesia

*Corresponding author: trisyuliana@gmail.com

**Abstract.** The development and validation of an analytical method for the simultaneous determination of Dextromethorphan Hydrobromide, Chlorpheniramine Maleate, and Potassium Sorbate in cough syrup using the High-Performance Liquid Chromatography (HPLC) method have been conducted. The determination was carried out on the simulation sample consisting of three substances and prepared five different concentrations, with three replications for each concentration. The optimum HPLC conditions were obtained using the C18 column, with UV detection at 230 nm. The mobile phase of acetonitrile & phosphate buffer pH 2.5 (containing sodium-heptane sulfonate as ion-pair), with a gradient elution system, at a 1 mL/minute, while the flow rate column temperature is set at 35oC. The results showed that the method has met the parameters set in the validation test with the recovery ranged from 98% - 102%, the precision test results (RSD) < 2%, and the linearity (r) ≥ 0.98. The method obtained is quite selective by confirmed identity with the maximum wavelength scanning of each analyte peak by PDA detector. The limit of detection (LoD) for Dextromethorphan-HBr, Chlorpheniramine maleate and Potassium sorbate were 3.13 µg/mL, 0.94 µg/mL and 0.12 µg/mL respectively and the limit of quantitation (LoQ) were 9.49 µg/mL, 2.85 µg/mL and 0.38 µg/mL respectively.

## 1. Introduction

Dextromethorphan-HBr (DTM) is chemically (+)-3-methoxy-17-methylmorphinan hydrobromide. The molecular formula is $C_{18}H_{25}NO.HBr$ and molecular weight is 370.3 g/mol. DTM is an antitussive cough medicine used to suppress coughs caused by irritation of the throat and bronchial airways. Chlorpheniramine maleate (CTM) is chemically (3-(4-Chlorophenyl)-N,N-dimethyl-3-pyridine-2-ylpropan-1-aminium (2Z)-3-carboxyprop-2-enoate). The molecular formula is $C_{16}H_{19}ClN_2.C_4H_4O_4$, and the molecular weight is 390.87 g/mol. CTM is an antihistamine that functions to treat hypersensitivity reactions. Potassium sorbate (PS) is chemically Potassium (2E,4E)-Hexa-2,4-dienoate. The molecular formula is C6H7KO2, and the molecular weight is 150,22 g/mol [1]. PS acts as a medicinal preservative, namely as an anti-bacterial and antifungal. The use of this preservative

IOP Conf. Series: Materials Science and Engineering 1115 (2021) 012035 doi:10.1088/1757-899X/1115/1/012035

must be regulated and controlled so that it does not cause harm such as poisoning and long-term use is a carcinogen [2]

Many methods have been described in the literature to determine DTM, CTM, and PS individually and in combination [3-7]. However, no HPLC method was reported for the simultaneous determination of these drugs in combined dosage forms. Fixed-dose combination containing DTM (5.0 mg), CTM (2.0 mg), and PS (10.0 mg) in the cough syrup and the HPLC method was validated following ICH Guidelines [8-9]. This study aimed to analyze the three substances simultaneously under the same conditions to simplify and shorten analysis time.

## 2. Methodology

### 2.1. Material and Reagents

Pure DTM from USP reference standard, pure CTM from Suriya Life Science, and PS from Celanese, Germany [10]. The cough syrup sample simulation consists DTM (5.0 mg), CTM (2.0 mg) & PS (10.0 mg) dissolved in sample matrix cough syrup. The reagents used were HPLC grade, KH2PO4, a reagent for ion pair chromatography (sodium 1-heptane sulfonate) from TCI, High purity aqua DM was prepared by using Millipore. Mobile phase and working solutions were filtered through 0.45 µm and degassed using a sonicator before use.

### 2.2. Apparatus

An HPLC system (Shimadzu LC-20AD), with LC solutions data handling system (Shimadzu LC Solution) and an autosampler (SIL-20AC), was used for the analysis. The detector consists of a photodiode array (SPD-M20A)). The column used was C-18, Inertsil ODS3, (15 cm x 4.6 mm i.d. x 5 µm) with UV detection at 230 nm.

### 2.3. Preparation & Standard stock solution

A combined standard stock solution containing DMP, CTM & PS were prepared by 50.0 mg, 20.0 mg, and 10 mg, respectively. All of them are transferred to a 50.0 ml volumetric flask and add the mixed solution of phosphate buffer pH 2,5: methanol (90:10 v/v) and sonicated in a bath sonicator for 15 minutes. After sonication, the volume was made up to the mark 50 mL with the same mixed solution. A solution was made five times dilutions for the working standard to obtain the final concentrations of DTM, CTM, and PS 200, 80, and 40 µg/mL, respectively.

### 2.4. Preparation sample solution

First, we weigh 200.0 mg DTM and 80.0 mg CTM in a 50 mL volumetric flask, then dissolved in a phosphate buffer: methanol mixture (90:10 v/v) up to the limit marked (solution A) and weighed separately 400 mg KS in a 50 mL volumetric flask then dissolved in the same mixed solution to the limit marked (solution B). We weighed 10 x density cough syrup matrix into 5 of 50mL volumetric flasks to make a simulation sample solution. Subsequently added 7, 9, 10, 11 and 13 mL solution A and 4, 9, 10, 11, 13 mL solution B, respectively. Then diluted with a mixed solution of phosphate buffer pH 2.5: methanol (90:10 v/v) and after sonicated for 15 minutes, volume was made up to 50 mL by the same mixed solution. For PS determination, it is necessary to make a 10x dilution of each simulated solution.

# 3. Results and Discussion

## 3.1 Method Development

Firstly, we tested several mobile phase compositions for effective separation of dextromethorphan hydrobromide, chlorpheniramine maleate, and potassium sorbate. The mobile phase containing acetonitrile and phosphate buffer pH 2.5 (containing sodium-heptane sulfonate as ion-pair) (5:95, v/v) with gradient elution system up to 30% acetonitrile in 15minute. The flow rate of 1 mL/min, the injection volume of 20 μL, the column temperature 35°C, and a detection wavelength of 230 nm afforded the best separation of these analytes. (see Figure 1 and Table 1)

## 3.2. Validation Method

### *3.2.1. Linearity, Limit of Detection, and Limit of Quantification*

To evaluate the linearity of the method, different concentrations of the three analytes in the range of 140.0–260.0 μg/mL for DTM, 56.0–104.0 μg/mL for CTM, 16.0–52.0 μg/mL for PS were analyzed, and the linearity between the peak-area and the concentration was examined for each analyte. The results obtained show that the current method is linear for the three analytes in the range specified above with a correlation coefficient of better than 0.99, see Table 2.

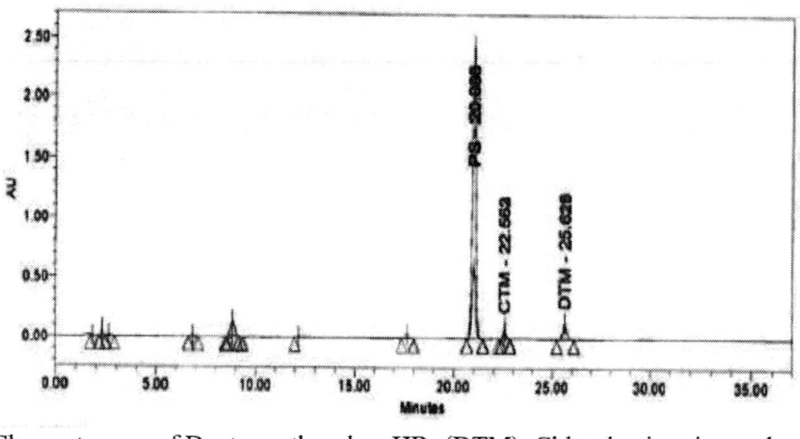


**Figure 1.** Chromatogram of Dextromethorphan HBr (DTM), Chlorpheniramine maleate (CTM) and Potassium Sorbate (PS) in cough syrup formulation

**Table 1.** Chromatographic parameters of the separated analytes

| Analyte | Retention time (minute) | Resolution | Asymmetry |
|---|---|---|---|
| Dextromethorphan HBr | 25.622 | 9.76 | 1.20 |
| Chlorpheniramine maleate | 22.557 | 5.37 | 1.09 |
| Potassium Sorbate | 20.982 | 31.12 | 1.06 |

*3.2.2 Accuracy (%Recovery)*

The method's accuracy was studied by preparing a sample matrix of the syrup cough formulation according to the formulation procedure. The required quantity of placebo, a known amount of the three active ingredients (DTM, CTM & PS) with the same proportion as in the cough syrup formulation, was added to get simulated drug formulation. Results have shown that the assay's mean recovery is (100.64 ± 0.63)% for three ingredients, and the RSD is lower than 1.0%, see Table 3.

**Table 2.** Linearity, LoD, and LoQ

| Analyte | Equation | Coef. Correl (r) | LoD (μg/mL) | LoQ (μg/mL) |
|---|---|---|---|---|
| Dextromethorphan HBr | Y = 19091X + 105144 | 0.9998 | 3.13 | 9.49 |
| Chlorpheniramine maleate | Y = 24693X + 17506 | 0.9999 | 0.94 | 2.85 |
| Potassium Sorbate | Y = 42941X + 17782 | 1.0000 | 0.12 | 0.38 |

*3.2.3 Precision*

This method's precision was evaluated by calculating the RSD of five simulation samples with different concentration levels and three replicates for each concentration. The result show %RSD ranged is (0.05 – 0.16)%, (0.02 – 0.26)% and (0.01 – 0.20)% for DTM, CTM and PS respectively. It was found to be less than RSD repeatability by Horwitz. See Table 3. These results show that the current method is repeatable.

**Table 3.** Percentage recovery & precision data of DTM, CTM, and PS in cough syrup formulation

| **Analyte** | **%Recovery (n = 5)** | **% RSD (n = 5)** |
|---|---|---|
| Dextromethorphan HBr | 101.3 ± 0.7 | 0.10 |
| Chlorpheniramine maleate | 100.1 ± 0.3 | 0.13 |
| Potassium Sorbate | 100.5 ± 0.9 | 0.11 |

*3.2.4 Selectivity*

Analytical selectivity relates to the extent to which the method can be used to determine particular analytes in mixtures or matrices without interferences from other components of similar behavior. On the other hand, an analytical approach is selective if only one component can be determined independently from all the different elements that give no analytical signal (9-10). The current method's selectivity was demonstrated by the excellent separation of the three analytes. (See Figure 1 and confirmed by scanning the wavelength for each analyte as shown in the Figure. 2, 3, and 4). The method's selectivity is strengthened in the absence of other compounds' coelution, which is indicated by the value of the purity angle less than the purity threshold and the resolution value greater than 1.5 as shown in Table 4.

IOP Conf. Series: Materials Science and Engineering 1115 (2021) 012035 doi:10.1088/1757-899X/1115/1/012035

**Table 4.** Comparing purity Angle vs. Purity Threshold

| Analyte | Resolution | Purity Angle | Purity Threshold |
|---|---|---|---|
| DTM standard | 11.11 | 0.205 | 0.232 |
| DTM in sample | 4.39 | 1.261 | 1.522 |
| CTM standard | 5.50 | 0.211 | 0.226 |
| CTM in sample | 5.78 | 0.190 | 0.566 |
| PS standard | 39.64 | 0.065 | 0.250 |
| PS in sample | 11.43 | 0.110 | 0.249 |

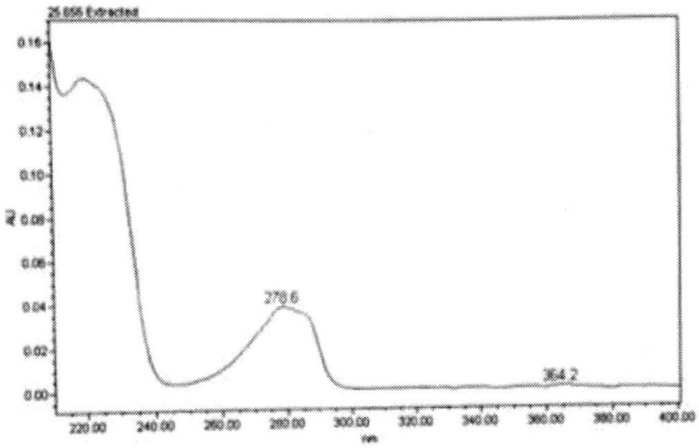


**Figure 2.** Dextromethorphan HBr spectrum

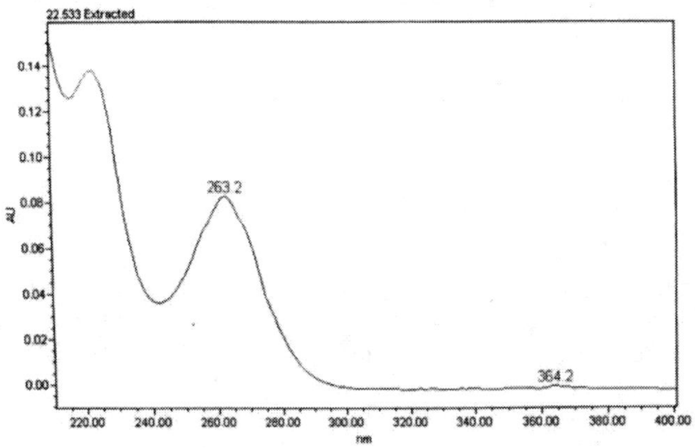


**Figure 3.** Chlorpheniramine maleate spectrum

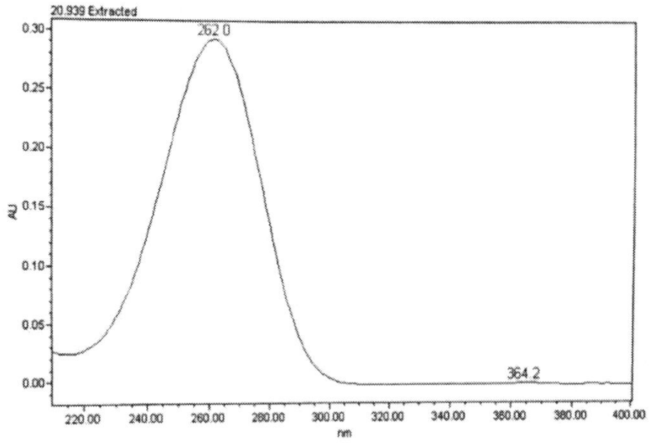


**Figure 4.** Potassium sorbate spectrum

*3.2.5 Robustness*

The robustness test of the analytical method measures its capacity to remain unaffected by small but deliberate method parameters and indicates its reliability during normal usage. Robustness of the current approach was investigated by analyzing samples of the cough syrup using the same chromatographic conditions specified in method development but with a small change in the following chromatographic parameters: (a) pH of phosphate buffer: 2.0 and 3.0 instead of 2.5, (b) detection wavelength: 228 nm and 232 nm instead of 230 nm, (c) column temperature: 30oC and 40oC instead of 35oC. All variation parameters three-time repeatedly. RSD of dextromethorphan HBr, chlorpheniramine maleate, and potassium sorbate assay under these conditions is calculated to be less than 1%, except phosphate buffer pH 2, which is obtained > 2% (see Table 5).

**Tabel 5.** Robustness testing

| Parameter | | Assay | | |
|---|---|---|---|---|
| | | Dextromethorphan HBr | Chlorpheniramine maleate | Potassium sorbate |
| pH phosphate buffer | 2.0 | 100.91 ± 0.07 | 88.44 ± 4.58 | 99.91 ± 0.92 |
| | 2.5 | 100.73 ± 0.06 | 98.92 ± 0.17 | 99.83 ± 0.11 |
| | 3.0 | 100.56 ± 0.05 | 98.28 ± 0.07 | 99.63 ± 0.13 |
| Wavelength (nm) | 228 | 100.01 ± 0.16 | 98.94 ± 0.10 | 100.05 ± 0.10 |
| | 230 | 100.91 ± 0.06 | 99.03 ± 0.05 | 100.18 ± 0.02 |
| | 232 | 100.32 ± 0.12 | 99.00 ± 0.05 | 100.35 ± 0.25 |
| Column temperature (°C) | 30 | 99.08 ± 0.22 | 98.72 ± 0.18 | 99.95 ± 0.14 |
| | 35 | 100.73 ± 0.18 | 98.92 ± 0.02 | 100.26 ± 0.14 |
| | 40 | 99.45 ± 0.15 | 98.37 ± 0.16 | 100.27 ± 0.15 |

## 4. Conclusion

A simple, accurate, precise HPLC method is developed and validated for simultaneous determination of dextromethorphan-HBr, chlorpheniramine maleate, and potassium sorbate simultaneously in cough syrup formulations as long as the pH more than 2.

## Acknowledgments

We would like to thank Combiphar PT gratefully for their support and for providing the necessary instruments/apparatus to perform this study.

## References

[1] Moffat A C, Osselton M D, Widdop B and Watts J 2011 *Clarke's analysis of drugs and poisons* London: Pharmaceutical press
[2] Colaco, Ramesh S, Rajan N S and Meyyanathan S 2017 Pharmacokinetic Evaluation of Once-Daily Sustained Release Dextromethorphan Hydrobromide (DM) and Immediate Release DM in Healthy Human Volunteers. *World Journal of Pharmaceutical Research* 7 (1) 879-890
[3] Rauha J P, Salomies H and Aalto M 1996 Simultaneous Determination of Bromhexine Hydrochloride and Methyl and Propyl P-Hydroxybenzoate and Determination of Dextromethorphan Hydrochloride in Cough-Cold Syrup by High-Performance Liquid Chromatography *Journal Pharmaceutical Biomedical Analysis* **15** (2) 287-293
[4] Sharma V J M C 2016 Validated RP-HPLC Method for determination of Bromhexine HCl, Chlorpheniramine Maleate, Dextromethorphan HBr and Guaiphenesin in Pharmaceutical Dosage Forms *Journal of Taibah University for Science* **10** 38-45
[5] Caglar H and Buyuktuncel S E 2014 HPLC method development and validation: simultaneous determination of active ingredients in cough and cold pharmaceuticals *International journal of pharmacy and pharmaceutical sciences* **6** (10) 421-428
[6] Byrnea J, Velasco T and Reinhardta R 2014 Development and validation of a novel stability-indicating HPLC method for the simultaneous assay of betamethasone-17-valerate, fusidic acid, potassium sorbate, methylparaben and propylparaben ina topical cream preparation, *Journal of Pharmaceutical and Biomedical Analysis* **96** 111 – 117
[7] Pylypiw H M and Grether M T 2000 Rapid high-performance liquid chromatography method for the analysis of sodium benzoate and potassium sorbate in foods *Journal of Chromatography A*. **883** (1-2) 299–304
[8] Bertil M and Örnemark U 2014 *The fitness for purpose of analytical methods* UK: A laboratory guide to method validation and related topics
[9] Danzer K 2001 Selectivity and specificity in analytical chemistry. General considerations and attempt of a definition and quantification *Fresenius Journal Analytical Chemistry* **369** 397 – 402
[10] Englert C, Brendel J C, Majdanski T C, Yildirim T, Schubert S, Gottschaldt M, Windhab N and Schubert U S 2018 Pharmapolymers in the 21st century: Synthetic polymers in drug delivery applications. Progress in Polymer Science **87** 107-164

IOP Conf. Series: Materials Science and Engineering 1115 (2021) 012036 doi:10.1088/1757-899X/1115/1/012036

# Implementation transmittal solutions at capital project to increase the effectiveness

**A Herdiansah[1,*], D Nurnaningsih[1], Y Sugiyani[1] and Handayani[2]**

[1]Informatics Engineering Department, University of Muhammadiyah Tangerang, Indonesia
[2]Informatics Department, University of Indraprasta PGRI, Indonesia

*Corresponding author: arief_herdiansah@umt.ac.id

**Abstract.** The capital project is a construction project which has a large scale of documentation. Transmittal solutions as part of Enterprise Document Management Solution (EDMS) could be offered as the solution of document management inefficiency in the capital projects. EDMS processed the document content in a more detail way and had simple integration with another internal system. EDMS used for the implementation of the transmittal process during document distribution in every tier to improve the system performance to be faster based on the design configuration. This study focused on the implementation of transmittal solutions in capital projects.

## 1. Introduction

The capital project is a project planning, engineering, procurement, construction and operation of a petroleum refinery building, plant, facility and infrastructure [1]-[5]. Organizations who are engaged in the design, construction, and petroleum mining industries, aside from facing increasingly fierce competition, they must also be able to increase HSE (Health, Safety, and Environmental) supervision. In the six life cycle phases of the facility (Feasibility, Design, Construction, Operations, Renewal and Deactivation) such as petroleum refining, significant content and processes are shared between individuals and internal and external organizations. As a result, the organization faces several critical information problems. Information is often lost between Project phases, which delays completion and increases overall Project costs because of the lack of integration between specially created systems, thereby increasing IT costs. The risks and costs of non-compliance with HSE regulations are also very high.

Implementation of Transmittal solution is part of the EDMS (Enterprise Document Management System) solution that obliged to ensure compliance with applicable oil and gas industry regulations to avoid penalties from the government [6]-[7]. This solution provides safe and efficient access to AFC and Enterprise and Project-wide "As-Built" engineering drawings, as well as across Company boundaries, streamlining supplier's interactions with internal Business Processes that rely on efficient access to Engineering content. As part of EDMS solutions, transparent solutions can enable interoperability between EDMS and other systems such as AutoCAD, Factory Maintenance and SAP.

Essentially the Transmittal process starts when the part that has the responsibility of controlling project documents (Document Controller) registers a new Incoming Transmittal and automatically bulk-loads the capital project documents sent by the Owner/Operators, Sub-Contractors, and the like. This is performed against the Transmittals Coversheet/Index the Owner/Operator, for example, sends. Documents associated with a Transmittal can be added directly from the EDM Repository (where relevant). This process is carried out on the Transmission Delivery Sheet/Index the Owner/Operator, for example, sends. EDMS permit the knowledge and the monitoring of the value of information and to decide the receiver (acceptor) and to manage the transmission of those data.

This research is an observation and analysis of the process of implementing a digital transmittal document solution in one company that has a capital project.

## 2. Literature study

### 2.1. The capital projects

Capital projects in the process industries involve the construction of physical plant facilities and materials processing equipment to produce a new product for expected profit or alternatively to maintain or develop operating-level capabilities. The construction industry involves diverse and complex information that flows between the various participants, much of which is conveyed using documents.

### 2.2. Document controller

The document controller is an employee of the project owner, lead designer, or lead contractor organization, or may work as an independent consultant/contractor. The document controller is responsible for reporting to the owner's project representative and work with an information management committee consisting of project managers and information specialists from key project participants.

### 2.3. Transmittal solution

Registration of messages transmission and incoming/outgoing messages within the enterprise provided by the electronic document management system has grown in organizations. Transmission of e-mail messages, logging incoming/outgoing, as well as documents created by the user, control the movement of documents within the organization and storing information in databases. It is possible to consider with a high level of confidence that the transmitted data is authentic.

### 2.4. Document management system

Document management responsible for the efficient and systematic control of the creation, receipt, maintenance, use and disposition of records, including processes for capturing and maintaining evidence of and information about business activities and transactions in the form of records. Most of the documents are pdf files according to ICH. Other formats are possible: xml, gif, jpeg, etc.

### 2.5. EDMS

EDMS contains enhanced features related to the life-cycle and versioning of particular classes of documents [8]-[10]. EDMS has multiple roles of storage, archiving, management, approbation and flow-control of Integrated Management System documents to facilitate organizational workflows.

# 3. Methodology

## 3.1. Gathering user requirement: user the initial stage of the development of transmittal solution

The problem with a capital project is how information can be shared with each team involved, taking into account the effectiveness, efficiency and security of the information data. Information that needs to be distributed includes asset images, specifications, calculations and other information, usually for review, commentary, or approval purposes.

**Table 1.** User requirement

| Functionalities | Non Functionalities |
|---|---|
| Auto number generation when uploaded document loading file. | Could have interfaced with another system |
| Manage all common file formats ex: Autocad | The data location is not allowed to store on the cloud |
| Transmittal process, i.e:<br>a. Manage, create, send, receive, and view transmittals.<br>b. Apply distribution matrices.<br>c. Acknowledge receipt capture.<br>d. Hyperlinks to items on a transmittal from the transmittal document | Usability, i.e:<br>Provide an intuitive, friendly GUI, Microsoft 'Look and Feel' |
| Distribution lists, i.e:<br>a. Create and maintain distribution lists.<br>b. Security to distribution lists | Scalability, i.e:<br>Scalable to encompass larger volumes and different types of documents into the future. |
| Workflow, i.e:<br>a. Could have tasks presented in an 'inbox' or 'work area'.<br>b. Monitor the progress/status of a workflow.<br>c. Send bulk documents on workflow and approve bulk workflows.<br>d. Email notification when the workflow stopped. | Performance, i.e:<br>Could be accessible from all geographic locations with acceptable response time. |
| Mark-up & commenting, i.e:<br>a. Review comments in real-time.<br>b. Ability to do side by side overlay compare.<br>c. Ability for multiple user to edit a document in a native format and add review comments. | Security |
| Searching, indexing & retrieval, i.e:<br>a. Search document based on its content, metadata, and advanced criteria.<br>b. Search for textual data using a full-text engine.<br>c. Search across MS.Office documents, CAS drawings, PDF Files.<br>d. Ability to save and share searches. | Security |

The initial stage of the transmittal solution implementation process is conducting an assessment relating to the features needed by the user in the process of distributing the project documents undertaken. It is discuss related how the features of a transmittal solution can be according to user needs and still be able to maintain good data security standards.

From the results of the assessment conducted, the user requires several functions available in a transmittal solution, namely functional and non-functional as shown in Table 1.

### 3.2. Transmittal solution design: the speed increasing of transmittal delivery

To be able to provide increased delivery of correspondence and information to users who need it, design an infrastructure architecture as a basic EDMS deployment that typically consists of the following tiers:

a) Client Tier: Browser/Client Machine
b) Application Tier: Application Server
c) Content Tier: Content Server
d) Service Tier: LDAP System, 3rd Party System, Library/Workflow Services, SMTP System
e) Storage Tier: Storage Server, Database Server, Full Text Index Server.

The transmittal solution high-level design application can be described in Figure 1.

| Transmittal Solution | | | | |
|---|---|---|---|---|
| Transmittal | Review & Approval | Progress Tracking | Bulk Loading | Distribution Matrics |
| Numbering & Versioning | Export Library | Project Configuration | Search & Reporting | Role Security |

**Figure 1.** Transmittal solution high level design

Transmittal solution provide the standard basic feature as shown on the image above. If the customization required, solution provide the SDK file which can be used by the solution administrators/developers to start customizing the system.

For the integration part, this solution provides some extensions with the 3rd party application which can be used based on customer's requirement.

Based on available information, a logical scenario is designed from a transmittal solution, the logical scenario for designing the appropriate transmittal solution is shown in Figure 2.

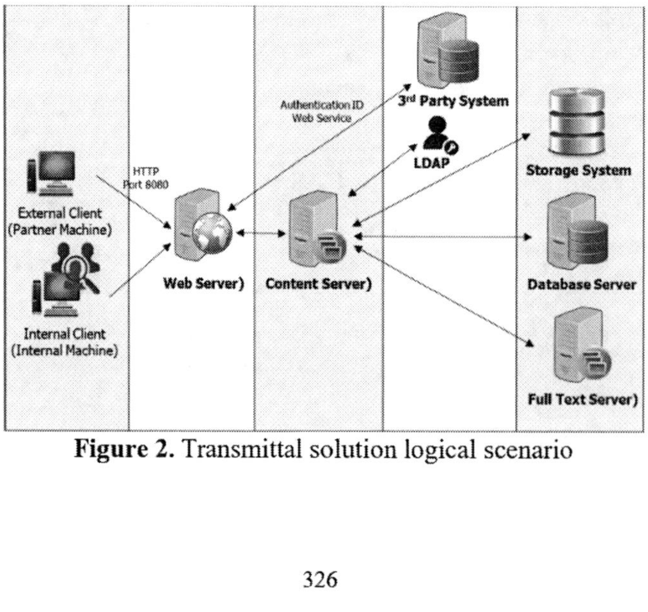


**Figure 2.** Transmittal solution logical scenario

Logical scenario description:

a) Clients will access a web server using a framework
b) Web will respond request from the client and interconnected with the content server.
c) Note: content server is a server that stores soft copies of all project documents classified by the project.
d) The content consists of a content server in which organization can store, manage, and deploy all types of content, including HTML & XML, graphic, multimedia, other types of rich media and traditional documents.
e) Service, at this level there are the application servers that provide web services interface, example integration with other software solution. This solution will be deployed on the application server (application entry point for users).
f) Storage, these solutions integrated networked storage technologies, storage systems, software and services

### 3.3. Environments Specification

To get a good performance solution based on Best Practice, we need the following environment requirements (see Table 2).

**Table 2.** Environment specification (best practice)

| Server Microsoft Windows Server 2016 | | | |
|---|---|---|---|
| Machine | CPU | RAM | HDD |
| Database Server | 12 Core | 16 GB | 250 GB |
| Content Server | 12 Core | 16 GB | 250 GB |
| Full Text Server | 8 Core | 8 GB | 250 GB |
| Web Server | 12 Core | 16 GB | 150 GB |
| Total | 44 Core | 56 GB | 900 GB |

## 4. Results and Discussion

The implementation of the distribution system in the section that handles capital projects requires all the teams involved in managing the capital project.

### 4.1. Use Case Transmittal Solution

Figure 3 is the use case diagram for general system functionality in transmittal solution.

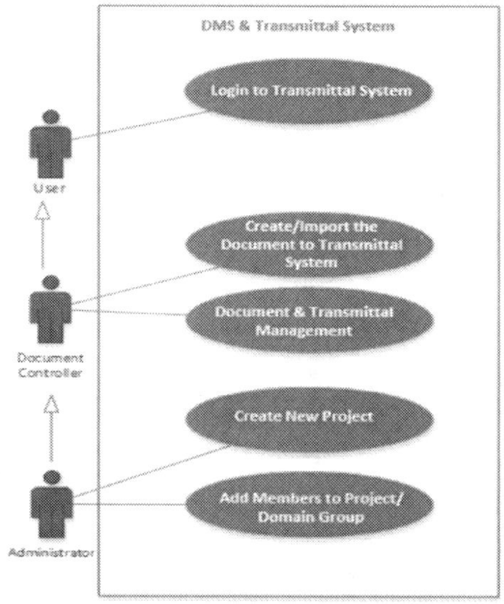


**Figure 3.** Use case general system transmittal solution

The following is an explanation of the DMS & Transmittal Systems use case:

*4.1.1. Login*

**Table 3.** Login to transmittal system group use case description

| **Name of Use Case:** | **Login to Transmittal System** |
|---|---|
| **Created By:** | **Herdiansah & Team** |
| **Date Created:** | **21-Jan-2020** |
| Description: | User who has privileged logs in to the system |
| Actors: | User, System |
| Pre-conditions: | User who logs in to system has to be registered in the system |
| Post-conditions: | User successfully logged in to the system |
| Flow: | 1. User accesses the link to log in to the system.<br>2. Login page of the system will be displayed.<br>3. User enters name and password then clicks on "Sign In" button.<br>4. Home of the system will be displayed. |
| Alternative Flows: | - |
| Exceptions: | - |
| Requirements: | - |

*4.1.2. Create new project domain*

**Table 4.** Create new project group use case description

| **Name of Use Case:** | **Create New Project/Domain** |
|---|---|
| **Created By:** | **Herdiansah & Team** |

| | |
|---|---|
| **Date Created:** | **21-Jan-2020** |
| Description: | Administrator creates a new project/domain by 'copying' project template. |
| Actors: | Administrator, System |
| Pre-conditions: | 1. Administrator already logged in to the system.<br>2. Administrator has the capability to create new project/domain. |
| Post-conditions: | Administrator successfully creates a new project/domain. |
| Flow: | 1. Administrator selects an existing project and then selects **Tools > Actions.**<br>2. System will show a page to initiate new project.<br>3. Administrator clicks on **Initiate New Project** button and must fill out Project Reference and Project Name field. Administrator optionally enter values for other fields.<br>4. Administrator clicks on **Perform Action & Close** button.<br>5. New project folder structure will be generated automatically. |
| Alternative Flows: | - |
| Exceptions: | - |
| Requirements: | - |

*4.1.3. Add member to project/domain group*

**Table 5.** Add member to project/domain group

| **Name of Use Case:** | **Add Members to Project/Domain Group** |
|---|---|
| **Created By:** | **Herdiansah & Team** |
| **Date Created:** | **21-Jan-2020** |
| Description: | Administrator adds members to project/domain group. |
| Actors: | Administrator, System |
| Pre-conditions: | 1. Administrator already logged in to the system.<br>2. Project folder structure is successfully initiated. |
| Post-conditions: | The new members are successfully added to the new project/domain group. |
| Flow: | 1. Administrator selects Project Groups node first on the left navigation then select the project group name.<br>2. Administrator double click on the roles from the selected project group then click on **File** >> **Add Member(s)**.<br>3. System will show a page to choose the user.<br>4. Administrator chooses the user to be added then click OK button.<br>5. New member was successfully added to the group. |
| Alternative Flows: | - |
| Exceptions: | - |
| Requirements: | - |

IOP Conf. Series: Materials Science and Engineering 1115 (2021) 012036 doi:10.1088/1757-899X/1115/1/012036

#### *4.1.4. Document number generation from creating/importing document use case description*

**Table 6.** Document number generation from creating/importing document

| **Name of Use Case:** | **Document Number Generation from Creating/Importing Document** |
|---|---|
| **Created By:** | **Herdiansah & Team** |
| **Date Created:** | **21-Jan-2020** |
| Description: | The document controller creates/imports the document to the system, and the document number will be generated automatically. |
| Actors: | The document controller |
| Pre-conditions: | 1. The document controller already logged in to the system. |
| Post-conditions: | The system successfully generates the document number. |
| Flow: | 1. The document controller clicks on **File** >> **New** >> **Document**.<br>2. **New Document: Create** a webpage dialogue will be displayed.<br>3. Select a document type from the provided list. Field name (Auto numbered) will be filled automatically.<br>4. Click on **Next** button.<br>5. Fill out some of the required metadata to generate the document number following the standard document numbering.<br>6. Click on **Finish** button.<br>7. The document is successfully created in the system with the generated document number following standard document numbering. |
| Alternative Flows: | - |
| Exceptions: | - |
| Requirements: | - |

### 4.2. Administrator Module

Transmittal solution system administrators can perform several functions, including:

a. User management
b. Documents configuration
c. Document loading configuration
d. Distribution matrices configuration distribution matrices is a function of managing users access rights to documents distributed through the system. Figure 4 shows the distribution matrices module.

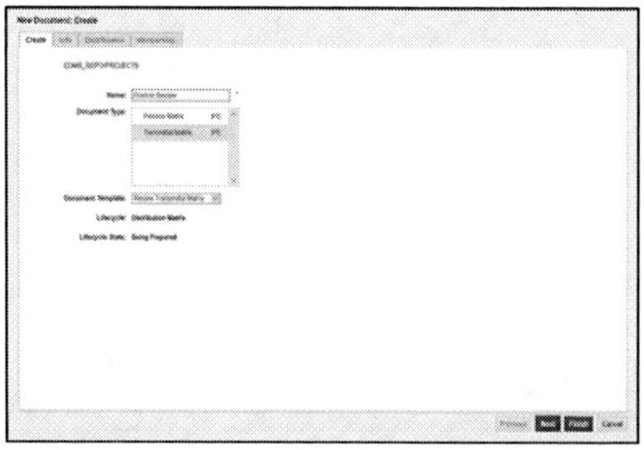

**Figure 4.** Create distribution matric menu

e. Comment sheet and cover template configuration.

### 4.3. Document Controller Module

The document management standard features and some features related to the shipping management process are explained in Table 7.

**Table 7.** Rules and responsibility document control user

| No | Role in System | Responsibility |
|---|---|---|
| 1 | Document Controller | - Distribution Matrices Management<br>- Transmittal Management<br>- Document Management<br>- Project Management |
| 2 | Contractor | - Transmittal Management<br>- Document Management |
| 3 | Reviewer | - Commenting and Mark-up Document<br>- Document Management |
| 4 | Responsible | - Consolidate Comments from Reviewers<br>- Document Management |
| 5 | Approver | - Approve Document<br>- Document Management |

Document Control is the user in charge of the initial distribution of project documents to all teams involved following the permissions of the user. Figure 5 shows the menu of a transmittal review process.

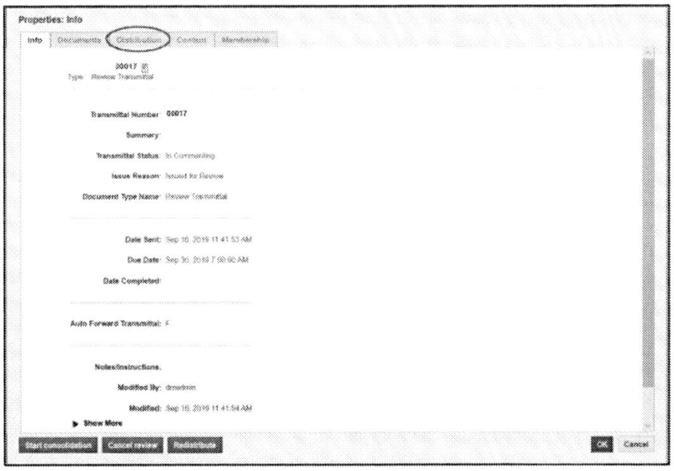


**Figure 5.** Create Transmittal Review Process

### 4.4. User Module

Transmittal system users can perform several functions ranging from reviewing project/transmittal documents, giving annotations to the reviewed documents, distributing project/transmittal documents to other users, to searching documents stored in the system. Figure 6 is a menu display for the project/transmittal document search process stored in the system.

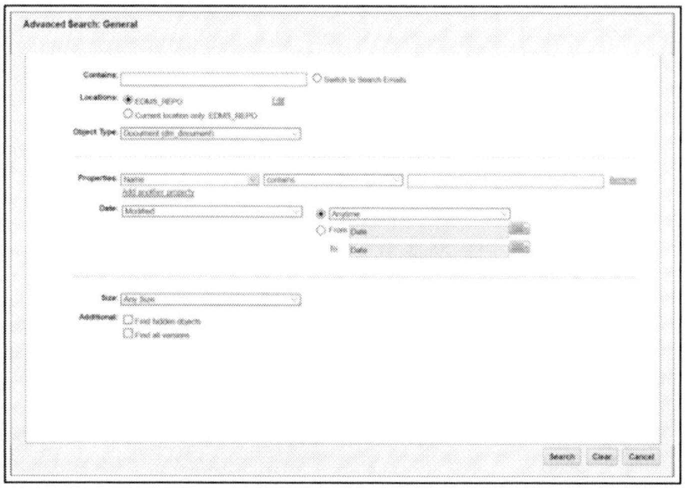


**Figure 6.** Searching module

## 5. Conclusion

This study tries to share experiences in implementing the use of transmittal solutions in a company that has a capital project. The use of this transmittal solution can also be implemented in companies that handle large-scale of projects where the involvement of users in the project team is quite large, and the process of document distribution/transmittal is quite high, thus requiring a system to regulate the traffic distribution of project documents.

The implementation using transmittal solution as EDMS deployment solution is a problem-solving solution for the capital project. The author concludes that transmittal solution has created distribution, revision process and information delivery to everyone in the team become more effective, efficient and controlled, specifically from the data security.

## References

[1] Yun S, Choi J, de Oliveira D P and Mulva S P 2016 Development of performance metrics for phase-based capital project benchmarking *International Journal of Project Management* **34** (3) 389-402

[2] Sanchez B and Haas C 2018 Capital project planning for a circular economy. *Construction Management and Economics* **36** (6) 303-312

[3] Chen, H. L. (2015). Performance measurement and the prediction of capital project failure. *International Journal of Project Management* **33** (6) 1393-1404

[4] Ahmed S M, Sang L P and Torbica Ž M 2003 Use of quality function deployment in civil engineering capital project planning *Journal of Construction Engineering and Management* **129** (4) 358-368

[5] Ness E C and Tandberg D A 2013 The determinants of state spending on higher education: How capital project funding differs from general fund appropriations *The Journal of Higher Education* **84** (3) 329-362

[6] Huang S and Meng X 2020 Research and Application of Integration Solution for Enterprise-Level Heterogeneous Document Management System *Journal of Physics: Conference Series* **1621** (1) 012109

[7] Obukhov A, Krasnyanskiy M and Nikolyukin M 2020 Algorithm of adaptation of electronic document management system based on machine learning technology *Progress in Artificial Intelligence* 1-17

[8] Cheung S C and Chiu D K 2003 A watermarking infrastructure for enterprise document management *36th Annual Hawaii International Conference on System Sciences* 1-10
[9] Tarkhanov I 2015 Policy algebra for access control in enterprise document management systems *9th International Conference on Application of Information and Communication Technologies* 225-228
[10] Maican C and Lixandroiu R 2016 A system architecture based on open source enterprise content management systems for supporting educational institutions *International Journal of Information Management* **36** (2) 207-214

# Slaughterhouse Wastewater Treatment by Electrocoagulation Process

**F A Nugroho, A Z Arif, G Z M Sabila and P T P Aryanti***

Chemical Eng. Dept., Universitas Jenderal Achmad Yani, Jl. Ters. Jend. Sudirman, Cimahi

*Corresponding author: p.teta@lecture.unjani.ac.id

**Abstract.** Slaughterhouse wastewater contains a high concentration of colloids, organic compounds, and suspended solids, which cannot be directly discharged to the environment. In this research, the electrocoagulation process was used to treat slaughterhouse wastewater. The pH of the solution was neutralized by adding $Ca(OH)_2$. The influence of current density (933 $mA/dm^2$ and 1400 $mA/dm^2$) and electrode configuration (3 anodes 3 (three) cathodes and 4 (four) anodes 2 (two) cathodes) was investigated. The experimental results showed that the treated wastewater at a current density of 1400 $mA/dm^2$ produced a lower contaminant concentration than the current density of 933 $mA/dm^2$. The current density of 1400 $mA/dm^2$ and electrode configuration of 4 anodes-2 cathodes reduced BOD by 56.4%, TDS by 20.25%, and TSS by 99.47%.

## 1. Introduction

The slaughterhouse wastewater contains a high concentration of biodegradable organic compounds (such as proteins and blood), colloidal (such as fat, oil, and grease), suspended solids (such as grit, manure, and undigested feeds), nitrogen, phosphorus, and pathogenic bacteria [1]. These contaminants lead to a high value of BOD, COD, and turbidity, which must be treated before being discharged into the environment [2]. Coagulation/flocculation [3-5], trickling filtration [6], and biological process [7-9] are standard techniques used for wastewater treatment. Among these technologies, biological treatment, either anaerobic or aerobic treatment [7, 10, 11], has been widely chosen due to its high removal efficiency to remove the biological contaminants in slaughterhouse wastewater. Some parameters, such as long hydraulic retention time, the high energy requirement for aeration, large area, and high concentration of sludge produced during the process, become a problem in extensive scale application of biological processes [12].

Electrocoagulation (EC) has recently been used as an alternative process for slaughterhouse wastewater treatment [13, 14]. The EC is an electrochemical process, which combines coagulation, adsorption, precipitation, and flotation in a one-stage procedure. This process involves two electrodes, namely anode, and cathode, which are arranged in pairs. When a direct current is applied to the electrodes, soluble metal cations are released from the anode, neutralizing the stable charge on the subsequent pollutants and destabilizing the contaminants to form flocs. Meanwhile, hydrogen gas and hydroxyl ions are released from the cathode. The hydrogen gas lifts the pollutants to the top of the solution. Then, they are collected and removed by a skimmer. The EC process's chemical reactions have been explained in kinds of literature [15, 16]. When using aluminum (Al) as an anode, $Al^{3+}$ ions are released from the anode. Besides destabilizing the charge of pollutants, the $Al^{3+}$ reacts with hydroxyl ($OH^-$) ion to form coagulants, $Al(OH)_3$, and enlarge the agglomeration of flocs.

Several studies have been conducted using the EC process to treat slaughterhouse wastewater, mostly carried out in a batch mode. Cruz et al. [17] compared three types of electrodes, namely pure aluminum (Al), pure iron (Fe), and a combination of Al and Fe. They found that the Al electrode provided higher COD removal, up to 97%, compared to other electrodes. Paulista et al. [18] used aluminum and graphite electrode for poultry slaughterhouse treatment. High-efficiency removal of COD was achieved when the coagulant load was more than 51 mg/L, and the gas load was more than 60 NmL/L. The COD removal was up to 85%, while the turbidity removal was up to 99%. Kobya et al. [13] used Al and Fe electrodes for treating poultry slaughterhouse waste with COD of 26.000 – 29.000 mg/L. Around 93% removal of COD was reached in 25 minutes of EC process when the applied current density was 150 A/m$^2$. It was estimated that the total operating cost of the slaughterhouse wastewater treatment by EC process was varied from 0.015 US$/kg COD removed (Fe electrode) to 0.027 US$/kg COD removed (Al electrode) or 0.4 US$/m$^3$ (Fe electrode) to 0.7 US$/m$^3$ (Al electrode) [19]. Potrich et al. [20] also compared Al and Fe electrodes' performances in EC reactor to remove total nitrogen and phosphorus in poultry slaughterhouse wastewater. The EC process was conducted in batch-mode. The optimum result was achieved at 20 minutes by using Al electrode at operating treatment conditions of 30 mA/cm$^2$ current density and initial pH of 8. The lowest cost was 3.89 US$/m$^3$. In this study, the slaughterhouse wastewater was treated using a continuous EC process. The EC reactor was equipped with a turbine impeller and six Al electrodes as baffles. The influence of electrode configurations and applied current densities was investigated.

## 2. Material and Methods

The experimental set-up and method were referred to in our previous work [21, 22]. A cylindrical EC reactor was used with a diameter of 25 cm and a height of 30 cm. Six Al electrodes were used as baffles, which had length, width, and thickness of 33 cm, 4 cm, and 3 mm, respectively. These electrodes were connected to a DC power supply to provide a current density of 933 mA/dm$^2$ (10A) and 1400 mA/dm$^2$ (15 A). The EC reactor's mixing process was conducted using a turbine impeller at a stirring rate of 100 rpm.

Meanwhile, the feed flow rate was maintained at 233,3 mL/min. The experimental apparatus used in this research is shown in Figure 1. The wastewater was obtained from one of the cattle slaughterhouse industries in Cimahi, West Java, Indonesia. The influent's pH was adjusted to 7 (neutral) by adding $Ca(OH)_2$. The concentration of TDS, turbidity, pH, and COD of the effluent was measured after 120 EC process minutes. The analysis methods refer to our previous research [21, 23].

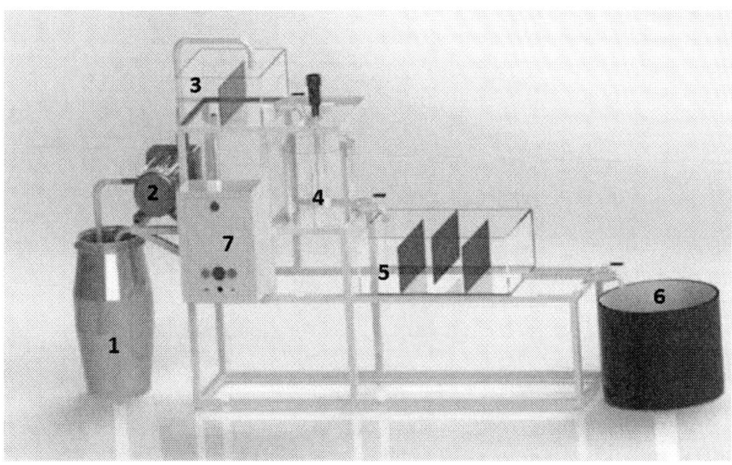


**Figure 1.** Experimental set-up: (1) feed tank, (2) feed pump, (3) stabilizer tank, (4) EC reactor equipped with a turbine impeller and baffles, (5) sedimentation tank, (6) effluent tank, and (7) control panel.

## 3. Result and Discussion

### 3.1. The influence of operating parameters on TDS, turbidity, and pH of the slaughterhouse wastewater

Figure 2 shows the influence of operating parameters on effluent qualities, including TDS, turbidity, and pH of the solution. It shows that higher TDS removal was achieved when four anodes and two cathodes (4A-2C) were used in the EC reactor compared to 3 anodes and three cathodes (3A-3C) configuration. When the current density was increased from 933 mA/dm$^2$ (10A) to 1400 mA/dm$^2$ (15A) the TDS concentration was reduced by 11.81% (from 2370 to 2090 mg/L) when using 3A-3C configuration. The TDS reduction was improved to 17.72% when using the 4A-2C configuration at a current density of 933 mA/dm$^2$ and 20.25% at a current density of 1400 mA/dm$^2$ (Figure 2A). In this research, more than 98% reduction in turbidity can be achieved in the whole experiments. It has been reported that the amount of applied current was related to the coagulant dose produced by the anode [24]. The increase of applied current density enhanced the amount of ion $Al^{3+}$ released from the anode. The $Al^{3+}$ ions form complex compounds that act as coagulants to destabilize the contaminants, allowing contaminants to agglomerate and form flocs. The amount of $Al^{3+}$ ions can be enhanced by increasing the number of anodes or increasing the applied current density value. Besides, the reaction of $Al^{3+}$ with hydroxide ions ($OH^-$) in solution resulted in $Al(OH)_3$, which was able to adsorb contaminants and form larger flocks, then settled to the bottom of the EC reactor. The $Al^{3+}$ ions and $Al(OH_3)$ were easier to destabilize colloids and particulates than ionic contaminants, and therefore, higher removal in turbidity was resulted compared to TDS.

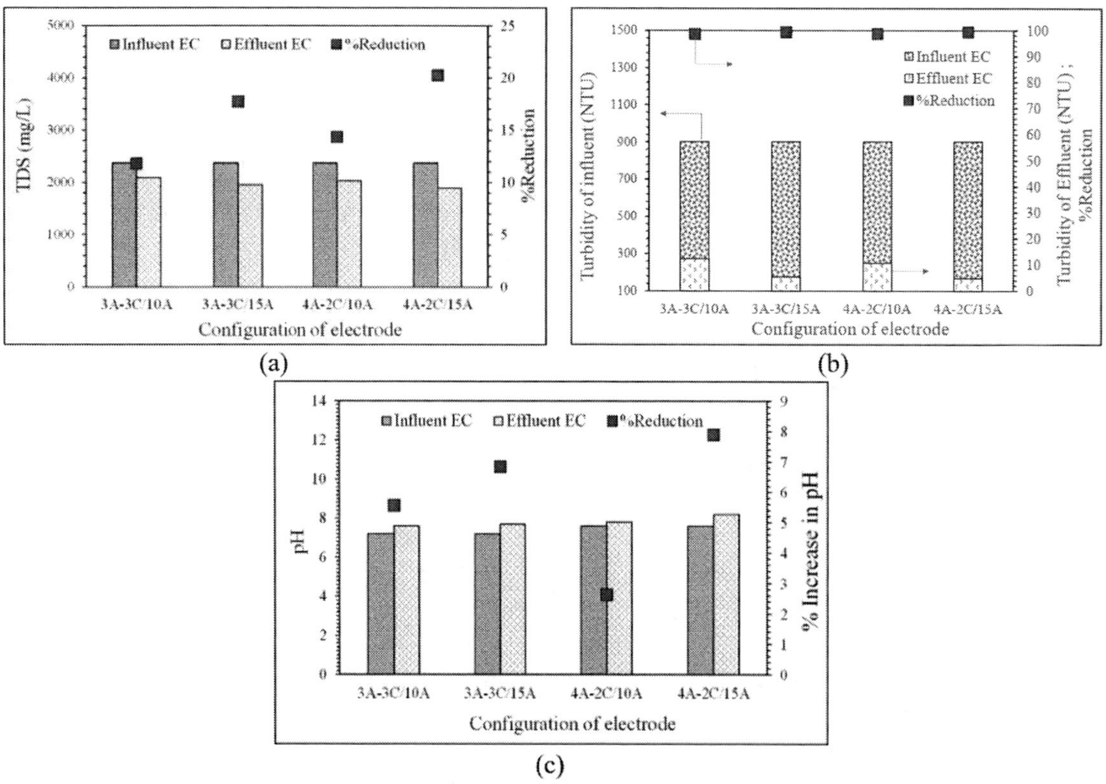


**Figure 2**. The influence of operating parameters on reducing: (a) TDS, (b) turbidity, and (c) pH of slaughterhouse wastewater.

Figure 2c shows the change in pH of the solution during the EC process, where the influent's initial pH was adjusted to 7. In general, the solution's pH will be raised during the EC process due to OH- ions in the cathode. The electrode configuration of 4A-2C showed a more significant change in pH compared to 3A-3C. The lowest change in pH (i.e., 2.63%) has occurred at the electrode configuration of 4A-2C and a current density of 933 mA/dm$^2$. It was suggested that the $OH^-$ ions in the solution was mostly reacted to $Al^{3+}$ ions to form $Al(OH)_3$. Therefore, the concentration of OH- was low, and the pH of the solution was slightly changed. The increase in current density led to a more significant change in pH. As the charge density increased to 1400 mA/dm$^2$, many OH- ions were produced. Consequently, the pH of the solution was changed by 7.79% (from 7.6 to 8.2). Several research studies have been conducted and reported that the slaughterhouse wastewater treatment using the EC process was optimum when the treatment process was completed at pH 3.

### 3.2. The influence of operating parameters on the decrease in BOD by the EC system

The influence of the configuration of electrodes and current density on BOD removal is presented in Figure 3. There was no significant change in BOD removal by increasing the number of anodes in the EC reactor. In both electrode configurations, i.e. 3A-3C and 4A-2C, the BOD removal was by 37.97% (from 710.60 to 440.80 mg/L) and 39.38% (from 710.60 to 430.80), respectively. However, BOD removal was improved to 56% by increasing the current density. It has been mentioned that the current per area of electrodes in the EC reactor was proportional to the number of metal ions or coagulant produced from the electrodes. As the current density increased, a larger amount of coagulant was produced to destabilize the wastewater contaminants. The increase of current density reduced the bubble size of $H_2$ gas in the cathode, which enhanced the bubble densities to upward the contaminants to the top of the EC reactor. The smaller size of bubbles improved particle attachment's surface area, and thus higher separation efficiency could be achieved [25]. However, the percentage of BOD reduction was still below 60%. Therefore, further experimental by varying the current density and other operating parameters are still required to improve the BOD removal.

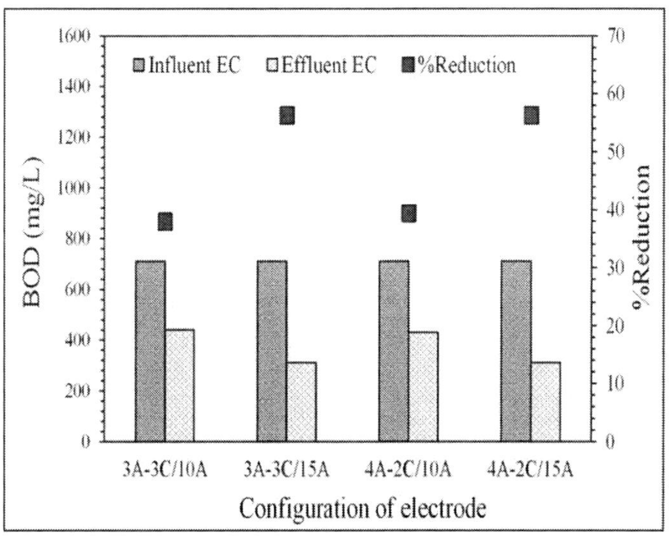


**Figure 3**. The influence of operating parameters on reducing BOD of slaughterhouse wastewater

## 4. Conclusion

In this research, A continuous electrocoagulation (EC) process is used to treat cattle slaughterhouse wastewater that contains a high concentration of colloids, organic compounds, and suspended solids. The cylindrical EC reactor is designed to combine mixing and electrocoagulation in a one-stage process and is conducted using a six-blade turbine impeller. Six Al electrodes are used as baffles to generated turbulence during the EC process. The pH of the solution is neutralized by adding $Ca(OH)_2$. The influence of current density (933 $mA/dm^2$ and 1400 $mA/dm^2$) and electrode configuration (3 anodes 3 (three) cathodes and 4 (four) anodes 2 (two) cathodes) is investigated. The experimental results showed that the treated wastewater at a current density of 80 $mA/cm^2$ produced a lower contaminant concentration than the current density of 50 $mA/cm^2$. The current density of 80 $mA/cm^2$ and electrode configuration of 4 anodes-2 cathodes reduced BOD by 56.4%, TDS by 20.25%, and TSS by 99.47%.

## Acknowledgment

LPPM Universitas Jenderal Achmad Yani supported this work through "Internal Competitive Research Program 2020" (letter no. SKEP/129/UNJANI/V/2020).

## References

[1] Aluoch E O 2015 *Use of enzymes in anaerobic sequencing batch reactor (ASBR) treatment of slaughterhouse wastewater* Doctoral Dissertation University of Nairobi

[2] Olayinka O, Adedeji O and Oladeru I 2013 Water Quality and Bacteriological Assessment of Slaughterhouse Effluent on Urban River In Nigeria *Journal of Applied Sciences in Environmental Sanitation* **8** (4) 277-286

[3] Al-Mutairi N, Hamoda M and Al-Ghusain I 2004 Coagulant selection and sludge conditioning in a slaughterhouse wastewater treatment plant. *Bioresource technology* **95** (2) 115-119

[4] Aguilar M, Saez J, Llorens M, Soler A and Ortuno J 2003 Microscopic observation of particle reduction in slaughterhouse wastewater by coagulation-flocculation using ferric sulphate as coagulant and different coagulant aids *Water Research* **37** (9) 2233-2241

[5] Sanchis M I A Sáez J, Lloréns M, Soler A, and Ortuño J F 2003 Particle size distribution in slaughterhouse wastewater before and after coagulation-flocculation *Environmental progress* **22** (3) 183-188.

[6] Kharat D S 2019 Pollution Control in Meat Industry *Current Environmental Engineering* **6** (2) 97-110.

[7] Aziz A, Basheer F, Sengar A, Khan S U and Farooqi I H 2019 Biological wastewater treatment (anaerobic-aerobic) technologies for safe discharge of treated slaughterhouse and meat processing wastewater *Science of the total environment* **686** 681-708

[8] Salehiyoun A R, Di Maria F, Sharifi M, Norouzi O, Zilouei H and Aghbashlo M 2020 Anaerobic co-digestion of sewage sludge and slaughterhouse waste in existing wastewater digesters *Renewable Energy* **145** 2503-2509

[9] Vidal J, Huiliñir C and Salazar R 2016 Removal of organic matter contained in slaughterhouse wastewater using a combination of anaerobic digestion and solar photoelectro-Fenton processes *Electrochimica Acta* **210** 163-170

[10] Salminen E and Rintala J 2002 Anaerobic digestion of organic solid poultry slaughterhouse waste–a review *Bioresource technology* **83** (1) 13-26

[11] Ardestani F and Abbasi M 2019 Poultry slaughterhouse wastewater treatment using anaerobic fluid bed reactor and aerobic mobile-bed biological reactor *International Journal of Engineering* **32** (5) 634-640

[12] Leicester D D, Amezaga J M, Moore A and Heidrich E S 2020 Optimising the Hydraulic Retention Time in a Pilot-Scale Microbial Electrolysis Cell to Achieve High Volumetric Treatment Rates Using Concentrated Domestic Wastewater *Molecules* **25** (12) 2945

[13] Kobya M, Senturk E and Bayramoglu M 2006 Treatment of poultry slaughterhouse wastewaters by electrocoagulation *Journal of hazardous materials* **133** (1-3) 172-176

[14] Reilly M, Cooley A P, Tito D, Tassou S A, and Theodorou M K 2019 Electrocoagulation treatment of dairy processing and slaughterhouse wastewaters *Energy Procedia* **161** 343-351

[15] Hakizimana J N, Gourich B, Chafi M, Stiriba Y, Vial C, Drogui P and Naja J 2017 Electrocoagulation process in water treatment: A review of electrocoagulation modeling approaches *Desalination* **404** 1-21

[16] Bayramoglu M, Eyvaz M and Kobya M 2007 Treatment of the textile wastewater by electrocoagulation: economical evaluation *Chemical Engineering Journal* **128** (2-3) 155-161

[17] Cruz K D, Francisco J T J, Mellendrez K J M and Pineda J M F Electrocoagulation treatment of swine slaughterhouse wastewater: effect of electrode material *E3S Web of Conferences*

[18] Paulista L O, Presumido P H, Theodoro J D P and Pinheiro A L N 2018 Efficiency analysis of the electrocoagulation and electroflotation treatment of poultry slaughterhouse wastewater using aluminum and graphite anodes *Environmental Science and Pollution Research* **25** (20) 19790-19800

[19] Bayramoglu M, Kobya M, Eyvaz M and Senturk E 2006 Technical and economic analysis of electrocoagulation for the treatment of poultry slaughterhouse wastewater *Separation and Purification Technology* **51** (3) 404-408

[20] Potrich M C, Duart E S A, Sikora M S and Costa da Rocha R D 2020. Electrocoagulation for nutrients removal in the slaughterhouse wastewater: comparison between iron and aluminum electrodes treatment *Environmental Technology* 1-15

[21] Nugroho F A, Aryanti P T P, Nurhayati S and Muna H M 2019 A combined electrocoagulation and mixing process for contaminated river water treatment *AIP Conference Proceedings* **2097** (1) 030017

[22] Nugroho F, Aryanti P T P, Irawan B, Kurnia M and Prasetyo T 2019 Integrated Electrocoagulation and Tight Ultrafiltration Membrane for Wastewater Reclamation and Reuse *Reaktor* **18** (4) 209-215

[23] Nugroho F A, Sani M M, Apriyanti F and Aryanti P T P 2020 The Influence of Applied Current Strength and Electrode Configuration in Laundry Wastewater Treatment by Electrocoagulation. *Journal of Physics: Conference Series* **1477** (5) 052018

[24] Zini L P, Longhi M, Jonko E and Giovanela M 2020 Treatment of automotive industry wastewater by electrocoagulation using commercial aluminum electrodes *Process Safety and Environmental Protection* **142** 272-284

[25] Chen G 2004 Electrochemical technologies in wastewater treatment *Separation and purification Technology* **38** (1) 11-41

# Hoshin Kanri's Strategic Planning Methodology through Dijkstra Algorithm within Industrial Engineering and Stakeholder Perspectives

**H C Ho* and K E N Soebandrija**

Industrial Engineering Department, Bina Nusantara University, Jakarta, Indonesia 11480

*Corresponding author: hhchie@binus.edu

**Abstract.** Hoshin Kanri is defined as strategic planning methodology to reinforce strategic work. This paper elaborates Hoshin Kanri's strategic planning methodology through Dijkstra algorithm within industrial Engineering and stakeholder perspectives. Dijkstra's algorithm is categorized as the single source shortest path algorithm. This algorithm is beneficial to measure the shortest distance on a directed and undirected path within delivery optimization. This paper objective is to evaluate strategic planning in term of delivery optimization and scheduling. Subsequently, by implementing the strategic planning, it is expected to improve existing distribution process in supply chain. Other than that, it is expected to reduce and optimize delivery's cost within its scheduling process through lens of industrial engineering and stakeholder priority. The research methodology in this paper refers to the quantitative approaches. These approaches are within sales performance's data collection in modern market and its customer address. Furthermore, the data analysis proceeds to purchasing power of each customer. Ultimately, the mentioned analysis scheduling method through routing planning is using Dijkstra's Algorithm. The mentioned Dijkstra Algorithm's delivery optimization and scheduling is through lens of industrial engineering and Stakeholder. The result and discussion of this paper elaborates analysis and its decision making toward strategic planning in order to accommodate stakeholder perspectives.

## 1. Introduction

Hoshin Kanri is defined as strategic planning methodology to reinforce strategic work. This paper elaborates Hoshin Kanri's strategic planning methodology through Dijkstra algorithm within industrial Engineering and stakeholder perspectives.

Dijkstra's algorithm is categorized as the single source shortest path algorithm. This algorithm is beneficial to measure the shortest distance on a directed and undirected path within delivery optimization.

**Table 1.** Hoshin Kanri Methodologies Involvement in Selected Scholar Works

| | |
|---|---|
| [1] Saraph, Benson, and Schroeder (1989) | [12] Dow, Sampson, and Ford (1999) |
| [2] Anderson, Schroeder, and Deverage (1995) | [13] Wilson and Collier (2000) |
| [3] Choi and Liker (1995) | [14] Cua, McKone, and Schroeder (2001) |
| [4] Flynn, Schroeder, and Sakibara (1995) | [15] Ghobadian and Gallear (2001) |
| [5] Powell (1995) | [16] Kaynak (2003) |
| [6] Black and Porter (1996) | [17] Sebastianelli and Tamimi (2003) |
| [7] Adam et al. (1997) | [18] Taylor and Wright (2003) |
| [8] McLachlin (1997) | [19] Achanga, Shehab, and Nelder (2006) |
| [9] Ahire and O'Shaughnessy (1998) | [20] Ahrens (2006) |
| [10] Zbaricki (1998) | [21] Nair (2006) |
| [11] Dale, Broaden, Wilcox, and McQuarter (1999) | [22] Shah and Ward (2007) |

Hoshin Kanri (HK) as strategic planning encompass wide spectrum of strategic planning; Leadership and management involvement; attitudes and communication; involvement, commitment and responsibility; goals and plans [23].

Hoshin Kanri is a methodology that was coined in Japan. Currently this methodology constitutes indispensable strategic management systems that are integrated in several systems. The integrated systems refer to the following but not limited to Total Quality Management (TQM), Lean Management, Six Sigma and Balanced Scorecard (BSC). Its implemented has produced successful implementation in wide spectrum of institutions [24].

This paper focuses the HK's strategic planning methodology through Djikstra algorithm within Industrial Engineering and Stakeholder Perspectives.

Searching algorithms or query process in probabilistic roadmaps are selected through the scholar work representative. The implementations of aforementioned searching algorithms are referring scholar work. Precisely, the implementation of, A* and Dijkstra algorithms are merely within the specified circumstances.

Those circumstances refer to the problem solving process, in which A*, Dijkstra algorithm or heuristic-based algorithms are frequently capitalized. To some extent, the optimal approaches are capitalized for problem solving process within coordinate system. There are wide spectrum and extensive research that elaborates searching algorithms for several circumstances in reasonable time. The algorithms, several wide spectrum of research, are categorized into its capacity as classic algorithms or heuristic-based algorithms [25, 26, 27].

## 2. Materials

This paper objective, through Hoshin Kanri, is to evaluate strategic planning in term of delivery optimization and scheduling. Subsequently, by implementing the strategic planning, it is expected to improve existing distribution process in supply chain.

Other than that, it is expected to reduce and optimize delivery's cost within its scheduling process through lens of industrial engineering and stakeholder priority, through CEO and Leadership Team of Unit Analysis of Organization in Figure 1 [28].

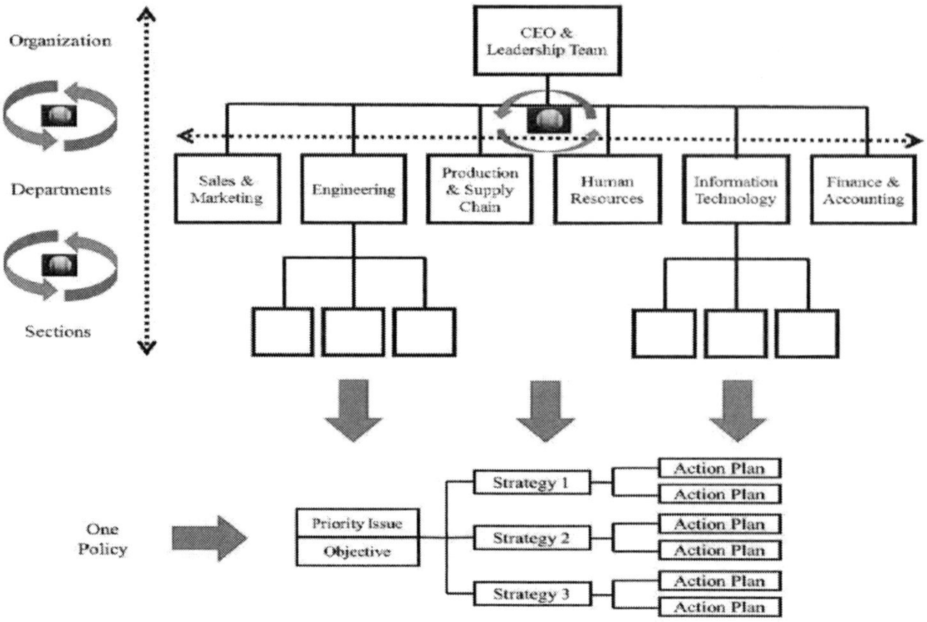


**Figure 1.** Hoshin Kanri and CEO Leadership for Strategic Planning

## 3. Methods

The research methodology in this paper refers to the quantitative approaches. In these approaches, Hoshin Kanri and CEO Leadership for Strategic Planning play vital role in its one policy through priority issue and objectives through strategy and action plan. These approaches are within sales performance's data collection in modern market and its customer address. Furthermore, the data analysis proceeds to purchasing power of each customer [29], [30], [31], [32].

## 4. Result and Discussion

This algorithm is beneficial to measure the shortest distance on a directed and undirected path within delivery optimization. This paper objective is to evaluate strategic planning in term of delivery optimization and scheduling. Subsequently, by implementing the strategic planning, it is expected to improve existing distribution process in supply chain. Other than that, it is expected to reduce and optimize delivery's cost within its scheduling process through lens of industrial engineering and stakeholder priority.

In term of classic algorithms; Dijkstra Algorithm is deemed as capitalized algorithm toward problem solving [33] This algorithm provides the network and its optimal outcome, during the situation in which 100% of its distances are positive. The A* algorithm is capitalized toward the problem solving in finding the shortest path problem on network [34].

Furthermore, in term of heuristic based algorithm; There are many heuristic-based methods for the collision-free path planning (CFPP) problem. Wavefront algorithm is one of the heuristics algorithms that are originated from potential field ideas within its wide spectrum specifications. Precisely, the aforementioned Wavefront algorithm is capitalized toward problem solving with 3D normal distributions transformations [35]. Wide spectrum of the metaheuristic are implemented toward the collision-free path planning (CFPP) problem within its algorithm of either network or coordinate system.

Evolutionary algorithms, such as the Genetic algorithm (GA), are frequently scrutinized within network system [36] that are relevant CFPP. In addition to the other circumstances, the wide spectrum of algorithms are deployed under coordinate system conditions [37]. Some novel algorithms within Genetic algorithm are generated from related scholars outcomes [38].

## 5. Conclusion

Hoshin Kanri is defined as strategic planning methodology to reinforce strategic work. This paper elaborates Hoshin Kanri's strategic planning methodology through Dijkstra algorithm within industrial Engineering and stakeholder perspectives. Dijkstra's algorithm is categorized as the single source shortest path algorithm.

This algorithm is beneficial to measure the shortest distance on a directed and undirected path within delivery optimization. This paper objective is to evaluate strategic planning in term of delivery optimization and scheduling. Subsequently, by implementing the strategic planning, it is expected to improve existing distribution process in supply chain. Other than that, it is expected to reduce and optimize delivery's cost within its scheduling process through lens of industrial engineering and stakeholder priority.

The research methodology in this paper refers to the quantitative approaches. These approaches are within sales performance's data collection in modern market and its customer address. Furthermore, the data analysis proceeds to purchasing power of each customer.

Ultimately, the mentioned analysis scheduling method through routing planning is using Dijkstra's Algorithm. The mentioned Dijkstra Algorithm's delivery optimization and scheduling is through lens of industrial engineering and Stakeholder. The result and discussion of this paper elaborates analysis and its decision making toward strategic planning in order to accommodate stakeholder perspectives.

In term of classic algorithms; Dijkstra Algorithm is deemed as capitalized algorithm toward problem solving. This algorithm provides the network and its optimal outcome, during the situation in which 100% of its distances are positive. The A* algorithm is capitalized toward the problem solving in finding the shortest path problem on network.

Furthermore, in term of heuristic based algorithm; There are many heuristic-based methods for the collision-free path planning (CFPP) problem. Wavefront algorithm is one of the heuristics algorithms that are originated from potential field ideas within its wide spectrum specifications. Precisely, the aforementioned Wavefront algorithm is capitalized toward problem solving with 3D normal distributions transformations. Wide spectrum of the metaheuristic are implemented toward the collision-free path planning (CFPP) problem within its algorithm of either network or coordinate system.

## References

[1] Saraph J, Benson G and Schroeder R 1989 An instrument for measuring the critical factors of quality management *Decision Sciences* **20** 810 – 829

[2] Anderson J C, Rungtusanatham M, Schroeder R G and Devaraj S 1995 A path analytic model of a theory of quality management underlying the Deming management method: Preliminary analytical findings *Decision Sciences* 26 (5) 637 – 658

[3] Choi T and Liker J 1995 Bringing Japanese continuous improvement approaches to US manu- facturing: The role of process orientation and communications. *Decision Sciences,* **26** (5) 589 – 620

[4] Flynn B B, Schroeder R G and Sakibara S 1995 The impact of quality management practices on performance and competitive advantage *Decision Sciences* 26 659 – 691

[5] Powell T 1995 Total quality management as competitive advantage: A review and empirical study *Strategic Management Journal* **16** 15 – 37

[6] Black S and Porter L 1996 Identification of the critical factors of TQM *Decision Sciences* **27** 1–21
[7] Adam E E, Corbett L M, Flores B E, Harrison N J, Lee T S, Rho B H and Westbrook R 1997. An international study of quality improvement approach and firm performance *International Journal of Operations and Production Management* **17** 842–873
[8] McLachlin R 1997 Management initiatives and just-in-time manufacturing *Journal of Operations Management* **15** 271 – 292
[9] Ahire S L and O'Shaughnessy K C 1998 The role of top management commitment in quality management: An empirical analysis of the auto parts industry *International Journal of Quality Management* **3** (1) 5-37
[10] Zbaricki M 1998 Rhetoric and reality of total quality management *Administrative Science Quarterly* **42** 366-394
[11] Dale B G, Broaden R J, Wilcox M and McQuarter R E 1999 Sustaining continuous improvement: What are the key issues? *Quality Engineering* **11** 369 – 377
[12] Dow D, Samson D and Ford S 1999 Exploding the myth: Do all quality management practices contribute to superior performance? *Production and Operations Management* **8** 1 – 27
[13] Wilson D and Collier G 2000 An empirical investigation of the Malcolm Baldrige National Quality award causal model *Decision Sciences* **31** 361-383
[14] Cua K, McKone K and Schroeder R 2001 Relationships between implementation of TQM, JIT, and TPM and manufacturing performance *Journal of Operations Management* **19** 675 – 694
[15] Ghobadian A and Gallear D 2001 TQM implementation: An empirical examination and pro- posed generic model *Omega* **29** 343 – 359
[16] Kaynak H 2003 The relationship between total quality management practices and their effects on firm performance *Journal of Operations Management* **21** 405 –435
[17] Sebastianelli R and Tamimi N 2003 Understanding the obstacles to TQM Success *Quality Management Journal* 10 (3) 45 – 54
[18] Taylor W and Wright G 2003 A longitudinal study of TQM implementation: Factors influencing success and failure *Omega* **31** 97 – 111
[19] Achanga P, Shehab E, Roy R and Nelder G 2006 Critical success factors for lean implementation within SMEs *Journal of Manufacturing Technology Management* **17** (4) 460-471
[20] Ahrens T 2006 *Lean production: Successful implementation of organisational change in oper- ations instead of short term cost reduction efforts. Seefeld* Germany: Lean Alliance GmbH.
[21] Nair A 2006 Meta-analysis of the relationship between quality management practices and firm performance *Journal of Operations Management* **24** 948 – 975
[22] Shah R and Ward P 2007 Defining and developing measures of lean production *Journal of Operations Management* **25** 785 – 805
[23] Nicholas J 2014 Hoshin kanri and critical success factors in quality management and lean production *Total Quality Management and Business Excellence* **27** (3-4) 250-264
[24] Ahmed H O 2016 A Proposed Systematic Framework for Applying Hoshin Kanri Strategic Planning Methodology in Educational Institutions *European Scientific Journal* **12** (16) 158-194
[25] Masehian E and Sedighizadeh D 2007 Classic and Heuristic Approaches in Robot Motion Planning–A Chronological Review *World Acad. Sci. Eng. Technol* **29** 101–106
[26] Mac T T, Copot C, Tran D T and De Keyser R 2016 Heuristic approaches in robot path planning: A survey. *Robot. Auton. Syst.* **86** 13–28
[27] Mohant P and Parhi D 2013 Controlling the Motion of an Autonomous Mobile Robot Using Various Techniques: A Review. *J. Adv. Mech. Eng.* **1** 24–39

[28] Liedtke C A 2017 *Big Data in Hoshin Kanri* Singapore: Asian Network for Quality Congress.
[29] Basias N and Pollais, Y 2018 Quantitative and Qualitative Research in Business & Technology: Justifying a Suitable Research Methodology *Review of Integrative Business and Economics Research* 7 91-105
[30] Goertz G and Mahoney J 2012 *A Tale of Two Cultures: Qualitative and Quantitative Research in the Social Sciences* New York: Princeton University Press.
[31] Martin, W and Bridgmon, K 2012 *Quantitative and Statistical Research Methods. From Hypothesis to Results* New York: John Wiley & Sons
[32] Yin, R 2011 *Applications of Case Study Research* New York: SAGE Publications
[33] Dijkstra E W 1959 A note on two problems in connexion with graphs *Numer. Math*, **1** 269–271
[34] Hart P E, Nilsson N J and Raphael B 1968 A Formal Basis for the Heuristic Determination of Minimum Cost Paths *IEEE Trans. Syst. Sci. Cybern.* **4** 100–107
[35] Stoyanov T, Magnusson M, Andreasson H, Lilienthal A J 2010 Path planning in 3D environments using the Normal Distributions Transform *Proceedings of the 2010 IEEE/RSJ International Conference on Intelligent Robots and Systems* 3263–3268
[36] Jiang A, Yao X and Zhou J 2018 Research on path planning of real-time obstacle avoidance of mechanical arm based on genetic algorithm *Journal Engineering* 1579–1586
[37] Yang Q and Yoo S J 2018 Optimal UAV Path Planning: Sensing Data Acquisition Over IoT Sensor Networks Using Multi-Objective Bio-Inspired Algorithms *IEEE Access* **6** 13671–13684
[38] Lee H-Y, Shin, H and Chae J 2018 Path Planning for Mobile Agents Using a Genetic Algorithm with a Direction Guided Factor *Electronics* **7** 212

# Implementing Cloud Computing Technology on Restaurant's Expenses Monitoring System

**F N Hasanah***, **F Renaldi and F R Umbara**

Science and Informatics Faculty, Dept of Informatics Engineering, Universitas Jenderal Achmad Yani, Cimahi, West Java, Indonesia

*Corresponding email: fnurulhasanah@gmail.com

**Abstract.** Cloud computing offers the advantage to increase business productivity because it combines computer technology and internet-based development that is currently needed in various fields, including the restaurant industry. The application of cloud computing in the restaurant industry that has been carried out includes building payment transaction systems, restaurant reservation systems, and customer satisfaction systems. Although there have been many studies on cloud computing applications in the restaurant industry, there is still minimal research that discusses the consolidation of many restaurant branches' expenses. Business activities in the restaurant industry require payments that need to be monitored appropriately to avoid waste, providing information on the cost-efficiency of costs quickly and accurately, allowing easy observation of expenditure data movement. The method used is by collecting expenditure data from all restaurant branches then processing it by comparing the income data obtained to produce output in the form of the level of efficiency achieved by each store. For three months, the sample data were obtained from the Cekeran Midun restaurant, which has seven branches in Bandung and Cimahi, Indonesia. The user acceptance test results show that all functions work well and have an overall acceptance rate of 83.33%, with most comments on the improvement being on the convenience of use and the system's layout. Further research combines the calculation of costs in a restaurant. It allows the development of a platform that provides cloud computing-based online services that can be used collectively by various restaurants (software as a service).

## 1. Introduction

Information technology has been widely applied in various fields, one of which is the restaurant industry. The restaurant industry's application of information technology started more than two decades ago [1]. Different kinds of information technology are used, including point-of-sale systems to assist payment transactions. They also used employee management systems, stock monitoring or inventory systems, production-support systems to support production functions, and profit and loss calculations in the accounting system [1]. Information technology can provide easier access to penetrate new or existing markets, provide cost-effectiveness, and present goods and services more effectively [2]. This condition proves that restaurant industry owners need to adapt to ever-evolving technology.

Business activities in the restaurant industry certainly require operational costs. They have an essential role in achieving its success, namely obtaining profits or operating profits [3]. The restaurant industry's operating activities' expenses need to be monitored to avoid waste [4]. This condition also applies to restaurants that have many branches. Monitoring activities are carried out to obtain information continuously from a specific action to become a positive result following the objectives set [5]. Monitoring will provide information about the status carried out repeatedly from time to time, up to an ongoing process check [6]. The application of monitoring information systems has been carried out in many research fields, including textile companies for monitoring production [5] [6]. Post offices need an information system for monitoring transactions between agents [7]. In small and micro enterprises (SME) newspaper agents, the monitoring model improves management functions [8].

Meanwhile, for monitoring activities or activities, research has been carried out in student affairs to monitor student activities [9] and monitor humanitarian activities carried out in Zimbabwe [10]. Monitoring information systems are also needed in the banking sector [11]. The monitoring process can also be related to inventory activities in the procurement and logistics department [12]. Another study created a monitoring information system in the Public Works Office by applying the yield value concept approach [13]. In Small and Medium Industries (SMI), research is carried out to build a system capable of accelerating the monitoring process. The system can be accessed in real-time, making it easier for policymakers to obtain the required data [14]. The latest technology is applied to other restaurant industry research to build an integrated system that can monitor all revenue from restaurant sales in real-time. The system design is made by implementing the RESTFul API architecture with Token security access and Web Services' use for data integration and processing in each information system [15]. The technology that is well known today is Cloud Computing, which can help create an application that regulates a restaurant system's management or management well [16]. Also, the application of Cloud Computing is needed by Small and Medium Enterprises to carry out global marketing [17].

In contrast to previous studies, this study aims to monitor expenditure cost data spread across many restaurant branches so that data management can be carried out centrally. Restaurant business activities are carried out every day and result in increasing expenditure data. If it is not well organized, reporting expenditure data to the top management level will experience difficulties. Therefore, it is crucial to conduct current research to help the company's progress and overcome problems in disseminating data and information in many places with information technology. This study's cost data were obtained from the Cekeran Midun restaurant, which has seven branches spread across Bandung and Cimahi cities. Issues arose when the management tried to monitor waste in each branch due to data spread in many places. This situation resulted in slow reporting so that the company suffered losses due to the absence of early handling of the waste. Based on the background description, it is deemed necessary to research the information system for monitoring expenditures on the Cekeran Midun.

## 2. Research Method

This study uses several techniques for data collection consisting of interviews with relevant parties and company observations. We conduct the interview processes at the three Cekeran Midun Outlet located in Cimahi and Padjajaran, Indonesia. The first visit was made on the 5th of October 2020 to meet with the head outlet and ask permission to conduct interviews to determine the workflow on Cekeran Midun. More meetings were held on the 6th of October 2020 to meet with the financial staff in head office and began speaking about the business process, the actors involved, and thoroughly explaining the company's activities. In addition to the interview, the observation process also took place from 23rd to the 24th of October 2020. During the observation, the activities carried out were to look at the company's events, see who the actors were there, and record the observations' results.

### 2.1. Business Process Identification and System Objective

The first stage in software development identifies the business flow to create the system's objective [18]. The business process related to expense consolidation expenses that run on Cekeran Midun starts with ordering raw material made by each branch. After the production receives a list of orders for raw materials from each branch, the production staff prepares the raw materials to be sent by the driver and writes a list of orders in a road letter. The road letter is useful for processing by the financial staff in calculating the expenses of each branch. Every day, there is daily closing to calculate the revenue earned by each branch. Expense cost consolidation is complete after obtaining efficiency level results. The efficiency level is obtained from calculating the percentage of costs, which divides the income and expenses.

We identify the gap or problems within the business process identification, as depicted in Table 1.

**Table 1.** Problems Found during Business Process Identification

| No | Problems Found | Description |
|---|---|---|
| 1 | Expenses data spread across many branches | The company is growing rapidly but not followed by technological advances. Ultimately resulting in the spread of data that cannot be appropriately consolidated. |
| 2 | It took a long time to collect expense data | Due to the spread of expenses data in many places, as a result, it takes a long time to collect expenses data for processing and monitoring. |
| 3 | Monitoring branches experiencing waste is difficult | It is difficult for stakeholders to know which branches are experiencing waste in managing their expenses. |
| 4 | Data on expenses consolidation are not mutually integrated | Data that is not integrated makes it less efficient the time needed to check. |
| 5 | Efficiency level information is not accurately recorded | Information regarding efficiency level is challenging to monitor by the CEO |
| 6 | Unstandardized report | Reports that have been done were known to have no standards. |

From this problem found during business process identification, we set up objectives using the SMART approach [19]. It is a framework of objectives used to help guide goal setting. SMART is an acronym that stands for Specific, Measurable, Achievable, Realistic, and Time-Bound. Therefore, a SMART goal incorporates all these criteria to help focus on making a sound system and increasing the chances of achieving that goal. SMART objectives are well known in the education sector [20] [21], although we also can find the framework being implemented in health [22] and even national defense [23]. Our interpretation of objectives based on the SMART approach can be seen in Table 2.

**Table 2.** SMART Objectives of the Restaurant's Expenses Monitoring System

| No | Objective Type | Objective Description | Solving Problem No- |
|---|---|---|---|
| 1 | Specific | The purpose of this system is to create clear and transparent information on the consolidated expense data of various branches | 1, 3, 4, 5 |
| 2 | Measurable | To collect and process expense data in one cloud-based platform and can be accessed from many places | 1, 2, 3, 5 |
| 3 | Achievable | To manage and monitor the company's operational activities in the expense consolidation process | 1, 4, 5, 6 |
| 4 | Realistic | All stakeholders can access the web-based system at the middle management and top management levels for monitoring purposes | 1, 2, 3, 4, 5, 6 |
| 5 | Time-Bound | To be used in the 1st semester of 2021 and measured after 60-days of use | - |

### 2.2. System Design

System design deals with planning information systems development by understanding and specifying what a system should do and how its components should be implemented and work together [24]. System analysts solve business problems by analyzing information systems' requirements and designing such systems by applying analysis and design techniques. The design emphasizes a *conceptual solution* that fulfills the conditions rather than its implementation [25]. For example, in this study, a description of a table structure of order and road letter (2 of our most important table in the system) is presented in Table 3 and Table 4.

**Table 3.** Table of table_order

| No | Field Name | Type Field | Size Field | Key | Ref Table | Remarks |
|---|---|---|---|---|---|---|
| 1. | id_order | Number | - | Primary Key | - | Not Null Integer [0-9] |
| 2. | id_road_letter | Number | - | Foreign Key | table_road_letter | Not Null Integer [0-9] |
| 3. | id_raw_material | Number | - | Foreign Key | table_raw_material | Not Null Integer [0-9] |
| 4. | name_raw_material | String | 40 | - | - | Not Null Varchar [Xx] |
| 5. | price | Number | - | - | - | Not Null Integer [0-9] |
| 6. | unit | String | 15 | - | - | Not Null Varchar [Xx] |
| 7. | category | String | 20 | - | - | Not Null Varchar [Xx] |

Table 3 explains the raw material ordering data used for the calculation of expenses. The actors who use this data are the head outlets from all branches and the production head. Table analysis produces two foreign keys, namely id_road_letter and id_raw_material, because later, the information is related to other tables.

**Table 4.** Table of table_road_letter

| No | Nama Field | Type Field | Size Field | Key | Ref Table | Remarks |
|---|---|---|---|---|---|---|
| **1.** | id_road_letter | Number | - | Primary Key | - | Not Null Integer [0-9] |
| **2.** | id_outlet | Number | - | Foreign Key | table_outlet | Not Null Integer [0-9] |
| **3.** | letter_number | Number | - | - | - | Not Null Integer [0-9] |
| **4.** | date_actual | Date | - | - | - | Not Null Date [ddmm-yyyy] |
| **5.** | total_expenses | Number | - | - | - | Not Null Integer[0-9] |
| **6.** | total_sales | Number | - | - | - | Not Null Integer [0-9] |
| **6.** | status | String | 10 | - | - | Null Varchar[Xx] |

Table 4 describes the road letter data that stores information about the id_outlet, the foreign key from table_outlet, letter_number, date_actual, total_expenses from the total of raw material orders total_sales of the total sales on that day. From these data, each branch's amount of profit can be seen by reducing total_sales by total_expenses.

*2.2.1. Actor's Identification.*

Identifying actors is one of the first steps in use case analysis [26]. An actor represents each type of external entities with which the system must interact. We define actors within our systems, and this can be seen in Table 5.

**Table 5.** Actor Identification of Restaurant's Expenses Monitoring System

| No. | Actors | Descriptions |
|---|---|---|
| **1.** | Head Outlet | access rights to adding raw material orders as the initial monitoring activities and adding sales data when restaurant operating hours are over. |
| **2.** | Head of Production | access rights to view details of raw material orders, change raw material orders according to available stock and provide confirmation of raw material orders. |
| **3.** | CEO | access rights to view expenses monitoring reports in graphs and tables |

*2.2.2. Functional Analysis*

Modules, functions, and features of a system are derived from the interpretation of objectives. We can find depicted in Table 6 the analysis and its correlation to the goals.

**Table 6.** Functional Analysis of the Restaurant's Expenses Monitoring System

| No. | Module(s) | Description of Functionalities | Objective No- |
|---|---|---|---|
| 1. | Manage raw material orders | Manage activities: create activities, update activities, and delete actions. | 1, 2, 4, 5 |
| 2. | Manage sales data | To manage the sales data required in calculating the level of efficiency | 1, 2, 4, 5 |
| 3. | Calculation of level efficiency | As a result of the percentage rate of expenses utilization. | 2, 3, 5 |
| 4. | Manage reports | Manage monthly expense reports from all branches, which can be presented in graphs or tables | 1, 3, 4, 5 |

## 2.3. System Development

In this study, we implemented the web-based software based on the PHP programming language with the editor is Visual Studio Code. It uses the MySQL database and Apache Web Server in the XAMPP application, Google Chrome / Mozilla Firefox as a Web Browser media, and Laravel Framework for PHP as a development restaurant's expenses monitoring system.

## 2.4 Implementing Cloud Computing Technology

After the system development stage is complete, then apply cloud computing technology to the system. The hosting method is carried out so that the restaurant's expenses monitoring system can be accessed via the link http://cekmid-monitoring.digital. This platform aims to provide easy access for users who want to use the system. Even though the design is made web-based, it is still compatible when accessed via a mobile screen.

# 3. Result and Discussion

The design and manufacture of software available in this study were completed in four months. In the testing phase, there are two tactics, namely system testing and user testing. The test will determine the quality, success, testing stages, and conclusions of the study.

## 3.1 Restaurant's Expenses Monitoring System

Implementation of the software interface results from the design concept or design created in the previous stage [27]. Interface design or mock-up helps developers make good designs to build monitoring systems for the restaurant's expenses. To enter the system based on the user level, namely the head outlet, the head of the production, and the CEO. Each user must first fill in their username and password, as depicted in Figure 1.

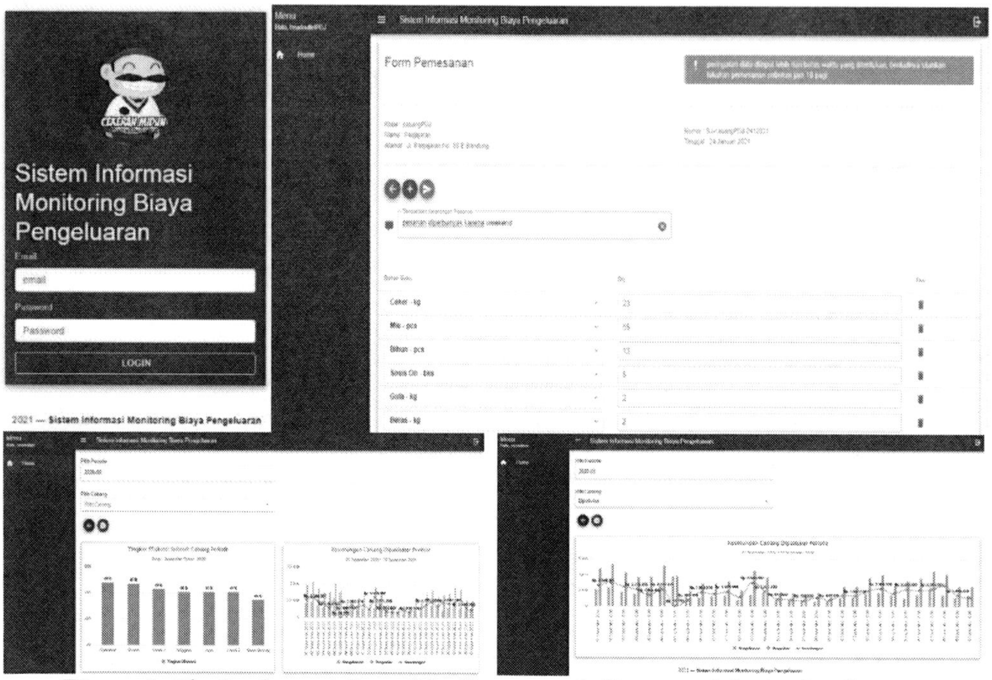


**Figure 1.** The Implementation of the Restaurant's Expenses Monitoring System

There is a raw material order management function in the system to input orders to fulfill the branch's needs. After the order of raw materials is confirmed, the cost data is recorded in the system., sales data are inputted to enter the day's total revenue. Furthermore, the cost percentage is calculated by dividing the total income by the total expenses to produce a branch efficiency level. In Table 7, a test case of filling out the raw material order form.

**Table 7.** Use Case Test of filling out raw material order form scenario

| **Use Case ID** | UC-1 |
|---|---|
| **Use Case Name** | Manage Raw Material Order |
| **Test Scenario** | Filling out Raw Material Order |
| **Test Case** | The filling outfield of order description, raw material, and quantity of raw material |
| **Pre-Condition** | Raw material order = empty |
| **Test Steps** | 1. Select the raw material ordering menu<br>2. Start filling out the data in all fields<br>3. Click "save." |
| **Test Data** | 1. Fill out order description <entering the payday date; orders are reproduced><br>2. Choose the raw material <Ceker, Mie, Bihun, Sosis><br>3. Fill out quantity according to the selected raw material <30, 50, 25, 6> |
| **Expected Result** | Filling out raw material order succeeded without fail and show a message about details raw material order then the message "Data has been successfully saved." |
| **Post Condition** | "Data has been successfully saved" |
| **Status (Pass/Fail)** | PASS |

**Actual Result**

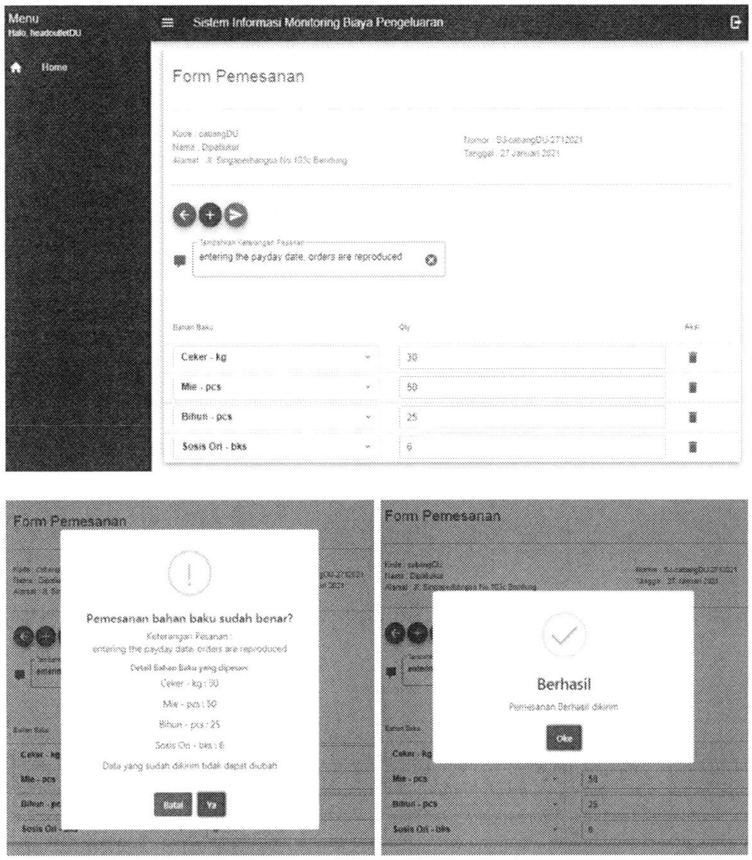


The activity of ordering raw materials from each branch is essential in monitoring expenditure costs because containing raw materials produces data on each branch's total raw materials. This natural material is processed in branches and has food sold to consumers and creates added value, which is its profit. Savings or wasteful of a branch in using expenses, seen from the level of efficiency obtained.

### 3.2. The Acceptance Test

To find out that the needs of users have been fulfilled, we conducted a user acceptance test. The test was made with 27 test scenarios for three types of users: (1) Head Outlet; (2) Head of Production; (3) CEO. User Acceptance Test results can be seen in Table 8.

**Table 8.** User Acceptance Test of Restaurant's Expenses Monitoring System

| No | User/Tester | Acceptance Rate | Notable comments | User Description |
|---|---|---|---|---|
| **1.** | Head Outlet | (10 out of 12) 83.3% | *"felt helped because it made the work easier, the process of adding raw material orders was fast."* | - Seven user<br>- Gender: man<br>- High school and vocational school graduate |
| **2.** | Head of Production | (8 out of 12) | *"a little trouble with the* | - One user |

| No | User/Tester | Acceptance Rate | Notable comments | User Description |
|---|---|---|---|---|
| | | 66.7% | *layout of the buttons, but it's a nice system."* | - Gender: man<br>- High school graduate |
| **3.** | CEO | (3 out of 3)<br>100% | *"easy to use, data can be well visualized and accurate."* | - One user<br>- Gender: man<br>- Bachelor degree in economics |
| | **Average Acceptance** | **83.33 %** | | |

## 4. Conclusions

This study is based on the spread of expense data in many restaurant branches, challenging to monitor. For these problems, the solutions presented with the help of technology make data scattered in many places that can be well organized and provide reports on each outlet's efficiency level that are accurate and can be accessed easily. Taking advantage of the advantages presented by Cloud Computing in the form of the use of computer technology and internet-based development, the expense monitoring system can be built into a platform as a service.
Further research combines the calculation of costs in a restaurant. It allows the development of a platform that provides Cloud Computing-based online services that can be used collectively by various restaurants (software as a service).

## References

[1] Ansel D and Dyer C A 1999 Framework for Restaurant Information Technology *Cornell Hotel and Restaurant Administration Quarterly* **40** (3) 74–84

[2] Huo Y H 1998 Information Technology and the Performance of the Restaurant Firms Journal of Hospitality & Tourism Research **22** (3) 239-251

[3] Frazer L 2012 The Effect Of Internal Control On The Operating Activities Of Small Restaurants *Journal of Business & Economics Research* **10** (6) 361-374

[4] Mun S G and Jang S C 2018 Restaurant operating expenses and their effects on profitability enhancement *Int. J. Hosp. Manag.*, **71** (Sept) 68–76

[5] Trisdiyanti R, Renaldi F, Umbara F R and Bon A T 2020 Implementation of Production Monitoring Systems in A Small Textile Company *Proceedings of the 5th NA International Conference on Industrial Engineering and Operations Management* 2980–2987

[6] Rohman H A, Renaldi F, Umbara F R and Bon A T 2020 Production Monitoring Information System : Managing Supply and Demand in the Textile Factory *Proceedings of the 5th NA International Conference on Industrial Engineering and Operations Management* 1902–1911

[7] Leonov S, Yarovenko H, Boiko A and Tetiana D 2019 *Proceedings of the Selected Papers of the 8th International Conference on Monitoring, Modeling & Management of Emergent Economy* **2422** 2980-2987

[8] Husni I and Amin A 2014 Information Model of Inventory and Finance Monitoring to Fix Managerial Function in Newspaper Agency **19** (1) 46-58

[9] Ariefni D F and Legowo M B 2018 Implementation of Monitoring and Evaluation in the Student Activity Information System at Perbanas Institute Jakarta *Jurnal Teknik Informatika dan Sistem Informasi* **4** (3) 422-432

[10] Hardlife Z and Zhou G 2013 Utilisation of monitoring and evaluation systems by development agencies: The case of the UNDP in Zimbabwe *Am. Int. J. Contemp. Res.* **3** (3) 70-83
[11] Legowo M B 2017 Monitoring and Evaluation Information System Modeling for Banking Credits *International Journal of Computer Science Issues* **14** (5) 21–30
[12] Sutedi S 2011 Designing Information System for Monitoring the Progress of Goods in the Procurement and Logistics Section of the Darmajaya Institute of Information and Business *J. Inform. Darmajaya* **11** (2) 41–52
[13] Ibnu D S K I M, Rahman, Abdul R 2016 Design of Public Works Monitoring Information System with Result Value Concept Approach *TRANSISTOR Elektro dan Inform.* **1** (1) 13–21
[14] Wahyuniardi R, Afrianti L H, Nurjaman S and Gusdya W 2016 Development of a WEB-Based Information System for Monitoring and Evaluation of Small and Medium Industry Centers in West Java **4** (1) 9-19
[15] Adianto I K, Informatika J T, and Service W 2019 Restaurant Billing System (Ebill Resto) by Implementing Billing Data Synchronization at Company Branches to the Parent Company Based on Rest API *Semin. Nas. Hum. Apl. Teknol. Inf.* **5** (1) 1–6
[16] Mahardika I G N A 2014 Cloud-Based Restaurant Management Back End Application *J. Merpati* **2** (1) 98-105
[17] Ghozali M I, Sugiharto W H, Afifi Z 2019 Cloud Computing as an Optimization Strategy for Expanding the Target Market for Small and Medium Enterprises in Facing the Industrial Revolution 4.0 *J. Mantik Penusa* **3** (1) 104-109
[18] Seqerloo Y, Arezoo, Amiri M J, Parsa S and Koupaee M Automatic Test Cases Generation from Business Process Models *Requir. Eng.* **24** (1) 119-132
[19] Bjerke M B and Renger R 2017 Being smart about writing SMART objectives *Eval. Program Plann.* **61** 125-127
[20] Jung L A 2007 Writing SMART Objectives and Strategies That Fit the ROUTINE *Teach. Except. Child.* **39** (4) 54-58
[21] Tofade T, Khandoobhai A and Leadon K 2012 Use of SMART learning objectives to introduce continuing professional development into the pharmacy curriculum *Am. J. Pharm. Educ.* **76** (4) 1-7
[22] Kollef M 2008 SMART approaches for reducing nosocomial infections in the ICU *Chest* **134** (2) 447-456
[23] McNerney M, Marquis J, Zimmerman S and Klein A 2016 *SMART Security Cooperation Objectives: Improving DoD Planning and Guidance* California: Santa Monica
[24] Vareilles E, Coudert T, Aldanondo M, Geneste L and Abeille J 2012 Coupling system design and project planning: Discussion on a bijective link between system and project structures *IFAC Proceedings Volumes* **45** (6) 23-25
[25] Stachtiari E, Mavridou A, Katsaros P, Bliudze S and Sifakis J 2018 Early validation of system requirements and design through correctness-by-construction *Journal of Systems and Software* **145** 1-28
[26] Mantzana V, Themistocleous M, Irani Z and Morabito V 2007 Identifying healthcare actors involved in the adoption of information systems *European Journal of Information Systems* **16** (1) 91-102
[27] Henderson-Sellers B and Edwards J M 1990 The object-oriented systems life cycle *Commun. ACM* **33** (9) 142-159

IOP Conf. Series: Materials Science and Engineering 1115 (2021) 012040 doi:10.1088/1757-899X/1115/1/012040

# Implementing Online Analytical Processing in Hotel Customer Relationship Management

**M Taufik***, **F Renaldi and F R Umbara**

Department of Informatics, Universitas Jenderal Achmad Yani, Cimahi – Indonesia

*Email: mtaufikhidayat16@if.unjani.ac.id

**Abstract.** Online Analytical Processing (OLAP) is increasingly being used by applying technology concepts that use a multidimensional view of grouping data to provide quick access to strategic information for analysis purposes. In the tourism industry, especially hospitality, this is very useful, especially in processing hotel operational data. Although OLAP technology has been widely applied in the hotel industry in reporting business sales, marketing, and reporting management analysis, there is still little research that discusses customer activity analysis. This study develops an OLAP-based analytical CRM system to analyze customer data and classify it into two main segments: geographically and demographically. Transactional data for hotels that have existed for the last three years are converted into a data warehouse, including the ETL process from the original database to a star schema database. It has a fact table and dimensions. Furthermore, OLAP cube operations are performed and generate customer reports. Testing in this study was carried out by calculating the total data's access speed and accuracy from 100 to 5000 customers, and the estimated value was 45.50 to 80 milliseconds in 30 experiments. For further research, it can be developed by integrating data from the hotel transactional system so that the analytical process can run in real-time.

## 1. Introduction

Online Analytical Processing (OLAP) is a process used to query data in complex forms and analyze large volumes of data, trend analysis, and structured multidimensional modeling [1]. OLAP capabilities can access various data sources, supports faster and more consistent Query Response Time and data aggregation in the Data Warehouse capable of navigating data in the form of Drill Down and Roll Up [2]. The use of OLAP technology has grown in popularity since 1991. The increase in data volume and the use of business analytics value for enterprises is enormous [3]. Transactions that occur every day must follow the business processes carried out, recorded in a system, and produce data with immense size and are increasing every day [4]. It is stored in a data warehouse, an integrated database architecture archives for processing the data. It analyzes historical data such as customers, sales, and other daily operations information [5]. The extraction process is carried out, and the transformation is carried out by applying various formulas, aggregations, and validations to obtain fact and dimension tables. Then, the data cube was generated by the data warehouse with Multidimensional features [6]. With this, you can see the data conceptually with slicing and dicing techniques to facilitate data analysis inside.

One of the OLAP technologies that have been carried out by many scientists in Analytical Customer Relationship Management (CRM) is using data warehouse storage of activities obtained from operational data designed to determine the best customer [7]. And other studies examine the data warehouse model designed to manage consumer and purchasing data. The design uses a nine-step methodology with a fact constellation schema modeling consisting of two fact tables: a customer fact table and a purchase fact table [8]. Previous research regarding customer licensing reporting using the OLAP method and tested with 500 data records can access at a speed of 90.70md, resulting in purchasing patterns on customer behavior in choosing services [9][10]. One of the main factors to increase business growth in a company is that sometimes there is a lot of unknown information from potential customers to the company, resulting in the company being unable to analyze the desired potential [11]. Over time, this analysis also helps to refine the prospect's previous business decisions by assessing a customer loyalty for purchasing a product or service [12]. Customer loyalty is one of the keys to achieving maximum profit in the company [13]. Many things that affect customer loyalty include customer satisfaction, such as price [14] and company location [15]. The role of analytical CRM is used to determine the profitability and segmentation of which ones are most profitable over time [16]. Customer segmentation aims to group customers into classes based on their behavior and needs in buying a product or service [17]. With segmentation, business people can more easily identify target consumers [18].

In the tourism industry, especially hospitality, this is very useful, especially in processing hotel operational data. In general, OLAP technology has been widely applied in the hotel industry in sales business reporting, reporting management [19]. However, there is still minimal research that discusses the analysis of customer activity. This study creates an Analytical CRM System with OLAP implementation where data processing is taken from customer data and analyzed to make information that is very useful for the manager.

## 2. Research Method

The data analysis stages at Hotel Simply Valore to get the software engineering process requirements can be done to see the source and current data flow as shown in Figure 1.

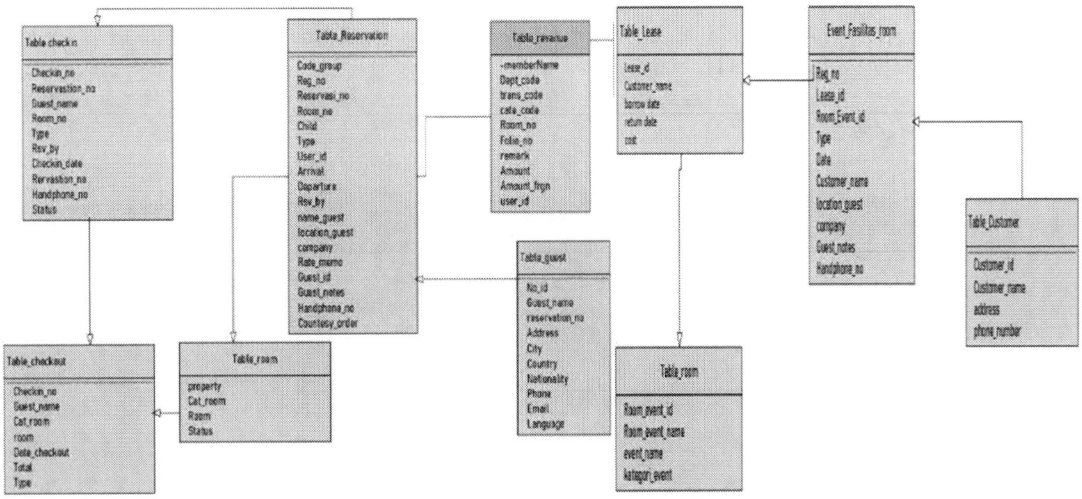


**Figure 1.** Physical Data Model

IOP Conf. Series: Materials Science and Engineering 1115 (2021) 012040 doi:10.1088/1757-899X/1115/1/012040

After determining the hotel's data source, the next stage will be carried out in Figure 2.

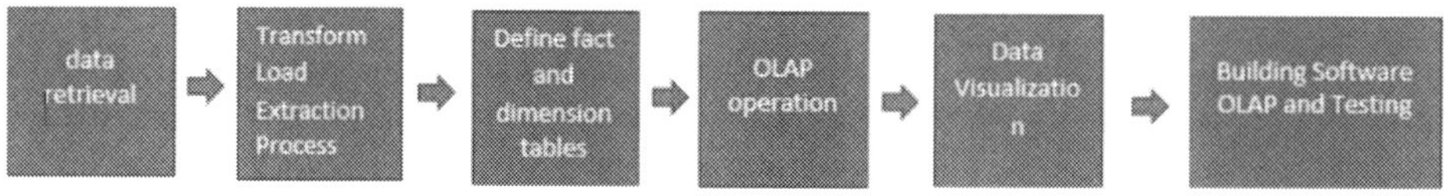


**Figure 2.** Methodology

## 2.1 Data Retrieval

Data is taken from hotel transactional activities, which have different sources. The data used are from 2018 to 2020, starting from customer activities such as guest reservation data, event room rentals, and restaurant use in hotels with the xls data format (see Figure 3).

| ARRIVAL | GUEST LIS | T (01/ | 05/18 - 30/12/18) | | | | | | | | |
|---|---|---|---|---|---|---|---|---|---|---|---|
| No | RmNo R | esNo | GuestName | RmCat | A/C | Co | NAT Company/TA | Arrival | Departure | Stay Email Address | Remark |
| 40 | 520 | 36322 | Zatadini Nabilah, | DLD3 | 2/0 | 0 | INA Booking.Com, | TOUR 26/05/18 | 27/05/2018 | 1 | 0856 9359 0915 |
| 41 | 618 | 36367 | Farhan Ilham, | FMS3 | 2/0 | 0 | INA Expedia, | 26/05/2018 | 27/05/2018 | 1 | 0895 3979 42509 |
| 42 | 650 | 36360 | Monika, | DLD3 | 2/0 | 0 | INA PEGIPEGI.COM, | 26/05/2018 | 27/05/2018 | 1 | |
| 43 | 1901 | 36355 | LENI, | SPR | 2/0 | 0 | INA | 26/05/2018 | 27/05/2018 | 1 | 081 387 723 378 |
| 44 | 331 | 36355 | LENI, | DLXS3 | 2/0 | 0 | INA | 26/05/2018 | 27/05/2018 | 1 | 081 387 723 378 |
| 45 | 350 | 36355 | LENI, | DLXS3 | 2/0 | 0 | INA | 26/05/2018 | 27/05/2018 | 1 | 081 387 723 378 |
| 46 | 352 | 36355 | LENI, | DLXS3 | 2/0 | 0 | INA | 26/05/2018 | 27/05/2018 | 1 | 081 387 723 378 |
| 47 | 356 | 36355 | LENI, | DLXS3 | 2/0 | 0 | INA | 26/05/2018 | 27/05/2018 | 1 | 081 387 723 378 |
| 48 | 322 | 36362 | Aaus S. | DLXS3 | 2/0 | 0 | INA Traveloka. | 26/05/2018 | 27/05/2018 | 2 | 0852 202 700 93 |

**Figure 3.** Transactional Hotel Data

## 2.2. Extract Transform Load (ETL)

We perform the ETL process so that data does not experience redundancy, duplicates, inconsistent data, data entry errors [20]. It is selecting and retrieving data from one or several sources (such as databases). Then, we start the data cleaning process by identifying empty (null) data, containing noise, and inconsistent due to the data entry process. Data cleaning waiting time is filled with the maximum appearance value (see Figure 4).

| No | NoRoom | NoRes | GuestName | RoomCa | A/C | NAT | Arrival | Departure | Stay | Remark | address |
|---|---|---|---|---|---|---|---|---|---|---|---|
| 1 | 626 | 36203 | Achmad Nuzulrullah | FMS3 | 2/0 | INA | 16/05/2018 | 25/05/2018 | 1 | 085697258758 | Kecamatan Batununggal, Kota Bandung, Jawa Barat |
| 2 | 522 | 36208 | Yongjun Cai, | DLD3 | 2/0 | INA | 20/05/2018 | 25/05/2018 | 1 | 087851423695 | Kemayoran,Jakarta Pusat,DKI Jakarta |
| 3 | 916 | 36149 | Yuriko kayamoto, | DLD3 | 2/0 | INA | 23/05/2018 | 25/05/2018 | 2 | 085736262372 | Kalideres,Jakarta Barat,DKI Jakarta |
| 4 | 506 | 36318 | Benediktus dicki kristi | DLD3 | 2/0 | INA | 24/05/2018 | 25/05/2018 | 4 | 082330602071 | Kecamatan Buahbatu, Kota Bandung, Jawa Barat |
| 5 | 338 | 36315 | Budi Budiman, | DLXS3 | 3/0 | INA | 24/05/2018 | 25/05/2018 | 1 | 083871576645 | Kecamatan Cibiru, Kota Bandung, Jawa Barat |

**Figure 4.** Extract data

Data transformation is the process of retrieving raw data that cannot be stored in a data warehouse. Therefore, the data must conform to the data warehouse's predetermined standard structure so that it can be stored in a data warehouse (see Figure 5).

IOP Conf. Series: Materials Science and Engineering 1115 (2021) 012040 doi:10.1088/1757-899X/1115/1/012040

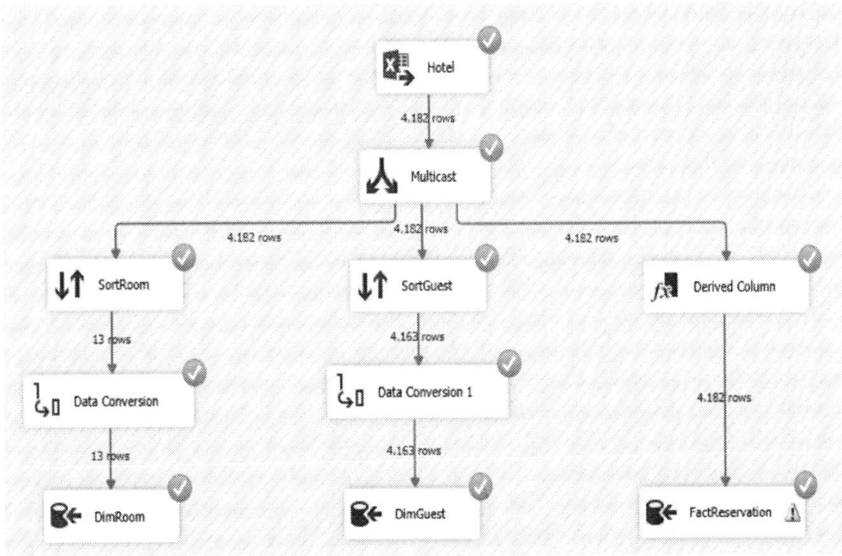


**Figure 5.** ETL Processes

Load data has been adjusted into the data format made previously, sending data from the Recordset to the destination table in the prepared Data Warehouse. This data can be in the form of dimension data or factual data. Then proceed with the process of filling in the Integration Log data. To be forwarded to the original face and services, where users can access them. Users access the output from this Loading stage in the form of data and information for reporting.

### 2.3. Define Fact and Dimension Table

As a physical data model, its structure can be seen from various points of view, with three parts (rows, columns, and layers), with multidimensional data objects in it whose attributes are divided into dimensional attributes and measurement attributes (facts) [21]. For dimensional objects in the form of time, which includes the year and month, store customer data in a month or year (see Figure 6). For fact table and dimension, the table can be seen in Table 1 and Table 2.

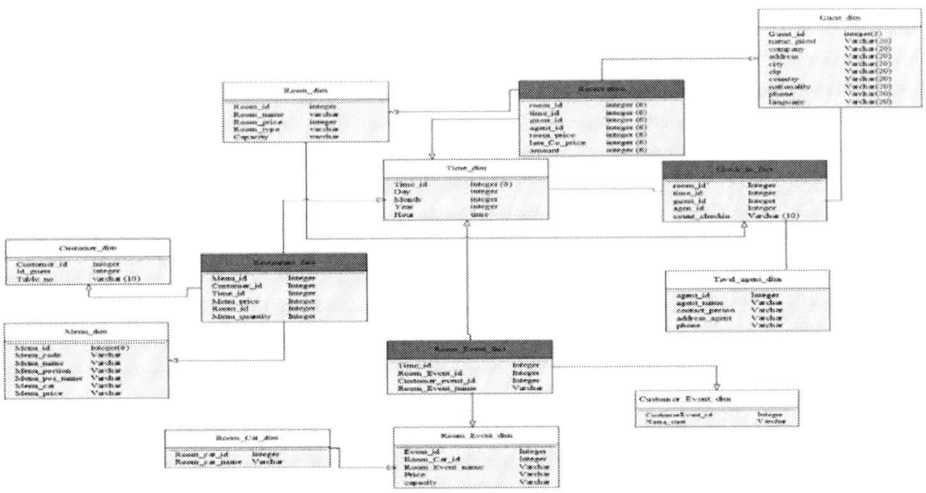

**Figure 6.** Star Schema

**Table 1.** Fact Table

| Table Fact | Description |
|---|---|
| Reservation Fact | Serves to see customer characteristics, location, and budget for hotel room selection. |
| Event Fact | Serves to see event room users in hotels by customers (meeting rooms, weddings, and big conferences). |
| Agent Travel Fact | Serves to see customers' characteristics, in making reservations to hotels in various ways such as to hotels directly or through applications. |
| Restaurant Hotel Fact | Serves to see the characteristics of a restaurant in making reservations in various ways, such as to hotels directly or through applications |

**Table 2.** Dimension Table

| Table Dimension Name | Attribute | Description |
|---|---|---|
| Dimension Guest | GuestId, GuestName, Address, PhoneNumber, City, Region, Gender | This dimension contains information relating to hotel guests |
| Dimension Room | Roomid, RoomCat, Price, TotalRoom | This dimension describes the room specifications provided by the hotel |
| Dimension Time | TimeId, Date, Day, month, year | This dimension contains the division of time by year and month. This dimension is useful when management wants to view reports based on a specific time |
| Dimension CustEvent | CustId, Custname, TypePromosi, CostPromosi | This dimension contains information relating to customers who rent unique meeting rooms |
| Dimension EventRoom | RoomId, RoomName, Capacity, Price | This dimension describes the specifications of the meeting room facilities provided by the hotel |
| Dimension Agent | Agent_Id, Agent_name, Agent_email, Agent_number | This dimension contains travel agent information that describes customer transactions when making a reservation |
| Dimension Restaurant | Restaurant_id, Restaurant_name | This dimension describes the restaurant specifications provided by the hotel |
| Dimension Menu | Menu_id, Category, Item_name, Price | dimension describes information about the name of the food in the restaurant |

### 2.4. Operation OLAP

*Slicing* and *Dicing* operations to view data as a visualization of the cube. By slicing and dicing, users can view data from several perspectives. Users can extract part of the aggregated data and can check in detail based on the desired dimensions. Aggregated data is pre-calculated data (pre-calculated). Furthermore, drill-down data aggregation users can do by drill-down and consolidation. Display of more detailed data such as customer information will display customer details.

### 2.5. Visualization Data

OLAP reporting results are visualized in the form of diagrams or graphs so that the information obtained is easy to understand by displaying information on sales of guest reservations, room rentals, and use of restaurants at Hotel Simply Valore.

### 2.6. Building Software

Software built for OLAP Analytical CRM reporting using the tools Microsoft SQL Studio and the PHP programming language.

## 3. Results and Discussion

In this study, an analytical CRM system was built with OLAP implementation to analyze customer data, where there are customer segmentation reporting results in Geography, Demography, and customer travel agents to carry out specific promotions for some hotel customers.

### 3.1. OLAP In CRM System of Hotel Simply Valore

There is a reservation data analysis menu on the front office dashboard to browse data and determine the number of sales and promotion locations with the most data for management. The decline is based on the data that customers do a lot of activity and recruit new potential customers during hotel promotions. The query response time testing can be seen in Table 3.

In Figure 7, the analytical CRM system displays the results of several facts and customer dimensions, including customer reporting each year with this information. The hotel is more aware of each customer's reports every month.

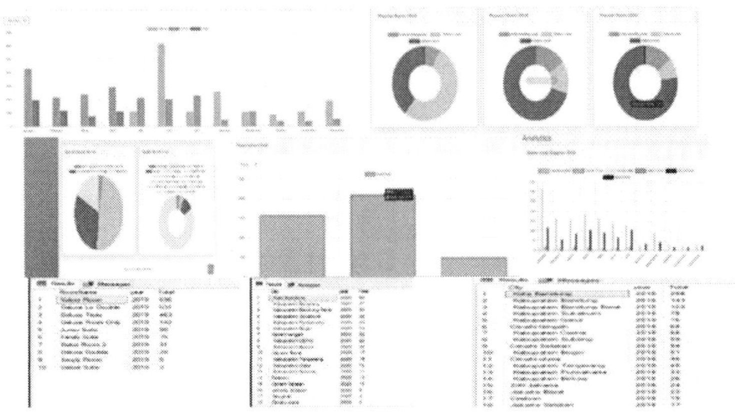

**Figure 7.** Analytical CRM System

IOP Conf. Series: Materials Science and Engineering 1115 (2021) 012040 doi:10.1088/1757-899X/1115/1/012040

**Table 3** Query respond time testing

| No | Description | Total Data | Method | Experiments | Average speed (milliseconds) OLAP |
|---|---|---|---|---|---|
| 1 | Testing Report on the Number of Hotel Guests (2018-2020) | 100 | Get | 30 Times | 44.90 |
| | | 3000 | Get | 30 Times | 65.77 |
| | | 5000 | Get | 30 Times | 79.91 |
| 2 | Testing Report on the Number of Travel Agents (2018-2020) | 5000 | Get | 30 Times | 75.00 |
| 3 | Testing Report Customer Event Room Hotel Facilities (2018-2020) | 700 | Get | 30 Times | 49.12 |
| 4 | Hotel Restaurant Testing Report (2018-2020) | 1000 | Get | 30 Times | 53.42 |
| | Result | 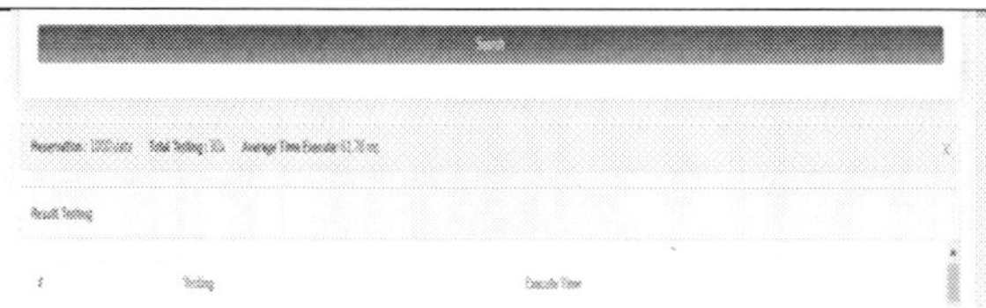 | | | |

The results can be obtained from the query response for data access speeds of less than 80ms to display information from the data in 30 times the experience. With this, the software is feasible to implement. This output is shown by fast data execution and precise accuracy, and a drill-down and roll-up process. Able to analyze more deeply and offer more detailed customer reporting.

## 4. Conclusions

This study presents an Analytical CRM system in implementing OLAP by displaying information and result data in the form of web-based operable graphs. Olap serves as a useful and essential tool used for analysis needs because of this star schema design. It can analyze data from various dimensional perspectives to quickly capture customer needs. With the use of OLAP technology, data from conventional databases are converted into a data warehouse. Data storage on a large scale so that it can be easily accessed by managers to make decisions. The research found that data access speeds were less than 80ms in 30 times of experience and showed that this software was feasible to implement. This condition was indicated by fast data execution and precise accuracy, and then drill down and roll-up. Able to analyze more deeply. This study produces OLAP Reporting, which can analyze large data sizes and minimize previous software gaps. Faster query response times. Further research can be developed by integrating data from the hotel transactional system so that the analysis process can run in real-time.

## References

[1] Bayyinah A Optimization Online Analytical Processing (OLAP) Data Sales Door Case Study CV Adilia Lestari *Int. J. Comput. Tech.* **5** (2) 16–22

[2] Girsang A, Isa S M, Susilo B, Maxlian, Satya D, Fariz S A and Ramdani D 2017 Decision support system using data warehouse for hotel reservation system *Proceeding International Conference on Sustainable Information Engineering and Technology* 369–373

[3] Chaudhuri S, Dayal U and Narasayya V 2011 An Overview of Business Intelligence Technology *Communications of the ACM* **54** (8) 88-98.

[4] Chavan R and Nehe R 2017 Analysis of Data Warehousing and OLAP Technology *International Conference on Innovations in IT and Management* **2** 225–231

[5] Dewan S, Aggarwal Y and Tanwar S 2013 Review on Data Warehouse, Data Mining and OLAP Technology : As Prerequisite aspect of business decision- making activity **1** (10) 30-39

[6] Micu A, Micu A-E and Capatina A 2009 Innovative Analysis of a CRM Database using Online Analytical Processing (OLAP) Technique in Value Chain Management Approach *proceedings of the 10th wseas Int. Conference on mathematics and computers in business and economics* 158-164

[7] Khodakarami F and Chan Y E 2014 Exploring the role of customer relationship management (CRM) systems in customer knowledge creation *Inf. Manag.* **51** (1) 27-42

[8] Case A S, Armada P T, and Motor I 2015 Analytical Customer Relationship Management for Garage Services Recommendation Using the Generalized Sequential Pattern Method *Journal of Computer Science and Software Engineering (IJCSSE)* **4** (4) 95-101

[9] Burez J and Poel D V D 2007 CRM at a pay-TV company : Using analytical models to reduce customer attrition by targeted marketing for subscription services *Expert Systems with Applications* **32** (2) 277–288

[10] Farquad M A H, Ravi V and Raju S B 2014 Churn prediction using comprehensible support vector machine: An analytical CRM application *Applied Soft Computing* **19** 31-40

[11] Hananto V R, Churniawan A D and Wardhanie A P 2017 Designing Analytical CRM to Support Customer Segmentation in Educational Institutions *J. Ilm. Teknol. Inf. Asia*, **11** (1) 79

[12] Josiah A, Ikenna O, Jennifer A, Chinaedum I, Justina R and Nnamonso A "The Relevance of Analytical CRM and Knowledge Management in an Organisation : A Data Mining Structure," vol. 4, no. 2, pp. 208–215, 2015.

[13] Faculty I 2020 Increasing Loyalty Using Customer Relationship Management : A Case on Co-working Space Company *Proceedings of the International Conference on Industrial Engineering & Operations Management* 1502–1512

[14] Winata A and Fiqri I A 2017 Emersia Hotel Customer Loyalty in Bandar *Journal Darmajaya* **3** (2) 133–149

[15] Hallowell R 1996 The relationships of customer satisfaction, customer loyalty, and profitability: An empirical study *Int. J. Serv. Ind. Manag.* **7** (4) 27–42

[16] Ersi D Y and Semuel H 2014 Crm Analysis, Customer Satisfaction and Product Loyalty *J. Manaj. Pemasar.***8** (1) 1–8

[17] Mardiani I 2014 Segmentation Analysis and Targetting of Students of the Faculty of Economics, Esa Unggul University *J. Ekon. Univ. Esa Unggul* **5** (2) 69–79

[18] Irawan D and Japarianto E 2013 Analysis of the Effect of Product Quality on Loyalty through Satisfaction as an Intervening Variable in Por Kee Restaurant Customers in Surabaya *J. Manaj. Pemasar.* **1** (2) 1–8

[19] Hakim D K 2011 Development of Licensing Reporting Applications with Online Analytical Processing ( OLAP ) ( Development of License Report Application with Online Analytical Processing ( OLAP )) *JUITA* **1** (3) 105–111

[20] Sidi E, El M and Amin E 2016 Star Schema Advantages on Data Warehouse: Using Bitmap Index and Partitioned Fact Tables *Int. J. Comput. Appl.* **134** (13) 11-13
[21] Maliappis M and Kremmydas D 2015 An online analytical processing (OLAP) database for agricultural policy data: A Greek case study *CEUR Workshop Proc.* **1498** 214-225

# Industry X.0 and Strategic Ambidextrous Trilogy of Professional Engineering, Industrial Engineering and Business Engineering

**K E N Soebandrija***

Industrial Engineering Department, Faculty of Engineering, Bina Nusantara University, Jakarta, Indonesia 11480

*Corresponding author: Knugroho@Binus.edu

**Abstract.** Industry X.0 colligates agile, nimble and breakthrough smart technologies in digital transformation of industry era. In this paper, the mentioned Industry X.0 synergizes strategic ambidextrous trilogy of professional engineering, industrial engineering and business engineering. The objective of this paper is to enhance the theoretical perspectives within industrial and managerial implementation in Indonesia and Global Situation. The enhancement is intended to accelerate corporate operational efficiency and corporate wide growth. The research methodology in this paper refers to quantitative strategic ambidextrous approach that requires exploration and exploitation trilogy. This trilogy enhances every single advantage from unison of professional engineering, industrial engineering and business engineering. The result and discussion of this paper refers to 4 indispensable corporate transformations through both exploration and exploitation within unison of incremental and radical innovation, known as ambidexterity. Those 4 corporate transformations are characterized by the strategic ambidextrous trilogy toward: a. manufacturing and production; b. design and product engineering; c. innovation and growth; d. support and services. Ultimately both Industry X.0 and Strategic Ambidextrous Trilogy synergize into sustainable transformation to enhance theoretical perspective within industrial and managerial implementation through several benchmarks. Those benchmarks refer to Industry 4.0 (IR 4.0), Making Indonesia 4.0 (MI 4.0), Society 5.0, and Hallyu 2.0.

## 1. Introduction

Industry X.0 colligates agile, nimble and breakthrough smart technologies in digital transformation of industry era. In this paper, the mentioned Industry X.0 synergizes strategic ambidextrous trilogy of professional engineering, industrial engineering and business engineering.

There are several backgrounds surrounding the decision making to prioritize Ambidexterity as recommended strategic implementation in Industry X.0. The aforementioned strategic implementation is useful for the turbulence of volatility, uncertainty, complexity and ambiguity of the situation in the digital transformation of industry era. Therefore, there are still open rooms for ambidexterity in the strategic implementation in the corporate.

## 2. Materials

The objective of this paper is to enhance the theoretical perspectives within industrial and managerial implementation in Indonesia and Global Situation. The enhancement is intended to accelerate corporate operational efficiency and corporate wide growth.

The implementation of the aforementioned industrial and managerial implementation refers to the productive publication of scholar works from the Institute of Industrial and Systems Engineering (IISE), within Industrial Engineering, as illustrated in Table 1 [2].

**Table 1.** Publication of Institute of Industrial and Systems Engineering (IISE) within Ambidextrous Organization

| No | Total Citation | Publication Title | Author(s) | Year |
|---|---|---|---|---|
| 1 | 775 | Economic production cycles with imperfect production processes | Rosenblatt M.J., Lee H.L[3] | 1986 |
| 2 | 510 | An EOQ model for items with Weibull distribution deterioration | Covert R.P., Philip G.C[4] | 1973 |
| 3 | 458 | A review of multivariate control charts | Lowry C.A., Montgomery D.C [5] | 1995 |
| 4 | 458 | Facility location under uncertainty: A review | Snyder L.V [6] | 2006 |
| 5 | 449 | An algorithm for the computer control of a flexible manufacturing system | Kimemia J., Gershwin S.B [7] | 1983 |
| 6 | 427 | A Review of Production Planning and Scheduling Models in the Semiconductor Industry Part I: System Characteristics, Performance Evaluation and Production Planning | Uzsoy R., Lee C.-Y [8] | 1992 |
| 7 | 412 | Residual-life distributions from component degradation signals: A Bayesian approach | Gebraeel N.Z., Lawley M.A., Li R., Ryan J.K [9] | 2005 |
| 8 | 381 | Appointment scheduling in health care: Challenges and opportunities | Gupta D., Denton B. [10] | 2008 |
| 9 | 329 | Travel-time models for automated storage/retrieval systems | Bozer Y.A., White J.A., Sr.[11] | 1994 |
| 10 | 323 | On how to prioritize design requirements during the QFD planning process | Wasserman G.S. [12] | 1993 |
| 11 | 312 | Strategies for robust multiresponse quality engineering | Pignattello J.J., Jr.[13] | 1993 |
| 12 | 292 | Application of the similarity coefficient method in group technology | Seifoddini H., Wolfe P.M., Sr.[14] | 1986 |
| 13 | 291 | A review of production planning and scheduling models in the semiconductor industry part h: Shop-floor control | Uzsoy R., Lee C.-Y., Martin-Vega L.A. [15] | 1994 |
| 14 | 280 | Reliable preventive maintenance scheduling | Malik M.A.K. [16] | 1979 |
| 15 | 279 | Determining the importance weights for the customer requirements In QFD using a fuzzy AHF with an extent analysis approach | Kwong C.K., Bai H.[17] | 2003 |
| 16 | 256 | A review of integrated analysis of production-distribution systems | Sarmiento A.M., Nagi R.[18] | 1999 |
| 17 | 234 | Design of automatic guided vehicle systems | Maxwell W.L.[19] | 1982 |
| 18 | 230 | Controlling factor weights in data envelopment analysis | Roll Y., Cook W.D., Golany B.[20] | 1991 |
| 19 | 226 | Robustness measures and robust scheduling for job shops | Leon V.J., Wu S.D., Storer R.H.[21] | 1994 |
| 20 | 220 | An AHP framework for prioritizing customer requirements in QFD: An industrialized housing application | Armacost R.L., Componation P.J., Mullens M.A., Swart W.W.[22] | 1994 |

## 3. Methods

The research methodology in this paper refers to quantitative strategic ambidextrous approach that requires exploration and exploitation trilogy [23], [24], [25], [26]. The ambidextrous trilogy enhances every single advantage from unison of professional engineering, industrial engineering and business engineering that require corporate transformation. In this situation, corporate transformations are characterized by the strategic ambidextrous trilogy toward: a. manufacturing and production; b. design and product engineering; c. innovation and growth; d. support and services.

## 4. Result and Discussion

The result and discussion of this paper refers to 4 indispensable corporate transformations through both exploration and exploitation within unison of incremental and radical innovation, known as ambidexterity.

Those 4 corporate transformations are characterized by the strategic ambidextrous trilogy toward: a. manufacturing and production; b. design and product engineering; c. innovation and growth; d. support and services. Ultimately both Industry X.0 and Strategic Ambidextrous Trilogy synergize into sustainable transformation to enhance theoretical perspective within industrial and managerial implementation through several benchmarks. Those benchmarks refer to Industry 4.0 (IR 4.0), Making Indonesia 4.0 (MI 4.0), Society 5.0, and Hallyu 2.0.

Agile, nimble and breakthrough smart technologies constitute characteristics of Industry X.0. Therefore, the roles of ambidextrous characteristics are indispensable, including its trilogy of professional engineering, industrial engineering and business engineering. The Industry X.0, according to Accenture, has its unique path within the ambidextrous organization toward digital transformation in digital era. In term of Innovation and Growth, corporate is required to synergize the decision making toward their disruptive business model innovation within products and services in order to provide customers the new excellence level.

Furthermore, in term of Manufacturing and Production, Industry X.0 is established based upon individual customer requirements within design phase. To some extent, this requirement is further followed up into manufacturing process using cutting edge technologies toward cost reductions while simultaneously augment the world class quality standards. Eventually, in term of Service and Support, Industry X.0 is personalized to cater each unique customer' requirements. The agile, nimble and breakthrough smart technologies of Industry X.0 is enabled through the Artificial Intelligence software and networks.

## 5. Conclusion

Industry X.0 colligates agile, nimble and breakthrough smart technologies in digital transformation of industry era. In this paper, the mentioned Industry X.0 synergizes strategic ambidextrous trilogy of professional engineering, industrial engineering and business engineering.

The objective of this paper is to enhance the theoretical perspectives within industrial and managerial implementation in Indonesia and Global Situation. The enhancement is intended to accelerate corporate operational efficiency and corporate wide growth. The research methodology in this paper refers to quantitative strategic ambidextrous approach that requires exploration and exploitation trilogy.

This trilogy enhances every single advantage from unison of professional engineering, industrial engineering and business engineering. The result and discussion of this paper refers to 4 indispensable corporate transformations through both exploration and exploitation within unison of incremental and radical innovation, known as ambidexterity.

Those 4 corporate transformations are characterized by the strategic ambidextrous trilogy toward: a. manufacturing and production; b. design and product engineering; c. innovation and growth; d. support and services. Ultimately both Industry X.0 and Strategic Ambidextrous Trilogy synergize into sustainable transformation to enhance theoretical perspective within industrial and managerial implementation through several benchmarks. Those benchmarks refer to Industry 4.0 (IR 4.0), Making Indonesia 4.0 (MI 4.0), Society 5.0, and Hallyu 2.0.

The Industry X.0 has its unique path within the ambidextrous organization toward digital transformation in digital era. In term of Innovation and Growth, corporate is required to synergize the decision making toward their disruptive business model innovation within products and services in order to provide customers the new excellence level.

Furthermore, in term of Manufacturing and Production, Industry X.0 is established based upon individual customer requirements within design phase. To some extent, this requirement is further followed up into manufacturing process using cutting edge technologies toward cost reductions while simultaneously augment the world class quality standards.

## References

[1] Lis, A, Józefowicz, B, Tomanek, M, Gulak-Lipka, P 2018 The Concept of the Ambidextrous Organization: Systematic Literature Review *International Journal of Contemporary Management* **17** 77–97
[2] Hu, S 2019 Fifty years of IISE Transactions: A bibliometric overview. *IOSR Journal of Mechanical and Civil Engineering* **16** 76-87
[3] Rosenblatt, M J and Lee, H L 1986 Economic production cycles with imperfect production processes *IIE Transactions* **18** 48–55
[4] Covert R P and Philip G C 1973 An EOQ model for items with Weibull distribution deterioration *AIIE Transactions* **5** (4) 323–326
[5] Laengle S, Modak N M, Merigó J M and De La Sotta C 2018 Thirty years of the International Journal of Computer Integrated Manufacturing: A bibliometric analysis. *International Journal of Computer Integrated Manufacturing* **31** (12) 1247– 1268
[6] Snyder L V 2006 Facility location under uncertainty: A review *IIE Transactions* **38** (7) 547–564
[7] KchaouBoujelben M and Boulaksil Y 2018 Modeling international facility location under uncertainty: A review, analysis, and insights *IISE Transactions* **50** (6) 535–551
[8] Vieira E S, Cabral J A S and Gomes J A N F 2014 Definition of a model based on bibliometric indicators for assessing applicants to academic positions *Journal of the American Society for Information Science &Technology* **65** (3) 560–577
[9] Fu H, Fang K and Fang C 2018 Characteristics of scientific impact of Resources Conservation and Recycling in the past 30 years *Resources, Conservation & Recycling* **137** 251–259
[10] Gupta D and Denton B 2008 Appointment scheduling in health care: Challenges and opportunities *IIE Transactions* **40** (9) 800–819
[11] Bottani E and Vignali G 2019 Augmented reality technology in the manufacturing industry: A review of the last decade *IISE Transactions* **51** (3) 284–310
[12] Wang C, Lim M K and Lyons A 2018 Twenty years of the International Journal of Logistics Research and Applications: A bibliometric overview *International Journal of Logistics Research and Applications* **22** (3) 304–323
[13] Noyons E C M, Moed H F and Luwel M 1999 Combining mapping and citation analysis for evaluative bibliometric purposes: A bibliometric study *Journal of the American Society for Information Science* **50** (2) 115–131
[14] Sci$^2$ Team. 2009 . Science of Science (Sci$^2$) Tool. Indiana University and SciTech Strategies. Retrieved from http://sci.slis.indiana.edu.
[15] Zupic I and Čater T 2015 Bibliometric methods in management and organization. *Organizational Research Methods* **18** (3) 429–472
[16] Merigó J M, Pedrycz W, Weber R and de la Sotta C 2018. Fifty years of Information Sciences: A bibliometric overview *Information Sciences* **432** 245–268
[17] Kessler M M 1963 Bibliographic coupling between scientific papers. *American*

*Documentation* **14** (1) 10–25
[18] Small H 1973 Co-citation in the scientific literature: a new measure of the relationship between two documents *Journal of the American Society for Information Science* **24** (4) 265–269
[19] Merigó J M, Pedrycz W, Weber R and de la Sotta C 2018 Fifty years of Information Sciences: A bibliometric overview *Information Sciences* **432** 245–268
[20] Bottani E and Vignali G 2019 Augmented reality technology in the manufacturing industry: A review of the last decade *IISE Transactions* **51** (3) 284–310
[21] Laengle S, Modak N M, Merigó J M and De La Sotta C 2018 Thirty years of the International Journal of Computer Integrated Manufacturing: A bibliometric analysis *International Journal of Computer Integrated Manufacturing* 31 (12) 1247– 1268
[22] Archambault É, Campbell D, Gingras Y and Larivière V 2009 Comparing bibliometric statistics obtained from the web of science and Scopus *Journal of the American Society for Information Science and Technology* **60** (7) 1320–1326
[23] Basias N and Pollais Y 2018 Quantitative and Qualitative Research in Business & Technology: Justifying a Suitable Research Methodology *Review of Integrative Business and Economics Research* **7** 91-105
[24] Goertz G and Mahoney J 2012 *A Tale of Two Cultures: Qualitative and Quantitative Research in the Social Sciences* New York: Princeton University Press.
[25] Martin W and Bridgmon K 2012 *Quantitative and Statistical Research Methods. From Hypothesis to Results* New York: John Wiley & Sons
[26] Yin R 2011 *Applications of Case Study Research* New York: SAGE Publications

IOP Conf. Series: Materials Science and Engineering 1115 (2021) 012042 doi:10.1088/1757-899X/1115/1/012042

# A Research to Design, Develop and Implementation of Android Application System for Waste Bank Sharia Community at Kampung Hijau Kemuning

**R Destriana*, N Handayani, S M Husain and A T P Siswanto**

Department of Information, University of Muhammadiyah Tangerang, Banten, Indonesia 15118

*Corresponding author: rachmat.destriana@ft-umt.ac.id

**Abstract.** The problem of garbage is a behavioral problem, therefore the need for a change in public behavior to discipline throwing garbage in place, and the problem of garbage becomes one of the problems in major cities and other metropolitan cities, including in Tangerang area. Household waste is one type of garbage that can cause increasing problems facing the government in dealing with it. Kampung Hijau Kemuning is one of the residential settlements in Binong Village, Curug Sub-District that manages sharia-based waste banks aimed at encouraging environmental rescue to be green, clean, reduce waste and become an additional source of income for the residents of Kampung Hijau Kemuning. In the process of waste bank management requires the development of either the use of computer applications based on android or in the process of financial management that is clean and free from usury in accordance with islamic (sharia) rules. The purpose of this research is to build an Android-based Sharia Waste Bank application with a user-friendly view using the Waterfall method. The waterfall model provides a sequential or sequenced approach to software life flow. The results of this research will help the process of managing innovative waste bank activities and become input to increase the empowerment of micro-business communities in an activity, especially in Tangerang area.

## 1. Introduction

The issue of garbage has become a global concern as all countries face such problems [1]. The problem of garbage is a behavioral problem, hence the need for a change in public behavior to discipline throwing garbage in place, and the problem of garbage becomes one of the problems in major cities and other metropolitan cities, including in Tangerang area. Household waste is one type of garbage that can cause increasing problems facing the government in dealing with it [2]. The Regent of Tangerang, stated that it is increasingly difficult to control and manage garbage. In Tangerang Regency, the volume of household waste every day is around 1,500 tons.

Law No. 18 of 2008 on Waste Management and Government Regulation No. 81 of 2012 mandates the need for a fundamental paradigm change in waste management, namely from the paradigm of garbage collection to processing that focuses on waste reduction and waste management, namely by implementing waste arising restriction activities, recycling and waste reuse or better known as

Reduce, Reuse and Recycle (3R) through smart, efficient and programmatic efforts. People are starting to move to take advantage of the garbage. Both dry and wet garbage.

Rahmadi and Mulyani (2016) [3] presented Sharia-based waste banks as methods or means of waste management in residential areas or residential complexes. Managed at the Waste Bank only covers dry waste and can still be recycled into other goods, such as plastics, bottles, cardboard and other goods that each have economic value and in accordance with religious rules. The Financial Management System of Waste Bank is sharia-based, due to the increasing public demand for a financial system that conforms to religious rules because the majority of Indonesians are Muslims. Society demands a clean and usury-free financial or banking system.

Kampung Hijau Kemuning is one of the residential settlements in Binong Village, Curug Sub-District that manages sharia-based waste banks aimed at encouraging environmental rescue to be green, clean, reduce waste and become an additional source of income for the residents of Kampung Hijau Kemuning. The processed waste comes from household waste (organic or non-organic) that has been sorted and then deposited into garbage collectors and some waste is made from recycled waste. Proceeds from the sale of non- organic waste in the form of customer savings with an agreement of 85% for customer savings and 15% for environment and operational costs [4]. The current waste bank management system is still conventional, namely the process of recording customer data, garbage data, savings data from garbage sales is still recorded in the book.

The increasing demands of the people of Kampung Hijau Kemuning in the process of managing waste banks so that it requires the development of either the use of android-based computer applications or in the process of financial management that is clean and free from usury in accordance with islamic rules. The need for android-based computer applications is urgently needed in managing the administration of sharia waste banks both in the process of recording junk data, customer data to know savings balances that can be accessed anywhere and anytime using smartphones. The purpose of this research is to build an Android-based Sharia Waste Bank application with a user- friendly view and can assist in the process of managing innovative waste bank activities and become an input to increase the empowerment of micro-enterprises in a community in an activity especially in Tangerang area.

Empowerment is essential for the integral economy of the people who have an important position and must be held thoroughly, optimally and continuously through conducive climate development, the provision of opportunities to strive, support the protection and development of the widest range of businesses, in order to increase the position, role, and potential of Micro Enterprises. In realizing economic growth, and can increase people's incomes, create jobs and reduce poverty

## 2. Literature Review

Ogedebe (2002) found that in the analysis and design of the system, especially for the transaction process, where the dialog displayed is easier to understand [5]. The greater the interaction between the computer and the user, the greater the benefits gained when the information system development process will be faster and make the user more interactive in the development process [6]-[8]. The systematic approach in waterfall, as follows [9-10]:

1. Communication

This step is an analysis of the needs of the software, and the stage of conducting data collection by conducting meetings with garbage banks, or collecting additional data in journals, articles, and the internet.

2. Planning

The planning process is a continuation of the communication (analysis requirement) process [8]. This stage generates a user requirement document which can be said as data about the user's desire in the creation of android-based applications, including plans to be carried out. Where the data can be in the form of junk bank savings book as well as data on garbage weighing transactions that will be processed by the garbage bank admin.

3. Modeling

This modeling process will translate the data that has been obtained in the previous phase to be processed into a software engineering design. The process focuses on structured design of data, software architecture, user interface representation, and procedural details. This stage will generate a document called software requirement.

4. Construction

Construction is the process of creating an application through the Coding process. Coding or coding is a translation of a design in a language recognized by a computer. The programmer translates the transaction requested by the user. This stage is the real stage in working on the creation of software, meaning the use of computers can be maximized in this stage. Once the encoding is complete, the next test will be tested against the application that has been created. The purpose of testing is to test all the functions in the application to be improved.

5. Deployment

This stage can be said to be the final stage in the creation of a software or application. After performing analysis, design, coding then the finished system will be used by the user i.e. the garbage bank and finally the software should be maintained periodically.

## 3. Methodology

### 3.1 Current System Analysis

From the results of interviews and observations that have been conducted in Kampung Hijau Kemuning, in accordance with the discussion taken about the management of waste bank data in Kampung Hijau Kemuning such as data collection of garbage transactions, sorting garbage until savings still use semi computer i.e. using Microsoft Excel. The current environment activity and current UseCase can be seen in Figure 1 and Figure 2.

The needs that users need are obtained from the results of interviews conducted by researchers to the admin of Kampung Hijau Kemuning, namely to analyze the needs of the system to be made, researchers get the following conclusions:

1. It is necessary an information system that can know the update of junk data as well as the savings balance of each customer.
2. A system is needed that can facilitate the documentation of garbage data calculations, garbage transactions to the savings balance of each customer.
3. Can provide the necessary reports quickly, accurately and accurately to admins and customers.

IOP Conf. Series: Materials Science and Engineering 1115 (2021) 012042 doi:10.1088/1757-899X/1115/1/012042

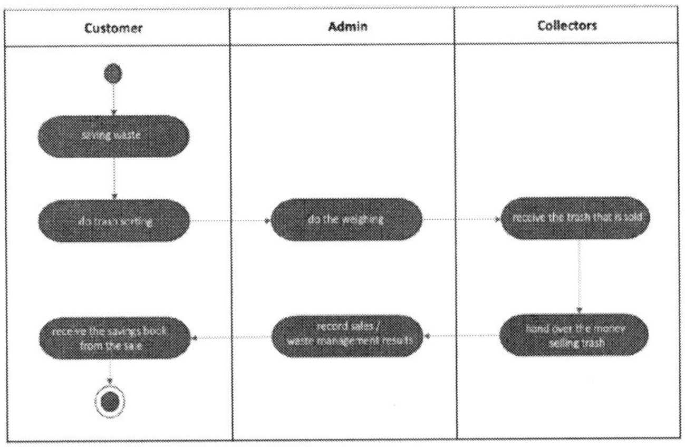


**Figure 1.**Current Environment Activity

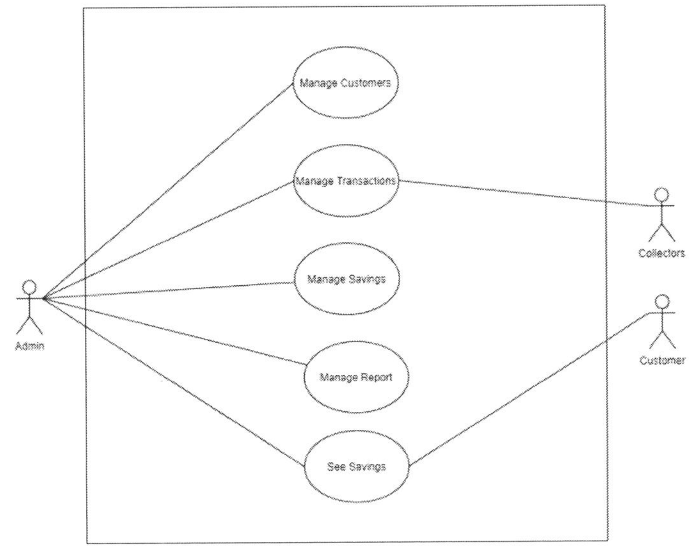


**Figure 2.** Current UseCase

## 4. Results and Discussion

### 4.1 Proposed System Program Design

Visual modeling that enables the development of the system makes blue prints of its vision in its default form by creating use case Diagrams, Activity Diagrams, Sequence Diagrams and Class Diagrams (see Figures 3-5).

IOP Conf. Series: Materials Science and Engineering 1115 (2021) 012042 doi:10.1088/1757-899X/1115/1/012042

1. Use Case Diagram application of Sharia Waste Bank Kemuning

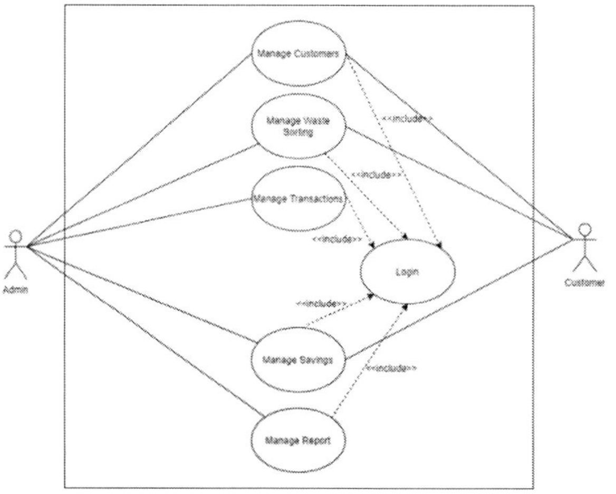


**Figure 3.** Future Design for UseCase Diagram

2. Activity Diagram application of Sharia Waste Bank Kemuning.

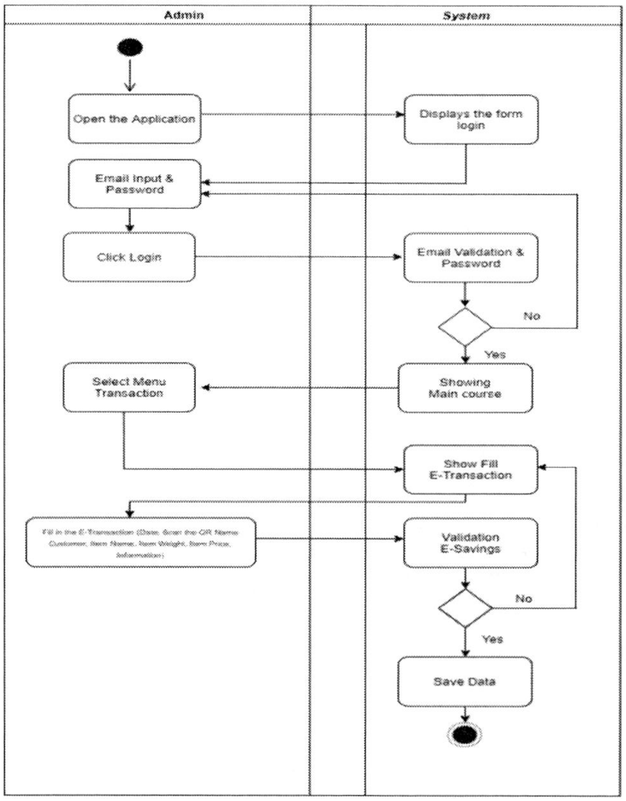


**Figure 4.** Future Design for Activity Diagram

3. Sequence Diagram application of Sharia Waste Bank Kemuning.

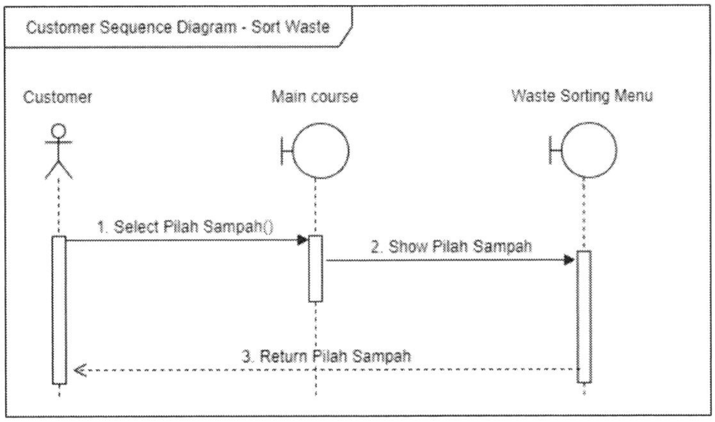


**Figure 5.** Future Design for Sequence Diagram

### 4.2 The Development of the Sharia Kemuning Waste Bank - Android based.

Part of the system design is the concept of interface design, process and data with the aim of producing system specifications that suit the needs. Here's a look at the design of Android-based Sharia Kemuning Waste Bank application.

This is the page for everyone who want to go to the application at the first time (see Figures 6).

**Figure 6.** Login Menu

This is a menu for users when they want to use the menus in the garbage bank application, there are several menus such as the customer menu, the waste sorting menu, the transaction menu, the savings menu and the reports menu.

This is the main menu when we are going to make transactions, whether it's sorting waste or saving transactions (see Figures 7 and Figure 8).

IOP Conf. Series: Materials Science and Engineering 1115 (2021) 012042 doi:10.1088/1757-899X/1115/1/012042

**Figure 7.** Customer Menu

**Figure 8.** Transaction Menu

This is the report menu for the waste bank transaction and saving report transaction (see Figures 9 and Figure 10).

**Figure 9.** Report Menu of waste bank

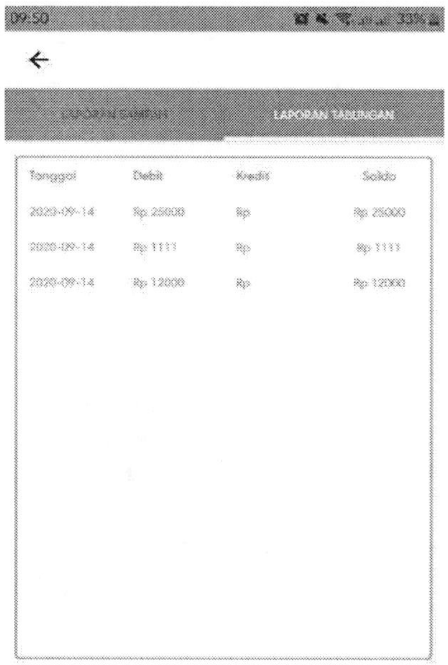


**Figure 10.** Report Menu saving

### 4.3 System Testing

At this testing stage, it is carried out using black box testing for all functions of the Android-based waste bank application carried out by the waste bank. The results of testing using the black box method can be seen in Table 1.

**Table 1**. The results of testing using the black box method

| No | Input | Expected Output | Actual output | Conclusion |
|---|---|---|---|---|
| 1 | Login | Application showing form login | showing login form | Valid |
| 2 | Existing email and password | Application showing HomeScreen | showing Homscreen | Valid |
| 3 | Vies Customer | Application showing List Nasabah | showing List of Customer | Valid |
| 4 | View waste Item | Application showing Menu Sorting Trash | showing Menu of waste collecting | Valid |
| 5 | Saving transaction | Application showing Menu of saving transaction and Input saving transaction | showing Menu of saving transaction and saving | Valid |
| 6 | Saving history | Application showing History | showing History of saving transaction | Valid |
| 7 | waste transaction menu | Application showing Menu of saving transaction and Input waste transaction | showing Menu of saving transaction and Input waste | Valid |
| 8 | waste transaction | Application showing History of waste transaction | showing History of waste transaction | Valid |
| 9 | waste transaction | Application showing report of waste transaction | showing report of waste transaction | Valid |

## 5. Conclusion

The research conclusions of Sharia Waste Bank based on Android Kampung Hijau Kemuning are as follows:

a. Waste bank data management, savings data from garbage sales and sales of craft products from garbage that can be used as value is input to the Android-based Sharia Waste Bank Application that very use full for the people on that area.

b. As result for user, to see the proceeds of garbage sales and the sale of craft works from garbage, customers can directly access to the android-based sharia waste bank application on their smartphone with access as a customer. The information displayed on the application can see the customer's savings balance, savings transactions, and data related to the activities of sharia waste bank kampung hijau kemuning.

IOP Conf. Series: Materials Science and Engineering 1115 (2021) 012042 doi:10.1088/1757-899X/1115/1/012042

## References

[1] Bezama A, Agamuthu P 2019 Addressing the big issues in waste management *Waste Management & Research* **37** (1) 1-3

[2] Dahlén L and Lagerkvist A 2009 Evaluation of recycling programmes in household waste collection systems *Waste Management & Research* **28** (7) 577-586

[3] Rahmadi and Mulyani D 2016 Model Sistem Informasi Keuangan Bank Sampah Sharia Jutisi: Jurnal Ilmiah Teknik Informatika dan Sistem Informasi **5** (3) 1199-1204

[4] Amaral R E C, Brito J, Buckman M, Drake E, Ilatova E, Rice P, Sabbagh C, Voronkin S and Abraham Y S 2020 Waste Management and Operational Energy forSustainable Buildings: A Review *Sustainability* **12** 1-21

[5] Ogedebe P M and Jacob B P 2012 Software Prototyping: A Strategy Touse when User Lacks Data Processing Experience *ARPN Journal of Systems and Software* **2** (6) 219-224

[6] Destriana R, Permana A, Legawa S, Yanuardi and Irawan H 2019 Security system development for vehicle using the method of "mail notification" at villa Rizki Ilhami Tangerang residential *IOP Conference Series: Materials Science and Engineering* **508** 012124

[7] Petersen K, Wohlin C and Baca D 2009 The Waterfall Model in Large-Scale Development *Lecture Notes in Business Information Processing* **32** 86-400

[8] Costabile M F 2001 Usability in the software life cycle *Handbook of Software Engineering and Knowledge Engineering* **1** 179-192

[9] Bassil Y 2012 A Simulation Model for the WaterfallSoftware Development Life Cycle *International Journal of Engineering& Technology(iJET)* **2** (5) 1-7

[10] Kulkarni R H, Padmanabham P and Basee K K 2015 Critical Review of Extended Waterfall Model *International Journal of Scientific & Engineering Research* **6** (11) 425-432

IOP Conf. Series: Materials Science and Engineering 1115 (2021) 012043 doi:10.1088/1757-899X/1115/1/012043

# Parameter Estimation and Hypothesis Testing of Geographically Weighted Bivariate Zero Inflated Poisson Inverse Gaussian Regression Models

**M S Nur, Purhadi* and A Choiruddin**

Department of Statistics, Faculty of Mathematics, Computing and Data Science, Institut Teknologi Sepuluh Nopember, Surabaya, Indonesia

*Corresponding author: purhadi@statistika.its.ac.id

**Abstract.** Count data is data that can be counted within a certain range of time. The count data itself can be accommodated in the Poisson distribution which is always positive value. Poisson regression is often used to model count data, but Poisson regression requires equality in means and variance (equidispersion). The count data often contains many zero values and is unable to meet the equidipersion assumption. Zero Inflated Poisson Inverse Gaussian Regression (ZIPIGR) is a mix poisson regression that can accommodate under/overdispersion and also the number of zero values in the data. ZIPIGR is global regression model, so the interpretation of the model obtained is global in all observational data. The ZIPIGR method that accommodate the location factor is the Geographically Weighted ZIPIGR (GWZIPIGR). Bivariate GWZIPIGR (GWBZIPIGR) can be used when there are two response variables that correlate each other. This paper will discuss parameter estimation of GWBZIPIGR using Maximum Likelihood Estimation (MLE) and numerical iteration of Berndt Hall Hall Hausman (BHHH) is used when the result of estimation equation does not close form, Meanwhile Maximum Likelihood Ratio Test (MLRT) is used to test the parameters where simultaneous hypothesis testing used the $G^2$ test statistic at large sample with $\chi^2$ distribution.

## 1. Introduction

Count data is a type of data whose event can be counted as integers and has a non-negative value, this event can be accommodated into Poisson Distribution. The use of Poisson Distribution in regression modeling is Poisson Regression, however Poisson regression requires equidispersion assumption, meaning that the mean and variance must be equal [1]. In count data, it often has a large value variance due to a large number of zeros [2], and causes Poisson regression to be unsuitable for use when the count data has excess zeros and overdispersion. The development of Poisson Regression to accommodate excess zeros and overdispersion on the response variable was carried out through the Zero-Inflated Poisson (ZIP) model [3]. Zero-Inflated Poisson Inverse Gaussian Regression (ZIPIGR) Model is another alternative to the regression model for overdispersed Poisson data. ZIPIGR combines the Zero Inflated (ZI) model which can accommodate excess zeros with the Poisson Inverse Gaussian (PIG) distribution [4]. One statistical technique commonly used to addressing spatial effects and characteristic is geographical weighted regression model (GWR) [5]. The ZIPIGR model produces parameter estimates that apply globally, meaning that the parameters generated apply to all observation units. Often the observational data not only contains the measured value but also contains location characteristics inherent in the observational data. several research on local bivariate model for ZIP has been conducted by [6][7], but the resulting model is less able to handle the presence of overdispersion in the correlated response variables. In this study, a Geographically Weighted Bivariate

Zero Inflated Poisson Inverse Gaussian Regression model was developed through parameter estimation using the Maximum Likelihood Estimation (MLE) and Hypothesis Testing using the Maximum Likelihood Ratio Test (MLRT).

## 2. Theoritical Review

### 2.1 Zero Inflated Poisson Inverse Gaussian Regression (ZIPIGR)

Suppose $Y$ has a ZIPIG distribution with parameters $\lambda, \tau,$ and $p$, where $\lambda$ is the mean, $\tau$ is the dispersion parameter, and $p$ is the zero-inflated parameter then the probabillity density function of ZIPIG can be written as follows:

$$P(Y=y\mid\lambda,\tau,p)=\begin{cases} p+(1-p)P\ (Y=0\mid\tau,p) & ;y=0 \\ (1-p)P\ (Y=y\mid\tau,p) & ;y=1,2,3,... \end{cases} \quad (1)$$

where,

$$P(Y=0\mid\lambda,\tau,p)=p+(1-p)e^{\frac{1}{\tau}}\left(\frac{2}{\pi\tau}\right)^{\frac{1}{2}}(2\lambda\tau+1)^{\frac{1}{4}}\left(\frac{\pi}{2}\right)^{\frac{1}{2}}\left(\frac{1}{\tau}\sqrt{(2\lambda\tau+1)}\right)^{-\frac{1}{2}}\exp\left(-\frac{1}{\tau}\sqrt{(2\lambda\tau+1)}\right)$$

$$P(Y=y\mid\lambda,\tau,p)=(1-p)\frac{\lambda^{y}e^{\frac{1}{\tau}}}{y!}\left(\frac{2}{\pi\tau}\right)^{\frac{1}{2}}(2\lambda\tau+1)^{-\frac{\left(y-\frac{1}{2}\right)}{2}}\left(\frac{\pi}{2}\right)^{y-\frac{1}{2}}\left(\frac{1}{\tau}\sqrt{(2\lambda\tau+1)}\right)^{\frac{1}{2}+y}\exp\left(-\frac{1}{\tau}\sqrt{(2\lambda\tau+1)}\right)$$

and

$$p=\frac{\exp\left(-\gamma\mathbf{x}^{T}\boldsymbol{\beta}\right)}{1+\exp\left(-\gamma\mathbf{x}^{T}\boldsymbol{\beta}\right)} \qquad (1-p)=\frac{1}{1+\exp\left(-\gamma\mathbf{x}^{T}\boldsymbol{\beta}\right)}$$

The ZIPIGR model can be written as follows:

$$\lambda=\exp\left(\mathbf{x}^{T}\boldsymbol{\beta}\right) \qquad \ln(\lambda)=\mathbf{x}^{T}\boldsymbol{\beta} \quad (2)$$

The estimation method used in the ZIPIGR model is the Maximum Likelihood Estimation, while the MLRT is used to test the hypothesis.

### 2.2 Bivariate Zero Inflated Poisson Inverse Gaussian Regression (BZIPIGR)

Let $Y_1$ and $Y_2$ is a random variable with BZIPIG Distribution $(Y_{1j},Y_{2j})\overset{iid}{\sim}BZIPIG(\lambda_1,\lambda_2,\lambda_0,p_1,p_2,\tau)$, then the probabillity density function of ZIPIGR can be witten as follows [8][9]:

$$P\left(Y_{1j},Y_{2j}\right)=\begin{cases} P\ \left(Y_{1j}=0,Y_{2j}=0\mid\lambda,\tau,p\right) \\ P\ \left(Y_{1j}>0,Y_{2j}=0\mid\lambda,\tau,p\right) \\ P\ \left(Y_{1j}=0,Y_{2j}>0\mid\lambda,\tau,p\right) \\ P\ \left(Y_{1j}=0,Y_{2j}=0\mid\lambda,\tau,p\right) \end{cases} \quad (3)$$

where

$$\begin{aligned} &P\ \left(Y_{1j}=0,Y_{2j}=0\mid\lambda,\tau,p\right) \\ &=p_1(\mathbf{x}_j)p_2(\mathbf{x}_j)+p_1(\mathbf{x}_j)(1-p_2(\mathbf{x}_j))\left(\frac{2}{\pi\tau}\sqrt{2\lambda_{2j}\tau+1}\right)^{\frac{1}{2}}\exp(\tau^{-1})K_{-\frac{1}{2}}\left(\tau^{-1}\sqrt{2\lambda_{2j}\tau+1}\right) \\ &+p_2(\mathbf{x}_j)(1-p_1(\mathbf{x}_j))\left(\frac{2}{\pi\tau}\right)^{\frac{1}{2}}\exp(\tau^{-1})(2\lambda_{1j}\tau+1)^{\frac{1}{4}}K_{-\frac{1}{2}}\left(\tau^{-1}\sqrt{2\lambda_{1j}\tau+1}\right) \\ &+(1-p_1(\mathbf{x}_j))(1-p_2(\mathbf{x}_j))\left(\frac{2}{\pi\tau}\right)^{\frac{1}{2}}\exp(\tau^{-1})\left(2(\lambda_{1j}+\lambda_{2j})\tau+1\right)^{\frac{1}{4}}K_{-\frac{1}{2}}\left(\tau^{-1}\sqrt{2(\lambda_{1j}+\lambda_{2j})\tau+1}\right) \end{aligned}$$

$$P\left(Y_{1j}>0,Y_{2j}=0\mid\lambda,\tau,p\right)$$
$$=\left(1-p_1\left(\mathbf{x}_j\right)\right)p_2\left(\mathbf{x}_j\right)\left(\frac{2}{\pi\tau}\sqrt{2\lambda_{1j}\tau+1}\right)^{\frac{1}{2}}e^{\frac{1}{\tau}}\frac{\lambda_{1j}^{y_{1j}}K_{y_{1j}-\frac{1}{2}}\left(\tau^{-1}\sqrt{2\lambda_{1j}\tau+1}\right)}{\left(\sqrt{2\lambda_{1j}\tau+1}\right)^{y_{1j}}y_{1j}!}$$
$$+\left(1-p_1\left(\mathbf{x}_j\right)\right)\left(1-p_2\left(\mathbf{x}_j\right)\right)\left(\frac{2}{\pi\tau}\sqrt{2\left(\lambda_{1j}+\lambda_{2j}\right)\tau+1}\right)^{\frac{1}{2}}e^{\frac{1}{\tau}}\frac{\lambda_{1j}^{y_{1j}}K_{y_{1j}-\frac{1}{2}}\left(\tau^{-1}\sqrt{2\left(\lambda_{1j}+\lambda_{2j}\right)\tau+1}\right)}{\left(\sqrt{2\left(\lambda_{1j}+\lambda_{2j}\right)\tau+1}\right)^{y_{1j}}y_{1j}!}$$

$$P\left(Y_{1j}=0,Y_{2j}>0\mid\lambda,\tau,p\right)$$
$$=\left(1-p_2\left(\mathbf{x}_j\right)\right)p_1\left(\mathbf{x}_j\right)\left(\frac{2}{\pi\tau}\sqrt{2\lambda_{2j}\tau+1}\right)^{\frac{1}{2}}\frac{\lambda_{2j}^{y_{2j}}e^{\frac{1}{\tau}}K_{y_{2j}-\frac{1}{2}}\left(\tau^{-1}\sqrt{2\lambda_{2j}\tau+1}\right)}{\left(\sqrt{2\lambda_{2j}\tau+1}\right)^{y_{2j}}y_{2j}!}$$
$$+\left(1-p_1\left(\mathbf{x}_j\right)\right)\left(1-p_2\left(\mathbf{x}_j\right)\right)\left(\frac{2}{\pi\tau}\sqrt{2\left(\lambda_{1j}+\lambda_{2j}\right)\tau+1}\right)^{\frac{1}{2}}\frac{e^{\frac{1}{\tau}}\lambda_{2j}^{y_{1j}}K_{y_{2j}-\frac{1}{2}}\left(\tau^{-1}\sqrt{2\left(\lambda_{1j}+\lambda_{2j}\right)\tau+1}\right)}{\left(\sqrt{2\left(\lambda_{1j}+\lambda_{2j}\right)\tau+1}\right)^{y_{2j}}y_{2j}!}$$

$$P\left(Y_{1j}>0,Y_{2j}>0\mid\lambda,\tau,p\right)$$
$$=\left(1-p_1\left(\mathbf{x}_j\right)\right)\left(1-p_2\left(\mathbf{x}_j\right)\right)\left(\frac{2}{\pi\tau}\sqrt{2\left(\lambda_{1j}+\lambda_{2j}\right)\tau+1}\right)^{\frac{1}{2}}e^{\frac{1}{\tau}}\frac{\lambda_{1j}^{y_{1j}}\lambda_{2j}^{y_{2j}}K_{y_{1j}+y_{2j}-\frac{1}{2}}\left(\tau^{-1}\sqrt{2\left(\lambda_{1j}+\lambda_{2j}\right)\tau+1}\right)}{\left(\sqrt{2\left(\lambda_{1j}+\lambda_{2j}\right)\tau+1}\right)^{y_{1j}+y_{2j}}y_{1j}!y_{2j}!}$$

The probability function for the BZIPIGR model can be formed by defining $\lambda_{1j},\lambda_{2j},p_1\left(x_j\right),p_2\left(x_j\right)$ into the BZIPIG probability function equation. Where:

$$\lambda_{1j}=\left(\mathbf{x}_j^T\boldsymbol{\beta}_1\right)\quad\lambda_{2j}=\left(\mathbf{x}_j^T\boldsymbol{\beta}_2\right)$$
$$p_1\left(x_j\right)=\frac{\exp\left(-\gamma_1\mathbf{x}_j^T\boldsymbol{\beta}_1\right)}{1+\exp\left(-\gamma_1\mathbf{x}_j^T\boldsymbol{\beta}_1\right)}\quad 1-p_1\left(x_j\right)=\frac{1}{1+\exp\left(-\gamma_1\mathbf{x}_j^T\boldsymbol{\beta}_1\right)}\tag{4}$$
$$p_2\left(x_j\right)=\frac{\exp\left(-\gamma_2\mathbf{x}_j^T\boldsymbol{\beta}_2\right)}{1+\exp\left(-\gamma_2\mathbf{x}_j^T\boldsymbol{\beta}_2\right)}\quad 1-p_2\left(x_j\right)=\frac{1}{1+\exp\left(-\gamma_2\mathbf{x}_j^T\boldsymbol{\beta}_2\right)}$$

MLE is used to estimate the parameters of the BZIPIGR model, but the first derivative of the ln likelihood function of the BZIPIGR model does not have a close form equation so that the BHHH iteration used to obtain parameter estimates. MLRT is used for hypothesis testing of BZIPIGR parameters.

### 2.3 Geographically Weighted Bivariate Zero Inflated Poisson Inverse Gaussian Regression (GWBZIPIGR)

GWBZIPIGR is the local model of BZIPIGR. GWBZIPIGR defines each observation on the response variable taken from the observation location in the format of latitude and longitude coordinates $\left(u_j,v_j\right)$. Based on the BZIPIGR probabillity function in the previous section, the GWBZIPIGR model can be written as follows:

$$\lambda_{1j}=\exp\left(\mathbf{x}_j^T\boldsymbol{\beta}_1\left(u_j,v_j\right)\right)\quad\lambda_{2j}=\exp\left(\mathbf{x}_j^T\boldsymbol{\beta}_2\left(u_j,v_j\right)\right)$$
$$p_1\left(x_j\right)=\frac{\exp\left(-\gamma_1\left(u_j,v_j\right)\mathbf{x}_j^T\boldsymbol{\beta}_1\left(u_j,v_j\right)\right)}{1+\exp\left(-\gamma_1\left(u_j,v_j\right)\mathbf{x}_j^T\boldsymbol{\beta}_1\left(u_j,v_j\right)\right)}\quad 1-p_1\left(x_j\right)=\frac{1}{1+\exp\left(-\gamma_1\left(u_j,v_j\right)\mathbf{x}_j^T\boldsymbol{\beta}_1\left(u_j,v_j\right)\right)}\tag{5}$$
$$p_2\left(x_j\right)=\frac{\exp\left(-\gamma_1\left(u_j,v_j\right)\mathbf{x}_j^T\boldsymbol{\beta}_2\left(u_j,v_j\right)\right)}{1+\exp\left(-\gamma_1\left(u_j,v_j\right)\mathbf{x}_j^T\boldsymbol{\beta}_2\left(u_j,v_j\right)\right)}\quad 1-p_2\left(x_j\right)=\frac{1}{1+\exp\left(-\gamma_2\left(u_j,v_j\right)\mathbf{x}_j^T\boldsymbol{\beta}_2\left(u_j,v_j\right)\right)}$$

MLE will be used to estimate the parameters of the GWBZIPIGR model, but the first derivative of the ln likelihood function of the GWBZIPIGR model does not have a close form equation so that the BHHH iteration used to obtain parameter estimates. MLRT is used for hypothesis testing of GWBZIPIGR parameters.

## 3. Result and Discussion

This section provides the result and discussion about parameter estimation and hypothesis testing of GWBZIPIGR model.

### 3.1 Parameter Estimation of GWBZIPIGR

The first step to obtaining the GWBZIPIGR model estimate is to determine the likelihood function based on the probability density function of the GWBZIPIG distribution in the previous section. Suppose $n$ random sample $(u_j,v_j), j=1,2,\dots n$ are taken independently, then the likelihood function of location $j$ can be written as follows:

$$\begin{aligned} L\left(\boldsymbol{\beta}_1(u_j,v_j),\boldsymbol{\beta}_2(u_j,v_j),\boldsymbol{\gamma}_1(u_j,v_j),\boldsymbol{\gamma}_2(u_j,v_j),\tau\right) &= \prod_{j=1}^{n} f\left(y_{1j},y_{2j}\right)^{w_{ij}^*} \\ &= \prod_{j=1}^{n}\left(\left(A_j\right)^{1-b_i-c_i-d_i}\left(B_j\right)^{b_i}\left(C_j\right)^{c_i}\left(D_j\right)^{d_i}\right)^{w_{ij}^*} \end{aligned} \tag{6}$$

where,

$$A_j = \frac{\exp\left(-\gamma_1(u_j,v_j)\mathbf{x}_j^T\boldsymbol{\beta}_1(u_j,v_j)\right)}{1+\exp\left(-\gamma_1(u_j,v_j)\mathbf{x}_j^T\boldsymbol{\beta}_1(u_j,v_j)\right)}\frac{\exp\left(-\gamma_2(u_j,v_j)\mathbf{x}_j^T\boldsymbol{\beta}_2(u_j,v_j)\right)}{1+\exp\left(-\gamma_2(u_j,v_j)\mathbf{x}_j^T\boldsymbol{\beta}_2(u_j,v_j)\right)}+O_1+O_2+O_3$$

$$O_1 = \frac{\exp\left(-\gamma_1(u_j,v_j)\mathbf{x}_j^T\boldsymbol{\beta}_1(u_j,v_j)\right)}{1+\exp\left(-\gamma_1(u_j,v_j)\mathbf{x}_j^T\boldsymbol{\beta}_1(u_j,v_j)\right)}\frac{1}{1+\exp\left(-\gamma_2(u_j,v_j)\mathbf{x}_j^T\boldsymbol{\beta}_2(u_j,v_j)\right)}\left(\frac{2}{\pi\tau}\sqrt{2\lambda_{2j}\tau+1}\right)^{\frac{1}{2}} \exp\left(\tau^{-1}\right)K_{-\frac{1}{2}}\left(\tau^{-1}\sqrt{2\lambda_{2j}\tau+1}\right)$$

$$O_2 = \frac{\exp\left(-\gamma_2(u_j,v_j)\mathbf{x}_j^T\boldsymbol{\beta}_2(u_j,v_j)\right)}{1+\exp\left(-\gamma_2(u_j,v_j)\mathbf{x}_j^T\boldsymbol{\beta}_2(u_j,v_j)\right)}\frac{1}{1+\exp\left(-\gamma_1(u_j,v_j)\mathbf{x}_j^T\boldsymbol{\beta}_1(u_j,v_j)\right)}\left(\frac{2}{\pi\tau}\right)^{\frac{1}{2}}\exp\left(\tau^{-1}\right) \left(2\lambda_{1j}\tau+1\right)^{\frac{1}{4}}K_{-\frac{1}{2}}\left(\tau^{-1}\sqrt{2\lambda_{1j}\tau+1}\right)$$

$$O_3 = \frac{1}{1+\exp\left(-\gamma_1(u_j,v_j)\mathbf{x}_j^T\boldsymbol{\beta}_1(u_j,v_j)\right)}\frac{1}{1+\exp\left(-\gamma_2(u_j,v_j)\mathbf{x}_j^T\boldsymbol{\beta}_2(u_j,v_j)\right)}\left(\frac{2}{\pi\tau}\right)^{\frac{1}{2}}\exp\left(\tau^{-1}\right) \left(2\tau\left(\lambda_{1j}+\lambda_{2j}\right)+1\right)^{\frac{1}{4}}K_{-\frac{1}{2}}\left(\tau^{-1}\sqrt{2\tau\left(\lambda_{1j}+\lambda_{2j}\right)+1}\right)$$

$$B_j = \frac{1}{1+\exp\left(-\gamma_1(u_j,v_j)\mathbf{x}_j^T\boldsymbol{\beta}_1(u_j,v_j)\right)}\frac{\exp\left(-\gamma_2(u_j,v_j)\mathbf{x}_j^T\boldsymbol{\beta}_2(u_j,v_j)\right)}{1+\exp\left(-\gamma_2(u_j,v_j)\mathbf{x}_j^T\boldsymbol{\beta}_2(u_j,v_j)\right)}P_1+O_4$$

$$P_1 = \left(\frac{2}{\pi\tau}\sqrt{2\lambda_{1j}\tau+1}\right)^{\frac{1}{2}}\frac{\exp\left(\lambda_{1j}\right)^{y_1}\exp\left(\tau^{-1}\right)K_{y_1-\frac{1}{2}}\left(\tau^{-1}\sqrt{2\exp\lambda_{1j}\tau+1}\right)}{\left(\sqrt{2\exp\lambda_{1j}\tau+1}\right)^{y_1}y_1!}$$

$$O_4 = \frac{1}{1+\exp\left(-\gamma_1(u_j,v_j)\mathbf{x}_j^T\boldsymbol{\beta}_1(u_j,v_j)\right)}\frac{1}{1+\exp\left(-\gamma_2(u_j,v_j)\mathbf{x}_j^T\boldsymbol{\beta}_2(u_j,v_j)\right)}\left(\frac{2}{\pi\tau}\sqrt{2\tau\left(\lambda_{1j}+\lambda_{2j}\right)+1}\right)^{\frac{1}{2}} \frac{\exp\left(\left(\lambda_{1j}\right)\right)^{y_1}\exp\left(\tau^{-1}\right)K_{y_1-\frac{1}{2}}\left(\tau^{-1}\sqrt{2\left(\lambda_{1j}+\lambda_{2j}\right)\tau+1}\right)}{\left(\sqrt{2\left(\lambda_{1j}+\lambda_{2j}\right)\tau+1}\right)^{y_1}y_1!}$$

$$C_j = \frac{1}{1+\exp\left(-\gamma_2(u_j,v_j)\mathbf{x}_j^T\boldsymbol{\beta}_2(u_j,v_j)\right)}\frac{\exp\left(-\gamma_1(u_j,v_j)\mathbf{x}_j^T\boldsymbol{\beta}_1(u_j,v_j)\right)}{1+\exp\left(-\gamma_1(u_j,v_j)\mathbf{x}_j^T\boldsymbol{\beta}_1(u_j,v_j)\right)}P_2 + O_5$$

$$P_2 = \left(\frac{2}{\pi\tau}\sqrt{2\lambda_{2j}\tau+1}\right)^{\frac{1}{2}}\frac{(\lambda_{2j})^{y_2}\exp(\tau^{-1})K_{y_2-\frac{1}{2}}\left(\tau^{-1}\sqrt{2\lambda_{2j}\tau+1}\right)}{\left(\sqrt{2\lambda_{2j}\tau+1}\right)^{y_2}y_2!}$$

$$O_6 = \frac{1}{1+\exp\left(-\gamma_1(u_j,v_j)\mathbf{x}_j^T\boldsymbol{\beta}_1(u_j,v_j)\right)}\frac{1}{1+\exp\left(-\gamma_2(u_j,v_j)\mathbf{x}_j^T\boldsymbol{\beta}_2(u_j,v_j)\right)}\left(\frac{2}{\pi\tau}\sqrt{2\tau(\lambda_{1j}+\lambda_{2j})+1}\right)^{\frac{1}{2}}$$
$$\frac{(\lambda_{2j})^{y_2}\exp(\tau^{-1})K_{y_2-\frac{1}{2}}\left(\tau^{-1}\sqrt{2(\lambda_{1j}+\lambda_{2j})\tau+1}\right)}{\left(\sqrt{2(\lambda_{1j}+\lambda_{2j})\tau+1}\right)^{y_2}y_2!}$$

$$D_j = \frac{e^{\frac{1}{2}}\left(\frac{2}{\pi\tau}\sqrt{2\tau(\lambda_{1j}+\lambda_{2j})+1}\right)^{\frac{1}{2}}}{\left(1+\exp\left(-\gamma_1(u_j,v_j)\mathbf{x}_j^T\boldsymbol{\beta}_1(u_j,v_j)\right)\right)\left(1+\exp\left(-\gamma_2(u_j,v_j)\mathbf{x}_j^T\boldsymbol{\beta}_2(u_j,v_j)\right)\right)}P_3$$

$$P_3 = \frac{\lambda_{1j}^{y_1}\lambda_{2j}^{y_2}K_{y_1+y_2-\frac{1}{2}}\left(\tau^{-1}\sqrt{2(\lambda_{1j}+\lambda_{2j})\tau+1}\right)}{\left(\sqrt{2(\lambda_{1j}+\lambda_{2j})\tau+1}\right)^{y_1+y_2}y_1!y_2!}$$

To simplify the derivative process of the likelihood function above respect to each parameter, then the ln likelihood function is formed as follows:

$$l = \ln L\left(\boldsymbol{\beta}_1(u_j,v_j),\boldsymbol{\beta}_2(u_j,v_j),\boldsymbol{\gamma}_1(u_j,v_j),\boldsymbol{\gamma}_2(u_j,v_j),\tau\right) = \ln\left(\prod_{j=1}^{n}\left((A_j)^{1-a_j-b_j-c_j-d_j}(B_j)^{b_j}(C_j)^{c_j}(D_j)^{d_j}\right)^{w_{ij^*}}\right)$$
$$l = \sum_{j=1}^{n}w_{ij^*}(1-b_j-c_j-d_j)\ln A_j + \sum_{j=1}^{n}w_{ij^*}(b_j)\ln B_j + \sum_{j=1}^{n}w_{ij^*}(c_j)\ln C_j + \sum_{i=j}^{n}w_{ij^*}(d_j)\ln D_j \quad (7)$$

The ln likelihood function obtained above then derived respect to each parameter as below:

$$\frac{\partial l}{\partial\boldsymbol{\beta}_1(u_j^*,v_j^*)} = \sum_{j=1}^{n}(1-b_j-c_j-d_j)\frac{1}{A_j}(A_{11}+A_{12}+A_{13}+A_{14}) + \sum_{j=1}^{n}(b_j)\left(\sum_{j=1}^{n}B_{11}+\sum_{j=1}^{n}B_{12}+\sum_{j=1}^{n}B_{13}+\sum_{j=1}^{n}B_{14}\right)$$
$$+\sum_{j=1}^{n}(c_j)\left(\sum_{j=1}^{n}C_{11}+\sum_{j=1}^{n}C_{12}\right)+\sum_{j=1}^{n}(d_j)\left(\sum_{j=1}^{n}D_{11}+\sum_{j=1}^{n}D_{12}\right) \quad (8)$$

where,

$$A_j = \frac{\exp\left(-\gamma_1(u_j^*,v_j^*)\mathbf{x}_j^T\boldsymbol{\beta}_1(u_j^*,v_j^*)\right)}{1+\exp\left(-\gamma_1(u_j^*,v_j^*)\mathbf{x}_j^T\boldsymbol{\beta}_1(u_j^*,v_j^*)\right)}\frac{\exp\left(-\gamma_2(u_j^*,v_j^*)\mathbf{x}_j^T\boldsymbol{\beta}_2(u_j^*,v_j^*)\right)}{1+\exp\left(-\gamma_2(u_j^*,v_j^*)\mathbf{x}_j^T\boldsymbol{\beta}_2(u_j^*,v_j^*)\right)}$$
$$+\frac{\exp\left(-\gamma_1(u_j^*,v_j^*)\mathbf{x}_j^T\boldsymbol{\beta}_1(u_j^*,v_j^*)\right)\left(\frac{2}{\pi\tau}\sqrt{2\lambda_{2j}\tau+1}\right)^{\frac{1}{2}}\exp(\tau^{-1})K_{-\frac{1}{2}}\left(\tau^{-1}\sqrt{2\lambda_{2j}\tau+1}\right)}{\left(1+\exp\left(-\gamma_1(u_j^*,v_j^*)\mathbf{x}_j^T\boldsymbol{\beta}_1(u_j^*,v_j^*)\right)\right)\left(1+\exp\left(-\gamma_2(u_j^*,v_j^*)\mathbf{x}_j^T\boldsymbol{\beta}_2(u_j^*,v_j^*)\right)\right)}$$
$$+\frac{\exp\left(-\gamma_2(u_j,v_j)\mathbf{x}_j^T\boldsymbol{\beta}_2(u_j,v_j)\right)\left(\frac{2}{\pi\tau}\right)^{\frac{1}{2}}\exp(\tau^{-1})(2\exp\lambda_{1j}\tau+1)^{\frac{1}{4}}K_{-\frac{1}{2}}\left(\tau^{-1}\sqrt{2\lambda_{1j}\tau+1}\right)}{\left(1+\exp\left(-\gamma_1(u_j^*,v_j^*)\mathbf{x}_j^T\boldsymbol{\beta}_1(u_j^*,v_j^*)\right)\right)\left(1+\exp\left(-\gamma_2(u_j^*,v_j^*)\mathbf{x}_j^T\boldsymbol{\beta}_2(u_j^*,v_j^*)\right)\right)}$$
$$+\frac{\left(\frac{2}{\pi\tau}\right)^{\frac{1}{2}}\exp(\tau^{-1})\left(2\tau(\lambda_{1j}+\lambda_{2j})+1\right)^{\frac{1}{4}}K_{-\frac{1}{2}}\left(\tau^{-1}\sqrt{2\tau(\lambda_{1j}+\lambda_{2j})+1}\right)}{\left(1+\exp\left(-\gamma_1(u_j^*,v_j^*)\mathbf{x}_j^T\boldsymbol{\beta}_1(u_j^*,v_j^*)\right)\right)\left(1+\exp\left(-\gamma_2(u_j^*,v_j^*)\mathbf{x}_j^T\boldsymbol{\beta}_2(u_j^*,v_j^*)\right)\right)}$$

$$A_{11} = \frac{\exp\left(-\gamma_2(u_j^*,v_j^*)\mathbf{x}_j^T\boldsymbol{\beta}_2(u_j^*,v_j^*)\right)}{1+\exp\left(-\gamma_2(u_j^*,v_j^*)\mathbf{x}_j^T\boldsymbol{\beta}_2(u_j^*,v_j^*)\right)}\frac{-\gamma_1(u_j^*,v_j^*)\mathbf{x}_j^T\left(\exp\left(-\gamma(u_j^*,v_j^*)\mathbf{x}_j^T\boldsymbol{\beta}_1(u_j^*,v_j^*)\right)\right)}{\left(1+\exp\left(-\gamma_1(u_j^*,v_j^*)\mathbf{x}_j^T\boldsymbol{\beta}_1(u_j^*,v_j^*)\right)\right)^2}$$

$$A_{12} = \frac{1}{1+\exp\left(-\gamma_2\left(u_j^*,v_j^*\right)\mathbf{x}_j^T\boldsymbol{\beta}_2\left(u_j^*,v_j^*\right)\right)}\left(\frac{2}{\pi\tau}\sqrt{2\lambda_{2j}\tau+1}\right)^{\frac{1}{2}}\exp\left(\tau^{-1}\right)\left(\left(\frac{\pi}{2}\tau\right)^{\frac{1}{2}}\left(1+2\tau\lambda_{2j}\right)^{-\frac{1}{4}}\exp\left(\left(\tau^{-1}\right)\sqrt{1+2\tau\lambda_{2j}}\right)\right)$$
$$\frac{-\gamma_1\left(u_j^*,v_j^*\right)\mathbf{x}_j^T\left(\exp\left(-\gamma_1\left(u_j^*,v_j^*\right)\mathbf{x}_j^T\boldsymbol{\beta}_1\left(u_j^*,v_j^*\right)\right)\right)}{\left(1+\exp\left(-\gamma_1\left(u_j^*,v_j^*\right)\mathbf{x}_j^T\boldsymbol{\beta}_1\left(u_j^*,v_j^*\right)\right)\right)^2}$$

$$A_{13} = \frac{\exp\left(-\gamma_2\left(u_j^*,v_j^*\right)\mathbf{x}_j^T\boldsymbol{\beta}_2\left(u_j^*,v_j^*\right)\right)}{1+\exp\left(-\gamma_2\left(u_j^*,v_j^*\right)\mathbf{x}_j^T\boldsymbol{\beta}_2\left(u_j^*,v_j^*\right)\right)}\exp\left(\tau^{-1}\right)$$
$$\Bigg(\Bigg(\left(\frac{2}{\pi\tau}\sqrt{2\lambda_{1j}\tau+1}\right)^{\frac{1}{2}}\left(\frac{\pi}{2}\tau\right)^{\frac{1}{2}}\left(1+2\tau\lambda_{1j}\right)^{-\frac{1}{4}}\exp\left(\left(\tau^{-1}\right)\sqrt{1+2\tau\lambda_{1j}}\right)$$
$$\left.\frac{-\gamma_1\left(u_j^*,v_j^*\right)\mathbf{x}_j^T\left(\exp\left(-\gamma_1\left(u_j^*,v_j^*\right)\mathbf{x}_j^T\boldsymbol{\beta}_1\left(u_j^*,v_j^*\right)\right)\right)}{\left(1+\exp\left(-\gamma_1\left(u_j^*,v_j^*\right)\mathbf{x}_j^T\boldsymbol{\beta}_1\left(u_j^*,v_j^*\right)\right)\right)^2}\right)$$
$$+\left(\frac{1}{1+\exp\left(-\gamma_1\left(u_j^*,v_j^*\right)\mathbf{x}_j^T\boldsymbol{\beta}_1\left(u_j^*,v_j^*\right)\right)}\left(\frac{\pi}{2}\tau\right)^{\frac{1}{2}}\left(1+2\tau\lambda_{1j}\right)^{-\frac{1}{4}}\exp\left(\left(\tau^{-1}\right)\sqrt{1+2\tau\lambda_{1j}}\right)\left(\frac{2}{\pi\tau}\right)^{\frac{1}{2}}\frac{\tau}{2}\left(1+2\tau\lambda_{1j}\right)^{-\frac{3}{4}}\lambda_{1j}\mathbf{x}_j\right)$$
$$+\left(\frac{1}{1+\exp\left(-\gamma_1\left(u_j^*,v_j^*\right)\mathbf{x}_j^T\boldsymbol{\beta}_1\left(u_j^*,v_j^*\right)\right)}\left(\frac{\pi}{2}\tau\right)^{\frac{1}{2}}\left(\frac{2}{\pi\tau}\sqrt{1+2\tau\lambda_{1j}}\right)^{\frac{1}{2}}\left(\left(\frac{-\tau}{2}\left(1+2\tau\lambda_{1j}\right)^{-\frac{5}{4}}\lambda_{1j}\mathbf{x}_j\exp\left(\left(\tau^{-1}\right)\sqrt{1+2\tau\lambda_{1j}}\right)\right)\right.$$
$$\left.\left.\left.+\left(\left(1+2\tau\lambda_{1j}\right)^{\frac{1}{4}}\exp\left(\left(-\tau^{-1}\right)\sqrt{1+2\tau\lambda_{1j}}\right)\lambda_{1j}\mathbf{x}_j\right)\right)\right)\right)$$

$$A_{14} = \frac{1}{1+\exp\left(-\gamma_2\left(u_j^*,v_j^*\right)\mathbf{x}_j^T\boldsymbol{\beta}_2\left(u_j^*,v_j^*\right)\right)}\exp\left(\tau^{-1}\right)$$
$$\Bigg(\Bigg(\left(\frac{2}{\pi\tau}\sqrt{1+2\tau\left(\lambda_{1j}+\lambda_{2j}\right)}\right)^{\frac{1}{2}}\left(\frac{\pi}{2}\tau\right)^{\frac{1}{2}}\left(1+2\tau\left(\lambda_{1j}+\lambda_{2j}\right)\right)^{-\frac{1}{4}}\exp\left(-\left(\tau^{-1}\right)\sqrt{1+2\tau\left(\lambda_{1j}+\lambda_{2j}\right)}\right)$$
$$\frac{-\gamma_1\left(u_j^*,v_j^*\right)\mathbf{x}_j^T\left(\exp\left(-\gamma_1\left(u_j^*,v_j^*\right)\mathbf{x}_j^T\boldsymbol{\beta}_1\left(u_j^*,v_j^*\right)\right)\right)}{\left(1+\exp\left(-\gamma_1\left(u_j^*,v_j^*\right)\mathbf{x}_j^T\boldsymbol{\beta}_1\left(u_j^*,v_j^*\right)\right)\right)^2}$$
$$+\left(\frac{1}{1+\exp\left(-\gamma_1\left(u_j^*,v_j^*\right)\mathbf{x}_j^T\boldsymbol{\beta}_1\left(u_j^*,v_j^*\right)\right)}\left(\frac{\pi}{2}\tau\right)^{\frac{1}{2}}\left(1+2\tau\left(\lambda_{1j}+\lambda_{2j}\right)\right)^{-\frac{1}{4}}\right.$$
$$\left.\exp\left(-\left(\tau^{-1}\right)\sqrt{1+2\tau\left(\lambda_{1j}+\lambda_{2j}\right)}\right)\left(\frac{2}{\pi\tau}\right)^{\frac{1}{2}}\frac{\tau}{2}\left(1+2\tau\lambda_{1j}\right)^{-\frac{3}{4}}\lambda_{1j}\mathbf{x}_j\right)$$
$$+\left(\frac{1}{1+\exp\left(-\gamma_1\left(u_j^*,v_j^*\right)\mathbf{x}_j^T\boldsymbol{\beta}_1\left(u_j^*,v_j^*\right)\right)}\left(\frac{\pi}{2}\tau\right)^{\frac{1}{2}}\left(\frac{2}{\pi\tau}\sqrt{1+2\tau\left(\lambda_{1j}+\lambda_{2j}\right)}\right)^{\frac{1}{2}}\left(\frac{-\tau}{2}\left(1+2\tau\left(\lambda_{1j}+\lambda_{2j}\right)\right)^{-\frac{5}{4}}\lambda_{1j}\mathbf{x}_j\right.\right.$$
$$\left.\left.\exp\left(\left(\tau^{-1}\right)\sqrt{1+2\tau\left(\lambda_{1j}+\lambda_{2j}\right)}\right)+\left(\left(1+2\tau\left(\lambda_{1j}+\lambda_{2j}\right)\right)^{\frac{1}{4}}\exp\left(\left(-\tau^{-1}\right)\sqrt{1+2\tau\left(\lambda_{1j}+\lambda_{2j}\right)}\right)\lambda_{1j}\mathbf{x}_j\right)\right)\Bigg)\Bigg)$$

$$B_{11} = \sum_{j=1}^{n}\left(\frac{\gamma_1\left(u_j^*,v_j^*\right)\mathbf{x}_j^T\left(\exp\left(-\gamma_1\left(u_j^*,v_j^*\right)\mathbf{x}_j^T\boldsymbol{\beta}_1\left(u_j^*,v_j^*\right)\right)\right)}{1+\exp\left(-\gamma_1\left(u_j^*,v_j^*\right)\mathbf{x}_j^T\boldsymbol{\beta}_1\left(u_j^*,v_j^*\right)\right)}+\frac{1}{2\left(2\tau\lambda_{1j}+1\right)^{\frac{1}{2}}}\left(\frac{1}{2}\left(2\tau\lambda_{1j}+1\right)^{-\frac{1}{2}}2\tau\lambda_{1j}\mathbf{x}_j\right)+y_{1j}\left(\mathbf{x}_j^T\boldsymbol{\beta}_1\left(u_j^*,v_j^*\right)\right)\mathbf{x}_j\right)$$

$$B_{12} = \sum_{j=1}^{n}\left(\left(-K_{y_1-\frac{1}{2}}\left(\tau^{-1}\sqrt{2\lambda_{1j}\tau+1}\right)+\frac{\tau\left(y_1-\frac{1}{2}\right)K_{y_1-\frac{1}{2}}\left(\tau^{-1}\sqrt{2\lambda_{1j}\tau+1}\right)}{\left(\sqrt{2\tau\lambda_{1j}+1}\right)}\right)\left(\left(2\tau\lambda_{1j}+1\right)^{-\frac{1}{2}}\lambda_{1j}\right)-\tau y_{1j}\frac{1}{1+2\lambda_{1j}\tau}\lambda_{1j}\mathbf{x}_j\right)$$

$$B_{13}=\sum_{j=1}^{n}\left(\frac{\gamma_1\left(u_j^*,v_j^*\right)\mathbf{x}_j^T\left(\exp\left(-\gamma_1\left(u_j^*,v_j^*\right)\mathbf{x}_j^T\boldsymbol{\beta}_1\left(u_j^*,v_j^*\right)\right)\right)}{1+\exp\left(-\gamma_1\left(u_j^*,v_j^*\right)\mathbf{x}_j^T\boldsymbol{\beta}_1\left(u_j^*,v_j^*\right)\right)}+\frac{1}{2\left(2\tau\left(\lambda_{1j}+\lambda_{2j}\right)+1\right)^{\frac{1}{2}}}\left(\frac{1}{2}\left(2\tau\left(\lambda_{1j}+\lambda_{2j}\right)+1\right)^{-\frac{1}{2}}2\tau\lambda_{1j}\mathbf{x}_j\right)+y_{1j}\left(\mathbf{x}_j^T\boldsymbol{\beta}_1\left(u_j^*,v_j^*\right)\right)\mathbf{x}_j\right)$$

$$B_{14}=\sum_{j=1}^{n}\left(-K_{y_1-\frac{1}{2}}\left(\tau^{-1}\sqrt{2\left(\lambda_{1j}+\lambda_{2j}\right)\tau+1}\right)+\frac{\tau\left(y_1-\frac{1}{2}\right)K_{y_1-\frac{1}{2}}\left(\tau^{-1}\sqrt{2\left(\lambda_{1j}+\lambda_{2j}\right)\tau+1}\right)}{\left(\sqrt{2\left(\lambda_{1j}+\lambda_{2j}\right)\tau+1}\right)}\right)\left(\left(2\tau\left(\lambda_{1j}+\lambda_{2j}\right)+1\right)^{-\frac{1}{2}}\lambda_{1j}\right)-\tau y_{1j}\frac{1}{1+2\lambda_{1j}\tau+2\lambda_{2j}\tau}\lambda_{1j}\mathbf{x}_j$$

$$C_{11}=\sum_{j=1}^{n}\left(\frac{-\gamma_1\left(u_j^*,v_j^*\right)\mathbf{x}_j^T}{1+\exp\left(-\gamma_1\left(u_j^*,v_j^*\right)\mathbf{x}_j^T\boldsymbol{\beta}_1\left(u_j^*,v_j^*\right)\right)}\right)$$
$$+\sum_{j=1}^{n}\left(\frac{\gamma_1\left(u_j^*,v_j^*\right)\mathbf{x}_j^T\left(\exp\left(-\gamma_1\left(u_j^*,v_j^*\right)\mathbf{x}_j^T\boldsymbol{\beta}_1\left(u_j^*,v_j^*\right)\right)\right)}{1+\exp\left(-\gamma_1\left(u_j^*,v_j^*\right)\mathbf{x}_j^T\boldsymbol{\beta}_1\left(u_j^*,v_j^*\right)\right)}+\frac{1}{2\left(2\tau\left(\lambda_{1j}+\lambda_{2j}\right)+1\right)^{\frac{1}{2}}}\left(\frac{1}{2}\left(2\tau\left(\lambda_{1j}+\lambda_{2j}\right)+1\right)^{-\frac{1}{2}}2\tau\lambda_{1j}\mathbf{x}_j\right)\right)$$

$$C_{12}=\sum_{j=1}^{n}\left(-K_{y_2-\frac{1}{2}}\left(\tau^{-1}\sqrt{2\left(\lambda_{1j}+\lambda_{2j}\right)\tau+1}\right)+\frac{\tau\left(y_2-\frac{1}{2}\right)K_{y_2-\frac{1}{2}}\left(\tau^{-1}\sqrt{2\left(\lambda_{1j}+\lambda_{2j}\right)\tau+1}\right)}{\left(\sqrt{2\left(\lambda_{1j}+\lambda_{2j}\right)\tau+1}\right)}\right)\left(\left(2\tau\left(\lambda_{1j}+\lambda_{2j}\right)+1\right)^{-\frac{1}{2}}\lambda_{1j}\right)-\tau y_{2j}\frac{1}{1+2\lambda_{1j}\tau+2\lambda_{2j}\tau}\lambda_{1j}\mathbf{x}_j$$

$$D_{11}=\sum_{j=1}^{n}\left(\frac{\gamma_1\left(u_j^*,v_j^*\right)\mathbf{x}_j^T\left(\exp\left(-\gamma_1\left(u_j^*,v_j^*\right)\mathbf{x}_j^T\boldsymbol{\beta}_1\left(u_j^*,v_j^*\right)\right)\right)}{1+\exp\left(-\gamma_1\left(u_j^*,v_j^*\right)\mathbf{x}_j^T\boldsymbol{\beta}_1\left(u_j^*,v_j^*\right)\right)}+\frac{1}{2\left(2\tau\left(\lambda_{ij}+\lambda_{2j}\right)+1\right)^{\frac{1}{2}}}\left(\frac{1}{2}\left(2\tau\left(\lambda_{ij}+\lambda_{2j}\right)+1\right)^{-\frac{1}{2}}2\tau\lambda_{ij}\mathbf{x}_j\right)+y_{1j}\left(u_j^*,v_j^*\right)\right)$$

$$D_{12}=\sum_{j=1}^{n}\left(-K_{y_1+y_2-\frac{1}{2}}\left(\tau^{-1}\sqrt{2\left(\lambda_{ij}+\lambda_{2j}\right)\tau+1}\right)+\frac{\tau\left(y_1+y_2-\frac{1}{2}\right)K_{y_1+y_2-\frac{1}{2}}\left(\tau^{-1}\sqrt{2\left(\lambda_{ij}+\lambda_{2j}\right)\tau+1}\right)}{\left(\sqrt{2\left(\lambda_{ij}+\lambda_{2j}\right)\tau+1}\right)}\right)$$
$$\left(\left(2\tau\left(\lambda_{ij}+\lambda_{2j}\right)+1\right)^{-\frac{1}{2}}\lambda_{ij}\right)-\tau\left(y_{1j}+y_{2j}\right)\frac{1}{1+2\lambda_{ij}\tau+2\lambda_{2j}\tau}\lambda_{ij}\mathbf{x}_j$$

Using the same way, we can obtained the first derivative of ln likelihood function respect to $\boldsymbol{\beta}_2(u_j^*,v_j^*)$ as below:

$$\frac{\partial l}{\partial\boldsymbol{\beta}_2(u_j^*,v_j^*)}=\sum_{j=1}^{n}\left(1-b_j-c_j-d_j\right)\frac{1}{A_j}\left(A_{21}+A_{22}+A_{23}+A_{24}\right)+\sum_{j=1}^{n}\left(b_j\right)\left(\sum_{j=1}^{n}B_{21}+\sum_{j=1}^{n}B_{22}\right)$$
$$+\sum_{j=1}^{n}\left(c_j\right)\left(\sum_{j=1}^{n}C_{21}+\sum_{j=1}^{n}C_{22}+\sum_{j=1}^{n}C_{23}+\sum_{j=1}^{n}C_{24}\right)+\sum_{j=1}^{n}\left(d_j\right)\left(\sum_{j=1}^{n}D_{21}+\sum_{j=1}^{n}D_{22}\right) \quad (9)$$

The first derivative of ln likelihood function respect to $\gamma_1(u_j^*,v_j^*)$ can be written as follow:

$$\frac{\partial l}{\partial\gamma_1(u_j^*,v_j^*)}=\sum_{j=1}^{n}\left(1-b_j-c_j-d_j\right)\frac{1}{A_j}\left(A_{31}+A_{32}+A_{33}+A_{34}\right)+\sum_{j=1}^{n}\left(b_j\right)\left(\sum_{j=1}^{n}B_{31}+\sum_{j=1}^{n}B_{32}\right)$$
$$+\sum_{j=1}^{n}\left(c_j\right)\left(\sum_{j=1}^{n}C_{31}+\sum_{j=1}^{n}C_{32}\right)+\sum_{j=1}^{n}\left(d_j\right)\left(\sum_{j=1}^{n}D_{31}\right) \quad (10)$$

$$A_{31}=\frac{\exp\left(-\gamma_2(u_j^*,v_j^*)\mathbf{x}_j^T\boldsymbol{\beta}_2(u_j^*,v_j^*)\right)}{1+\exp\left(-\gamma_2(u_j^*,v_j^*)\mathbf{x}_j^T\boldsymbol{\beta}_2(u_j^*,v_j^*)\right)}\frac{-\mathbf{x}_j^T\boldsymbol{\beta}_1(u_j^*,v_j^*)\left(\exp\left(-\gamma_1(u_j^*,v_j^*)\mathbf{x}_j^T\boldsymbol{\beta}_1(u_j^*,v_j^*)\right)\right)}{\left(1+\exp\left(-\gamma_1(u_j^*,v_j^*)\mathbf{x}_j^T\boldsymbol{\beta}_1(u_j^*,v_j^*)\right)\right)^2}$$

$$A_{32}=\frac{1}{1+\exp\left(-\gamma_2(u_j^*,v_j^*)\mathbf{x}_j^T\boldsymbol{\beta}_2(u_j^*,v_j^*)\right)}\left(\frac{2}{\pi\tau}\sqrt{2\lambda_{2j}\tau+1}\right)^{\frac{1}{2}}\exp\left(\tau^{-1}\right)$$
$$\left(\left(\frac{\pi}{2}\tau\right)^{\frac{1}{2}}\left(1+2\tau\lambda_{2j}\right)^{-\frac{1}{4}}\exp\left(\left(\tau^{-1}\right)\sqrt{1+2\tau\lambda_{2j}}\right)\right)\frac{-\mathbf{x}_j^T\boldsymbol{\beta}_1(u_j^*,v_j^*)\left(\exp\left(-\gamma_1(u_j^*,v_j^*)\mathbf{x}_j^T\boldsymbol{\beta}_1(u_j^*,v_j^*)\right)\right)}{\left(1+\exp\left(-\gamma_1(u_j^*,v_j^*)\mathbf{x}_j^T\boldsymbol{\beta}_1(u_j^*,v_j^*)\right)\right)^2}$$

$$A_{33}=\frac{\exp\left(-\gamma_2(u_j^*,v_j^*\mathbf{x}_j^T\boldsymbol{\beta}_2(u_j^*,v_j^*)\right)}{1+\exp\left(-\gamma_2(u_j^*,v_j^*)\mathbf{x}_j^T\boldsymbol{\beta}_2(u_j^*,v_j^*)\right)}\exp\left(\tau^{-1}\right)\left(\frac{2}{\pi\tau}\sqrt{2\lambda_{1j}\tau+1}\right)^{\frac{1}{2}}\left(\frac{\pi}{2}\tau\right)^{\frac{1}{2}}\left(1+2\tau\lambda_{1j}\right)^{-\frac{1}{4}}$$
$$\exp\left(\left(\tau^{-1}\right)\sqrt{1+2\tau\lambda_{1j}}\right)\frac{-\mathbf{x}_j^T\boldsymbol{\beta}_1(u_j^*,v_j^*)\left(\exp\left(-\gamma_1(u_j^*,v_j^*)\mathbf{x}_j^T\boldsymbol{\beta}_1(u_j^*,v_j^*)\right)\right)}{\left(1+\exp\left(-\gamma_1(u_j^*,v_j^*)\mathbf{x}_j^T\boldsymbol{\beta}_1(u_j^*,v_j^*)\right)\right)^2}$$

$$A_{34}=\frac{1}{1+\exp\left(-\gamma_2(u_j^*,v_j^*)\mathbf{x}_j^T\boldsymbol{\beta}_2(u_j^*,v_j^*)\right)}\exp\left(\tau^{-1}\right)\left(\frac{2}{\pi\tau}\sqrt{1+2\tau\left(\lambda_{1j}+\lambda_{2j}\right)}\right)^{\frac{1}{2}}\left(\frac{\pi}{2}\tau\right)^{\frac{1}{2}}\left(1+2\tau\left(\lambda_{1j}+\lambda_{2j}\right)\right)^{-\frac{1}{4}}$$
$$\exp\left(-\left(\tau^{-1}\right)\sqrt{1+2\tau\left(\lambda_{1j}+\lambda_{2j}\right)}\right)\frac{-\mathbf{x}_j^T\boldsymbol{\beta}_1(u_j^*,v_j^*)\left(\exp\left(-\gamma_1(u_j^*,v_j^*)\mathbf{x}_j^T\boldsymbol{\beta}_1(u_j^*,v_j^*)\right)\right)}{\left(1+\exp\left(-\gamma_1(u_j^*,v_j^*)\mathbf{x}_j^T\boldsymbol{\beta}_1(u_j^*,v_j^*)\right)\right)^2}$$

$$B_{31}=\sum_{j=1}^{n}\left(\frac{\mathbf{x}_j^T\boldsymbol{\beta}_1(u_j^*,v_j^*)\left(\exp\left(-\gamma_1(u_j^*,v_j^*)\mathbf{x}_j^T\boldsymbol{\beta}_1(u_j^*,v_j^*)\right)\right)}{1+\exp\left(-\gamma_1(u_j^*,v_j^*)\mathbf{x}_j^T\boldsymbol{\beta}_1(u_j^*,v_j^*)\right)}\right)$$

$$B_{32}=\sum_{j=1}^{n}\left(\frac{\mathbf{x}_j^T\boldsymbol{\beta}_1(u_j^*,v_j^*)\left(\exp\left(-\gamma_1(u_j^*,v_j^*)\mathbf{x}_j^T\boldsymbol{\beta}_1(u_j^*,v_j^*)\right)\right)}{1+\exp\left(-\gamma_1(u_j^*,v_j^*)\mathbf{x}_j^T\boldsymbol{\beta}_1(u_j^*,v_j^*)\right)}\right)$$

$$C_{31}=\sum_{j=1}^{n}\left(\frac{-\mathbf{x}_j^T\boldsymbol{\beta}_1(u_j^*,v_j^*)}{1+\exp\left(-\gamma_1(u_j^*,v_j^*)\mathbf{x}_j^T\boldsymbol{\beta}_1(u_j^*,v_j^*)\right)}\right)$$

$$C_{32}=\sum_{j=1}^{n}\left(\frac{\mathbf{x}_j^T\boldsymbol{\beta}_1(u_j^*,v_j^*)\left(\exp\left(-\gamma_1(u_j^*,v_j^*)\mathbf{x}_j^T\boldsymbol{\beta}_1(u_j^*,v_j^*)\right)\right)}{1+\exp\left(-\gamma_1(u_j^*,v_j^*)\mathbf{x}_j^T\boldsymbol{\beta}_1(u_j^*,v_j^*)\right)}\right)$$

$$D_{31}=\sum_{j=1}^{n}\left(\frac{\mathbf{x}_j^T\boldsymbol{\beta}_1(u_j^*,v_j^*)\left(\exp\left(-\gamma_1(u_j^*,v_j^*)\mathbf{x}_j^T\boldsymbol{\beta}_1(u_j^*,v_j^*)\right)\right)}{1+\exp\left(-\gamma_1(u_j^*,v_j^*)\mathbf{x}_j^T\boldsymbol{\beta}_1(u_j^*,v_j^*)\right)}\right)$$

Using the same way, we can obtained the first derivative of ln likelihood function respect to $\gamma_2(u_j^*,v_j^*)$ as below:

$$\frac{\partial l}{\partial\gamma_2(u_j^*,v_j^*)}=\sum_{j=1}^{n}\left(1-b_j-c_j-d_j\right)\frac{1}{A_j}\left(A_{41}+A_{42}+A_{43}+A_{44}\right)+\sum_{j=1}^{n}\left(b_j\right)\left(\sum_{j=1}^{n}B_{41}+\sum_{j=1}^{n}B_{42}\right)$$
$$+\sum_{j=1}^{n}\left(c_j\right)\left(\sum_{j=1}^{n}C_{41}+\sum_{j=1}^{n}C_{42}\right)+\sum_{j=1}^{n}\left(d_j\right)\left(\sum_{j=1}^{n}D_{41}\right) \quad (11)$$

The first derivative of ln likelihood function respect to $\tau$ can be written as follow:

$$\frac{\partial l}{\partial\tau}=\sum_{j=1}^{n}\left(1-b_j-c_j-d_j\right)\frac{1}{A_j}\left(A_{52}+A_{45}+A_{54}\right)+\sum_{j=1}^{n}\left(b_j\right)\left(\sum_{j=1}^{n}B_{51}+\sum_{j=1}^{n}B_{52}+\sum_{j=1}^{n}B_{53}+\sum_{j=1}^{n}B_{54}\right)$$
$$+\sum_{j=1}^{n}\left(c_j\right)\left(\sum_{j=1}^{n}C_{51}+\sum_{j=1}^{n}C_{52}+\sum_{j=1}^{n}C_{53}+\sum_{j=1}^{n}C_{54}\right)+\sum_{j=1}^{n}\left(d_j\right)\left(\sum_{j=1}^{n}D_{51}+\sum_{j=1}^{n}D_{52}\right) \quad (12)$$

$$A_{52}=\frac{\exp\left(-\gamma_2(u_j^*,v_j^*)\mathbf{x}_j^T\boldsymbol{\beta}_2(u_j^*,v_j^*)\right)}{1+\exp\left(-\gamma_2(u_j^*,v_j^*)\mathbf{x}_j^T\boldsymbol{\beta}_2(u_j^*,v_j^*)\right)}\frac{1}{1+\exp\left(-\gamma_1(u_j^*,v_j^*)\mathbf{x}_j^T\boldsymbol{\beta}_1(u_j^*,v_j^*)\right)}$$
$$\left(\left(\left(\frac{2}{\pi\tau}\sqrt{2\lambda_{1j}\tau+1}\right)^{\frac{1}{2}}\left(\frac{\pi}{2}\tau\right)^{\frac{1}{2}}\left(1+2\tau\lambda_{1j}\right)^{-\frac{1}{4}}\exp\left(-\left(\tau^{-1}\right)\sqrt{1+2\tau\lambda_{1j}}\right)\left(-\frac{\exp\left(\tau^{-1}\right)}{\tau^{-2}}\right)\right)\right.$$
$$+\left(\exp\left(\tau^{-1}\right)\left(\frac{\pi}{2}\tau\right)^{\frac{1}{2}}\left(1+2\tau\lambda_{1j}\right)^{-\frac{1}{4}}\exp\left(-\left(\tau^{-1}\right)\sqrt{1+2\tau\lambda_{1j}}\right)\left(\frac{2}{\pi\tau}\right)^{\frac{1}{2}}\frac{\frac{\lambda_{1j}}{2}\left(1+2\tau\lambda_{1j}\right)^{-\frac{3}{4}}(\tau)^{\frac{1}{2}}-\left(1+2\tau\lambda_{1j}\right)^{\frac{1}{4}}\frac{1}{2}(\tau)^{\frac{1}{2}}}{\tau}\right)$$
$$+\left(\left(\frac{2}{\pi\tau}\sqrt{2\lambda_{1j}\tau+1}\right)^{\frac{1}{2}}\exp\left(\tau^{-1}\right)\right)\left(\left(\frac{1}{2}\right)^{\frac{1}{2}}\frac{\pi}{2}(\pi\tau)^{-\frac{1}{2}}\left(1+2\tau\lambda_{1j}\right)^{-\frac{1}{4}}\exp\left(-\left(\tau^{-1}\right)\sqrt{1+2\tau\lambda_{1j}}\right)+\left(\frac{\pi}{2}\tau\right)^{\frac{1}{2}}\left(-\frac{\lambda_{1j}}{2}\right)\right.$$
$$\left.\left.\left(1+2\tau\lambda_{1j}\right)^{-\frac{5}{4}}\exp\left(-\left(\tau^{-1}\right)\sqrt{1+2\tau\lambda_{1j}}\right)+\left(\frac{\pi}{2}\tau\right)^{\frac{1}{2}}\left(1+2\tau\lambda_{1j}\right)^{-\frac{1}{4}}\left(-\left(\tau^{-1}\right)\sqrt{1+2\tau\lambda_{1j}}\right)\right)\right)$$

$$A_{53}=\frac{\exp\left(-\gamma_1(u_j^*,v_j^*)\mathbf{x}_j^T\boldsymbol{\beta}_1(u_j^*,v_j^*)\right)}{1+\exp\left(-\gamma_1(u_j^*,v_j^*)\mathbf{x}_j^T\boldsymbol{\beta}_1(u_j^*,v_j^*)\right)}\frac{1}{1+\exp\left(-\gamma_2(u_j^*,v_j^*)\mathbf{x}_j^T\boldsymbol{\beta}_2(u_j^*,v_j^*)\right)}$$
$$\left(\left(\left(\frac{2}{\pi\tau}\sqrt{2\lambda_{2j}\tau+1}\right)^{\frac{1}{2}}\left(\frac{\pi}{2}\tau\right)^{\frac{1}{2}}\left(1+2\tau\lambda_{2j}\right)^{-\frac{1}{4}}\exp\left(-\left(\tau^{-1}\right)\sqrt{1+2\tau\lambda_{2j}}\right)\left(-\frac{\exp\left(\tau^{-1}\right)}{\tau^{-2}}\right)\right)\right.$$
$$+\left(\exp\left(\tau^{-1}\right)\left(\frac{\pi}{2}\tau\right)^{\frac{1}{2}}\left(1+2\tau\lambda_{2j}\right)^{-\frac{1}{4}}\exp\left(-\left(\tau^{-1}\right)\sqrt{1+2\tau\lambda_{2j}}\right)\left(\frac{2}{\pi\tau}\right)^{\frac{1}{2}}\frac{\frac{\lambda_{2j}}{2}\left(1+2\tau\lambda_{2j}\right)^{-\frac{3}{4}}(\tau)^{\frac{1}{2}}-\left(1+2\tau\lambda_{2j}\right)^{\frac{1}{4}}\frac{1}{2}(\tau)^{\frac{1}{2}}}{\tau}\right)$$
$$+\left(\left(\frac{2}{\pi\tau}\sqrt{2\lambda_{2j}\tau+1}\right)^{\frac{1}{2}}\exp\left(\tau^{-1}\right)\right)\left(\left(\frac{1}{2}\right)^{\frac{1}{2}}\frac{\pi}{2}(\pi\tau)^{-\frac{1}{2}}\left(1+2\tau\lambda_{2j}\right)^{-\frac{1}{4}}\exp\left(-\left(\tau^{-1}\right)\sqrt{1+2\tau\lambda_{2j}}\right)+\left(\frac{\pi}{2}\tau\right)^{\frac{1}{2}}\left(-\frac{\lambda_{2j}}{2}\right)\right.$$
$$\left.\left.\left(1+2\tau\lambda_{2j}\right)^{-\frac{5}{4}}\exp\left(-\left(\tau^{-1}\right)\sqrt{1+2\tau\lambda_{2j}}\right)+\left(\frac{\pi}{2}\tau\right)^{\frac{1}{2}}\left(1+2\tau\lambda_{2j}\right)^{-\frac{1}{4}}\left(-\left(\tau^{-1}\right)\sqrt{1+2\tau\lambda_{2j}}\right)\right)\right)$$

$$A_{54}=\frac{\exp\left(-\gamma_2(u_j^*,v_j^*)\mathbf{x}_j^T\boldsymbol{\beta}_2(u_j^*,v_j^*)\right)}{1+\exp\left(-\gamma_2(u_j^*,v_j^*)\mathbf{x}_j^T\boldsymbol{\beta}_2(u_j^*,v_j^*)\right)}\frac{1}{1+\exp\left(-\gamma_1(u_j^*,v_j^*)\mathbf{x}_j^T\boldsymbol{\beta}_1(u_j^*,v_j^*)\right)}$$
$$\left(\left(\left(\frac{2}{\pi\tau}\sqrt{2\lambda_{1j}\tau+1}\right)^{\frac{1}{2}}\left(\frac{\pi}{2}\tau\right)^{\frac{1}{2}}\left(1+2\tau\lambda_{1j}\right)^{-\frac{1}{4}}\exp\left(-\left(\tau^{-1}\right)\sqrt{1+2\tau\lambda_{1j}}\right)\left(-\frac{\exp\left(\tau^{-1}\right)}{\tau^{-2}}\right)\right)\right.$$
$$+\left(\exp\left(\tau^{-1}\right)\left(\frac{\pi}{2}\tau\right)^{\frac{1}{2}}\left(1+2\tau\lambda_{1j}\right)^{-\frac{1}{4}}\exp\left(-\left(\tau^{-1}\right)\sqrt{1+2\tau\lambda_{1j}}\right)\left(\frac{2}{\pi\tau}\right)^{\frac{1}{2}}\frac{\frac{\lambda_{1j}}{2}\left(1+2\tau\lambda_{1j}\right)^{-\frac{3}{4}}(\tau)^{\frac{1}{2}}-\left(1+2\tau\lambda_{1j}\right)^{\frac{1}{4}}\frac{1}{2}(\tau)^{\frac{1}{2}}}{\tau}\right)$$
$$+\left(\left(\frac{2}{\pi\tau}\sqrt{2\lambda_{1j}\tau+1}\right)^{\frac{1}{2}}\exp\left(\tau^{-1}\right)\right)\left(\left(\frac{1}{2}\right)^{\frac{1}{2}}\frac{\pi}{2}(\pi\tau)^{-\frac{1}{2}}\left(1+2\tau\lambda_{1j}\right)^{-\frac{1}{4}}\exp\left(-\left(\tau^{-1}\right)\sqrt{1+2\tau\lambda_{1j}}\right)+\left(\frac{\pi}{2}\tau\right)^{\frac{1}{2}}\left(-\frac{\lambda_{1j}}{2}\right)\right.$$
$$\left.\left.\left(1+2\tau\lambda_{1j}\right)^{-\frac{5}{4}}\exp\left(-\left(\tau^{-1}\right)\sqrt{1+2\tau\lambda_{1j}}\right)+\left(\frac{\pi}{2}\tau\right)^{\frac{1}{2}}\left(1+2\tau\lambda_{1j}\right)^{-\frac{1}{4}}\left(-\left(\tau^{-1}\right)\sqrt{1+2\tau\lambda_{1j}}\right)\right)\right)$$

$$B_{51}=\sum_{j=1}^{n}\left(\frac{1}{2}\left(-\frac{1}{\tau}+\frac{1}{\left(2\tau\lambda_{1j}+1\right)^{\frac{1}{2}}}\left(\frac{1}{2}\left(2\tau\lambda_{1j}+1\right)^{-\frac{1}{2}}2\lambda_{1j}\right)+\left(-\tau^{-2}\right)\right)\right)$$

$$B_{52}=\sum_{j=1}^{n}\left(\left(\left(-K_{y_1-\frac{1}{2}}\left(\tau^{-1}\sqrt{2\lambda_{1j}\tau+1}\right)+\frac{\tau\left(y_1-\frac{1}{2}\right)K_{y_1-\frac{1}{2}}\left(\tau^{-1}\sqrt{2\lambda_{1j}\tau+1}\right)}{\left(\sqrt{2\tau\lambda_{1j}+1}\right)}\right)\frac{\frac{\lambda_{1j}}{2}\left(2\tau\lambda_{1j}+1\right)^{-\frac{3}{4}}(\tau)^{\frac{1}{2}}-\left(2\tau\lambda_{1j}+1\right)^{\frac{1}{4}}\frac{1}{2}(\tau)^{-\frac{1}{2}}}{\tau}-\frac{1}{2}y_{1j}\frac{1}{1+2\lambda_{1j}\tau}2\lambda_{1j}\right)$$

$$B_{53}=\sum_{j=1}^{n}\left(\frac{1}{2}\left(-\frac{1}{\tau}+\frac{1}{\left(2\tau\left(\lambda_{1j}+\lambda_{2j}\right)+1\right)^{\frac{1}{2}}}\left(\frac{1}{2}\left(2\tau\left(\lambda_{1j}+\lambda_{2j}\right)+1\right)^{-\frac{1}{2}}2\left(\lambda_{1j}+\lambda_{2j}\right)\right)+\left(-\tau^{-2}\right)\right)\right)$$

$$B_{54}=\sum_{j=1}^{n}\left(\left(-K_{y_1-\frac{1}{2}}\left(\tau^{-1}\sqrt{2\left(\lambda_{1j}+\lambda_{2j}\right)\tau+1}\right)+\frac{\tau\left(y_1-\frac{1}{2}\right)K_{y_1-\frac{1}{2}}\left(\tau^{-1}\sqrt{2\left(\lambda_{1j}+\lambda_{2j}\right)\tau+1}\right)}{\left(\sqrt{2\left(\lambda_{1j}+\lambda_{2j}\right)\tau+1}\right)}\right)\right.$$
$$\left.\left(\frac{\frac{\left(\lambda_{1j}+\lambda_{2j}\right)}{2}\left(2\tau\left(\lambda_{1j}+\lambda_{2j}\right)+1\right)^{-\frac{3}{4}}(\tau)^{\frac{1}{2}}-\left(2\tau\left(\lambda_{1j}+\lambda_{2j}\right)+1\right)^{\frac{1}{4}}\frac{1}{2}(\tau)^{-\frac{1}{2}}}{\tau}\right)-\frac{1}{2}y_{1j}\frac{1}{1+2\lambda_{1j}\tau+2\lambda_{2j}\tau}2\lambda_{1j}+2\lambda_{2j}\right)$$

$$C_{51}=\sum_{j=1}^{n}\left(\frac{1}{2}\left(-\frac{1}{\tau}+\frac{1}{\left(2\tau\lambda_{2j}+1\right)^{\frac{1}{2}}}\left(\frac{1}{2}\left(2\tau\lambda_{2j}+1\right)^{-\frac{1}{2}}2\lambda_{2j}\right)+\left(-\tau^{-2}\right)\right)\right)$$

$$C_{52}=\sum_{j=1}^{n}\left(\left(\left(-K_{y_2-\frac{1}{2}}\left(\tau^{-1}\sqrt{2\lambda_{2j}\tau+1}\right)+\frac{\tau\left(y_2-\frac{1}{2}\right)K_{y_2-\frac{1}{2}}\left(\tau^{-1}\sqrt{2\lambda_{2j}\tau+1}\right)}{\left(\sqrt{2\tau\lambda_{2j}+1}\right)}\right)\frac{\frac{\lambda_{2j}}{2}\left(2\tau\lambda_{2j}+1\right)^{-\frac{3}{4}}(\tau)^{\frac{1}{2}}-\left(2\tau\lambda_{2j}+1\right)^{\frac{1}{4}}\frac{1}{2}(\tau)^{-\frac{1}{2}}}{\tau}-\frac{1}{2}y_{1j}\frac{1}{1+2\lambda_{2j}\tau}2\lambda_{2j}\right.\right.$$

$$C_{53}=\sum_{j=1}^{n}\left(\frac{1}{2}\left(-\frac{1}{\tau}+\frac{1}{\left(2\tau\left(\lambda_{1j}+\lambda_{2j}\right)+1\right)^{\frac{1}{2}}}\left(\frac{1}{2}\left(2\tau\left(\lambda_{1j}+\lambda_{2j}\right)+1\right)^{-\frac{1}{2}}2\left(\lambda_{1j}+\lambda_{2j}\right)\right)+\left(-\tau^{-2}\right)\right)\right)$$

$$C_{54}=\sum_{j=1}^{n}\left(\left(-K_{y_1-\frac{1}{2}}\left(\tau^{-1}\sqrt{2\left(\lambda_{1j}+\lambda_{2j}\right)\tau+1}\right)+\frac{\tau\left(y_1-\frac{1}{2}\right)K_{y_1-\frac{1}{2}}\left(\tau^{-1}\sqrt{2\left(\lambda_{1j}+\lambda_{2j}\right)\tau+1}\right)}{\left(\sqrt{2\left(\lambda_{1j}+\lambda_{2j}\right)\tau+1}\right)}\right)\right.$$
$$\left.\left(\frac{\frac{\left(\lambda_{1j}+\lambda_{2j}\right)}{2}\left(2\tau\left(\lambda_{1j}+\lambda_{2j}\right)+1\right)^{-\frac{3}{4}}(\tau)^{\frac{1}{2}}-\left(2\tau\left(\lambda_{1j}+\lambda_{2j}\right)+1\right)^{\frac{1}{4}}\frac{1}{2}(\tau)^{-\frac{1}{2}}}{\tau}\right)-\frac{1}{2}y_{1j}\frac{1}{1+2\lambda_{1j}\tau+2\lambda_{2j}\tau}2\lambda_{1j}+2\lambda_{2j}\right)$$

$$D_{51}=\sum_{j=1}^{n}\left(\frac{1}{2}\left(-\frac{1}{\tau}+\frac{1}{\left(2\tau\left(\lambda_{1j}+\lambda_{2j}\right)+1\right)^{\frac{1}{2}}}\left(\frac{1}{2}\left(2\tau\left(\lambda_{1j}+\lambda_{2j}\right)+1\right)^{-\frac{1}{2}}2\left(\lambda_{1j}+\lambda_{2j}\right)\right)+\left(-\tau^{-2}\right)\right)\right)$$

$$D_{52} = \sum_{j=1}^{n} \left( -K_{y_1+y_2-\frac{1}{2}} \left( \tau^{-1} \sqrt{2(\lambda_{1j}+\lambda_{2j})\tau+1} \right) + \frac{\tau\left(y_1+y_2-\frac{1}{2}\right) K_{y_1+y_2-\frac{1}{2}} \left( \tau^{-1} \sqrt{2(\lambda_{1j}+\lambda_{2j})\tau+1} \right)}{\left(\sqrt{2(\lambda_{1j}+\lambda_{2j})\tau+1}\right)} \right)$$

$$\left( \frac{\frac{(\lambda_{1j}+\lambda_{2j})}{2}\left(2\tau(\lambda_{1j}+\lambda_{2j})+1\right)^{-\frac{3}{4}} (\tau)^{\frac{1}{2}} - \left(2\tau(\lambda_{1j}+\lambda_{2j})+1\right)^{\frac{1}{4}} \frac{1}{2}(\tau)^{-\frac{1}{2}}}{\tau} \right)$$

$$-\frac{1}{2}(y_{1j}+y_{2j})\frac{1}{1+2\lambda_{1j}\tau+2\lambda_{2j}\tau} 2\lambda_{1j} + 2\lambda_{2j}$$

The results obtained from the first derivative of the ln likelihood function for each parameter are explicit derivatives, so the BHHH algorithm is used to obtain the parameter estimators based on the first derivative of the likelihood function. The steps of the BHHH algorithm can be written as follows[10]:

1. Determining the initial value of $\hat{\boldsymbol{\theta}}$ and m=0 with $\varepsilon > 0$ value is the convergence limit.

   $$\hat{\boldsymbol{\theta}} = \left[ \hat{\boldsymbol{\gamma}}_{1(0)}^T(u_i,v_i) \ \hat{\boldsymbol{\gamma}}_{2(0)}^T(u_i,v_i) \ \hat{\boldsymbol{\beta}}_{1(0)}^T(u_i,v_i) \ \hat{\boldsymbol{\beta}}_{2(0)}^T(u_i,v_i) \ \hat{\tau} \right]^T$$

   The Initial value of $\hat{\boldsymbol{\gamma}}_{1(0)}^T(u_i,v_i), \hat{\boldsymbol{\gamma}}_{2(0)}^T(u_i,v_i)$, $\hat{\boldsymbol{\beta}}_{1(0)}^T(u_i,v_i)$, $\hat{\boldsymbol{\beta}}_{2(0)}^T(u_i,v_i)$ parameters obtained from estimators of the ZIPR model while the initial value for $\hat{\tau}$ parameters obtained from the dispersion value of the ZIPIGR model.
2. Determine the gradient vector $\mathbf{g}(\hat{\boldsymbol{\theta}}_m)$

   $$\mathbf{g}(\hat{\boldsymbol{\theta}}_r)^T = \left[ \frac{\partial L(\bullet)}{\partial \hat{\boldsymbol{\gamma}}_{1(0)}^T(u_i,v_i)} \ \frac{\partial L(\bullet)}{\partial \hat{\boldsymbol{\gamma}}_{2(0)}^T(u_i,v_i)} \ \frac{\partial L(\bullet)}{\partial \hat{\boldsymbol{\beta}}_{1(0)}^T(u_i,v_i)} \ \frac{\partial L(\bullet)}{\partial \hat{\boldsymbol{\beta}}_{2(0)}^T(u_i,v_i)} \ \frac{\partial L(\bullet)}{\partial \hat{\tau}} \right]^T$$

   With $L(\bullet) = \ln L(\boldsymbol{\gamma}_1(u_i,v_i), \boldsymbol{\gamma}_2(u_i,v_i), \boldsymbol{\beta}_1(u_i,v_i), \boldsymbol{\beta}_2(u_i,v_i), \tau)$
3. Determine the Hessian matrix

   $$\mathbf{H}(\hat{\boldsymbol{\theta}}_r) = -\sum_{i=1}^{n} \mathbf{g}_i(\hat{\boldsymbol{\theta}}_m)\mathbf{g}_i(\hat{\boldsymbol{\theta}}_m)^T$$
4. Substituting value of $\hat{\boldsymbol{\theta}}_m$ into gardient vector element $\mathbf{g}(\hat{\boldsymbol{\theta}}_m)$ and Hessian matrix.
5. Iterate the BHHH equation started when m=0,

   $$\hat{\boldsymbol{\theta}}_{m+1} = \hat{\boldsymbol{\theta}}_m - \mathbf{H}^{-1}(\hat{\boldsymbol{\theta}}_m)\mathbf{g}(\hat{\boldsymbol{\theta}}_m)$$

   the iteration stopped when $\left\|\hat{\boldsymbol{\theta}}_{m+1} - \hat{\boldsymbol{\theta}}_m\right\| \le \varepsilon$ with $\varepsilon$ e is a small positive number close to 0
6. Repeat steps number 2 and so on with m=m+1.

### 3.2 Hypothesis Testing of GWBZIPIGR

Hypothesis testing of the GWBZIPIGR parameters consists of two types of hypothesis testing, namely simultaneous and partial. Tests were conducted to determine the significance of each of the parameters obtained. In this paper, the Maximum Likelihood Ratio Test (MLRT) is used to test the hypothesis.

#### *3.2.1 Simultaneous Test of $\boldsymbol{\beta}_1(u_j^*,v_j^*), \boldsymbol{\beta}_2(u_j^*,v_j^*)$ parameters*

The $\boldsymbol{\beta}_1(u_j^*,v_j^*)$and $\boldsymbol{\beta}_2(u_j^*,v_j^*)$ parameters were tested simultaneously with the following hypothesis:

$H_0 : \beta_{l1}(u_j^*,v_j^*) = \beta_{l2}(u_j^*,v_j^*) = \ldots = \beta_{lr}(u_j^*,v_j^*) = 0, l = 1,2; r = 1,2,\ldots,q; j = 1,2,\ldots,n$

$H_1$ : there is at least one $\beta_{lr}(u_j^*,v_j^*) \neq 0, l = 1,2, r = 1,2,\ldots,q, j = 1,2,\ldots,n$

To test the hypothesis, it is needed to determine the likelihood function for the parameter set under the population $L(\hat{\Omega})$ and the parameter set under the null hypothesis $L(\hat{\omega})$. The likelihood function for the parameter set under the population $L(\hat{\Omega})$ is the likelihood function at section 3.1 $L(\Omega)$, meanwhile the likelihood function for the parameter set under the null hypothesis can be written:

$$L(\omega)=\left\{\beta_{10}\left(u_j^*,v_j^*\right),\beta_{20}\left(u_j^*,v_j^*\right),\gamma_{10}\left(u_j^*,v_j^*\right),\gamma_{20}\left(u_j^*,v_j^*\right),\tau;j=1,2,...,n\right\}$$

$$L\left(\boldsymbol{\beta}_{10}\left(u_j^*,v_j^*\right),\boldsymbol{\beta}_{20}\left(u_j^*,v_j^*\right),\boldsymbol{\gamma}_{10}\left(u_j^*,v_j^*\right),\boldsymbol{\gamma}_{20}\left(u_j^*,v_j^*\right),\tau\right)=\prod_{j=1}^{n}\left(A_j^*\right)+\prod_{j=1}^{n}\left(B_j^*\right)+\prod_{j=1}^{n}\left(C_j^*\right)+\prod_{j=1}^{n}\left(D_j^*\right)$$

With the GWBZIPIGR model without the dependent variable can be written as follows:

$$\lambda_{1j}=\exp\left(\boldsymbol{\beta}_{10}\left(u_j^*,v_j^*\right)\right)\quad \lambda_{2j}=\exp\left(\boldsymbol{\beta}_{20}\left(u_j^*,v_j^*\right)\right)$$

$$p_{1j}=\frac{\exp\left(-\gamma_{10}\left(u_j^*,v_j^*\right)\boldsymbol{\beta}_{10}\left(u_j^*,v_j^*\right)\right)}{1+\exp\left(-\gamma_{10}\left(u_j^*,v_j^*\right)\boldsymbol{\beta}_{10}\left(u_j,v_j^*\right)\right)}\quad 1-p_{1j}=\frac{1}{1+\exp\left(-\gamma_{10}\left(u_j^*,v_{j^*}\right)\boldsymbol{\beta}_{10}\left(u_j^*,v_j^*\right)\right)} \tag{13}$$

$$p_{2j}=\frac{\exp\left(-\gamma_{10}\left(u_j^*,v_j^*\right)\boldsymbol{\beta}_{20}\left(u_j^*,v_j^*\right)\right)}{1+\exp\left(-\gamma_{10}\left(u_j^*,v_j^*\right)\boldsymbol{\beta}_{20}\left(u_j^*,v_j^*\right)\right)}\quad 1-p_{2j}=\frac{1}{1+\exp\left(-\gamma_{20}\left(u_j^*,v_j^*\right)\boldsymbol{\beta}_{20}\left(u_j^*,v_j^*\right)\right)}$$

The likelihood function $L(\omega)$ formed into its ln likelihood function, then derived respect to the parameter $\beta_{10}(u_j,v_j),\beta_{20}(u_j,v_j),\gamma_{10}(u_j,v_j),\gamma_{20}(u_j,v_j),\tau$ equated to zero, and solved by the BHHH algorithm as in section 3.1.

The following is the lambda ratio $\Lambda$ which is formed from the likelihood function for the parameter set under the population $L(\hat{\Omega})$ and the parameter set under the null hypothesis $L(\hat{\omega})$

$$\Lambda=\frac{L(\omega)}{L(\Omega)}$$

With the distribution of the statistical test for large sampel (n) can be written:

$$\Lambda<\Lambda_0$$
$$\Lambda^2<\Lambda_0^{\ 2}$$
$$\ln\Lambda^2<\ln\Lambda_0^{\ 2}$$
$$-\ln\Lambda^2>-\ln\Lambda_0^{\ 2}$$
$$G^2>-\ln(\Lambda^2)$$

Where $G^2>c$ and $\alpha=P\left(G^2>c\mid H_0\text{ true}\right)$ $c$ is obtained from when the distribution of $G^2$ is known, so the simultaneous test can be written as follows:

$$G^2=-\ln\left[\Lambda^2\right]=-2\ln\left[\frac{L(\hat{\omega})}{L(\hat{\Omega})}\right] \tag{14}$$

The $G^2$ will be compared to $\chi_{\alpha;(a-b)}$ with $a$ is the number of parameters under population and $b$ is the number of parameters under $H_0$, meanwhile the rejection area of $H_0$ is if $\left|G^2\right|>\chi_{\alpha;(a-b)}$.

### *3.2.2 Simultaneous Test of $\gamma_1\left(u_j^*,v_j^*\right),\gamma_2\left(u_j^*,v_j^*\right)$ parameters*

The $\gamma_1\left(u_j^*,v_j^*\right)$ and $\gamma_2\left(u_j^*,v_j^*\right)$ parameters are tested simultaneously with the following hypothesis:

$H_0:\gamma_{l1}\left(u_j^*,v_j^*\right)=\gamma_{l2}\left(u_j^*,v_j^*\right)=...=\gamma_{lq}\left(u_j^*,v_j^*\right)=0, l=1,2, r=1,2,...,q$

$H_1$: there is at least one $\gamma_{lr}\left(u_j^*,v_j^*\right)\neq 0, l=1,2, r=1,2,...,q$

To test the hypothesis, it is needed to determine the likelihood function for the parameter set under the population $L(\hat{\Omega})$ and the parameter set under the null hypothesis $L(\hat{\omega})$. The likelihood function for the parameter set under the population $L(\hat{\Omega})$ is the likelihood function at section 3.1 $L(\Omega)$, meanwhile the likelihood function for the parameter set under the null hypothesis can be written:

$$L(\omega)=\{\gamma_{10}(u_j^*,v_j^*),\gamma_{20}(u_j^*,v_j^*),\beta_{10}(u_j^*,v_j^*),\beta_{20}(u_j^*,v_j^*),\tau; j=1,2,...,n\}$$

$$L(\gamma_{10}(u_j^*,v_j^*),\gamma_{20}(u_j^*,v_j^*),\boldsymbol{\beta}_1(u_j^*,v_j^*),\boldsymbol{\beta}_2(u_j^*,v_j^*),\tau)=\prod_{j=1}^{n}(A_j^*)+\prod_{j=1}^{n}(B_j^*)+\prod_{j=1}^{n}(C_j^*)+\prod_{j=1}^{n}(D_j^*)$$

With the GWBZIPIGR model without the dependent variable can be written as follows:

$$\lambda_{1j}=\exp\left(\mathbf{x}_j^T\boldsymbol{\beta}_1(u_j^*,v_j^*)\right) \quad \lambda_{2j}=\exp\left(\mathbf{x}_j^T\boldsymbol{\beta}_2(u_j^*,v_j^*)\right)$$

$$p_1(x_j)=\frac{\exp\left(-\gamma_1(u_j^*,v_j^*)\mathbf{x}_j^T\boldsymbol{\beta}_1(u_j^*,v_j^*)\right)}{1+\exp\left(-\gamma_{10}(u_j^*,v_j^*)\mathbf{x}_j^T\boldsymbol{\beta}_1(u_j^*,v_j^*)\right)} \quad 1-p_1(x_j)=\frac{1}{1+\exp\left(-\gamma_{10}(u_j^*,v_j^*)\mathbf{x}_j^T\boldsymbol{\beta}_1(u_j^*,v_j^*)\right)} \tag{15}$$

$$p_2(x_j)=\frac{\exp\left(-\gamma_{10}(u_j^*,v_j^*)\mathbf{x}_j^T\boldsymbol{\beta}_2(u_j^*,v_j^*)\right)}{1+\exp\left(-\gamma_{10}(u_j^*,v_j^*)\mathbf{x}_j^T\boldsymbol{\beta}_2(u_j^*,v_j^*)\right)} \quad 1-p_2(x_j)=\frac{1}{1+\exp\left(-\gamma_{20}(u_j^*,v_j^*)\mathbf{x}_j^T\boldsymbol{\beta}_2(u_j^*,v_j^*)\right)}$$

The likelihood function $L(\omega)$ formed into its ln likelihood function, then derived respect to the parameter $\gamma_{10}(u_j,v_j),\gamma_{20}(u_j,v_j),\beta_1(u_j,v_j),\beta_2(u_j,v_j),\tau$ equated to zero, and solved by the BHHH algorithm as in section 3.2.

The following is the lambda ratio $\Lambda$ which is formed from the likelihood function for the parameter set under the population $L(\hat{\Omega})$ and the parameter set under the null hypothesis $L(\hat{\omega})$

$$\Lambda=\frac{L(\omega)}{L(\Omega)}$$

With the distribution of the statistical test for large sampel (n) can be written:

Where $G^2>c$ and $\alpha=P(G^2>c\mid H_0 \text{ true})$ $c$ is obtained from when the distribution of $G^2$ is known, so the simultaneous test can be written as follows:

$$G^2=-\ln\left[\Lambda^2\right]=-2\ln\left[\frac{L(\hat{\omega})}{L(\hat{\Omega})}\right] \tag{16}$$

The $G^2$ will be compared to $\chi_{\alpha:(a-b)}$ with $a$ is the number of parameters under population and $b$ is the number of parameters under $H_0$, meanwhile the rejection area of $H_0$ is if $|G^2|>\chi_{\alpha:(a-b)}$.

### *3.2.3 Partial Test of parameters*

If the result of simultaneous hypothesis testing is rejecting $H_0$, then partially testing is carried out using the Wald test. The hypothesis of these test can be written as follows:

1. $H_0:\beta_{lr}(u_j,v_v)=0$

   $H_1:\beta_{lr}(u_j,v_j)\neq 0$
2. $H_0:\gamma_{lr}(u_j,v_j)=0$

   $H_1:\gamma_{lr}(u_j,v_j)\neq 0$
3. $H_0:\tau=0$

   $H_1:\tau\neq 0$

The test statistic can be written:

$$Z_\beta = \frac{\hat{\beta}_{lr}}{se(\hat{\beta}_{lr})}, Z_\gamma = \frac{\hat{\gamma}_{lr}}{se(\hat{\gamma}_{lr})}, Z_\tau = \frac{\hat{\tau}}{se(\hat{\tau})} \quad (17)$$

with $se(\hat{\beta}_{lr}) = \sqrt{\hat{\text{var}}(\hat{\beta}_{lr})}$, $se(\hat{\gamma}_{lr}) = \sqrt{\hat{\text{var}}(\hat{\gamma}_{lr})}$, $se(\hat{\tau}) = \sqrt{\hat{\text{var}}(\hat{\tau})}$

The value of $\sqrt{\hat{\text{var}}(\hat{\beta}_{lr})}$, $se(\hat{\gamma}_{lr}) = \sqrt{\hat{\text{var}}(\hat{\gamma}_{lr})}$, and $se(\hat{\tau}) = \sqrt{\hat{\text{var}}(\hat{\tau})}$ obtained from the main diagonal element of the Hessian matrix. Reject $H_0$ if $|Z| > Z_{\frac{\alpha}{2}}$

## 4. Conclusion

Based on the estimates that have been made in the results section, it can be concluded that the estimation of the GWBZIPIGR model parameters using MLE results in the first derivative of the ln likelihood function that cannot be solved mathematically so the estimation is solved using the BHHH iteration. The simultaneous hypothesis test was carried out using the MLRT for $\beta(u_j,v_j)$ and $\gamma(u_j,v_j)$ parameters, and the Wald test was used to test the parameters partially.

## References

[1] Winkelmann R 2008 *Economic Analysis of Count Data* Berlin: Springer

[2] Hu M C, Pavlicova and Nunes E V 2011 Zero-Inflated and Hurdle Models of Count Data with Extra Zeros: Examples from an HIV-Risk Reduction Intervention Trial *The American Journal of Drug and Alcohol Abuse* **37** (5) 367-375

[3] Lambert D 1992 Zero-Inflated Poisson Regression With an Application to Defects in Manufacturing *Technometrics* **34** (1) 1-14

[4] Hilbe J M 2014 *Modeling Count Data* New York: Cambridge University Press

[5] Krishna P M and Tukaram S D 2011 Bivariate Zero-Inflated Power Series Distribution *Applied Mathematics* **2** 824–829

[6] Pangulimang J, Purhadi and Sutikno 2016 Parameter Estimation and Hypothesis Testing of Geographically Weighted Bivariate Zero-Inflated Poisson *Proceeding of 3rd International Conference on Research, Implementation and Education of Mathematics and Science* 73-78

[7] Lubis S M, Pramoedyo H and Astutik S 2019 Mapping Leprosy Distribution with Geographically Weighted Bivariate Zero Inflated Poisson Regression *International Journal of Recent Technology and Engineering* **8** 32-39

[8] Fotheringham A S, Brundson C and Charlton M 2002 *GeographicallyWeighted Regression: The analysis of spatially varying relationship* London: JohnWiley & Son Ltd

[9] Yuen H K, Chow S C and Tse S K 2015 On Statistical Tests for Homogeneity of Two Bivariate Zero-Inflated Poisson Populations *Journal of Biopharmaceutical Statistics* **25** 44-53

[10] Cameron C and Ttrivedi P K 2005 *Microeconometric Methods and applications* UK: Cambridge University Press

IOP Conf. Series: Materials Science and Engineering 1115 (2021) 012044 doi:10.1088/1757-899X/1115/1/012044

# Image Processing Technology for Edge Detection Based on Vision and Raspberry Pi

**L Patria**[1,*] **and A Sambas**[2]

[1]Department of Information, Universitas Terbuka, Indonesia.
[2]Department of Mechanical Engineering, Universitas Muhammadiyah Tasikmalaya, Indonesia

*Email: lintang@ecampus.ut.ac.id

**Abstract** Rice (*Oryza sativa L)* is the main food crop in Indonesia because most of Indonesia's population consumes rice as a staple food. In this paper, we present a system that functions to detect weeds in rice plants using imaging analysis with a camera device and a Raspberry Pi. The Raspberry Pi got a picture of the field from camera that connected through USB connection and send it to the IoT server. The method used in the analysis is to combine the vertical projection method and the linear scanning method. The results showed that our system was very effective in detecting edges in rice plants.

## 1. Introduction

There are some applications that use image processing in agriculture such as agricultural plant leaf disease detection [1], classification of agricultural products [2], weed detection [3], assessing sugar beet leaf nitrogen status [4], apple classification [5], plant detection of intact tomato fruits [6], non-destructive crop measurements [7], sorting systems in agricultural [8], estimating antioxidant activity and anthocyanin content of sweet cherry [9], extraction of hedgerows [10], quality and shelf-life prediction of cauliflower [11], recognition and classification of paddy leaf diseases [12].

Various important studies involve the edge detection. Vincent and Folorunso (2009) presented a descriptive algorithm for Sobel image edge detection [13], Revathi and Hemalatha (2012) conducted classification of cotton leaf spot diseases using image processing edge detection techniques [14]. Ezzaki et al (2018) presented a comparison of edge detectors performance through use in agricultural robotics [15]. The result shows that the good performance of Bouda's and Kayyali's detector. Qul'am et al (2019) created online image processing of a vision guide pick and place robot by using edge detection [16]. Danton et al (2020) developed a spraying robot for precision agriculture based on the edge approach [17].

In this work, we proposed a low-cost Raspberry Pi based image processing system for identifying edge from rice field. The Raspberry Pi got picture of the field from camera that connected through USB connection. The Raspberry Pi analyzes the picture using vertical projection and linear scanning method and send the result to IoT server. Finally, edge detection in rice plant can be analyzed comprehensively

## 2. Architecture

The system consisted of Microcomputer (based on Raspberry Pi 4, 2 GB of RAM and 16 GB of memory card), camera, modem, and IoT server. The Raspberry Pi was a central system of all devices, that perform calculating, processing and converting. The Raspberry Pi processed image of field based on its rows in certain interval (one shoot for every day) with camera via USB 2.0 / 3.0 connector. The taken image was processed inside the Raspberry Pi using image processing algorithm without delayed after the camera shot a row of the field.

The processed image was sent to IoT server based on HTTP request through internet using the Modem (GSM Module). The IoT Server built from Virtual Private Server which could be adding an Operating System (OS) and several image processing resource independently (not a sharing hosting). The IoT Server saved the sent image inside server drive with unique name (based on time) and extension as an image file.

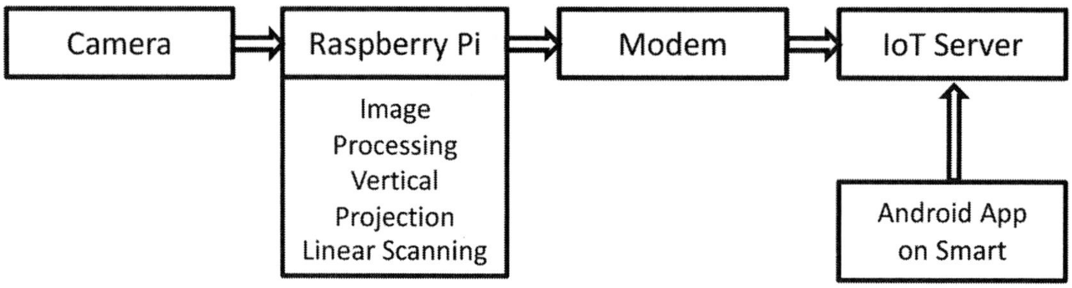


**Figure 1.** Image processing scheme based Raspberry Pi

**Figure 2.** Camera for edge detection

# 3. Methodology

## 3.1 Vertical projection method

The vertical projection of the image is a graph, which is the overall magnitude of the figure according to the y-axis. If we compute the vertical projection of the image after applying the vertical edge detection filter, the magnitude of a certain point represents the occurrence of the vertical edges at that time. Then, the vertical projection will be processed so that the image changes can then be used for vertical localization of the license plate. The vertical projection is detected by obtaining the pixel quantization value based on the image height (line). At this stage the image will be cut based on the graph of the vertical projection. The equation model used is as follows

$$P_x(i) = \sum_{j=1}^{W} bw(i,j), \dots i = 1,2, \dots H \quad (1)$$
$$P_y(j) = \sum_{i=1}^{H} bw(i,j), \dots j = i,2, \dots W \quad (2)$$

W denotes the width of images, $H$ means the height of images, $P_x(i)$ represents all the non-zero pixels of the i column in the horizontal projection direction; $P_y(j)$ denotes all the non-zero pixels of the j row in the vertical projection direction.

## 3.2 Linear scanning algorithm

The vertical projection method generates the number of crop lines and the coverage of each crop line, with n denoted the number of the crop rows, $P_{iTL}$ mean the left top boundary of i column of crop rows, $P_{iTR}$ represented the right top boundary of i column of crop rows,PiBL denoted the left bottom boundary of i column of crop rows, $P_{iBR}$ designated the right bottom boundary of i column of crop rows [18]. In this method, we use ref from [18], Linear scanning algorithm flow chart is shown in Figure 3.

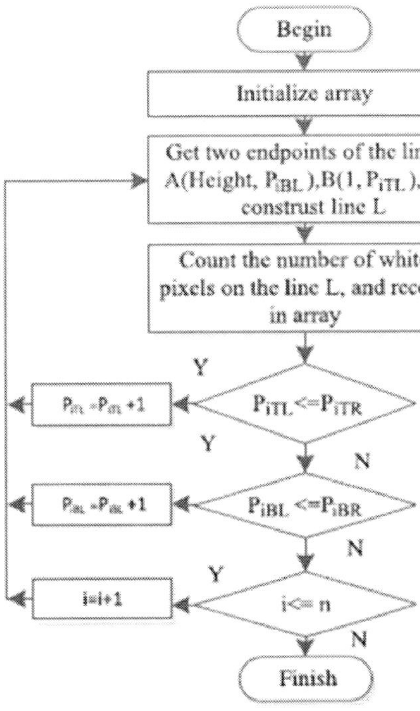


**Figure 3** Linear scanning crop line detection algorithm [18].

## 4. Result and Discussion

This research was conducted in Bojonggambir, Tasikmalaya. The first stage of this research was to create a manual outline for rice plants. The aim was to facilitate analysis of the rice plant row structure. The resulting pattern could be compared with the processed image. Manual outline creation can be seen in Figure 4.

By using the vertical projection method and the linear scanning method, we had processed the image on the rice plant to detect the edges of the plant. The goal was to see the regularity pattern of rice plants. If the pattern was found to be irregular, then there were other plants which called "weeds".

The next step was to detect plants in green and convert them to black and white. If we applied this model to rice plants that grew regularly, the results obtained were edge detection that corresponded to the structure of rice plants (See Figure 5). But if we applied this model to rice plants that growth irregularly, the results showed irregular edge detection. With this method, we could conclude that plants that the irregular growth plants were weeds (See Figure 6).

**Figure 4** Manual outline creation

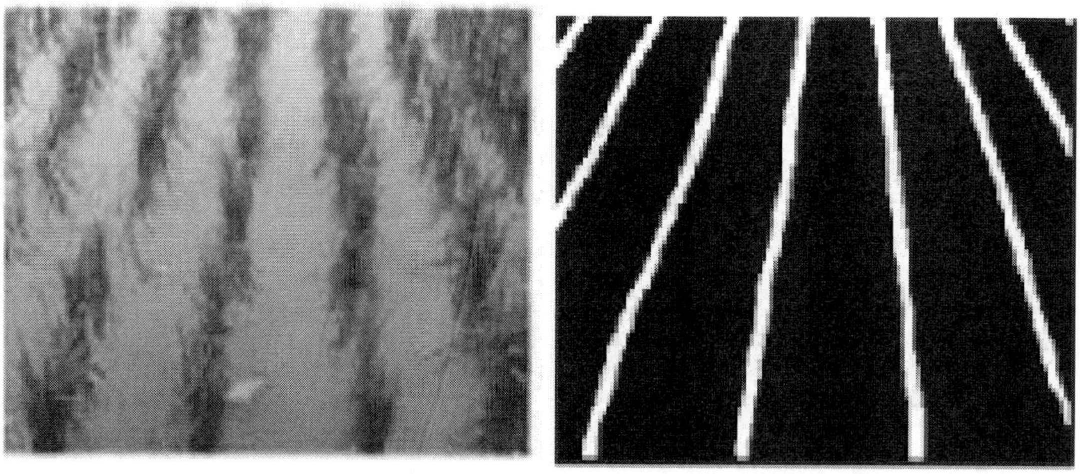

**Figure 5** Detect green plants and convert them to black and white on regular rice plants

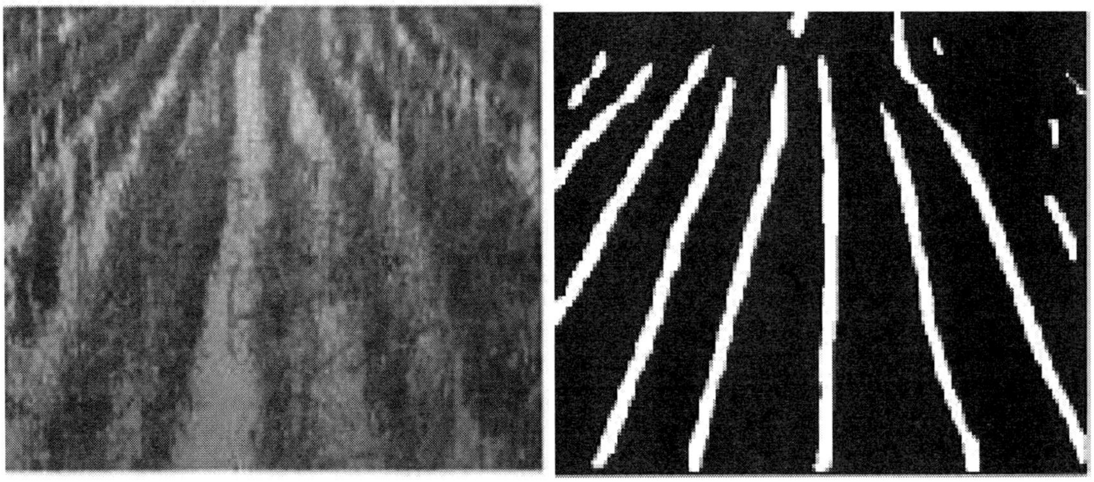

**Figure 6** Detect green plants and convert them to black and white on irregular rice plants

## 5. Conclusions

This paper combined vertical projection method with linear scanning method to identify the center line of the crop rows, and divided the grid cells. Also, we proposed a low-cost Raspberry Pi based image processing system for identifying edge from rice field. The Raspberry Pi processed pictures of the field from USB connected camera. The results showed that our system was very effective in detecting edges in rice plants. The next research is to detect weeds using the convolutional neural network method.

## Acknowledgment

Authors thanks to Universitas Terbuka for the financial support to through Project No. 15970/UN31.LPPM/PT.01.03/2020 (Penelitian Terapan).

## References

[1] Jagtap S B and Hambarde M S M 2014 *IOSR J VLSI Sig Proc* **4** 24-30
[2] Tripathi M K and Maktedar D D 2016 In *2016 International Conference on Computing Communication Control and automation* 1-6
[3] Wang A, Zhang W and Wei X 2019 *Computers and electronics in agriculture* **158** 226-240
[4] Moghaddam P A, Derafshi M H and Shayesteh M 2010 *Journal of Food, Agriculture and Environment* **8** 485-489
[5] Shahin M A, Tollner E W, McClendon R W and Arabnia H R 2002 *Transactions of the ASAE* **45** 1619
[6] Yamamoto K, Guo W, Yoshioka Y and Ninomiya S 2014 *Sensors* 14 12191-12206
[7] Van Henten E J and Bontsema J 1995 *Journal of agricultural engineering research* **61** 97-105
[8] Raji A O and Alamutu A O 2005 *Agricultural Engineering International* **7** 1-12
[9] Taghadomi-Saberi S, Omid M, Emam-Djomeh Z and Ahmadi H 2014 *Journal of the Science of Food and Agriculture* **94** 95-101
[10] Tansey K, Chambers I, Anstee A, Denniss A and Lamb A 2009 *Applied geography* **29** 145-157
[11] Alden K M, Omid M, Rajabipour A, Tajeddin B and Firouz M S 2019 *Computers and*

*Electronics in Agriculture* **163** 104861
[12] Ramesh S and Vydeki D 2020 *Information processing in agriculture* **7** 249-260
[13] Vincent O R and Folorunso O 2009 *Proceedings of Informing Science & IT Education Conference* **40** 97-107
[14] Revathi P and Hemalatha M 2012 In *2012 International Conference on Emerging Trends in Science, Engineering and Technology* 169-173
[15] Ezzaki A, Masmoudi I & Mohamed E A (2018) In *The International Conference on Micro and NanoSatellites* 1-5
[16] Qul'am H M, Dewi T, Risma P, Oktarina Y and Permatasari D 2019 In *2019 International Conference on Electrical Engineering and Computer Science* 102-106
[17] Danton A, Roux J C, Dance B, Cariou C and Lenain R 2020 In *2020 IEEE Conference on Control Technology and Applications* 267-272
[18] Tang J L, Chen X Q, Miao R H and Wang D 2016 *Computers and Electronics in Agriculture* **122** 103-111

# Vulnerability Analysis of Internet Devices from Indonesia Based on Exposure Data in Shodan

**B Novianto*, Y Suryanto and K Ramli**

Department of Electrical Engineering, Faculty of Engineering, University of Indonesia, Depok, Indonesia

*Corresponding author: bambang.novianto81@ui.ac.id

**Abstract.** The growth of internet-enabled devices has increased interest in cybersecurity. In 2014, Project SHINE (SHodan INtelligence Extraction) published a report of large-scale security assessments for devices connected to the Internet. However, the number of IP addresses harvested from Indonesia in 2014 is very small. There were 7.182 IP address from Indonesia. It was about 0,0032% from the total 2.186.971 IP addresses. In this paper, we propose an initiative to gather all information for all Autonomous System Number (AS Number) from Indonesia in Shodan. We have gathered a dataset about all information of AS Numbers in Indonesia such as 12.787 unique ports, 79 unique operating systems, 409 unique products, 3.634 unique domains, 145.543 unique IP addresses, and 790 unique organizations. We use the K-Means algorithm to cluster all AS Numbers into several classes according to the exposure level in shodan. Based on the result, we have 4 classes of AS Numbers. There are 1.075 AS Numbers in class:0 (no information in Shodan yet), 614 AS Numbers in class:1 (exposure level = low), 9 AS Numbers in class:2 (exposure level = medium), and 1 AS Number in class:3 (exposure level = high). This information can be used to warn the organizations that manage AS Numbers in Indonesia to be aware of the security and the threats to their systems.

## 1. Introduction

Nowadays, the Internet has been one of the most transformative and fast-growing technologies. The growth of internet-enabled devices has increased interest in the security of data. To collect data from internet-enabled devices, Project SHINE was created. Project SHINE collects data by using the Shodan search engine. Shodan is a search engine that scans the internet to find open ports. Shodan is a reference to "Sentient Hyper-Optimized Data Access Network" [1]. Shodan's information is considered valuable when attempting to discover critical systems that are directly connected to the Internet. There will be a positive or negative aspect of Shodan. It depends on its user's intent. Shodan can be used as either an assessment portal or as a weapon for hackers to scan targets passively.

Based on Project SHINE reports in 2014, there were 211 countries identified based on the IP addresses harvested. The highest countries are the United States with 616.994 IP addresses (33.75%), Germany with 280.248 IP addresses (15.33%), China with 112.114 IP addresses (6.13%), Korea with 99.856 IP addresses (5.46%) and United Kingdom with 66.234 IP addresses (3.62%). Unfortunately, the number of IP addresses harvested from Indonesia in 2014 is very small. There are 7.182 IP address (0,0032% from the total 2.186.971 IP addresses).

In this paper, we propose an initiative to gather all information for all Autonomous System Numbers (AS Numbers) in Indonesia from Shodan. An autonomous system is a group of connected Internet Protocol under the control from a single administrative entity. We use a clustering algorithm to categorize all AS Numbers in Indonesia. Clustering is one of the most widely studied problems in data mining [2]. Clustering is the process of grouping a set of unlabeled objects into classes whose members are similar in some way [3]. There are many methods for clustering. The most common method is partitioning-based clustering. This method divides the large dataset into a number K of groups by using a distance as a measure of similarity, where each group represents a cluster [4]. K-Means [5], K-Medoids [6], K-Modes [7], PAM [8], CLARA [9], CLARANS [10] are the most algorithms that use this method. We use K-Means clustering because it is the most popular distance-based partitional clustering algorithm that is simple yet efficient [11]. The result of this study can be used to warn all organizations in Indonesia that manage AS Number to be aware of potential vulnerabilities that can be exposed from Shodan.

The structure of the rest of this paper is as follows. We provide related works in Section II. In Section III, we show the methodology that we used. Section IV, we present the result and discussion of this research. In Section V, we present the conclusion.

## 2. Related Works

Our first reference is the SHINE project [1]. SHINE project conducted a large-scale security assessment for devices connected to the Internet. There were 211 countries identified based on the IP addresses discovered. The highest country was the United States with 616.994 IP address (33.75%). The number of IP addresses harvested from Indonesia in 2014 is very small. There were 7182 IP address only. It is about 0,0032% from the total 2.186.971 IP address discovered.

Ercolani et al. [12] visualized IP addresses and open ports from Shodan. The dataset was derived from the set of all scans from the 1st through the 14th of September 2015. They extracted all IP addresses that were identified from Iran. Ali et al. [13] used Shodan to present vulnerability scanning for IoT devices in Jordan. They wanted to warn the community about IoT security issues and could be exposed. They pointed-out the critical cyber-attack against IoT and presented statistics about vulnerable devices in Jordan. Rae et al. [14] surveyed methods of assessment using Shodan. They presented the impact of Shodan to identify Internet-connected devices for both good and bad purposes. Williams et al. [15] assessed IoT devices exposed on the Internet with Shodan. However, their focus were on the vulnerabilities of webcams, smart TV, and printers. They collected 8,127 data of Webcams, 1,437 data of Smart TV, and 10,675 data of Printers.

There are many different kinds of research to assess internet-enabled devices with Shodan. However, none that focus to analyze and gather all information for all AS Numbers from Indonesia in Shodan. Ercolani et al. [12] and Ali et al. [13] did their research in their own countries such as Iran and Jordan. SHINE project [1] did a global-scale assessment, but they were not focused on Indonesia. The number of IP addresses from Indonesia harvested by the SHINE project is very small. Williams et al. [15] just focused their assessment on webcams, smart TV, and printers.

## 3. Methodology

The main purpose to study the exposure of all internet-enabled devices in Shodan is to be able to profile vulnerabilities in Indonesia. Because there are a huge number of IP address from Indonesia, we cluster all IP address based on the AS Numbers. These are our methodology for this study:

### 3.1. Data Collection

In this step, we will create a dataset by collecting all information in Shodan for all AS Numbers from Indonesia. First of all, we collect all AS Numbers in Indonesia. We can get information about the list of AS Numbers in Indonesia from ipinfo.io [16]. Then, we gather all subnets for each AS Numbers. We can look up all subnets for each AS Number by using API from hackertarget.com. We create bash scripting to collect all subnet for each AS Number into CSV files. The fields of the data are as follow.

**Table 1.** The fields of data from ipinfo.io and hackertarget.com

| Fields | Information |
|---|---|
| AS Number | Autonomous System Number |
| Subnet | List of subnets for each AS Number |

We scan all subnets for each AS Number in Shodan. We register a business account to get the API Key of Shodan. We create bash scripting to automatically scan all subnets for each AS Number from Indonesia in Shodan. Daily downloads of scans have been parsed and inserted into CSV files. The dataset was derived from December 2019 until January 2020. We observed 1.699 AS Numbers in Indonesia. These are the fields that we collect from Shodan.

**Table 2.** The fields of data from shodan

| Fields | Information |
|---|---|
| data | Information with raw data |
| domains | Information about Domain Name System |
| hash | Hash value of information |
| host | Information about host or user |
| hostnames | Information about hostnames |
| info | Additional information |
| ip_str | IP address of the target |
| isp | Information about Internet Service Provider |
| org | Information about the organization |
| os | Information about operating system |
| port | Information about port |
| product | Information about product of services |
| tags | Tags of information |
| timestamp | Timestamp of information |
| transport | Information about protocol transport |
| version | Information about the version of services |

### 3.2. Data Cleaning and Integration

We use Splunk Data Analytics for indexing data in a searchable repository that can generate reports, dashboards, and visualizations. Before we input our dataset into Splunk, we need to clean our data from blank spaces. We integrate all CSV files for each AS Number in Shodan into a single file CSV. Then, we convert semicolon CSV into a comma-delimited CSV for input to Splunk.

### 3.3. Data Featuring

For the efficiency of processing, we need to keep the number of feature space as low as possible but should present host behavior carrying rich enough information. We found the following features are the most important to represent the exposure quantity in shodan:

- Number of queries
  This feature represents the number of queries for each AS Number in Shodan. If there are many queries that we collect from Shodan, it indicates that the exposure level for that AS Number is high. If there are few queries that we collect from Shodan, it indicates that the exposure level for that AS Number is low.
- Size (in bytes)
  This feature represents the numbers of information in bytes. If the size of the data that we collect from Shodan is big, it indicates that the exposure level for that AS Number is high. If the size of the data that we collect from Shodan is small, it indicates that the exposure level for that AS Number is low.

### 3.4. Data Mining

We use K-Means Algorithm to cluster all AS Numbers based on the level of exposure fin Shodan. K-Means Algorithm is the most commonly used clustering algorithm. The distance measure it generally uses is Euclidean distance, given by:

$$\sqrt{\Sigma_{i=1}^{d}(x_1 - y_1)^2} \quad (1)$$

where $x_i$ and $y_i$ are two points in a d-dimensional Euclidean space. The objective function to be minimized is called Sum of Squared Error (SSE), given by the following equation:

$$SSE = \Sigma_{i=1}^{d} \Sigma_{x_{i \in c_k}} (x_i - c_k)^2 \quad (2)$$

The cluster centroid $c_k$ can be updated as:

$$c_k = \frac{\Sigma_{x_{i \in c_k}} x_i}{c_k} \quad (3)$$

Here is the algorithm of K-Means clustering [17].

| K-Means Clustering Algorithm |
|---|
| **Input:** *k* (number of clusters),<br>*D* (dataset)<br>**Output:** a set of k clusters<br>**Method:**<br>Arbitrarily choose *k* objects from dataset *D* as an initial cluster;<br>**Repeat:**<br>1: reassign each object to the cluster that most similar based on the mean value in the cluster<br>2: update the cluster means by calculating the means value of the objects in the cluster<br>**Until** no update |

### 3.5. Evaluation

We evaluate vulnerabilities of AS Numbers from Indonesia based on the exposure level in Shodan. In this paper, we provide several visualization and statistics about ports, products/services, operating systems, and domains. Information detail about the name of organizations and all IP address that exposed in Shodan is not presented in this paper to protect the security of the organizations and IP address.

## 4. Numeric Examples

We observe 1.699 AS Numbers in Indonesia. There are 624 AS Numbers information that we collect from Shodan such as IP address, ports, product services, domains, operating systems, and organizations. We gather 289,715 queries of Shodan into a single file. The size of the file is 237 MB. These are all the information that we collect from Shodan.

### 4.1. Ports

There are 272.457 information about ports. From 272.457 information, there are 12.787 unique ports. Port 80 is the highest port from Indonesia in our dataset. There are 380.008 information about port 80 (13,95%). Figure 1 provides us information about the statistics about ports from Indonesia in Shodan.

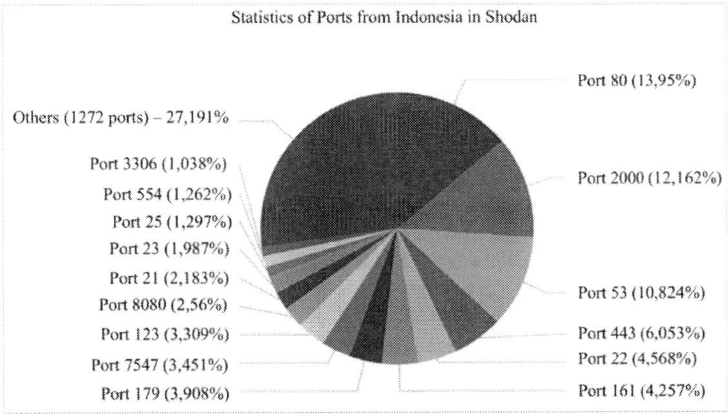


**Figure 1.** Statistics of ports from Indonesia in Shodan

### 4.2 Operating systems

There is 2.141 information about operating systems. From 2.141 information, there are 79 unique operating systems. Linux 3.0 is the highest operating system in this dataset. There are 768 information about Linux 3.x (35,871%). Figure 2 provides us information about the statistics of operating systems from Indonesia in Shodan.

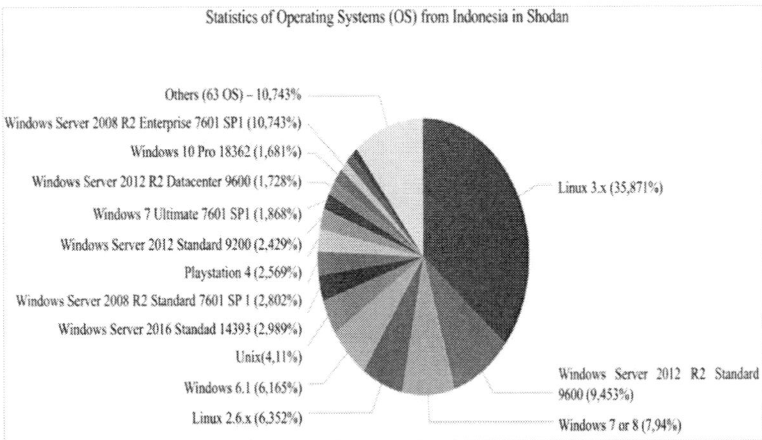


**Figure 2.** Statistics of operating systems from Indonesia in Shodan

### 4.3 Product/Services

There are 98.152 information about product services. From 98.152 information, there are 409 unique products. "MikroTik bandwidth-test server" is the highest product in this dataset. The information about product/services are MikroTik bandwidth-test server (32.759 information), apache httpd (16.236 information), Open SSH (7.638 information), nginx (5.183 information), ntpd (3.343 information), Mikrotik router ftpd, (3.118 information), Dropbear sshd (3.088 information), MySQL (2.456 information), Postfix smtpd (2.391 information), etc. Figure 3 provides us information about the statistics of products/services from Indonesia in Shodan.

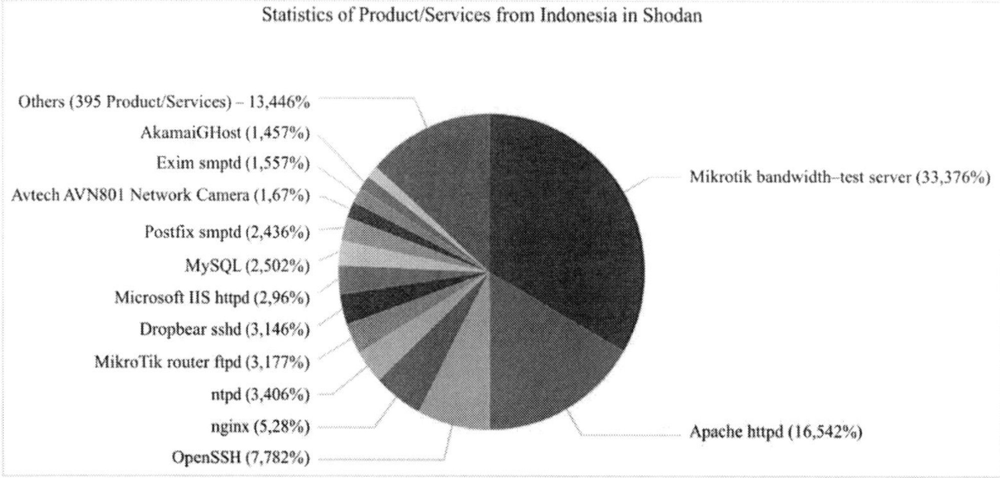


**Figure 3.** Statistics of products/services from Indonesia in Shodan

### 4.4 Domains

There are 111.732 information about domains. From 111.732 information, there are 3.350 unique domains. The highest top-level domain is *.id. Table 3 gives us information about the Top-Level Domains from Indonesia in Shodan.

**Table 3.** Top-Level Domains from Indonesia in Shodan

| Top-Level Domains | Number of Unique Domains | Total Information |
|---|---|---|
| *.id | 1.881 | 85.238 |
| *.com | 1.242 | 17.498 |
| *.net | 190 | 8.190 |
| *.arpa | 15 | 666 |
| *.tv | 9 | 91 |
| *biz | 13 | 49 |

### 4.5 IP addresses

There are 272.461 information about IP addresses. From 272.461 information, there are 145.543 unique IP addresses. Unfortunately, for security and secrecy reasons, we cannot publish all specific IP addresses in this paper because the information will refer to specific organizations in Indonesia.

### 4.6 Organizations

There are 271.731 information that includes the name of the organizations such as Internet Service Providers, agencies, and universities. From 271.731 information, there are 790 unique organizations. Unfortunately, we can not publish the name of organizations in this paper for security and secrecy reasons. The highest organizations in Shodan has 64.505 information (23,738%).

### 4.7 Clustering for AS Number in Indonesia

There are 1.699 AS Numbers in Indonesia. We create a dataset about all information of 624 AS Numbers that exposed in Shodan. There are 1.075 AS Numbers that are not exposed to Shodan. 1.075 AS Numbers will be categorized as class:0 that indicates that there is no information in Shodan yet. According to the number of information in Shodan, we will cluster 624 AS Numbers to be several classes with the K-Means algorithm. We use two features for the K-Means algorithm. The features are the "number of the query" and "size (in bytes)". These features represent the quantity of information that we collect from Shodan including ports, IP addresses, products, operating systems, domains, and organizations. We want to cluster the exposure level to be 3 classes such as: low, medium and high. These are the result of the K-Means algorithm with value of centroid K=3.

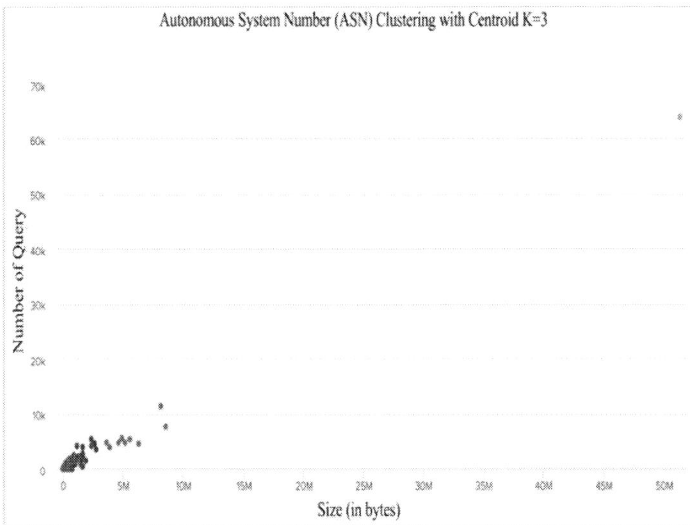


**Figure 4.** The result of K-Means clustering algorithm

Figure 4 shows us that there are many blue nodes (614 nodes) as class:1, several green nodes (9 nodes) as class:2 and a single orange node as class:3. We have tried for centroid K>3. The result is not better than centroid K=3 because there is no clear boundary for the blue nodes to be separated into more classes.

## 5. Conclusion

We successfully create a dataset about all information of AS Numbers from Indonesia in Shodan. From our dataset, there are 272.457 information about ports (with 12.787 unique ports), 2.141 information about operating systems (with 79 unique operating system), 98.152 information about product services (with 409 unique products), 113.358 information about domains (with 3.634 unique domains), 272.461 information about IP addresses (with 145.543 unique IP addresses), and 271.731 information about the name of organization such as Internet Service Providers, agencies, and universities with (790 unique organizations). Based on the result of the clustering algorithm, we get four classes of AS Number related to the level of exposure in Shodan such as :

- Class:0.
  This class indicates that there is no information about the AS Number in Shodan yet. There are 1.075 AS Numbers in this class.
- Class:1.
  This class indicates that there is low information about AS Number in Shodan. There are 614 AS Numbers in this class.
- Class:2.
  This class indicates that there is medium information about AS Number in Shodan. There are 9 AS Numbers in this class.
- Class:3 :
  This class indicates that there is high information about AS Number in Shodan. There is 1 AS Number in this class.

This information is important to warn organizations that manage the AS Number to be aware of the security of their systems.

## Acknowledgments

This paper's publication is supported by Hibah PITTA UI 2020 funded by the University of Indonesia.

### References

[1] Koblentz G. D 2020 Emerging technologies and the future of CBRN terrorism. The Washington Quarterly **43** (2) 177-196
[2] Chehouri A, Younes R, Khoder J, Perron J and Ilinca A 2017 A selection process for genetic algorithm using clustering analysis. Algorithms **10** (4) 123
[3] Jain A K, Murty M N and Flynn P J 1999 Data clustering: a review ACM computing surveys (CSUR) **31** (3) 264-323
[4] Kumar K M and Reddy A R M 2016 A fast DBSCAN clustering algorithm by accelerating neighbor searching using Groups method Pattern Recognition **58** 39-48
[5] Liu X, Zhu X, Li M, Wang L, Zhu E, Liu T and Gao W 2019 Multiple kernel $ k $ k-means with incomplete kernels IEEE transactions on pattern analysis and machine intelligence **42** (5) 1191-1204
[6] Sammut C and Webb G I 2017 Encyclopedia of machine learning and data mining Springer
[7] Jiang F, Liu G, Du J and Sui Y 2016 Initialization of K-modes clustering using outlier detection techniques Information Sciences **332** 167-183
[8] Ng R T and Han J 2002 CLARANS: A method for clustering objects for spatial data mining IEEE transactions on knowledge and data engineering **14** (5) 1003-1016
[9] Gupta T and Panda S P 2018 A comparison of k-means clustering algorithm and clara clustering algorithm on iris dataset International Journal of Engineering & Technology **7** (4) 4766-4768
[10] Matioli L C, Santos S R, Kleina M and Leite E A 2018 A new algorithm for clustering based on kernel density estimation Journal of Applied Statistics **45** (2) 347-366
[11] Jain A K 2010 Data clustering: 50 years beyond K-means Pattern recognition letters **31** (8) 651-666
[12] Fernández-Caramés T M and Fraga-Lamas P 2020 Teaching and learning iot cybersecurity andvulnerability assessment with shodan through practical use cases Sensors **20** (11) 3048
[13] Yu M, Zhuge J, Cao M, Shi Z and Jiang L 2020 A survey of security vulnerability analysis discovery detection and mitigation on IoT devices Future Internet **12** (2) 27

[14] Darlaston J and Wintle J 2007 Safety factors in the design and use of pressure equipment Engineering failure analysis **14** (3) 471-480
[15] Alshehri M D, Hussain F K and Hussain O K 2018 Clustering-driven intelligent trust management methodology for the internet of things (CITM-IoT) Mobile networks and applications **23** (3) 419-431
[16] Rubner Y, Puzicha J, Tomasi C and Buhmann J M 2001 Empirical evaluation of dissimilarity measures for color and texture Computer vision and image understanding **84** (1) 25-43

# Air Quality Monitoring System in South Tangerang Based on Arduino Uno: From Analysis to Implementation

**S Listyarini[1,*], L Warlina[2] and A Sambas[3]**

[1,2]Department of Environment, Universitas Terbuka, Indonesia.
[3]Department of Mechanical Engineering, Universitas Muhammadiyah Tasikmalaya, Indonesia

*Email: listyarini@ecampus.ut.ac.id

**Abstract.**
Pollution is a major factor in reducing air quality. A decrease in air quality can cause many problems such as global climate change, environmental damage, asthma and lung cancer. In this paper, we present an air quality monitoring system based on the Arduino Uno at Tangerang City. Air quality measured was carbon monoxide (CO), butane ($C_4H_{10}$), particulate matter ($PM_{2.5}$) and ozone ($O_3$). The types of sensors used were MQ-131 for the ozone sensor ($O_3$), MQ-7 for carbon monoxide (CO), GP2Y1010AU0F for the particulate matter ($PM_{2.5}$), and MQ-2 for butane ($C_4H_{10}$). This tool was built using Arduino and data from the sensor was sent to a Liquid Crystal Display (LCD) which visualized in numeric form. Data collection was carried out at Jalan Cabe Raya Pondok Cabe, South Tangerang. The test results showed the carbon monoxide (CO) had an average concentration of 2.79 ppm, the ozone gas concentration ($O_3$) was 0.08 ppm, particulate matter concentration ($PM_{2.5}$) was 35.51 $\mu g/m^3$ and butane average concentration ($C_4H_{10}$) was 35.05 ppm. Based on the data obtained, South Tangerang, Indonesia had a high level of pollution in particulate matter ($PM_{2.5}$).

## 1. Introduction

Air pollution is a condition in which the presence of one or more chemical, physical or biological substances in the atmosphere in dangerous quantities [1]-[2]. It harms human, animal and plant health, disturb aesthetics and comfort, or damage property. Air pollution is one type of environmental pollution besides soil pollution, water pollution, and sound pollution.

Among the impacts of air pollution are infant mortality [3], pregnancy complications [4], chronic respiratory diseases [5], climate change [6], allergic diseases [7], asthma and rhinitis [8], neuroinflammation and neurodegeneration [9], central nervous system pediatric impact [10], tuberculosis [11], oxidative DNA damage [12], morphological characteristics of Eucalyptus camaldulensis [13], cardiovascular disease [14], bronchiolitis obliterans syndrome and mortality after lung transplantation [15].

Various important studies involve the quality air monitoring system based on Arduino uno microcontroller. Karami et al conducted a toolbox which provides low-cost for indoor environmental quality (IEQ) monitoring based Arduino [16]. Kumar and Jasuja developed a real-time standalone air quality monitoring system which includes various parameters: $PM_{2.5}$, carbon monoxide, carbon dioxide, temperature, humidity and air pressure [17]. They built the system with Raspberry Pi and

Internet of Thing. Gunawan et al presented a cost effective and portable air quality measurement system using Arduino Uno microcontroller and four low-cost sensors. It is capable to measure the concentration of carbon monoxide (CO), ground level ozone ($O_3$) and particulate matters ($PM_{10}$ & $PM_{2.5}$) [18]. Husain et al constructed air quality monitoring based on Arduino Uno microcontroller and android [19]. However, Further research related to air quality monitoring should be developed as a literature on improving equipment quality.

In this work, we proposed a low-cost Arduino uno microcontroller-based air quality monitoring system. The Arduino is connected to four sensors and the measured data is displayed on the LCD. The sensors used are Mq-131 for the ozone sensor ($O_3$), Mq-7 for carbon monoxide (CO), GP2Y1010AU0F for the particulate matter ($PM_{2.5}$), and MQ-2 for butane ($C_4H_{10}$).

## 2. Architecture

The Microcontroller reads all sensors using Analog Digital Converter (ADC) based on 8 bits digital data (0 to 255 of digital data) [20]. ADC hardware was not needed in the architecture because the ADC feature included inside the Microcontroller's features. The Microcontroller converted the ADC data to voltage value using general equation. The voltage data was added to sensor equation using sensor datasheet and sensor company equation that implemented to Microcontroller's program (See Figure 1). The hardware of the air quality monitoring system can be seen in Figure 2.

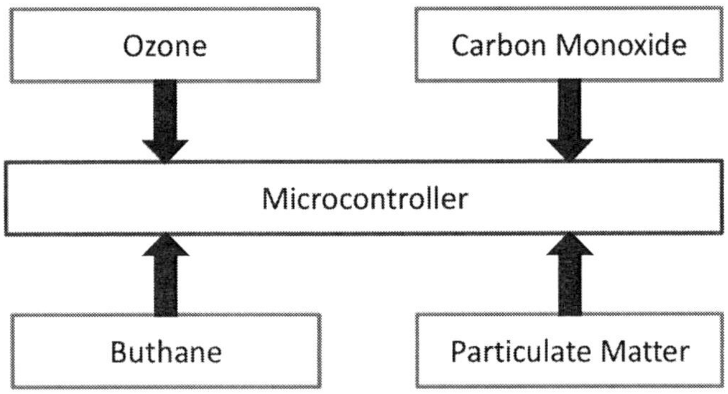


**Figure 1.** Microcontroller based air quality monitoring block diagram

**Figure 2.** Architecture Hardware air quality monitoring system

## 3. Methodology

We have created an air quality monitoring system that is low cost and easy to use. **GP2Y1010AU0F** is a PM sensor connected to digital pin A5 Arduino, MQ-7 is a carbon monoxide sensor connected to the Arduino A2 digital pins, MQ-131 (ozone sensor) and MQ-2 (butane sensor) connected to analog pins A4 and A0 of the Arduino.

We connect 4 sensors to the Arduino uno board. The data read from the sensor will be processed by the Arduino Uno microprocessor, then the data will be displayed on an LCD in the form of numbers. The results obtained will be analyzed according to the measured air quality threshold. The steps of this research can be seen in Figure 3.

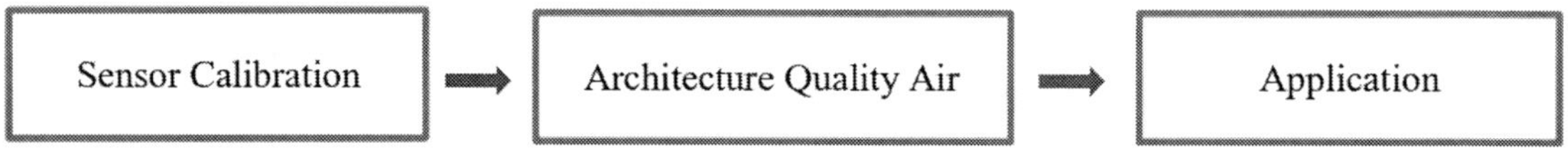


**Figure 3.** Methodology of our research

## 4. Result and Discussion

Data collection was carried out at Cabe Raya Pondok Cabe, Pamulang subdistrict, South Tangerang City. Data were collected in the morning (10.00 AM) with a large volume of vehicles. In addition, we took 10 samples of data on each sensor. The results of air quality monitoring can be seen in Table 1.

**Table 1.** Result of Monitoring Quality air in Tangerang

| Data | Carbon Monoxide (CO) (ppm) | Butane ($C_4H_{10}$) (ppm) | particulate matter ($PM_{2.5}$) ($\mu g/m^3$) | ozone ($O_3$) (ppm) |
|---|---|---|---|---|
| 1 | 2.40 | 35.42 | 29.20 | 0.02 |
| 2 | 3.20 | 35.20 | 35.10 | 0.04 |
| 3 | 3.40 | 34.21 | 38.25 | 0.08 |
| 4 | 2.52 | 35.23 | 39.42 | 0.20 |
| 5 | 1.54 | 35.24 | 39.38 | 0.14 |
| 6 | 2.84 | 35.64 | 40.68 | 0.16 |
| 7 | 2.80 | 35.19 | 34.87 | 0.08 |
| 8 | 1.40 | 35.11 | 33.25 | 0.04 |
| 9 | 3.60 | 34.12 | 32.40 | 0.06 |
| 10 | 3.20 | 35.14 | 32.54 | 0.04 |
| **Average** | **2.79** | **35.05** | **35.51** | **0.08** |

The carbon monoxide (CO) had an average concentration of 2.79 ppm. At this value, conditions in the city of South Tangerang were good because it is still below the predetermined threshold. For butane ($C_4H_{10}$) and ozone ($O_3$) were 35.05 ppm and 35.51 ppm, respectively. Finally, particulate matter ($PM_{2.5}$) was 35.51 $\mu g/m^3$. This value is close to the threshold of particulates. Therefore, South Tangerang had a high level of pollution in particulate matter ($PM_{2.5}$).

## 5. Conclusions

This paper discusses the process of air quality monitoring based on the Arduino Uno microcontroller. The sensors used were MQ-131 for the ozone sensor ($O_3$), MQ-7 for carbon monoxide (CO), GP2Y1010AU0F for the particulate matter ($PM_{2.5}$), and MQ-2 for butane ($C_4H_{10}$). The results showed that there was a significant value in the particulate matter ($PM_{2.5}$) of 35.51 $\mu g/m^3$. This indicated that South Tangerang had a high level of particulate matter. So, we will focus on monitoring particulate matter for further research.

## References

[1] Brunekreef B and Holgate S T 2002 *The lancet* **360** 1233-1242
[2] Landrigan P J 2017 *The Lancet Public Health* **2** e4-e5
[3] Chay K Y and Greenstone M 2003 *The quarterly journal of economics* **118** 1121-1167
[4] Lee P C, Roberts J M, Catov J M, Talbott E O and Ritz B 2013 *Maternal and child health journal* **17** 545-555
[5] Guan W J, Zheng X Y, Chung K F and Zhong N S 2016 *The Lancet* **388** 1939-1951
[6] Tagaris E, Liao K J, DeLucia A J, Deck L, Amar P and Russell A G 2009 *Environmental science & technology* **43** 4979-4988
[7] Takizawa H 2011 *The Korean journal of internal medicine* **26** 262-270
[8] Bousquet J, Anto J M, Annesi-Maesano I, Dedeu T, Dupas E, Pépin J L and Benveniste S 2018 *Clinical and translational allergy* **8** 1-13
[9] Calderón-Garcidueñas L, Leray E, Heydarpour P, Torres-Jardón R and Reis J 2016 *Revue neurologique* **172** 69-80
[10] Calderón-Garcidueñas L, Kulesza R J, Doty R L, D'Angiulli A and Torres-Jardón R 2015 *Environmental research* **137** 157-169
[11] Hwang S S, Kang S, Lee J Y, Lee J S, Kim H J, Han S K and Yim J J 2014 *The Korean journal of internal medicine* **29** 183-190
[12] Ambroz A, Vlkova V, Rossner Jr P, Rossnerova A, Svecova V, Milcova A and Sram R J 2016 *International journal of hygiene and environmental health* **219** 545-556
[13] Assadi A, Pirbalouti A G, Malekpoor F, Teimori N and Assadi L 2011 *Journal of Food, Agriculture and Environment* **9** 676-679
[14] Rajagopalan S, Al-Kindi S G and Brook R D 2018 *Journal of the American College of Cardiology* **72** 2054-2070
[15] Nawrot T S, Vos R, Jacobs L, Verleden S E, Wauters S, Mertens V and Dupont L J 2011 *Thorax* **66** 748-754
[16] Karami M, McMorrow G V and Wang L 2018 *Journal of Building Engineering* **19** 412-419
[17] Kumar S and Jasuja A 2017 In *2017 International Conference on Computing, Communication and Automation,* 1341-1346
[18] Gunawan T S, Munir Y M S, Kartiwi M and Mansor H 2018 *International Journal of Electrical and Computer Engineering* **8** 280-290
[19] Husain A M, Rini T H, Haque M I and Alam M R 2016 *Journal of Modern Science and Technology* **4** 86-96
[20] Kuncoro A H, Mellyanawaty M, Sambas A, Maulana D S, Subiyanto and Mamat M 2020 *Journal of Advance Research in Dynamical & Control Systems* **12** 2473-2479

IOP Conf. Series: Materials Science and Engineering 1115 (2021) 012047 doi:10.1088/1757-899X/1115/1/012047

# Build Website-Based Equipment Inventory Information System in Pesantren Al Ma'soem

**S F Nurrohmah[*] and M F Nugraha**

Informastion System, Universitas Ma'soem, Sumedang, Indonesia

[*]Email: s.fatimah1261@gmail.com

**Abstract.** Pesantren Siswa Al Ma'soem (PSAM) is an educational institution that has more than 500 equipment's used to support operational activities. But in the management of equipment inventory is still done semi-computerized, namely using Microsoft excel. Thus, causing problems such as officers' difficulty in performing physical checks of equipment with existing data, difficulty knowing the information of equipment in the PSAM, limited working hours to access equipment information and allow data loss and duplication of data. The purpose of this research is to overcome problems arising from the current system. The purpose of this research is to facilitate officers in performing physical checks of equipment with existing data, make it easier to know the information of equipment located in branch offices, equipment information can be accessed without limited working hours during the internet network and minimize the occurrence of data loss and duplication of data. The analysis carried out in this study includes analysis of the current system and analysis of the proposed system. The system design used is by approaching the OOAD (Object Oriented Analysis and Design) method with the Unified Modeling Language (UML) language. The design of this information system is expected to facilitate officers in conducting equipment inventory, presenting necessary equipment information anywhere and anytime, and minimizing the occurrence of data loss and duplication of data.

## 1. Introduction

The development of information technology at this time has brought very important benefits for the advancement of human life. The development of information technology can now really be felt benefits in human life, it is undeniable that the development of information technology provides prosperity to human life, ease in doing everything and can make the relationship between humans is no longer limited by space and time.

At this time the development of information technology is growing with the existence of the internet. Internet is an electronic communication network that connects one media with another. With the internet all things can be done, one of them is like looking for the necessary information. Usually the media that provides about this information is the website. Website is a communication media connected to the internet network where users of information can access all information contained in it.

The development of information technology at this time has also penetrated into various fields such as education, industry and others. With the existence of information technology applied to a particular field will provide convenience in providing the information needed. In addition, information technology must be continuously developed in accordance with the needs of an institution or organization or institution. One of them is in providing information about equipment inventory in an educational institution, namely Al Ma'soem Student Boarding School.

This research was conducted at PSAM which is located at Jalan Raya Cipacing No.22 RT 01 RW 14, Jatinangor District, Sumedang, West Java 40394. PSAM has more than 500 equipment's used to support its operations. The equipment includes whiteboards, desks, chairs, computers, cabinets, CCTV and so on. Inventory system equipment that is now running is semi-computerized, namely by using Microsoft excel. Equipment data on the procurement of equipment until inventory of this equipment is carried out by the Management of Facilities and Infrastructure in PSAM managed by less than 3 employees.

Inventory management is carried out independently based on data that has been done previous equipment inspection and the cost of equipment is borne by pesantren itself. The procedure for the procurement of equipment on the current system is first, the manager of facilities and infrastructure makes a submission for the procurement of equipment to the director. Second, after the submission is approved by the director then proceeded to the advice section to be carried out the realization of equipment procurement in accordance with the agreed and lastly done bookkeeping by the recording of facilities and infrastructure. Submissions and creation of equipment inventory reports are currently made on an annual basis and are only used for reports. Report creation using Microsoft Excel tools. Thus, with the running of the current system, can be identified several weaknesses, including:

1. Some equipment has not been given an identity, so it is difficult to match the data on the field with the data recorded in the inventory list.
2. It is difficult to know the information of equipment in the PSAM environment because it can only be known in a few rooms.
3. To find out equipment information can only be known during business hours only
4. Possible occurrence of data loss and duplication of data.

Based on the identification of the weaknesses above, it is expected that there is a design of information system that can be used to overcome these weaknesses in order to facilitate the process of inventorying equipment in Al Ma'soem Student Boarding School.

## 2. Methodology

### 2.1. System Development Methods

The approach technique used in the design of this website-based equipment inventory information system is to use the OOAD approach. OOAD is a method used for analysis and design of systems with an object-oriented approach [1]. This object can be interpreted as an entity that has identity, behavior and existence. The system development method used is Rational Unified Process (RUP). This RUP is a system development method with object-oriented concept with UML modeling language. RUP (Rational Unified Process) is an iterative process for software development; originally proposed in 1988 as the Rational Objectory Process. Then proposed the Unified Process [2].Based on the picture below, it can be seen that the RUP has two dimensions, namely:

1. The first dimension is depicted horizontally representing the dynamic aspects of software development. This aspect is described in the development stage or phase. Each phase will have a major milestone that signifies the end of the beginning of the next phase. Each phase can consist of one or more iterations. This dimension consists of inception, elaboration, construction and transition.
2. The second dimension is depicted vertically representing the static aspects of the software development process grouped into several disciplines. The software development process described into several disciplines consists of four important elements, namely who is doing, what, how and when. This dimension consists of Business Modeling, Requirements, Analysis and Design,

Implementation, Test, Deployment, Configuration and Change Management, Project Management and Environment.

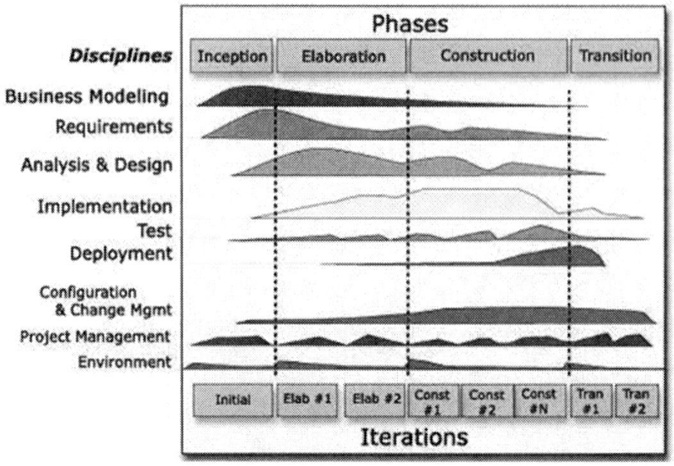


**Figure 1.** Fase RUP

The system development devices used are Flow map, Use Case Diagram, Class Diagram, and Activity Diagram.

### 2.2 Flow Map

Flow map is a stand-alone software package dedicated to the analysis of spatial interaction data. In this context, spatial interaction data is considered as all data that describe the movement of people, goods, energy or information through space on a regular or irregular basis usually via some form of transport network[3].

### 2.3 Use Case Diagram

Use Case Diagrams are artifacts created during early phases of software development that are significant for formulating, corroborating and documenting the system behaviour [4]. Use cases describe a set of sequence of actions, including the variants that a system performs to generate a visible result of value to an actor. Use Case diagrams are central to modelling the behavior of the system or a sub-system. Each one of these represents a set of Use Cases, actors and their relationship [5].

### 2.4 Class Diagram

Class diagrams are static structures that provide an overview of the system by specifying classes and the relationships between them. They are used for a variety of purposes such as understanding requirements, modeling the domain-specific data structure, and describing detailed design of the target system[6].

Class diagrams are static structures that provide an overview of the system by specifying classes and the relationships between them. They are used for a variety of purposes such as understanding requirements, modeling the domain-specific data structure, and describing detailed design of the target system [7].

### 2.5 Activity Diagram

Activity Diagram (AD) is an important diagram for modeling the dynamic aspects of a system[8]. Activities and the corresponding activity diagrams belong to the basic concepts of the Unified Modeling Language (UML) which has become the standard modeling language for specifying, visualizing and documenting software systems. Activity diagrams are used for modeling behavioral logic like business processes or workflow. In the modern UML 2, they experienced an increasing importance[9].

## 3. System Analysis and Design

### 3.1 System Analysis

*3.1.1 Flow map*

This flow map serves to explain the work program for the development of the new system.

1. New Equipment Submission Procedure
   a. The Facility Manager submits the procurement of new equipment by inputting the data of the procurement submission into the system and then printing the procurement sheet. After that, the procurement sheet is submitted to the facilities and infrastructure
   b. The facilities and infrastructure department receives a procurement sheet and notification of procurement submissions from the system. Then, check the proposed price, the background of the procurement submission and the physical check of the equipment. If it is appropriate, the procurement sheet is approved and then submitted to the director. If it is not appropriate then the procurement sheet is not approved and returned to the manager.
   c. The Director double-checks the proposed price and background of the submission if it is appropriate then the procurement sheet is approved by the director.
   d. If approved by the director, the facilities and infrastructure section will realize the proposed equipment. Then the purchase note and procurement sheet are submitted to the support/accounting section to be inputted into the system.

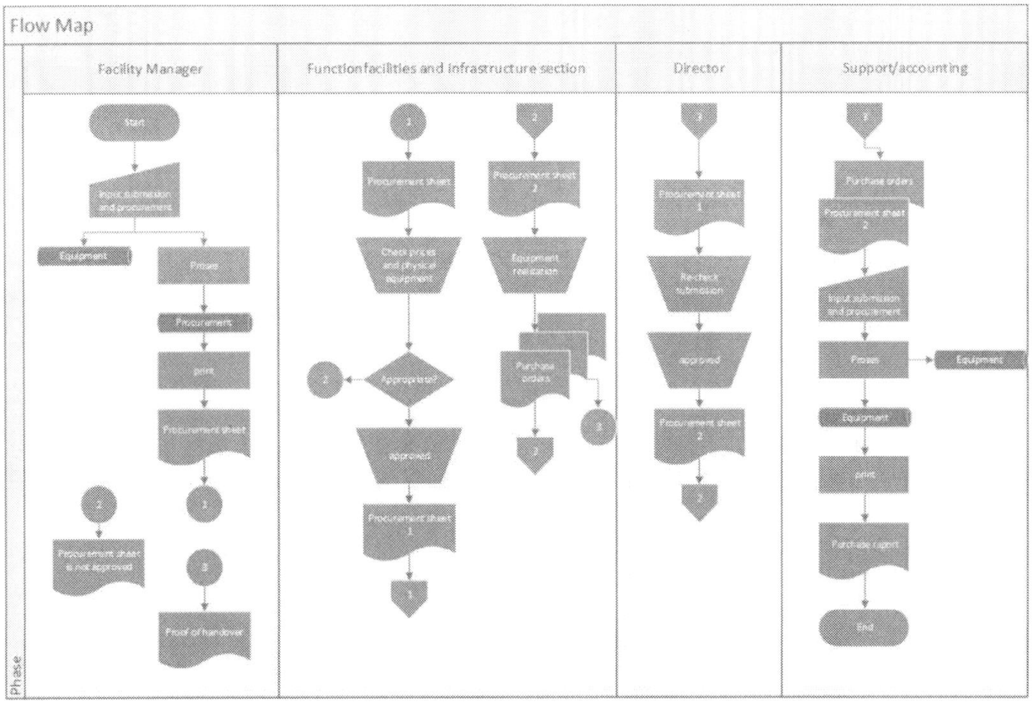


**Figure 2.** Flow map Proposed New Equipment Submission

### *3.1.2 Use Case Diagram*

Use Case Diagram serves to describe how the interaction between actors and the system. Here is a use case diagram of equipment inventory information system in PSAM.

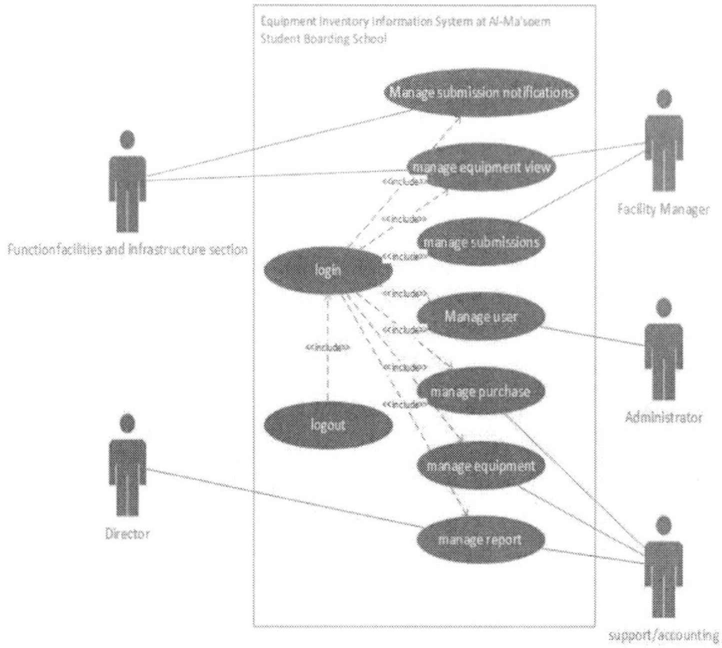


**Figure 3.** Use Case Diagram

**Activity Diagram**

1. Activity Diagram Administrator

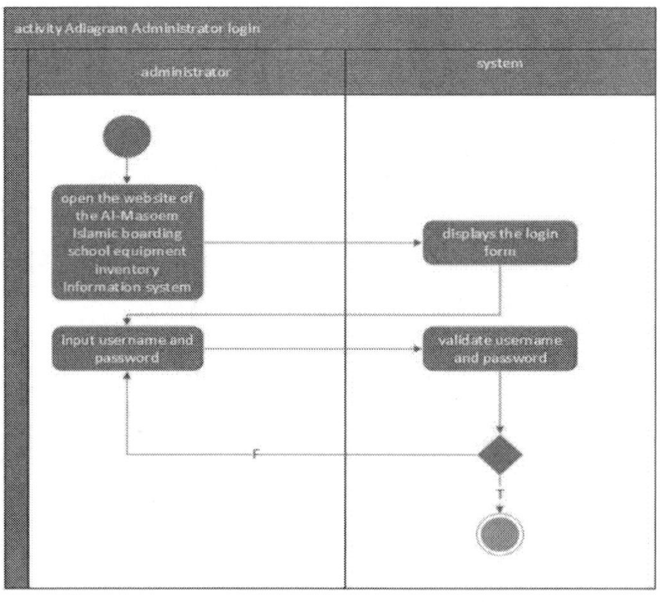


**Figure 4.** Activity Diagram Administrator Login

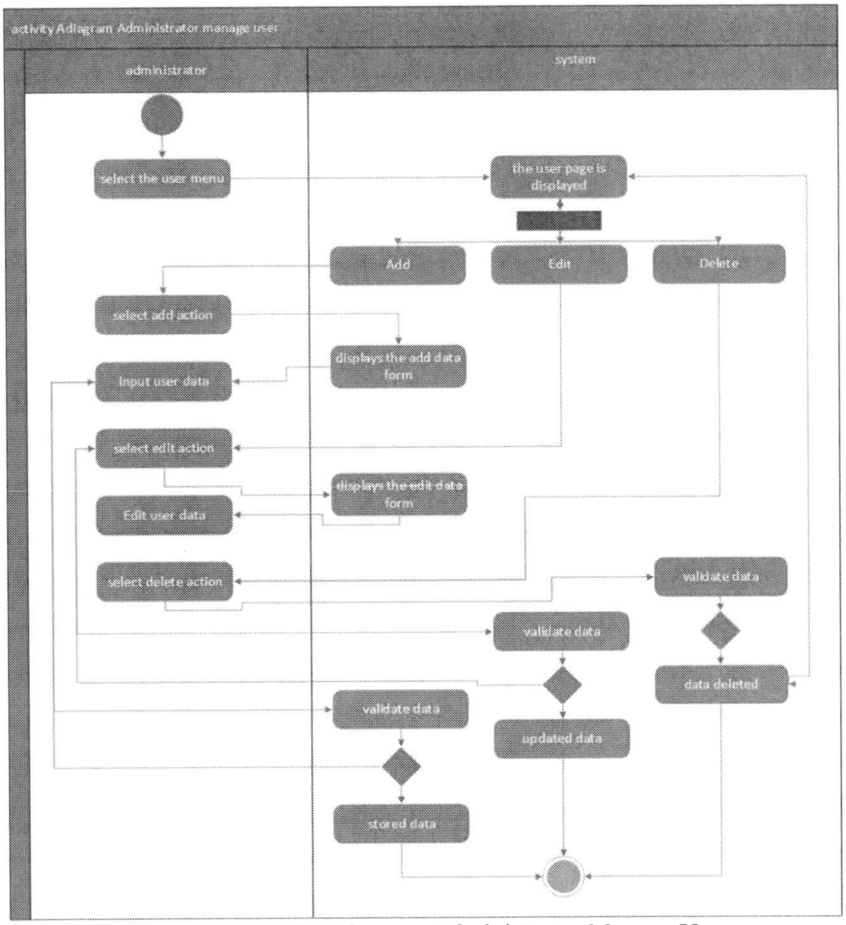


**Figure 5.** Activity Diagram Administrator Manage User

2. Activity Diagram Infrastructure Department

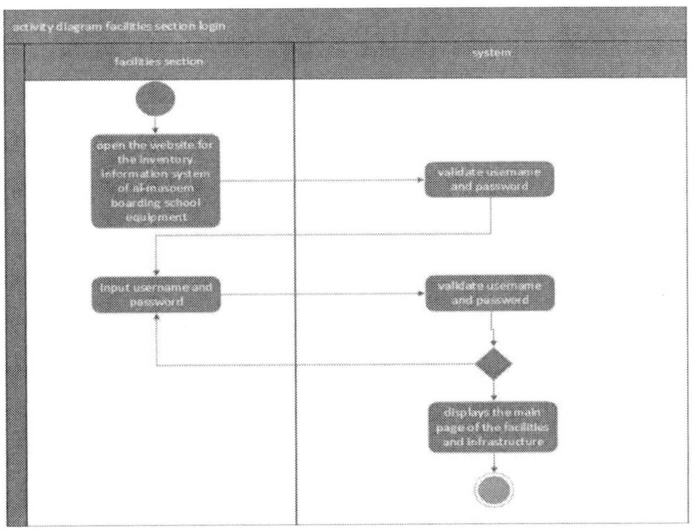


**Figure.6** Activity Diagram Facilities Section Login

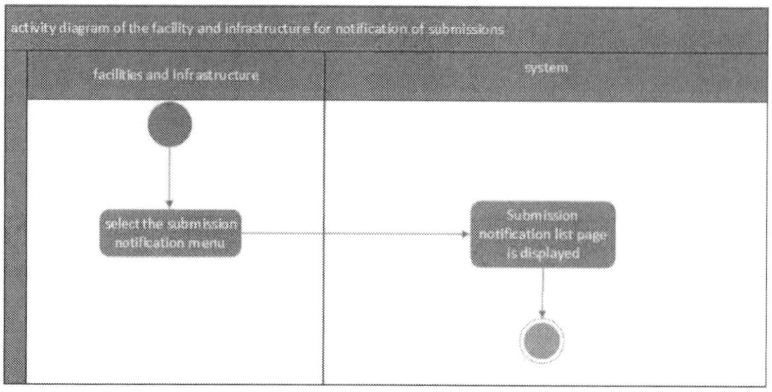


**Figure 7.** Activity Diagram Facilities Section Submission Notification

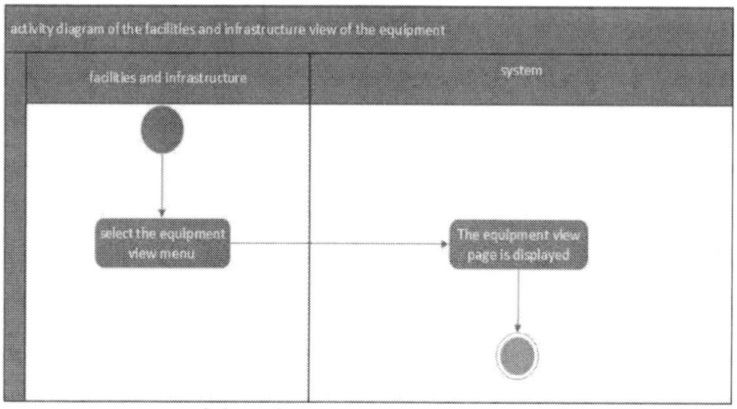


**Figure 8.** Activity Diagram Facilities Section View Tools

3. Activity Diagram Bag. Support/Accounting

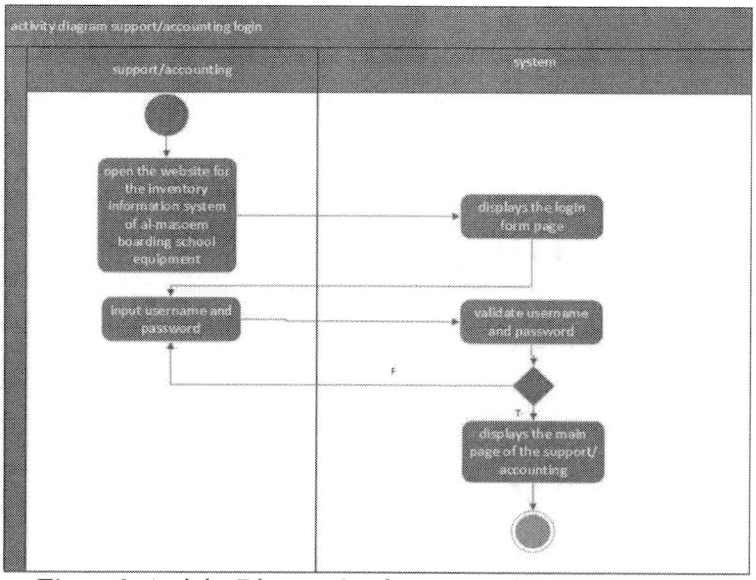


**Figure 9.** Activity Diagram Section Support/Accounting Login

IOP Conf. Series: Materials Science and Engineering 1115 (2021) 012047 doi:10.1088/1757-899X/1115/1/012047

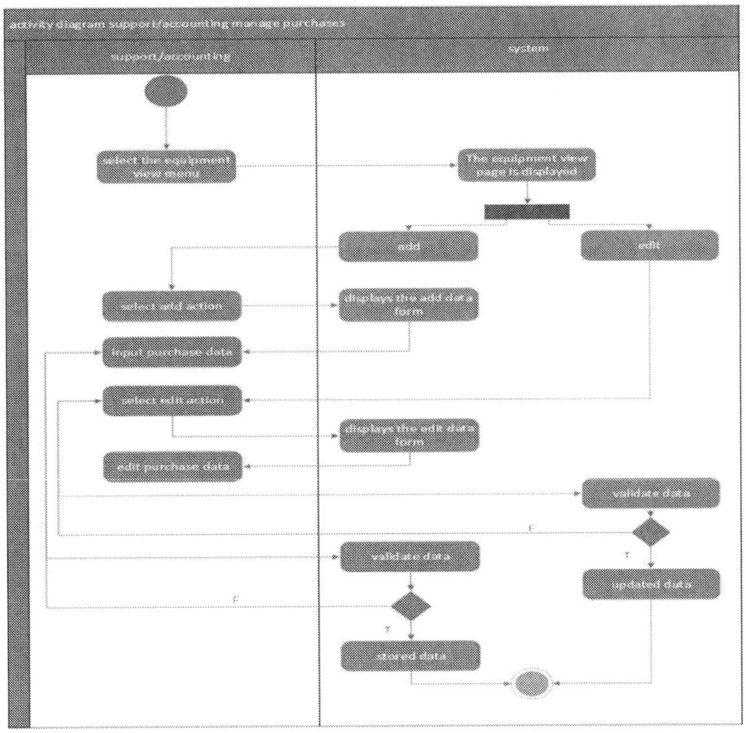


**Figure 10.** Activity Diagram Section Support/Accounting Manage Purchases

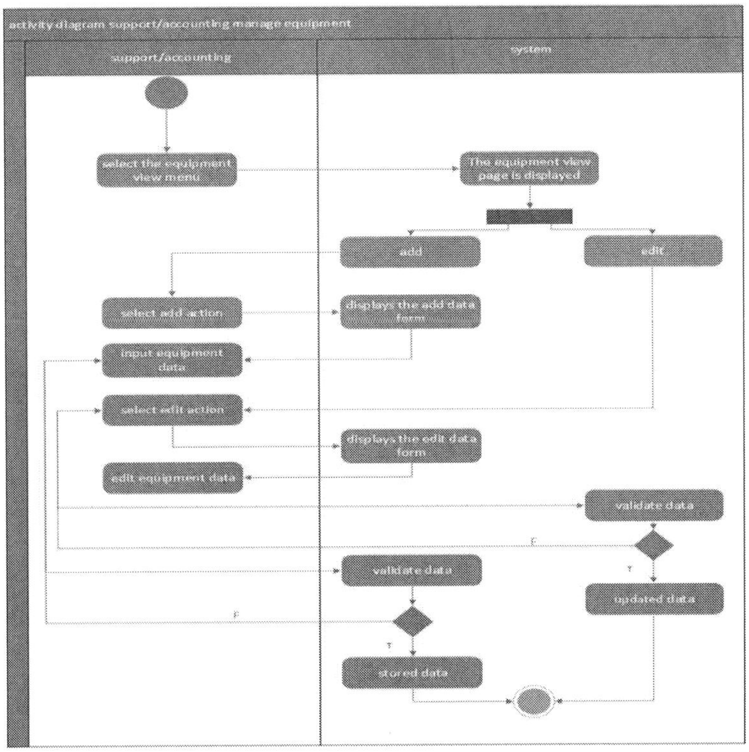


**Figure 11.** Activity Diagram Section Support/Accounting Manage Tools

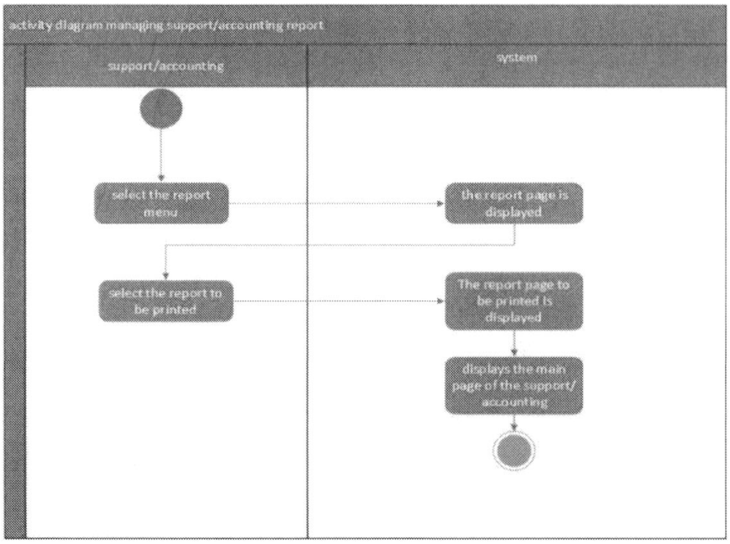


**Figure 12.** Activity Diagram Section Support/Accounting Manage Reports

4. Activity Diagram Facility Manager

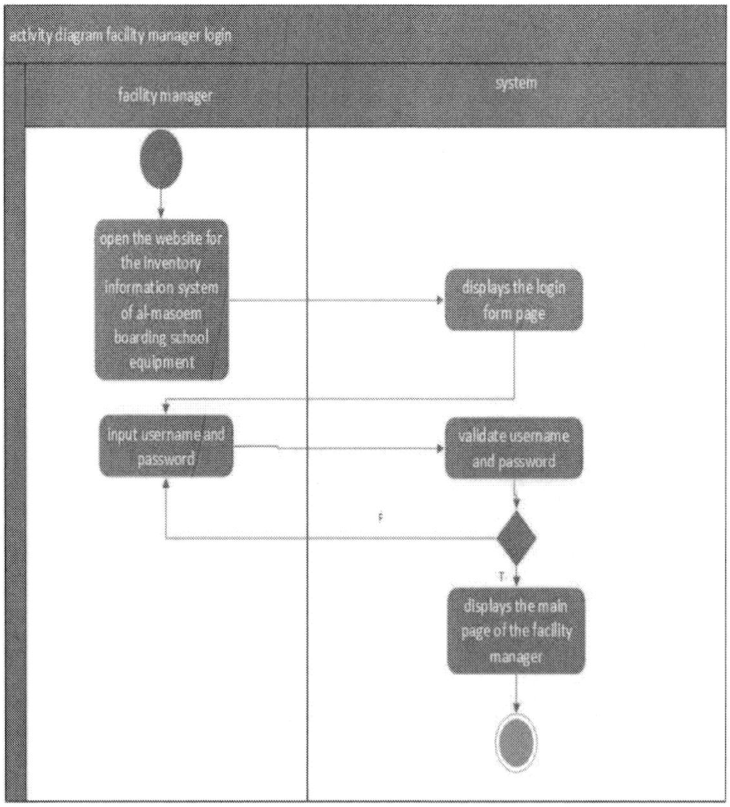


**Figure 13.** Activity Diagram Facilities Manager Login

IOP Conf. Series: Materials Science and Engineering 1115 (2021) 012047 doi:10.1088/1757-899X/1115/1/012047

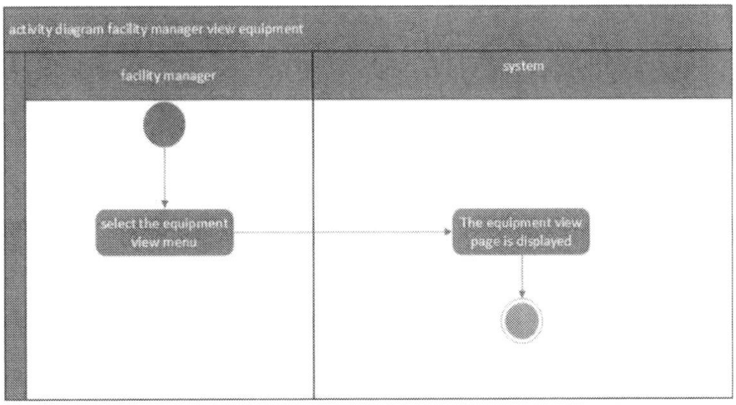


**Figure 14.** Activity Diagram Facilities Manager View Tools

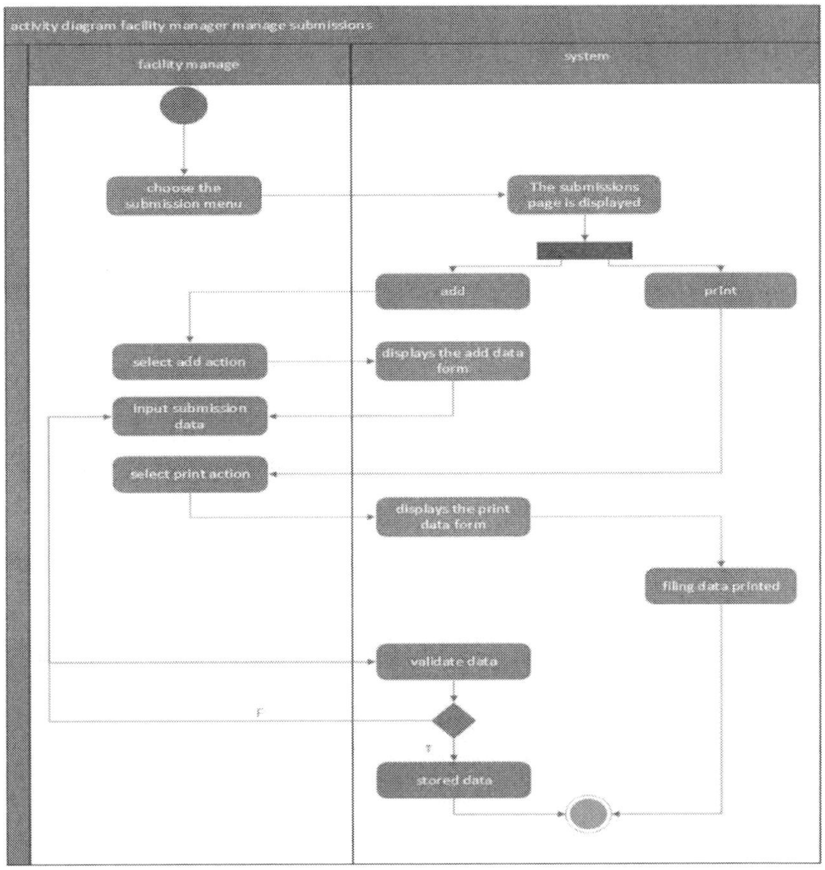


**Figure 15.** Activity Diagram Facilities Manager Manage Submissions

IOP Conf. Series: Materials Science and Engineering 1115 (2021) 012047 doi:10.1088/1757-899X/1115/1/012047

5. Activity Diagram Director

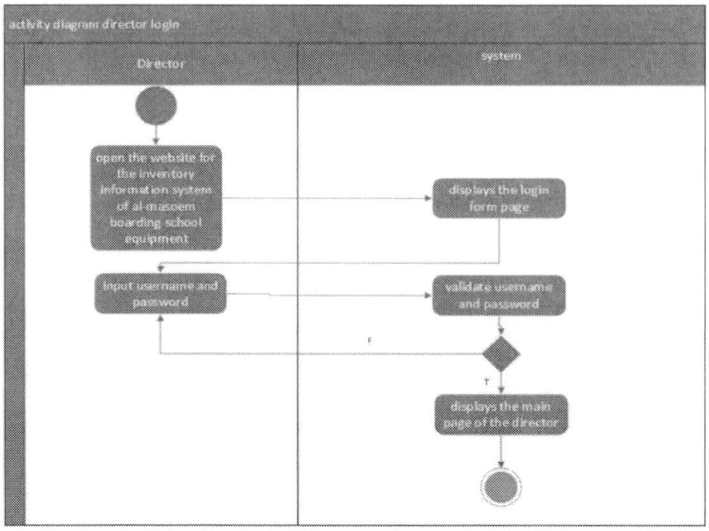


**Figure 16.** Activity Diagram Director Login

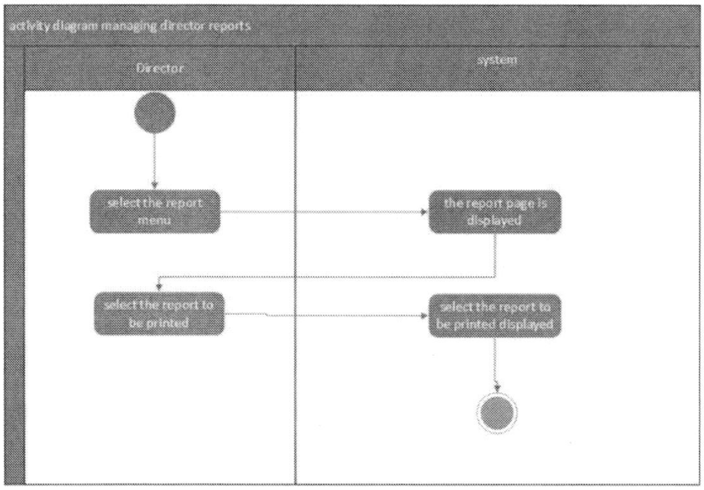


**Figure 17.** Activity Diagram Director Manage Report

## 3.2 System Design

### *3.2.1 Class Diagram*

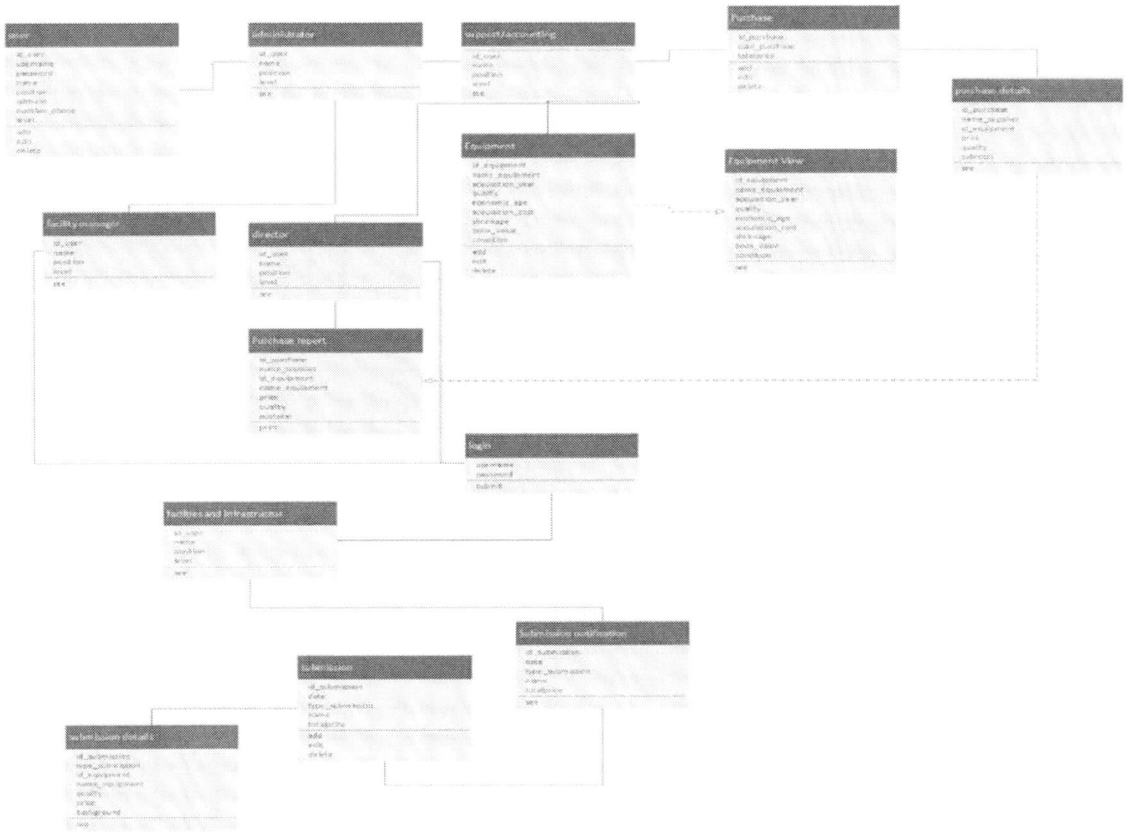

**Figure 18.** Class Diagram

### *3.2.2 User Interface*

1. Administrator Interface

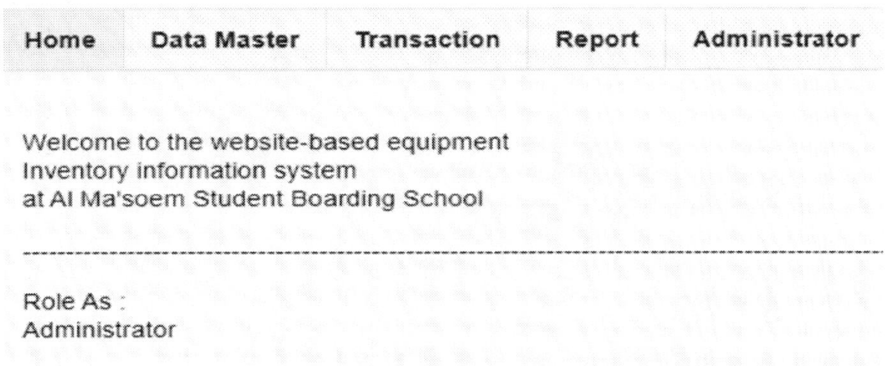


**Figure 19.** Administrator Interface

2. Infrastructure Advice Section Interface

IOP Conf. Series: Materials Science and Engineering 1115 (2021) 012047 doi:10.1088/1757-899X/1115/1/012047

**Figure 20.** Infrastructure Advice Section Interface

3. Facilities Manager Interface

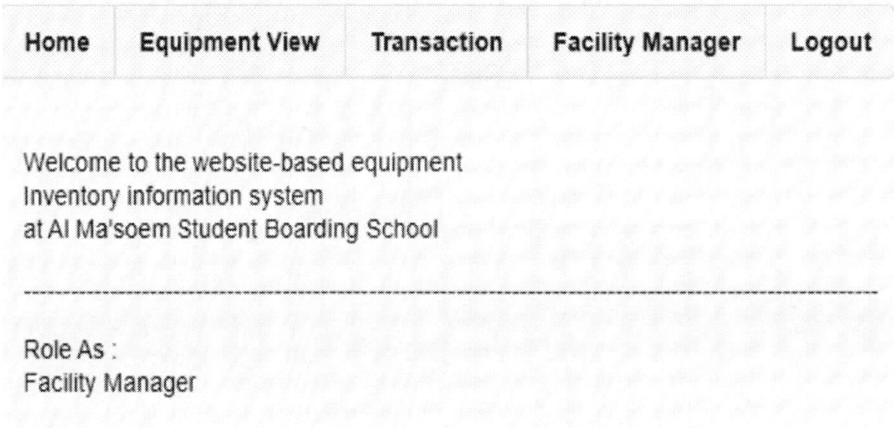


**Figure 21.** Facilities Manager Interface

4. Section Support/Accounting Interface

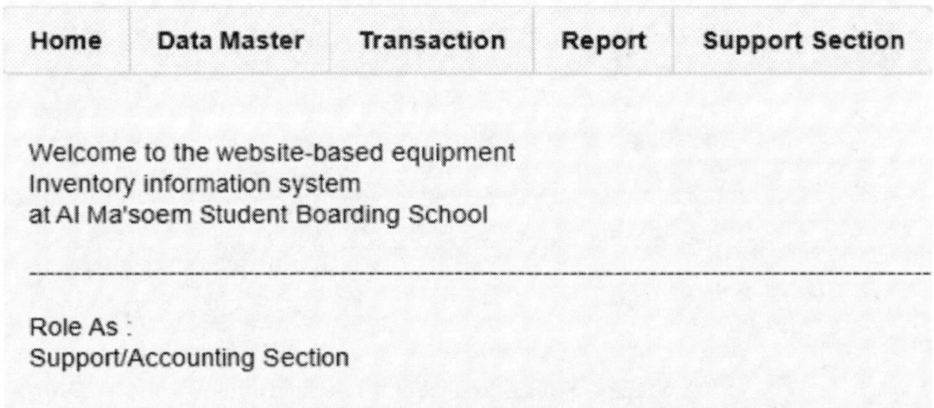


**Figure 22.** Interface Section Support/Accounting

5. Director Interface

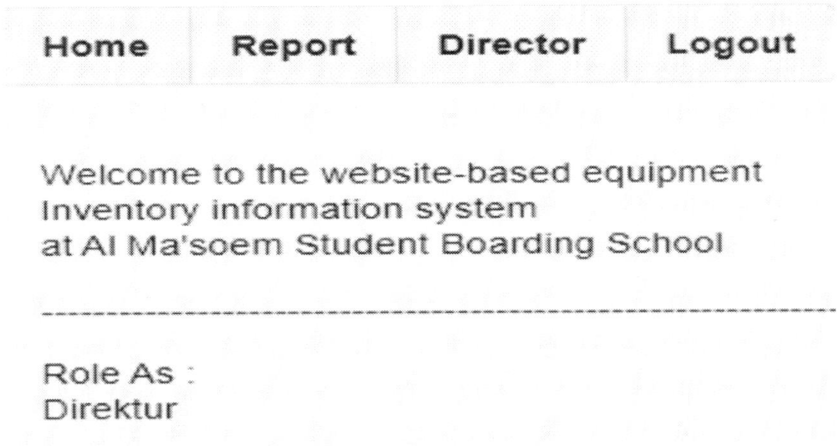


**Figure 23.** Director Interface

## 4. Conclusion

With the design of information system inventory equipment Based PSAM, it can be concluded as follows:

1. This system will make it easier for officers in performing physical checks of equipment in the field.
2. This system will make it easier to know the information of equipment in the environment around PSAM.
3. This system can be accessed without being limited by working hours as long as there is an internet network.
4. This system will minimize the possibility of data loss and duplication of data

## References

[1] Purwaningtias F P 2018 E-Commerce Penjualan Berbasis Metode Ooad. *Jurnal Cendikia*, **16** 1-5
[2] Anwar A 2014 A review of rup (rational unified process) *International Journal of Software Engineering (IJSE)* **5** 12-19
[3] Geertman S, de Jong T and Wessels C 2003 Flowmap: A support tool for strategic network analysis *Planning support systems in practice* 155-175
[4] Sabharwal S, Kaur P and Sibal R 2017 Empirical and Theoretical Validation of a Use Case Diagram Complexity Metric *International Journal of Information Technology and Computer Science* **9** 35-47
[5] Zheng J, Feng Y and Zhao Y 2014 A unified modeling language-based design and application for a library management information system *Cybernetics and Information Technologies* **14** 129-144
[6] Al-Fedaghi, S. (2017). Diagramming the class diagram: toward a unified modeling methodology. *arXiv preprint arXiv:1710.00202*
[7] Riaz Ahamed S S 2010 Review and Analysis of The Issues of Unified Modeling Language for Visualizing, Specifying, Constructing and Documenting the Artifacts of a Software-Intensive System. *arXiv e-prints*, arXiv-1001

[8] Ahmad T, Iqbal J, Ashraf A, Truscan D and Porres I 2019 Model-based testing using UML activity diagrams: A systematic mapping study *Computer Science Review* **33** 98-112
[9] Siebenhaller M and Kaufmann M 2006 Drawing activity diagrams *Proceedings of the 2006 ACM symposium on Software visualization* 159-160
[10] Bagir H and Putro B E 2018 Analisis Perancangan Sistem Informasi Pergudangan di CV. Karya Nugraha *Jurnal Media Teknik dan Sistem Industri* **2** 30-38

IOP Conf. Series: Materials Science and Engineering 1115 (2021) 012048 doi:10.1088/1757-899X/1115/1/012048

# Design of website-based library information systems at SMK Pasundan Jatinangor

**T I Yani[1*] and F Hamdani**
Accounting Computer, Ma'soem University, Cipacing Street No. 22, Jatinangor District, 45363, Indonesia

*Email: Tamiindriani28@gmail.com

**Abstract.** SMK Pasundan Jatinangor is a private vocational high school located in Jatinangor District. Currently the system in effect in the place under study is semi-computerized based, when a book is borrowed in the library the data is inputted in Ms. Excel, as well as member data, book data, date of borrowing and date of return are inputted manually which resulted in the calculation of fines being calculated manually. So that some significant obstacles are found, including 1.can be lost because the book does not automatically update 2.When there is a loan, there is an error recording book data and member data 3.Lower handling of services in the library 4. A error in late fines for returning books 5. requires a long process of inputting books, members and borrowing. In this study, the authors use a system development tool using UML (Unified Modeling Language) which consists of: Use Case Diagrams, Activity Diagrams and Class Diagrams. With the design of this library information system, it is hoped that it can make it easier for officers / librarians to monitor the availability of books, add books, borrow and return books, so that they can assist in library services and management.

## 1. Introduction

Information technology is the technology needed to process information using electronic computers, the development of information technology is very rapid, of course, providing benefits to human life. Various areas of human life are now starting to implement various new technological devices to make it easier for humans [1]-[4].

SMK Pasundan Jatinangor which is a private vocational high school located in the Jatinangor District area. More precisely, having his address at Jalan Colonel Ahmad Syam No.46, Cikeruh Village, Jatinangor District. This school is under the auspices of the Pasundan Elementary and Secondary Education Foundation (YPDM). It has 5 departments, namely: Computer Network Engineering, Motorcycle Engineering, Light Vehicle Engineering, Accounting, and Office Administration.

One of the facilities at SMK Pasundan Jatinangor is a library which is often used every day to read and borrow books. The system that the author wants to design is a Library Information System which is an information system created to provide information on a list of available books that students can borrow and make it easier for officers / librarians to monitor book availability, add books, borrow and return books, so that they can assist in service and management library.

Currently the library system process is semi-computerized based, namely using the Ms. Excel application, when a book is borrowed in the library the data is inputted in Ms. Excel, as well as member data, book data, date of borrowing and date of return are manually inputted resulting in calculations.

Based on the description above, the authors conclude this research will focus on designing a website-based library information system at SMK Pasundan Jatinangor. The writer hopes that this system design can facilitate all library services reduce the work time of officers to be faster and more effective, minimize the occurrence errors in inputting book and member data, work that was originally done in a manner manual becomes automatic especially facilitate the work of library management staff.

## 2. System Devices

2.1 System Development Tools

In an information system design, the system development tool used is UML [5]. UML is a set of tools used to perform abstraction of an object-based system or software. UML stands for Unified Modeling Language. UML is also one way to facilitate sustainable application development. The tools used in this information system model are:

a. Use Case Diagram

Use Case is a technique for recording the functional requirements of a system [6]. Use case diagrams are UML diagram models that are used to describe the expected functional requirements of a system or can be said to briefly describe who uses the system and what they can do.

a. Activity Diagram

Activity Diagram is a technique for describing procedural logic, business processes, and work lines. In some ways, activity diagrams play a role similar to flowcharts, but the principle difference between flow chart notations is that they support parallel behaviors [7]-[8]. The nodes in an activity diagram are called actions, so the diagram shows an activity composed of actions.

c. Class Diagram

Class Diagram is a collection of similar objects [9]-[10]. Class diagram is a type of diagram that is useful in UML which can map a particular system structure by modeling classes, attributes, operations and relationships between objects.

## 3. System Analysis

### 3.1 Use Case Diagram

The use case diagram of the website-based library information system at SMK Pasundan Jatinangor has visitor use cases and librarian and admin use cases (See Figure 1 and Figure 2).

### 3.2 Activity Diagram

The activity diagrams in the website-based library information system at SMK Pasundan Jatinangor include visitor activity diagrams, admin activity diagrams and librarian activity diagrams (See Figures 3-7).

### 3.3 Class Diagram

The class diagram of the website-based library information system at SMK Pasundan Jatinangor is as follows:

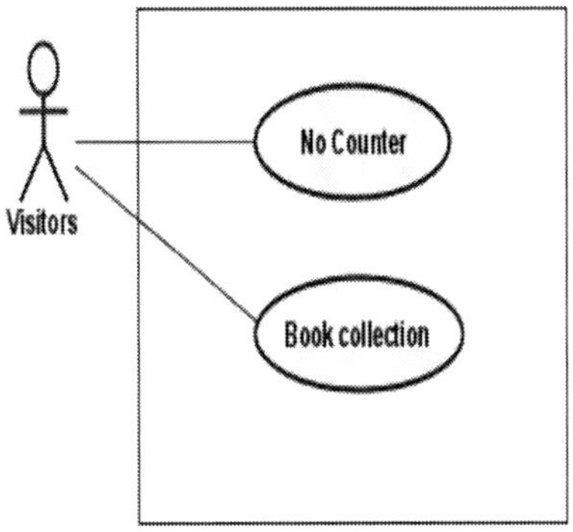


**Figure 1.** Visitors use case

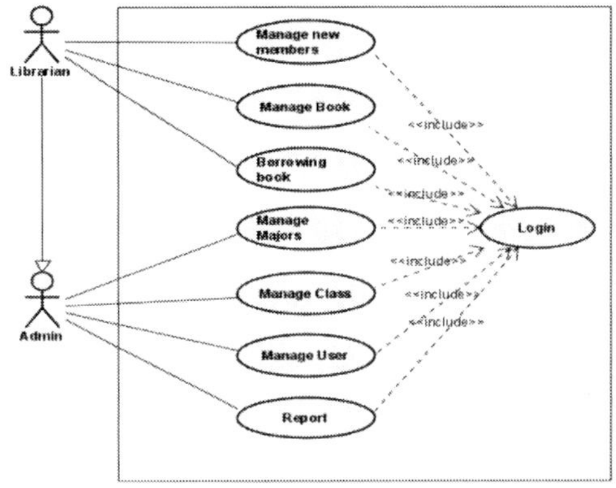


**Figure 2.** Librarian and Admin Use Case

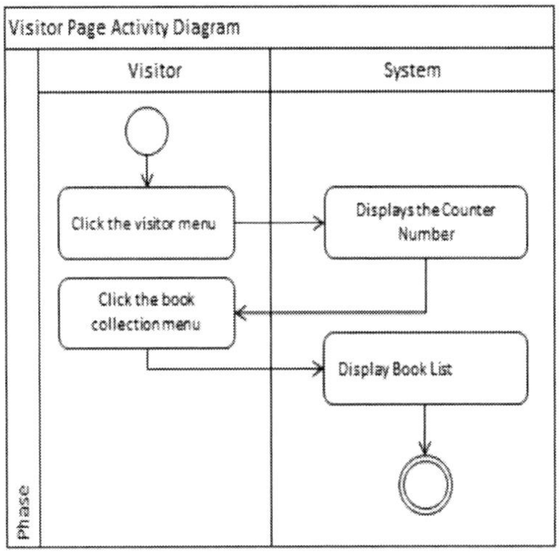


**Figure 3.** Visitor Activity Diagram

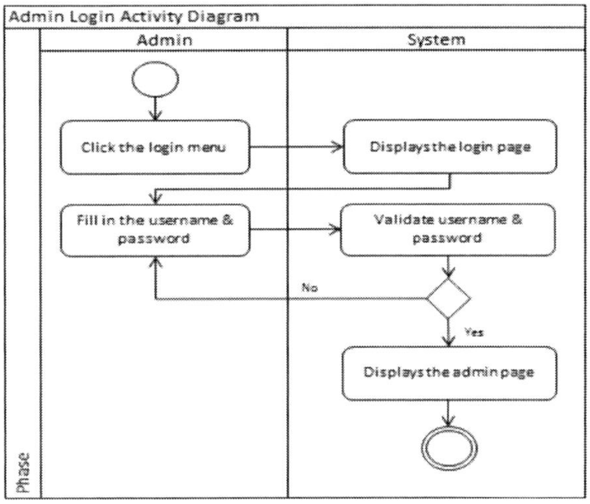


**Figure 4.** Admin Login Activity Diagram

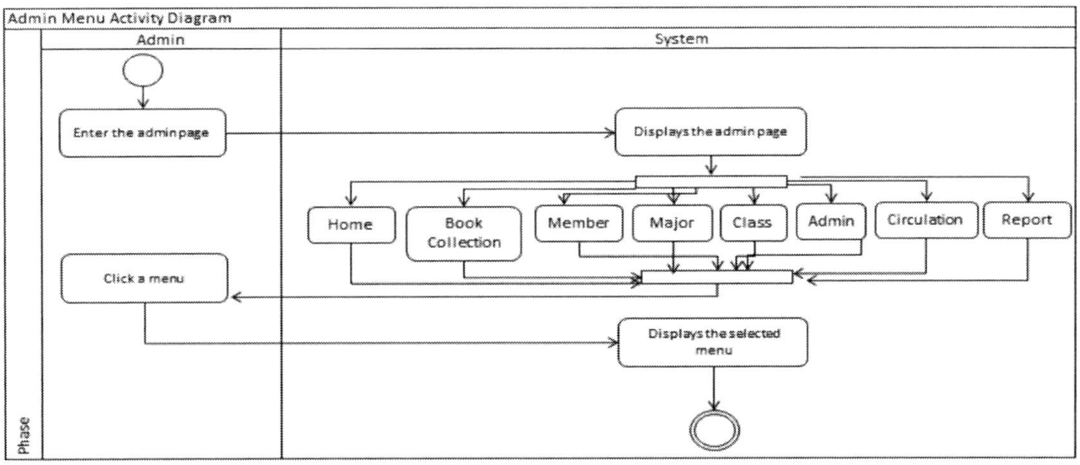


**Figure 5.** Admin Menu Activity Diagram

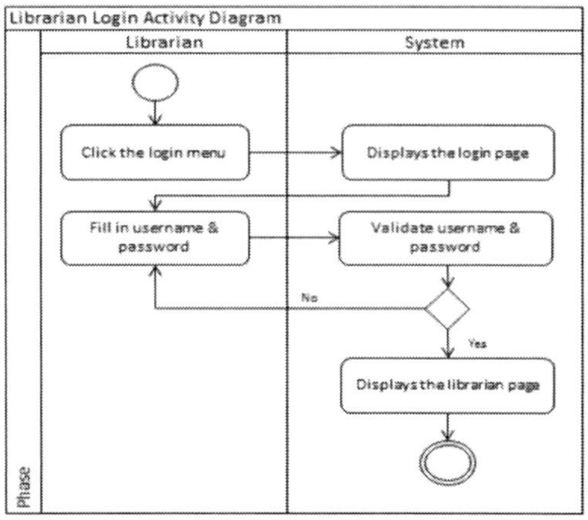


Figure 6. Librarian Login Activity Diagram

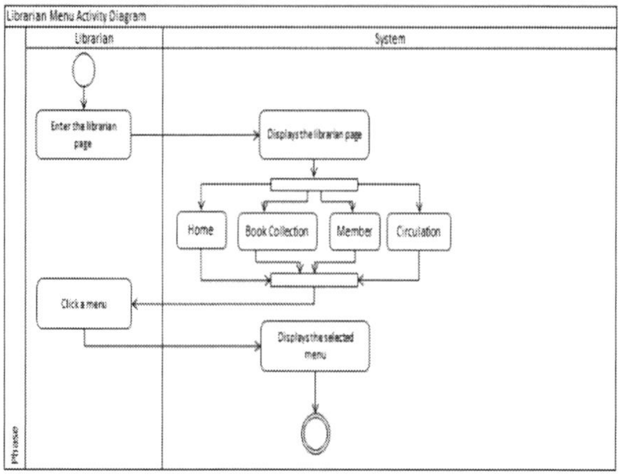


Figure 7. Librarian Menu Activity Diagram

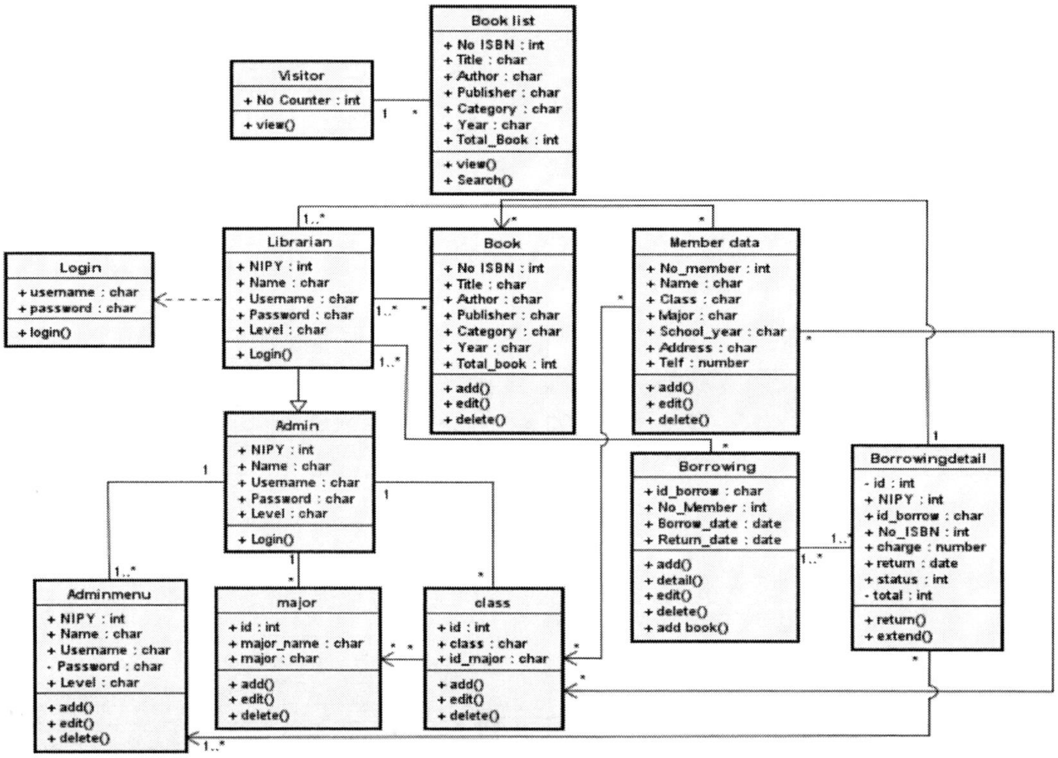


Figure 8. Class Diagram

### 3.4 User Interface

To facilitate interaction between users and the system, it is necessary to design a user interface. User interface can be interpreted as a means or media used to connect microprocessor devices and then communicate with the user [11]-[12]. This user interface serves to provide new knowledge, display system explanations and provide guidance on system usage as a whole so that the user understands the workings of a system that provides easy use of the system and interactive ease. The user interface design for the website-based library information system at SMK Pasundan Jatinangor can be seen in Figures 9-13.

Figure 9. Librarian Login Activity Diagram

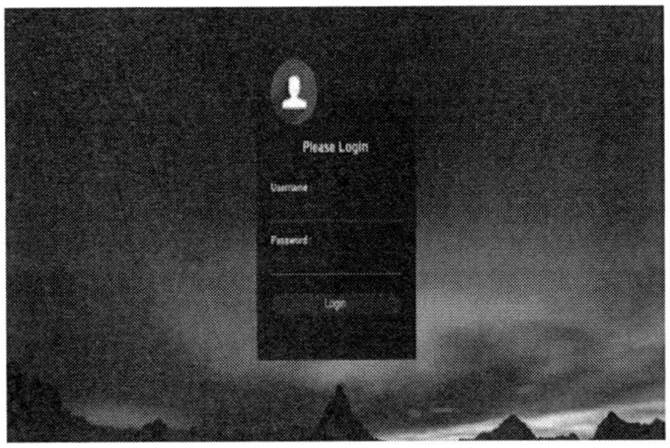


**Figure 10.** Librarian Menu Activity Diagram

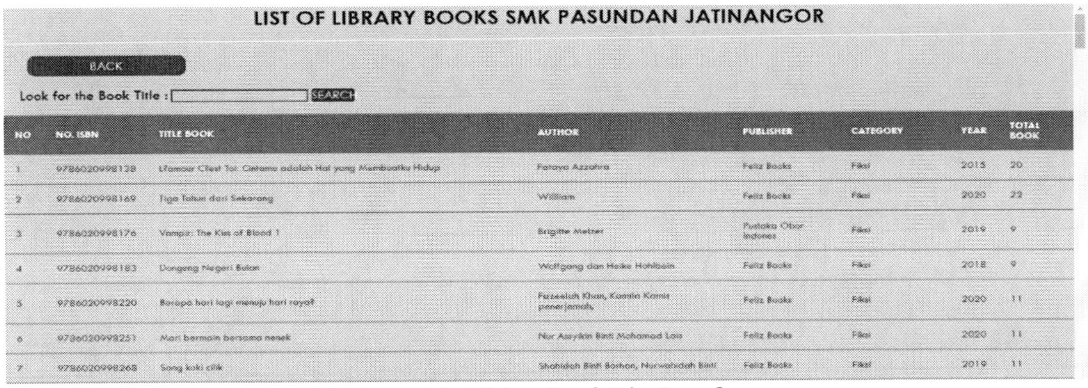

LIST OF LIBRARY BOOKS SMK PASUNDAN JATINANGOR

BACK

Look for the Book Title : SEARCH

| NO | NO. ISBN | TITLE BOOK | AUTHOR | PUBLISHER | CATEGORY | YEAR | TOTAL BOOK |
|---|---|---|---|---|---|---|---|
| 1 | 9786020998138 | L'amour C'est Toi: Cintamu adalah Hal yang Membuatku Hidup | Fataya Azzahra | Feliz Books | Fiksi | 2015 | 20 |
| 2 | 9786020998169 | Tiga Tahun dari Sekarang | William | Feliz Books | Fiksi | 2020 | 22 |
| 3 | 9786020998176 | Vampir: The Kiss of Blood 1 | Brigitte Melzer | Pustaka Obor Indones | Fiksi | 2019 | 9 |
| 4 | 9786020998183 | Dongeng Negeri Bulan | Wolfgang dan Heike Hohlbein | Feliz Books | Fiksi | 2018 | 9 |
| 5 | 9786020998220 | Berapa hari lagi menuju hari raya? | Fazeelah Khan, Kamila Kamis penerjemah, | Feliz Books | Fiksi | 2020 | 11 |
| 6 | 9786020998251 | Mari bermain bersama nenek | Nur Asyikin Binti Mohamad Lais | Feliz Books | Fiksi | 2020 | 11 |
| 7 | 9786020998268 | Sang koki cilik | Shahidah Binti Borhan, Nurwahidah Binti | Feliz Books | Fiksi | 2019 | 11 |

**Figure 11.** Visitor Book List Interface

**Figure 12.** Admin Menu Interface

**Figure 13.** Librarian Menu Interface

## 4. Conclusions

Based on the design results of the Library Information System at SMK Pasundan Jatinangor, it can be concluded as follows:

a. Can facilitate computerized input of new books and reduce errors in inputting.
b. Reducing the processing time in calculating the availability of books in the library.
c. Provide new changes for staff and teachers in keeping up with the times with a computerized system.
d. Reducing the difficulty in calculating fines that were originally calculated manually.
e. Minimizing the loss of books in borrowing.

## References

[1] Yusuf M O 2005 Information and communication technology and education: Analysing the Nigerian national policy for information technology. *International education journal* **6** (3) 316-321.

[2] Volman M and Van Eck E 2001 Gender equity and information technology in education: The second decade *Review of educational research* **71** (4) 613-634

[3] Alavi M, Yoo Y and Vogel D R 1997 Using information technology to add value to management education *Academy of management Journal* **40** (6) 1310-1333

[4] Vaidyanathan S, Sambas A, Mamat M and Sanjaya W S M 2017 Analysis, synchronisation and circuit implementation of a novel jerk chaotic system and its application for voice encryption *International Journal of Modelling, Identification and Control* **28** (2) 153-166

[5] Achour S L, Dojat M, Rieux C, Bierling P and Lepage E 2001 A UMLS-based knowledge acquisition tool for rule-based clinical decision support system development *Journal of the American Medical Informatics Association* **8** (4) 351-360

[6] Aleryani A Y 2016 Comparative study between data flow diagram and use case diagram *International Journal of Scientific and Research Publications* **6** (3) 124-126

[7] Swain R K, Panthi V and Behera P K 2013 Generation of test cases using activity diagram *International journal of computer science and informatics* **3** (2) 1-10

[8] Aguilar-Saven R S 2004 Business process modelling: Review and framework *International Journal of production economics* **90** (2) 129-149

[9] Ben Abdessalem Karaa W, Ben Azzouz Z, Singh A, Dey N, S Ashour A and Ben Ghazala H 2016 Automatic builder of class diagram (ABCD): an application of UML generation from functional requirements *Software: Practice and Experience* **46** (11) 1443-1458

[10] Ali N H, Shukur Z and Idris S 2007 Assessment system for UML class diagram using notations extraction *International Journal on Computer Science Network Security* **7** 181-187

[11] Bowman D A, Kruijff E, LaViola Jr J J and Poupyrev I 2001 An introduction to 3-D user interface design *Presence: Teleoperators & Virtual Environments* **10** (1) 96-108

[12] Mosier J N and Smith S L 1986 Application of guidelines for designing user interface software *Behaviour & information technology* **5** (1) 39-46

# Performance Analysis of Hydraulic Machines Using Impellers Made from Conventional Steel and 3D Printing Co-polymer

**D Sitte and H Sutanto***

[1]Department of Mechanical Engineering, Faculty of Engineering, Czech University of Life Sciences Prague, Prague, Czech Republic
[2]Department of Mechanical Engineering, Faculty of Engineering, Atma Jaya Catholic University of Indonesia, Jakarta, Indonesia

*Corresponding author: hadi.sutanto@atmajaya.ac.id

**Abstract.** The research on hydraulic machines in industries is appropriate for many reasons. This paper deals with the creation of 3D prototypes of impellers of hydraulic machines printed from ABS plastic and verification of operating characteristics compared to conventional steel impellers of pump and also in turbine mode. The work includes properties, composition and verification of mechanical properties of ABS for different printing methods. Practical measurements of the specific geometry of the different 3D variants of the radial one-stage centrifugal impeller from ABS compared to metal wheels in the real hydraulic circuit are described. The results of the development of experiments demonstrate that the performance characteristics of the printed impellers of all variants have very similar functional values and very identical trends of functional curves of mechanical power and overall mechanical efficiency. At the same optimal speed, the highest performance and efficiency also as the metal wheels have been produced. It is clear from the results of work that prototypes produced by 3D printing achieve optimal performance parameter values in hydraulic testing and it is possible to successfully use this advanced technology in the verification of newly developed pumps, turbines and pumps turbines.

## 1. Introduction

Hydraulic machines, such as pumps and turbines are widely used in industrial and agricultural fields for irrigation and water supply [1]. In general, they have four different parts as suction and exit pipes, impellers and volute. The impeller is the core part and it converts the mechanical energy into pressure energy [2], which directly determine the transport capacity and the hydraulic performance of centrifugal pumps. It means that optimized design of impeller is essential and significant for the efficient operation of a centrifugal pump [3].

The use of 3D-CAD graphics systems is now a great tool for experiment, research and manufacturing activities in engineering. The basic principle is a parameterized graphical three-dimensional model in CAD software and now it can be used like CATIA V5, Creo, Inventor or SOLIDWORKS. Rapid Prototyping technology is one of the systems for creating faster and now affordable prototypes from 3D printers. The results are also the quality, possibilities and speed of 3D printed components to begin for replacing conventional ways of production itself [4].

Manufacturing companies offer a wide range of amateur and industrial 3D printers. In practice, the nomenclature of printing technology is based on the type of printing material such as powdered BJ

(Binder Jetting) and SLS (Laser Sintering), liquid SLA (Stereo Lithography) and rigid FDM (Fused Deposition Modelling). A common principle of production of a 3D printing component is the gradual addition of material by layer, as opposed to classical machining when the material is taken away by the subtractive processes. FDM technology was developed by Stratasys and it is the most widely used method of 3D printing of plastic components in engineering. For its positive properties, it was used in this work for making all functional prototypes on the 3D printer Dimension Elite Printer 180-00105 [5], [6].

Acrylonitrile-butadiene-styrene or ABS material, is an amorphous thermoplastic industrial co-polymer with a wide range of uses because of its universal properties. It is very well resistant to mechanical damage, impact and strain at low and high temperatures. ABS has considerable toughness and very good machinability. It is stable and resistant to UV radiation for a long time. ABS material is also low-soak, health-resistant and easy to withstand acids, hydrocarbons, hydroxides, oils and fats. On the contrary, it is very well soluble in acetone or in 1.2 dichloromethane. The ABS plus (ABS-P430 Model) variant, which has 40% higher strength than standard ABS, was used for printing in this particular case [7], [8], [9].

Verification of performance characteristics and operating parameters in the development, manufacture and supply of pumps is specified in ISO 9906. The standard specifies hydraulic power tests for all rotor dynamic pumps. And it can be used for pumps of all sizes and on which wheels pumped liquids that behave like pure cold water. Testing of the impellers used in the experiment was based on this standard [10].

The main purpose of this work is to verify the functional parameters and utility properties of a specific geometry of 3D printing plastic impellers prototypes compared to the parameters of the identical geometry of metal impellers produced by conventional technologies, such as metal casting, for pump and turbine mode during the dynamic load in the real hydraulic circuit.

## 2. Background Theory

### 2.1. Preparing a 3D model

The process of fast prototyping begins with creation of a 3D digital model in a CAD modelling program, here Inventor and Creo. Preparing a 3D model for printing begins with checking and editing geometry based on experience with a certain percentage of the shrinking of print dimensions, on modelling the necessary technological additions for clamping and machining. For functional impeller surfaces a machining allowance of 1 mm per radius was chosen, as seen in Figure 1. The following step was the conversion of a 3D digital model to Standard Tessellation Language (STL). The purpose was to create a sufficiently fine triangulation network that determines the resulting quality of the printed prototype, see Figure 2.

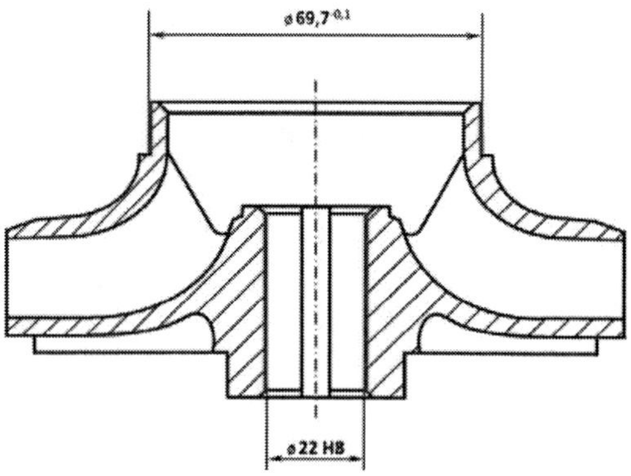


**Figure 1.** Working areas of the impeller.

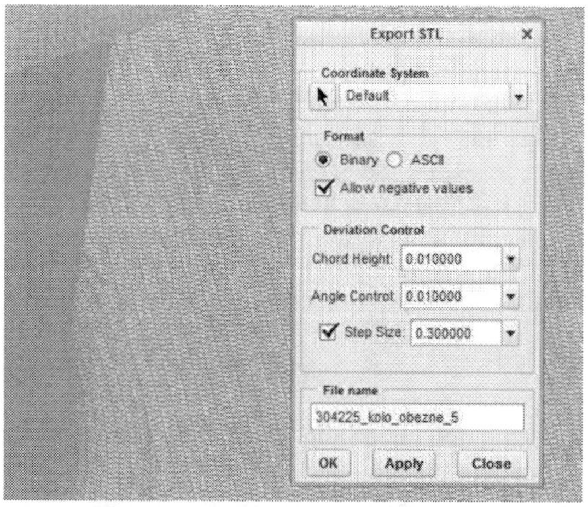


**Figure 2.** Creation of STL network.

The STL model was load into Grab-CAD Print. The orientation of the prototype on the print mat has been chosen, since the tensile strength is better in the direction of printed fibers than the strength perpendicular to the printed layers. The orientation of the part also affects the construction and consumption of supporting material (Figure 3). The position of the prototype on the mat affects the print temperature, which is important for test samples. In the following step, the print parameters were selected. The construction of the carrier material was related to the value of automatic construction (smart). The nature of the prototype construction was chosen depending on the printed variant, either for full material (solid) or shell structure with a honeycomb filling inside. The thickness of the print layers has always been set to 178 micrometers for softer and more precise printing. Figure 4 shows the simulation of automatic construction of the model and support material [11]. This process was followed by printing processes with parameters as in Table 1.

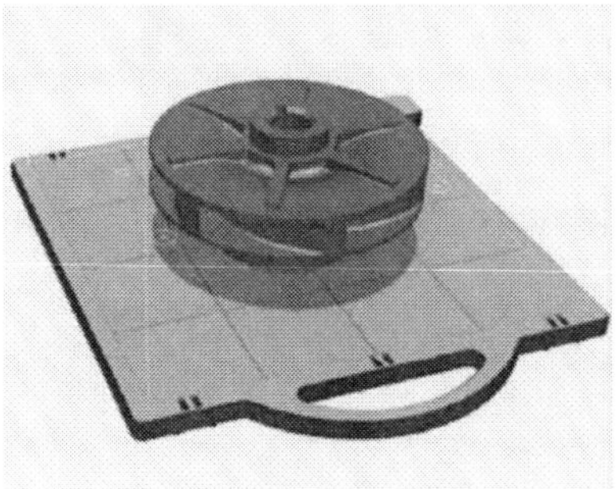

**Figure 3.** Position of the impeller on the mat [9]

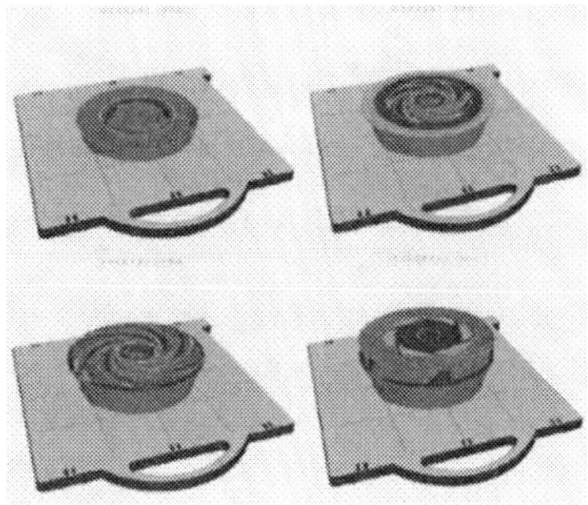

**Figure 4.** Simulation of the print job [9]

**Table 1.** The printing parameters [12].

| Print job | Time (h:min:s) | ABS | Amount (cm$^3$) | Support | Amount (cm$^3$) |
|---|---|---|---|---|---|
| ABS full impeller | 32:29:11 | ABS-P430 | 187.5 | P400SR | 69.9 |
| Composite impeller | 30:57:01 | ABS-P430 | 159.4 | P400SR | 72.7 |
| ABS shell impeller | 23:21:52 | ABS-P430 | 87.9 | P400SR | 69.9 |

After printing process, the component was removed from the printer and left to cool naturally to room temperature. This was followed by rough cleaning of the supporting material, dimensional and visual inspection, softly cleaning in the SCA-1200 printer accessories in lye at 70°C. The cleaned part was washed with 70°C water and put in an ultrasonic bath, then was naturally dried at room temperature of 20°C. Now the prototype is in terms of printing completely finished and it can be proceed to the advanced machining. After machining, the process of component is already finish, and it remains to verify the appearance and required dimensions [8], [13].

## 2.2. Mechanical tests of ABS-P430 plastic

To verify the mechanical properties of printed ABS-P430 plastic, static load tensile tests have been established according to ISO 527-1 and ISO 527-2. The three sets of type A1 samples were tested after five pieces printed with different directions of printing fibers, see Figure 5. Samples were printed with consistent compliance using the same printing parameters as prototypes of the impellers were produced. The tests were carried out on the Instron 3382 test engine, see Figure 6. The test was controlled by a feed of 1 mm. min$^{-1}$ and it was completed at the time of the sample infringement. Figure 7 shows a test sample which is captured in the detail of the breach [14], [15].

IOP Conf. Series: Materials Science and Engineering 1115 (2021) 012049 doi:10.1088/1757-899X/1115/1/012049

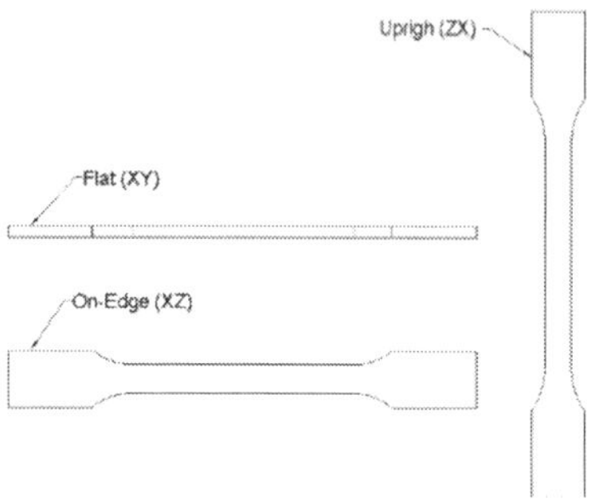


**Figure 5.** Test samples A1 [14].

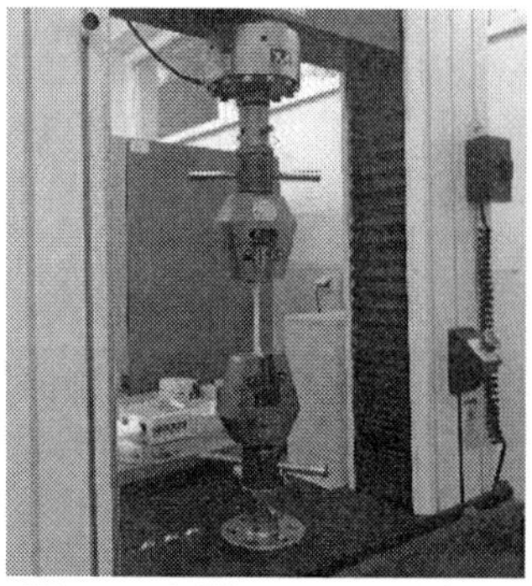

**Figure 6.** The test machine.

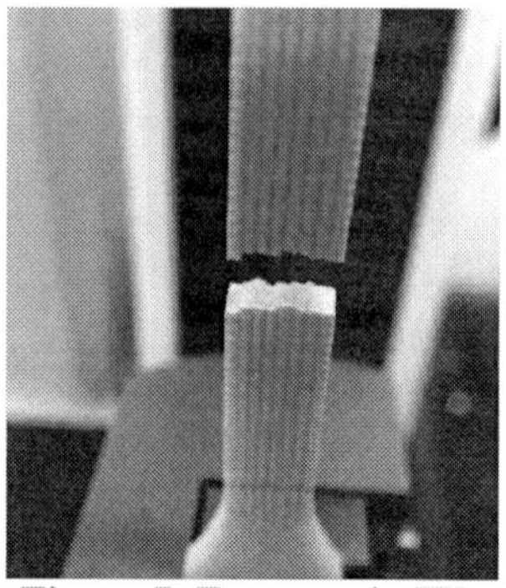

**Figure 7.** Test sample ZX.

## 2.3. Experimental verification of turbine and pump performance parameters

For laboratory verification, the geometry of the impeller was selected, which is displayed in the spiral cabinet together with the pump parameters as specified in Figure 8.

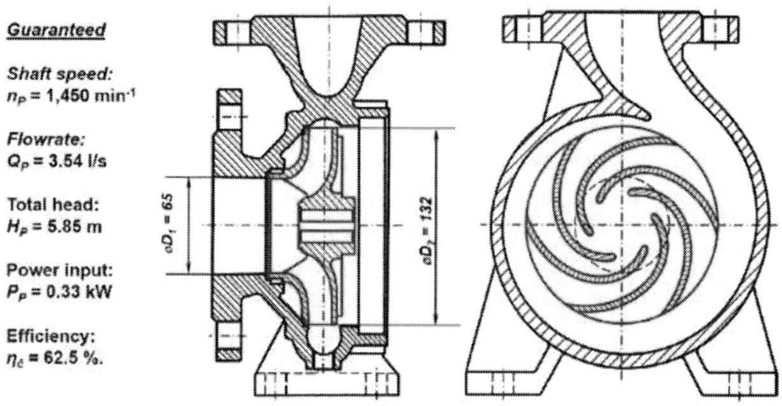


**Figure 8.** Metal impeller in spiral housing and pump parameters [15].

The radial one-stage centrifugal pump impellers produced by conventional casting methods (cast iron, stainless steel) and Rapid Prototyping technology with 3D printing technology have been tested. These wheels were tested in pump operation at 1,450 rpm.min$^{-1}$ and 2,950 rpm.min$^{-1}$ as well as in a turbine mode.

Impellers produced by Rapid Prototyping technology have been printed in three 3D printing variants, while consistently maintaining the conformity of all STL formation parameters, orientation, location, layer thickness and support creation. The experiments using the following 5 types of impellers:

- Steel impeller, **Figure 9**.
- Cast iron impeller, **Figure 10.**
- ABS full impeller made of solid, **Figure 11.**
- ABS shell impeller, solid surface and porous (honeycomb) interior. **Figure 12.**
- Composite impeller using ABS solid with glued metal disc, **Figure 13.**

**Figure 9**. Steel impeller

**Figure 10.** Cast iron impeller

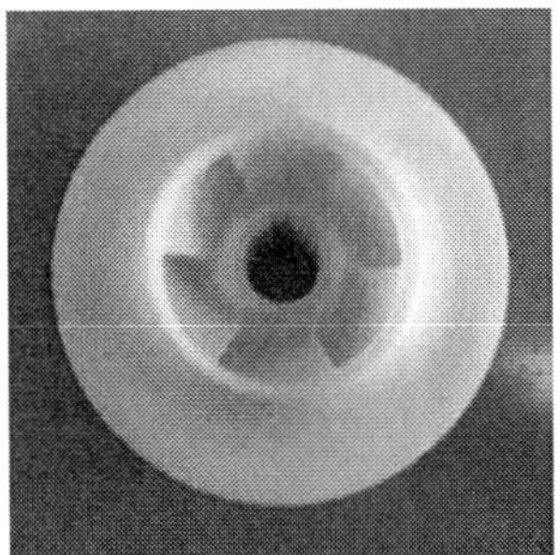

**Figure 11.** ABS full impeller

**Figure 12.** ABS shell impeller

**Figure 13.** Composite impeller

### 2.4. Hydraulic test circuit

A hydraulic test circuit was used to test the functional parameters of radial single-stage centrifugal impellers in pump and turbine mode. The diagram of the circuit for Figure 14.

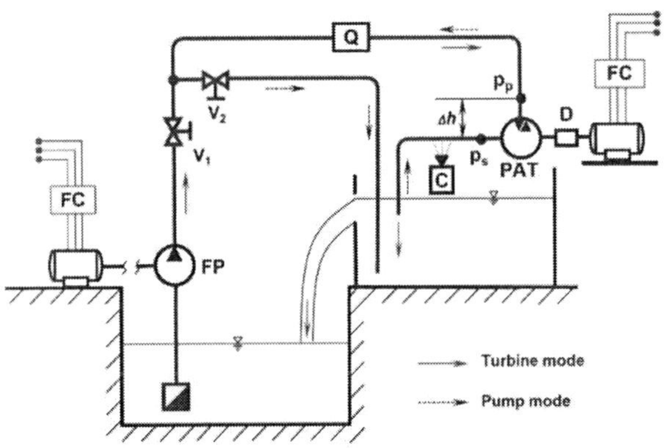


*Q – flow meter, FP – power pump, PAT – tested pump or pump in turbine mode, V1, V2 – control valves, Pp, Ps – pressure sensors, D – dynamometer, FC – frequency converter, C – camcorder.*

**Figure 14.** Hydraulic circuit scheme [16].

The hydraulic test circuit consists of two tanks, connecting pipes, control and measuring elements and the pump tested (P/T). By closing the V1 valve and regulating the V2 valve, water flows in the direction of the full arrows and forms a closed circuit for the tested pump. The pump was powered by a three-phase asynchronous electric motor with continuous shaft speed control using the LSLV0055s100- 4EOFNS frequency converter. The torque on the pump shaft was measured by the torque sensor (D) Magtrol TMB 307/41 (accuracy 0.1%). The water flow rate was measured using the electromagnetic flow meter (Q) SITRANS F M MAG 5100 W (accuracy 0.5%). Pressures in extrusion ($p_P$) and intake manifold ($p_S$) were measured by pressure sensors HEIM3340 (accuracy 0.5 %) installed according to the measurement requirement 1. accuracy class (ISO 9906).

The measurements were carried out with two constant speeds (1,450 rpm.min$^{-1}$ or 2,950 rpm.min$^{-1}$) and set by the frequency converter. Gradually closing of the V2 valve increased the resistance of the discharge pipe. During the measurement, all values of the flow, engine speed, torque on the engine shaft, fluid pressures at the pump inlet and outlet were recorded.

The turbine operation of the pump took place on the same test line as follows. By closing valve V2 at the open valve V1, water flowed in the direction of intermittent arrows, and the power pump (FP) created hydro technical potential for the turbine. The measurement was carried out when the power pump (FP) was set up by the frequency converter to the required hydro technical potential. With this setting, the turbine was gradually braked using a three-phase electric motor with a frequency converter and braking resistors. The values of the flow, engine speed, torque on the engine shaft, fluid pressures at the pump inlet and outlet were measured and recorded. Declining hydro technical gradient with increasing turbine speeds was due to constant speeds of the power pump. To maintain a constant gradient, the power pump speed needs to be controlled, for example, by a frequency converter. The downward gradient was the same for all turbine variants and did not affect the measurement results. The same instruments were used for measurements as in pump operation, see Figure 15, Figure 16 and Figure 17 [10], [16].

**Figure 15.** Hydraulic circuit.

**Figure 16.** Detail of hydraulic circuit.

**Figure 17.** Composite impeller.

### 2.5. Calculation for pump and turbine

Pump mode:

- From Bernoulli's equation between intake ($p_s$) and displacement ($p_p$) of the pump and the continuity equation, the pump's variable energy was calculated as,

$$0 + \frac{p_s}{\rho} + \frac{v_s^2}{2} + Y_p = g.h + \frac{p_p}{\rho} + \frac{v_p^2}{2} \quad [J.kg^{-1}] \tag{1}$$

$$Q = v.S \quad [m^3.s^{-1}] \tag{2}$$

$$Y_p = \frac{p_p - p_s}{\rho} + \frac{v_p^2 - v_s^2}{2} + g.h \quad [J.kg^{-1}] \tag{3}$$

Turbine mode:

- From Bernoulli's equation between intake ($p_{st}$) and displacement ($p_{pt}$) of the turbine and continuity equation, the turbine's variable energy was calculated as,

$$0 + \frac{p_{st}}{\rho} + \frac{v_{st}^2}{2} + Y_t = g.h + \frac{p_{pt}}{\rho} + \frac{v_{pt}^2}{2} \quad [J.kg^{-1}] \tag{4}$$

$$Q_t = v_t.S \quad [m^3.s^{-1}] \tag{5}$$

$$Y_t = \frac{p_{pt} - p_{st}}{\rho} + \frac{v_{pt}^2 - v_{st}^2}{2} + g.h \quad [J.kg^{-1}] \tag{6}$$

**Nomenclature**

| | |
|---|---|
| $p$ – liquid pressure at a given location | [Pa] |
| $\rho$ – specific mass of liquid | $[\mathrm{kg \cdot m^{-3}}]$ |
| $v$ – medium fluid speed at a given location | $[\mathrm{m \cdot s^{-1}}]$ |
| $Y_p$ – specific pump energy | $[\mathrm{J \cdot kg^{-1}}]$ |
| $Y_t$ – specific energy of the turbine | $[\mathrm{J \cdot kg^{-1}}]$ |
| $g$ – gravitational acceleration | $[\mathrm{m \cdot s^{-2}}]$ |
| $h$ – vertical lightness of pressure collection | [m] |
| $Q$ – volumetric flow rate | $[\mathrm{l \cdot s^{-1}}]$ |
| $S$ – flow cross-section at a given location | $[\mathrm{m^2}]$ |
| $d$ – the luminosity of the pipe at a given location | [m] |

### 2.6. Test circuit input parameters

Pump mode:
- Intake manifold luminosity: $d_s$ = 65 mm
- Extrusion pipe luminosity: $d_p$ = 50 mm
- Vertical pressure relief: h = 290 mm

Turbine mode:
- Intake manifold luminosity: $d_s$ = 50 mm
- Extrusion pipe luminosity: $d_p$ = 65 mm
- Vertical pressure relief: h = 290 mm

## 3. Results and discussion

Test samples ABS-P430, see Figure 5, were tested in five pieces for each of the three directions of the printed fibers and reached the average tensile strength values in ZX = 14.8 MPa., XY= 26.8 MPa., and XZ = 29.9 MPa.

The measured and calculated values of pump and turbine mode have compiled with the performance characteristics of all impeller variants as shown in Figure 18 – 25. The pump mode experiments used two pump operation at speed 1,450 rpm.min$^{-1}$ (see Figure 18 – 20) and 2,950 rpm.min$^{-1}$ (see Figure 21– 23). Pump mode power characteristics at 1,450 rpm.min$^{-1}$

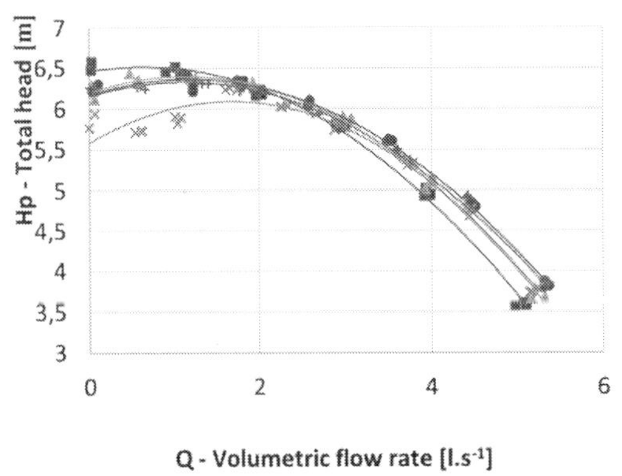


**Figure 18.** Total head vs flow rate

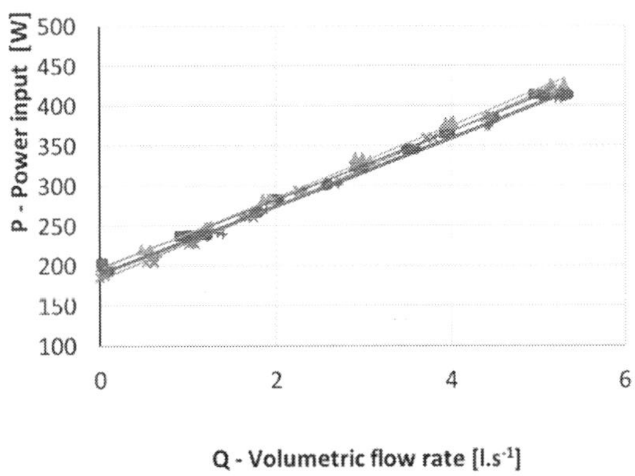


**Figure 19.** Power input vs flow rate

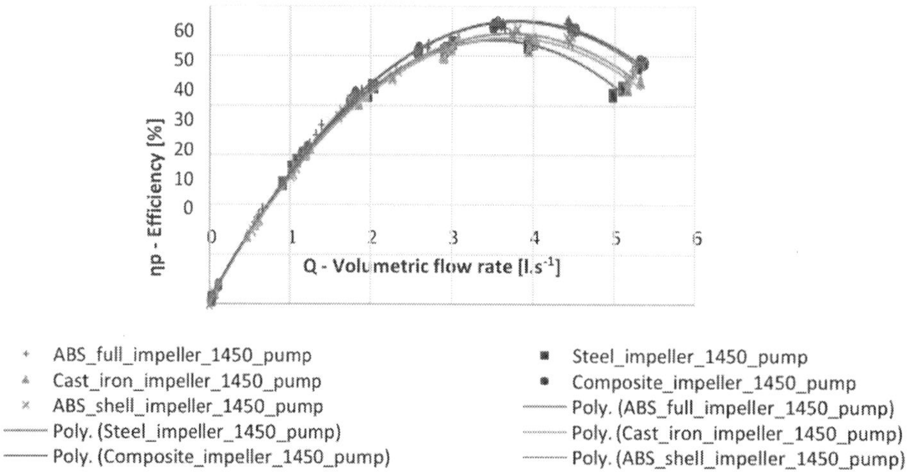


**Figure 20.** Pump efficiency vs flow rate

Pump mode power characteristics at 2,950 rpm.min$^{-1}$ :

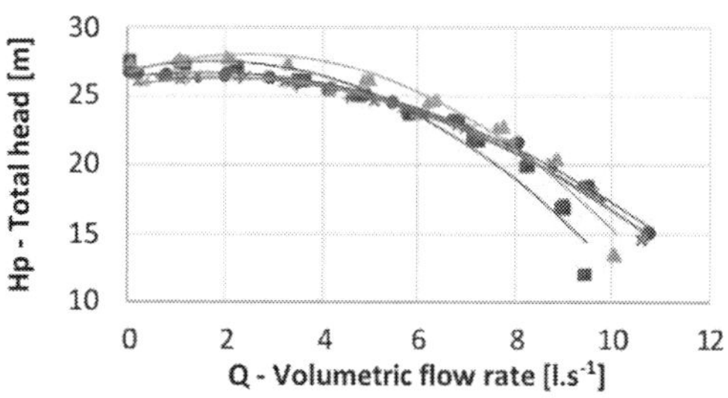


**Figure 21.** Total head vs flow rate

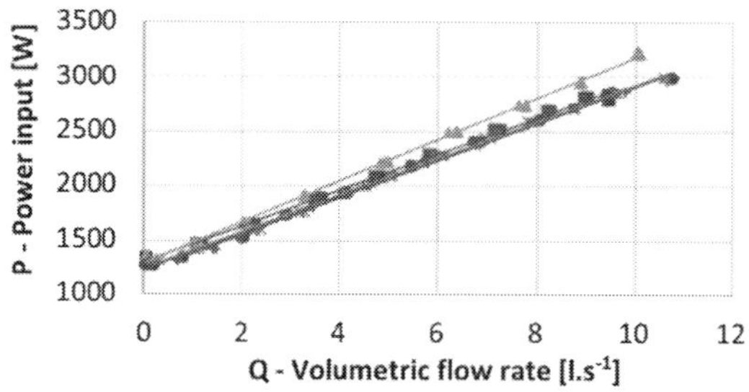


**Figure 22.** Power input vs flow rate

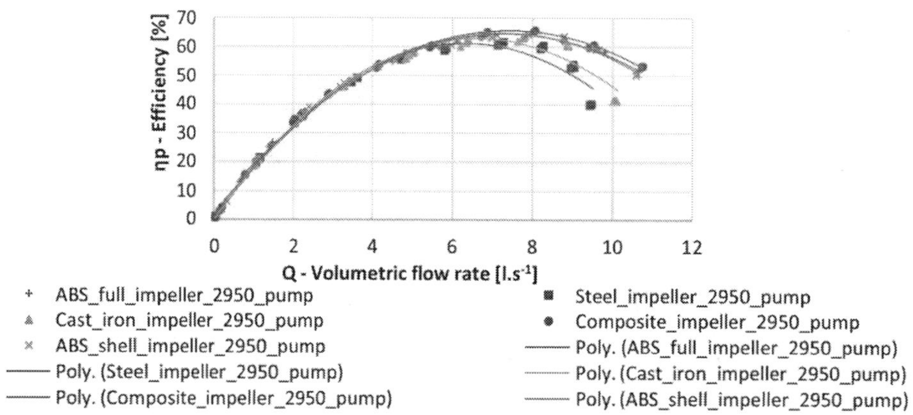


**Figure 23.** Pump efficiency vs flow rate

Turbine mode power characteristics:

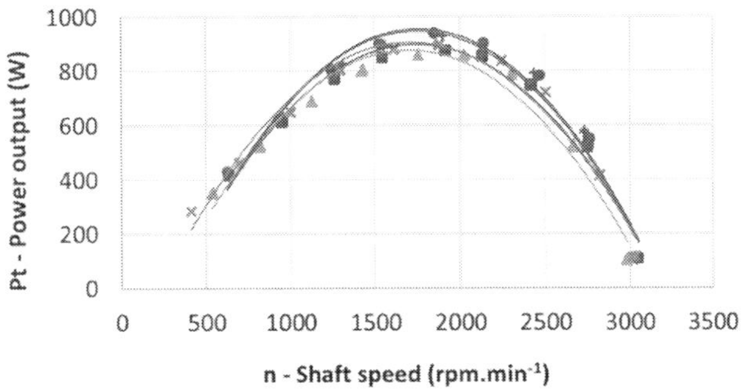


**Figure 24.** Mechanical power of the turbine vs shaft speed

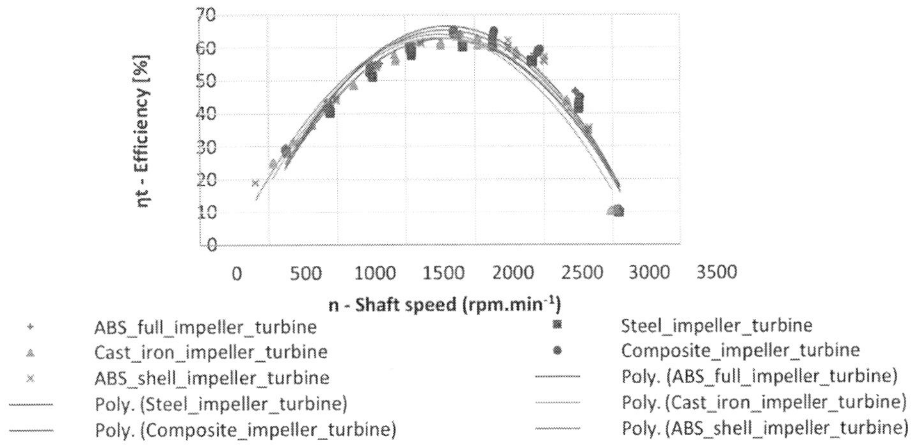


**Figure 25.** Turbine efficiency vs shaft speed

As in the reference, there is a good experience in achieving a higher quality (enamel) surface of prototypes during short-term soaking or exposure to the effects of acetone vapor or 1.2 dichloromethane. This often happens at the cost of reducing the strength of the component or losing shape and dimensional accuracy. The original intention was also to incorporate the adjustment in the production of impellers, but due to the shape-intensive geometry, the significantly different thicknesses of the components walls and the significant load in the verification tests in the hydraulic circuit. It was surrounded from this intention after several attempts. This is an issue for further independent work and for the repeatable use of these methods it is necessary to have the process properly tuned. The positive benefit of this treatment could be possible to prevent savories into the cavities of the wheels with liquid and thus avoid any mass imbalance. In the stress tests, it has been shown that although the wheels are capable of partially soaked with liquid, tests are of a short-term nature and minor intake had no effect on the course and test results. All variants of tested wheels have always endured high stress at full load in turbine and pump operation at 1,450 rpm.min$^{-1}$ and 2,950 rpm.min$^{-1}$ [7], [8].

Material tests of printed ABS at static tests on the tearing machine according ISO 527-1 and ISO 527-2 indicate a significant different of strong characteristic of samples printed in different directions of printing fibers. Strength properties should be taken into account already when designing the prototype itself, appropriately set the printing parameters due to the necessary time to print, consume materials and the purpose of the prototype, especially for mechanically strained prototypes.

## 5. Conclusions

The performance characteristics of the impellers tested and measured on the hydraulic circuit for the pump mode showed that all variants have very similar functional values to each other and very identical trends of functional curves of total head, power consumption and overall mechanical efficiency depending on flow compared to two operating conditions at 1,450 rpm.min$^{-1}$ and 2,950 rpm.min$^{-1}$. The shapes of individual curves are almost identical, and values of the transport height and power correspond to the multiple of the speed of the operation. The total efficiency values of pumps in operation at a speed of 1,450 rpm.min$^{-1}$ are around 52 % and in operation at a speed of 2,950 rpm.min$^{-1}$ are around 62 %. The power characteristics of the impellers for the turbine mode show that all variants have also very similar functional values and very identical trends in functional curves of mechanical power and overall mechanical efficiency depending on speed. The values of the total mechanical efficiency of turbines are around 62 %, which corresponds to the type of operation. The characteristics of this turbine can be read from the measured values, where, at increasing speeds, the flow and overall efficiency grow up to the maximum of power and, with further increasing speed, the flow and efficiency decrease. In conclusion, the turbine achieves the highest performance and efficiency at optimum speeds.

Based on the experiments, prototypes produced by Rapid Prototyping technology achieve optimal values in hydraulic testing compared to conventionally produced impellers in both pump and turbine mode. While the choice of parameters in the processing of the virtual model, the quality of the material used and above all, the quality of 3D printing, plays a significant role. Following these boundaries conditions, this advanced technology can be successfully used in the verification of newly developed pumps, turbines and pump turbines.

## Acknowledgements

The authors thanks to (a) the Internal Grant Agency 2020 of the Czech University of Life Sciences Prague for funding from the project of liquid flow analysis in hydrodynamic pump and pump in turbine mode, and (b) Atma Jaya Catholic University of Indonesia for the collaboration of the student exchange program.

### References

[1] Pérez-Sánchez M, Sánchez-Romero F J, Ramos H M and López-Jiménez P A 2017 Energy recovery in existing water networks: Towards greater sustainability *Water* **9** (2) 97

[2] Li X, Chen B, Luo X and Zhu Z 2020 Effects of flow pattern on hydraulic performance and energy conversion characterisation in a centrifugal pump *Renewable Energy* **151** 475-487

[3] Han X, Kang Y, Li D and Zhao W 2018 Impeller optimized design of the centrifugal pump: A numerical and experimental investigation *Energies* **11** (6) 1444

[4] Mohan N, Senthil P, Vinodh S and Jayanth N 2017 A review on composite materials and process parameters optimisation for the fused deposition modelling process *Virtual and Physical Prototyping* **12** (1) 47-59

[5] Wickramasinghe S, Do T and Tran P 2020 FDM-based 3D printing of polymer and associated composite: A review on mechanical properties defects and treatments *Polymers* **12** (7) 1529

[6] Ramya A and Vanapalli S L 2016 3D printing technologies in various applications *International Journal of Mechanical Engineering and Technology* **7** (3) 396-409

[7] Kandananond K 2020 Optimization of fused filament fabrication system by response surface method *International Journal of Metrology and Quality Engineering* **11** 4

[8] Dave H K, Rajpurohit S R, Patadiya N H, Dave S J, Sharma K S, Thambad S S and Sheth K V 2019 Compressive strength of PLA based scaffolds: effect of layer height infill density and print speed *Int J Mod Manuf Technol* **11** (1) 21-7

[9] Höller S, Benigni H and Jaberg H 2019 Numerical and experimental investigation of the 4-Quadrant behavior of different mixed flow diffuser pumps *International Journal of Turbomachinery Propulsion and Power* **4** (1) 3
[10] Özkil A G 2017 Collective design in 3D printing: A large scale empirical study of designs designers and evolution *Design Studies* **51** 66-89
[11] Costa S F, Duarte F M and Covas J A 2015 Thermal conditions affecting heat transfer in FDM/FFE: a contribution towards the numerical modelling of the process: This paper investigates convection conduction and radiation phenomena in the filament deposition process *Virtual and Physical Prototyping* **10** (1) 35-46
[12] Garg A, Bhattacharya A and Batish A 2016 On surface finish and dimensional accuracy of FDM parts after cold vapor treatment *Materials and Manufacturing Processes* **31** (4) 522-529
[13] Stergiou G S, Alpert B, Mieke S, Asmar R, Atkins N, Eckert S and O'Brien E 2018 A universal standard for the validation of blood pressure measuring devices: Association for the Advancement of Medical Instrumentation/European Society of Hypertension/International Organization for Standardization Collaboration Statement Hypertension **71** (3) 368-374
[14] Iso E 2012 *Plastics—Determination of tensile properties—Part 1: General principles* Geneva Switzerland: International Organization of Standardization
[15] Polák M 2017 Experimental evaluation of hydraulic design modifications of radial centrifugal pumps *Agron. Res* **15** 1189-1197
[16] Polák M 2019 The influence of changing hydropower potential on performance parameters of pumps in turbine mode *Energies* **12** (11) 2103

# Analysis of water-ethanol-gasoline (RON 88) compositions in one phase substance-using triangular graph

**H F Sangian[1*], B H Manialup[1], A As'ari[1], A R Ibrahim[1], J R Rombang[1], J A Lametige[1], H I R Mosey[1], A Tanauma[1], G Pasau[1], S H J Tongkukut[1], R Purwadi[2], G F Neonufa[3], J A Rorong[4], M M Sangi[4], S Gunawan[5], A Widjaja[5], A Tuhuloula[6], R Thahir[7], B A Sadjab[8], V A J Masinambow[9], T Sriana[10], Z Furqon[10], S Y Agnesty[10] and S B Rondonuwu[11]**

[1]Department of Physics, Sam Ratulangi University, Manado, Indonesia
[2]Department of Chemical Engineering, Institut Teknologi Bandung, Indonesia
[3]Department of Agriculture Product Technology, UKAW Kupang, Indonesia
[4]Department of Chemistry, Sam Ratulangi University, Manado, Indonesia
[5]Department of Chemical Engineering, Institut Teknologi Sepuluh Nopember, Surabaya, Indonesia
[6]Department of Chemical Engineering, Lambung Mangkurat University, Banjarbaru, Indonesia
[7]Department of Chemical Engineering, Politeknik Negeri Samarinda, Indonesia
[8]Department of Physics, Halmahera University, North Maluku, Tobelo, Indonesia
[9]Department of Economic Development, Sam Ratulangi University, Manado, Indonesia
[10]Department of Oil and Gas Processing, Energy and Minerals Polytechnics, Cepu Blora, Indonesia
[11]Department of Biology, Sam Ratulangi University, Manado, Indonesia

[*]Corresponding author: hannysangian@yahoo.co.id

**Abstract.** The present work is aimed at analysing the compositions of the water, ethanol, and gasoline which has *Research Octane Number*/RON was 88 forming a stable emulsion (one phase) employing a ternary graph. When the mixture process, the blended fuels consisted of water-ethanol-gasoline were successfully prepared in which they were formed in one phase. Ethanol was derived from *Arenga pinnata* liquor which is locally called *cap tikus* using a home-made reflux distillation filled by packing materials. Ethanol obtained were differing their concentration that depended on the column temperature set. It was found that the purities were ranged from 80 to 96% and the higher column temperature was chosen the lower concentration was obtained. Each aqueous ethanol was blended with gasoline to obtain a homogenous solution. For ethanol 80%, compositions of water, pure ethanol, and gasoline were observed at 18, 74, and 8 (%v/v), and 22, 70, and 7 (%w/w). While ethanol 96%, the compositions ratios were 1:22:77 (%v/v) and 1:23:76 (%w/w). The ranges of pure ethanol, gasoline, and water in which they formed one phase solution were recorded at 23-70%, 7-76%, and 1-22%. The work found that substance was in one phase if the wet ethanol keeps being added. When the ethanol composition has decreased the substance was separated into wet ethanol and gasoline. The minimum ethanol dissolved completely into gasoline was of 80%.

## 1. Introduction

The world is facing a difficult time caused by the dwindling of the petroleum reserved and the increasing of the atmospherically temperature called (*global warming*). The utilization of renewable resources which are processed into biofuels must be developed massively around the world. The biomass (lignocellulose and starch) have been attracting many scientists to improved becoming modified materials and food [1-7]. The treated biomass as stated previously were more easy to convert them into sugars [8] and then were fermented into alternative energies such as the bioethanol [9], hydrogen gas [10], and solid fuel [11]. The other green fuel derived from a renewable resource and had been produced at the industry scale was biodiesel [12].

Indonesia still uses fossil-based fuels as the main energy source in the transportation sector, electrical generations, and industrial sector. On the other hand, their reserves are decreasing and will be run out in the coming years because are not renew able. The blended fuels are an alternative solution to substitute a part of petroleum.

The blended fuels have been investigated for years by authors and reported by inventors previously as printed in journals and patent documents. The study of diethyl ether and ethanol added to biodiesel/diesel blends was evaluated on performance, emission, and combustion characteristics. The result showed that CO emission was lower than the non-blended fuels [13]. The feasibility study of biodiesel/diesel blends in which the biodiesel ratio was studied. The work showed that the increment of 10% biodiesel, the energy decreased at 1.42%. The engine test, however, there was no significant difference if compared to the low ratio [14].

The smoke characteristics of the four-stroke engine employing an ethanol-gasoline fuel which ethanol part was 10% (v/v), E10, was analyzed. It was found the CO and HC emission was lower than that of pure gasoline. The smoke analysis showed the ethylene, acetaldehyde, and ethanol emissions were bigger compared to than unleaded gasoline engine [15]. The effect of ethanol-gasoline blends for SI engine especially on cold-start emission. The results suggested that the 5, 10, 20, and 30 ethanol blends with gasoline, the engine was started well and HC and CO emissions were decreasing. At E40 the engine work faced a problem because the air-fuel mixture was not proportional [16].

The blended fuels, biodiesel-diesel, biodiesel-ethanol, and ethanol-gasoline also were reported in many document patents. The fuel blends, diesel fuel oil and a $C_3$ (excluding n-propanol)-$C_{32}$ organic alcohol, were disclosed as stated in US Patent No. 5,720,784. The blend fuels of diesel, ethanol and a coupling agent were discovered documented in US Patent No. 7,357,819 B2. The invention revealed the coupling agent comprises: 2-ethyl hexanol; fatty alkanolamides; and one or more fatty acids; wherein one or more fatty acids are present in the coupling agent.

Dissolution of water and pure ethanol in gasoline was investigated whereby ethanol was derived from Arenga pinnata. The work investigated the soluble state of water-pure ethanol-gasoline in normal conditions [17]. However, the previous work did not investigate the compositions of three components in one phase substance. As mentioned previously in references survey was mainly describing the blend fuels in which the purity ethanol was close to un-hydrous alcohol and the ethanol-fossil fuel bonds must be other agents, surfactants and coupling agents.

The present study aims to study the blend fuels of water-ethanol-gasoline in one phase whereby alcohol concentrations were ranged from 80 to 96%. The composition of water was taken from aqueous ethanol by multiplying the purity with a total volume and divided by volume of three components. The procedure was briefly summarized: The preparation of palm sap, the ethanol yeasting, and distillation.

## 2. Materials and Methods

Ethanol used in this study was derived from liquor which was produced by the farmer. The sap consisting of sugar 14% was tapped from *Arenga pinnata* and kept in a plastic container in which it occurred a fermentation for 3-5 days. The yeasted solution that its ethanol concentration was of 7% was poured inside the boiler and heated until its boiling temperature. The vapor was channeled into a simple distillation and condensed into the liquid phase and the ethanol average purity of liquor obtained was of 40%.

The liquor was purchased and then continued to increase its purity ranging 80 to 96% employing a reflux distillation as shown in Figure 1. The liquor was poured inside the boiler and the temperature was increased until its boiling. The liquor volume filled in the boiler was of 0.75 of total volume, while the packing materials were 80% of column volume. The length of a reflux column was 2 meters and diameter was of 2 inches. The purity of ethanol produced was depending on the column temperature, ethanol content in liquor, and packing materials. The vapor rose to the reflux column and was occurred a separation between ethanol and water and then ethanol dropped into a flask and was isolated.

The blending procedures were the following: The gasoline denoted with gas was measured at 15 mL and poured in Erlen-Meyer glass. In the industrial scale, the unit can be extended into either of a liter or kiloliter. Firstly, the aqueous ethanol assigned by Aq. Et with purity 80% was mixed with gasoline. In a similar volume, two components were separated and when the wet ethanol was added continually the gasoline was starting dissolved. The components were not distinguished at wet ethanol added was of 75 mL. The volume of pure ethanol was obtained by employing a formula %x volume of Aq. ethanol, while pure water using a relation of (100%-% Aq. ethanol) x volume of Aq. ethanol. The composition was calculated by ratio equation of volume of the certain component divided by total volumes. In this study also described the composition in the triangular graph.

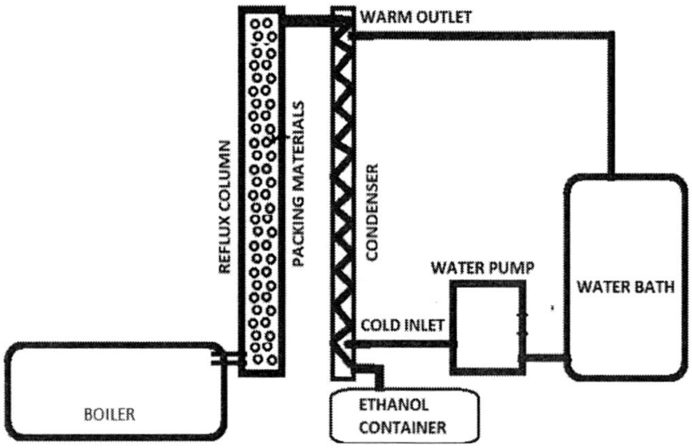


Figure-1. The reflux distillation design

## 3. Results and Discussion

Table 1 shows the composition of one phase of water (Wat), ethanol (Et), and gasoline (Gas) which has RON 88. Three components could be forming one phase substance but the composition must be exact. The 175 mL of ethanol 80% could be dissolved completely with 15 mL gasoline. The composition of aqueous ethanol and gasoline in one phase was recorded at 92.11 and 7.89%. The composition in %w/w is shown in the right column as displayed in the table and in the text just described. The ethanol 81% formed a homogenous substance in which percentages of water, pure ethanol and gasoline were of 17.42, 74.25, and 8.33, while at 85% compositions were of 13.33, 75.56, and 11.11%. It was discovered that the higher ethanol concentration employed was the higher gasoline-ethanol used.

If ethanol purity was increased to 90% the volume needed of three components was changed widely. To form one phase was needed gasoline 10 mL, wet ethanol 30 mL, or water 3 mL or in percentage compositions were of 7.50% water, 67.50% pure ethanol, and 25.50% gasoline, meanwhile the compositions (in %w/w) were of 9.00, 67.00, and 24.00%. When ethanol increased slightly to 92%, the volume ratio of gasoline and aqueous ethanol was of 10:21 which were similar to compositions ratio 69:31 (w/w).

The compositions of water, pure ethanol and gasoline were recorded at 3.82, 50.73, and 45.45% if wet ethanol was of 93% that was equivalent to volume and weight ratios observed at 0.85:11.16:10 and 5.00:51.00:44.00. The weight and volume ratios changed to 4.00:45.00:51.00 (%w/w) and 0.54:8.46:10 when ethanol concentration went up to 94%. While the compositions were noted at 2.84:44.53:52.63 (%v/v).

**Table 1.** Compositions of water, ethanol, and gasoline (RON 88) in one phase substances

| Et Con (%) | Volume (mL) | | | | Composition (%) | | | | | |
|---|---|---|---|---|---|---|---|---|---|---|
| | | | | | (V/V) | | | (W/W) | | |
| | Gas | Aq.Et | Wat in Aq.E | Pure Et | Wat | Et | Gas | Wat | Et | Gas |
| 80 | 15 | 175 | 35.00 | 140.00 | 18.42 | 73.68 | 7.89 | 22.00 | 70.00 | 7.00 |
| 81 | 10 | 110 | 20.90 | 89.10 | 17.42 | 74.25 | 8.33 | 21.00 | 71.00 | 8.00 |
| 82 | 10 | 100 | 18.00 | 82.00 | 16.36 | 74.55 | 9.09 | 20.00 | 72.00 | 8.00 |
| 83 | 10 | 90 | 15.30 | 74.70 | 15.30 | 74.70 | 10.00 | 19.00 | 72.00 | 9.00 |
| 84 | 10 | 85 | 13.60 | 71.40 | 14.32 | 75.16 | 10.53 | 18.00 | 73.00 | 10.00 |
| 85 | 10 | 80 | 12.00 | 68.00 | 13.33 | 75.56 | 11.11 | 16.00 | 73.00 | 10.00 |
| 86 | 10 | 65 | 9.10 | 55.90 | 12.13 | 74.53 | 13.33 | 15.00 | 73.00 | 12.00 |
| 87 | 10 | 60 | 7.80 | 52.20 | 11.14 | 74.57 | 14.29 | 14.00 | 73.00 | 13.00 |
| 88 | 10 | 55 | 6.60 | 48.40 | 10.15 | 74.46 | 15.38 | 13.00 | 73.00 | 14.00 |
| 89 | 10 | 42 | 4.62 | 37.38 | 8.88 | 71.88 | 19.23 | 11.00 | 71.00 | 18.00 |
| 90 | 10 | 30 | 3.00 | 27.00 | 7.50 | 67.50 | 25.00 | 9.00 | 67.00 | 24.00 |
| 91 | 10 | 28 | 2.52 | 25.48 | 6.63 | 67.05 | 26.32 | 8.00 | 67.00 | 25.00 |
| 92 | 10 | 21 | 1.68 | 19.32 | 5.42 | 62.32 | 32.26 | 7.00 | 62.00 | 31.00 |
| 93 | 10 | 12 | 0.84 | 11.16 | 3.82 | 50.73 | 45.45 | 5.00 | 51.00 | 44.00 |
| 94 | 10 | 9 | 0.54 | 8.46 | 2.84 | 44.53 | 52.63 | 4.00 | 45.00 | 51.00 |
| 95 | 10 | 4 | 0.20 | 3.80 | 1.43 | 27.14 | 71.43 | 2.00 | 28.00 | 70.00 |
| 96 | 20 | 6 | 0.24 | 5.76 | 0.92 | 22.15 | 76.92 | 1.00 | 23.00 | 76.00 |

The ethanol 95 and 96% were obtained more difficult compared to that of below 94% since the column temperature should be controlled tightly. In the blending process, however, those are very easy to dissolve into gasoline. The ethanol needed for 95 and 96% were only 4 and 6 mL compared to gasoline recorded at 10 and 20 mL. The percentage of water, pure ethanol, and gasoline were of 1.43, 27.14 and 71.43 (%v/v) and 2.00, 28.00, and 70.00 (%w/w). The compositions varied when ethanol purity to 96% and was observed at 0.92, 22.15, and 76.92 and 1.00, 23.00, and 76.00.

Fig. 2 shows the triangular diagram of substances, water-ethanol-gasoline (RON 88) in one phase in which their compositions are assigned with solid circles and the data included in a graph were only compositional in %v/v. The solid circle as shown was the compositions of water, ethanol, and gasoline whereby they were just formed a soluble liquid. The region located on the right line (in front of the reader) was one phase whereby three components were in one phase. On the other hand, the region situated on the left line was two phases aqueous ethanol and gasoline was separated. It was found that the separation was gradually depending on ethanol composition.

The study showed the one phase fuel consisting of the polar (water/ethanol) and nonpolar (gasoline) substance was possible to be formed. The ethanol role is very important to connect between water and gasoline. As known the ethanol consists of two parts, R ($C_2H_5$-) and (-OH that could bond nonpolar in fossil fuel and polar branches of water.

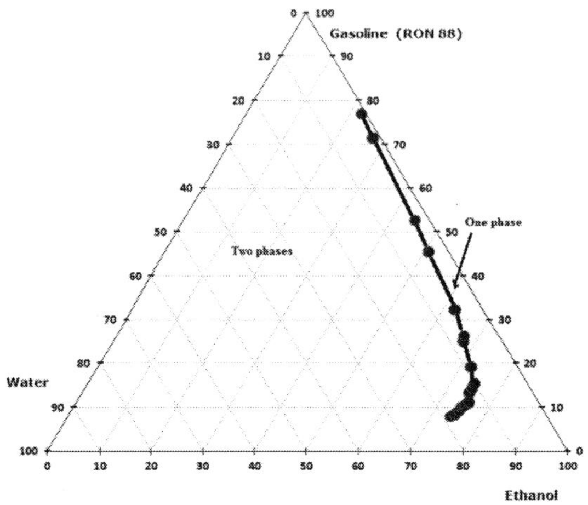


Figure 2. The triangular diagram of water-ethanol-gasoline (RON 88) in one phase in normal condition

For it has two parts as previously described, the ethanol could be functioning as a surfactant which was a bridge connecting gasoline and water. The bond is mainly coming from electrical interaction between positive- and negative charges. When the gasoline composition was larger than ethanol the components were separated since the bond strength could be weaker than the oil weight. The most important findings in this study are the composition is the critical point in the successful blending. The previous work which was comparable with the present study was carried to blend a gasoline and a technical ethanol 96% as reported by authors [18]. It was indicative that the aqueous ethanol could be blended with gasoline to form one phase substance as alternative energy (see Figure 2).

## 4. Conclusion

The aqueous one phase emulsion of the water, pure ethanol and gasoline (RON 88) were successfully blended in which ethanol was produced from *Arenga pinnata* sap. The palm juice was yeasted into a liquor which consisted of ethanol and water. The ethanol was separated using a reflux distillation and produced ethanol with concentration 80-96% depending on a column temperature. It was found that aqueous ethanol which still had water could dissolve easily with gasoline in a certain composition. The ethanol 80% was a minimum concentration to form one phase solution with gasoline. The data showed that the higher ethanol purity added into the mixture was the less ethanol needed to mix with gasoline forming one phase solution. It was discovered if ethanol composition increased continually after one phase formed the components were not separated into two phases.

## Acknowledgments

Authors wishing to acknowledge the Ministry of Higher Education Department of the Indonesian Government and Sam Ratulangi University for financial support through the PNBP program.

## References

[1] Gunawan S, Widjaja T, Zullaikah S, Ernawati L, Istianah N, Aparamarta H W and Prasetyoko D 2015 Effect of fermenting cassava with Lactobacillus plantarum, Saccharomyces cereviseae, and Rhizopus oryzae on the chemical composition of their flour *International Food Research Journal* **22** (3) 1280-1287

[2] Sangian H F, Kristian J, Rahma S, Agnesty S Y, Gunawan S and Widjaja A 2015 Comparative study of the preparation of reducing sugars hydrolyzed from high-lignin lignocellulose pretreated with ionic liquid, alkaline solution and their combination *Journal of Engineering and Technological Sciences* **47** (2) 137-148

[3] Sangian H F and Widjaja A 2017 Effect of Pretreatment Method on Structural Changes of Coconut Coir Dust *BioResources* **12** (4) 8030-8046

[4] Muharja M, Junianti F, Ranggina D, Nurtono T and Widjaja A 2018 An integrated green process: subcritical water, enzymatic hydrolysis, and fermentation, for biohydrogen production from coconut husk *Bioresource Technology* **249** 268-275

[5] Sangian H F, Sehe M R, Tamuntuan G H and Zulnazri Z 2018 Utilization of Saline Solutions in the Modification of Lignocellulose from Champaca Wood *J. Korean Wood Sci. Technol.* **46** (4) 368-379

[6] Muharja M, Umam D K, Pertiwi D, Zuhdan J, Nurtono T and Widjaja A 2019 Enhancement of sugar production from coconut husk based on the impact of the combination of surfactant-assisted subcritical water and enzymatic hydrolysis *Bioresource Technology* **274** 89-96

[7] Sangian H F, Aruan I, Tamuntuan G H, Bobanto M D, Sadjab B A, Purwadi R, Zulnazri Z, Masinambow V J and Gunawan S 2019 A Comparative Study of the Structures, Crystallinities, Miller Indices, Crystal Parameters, and Particle Sizes of Microwave-and Saline Water-Treated Cassava Starch *BioResources* **14** (4), 8212-8228

[8] Ebrahimi M, Caparanga A R and Ordono E E 2017 Evaluation of organosolv pretreatment on the enzymatic digestibility of coconut coir fibers and bioethanol production via simultaneous saccharification and fermentation *Renewable Energy* **109** (Supplement C) 41–48

[9] Ebrahimi M, Caparanga A R and Villaflores O B 2018 Weak base pretreatment on coconut coir fibers for ethanol production using a simultaneous saccharification and fermentation process *Biofuels* 2018 1-7

[10] Widjaja A, Agnesty SY, Sangian H F and Gunawan S 2015 Application of ionic liquid [DMIM] DMP pretreatment in the hydrolysis of sugarcane bagasse for biofuel production *Bulletin of Chemical Reaction Engineering & Catalysis* **10** (1) 70-77

[11] Pestaño L D B and Jose W I 2016 Production of Solid Fuel by Torrefaction Using Coconut Leaves as Renewable Biomass *International Journal of Renewable Energy Development* **5** (3) 187 -197

[12] Hoang T A and Le V V 2017 The Performance of A Diesel Engine Fueled With Diesel Oil, Biodiesel and Preheated Coconut Oil *International Journal of Renewable Energy Development* **6** (1) 1-7

[13] Qi D H, Chen H, Geng L M and Bian Y Z 2011 Effect of diethyl ether and ethanol additives on the combustion and emission characteristics of biodiesel-diesel blended fuel engine *Renewable Energy* **36** 1252-1258

[14] Ali O M, Mamat R, Abdullah N R and Abdullah A A 2016 Analysis of blended fuel properties and engine performance with palm biodiesel–diesel blended fuel *Renewable Energy* **86** 59-67

[15] Jia L W, Shen M Q, Wang J and Lin M Q 2005 Influence of ethanol–gasoline blended fuel on emission characteristics from a four-stroke motorcycle engine *Journal of Hazardous Materials* **123** (1-3) 29-34

[16] Chen R H, Chiang L B, Chen C N and Lin T H 2011 Cold-start emissions of an SI engine using ethanol–gasoline blended fuel *Applied Thermal Engineering* **31** (8-9) 1463-1467

[17] Sangian H F, Tamuntuan G H, Mosey H I, Suoth V and Manialup B H 2017 The utilization of arenga pinnata ethanol in preparing one phase-aqueous gasohol *ARPN Journal of Engineering and Applied Sciences* **12** (24) 7039-7046
[18] Murachman B, Pranantyo D and Putra E S 2014 Study of gasohol as alternative fuel for gasoline substitution: characteristics and performances *International Journal of Renewable Energy Development* **3** (3) 175-183

# IoT based Two Levels Feeding System for Koi Fish Pond

**F Lekswina[1] and D Widjaja[2,]***

Department of Electrical Engineering, Faculty of Sciences and Technology, Sanata Dharma University, Yogyakarta, Indonesia

*Corresponding author: damar@usd.ac.id

**Abstract.** Raising fish in a pond at home can be a hobby and add to the aesthetic value of the owner's residence. Sometimes people doesn't aware with the routine of feeding due to busy life that takes up a lot of time. The system that is designed in this work will help people who raise fish to monitor and control feeding activity and the number of pellets to be given remotely. The App Inventor is used to control and monitor an Internet of Things (IoT) based system. The system can display the pellets weight from the load cell sensor measurements on the container. The system is able to send notification if the feed weight is less than 300 grams. Users can choose four feeding settings, namely: S size, M size, L size, and customize. The servo motor will open the valve on the feeder and container containing pellets. The results of this work indicate that the two level feeding system for koi fish is working well. Sending and receiving data from and the webserver works well. The weight that is read by the load cell is displayed correctly.

## 1. Introduction

Raising fish in a pond at home can be a hobby or add to the aesthetic value of the residence owner. Sometimes people with a lot of daily routines ignore the food feeding activity to their fish. Feeding koi fish needs to be done carefully, especially in providing the portion or amount of the pellets to be given [1]-[5]. The impact of overfeeding causes koi fish release more manure pile. This will degrade the quality of the fish pond water.

Weku et al. (2015) designed Automatic Fish Feeding Device using Microcontroller [6]. This device will automatically feed the fish in certain time according programmed schedule with predetermined amount of pellets. When feeding process is done and the pellets container is empty, the device will send information via Short Message Service (SMS).

Another work has been done by Firdaus et al (2016), He made automatic fish feeding and pH controlling device [7]. This device is equipped with Liquid Crystal Display (LCD) to display feeding time schedule and pH index at that time.

In this research, a control device designed to use remote control through the IoT platform application on an Android based smartphone, which is App Inventor that is managed by MIT (Massachusetts Institute of Technology) [8]. This device allows fish feeding process is done according to predetermined standard portions regularly or customize timing. There are two pellets containers and the opening of each container outlet is controlled by using servo motors. The system provides 4 feeding options with different size of pellets that are size S, M, L, and customize size. One container has bigger size than the other so that it can be filled with more pellets. Therefore, this system is called 2 levels system.

When the small container (fish pond pellet feeder) run out of pellet, the bigger container will supply more pellets. Containers are equipped with a load cell sensor for measuring the weight of the pellets, as well as monitoring the amount of remaining available pellets. If the pellets in the big container less than 300 grams, sensor subsystem will send information to smartphone to warn the user anywhere and anytime.

## 2. System Model

System model that is used in this research is depicted in Figure 1. Arduino Mega is used as a microcontroller and connected with ESP8266 Wi-Fi module using serial communication to send and receive data [9]-[12]. The ESP8266 Wi-Fi module is an additional module in order to connect the sub system to the internet and send data on the clouds (web server). The user selects the pellets size via the application on the smartphone. Servo motor on the containers will move to open the outlets and release the pellets. Feeding time can be set from the application. The weight sensor will measure the pellets weight in the containers every time after pellets releasing.

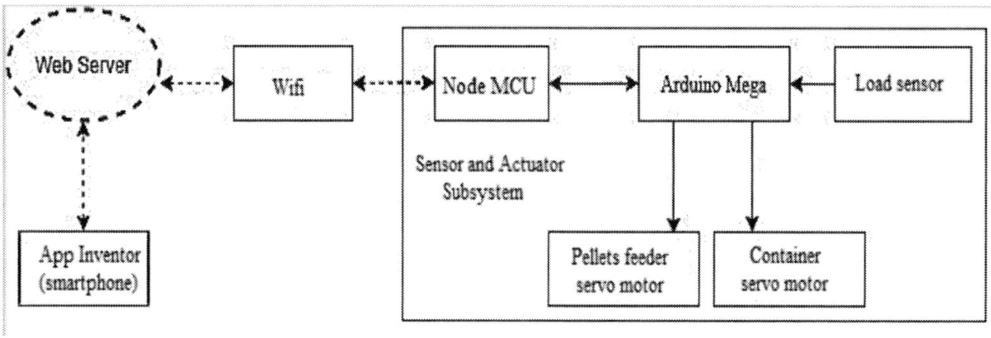


**Figure 1.** System Model

### 2.1. Activity Diagram

General flow diagram can be seen in Figure 2. The Figure 2.(a) shows the flowchart of the connection procedure and overall system work process. At the beginning of the process, the system will search for a Wi-Fi connection. If it is connected, then it will proceed to process monitoring and controlling procedure. The load cell sensor then will initialize the input port of microcontroller in figure 2.(b) and it will read the analogue data value of the sensor. Microcontroller which is connected to Wi-Fi, will send the data to App Inventor IoT platform webserver and display the data on the smartphone.

Figure 2(c) shows the controlling procedure of the automatic feeding system. First, user selects the pellets size from the App Inventor interface on the smartphone. If small size pellets (Size S) is chosen, then the interface will give 2 options, whether user can do the feeding process automatically or manually. If user selects automatic process, the system will give command to feeder and container outlets to open for 3 seconds every 12 hours.

Every time the feeder gives pellets to the pond for certain time period, it will also be given the same amount of pellets from the container. The procedure will have the same process for size M and size L, except that the opening time of feeder and container outlets become 5 seconds and 7 seconds. If user selects manual procedure, then every time user selects this option, system will do the feeding procedure without waiting for 12 hours.

Different procedure will be done if the user chooses "customize" option. User will not select the size of the pellets and the opening time of feeder and container outlets. However, user needs to determine how many fishes that has to be fed. According to this number of fish, system will automatically determine the opening time of the outlets.

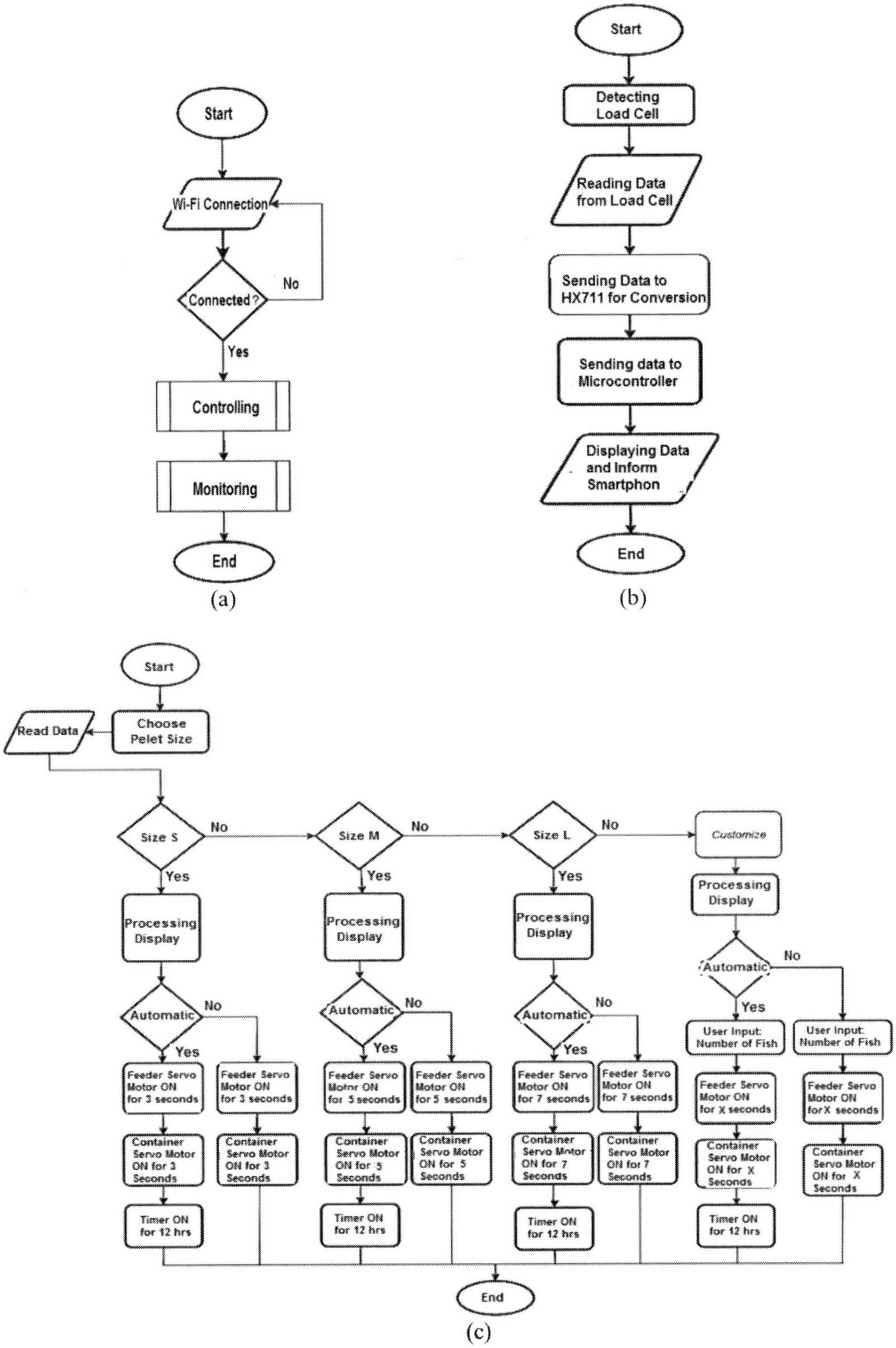


**Figure 2.** System Flowchart
(a). Connection procedure. (b). Monitoring procedure. (c) Controlling procedure.

IOP Conf. Series: Materials Science and Engineering 1115 (2021) 012051 doi:10.1088/1757-899X/1115/1/012051

Figure 3 shows the model construction of this two level feeding system. There are two containers. The lower container is pellets feeder and the outlets pipe directly leads the pellets to the fish pond. The upper container is the backup container which has bigger size and can be filled with larger amount of pellets. When the feeder takes out the pellets, the big container automatically supply more pellets to the feeder.

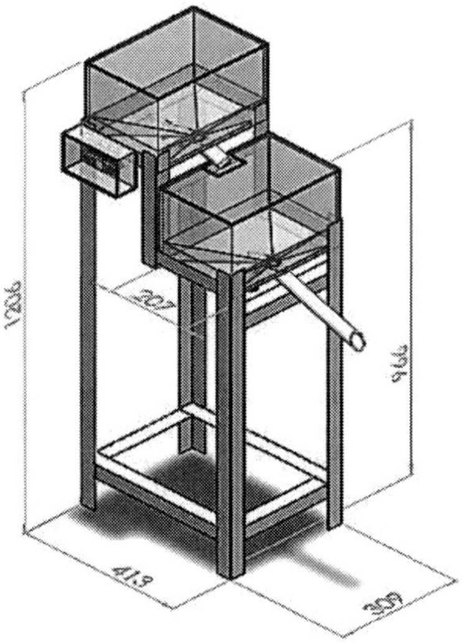

**Figure 3.** Design of the Hardware Model

## 3. Results and Discussion

The test results are shown in Table 1 and Table 2. Tests are done for S size, M size, L size, and customize size with manual and automatic settings. It can be seen that the two servo motors can open the containers outlets according to the setting in application on the smartphone. The number of pellets in Table 1 increase from around 50 grams, 70 grams, to 100 grams when the size of pellets increase from S, M, to L. It also increases according to the number of fish in customize option.

**Table 1.** Manual Setting

| | Trial | Servo feeder | Servo container | Pellets weight (gram) | Note |
|---|---|---|---|---|---|
| S | 1 | Open | Open | 46 | - |
| | 2 | Open | Open | 51 | - |
| | 3 | Open | Open | 50 | |
| | 4 | Open | Open | 47 | - |
| M | 1 | Open | Open | 77 | - |
| | 2 | Open | Open | 74 | - |
| | 3 | Open | Open | 72 | |
| | 4 | Open | Open | 75 | - |
| L | 1 | Open | Open | 101 | - |
| | 2 | Open | Open | 103 | - |
| | 3 | Open | Open | 101 | |
| | 4 | Open | Open | 102 | - |
| Customize | 1 | Open | Open | 30 | 1 fish |
| | 2 | Open | Open | 89 | 6 fish |
| | 3 | Open | Open | 147 | 10 fish |
| | 4 | Open | Open | 228 | 15 fish |

When automatic setting is selected as shown in Figure 3, the same condition is also apply. Time setting for automatic feeding is 12 hours. However, data in Figure 3 is collected once in a day, particularly in midnight time.

**Table 2.** Automatic Setting

| | Trial | Time-1 | Servo feeder | Servo container | Pellets weight time-1 (gram) | Time-2 | Servo feeder | Servo container | Pellets weight time-2 (gram) | Note |
|---|---|---|---|---|---|---|---|---|---|---|
| S | 1 | 23. 50 | Open | Open | 47 | 23.55 | Open | Open | 48 | - |
| | 2 | 24.34 | Open | Open | 46 | 24.38 | Open | Open | 54 | - |
| | 3 | 1.10 | Open | Open | 51 | 1.15 | Open | Open | 51 | - |
| | 4 | 1.45 | Open | Open | 54 | 1.50 | Open | Open | 50 | - |
| M | 1 | 24.05 | Open | Open | 75 | 24.1 | Open | Open | 74 | - |
| | 2 | 24.45 | Open | Open | 72 | 24.50 | Open | Open | 76 | - |
| | 3 | 1.20 | Open | Open | 76 | 1.25 | Open | Open | 73 | - |
| | 4 | 1.55 | Open | Open | 72 | 1.59 | Open | Open | 72 | - |
| L | 1 | 24.15 | Open | Open | 102 | 24.20 | Open | Open | 101 | - |
| | 2 | 24.55 | Open | Open | 103 | 24.59 | Open | Open | 102 | - |
| | 3 | 1.30 | Open | Open | 102 | 1.35 | Open | Open | 104 | - |
| | 4 | 2.50 | Open | Open | 101 | 2.10 | Open | Open | 101 | - |
| Customize | 1 | 2.25 | Open | Open | 29 | 2.30 | Open | Open | 30 | 1 fish |
| | 2 | 2.33 | Open | Open | 88 | 2.38 | Open | Open | 87 | 6 fish |
| | 3 | 2.45 | Open | Open | 146 | 2.50 | Open | Open | 147 | 10 fish |
| | 4 | 3.05 | Open | Open | 227 | 3.10 | Open | Open | 229 | 15 fish |

Load cell sensor testing can be seen in Table 3. Based on the test results, the load cell sensor can work properly. The App Inventor as an IoT platform also work well as it was designed. When the pellets weight is < 300 grams, a notification in smartphone will be "Pellets is running low", and if the pellets weight is > 300, notification is "Pellets is available".

**Table 3.** Load Cell Testing

| Weight reading | | Picture | | Error (%) |
|---|---|---|---|---|
| Scale (gram) | Load Cell (gram) | Scale | Load Cell | |
| 145 | 147 | 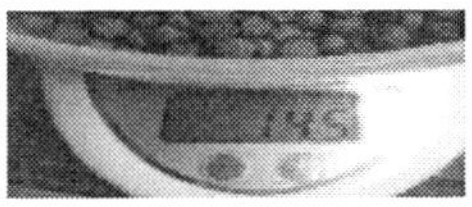 | 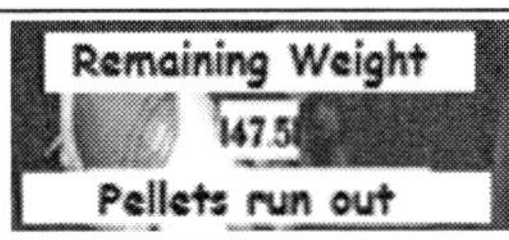 | 1.379 |
| 249 | 250 |  | 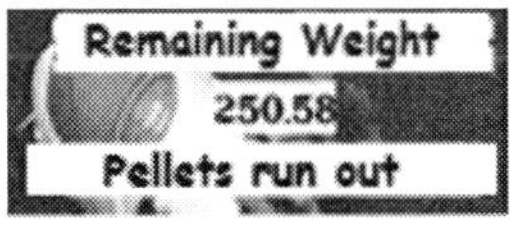 | 0.402 |
| 582 | 584 | 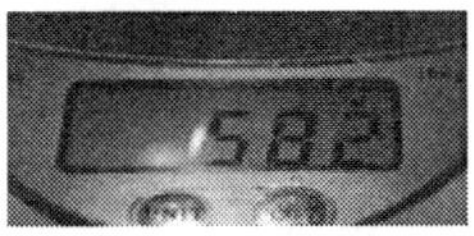 | 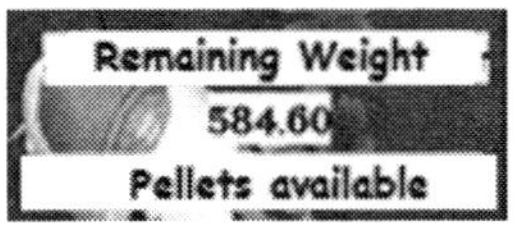 | 0.344 |
| 850 | 853 | 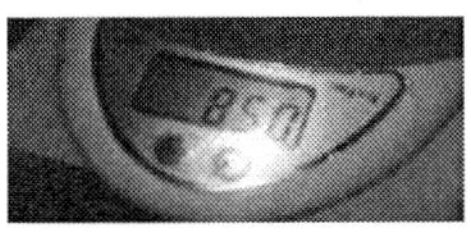 | 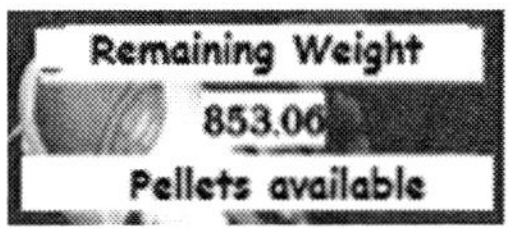 | 0.353 |
| 1534 | 1537 | 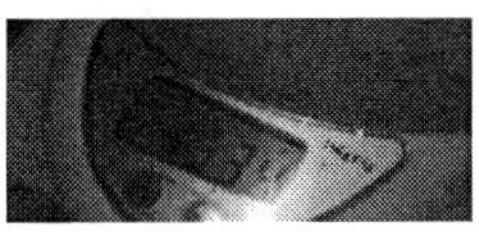 | 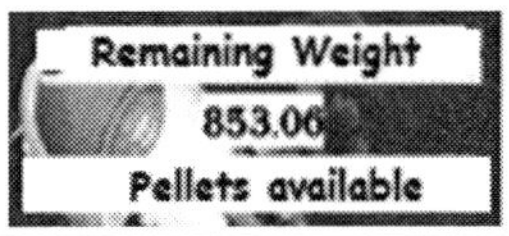 | 0.196 |

## 4. Conclussion

The system works well as it was designed. Systems can send and receive data to and from App Inventor, as IoT platform, correctly. The amount of pellets that are fed to the pond according to the pellet size or the number of fish selected. The weight of remaining pellets in the feeder and container can be read and display on the smartphone correctly.

## Acknowledgments

Thanks to the Institute of Research and Community Service (Lembaga Penelitian dan Pengabdian kepada Masyarakat - LPPM) Sanata Dharma University for funding and facilitating this research.

## References

[1] Bhaskar P, Pyne S K and Ray A K 2015 Growth performance study of Koi fish, Anabas testudineus (Bloch) by utilization of poultry viscera, as a potential fish feed ingredient, replacing fishmeal *International Journal of Recycling of Organic Waste in Agriculture* **4** (1) 31-37

[2] Kusku H, Ergun S, Yilmaz S, Guroy B and Yigit M 2019 Impacts of urban noise and musical stimuli on growth performance and feed utilization of koi fish (Cyprinus carpio) in recirculating water conditions *Turkish Journal of Fisheries and Aquatic Sciences* **19** (6) 513-523

[3] Hancz C, Magyary I, Molnar T, Sato S, Horn P and Taniguchi N 2003 Evaluation of color intensity enhanced by paprika as feed additive in goldfish and koi carp using computer-assisted image analysis *Fisheries science* **69** (6) 1158-1161

[4] Sarker M A A, Arshad F M, Alam M F, Mohamed Z A and Khan M A 2016 Feed technology and production performance of Thai koi (Anabas testudeneus) in Bangladesh *Journal of Applied Aquaculture* **28** (2) 76-91

[5] Sudirman D A, Arief M and Fasya A H 2020 Addition different algae (Spirulina) flour to artificial feed on color quality and growth of Koi fish (Cyprinus carpio-Koi). *IOP Conference Series: Earth and Environmental Science* **441** (1) 012086

[6] Weku H S, Poekoel V C and Robot R F 2015 Rancang Bangun Alat Pemberi Pakan Ikan Otomatis Berbasis Mikrokontroler *Jurnal Teknik Elektro dan Komputer* **4** (7) 54-64

[7] Firdaus B A, Kridalukmana R and Widianto E D 2016 Pembuatan Alat Pemberi Pakan Ikan Dan Pengontrol PH Otomatis *Jurnal Teknologi dan Sistem Komputer* **4** (1) 133-138

[8] Aris A E 2016 *Belajar dan Bermain dengan MIT App Inventor* Surabaya: MB Dev

[9] Kuncoro A H, Mellyanawaty M, Sambas A, Maulana D S, Subiyanto and Mamat M 2020 Air Quality Monitoring System in the City of Tasikmalaya based on the Internet of Things (IoT) *Jour of Adv Research in Dynamical & Control Systems* **12** (2) 2473-2479

[10] Lee C and Wang Y J 2020 Development of a cloud-based IoT monitoring system for Fish metabolism and activity in aquaponics *Aquacultural Engineering* **90** 102067

[11] Jiangbin Z, Rui C, Hao Z, Lei Z and Chunyan M 2018 Intelligent Fish Tank Control System Based on Internet of Things Cloud Computing Platform *Proceedings of the 2nd International Symposium on Computer Science and Intelligent Control* 1-6

[12] Ma Y and Ding W 2018 Design of intelligent monitoring system for aquaculture water dissolved oxygen *2018 IEEE 3rd Advanced Information Technology, Electronic and Automation Control Conference* 414-418

IOP Conf. Series: Materials Science and Engineering 1115 (2021) 012052 doi:10.1088/1757-899X/1115/1/012052

# Study of SEM, XRD, TGA, and DSC of Cassava Bioplastics Catalyzed by Ethanol

**H F Sangian[1*], E Maneking[1], S H J Tongkukut[1], H I R Mosey[1], V Suoth[1], H Kolibu[1], A Tanauma[1], G Pasau[1], A As'ari[1], V A J Masinambow[2], B A Sadjab[3], M M Sangi[4] and S B Rondonuwu[5]**

[1]Department of Physics, Sam Ratulangi University, 95115 Manado, Indonesia
[2]Department of Economic Development, Sam Ratulangi University, 95115 Manado, Indonesia
[3]Department of Physics, Halmahera University, 97762 Tobelo, Indonesia
[4]Department of Chemistry, Sam Ratulangi University, Manado, Indonesia
[5]Department of Biology, Sam Ratulangi University, Manado, Indonesia

[*]Corresponding author: hannysangian@yahoo.co.id

**Abstract.** This study aims at preparing and characterizing of bioplastics that utilized cassava biomass which was available abundantly in Indonesia. The bioplastic synthesis was conducted with two variations of the reactant, namely cassava + starch + glycerol + water + acetic acid (vinegar) and cassava + starch + glycerol + water + acetic acid + ethanol. Bioplastic obtained was characterized by employing scanning electron microscopic (SEM), X-Ray diffraction (XRD), Fourier transform infra-red (FTIR), thermogravimetric analysis (TGA), and differential scanning calorimetric (DSC). The results show that bioplastics with the second combination have a high degree of degradation, whereby they were consistent with XRD analysis, appearing a low crystallinity value. The functional groups have shown IR spectra presented the existence of C-H alkanes, C = O esters, and C-H alkene groups. While the surface morphology displayed a flat surface, which was relatively comparable to both samples and the reduction of sample mass during heating was at 2.32 mg.

## 1. Introduction

Plastic is one type of material formed by the polymerization process, which is the process of combining several simple molecules (monomers) through a chemical process into more complex molecules [1]. This material is widely used in modern society for has several advantages such as durable, lightweight, waterproof, and the price is relatively low. Every year around 265,000,000 tons of plastic are produced and used worldwide for industrial and household [2]. However, plastic currently that massively is utilized a synthetic polymer which cannot be decomposed in a short time, causing environmental problems. Plastic waste tends to create a buildup of soil pollution, causing land degradation, environmental damage, and floods [3].

Recently, scientists have been giving more attention to the development of renewable resources, such as starch, lignocellulose, and ethanol [4-7]. Those materials could be improved, becoming the bioplastics that can be decomposed by microorganisms [8]. The development of bioplastics has focused on starch, which is available abundantly in nature [9]. The potential for bioplastic development is immense because the raw material is available in the environment. The bioplastics have been reported by authors in which they were processed from great range sources, such as cassava [10].

The present study utilized the cassava (*Manihot esculenta*) as the raw material with some considerations. Indonesia is the third-largest cassava producing country in the world with production capacity in 2014, reaching 26 million tons, meanwhile based on the FAO report in 2017, the world capacity attained at 278.0 Million Tons. The huge cassava is an excellent opportunity to increase its value, becoming bioplastic.

As described previously, that bioplastics had been prepared from many biomass. It was known generally that bioplastic was obtained by reacting reactants, cassava, glycerin, which is enhanced by a catalyst. Meanwhile, the properties and improvement of bioplastic also were reported by many authors. The deficiencies of this material from starch were low mechanical properties (tensile strength, strain, and young modulus) and hydrophilic nature. The increase of bioplastic quality was the addition of biopolymers such as glycerol and sorbitol. The addition of glycerol provides a higher solubility in plastic films than sorbitol. Bioplastics made from cassava starch and glycerol plasticizers are transparent, bright, homogeneous, flexible, and easy to handle [11].

Based on the reference survey as described above, the preparation and characterization of plastic films synthesized with a combination of cassava starch + glycerol + water + acetic acid and cassava starch + glycerol + water + acetic acid + alcohol are needed to be studied and analyzed. Bioplastics were characterized by using instruments. SEM was to observe the surface morphology; XRD was to determine the crystallinity; FTIR was to examine the functional groups formed; TGA and DSC were to identify the thermal properties of plastic films.

## 2. Materials and Methods

### 2.1 Chemicals

The ethanol employed in this work was fermented from Arenga pinnata sap and distilled using reflux distillation. Glycerol was purchased from a drug store and produced by One Med Health Care, Jayamas Medica Industry, Indonesia. Acetic acid used an industrial-grade with purity was 99.8%, and the solvent water was obtained through simple distillation. The cassava was obtained from a farmer in Minahasa Regency and then was milled until particle sizes were ranged 100 – 120 mesh. Before using, the cassava was dried under sunlight until the weight was constant in which the water content was reduced significantly.

### 2.2 Instruments

The tools used were an electrical stove equipped by magnetic stirring, aluminum pan 100mL. Apparatuses XRD and FTIR used were the Analytical X'pert Pro (Almelo, Netherlands) and Shimadzu PRESTIGE 21 (Tokyo, Japan), respectively.

SEM measured the characterization of surface morphology with specification FEI Inspect S50 SEM (Tokyo, Japan), which is available at Central Laboratory, Universitas Negeri Malang (UM), East Java, Indonesia.

The XRD procedures followed the steps as described below. The degree measurements were set from $2\theta = 10.01$ until 89.9° with beam wavelengths were $K\alpha$ ($\lambda = 1.54$ Å) and $K\beta$ ($\lambda = 1.39$ Å) at 25 °C. The operating current and potential differences were set at 35 mA and 40 kV, respectively. The functional vibrations of the substrates were detected by instrument FTIR, whose frequencies were in the range of 400 to 4000 $cm^{-1}$.

The SEM pictures were taken employing an electrical potential for accelerating operating at 15 kV, and picture magnifications were set at 5000x. The working distance (WD) and the spot calibration were 10.8 mm, and 5.5, respectively.

### 2.3 Procedures

The work was conducted in two phases, such as the manufacturing of bioplastic films and characterizing the sample. The preparation employed two combinations: The 50 grams of cassava starch, 25 ml of glycerol, 25 ml of acetic acid (vinegar), and 50 ml of water; the 50 grams of cassava starch, 25 ml of glycerol, 25 ml of acetic acid (vinegar), 50 ml of water and the addition of 25 ml of alcohol.

The ingredients were mixed inside a flask equipped by a reflux column and heated at 80 °C under gentle stirring. The heating and stirring are done until the liquid thickened and resembled as glue. After that, the mixture was cooled down for minutes and printed on a wooden board laid on the aluminum foil. The biodegradable plastic mold was dried under sunlight for hours. After drying, the plastic was removed from the frame; then, the process was repeated for variation 2.

For characterization purposes, the film was cut into small pieces with a size of 2 × 2 $cm^2$. The characteristics of films were analyzed using instruments XRD, SEM, FTIR, TGA, and DSC. The measurement employing TGA and DSC used the device manufactured by Toledo Mettler at Energy Laboratory, ITS, in Surabaya.

The crystallinity value (Cr) was calculated using the Herman method, which was followed from the previous report [12] as presented in Eq.(1) and employing the Origin Lab and Microsoft Excel software. Firstly, the value of high intensity was marked (for crystal region), and the second, the amorphous intensity was obtained by searching for the crystal intensity and was marked by the valley before the crystal region. To determine the crystallinity applying equation (1) as follows

$$Cr = 100 \frac{I_{crystal}}{I_{amorphous+} \quad Crystal} \% \qquad (1)$$

## 3. Results and Discussion

### 3.1 Surface Morphology

The bioplastics grown was a printed sheet whose properties were a light brown, transparent, and elastic, as shown in Figure 1. The investigation of surface morphology was carried out at determining the surface structure, cracks, and surface smoothness. The study was intended to study a comparison of two combinations, namely cassava starch +glycerol+water+acetic acid and cassava starch+glycerol+water+acetic acid+ alcohol. Figure 2 show the surface morphology of the samples grown employing two combinations cassava starch+glycerol+water+acetic acid (**A**), cassava starch+glycerol+water+acetic acid+ethanol (**B**) and images are magnified 50,000x.

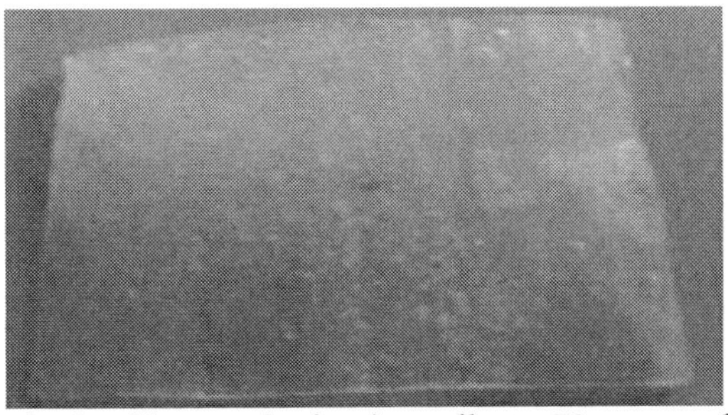

**Figure 1.** The bioplastic obtained by employing ingredients 50 grams of cassava starch, 25 ml of glycerol, 25 ml of acetic acid (vinegar), 50 ml of water, and 25 ml of ethanol

The image showed that the surface structures were fractured and cracked both samples, but the ethanol addition appeared smoother compared to the first variation. The cracks might be caused by the starch bonds α-1,4-glucosidic connection, which was more amorphous. Another reason could be explained that the size of amylopectin was large in which the particles were not dense in the starch matrix. The addition of ethanol influenced surface morphology, which was more compact and homogeny.

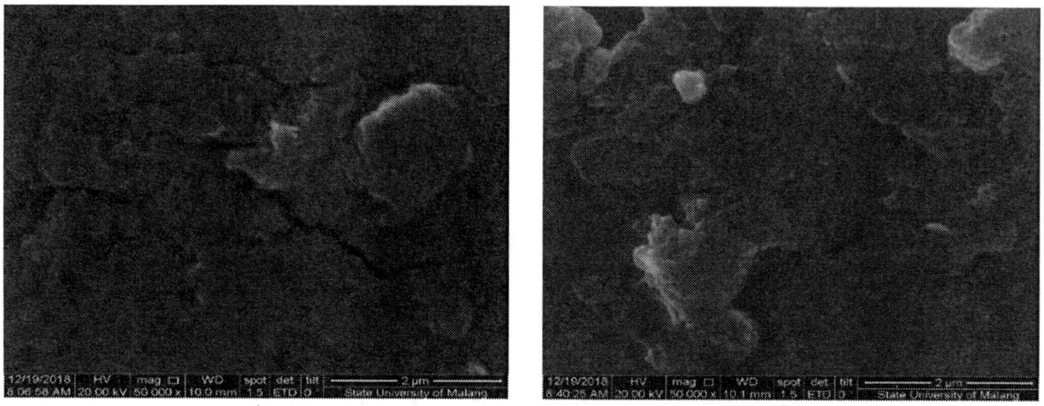

A B

**Figure 2.** The surface morphology of bioplastic prepared by using first combination cassava starch+glycerol+water+acetic acid (A); cassava starch+glycerol+water+acetic acid+ethanol (B) and the images were magnified at 50000x

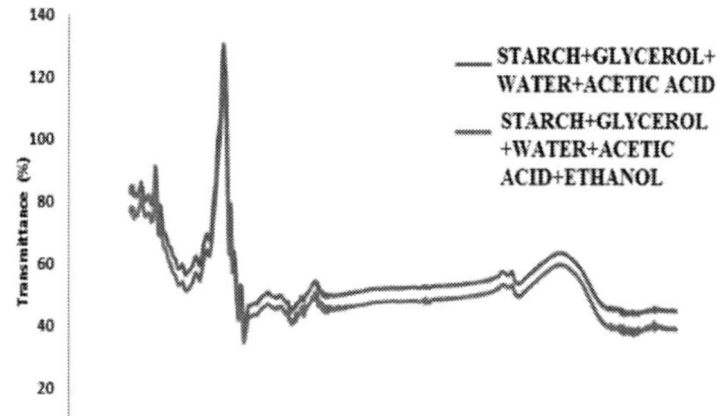


**Figure 3.** The fingerprint of the bioplastics reacted from cassava starch+glycerol+water+acetic acid (red) and cassava starch+glycerol+water+acetic acid+ethanol (blue)

### 3.2 Functional Groups Analysis

The FTIR (fingerprint) is a characterization based on the vibrations of atoms or molecules by passing infrared radiation through a material or sample in which the energy corresponds to the wave frequency. The absorbance bands of the functional groups in the material are used to identify the substance properties. Figure 3 is the FTIR spectra of the bioplastics prepared by reacting ingredients cassava starch+glycerol+water+acetic acid (red) and cassava starch+glycerol+water+acetic acid+ethanol (blue).

Both fingerprints showed similar peaks and valley, but the transmittances or absorbances were significantly different. Generally, the transmittance of bioplastic derived from the second variation was higher compared to that of the first sample. It was indicative that the first substrate which was prepared from in which ethanol did not exist was a different structure with the second sample. It was also supported by SEM images and XRD spectra, as described in the next section.

The spectra showed that the wave absorbed profoundly in the range of 675-995 $cm^{-1}$ and 2966.52 $cm^{-1}$ positioned in the variety of 2850 - 2970 $cm^{-1}$ belonged to the C-H alkene groups. According to work conducted by [13] showed the C-C, C-O, C-O (ester), and C-O-H (carbonic acid) bonds appeared at energy bands between 800 and 1300 $cm^{-1}$ which was relatively comparable with present spectra in the range of 1050 - 1300 $cm^{-1}$. The absorbance whose wavelengths were of 1000 - 300 $cm^{-1}$ were typical of the C-O (ester group), which was an indication of the ability to decompose. The bonds also are hydrophilic groups whereby water can cause microorganisms entering the bioplastic matrix for decomposing [14].

### 3.3 Crystallinity

The XRD was employed to investigate the crystalline structures and particle size of samples. Fig. 4 is the XRD spectra of bioplastics grown from combining of cassava starch+glycerol+water+acetic acid (red) and cassava starch+glycerol+water+acetic acid+ethanol (blue). The XRD patterns of each sample, as shown, are the characteristics of bioplastic and have peaks, which are known as a crystalline region [15].

The vivid peak of the first combination was at 22.30° and shifted slightly to 22.27° of the second variation in which ethanol was added. Meanwhile, the valley of both samples was located at 12.5°. The present result has a similarity to the previous study about lignocellulose crystallinity, as published by researchers [16].

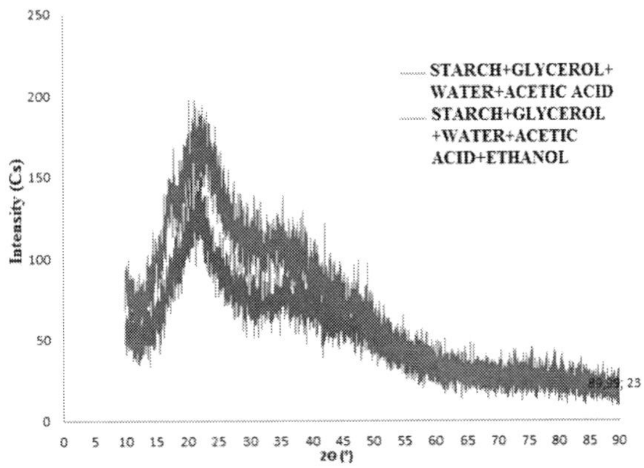


**Figure 4**. The XRD spectra of bioplastics obtained from reacting of cassava starch+glycerol+water+acetic acid (red) and cassava starch+glycerol+ water+aceticacid+ethanol (blue)

Table 1 presents the crystallinity values of substrates grown from cassava starch + glycerol + water + acetic acid (Bioplastic 1) and cassava starch + glycerol + water + acetic acid + ethanol (Bioplastic 2). Based on the results obtained, that crystallinity of the bioplastic employing Eq. (1) was of 0.67, which was higher than that of the second bioplastic observed at 0.63. Even though the surface morphology of the second bioplastic characterized by SEM was smoother than the first substrate, its crystallinity was less compared to that of the first bioplastic as resulted. For the second bioplastic was more amorphous than the first one, so it is better to be decomposed.

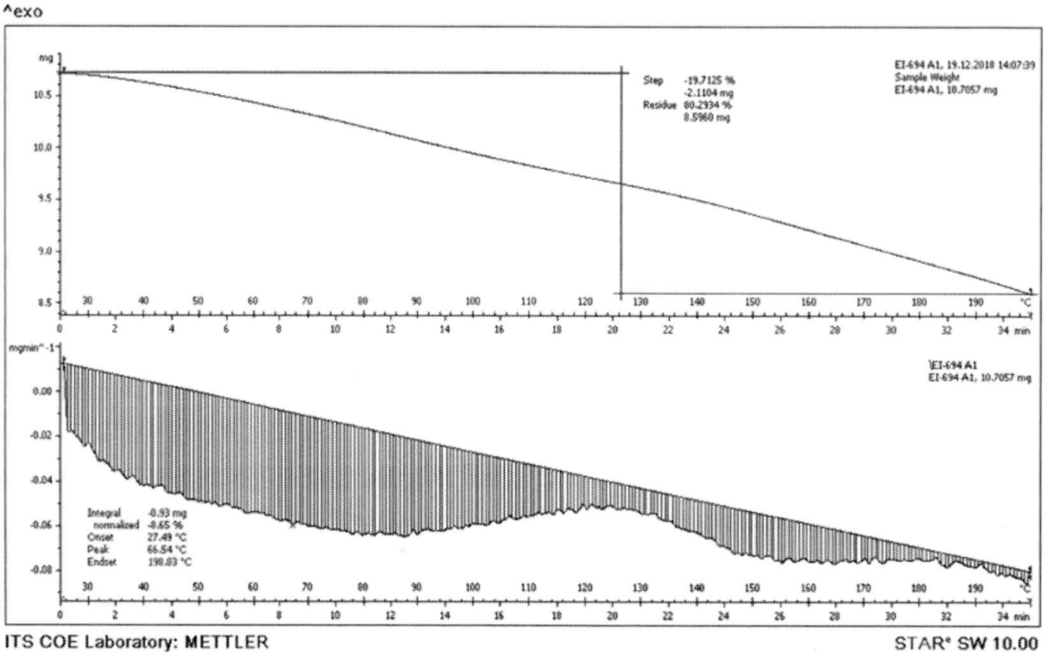


**(A)**

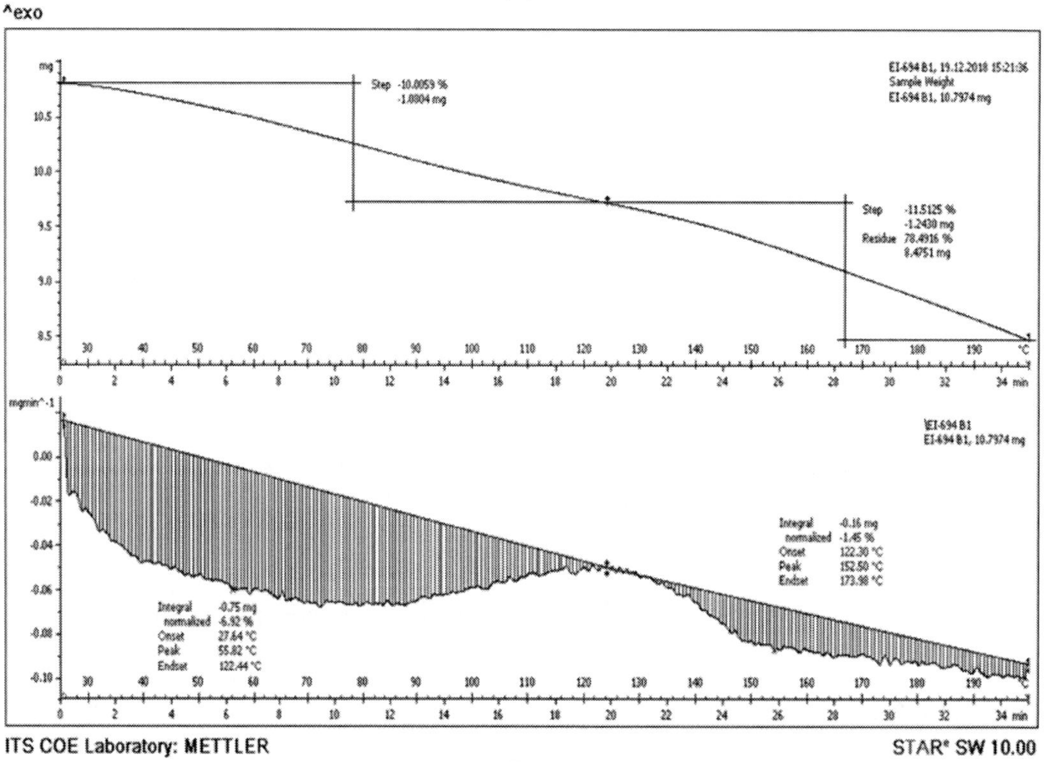


**(B)**

**Figure 5.** The TGA diagram of the bioplastic derived from cassava starch+glycerol+water+acetic acid (A) and cassava starch+glycerol+water+acetic acid+ethanol (B)

Table 1. The crystallinity values of bioplastics prepared from cassava starch + glycerol + water + acetic acid (Bioplastic 1) and cassava starch + glycerol + water + acetic acid + ethanol (Bioplastic 2)

| Substrate | Crystal intensity (Cs) | Amorphous intensity (Cs) | Crystallinity |
|---|---|---|---|
| Bioplastic 1 | 134 | 67 | 0.67 |
| Bioplastic 2 | 182 | 108 | 0.63 |

### 3.4 Thermo Gravimeter Analysis (TGA)

To investigate the thermal property of bioplastic concerning temperature was measured by employing the methods *Thermo Gravimeter Analysis* (TGA) and *Differential Scanning Calorimetric* (DSC).

This work analyzed the change of weight of both substrates due to the influence of temperature on the material [17]. The 10 mg each sample was heated in a temperature range from 27 °C to 190 °C at a rate of 10 °C / min, as presented in Figures 5a and 5b.

When the heat was supplied into the bioplastic, the temperature increased as time inclined. It was discovered that the weight of bioplastic decreased as the temperature went up, which was comparable to the investigation of the authors.

The first substrate occurred the thermal decomposition at 27.49 - 122.44 °C, and the mass decreased by around 19.71%, which was similar to 2.11 mg. Meanwhile, the mass decline of the second sample with the range of temperature 27.67 - 173.98 °C was 21.51%, which was of 2.32 mg. These results were comparable to those that were obtained from XRD in which the second bioplastic was more amorphous than the first one.

### 3.5 Differential Scanning Calorimetric (DSC)

The DSC method is at analyzing the caloric value required entering the material as temperature inclines. Figure 5 shows the DSC diagram of the bioplastic derived from cassava starch + glycerol + water + acetic acid (A) and cassava starch + glycerol + water + acetic acid + ethanol (B). The diagrams give information about critical temperatures. Both diagrams presented that the bioplastics prepared were semicrystalline forms since many peaks were appearing do the glass transition was difficult to be determined. The glass transitions (Tg) the first- and the second substrate could be relatively similar at 62 °C. The second diagram representing the second sample displays more amorphous compared to the first one, whereby it is comparable with the XRD spectra. The temperatures which are points to valleys and peaks correlate to the crystallization and melting points of the material. The diagrams showed that the first bioplastic was decomposed with temperatures which are higher than 185 °C, while the second was destroyed above 195 °C. The first material showed the melting point was at 175 °C; meanwhile, the second sample was at 180 °C, respectively.

The samples, as shown in the diagram, were occurred multiple phase transformations in which it could be seen from peaks and valleys formed. The least valley gave information about crystallization, while the peaks showed that the material transformed from solid to liquid and finally was destroyed at high temperatures. The findings verified the investigations on the bioplastics, as reported previously [18-22]. The present work is an alternative technique to improve the quality of bioplastics. It is crucial to change the synthetic plastics which have been utilized by people for a hundred years. It is possible since the raw material; cassava can be planted in most tropical countries. The urgent action now is to overcome the damage of the environment because of plastics, especially oceans. Much big fish had been found dead whose bellies contain plastics. The most problematic to develop bioplastic is the price of cassava, which is higher than fossil oils, and the properties of bioplastics are not fulfilling the requirements of a daily application for people.

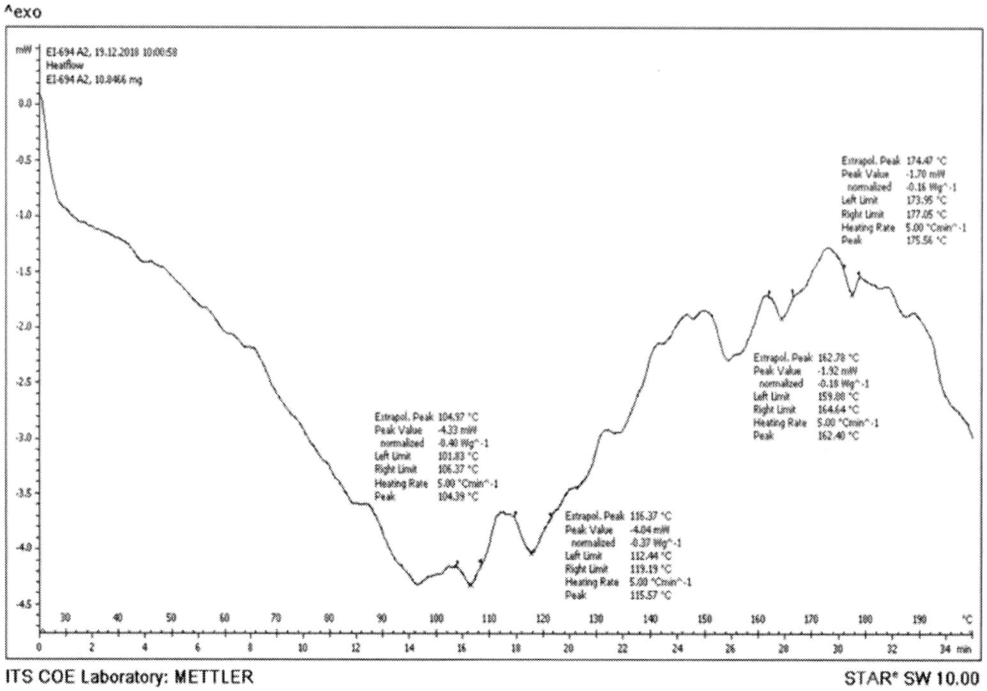


(A)

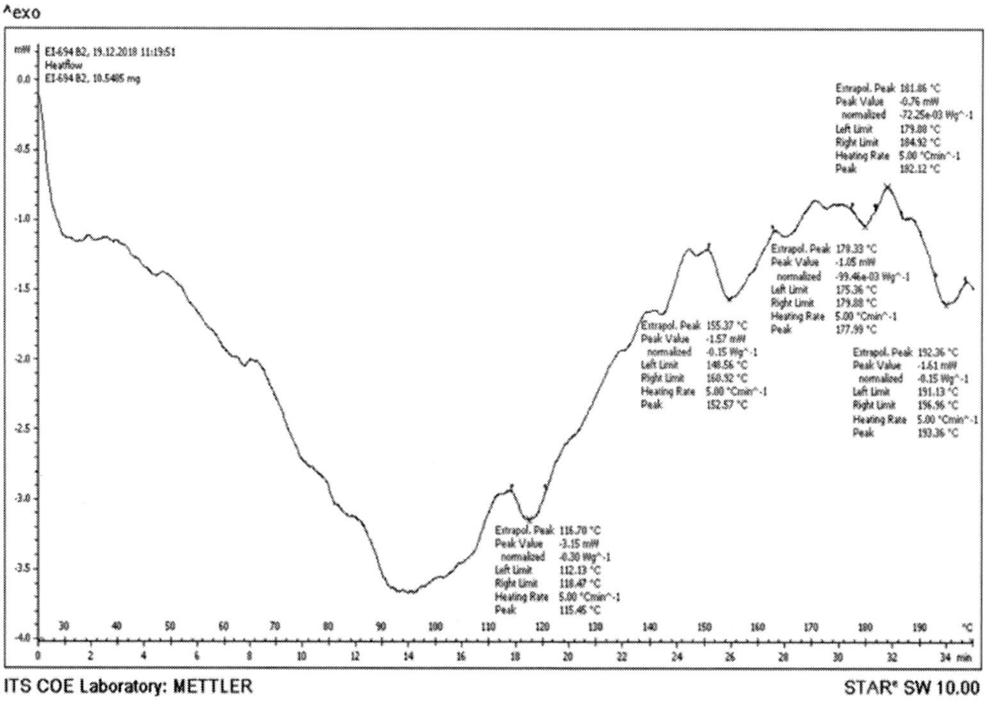


(B)

**Figure 6.** The DSC diagram of the bioplastic derived from cassava starch+glycerol+water+acetic acid (A) and cassava starch+glycerol+water+acetic acid+ethanol (B)

## 4. Conclusion

The bioplastics which were prepared from a combination of cassava starch+glycerol+water+acetic acid and cassava starch+glycerol+water+acetic acid+ethanol were synthesized successfully. The SEM images showed that the surface morphology of the second bioplastic was smoother than that of the first one. The XRD presented that the bioplastic employing the second combination was more amorphous in which TGA and DSC measurements supported the claim. The Cr value of the first sample was 0.66, while the second was recorded at 0.63. The biodegradable material was established since the existence of the functional ester, which was easily trimmed by micro-organisms. The bioplastic obtained was semi-crystalline structure and was occurred multiple phase changes.

## Acknowledgments

The authors would like to acknowledge the Higher Education Department of the Indonesian Government and Sam Ratulangi University (UNSRAT Manado) for financial support. We thank Professor Benny Pinontoan as a Dean of Fac. Mathematics and Sciences who provided the facility and instrument and assisted the authors in work.

## References

[1] Kumar A A, Karthick K and Arumugam K P 2011 Properties of biodegradable polymers and degradation for sustainable development. *International Journal of Chemical Engineering and Applications* **2** 164-167
[2] Kipngetich T E and Hillary M 2013 A blend of green algae and sweet potato starch as a potential source of bioplastic production and its significance to the polymer industry *International Journal of Green and Herbal Chemistry* **2** (1) 15-19
[3] Davis G and Song J H 2006 Biodegradable packaging based on raw materials from crops and their impact on waste management. *Industrial crops and products* **23** 147-161
[4] Sangian H F, Aruan I, Tamuntuan G H, Bobanto M D, Sadjab B A, Purwadi R, Zulnazri Z, Masinambow V J and Gunawan G 2019 A Comparative Study of the Structures, Crystallinities, Miller Indices, Crystal Parameters, and Particle Sizes of Microwave-and Saline Water-Treated Cassava Starch *BioResources* **14** (4) 8212-8228
[5] Widjaja A, Agnesty S Y, Sangian H F and Gunawan S 2015 Application of ionic liquid [DMIM] DMP pretreatment in the hydrolysis of sugarcane bagasse for biofuel production. *Bulletin of Chemical Reaction Engineering & Catalysis* **10** (1) 70
[6] Sangian H F, Telleng R, Aruan I ,Mosey H I R and Tamuntuan G H 2018 The structural modification of cassava starch using a saline water pretreatment *Food Science and Technology* **38** 215-220
[7] Sangian H F, Tamuntuan G H, Mosey H I R, Suoth V and Manialup B H 2017 The utilization of arenga pinnata ethanol in preparing one phase-aqueous gasohol ARPN Journal of Engineering and Applied Sciences **12** (24) 7039-7046
[8] Bourtoom T 2008 Edible films and coatings: characteristics and properties *International food research journal* **15** (3) 237-248
[9] Liu L, Kerry J F and Kerry J P 2005 Selection of optimum extrusion technology parameters in the manufacture of edible/biodegradable packaging films derived from food-based polymers *Journal of Food Agriculture and Environment* **3** (3/4) 51-64
[10] Ezeoha S L and Ezenwanne J N 2013 Production of biodegradable plastic packaging film from cassava starch *IOSR Journal of Engineering* **3** (10) 14-20
[11] Phan D, Debeaufort F. Luu D and Voilley A 2005 Functional properties of edible agar-based and starch-based films for food quality preservation *Journal of Agricultural and Food Chemistry* **53** (4) 973-981

[12] Bansal P, Hall, M.J, Realff J H, Lee A S, and Bommarius 2010 Multivariate statistical analysis of X-ray data from cellulose: a new method to determine degree of crystallinity and predict hydrolysis rates. *Bioresource technology* **101** (12) 4461-4471

[13] Huang C B, Jeng R, Sain M, Saville B A and Hubbes M 2006 Production, characterization, and mechanical properties of starch modified by Ophiostoma spp *BioResources* **1** (2) 257-269

[14] Harry T F, Joel O F, Ademiluyi F T, and Oduola K 2016 Performance evaluation of local cassava starches with imported starch for drilling fluid *American Journal of Engineering Research* **5** (10) 111-120

[15] Sangian H. F and Widjaja A. 2017 Effect of pretreatment method on structural changes of coconut coir dust *BioResources* **12** (4) 8030-8046

[16] Zhaosheng Y, Xiaoqian M and Ao L 2008 Kinetic studies on catalytic combustion of rice and wheat straw under air-and oxygen-enriched atmospheres, by using thermogravimetric analysis *Biomass and bioenergy* **32** (11) 1046-1055

[17] Agustin M B, Ahmad B, Alonzo S M M and Patriana F M 2014 Bioplastic based on starch and cellulose nanocrystals from rice straw *Journal of Reinforced Plastics and Composites* **33** (24) 2205-2213

[18] Edhirej E, Sapuan S M, Jawaid M and Sahari N I 2015 Bioplastic based on starch and cellulose nanocrystals from rice straw *Journal of Reinforced Plastics and Composites* **33** (24) 2205-2213

[19] Liu H, Sale K L, Holmes B M, Simmons B A, Singh S 2010 Understanding the interactions of cellulose with ionic liquids: a molecular dynamics study *The Journal of Physical Chemistry B* **114** (12) 4293-4301

[20] Singh V, Ali S Z, Somashekar R, Mukherjee P S 2006 Nature of crystallinity in native and acid modified starches *International Journal of Food Properties* **9** (4) 845-854

[21] Slopiecka K, Bartocci P and Fantozzi F 2012 Thermogravimetric analysis and kinetic study of poplar wood pyrolysis *Applied Energy* **97** 491-497

[22] Zuraida A, Hazleen A and Yusliza Y 2011 The study of biodegradable thermoplastics sago starch *Key Engineering Materials* **471** 397-402

# IoT based Trash Can Monitoring System for Smart Garden Cleanliness

**F D Moris and D Widjaja***

Electrical Engineering Department, Faculty of Science and Technology, Sanata Dharma University, Yogyakarta, Indonesia

*Corresponding author: damar@usd.ac.id

**Abstract.** Garbage is always produced every day, including household trash and many kind of waste. IoT based Trash Can Monitoring System was designed to automatically maintain the garden cleanliness and help to ease the cleaning staff work. This system uses two controlled trash bins with different location identity (ID). There are many devices to build this system, such as NodeMCU microcontroller, Thingspeak as an IoT platform, two ultrasonic sensors to detect the height of trash and approached object, two servo motors to control can cover and disposal outlet, and solenoid as a locking part. The results of this work show that the system designed works well. Location ID of the two trash cans, approached object detection, measurement of garbage height inside the trash can, opening and locking of disposal outlet and cover, work with 100% success rate and all data can be sent to Thingspeak without any error.

## 1. Introduction

Garbage is produced every day, both waste from organic substances (vegetables, meat, leaves, etc.) or from inorganic substances (plastic, glass, paper, rubber, metal and so on) [1]-[4]. At this time, disposing of garbage is still a problem that is concerning the government and people who care about environmental cleanliness. Lack of awareness to dispose of garbage in its place and uncoordinated disposal from many trash cans also need attention. These problems make people tend to carelessly litter because when they want to throw the trash, it turns out that the trash can is full.

Nainggolen (2018) has designed smart trash can that only can measure the garbage level inside the trash can [5]. When the trash can is full, the cover will automatically locked and no one can put more trash to the can. This system is also capable of sending trash can data to the cleaning staff via Short Message Service (SMS) when the trash can is full. Memon et al. (2019) built automatic trash can which can simply open the cover using ultrasonic sensor [6]. Some technologies related to IoT can be seen in ref [7]-[12].

This work on remote monitoring of trash can and automatic trash can lid opener aims to increase public awareness in making trash in its place and make it easier for cleaning staff to monitor the garbage inside the trash can. This system uses two controlled trash bins to differentiate the location, NodeMCU microcontroller, Thingspeak IoT platform as monitoring server, two ultrasonic sensors to detect the height of the garbage and the approached object, two servo motors to control top cover and automatic disposal cap, and the solenoid as a disposal cap locking system.

The remainder of this paper is organized as follows. Section 2 provides the description of system model used in this work. Section 3 presents results and discussion. It will describe the system performance based on the system success rate. Section 4 provides the conclusion of the work.

## 2. System Model

Figure 1 shows the system model of this study. The system input comes from the ultrasonic sensor for object detection and the height of the garbage and then processed by NodeMCU. The data read by NodeMCU is then used to drive the actuator. The actuator consists of a top servo motor to drive the top cover of the trash can based on object detection sensor data. Then the data on the height of the garbage is used to drive the servo motor when the garbage is full for 3 minutes and no cleaning staff arrives. Red and green LEDs are used for the indicator of the garbage height. Garbage height data will be sent and displayed on Thingspeak for monitoring.

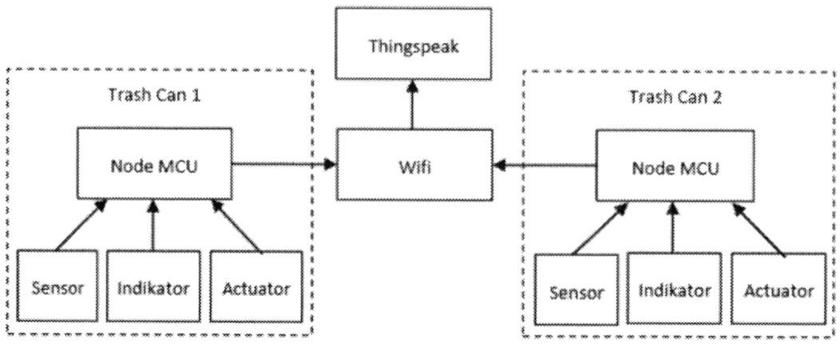


**Figure 1.** System Model

The model that will be made in this work can be seen in Figure 2. It consist of 2 trash cans that supossed to be separated in different locations. However, it location is simplified into one place with different location ID. Below the two trash cans, it has common disposal container model for disposing the garbage from the full trash can that is not visited by the cleaning staff after the can is full for 3 mi nutes. Each can has two LED indicators. When red LED is on, it means the trash can is full and the cover will be locked.

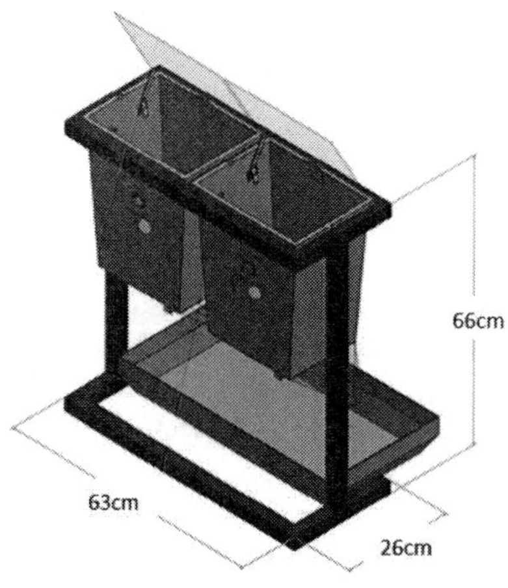


**Figure 2.** Design of the Hardware Model

The system flowchart can be seen in Figure 3. This flowchart will describe an outline from the beginning to the end process. At the beginning of the process, the top cover is closed, a solenoid condition is OFF (locking the bottom cover), and the system is connected to the internet. Then the ultrasonic sensor checks the load (garbage) of the trash can. When the load in the trash is full, the full indicator LED will be ON, while the empty indicator LED will OFF. After that, the data on the trash can will be sent to the Thingspeak IoT platform. After 3 minutes, unattended full trash can will dispose of the load automatically to the common disposal container. A time delay of 3 minutes is used for cleaning staff to manually dispose of garbage.

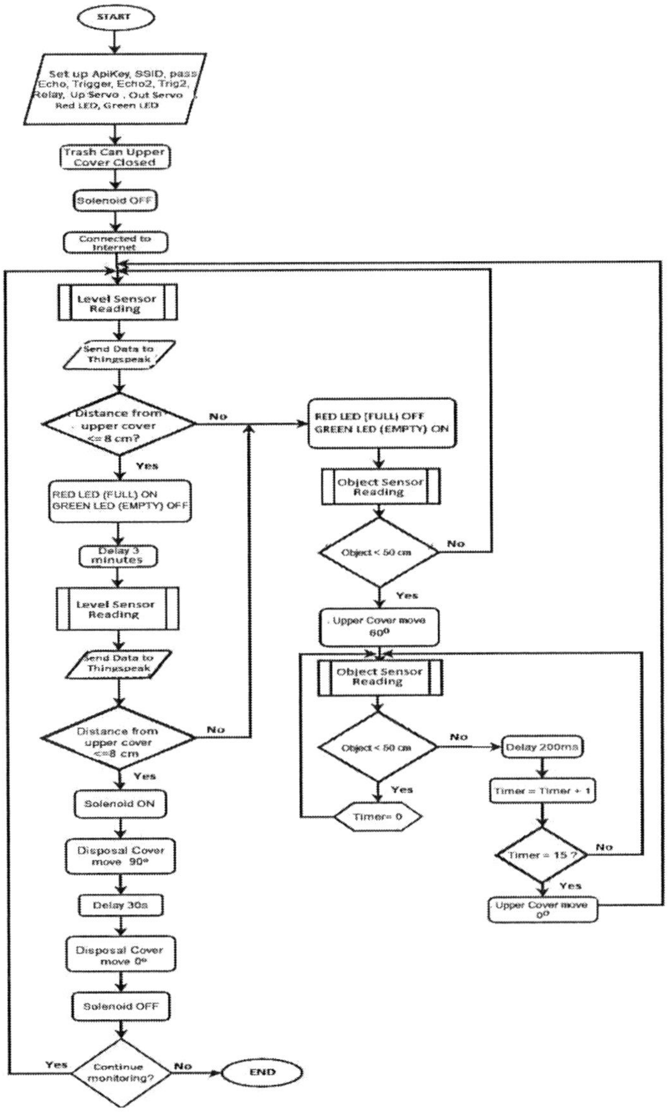


**Figure 3.** General Flowchart System

## 3. Result and Discussion

Table 1 shows the test result data of the whole system. This test is used to measure the success rate of the system. Testing the whole system is done by filling the trash can from empty to full, and testing the top cover opening. The testing includes opening the top cover, manual and automatic disposal covers, and finally testing the data transmission to Thingspeak. All the experiments show that the system was able to work in every condition as designed.

Table 2 shows testing result of garbage height at location 1. The measured garbage height value is the same as the value read by the sensor. However, when the trash was around 49cm high, the sensor failed to read the data. Based on the characteristics, the ultrasonic sensor has the ability to read a minimum distance of 2cm, therefore when the distance between the sensor and the garbage is too close the value read by the sensor will have error. This problem is also happened at location 2.

**Table 1.** Overall System Test Result

| **Trash Can Indicator** | | **Object Status** | | **Automatic Disposal Cover** | | **Upper Cover** | | **Send Data** | |
|---|---|---|---|---|---|---|---|---|---|
| Location 1 | Location 2 | Location 1 | Location 2 | Location 1 | Location 2 | Location 1 | Location 2 | Location 1 | Location 2 |
| Full | Not Full | Not Detected | Not Detected | Yes | No | Closed | Closed | Yes | Yes |
| Full | Not Full | Not Detected | Detected | Yes | No | Closed | Opened | Yes | Yes |
| Not Full | Full | Not Detected | Not Detected | No | Yes | Closed | Closed | Yes | Yes |
| Not Full | Full | Detected | Not Detected | No | Yes | Opened | Closed | Yes | Yes |
| Not Full | Not Full | Detected | Not Detected | No | No | Opened | Closed | Yes | Yes |
| Not Full | Not Full | Not Detected | Detected | No | No | Closed | Opened | Yes | Yes |
| Not Full | Full | Detected | Detected | No | No | Opened | Opened | Yes | Yes |
| Full | Not Full | Detected | Detected | No | No | Opened | Opened | Yes | Yes |
| Full | Full | Not Detected | Not Detected | Yes | Yes | Closed | Closed | Yes | Yes |
| Full | Full | Not Detected | Not Detected | Yes | Yes | Closed | Closed | Yes | Yes |

**Table 2.** Testing of Garbage Height, Location 1

| **Garbage Height (cm)** | **Sensor read (cm)** | **Trash Can Indicator** | **Note** |
|---|---|---|---|
| 0 | 47 | Not Full | Succesfull |
| 10 | 37 | Not Full | Succesfull |
| 16 | 31 | Not Full | Succesfull |
| 18 | 29 | Not Full | Succesfull |
| 29 | 18 | Not Full | Succesfull |
| 32 | 15 | Not Full | Succesfull |
| 34 | 13 | Not Full | Succesfull |
| 40 | 7 | Full | Succesfull |
| 42 | 5 | Full | Succesfull |
| 49 | 2512 | Not Full | Not Succesfull |

Table 3 shows the object detection distance at location 1. This object is representation of the hand of cleaning staff. When the object distance is less than 50 cm, the top cover will open until sensor detects no object around it. When the object distance is more than 50 cm, the top cover will remain close. When the trash can is full, object detection will not work. The same test result also occurs at location 2.

**Table 3.** Object Detection Distance Test

| Object Status | Distance (Cm) | Trash Can Indicator | Trash Can Cover | Upper Servo Motor | Note |
|---|---|---|---|---|---|
| Not Detected | 100 | Not Full | Close | 90° | Succesfull |
| Not Detected | 70 | Not Full | Close | 90° | Succesfull |
| Detected | 50 | Not Full | Open | 0° | Succesfull |
| Detected | 45 | Not Full | Open | 0° | Succesfull |
| Detected | 40 | Not Full | Open | 0° | Succesfull |
| Detected | 35 | Not Full | Open | 0° | Succesfull |
| Detected | 30 | Not Full | Open | 0° | Succesfull |
| Detected | 25 | Not Full | Open | 0° | Succesfull |
| Detected | 20 | Not Full | Open | 0° | Succesfull |
| Detected | 15 | Not Full | Open | 0° | Succesfull |
| Detected | 10 | Not Full | Open | 0° | Succesfull |
| Detected | 5 | Not Full | Open | 0° | Succesfull |
| Detected | 2 | Not Full | Open | 0° | Succesfull |

## 4. Conclusion

The system work well as it was designed. The two trash cans with different location ID successfully send all monitoring data to Thingspeak. Ultrasonic sensor can sense the garbage height correctly, except when the distance of the garbage to the sensor is less than 2 cm. Automatic disposal work as it was designed. It disposes garbage when the trash can is full and there is no cleaning staff arrive within 3 minutes. All the motor servo and solenoid work well as it can automatically open and close cover and locking system. Overall, the system works well with 100% success rate.

## Acknowledgments

Thanks to the Institute of Research and Community Service (Lembaga Penelitian dan Pengabdian kepada Masyarakat) Sanata Dharma University for funding and facilitating this research.

## References

[1] Rahmayanti H, Ichsan I Z, Azwar S A, Oktaviani V, Ladesi V K and Pertiwi N 2020 Garbage Sorting Games, DIFMOL, and ILMIZI: Technology Innovation for Environmental Learning of Disaster Mitigation *International Journal of Advanced Science and Technology* **29** (5) 11255-11265

[2] Bolzonella D, Pavan P, Battistoni P and Cecchi F 2003 The under sink garbage grinder: a friendly technology for the environment *Environmental technology* **24** (3) 349-359

[3] Yufeng Z, Na D, Jihong L and Changzhong X 2003 A new pyrolysis technology and equipment for treatment of municipal household garbage and hospital waste *Renewable Energy* **28** (15) 2383-2393

[4] Chen G, Wang H and Zheng J 2019 Application of Image Recognition Technology in Garbage Classification Education Platform *2019 5th International Conference on Control, Automation and Robotics* 290-294

[5] Nainggolan O D 2018 *GSM based Trash Can Monitoring System* Magister Thesis, Sanatha Dharma University, Yogyakarta
[6] Memon S K, Shaikh F K, Mahoto N A and Memon A A 2019 IoT based smart garbage monitoring & collection system using WeMos & Ultrasonic sensors *2019 2nd International Conference on Computing, Mathematics and Engineering Technologies* 1-6
[7] Satria D and Hidayat T 2019 Implementation of wireless sensor network (WSN) on garbage transport warning information system using GSM module *Journal of Physics: Conference Series* **1175** (1) 012054
[8] Putra R H, Kusuma F T, Damayanti T N and Ramadan D N 2019 IoT: smart garbage monitoring using android and real time database *Telkomnika* **17** (3) 1483-1491
[9] Suryawanshi S, Bhuse R, Gite M and Hande D 2018 Waste management system based on IoT. *Waste Management* **5** (3) 1-3
[10] Addabbo T, Fort A, Mecocci A, Mugnaini M, Parrino S, Pozzebon A and Vignoli V 2019 A lora-based iot sensor node for waste management based on a customized ultrasonic transceiver *2019 IEEE Sensors Applications Symposium* 1-6
[11] Kuncoro A H, Mellyanawaty M, Sambas A, Maulana D S, Subiyanto and Mamat M 2020 Air Quality Monitoring System in the City of Tasikmalaya based on the Internet of Things (IoT) *Jour of Adv Research in Dynamical & Control Systems* **12** (2) 2473-2479
[12] Bajaj A and Reddy S 2017 Garbage monitoring system using IOT *International Journal of Pure and Applied Mathematics* **114** (12) 155-161

# Mixed Truncated Spline and Kernel Nonparametric Regression Model on Population Growth Rate in West Nusa Tenggara Province

**N Fitriyani*, M Hadijati, L Harsyiah and M S Sauri**

Department of Mathematics, Faculty of Mathematics and Natural Sciences, Universitas Mataram, Jl Majapahit 62, Mataram, Indonesia

*Corresponding author: nurul.fitriyani@unram.ac.id

**Abstract.** This study was conducted by considering the data pattern that differs from each independent variable to the dependent variable. If only one estimator is used to estimate the nonparametric regression curve, the resulting estimator does not match the data pattern, less precise, and tends to produce large errors. Therefore, this study aimed to model the mixed truncated spline and kernel nonparametric regression on data and predict the population growth rate in West Nusa Tenggara Province by considering the Mean Absolute Percentage Error (MAPE). Based on the study conducted, relationships generated by $x_1, x_2, x_7, x_8$ formed specific patterns and were indicated to follow the spline approach's characteristics. In contrast, the pattern generated by $x_3, x_4, x_5, x_6, x_9$ showed the data distribution with no specific pattern and were addressed to follow the kernel approach's characteristics. The best model was obtained with the optimal bandwidth for each variable and 3 points of optimal knots. The mixed truncated spline and kernel nonparametric regression model was suitable for modeling and predicting population growth rate data in West Nusa Tenggara Province with a relatively high proportion and a high degree of accuracy.

## 1. Introduction

Regression analysis is a statistical analysis to estimate the relationship pattern between predictor variables and the response variable. The relationship can be explained by utilizing a regression curve, using parametric and nonparametric regression estimation approaches. If the shape of the regression curve is known, then the estimation approach is carried out with parametric regression. However, if no shapes are ruled out for the regression function and initial information regarding the regression curve is limited, the estimation approach is carried out with nonparametric regression [1], [2].

Nonparametric regression provides estimates with great flexibility in the possible form of the regression curve. It generally only assumes that the regression curve belongs to some infinite-dimensional collection of functions without any parametric form assumptions [2]. There are several model approaches for estimating nonparametric regression curves. Spline truncated is one of the models with very specific and good statistical interpretation and visual interpretation among nonparametric models. It also has high flexibility and a good ability to handle data with changeable behaviors on certain sub-intervals [1], [2], [3], [4]. Besides, kernel estimator has a good ability to model data with no specific pattern [3].

One indicator that is often used in describing an area's population conditions is the population growth rate [5]. One way to monitor the population growth rate is by modeling, to check the relationship between population growth rate and its factors. In general, the population growth rate pattern tends to change at certain intervals, so that if it is estimated using parametric regression, it will produce inaccurate results. Moreover, in some other cases, data pattern does not follow a particular pattern, such as pattern data related to children growth [6], fertility [4], [7], and rice production [8], so they were more advisable to used nonparametric regression approach.

There are primary assumptions that need to be regarded in the nonparametric regression. The pattern in each independent multivariable is supposed to have the same pattern, and the researcher only uses one form of the model estimator for each independent variable. In its application in several cases, the resulting data patterns often differ from each independent variable. Therefore, if only one estimator is used to estimate the nonparametric regression curve, the resulting estimator does not match the data pattern. As a result, the resulting regression model estimation is less precise and tends to produce large errors [1].

This research was expected to produce the best model of the mixed truncated spline and kernel nonparametric regression to describe the population growth rate data in West Nusa Tenggara Province. This model was also required to provide predictive values for population growth rate in West Nusa Tenggara Province by considering the Mean Absolute Percentage Error (MAPE).

## 2. Methods

This study used secondary data obtained from the Central Bureau of Statistics, West Nusa Tenggara Province. A total of $n = 32$ sub-district sample data from all districts were taken, including population growth rate $(y)$, and nine independent variables, i.e., Number of Births $(x_1)$, Number of Deaths $(x_2)$, Number of Reproductive Age Couples $(x_3)$, Number of Contraceptive Users $(x_4)$, Number of Health Workers $(x_5)$, Number of Health Facilities $(x_6)$, Population $(x_7)$, and Population Density $(x_8)$, and Number of Educational Facilities $(x_9)$. The research on population growth rate data was carried out in several steps, including:

a. Collecting data and determining variables.
b. Constructing scatter plots between variables and determining the appropriate approach. Given the data pairs $(x_{i1}, x_{i2}, \cdots, x_{i9}, y_i)$, $i = 1,2,3,....,32$, they were approached with the following regression model.

$$y_i = \mu(x_{i1}, x_{i2}, \cdots, x_{i9}) + \varepsilon_i \quad (1)$$

Random errors $\varepsilon_i$ were assumed to be independent and normally distributed with zero mean and constant variance. The regression curve was assumed to follow the following nonparametric mixed model truncated spline and additive kernel.

$$\mu(x_{i1}, x_{i2}, \cdots, x_{i9}) = \sum_j f_j(x_{ij}) + g(x_{ij}) \quad (2)$$

The function $f_j(x_{ij})$ and $g(x_{ij})$ were assumed to be smooth, in the sense that they are continuous and differentiable. Functions $f_j(x_{ij})$ were approached by truncated spline function, while $g(x_{ij})$ was approached by Nadaraya-Watson kernel. Therefore Equation (2) can be written in the following form.

$$\mu(x_{i1}, x_{i2}, \cdots, x_{i9}; \mathbf{k}, \mathbf{h}) = \sum f_j(x_{ij}; \mathbf{k}, \mathbf{h}) + g(x_{ij}; \mathbf{h}) \quad (3)$$

Vector $\mathbf{k}$ and $\mathbf{h}$ represented knot and bandwidth vectors, respectively.

c. Modeling related variables using kernel approach and determining the optimum bandwidth based on the minimum value of Generalized Cross-Validation (GCV). In matrix form, the estimator for mixed nonparametric regression truncated spline and kernel model shown in Equation (3) was given as follows, estimated by using Least Square Method [9].

$$\hat{\mu}(\mathbf{x};\ \mathbf{k},\mathbf{h}) = \mathbf{F}(\mathbf{x};\ \mathbf{k},\mathbf{h})\mathbf{y} + \mathbf{G}(\mathbf{x};\ \mathbf{h})\mathbf{y} \tag{4}$$

Vector $\mathbf{y}$ revealed the dependent variable vector and matrix $\mathbf{G}(\mathbf{x};\ \mathbf{h})$ approach using a nonparametric method that is generally used, namely multivariate Nadaraya-Watson kernel density estimator. Therefore, the estimator form was given as the following [10], [11].

$$\hat{g}(x_{i1}, x_{i2}, \cdots, x_{i9}; \mathbf{h}) = \frac{\frac{1}{n}\sum_{i=1}^{n=32}\frac{1}{\mathbf{h}}K\left(\frac{x_1 - X_{i1}}{h_1}, \frac{x_2 - X_{i2}}{h_2}, \cdots, \frac{x_9 - X_{i9}}{h_9}\right)Y_i}{\frac{1}{n}\sum_{i=1}^{n=32}\frac{1}{\mathbf{h}}K\left(\frac{x_1 - X_{i1}}{h_1}, \frac{x_2 - X_{i2}}{h_2}, \cdots, \frac{x_9 - X_{i9}}{h_9}\right)}$$

Vector $\mathbf{h} = (h_1, h_2, \cdots h_9)^T$ , $h_j$ indicated a variable bandwidth. The easiest solution for multivariate kernel function was generally relying on multiplicative (product) kernel form, given as [12], [13].

$$\mathrm{K}(u_1, u_1, \cdots, u_9) = \mathrm{K}(u_1)\cdot \mathrm{K}(u_2)\cdot \cdots \cdot \mathrm{K}(u_9)$$

Hence, the multivariate Nadaraya-Watson kernel estimator could be written as the following form.

$$\hat{g}(x_{i1}, x_{i2}, \cdots, x_{i9}; \mathbf{h}) = \frac{\frac{1}{n}\sum_{i=1}^{32}\left\{\prod_j \frac{1}{h_j}K\left(\frac{x_j - X_{ij}}{h_j}\right)\right\}Y_i}{\frac{1}{n}\sum_{i=1}^{32}\left\{\prod_j \frac{1}{h_j}K\left(\frac{x_j - X_{ij}}{h_j}\right)\right\}}$$

Function $\mathrm{K}(\cdot)$ denoted a kernel function. This study used the quartic kernel function with the following form [2].

$$\mathrm{K}(u) = \frac{15}{16}(1 - u^2)^2\ ,\ I(|u| \le 1) \tag{5}$$

Nonparametric regression curve estimation using the Nadaraya-Watson kernel approach was very dependent on bandwidth. Therefore, to get the best results, it was necessary to select the optimal bandwidth parameter. The selection of the optimum bandwidths $h_j$ for each variable $j$ were carried out by minimizing the GCV criterion given by the following formula.

$$GCV(\mathbf{h}) = \frac{n^{-1}\left\| y_i - \mathbf{G}(\mathbf{x};\ \mathbf{h})\mathbf{y} \right\|^2}{\left(n^{-1} trace[\mathbf{I} - \mathbf{G}(\mathbf{x};\ \mathbf{h})]\right)^2} \tag{6}$$

The matrix $\mathbf{I}$ indicated an identity matrix.

d. Modeling related variables $j$ using truncated spline functions $f_j(x_{ij})$, as the following [14].

$$f_j(x_{ij}) = \beta_0 + \beta_j x_{ij} + \sum_{l=1}^{m}\theta_{lj}(x_{ij} - k_{lj})_+\ ,\ j = 1,2,\cdots,9$$

Constants $\beta_0, \beta_j, \theta_{lj}$ are unknown truncated spline parameters and $l = 1,2,\cdots,m$ represent the number of knots used. The linear truncated function was defined as follows.

$$(x_{ij} - k_{lj})_+ = \begin{cases} (x_{ij} - k_{lj}) & ,\ x_{ij} \ge k_{lj} \\ 0 & ,\ x_{ij} < k_{lj} \end{cases}$$

In this study, 1-knot point $(l = 1)$, 2-knot points $(l = 2)$, and 3-knot points $(l = 3)$ were used, which then determined the optimal knot points' number and location using GCV criteria, based on the optimal bandwidth previously obtained. The estimator of mixed truncated spline and kernel nonparametric regression curve was given by Equation (4), with $\mathbf{F}(\mathbf{x};\ \mathbf{k},\mathbf{h})$ given by [9].

$$\mathbf{X}(\mathbf{k},\mathbf{h})\left(\mathbf{X}(\mathbf{k},\mathbf{h})^T\mathbf{X}(\mathbf{k},\mathbf{h})\right)^{-1}\mathbf{X}(\mathbf{k},\mathbf{h})^T(\mathbf{I} - \mathbf{G}(\mathbf{h}))$$

Mixed modeling was acquired based on the optimum bandwidth obtained and determining the optimum number and location of knot points by minimizing the GCV criterion given as follows.

$$G(\mathbf{k},\mathbf{h}) = \frac{n^{-1}\left\| y_i - \hat{\mu}(\mathbf{x};\, \mathbf{k},\mathbf{h}) \right\|^2}{\left(n^{-1} trace\left[\mathbf{I} - \mathbf{F}(\mathbf{x};\, \mathbf{k},\mathbf{h}) - \mathbf{G}(\mathbf{x};\, \mathbf{h})\right]\right)^2} \tag{7}$$

Vector $\mathbf{k} = (k_{l1}, k_{l2}, \cdots, k_{l9})^T$ , $l = 1,2,3$ represented the knot vector.

e. Measuring the proportion of total variation about the mean of data $\bar{y}$ explained by the regression, using an $R^2$ statistic, with the following formula [15].

$$R^2 = \frac{\sum_{i=1}^{n} \left(\hat{\mu}(\mathbf{x};\, \mathbf{k},\mathbf{h}) - \bar{y}\right)^2}{\sum_{i=1}^{n} (y_i - \bar{y})^2} \tag{8}$$

f. Checking whether the model residuals have fulfilled the model assumptions [15].

g. Predicting data and calculating the MAPE measurement for prediction accuracy using the following Equation [16].

$$MAPE = \frac{1}{n}\sum_{i=1}^{n} \frac{\left| y_i - \hat{\mu}(\mathbf{x};\, \mathbf{k},\mathbf{h}) \right|}{y_i} \times 100\% \tag{9}$$

h. Drawing a conclusion.

## 3. Results and Discussion

The appropriate approach can be determined based on the scatter plots between each of the nine independent variables $(x_i)$ and the dependent variable $(y)$. The relationship patterns produced in regards to choose the approaches that will be used.

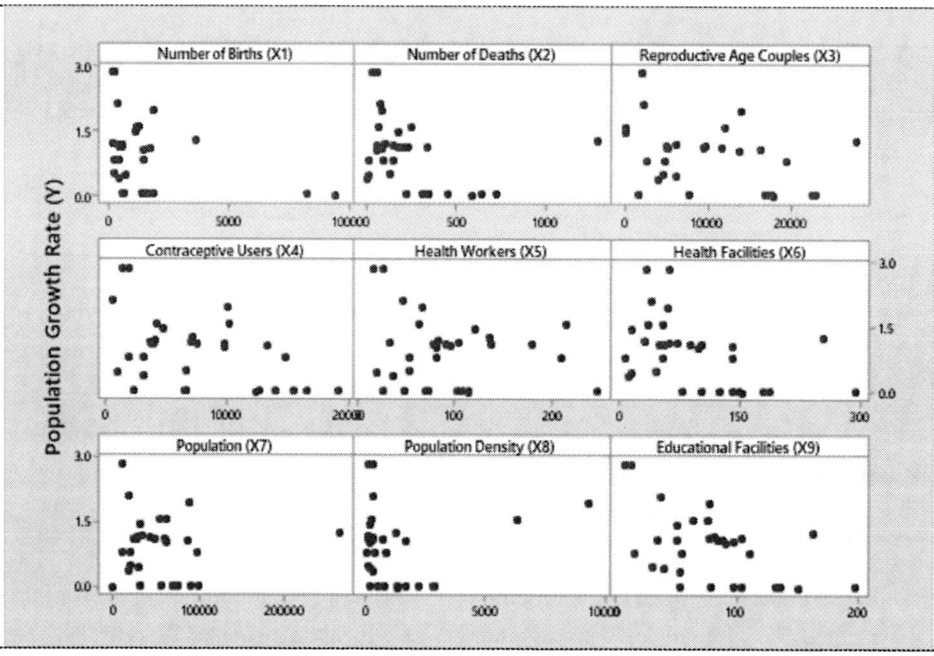


**Figure 1.** Scatter plot between population growth rate $(y)$ and predictor variables $(x_i)$.

Based on Figure 1, relationships generated by $x_1, x_2, x_7, x_8$ formed specific patterns, while patterns generated by $x_3, x_4, x_5, x_6, x_9$ showing the data distribution with no form or particular patterns. Based on the initial analysis of the predictors' relationship with the population growth rate $(y)$, four variables, i.e., Number of Births $(x_1)$, Number of Deaths $(x_2)$, Population $(x_7)$, and Population Density $(x_8)$, are indicated to follow the characteristics of the spline approach. In contrast, five variables, i.e., Number of Reproductive Age Couples $(x_3)$, Number of Contraceptive Users $(x_4)$, Number of Health Workers $(x_5)$, Number of Health Facilities $(x_6)$, and Number of Educational Facilities $(x_9)$, are addressed to follow the kernel approach characteristics.

Based on the initial analysis using a scatter plot, the truncated spline function approaches the functions $f_j(x_{ij})$, $j = 1,2,7,8$, while the Nadaraya-Watson kernel approaches the function $g_{\mathbf{h}}(x_{ij})$, $j = 3,4,5,6,9$. Nonparametric regression curve estimation using mixed truncated spline and kernel is highly dependent on knot points and bandwidth parameters. To get the best model, it is necessary to select the optimal knot point and bandwidth parameters. Modeling several variables $(x_3, x_4, x_5, x_6, x_9)$ related to the kernel nonparametric statistical approach is done with the estimator form given as the following.

$$\hat{g}_{\mathbf{h}}(x_3, x_4, x_5, x_6, x_9) = \frac{\frac{1}{n}\sum_{i=1}^{32}\left\{\prod_{j=3,4,5,6,8}\frac{1}{h_j}K\left(\frac{x_j - X_{ij}}{h_j}\right)\right\}Y_i}{\frac{1}{n}\sum_{i=1}^{32}\left\{\prod_{j=3,4,5,6,8}\frac{1}{h_j}K\left(\frac{x_j - X_{ij}}{h_j}\right)\right\}}$$

Function $K(\cdot)$ denotes a quartic kernel function in Equation (5).

Nonparametric regression curve estimation using the Nadaraya-Watson kernel approach is very dependent on bandwidth. Therefore, to get the best results, it is necessary to select the optimal bandwidth parameter. The selection of the optimum bandwidths $h_j$ for each variable $j$ is carried out by considering the minimum GCV based on Equation (6). The optimum bandwidths obtained are listed in Table 1.

**Table 1.** Optimum bandwidths for independent variables.

| | | |
|---|---|---|
| Optimum Bandwidth $(h)$ | $h_3$ | 0.2871 |
| | $h_4$ | 0.9754 |
| | $h_5$ | 1.1456 |
| | $h_6$ | 0.8856 |
| | $h_9$ | 0.6788 |
| Generalized Cross-Validation (GCV) | | 0.8577258 |

Furthermore, the population growth rate data modeling is carried out using a mixed regression model of nonparametric spline truncated and kernel, selecting optimal knot points for the related variables $(x_1, x_2, x_7, x_8)$. Regression curves $f_j(x_{ij})$ are approached by truncated spline linear function with knots $k_{lj}$. In this study, 1-knot point $(l = 1)$, 2-knot points $(l = 2)$, and 3-knot points $(l = 3)$ are used, which then determined the optimal knot points' number and location by considering GCV criteria, based on the optimal bandwidth previously obtained.

Based on what has been explained previously, the nonparametric regression curve's estimation using the mixed truncated spline and kernel approach is highly dependent on the knot point and bandwidth parameters. Selection of the optimum knots $k_{lj}$ for each variable $j$, related to the optimum bandwidths obtained, is also carried out by considering the minimum GCV. This method is used to choose the knots' number and locations for the estimator, with the following results (see Table 2).

**Table 2.** Optimum knots for independent variables.

| Knots | | Variables $(j)$ | | | | Generalized Cross-Validation (GCV) | $R^2$ statistic |
|---|---|---|---|---|---|---|---|
| | | $x_1$ | $x_2$ | $x_7$ | $x_8$ | | |
| 1 knot | $k_{1j}$ | 5413.69 | 725.18 | 148933.30 | 5246.49 | 0.8198 | 50.63 % |
| 2 knots | $k_{1j}$ | 393.08 | 34.87 | 6845.10 | 273.79 | 0.5177 | 72.95 % |
| | $k_{2j}$ | 871.23 | 100.62 | 20377.31 | 747.38 | | |
| 3 knots | $k_{1j}$ | 393.08 | 34.87 | 6845.10 | 273.79 | 0.2293 | 83.32 % |
| | $k_{2j}$ | 871.23 | 100.62 | 20377.31 | 747.38 | | |
| | $k_{3j}$ | 1349.38 | 166.36 | 33909.51 | 1220.97 | | |

Based on Table 2, it can be seen that the minimum GCV value is obtained by using three knot points. Furthermore, the population growth rate data modeling is carried out using the knots point and the optimal bandwidth obtained. The unknown truncated spline parameters $\left(\beta_0, \beta_j, \theta_{lj}\ ,\ j = 1,2,7,8\ ;\ l = 1,2,3\right)$ for $x_1, x_2, x_7, x_8$ were then estimated with the following results.

**Table 3.** Parameter estimates for the truncated spline approach.

| Variables | Parameter estimates for $x_j$ | | | | |
|---|---|---|---|---|---|
| | $\beta_j$ | $\theta_{1j}$ | $\theta_{2j}$ | $\theta_{3j}$ | $\beta_0$ |
| $x_1$ | $-6.78 \times 10^{-05}$ | $2.73 \times 10^{-05}$ | $1.82 \times 10^{-05}$ | $9.08 \times 10^{-06}$ | 0.5611 |
| $x_2$ | $-0.0015$ | $5.42 \times 10^{-07}$ | $3.61 \times 10^{-07}$ | $1.81 \times 10^{-07}$ | |
| $x_7$ | $-5.97 \times 10^{-06}$ | 0.0001 | $7.44 \times 10^{-05}$ | $3.72 \times 10^{-05}$ | |
| $x_8$ | $2.64 \times 10^{-05}$ | 0.0009 | 0.0006 | 0.0003 | |

The nonparametric regression curve's estimator using mixed truncated spline and kernel for multivariable predictors that depend on the spline knots and kernel bandwidths is as follows.

IOP Publishing
doi:10.1088/1757-899X/1115/1/012054

$$\hat{y}_i = 0.5611 - 6.78\times10^{-5}x_{i1} - 0.0015x_{i2} - 5.97\times10^{-6}x_{i7} + 2.64\times10^{-5}x_{i8}$$
$$+2.73\times10^{-5}(x_{i1}-393.08)_+^1 + 1.82\times10^{-5}(x_{i1}-871.23)_+^1 + 9.08\times10^{-6}(x_{i1}-1349.38)_+^1$$
$$+5.42\times10^{-7}(x_{i2}-34.87)_+^1 + 3.61\times10^{-7}(x_{i2}-100.62)_+^1 + 1.81\times10^{-7}(x_{i2}-166.36)_+^1$$
$$+0.0001(x_{i7}-6845.10)_+^1 + 7.44\times10^{-5}(x_{i7}-20377.31)_+^1 + 3.72\times10^{-5}(x_{i7}-33909.51)_+^1$$
$$+0.0009(x_{i8}-273.79)_+^1 + 0.0006(x_{i8}-747.38)_+^1 + 0.0003(x_{i8}-1220.97)_+^1$$
$$+\frac{\sum_{i=1}^{32}\left(K\left(\frac{x_3-X_{i3}}{0.2871}\right)\cdot K\left(\frac{x_4-X_{i4}}{0.9754}\right)\cdot K\left(\frac{x_5-X_{i5}}{1.1456}\right)\cdot K\left(\frac{x_6-X_{i6}}{0.8856}\right)\cdot K\left(\frac{x_9-X_{i9}}{0.6788}\right)\right)y_i}{\sum_{i=1}^{32}\left(K\left(\frac{z_1-Z_{i1}}{0.2871}\right)\cdot K\left(\frac{z_2-Z_{i2}}{0.9754}\right)\cdot K\left(\frac{z_3-Z_{i3}}{1.1456}\right)\cdot K\left(\frac{z_4-Z_{i4}}{0.8856}\right)\cdot K\left(\frac{z_5-Z_{i5}}{0.6788}\right)\right)}$$

Based on Equation (8), it is known that the mixed model has an $R^2$ statistic of 83.32%, which is quite a large proportion. This means that the fitted regression $\hat{y}_i$ explains 83.32% of the total variation in the data's average population growth rate data.

Moreover, checking the residual assumptions of the truncated spline and kernel mixture models obtained. The results obtained show that the residual fulfilled the assumptions about the errors, that the errors are independent, have a zero mean and constant variance, and follow the normal distribution so that the model is suitable for prediction. Figure 2 shows the prediction from population growth rate data using the best model obtained.

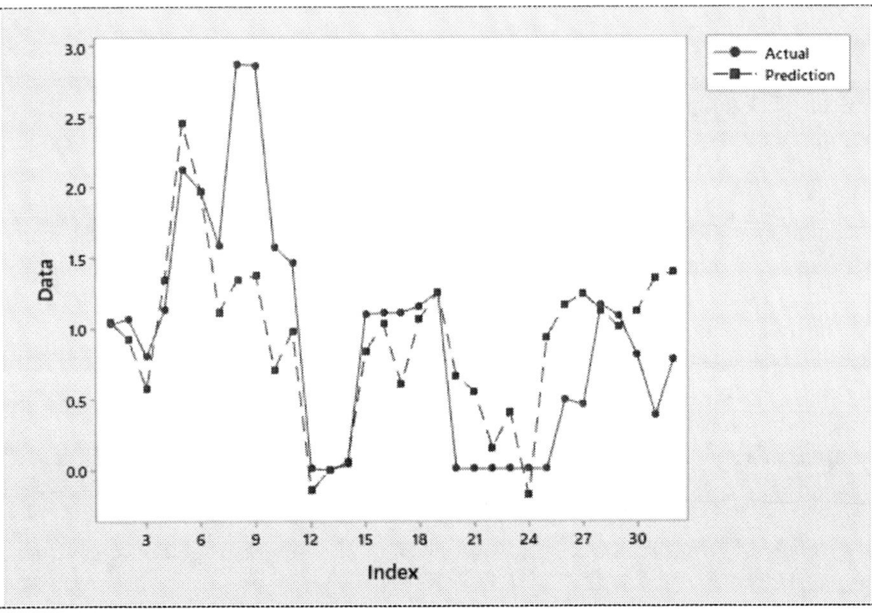


**Figure 2.** Actual versus prediction data.

The population growth rate prediction is then tested for accuracy using the Mean Absolute Percentage Error (MAPE) measurement according to Equation (9). The accuracy of data prediction obtained by MAPE value was 9.79 %. This value is below 10%, so that the forecasting has an excellent forecast.

## 4. Conclusion

The best model is obtained with the optimal bandwidth for each variable, and 3 points of optimal knots from the options tried, taking into account the smallest GCV value. The truncated spline and kernel nonparametric mixed regression model is suitable for modeling and predicting population growth rate data in West Nusa Tenggara Province with a relatively high proportion and a high degree of accuracy.

## Acknowledgments

This work was supported by the Statistics Areas of Expertise, Universitas Mataram, by using PNBP Universitas Mataram year 2020 funding.

## References

[1] Budiantara I N, Ratnasari V, Ratna M, Zain I 2015 The combination of spline and kernel estimator for nonparametric regression and its properties *Applied Mathematical Sciences* **9** (122) p 6083-6094.
[2] Eubank R L 1999 *Nonparametric regression and spline smoothing* New York: Marcel Dekker Inc
[3] Hardle W 1994 *Applied nonparametric regression* Berlin: Humboldt-Universität zu Press
[4] Fitriyani N, Budiantara I N, Zain I, Ratnasari V 2016 Nonparametric regression spline in the estimation of the average number of children born alive per woman *Proceeding of the 1st ICST Mataram University* 169-172.
[5] Bongaarts J 2009 Human population growth and the demographic transition *Philosophical Transactions of the Royal Society B* **364** 2985-2990
[6] Chamidah N, Budiantara I N, Sunaryo, S, and Zain I 2012 Designing of child growth chart based on multi-response local polynomial modeling *Journal of Mathematics and Statistics* **8** (3) 342-347
[7] Fitriyani N and Budiantara I N 2014 Cross-validation and generalized cross-validation methods in the nonparametric regression spline (a case study of fertility data in East Java) *Proceeding of National Seminar on Science Education Innovation* 1089-1096
[8] Bidayani, Hadijati M, Fitriyani N 2019 Spline semiparametric regression model of rice production in east lombok regency *Eigen Mathematics Journal* **2** (1) 6-12
[9] Ratnasari V, Budiantara I N, Ratna M, and Zain I 2016 Estimation of nonparametric regression curve using mixed estimator of multivariable truncated spline and multivariable kernel *Global Journal of Pure and Applied Mathematics* **12** (6) 5047-5057
[10] Nadaraya E A 1964 On estimating regression *Theory of Probability and Application* **9** 141-142
[11] Watson G S 1964 Smooth regression analysis *The Indian Journal of Statistics, Series A* **26** (4) 359-372
[12] Scott D W 2004 Multivariate density estimation and visualization *Working Papers Humboldt-Universität zu Berlin* **16** 1-12
[13] Hardle W, Muller M, Sperlich S, and Werwatz A 2004 *Nonparametric and semiparametric models: an introduction* Germany: Springer
[14] Budiantara I N, Ratna M, Zain I, and Wibowo W 2012 Modeling the percentage of poor people in indonesia using spline nonparametric regression approach *International Journal of Basic & Applied Sciences IJBAS-IJENS* **12** (6) 119-124
[15] Draper NR and Smith H 1998 *Applied regression analysis* Canada: John Wiley & Sons
[16] Wei W W S 2006 *Time series analysis: univariate and multivariate methods* New York: Pearson Education

# The Effect of Human Development Index on Poverty Model in Indonesia using Penalized Basis Spline Nonparametric Regression.

**N Hasanah*, S Bahri and N Fitriyani**

Department of Mathematics, Faculty of Mathematics and Natural Sciences, Universitas Mataram, Jl. Majapahit 62 Mataram, Indonesia

*Corresponding author: nurlizahasanah07@gmail.com

**Abstract.** Poverty is an issue of special concern to various countries in the world, including Indonesia. One factor that can affect poverty is the Human Development Index (HDI). The aim of this study is to model of the problem of poverty based on HDI. Using nonparametric spline regression was used with the penalized basis spline (PB-Spline) approach, which can overcome the problem of selecting many knot points and the location of knots in spline regression. This study was obtained three models, i.e. model with one knot point, model with two knot points, and model with three knot points. Based on indicators of the value of Generalized Cross Validation (GCV) and the value of Mean Square Error (MSE), the best model was a model with three knots, with smoothing parameter value of 1000, 11.26236 value of GCV and 11.08420 value of MSE.

## 1. Introduction

Poverty is an issue of particular concern to various countries in the world, including Indonesia. Indonesia is a country with a high poverty rate, with more than 10% of Indonesia's population is categorized as poor [1]. Based on several previous studies, we can see that the Human Development Index (HDI) significantly affects poverty levels [2]. The impact of the HDI on poverty can be determined by analyzing the relationship between the two variables. A nonparametric regression approach can be used to estimate the model. This method is not bound by assumptions and has high flexibility because estimating the regression curve adjusts the data without being influenced by the researcher's subjectivity factor [3]. Some of the nonparametric regression approaches that have been developed are spline, kernel, MARS, Fourier series, wavelet, and others.

The spline is a polynomial slice that has segmented properties that are joined by knot points and can explain the character of the data [3]. The advantages of spline regression in study are that the model will tend to seek its estimation wherever the data moves. It can overcome fluctuating data patterns with knot points and produces relatively smooth curves [4]-[7]. The spline model's weakness during high order and the selection of a large number of knots can be overcome by using the B-splines basis, namely by building a base based on the knot point. The problem in creating the B-spline's basic function is the knots' determination and placement, namely the place where the polynomial pieces are connected to the B-splines [2].

P-splines contains a function that takes into account smoothing parameters [2]. The P-splines consists of two components namely B-splines component and the distinction penalty. The B-splines is developed into a P-splines that is finished using the B-splines with the same knot distance, called the PB-splines [8]. Optimizing the number and placement of knots is resolved by first determining the number of knots used on the B-splines. Furthermore, the knots' order is carried out with the concept of equal space knots, which is to adjust the knots' position so that the distance between the knots is the same. Then to increase the smoothness, a differentiating penalty is given [9].

This study aims to model the percentage of poverty based on human development index using PB-Spline nonparametric regression. So that, through this model the government is expected to make efforts to manage resources to improve the human development index in order to reduce the percentage of Poverty.

## 2. Method

The analysis steps were:

a. Preparing for the data obtained from central Bureau of Statistics about to the problem of poverty in Indonesia year 2018.

b. Determining the variables used with Human Development Index as a independent variabel and Poverty Percentage as a dependent variable and then input the dependent variable and the independent variable.

c. Determining a descriptive statistics of each variable and scatter the plot between the response variable and the predictor variable to find out the initial data pattern.

d. Determining knot point by defining a new predictor that contains the unique value of the predictor and then sort. Next determine the warts of the new predictor as the knot point by dividing the new predictor as much as the part.

e. Determining the base order of B-Spline function to find the base value of the B-Spline function on PB-Spine estimator by using the following Equation [10]

$$B_{j,p}(x) = \frac{x - k_j}{k_{j+p-1} - k_j} B_{j,p-1}(x) + \frac{k_{j+p} - x}{k_{j+p} - k_{j+1}} B_{j+1,p-1}(x) \tag{1}$$

Where $\alpha_j$ is a coefficient regression, $B_{j-p,p}(x)$ is a B-spline define over the knot, $p$ is a degree of B-spline function and $k_j$ is a knot points.

f. Modeling the dependent variable using linear PB-splines regression with knots points obtained from quantile samples

$$f(x_i) = \sum_{j=1}^{p+k} \alpha_j(\lambda; D) B(x_i)_{j-p,p} \tag{2}$$

g. Calculating the GCV value for each knot in the PB-spline linear regression model using the following equations

$$GCV = \frac{n^{-1}\sum_{i=1}^{n}(y_i - \hat{y}_i)^2}{n^{-1}\mathbf{trace}[\mathbf{I} - \mathbf{S}]} \tag{3}$$

h. Selecting of knot points and smoothing parameter based on the minimum GCV value of the PB-spline linear regression model for each knot points.

i. Modeling using the optimal knot points obtained.

j. Calculating the estimated parameter value.

$$\hat{\boldsymbol{\alpha}} = (\mathbf{B}^T\mathbf{B} + \lambda \mathbf{D}_d{}^T \mathbf{D}_d)^{-1}\mathbf{B}^T\mathbf{Y} \tag{4}$$

k. Determining the best model based on MSE using the following equation[8]

$$MSE = n^{-1}\sum_{i=1}^{n}(y_i - \hat{y}_i)^2 \tag{5}$$

l. Interpreting the models and calculating cofficient determinant $(R^2)$ using the following equation

$$R^2 = \frac{\sum_{i=1}^{n}(\hat{y}_i - \bar{y})^2}{\sum_{i=1}^{n}(y_i - \bar{y})^2} \tag{6}$$

## 3. Result and Discussion

Before further analysis or modeling using PB-Spline nonparametric regression, the first thing to do is investigate the pattern of the relationship between dependent variables and independent variables by creating graphs. The form pattern relationship between dependent variables and independent variables can be seen in Figure 1.

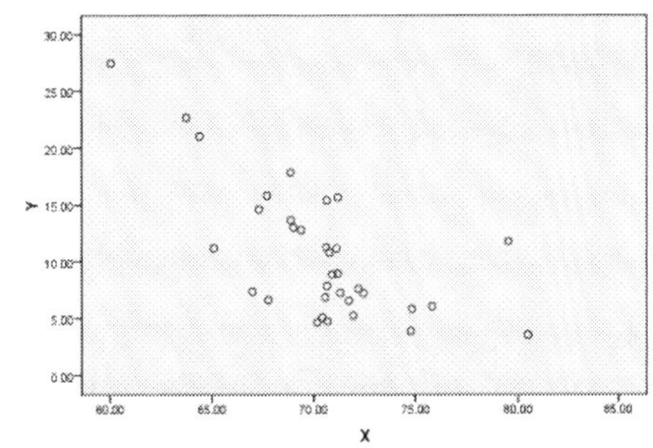


**Figure 1.** Scatter Plot between Human Development Index and Poverty Percentage..

The following scatter plot describes how data scatter patterns where it is known that data points have different patterns at certain intervals. Then the analysis is done using one of the regression analysis spline namely PB-Spline. Since $n = 34$ in this case and only trial value of many knots point $(K)$ less than $n - p - 1$ are used. And this study we used one, two and three knots for modeling poverty. The model is obtained by selecting the optimal finer parameters and knot points. Because in the selection of optimal knots and optimal finer parameters must meet the minimum GCV criteria. Figure 2 are three prediction models obtained by model using the optimal knots obtained.

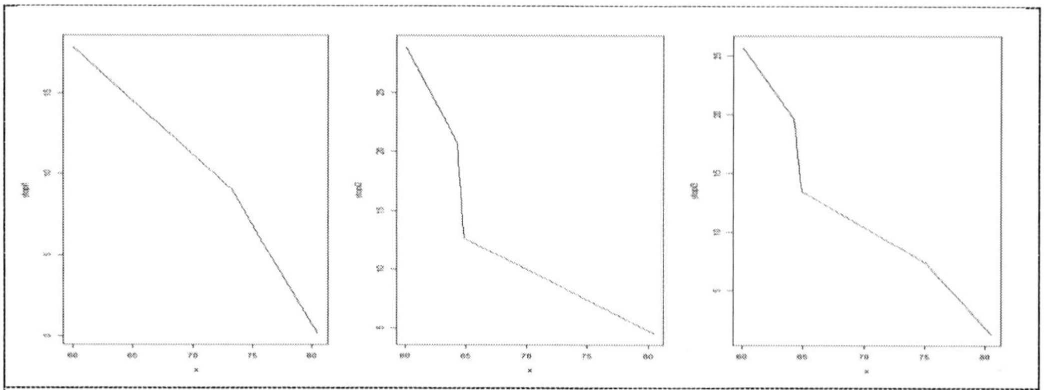

**Figure 2.** Prediction Models with One Knot Points, Two Knot Points and Three Knot Points.

To measure the accuracy of the model obtained on the estimation used MSE. Table 1 is a comparison of MSE values for the optimal one, two and three knot models obtained.

**Table 1.** A table comparison of MSE value

| Many Knot | Knot Points | MSE |
|---|---|---|
| 1 | $k_1 = 73.26647$ | 12.94277 |
| 2 | $k_1 = 64.26206$<br>$k_2 = 64.86235$ | 11.90754 |
| 3 | $k_1 = 64.26206$<br>$k_2 = 64.86235$<br>$k_3 = 75.06735$ | 11.08420 |

## 4. Conclusion

The study conducted three models is a model with a one-knot point, a model with two-knot points, and a model with three-knot points. Based on indicators of the value of Generalized Cross-Validation (GCV) and the amount of Mean Square Error (MSE), the best model was a model with three knots, with a 1000 value of smoothing parameter, 11.26236 value of GCV, 11.08420 value of MSE, and 56.1% value of the coefficient of determination $(R^2)$. The best model obtained is listed in Equation (7).

$$\begin{aligned}\hat{y} &= \hat{\alpha}_1 B_{-1,2} + \hat{\alpha}_2 B_{0,2} + \hat{\alpha}_3 B_{1,2} + \hat{\alpha}_4 B_{2,2} + \hat{\alpha}_5 B_{3,2} \\ &= 25{,}702932\, B_{-1,2} + 19{,}592885\, B_{0,2} + 13{,}484852\, B_{1,2} + 7{,}385023\, B_{2,2} + 1{,}295194\, B_{3,2}\end{aligned} \tag{7}$$

On a simplified base and formed the following function

$$\hat{y} = \begin{cases} \dfrac{474.9746852599 - 6.110047x}{4.20206}, 60.06 < x \le 64.26206 \\ \dfrac{404.2761960647 - 6.108033x}{0.60029}, 64.26206 < x \le 64.86235 \\ \dfrac{533.2621581981 - 6.099829x}{10.205}, 64.86235 < x \le 75.06735 \\ \dfrac{497.0460194941 - 6.089829x}{5.40265}, 75.06735 < x \le 80.47 \end{cases} \quad (8)$$

## Acknowledgments

This work was supported by the Statisticals Research Areas, Universitas Mataram.

## References

[1] Nansadiqa L, Masbar R and Majid M S A 2019 Does economic growth matter for poverty reduction in Indonesia *East African Scholars Journal of Economics, Business and Management* **2** (2) 46-52

[2] Novianto, S and Sudarsono H 2018 Analysis Of Poverty Level in Districts/Cities Of Central Java *Jurnal Ekonomi Pembangunan* **16** (1) 1-12

[3] Fahrika A I, Salam H and Buhasyim M A 2020 Effect of Human Development Index (HDI), Unemployment, and Investment Realization toward Poverty in South Sulawesi-Indonesia *The International Journal of Social Sciences World* **2** (2) 110-116

[4] Massaid A, Hanif M, Febrianti D and Chamidah N 2019 Modelling of Poverty Percentage of Non-Food Per Capita Expenditures in Indonesia Using Least Square Spline Estimator *IOP Conference Series: Materials Science and Engineering* **546** (5) 052044

[5] Sriliana I, Sunandi E and Rafflesia U 2018 The Poverty Modeling Using Small Area Estimation with Semiparametric P-spline *Asian Journal of Applied Sciences* **6** (4). 11-19

[6] Mariati N P A M, Budiantara I N and Ratnasari V 2019 Modeling Poverty Percentages in the Papua Islands using Fourier Series in Nonparametric Regression Multivariable *Journal of Physics: Conference Series* **1397** (1) 012071

[7] Dewanti P, Budiantara I N and Rumiati A T 2020 Modelling of SDG's Achievement in East Java Using Bi-responses Nonparametric Regression with Mixed Estimator Spline Truncated and Kernel. In *Journal of Physics: Conference Series* **1562** (1) 012016

[8] Iorio C, Frasso G, D'Ambrosio A and Siciliano R 2018 A P-spline based clustering approach for portfolio selection *Expert Systems with Applications* **95** 88-103

[9] Wagner J, Münnich R, Hill J, Stoffels J and Udelhoven T 2017 Non-parametric small area models using shape-constrained penalized B-splines *Journal of the Royal Statistical Society: Series A* **180** (4) 1089-1109

[10] Zamani P, Amirabadi-Farahani M, Aliarabi H and Malecky M 2016 Comparison of different Legendre and B-Spline random regression models of lamb's estimate variance components for average birth weight per lambing in Mehraban sheep *Iranian Journal of Animal Science* **46** (4) 407-415

# IoT based Soil Moisture and Sun Light Monitoring for Smart Planting

**Jonathan and D Widjaja**

Electrical Engineering Department, Faculty of Science and Technology, Sanata Dharma University, Yogyakarta, Indonesia

Corresponding author: damar@usd.ac.id

**Abstract.** Plant enthusiasts, entrepreneurs, and farmers usually have more than one plant that needs to be looked after every day. Lack of water and unproportioned sun light can be a serious problem if they have to leave the plant for a long period of time. An Internet of Things (IoT) based automatic remote plant sprinkler with sun light cover was designed to solve this problem. This system eases the users for monitoring the condition of plants without having to be in the field. The system model uses an ESP32 microcontroller, Light Dependant Resistor (LDR), relay, and L298N DC motor driver. In dry soil conditions, the water pump and sprinkler will turn on and water the soil. When the soil conditions are moist or wet the pump and sprinkler will not turn on. When there is intense sun light, the DC motor will pull the curtains to cover the plant. All the data can be monitored remotely via Wi-Fi connection. The results of this work shows that the system works well with 100% monitoring success rate.

## 1. Introduction

Plant enthusiasts, entrepreneurs, and farmers must have more than one plant that needs to be cared every day. One of the most important treatments is to provide water with sprinklers. Lack of watering can cause serious problem for plant if the owner have to leave the plant for a long time [1]. The other factor to provide a good environment for plant to grow is the sun light intensity [2]. Sufficient water and adequate sun light are important factors in carrying out photosynthetic activities. If this is not well considered, the plants will quickly wither and die. An IoT based automatic plant sprinkler and cover system that can be monitored and controlled remotely makes it easier for users to monitor the condition of these plants anywhere and anytime [3]-[7].

Gunawan (2018) has designed automatic sprinkler using soil moisture sensor [8]. This system will automatically water the plant whenever the plant need watering based on the soil condition. Prasojo et al. (2020) has also designed automatic sprinkler using Arduino Uno microcontroller [9]. Other references about IoT based automatic plant sprinkler can be seen in [10]-[12]. In this work, Android based smartphone receives and displays the soil condition data, whether it is dry, moist, or wet according to the sensor readings.

Based on the aforementioned problems, the development of automatic sprinkler combined with sun light plant cover was proposed in this work. Soil condition and sun light intensity will be sensed to drive the system whether it will water and cover the plant or not.

The remainder of this paper is organized as follows. Section 2 provides the description of system model used in this work. Section 3 presents results and discussion. It will describe the system performance based on the system success rate. Section 4 provides the conclusion of the work.

## 2. System Model

The system is using a remote control via the Blynk application as an IoT platform embedded in smartphone. This tool uses a soil moisture sensor to determine the moisture content in the soil and an LDR sensor to determine the intensity of sun light. DC motor is used as actuator to open and close of the curtains as plant cover from the intense sun light. DC water pump is used for watering the plants. The data that has been obtained from each sensor will be sent to the Blynk application. This system uses ESP32 microcontroller to control all actuators.

Figure 1 show the system model of this work. Soil moisture sensor will detect three condition; dry, humid and wet. If the soil is dry, when it gets an output with a range > 3500, the relay will turn on then the DC water pump will water the plants. If the soil is in a humid condition with a range > 1600 and < 3500 or wet with a range < 1600, the relay will not turn on and the DC water pump will not flush. The LDR sensor will work if the sunlight intensity is too hot and it results an output with a range < 1500 and the DC motor will close the plant curtains. Likewise, when the sunlight intensity is not too hot and an output has range > 1500, the DC motor will open the plant curtains.

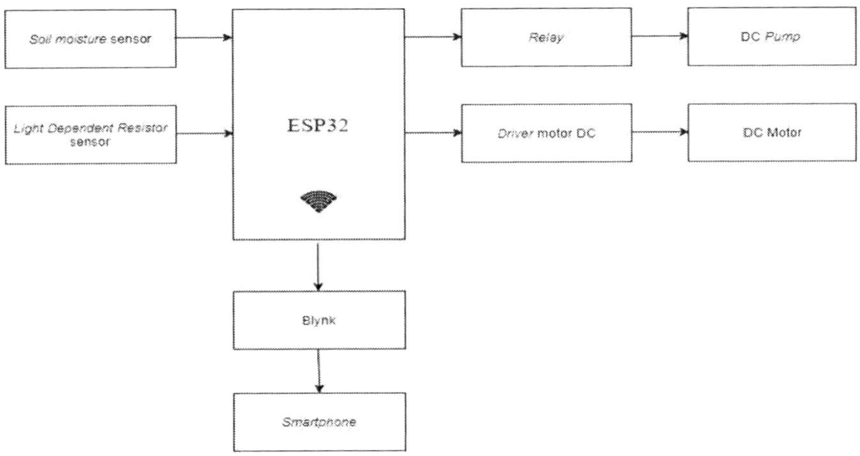


**Figure 1.** System Model

### 2.1. Activity Diagram

Figure 2 show the flowchart of the overall system work process. At the beginning of the process, the system will search for a Wi-Fi connection. If it is connected, then it will proceed to process each sensor reading. The soil moisture sensor then will initialize the input port of microcontroller and it will read the analogue data value of the sensor.

The soil is classified as dry soil when it gets an output sensor value with a range above 3500, humid soil when the value range below 3500 and above 1600, and wet soil when the value range below 1600. When the value of soil moisture sensor above 3500, then the relay and DC water pump will turn on. When the value of soil moisture sensor is below 3500, then the relay and DC water pump will not turn on. Data from sensor measurement results are then sent to Android smartphone via Blynk application.

The sensing and monitoring is done in the same process for LDR sensor. The LDR sensor will initialize the input of microcontroller and it will read the analogue data value of the sensor. When the analogue value of the LDR is below 1500, the plant curtains will open. When the LDR sensor value is above 1500, it will close the curtains and cover the plants. Data from sensor measurement results are then sent to Android smartphone via Blynk application.

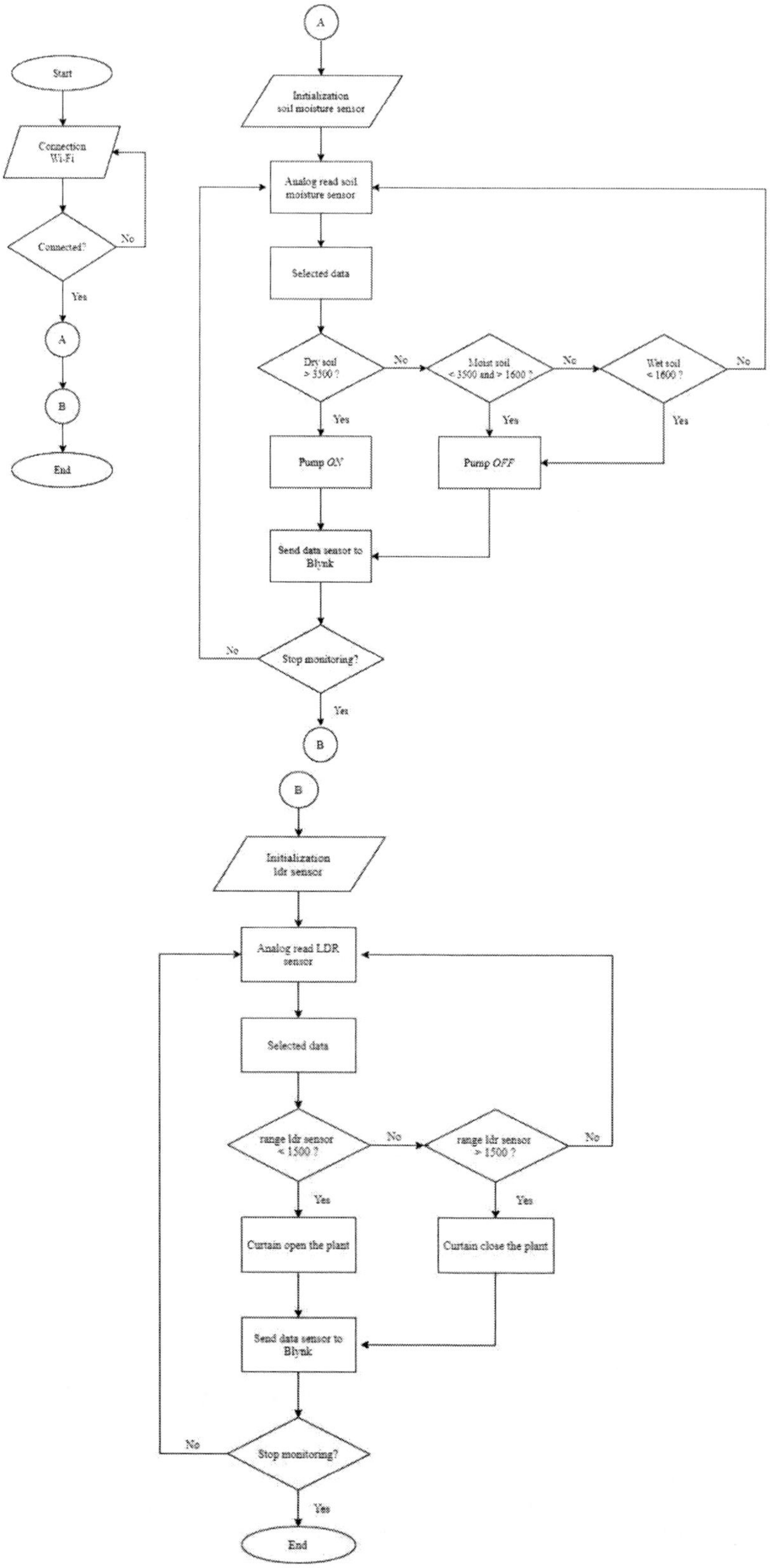


**Figure 2.** System Flowchart

The frame of this system is made using a metal hole angled and shown in Figure 3. The system consists of two identical sub systems. Each sub systems have different IP addresses. Moreover, it have two sensors, that are soil moisture and LDR sensor, and two actuators, that are water pump and DC motor to water the plan and push-pull the black curtain plant cover. The use of two identical sub systems is to test the performance of IoT platform based on the two different IP addresses.

(a). Front view (b). Back view

**Figure 3.** Design of the Hardware Model

## 3. Result and Discussion

Table 1 show the results of the data from the system testing used to measure the success rate. System testing is carried out in several parts, namely the sensor which is *soil moisture* able to detect soil, dry, moist and wet conditions. The LDR sensor is able to detect light intensity. The actuator DC motor is able to open and close plant curtains and a water pump can water plants. All of the data obtained is sent to Blynk successfully.

**Table 1.** Overall System Test Result

| No | Light Condition | DC Motor | Plant Curtain | Soil Condition | Water Pump | Send Data to Blynk | Note |
|---|---|---|---|---|---|---|---|
| 1 | Dark | OFF | CLOSE | DRY | ON | YES | Success |
| 2 | Lamp 5 Watt | OFF | CLOSE | | | YES | Success |
| 3 | Lamp 15 Watt | OFF | CLOSE | | | YES | Success |
| 4 | Lamp 25 Watt | OFF | CLOSE | MOIST | OFF | YES | Success |
| 5 | Lamp 40 Watt | OFF | CLOSE | | | YES | Success |
| 6 | Lamp 60 Watt | ON | OPEN | WET | OFF | YES | Success |
| 7 | Lamp 100 Watt | ON | OPEN | | | YES | Success |

**Table 2**. Test Result of Plant Watering Sub System 1

| No | Test # | Soil Moisture Sensor Value | ADC | Soil Condition | Analog reading | Digital reading | Pump ON (Second) | Note |
|---|---|---|---|---|---|---|---|---|
| | 1 | 4095 | 3.3V | DRY | DRY | DRY+ | 16.17 | Success |
| | 2 | 4095 | 3.3V | DRY | DRY | DRY+ | 16.92 | Success |
| 1 | 3 | 4095 | 3.3V | DRY | DRY | DRY+ | 20.58 | Success |
| | 4 | 4095 | 3.3V | DRY | DRY | DRY+ | 22.04 | Success |
| | 5 | 4095 | 3.3V | DRY | DRY | DRY+ | 31.16 | Success |
| | 1 | 1641 | 1.33V | MOIST | MOIST | NOR | 0 | Success |
| | 2 | 1759 | 1.42V | MOIST | MOIST | NOR | 0 | Success |
| 2 | 3 | 1821 | 1.47V | MOIST | MOIST | NOR | 0 | Success |
| | 4 | 2055 | 1.65V | MOIST | MOIST | NOR | 0 | Success |
| | 5 | 2352 | 1.9V | MOIST | MOIST | NOR | 0 | Success |
| | 1 | 1008 | 0.82V | WET | WET | WET+ | 0 | Success |
| | 2 | 1052 | 0.89V | WET | WET | WET+ | 0 | Success |
| 3 | 3 | 1103 | 0.92V | WET | WET | WET+ | 0 | Success |
| | 4 | 1133 | 0.92V | WET | WET | WET+ | 0 | Success |
| | 5 | 1112 | 0.9V | WET | WET | WET+ | 0 | Success |

**Table 3**. Test Result of Plant Watering Sub System 1

| No | Test # | Soil Moisture Sensor Value | ADC | Soil Condition | Analog reading | Digital reading | Pump ON (Second) | Note |
|---|---|---|---|---|---|---|---|---|
| | 1 | 4095 | 3.3V | DRY | DRY | DRY+ | 16.32 | Success |
| | 2 | 4095 | 3.3V | DRY | DRY | DRY+ | 16.85 | Success |
| 1 | 3 | 4095 | 3.3V | DRY | DRY | DRY+ | 20.16 | Success |
| | 4 | 4095 | 3.3V | DRY | DRY | DRY+ | 21.7 | Success |
| | 5 | 4095 | 3.3V | DRY | DRY | DRY+ | 30.29 | Success |
| | 1 | 1708 | 1.38V | MOIST | MOIST | NOR | 0 | Success |
| | 2 | 1762 | 1.42V | MOIST | MOIST | NOR | 0 | Success |
| 2 | 3 | 1835 | 1.48V | MOIST | MOIST | NOR | 0 | Success |
| | 4 | 2032 | 1.64V | MOIST | MOIST | NOR | 0 | Success |
| | 5 | 2384 | 1.92V | MOIST | MOIST | NOR | 0 | Success |
| | 1 | 1023 | 0.84V | WET | WET | WET+ | 0 | Success |
| | 2 | 1072 | 0.88V | WET | WET | WET+ | 0 | Success |
| 3 | 3 | 1141 | 0.92V | WET | WET | WET+ | 0 | Success |
| | 4 | 1210 | 0.98V | WET | WET | WET+ | 0 | Success |
| | 5 | 1125 | 0.94V | WET | WET | WET+ | 0 | Success |

Table 2 and Table 3 shows that analogue reading from sensors and digital value from ADC can detect soil conditions when it is dry, moist or wet. The tests for each condition were done 5 times. When the soil condition is dry, the analogue value is 4095 and ADC value is 3.3V. The water pump will turn on. When the condition is moist or wet, the water pump will turn off. The soil moisture sensor and the DC pump work well according to the design with 100 % success rate.

Table 4 and Table 5 show the test result of plant cover of sub system 1 and 2. When the lights are dark, 5 Watt, 15 Watt, 25 Watt and 40 Watt, the DC motor is OFF. When use the lamp of 60 Watt and 100 Watt, the DC motor is ON. The digital light meter shows the measurement of the lux value of each lamp condition. When the analogue sensor value increase, the lux value will decrease and vice versa. The LDR sensor and DC motor work well with a 100 % success rate.

**Table 4**. Test Result of Plant Cover Sub System 1

| No | Lamp Condition | LDR Value | ADC | Digital Light Meter AS803 | DC Motor | Note |
|---|---|---|---|---|---|---|
| 1 | Dark | 4095 | 3.3V | 0 lux | OFF | Success |
| 2 | 5 W | 2538 | 2.05V | 3 lux | OFF | Success |
| 3 | 15 W | 2384 | 1.92V | 5 lux | OFF | Success |
| 4 | 25 W | 2128 | 1.74V | 10 lux | OFF | Success |
| 5 | 40 W | 1606 | 1.3V | 22 lux | OFF | Success |
| 6 | 60 W | 1358 | 1.09V | 36 lux | ON | Success |
| 7 | 100 W | 974 | 0.78V | 78 lux | ON | Success |

**Table 5**. Test Result of Plant Cover Sub System 2

| No | Lamp Condition | LDR Value | ADC | Digital Light Meter AS803 | DC Motor | Note |
|---|---|---|---|---|---|---|
| 1 | Dark | 4095 | 3.3V | 0 lux | OFF | Success |
| 2 | 5 W | 2445 | 1.97V | 3 lux | OFF | Success |
| 3 | 15 W | 2346 | 1.88V | 5 lux | OFF | Success |
| 4 | 25 W | 2110 | 1.7V | 10 lux | OFF | Success |
| 5 | 40 W | 1685 | 1.36V | 22 lux | OFF | Success |
| 6 | 60 W | 1454 | 1.33V | 36 lux | ON | Success |
| 7 | 100 W | 1125 | 0.91V | 78 lux | ON | Success |

## 4. Conclusion

The system works well as it was designed. Both of sub systems can send data to the Blynk correctly. Sprinkler work well with 100% success rate. Soil moisture sensor and LDR can detect soil and sun light condition as it specification range. Water pump will water the soil when the condition is dry.

## Acknowledgments

Thank you to the Institute for Research and Community Service (Lembaga Penelitian dan Pengabdian kepada Masyarakat - LPPM) Sanata Dharma University for funding and facilitating this research.

## References

[1] Purnawarman R 2004 *AT89C51 Microcontroller Based Automatic Plant Watering Tool* Final Project, Sanata Dharma University, Yogyakarta.

[2] Pardede F O I and Simamora B A 2020 Designing Automatic Sprinklers and Web-Based Plant Monitoring Using a Microcontroller *Jurnal Mantik* **4** (3) 2232-2236

[3] Kumar J, Gupta N, Kumari A and Kumari S 2019 Automatic plant watering and monitoring system using NodeMCU *2019 9th International Conference on Cloud Computing, Data Science & Engineering* 545-550

[4] Bhardwaj S, Dhir S and Hooda M 2018 Automatic plant watering system using IoT *2018 Second International Conference on Green Computing and Internet of Things* 659-663

[5] Monica M, Yeshika B, Abhishek G S, Sanjay H A and Dasiga S 2017 IoT based control and automation of smart irrigation system: An automated irrigation system using sensors, GSM, Bluetooth and cloud technology *2017 International Conference on recent innovations in signal processing and embedded systems* 601-607

[6] Johar R, Bensenouci A and Bensenouci M A 2018 IoT based smart sprinkling system. In *2018 15th Learning and Technology Conference* 147-152

[7] Kuncoro A H, Mellyanawaty M, Sambas A, Maulana D S, Subiyanto and Mamat M 2020 Air Quality Monitoring System in the City of Tasikmalaya based on the Internet of Things (IoT) *Jour of Adv Research in Dynamical & Control Systems* **12** (2) 2473-2479

[8] Gunawan M S 2018 Automatic Water Sprinkler using Soil Moisture Sensor *Journal of Electrical Technology* **3** (1) 1-15

[9] Prasojo I, Maseleno A and Shahu N 2020 Design of automatic watering system based on Arduino *Journal of Robotics and Control* **1** (2) 59-63

[10] Noerhayati E, Dwisulo B and Rahmawati A 2020 Sprinkler irrigation design with microcontroller based on IoT *IOP Conference Series: Earth and Environmental Science* **456** (1) 012063

[11] Chowdhury B S and Raghukiran N 2017 Autonomous sprinkler system with Internet of Things *International Journal of Applied Engineering Research* **12** (16) 5430-5432

[12] Athawale S V, Solanki M, Sapkal A, Gawande A and Chaudhari S 2020 An IoT-Based Smart Plant Monitoring System *Smart Computing Paradigms: New Progresses and Challenges* 303-310

# Intervention Model of IDX Finance Stock for the Period May 2010 – May 2020 Due to the Effects of the Corona Virus

**R Utami[1], M Hadijati and I G A W Wardhana**

Department of Mathematic, Faculty of Mathematic and Natural Science, Universitas Mataram, Mataram, Nusa Tenggara Barat, Indonesia

Corresponding author: riyana.utami20@gmail.com

**Abstract.** This study aim to model IDX Finance data that was intervened by the corona virus. With the increasing spread of the corona virus, the consequence is that it is necessary to anticipate to prevent further losses. To model time series data that is intervened by an event, it is necessary to carry out intervention analysis to describe the changes that have occurred in the data and forecasting can be done for some future data. By looking at changes in patterns in IDX Finance data since January 2020 which are immediately felt and are prolonged in nature, it can be concluded that IDX Finance's stock closing price data has a permanent abrupt response pattern and intervention order $b = 0$, $s = 0$, $r = 0$. However, the final model that is produced does not have significant parameters that are generated by the maximum likelihood estimation method. Because there are several possibilities that the data does not contain intervention or data is used too little to identify the order of intervention.

## 1. Introduction

The term intervention is often found in various statistical analyzes, especially in data analysis related to time, also known as time series analysis. Intervention defined as the following, namely the act of interfering with the results or direction, especially of a condition or process (such as to prevent loss or increase function) or defined as the interference of a state in the affairs of another country to force it to take or refrain from taking specific actions. While the time series is a series of observations on a variable taken based on a fixed time sequence [1]. Time series data that are intervened by an event or certain events result in the data experiencing a changing pattern depending on the nature of the events that affect it, namely temporary (pulse) and prolonged (step) [2]. To describing the effect of an intervention that affects data, intervention analysis is used [3].

Since the coronavirus (covid-19) was declared a global pandemic by the WHO Direct-General. Covid-19 intervened in all fields, one of which was in the economic sector [4]-[7]. The quick spread of the coronavirus in China to various countries has sparked concern among market players. Indonesia as one of the affected countries feels the consequences. The domestic stock market has been hit by negative sentiment from the coronavirus (Corona effect), one of which is that the shares of the Indonesia Stock Exchange (IDX) in the financial sector have decreased.

So to describe the effect of the coronavirus intervening in IDX finance stock data, it is necessary to conduct an intervention analysis so that a model will obtain. The intervention model will be said to be appropriate to describe the corona effect that occurs in the data if the parameters of the model are significant and meet the model's assumptions. The basic assumption is that the residual is white noise (identically-independent distributed) and normally distributed. To obtaining significant parameters, it is necessary to estimate the parameters. The method of parameter estimation often used is the maximum likelihood method. It is because that the maximum likelihood method is closely related to numerical ability, especially in producing a solution point in an equation [8]. Maximum likelihood is a parameter estimation method that maximizes the likelihood function [9]-[10]. This method can be used to estimate parameters simultaneously from several parameters from a specific population. In this case, the parameter estimate value maximizes its likelihood function.

## 2. Methodology

The data used is secondary data obtained from the Investing.id website regarding the latest price data (or closing price) of Idx finance shares from May 2010 - May 2020. The initial prediction of the intervention variable is the step function because the intervention affects the stock closing price in the long term. The intervention variables are defined as follows:

$$S_t^{(T)} = \begin{cases} 0 \ , & t \neq T \ (\textit{before the decline in share prices due to the effects of the coronavirus}) \\ 1 \ , & t \geq T \ (\textit{during and after the decline in share prices due to the effects of the coronavirus} \end{cases} \quad (1)$$

Where $T$ is the time of intervention, which is the time when the effects of the coronavirus begin to affect the IDX finance stock data.

The analysis steps that will be carried out are:

1. Create a time series plot to see patterns of intervention responses and predict possible intervention variables.
2. Dividing the data into two groups of data, namely the pre-intervention data group and the post-intervention data group.
3. Create the ARIMA model from the pre-intervention data group so that an ARIMA model is produced with all significant parameters and fulfilling the residual assumptions, namely white noise (identical-independent) and normally distributed.
4. Predict several post-intervention data groups using the ARIMA model that is obtained from the pre-intervention data group.
5. Calculating the response residuals between the post-intervention data and the forecast results from the pre-intervention data.
6. Identifying patterns of intervention response through residual graphs of response to forecasting results, and identifying intervention orders, namely, $b$, $s$, and $r$.
7. Create an intervention model to produce a model with all significant parameters and satisfying the residual assumptions.
8. Make forecasts using the best intervention model.

## 3. Results and Discussion

To get an intervention model from the IDX stock closing price data, the first step that needs to be done is to create a time series plot to see changes in patterns that occur in data due to the impact of the coronavirus. The following is the result of the time series plot from the IDX Finance data in Figure 1.

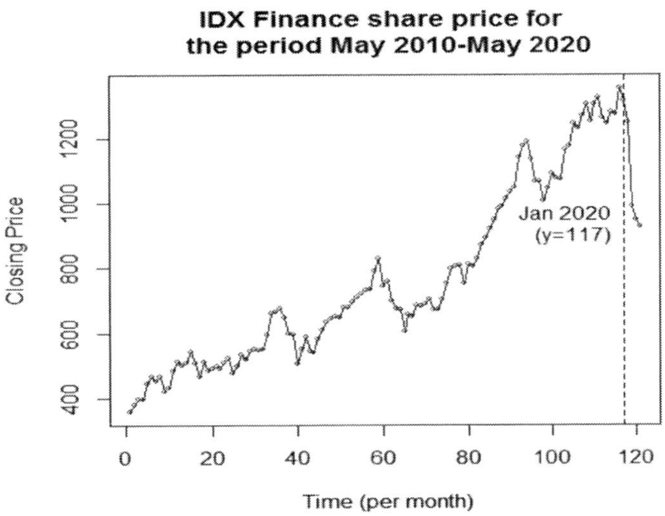


**Figure 1.** Time series plot of IDX Finance Period May 2010 - May 2020

Based on Figure 1, that can be concluded that from January 2020 to May 2020 the closing price of IDX shares has decreased, inversely proportional to the data in the period May 2010 to December 2019 which tends to increase every month. This decrease was due to intervention by the increasingly rapid spread of the coronavirus which resulted in many changes in all fields. This decline was seen in the 117th series.

### 3.1 Formation of ARIMA Model from Pre-Intervention Data Group

By looking at the pattern of changes in the time series plot in Figure 1, IDX Finance data is dividing into two data groups, namely the pre-intervention data group (pre-intervention) and the data group during and after the intervention (post-intervention). After obtaining the pre-intervention data group, namely the stock data from May 2010 to December 2019, the next step is to create an ARIMA model from the pre-intervention data group.

To create an ARIMA model, the data must first meet the stationary data, namely stationary to variance and average. By looking at the Box-Cox graph, a $\lambda$ value is $-0.35$ which means that the data needs to be transformed by the transfer function used, namely:

$$Z_t = \frac{Z_t^{\lambda} - 1}{\lambda} \tag{2}$$

Then the differencing is made once. After generating the stationary data, to identify the ARIMA model can use ACF and PACF plot.

The best ARIMA model chosen is ARIMA $(2,1,2)$ because its parameters are significant, satisfied all residual assumptions, and have the smallest RMSE, AICc, and AIC values compared to other ARIMA models as in Table 1.

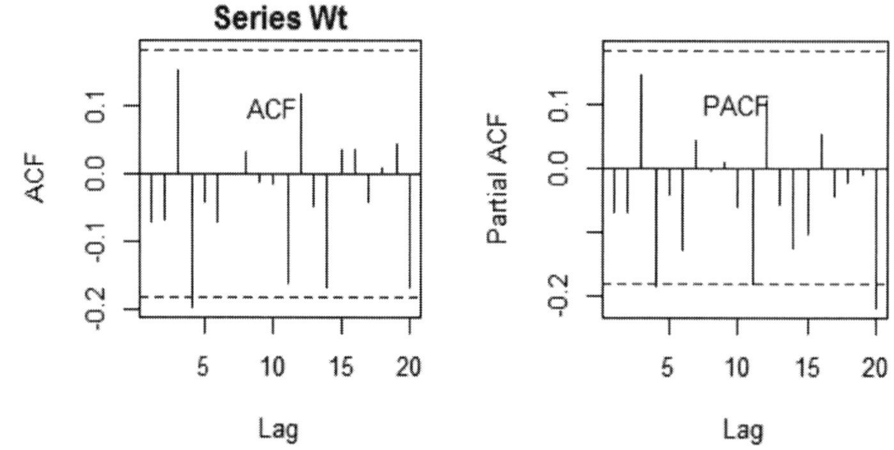


**Figure 2.** ACF and PACF plot of a pre-intervension data set

**Table 1** Comparative analysis

| d=1 | | MA(q) | | | |
|---|---|---|---|---|---|
| | | 0 | 1 | 2 | 3 |
| | 0 | ARIMA(0,1,0)<br>AICC = -1161.56<br>AIC = -1161.59<br>BIC = -1158.85<br>RMSE = 0.00542 | ARIMA(0,1,1)<br>AICC = -1159.56<br>AIC = -1159.66<br>BIC = -1154.17<br>RMSE = 0.00542 | ARIMA(0,1,2)<br>AICC = -1157.5<br>AIC = -1157.72<br>BIC = -1149.48<br>RMSE = 0.00542 | ARIMA(0,1,3)<br>AICC = -1160.25<br>AIC = -1160.61<br>BIC = -1149.63<br>RMSE = 0.00530 |
| | 1 | ARIMA(1,1,0)<br>AICC = -1159.55<br>AIC = -1159.66<br>BIC = -1154.17<br>RMSE = 0.00542 | ARIMA(1,1,1)<br>AICC = -1157.45<br>AIC = -1157.66<br>BIC = -1149.43<br>RMSE = 0.00542 | ARIMA(1,1,2)<br>AICC = -1158.58<br>AIC = -1158.95<br>BIC = -1147.97<br>RMSE = 0.00534 | ARIMA(1,1,3)<br>AICC = -1159.61<br>AIC = -1160.16<br>BIC = -1146.44<br>RMSE = 0.00527 |
| AR(p) | 2 | ARIMA(2,1,0)<br>AICC = -1157.52<br>AIC = -1157.74<br>BIC = -1149.5<br>RMSE = 0.00542 | ARIMA(2,1,1)<br>AICC = -1159.19<br>AIC = -1159.55<br>BIC = -1148.57<br>RMSE = 0.00533 | ARIMA(2,1,2)[a]<br>AICC = -1161.96<br>AIC = -1162.51<br>BIC = -1148.79<br>RMSE = 0.00519 | ARIMA(2,1,3)<br>AICC = -1157.99<br>AIC = -1158.77<br>BIC = -1142.3<br>RMSE = 0.00525 |
| | 3 | ARIMA(3,1,0)<br>AICC = -1159.6<br>AIC = -1159.96<br>BIC = -1148.98<br>RMSE = 0.00532 | ARIMA(3,1,1)<br>AICC = -1160.03<br>AIC = -1160.58<br>BIC = -1146.85<br>RMSE = 0.00526 | ARIMA(3,1,2)<br>AICC = -1157.85<br>AIC = -1158.63<br>BIC = -1142.16<br>RMSE = 0.00526 | ARIMA(3,1,3)<br>AICC = -1159.44<br>AIC = -1160.49<br>BIC = -1141.27<br>RMSE = 0.00515 |

[a]model has a significant parameters obtained from the maximum likelihood estimation method and satisfied the residual assumptions

The ARIMA (2,1,2) model generated above can be defined as follows:

$$(1-\hat{\phi}_1 B-\hat{\phi}_2 B^2)\,(1-B)\,Z_t = (1-\hat{\theta}_1 B-\hat{\theta}_2 B^2)\,a_t$$

$$(1+1.717584B+0.873211B^2)\,(1-B)\,Z_t = (1+1.836051B+0.950613B^2)\,a_t$$

or,

$$Z_t = Z_{t-1} - 1.717584 Z_{t-1} + 1.717584 Z_{t-2} - 0.873211 Z_{t-2} + 0.873211 Z_{t-3} + a_t + 1.836051 a_{t-1} + 0.950613 a_{t-2} \quad (3)$$

With the ARIMA model above, several post-intervention data groups that predicted to produce the following:

**Table 2.** Forecasting results of ARIMA (2,1,2)

| Month | Forecast |
|---|---|
| **January 2020** | 2.629425 |
| **February 2020** | 2.628386 |
| **March 2020** | 2.629071 |
| **April 2020** | 2.628801 |
| **May 2020** | 2.628667 |

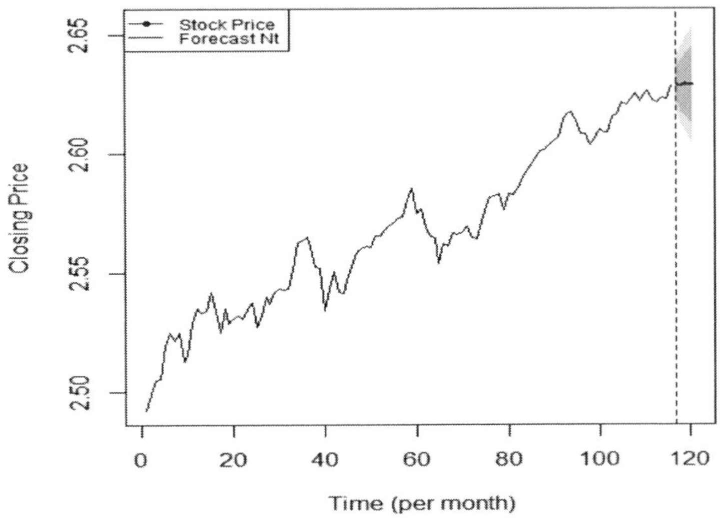


**Figure 3.** Comparison of Actual Value and Forecast Data of IDX Finance

### 3.2 Formation of an intervention model

Figure 4 is a residual graph obtained from the difference between the actual values from June 2019 - May 2020 that was used to identify the intervention orders, namely $b$, $s$ and $r$. Order $b$ represents the starting time of the impact of the intervention, order $s$ represented the delay time the data starts to stabilize calculated from the time the intervention occurs, and order $r$ represents the next time-lag after $b$ and $s$ when the data forms have a clear pattern.

Based on Figure 4, it is found that the residuals exceed the significance limit, which is three times the RMSE of the ARIMA (2,1,2) model, which is $0.00518973$. So it can be concluded that there are several intervention models with various possible intervention orders. With the initial assumption that the data was intervened by the coronavirus for a long term (step function), and by looking at Figure 1 it can be concluded that the coronavirus directly affects the data permanently, it can be concluded that the intervention response pattern is abrupt permanent. The abrupt permanent pattern is defined as follows:

$$\omega B^b S_t^{(T)} \tag{4}$$

Because the effect of direct intervention is immediate, the order $b=0$. Several possible intervention orders include $b=0, s=0, r=0$; $b=0, s=1, r=0$; Z; $b=0, s=2, r=0$; $b=0, s=3, r=0$ and $b=0, s=4, r=0$. From the several possible orders above, it is found that none of the intervention models with the intervention order has a significant parameter. And for the intervention model with order $s>1$, there are obstacles in running the R program, namely that the intervention model is not produced. So that the intervention model that is finally used is an intervention model with the order $b=0, s=0, r=0$ which is defined as follows:

$$Z_t = \hat{\omega}_0 S_t^{(T)} + \frac{\left(1-\hat{\theta}_1 B - \hat{\theta}_2 B^2\right)}{\left(1-\hat{\phi}_1 B - \hat{\phi}_2 B^2\right)(1-B)} a_t \tag{5}$$

$$Z_t = (-0.0048753)\, S_t^{(T)} + \frac{\left(1+0.0021066B-0.9785980B^2\right)}{\left(1-0.0936719B-0.8995401B^2\right)(1-B)} a_t \tag{6}$$

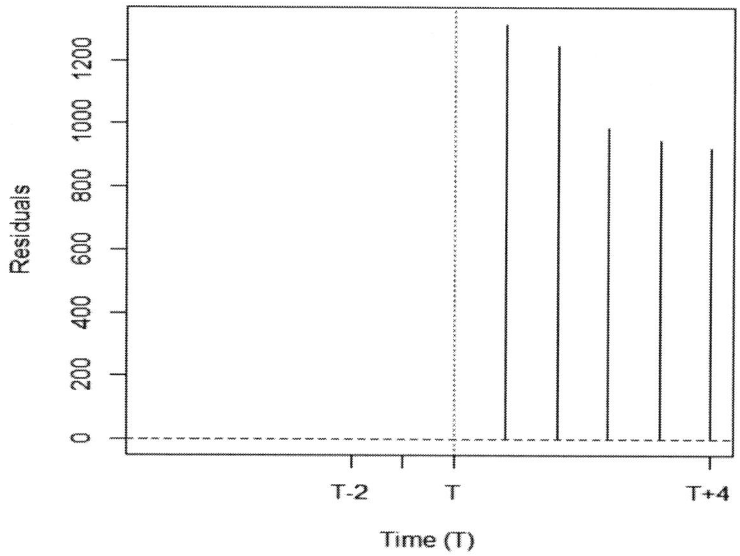


**Figure 4.**Residual Graph

## 4. Conclusion

By looking at the results of the intervention model obtained, namely the model that is not significant to the parameters. So it can be concluded that several possibilities resulted in this, including that there was no intervention in IDX Finance data or other words changes in IDX Finance data were not caused by the intervention of the coronavirus, and the data is used too little to identify the order of intervention.

## Acknowledgments

Special thanks to KPBI Statistika of Universitas Mataram for their support.

## References

[1] Alwan L C and Roberts H V 1988 Time-series modeling for statistical process control *Journal of business & economic statistics* **6** (1) 87-95
[2] Rehman A, Jingdong L, Chandio A A, Hussain I, Wagan S A and Memon Q U A 2019 Economic perspectives of cotton crop in Pakistan: A time series analysis (1970–2015)*Journal of the Saudi Society of Agricultural Sciences* **18** (1) 49-54
[3] Etuk E H and Eleki A G 2017Arima intervention analysis of monthly Xaf-Ngn Exchange rates occasioned by Nigerian economic recession *Noble International Journal of Economics and Financial Research* **2** (4) 76-81
[4] Caraka R E, Lee Y, Kurniawan R, Herliansyah R, Kaban P A, Nasution B I and PardameanB 2020Impact of COVID-19 large scale restriction on environment and economy in Indonesia *Global Journal of Environmental Science and Management* **6** 65-84
[5] Tanrıvermiş H 2020Possible impacts of COVID-19 outbreak on real estate sector and possible changes to adopt: A situation analysis and general assessment on Turkish perspective *Journal of Urban Management* **9** (3) 263-269
[6] Laing T 2020 The economic impact of the Coronavirus 2019 (Covid-2019): Implications for the mining industry *The extractive industries and society* **7** (2) 580-582
[7] Martin A, Markhvida M, Hallegatte S and Walsh B 2020 Socio-economic impacts of COVID-19 on household consumption and poverty *Economics of disasters and climate change* **4** (3) 453-479
[8] Durham G B and Gallant A R 2002 Numerical techniques for maximum likelihood estimation of continuous-time diffusion processes *Journal of Business & Economic Statistics* **20** (3) 297-338
[9] Johnstone D J 2011 Economic interpretation of probabilities estimated by maximum likelihood or score *Management Science* **57** (2) 308-314
[10] Solarin S A 2020 The effects of shale oil production, capital and labour on economic growth in the United States: A maximum likelihood analysis of the resource curse hypothesis *Resources Policy* **68** 101799

IOP Conf. Series: Materials Science and Engineering 1115 (2021) 012058 doi:10.1088/1757-899X/1115/1/012058

# Inflation modeling in Indonesia using hybrid autoregressive integrated moving average (ARIMA)-neural network (NN)

**D Pratiwi and M Hadijati**

Department of Mathematics, Faculty of Mathematics and Natural Sciences, Universitas Mataram, Mataram, Indonesia

*Email: denipratiwi26@gmail.com

**Abstract.** One of the macroeconomic indicators to see the stability of a country's economy is inflation. This study aims to model the value of monthly inflation in Indonesia from January 2003 to December 2019 using the ARIMA-NN hybrid. The data plot shows a non-linear pattern and trends, so that the differencing process is carried out and the model is built using ARIMA model. The best ARIMA model obtained is SARIMA $(1,1,0)(0,1,1)^{12}$ with a Root Mean Square Error (RMSE) of 0.01134. Furthermore, ARIMA residuals that do not satisfy white noise and normality are modeled using NN. The best structure obtained of NN model is $(3\times2\times1)$ with an RMSE of 0.023984. From the ARIMA and residual NN prediction results, the ARIMA-NN hybrid model is obtained to predict the value of monthly inflation in Indonesia for the next 12 months with the Mean Absolut Percentage Error (MAPE) value is 11.40873%. It means that the model result has high prediction accuracy.

## 1. Introduction

Inflation can briefly be interpreted as a tendency to increase the prices of goods and services generally and continuously [1]. Based on BPS [2], the month-to-month inflation rate in Indonesia on June 2019 was 0.55 percent with 138.16 of Consumer Price Index (CPI). The highest inflation occurred in Manado at 3.60 percent and the lowest occurred in Singaraja at 0.02 percent. Meanwhile, on August 2019, there was inflation of 0.12 percent. The highest inflation occurred in Kudus with 0.82 percent and the lowest occurred in Tasikmalaya, Madiun, and Pare-Pare with 0.04 percent in each. It can be seen that there has been an increase and decrease in the value of inflation in Indonesia, therefore inflation modeling is necessary. One of the statistical methods that can be used is time series. Time series is a series of observations collected based on the same time interval. One of the time series methods that can be used is Autoregressive Integrated Moving Average (ARIMA). ARIMA is able to model linear pattern. However, in reality the data does not only contain linear elements but still contains non-linear elements [3-5]. To overcome the limitations of ARIMA, the Neural Network (NN) method can be used. The NN method can model data without considering the linearity pattern of the data [6]. Therefore, the ARIMA-NN hybrid method is obtained which will be used to model monthly inflation data in Indonesia for the period January 2003 to December 2019.

Research on the ARIMA-NN hybrid model has been carried out by Zhang [6], who applied data on wolf's sunspot, Canadian lynx and dollar exchange rates. From that research, it is found that the ARIMA-NN hybrid model produced can be used as an effective way to improve forecasting performance. In the ARIMA-NN hybrid modeling the data is modeled using ARIMA, while the residuals of the ARIMA model are modeled using NN. To model the data with the hybrid ARIMA-

NN, initial testing is carried out using Tersvirta test to determine whether there was a non-linear pattern in the data. The accuracy of forecasting is measured based on the Mean Absolute Percentage Error (MAPE) value [7-8].

The paper is organized as follows. In the next section, it is explained the data set and methodology of research. Empirical results from the real data sets are reported in Section 3. Section 4 contains the concluding remarks.

## 2. Methodology

Control chart structure and log-normal transformation for CV statistic. The data in this study is secondary data obtained from the official website of Bank Indonesia https://bi.go.id which is the monthly data on inflation from Januari 2003 to December 2019. Data is divided to in sample data (from Januari 2003 to December 2018) and out sample data (Januari 2019 to December 2019).

The steps for modeling hybrid ARIMA-NN are as follows.

1. Collecting inflation data in Indonesia, then dividing the data into in sample data for modeling or training and out sample data to measure the accuracy of forecasting.
2. Testing the linearity data using Terasvirta test.
3. Testing the stationary of the data, then identify the ARIMA model against the stationary data and estimating the ARIMA parameters using Ordinary Least Square (OLS).
4. Testing the best ARIMA model residuals assumptions, including normality and white noise residuals.
5. Next, modeling ARIMA residuals using NN. Determining the NN input combination and conduct trial and error the NN paramaters used for training.
6. Calculate the value of out sample RMSE from the trial error results of the NN training. The NN structure with the smallest out sample RMSE value will be used for the NN testing step.
7. Perform the ARIMA residuals testing step and then forecasting using hybrid ARIMA-NN.

## 3. Result and Discussions

Hybrid is a combination of two or more systems in one function, in this case it is a combinations of ARIMA and NN. In general, the combinations of linear and non linear time series models can be written as follows

$$y_t = L_t + E_t,$$

where $L_t$ denotes a linear component and $E_t$ denotes a non linear component. There are two components that must be estimated from the data, thus are the ARIMA model is used to solve linear cases and the residuals of the linear model still containing non linear relationship information.

### 3.1 Results of the Terasvirta linearity test

To perform the Terasvirta test, it was carried out by regressing $X_t$ with the predictor variables $X_{t-1}, X_{t-2}, X_{t-3}$. Furthermore, from the regression results obtained residual ($\hat{u}_t$), then regressed with the initial predictor variable, 6 additional predictor variables, and the quadratic and cubic terms of the initial predictor variables, with a significance level of alfa 5% . The result is obtained p-value 0.000. So that it can be concluded to reject $H_0$ and it means that $x_t$ has a non linear pattern.

### 3.2. ARIMA Modeling

In an autoregressive integrated moving average model, the future value of a variable is assumed to be a linear function of several past observations and random errors. According to Wei [9], the general form of the AR model with the p-order or the ARIMA (p,0,0) is stated as follows

$$Z_t = \mu + \emptyset_1 Z_{t-1} + \emptyset_2 Z_{t-2} + \cdots + \emptyset_p Z_{t-p} + e_t.$$

The general form of the q-order MA model or ARIMA (0,0,q) can be written as follows

$$Z_t = \mu + e_t - \theta_1 e_{t-1} - \theta_2 e_{t-2} - \cdots - \theta_q e_{t-q}.$$

Therefore, the ARIMA Box-Jenkins or ARIMA (p,d,q) model is obtained as follows

$$\emptyset_p(B)(1-B)^d X_t = \theta_q(B)$$

where $X_t$ and $\varepsilon_t$ are the actual value and random error at time period $t$, respectively $\emptyset_p$ and $\theta_q$ are model parameters. The symbols p and q are integers and often referred to as orders of the model. Random errors ($\varepsilon_t$) are assumed to be independently and identically distributed witha mean of zero and a constant variance of $\sigma^2$. According to Aguilar et al. [10], the ARIMA model which contains seasonal patterns is called Seasonal ARIMA (SARIMA). In general, it can be written as the following equation

$$\varphi_p(L)\phi_p(B^d)\nabla^d\nabla_S^D X_t = \theta_q(B)\Theta_Q(B^S)a_t.$$

The SARIMA model can be used on data that shows a recurring pattern with a certain period and is a combination of non seasonal factors and seasonal factors. Figure 1 shows time series and Box cox plot $X_t$.

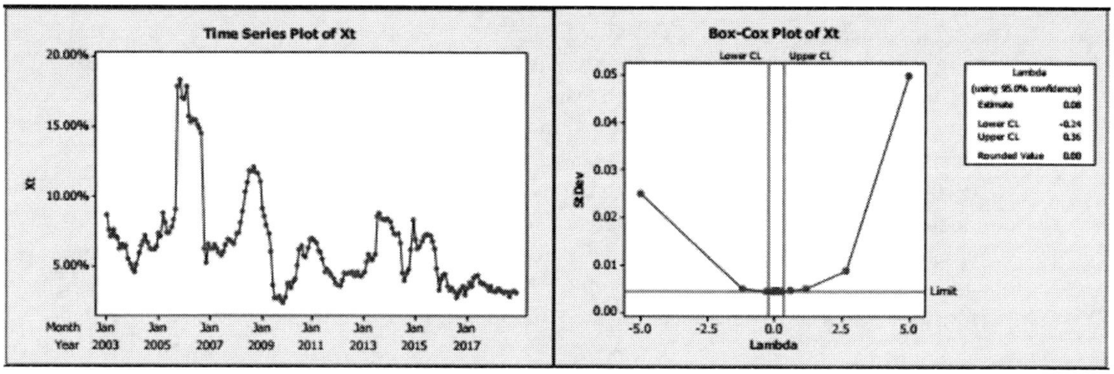


**Figure 1.** Time series and Box-cox plot $X_t$

Figure 1 shows the data is not stationary in the mean because it has trend pattern and Box-cox plot which shows a rounded value of 0 so that the data is not stationary in varian, so it is necessary to transform $\ln(X_t)$ and differencing 1st and 12th orders because the data forms a seasonal pattern every 12 months and then was obtained time series plot in Figure 2.

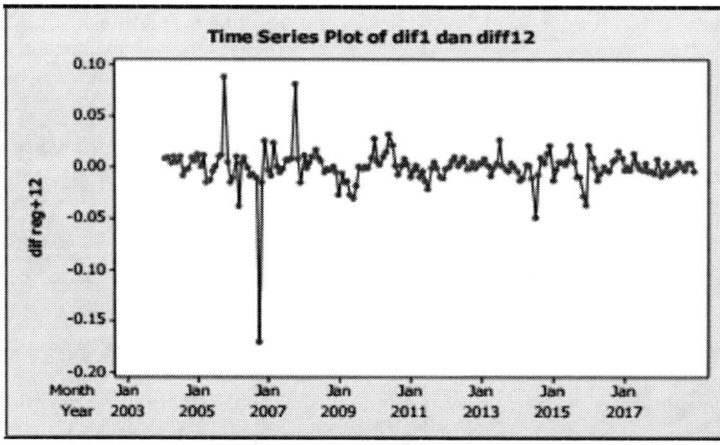


**Figure 2.** Time series plot stationery data

Figure 2 shows stationary time series plot of $X_t$. And then, identify the ARIMA model using ACF and PACF of the stationery data. The result is shown in Figure 3.

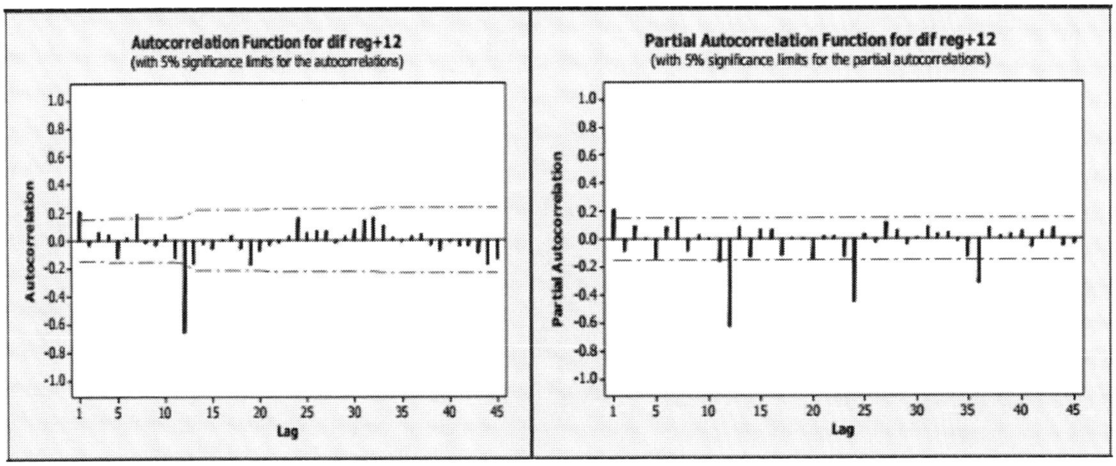


**Figure 3.** ACF and PACF stationery data

Based on the ACF and PACF in Figure 3, it can be seen that the seasonal pattern of ACF is cut off at lag 12, and PACF dies down, so that in the seasonal pattern it can be concluded that it contains ARIMA $(0,1,1)^{12}$. The ACF regular pattern is cut off pattern at lag 1 and 7, while PACF is cut off at lag 1. In the selection of the ARIMA model order can be done with the parsimony principle, meaning that selection a model with as few parameters as possible. And from all of the possible combinations of the regular and seasonal ARIMA model, it can be found the best model that is SARIMA $(1,1,0)(0,1,1)^{12}$ with RMSE 0.01134. After that the residual diagnostic checking is carried out using Kolomogorov-Smirnov test for testing normality residual and Ljung-Box test for testing white noise of residual ARIMA. With 5 percent significance level, it is concluded that the residuals do not satisfy normality and white noise residuals so the residual pattern still affected the SARIMA$(1,1,0)(0,1,1)^{12}$ models, therefore the residuals can still be modeled using NN and is expected to increase the level of accuracy of the prediction results.

### 3.3 Neural Network (NN) modeling

Neural Network (NN) is a connected network that mimics the working of nerve cells in the human brain. NN are one of such models that are able to approximate various nonlinearities in the data. NN are Dexible computing frameworks for modeling a broad range of nonlinear problems. One significant advantage of the NN models over other classes of nonlinear model is that NN are universal approximators which can approximate a large class of functions with a high degree of accuracy. Relationship of output $(y_t)$ with input $(y_{t-1}, \dots, y_{t-p})$ can be expressed as [6]

$$y_t = w_0 + \sum_{j=1}^{q} w_j g\left(w_{0j} + \sum_{i=1}^{p} w_{i,j} y_{t-i}\right) + e_t,$$

where $w_j (j = 1,2, \dots, q)$ and $w_{ij} (i = 0,1,2, \dots, p; j = 1,2, \dots, q)$ are the model parameters often called the connection weights; p is the number of input nodes and q is the number of hidden nodes. In this study, a combination of $e_{t-12}$, $X_t$, and $X_t$ which were stationary both in varians and mean. Thus, the number of input neurons used in this research is 3 neurons with the parameters and structure of the NN used for training determined by trial error. Figure 4 shows the NN structure that produces the smallest out sample RMSE 0.023984.

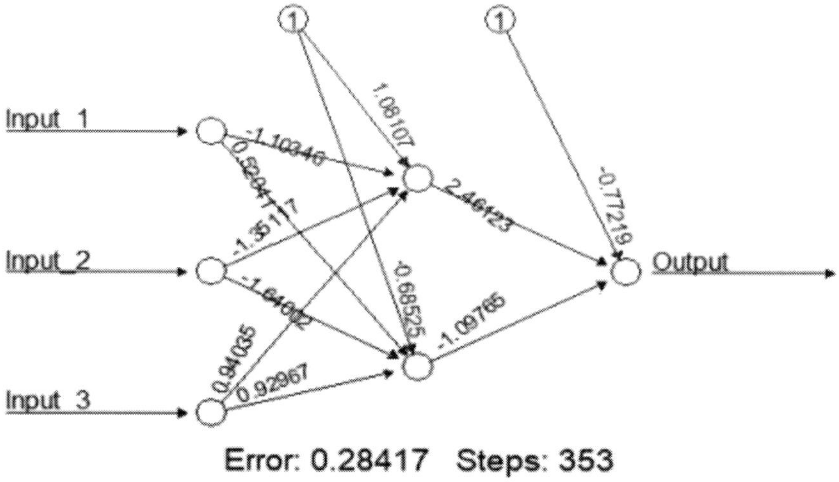


**Figure 4.** Structure NN (3×2×1) for training

Then performed ARIMA residual testing using NN with best structure NN in Figure 4. The testing result shows that the residual prediction by the best structure NN obtained is not much different with the actual ARIMA residual. So that it can be continued with the prediction of inflation based on the ARIMA-NN hybrid model by adding the ARIMA prediction of inflation and the residual NN prediction. The result of data prediction of inflation based on the ARIMA-NN hybrid model is shown in Table 1.

**Table 1.** Comparison table of actual and prediction data using hybrid ARIMA-NN

| Period | Actual | Prediction | Residual |
|---|---|---|---|
| Januari | 0.0282 | 0.0332 | -0.0050 |
| Februari | 0.0257 | 0.0325 | -0.0068 |
| Maret | 0.0248 | 0.0322 | -0.0074 |
| April | 0.0283 | 0.0347 | -0.0064 |
| Mei | 0.0332 | 0.0353 | -0.0021 |
| Juni | 0.0328 | 0.0346 | -0.0018 |
| Juli | 0.0332 | 0.0333 | -0.0001 |
| Agustus | 0.0349 | 0.0332 | 0.0017 |
| September | 0.0339 | 0.0331 | 0.0008 |
| Oktober | 0.0313 | 0.0309 | 0.0004 |
| November | 0.0300 | 0.0303 | -0.0003 |
| Desember | 0.0272 | 0.0322 | -0.0050 |

Figure 5 shows comparison graph of actual and prediction data of inflation rate in Indonesia period Januari 2019 to December 2019 by using hybrid ARIMA-NN.

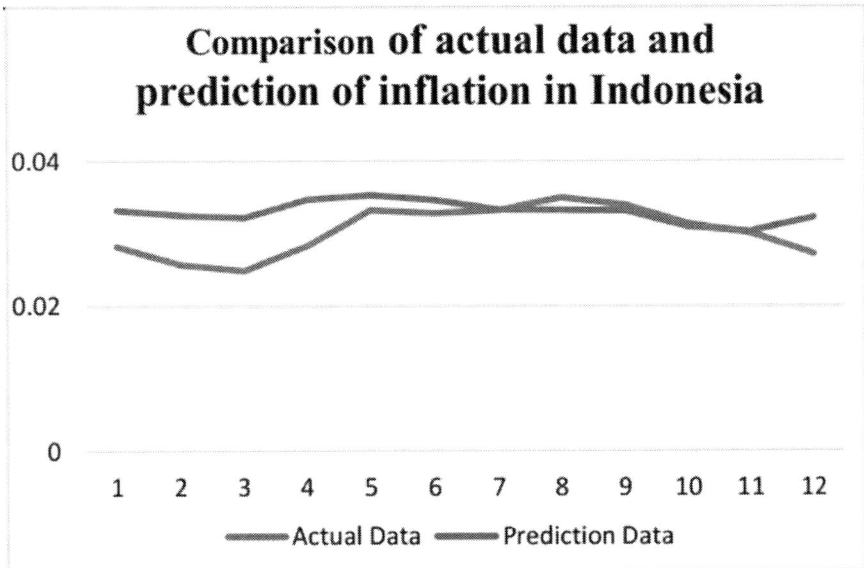


**Figure 5.** Comparison graph of actual and prediction data using ARIMA-NN

Based on Figure 5, it can be seen that the dashed red line shows the prediction data, while the blue line is the actual data. The graph shows that the prediction data pattern is close to the actual data pattern. This is supported by the result of calculation of MAPE value of 11.40873% that it means the prediction of inflation rate using ARIMA-NN hybrid model is relatively good with high prediction accuracy.

## 4. Conclusion

Based on the result and discussion, it can be concluded that:

a. Monthly inflation data Indonesia from January 2003 to December 2019 still contain a non linear pattern so it is suitable to be modeled using hybrid ARIMA-NN method. In addition, inflation data in Indonesia also has seasonal pattern every 12 periods, so that the best model is obtained using ARIMA method, SARIMA $(1,1,0)(0,1,1)^{12}$ with RMSE value 0.01134 while residuals SARIMA $(1,1,0)(0,1,1)^{12}$ which still contains non linear patterns modeled using NN obtained the best structure, namely NN $(3\times2\times1)$ with RMSE of 0.023984.

b. From the ARIMA-NN hybrid model that has been obtained, it can be seen that the accuracy of the forecasting value based on the MAPE value is 11.408730%, meaning that the resulting model has high prediction accuracy.

## Acknowledgments

Special thanks to Statisctical Research Group (KPBI Statistika) of Universitas Mataram for their support.

## References

[1] Suseno, Astiyah S 2009 *Seri Kebanksentralan* Retrieved from Bank Indonesia: http://www.bi.go.id
[2] Badan Pusat Statistik (BPS) 2019 *Berita Resmi Statistik* NTB: Badan Pusat Statistik
[3] Bakar N A, Rosbi S 2017 Autoregressive Integrated Moving Average (ARIMA) Model for Forecasting Cryptocurrency Exchange Rate in High Volatility Environment: A New Insight of

Bitcoin Transaction *International Journal of Advanced Engineering Research and Science* **4** (11) 130-137

[4] Hatta H H M, Daud F M, Mohamad N 2018 An Application of Time Series ARIMA Forecasting Model for Predicting the Ringgit Malaysia-Dollar Exchange Rate *Journal of Data Analysis* **1** (1) 42-48

[5] Unggara I, Musdholifah A, Sari A K 2019 Optimization of ARIMA Forecasting Model using Firefly Algorithm *Indonesian Journal of Computing and Cybernetics Systems* **13** (2) 127-136

[6] Zhang G P 2003 Time Series Forecasting Using a Hybrid ARIMA and Neural Network Model. *Neurocomputing* **50** 159-175

[7] Estiko F, Wahyuddin S 2019 Analysis of Indonesia's Inflation Using ARIMA and Artificial Neural Network *Economics Development Analysis Journal* **8** (2) 151-162

[8] Qadrini L, Asrirawan, Mahmudah N, Fahmuddin M S, Amri I F 2021 Forecasting Bank Indonesia Currency Inflow and Outflow Using ARIMA, Time Series Regression (TSR), ARIMAX, and NN Approaches in Lampung *Jurnal Matematika, Statistika & Komputasi* **17** (2) 166-177

[9] Wei W S 2006 *Time Series Analysis: Univariate and Multivariate Methods* (2nd Edition) Boston: Addison Wesley

[10] Aguilar J J, Turias I J, Come M J 2014 Hybrid Approaches Based on SARIMA and Artificial Neural Networks for Inspection Time Series Forecasting *Transportation Research Part E* **67** 1-13

IOP Conf. Series: Materials Science and Engineering 1115 (2021) 012059 doi:10.1088/1757-899X/1115/1/012059

# RSEM Modelling for Eruption Precursors Investigation on Lokon Volcano

**V Kaendung, D P Pandara*, G Pasau, Ferdy, M Bobanto, and R Telleng**

Department of Physics, Sam Ratulangi University, Indonesia.

*Coresponding author: dpandara_fisika@unsrat.ac.id

**Abstract.** Acceleration of volcanic activity can be detected from changes in the amplitude and energy of the seismic signals recorded around the volcano. The acceleration can be described from RSEM modelling. RSEM is useful for determining eruption precursors on volcanoes. In this study, the calculation of the cumulative RSEM value on Lokon Volcano used a very powerful software, namely ObsPy. Based on the RSEM model, the eruption of Mount Lokon on 12 September 2014 was different in character from the eruption on 20 May 2015. One was dominated by hybrid earthquakes and the other was triggered by volcanic earthquakes. These result showed that RSEM modelling can be used to investigate precursor of eruption.

## 1. Introduction

Seismic signals emitted by sources of volcanic activity contain a number of energies which can be important indicators in studying the behavior of volcanic eruptions. The real-time seismic energy measurement (RSEM) is a parameter that can describe the rate of seismic energy release during periods of intense volcanic activity [1]. RSEM is the standard deviation of the squared amplitude of the seismic signal over a certain time interval divided by the number of measurements [2]. Based on the cumulative RSEM model, investigations of the dominant seismic precursor triggered an eruption were carried out. RSEM modeling of seismic signal data can be done effectively using the ObsPy device which is designed to access and process seismic waveform data and metadata based on the Python programming language [2] [3] [4].

RSEM analysis using ObsPy will be useful to identify kind of frequencies associated with earthquakes on volcano. Thus, type of earthquake related with acceleration of volcanic activity can be determined. In this study, modelling of RSEM implemented on seismic data of Lokon Volcano and connected to eruption at September 12$^{nd}$, 2014 and May 20$^{th}$, 2015.

IOP Conf. Series: Materials Science and Engineering 1115 (2021) 012059 doi:10.1088/1757-899X/1115/1/012059

## 2. Material and Method

The data used in this research are seismic recording data in Seisan format of Empung Station during June 2014 until September 2014 dan during January until May 2015. These data obtained from KKVO at Tomohon. Data were converted into MSEED format and stored in stream of the ObsPy. In the calculation of RSEM and their cumulative, seismic signals were divided into 5 frequency categories, namely 1-16 Hz, 1-3 Hz, 3-5 Hz, 5-10 Hz and 10-16 Hz. Then, cumulative RSEM were plotted using matplotlib modul in Python 2. Based on the graph of cumulative RSEM, eruption precursor were investigated.

## 3. Results and Discussion

RSEM modeling in this study were applied to seismic signal data related to the Mount Lokon eruption on September 12[nd], 2014 at 19:00 UTC (see Figure 1) and May 20[th] 2015 at 07:20 UTC (Figure 2). The cumulative RSEM model in Figure 1 shows that a seismic signal with a frequency between 1-3 Hz has the same pattern as an overall seismic signal with a frequency of 1-16 Hz. The accelerated RSEM increase of the 1-3 Hz seismic signal indicated by the red dashed elliptical curve in Figure 1 was observed on 18-22 June 2014 and about 3 days before the eruption. Seismic signals with a frequency of 1-3 Hz were very dominant prior to the eruption. The seismic signal correlates with long period (LP) earthquakes and hybrid earthquakes [5] Hybrid earthquakes are a combination of signal and frequency characteristics between LP earthquakes and shallow volcanic earthquakes (VB), while LP earthquakes are gusts of high energy gas which are locally called gusts of earthquake [6]. The increase in LP earthquakes correlates with an increase in degassing or the release of water vapor due to high pressure around the crater [6]. Increasing of hybrid earthquakes is in line with the increased flow of fluids or gases that break through rock fractures, causing vibrations [7].

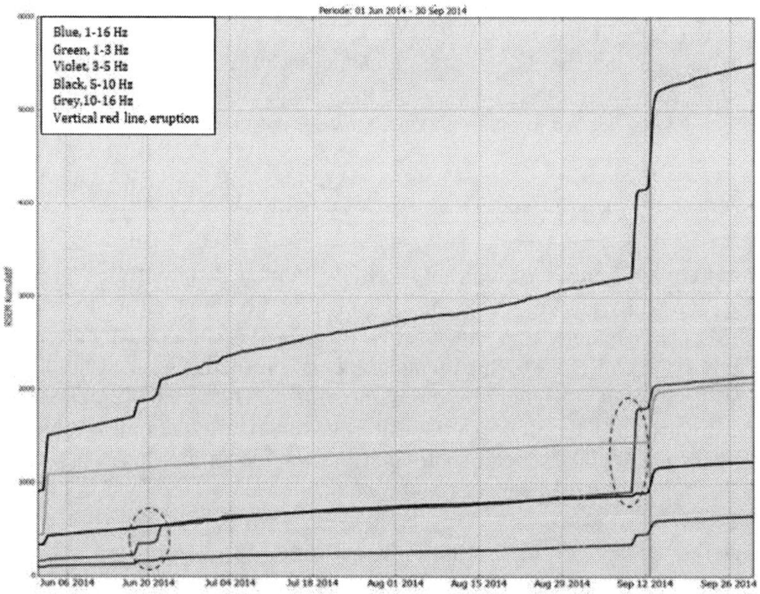


**Figure 1.** Graph of Cumulative RSEM June until September, 2014 (Vertical red line state eruption at September 12[nd], 2014 19:00 UTC).

The cumulative RSEM model in Figure 2 shows that the eruption that occurred on 20 May 2015 at 07:20 UTC was marked by a significant increase in seismic signals with a frequency of 1-3 Hz (dotted ellipse light blue) followed by an acceleration of 10-16 Hz seismic signals. (marked with a dashed ellipse in red). Seismic signals with a frequency of 10-16 Hz are associated with shallow volcanic-tectonic earthquakes (VB) and an increase in VB earthquakes is a marker for explosive eruptions of the volcanic type [8]. The volcanic type eruption was also confirmed through the microstructure analysis of the andesite and vesicular ash [9]. The acceleration of the number of VB earthquakes is related to the increase in scouring of the magma plug due to increased gas pressure [8]. The presence of a magma plug can be confirmed by the eruption product in the form of a volcanic bomb [10].

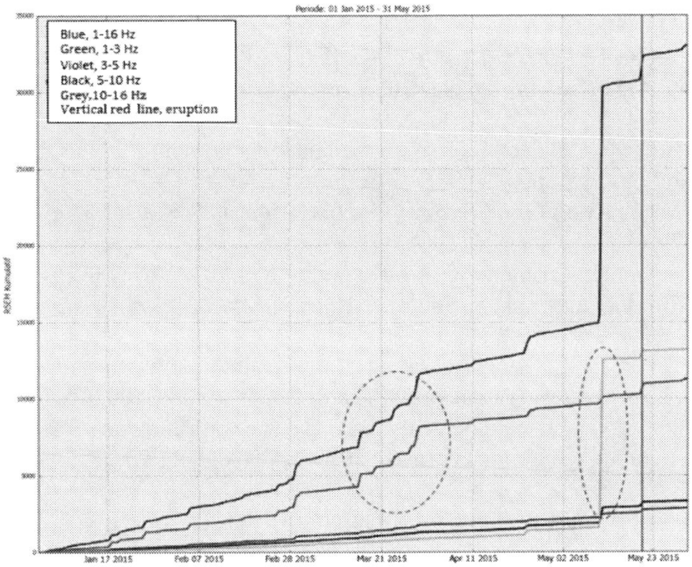


**Figure 2.** Graph of Cumulative RSEM January until May, 2015 (Vertical red line state eruption at May 20th, 2015).

The Geological Agency confirmed that the eruption on September 12, 2014 at 19:00 UTC was a small, phreatic type eruption with an eruption column height of about 500 m [5]. Phreatic eruptions are associated with an ash emission process that begins with the migration of low frequency seismic signals from 0.3 to 3 Hz [11] The eruption on May 20 2015 at 07:20 UTC was an eruption accompanied by a bang with a black eruption column and was a small volcanic type. The two eruptions were clearly of different character and these could be inferred from the RSEM model.

## 4. Conclusion

The significance of the differences between the two eruption types is indicated by differences in the cumulative RSEM model as well as differences in the types of precursors and differences in energy intensity. The precursors associated with the eruption of Mount Lokon on 12 September 2014 were hybrid earthquakes, while the precursors for the eruption on 20 May 2020 were volcanic earthquakes.The results of this research indicate that seismic precursors can be detected from the cumulative RSEM model.

## Acknowledgment

We grateful thanks to LPPM Unsrat for funding this research through PDUPT 2019 Grant.

### References

[1] De la Cruz-Reyna S and Reyes-Dávila G A 2001 A model to describe precursory material-failure phenomena: applications to short-term forecasting at Colima volcano, Mexico *Bulletin of Volcanology* **63** (5) 297-308

[2] Krischer L, Megies T, Barsch R, Beyreuther M, Lecocq T, Caudron C and Wassermann J 2015 ObsPy: A bridge for seismology into the scientific Python ecosystem *Computational Science & Discovery* **8** (1) 014003

[3] Bueno A, Zuccarello L, Díaz-Moreno A, Woollam J, Titos M, Benítez C and De Angelis S 2020 PICOSS: Python Interface for the classification of seismic signals *Computers & Geosciences* **142** 104531

[4] Megies T, Beyreuther M, Barsch R, Krischer L and Wassermann J 2011 ObsPy–What can it do for data centers and observatories? *Annals of Geophysics* **54** (1) 47-58

[5] Pandara D P, Ferdy F, Pasau G and Suoth V A 2020 Analisis Prekursor Erupsi Gunung Lokon *Jurnal MIPA* **9** (2) 47-50

[6] Haerani N, Gunawan H, Kristianto K, Kushendratno K and Wittiri S R 2010 Studi terpadu seismik dan deformasi di Gunung Lokon, Sulawesi Utara *Jurnal Lingkungan dan Bencana Geologi* **1** (3) 151-164

[7] Shakirova A and Firstov P 2019 Observation of the seismic mode «drumbeats» on volcanoes of the world and Kizimen volcano (Russia) *E3S Web of Conferences* **127** 03004

[8] Kristianto K, Gunawan H, Haerani N, Mulyana I, Basuki A, Primulyana S and Bina F U 2012 Gejala awal letusan Gunung Lokon Februari 2011-Maret 2012 *Jurnal Lingkungan dan Bencana Geologi* **3** (3) 151-168

[9] Padara D P 2017 Analisis Tipikal Erupsi Gunung Lokon Periode Erupsi 2012-2013 Berdasarkan Karakterisasi Mikrostruktur Abu Vulkanik *Jurnal MIPA* **6** (2) 36-41

[10] Wright H M, Cashman K V, Rosi M and Cioni R 2007 Breadcrust bombs as indicators of Vulcanian eruption dynamics at Guagua Pichincha volcano, Ecuador *Bulletin of Volcanology* **69** (3) 281-300

[11] Woitischek J, Woods A W, Edmonds M, Oppenheimer C, Aiuppa A, Pering T D and Garaebiti E 2020 Strombolian eruptions and dynamics of magma degassing at Yasur Volcano (Vanuatu) *Journal of Volcanology and Geothermal Research* **398** 106869

IOP Conf. Series: Materials Science and Engineering 1115 (2021) 012060 doi:10.1088/1757-899X/1115/1/012060

# Grain Shape Analysis of the Volcanic Ash Particle from Soputan Volcano to Investigate Eruption Type

**D Tampenawas*, M Bobanto, D Pandara, Ferdy, G Pasau, GAbidjulu and V Kaendung**

Department of Physics, Sam Ratulangi University, Manado, Indonesia

*Corresponding author: mariabobanto5@unsrat.ac.id

**Abstract.** Volcanic ash showcases a diverse spectrum of shapes, each with different physical properties and behaviours.Those properties and behaviours are often linked with fragmentation process of the magma inside the volcano. Analyzing the shape of a volcanic ash can give us valuable information behind the magma formation and the eruption style of a volcano. This research are conducted in hope to find implication between volcanic ash grain size and the eruption style of Mount Soputan. Sample from Soputan volcano eruption are choosed because of the danger the mountain holds to the populated area that surrounds it. Samples from three different eruption date are collected and being examined by ImageJ to determine the shape parameters needed in order to find the eruption type. Manual observation are also conducted to support the result from software analysis.

## 1. Introduction

Volcanic eruptions are one of the natural disasters with the greatest potential danger in Indonesia. Apart from the high number of casualties, the eruption disaster was also followed by large-scale damage to infrastructure and long-term social and economic impacts [1]-[3]. North Sulawesi is one of the areas in Indonesia that is vulnerable to the threat of volcanic eruptions. One of the most active volcanoes with high potential hazards is Mount Soputan [4]. The danger aspect of Mount Soputan is due to its typical eruption in the form of an explosive eruption with a VEI index of 2-3 which can eject pyroclastic material such as ash, sand, lapilli, and bombs, as well as incandescent lava avalanches [5]. Given the potential dangers, early warning efforts of an eruption must be carried out. One of the important early warning efforts in anticipating the dangers of volcanic eruptions is through understanding the character of volcanic eruptions.

Volcano-inspired ash can provide significant information because the ash comes from magma rising from the depths of the mountain body. This volcanic ash is the result of magma interaction with various processes ranging from magma sacs to oxidation in the atmosphere. Characterization and analysis of the microstructure and geochemical composition of volcanic ash will be able to provide an explanation of the eruption process that occurred in aspects of mineralization, crystallization, vesiculation and fragmentation [6]-[7]. Heiken, 1972 shows that in-depth studies related to the morphology of ash have a relationship with magma composition and type of eruption [8]. Therefore, by analyzing the form of ash, we can get a deeper understanding of the physical mechanisms of volcanic eruptions that can help mitigate disasters due to volcanic eruptions [9]-[10].

Thus, characterization and analysis of the microstructure and geochemistry of the ash from the eruption of Mount Soputan can answer problems related to the mechanism and typical eruptions so as to provide basic information that is important for future mitigation efforts.

## 2. Methodology

This research uses *Scanning Electron Microscopy* technique to get the micro structure of the ash particle and then we uses image processing software *Image-J* to get the shape parameter needed. The research workflows follows the following steps:

*a. Preparation*
At the preparation stage, the sample will be separated into three fractions, namely fine fraction (<75 μm), medium fraction (70 - 125 μm) and coarse fraction (> 125 μm) with variations in sieve arrangement.

*b. Microstructure Analysis*
The eruption ash microstructure analysis will be carried out using SEM. The SEM image will be analyzed with Image J software to observe grain size and vesicle size. Morphological parameters of the microstructure of the ash to be studied are grain size and vesicularity.

*c. Conclusion*
The results of the microstructure analysis of the eruption ash using SEM in the form of parameters of grain size, form factor, roundness, compactness, aspect ratio, will be used to observe the eruption processes and determine the type of eruption.

### 3. Results and Discussion

Results from the research conducted by using image analysis software ImageJ produces number of shape perimeters essential for determining eruption style such as area of particle, aspect ratio, compactness, roundness.

The area size obtained from ImageJ processing showcases a wide range of size varying from the 15720.23 to 23749.8 $\mu m^2$ as shown in Table 1, Table 2 and Table 3. The area size parameter obtained itself cannot be the sole parameter to determine the type of eruption and thus needs to be compared with other parameters to be able to determine the type of eruption better. The numbers obtained from the shape parameter analysis and result from manual observation shows that most of the ash particle have blocky and blocky vesicular shape as seen in Figure 1 which implicates a phreatomagmatic eruption [9]. Phreatomagmatic are driven from thermal contraction of magma when it comes in contact with water. This temperature difference between the two causes violent water-lava interactions that make up the eruption. The products of phreatomagmatic eruptions are believed to be more regular in shape and finer grained than the products of magmatic eruptions because of the differences in eruptive mechanisms [8].

Variety in both compactness between all three samples doesn't shows much difference with one another except for sample collected from October 2018 eruption as shown in Figure 2, Figure 3 and Figure 4. This may have caused by difference in eruption style between eruption on October 2018 and the other two eruption. Furthermore, deeper investigation regarding the chemical parts of the ash can shows a lot more details as to the difference in eruption style but overall, it conforms to the theory and previous research that the numbers shown which ranges from 0.56-0.92 to the highest shows the almost complete level of magmatic process inside the volcanic chamber that might helps shape the ash into a more regular shape.

IOP Conf. Series: Materials Science and Engineering 1115 (2021) 012060 doi:10.1088/1757-899X/1115/1/012060

**Table 1**. Shape parameters obtained from ImageJ on sample collected from Mt. Soputan eruption on October 2018 size 75-125 ( μm ) under 500x magnification.

| Count | Area | AR | Comp | Round | Form Factor |
|---|---|---|---|---|---|
| 55 | 23749.8 | 3.21 | 0.56 | 0.31 | 0.27 |
| 39 | 22311.62 | 2.75 | 0.6 | 0.36 | 0.4 |
| 40 | 19850.02 | 1.24 | 0.9 | 0.81 | 0.56 |
| 1 | 15720.23 | 1.69 | 0.77 | 0.59 | 0.35 |

**Table 2**. Shape parameters obtained from ImageJ on sample collected from Mt. Soputan eruption on February 2016 size 75-125 ( μm ) under 500x magnification.

| Count | Area | AR | Comp | Round | Form Factor |
|---|---|---|---|---|---|
| 1 | 19444.79 | 1.2 | 0.91 | 0.84 | 0.52 |
| 3 | 20167.78 | 1.29 | 0.88 | 0.77 | 0.35 |
| 8 | 20661.89 | 1.18 | 0.92 | 0.84 | 0.41 |
| 49 | 13981.24 | 1.41 | 0.84 | 0.71 | 0.51 |

**Table 3**. Shape parameters obtained from ImageJ on sample collected from Mt. Soputan eruption on January 2016 size 75-125 ( μm ) under 500x magnification.

| Count | Area | AR | Comp | Round | Form Factor |
|---|---|---|---|---|---|
| 148 | 17319.75 | 1.21 | 0.91 | 0.83 | 0.48 |
| 101 | 19985.53 | 1.33 | 0.87 | 0.75 | 0.38 |
| 115 | 19735.76 | 1.34 | 0.86 | 0.75 | 0.58 |
| 59 | 18990.51 | 1.26 | 0.89 | 0.79 | 0.58 |

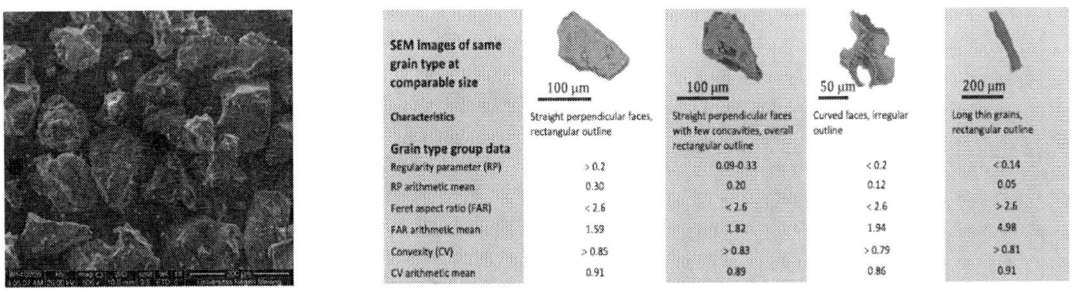


**Figure 1**. Comparison between ash collected from Mt.Soputan eruption on January 2016 (Left ) and Grain type morphology overview from Liu,2015 ( Right )

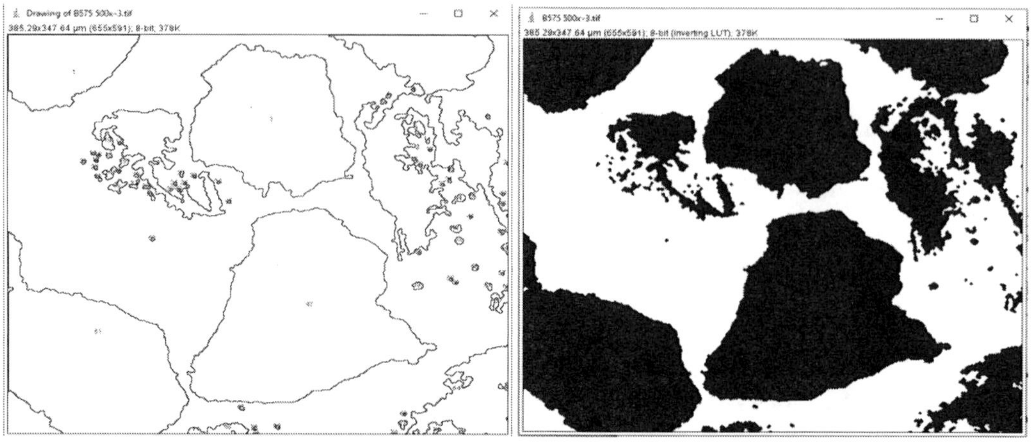

**Figure 2.** A Sample of the ash collected from October 3rd, 2018 eruption after thresholding_1

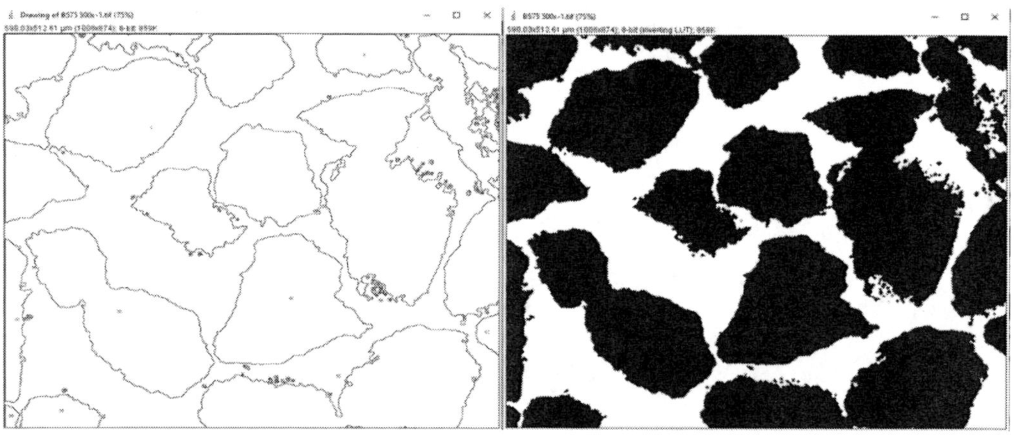

**Figure 3.** A Sample of the ash collected from October 3rd, 2018 eruption after thresholding_2

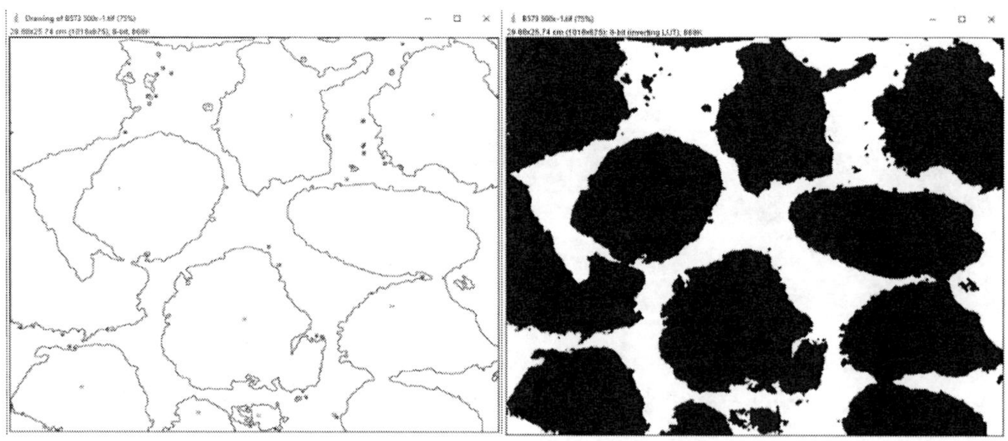

**Figure 4.** A Sample of the ash collected from October 3rd, 2018 eruption after thresholding_3

## 4. Conclusion

The shape parameters obtained from ImageJ conforms to previous research that the numbers shown from examining all three samples shows blocky to blocky vesicular figure which usually produced by explosive type eruption. Further investigation on other aspects of the material may provide a more detailed information about the eruption type but as for general information, shape parameters analysis can bring sufficient information

## Acknowledgments

We would like to thanks to the Rector of Samratulangi University who support us, so that this research was fully funded by PNBP Sam Ratulangi University through RDTPU 2020 Research Grant to MB & FF.

## References

[1] Brenguier F, Shapiro N M, Campillo M, Ferrazzini V, Duputel Z, Coutant O and Nercessian A 2008 Towards forecasting volcanic eruptions using seismic noise *Nature Geoscience* **1** (2) 126-130
[2] Robock A and Mao J 1992 Winter warming from large volcanic eruptions *Geophysical Research Letters* **19** (24) 2405-2408
[3] Linde A T and Sacks I S 1998 Triggering of volcanic eruptions *Nature* **395** (6705) 888-890
[4] Sparks R S J 2003 Forecasting volcanic eruptions *Earth and Planetary Science Letters* **210** (1-2) 1-15
[5] Pallister J S, Bina F R, McCausland W, Carn S, Haerani N, Griswold J and Keeler R 2012 Recent explosive eruptions and volcano hazards at Soputan volcano—a basalt stratovolcano in north Sulawesi, Indonesia *Bulletin of volcanology* **74** (7) 1581-1609
[6] Minami Y, Imura T, Hayashi S and Ohba T 2016 Mineralogical study on volcanic ash of the eruption on September 27, 2014 at Ontake volcano, central Japan: correlation with porphyry copper systems *Earth, Planets and Space* **68** (1) 1-11
[7] Padara D P 2017 Analisis Tipikal Erupsi Gunung Lokon Periode Erupsi 2012-2013 Berdasarkan Karakterisasi Mikrostruktur Abu Vulkanik *Jurnal MIPA* **6** (2) 36-41
[8] Zimanowski B, Wohletz K, Dellino P and Büttner R 2003 The volcanic ash problem *Journal of Volcanology and Geothermal Research* **122** (1-2) 1-5
[9] Shoji D, Noguchi R, Otsuki S and Hino H 2018 Classification of volcanic ash particles using a convolutional neural network and probability *Scientific reports* **8** (1) 1-12
[10] Schmith J, Höskuldsson Á and Holm P M 2017 Grain shape of basaltic ash populations: implications for fragmentation *Bulletin of Volcanology* **79** (2) 14-27

# A discrete choice model of voters' decision in the general election 2019 using LASSO regression

**D Irvani[1,2], K A Notodiputro[2,] * and B Sartono[2]**

[1]Saiful Mujani Research and Consulting (SMRC), Jakarta, Indonesia
[2]Departement of Statistics, Institut Pertanian Bogor (IPB), Bogor, Indonesia

*Email: khairil@apps.ipb.ac.id

**Abstract.** Statistical modelling faces problems when too many variables are involved in the model. Using all variables in the model will result in overfitting which in turn produces unstable predictions. LASSO (Least Absolute Shrinkage and Selection Operator) is a method that can be used to overcome this problem. LASSO selects more important variables with the goal of improving the prediction accuracy and interpretability. This paper identifies factors those influenced voters in 2019 presidential election using LASSO penalized logistic regression. This logistic model is also known as a discrete choice model. The data used in this study is the result of an exit poll conducted during the election day with a total sample of 2,289 respondents. The response variable in this study is a vote choice. The independent variables include sociological, psychological, political economy, and campaign variables. The results showed that there were 17 of 27 variables having nonzero coefficients. These variables include candidates' personal qualities which show the largest effect, followed by the incumbent performance variable. On the other hand, variables such as religion, education, and ethnicity have much less effects. This implies that the psychological and political economy factors played more important role than the sociological factor in determining voters' decision. Based on this finding the prospective candidate in the future election should focus their campaign strategy on enhancing positive images (empathy, integrity and capability) as well as offering better programs to convince voters.

## 1. Introduction

Statistical modeling faces problems when too many variables are involved in the model. Using all variables in the model will result in overfitting which in turn produces unstable predictions. LASSO (Least Absolute Shrinkage and Selection Operator) is a method that can be used to overcome this problem.

LASSO was introduced by Tibshirani [1] as a method for shrinking coefficients and selecting more important variables with the goal of improving the prediction accuracy and interpretability. In least squares method LASSO shrinks the coefficients $\beta = (\beta_1, \dots, \beta_p)^T$ toward zero according to a penalty $\lambda \sum_{j=1}^{p} |\beta_j|$ in the objective function $\min_{(\beta_0,\beta)\in\mathbb{R}^{p+1}} \left[\frac{1}{2n}\sum_{i=1}^{n}\left(y_i - \beta_0 - x_i^T\beta\right)^2 + \lambda\sum_{j=1}^{p}|\beta_j|\right]$, $\lambda > 0$, which is equivalent to $\min_{(\beta_0,\beta)\in\mathbb{R}^{p+1}} \left[\frac{1}{2n}\sum_{i=1}^{n}\left(y_i - \beta_0 - x_i^T\beta\right)^2\right]$ subject to $\sum_{j=1}^{p}|\beta_j| \le t$, $t \ge 0$. While in logistic regression LASSO shrinks the coefficients toward zero based on the objective function $\min_{(\beta_0,\beta)\in\mathbb{R}^{p+1}}\left[-\frac{1}{n}\sum_{i=1}^{n}\left\{y_i\left(\beta_0+x_i^T\beta\right)-\log(1+e^{\beta_0+x_i^T\beta})\right\}+\lambda\sum_{j=1}^{p}\left|\beta_j\right|\right]$ [2]. The LASSO coefficients can be obtained using coordinate descent algorithm [3].

LASSO has been applied in a number of previous studies. Kristensen et al. (2017) applied a LASSO-penalized multinomial logit model to predict party choice [4]. Cantoni and Pons (2019) used LASSO to select more important variables associated with voter turnout [5]. Meanwhile, Chen et al. (2019) applied LASSO to identify factors associated with tuberculosis in West Java [6].

This paper identifies factors that influence voters in 2019 presidential election using LASSO penalized logistic regression. This logistic model is also known as a discrete choice model.

## 2. Research Method

### 2.1. Data

The data used in this study is the result of an exit poll conducted during the election day with a total sample of 2,289 respondents. The response variable in this study is a vote choice (1 = Candidate A, 0 = Candidate B). The independent variables include sociological, psychological, political economy, and campaign variables, those are 27 variables in total. Overall, the independent variables used in this study are explained in Appendix.

### 2.2. Data Analysis

The data analysis in this study is divided into 3 parts:

a. Descriptive analysis to describe the relationships between the response variable and each independent variable.

b. 10-fold cross validation 10-folds cross-validation to determine the optimum hyperparameter $\lambda$ ($\lambda_{opt}$). This cross-validation is as follows. The dataset is divided at random into 10 groups ($g = 1, \ldots, 10$), where one group is fixed as the test set and the remaining 9 groups are designated as the training set. LASSO logistic regression with a value of $\lambda$ is applied to the training set (9 groups beside the $g^{\text{th}}$ group) then the model is used to calculate the deviance of the testing set (the $g^{\text{th}}$ group) $d_\lambda^{(g)} = \frac{1}{n_g}\left\{\sum_{i=1}^{n_g} -2[y_i \log\tilde{p}_i + (1-y_i)\log(1-\tilde{p}_i)]\right\}$, where $\tilde{p}_i = \frac{\exp\left(\tilde{\beta}_0^{(g)}+x_i^T\tilde{\beta}^{(g)}\right)}{\left[1+\exp\left(\tilde{\beta}_0^{(g)}+x_i^T\tilde{\beta}^{(g)}\right)\right]}$ is the predicted probability for observation $i$ in the testing set, $\tilde{\beta}_0^{(g)}$ and $\tilde{\beta}^{(g)}$ are the LASSO coefficients based on the training set, and $n_g$ is the number of observations in the testing set. This procedure is repeated 10 times using different testing and training sets to obtain deviances for each testing set ($d_\lambda^{(1)}, \ldots, d_\lambda^{(10)}$), and the model performance is measured by CV-Dev$_\lambda$ $= \frac{1}{10}\sum_{g=1}^{10} d_\lambda^{(g)}$. This cross validation is repeated for different $\lambda$ values. The $\lambda_{opt}$ is determined by a criterion $\lambda_{opt} = \begin{cases}\lambda_{min}, & \text{if } nonzero_{\lambda_{min}} = nonzero_{\lambda_{1se}} \\ \lambda_{1se}, & \text{if } nonzero_{\lambda_{min}} > nonzero_{\lambda_{1se}}\end{cases}$, where $nonzero_\lambda$ is the number of variables having nonzero coefficients based on a $\lambda$, $\lambda_{min}$ is the value of $\lambda$ which produces the minimum CV-Dev$_\lambda$ (CV-Dev$_{\lambda_{\min}}$), and $\lambda_{1se}$ is the largest value of $\lambda$ which produces CV-Dev$_\lambda$ within 1 standard error of CV-Dev$_{\lambda_{\min}}$.

c. Estimate the LASSO panelized logit regression model by using the full dataset and the $\lambda_{opt}$. This final model performance is assessed by the fraction of deviance explained $D_\lambda^2 = \frac{\text{Dev}_{null}-\text{Dev}_\lambda}{\text{Dev}_{null}}$, where Dev$_{null}$ is the deviance with a constant model (mean), Dev$_\lambda = -2 \times$ loglik$(\tilde{\beta}_{0\lambda}, \tilde{\beta}_\lambda)$ is the deviance of the model using $(\tilde{\beta}_{0\lambda}, \tilde{\beta}_\lambda)$, and $(\tilde{\beta}_{0\lambda}, \tilde{\beta}_\lambda)$ are the LASSO coefficients estimates. The standard error of each coefficient is estimated using bootstrap method with 1000 replications [2], [7].

The LASSO coefficients estimates are calculated using *glmnet* package in R [3].

## 3. Results and Discussion

### 3.1. Descriptive Analysis

Support for a candidate appeared to vary according to values of the independent variables. In terms of demographics, differences in support can be seen especially in terms of religion, ethnicity, education and region. Candidate A tended to get a bigger support from the non-Muslim voters, Javanese, voters with lower education, and voters who live in areas outside Sumatera. Candidate A also seemed to have greater support from voters who assessed economic condition as better and voters who were satisfied with the incumbent performance. Party identification and candidates' personal qualities appeared to have positive relationships with the electability of the candidates concerned, whereas belief in negative issues had negative relationship with the electability. Voters who were more exposed to candidate's campaign in various media appeared to be more supportive of the candidate (Figure 1).

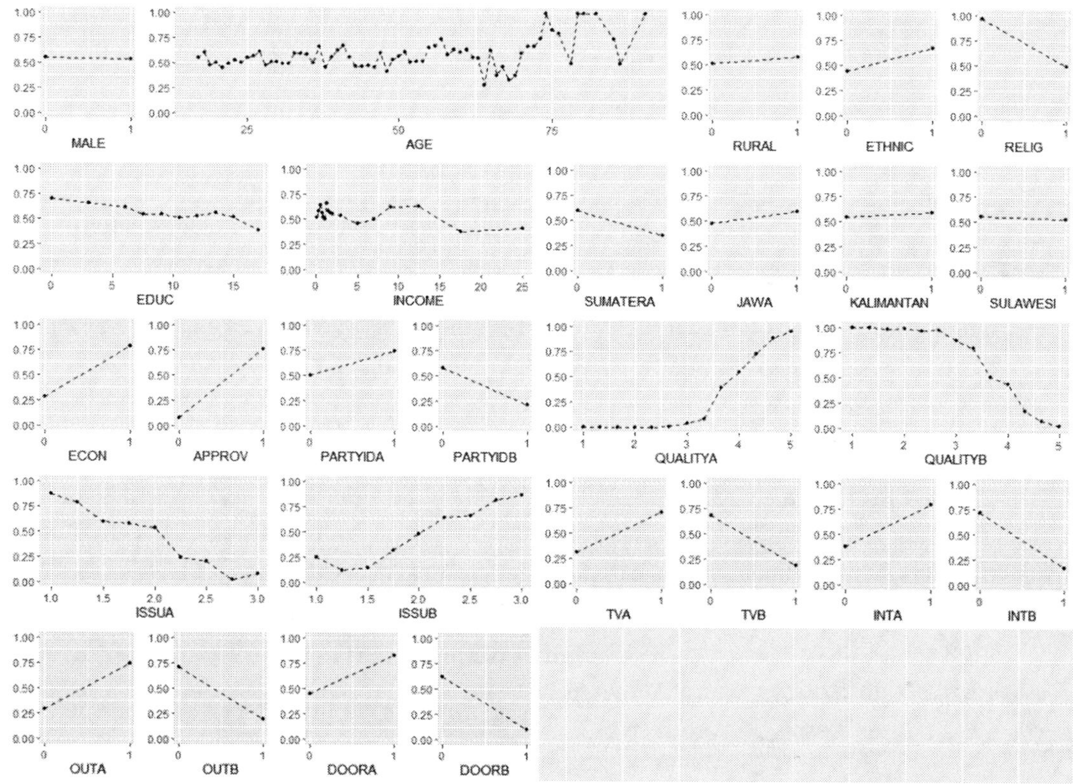


**Figure 1.** Proportion of support for Candidate A (vertical axis) according to explanatory variables (horizontal axis).

### 3.2. Cross-Validation and Optimum-$\lambda$

The cross-validation analysis showed that $\lambda_{min}$ =0.002466 resulted in a CV-Dev$_\lambda$ of 0.3184, while $\lambda_{1se}$ = 0.009069 resulted in CV-$Dev_\lambda$ of 0.3362. The LASSO logistic regression with $\lambda_{min}$ shrinks the coefficients of 7 independent variables to zero, while $\lambda_{1se}$ shrinks the coefficients of 10 independent variables to zero (Figure 2a). Since using $\lambda_{1se}$ shrinks more coefficients to zero than $\lambda_{min}$ then $\lambda_{1se}$ is chosen as the optimum-λ.

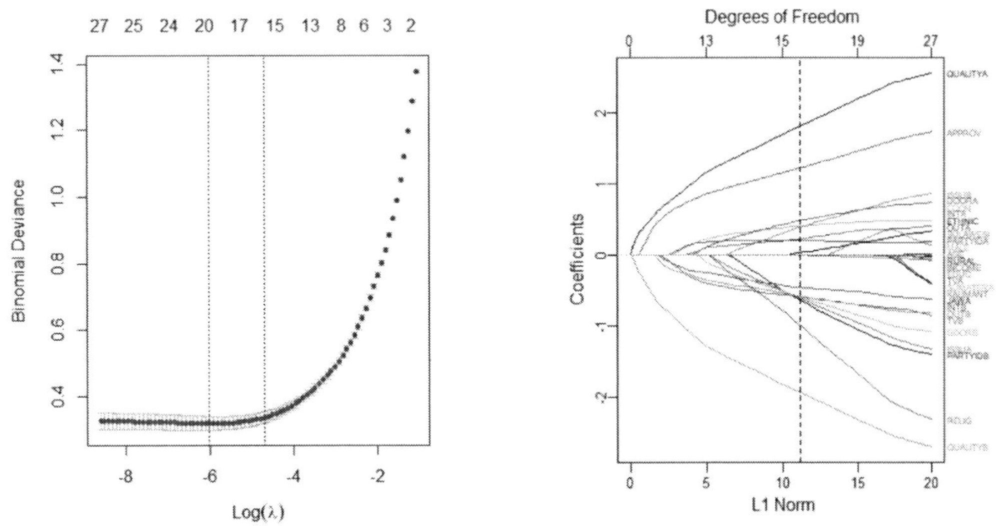


(a) (b)

**Figure 2.** (a) 10-folds cross-validation, the vertical dash lines are the locations of $\lambda_{min}$ and $\lambda_{1se}$; (b) The coefficient path for LASSO penalized logistic regression with all data, plotted versus the $\ell_1$-norm of the coefficient vector relative to the $\ell_1$-norm of the unrestricted estimate, the vertical dash line is the location of $\ell_1$-norm corresponding to $\lambda_{1se}$

### 3.3. LASSO Penalized Logistic Regression Model

The LASSO logistic regression model by using the full dataset and the optimum-λ resulted in a deviance ratio of $D_\lambda^2$ = 76.4%, which means that the model has a fairly good performance. This model shrinks 10 variable coefficients to zero: MALE, RURAL, AGE, INCOME, SUMATERA, JAVA, KALIMANTAN, SULAWESI, PARTYIDA, and TVA. This shows that gender, rural-urban, income, region, closeness to parties those supported Candidate A, and Candidate A's TV commercial exposure were not important factors in determining the vote choice. Then the explanation for voting behavior can be focused more on the remaining 17 variables (Table 1).

The plot of coefficient paths visualizes the most to the least important variable (Figure 2b). By observing each coefficient path from left to right, it can be seen how each variable appears as nonzero (having nonzero coefficient) as the budget $t$ increases, or as the hyperparameter $\lambda$ decreases (the value of λ is getting smaller as it moves to the right).

If we use a very large penalty then all variables having zero coefficients. When the penalty is reduced slightly, QUALITYA (Candidate A's personal quality) and QUALITYB (Candidate B's personal quality) appear almost simultaneously as nonzero. If the penalty is reduced again then APPROV variable (the incumbent performance) becomes nonzero. Furthermore, the gradual decreasing of the value of hyperparameter $\lambda$ until $\lambda_{1se}$ makes the following variables appear as nonzero in sequence: OUTB, INTB, TVB, OUTA, ECON, INTA, DOORA, DOORB, RELIGION, ISSUA, PARTYIDB, ISSUB, EDUC, ETHNIC.

Candidates' personal qualities were the most important variables, followed by the incumbent performance. Meanwhile, ethnicity had the lowest effect. In general, this study confirms the results of previous studies that voting behavior can be explained through psychological, political economy, and sociological approaches [8]. In addition, this study also confirms previous research that campaign and disinformation are also factors that can influence voting behavior [9-10].

**Table 1.** LASSO-Penalized Logistic Regression Model

| | Coefficient | Standard Error | Z | P(>\|Z\|) | Odds Ratio |
|---|---|---|---|---|---|
| (Intercept) | 0.490 | 0.837 | 0.59 | 0.558 | - |
| MALE | - | - | - | - | - |
| RURAL | - | - | - | - | - |
| AGE | - | - | - | - | - |
| EDUC | -0.002 | 0.011 | -0.20 | 0.840 | 1.0 |
| INCOME | - | - | - | - | - |
| ETHNIC | 0.009 | 0.086 | 0.10 | 0.921 | 1.0 |
| RELIG | -0.900 | 0.245 | -3.68 | 0.000 | 0.4 |
| SUMATERA | - | - | - | - | - |
| JAWA | - | - | - | - | - |
| KALIMANTAN | - | - | - | - | - |
| SULAWESI | - | - | - | - | - |
| ECON | 0.393 | 0.149 | 2.63 | 0.008 | 1.5 |
| APPROV | 1.196 | 0.200 | 5.99 | 0.000 | 3.3 |
| PARTYIDA | - | - | - | - | - |
| PARTYIDB | -0.582 | 0.279 | -2.09 | 0.037 | 0.6 |
| QUALITYA | 1.766 | 0.135 | 13.08 | 0.000 | 5.8 |
| QUALITYB | -1.897 | 0.124 | -15.36 | 0.000 | 0.2 |
| ISSUA | -0.563 | 0.170 | -3.31 | 0.001 | 0.6 |
| ISSUB | 0.354 | 0.185 | 1.92 | 0.055 | 1.4 |
| TVA | - | - | - | - | - |
| TVB | -0.579 | 0.180 | -3.21 | 0.001 | 0.6 |
| INTA | 0.208 | 0.150 | 1.38 | 0.167 | 1.2 |
| INTB | -0.452 | 0.193 | -2.34 | 0.019 | 0.6 |
| OUTA | 0.202 | 0.171 | 1.18 | 0.237 | 1.2 |
| OUTB | -0.568 | 0.220 | -2.58 | 0.010 | 0.6 |
| DOORA | 0.456 | 0.185 | 2.47 | 0.013 | 1.6 |
| DOORB | -0.562 | 0.237 | -2.37 | 0.018 | 0.6 |

### *3.3.1. Psychological Factor*

Within the psychological variables, PARTYIDA having zero coefficient. The closeness of the voters to parties those supported Candidate A (compared to not closed to any party) didn't seem to have an important role in the choice. However, this doesn't mean that party identification is not important at all. The model shows that the closeness of the voters to the parties those supported Candidate B reduced the tendency to vote for Candidate A, and this effect is significant at 95% confidence level. Party identification, however, remains important despite its limited influence.

The most important psychological variable is Candidates' personal qualities. Every addition of 1 point of score on the assessment of Candidate A's personal quality increased 5.8 times the odds for choosing Candidate A. Conversely, every addition of 1 point of score on Candidate B's personal quality decreased the odds of choosing Candidate A by 0.2 times.

Negative issues regarding candidates contributed to voters' decision. The more voters were exposed to and believe in negative issues about a candidate, the lower the chance of the voters choosing the candidate. Considering that some of the negative issues during the campaign period were a form of disinformation or hoaxes, this finding needs to be of concern to various parties. Voters' political decisions should be made based on correct information. The spread of fake news during campaign period would ultimately reduce the quality of election.

### *3.3.2. Political Economy Factor*

Evaluation on the economic condition and incumbent performance have positive coefficients. This shows that voters are basically rational. Voters who were satisfied with the incumbent performance and thought that the economic condition was better than the last year had a tendency to prefer Candidate A (as the incumbent) over voters who had the opposite opinion. The relative tendency to vote for Candidate A among voters who were satisfied with the incumbent performance was 3.3 times greater than among those who were dissatisfied. Meanwhile, voters who considered the economic condition was better than before tended to be more likely to vote for Candidate A 1.5 times greater than voters who thought otherwise.

### *3.3.3. Sociological Factor*

The sociological variables those having nonzero coefficients are religion, education, and ethnicity. Religion and education have negative coefficients while ethnicity has a positive coefficient. However, the effects of education and ethnicity less convincing, not significant at 95% confidence level. The more convincing sociological variable is religion. The tendency of voting for Candidate A (versus Candidate B) among Muslim voters is 0.4 times (lower) than that of non-Muslim voters. The influence of religion on the choice is interesting, since Candidate A and Candidate B have the same religious background (Islam). Muslim voters might think Candidate A was less representative of their interests, whereas non-Muslim voters thought Candidate A was more representative of their aspirations.

### *3.3.4. Campaign Factor*

Campaigns played important role in voting. Candidate B's campaign appeared to have stronger influence than Candidate A's campaign. The effect of Candidate A's campaign through TV commercials has even shrunk to zero. This might be related to the different position of the two candidates: Candidate A as the incumbent and Candidate B as the challenger. Most voters probably already knew and could evaluate Candidate A's programs during his 5 years in office. Therefore, the campaign conducted by Candidate A during the campaign period had no significant effect. At 95% confidence level, only campaigns through volunteers' visits could increase Candidate A's electability.

Unlike Candidate A, voters generally only knew about Candidate B's program when they entered the campaign period. Therefore, the campaign conducted by Candidate B played an important role in providing information and encouraging voters to vote. All campaign forms conducted by Candidate B had significant effects. Voters who were more exposed to any forms of Candidate B's campaign lowered the tendency to vote for Candidate A to 0.6 times compared to voters who had never been exposed to any candidate campaign.

## 4. Conclusion

The LASSO penalized logistic regression model with a λ-optimum in this study has a ratio of deviance of $D_\lambda^2 = 76.4\%$. Based on this model, there were 17 of the 27 variables having nonzero coefficients. These variables were candidate's personal qualities (QUALITYA and QUALITYB), incumbent performance (APPROV), campaigns through TV commercials (TVB), campaigns through advertisements on the internet or social media (INTA and INTB), campaigns through outdoor media (OUTA and OUTB), campaigns through volunteers' visits (DOORA and DOORB), economic condition (ECON), negative issues about candidates (ISSUA and ISSUB), religion (RELIG), education (EDUC), and ethnicity (ETHNIC).

Candidates' personal qualities show the largest effect, followed by the incumbent performance variable. On the other hand, variables such as religion, education, and ethnicity have much less effects. This implies that the psychological and political economy factors played more important role than the sociological factor in determining voters' decision. Based on this finding the prospective candidate in the future election should focus their campaign strategy on enhancing positive images (empathy, integrity and capability) as well as offering better programs to convince voters.

## Acknowledgments

The authors would like to thank SMRC, IPB University, and all persons who contributed to the improvement of this paper: Muhlis Ardiansyah, Etis Sunandi, Asep Rusyana, Alona Dwinata, Embay Rohaeti.

## References

[1] Tibshirani R 1996 Regression shrinkage and selection via the *LASSO Journal of the Royal Statistical Society, Series B (Methodological)* **58** (1) 267-288

[2] Hastie T, Tibshirani R, Wainwright M 2016 *Statistical Learning with Sparsity* Florida: CRC Press

[3] Friedman J, Hastie T, Tibshirani R 2010 Regularization Paths for Generalized Linear Models via Coordinate Descent *Journal of Statistical Software* **33** (1) 1-22

[4] Kristensen J B, Albrechtsen T, Dahl-Nielsen E, Jensen M, Skovrind M, Bornakke T 2017 Parsimonious data: How a single Facebook like predicts voting behavior in multiparty systems *PLoS ONE* **12** (9) e0184562

[5] Cantoni E, Pons V 2019 Does Context Trump Individual Drivers of Voting Behavior? Evidence from U.S. Movers *Harvard Business School Working Paper* 19-25

[6] Chen S, Notodiputro K A, Rahardiantoro S 2019 Penerapan Analisis LASSO dan Group LASSO dalam Mengidentifikasi Faktor-faktor yang Berhubungan dengan Tuberkulosis di Jawa Barat *Indonesian Journal of Statistics and Its Applications* **4** (1) 39-54

[7] Efron B, Tibshirani R 1993 *An Introduction to the Bootstrap* London: Chapman & Hall

[8] Mujani S, Liddle R W, Ambardi K 2018 *Voting behavior in Indonesia since democratization: critical democrats* Cambrige: Cambridge University Press

[9] Peterson D A M 2009 Campaign Learning and Vote Determinant *American Journal of Political Science* **53** (2) 445-460

[10] Zimmermann F, Kohring M 2019 Mistrust, Disinforming News, and Vote Choice: A Panel Survey on the Origins and Consequences of Believing Disinformation in the 2017 German Parliamentary Election *Political Communication* **37** (2) 215-237

**Appendix.** Independent Variables and The Explanation.

1. Gender (MALE): 1 = male, 0 = female.
2. Rural-Urban (RURAL): 1 = rural, 0 = urban.
3. Age in years (AGE).
4. Education: length of schooling in years (EDUC).
5. Income per month in million rupiahs (INCOME).
6. Ethnicity (ETHNIC): 1 = Javanese, 0 = others.
7. Religion (RELIG): 1 = Islam, 0 = others.
8. Region, represented by the four dummy variables:

| Region | Dummy Variables | | | |
|---|---|---|---|---|
| | SUMATERA | JAVA | KALIMATAN | SULAWESI |
| Sumatera | 1 | 0 | 0 | 0 |
| Java | 0 | 1 | 0 | 0 |
| Kalimantan | 0 | 0 | 1 | 0 |
| Sulawesi | 0 | 0 | 0 | 1 |
| Others | 0 | 0 | 0 | 0 |

9. Evaluation of current national economic condition compared to last year (ECON): 1 = better, 0 = worse or no change.
10. Satisfaction with the incumbent's performance (APPROV): 1 = satisfied, 0 = dissatisfied.
11. Identity of party A (PARTYIDA): 1 = feeling closer to the party that supports Candidate A, 0 = does not feel close to the party that supports Candidate A.
12. Identity of party B (PARTYIDB): 1 = feeling closer to the party that supports Candidate B, 0 = does not feel close to the party that supports Candidate B.
13. Assessment of the personal quality of candidate A (QUALITYA): a composite index of three items on the quality of Candidate A, (1) attentive/care for the people, (2) able to lead in overcoming national problems, (3) trustworthiness; where each item is measured on a 5-point scale (1 = very inattentive/unable to lead/untrustworthy, 5 = very attentive/able to lead/trustworthy). The three items are averaged to form a score on a scale of 1-5, where 1 means very bad and 5 means very good.
14. Assessment of the personal quality of Candidate B (QUALITYB): a composite index of three items on the quality of Candidate B: (1) attentive/care for the people, (2) able to lead in overcoming national problems, (3) trustworthiness; where each item is measured on a 5-point scale (1 = very inattentive/unable to lead/untrustworthy, 5 = very attentive/able to lead/trustworthy). The three items are averaged to form a score on a scale of 1-5, where 1 means very bad and 5 means very good.
15. Attitudes towards negative issues for Candidate A (ISSUA): a combined index of four negative issue items for Candidate A: (1) related to PKI, (2) an accomplice to PRC, (3) anti-Islam, (4) authoritarian; where each item is measured on a 3-point scale (1 = don't believe, 2 = never heard of, 3 = believe). The four items are averaged to form a score on a scale of 1-3, where 1 means do not believe in all issues and 3 means believe with all issues.
16. Attitudes towards negative issues for Candidate B (ISSUB: a composite index of four negative issue items for Candidate B: (1) Christian/Catholic family, (2) involved in activist kidnapping events, (3) fired from the army, (4) temperamental/has mental instability; where each item is measured on a 3-point scale (1 = don't believe, 2 = never heard of, 3 = believe). The four items are averaged to form a score on a scale of 1-3, where 1 means don't believe in all issues and 3 means believe in all issues.
17. Candidate advertisements that are seen more frequently on TV, which are represented by two dummy variables with the reference category "balanced/never seen":

| Candidate ads that are seen more frequently on TV | Dummy Variables | |
|---|---|---|
| | TVA | TVB |
| Ads of Candidate A | 1 | 0 |

| | | |
|---|---|---|
| Ads of Candidate B | 0 | 1 |
| Balanced/Never seen | 0 | 0 |

18. Candidate advertisements that are seen more frequently on internet or social media, which are represented by two dummy variables with the reference category "balanced/never seen":

| Candidate ads that are seen more frequently on internet | Dummy Variables | |
|---|---|---|
| | INTA | INTB |
| Ads of Candidate A | 1 | 0 |
| Ads of Candidate B | 0 | 1 |
| Balanced/Never seen | 0 | 0 |

19. Candidate advertisements that are seen more frequently on outdoor media (banners, stickers, flags, etc.), represented by two dummy variables with the reference category "balanced/never seen":

| Candidate advertisements that are seen more frequently on outdoor media (banners, stickers, flags, etc.) | Dummy Variables | |
|---|---|---|
| | OUTA | OUTB |
| Outdoor Media of Candidate A | 1 | 0 |
| Outdoor Media of Candidate B | 0 | 1 |
| Balanced/Never seen | 0 | 0 |

20. The candidate's campaign team that visited the most frequently, represented by two dummy variables with the reference category "balanced/none":

| The candidate's campaign team that visited the most frequently | Dummy Variables | |
|---|---|---|
| | DOORA | DOORB |
| Campaign team of Candidate A | 1 | 0 |
| Campaign team of Candidate B | 0 | 1 |
| Balanced/None | 0 | 0 |

IOP Conf. Series: Materials Science and Engineering 1115 (2021) 012062 doi:10.1088/1757-899X/1115/1/012062

# Analysis of Ionosphere disturbance caused by the Lokon Volcano Eruption using GPS TEC data

**D P Pandara[1,*], B Muslim[2], B Sunardi[3], Ferdy[1], G Pasau[1], M Mananohas[4] and Ch Ango[1]**

[1]Physics Department, FMIPA, Sam Ratulangi University, INDONESIA
[2]Space Science Center, National Institute of Aeronautics and Space (LAPAN), Bandung, INDONESIA
[3]Research and Development Centre of BMKG, INDONESIA
[4]Mathematics Department, FMIPA, Sam Ratulangi University, INDONESIA.

*Corresponding author: dpandara_fisika@unsrat.ac.id

**Abstract.** Eruption of Mount Lokon has the potential to cause danger to the community. The study of the ionospheric disturbances triggered by Lokon activity is very important to determine eruptive behavior. TEC anomaly before eruption is succesfully detected using correlation methods. Based on GPS TEC data from the GNSS satellite it was concluded that the TEC anomaly occurred three days before the eruption on July 14$^{th}$, 2011. Using leveling method and polynomial technique, eruption effects on ionosphere were be determined. Eruption on July 2011 caused ionospheric disturbance. These results will be important considerations in decision making related to disaster mitigation

## 1. Introduction

Volcanic activity is very potential to encourage ionospheric disturbance through explosion or infrasound[1]. Volcanic eruptions and shallow seismic activity can trigger the propagation of gravitational and acoustic waves upward into the ionosphere[2]. These waves produced significant disturbances on the total electron content (TEC) in ionosphere[3]. The TEC variation and anomalies on some volcanoes happened 6-17 days before eruptions [3] dan some minutes to hour after eruption [4]. The pattern of TEC disturbances can provide important information about the characteristics of a volcano [5]. Analysis of ionospheric disturbances related volcano eruption were done using TEC (Total Electron Content) data from dual-frequency Geographic Positioning Systems (GPS) observations[5][6]. GPS TEC from Global Navigation Satellite System (GNSS) is a useful tool to observe the ionosphere disturbance caused by volcanic activities [7].

Lokon volcano is one of the most active volcanoes in Indonesia. Its eruption could endanger the people living in Tomohon City, North Sulawesi Province and potentially cause disruption on ionosphere. Research of TEC disturbation related to Lokon eruptions has never been carried out even though it is very important as a clue about an eruption behavior. Compared with previous research, this research combined RSAM data and GPS TEC data to determine the effect of eruptions activities on the ionosphere. Aims of the paper investigates TEC disruption before and after eruption of Lokon Volcano at July 14th, 2011. This study is one of the keys to optimizing mitigation efforts in the future.

IOP Conf. Series: Materials Science and Engineering 1115 (2021) 012062 doi:10.1088/1757-899X/1115/1/012062

## 2. Materials and Methods

### 2.1. Materials

The data analyzed in this study consisted of TEC derived from the Global Ionosphere Map (GIM TEC), TEC calculated from GPS (GPS TEC), Disturbance storm time index (Dst), and Realtime Seismic Amplitude Measurement (RSAM). GIM TEC data were taken from Center for Orbit Determination in Europe (CODE; ftp://ftp.unibe.ch/aiub/CODE/). GIM TEC is generated every day using GPS data from TEC modeled in the reference frame of geomagnetic sun by spherical harmonic expansion based on a single layer model. The GIM TEC is arranged in $2.5^0$ x $5.0^0$ (latitude versus longitude) grids with a temporal resolution of 2 hours. GPS TEC data were computed from GPS observation data of high resolution in Rinex format (Rinex GPS data), GPS satellite orbit data, GPS station coordinate data and Differential Code Bias (DCB). Rinex GPS data (*.* O) and GPS satellite orbit data (*.sp3) were downloaded from ftp://cddis.nasa.gov/, DCB data for calibration of TEC computational were downloaded from ftp: //ftp.unibe. ch / aiub / CODE/ and GPS station coordinate data were downloaded from ftp://igscb.jpl.nasa.gov/pub/station/general/igs.snx. Dst index obtained in real time from the World Data Center for Geomagnetism, Kyoto (http://wdc.kugi.kyoto-u.ac.jp/dst_real time/index.html). Dst index is a measure of the strength of magnetic storms. GIM TEC and Dst index were used to analyze ionosphere anomalies before the Mount Lokon eruption, while GPS TEC data were used to analyze the ionosphere fluctuation after eruptions. Furthermore, the seismic activity that occurs in the body of the mountain were described by RSAM. Seismic signals in the SEISAN form were obtained from Center of Volcanology and Geology Hazard Mitigation Bandung (CVGHM) database.

### 2.2. Methods

To investigate TEC anomaly before Lokon eruption on July 14th 2011, TEC from GIM TEC were processed using correlation techniques which consists of auto correlation and cross correlation [8]. Auto correlation process between TEC daily value and diurnal variation of TEC of a station was carried out. To represent normal conditions, TEC data were selected in 27 days back so that there were 27 TEC data. At each observation time starting at 00:00 to 22:00 with an interval of 2 hours, the average TEC were calculated at a certain hour for 27 days so that the monthly average of TEC diurnal variation were obtained. Furthermore, auto correlation analysis was performed between diurnal variation of TEC on a particular day for 27 days with a diurnal variation of average monthly of TEC. The process produced 27 correlation coefficients. From this step, the average of 27 coefficient correlation and standard deviation of coefficient correlation (SDCC) were calculated. Each coefficient correlation values subtracted with the average would determine deviation of coefficient correlation (DCC). Anomaly of TEC is indicated when the ratio is DCC/SDCC $\leq$ -1. To determine whether the TEC anomaly occurs locally or globally, TEC data obtained from the GIM at the closest latitude to the monitoring point were cross-correlated with the TEC data at the next latitude (latitude north and south). This fact is expressed by the variability index. To show range of ionosphere disturbances were calculated DTEC. DTEC is the logarithmic value of the comparison of the TEC at a particular hour with the TEC value at the same hour representing median value of the last 27 days from the day [9]. Based on DTEC value, ionosphere condition divided into four categories and stated in W index. To see whether TEC fluctuations were influenced by magnetic effects or not, Dst index were analyzed. Dst index are usually expressed with a negative value that will indicate a decrease in the Earth's magnetic field. The TEC anomaly considered as precursory signal of earthquakes or volcano activities while the geomagnetic level characterized as quiet day. The quiet day represents that TEC fluctuations are not influenced by magnetic storms which ionospheric compared with $TEC_{model}$ data. $TEC_{model}$ data were constructed using polynomial order 6 for diurnal variations, while GPS TEC data were constructed from two L-band signals transmitted by GNSS satellites [10].

GPS TEC data near real time were carried out by processing RINEX GPS data, GPS satellite orbit data and DCB data into Slant TEC (STEC) data [7]. Rinex GPS data and satellite orbit data were extracted to obtain code and phase data, receiver station coordinates, GPS observation time, and GPS satellite orbit. Based on the coordinates of the receiving station, GPS observation time and the orbit of the GPS satellites, Ionospheric Pierce Point (IPP) was calculated. IPP includes the elevation angle, latitude and longitude. Then, the code data and phase data were combined at multiple frequencies of the GPS signal to obtain the estimated $STEC_{code}$ and $STEC_{phase}$. The $STEC_{code}$ and $STEC_{phase}$ were calibrated with DCB data. Due to the higher noise in the $STEC_{code}$, we used the $STEC_{phase}$ data. These data stored in ASCII format with a matrix structure. The matrix column shows the $STEC_{phase}$ of 32 GNSS satellites and the matrix row shows the observation time every 30 seconds in 24 hours. To analyze the effects of the Lokon eruption, the $STEC_{phase}$ provided fairly comprehensive data related with eruption were selected from the matrix. $STEC_{phase}$ data and $STEC_{model}$ data at the day of eruption were plotted in one graph. From this graph, we examine the occurrence of disturbances. RSAM are estimated from seismic signals based on Endo and Murray Work's[11]. RSAM were calculated using ObsPy software [12].

## 3. Results and Discussion

### 3.1. Anomaly Before The July 14th, 2011 Eruption

The characteristics of the ionosphere and geomagnetic interval of 24 days before the first magmatic eruption of Lokon volcano on July 14th, 2011 at 15:31:52 UTC is shown in Figure 1.

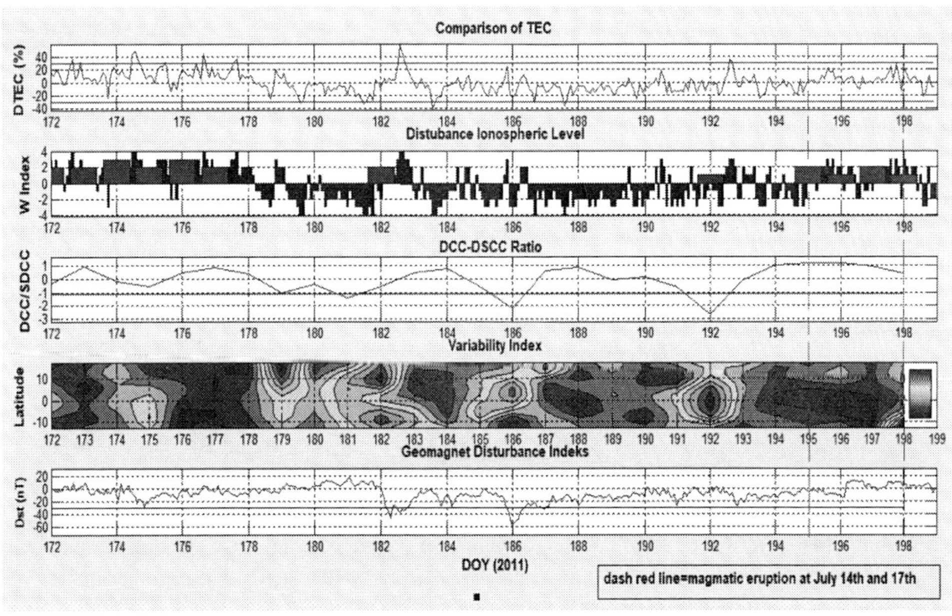


**Figure 1**. Ionosphere Anomaly Related Lokon Eruption July 14th, 2011

The data shows that on July 1st, 2011 (the 181th day) the DCC / SDCC ratio is below -1 and the ionosphere disturbance index is at the level of -2. This means that there is a fairly small disturbance in the ionosphere. The disturbance did not originate from magnetic storms because the DST indexes were not significant and variability index was quite low. Between 182nd days and 186th, the DCC / SDCC ratio is below minus 1 so that it only causes minor disturbances as indicated by the W index.

Ionosphere activities during three days were dominated by interactions with magnetic storms. Ionospheric disorders increased on July 11th, 2011 (day 192nd) with the variability of the disturbance to be very local and followed by a fluctuation in the DTEC ratio with big amplitude. Fluctuation in DTEC is significant with the condition of the ionosphere which has been disturbed at a moderate level.

Based on the performance of the DTEC, the W index, the DCC / DSCC ratio and the DST index in Figure 1, it can be concluded that there is a TEC anomaly which happened on July 11th, 2011 [13]. Hasil sesuai dengan analisis terhadap data TEC dari stasiun GPS terdekat yaitu stasiun PIMO di Luzon, Filipina. This anomaly occurred 3 days before the eruption of July14th, 2011 or 8 days before the eruption of 17 July 2011. Ionosphere anomalies occured before volcanic eruptions can be caused by rising radon gas to the surface along with rising magma [3]. They caused the atmospheric ionization process and changes in air conductivity. TEC anomaly associated also with a decrease in electron density in the ionosphere [3]. Hot magma that moves up from the magma chamber to the surface will trigger the release of positive ions into the ionosphere. The recombination of positive ions with free electrons causes a decreasing of the TEC [13].

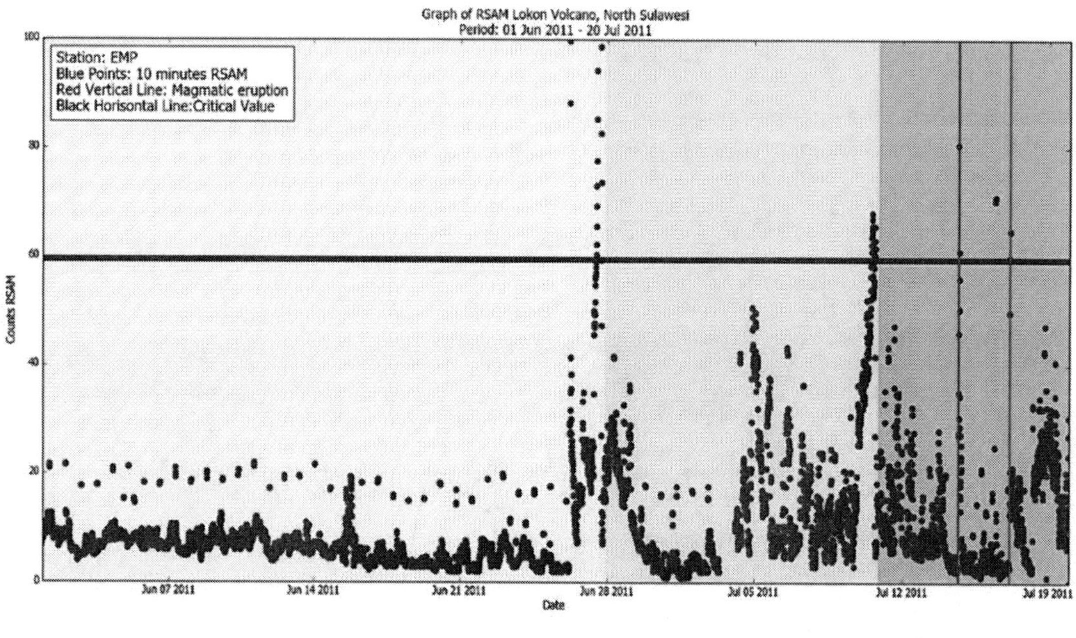


**Figure 2.** RSAM Lokon Volcano during on June until July, 2011

Figure 2 describes pattern of RSAM on Lokon Volcano. RSAM inform the state of volcanic activity. Since the emergence of a significant tremor on June 28-29th 2011, the increasing of volcanic activity has been very intensive in the Lokon volcano. From July 4th, 2011 until July 9th, 2011, RSAM increase. RSAM increasing indicates fluid flow in the volcano conduit and it cause the conduit vibrates. RSAM values reach a critical values at July 11st, 2011 where tremor earthquake was very dominant. Tremors have been thought to be related to the supply of hot magma into the magma chamber [14] and followed by increased pressure in the conduit due to the degassing process [15]. Hot magma flow would change magnetic's field around volcano. Therefore, the TEC anomaly was allegedly associated with volcanic activity related tremor increasing.

### 3.2.Ionosphere Disturbance After Eruption

Figure 3 shows that pertubation of TEC were recorded by satellite 2 for BAKO and XMIS stations with fluctuations amplitude of 1.5 TECU and 1 TECU respectively. The fluctuations at BAKO and XMIS stations occurred each 40 and 50 minutes after eruption. The eruption occurred at 15:31 UTC which is marked by a pink vertical line. The polynomial line is the value of TEC smoothing and serves as a comparison in determining the presence of interference with TEC. The time difference is influenced by the distance of the station from the source of the eruption and the propagation speed between acoustic and gravity waves [16]. Gravitational and acoustic waves are two forms of Rayleigh waves arising due to upward pressure from the force of the eruption [4]. Because of the difference in propagation time, these waves affect the ionosphere at different time periods and detected by GNSS satellites [17].

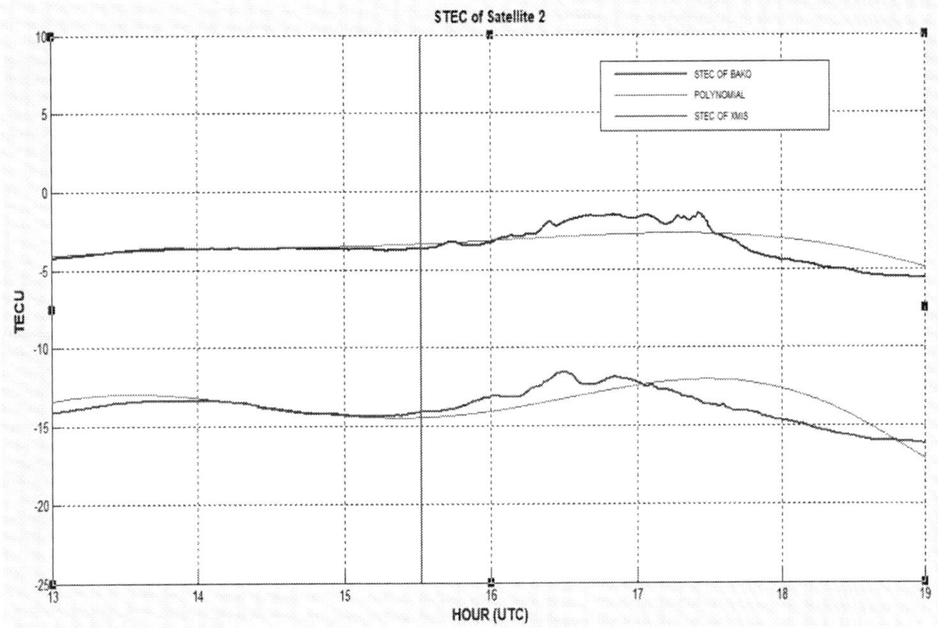


**Figure 3**. TEC fluctuations on July 14th, 2011 recorded by satellite 2 at BAKO and XMIS stations

On the day of the eruption, there were not magnetic storm, significant earthquakes and volcanic eruption other than the only eruption from the Mount Lokon so the TEC fluctuations observed in Figure 3 are related to the eruption of Lokon volcano. From figure 3 it can also be seen that the fluctuation is irregular. This irregularity occurs because of Indonesia's position in the equatorial area as known scintillation effect. It influenced temporal variations of TEC in the ionosphere and caused the signal phase different in the equatorial region. Thus, Mount Lokon's eruption at July 14th, 2011 affected the ionosphere layer.

## 4. Conclusions

The cross-correlation method applied to the GPS TEC data can detect anomalies in the ionosphere. Ionosphere disturbances after Lokon volcano eruption can be described using leveling and polynomial methods. Precursor eruption on July 2011 is tremor increasing related hot magma data GPS flow.

Magma flow triggered TEC anomaly so explosion activities disrupted the ionosphere above Lokon volcano. Acoustic shock waves and gravity wave caused fluctuation of the GPS TEC. These results indicate that the interaction between Lokon activities and the ionosphere which can be observed. This research must be conducted on many eruptions so that the character of volcano is more understandably. So that, studies on ionosphere disturbance are more accurate to GPS TEC data that must be taken from the nearest GPS station.

## Acknowledgments

We would like to thanks to the Rector of Samratulangi University who support us, so that this research was fully funded by PNBP Sam Ratulangi University through RDUU 2019 Research Grant to DP, F & MM.

## References

[1] Matoza R, Fee D, Green D, and Mialle P 2019 Infrasound monitoring for atmospheric studies Berlin: Springer

[2] Panda D, Senapati B, Tyagi B and Kundu B 2019 Effects of Rayleigh-Taylor instability and ionospheric plasma bubbles on the global navigation satellite System signal Journal of Asian Earth Sciences **170** 225-233

[3] Li W, Guo J, Yue J, Shen Y, and Yang Y 2016 Total electron content anomalies associated with global VEI4+ volcanic eruptions during 2002–2015. *Journal of Volcanology and Geothermal Research* **325** 98-109

[4] Nakashima Y, Heki K, Takeo A, Cahyadi M.N, Aditiya A, and Yoshizawa K 2016 Atmospheric resonant oscillations by the 2014 eruption of the Kelud volcano, Indonesia, observed with the ionospheric total electron contents and seismic signals *Earth and Planetary Science Letters* **434** 112-116

[5] Cahyadi M N, Rahayu R W, Heki K and Nakashima Y 2020 Harmonic ionospheric oscillation by the 2010 eruption of the Merapi volcano, Indonesia, and the relevance of its amplitude to the mass eruption rate *Journal of Volcanology and Geothermal Research* **405** 107047

[6] Calais E, Bernard Minster J, Hofton M and Hedlin M 1998 Ionospheric signature of surface mine blasts from Global Positioning System measurements *Geophysical Journal International* **132** (1) 191-202

[7] Cahyadi M N, Rahadyan A N and Muslim B 2019 Analysis of Total Electron Content (TEC) Near Real Time Using Dual Frequency GPS Data (Study Case: Surabaya) *E3S Web of Conferences* **94** 05005

[8] Sunardi B, Muslim B, Sakya A E, Rohadi S and Murjaya J 2018 Ionospheric earthquake effects detection based on Total Electron Content (TEC) GPS Correlation *IOP Conference Series: Earth and Environmental Science* 132 (1) 012014

[9] Stanislawska I, Gulyaeva T and Jin S 2015 Ionospheric W index based on GNSS TEC in the operational use for navigation systems *Satellite positioning—methods, models and applications* 131-148

[10] Muslim B, Haralambous H and Oikonomu C 2019 Ionospheric earthquake signatures on GPS TEC *Journal of Physics: Conference Series* **1170** (1) 012070

[11] Endo E T and Murray T 1991 Real-time seismic amplitude measurement (RSAM): a volcano monitoring and prediction tool *Bulletin of Volcanology* **53** (7) 533-545

[12] Krischer L, Megies T, Barsch R, Beyreuther M, Lecocq T, Caudron C and Wassermann J 2015 ObsPy: A bridge for seismology into the scientific Python ecosystem *Computational Science & Discovery* **8** (1) 014003

[13] Ango C I, Pandara D P, Ferdy F and Tongkukut S H 2020 Investigasi Anomali TEC-Ionosfer

Sebelum Letusan Gunung Lokon 14 Juli 2011 Menggunakan Metode Sliding Interquartile *Jurnal MIPA* **9** (1) 28-32
[14] Sgroi T, Di Grazia G and Favali P 2019 Volcanic Tremor of Mt. Etna (Italy) Recorded by NEMO-SN1 Seafloor Observatory: A New Perspective on Volcanic Eruptions Monitoring *Geosciences* **9** (3) 115-124
[15] Pandara D P, Ferdy F, Pasau G and Suoth V A 2020 Analisis Prekursor Erupsi Gunung Lokon Tanggal 12 September 2014 *Jurnal MIPA* **9** (2) 47-50
[16] Holding D 2017 Multi-GNSS vertical total electron content estimates: Data analysis and machine learning with python to evaluate ionospheric perturbations from earthquakes *Hold. Eng. Consult* **1** 5-8
[17] Cahyadi M N, Dermawan,A A A and Muslim B 2018 Studi Perubahan Ionosfer Akibat Letusan Gunung Berapi Dengan Pengamatan Total Electron Content GNSS *Geoid* **13** (2) 196-199

# Geographically and Temporally Weighted Regression Model with Gaussian Kernel Weighted Function and Bisquare Kernel Weighted Function

**Harianto*, W H Nugroho and E Sumarminingsih**

Department of Statistic, Malang Brawijaya University, Indonesia

*Corresponding author: harianto@student.ub.ac.id

**Abstract.** The Geographically and Temporally Weighted Regression (GTWR) model is a development of the Geographically Weighted Regression (GWR) model. Estimation of parameters in the GTWR model uses Weighted Least Square (WLS). The purpose of this study is to build two GTWR models using the Gaussian kernel weight function and the Bisquare kernel weight function on the percentage of poor people in East Java Province in 2010-2015. Each weighted function produces 228 different models at each location and time. The GTWR method with the Gaussian Kernel weight function produces parameter estimates at the observation location in the range between 0.7796 to 48.1435. Meanwhile, the parameter estimation of the GTWR model with the Kernel Bisquare weight function ranges from 0.7828 to 48.1444. The two weight functions can model the percentage of poor people in East Java Province in 2010-2015, with $R^2$ values of 98.994% and 98.987%, respectively.

## 1. Introduction

The Geographically and Temporally Weighted Regression (GTWR) method is a development of the Geographically Weighted Regression (GWR) method with the aim of being able to simultaneously handle spatial and temporal non-stationary data [1]. Furthermore, based on ref [2], the GTWR model has the advantage that the resulting model is local at each location and time, this makes the resulting model more representative in terms of location and time. Several previous studies on GTWR have been conducted in ref [3], for modeling economic growth using R and in ref [4], for modeling the proportion of poor people in districts and cities in Central Java Province. Also, in ref [5] for determining the optimum bandwidth result of the GTWR model using the cross validation method. Dwi et al. (2016) modeling of air pollutant elements $No_2$ using GTWR [6]. One of the important elements in the GTWR method is to construct a weighted matrix with data containing spatial and temporal information that can identify spatial and temporal diversity. This weighted idea is based on Tobler's I law which says "All things are related to each other, but something closer affects something far away" [7]. A weighted function that can be used either by using kernel functions. This research uses the Gaussian kernel function and the Bisquare kernel function to build the model.

Yusuf (2013) implemented Gaussian kernel functions and Bisquare kernel functions to build a GWR model [8]. The weighting function is used because both involve the element of distance between the observed locations whose value is continuous in constructing the weighting matrix, so that each location will receive a weight according to the distance between the location and the observed location. The purpose of this research is to apply the Gaussian kernel weight function and the Bisquare kernel function in building the GTWR model with a case study of the percentage of poor people in East Java Province in 2010-2015.

## 2. Literature Review

### 2.1. Geographically Weighted Regression (GWR)

One of the methods of regression that incorporates spatial elements in building a model is the Geographically Weighted Regression (GWR) method. GWR was developed from a global regression model based on non-parametric regression [9]. GWR is a method aimed at building a regression model that is local. The estimation of the regression parameters in the GWR method is carried out at each location by involving the surrounding information that has been weighted by geographical distance [10]. The GWR model can generally be written as follows:

$$Y_i = \beta_o(u_i, v_i) + \sum_{k=1}^{p} \beta_k(u_i, v_i)x_{ik} + \varepsilon_i \text{ ; } k = 1,2, \dots p \text{ ; } i = 1,2, \dots, n \quad (1)$$

The parameter estimation on GWR model using Weighted Least Square (WLS) Method by providing different weighteds for each location where data is observed. Here are the following formulation of WLS estimator:

$$\hat{\beta}(u_i, v_i) = [X^T W(u_i, v_i)X]^{-1} X^T W(u_i, v_i)y \quad (2)$$

with

$$\boldsymbol{W}(u_i, v_i) = \begin{bmatrix} \boldsymbol{w}_{i1} & 0 & \dots & 0 \\ 0 & \boldsymbol{w}_{i2} & \dots & 0 \\ \vdots & \vdots & \ddots & \vdots \\ 0 & 0 & \dots & \boldsymbol{w}_{in} \end{bmatrix}$$

In equation (2), estimator $\hat{\beta}(u_i, v_i)$ is an unbiased and consistent estimator of $\beta(u_i, v_i)$ [4].

### 2.2. Spatial Weight Function

Regression involving spatial effects, one that determines the results of the analysis is the choice of weighting function [11]. In 2013, Yusuf [8] applied the two weight functions to the GWR method. Meanwhile, in this study, the Gaussian Kernel weighted function and the Bisquare Kernel weight function will be used in constructing the weight matrix in the GTWR model.

#### *2.2.1. Gaussian Kernel function*

$$\boldsymbol{w}_{ij} = exp\left(-\frac{1}{2}\left(\frac{d_{ij}}{h}\right)^2\right) \quad (3)$$

where $d_{ij}$ is the euclid distance from location-*i* to location-*j*, and *b* are the optimum window width. In the gaussian kernel weight function, the weight $(w_j(i))$ will approach one along the closer the distance between the location-*i* with location-*j* and weighted value $(w_j(i))$ decreases as the distance grows between location-*i* and location-*j*.

#### *2.2.2. Bisquare Kernel function*

$$\boldsymbol{w}_{ij}(u_i, v_i) = \begin{cases} \left(1 - \left(\frac{d_{ij}}{h}\right)^2\right) & ; d_{ij} < h \\ 0 & ; \text{another} \end{cases} \quad (4)$$

where $d_{ij}$ is the euclid distance from location-*i* to location-*j*, and *b* is the optimum window width, namely the optimum distance to a location still give influence to the location is being observed. In the bisquare kernel weighting function, if the distance between location-*i* and location-*j* greater than or equal to the window width, then the location will be given a zero weight. While the distance between locations is less than the width of the window will be weighted close to one as the distance between location-*i* and location-*j* gets closer.

While, to determine the optimum bandwidth using the Cross Validation (CV) method, mathematically it can be written in equation 5 [11].

$$CV(h) = \sum_{i=1}^{n}(y_i - \hat{y}_{\neq i}(h))^2 \quad (5)$$

## 2.3. Geographically and temporally Weighted Regression (GTWR)

The GWR model only considers spatial effects, while GTWR combines spatial and temporal data information in a weighted matrix to identify spatial and temporal heterogeneity [4]. The GTWR model for the independent variable p with the dependent variable at the location $\{u_i, v_i, t_i\}$ for each observation is written in equation (6):

$$Y_i = \beta_o(u_i, v_i, t_i) + \sum_{k=1}^{p}\beta_k(u_i, v_i, t_i)\, x_{ik} + \varepsilon_i \quad (6)$$

Estimation of the regression coefficient parameter $\hat{\beta}(u_i, v_i, t_i)$ at point $i$ can be obtained using the Weighted Least Square (WLS) method as follows:

$$\hat{\beta}(u_i, v_i, t_i) = [\boldsymbol{X^T W}(u_i, v_i, t_i)\boldsymbol{X}]^{-1}\boldsymbol{X^T W}(u_i, v_i, t_i)\boldsymbol{Y} \quad (7)$$

Where $\boldsymbol{W}(u_i, v_i, t_i) = diag\ (w_{i1}, w_{i2}, \dots, w_{in})$ is a matrix on the observation location $(u_i, v_i)$ and time $t_i$. Diagonal element $\boldsymbol{w_{ij}}(1 \le j \le n)$ is a spatial-temporal distance function at the observation point $(u_i, v_i, t_i)$. The built-in spatial-temporal distance is a combination of spatial distance function and temporal distance function [12]. In 2010 Huang *et al* [2] given Illustration of spatial-temporal distance uses an ellipsoidal coordinate system to measure the closeness between the regression point and the observation point that surrounds it. The following illustration of ellipsoidal coordinates is shown in Figure 1.

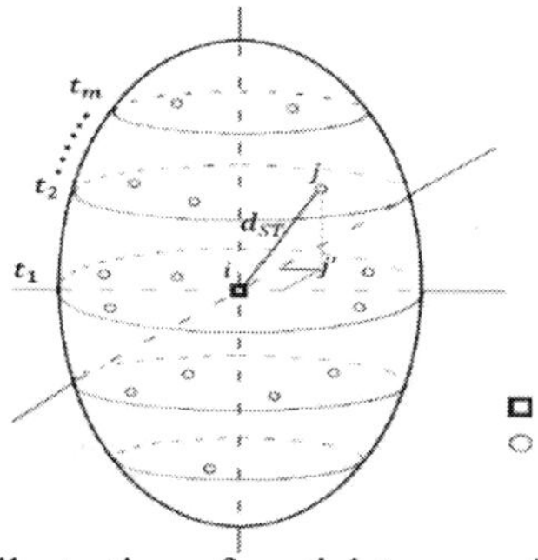


**Figure 1.** illustration of spatial-temporal distance at ellipsoidal coordinates [2]

Spatial-temporal distance function $(d_{ij}^{ST})$ is a combination of spatial distance function $(d_{ij}^{S})$ and temporal distance function $(d_{ij}^{T})$, which are written as follows [2]:

$$(d^{ST})^2 = \lambda(d^S)^2 + \mu(d^T)^2 \quad (8)$$

where

$$\begin{cases} (d_{ij}^S)^2 = (u_i - u_j)^2 + (v_i - v_j)^2 \\ (d_{ij}^T)^2 = (t_i - t_j)^2 \\ (d_{ij}^{ST})^2 = \lambda\left[(u_i - u_j)^2 + (v_i - v_j)^2\right] + \mu\left[(t_i - t_j)^2\right] \end{cases} \quad (9)$$

where λ and μ are used as a balancing parameter between the influence of location and time on the measurement of spatial-temporal distances. Based on the equation above obtained:

$$w_{ij} = exp\left\{-\left(\frac{\lambda[(u_i-u_j)^2+(v_i-v_j)^2]+\mu[(t_i-t_j)^2]}{h_{ST}^2}\right)\right\} \quad (10)$$

$$= exp\left\{-\left(\frac{\left[(u_i - u_j)^2 + (v_i - v_j)^2\right]}{h_S^2} + \frac{\left[(t_i - t_j)^2\right]}{h_T^2}\right)\right\}$$

For example $h_S^2 = \frac{h_{ST}^2}{\lambda}$ and $h_T^2 = \frac{h_{ST}^2}{\mu}$, it is obtained:

$$w_{ij} = exp\left\{-\left(\frac{(d_{ij}^S)^2}{h_S^2} + \frac{(d_{ij}^T)^2}{h_T^2}\right)\right\}$$

$$= exp\left\{-\left(\frac{(d_{ij}^S)^2}{h_S^2}\right)\right\} \times exp\left\{-\left(\frac{(d_{ij}^T)^2}{h_T^2}\right)\right\}$$
$$= w_{ij}^S \times w_{ij}^T$$

Where $w_{ij}^S = exp\left\{-\left(\frac{(d_{ij}^S)^2}{h_S^2}\right)\right\}$ and $w_{ij}^T = exp\left\{-\left(\frac{(d_{ij}^T)^2}{h_T^2}\right)\right\}$

As for the parameter values $\tau$, for example $\tau$ is the ratio parameter of $\tau = \frac{\mu}{\lambda}$ with $\lambda \neq 0$, the equation is obtained [12]:

$$\frac{(d_{ij}^{ST})^2}{\lambda} = \left[(u_i - u_j)^2 + (v_i - v_j)^2\right] + \tau\left[(t_i - t_j)^2\right] \quad (11)$$

For example $\lambda = 1$, which aims to reduce unknown parameters. There is one unknown parameter, which is $\tau$. The parameter $\tau$ is useful to balance the effect of temporal distance on spatial distance. This parameter is obtained from the minimum Cross Validation (CV) method through the initial value $\tau$ written in equation (12):

$$CV(\tau) = \sum_{i=1}^{n}\left(y_i - \hat{y}_{\neq i}(\tau)\right)^2 \quad (12)$$

In this study, the model accuracy test (Goodness of fit) uses the coefficient of determination ($R^2$). The coefficient of determination can explain the large variance of response variables that can be explained by the predictor variable. The value of $R^2$ is obtained through a mathematical equation as follows [11]:

$$R^2(u, v, t) = \frac{JKR_w}{JKT_w}\frac{\sum_{j=1}^{P} W_{ij}(y_j - \hat{y}_j)^2}{\sum_{j=1}^{P} W_{ij}(y_j - \bar{y})^2} \quad (13)$$

## 3. Result and Discussion

This study uses secondary data sourced from the Central Statistics Agency (BPS) of East Java Province during 2010-2015. Spatial units used in this study were 38 regencies / cities in East Java Province consisting of 29 regencies and 9 cities. The response variable in this study was the Percentage of poor population ($Y$). While the Predictor variables used were the School Participation Rate aged 16-18 years ($x_1$), Morbidity ($x_2$) and Unemployment Rate ($x_3$). The analysis Steps are as follows: (1) Describe the research variables. (2) Conduct a test of spatial and temporal heterogeneity. (3) Calculating the Euclidean Spatial-temporal distance at coordinates $(u_i, v_i, t_i)$ Include: determine the parameters of the spatial-temporal ratio ($\tau$), determine spatial parameters ($\lambda$) and temporal parameters ($\mu$) and Calculate spatial-temporal bandwidth ($h_{ST}$). (4) Determine the weighted value using the Gaussian kernel function and the Bisquare kernel function. (5) Estimated parameters $\hat{\beta}(u_i, v_i, t_i)$. (6) Calculating the value of $R^2$ in each model with equation (13). The statistical description of the research data is presented in Table 1

**Table 1.** Data Description

| Variable | Min | Max | Mean |
|---|---|---|---|
| Percentage of poor people (Y) | 4.45 | 32.47 | 23.12 |
| School participation rate Age 16-18 years ($X_1$) | 38.61 | 93.75 | 66.38 |
| Morbidity ($X_2$) | 7.02 | 27.72 | 15.34 |
| unemployment rate ($X_3$) | 0.87 | 10.62 | 4.52 |

Table 1 shows that in the period of 6 years the average percentage of poor people in East Java province amounted to 23.12% with the highest percentage of poverty at 32% and the lowest 4.45%. As for predictor variables, the 16-18 years old school participation rate has the highest number of 93.75%, the morbidity variable has the lowest value of 7.02%. While the unemployment rate variable for a period of 6 years has an average of 4.52% with the lowest number of 0.87%. The relationship variable responses with each of the predictor variables in a linear can be described by using a Pearson correlation. The relationship between response variable and predictor variables can be seen in Table 2.

**Table 2.** Correlation coefficient of the response variable and the predictor variables

| Variable | *Pearson* Correlation | *P-value* |
|---|---|---|
| $Y$ with $X_1$ | -0.5849776 | $2.2 \times 10^{-16}$** |
| $Y$ with $X_2$ | 0.121656 | 0.0667* |
| $Y$ with $X_3$ | -0.472064 | $4.692 \times 10^{-14}$** |

(* *) significant at a level of 5%, (*) significant at a level of 10%

Table 2 shows the relationship between variables $Y$ with variables $X_1$ and $X_3$ which are negative or have an inverse relationship with correlation coefficient values of 58% and 47% at the 95% confidence level. As for the variable $Y$ with the variable $X_2$ has a positive sign or has a direct relationship with a coefficient of 12% at a 90% confidence level. Furthermore the relationship between predictor variables needs to be seen to meet the assumption of non-multicollinearity. linear correlation between predictor variables can be tested with multicollinearity assumption test.

**Table 3.** VIF value on each predictor variable

| Years | $X_1$ | $X_2$ | $X_3$ |
|---|---|---|---|
| Full (2010-2015) | 1.661568 | 1.031367 | 1.539207 |
| 2010 | 2.829173 | 1.113492 | 3.278216 |
| 2011 | 2.268411 | 1.163203 | 2.501853 |
| 2012 | 1.667146 | 1.075252 | 1.526519 |
| 2013 | 1.565432 | 1.025933 | 1.354471 |
| 2014 | 1.436456 | 1.107825 | 1.545898 |
| 2015 | 1.730509 | 1.076934 | 1.514770 |

Table 3 shows the value of the Variance Inflation Factor (VIF) of the multicollinearity assumption test obtained from each predictor variable entirely under 10, meaning that the risk of multicollinearity is relatively low. In the spatial heterogeneity test showed the p-value during 2010-2015 was $3.027 \times 10^{-11}$. These results can be concluded that there is spatial heterogeneity. The following Breusch-pagan test results can be seen in Table 4.

**Table 4.** Breusch-Pagan Test

| Years | Breusch-pagan value | p-value |
|---|---|---|
| Full (2010-2015) | 55.147 | $3.027 \times 10^{-11}$ ** |
| 2010 | 9.2353 | 0.05548 * |
| 2011 | 9.304 | 0.05393 * |
| 2012 | 14.022 | 0.007225 ** |
| 2013 | 9.5494 | 0.04874 ** |
| 2014 | 10.184 | 0.03744 ** |
| 2015 | 10.949 | 0.02715 ** |

(* *) significant at a level of 5%, (*) significant at a level of 10%

While from the temporal dimension, the percentage of poor people in the province of East Java tends to decrease every year. The differences in each year indicate temporal heterogeneity. The following temporal heterogeneity with boxplot can be seen in Figure 2.

Boxplot describe temporal diversity of data. Figure 2 above shows that each year the size of the boxplot tends to narrow. Outliers in boxplot also appear to be decreasing every year. This can be explained that there is one regenci / city that has the highest percentage of poverty compared to other regencies / cities in a period of 6 years.

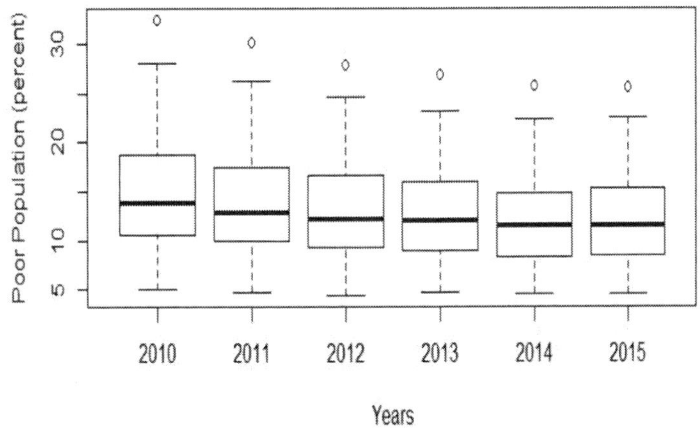


**Figure 2.** Boxplot of temporal heterogeneity of the percentage of poor people in East Java province in 2010-2015

Estimated parameter GTWR model on the percentage of poor people in East Java Province covering 38 regencies / cities in the period of 2010-2015, produced as many as 228 different models for each weighted function. Descriptions of the estimated parameters can be seen in Table 5.

Table 5 explains the parameter estimation description on GTWR by weight using the Gaussian Kernel function and the Bisquare Kernel function. The estimated parameters of the two weight functions produce almost the same value. Parameter estimation with Gaussian Kernel weight function at the observation location is in the range between 0.7796 to 48.1435. While the estimated parameters with the Bisquare Kernel weight function range from 0.7828 to 48.1444. In the predictor variable, the two weight functions have the same average in the school participation rate of 16-18 years old has a negative average value. This means that every time there is a decrease in the participation rate of schools aged 16-18 years will cause an increase in the percentage of the poor population in East Java. while the average morbidity and unemployment rate has a positive average value which means that each increase in morbidity and unemployment will cause an increase in the percentage of poor people in East Java. The value of $R^2$ on the GTWR model using the two weight functions is also almost the same. The Gaussian Kernel weight function has an $R^2$ value of 98.994% and the Bisquare Kernel weight function has an $R^2$ value of 98.987%.

**Table 5.** Description of Parameter Estimation Models

| Weight | Parameter | Min | Max | Mean | Std. Deviation |
|---|---|---|---|---|---|
| Kernel Gaussian | $\hat{\beta}_0$ | 0.7796 | 48.1435 | 19.2562 | 11.8940 |
| | $\hat{\beta}_1$ | -0.6466 | 0.06396 | -0.1241 | 0.1549 |
| | $\hat{\beta}_2$ | -0.2708 | 0.8837 | 0.0604 | 0.2294 |
| | $\hat{\beta}_3$ | -1.5191 | 1.9152 | 0.0473 | 0.7399 |
| Kernel Bisquare | $\hat{\beta}_0$ | 0.7828 | 48.1444 | 19.256 | 11.8947 |
| | $\hat{\beta}_1$ | -0.647 | 0.0639 | -0.1242 | 0.1549 |
| | $\hat{\beta}_2$ | -0.2708 | 0.8836 | 0.0604 | 0.2295 |
| | $\hat{\beta}_3$ | -1.5190 | 1.915 | 0.0474 | 0.7401 |

## 4. Conclusion

GTWR modeling in poverty cases in 38 regencies / cities in East Java Province during 2010-2015 using the Gaussian Kernel weight function and the Bisquare Kernel Weight function each produce 228 models in each location and time. GTWR modeling uses the Gaussian Kernel weight function and the Bisquare Kernel weight function in poverty cases in East Java Province, resulting in almost the same estimated parameters. The two weight functions can model the percentage of poor people in East Java Province in 2010-2015, with $R^2$ values of 98.994% and 98.987%, respectively. This study is limited to the Gaussian kernel function and Bisquare kernel function in constructing a weight matrix.

## References

[1] Wang P 2006 *Exploring spatial effects on housing price: the case study of the city of Calgary*, Master dissertation, University of Calgary, Canada
[2] Huang B, Wu B and Barry M 2010 Geographically and Temporally Weighted Regression for Modeling Spatio-Temporal Variation in House Prices *International Journal of Geographical Information Science* **24** (3) 383-401
[3] Sholihin M, Soleh AM and Djuraidah A 2017 Geographically and Temporally Weighted Regression (GTWR) for Modeling Economic Growth using R *International Journal of Computer Science and Network* 800-805
[4] Widiyanti KY, Yasin H and Sugito 2014 Pemodelan Proporsi Penduduk Miskin Kabupaten dan Kota di Provinsi Jawa Tengah Menggunakan Geographically and Temporally Weighted Regression *Jurnal Gaussian* **3** (4) 691-700
[5] Farihah I 2017 *Penentuan Hasil Bandwidth Optimum Model Geographically and Temporally Weighted dengan metode Cross Validation* Master dissertation UIN Malang, Indonesia.
[6] Dewi I, Yasin H and Hoyyi A 2016 Pemodelan Unsur Pencemar Udara No2 Menggunakan Geographically and Temporally Weighted Regression (GTWR) *Prosiding Seminar Nasional Variansi Universitas negeri Makasar Forum Pendidikan Tinggi Statistika Indonesia* 1-10
[7] Miller H J 2004 Tobler's First Law and Spatial Analysis Annals of the Association of America *Geographers* **94** (2) 284-289
[8] Yusuf L M 2013 *Regresi Terboboti Geografis dengan Fungsi Pembobot Kernel Gaussian dan Kernel Bisquare pada angka Harapan Hidup Studi kasus: Angka Harapan Hidup Kabupaten/Kota Provinsi Jawa Timur* Master dissertation Institut Pertanian Bogor
[9] Mei C I 2005 *Geographically Weighted Regression Technique for Spatial Data Analysis* Master dissertation Xi'an Jiaotong University
[10] Brunsdon C, Fotheringham AS and Charlton M 1999 Some Notes on Parametric Significance Test for Geographically Weighted Regression *Journal of Regional Science* **39** (3) 497-524
[11] Fotheringham A S, Brunsdon C and Chartlon M 2002 *Geographically Weighted Regression, The Analysis of Spatially Varying Relationships* England: John Wiley & Sons
[12] Liu J, Zhao Y, Yang Y, Xu S, Zhang F, Zhang X, Shi L and Qiu A 2017 A Mixed Geographically and Temporally Weighted Regression: Exploring Spatial-Temporal Variations from Global and Local Perspectives *Entropy* **53** (19) 1-20

# Mental Model of Prospective Chemistry Teachers on Electrolyte and Nonelectrolyte Solutions

**I W Suja*, I K Sudiana, I W Redhana and I B N Sudria**

Chemistry Department, Universitas Pendidikan Ganesha, Indonesia.

*Corresponding author: wayan.suja@undiksha.ac.id

**Abstract**. This study aims to describe and explain the understanding of the three levels of chemistry and the mental model of prospective chemistry teachers on electrolyte and nonelectrolyte solutions. The research population was all 140 students of the Chemistry Education Study Program of the Faculty of Mathematics and Natural Sciences Undiksha on the odd semester of 2020/2021 academic year, with a sample of 114 students. Sampling was done by proportionate stratified random sampling. Data were collected by means of three-level multiple-choice objective tests, each level sequentially to measure understanding of the macroscopic, submicroscopic, and symbolic levels. Data analysis was carried out descriptively. The results showed that students' understanding of the macroscopic level was 34.21%, the submicroscopic level was 27.54%, and the symbolic level was 29.47%. Only 12.46% of mental model profiles are considered conceptual models, and the remaining 87.54% are classified as alternative mental models, which include 38.95% partially correct and 48.59% specific misconceptions. The data shows that students' understanding of the three levels of chemistry is very low and most of their mental models on electrolyte and nonelectrolyte solutions are classified as alternative mental models.

## 1. Introduction

Meaningful learning in chemistry classes, including in college, involves forming mental models of scientifically correct chemical concepts. Mental models are ideas in the mind that are used to describe, explain, and predict phenomena [1]-[2]. The correct mental model (conceptual model) will be formed in the minds of students if learning involves studying the macroscopic, submicroscopic, and symbolic levels of chemistry, as well as the interconnection of the three levels [3]-[5].

In chemistry lectures, lecturers often assume that students are able to transfer their knowledge from one level to another. This assumption underlies the tendency of teaching chemical concepts directly from the macroscopic to the symbolic level without relating it to the submicroscopic level of understanding [6]. This condition causes students to experience difficulty in providing explanations at the submicroscopic level related to the macroscopic level phenomena [7]-[8]. Meanwhile, prospective chemistry teachers are the spearhead of the implementation of chemistry education in the future so that their chemical mental models must be scientifically correct.

As prospective teachers, students of Chemistry Education Study Program must have the correct mental model related to chemical concepts in the school curriculum. However, in reality, the research findings of ref [9], show that chemistry concepts taught in senior high school are only 8.37% understood by prospective chemistry teachers in the form of scientific mental models; while the rest are classified as alternative mental models (not scientifically correct). The same situation is also shown in ref [10], that the understanding of senior high school students about the basic concepts of chemistry taught in junior high school, is only 6.64% scientifically correct; while the remaining 93.36% is in the form of alternative mental models, including 44.22% classified as misconceptions.

One of the materials for chemical studies that involves understanding three levels of chemistry as a whole is electrolyte and nonelectrolyte solutions. The study material is studied in senior high school in the even semester of class X, then in Collage in Basic Chemistry, Senior High School Chemistry Study, and in Learning Strategy and Design. The contents of electrolyte and nonelectrolyte solutions have a relationship with solution subject, matter particles, the concept of ionization, acid-base and salt theory, colligative properties of solutions, and electrolysis. The study material is strategic for teaching chemistry mental models as a whole and can be used as a basis for getting used to three levels of thinking. For this reason, it is necessary to measure the mental models of prospective chemistry teacher regarding electrolyte and nonelectrolyte solutions. Selection of prospective chemistry teachers as research subjects because they will spearhead the implementation of chemistry education in the future so that their chemistry mental models must be scientifically correct.

In connection with the above problems, this study was carried out with the aim of describing and explaining the understanding of prospective chemistry teachers on the three levels of chemistry and their mental model profiles on electrolyte and nonelectrolyte solutions. The results of this study are indispensable for students and lecturers to reflect and remediate understanding the three levels of chemistry and build mental models of students by properly interconnecting three levels of chemistry.

## 2. Methodology

The research was carried out in the odd semester of the 2020/2021 academic year at the Undiksha Chemistry Education Study Program. Data collection involved a population of 140 students, with a sample of 114 people. Sampling was done by proportionate stratified random sampling. Data were collected by means of a three-tier multiple-choice objective test, sequentially to measure understanding at the macroscopic, submicroscopic, and symbolic levels. Data analysis was performed in a descriptively.

The test used was developed by ref [9]. The test equipment has been validated by a team of experts and tested in the field. The results of the trial which consisted of 10 items showed that all items were classified as valid ($r_{xy}$ = 0.445 - 0.764). Calculations using the Cronbach alpha formula show that the reliability of the test is high with a reliability coefficient of 0.723 so it is suitable for measuring students' mental models of electrolyte and nonelectrolyte solutions.

Data analysis was performed descriptively. Students' mental models are classified into the following three categories [11].

a. Scientifically correct (SC), if the answer to the three levels of chemistry is correct.
b. Partially correct (PC) if the answers to all three levels there are true and false.
c. Specific misconceptions (SM), if the answers to all three levels of chemistry are wrong.

The first type of mental model is called a conceptual model, while the other types of mental models are classified as alternative mental models.

## 3. Results and Discussion

The results of this study are related to the understanding of three levels of chemistry and the mental model profiles of prospective chemistry teachers at the Chemistry Education Study Program on electrolyte and nonelectrolyte solutions. The results of research on understanding the three levels of chemistry and mental models of prospective chemistry teacher are shown in Table 1.

The data in Figure 1 shows that students' understanding of the three levels of chemistry is as follows. Understanding the macroscopic level only reached 34.21%, the submicroscopic level was 27.54%, and the symbolic level was 29.47%. The data shows that the understanding of prospective chemistry teachers on three levels of chemistry in relation to electrolyte and nonelectrolyte solutions is classified as very low, and the lowest is mastery of the submicroscopic level.

**Table 1.** The Understanding Three Levels of Chemistry and Students' Mental Models on Electrolyte and Nonelectrolyte Solutions (N = 114)

| **Indicators** | **Item** | **Chemistry Level** | | | **Mental Model** | | |
|---|---|---|---|---|---|---|---|
| | | **Ma(%)** | **Sub(%)** | **Sym(%)** | **SC(%)** | **PC(%)** | **SM(%)** |
| 3.8.1. Identify the characteristics of an electrolyte solution when an electric current is applied. | 1 | 54 (47.37) | 50 (43.86) | 54 (47.37) | 20 (17.54) | 74 (64.91) | 20 (17.54) |
| 3.8.2. Determine electrolyte or nonelectrolyte compounds based on their electrical conductivity. | 2 | 60 (52.63) | 56 (49.12) | 50 (43.86) | 32 (28.07) | 48 (42,11) | 34 (29,82) |
| 3.8.3. Predicting the electrical conductivity of a liquid ionic or polar covalent compound. | 3 | 6 (5.26) | 16 (14.04) | 10 (8.77) | 2 (1.75) | 20 (17.54) | 92 (80.70) |
| 3.8.3. Predicting the electrical conductivity of a liquid ionic or polar covalent compound. | 4 | 38 (33.33) | 34 (29.82) | 44 (38.60) | 30 (26.32) | 20 (17.54) | 64 (56.14) |
| 3.8.4. Analyzing the electrical conductivity of a solution containing several solute components. | 5 | 20 (17.54) | 4 (3.51) | 18 (15.79) | 2 (1.75) | 32 (28.07) | 80 (70.18) |
| 4.8.2. Explaining the phenomenon that occurs in the electrical conductivity test of the solution. | 6 | 8 (7.02) | 32 (28.07) | 22 (19.30) | 4 (3.51) | 40 (35.09) | 70 (61.40) |
| 4.8.1 Concluding the type of solution based on the electrical conductivity and the degree of acidity | 7 | 80 (70.18) | 52 (45.61) | 56 (49.12) | 18 (15.79) | 74 (64.91) | 22 (19.30) |
| 3.8.5. Describing the electrical conductivity of salt crystals. | 8 | 28 (24.56) | 28 (24.56) | 28 (24.56) | 20 (17.54) | 18 (15.79) | 76 (66.67) |
| 4.8.2. Explaining the phenomenon that occurs in the electrical conductivity test of the solution. | 9 | 36 (31.58) | 32 (28.07) | 32 (28.07) | 10 (8.77) | 46 (40.35) | 58 (50.88) |
| 4.8.2. Explaining the phenomenon that occurs in the electrical conductivity test of the solution. | 10 | 60 (52.63) | 10 (8.77) | 22 (19.30) | 4 (3.51) | 72 (73.16) | 38 (33.33) |
| **Average** | | 39.00 (34.21) | 31.40 (27.54) | 33.60 (29.47) | 14.20 (12.46) | 44.40 (38.95) | 55.40 (48.59) |

Information, **Ma** = understanding at the macroscopic level, **Sub** = understanding at the submicroscopic level, Sym = understanding at the symbolic level, SC = scientifically correct, PC = partially correct, and SM = specifically misconceptions.

Students' understanding of the three levels of chemistry according to the data in Table 1 is shown in Figure 1.

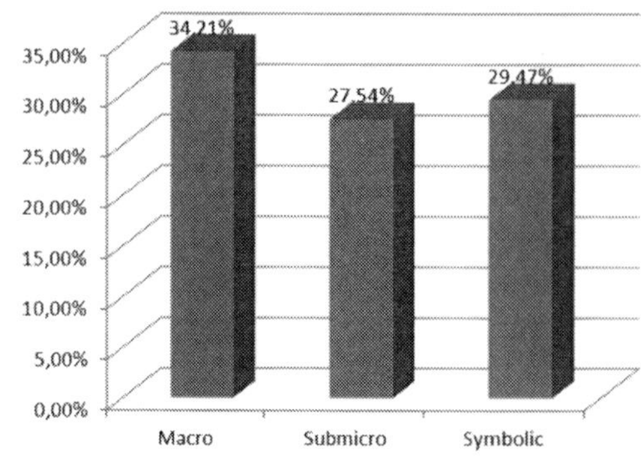


**Figure 1.** Percentage of Understanding Three Levels of Chemistry

Students' understanding on material particles (submicroscopic level) related to electrolyte and nonelectrolyte properties was the lowest among the three chemical levels, which only reached 27.54%. The low understanding of students about the submicroscopic level of chemistry is due to the learning carried out by educators only introducing the macroscopic level supported by the symbolic level [12]. Learning chemistry without touching the submicroscopic level causes students to be unable to explain chemical macroscopic phenomena based on the structure of the material particles using verbal language or chemical symbolic language [6]-[7]. Thus, according to Cooper et al. (2013), the introduction of the submicroscopic level is at the core of chemistry learning, but is often ignored by educators [13].

The ability of students to build interconnections of the three levels of chemistry is related to their mental models. For each item, if the three levels of chemistry can be answered correctly, then the student has a scientifically correct mental model (conceptual model). If all three levels are wrong, then it is classified as a specific misconceptions. Furthermore, if the three levels are true and false, then it is classified as a partially correct mental model. The profiles of students' mental models of electrolyte and nonelectrolyte solutions in general are shown in Figure 2.

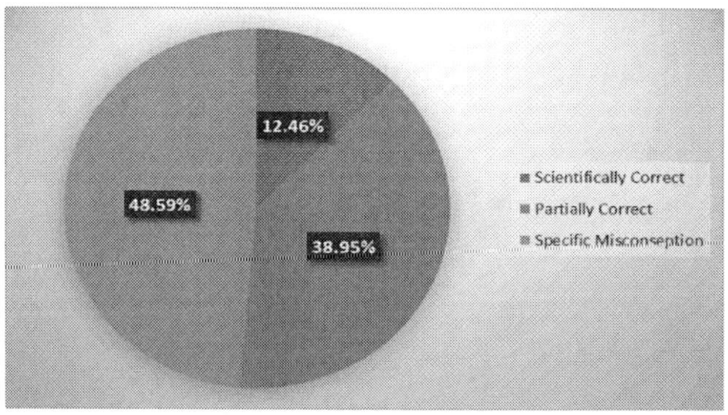


**Figure 2.** Profile of a student's mental model

The data in Figure 2 shows that only 12.46% of the prospective chemistry teachers' mental models on electrolyte and nonelectrolyte solutions are classified as conceptual models, the truth of which can be scientifically recognized. The rest, as much as 87.54% are classified as alternative mental models, consisting of 38.95% partially correct mental models and 48.59% classified as specific misconception mental models. These data show that most of the concepts of electrolyte and nonelectrolyte solutions are understood by prospective chemistry teachers in the form of specific misconceptions.

The mental models of prospective chemistry teachers about electrolyte and nonelectrolyte solutions include 12.46% classified as conceptual models, 38.95% partially correct, and 48.59% classified as specific misconceptions. These data show that most of the concepts of electrolyte and nonelectrolyte solutions are understood by prospective chemistry teachers in the form of specific misconceptions. This means that most students experience errors in choosing the option of the macroscopic, submicroscopic, and symbolic levels that are related between one level and another regarding the phenomena (cases) given in the questions.

The students' inability to think analysis-synthesis about the correlation of macroscopic properties and the types of material particles at the submicroscopic level, as well as imaginative thinking related to the symbolic level will cause students to experience difficulty in constructing mental models on three chemistry levels [14]. As a result, the mental models formed in students' minds are alternative mental models, which include partially correct and specific misconceptions.

Internal factors that have the potential to lead to the formation of alternative mental models in students' thoughts about electrolyte and nonelectrolyte solutions are as follows.

a. Lack of information related to the problems at hand. Information related to the formation of gas bubbles and the presence of ions as conductors of electric current in the electrolyte solution test has been obtained by students in the even semester of class X senior high school. Class X chemistry books do not explain the relationship between the electrical conductivity of the electrolyte solution and the gas bubbles formed at the electrodes. There is no information about the formation of gas bubbles at either electrode or at both electrodes. The low curiosity of students caused them to remain unaware of the type of gas produced at each electrode.
b. Errors due to over generalization. When learning about electrolyte and nonelectrolyte solutions, students usually define electrolyte compounds as compounds which in the form of a solution or their melt can conduct electric current. This definition actually only applies to ionic compounds, and not to electrolyte compounds containing polar covalent bonds. Those who have tested several strong acid solutions, such as HCl, $HNO_3$, and $H_2SO_4$ as strong electrolytes, view these strong acid compounds as capable of conducting electric current. Meanwhile, the material particles of these compounds are molecules so that in their pure state they are unable to conduct electric current. An overly broad generalization, without considering the physical condition of the compound, also occurs in the opinion that KCl crystals are capable of conducting electricity because they contain ions, whereas ions in solid crystals are not able to move freely.
c. Inability to apply concepts to solve the problem at hand. The explanation relates to the type of gas formed during the electric conductivity test of the solution related to the electrolysis concept taught in class XII. As prospective chemistry teachers, they should have understood that electrolysis of water produces hydrogen gas (at the cathode, negative pole) and oxygen gas (at the anode, positive pole) or oxidation of electrolyte anions to produce certain gases according to the type of anion. The inability to apply the concept of electrolysis causes students to be unable to explain the formation of gas bubbles in the electric conductivity test of electrolyte and nonelectrolyte solutions.
d. The fallacy of reasoning. The electrical conductivity of a solution does not combine the properties of the constituent compounds. The solution contains a mixture of cane sugar and table salt as much, does not cause it to become a weak electrolyte solution. The reasoning for mixing strong electrolyte and nonelectrolyte compounds into weak electrolytes cannot be explained scientifically. The movement of sodium ions and chloride ions which deliver electric currents is not inhibited by sucrose molecules (sugar cane) because there is no interaction between ions from table salt and sucrose molecules. Thus, the solution will be a strong electrolyte because it contains many ions from the ionization of the sodium chloride salt.
e. Lack of attention to process at the submicroscopic level. Most of the students who had partially correct mental models were due to their ignorance of the submicroscopic level. They are able to determine the phenomenon that is observed in the electrical conductivity of a pure solution or compound, but are unable to explain this phenomenon at the particle level of matter. This condition is still far from the expectations of ref [15], who view chemistry as a submicroscopic science. These findings are in line with the results of research in ref

[16], which state that students have difficulty providing explanations at the submicroscopic level of observable macroscopic phenomena.

The findings of this study have implications for chemistry learning in college, especially for prospective chemistry teachers. As the spearhead of chemistry education in the future, they must have a conceptual model related to the chemical concepts they will teach students, including the concepts taught in the subject matter of electrolyte and nonelectrolyte solutions. Learning chemical concepts must cover the three levels of chemistry, namely the macroscopic, submicroscopic and symbolic levels as a whole and build the interconnection of the three levels in the learners' memory in the form of mental models.

Students' mental models are built through perception, imagination and understanding of scientific discourse [1]. The same thing was also conveyed in ref [17], who views that students' mental models are built through observation, interpretation and explanation which are used to describe their understanding of the submicroscopic level of chemistry. Based on ref [5], in the syntax of the TripleChem model, perception arises from observing, the interpretation of the submicroscopic level is carried out by reasoning, imagination is carried out to visualize the submicroscopic level, and understanding of the three chemical levels can be built by interconnecting the three chemical levels with explanation. Thus, the formation of a scientific mental model can be done through the stages of observing, reasoning, modeling, and explanation. Mental models are built not once, but through the process of becoming because of its dynamic nature and continuously modified according to new information entered into it [18].

## 4. Conclusion

The conclusions of this research can be formulated as follows. First, students' understanding of prospective chemistry teachers about the three levels of chemistry each reached the macroscopic level of 34.21%, the submicroscopic level of 27.54%, and the symbolic level of 29.47%. This data shows that the understanding of prospective chemistry teachers about three levels of chemistry in relation to electrolyte and nonelectrolyte solutions is classified as very low, with mastery of the submicroscopic level being the lowest. Second, only 12.46% of students' mental model profiles are classified as conceptual models (scientifically correct) and the remaining 87.54% are classified as alternative mental models, which include 38.95% classified as partially correct and 49.59% as specific misconceptions.

## Acknowledgements

The researcher would like to thank the Rector of Universitas Pendidikan Ganesha for funding this research in the form of a Basic Research Grant in the 2020 fiscal year with a research contract number: 575/UN48.16/LT/2020.

## References

[1] Jansoon N, Coll R K and Somsook E 2009 Understanding mental models of dilution in Thai students *International Journal of Environmental & Science Education* **4** (2) 147-168
[2] Wang Ch Y 2007 *The role of mental-modelling ability, content knowledge, and mental model in general chemistry students' understanding about molecular polarity* Dissertation University of Missouri, Columbia
[3] Johnstone A H 2006 Chemical education research in glasgow in perspective *Chemistry Education and Practice* **7** (2) 49-63
[4] Talanquer V 2011 Macro, submicro, and symbolic: The many faces of the chemistry triplet. *International Journal of Science Education* **33** (2) 179-195
[5] Suja I W 2018 *Model pembelajaran TripleChem untuk mengembangkan kompetensi mahasiswa calon guru kimia* Dissertation Universitas Negeri Surabaya, Indonesia

[6] Tasker D and Dalton R 2006 Research into practice: visualization of the molecular world using animations *Chemistry Education Research and Practice* **7** (1) 141-159
[7] Sirhan G 2007 Learning difficulties in chemistry: An overview *Journal of Turkhis Science Education* **4** (2) 2-20
[8] Suja I W and Nurlita F 2016 Persepsi dan model mental mahasiswa tahun pertama terhadap pembelajaran kimia organik di SMA *Prosiding Seminar Nasional Riset Inovatif ke-4* 355-361
[9] Suja I W and Selamat I N 2018 *Pengembangan perangkat tes model mental siswa tentang materi kimia kelas X sekolah menengah atas* Singaraja: Universitas Pendidikan Ganesha
[10] Yoni A A S, Suja I W and Karyasa I W 2018 Profil model mental siswa SMA kelas X tentang konsep-konsep dasar kimia pada kurikulum sains SMP *Jurnal Pendidikan Kimia Indonesia* **2** (2) 64-69
[11] Sendur G, Toprak M and Pekmez E 2010 Analyzing of students' misconceptions about chemical equilibrium *International Conference on New Trends in Education and Their Implications* 1-15
[12] Ben-Zvi R, Eylon B and Silberstein J 1987 Is an atom of copper malleable? *Journal of Chemical of Education* **63** (1) 64-66
[13] Cooper M M, Corley L M and Underwood S M 2013 An investigation college chemistry students' understanding of structure-property relationships *Journal of Research in Science Teaching* **50** (6) 669-721
[14] Treagust D F and Chandrasegaran A L 2009 *Linking the macroscopic and sub-microscopic level: Diagrams* Scotland: Springer
[15] Wu H K 2002 Linking the microscopic view of chemistry to real life experience: Intertextuality in a high school science class room *Journal Chemical Education* **9** (1) 1-48
[16] Ardac D and Akaygun S 2004 Effectiveness of multimedia-based instruction that emphasizes molecular representations on students' understanding of chemical change *Journal of Research in Science Teaching* **41** (1) 317-337
[17] Chittleborough G 2004 *The role of teaching models and chemical representations in developing student's mental model of chemical phenomena* Dissertation Curtin University of Technology
[18] Chittleborough G and Treagust D F 2008 Correct interpretation of chemical diagrams requires transforming from one level of representation to another *Research in Science Education* **38** (1) 463-482

IOP Conf. Series: Materials Science and Engineering 1115 (2021) 012065 doi:10.1088/1757-899X/1115/1/012065

# Modeling Spatial Variation of Money Laundering Crime in Indonesia Using Geographically Weighted Multinomial Logistic Regression

**A Yulia*, S Astutik and U Sa'adah**

Faculty of Mathematics and Natural Sciences, Brawijaya University, Indonesia

*Corresponding author: anisyulia7@gmail.com

**Abstract.** This document presents the mapping of crime in Indonesia. Crime is a real threat that needs to be optimally handled which increases significantly and it is very detrimental to the country's economy. The crime that occurred created another more serious crime, namely Money Laundering Crime. The impact of Money Laundering Crime is the flow of illicit fund through the payment system to financial service providers as well as providers of goods and services. The perpetrator of Money Laundering crime is a player who is shrewd and familiar with the state financial system, so that the flow of funds does not involve a region but will expand to other areas in a country. This study analyses the spatial variations in the factors causing the Money Laundering crime in Indonesia by using a Geographically Weighted Multinomial Logistic Regression (GWMLR) technique. In this document, the GWMLR method is used to model the Money Laundering Crime and its explanatory variables by considering the effect of geographic location using the Adaptive Kernel weighting function. The response variables were divided into four categories with nominal scales by combining the general criminal case and money laundering crime.

## 1. Introduction

Regression analysis is a statistical analysis tool that utilizes the relationship between two or more variables [1]-[4]. The aim is to make unreliable estimates (predictions) for the value of a variable (response variable) if the value of another related variable (variable prediction) is known. In spatial data, regression analysis began to be developed by considering the effect of location or geographic location which was then called the model called Geographically Weighted Regression (GWR) [5]-[7]. GWR is starting to develop and not only uses simple spatial regression analysis, but can be applied to many other regression models, such as a logistic regression model called Geographically Weighted Multinomial Logistic Regression (GWMLR) [8]-[10].

Money laundering is a dual crime. By handling cases thoroughly in the existing of two crimes, namely predicate crime and money laundering crime, should be carried out more efficiently and effectively. In general, every perpetrator of a criminal act will try to hide or disguise the origin of assets that are the result of a crime in various ways so that the assets resulting from the crime are difficult to trace by law enforcement officials, so that he can freely use these assets for legal or illegal activities. Therefore Indonesian Law no. 8/2010 concerning the Prevention and Eradication of the Crime of Money Laundering states that the crime of money laundering not only threatens the stability and integrity of the economic system and the financial system, but can also endanger the joints of society, nation and state based on the 1945 Constitution of the Republic of Indonesia.

The impact of the crime of money laundering is the flow of illicit fund that is not only involves a region but also will develop into other areas in a country. Therefore, an area map is needed to find out criminal acts in each region in Indonesia to facilitate handling of crimes that occur.

## 2. Research Model

### 2.1 Data Source

In this research, the data used are data on factors that influence criminal acts obtained from the Central Statistics Agency (BPS) in 34 Provinces in Indonesia in 2019. The variable used in this research consists of one response variable ($Y$) with four categories with a scale nominal and five predictor variables ($X$). The response variable ($Y$) used is the number of cases of predicate crime and money laundering in the province. While the predictor variables used are Poverty Percentage ($X_1$), Open Unemployment Rate ($X_2$), Population Density ($X_3$), Higher Education Enrolment Rates ($X_4$), Per Capita Expenditure ($X_5$), Human Development Index ($X_6$), and Percentage of Households Have Access to Decent Housing ($X_7$).

### 2.2 Analysis Method

The method used in this research is Geographically Weighted Multinomial Logistic Regression using the best weight based on the smallest AIC value. The conducted assumption test is dependency spatial test and heterogeneity spatial test.

## 3. Result and Discussion

### 3.1 Descriptive Analysis

The variable dependent used in this research is criminal cases in Indonesia which are divided into 4 (four) categories. Descriptively, the causes of this crime could be explained in Figure 1.

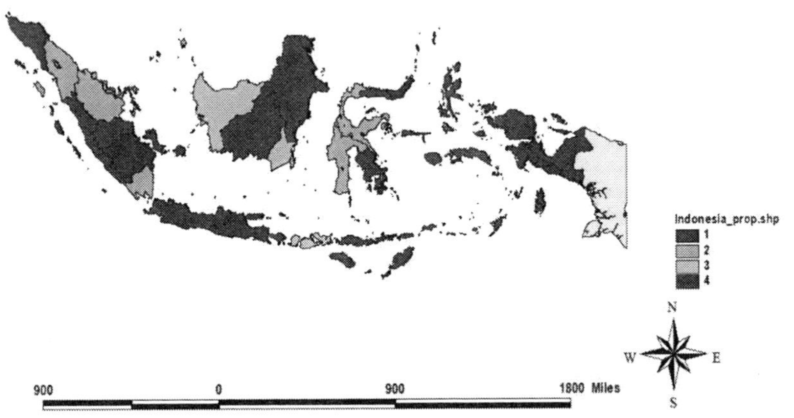


**Figure 1.**Distribution of Criminal Cases in Indonesia.

Based on the distribution of criminal cases in Indonesia, it can be seen that as many as 7 provinces (20.6%) are included in category 1, namely the number of cases of Predicate Crime and Money Laundering in the Province which is greater than the average, 1 province (2.9%) which is included in category 2, namely the number of cases of predicate crime in large provinces and money laundering in small provinces, 4 provinces (11.8%) are included in category 3, namely the number of cases of predicate offense in small provinces. Small regencies / cities and money laundering in big regencies / cities, and 22 provinces (64.7%) are included in category 4, namely the number of cases of predicate crime and money laundering in districts / cities that are small. Because there is only a fairly far comparison of the number of provinces between categories, in this study criminal cases are divided into 2 categories. Descriptively, the research variable, the independent variable could be explained in Table 1.

**Table 1.** Descriptive Independent Variable.

| Variable | Minimum | Maximum | Average | Deviation Standard |
|---|---|---|---|---|
| Poverty Percentage ($X_1$) | 3.42 | 26.55 | 10.24 | 5.52 |
| Open Unemployment Rate($X_2$) | 1.52 | 8.11 | 4.77 | 1.51 |
| Population Density($X_3$) | 9.00 | 15900.00 | 742.03 | 2708.76 |
| Higher Education Enrolment Rates($X_4$) | 14.27 | 73.14 | 33.59 | 10.75 |
| Per Capita Expenditure($X_5$) | 7336.00 | 18527.00 | 10915.59 | 2213.47 |
| Human Development Index($X_6$) | 60.84 | 80.76 | 71.04 | 3.91 |
| Percentage of Households Have Access to Decent Housing($X_7$) | 26.16 | 81.61 | 54.14 | 12.42 |

### 3.2 Assumption Test

The assumption test that needs to be done is dependency spatial test and heterogeneity spatial test.

#### *3.2.1 Dependency Spatial Test*

The assumption dependency test that is carried out to determine whether the residual variable respond in a region is related to the surrounding region which is carried out by using Moran's I test.

**Table 2.** Moran's I test

| | |
|---|---|
| **Observed** | 0.1062 |
| **Expected** | -0.02702 |
| **Sd** | 0.051193 |
| **P-Value** | 0.009256 |

Based on Table 2, it is obtained a *p-value* of 0.009256 or less than $\alpha$,so that $H_0$ is rejected and it can be concluded that the dependency assumption is met.

#### *3.2.2 Heterogeneity Spatial Test*

Testing of the heterogeneity spatial assumption is carried out to determine variance of the residuals of different respond variable in each location or there is one observation location that has different residual variance. Testing of the Spatial Heterogeneity is carried out by using Breusch Pagan test.

**Table 3.** Breusch-Pagan Test.

| | |
|---|---|
| **Breusch Pagan** | 13.728 |
| **df** | 6 |
| **P-Value** | 0.0362 |

The critical area rejects $H_0$ $if$ the value of $BP > \chi^2(\alpha, p)$. Obtained chi-square value $\chi^2(0{,}05, 6)$of 3.325 and *p-value* is less than $\alpha$. So, it can be concluded that there is $\sigma_i{}^2 \neq \sigma^2$or heterogeneity occurs. So that the assumption of spatial heterogeneity is fulfilled.

## 3.3 The testing result of Geographically Weighted Multinomial Logistics Regression

### 3.3.1 Selection of the best weighted

The first step in modeling a criminal case is to determine the weights to be used with the kernel function weights. In determining the best weight, what must be done is to determine the smallest AIC.

**Table 4.**Comparison of AIC Value Each Weighting Kernel Function

| Kernel Weighted | AIC | $R^2$ |
|---|---|---|
| Adaptive Gaussian Kernel | 268.562 | 0.688 |
| Adaptive Bisquare Kernel | 261.473 | 0.778 |

Based on Table 4, it is found that the appropriate weight is the Adaptive Bisquare Kernel because this weight has the smallest AIC value which is 261.473 with the highest$R^2$value among other weights namely 77,8%. The estimation of the criminal act model will be carried out using the GWMLR approach with adaptive weighting Bisquare Kernel.

### 3.3.2 Parameter Testing

Furthermore, the variables that have a significant effect on criminal cases will be classified based on significant variables in each province in Indonesia according to Table 5.

**Table 5.** Grouping of Provincial Based on Predictor Variables with Significant Influence.

| Province | Significant Variable |
|---|---|
| Aceh, West Sumatera, Riau | 1. $X_3$(Population Density)<br>2. $X_5$(Per Capita Expenditure)<br>3. $X_7$(Percentage of HouseholdsHave Access to Decent Housing) |
| North Sumatera, West Nusa Tenggara, South Kalimantan | 1. $X_3$(Population Density) |
| Jambi, South Sumatera, Bangka Belitung Islands, Central Java, Yogyakarta, Bali, East Nusa Tenggara, West Kalimantan, Central Kalimantan, North Kalimantan, North Sulawesi, West Sulawesi, Maluku | 1. $X_3$(Population Density)<br>2. $X_7$(Percentage of HouseholdsHave Access to Decent Housing) |
| Bengkulu, Lampung, Riau Islands, DKI Jakarta, West Java, East Java, Banten | 1. $X_2$(Open Unemployment Rate)<br>2. $X_3$(Population Density)<br>3. $X_7$(Percentage of HouseholdsHave Access to Decent Housing) |
| Central Sulawesi, South Sulawesi, Southeast Sulawesi, Gorontalo, North Maluku, West Papua, Papua | 1. $X_3$(Population Density)<br>2. $X_6$(Human Development Index)<br>3. $X_7$(Percentage of HouseholdsHave Access to Decent Housing) |

IOP Conf. Series: Materials Science and Engineering 1115 (2021) 012065 doi:10.1088/1757-899X/1115/1/012065

The final step is to make a description of the distribution of predictor variables that have a significant effect on criminal cases in each province in Indonesia using thematic maps. The use of this thematic map will make it easier to visually describe groupings of regions. The distribution of areas according to significant variables in each province can be presented in Figure 2, as follows.

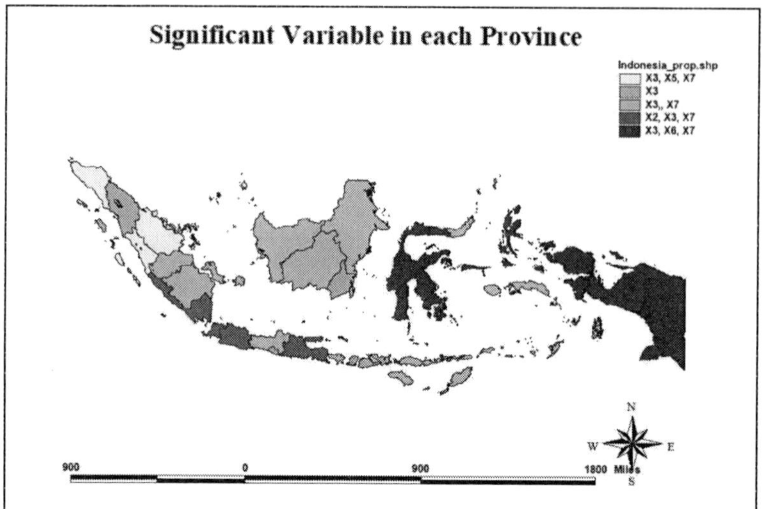


**Figure 2.** The Factors that Effect Significantly to the Criminal Act in each Province.

Figure 2 shows the distribution of the predictor variables that have a significant effect on each province, areas that tend to be visible, with the same factors. Provincial grouping that shows the existence of significant influencing factors, for example occurs in Central Sulawesi, South Sulawesi, Southeast Sulawesi, Gorontalo, North Maluku, West Papua, and Papua, with factors that have a significant effect are population, Human Development Index, and the Percentage of Households Have Access to Decent Housing.

With parameter estimation that obtained, the Model GWMLR formed. One of the Model GWMLR in East Java Province as follows:

$$g_1(x) = 1{,}231 + 0{,}77X_2 + 4{,}029X_3 - 4{,}136X_7$$

$$g_2(x) = 2{,}231 + 0{,}623X_2 + 2{,}098X_3 - 3{,}222X_7$$

$$g_3(x) = 0{,}445 + 0{,}854X_2 + 3{,}097X_3 - 3{,}109X_7$$

Criminal cases in East Java Province are influenced by three factors, that is Open Unemployment Rate ($X_2$), Population Density ($X_3$), and Percentage of HouseholdsHave Access to Decent Housing ($X_7$). With calculating the odd ratio, interpretation of the first Logit Model is the higher the open unemployment rate in the area, the chances of the high number of cases of predicate crime and money laundering will increase 2.16 times compared to the chances of having a low number of predicate crimes and money laundering cases.

## 4. Conclusion

This study examines whether the Poverty Percentage ($X_1$), Open Unemployment Rate ($X_2$), Population Density ($X_3$), Higher Education Enrolment Rates ($X_4$), Per Capita Expenditure ($X_5$), Human Development Index ($X_6$), and Percentage of Households Have Access to Decent Housing ($X_7$) spatially affects the number of cases of predicate crime and money laundering in the province ($Y$). The results of the analysis using the Geographically Weighted Multinomial Logistics Regression show that the Population Density variable ($X_3$) affects the number of cases of Origin and Money Laundering in the Province ($Y$). While the variable Open Unemployment Rate ($X_2$), Per Capita Expenditure ($X_5$), Human Development Index ($X_6$), and the Percentage of Households who Have Access to Decent Housing($X_7$) have an effect on several provinces in Indonesia.

## References

[1] Stancu I and Rece D 2009 The relationship between economic growth and money laundering–a linear regression model *Theoretical and Applied Economics* **9** (9) 20-29
[2] Stanley T D and Jarrell S B 2005 Meta-regression analysis: a quantitative method of literature surveys *Journal of economic surveys* **19** (3) 299-308
[3] Kaneko H, Arakawa M and Funatsu K 2008 Development of a new regression analysis method using independent component analysis *Journal of chemical information and modeling* **48** (3) 534-541
[4] Chao S H and Chen Y H 2019 A novel regression analysis method for randomly truncated strong-motion data *Earthquake Spectra* **35** (2) 977-1001
[5] Brunsdon C, Fotheringham A S and Charlton M E 1996 Geographically weighted regression: a method for exploring spatial nonstationarity *Geographical analysis* **28** (4) 281-298
[6] Yu H, Fotheringham A S, Li Z, Oshan T, Kang W and Wolf L J (2020) Inference in multiscale geographically weighted regression *Geographical Analysis* **52** (1) 87-106
[7] Mennis J 2006 Mapping the results of geographically weighted regression *The Cartographic Journal* **43** (2) 171-179
[8] Luo J and Kanala N K 2008 Modeling urban growth with geographically weighted multinomial logistic regression *Geoinformatics 2008 and Joint Conference on GIS and Built Environment: The Built Environment and Its Dynamics* **7144** 71440
[9] Mustafa A, Heppenstall A, Omrani H, Saadi I, Cools M and Teller J 2018 Modelling built-up expansion and densification with multinomial logistic regression, cellular automata and genetic algorithm *Computers, Environment and Urban Systems* **67** 147-156
[10] Laili I 2018 *Estimasi parameter model mixed Geographically Weighted Multinomial Logistic Regression (GWMLR) dengan Pembobot Adaptive Bisquare Kernel* Doctoral dissertation Universitas Islam Negeri Maulana Malik Ibrahim,Indonesia

# COVID-19 Mathematical Epidemic Model for Impact Analysis of Large Scale Social Restriction: The Case Study of Indonesia

**U M Rifanti[1,*], A R Dewi[2], Nurlaili[1] and S T Hapsari[1]**

[1]Department of Telecommunication and Electrical Engineering, Institut Teknologi Telkom Purwokerto, D.I Panjaitan 128 Purwokerto, Indonesia
[2]Department of Informatics Engineering, Institut Teknologi Telkom Purwokerto, D.I Panjaitan 128 Purwokerto, Indonesia

*Corresponding author: marina@ittelkom-pwt.ac.id

**Abstract.** Coronavirus disease 2019 (COVID-19) is an infectious disease caused by Severe Acute Respiratory Syndrome Coronavirus 2. There have been 218,382 confirmed cases of COVID-19, including 8,723 deaths in Indonesia till September 15, 2020. To fight Covid-19, the Indonesian Government modified the lockdown policy to become a large-scale social restriction (LSSR) according to the pandemic level in each region. The LSSR was implemented in DKI Jakarta on April 10, 2020, the epicenter of the Covid-19 outbreak in Indonesia. In this article, we propose the Susceptible-Infected-Hospitalized-Recovered (SIHR) epidemic model for impact analysis of LSSR on epidemic conditions and predict the long-term dynamic COVID-19 in Indonesia. This study extends the SIR model by adding a hospitalized compartment. Finally, we analyze the results of our mathematical model, estimate the model of essential parameters, and predict the disease by considering the real Indonesian cases from March 2 to September 6, 2020. Based on the result, LSSR has a good enough impact on solving the pandemic in Indonesia. One of the factors is because there is a significant difference in transmission rate ($\beta$) values. The new normal transmission rate is 75% higher than the LSSR's.

## 1. Introduction

Coronavirus disease 2019 (COVID-19) from Wuhan, China, is an infectious disease caused by Severe Acute Respiratory Syndrome Coronavirus 2 (SARS-CoV-2) [1]. A person can catch this virus from other humans who have been infected. It has spread widely to 215 countries in the world. The spread of this virus can be through cough or breath droplets from someone infected with COVID-19. These droplets can settle on surrounding objects. Someone can become infected if they touch objects with droplets, and then they touch their eyes, nose, or mouth. If a person inhales these droplets, he or she can also be infected [2].

Every country has a different epidemic level. In most countries where the spread of the virus has led to high outbreaks, physical distancing and movement restrictions, known as lockdowns, are imposed to slow the spread of the COVID-19 outbreak. However, several countries have successfully handled the COVID-19 attack without a lockdown, such as South Korea. Instead of locking up entire cities and cutting direct contact, South Korea focuses on rapid and extensive testing and tracing of all exposed contacts. Testing and tracking have enabled countries to blunt the exponential spread of disease without having to stop all internal movement and access between cities [3]. Early action and implementation of comprehensive public health measures - such as rapid case identification, rapid testing, complete contact tracing, isolation of positive cases, and subnational quarantine have kept the spread of COVID-19 below the threshold [4].

Till September 15, 2020, there have been 28,918,900 confirmed cases of COVID-19, including 922,252 deaths, reported to WHO. There have been 218,382 confirmed cases of COVID-19, including 8,723 deaths in Indonesia [5]. The first confirmed positive COVID-19 case in Indonesia was announced on March 2, 2020, by President Joko Widodo. Two Indonesian citizens living in Depok, West Java, have tested positive for COVID-19. They have interacted with Japanese citizens who were positive for COVID-19 before them [6]. At that time, the Indonesian Government implemented a lockdown policy under Law Number 6 of 2018 on Health Quarantine in anticipation of the spread of COVID-19 [7]. To fight Covid-19, the Indonesian Government modified the lockdown policy to become a large-scale social restriction (LSSR) according to the pandemic level in each region. The LSSR was implemented in DKI Jakarta on April 10, 2020, the epicenter of the Covid-19 outbreak in Indonesia. Technically, the types of community activities regulated in the LSSR Guidelines for the acceleration of handling COVID-19 include dismissing schools (school from home) and workplaces (work from home), restrictions on religious, public places, and socio-cultural activities, and restrictions on transportation modes. After three months of passing through the emergency response period and LSSR, the Indonesia Government began to explore implementing a new normal life and loosening the LSSR [8]. Indonesia is a vast country with different outbreak intros so that the management of epidemics between regions is not the same. Therefore data on confirmed COVID-19 cases is also fluctuating and difficult to predict. Until early September, the confirmed cases were still happening and tended to increase day by day.

Several studies have been conducted to study the spread of COVID-19. Mathematical models have been proposed by authors in their research in several countries, such as China [9][10][11], India [12][13], Italy [14], Mexico [15], Indonesia [16][17]. The simplest way to model the spread of disease in a population is to classify people into different compartments or groups. The compartment model uses a system of differential equations that tracks the population as a function of time and groups them into different groups based on infection status. In [18][19][20], the authors proposed a basic susceptible-infected-recovered (SIR) model. This study extends the SIR model by adding a hospitalized compartment, namely the SIHR model (susceptible-infected-hospitalized-recovered).

We proposed dynamical coefficients to quantify the variation of infectivity and intervention policies to adjust the pandemic's social responses in our model. Then we simulate the results about advancing or postponing, reinforcing, or weakening relevant systems. Through mapping the containment policies into measurable interval coefficients to observe their influence on the epidemic, the results of our model introduce statistical pieces of evidence that such systems can effectively suppress or even block the outbreak of COVID-19. Finally, we analyze the results of our mathematical model, estimate the model of important parameters, and predict the disease by considering the real Indonesian cases from 2 March to 6 September 2020.

## 2. Development of Model

In this study, we proposed SIHR (Susceptible-Infected-Hospitalized-Recovered) model to analyze the impact of LSSR on epidemic conditions and predict the long-term dynamic COVID-19 in Indonesia. This study extends the SIR model by adding a hospitalized. Several studies have been conducted to study the spread of COVID-19. Mathematical models have been proposed by authors in their research in several countries, such as China [11], India [13], Italy [14], Mexico [15], Indonesia [17] [18][19]. The compartment model uses a system of differential equations that tracks the population as a function of time and groups them into different groups based on infection status. The purpose of this study was to analyze the impact of LSSR on epidemic conditions and predict the long-term dynamic COVID-19 in Indonesia. To achieve this goal, we divide the human population into four categories based on their health status can be seen in Figure 1.

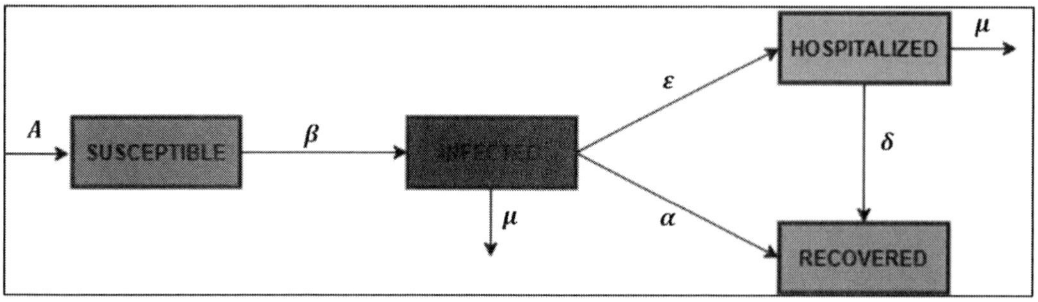


**Figure 1.** Transmission diagram of COVID-19

Susceptible ($S(t)$) are individuals who have not been infected but can become infected. Infected ($I(t)$) are individuals who are infected with COVID-19 and are able to infect other people. Hospitalized ($H(t)$) are individuals who have been hospitalized for COVID-19. Recovered ($R(t)$) are individuals who no longer have the COVID-19 virus. Therefore

$$N(t) = S(t) + I(t) + H(t) + R(t). \tag{1}$$

The Table 1 contains a description of each parameter used in the model.

**Table 1.** Description Of Parameters

| Notation | Description |
|---|---|
| $\beta$ | Transmission rate |
| $\mu$ | Death rate |
| $\varepsilon$ | Hospitalized rate |
| $\alpha$ | Recovery rate of I class |
| $\delta$ | Recovery rate of H class |
| $A$ | Recruitment rate |

We make the following assumptions for the formulation of the model.

1. Disease induced death only exists for individuals in $I$ and $H$.
2. Infected individuals in hospitals have a chance to recover faster than infected individuals who have not received medical treatment.
3. Recovered individuals have a long-term immunity to COVID-19. So that they don't return to class S.
4. The hospitalized rate of LSSR is higher than the new normal. When the number of infected patients increases day by day, the number of hospitals and medical personnel capable of handling COVID-19 patients is increasingly limited.

Using the transmission diagram given in figure 1 and the aforementioned assumptions, the model to describes the transmission of COVID-19 is expressed as the following systems of equation as follows.

$$\frac{dS}{dt} = A - \beta SI$$
$$\frac{dI}{dt} = \beta SI - (\mu + \alpha + \varepsilon)I$$

$$\frac{dH}{dt} = \varepsilon I - (\mu + \delta)H$$
$$\frac{dR}{dt} = \alpha I + \delta H \qquad (2)$$

with the initial conditions $S(t) > 0, I(t) > 0, H(t) \geq 0, R(t) \geq 0$.

## 3. Results and discussions

Indonesia is the largest archipelago country in the world, consisting of 17,504 islands [21]. Indonesia is ranked the fourth most populous country in the world with 274 billion people after China (1.44 billion people), India (1.38 billion people), and the United States (331 million people) [22]. Till September 15, 2020, there have been 28,918,900 confirmed cases of COVID-19, including 922,252 deaths, in the world. There have been 218,382 confirmed cases of COVID-19, including 8,723 deaths in Indonesia [5]. Based on this, Indonesia ranks 9th in Asia and 25th globally with the most COVID-19 infection cases [23].

In this section, we analyze the results of our mathematical model, estimate the model of essential parameters, and predict the disease by considering the real Indonesian cases from March 2 to September 6, 2020. The real data used in this article downloaded from the official website of the Indonesian Task Force for Handling COVID-19 (https://covid19.go.id/peta-sebaran). The LSSR was start implemented on April 10, 2020. Technically, the types of community activities regulated in the LSSR Guidelines for the acceleration of handling COVID-19 include dismissing schools (school from home) and workplaces (work from home), restrictions on religious, public places, and socio-cultural activities, and restrictions on transportation modes. So for about three months, most people carry out their activities from home. During that time, the community had meetings, studies, seminars, and online discussions using conference applications such as Zoom and Google Meeting. Restaurants only provide a takeaway service using a delivery service application such as Go-Food and Grab-Food. Religious places are also closed so that people worship from home. School and campus students learn from home by watching video tutorials from teachers and lecturers or having online discussions using conference applications. Several major cities have been locked down, so that inter-city transportation means are also restricted. The government has prohibited going to their hometown during Eid al-Fitr. Those policies were done to reduce the spread of COVID-19 in Indonesia. After three months of passing through the emergency response period and LSSR, on June 2, 2020, the Indonesia Government began to explore implementing a new normal life and loosening the LSSR.

The Figure 2 illustrates the differences in Indonesia's pandemic conditions between the LSSR period and the new normal when viewed from the number of confirmed and positive COVID-19 patients.

From Figure 2a, it can be seen that the number of confirmed increases increases rapidly in the new normal period. This is because the local government is increasingly aggressively conducting COVID-19 tests on the community. Furthermore, from Figure 2b it can be seen that in the new normal period there was an increase in the number of infected individuals which was higher than in the LSSR period. This is because the intensity of community contact in the new normal period is higher than the intensity of contact during the LSSR period. The easing of the LSSR also resulted in the relaxation of lockdowns in several regions. This has resulted in high mobilization from the red zone of COVID-19 to the green zone of COVID-19, and vice versa.

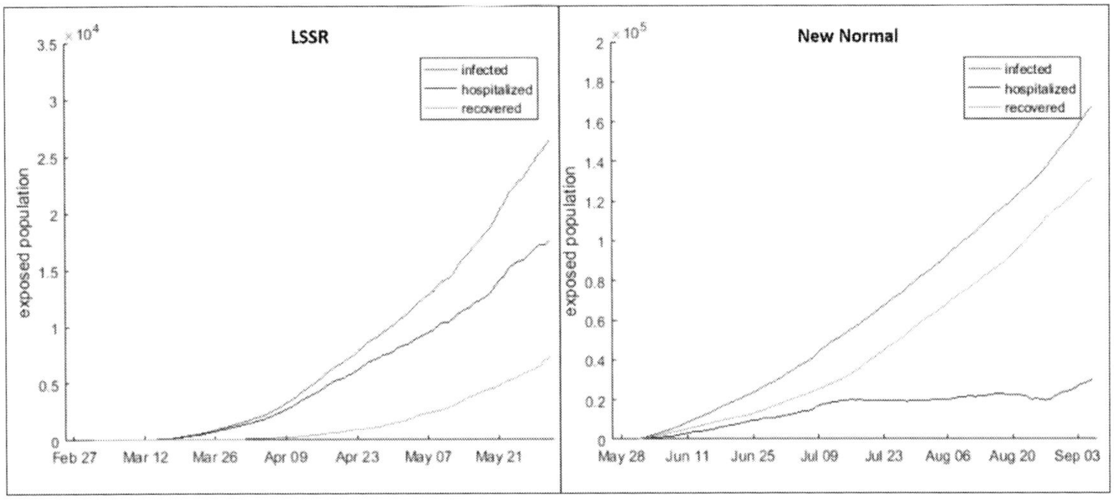


**Figure 3.** The comparison of real data between LSSR and New Normal condition

Based on Figure 3, in both periods, the number of infections increased day by day. However, the number of hospitalized patients during the LSSR period increased higher than the new normal period. During the LSSR period, the number of treatments in Indonesia could still accommodate almost all COVID-19 patients. However, in the new normal period, the number of treatments was limited due to the higher number of infections. The number of hospitalized patients increased in September because the number of recovered patients also increased quite rapidly so that over time the number of treatments increased as well.

From the previous explanation, it is known that the transmission rate ($\beta$) is an essential parameter in the COVID-19 mathematical model. Next, we calculate $\beta$ value using the simulation results of the previous model for each data set. Based on the first equation of system (1), we get

$$\beta = \frac{1}{SI}\left(A - \frac{dS}{dt}\right) \tag{3}$$

where $t$ denotes the time in days and

$$dS = \Delta S = S(t+1) - S(t). \tag{4}$$

Then, a significant difference in $\beta$ values is obtained from these calculations, as listed in the following table.

**Table 3.** The difference in $\beta$ values

| Parameter | LSSR | New Normal |
|---|---|---|
| $\beta$ | $5.214196498951317 \times 10^{-7}$ | $6.878624326294266 \times 10^{-7}$ |

Based on Figure 4, LSSR has a good enough impact on solving the pandemic in Indonesia. One of the factors is because there is a significant difference in transmission rate ($\beta$) values. Based on table 3, the new normal transmission rate is 75% higher than the LSSR's. This causes the effectiveness of transmission to be higher, and the disease will become more widespread. From the results of this study, it is predicted that the disease will disappear in October 2020 if the LSSR conditions are applied. However, due to the implementation of new normal conditions in June 2020, it is predicted that the disease will begin to disappear at the end of February 2021.

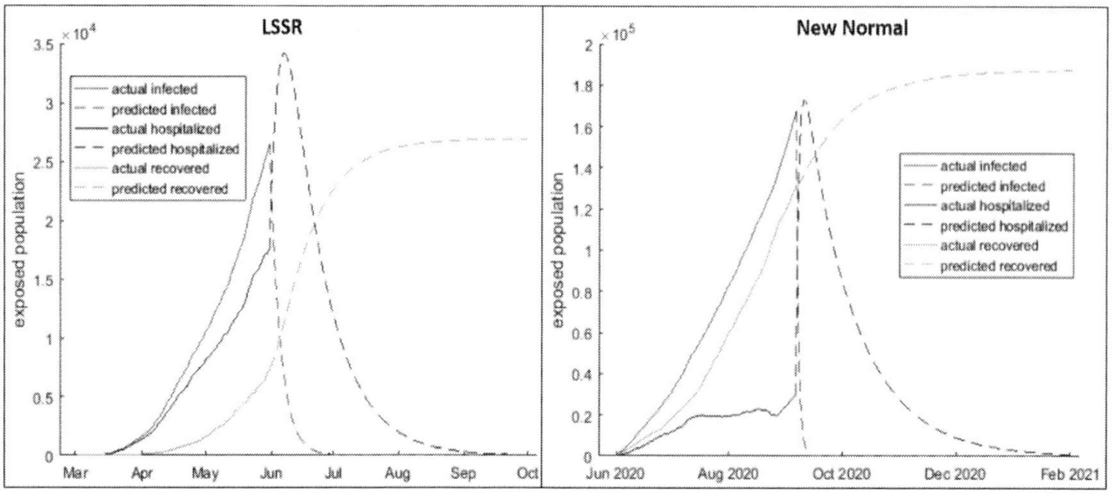


**Figure 4.** The comparison of predicted data between LSSR and new normal condition

## 4. Conclusions

Until September, the confirmed cases of COVID-19 in Indonesia were still happening and tended to increase day by day. Based on the result, LSSR has a good enough impact on solving the pandemic in Indonesia. One of the factors is because there is a significant difference in transmission rate parameter. The intensity of community contact in the new normal period is 75% higher than the intensity of contact during the LSSR period. This causes the disease will become more widespread. From the results of this study, it is predicted that the disease will disappear in October 2020 if the LSSR conditions are applied. However, due to the implementation of new normal conditions in June 2020, it is predicted that the disease will begin to disappear at the end of February 2021.

## Acknowledgments

This article is the one of the outcomes of our internal research at Telkom Institute of Technology Purwokerto (ITTP). So we would like to express our special gratitude to the Institute for Research and Community Service (LPPM) of ITTP for the spiritual and material support so that this research can run smoothly. Thanks also to all colleagues who helped this research, including our respective departments and expertise groups of research. Without your help, this research would not have been as smooth and profitable as this. Hopefully, we can collaborate on other, better studies. Thank you.

## References

[1] Hu T, Liu Y, Zhao M, Zhuang Q, Xu L and He Q 2020 A comparison of COVID-19, SARS and MERS *PeerJ* **8** e9725
[2] World Health Organization, "QA for public," 2020. [Online]. Available: https://www.who.int/indonesia/news/novel-coronavirus/qa-for-public. [Accessed: 15-Sep-2020]
[3] Lee D and Lee J 2020 Testing on the move: South Korea's rapid response to the COVID-19 pandemic *Transportation Research Interdisciplinary Perspectives* **5** 100111
[4] World Health Organization, "COVID-19 Strategy Up Date," Switzerland, 2020
[5] World Health Organization, "WHO Coronavirus Disease (COVID-19) Dashboard," 2020. [Online]. Available: https://covid19.who.int/table. [Accessed: 15-Sep-2020]

[6] Nuraini R *Kasus Covid-19 Pertama, Masyarakat Jangan Panik Portal Informasi Indonesia*, 2020. [Online]. Available: https://indonesia.go.id/narasi/indonesia-dalam-angka/ekonomi/kasus-covid-19-pertama-masyarakat-jangan-panik. [Accessed: 11-Sep-2020]
[7] Yunus N R and Rezki A 2020 Kebijakan pemberlakuan lock down sebagai antisipasi penyebaran corona virus Covid-19 *Salam: Jurnal Sosial dan Budaya Syar-i* 7 (3) 227-238
[8] Muhyiddin M 2020 Covid-19, New Normal, dan Perencanaan Pembangunan di Indonesia *Jurnal Perencanaan Pembangunan: The Indonesian Journal of Development Planning* **4** (2) 240-252
[9] Lin Q, Zhao S, Gao D, Lou Y, Yang S, Musa S S and He D 2020 A conceptual model for the coronavirus disease 2019 (COVID-19) outbreak in Wuhan, China with individual reaction and governmental action *International journal of infectious diseases* **93** 211-216
[10] Yang S, Cao P, Du P, Wu Z, Zhuang Z, Yang L and He D 2020 Early estimation of the case fatality rate of COVID-19 in mainland China: a data-driven analysis *Annals of translational medicine* **8** (4) 24-42
[11] Chen N, Zhou M, Dong X, Qu J, Gong F, Han Y and Zhang L 2020 Epidemiological and clinical characteristics of 99 cases of 2019 novel coronavirus pneumonia in Wuhan, China: a descriptive study *The lancet* **395** (10223) 507-513
[12] Sahoo B K and Sapra B K 2020 A data driven epidemic model to analyse the lockdown effect and predict the course of COVID-19 progress in India *Chaos, Solitons & Fractals* **139** 110034
[13] Mahajan A, Sivadas N A and Solanki R 2020 An epidemic model SIPHERD and its application for prediction of the spread of COVID-19 infection in India *Chaos, Solitons & Fractals* **140** 110156
[14] Ding Y and Gao L 2020 An evaluation of COVID-19 in Italy: A data-driven modeling analysis *Infectious Disease Modelling* **5** 495-501
[15] de León U A. P, Pérez Á. G and Avila-Vales E 2020 An SEIARD epidemic model for COVID-19 in Mexico: mathematical analysis and state-level forecast *Chaos, Solitons & Fractals* **140** 110165
[16] Setiati S and Azwar M K 2020 COVID-19 and Indonesia *Acta Medica Indonesiana* **52** (1) 84-89
[17] Aldila D, Khoshnaw S H, Safitri E, Anwar Y R, Bakry A R, Samiadji B M and Salim S N 2020 A mathematical study on the spread of COVID-19 considering social distancing and rapid assessment: The case of Jakarta, Indonesia *Chaos, Solitons & Fractals* **139** 110042
[18] Adi-Kusumo F, Susyanto N, Endrayanto I and Meliala A Model berbasis SIR dalam prediksi awal penyebaran covid-19 di daerah istimewa Yogyakarta (DIY) *Jurnal Matematika Thales* **2** (1) 1-10
[19] Sulaiman A 2020 On dynamical analysis of the data-driven sir model (covid-19 outbreak in indonesia) *medRxiv* 1-14
[20] Soldati G, Smargiassi A, Inchingolo R, Buonsenso D, Perrone T, Briganti D F and Demi L 2020 Proposal for international standardization of the use of lung ultrasound for patients with COVID-19: a simple, quantitative, reproducible method *Journal of Ultrasound in Medicine* **39** (7) 1413-1419
[21] Bappenas and BPS 2013 *Indonesia Population Projection* Jakarta: BPS
[22] Worldometer, Indonesia Population (LIVE), *Worldometer.info*, 2020. [Online]. Available: https://www.worldometers.info/world-population/indonesia-population/. [Accessed: 30-Oct-2020]
[23] Rizal J G 2020 Kasus Covid-19 Indonesia Peringkat 9 di Asia dan 25 di Dunia, *Kompas.com*, [Online]. Available: https://www.kompas.com/tren/read/2020/07/20/083200665/kasus-covid-19-indonesia-peringkat-9-di-asia-dan-25-di-dunia?page=all.

# Feature Extraction And Sound Pattern Analysis Of Wijaaksara Script

**I G A Gunadi**[1,2,*] **and D O Rahmawati**[3]

[1]Programe study Magister Computer Sciences, Universitas Pendidikan Ganesha, Singaraja, Bali, Indonesia
[2]Departement of Physiscs Education, Universitas Pendidikan Ganesha, Singaraja, Bali, Indonesia
[3]Departement of Physics Education , Universitas Pendidikan Ganesha, Singaraja, Bali, Indonesia

*Corresponding author: igedearisgunadi@undiksha.ac.id

**Abstract.** Wijaksara script is a sacred script in the Balinese script structure in Balinese Hindu culture, the sound of this script is believed to create vibrations that are spiritual and magical. In Balinese script literature, Wijaksara script is a combination of the Swalatita script with the Amsa script. Wijaksara script are categorized into several groups, namely (1) Eka Aksara (Om), (2) Dwi aksara (Ang, Ah), (3) Tri Aksara (Ang, Ung, Mang), (4) Panca Aksara (Nang, Mang , Sing, Wang, Yang), and (5) Dasa Aksara (Sang, Bang, Tang, Ang, Ing, Nang, Mang, Sing, Wang, Yang). In this research, several acoustic features of sound were extracted (1) constituent frequencies, (2) waveforms, (3) sound energy. Sound recordings from 20 samples were taken and then extracted to get the constituent frequency using FFT (*Fast Fourier Transform*). Based on the sound waveform, the waveform and sound energy are obtained. In this study, the acoustic character patterns were obtained, namely Eka, Dwi, Tri, Panca, and Dasa. Aksara is influenced by the sound character of the eka script. The main frequencies for the sounds of the letters eka, bi, tri, and dasa are: 4.37 kHz, 7.22 kHz, 3.97 kHz. 4.18 KHZ, 11.8 KHz, while the sound energy for normal sound pronunciation conditions are: 87.31 dB, 96.72 dB, 83.25 dB, and 80.29 dB, respectively.

## 1. Introduction

Balinese script is something that cannot be separated from Hindu culture. In India the use of certain sacred script always gives a certain sound in the Vedas. This is known as yantra, tantra and mantra. Related to the characters that produce sacred sounds, the first script to develop in India was the Karosti script [1]. In its development, the Karosti script developed into the Dewanagari script and the Pallawa script. The Dewanagari script is the script used to write Sanskrit, while the Pallawa script is used to write Pallawa. The two characters spreads throughout the archipelago in the form of the kawi script, which later developed into Javanese, Balinese, and other scripts [2].

In the development of script in the archipelago, there are differences and similarities in several things. Each region has its own character. For example, based on the form of Balinese script and Javanese script, it has similar forms, but when viewed functionally there are very significant differences. In Balinese script, there are still parts of the sacred script, namely wijaaksara and modre script. Both types of script are not found in Javanese script, or even other archipelago scripts. Wijaksara script, also known as the Bija aksara (Bija means seed), is a combination of the Swalalita script and the hamsa script [3].

The sound generated by the pronunciation of a letter symbol is believed to have a vibration that can be positive or negative effect, depending on how it is pronounced. Related to this, one of the wijaaksara script groups, Dwi Aksara consisting of two sounds (1) Ang, (2) Ah. According to [4], in his commentary on the Jnanasidhanta book, it states that the sound of (Ang Ah) is related to the sound of lifeness , on the contrary , if the pronunciation is (Ah Ang) it is related to deathness.

In Japan, in 1865 a Buddhist hermit named Mikao Usui developed a spiritual practice, as a form of alternative medicine, by cultivating natural energy. This spiritual practice is currently known as Reiki. Reiki has grown throughout the world, including Indonesia. In practice, a Reiki practitioner will clean and activate the 7 energy points in the human body. The 7 energy points (chakras) are the muladhara chakra, the svadhisthana chakra, the manipura chakra, the anahata chakra , the visuddhi chakra, the ajna chakra, the sahasrara chakra. One thing that is very interesting, according to reiki practitioners, is that to clean and activate these energy points ( Chakra), one of which is by pronunciation the sounds . some of that sounds are wijaaksara sounds. The activation of the muladra chakra with the pronunciation of the sound of " Vam" , the activation of the svadhista chakra with the pronunciation of the sound of "Ram" , the heart chakra (anahata) with the sound of "Yam", the throat chakra (visudhi) with the sound of "Ham", the ajna chakra with the sound of "Om", the activation of the sahasra chakra by using the sound of the "Ng" sound, [5],[6].

Wijaaksara script is a legacy of Balinese ancestors that is very extraordinary, practically its use has received recognition from many circles, even outside Balinese, and from different cultures. In this study, it tries to analyze more deeply related to the features of sound, seen from the side of acoustic physics, and sound patterns.

## 2. Theoritical Review

### 2.1 Review of Wijaaksara Script

Balinese script can be broadly divided into two categories, namely ordinary script and sacred script. Ordinary script consists of two groups, namely (1) Wreastra script, a common script used in everyday life, consisting of 18 characters. These commonly known characters consist of: a, na, ca, ra, ka, da. ta, sa, ... etc. (2) Swalalita script, is a kawi literary script consisting of 47 characters, consisting of 14 vowels and 33 wianjana (consonant) characters. Meanwhile, the sacred script also consists of two categories, the first (1) Wijaksara Akasara, which is a self-help script combined with the Hamsa script. (2) Modre script, is a combination of several characters and is generally a symbol of an object or thing that is spiritual and magical. Figure 1 shows the structure of the Balinese script, [2], [3].

Associated with Wijaksara script is swalalita script combined with the Hamsa script. Hamsa letters are characters associated with sound (O Ang, Ing, ah). The wijaksra script is a divine script that is believed to have magical powers. There are several categories of wijaksara characters, (1) Eka aksara (omkara), (2) Dwi aksara , (3) Tri aksara , (4) Panca aksara, and (5) Dasa aksara. Detail of wijaaksara script in Table 1

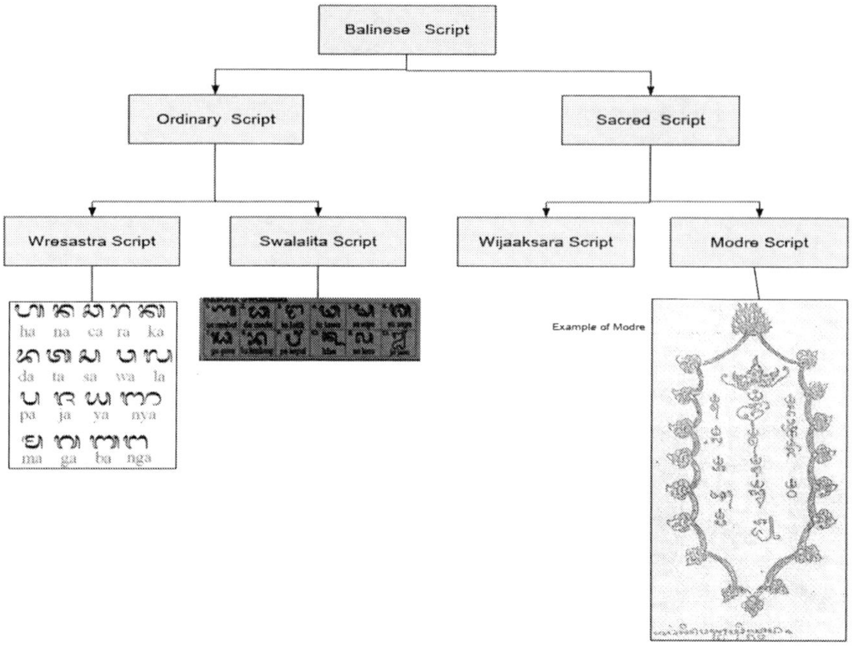


**Figure 1.** Classification of Balinese Script

**Table 1.** Detail of Wijaaksara Script

| Group Script | Symbol | Pronunciation |
|---|---|---|
| Eka Aksara |  | OM |
| Dwi Aksara |  | Ang , Ah |
| Tri Aksara |  | Ang, Ung, Mang |
| Panca Aksara | 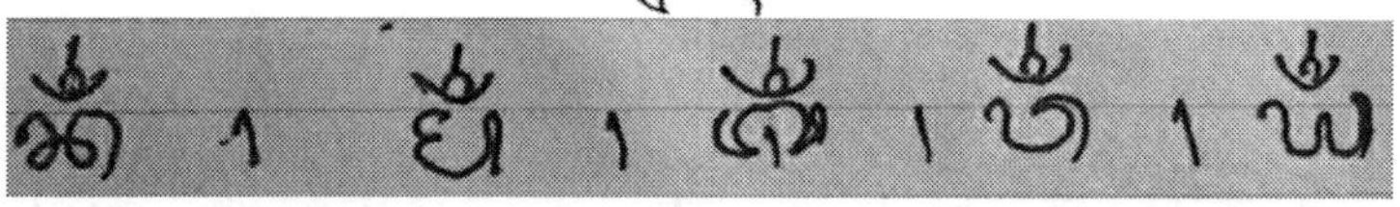 | Nang, Mang, Sing, Wang, Yang |
| Dasa Aksara | 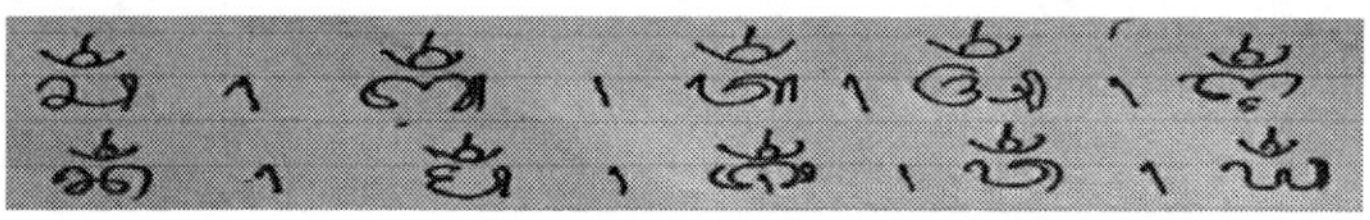 | Sang, Bang , Tang, Ang, Ing, Nang, Mang, Sing, Yang |

### 2.2 Feature Extraction Of Sound

In this study, 3 main features of sound are used which can be extracted from a sound data, frequency features, sound energy, and wave form. There are several methods that can be used to transform sound signals into frequency signals. FFT (Fast Fourier Transform) is one method to solve this problem, as was done in research [7], [8], [9]. In this study, to get the frequency feature and sound SPL, the FFT method will also be applied, the process illustration is shown in Figure 2.

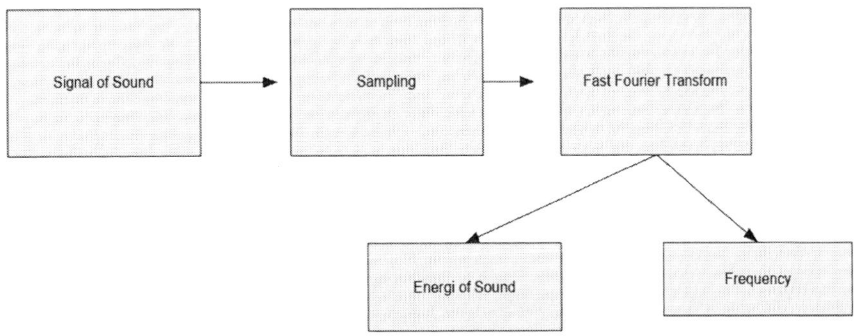


**Fig 2.** Feature Extraction of Sound Based On FFT ( Fast Fourier Transform)

By using FFT, the sound data transformation is carried out from the spatial (time) domain to the frequency domain. Thus it can be determined the main frequency, and several follower frequencies of a sound signal. Figure 3 illustrates an FFT result, the sound signal has been represented in the frequency domain.

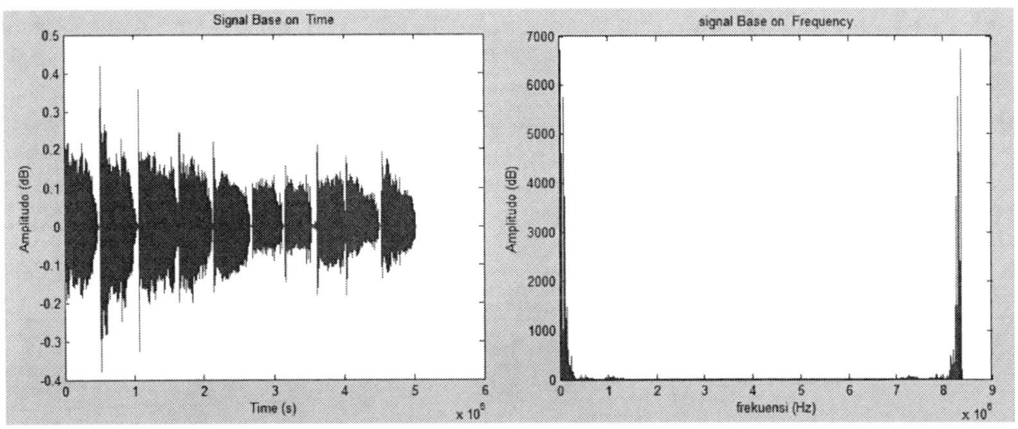


**Figure 3.** (a) Signal base on Spatial (time) Domain, (b) Signal Based on Frequency

The extraction of the waveform features is done by converting the sound signal to a wave image, which represents the change in amplitude with time. In general, a sound signal will definitely be represented with a distinctive wave shape. Sound waveforms can also provide information on another feature, namely the timing of the sound's reverberation. Figure 3 (a) illustrates waveform of signal sound.

IOP Conf. Series: Materials Science and Engineering 1115 (2021) 012067 doi:10.1088/1757-899X/1115/1/012067

## 3. Methodology of Reasearch

Research to determine the characteristics of wijakasara sound is carried out in several sequences, stated in Figure 4.

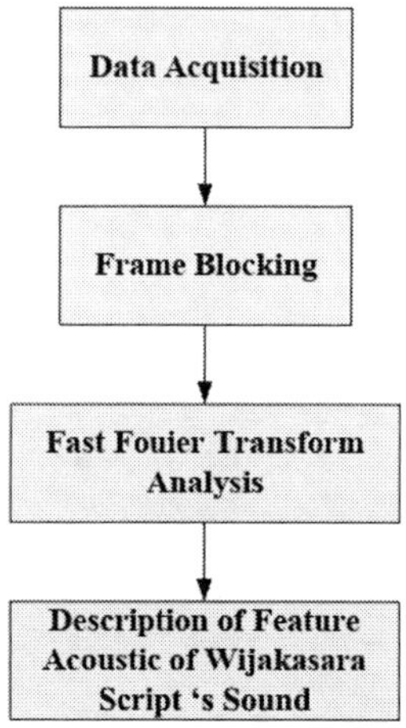


**Figure 4.** The Method Proposed for Extraction Sound Feature of Wijaaksara Script

### 3.1 Data Acquisition

Data collection of Wijaaksara sound samples was carried out by recording sounds on 20 respondent subjects with 10 male and 10 female. Recording is done for each particular sound wija script, adjusted to the length of breath of the subject. In this condition, the subject in the pronunciation is made in as comfortable a condition as possible. Respondents were drawn from students departement of physics education, Universitas Pendidikan Ganesha ( indonesia) with various backgrounds. The existence of this wijaaksara script sound is closely related to the Hindu religious belief system. However, in this study several subjects were taken from different belief backgrounds. Initially the presumption of acoustic properties / features obtained does not depend on matters of the person's background beliefs. The Process is illustrated in Figure 5.

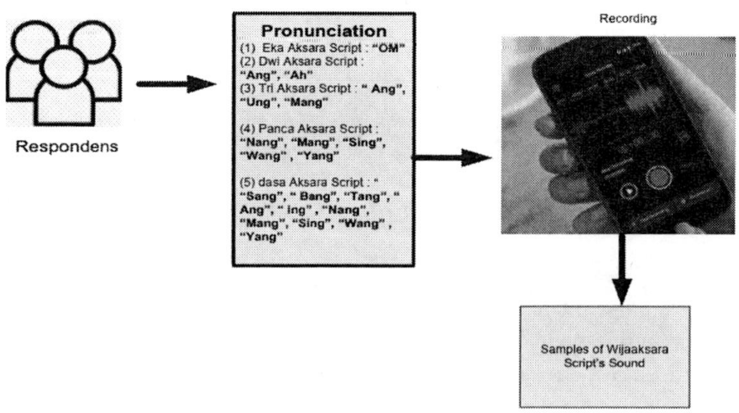


**Figure 5.** Data Acquisition

IOP Conf. Series: Materials Science and Engineering 1115 (2021) 012067 doi:10.1088/1757-899X/1115/1/012067

### 3.2 Frame Blocking

Frame blocking is cutting the signal, so that it gets 1 cycle of wijakasara sound. This process is manually marked, by marking the beginning of the cycle and the end of the chime cycle of the entire file. Figure 6 shows the sound waveform in a recording file (wav).

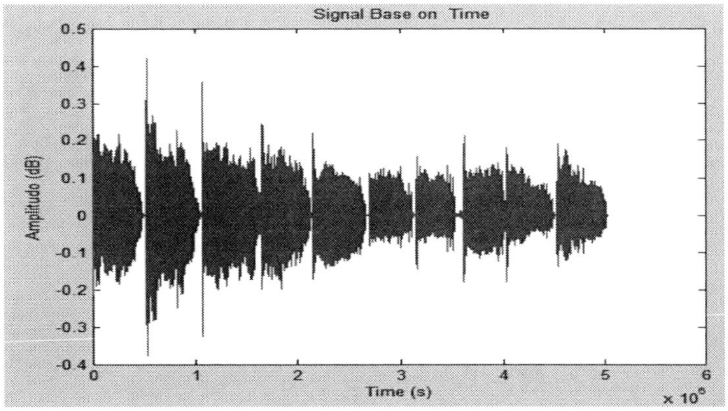


**Figure 6** The Cuting of Signal Sound

In a recording selected and cut in a certain part so that it gets 1 piece which is 1 cycle of wijaaksara sound.

### 3.3 Fast Fouier Transform Analysis

Feature extraction is done to get the basic frequency (F0) and prominent frequency, in this study limited to F1, F2, and F3. Based on that, in this study only 4 main frequencies were determined. For this purpose, FFT (Fast Fouier Transform) is used. FFT is a form of fourier transformation, which transforms the representation of signal data from the time domain into the frequency domain. The FFT algorithm is a development of the Discrete Fourier Transform algorithm, with a much faster computational process. DFT computing time is expressed O (N2) while FFT computing time is expressed N log N [10]. The definition of basic frequency and prominent frequency is illustrated in Figure 7.

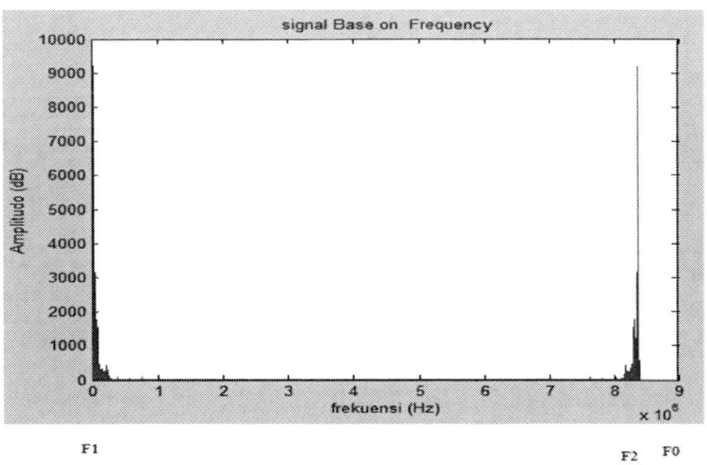


**Figure 7.** Basic Frequency and Prominent Frequency based on FFT

IOP Conf. Series: Materials Science and Engineering 1115 (2021) 012067 doi:10.1088/1757-899X/1115/1/012067

The determination of sound energy is determined by Equations 1 and 2. The form of sound waves consisting of positive and negative amplitude parts and is symmetrical, is taken only the positive part. The determination of the overall energy in t seconds is expressed in Equation 2. This energy is an approximation of the average sound energy for each time component. The average energy approach is shown in Figure 8.

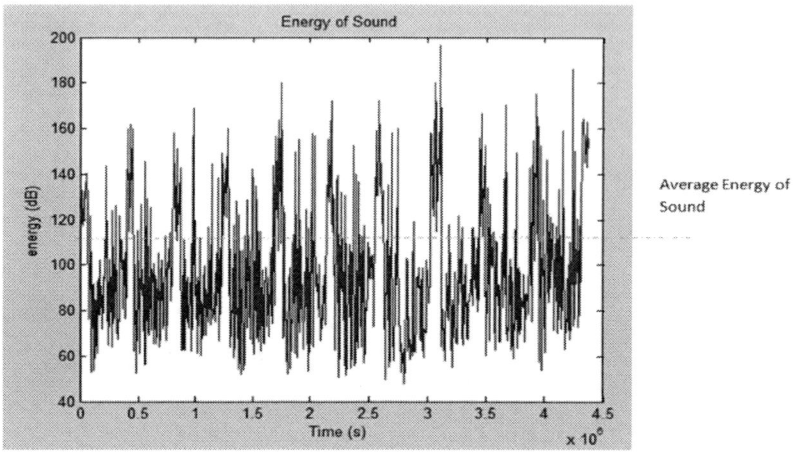


**Figure 8.** Average Approach to Determination of Sound Energy

Energy calculations using equation 1[11] .

$$EB = abs\ (\ 20 \log(A)) \tag{1}$$

Description of Notation formula :
EB : Energy of Sound (dB)
A : Amplitudo (dB)

In each time the sound signal has a certain amplitude (A), then at that time it also has energy (EB), based on Equation 1. Thus, overall in a period of t seconds, the total sound energy is calculated using the mean EB , Equation 2, [11].

$$ET = \overline{EB} \tag{2}$$

Description notation formula
ET : Total Energy in duration t seconds (dB)

## 4. Results and Discussion

The application developed in this research will be implemented using Matlab 2016. We do computations and simulations based on the FFT algorithm, to get the features of the frequency of the sound, and the average energy of the sound. To get the waveform, visualization of the signal in the time domain is performed. Table 2 summarizes the data patterns obtained in this study. For the waveform feature is taken from the dominant waveform or the most common pattern is taken from 20 samples, in each group of wijaaksara script. The sound frequency, consisting of the main frequency (F0), and the prominent frequency (F1, F2, F3), in this study used 4 constituent frequencies of sounds. This has been described in sub 3.3.

IOP Conf. Series: Materials Science and Engineering 1115 (2021) 012067 doi:10.1088/1757-899X/1115/1/012067

**Table 2.** Acoustic Feature of Wijaaksara Script

| **Part group of Wijaaksara Script** | **F0,F1,F2,F3( KHz)** | **Energy (dB)** | **Wave Form** |
|---|---|---|---|
| Eka aksara | 4.37 ; 3.92 ; 3.83 ; 3.81<br>Average Frequency : 3.98 KHz | 8.731 | 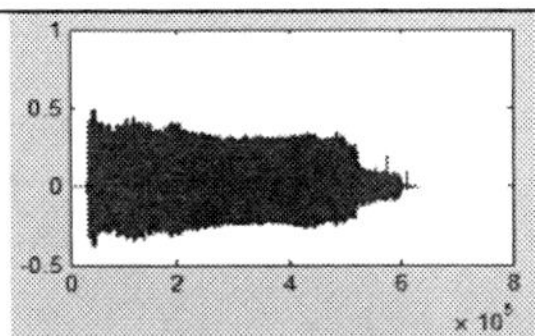 |
| Dwi Aksara | 8.91 ; 8.8 ; 4.22; 3.8<br>Average Frequency : 6.43 KHz | 9. 672 | 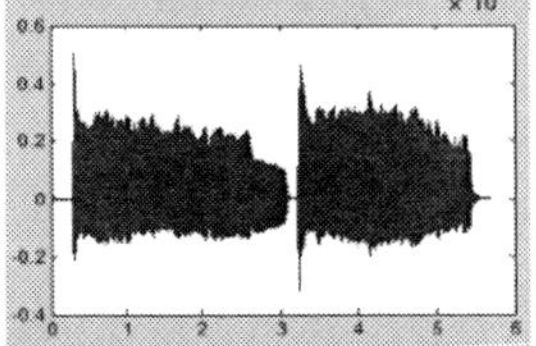 |
| Tri Aksara | 14.72 ; 7.89; 3.97; 3.88<br>Average Frequency : 7.62 KHz | 8.325 | 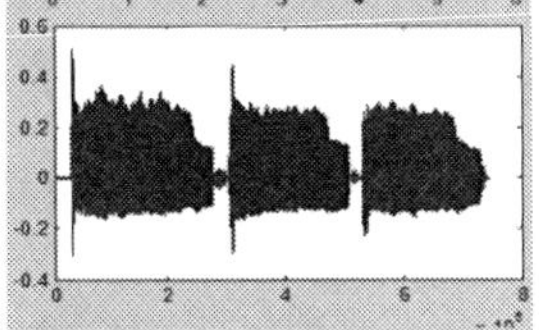 |
| Panca Aksara | 11.45 ; - ; -; 4.18<br>Average Frequency : 7.82 | 8.530 | 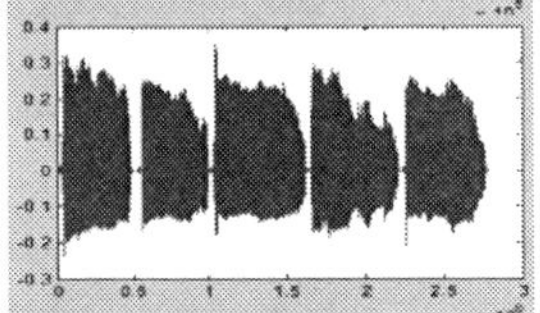 |
| Dasa Aksara | 20. 8 ;11.63 ; 11.82 ;-<br>Average Frequency : 14.72 KHz | 8.029 | 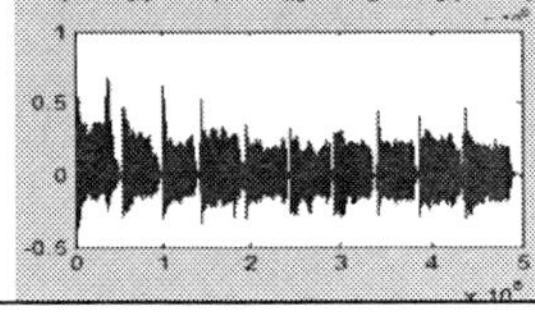 |

Related with the frequency feature, it is found that, in terms of the compiler frequency features, the values are different. There is a pattern to increase the frequency value from eka aksara script, dwi aksara, tri aksara, panca aksara, and dasa aksara script, as shown in Table 2 and Figure 9. Based on the waveforms, the eka aksara waveform is the basis for dwi aksara's waveform, tri aksara's waveform, panca aksara's waveform, and dasa aksara's waveform, Table 2. Another thing related to the waveform, it was found that in general there was no difference in shape, from 20 subjects, 19 subjects had the same waveform.

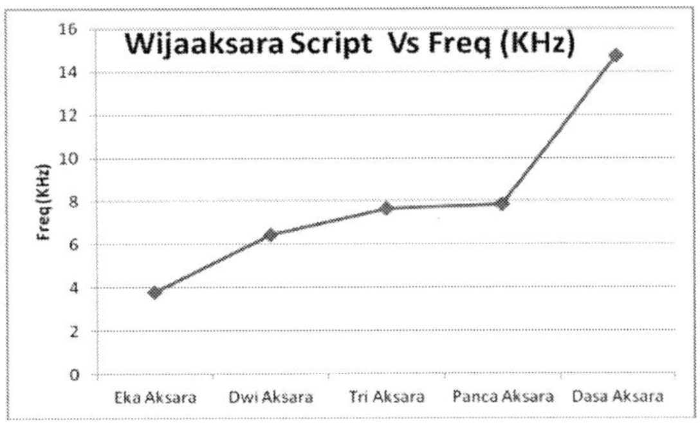


**Figure 9**. Frequency of Wijaaksara Script

## 5. Conclusions

In this paper, the sound produced by the pronounciation of the wijaaksara script with a variety characters of subjects has almost the same features, both in the form of waves, frequencies, and sound effects.

The Sound's frequencies have pattern, that pattern is value of frequency increase from eka aksara, dwi aksara, tri aksara, panca aksara and dasa aksara . Although that is still in a low frequency and in the audiosonic sound group. The Third based on the frequency and sound energy of the sound of the wijaaksara script, including the low frequency audiosonic sound, the sound energy also states the sound related to an atmosphere of calm (0- 20 dB). Based on this, it can be stated that the sound of wijaaksara is a sound related to relaxation.

## References

[1] Indrawan G, Paramarta I K, Agustini K and Sariyasa S 2018 Latin-to-Balinese script transliteration method on mobile application: A comparison *Indonesian Journal of Electrical Engineering and Computer Science* **10** (3) 1331-1342

[2] Kumar V and Sengar P K 2010 Segmentation of printed text in devanagari script and gurmukhi script *International Journal of Computer Applications* **3** (8) 30-33

[3] Mudiarta I M D R, Atmaja I M D S, Suharsana I K, Antara I W G S, Bharaditya I W P, Suandirat G A and Indrawan G 2020 Balinese character recognition on mobile application based on tesseract open source OCR engine *Journal of Physics: Conference Series* **1516** (1) 012017

[4] Haryati S 1985 *Jnanasiddhanta*. Jakarta: Djambatan

[5] Simpen I W, Udayana I N and Putra I B R The Caligraphy of Balinese Script as A Form of Bali Creative Industry Based on Balinese Script *Udayana Journal of Social Sciences and Humanities* **3** (1) 13-16

[6] Crisnapati P N, Novayanti P D, Indrawan G, Aryanto K Y E and Wibawa M S 2018 Accuracy Analysis of Pasang Aksara Bot using Finite State Automata Transliteration Method.*2018 6th International Conference on Cyber and IT Service Management* 1-6

[7] Gunadi I G A and Hartawan I G N Y 2019 Analysis of Frequency on Sound of Genta Based on Fast Fourier Transform Method *Proceedings of the 2nd International Conference on Software Engineering and Information Management* 185-188

[8] Zeng Y, Zhang M, Han F, Gong, Y and Zhang J 2019 Spectrum analysis and convolutional neural network for automatic modulation recognition *IEEE Wireless Communications Letters* **8** (3) 929-932

[9] Bozzo E, Carniel R and Fasino D 2010 Relationship between Singular Spectrum Analysis and Fourier analysis: Theory and application to the monitoring of volcanic activity *Computers & Mathematics with Applications* **60** (3) 812-820

[10] Efnusheva D, Tagasovska N, Tentov A and Kalendar M 2018 Efficiency Comparison of DFT/IDFT Algorithms by Evaluating Diverse Hardware-Implementations, Parallelization Prospectsand Possible Improvements *Proceedings of the 2nd International Conference on Applied Innovations* 81-89

[11] Yang L J, Zhang B H and Ye X 2004 Fast Fourier transform and its applications *Opto-Electronic Engineering* **31** 1-7

# Physical Properties of Bio-Composite Board Reinforced with Shell Particle and Coconut Fiber

**D O Rachmawati* and P Widiarini**

Department of Physic and Science Teaching, Faculty of Mathematics and Natural Science, the University of Ganesha Education, Singaraja-Bali, Indonesia

Corresponding author: dewioktofa.r@undiksha.ac.id

**Abstract.** The properties of the particles which have best binding capacity with the matrix can be combined with the properties of the fibers which give the composite strength. The use of shell particles and coconut fiber together as a composite reinforcement provides a variety of new physical properties that are different from its constituents. Therefore, the variation in weight fraction of the two fibers as reinforcement is interesting to study. The bio-composite board was made to use open mold process. The bio-composite board reinforced with shell particles and coconut fiber has a density of $0.938 \leq \rho \leq 1{,}020$ g/cm$^3$. The greatest water content reaches 5.43%. The immersion time of 24 hours, water absorption of this bio-composite board is in the range of 1.95% to 3.38%, and the largest thickness swelling reaches 2.86 %. Bio-composite board which is composed of 40% by weight of coconut fiber has a density of 1.111 g/cm$^3$. Water content reaches 5.90%%. Water absorption capacity and thickness swelling of 4.08% and 3.12%, respectively. While the bio-composite board which is composed of 40% by weight of shell particles has a smaller density, moisture content, absorption capacity and thickness expansion, namely 0.901 g/cm$^3$, 4.13%, 1.61% and 1.18%.

## 1. Introduction

Innovations in the field of composite materials make natural fibers attractive for researchers. Its use as a composite reinforce is very potential. One of these natural fibers is coconut shell and coir. Coconut shells and coir are renewable, renewable raw materials. It is environmentally friendly, biodegradable, abundant, renewable, inexpensive, and has a low density. Polymeric composites reinforced with natural fibres have been developed in recent years. It shows significant potential for various engineering applications due to their intrinsic sustainability, low cost, low weight and mechanical strength. The coconut coir reinforced composites were applied to noise-reducing materials, ebonite rubber, and technical material [1]-[3]. Also, the coconut shell particle can be applied as fillers for property development in natural rubber compound [4].

Coconut shell is a layer of coconut which layer sequence is after the fiber layer shell. Its volume fraction is about 15-19 % of the coconut volume fraction with 3-5 mm thickness. It has a density of 0.65 g/cm$^3$ and contains 8.00% water, 26.60% cellulose, and 29.40% lignin [5]. Coconut fiber is composed of 75% fiber and 25% cork. The fiber has a diameter of 0.1-1.5 mm, and contains 35% - 45% lignin and 23% -43% cellulose.The capacity of water holding reaches 3 times of its volume fraction. Coconut fiber is durable, high buoyancy, non-flammable and excellent insulation against temperature and sound [6]. Maximum water absorption in the coconut fiber reaches 180%. Its density is 1.2 g/cm$^3$ [7].

Several studies on the physical properties of coconut fiber composites or coconut shell powder have been conducted by previous researchers. The density of coconut shell particle reinforced composites with 200-800 μm size decreased with the increasing of particle volume fraction, namely 1.288-1.171 g / $cm^3$ [8]. Tensile property and modulus of elasticity reduces with the increase of wt% of shell particle namely 25.20 MPa and 654.00 MP for 35 wt% shell particles. Particles size, density and light absorbance of of uncarbonized coconut shell nanoparticles was studied in ref [9]. The results of these studies indicated that density of the uncarbonized coconut shell nanoparticles obtained at 70 hours of milling are 0.65 g/$cm^3$. Average particles size at 16, 46 and 70 hours of milling are 119.2 nm, 72.1 nm and 49.85 nm respectively. a decrease in particle size is accompanied by a decrease in the maximum absorbance wavelength when milling lasts up to 70 hours. Another case with the research conducted in ref [10], composites with coconut fiber reinforcement were successfully made with various fiber volume fractions. The tensile and impact strengths depend on the volume fraction of the fiber. Based on ref [11], the natural fibers as reinforcing polymer composites stated that the mechanical strength of coconut fibers can compete with the technical fibers. The results of these studies indicated that density, ultimate strength, modulus of elasticity and % elongation of particle composite depend on wt% of shell particle and particles size. Composite of coconut fiber can made in various fiber volume fractions.

Coconut fiber as a reinforcing element greatly determines the physical and mechanical properties of the composite. Fiber will be able to continue the load distributed by the matrix [12]. Shell particles and coconut fiber, which is combined with a polymer as a matrix, will produce alternative composites that are very beneficial to the industrial application. It is expected to produce maximum composite physical properties to support the application of alternative composites.

Composite physical properties is largely determined by the fiber, matrix and the volume fraction of its constituents [13]. The addition of a wetting agent is needed to wet the fibers. Interaction between particles or fibers can occur if there is contact between the wetting agent and the fiber surface. Wetting agent functions as a matrix. The matrix will bind the fibers into a structural unit, protect the fibers from external damage, transmit, or transfer external loads to the shear plane between the fibers and the matrix. Composite quality is influenced by the type of matrix used [14]. The unsaturated polyester resin matrix is a liquid. This resin is easy to handle, fast curing, perfect dimensional stability, good physical and electrical properties, easy to colour and modify. Unsaturated polyester resin has a density of 1.2 g / $cm^3$ to 1.5 g / $cm^3$, and the water absorption at immersion for 24 hours at temperature of $20^0$ ranges from 130% to 180%. Research conducted by BPPT Jakarta on the quality of tensile strength in matrix-reinforced composite structures (epoxy, polyester, vinyl ester) with natural fibers (abaca, hemp, kenaf, and pineapple fibers). The matrix with 40% -50% by volume fraction of unsaturated polyester resin has the highest tensile strength qualities. This resin is suitable for composites was made by casting method in open mould. Methyl Ethyl Ketone Peroxide (MEKP) is a hardener catalyst. MEKP is the result of the reaction of methyl ethyl ketone with hydrogen peroxide. The addition of this catalyst helps the resin hardening process. The time it takes for the resin to turn into plastic depends on the amount of catalyst mixed. The more catalyst you add, the faster the curing process. The addition of excessive catalyst results in a brittle material or burnt resin. The alkaline treatment of the fibers removes the ineffective fiber constituents, dirt and foam adhering to the fibers, increases the adhesion of the fiber and matrix interfaces and obtains mechanical bonding [15].

Physical testing is an initial test before further testing is carried out. Unlike the mechanical test, the physical test reflects properties that are not related to the influence of external forces. The result of physical testing has a big effect on the strength and appearance of the resulting composite board. Composite board density is defined as mass per unit volume. Density is determined by the mass of the fibers and the matrix as a constituent of the composite. Water content reflects the amount of water remaining in the cell cavity and voids trapped in the crystal structure. The amount of water contained in the fiber, resin, and humidity in the surrounding air affects the moisture content in the composite. The amount of water that can be absorbed by the cell cavities and voids is called water absorption capability.

Variations in composite research can done by making a combination of coconut shell particles and coconut fiber as a bio-composite reinforcement. This aims to get a new composite from the best combination of characteristics of each constituent component. This research is a preliminary study to describe the physical properties of the bio-composite board. The physical properties studied were density, water content, water absorption capacity, and thickness swelling.

## 2. Experiment

### 2.1 Preparation

Coconut shells and dried coconut fibers were chosen as bio-composite reinforcement. The separation of the coconut fiber from the cork is done using natural brushing techniques. The coconut shell is cleaned with sandpaper with the aim of removing the sticky cork. Next, immersion in 5% NaOH solution for 2 hours. Coconut shell particles with ≤ 74 μm size were obtained by filtering using a 200 mesh sieve. The fibers were chosen to be the same size and cut to a length of 3 cm.

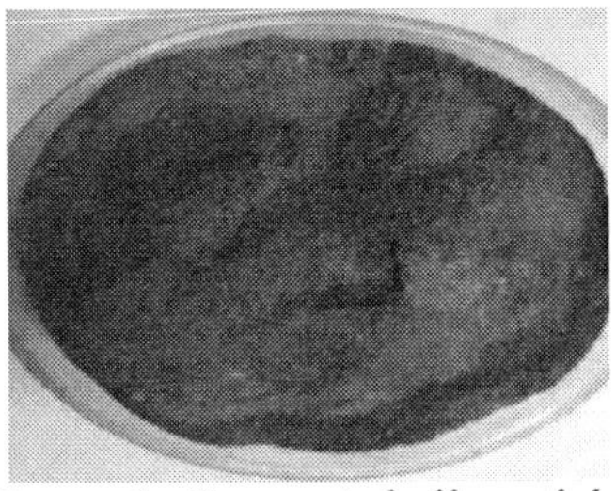

**Figure 1.** Coconut shell particles

**Figure 2.** Coconut fibre

### 2.2 The Making of Bio-composite

The bio-composite was made by lay-up hand method. The volume fraction ratio of fiber and matrix used is 40%:60%. The ratio of the weight fraction of the coconut shell particles and fiber are as follow, namely; 0%: 40%, 10%: 30%, 20%: 20%, 30%: 10%, and 40%: 0%. The unsaturated polyester resin matrix used was Yukalac type BQTN-EX 157. The density of this resin is 1,215 g/cm$^3$ with a water absorption capacity of 0.188%. The hardening of the resin using a MEKP catalyst was 3% of the volume of the resin used. Coconut fibers are arranged in a random fiber model in a mold. The coconut shell particles are mixed with unsaturated polyester resin.

**Figure 3.** Bio-composite molding process

Pressure is applied to the fibers until the thickness of the bio-composite board reaches 1 cm. The bio-composite board is allowed to dry. To get a stable condition, the bio-composite board is left to stand for 14 days. The test sample is 3x10x1 cm in size.

## 2.3 Testing

The test of bio-composite board density was carried out in dry air and volume conditions. The water content test was carried out by heating at $100^0$ C for 6 hours. The moisture content was determined by the mass under dry air conditions and the mass after heating. The water absorption of the test sample was carried out by immersion method for 24 hours. The water absorption capacity of the bio-composite board was determined by the mass under dry air conditions and the mass after immersion. The mass of the test sample was measured with the Ohaus digital balance.

**Figure 4.** Massa measurement

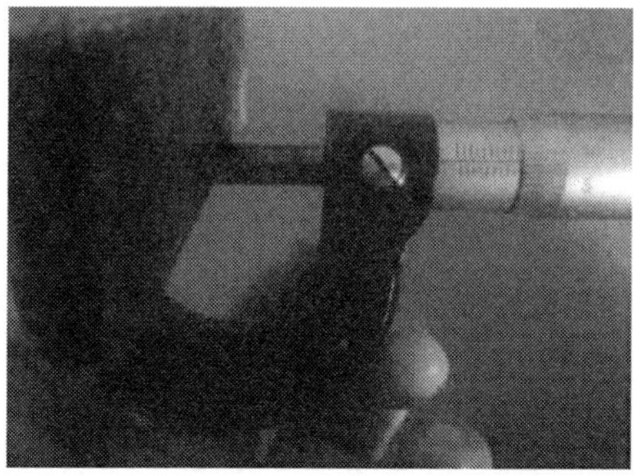

**Figure 5.** Thickness measurement

The thick swelling test used the immersion method. The test sample was immersed for 24 hours. The thickness swelling is determined from the thickness of the test sample at a certain point under air-dry conditions and the thickness after immersion. The thickness of the test sample was measured by using a measuring instrument in the form of a screw micrometer.

## 3. Results and Discussion

The 5% NaOH treatment for 2 hours changed the color of the shell and coconut husks from pale yellowish brown to dark brown. Research by M Leena [16] also showed similarity in color to coconut fibers after soaking with NaOH. The measured average density of coconut fiber was 1.125 g/cm$^3$.

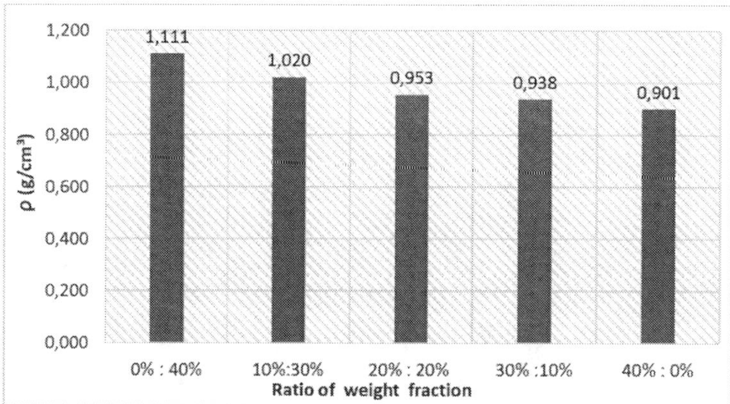


**Figure 6.** Density of bio-composite board

The bio-composites filled with only coconut shell particles had a density of 0.901 g/cm$^3$, and those filled with only coconut fiber had a density of 1.111 g/cm$^3$.

**Figure 7.** The Bio-composite

The combination of unsaturated polyester resin with a density of 1.215 g/cm$^3$, coconut shell particles with a density of 0.65 g/cm$^3$, and coconut fiber with a density of 1.125 g/cm$^3$resulted in a composite with a density of 0.938 $<\rho<$1.020 g/cm$^3$. This value is smaller than the density of composites made from coconut fiber alone or greater than the density of composites made from coconut shell particles. The difference in density between coconut shell particles, unsaturated polyester resin, and coconut fiber resulted in a combined density. The addition of 10% coconut shell particles weight fraction accompanied by the 10 % decrease in coconut fiber decrease the density of the composite. The decrease of density is shown in Figure 6. The greater the weight fraction of the high density fibers in the fixed composite volume, the higher the density of the composite. The density of the bio-composite board produced was $\geq$ 0.90 g/cm$^3$.

The bio-composite board filled only with coconut shell particles had the lowest water content, namely 4.13%. The highest water content, namely 5.90%, is on the bio-composite board which is only filled with coconut fiber. The bio-composite board with variations in the weight fraction of coconut shell and coconut shell particles has a water content of 4.71% to 5.43%, as shown in Figure 8. The water content of the bio-composite board decreases with increasing weight fraction of coconut shell particles on the constant volume of the bio-composite.

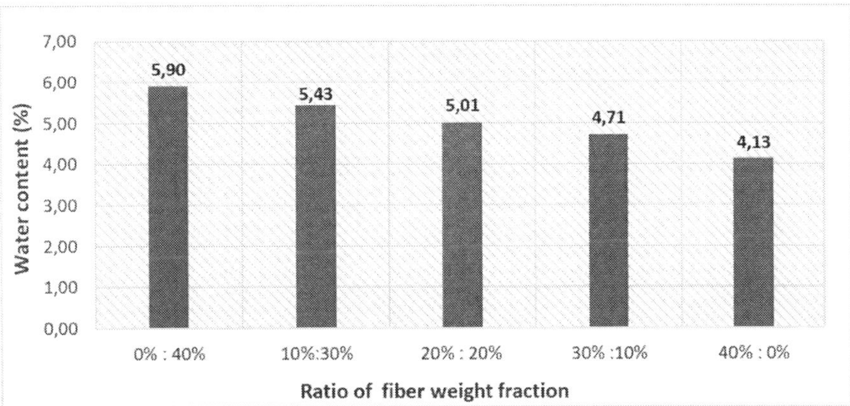


**Figure 8.** Water content of bio-composite board

The fiber weight fraction which is greater than shell particles contain a larger number of cell cavities and voids. The damaged fiber shape can create voids. Voids are also formed because air is trapped between fibers and particles that are not properly attached. Voids and cell cavities become free water reservoirs due to trapping of water vapour in them. The addition of coconut shell particles reduces the water content shown in Figure 8. The resin, the shell particles and fibers in small volume fraction can coats the composite perfectly. Shell particles are evenly distributed in the fiber arrangement. Water vapor from the outside environment cannot enter the composite.

By soaking for 24 hours, the bio-composite board with 10% weight fraction of coconut shell particles and 30% weight fraction of coconut fiber has a water absorption capacity of 3.38%. This value is smaller than the bio-composite board which is only filled with 40% coconut husk or greater than the bio-composite board which is only filled with coconut shell particles. The longer the immersion time, the greater the water absorption capacity of the bio-composite board, as shown in Table 1.

**Table 1.** The water absorption capacity of the bio-composite board

| Weight fraction (%) | | The water absorption capacity (%) | | | | | |
|---|---|---|---|---|---|---|---|
| Coconut shell particles | Coconut fibre | 2 h | 4 h | 6 h | 8 h | 10 h | 24 h |
| 0 | 40 | 0.51 | 0.90 | 1.28 | 1.98 | 3.23 | 4.08 |
| 10 | 30 | 0.40 | 0.80 | 1.34 | 1.86 | 2.34 | 3.38 |
| 20 | 20 | 0.37 | 0.77 | 1.10 | 1.49 | 1.90 | 2.55 |
| 30 | 10 | 0.34 | 0.75 | 1.04 | 1.35 | 1.49 | 1.95 |
| 40 | 0 | 0.30 | 0.63 | 0.78 | 0.93 | 1.01 | 1.61 |

The increase of water absorption becomes greater after 2 hours of immersion. In addition to the hygroscopic lignocellulose contained in the fiber, the resin can also absorb water. The ester group can bind water. Coconut fiber which has a lignocellulose content greater than coconut shell particles is able to absorb more water. When using 40% by weight fraction of coconut fiber, the resin does not completely wet or coat the entire surface of the fiber. The contact area between the fibers become smaller and adhesion of the interface between fiber and resin is also small. Increase in volume fraction of coconut shell particle, decrease in adhesion interface shell particles, fiber and resin. The particles and fibers are tightly bound. At 40% weigt fraction of coconut shell particles, the resin wets the entire the particles and fibers. In this condition, the surface of bio-composite was coated the resin completely. It is difficult for water to enter the fibers and voids. The water absorption capacity of this bio-composite board is 1.61%. The interfacial adhesion between natural fibres and polymeric matrices is critical to the composite performance [17-18]

With a soaking time of 24 hours, the thickness swelling of the bio-composite board with 10% weight fraction of coconut shell particles and 30% weight fraction of coconut fiber was 2.86%. This value is smaller for bio-composite boards that are only filled with 40% coconut fiber or greater on composite boards filled with only coconut shell particles. The longer the immersion time, the greater the thickness of the bio-composite board for all variations in the volume fraction of fibers.

**Table 2.** Thickness swelling of the bio-composite board

| Weight fraction (%) | | Thickness swelling (%) | | | | | |
|---|---|---|---|---|---|---|---|
| Coconut shell particles | Coconut fibre | 2 h | 4 h | 6 h | 8 h | 10 h | 24 h |
| 0 | 40 | 0.57 | 1.03 | 1.57 | 1.92 | 2.26 | 3.12 |
| 10 | 30 | 0.35 | 0.52 | 0.93 | 1.19 | 1.59 | 2.86 |
| 20 | 20 | 0.18 | 0.49 | 0.98 | 1.13 | 1.30 | 2.11 |
| 30 | 10 | 0.17 | 0.43 | 0.88 | 1.06 | 1.30 | 1.78 |
| 40 | 0 | 0.13 | 0.27 | 0.48 | 0.59 | 0.76 | 1.18 |

In high-density bio-composite boards, the fibers are subjected to greater stress during the pressing process. The fiber cell cavity is shrinking. The resin cannot enter the very small fiber cavities during the pressing process. Fibers that are not completely bonded to other fibers make it easier for the fibers to break away from the remaining tension. The absorption of water by the fibers causes the restoration of the fiber size to its original dimensions. The cell wall of the fiber or the size of the fiber cavity expands. After 2 hours of immersion, the expansion in cell wall size increased. Bio-composite boards with shell particles weight fraction greater than the weight fraction of fibers, resins perfectly bond particles and fibers. the adhesive force is large. The resin also coats the entire surface of the composite. As a result, it is difficult for water to enter the fiber, the recovery of the fiber size to its original dimension becomes quite small.

## 4. Conclusion

The density of the bio-composite board reinforced with shell particles and coconut fiber is in the range $0.938 \leq \rho \leq 1.020$ g/cm$^3$. The increase in weight fraction of coconut shell particles ($\leq 74$ μm in size) can reduce the density of the bio-composite board. The increase in the weight fraction of the resin can increase the adhesion of fibers and particles. Its use for furniture requires further testing, namely the screw strength

## Acknowledgement

Thank you to LPPM Ganesha University of Education, the basic physics laboratory of the S1 Physics Education study program, for the funds and facilities provided for this research. Special thanks to Iwan Suswandi for his helpful comments on this paper.

## References

[1] Lai C Y, Sapuan S M, Ahmad M, Yahya N and Dahlan K Z H M 2005 Mechanical and electrical properties of coconut coir fiber-reinforced polypropylene composites *Polymer-Plastic Technology and Engineering* **44** (4) 619-632

[2] Priya N A S, Raju P V and Naveen P N E 2014 Experimental testing of polymer reinforced with coconut coir fiber composites *International Journal of Emerging Technology and Advanced Engineering* **4** (12) 453-460

[3] Sudharsan B, Kumar M L and Kumar R M L 2014 The mechanical behaviour of eggshell and coconut coir reinforced composites *International Journal of Engineering Trends and Technology* **18** (1) 9-13

[4] Momoh F P, Hamza P A P and Nkeonye P Morphology And Crystallinity Of Modified Coconut Shell Powder In Natural Rubber Development *International Journal of Innovative Research and Advanced Studies* **4** 257-262

[5] Totla S K, Pillai A M, Chetan M, Warad C, Vinodkumar S K, Patil A Y and Kotturshettar B B 2020 Analysis of helmet with coconut shell as the outer layer *Materials Today: Proceeding* **32** 365-373

[6] Owolabi O, Czvikovszky T and Kovacs I 1985 Coconut-fiber-reinforced thermosetting plastics *Journal of applied polymer science* **30** (5) 1827-1836

[7] Naveen P N E and Prasad R V 2013 Evaluation of Mechanical Properties of Coconut Coir/Bamboo Fiber Reinforced Polymer Matrix Composites *Science and Engineering* **3** (4) 15-22

[8] Bhaskar J and Singh V K 2013 Physical and mechanical properties of coconut shell particle reinforced-epoxy composite *J. Mater. Environ. Sci* **4** (2) 227-232

[9] Bello S A, Agunsoye J O, Adebisi J A, Kolawole F O and Hassan S B 2016 Physical properties of coconut shell nanoparticles *journal of Science, Engineering and Technology* **12** 63-79

[10] Ali M, Liu A, Sou H and Chouw N 2012 Mechanical and dynamic properties of coconut fibre

reinforced concrete *Construction and Building Materials* **30** 814-825
[11] Taj S, Munawar M A and Khan S 2007 Natural fiber-reinforced polymer composites *Proceedings-Pakistan Academy of Sciences* **44** (2) 129-139
[12] Schwartz M M 1984 *Composite Materials Handbook* New York: McGraw-Hill
[13] Gibson R F 1994 *Principles Of Composite Material Mechanics* New York: McGraw-Hill
[14] Li Y and Ramesh K T 1998 Influence of particle volume fraction, shape, and aspect ratio on the behavior of particle-reinforced metal–matrix composites at high rates of strain *Acta Materialia* **46** (16) 5633-5646
[15] Alex S, Stanly J R and Ramachandran M 2015 A review on biodegradability of hybrid bamboo/glass fiber polymer composites *International Journal of Applied Engineering Research* **10** (11) 10565-10569
[16] Oksman K, Skrifvars M and Selin J F 2003 Natural fibres as reinforcement in polylactic acid (PLA) composites *Composites science and technology* **63** (9) 1317-1324
[17] Verma D, Gope P C, Shandilya A, Gupta A and Maheshwari M K 2013 Coir fibre reinforcement and application in polymer composites *J. Mater. Environ. Sci* **4** (2) 263-276
[18] Mulinari D R, Baptista C A R P, Souza J V C and Voorwald H J C 2011 Mechanical properties of coconut fibers reinforced polyester composites *Procedia Engineering* **10** 2074-2079

# Empirical Attenuation Model for Predicting Peak Ground Acceleration in North Arm Sulawesi, Indonesia

**G Pasau***, **G H Tamuntuan, A Tanauma, D P Pandara, F Ferdy, A As'ari and C Maramis**

Department of Physics, Sam Ratulangi University, Manado, Indonesia

*Corresponding author: guntur_pasau@unsrat.ac.id

**Abstract.** The North Arm of Sulawesi is one of the areas that has very active tectonic conditions. The earthquake source in this area comes from the activity of several tectonic plates such as the North Sulawesi Subduction, the Molluca Sea Collision, the Philippine Sea Plate, the Pacific Plate and several local faults such as the Palu Koro Fault, Gorontalo Fault, Bolmong Fault and Manado Fault. The purpose of this study was to determine the peak ground acceleration value using several empirical attenuation models such as the Fukushima-Tanaka, Donovan, McGuire, Campbell, and Crouse. The peak ground acceleration is the value of the greatest acceleration a place has ever occurred which describes the strength of the tremor or shock of an earthquake. Geographically, this research area is located at 118$^0$E to 127$^0$E and 0.5$^0$S to 4$^0$N. The data used are data from the ISC-EHB and USGS catalog during the observation interval 1964-2020. Data processing includes the uniformity of the magnitude scale into the Surface Magnitude (*Ms*) scale, determining the main shock, sorting earthquake data with a magnitude greater than M6.0 and a depth of less than 100 km, creating a grid measuring 0.05 x 0.05 degree. Next step, determine the peak ground acceleration using an empirical attenuation relationships. The analysis results show that the peak ground acceleration in the North Arm Sulawesi is as follows: McGuire Method 0.4295g, Fukushima-Tanaka 0.3595g method, Crouse method 0.314g, Donovan method 0.3149g, and Campbell method 0.2550g.

## 1. Introduction

Earthquake disasters are geological natural disasters that cause the most casualties to date. Thousands of victims were crushed under the rubble of the building due to the shock of the earthquake. One of the areas in Indonesia that has the potential and is prone to moderate to large earthquakes is North Arm of Sulawesi. The source of the earthquake in this area came from the activity of several tectonic plates such as the North Sulawesi Subduction, the Molluca Sea Collision, the Philippine Sea Plate and the Pacific Plate coupled with local faults [1]. Earthquakes always come suddenly even geologists cannot predict when they will occur. The only way to minimize casualties is through disaster mitigation efforts one of which is by determining or mapping the maximum horizontal acceleration value on the surface or Peak Ground Acceleration (PGA). The peak ground acceleration is the greatest acceleration value in a place that has ever occurred, which describes the strength of an earthquake vibration or shock [2]. The higher the ground acceleration in a place the higher the level of risk in that area. The purpose of this study was to determine the peak ground acceleration (PGA) in the North Arm Sulawesi region through an empirical attenuation relationships model. The attenuation function describes the relationship between ground motion intensity ($I$) and the magnitude of the earthquake ($M$) and the hypocenter distance ($R$) from a source point.

The empirical attenuation relationships used in modeling the peak ground acceleration are the McGuire, Campbell, Donovan, Crouse-Mc.Guire, and Fukushima-Tanaka Method (see Table 1) [3]-[5]. This method is used if the source of the earthquake is mostly from subduction activity and faults.

**Table 1**. The empirical attenuation relationship use

| ID | Source | Empirical Attenuation |
|---|---|---|
| 1 | Fukushima-Tanaka | $\log A = 0.41Ms - \log(R + 0.032x100.41Ms) - 0.0034R + 1.30$ |
| 2 | Crouse & Mc.Guire | $Ln\ Y = 2.4846 + 0.7334M - 0.0151M^2 - 0.5056Ln(R+1) - 0.0094R$ |
| 3 | Mc.Guire | $\alpha = \frac{472.38x10^{0.278M}}{(R+25)^{1.301}}$ |
| 4 | Donovan | $\alpha = \frac{1080x10^{0.5M}}{(R+25)^{1.32}}$ |
| 5 | Campbell | $\ln Y = -2.501 + 0.623Ml - \ln(R + 7.28)$ |

where: $A$, $Y$, $\alpha$ : acceleration (PGA value) in $g$
$Ms$: Magnitude Surface ($Ms$)
$R$: $(r^2 + h^2)^{1/2}$
$r$: hypocenter distance
$h$: focal depth

In this study we used data of good quality / high resolution. The data we have used are mostly from the ISC-EHB earthquake catalog (72%) where the data have been relocated. We also tried to use the mainshock data from each events earthquake.

The purpose of this study was to determine the peak ground acceleration (PGA) value in the North Arm of Sulawesi using the empirical attenuation relationship model. The distribution of PGA values is then plotted on a map that produces a peak ground acceleration map to find out which areas have tectonic conditions that are particularly vulnerable.

## 2. Materials and methods

### 2.1. Data Collection

The data used in this study is earthquake data that has been recorded occurs around the North Arm of Sulawesi. Recorded earthquakes were collected from the USGS (United Stated Geological Survey) and ISC-EHB (International Seismological Centre - Engdahl, van der Hilst & Buland) Bulletin during the observation period from 1964 - 2020 (see Figure 1)

From a number of earthquake data that have been collected, namely earthquakes with a magnitude greater than M5.0 and a depth of less than 100 km, then carried out the magnitude scale uniformity. Uniformity or conversion of various magnitude scales into Surface Magnitude ($Ms$) using the empirical model proposed by Scordilis:

$Ms = 0.85Mb + 1.02 \qquad 2.6 \leq \text{mb} \leq 5.3$ ..................(1)

$Mw = 0.65Ms + 2.20 \qquad 3.0 \leq Ms \leq 6.1$ ..................(2)

$Mw = 1.00Ms - 0.02 \qquad 6.2 \leq Ms \leq 8.0$ ..................(3)

The magnitude scale $Mb > 6.3$ is considered to be equivalent to the moment magnitude $Mw$ [6].

Data processed with a magnitude greater than M5.0 with maximum depth of 100 km. Earthquake data which has been collected from various catalogs have a different scale of magnitude therefore this is the first step to take what to do is conversion scale into the surface magnitude scale ($Ms$).

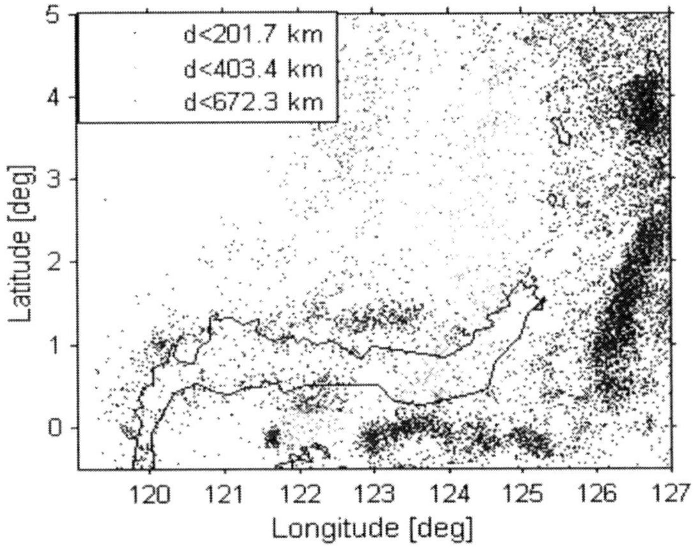


**Figure 1.** Plot earthquakes data period of 1964 - 2020.

Uniformity results later sorted or declustering catalogues using Gardner and Knopoff to determine mainshock apart from the foreshock and aftershock [7]. This uniformity process done based on the time window criteria and window spacing using empirical methods proposed by Gardner and Knopoff (1974). with the zmap software rocks. From the main earthquake the selected socks are then counted. Data at depths greater than 300 km do not contribute to ground acceleration on the surface.The results of the uniformity of earthquake data and earthquake data were selected as many as 11 data that caused the biggest shock impact in the North Arm of Sulawesi ( see Table 2).

**Table 2.** Earthquake data used in this study

| Lon ($^0$E) | Lat ($^0$N) | Origin Time (UTC) | | *Ms* | Depth (km) |
|---|---|---|---|---|---|
| 126.244 | 1.441 | 8/10/1968 | 02:07:4.15 | 7.6 | 20.9 |
| 119.685 | 0.091 | 8/14/1968 | 22:14:20.15 | 7.2 | 16.8 |
| 126.475 | 1.799 | 8/14/1986 | 19:39:14.65 | 7.1 | 32.3 |
| 122.811 | 1.196 | 4/18/1990 | 13:39:22.45 | 7.3 | 39.8 |
| 122.748 | 1.184 | 6/20/1991 | 05:18:53.12 | 7.0 | 26.3 |
| 119.908 | 0.698 | 1/1/1996 | 08:05:12.2 | 7.5 | 24 |
| 126.359 | 1.106 | 1/21/2007 | 11:27:44.48 | 7.3 | 22.2 |
| 122.098 | 1.308 | 11/16/2008 | 17:02:32.97 | 7.0 | 31 |
| 119.8462 | -0.2559 | 9/28/2018 | 10:02:45 | 7.5 | 20 |
| 126.1892 | 0.5126 | 7/7/2019 | 15:08:40 | 7.2 | 35 |
| 126.4156 | 1.6213 | 11/14/2019 | 16:17:40 | 7.1 | 33 |

### 2.2. Methods

Earthquake data collected from various catalogs usually have different magnitude scales. Therefore, the first step that must be done is to uniform the magnitude scale into the Surface Magnitude (*Ms*) scale. The results of uniformity are then sorted to determine the mainshock by applying the time window and distance window criteria using ZMAP software [8]-[10]. The uniformity results were then selected for large earthquakes with magnitudes greater than M6.0 and depths less than 100 km. The next step is to determine the amount of peak ground acceleration using the empirical attenuation relationships such as Fukushima & Tanaka, Campbell, Crouse, Donovan, and McGuire. Then map the spatial acceleration value by first creating a grid measuring 0.05 x 0.05 degree to determine which area has a significant value (greatest acceleration).

## 3. Result and Discussion

The results of calculations using the empirical attenuation relationships model show that the peak ground acceleration values in the North Arm of Sulawesi range from 0.001 g to 0.429590 g. The largest value is obtained from the Mc.Guire attenuation, which is 0.002244 g - 0.429590 g while the smallest value is obtained from the Crouse-Mc.Guire attenuation with a value of around 0.00022 g - 0.2254655 g (see Table 3).

**Table 3**. PGA value from several attenuation model

| No | Methods | Peak Ground Acceleration (g) | | |
|---|---|---|---|---|
| | | Min | Average | Max |
| 1 | Fukhusima-Tanaka | 0.0000082 | 0.010041 | 0.359595 |
| 2 | Mc.Guire | 0.002244 | 0.027128 | 0.429590 |
| 3 | Donovan | 0.0021811 | 0.021102 | 0.3149190 |
| 4 | Crouse & McGuire | 0.00022 | 0.01068 | 0.2254655 |
| 5 | Campbell | 0.001002 | 0.015608 | 0.2550339 |

The minimum value is generated at a place far from the epicenter earthquake while the maximum value is generated at the epicenter. This is the result of attenuation as a function of the distance, magnitude, and depth of the earthquake. Table 3 shows that the reasonable value is the value of the Mc.Guire method, the Donovan method and the Fukushima-Tanaka method. Meanwhile, the value generated by Crouse and Campbell attenuation is considered too small by considering the existing tectonic conditions.

The relationship between distance (km) versus PGA (g) for the M7.0 and M7.6 magnitudes is shown in Figure 2.

From Figure 2, we can see that the magnitude M7.0 shows that the graph of the acceleration value generated by the Crouse-Mc.Guire and Campbell method is almost the same (coincide). Meanwhile, the graph with magnitude M7.6 shows that the graphs produced by the two methods have separate values. The interesting thing about the two graphs above is that for the magnitude M≤7.0 the value of the ground acceleration generated by the two methods is almost the same while for M>7.0 the value of the ground acceleration is different.

The distribution of the peak ground acceleration values uses several attenuation relationship formulas as shown in Figure 3. In general, the North Arm of Sulawesi region has a high PGA value, this is due to its proximity to the earthquake source. The distribution of PGA values for each method shows the same pattern as the maximum values found in the western and northern regions. The distribution of PGA values for each method shows the same pattern as the maximum values found in the western region, namely Donggala and Tolitoli. Meanwhile, in the north, they are in the Buol, North Gorontalo and Boroko regions.

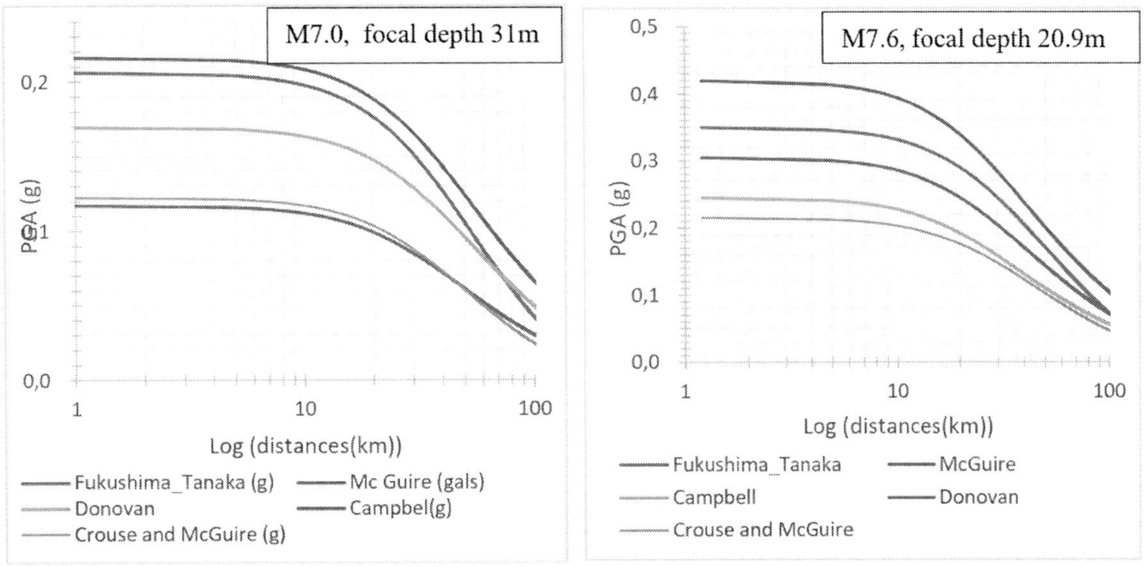


**Figure 2.** The PGA Attenuation Model Curve for Magnitude 7.0 *Ms* and 7.6 *Ms*

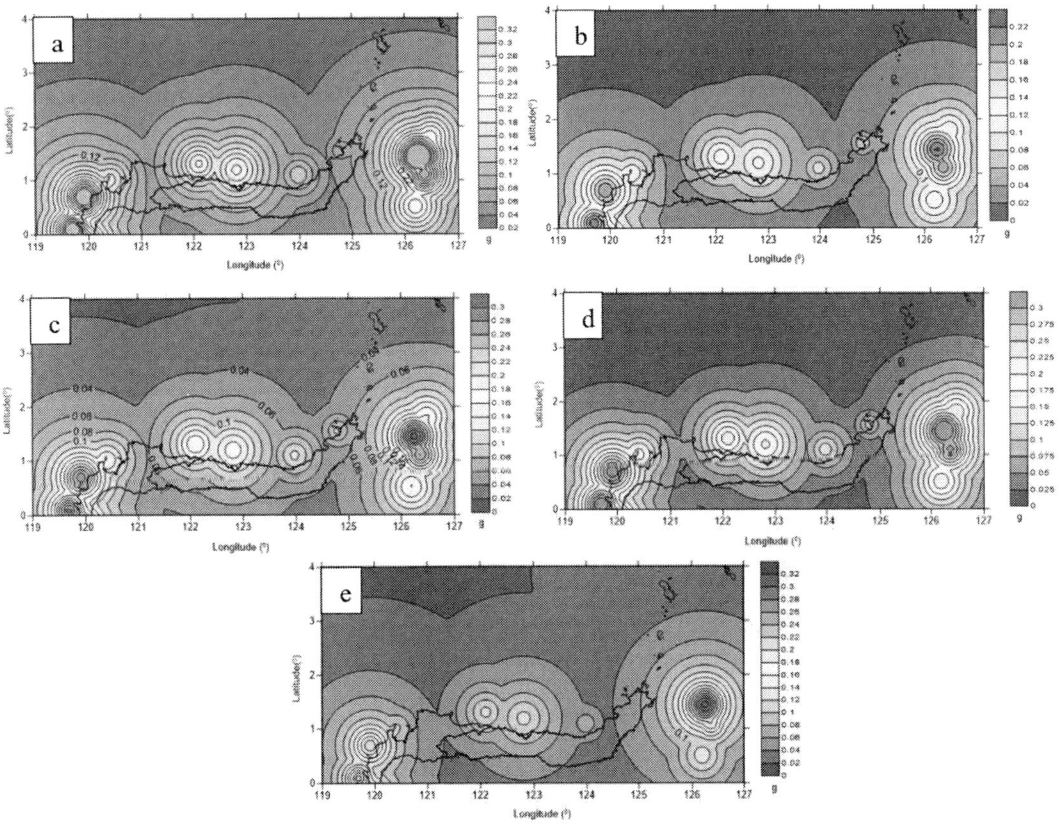


**Figure 3**. Distribution PGA value by using attenuation: (a) Mc.Guire (b) Crouse-McGuire (c) Donovan (d) Fukushima-Tanaka (e) Campbell

## 4. Conclusion

Prediction of peak ground acceleration using the attenuation relationship model has been carried out for the North Arm of Sulawesi. The analysis results show that the peak ground acceleration in this region ranges from 0.0000082 g - 0.429590 g. The greatest peak ground acceleration is obtained from the Mc.Guire attenuation relationship, while the smallest is obtained from the Crouse attenuation relationship. Generally, the greatest ground acceleration results from large magnitude earthquakes originating from subduction zones. The seismic vulnerability of a point depends on the proximity of an earthquake source such as the North Sulawesi Subduction and the Molluca Sea double collision.

## Acknowledgments

We express our sincere thanks to the Institute for Research and Community Service (LPPM) Unsrat for funding this research .

## References

[1] Pasau G, Tamuntuan G H, Tanauma A 2019 Numerical Modeling For Tsunami Wave Propagation (Case Study: Manado Bays) *IOP Conference Series: Material Science and Engineering* **567** 012005
[2] Pradhan B, Moneir A A A and Jena R 2018 Sand dune risk assessment in Sabha region, Libya using Landsat 8, MODIS, and Google Earth Engine images *Geomatics, Natural Hazards and Risk* **9** (1) 1280-1305
[3] Fukushima Y and Tanaka T 1990 A new attenuation relation for peak horizontal acceleration of strong earthquake ground motion in Japan *Bulletin of the seismological Society of America* **80** (4) 757-783
[4] Campbell K W 1997 Empirical near-source attenuation relationships for horizontal and vertical components of peak ground acceleration, peak ground velocity, and pseudo-absolute acceleration response spectra *Seismological research letters* **68** (1) 154-179
[5] Crouse C B, Vyas Y K and Schell B A 1988 Ground motions from subduction-zone earthquakes *Bulletin of the Seismological Society of America* **78** (1) 1-25
[6] Scordilis E M 2006 Empirical global relations converting MS and mb to moment magnitude *Journal of seismology* **10** (2) 225-236
[7] Gardner J K and Knopoff L 1974 Is the sequence of earthquakes in Southern California, with aftershocks removed, Poissonian? *Bulletin of the Seismological Society of America* **64** (5) 1363-1367
[8] Wiemar S 2001 A software package to analyze seismicity ZMAP *Seismological Research Letters* **72** (2) 373-382
[9] Öztürk S 2011 Characteristics of seismic activity in the Western, Central and Eastern parts of the North Anatolian Fault Zone, Turkey: Temporal and spatial analysis *Acta Geophysica* **59** (2) 209-238
[10] Zarifi Z., Nilfouroushan F and Raeesi M 2014 Crustal stress map of Iran: insight from seismic and geodetic computations *Pure and Applied Geophysics* **171** (7) 1219-1236

IOP Conf. Series: Materials Science and Engineering 1115 (2021) 012070 doi:10.1088/1757-899X/1115/1/012070

# Plastic Waste Conversion Reinforced with Rice Husk Ash and Red Stone Powder as Partial Sand Replacement for Paving Blocks Production

**I D K Sastrawidana**

Department of Chemistry, Faculty of Mathematic and Natural Sciences, Universitas Pendidikan Ganesa, Bali, Indonesia.

Corresponding author: ketut.sastrawidana@undiksha.ac.id

**Abstract.** The conversion of plastic waste into economically valuable materials is an innovative way to reduce plastic waste in the environment. At the present study aimed to assess the compressive strength and water absorption capacity of paving blocks from plastic waste combined with rice hausk ash as sand partial replacement and red stone powder as its natural colorant. The speciment were casted in a 6 x 6 x 6 cm mold and pressed with a load of 6 tons using a mechanical pressing machine. The compressive strength and water absorption of paving block at different content of plastic powder and rice husk ash were investigated. The study revealed that paving blocks made by replacing sand with 40% of plastic powder and 5% of rice husk ash pigmented with red stone at the ages of 28 days have compressive strengths of 12.7 MPa with water absorption value of 3.75%. We recommend this paving block was suitable for non-traffic areas such as pedestrians, landscapes, and public walkways.

## 1. Introduction

The use of plastic recently has become a problem for the environment because it is very difficult to degrade naturally and its presence in the environment has gradually increased year by year. It is estimated that it need of 10-15 years to naturally degrade plastics in the environment. Most of the plastic used is single-use plastic which is then disposed of as a solid waste. The existence of plastic waste in the soil has an impact on decreasing soil fertility, while in the sea it has the potential to kill marine animals. Indonesia, which is one of the largest contributors to plastic waste in the sea, has committed to reducing 70% of marine plastic waste by 2025. To realize this commitment, various efforts have been conducted, including prohibiting the plastic supply for supermarket consumers, degradation using the incineration method, and conversion to economic value products.

Plastics waste treatment through conventional incineration has releasing toxic gases into the environment such as Polychlorinated biphenyls, Dioxins, Furan and Mercury [1]. The use of plastic waste to liquid fuel [2], enhance hot mix asphalt performance [3], and wax recovery [4] seems to be a good alternative way to reduce environmental damage, but it requires high costs. Innovation of the conversion of plastic waste into a product or product mixture that is economical but low cost has also been developed, such as for the design and manufacture of furniture [5] and paving blocks [6].

According to Indonesian National Standard (SNI) 03-0691-1996 [7], paving blocks are defined the building materials made from a mixture of water, aggregate and Portland cement or other adhesives with or without additional materials that do not reduce the quality of the paving blocks. Two parameters commonly used to assess the suitability of using paving blocks are compressive strength and water absorption. The application of paving blocks for road, parking area, pedestrians and parks as well as other uses have a minimum average compressive strength of 35.0, 17.0, 12.5 and 8.5 MPa while the maximum average water absorption was 3, 6, 8 and 10%, respectively. Sand is one of the main materials in building and commonly used as fine aggregate in paving blocks production. The sand requirement is increases from year to year, but on the other hand the availability is tending to decrease. Therefore it is necessary to investigate other materials that can be used to sand replacement for building construction, including the paving blocks production. A numerous of recycled materials and by product of agriculture have been reported to have great potential to partially sand replacement in building materials without reducing their quality were sawdust ash [8], bagasse ash, and groundnut shell [9], sheet glass powder, crushed granite fine, and spent fire bricks [10]. In this study, eco-friendly concrete paving block production with the partial substitution of sand using plastic powder combined with rice husk ash as fine aggregate and inorganic pigments from red stone powder as natural inorganic pigment. Rice husk is a very reactive pozzolanic material, and can be used to improve the mechanical properties of paving blocks while red stone powder is used as a pigment to make the paving surface more attractive. Various strength parameters such as compressive strength and water absorption were investigated, and then compared to the strength of paving block according to SNI 03-0691 to assess its application area.

## 2. Materials and methods

### 2.1 Materials

The paving block material comprises river sand (passed in 5 mm of sieve), crushed stone (grain size of 5-10 mm), plastic powder, rice husk ash and red stone powder (passed in 1 mm sieve), Portland cement named as Tiga Roda and water. The plastic used was drinking water bottles, which is a type of polyethylene terephthalate (PET) obtained from plastic waste scavengers, while rice husk ash is collected from the rice milling industry in Bali, Indonesia. Performances of paving block materials are given in Figure 1, while physical properties of paving block materials are shown in Table 1.

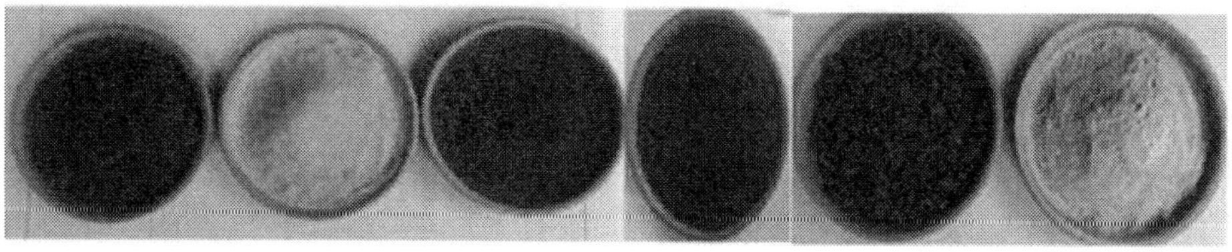

**Figure 1**. Paving block materials

**Table 1**. Physical properties of paving block materials

| Properties | Fine Aggregate | | | Coarse aggregate | Pigment |
|---|---|---|---|---|---|
| | River sand | Plastic powder | Rice husk | Crushed stone | Red stone |
| Specific gravity (kg/m$^3$) | 2.54 | 1.04 | 2.25 | 2.34 | 2.47 |
| Water absorption (wt. %) | 1.25 | 0.16 | 6.25 | 1.05 | 1.26 |

## 2.2 Plastics powder preparation

Plastic powder preparation was carried out by cutting the plastic into small pieces, and then put it in a drum container that already contain a little oil and heated while stirring manually until plastic slurry formed. The plastic slurry was poured into a plate container and allowed to cool at room temperature. Plastic solids are converted to plastic powder using a grinder machine completed with a 1 mm grain size sieve.

## 2.3 Paving block manufacture

Paving block was prepared in two steps. In the first step, paving blocks made with variation of plastic powder to sand ratio at other materials are constant. The mix proportion of paving block materials is shown in Table 2.

**Table 2.** Proportion of paving block materials for one sample tested.

| Mix | Fine aggregate (g) | | Coarse aggregate (g) | Binder (g) | Pigment(g) |
|---|---|---|---|---|---|
| | Sand | Plastic powder | Crushed stone | Portland cement | Red stone powder |
| 1 | 550 | 0 | 180 | 180 | 90 |
| 2 | 495 | 55 | 180 | 180 | 90 |
| 3 | 440 | 110 | 180 | 180 | 90 |
| 4 | 385 | 165 | 180 | 180 | 90 |
| 5 | 330 | 220 | 180 | 180 | 90 |
| 6 | 275 | 275 | 180 | 180 | 90 |
| 7 | 220 | 330 | 180 | 180 | 90 |

The paving block material was placed in a pan, then manually stirred while added amount water of 12% (v / w) from the total material weight used. The paving block material was placed in a pan, then manually stirred while added amount water of 12% (v / w) from the total material weight used. The mixture was casted in paving block mould with dimension of 6 x 6 x 6 cm and then pressed at load of 6 tons using a mechanical pressing machine. Paving blocks are cured for 28 days under atmospheric conditions. In the second step, paving blocks are made following the previous procedure, but sand usage gradually replaced with rice husk ash while the composition of other materials is constant. The material compositions for paving block preparation in the second steps were presented in Table 3.

**Table 3.** Proportion of paving block materials for one specimen.

| Mix. | Fine aggregate (kg) | | | Coarse aggregate (kg) | Binder (kg) | Pigment (kg) |
|---|---|---|---|---|---|---|
| | Sand | Plastic powder | Rice husk ash | Crushed stone | Portland cement | Red stone powder |
| 1 | 550 | 0 | 0.0 | 180 | 180 | 90 |
| 2 | 330 | 220 | 0.0 | 180 | 180 | 90 |
| 3 | 302.5 | 220 | 27.5 | 180 | 180 | 90 |
| 4 | 275 | 220 | 55.0 | 180 | 180 | 90 |
| 5 | 247.5 | 220 | 82.5 | 180 | 180 | 90 |
| 6 | 220 | 220 | 110.0 | 180 | 180 | 90 |
| 7 | 192.5 | 220 | 137.5 | 180 | 180 | 90 |
| 8 | 165 | 220 | 165.0 | 180 | 180 | 90 |

## 2.4 Testing

Paving blocks were assessed its quality after 28 days of cured with parameters test consist of the compressive strength and water absorption. The test result was compared to the compressive strength and water absorption which issued by SNI 03-0691-1996 to assess the usage feasibility. The compressive strength and water absorption values of paving blocks based on their application are listed in Table 4.

**Table 4.** Compressive strength and water absorption of paving blocks as maintained by Indonesian National Standard (SNI) 03-0691

| Class | Compressive strength (MPa) | | Water absorption (%) | Application |
|---|---|---|---|---|
| | Max. | Min. | Max. average | |
| A | 40 | 35 | 3 | Road |
| B | 20 | 17 | 6 | Parking area |
| C | 15 | 12.5 | 8 | Pedestrian |
| D | 10 | 8.5 | 10 | Parks and other users |

### *2.4.1 Compressive strength measurement*

Paving block test sample with dimension of 6 x 6x 6 cm was put in a hydraulic compressive strength machine and pressed slowly with a certain load until the test sample breaks. Compressive strength was calculated using the formula:

$$Compressive\,strength\,(MPa)\,\sigma = \frac{P}{A}$$

where σ = compressive strength (MPa), P = maximum pressing load (N) and A = surface area ($cm^2$). The tet-up for the compressive strength machine can be seen in Figure 2.

**Figure 2** Set-up for the compressive strength machine

*2.4.2 Water absorption measurement*

Water absorption of paving block was carried out according to SNI 03-0691.The paving block test sample was immersed in water for 24 hours, then rinsed and left for about 2 hours while removing the water bound to the surface of the paving block using a cloth and then weighed (Ww).Paving block was dried in an oven at 105°C for 24 hours. Paving blocks was dried in an oven for approximately 24 hours at a temperature of 105oC until their weight at two times of weighing is less than 0.2% from the previous weighing (Wd). Water absorption of sample was calculated using formula:

$$Water\,absorption\,(\%) = \frac{W_w - W_d}{W_d} x\,100\%$$

where, $W_w$ and $W_d$ are the weight of paving block in wet and dry condition.

## 3. Result and Discussion

In this study, paving block were prepared in two versions, namely paving block in cubic formed with a size of 6 x 6 x 6 cm for tested samples and a tri-hexagonal paving block with dimensions of 19.7 x 9.6 x 6 cm. All paving block were presented in Figure 3.

(a) Tested samples (b) Paving block with red stone powder (c) Paving block without red stone powder

**Figure 3.** Paving block

### 3.1 Compressive strength

Compressive strength is defined as the maximum load that causes the paving block to crack and is one of the parameters to assess its durability. Figure 4, show the relationship between the different substitution level of sand with powder plastic and the compressive strength of paving block. It is clear to observe that the compressive strength of a paving block decrease gradually when the amount of plastic was used to replace the sand increased. The decrease of compressive strength with increase rice husk ash addition caused by high porosity and low bulk density of rice husk ash. The compressive strength of paving block without sand addition was 25 MPa, and then decrease to 26.8 MPa, 19.3 MPa, 16.4 MPa, 14.9 MPa, 13.5 MPa, 12.9 MPa and 11.6 MPa when replacement level of sand by plastic powder were 10%, 20%, 30%, 40%, 50% and 60%, respectively. This finding is in line with ref [11]-[12], who reported that the compressive strength of paving block decreased with increasing plastic content.

.

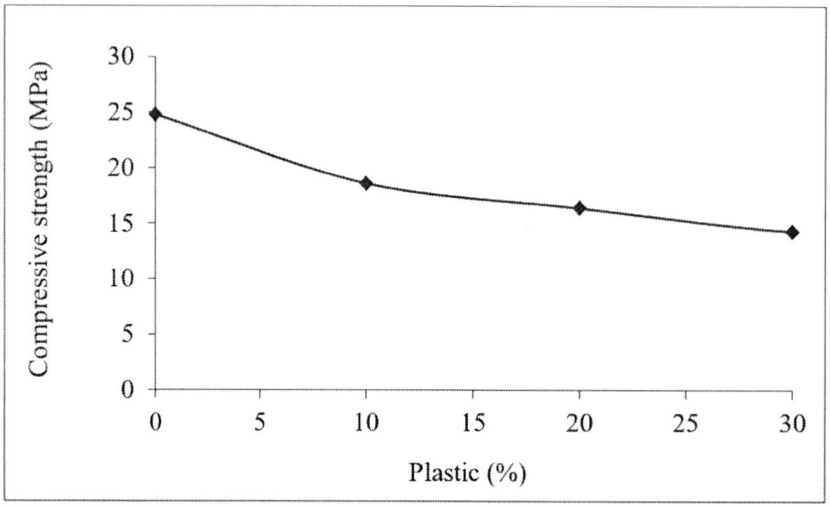


**Figure 4**. Effect of plastic waste substitution on compressive strength of paving block

According to SNI 03-0691: 1996, the compressive strength minimum value of paving block application for road, parking area, pedestrian and parks or other uses in successively were 35 MPa, 17 MPa, 12.5 MPa and 8.5 MPa. This indicates that the replacement of sand by plastic up to 50% results in paving blocks that meet the requirements used for pedestrians. A decrease in compressive strength can also be seen in the addition of rice husk ash as shown in Figure 5. It was observed that the compressive strength of paving block containing plastic powder 40% of total sand used was found to be 13.5 MPa in which decreased to 12.7 MPa, 10.1 MPa, 8.5 MPa, 6.4 MPa, 4.7 MPa, and 2.5 MPa with the replacement of 5%, 10%, 15%, 20%, 25% and 30% sand by rice husk ash. This finding consistent with earlier studies was carried out in ref [13], who reported that paving block without addition of rice husk ash have higher compressive strength than those with rice husk ash.

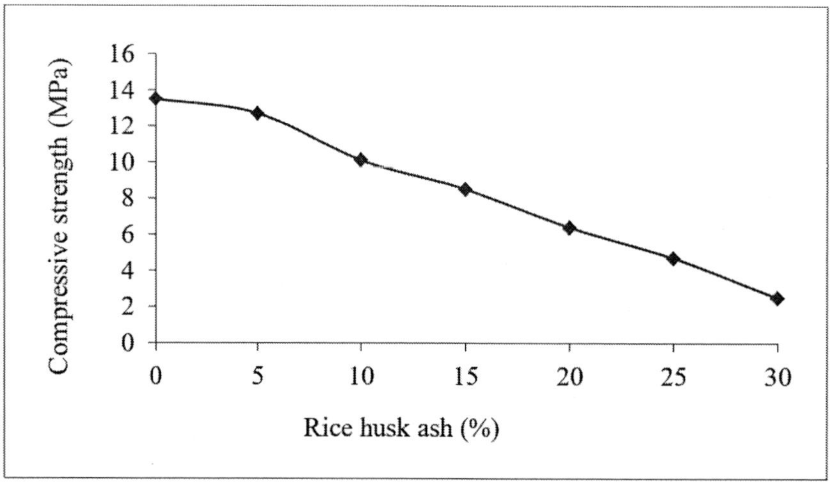


**Figure 5**. Effect of rice husk ash substitution level on compressive strength of paving block

According to SNI 03-0691, the application of paving block for parking area has minimum compressive strength of 17 MPa whereas for pedestrian 12.5 MPa. It was indicate that the replacement of sand up to 5% by rice husk ash at the plastic content of 40% resulted in paving blocks suitable for pedestrians.

### 3.2 Water absorption

Water absorption is defined as the ability of paving block to absorb water when immersed in it. The durability of concrete mainly depends on its permeability. The high permeability causes the penetration of water molecules, sulphate ions, chloride ions and others harmful substances easier into the inner part of paving block and destroys their stability. It was clear observed from Figure 6, that the water absorption value increased with increase the rice husk content within paving block. As presented in Table 1, rice husk ash has the highest water absorption compared to all aggregates used. This is a great contribute to the increase in water absorption of paving blocks. Water absorption of paving block was found to be 3.75%, 5.17%, 7.78%, 8.04%, 10.8%, 15.0% and 17.0% when sand replaced by rice husk ash at portion of 5%, 10%, 15%, 20%, 25% and 30%, respectively. This result in line with on ref [13], who reported that the water absorption properties of concrete increased with increasing percent replacement of sand with rice husk. In another study conducted by Djamaludin et al. (2020) who used tea waste ash, also found that water absorption from paving blocks increased with increasing tea ash content [14].

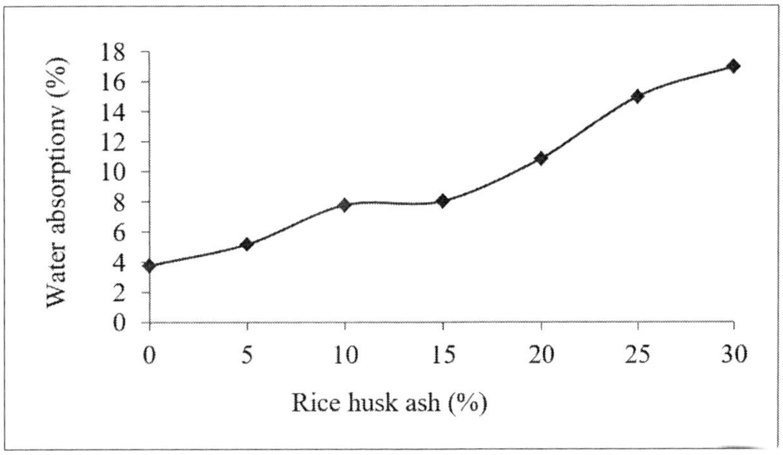


**Figure 6.** Effect of rice husk ash substitution on water absorption of paving block

According to SNI 03-0691, paving block with the maximum average water absorption equal to 8% are classified into C class (suitable for pedestrian or sidewalks application). Based on the compressive strength and water absorption test results, that paving block which was made by sand partial replacement by plastic powder (40%) and rice husk ash (5%) meet the requirements for pedestrian applications.

## 4. Conclusion

Plastic waste reinforced with rice husk ash has great potential to be used as fine aggregate to replace sand partially in paving blocks production. Paving blocks were made by replacing sand with 40% of plastic powder and 5% of rice husk ash at the ages of 28 days have compressive strengths of 12.7 MPa with water absorption value of 3.75%. We recommend this paving block was suitable for non-traffic areas such as pedestrians, landscapes, and public walkways.

## Acknowledgments

The author would like to thank Universitas Pendidikan Ganesha for the financial support provided (contract No. 577/UN48.16/LT/2020). Authors also appreciate the support of Chemistry Department in terms of laboratory facilities services..

## References

[1] Verma R, Vinoda K S, Papyreddy M and Gowda A N S 2016 Toxic pollutants from plastic waste-a review *Procedia Environmental Sciences* **35** 701-708.
[2] Alla M M G, Ahmed A I and Abdalla B K 2014 Conversion of plastic waste to liquid fuel *International Journal of Technical Research and Application* **2** (3) 29-31
[3] Awad A and All adday F 2017 Utilization of waste plastics to enhance the performance of modified hot mix asphalt *International Journal Geomate* **13** (40) 132-139
[4] Al-Salem S A, Yang Y, Wang J and Leeke G A 2020 Pyro-oil and wax recovery from reclaimed plastic waste in a continuous Auger pyrolysis reactor *Energies* **13** (8) 2040
[5] Nishikant K, Nachiket A, Avadhut I and Sangar A 2016 Manufacturing of concrete paving block by using waste glass material *International Journal of Scientific and Research Publications* **6** (6) 61-77
[6] Aegyman S, Obeng-Ahenkora N K, Assiamah S and Twumasi G 2019 Exploiting recycled plastic waste as an alternative binder for paving blocks production *Case Studies in Construction Materials* **11** e00246
[7] SNI 03-0691-1996: Indonesian National Standard Method for Paving Block Test
[8] Abdullahi A, Abubakar M and Afolayan A 2013 Partial replacement of sand with sawdust in concrete production *Proceedings of 3rd Biennial Engineering Conference, School of Engineering and Engineering Technology, Federal University Technology Minna, Niger State, Nigeria* 333-338
[9] Bahoria B V, Parbat D K and Naganaik P B 2013 Comprehensive literature review on use of waste product in concrete *International journal of application or innovation in Engineering & management* **2** (4) 387-394
[10] Keerthinarayana S and Srinivasan R 2010 Study on strength and durability of concrete by partial replacement of fine aggregate using crushed spent fire bricks *Buletinul Institutului Politehnic din Iasi Sectia Constructii Arhitectura* **56** (2) 51-59
[11] Karthikeyan M, Balamurali K, Kumar V B, Prabakar S M and Janarthanan R 2019 Utilization of waste plastic in concrete. *International Research Journal of Engineeringand Technol* **6** (4) 1400-1405
[12] Padhi R S and Mukharjee B B 2017 Influence of Rice Husk Ash and Recycled Coarse Aggregates on Properties of Concrete *International Journal of Engineering Technology Science and Research* **4** (6) 528-533
[13] Akinwumi I I, Awoyera P O, Olofinnade O M, Busari A A and Okotie M 2016 Rice husk as a concrete constituent: Workability, water absorption and strength of the concrete *Asian Journal of Civil Engineering* **17** (7) 887-898
[14] Djamaluddin A R, Caronge M A, Tjaronge M W, Lando A T and Irmawaty R 2020 Evaluation of sustainable concrete paving blocks incorporating processed waste tea ash *Case Studies in Construction Materials* **12** e00325

# Effect of Storage and Ageing Time on Sensorial Quality and Composition of Wine Phenolic Compounds Improved with Saccharomyces Cerevisiae Local Hybrid

**I N Tika*[1], I G A T Agustiana[2] and L P D Puspaningrat[3]**

[1]Department of Chemistry, Faculty of Mathematics and Natural Sciences, Universitas Pendidikan Ganesha-Indonesia
[2]Department of Elementery School education, Faculty of Education, Universitas Pendidikan Ganesha- Indonesia
[3]Departement of Farmacy, Institute of Health Science Buleleng Bali- Indonesia

*Corresponding author: nyoman.tika@undiksha.ac.id

**Abstract.** High quality wine can be produced by the perfect ageing process.The purpose of this study was to determine the effect of ageing process container on sensory quality and composition of phenolic compounds of wine fermented with local hybrid Sacharomyces cereviciae. The research design used a randomized block design with two factorials. The first factor is the materials for storing during the ageing process, namely plastic bottles, glass bottles and pottery barrels (from Banyuning). The second factor is ageing time, namely 1, 2, 3, 4 and 5 months. The experiment was carried out two times over so that 30 experimental units were obtained. The variables measured were phenolic compounds and sensory quality (aroma and taste, color, and turbidity level). The composition of the phenolic compounds was determined by gas chromatography-mass spectrometer (GC-MS) and sensory quality test using 15 trained fanelis. The results showed that the composition of the highest phenolic compound content was shown successively with containers: plastic bottles, glass and earthenware and the ageing time was 4 months. The content of phenolic compounds is Isoamyl alcohol (ISM), Isobutanol (ISB), n-propanol (PR), Acetaldehyde (ADE), 2,3 –Butanediol (BUT), Acetone (ACE), while the best sensory quality is using pottery. because it has better aroma and taste, lower turbidity level. Based on data analysis, it can be concluded that the length of time for the ageing process and the container during the ageing process has a significant effect on the quality and content of phenolic compounds in grape wine fermented with local hybrid Sacharomyces cereviciae.

## 1. Introduction

Wine is a fermented drink. The process of aging wine is an important process to produce high-quality wine [1]. Traditionally, the process of aging wine uses oak barrels. At least 150 types of oak are used to improve the quality of wine [2]. Various attempts were made to produce high quality wines in the aging process, namely using oak wood fragments [3] and microoxygenation applications [4]. Treatment through increasing levels of lees (yeast waste ) which is made in high concentrations [5] or the application of physical methods by replacing non-wooden containers [6]. Various modifications are still problematic, namely a long time being inefficient, unhygienic, yeast activity is not disturbed, causing microbial damage to grow, so that the quality of wine decreases, and the esterification process is disturbed [7]. Another breakthrough, using materials that have an orthodoxy high, allows for rapid and homogeneous esterification, and resistance to other microbial contaminants [8]. This condition can use hybrid yeast types that have stable properties, namely nutrient deficiencies and high alcohol content between yeast Saccharomyces cerevisiae and microbes [9].

This strategy has continued to be developed in the last decade, but the results have not been satisfactory [10].Yeast hybrid technology can be developed to produce wine, namely local hybrids of Saccharomyces cerevisiae [11] and has been applied to a structured fermentation method that can produce zero percent alcoholic wines [12]. Local hybrid Saccharomyces cerevisiae yeast in general medicine can produce wine with levels of 13% (unpublished data). Therefore, in this study it is necessary to know in research the ability of yeast to produce phenolic compounds and sensory qualities in the aging phase using various materials such as plastic bottles, glass bottles and pottery from Banyuning Buleleng Bali.

## 2. Materials and methods

The main material used in this study. Grapes from a people's garden in Banjar village, Buleleng, Bali. The grapes are black and ripe on the trees. The chemicals used are all in the proanalytic grade, 70% alcohol, glucose, citric acid, NaOH, and sodium metabisulfite, Nelson A's reagen, Nelson B's reagen, Arsenomolybdate solution, Ethanol, Methanol, and Penolphtalein. The main tools are a spectrophotometer, digital scale (Mettler Toledo AB 204), autoclave, laminar flow, centrifuge, refrigerator, pH meter (Istek), injector, stove (Rinai), and a distiller and a set of glassware.

### 2.1. Experimental design

This research was conducted with a factorial randomized block design. The first factor is the storage materials for plastic bottles, glass bottles and local ground barrels (Banyuning). The second factor was storage time from 1, 2, 3, 4 and 5 weeks. The experiment was carried out in two repetitions so that 20 experimental units were obtained. The variables measured were phenolic compounds and sensory quality (aroma and taste, color, and turbidity level). The data obtained were analyzed for their diversity and continued with the Duncan test if the treatment had a significant effect ($P < 0.05$).

### 2.2. Observed variables

The variables observed in this study were reducing sugar levels using the Nelson Somogyi method [13], ethanol content using the gas chromatography method [14], phenolic compounds using GC MS, and methanol content [14], total acid using the titration method, and the degree of acidity using a pH meter and a session test [15] by 15 trained panelists from UKM Amertha Nadi.

### 2.3. Experimental procedure

The minimum medium (MM) consists of: 1.34% basic yeast nitrogen without ammonium sulfate and amino acids (YNB, Difco, Detroit, USA) and 4% glucose supplied by Merck Farma Quimica (Barcelona, Spain). Wine Grapes should be treated with 0.4 gl 1 bentonite for 2 days at 48C to enhance sedimentation prior to fermentation, and be inoculated with 5% a local hybrid Sacharomyces cereviciae yeast preculture. Bottles with 250 ml MM of medium or wine should be covered with two layers of sterile gauze to avoid contamination, and fermentation carried out in duplicate at 18 ° C for 20 hours

### 2.4. Sequential fermentation

Mixed fermentation carried out with a sterile 250 ml aliquot should be maintained at 18C using 106 ufc / ml of P. fermentans for 2, 3, 4, 6 or 8 days. After each of these time intervals, 106 ufc / ml of S. cerevisiae was added to form the yeast mixture, which was then maintained for a total of 20 days in all cases. All of this sequential fermentation was made in duplicate. The resulting wines were centrifuged and stored at 20,8C, for a maximum of one month, for co-analysis.

### 2.5. Analysis using GC-MS

The volatile and alcohol compounds were analyzed, in triplicate, by direct injection of 1 microliter each of the final product into a gas chromatograph of Variant 3900 with a flame ionization detector (220 C) and CPWAX Column 57CB (50 0.25 mm). Both the injector and detector are operated at 220°C. The carrier gas $N_2$ (99.999%) at a flow rate of 36 ml / min. The injection is divided into 30 modes with an air flow of 300 ml / min. The column temperature was programmed from 40 ° C for 2 minutes to 120 ° C at an increment rate of 3 minutes and 4 minutes at 120 ° C. The samples were analyzed in triplicate. The identification and quantification of compounds is carried out by comparing the retention times and concentrations with standard solutions used in the alcohol industry.

## 3. Result and Discussion

The aging process was carried out for 1, 2, 3, 4, 5 months, carried out on plastic bottles, glass bottles, and pottery from Banyuning. The best storage time for the aging process is 4 months, with a score of 4.75. The lowest sensory quality was found in pottery within 2 months of treatment (see Figure 1).

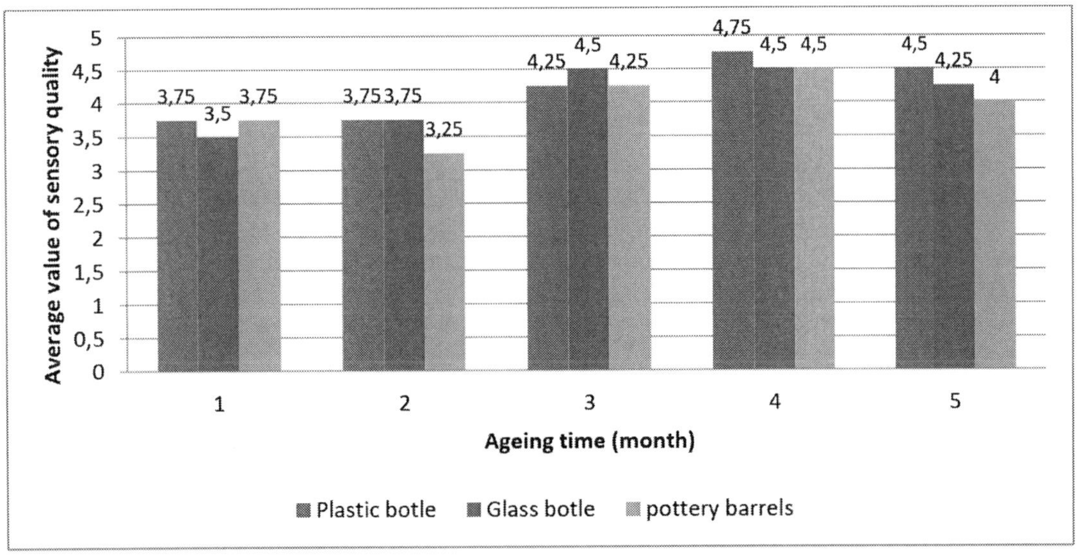


**Figure 1.** The sensory quality of the length of time in the agieng process.

Storage in plastic bottles, during the aging process, shows that the sediment that occurs during the aging process does not stick to the bottle walls. The same thing happens with glass bottles, making the wine appear clear. The tightly closed plastic bottle allows for the formation of aroma due to the better esterification reaction of glass and pottery containers, while the storage of good quality pottery from Banyuning village takes 1 month, then decreases in the 2nd month then increases again at 3 months storage and remains stagnant in the fourth month. Thus, the use of pottery is still relevant for short storage periods.

The content of wine polyphenols after the fermentation process in the aging phase is used to see the continued activity that occurs during the aging process. Some polyphenol compounds in storage containers such as plastic bottles, glass bottles and barrels or pottery from Banyuning are shown in Table 1. The phenolic compounds are Isoamil alcohol (ISM), Isobutanol (ISB), n-propanol (PR), acetaldehyde (ADE), 2,3 - Butanediol (TAPI), Acetone (ACE). The highest content is shown by plastic bottle containers.Storage in plastic bottles, in the aging process the sediment does not stick to the wall, and allows the wine to appear clear, and closes tightly, so the process of forming aroma for esterification is better than glass and earthenware containers.

**Table 1.** Polyphenol compounds in Sacharomyces cerevicae fermentation

| Polyphenol compounds | The content of polyphenol compounds (mg / L) in the aging process container | | |
|---|---|---|---|
| | Plastic botle | Glass botle | Pottrey barell from Banyuning Village |
| Isoamil alcohol (ISM) | 258 | 231 | 218 |
| Isobutanol (ISB) | 78 | 67 | 58 |
| n-propanol (PR) | 32 | 31 | 42 |
| Ethyl acetat (ETAC) | 54 | 65 | 43 |
| Acetaldehide (ADE) | 46 | 56 | 53 |
| 2,3 –Butanediol (BUT) | 630 | 562 | 541 |
| Aceton (ACE) | 25 | 21 | 35 |

This study also shows that wines are made from grapes from community gardens in Buleleng Bali, indicating that the aging process affects the quality of taste, aroma and color, as well as the level of clarity of the wine. This justifies modern alcoholic beverages undergoing an aging process in which compounds extracted from the container are used, such as wood, which can affect the chemical composition of the wine. The porous container contributes to the overall character of the wine [16]. Therefore, in making wine from Buleleng Bali grapes, choosing a storage place during the aging process is very important, because it can accelerate the chemical changes in wine [17].

Furthermore, aging for 4 months shows the highest sensory quality, this shows that the curing time affects sensory properties (taste, color, taste and clarity) and physicochemical (alcohol content, color and turbidity level), thus giving a change in the aroma and taste of wine [18]. The increased chemical content of wines preserved in plastic bottles produces a smoky aroma like toast, while wines preserved in glass bottles show the aroma of vanilla and cocoa, while wines using Banyuning gerbah containers are like toast and appear clearer.

## 4. Conclusion

The aging process in wines made from Bali Buleleng grapes fermented with local hybrid yeast Saccharomyces cerevisiae, which uses different places, actually affects the sensory qualities and the wine's polyphenolic compounds. Storage containers with plastic bottles work best for a storage period of 4 months. However, storage with pottery from Banyuning village is suitable for short term storage. The results of this study recommend the possibility of other ingredients for storage, so as to improve the wine quality of the Buleleng Bali grapes.

## Acknowledgments

Thanks are conveyed to ReCtor, Chair of LPPM Undiksha for the applied research grant given in 2020, with Number: SP DIPA - 023.17.2.677530/2020 revision 1 dated March 16, 2020 In accordance with Research Contract Number: 855 / UN48.16 / LT / 2020

## References

[1] Ugliano M 2013 Oxygen contribution to wine aroma evolution during bottle aging *Journal of agricultural and food chemistry* **61** (26) 6125-6136
[2] Dumitriu G D, Teodosiu C, Gabur I, Cotea V V, Peinado R A and de Lerma N L 2019 Evaluation of Aroma Compounds in the Process of Wine Ageing with Oak Chips *Foods* **8** (12) 662
[3] Tao Y, García J F and Sun D W 2014 Advances in wine aging technologies for enhancing wine quality and accelerating wine aging process *Critical reviews in food science and nutrition* **54** (6) 817-835
[4] Schmidtke L M, Clark A C and Scollary G R 2011 Micro-oxygenation of red wine: techniques, applications, and outcomes *Critical reviews in food science and nutrition* **51** (2) 115-131
[5] Loira I, Vejarano R, Morata A, Ricardo-da-Silva J M, Laureano O, González M C and Suárez-Lepe J A 2013 Effect of Saccharomyces strains on the quality of red wines aged on lees *Food Chemistry* **139** (1-4) 1044-1051
[6] Sanchez-Iglesias M, González-Sanjosé M L, Perez-Magarino S, Ortega-Heras M and Gonzalez-Huerta C 2009 Effect of micro-oxygenation and wood type on the phenolic composition and color of an aged red wine *Journal of agricultural and food chemistry* **57** (24) 11498-11509
[7] Juega M, Carrascosa A V and Martinez-Rodriguez A J 2015 Effect of short ageing on lees on the mannoprotein content, aromatic profile, and sensorial character of white wines *Journal of food science* **80** (2) M384-M388
[8] Martínez J M, Delso C, Maza M A, Álvarez I and Raso J 2019 Pulsed electric fields accelerate release of mannoproteins from Saccharomyces cerevisiae during aging on the lees of Chardonnay wine *Food Research International* **116** 795-801
[9] Moreno-Arribas V, Pueyo E and Polo M C 1996 Peptides in musts and wines. Changes during the manufacture of cavas (sparkling wines) *Journal of Agricultural and Food Chemistry* **44** (12) 3783-3788
[10] Contreras A, Hidalgo C, Schmidt S, Henschke P A, Curtin C and Varela C 2015 The application of non-Saccharomyces yeast in fermentations with limited aeration as a strategy for the production of wine with reduced alcohol content *International journal of food microbiology* **205** 7-15
[11] Tika I N 2017 *Widya Laksana* Singaraja: Undiksha Press
[12] Goold H D, Kroukamp H, Williams T C, Paulsen I T, Varela C and Pretorius I S 2017 Yeast's balancing act between ethanol and glycerol production in low-alcohol wines *Microbial biotechnology* **10** (2) 264-278
[13] Shao Y and Lin A H M 2018 Improvement in the quantification of reducing sugars by miniaturizing the Somogyi-Nelson assay using a microtiter plate *Food chemistry* **240** 898-903
[14] Cardenia V, Toschi T G, Scappini S, Rubino R C and Rodriguez-Estrada M T 2018 Development and validation of a Fast gas chromatography/mass spectrometry method for the determination of cannabinoids in Cannabis sativa L *journal of food and drug analysis* **26** (4) 1283-1292
[15] Volmer D A, Curbani L., Parker T A, Garcia J, Schultz L D and Borges E M 2017 Determination of titratable acidity in wine using potentiometric, conductometric, and photometric methods *Journal of Chemical Education* **94**( 9) 1296-1302

[16] Teodosiu C, Gabur I, Cotea V V, Peinado R A and López de Lerma N 2019 Evaluation of aroma compounds in the process of wine ageing with oak chips *Foods* **8** (12) 662
[17] Coppola F, Picariello L, Forino M, Moio L and Gambuti A 2021 Comparison of Three Accelerated Oxidation Tests Applied to Red Wines with Different Chemical Composition *Molecules* **26** (4) 815
[18] Liu Q, Weng P and Wu Z 2020 Quality and aroma characteristics of honey peach wines as influenced by different maturity *International Journal of Food Properties* **23** (1) 445-458

IOP Conf. Series: Materials Science and Engineering 1115 (2021) 012072 doi:10.1088/1757-899X/1115/1/012072

# Study on Generalized Prefer-Opposite (GPO) Algorithm for Constructing the De Bruijn Binary Sequence

**A N Setiawan*, M Astuti and Z M Mayasari**

Department of Mathematics, Faculty of Mathematics and Natural Science, University of Bengkulu, Indonesia.

*Corresponding author: aisyahnsetiawan@gmail.com

**Abstact** De Bruijn binary sequence $S_n$ of order $n$ is a string of bits $s_i \in \{0,1\} = \{s_1, \dots, s_{2^n}\}$, so that each string with length $n$, $\{a_1, \dots, a_n\} \in \{0,1\}^n$ appears exactly once. De Bruijn binary sequence can be built using the Generalized Prefer-Opposite (GPO) algorithm. The GPO algorithm is based on the Feedback Shift Register (FSR), which is a clock-regulated circuit with $n$ sequential storage units. In running the algorithm, we need input consisting of the feedback function $f(x_0, x_1, \dots, x_{n-1})$ and initial state $\boldsymbol{b} = b_0, b_1, \dots, b_{n-1}$. There are many combinations of feedback functions and initial states that can be used, then grouped into three families namely family one, family two and family three. The purpose of this research is studying the GPO algorithm which use the feedback function from family two $f(x_0, .., x_{n-1}) = 1 + \prod_{i=t}^{n-1} x_i$, $0 < t < n$, $t = 2$, $n = 4$ and initial state $\boldsymbol{b} = 1100$ in constructing the de Bruijn binary sequence. The result of this study indicate that the GPO algorithm with the feedback function and the initial state from the family two construct a de Bruijn binary sequence.

## 1. Introduction

In an era full of technological sophistication, many people are given the convenience of carrying out activities. One of them is the ease of obtaining information. This convenience, of course, requires security over the assets that are input and accessed, especially those concerning the secrets of a country, an institution or something that is private. This is the starting point for the development of a data security system.

Cryptography is the study of protecting information by encrypting the inputted data into a complex code, the code can be converted back into the original data through the description process by the recipient [1]-[2]. In the field of cryptography there is a sequence, called the de Bruijn binary sequence, which can be used to produce information that has a security quantity. This sequence combines the alphabet zero and one.

De Bruijn binary sequence $S_n$ of order $n$ is a string of bits $s_i \in \{0,1\} = \{s_1, \dots, s_{2^n}\}$, so that each string with length $n$, $\{a_1, \dots, a_n\} \in \{0,1\}^n$ appears exactly once. The de Bruijn binary sequence was introduced by De Bruijn in 1946 where each element used a binary alphabet of zeros and ones. Based on the article entitled "A Combinatorial Problem", De Bruijn stated that there are $2^{2^{n-1}-n}$ kinds of De Bruijn binary Sequences of order n. De Bruijn binary sequences can be constructed using the Generalized Prefer-Opposite (GPO) algorithm [3]-[4].

There are many combinations of feedback functions and initial states that can be formed which are grouped into three families, namely family one, family two and family three. However, it is difficult to confirm which functions are actually valid for the GPO algorithm. Therefore, this paper will discuss about the GPO algorithm using a feedback function and the initial state from the family two $f(x_0,..,x_{n-1}) = 1 + \prod_{i=t}^{n-1} x_i$, with $0 < t < n,$ ; $t = 2$ and $n = 4$ and *initial state* $\boldsymbol{b} = 1100$ in constructing the De Bruijn binary sequence. The purpose of paper is to study the feedback function and initial state of family two in building de Bruijn binary sequences.

## 2. Literature Review

### 2.1. Graph

A graph $G\ (V, E)$ is a discrete structure consisting of vertices and edges. According to Bondy and Murty in 1976, A graph $G$ is an ordered triple $(\ V(G)\ ,\ E(G),\ \boldsymbol{\psi_G})$ consisting of a nonempty set $V(G)$ of vertices, a set $E(G)$, disjoint from $V(G)$, of edges, and an incidence function $\boldsymbol{\psi_G}$ that associates with each edge of $G$ an unordered pair of (not necessarily distinct) vertices of $G$ [5]. If $e$ is an edge and u and v are vertices such that $\boldsymbol{\psi_G}\ (e) = uv$, then $e$ is said to join $u$ and $v$; the vertices $u$ and $v$ are called the end of $e$.

### 2.2. Tree

Tree is a connected graph that does not contain circuits. In a tree, there is always a path that connects the two nodes in the tree.

### 2.3. De Bruijn Binary Sequence

De Bruijn binary sequence $S_n$ of order $n$ is a string of bits $s_i \in \{0,1\} = \{s_1, \dots, s_{2^n}\}$, so that each string with length $n$, $\{a_1, \dots, a_n\} \in \{0,1\}^n$ appears exactly once. De Bruijn (1946) stated that there are $2^{2^{n-1}-n}$ kinds of De Bruijn binary Sequences of order n [6].

### 2.4. Feedback Shift Register (FSR)

The feedback shift registers are most useful in designing and generating such *pseudorandom* or *pseudonoise* (PN) sequences [7]-[8]. This is due to their simplicity of defining rules and their capability of generating sequences with much longer period. An $n$-stage shift register is a clock-regulated circuit with $n$ consecutive storage units. Each of the units holds a bit and, as the clock pulses, the bit is shifted to the next stage in line. The output is a new bit $\boldsymbol{s_n}$ based on the $n$ bits $\boldsymbol{s_0} = s_0, \dots, s_{n-1}$ called the initial state. The corresponding feedback function $f(x_0, \dots, x_{n-1})$ is the Boolean function that, on input $s_0$, outputs $s_n$.

### 2.5. State Graph

The state graph of the FSR with feedback function $f(x_0, x_1, \dots, x_{n-1})$ is a directed graph $G_f$ whose vertices are all of the $n$-stage states. There is a directed edge from a state $\boldsymbol{u} := u_0, u_1, \dots, u_{n-1}$ to a state $\boldsymbol{v} := u_1, u_2, \dots, u_{n-1}\ f(u_0, u_1, \dots, u_{n-1})$. We call $u$ a child of $v$ and $v$ the parent of $u$. We allow $u\ =\ v$, in which case $G_f$ contains a loop. A leaf in $G_f$ is a vertex with no child, i.e., a leaf has outdegree 1 and indegree 0 [9]-[10].

### 2.6. Generalized Prefer-Opposite (GPO) Algorithm

The following theorem and lemma will be used in the construction of the De Bruijn sequence :

**Lemma 1**
Given the state graph $G_f$ of the FSR with feedback function $f$, let the state $\boldsymbol{s}$ be a vertex with two children. Let the GPO Algorithm starts with an initial state $\boldsymbol{b} \neq \boldsymbol{s}$. By the time the algorithm visits $\boldsymbol{s}$ it must have visited both children of $\boldsymbol{s}$ [3].

**Theorem 2**
The GPO Algorithm, on input a function $f(x_0, x_1, \dots, x_{n-1})$ and an initial state $\boldsymbol{b} = b_0, b_1, \dots, b_{n-1}$, generates De Bruijn binary sequence of order $n$ if the state graph $G_f$ of the FSR satisfies the following two conditions [3]:
1) all of the states, except for the leaves, have exactly two children;
2) there is a unique directed path from any state $v$ to $b$.

**Table 1.** Generalized Prefer-Opposite Algorithm [3]

**Algorithm 1.** *Generalized Prefer-Opposite*

***Input*** : *Feedback function* $f(x_0, x_1, \dots, x_{n-1})$ dan *initial state* $\boldsymbol{b} = b_0, b_1, \dots, b_{n-1}$

***Output*** : Barisan biner

$\boldsymbol{c} = c_0, c_1, \dots, c_{n-1} \leftarrow \boldsymbol{b}$
**do**
*Print* $(c_0)$
$y \leftarrow f(c_0, c_1, \dots, c_{n-1})$
**if** $c_1, c_2, \dots, c_{n-1}, \tilde{y}$ belum muncul **then**
$\boldsymbol{c} \rightarrow c_1, c_2, \dots, c_{n-1}, \tilde{y}$
**else**
$\boldsymbol{c} \rightarrow c_1, c_2, \dots, c_{n-1}, y$
**while** $\boldsymbol{c} \neq \boldsymbol{b}$

### 2.7. Family Two

To contruct De Bruijn's sequence using Algorithm 1, it is enough to find a feedback function and initial state that satisfies Theorem 2. There are many combinations of feedback functions and initial state which are grouped into three families, one of which will be discussed, namely the family two that have been obtained [3].

**Teorema 3**

Given n and an integer $0 < t < n$, let the feedback function be [3]

$$f(x_0 x_1, \dots, x_{n-1}) := 1 + \prod_{i=t}^{n-1} x_i$$

and the initial state $b$ be any $n$-stage state of the sequence $(0, 1^{n-t})$. Then the GPO Algorithm generates the $f_2(n)$ family of de Bruijn sequences of order $n$.

## 3. Result and Discussion

In this paper, De Bruijn binary sequences with order 4. Based on the definition, the de Bruijn sequence $S_4$ is a string of bits $s_i \in \{0,1\} = \{s_1, .., s_{2^4}\}$, $i = 1,2, .. 2^4$, such that each string with $n = 4$, $\{a_1, \dots, a_4\} \in \{0,1\}^4$, occurs consecutively once. By using the Generalized Prefer-Opposite algorithm, the output of the De Bruijn binary sequence order 4 with a feedback function of the family two $f(x_0, .., x_3) = 1 + x_2 \cdot x_3$ and the initial state $b = 1100$ are as follows:

$$\boldsymbol{S_4} := (1100\ 0010\ 0111\ 1010)$$

The following sequence of states

b=1100→1000→0000→0001→0010→0100→1001→0011→0111→1111→1110→1101→1010→0101 →1011→0110→1100.

**Proof**:
The $x_0$ coefficient in $f$ is 0, for any non-leaf state $G_{f_2(4)}$ in figure 1 has exactly two children. The graph state $G_{f_2(4)}$ contains a directed cycle, namely $g_{f_2(4)}$ in figure 2 where each vertex is the $n$-stage state of $\boldsymbol{S_4}$. All states outside the cycle are vertices in the tree where each root is the state that is above the cycle. Because there is a unique path leading from any state $u$ to state $v$ in the cycle, as a result there is a unique path leading to any state in $G_{f_2(4)}$ to the initial state $b$. Both conditions in Theorem 2 are met. ■

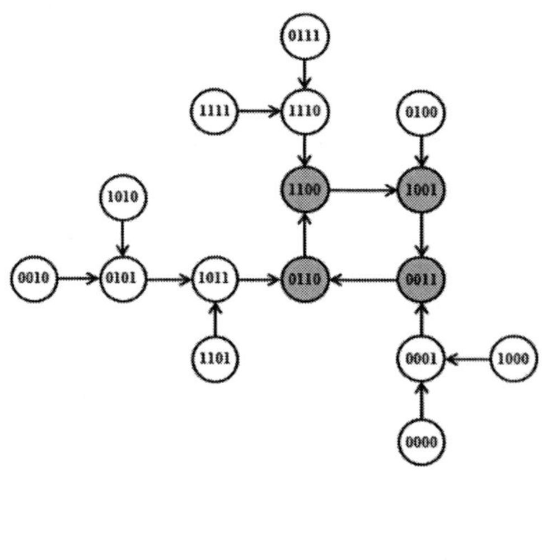

**Figure 1.** State Graph $G_{f_2(4)}$

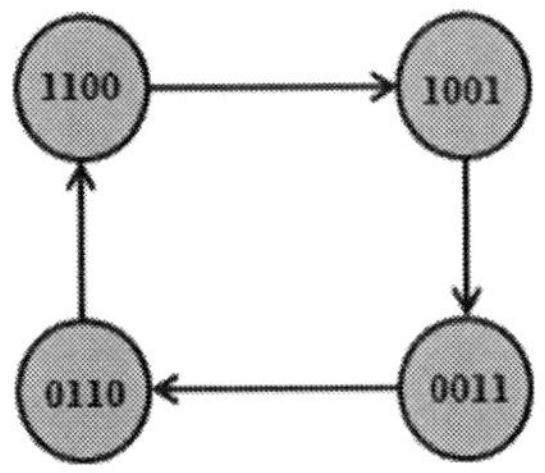


**Figure 2.** Directed *Cycle* $g_{f_2(4)}$

Furthermore, it is verified that the sequence obtained satisfies Lemma 1. For each non-leaf state $u$ has two children $v$ and $w$ in $G_{f_2(4)}$, it will be proven that the children of $u$, namely state $v$ and $w$, are first passed by the GPO algorithm. Suppose that any non-leaf state $u = 1110$ is chosen with the two children, $v = 0111$ and $w = 1111$. The GPO algorithm is run from the initial state $b \neq u$ onwards until it arrives at state $u$. It can be seen in sequence of states that state $v$ and $w$ are passed before state $u$. So, Lemma 1 is fulfilled.

## 4. Conclusion

Based on the results of the discussion of this journal, it can be concluded that the Generalized Prefer-Opposite (GPO) Algorithm with input feedback function and initial state of family two for $(x_0,..,x_{n-1}) = 1 + \prod_{i=t}^{n-1} x_i$, $0 < t < n, t = 2$ and $n = 4$ construct a de Bruijn binary sequence, namely $\boldsymbol{S_4} := (1100\ 0010\ 0111\ 1010)$. Each string with $n = 4$, $\{a_1, \ldots, a_4\} \in \{0,1\}^4$ in $\boldsymbol{S_4}$ appears exactly one time.

## Acknowledgments

We would like to say thanks to the Mathematics Department, Faculty of Mathematics and Natural Science, University of Bengkulu for the support of this paper.

## References

[1] Sambas A, Vaidyanathan S, Tlelo-Cuautle E, Abd-El-Atty B, Abd El-Latif A A, Guillén-Fernández, O, Hidayat Y, Gundara G 2020 A 3-D multi-stable system with a peanut-shaped equilibrium curve: Circuit design, FPGA realization, and an application to image encryption *IEEE Access* **8** 137116-137132

[2] Vaidyanathan S, Azar A T, Rajagopal K, Sambas A, Kacar S and Cavusoglu U 2018 A new hyperchaotic temperature fluctuations model, its circuit simulation, FPGA implementation and an application to image encryption *International Journal of Simulation and Process Modelling* **13** (3) 281-296

[3] Chang Z, Ezerman M F and Fahreza A A 2020 On greedy algorithms for binary de Bruijn sequences *Applicable Algebra in Engineering, Communication and Computing* 1-28

[4] Kang Y, Li J, Yuan Q and Sun S 2020 Adjacent vertex sum reducible edge coloring algorithm based on graph cryptography *2020 IEEE 9th Joint International Information Technology and*

*Artificial Intelligence Conference* **9** 1-4
[5] Bondy J A and Murty U S R 1976 *Graph theory with applications* London: Macmillan
[6] De Bruijn N G 1946 A combinatorial problem *Koninklijke Nederlandse Academie van Wetenschappen* **49** 758-764
[7] Woo K S, Wang Y, Kim Y, Kim J, Kim W and Hwang C S 2020 Hardware-Based Security: A Combination of a Volatile-Memristor-Based True Random-Number Generator and a Nonlinear-Feedback Shift Register for High-Speed Encryption *Advanced Electronic Materials* **6** (5) 2070022
[8] Zhong J and Lin D 2018 On minimum period of nonlinear feedback shift registers in grain-like structure *IEEE Transactions on Information Theory* **64** (9) 6429-6442
[9] Bonifati A, Holubová I, Prat-Pérez A and Sakr S 2020 Graph Generators: State of the art and open challenges *ACM Computing Surveys* **53** (2) 1-30
[10] Bi S and Zhang Y J A 2017 Graph-based cyber security analysis of state estimation in smart power grid *IEEE Communications Magazine* **55** (4) 176-183

# Utilization of Virgin Coconut Oil that has been Extracted in Phenolic Compounds as Resource of Diethanolamide Surfactants

**J H Mandei[*], Y F Assah, M Edam and N P Kumolontang**

Balai Riset dan Standardisasi Industri Manado, Manado, Sulawesi Utara, Indonesia

*Corresponding author: nenimandei@yahoo.com

**Abstact** One of the vegetable oils that can be used as a surfactant is extracts of phenolic compounds from virgin coconut oil (VCO) or VCO residue.The purpose of this study was to make surfactants from VCO residue and to obtain the correct ratio of reactants and the appropriate concentration of NaOH catalyst to produce diethanolamide surfactant.The materials used are methyl ester from VCO residue and dietanolamine with NaOH catalyst. The amidation process uses a mole ratio of dietanolamine and methyl ester (1:1, 1.25: 1, 1.5: 1, 1.75: 1 and 2: 1) at a temperature of 160 ° C for 3 hours, and a NaOH catalyst with a concentration of 0.25, 0.5, and 0.75%.The results showed that the treatment of the ratio of methyl ester and diethanolamine and the concentration of NaOH catalyst produced DEA surfactant with the same consistency, namely a semi-solid viscous liquid with a clear yellow color.The higher the concentration of catalyst, the yield tends to increase.. All DEA surfactant products have a pH of 10, with viscosities ranging from 252.47 - 509.89 cP, soaping numbers ranging from 35.22 - 52.95 mg KOH / g, and glycerol levels from 0.08 to 29%.

## 1. Introduction

A surfactant is a molecule that has both a hydrophilic group and a lipophilic group so that it can unite a mixture consisting of water and oil. Surfactant molecule has a polar part that likes water (hydrophilic) and non-polar parts like the oil (lipophilic) [1]. Surfactants are an ingredient needed in the manufacture of cleaning products such as soap and shampoo, as well as cosmetic and food products. Along with the increase in the population of Indonesia, there is also an increase in products that use surfactants.The surfactant industry in Indonesia is still limited, even though surfactants are needed in large quantities as a wetting agent, emulsifying agent and solubilizing agent in personal care products such as soap and shampoo, cosmetic products, and food products. The use of surfactants as an emulsifier aims to increase the stability of the emulsion by reducing the interface stress between the oil phase and the water phase. Surfactants are used either in the form of an oil-in-water emulsion or in the form of a water-in-oil emulsion. The potential to develop surfactants is still very large, so it is necessary to carry out research on surfactants more broadly. Surfactant production from vegetable raw materials has bright prospects in

Indonesia. Currently, from 95,000 tons of Indonesia's surfactant needs per year, about 45,000 tons are still imported [2].

Surfactants are generally synthesized from petroleum (petrochemicals). While the need for surfactants is increasing, the procurement of surfactants made from vegetable oil has been developed. Compared to surfactants made from petrochemical raw materials, surfactants made from vegetable oil raw materials are biodegradable so they are more environmentally friendly. In addition, the sustainability of its procurement is guaranteed because vegetable oil is a renewable natural resource.

One of the vegetable oils used as a surfactant is coconut oil. With the increasing awareness of good health and environment, the demand for biodegradable surfactants based on plants is also increasing. Therefore, a study is needed to obtain surfactants that have these two criteria, namely obtained from raw materials that can be renewed (renewable) and are degradative in nature so that they are ecologically acceptable. One of the surfactants that meet these two criteria is diethanolamide surfactant.

Coconut oil contains up to 48% lauric acid [3]. Meanwhile, virgin coconut oil (VCO) contains ± 90% saturated fatty acids which are dominated by lauric fatty acids around 47-53% and are believed to be bioactive components [4]. The high lauric acid content in virgin coconut oil has the potential to be made as a surfactant because of the nature of lauric acid, namely as a thickener, softener and moisturizer so that it can be used in industrial cleaning products such as soaps, shampoos and others [5]. In addition, because of its good antimicrobial properties, lauric acid is also used in the pharmaceutical industry. The use of coconut oil as a surfactant has several advantages, including being renewable (renewable resources), cleaner (cleaner) and purer than using petrochemical-based raw materials. Research on vegetable oil into surfactants has been carried out, especially palm oil and coconut oil, but research on virgin coconut oil that has been extracted by minor compounds or known as VCO residue has never been carried out. VCO is currently developing the extraction process of minor phenolic compounds contained therein and used as pharmaceutical and cosmetic preparations. In the extraction process, the amount of VCO taken (extracted) is around 20-25%, while the remaining 80-90% has not been utilized. One of the uses of the VCO residue is processing it into dietanolamide surfactant.

Dietanolamide surfactants are nonionic surfactants, namely surfactants that are not charged. The hydrophobic / lipophilic part of dietanolamide is the carbon chain of lauric acid as the most fatty acid component in coconut oil, while the hydrophilic / lipophobic part is a molecule that is uncharged [6]. The synthesis of dietanolamide is carried out in two ways, namely the reaction between methyl ester and dietanolamine with a by-product in the form of methanol or the reaction between fatty acids and dietanolamine with a by-product in the form of water. These two reactions are called the amidation reaction [7]. During the synthetic dietanolamide or the amidation process, the mole ratio of the reactants, temperature and reaction time affect the amount of fatty amides produced. The use of dietanolamine must be excessive, at least the required mole ratio of dietanolamine: methyl ester is 1.1: 1. This is because the higher the mole ratio of dietanolamine, the better the process [8]. However, excessive use of dietanolamine allows the presence of dietanolamine which does not react to form dietanolamide.

Based on this, a research was conducted on the manufacture of dietanolamide surfactants from coconut oil (crude coconut oil and VCO which had been extracted the phenolic compounds). The aim of this research is to make VCO surfactant that has been extracted from phenolic compounds and to obtain the correct ratio of reactants (dietanolamine: methyl ester) and concentration of NaOH catalyst to produce diethanolamide surfactant.

## 2. Methodology

### 2.1. Materials and Tools.

The main ingredients used are virgin coconut oil / VCO which has extracted its phenolic compounds (VCO residue), 96% methanol, sodium hydroxide catalyst, 10% H2SO4, and anhydrous Na2SO4 for the transesterification process. Meanwhile, the materials used for the amidation process are methyl ester from VCO and dietanolamine with sodium hydroxide as catalyst.

The tools used in this study were Pyrex 250 and 500 ml beaker, filter paper, 1000 ml Pyrex three-neck flask, reverse cooler, 5 cm magnetic stirrer, Biosan MM-1000overhead stirrer, Memmertwaterbath, statif, clamp, alcohol thermometer (0-100 ° C) and (0-250 ° C), Pyrex Erlenmeyer 250 and 500 ml, Pyrex 1000 ml separating funnel, Pyrex 500 ml measuring cup, pipette, stirrer, Sartorius analytical balance (0.0001 g) type BSA224S-CW, freezer and refrigerator, also Shimadzu Gas Chromatograph Mass Spectrometer with Auto-Injector & FID Detector type GCMS-QP2010 ultra

### 2.2. Methyl ester resulting from the transesterification reaction

The composition of the fatty acid methyl ester using Gas Chromatography was tested in the laboratory of the Manado Industrial Research and Standardization Center. The yield calculation, as well as the methyl ester quality test were tested in the SBRC testing laboratory of the Surfactant and Bioenergy Research Center of IPB Bogor.The density of the test method according to ASTM D4052-11.The viscosity of the test method according to the Brookfield IK-40 Rheometer. Saponation number, acid number, Iod number, total glycerol content according to SNI 7182: 2012 [9].

### 2.3. Dietanolamide surfactant

The viscosity of the test method according to IK-40 Rheometer Brookfield, saponation number and total glycerol content according to SNI 7182: 2012 and pH, were all tested in the SBRC testing laboratory of the Surfactants and Bioenergy Research Center of IPB Bogor.

## 3. Result and Discussion

### 3.1 VCO Residue Raw Material

Table 1 shows that the quality of the residual VCO raw material has a moisture content and free fatty acids that meet the quality requirements of SNI 7381: 2008 VCO [10], namely <0.2%. The quality of oil raw materials greatly affects the transesterification process of oil and alcohol into methyl esters. If the free fatty acids are <2%, the conversion of oil to methyl esters is carried out directly using the transesterification process, whereas if ALB> 2%, the esterification process must be carried out before the transesterification process, or there will be lathering. Bsed on ref [11], high levels of free fatty acid (FFA) oil in the process of making biodiesel or methyl ester cause a saponification reaction which will result in a decrease in FAME (fatty acid methyl ester) levels. One of the main factors determining the success of methyl ester production is the level of FFA. High levels of FFA in the production of methyl ester can trigger a saponification reaction which will result in a decrease in the resulting FAME levels. The FFA levels from oil above 1% will reduce the yield rate produced and increase soap formation, so that the separation process of biodiesel / methyl ester and glycerol becomes difficult. Therefore, oil FFA levels above 1% are not recommended to use alkaline catalysts (transesterification reaction) directly without reducing the FFA levels by using an acid catalyst (esterification reaction). The flowchart diethanolamide surfactant manufacturing process can be seen in Figure 1.

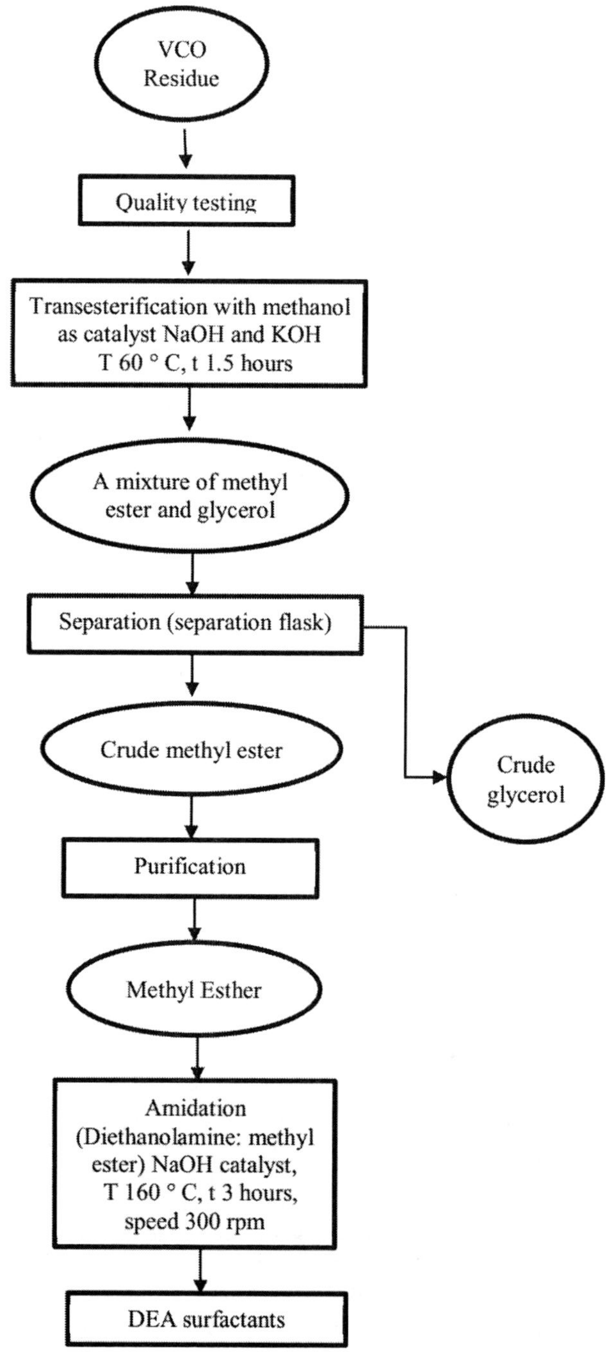


**Figure 1.** Flowchart diethanolamide Surfactant Manufacturing Process

**Table 1.** VCO Residue Test Results

| No. | Parameter Test | Unit | Result (Average) |
|---|---|---|---|
| 1. | Water content | % | 0.16 |
| 2. | Free Fatty Acids (FFA) | % | 0.19 |

From Table 2, it can be seen that the VCO residue still contains lauric acid which is quite high with an amount of 37.96% compared to the lauric acid content of VCO which is around 48-53%. Lauric acid is a thickener, softener, and moisturizer so that it can be used in the cleaning products industry such as soap, shampoo and others.

**Table 2.** Composition of VCO Residue

| Composition of VCO Residue | Amount (%) |
|---|---|
| Caprylic Acid (C8) | 6.02 |
| Capric Acid (C10) | 6.94 |
| Lauric Acid (C12) | 37.96 |
| Myristic Acid (C14) | 21.01 |
| Palmitic Acid (C16-0) | 12.52 |
| Stearate (C18-0) | 4.74 |
| Linoleic Acid (C18-2) | 2.12 |
| Oleic Acid (C18-1) | 8.69 |

### 3.2 Methyl Esther

From Table 3 it can be seen that the resulting methyl ester product has a density of 0.86 gr / cm3 which meets the quality requirements of SNI 7182: 2015, which is around 0.85-0.89 gr / cm3. This means that the methyl ester obtained from the transesterification results has been separated from glycerol because the presence of glycerol in biodiesel / methyl ester affects the density of biodiesel / methyl ester because glycerol has a fairly high density (1.26 g / cm3). If glycerol is not properly separated from biodiesel / methyl ester, the density of the methyl ester will increase [12]. The total glycerol content in the methyl ester product is 0.33%, slightly above the requirements of SNI 7182: 2015.

The viscosity of the methyl ester product is 1.39 cP (30°C). Viscosity is one of the factors that affect the speed of separation of glycerol from biodiesel in addition to density. High viscosity causes slow separation of methyl esters and glycerol. In addition, the viscosity of the methyl ester is related to the specific density [13]. The high specific density causes high viscosity, making it difficult for fuel to flow and will slow down the combustion process. The viscosity produced in this study was still below the SNI Biodiesel standard. This viscosity is slightly below biodiesel quality requirements which means it is slightly thinner.

The acid number of the transesterified methyl ester is higher than the quality requirements of SNI Biodiesel (Methyl Ester) because it ranges from 1.20-2.53 mgKOH / gr, while the requirement is a maximum of 0.5%. This happens because in the purification of the transesterified methyl ester, 10% H2SO4 is used to deactivate the NaOH catalyst. In the next washing process, warm water is used until the washing water has a neutral pH, but it turns out that there is still sulfuric acid bound to the methyl ester product [14].

**Table 3.** Results of Methyl Ester Quality Testing from VCO Residue

| Parameter | Results | Test method | SNI 7182:2015 |
|---|---|---|---|
| Density (gr / cm3) | 0.86 | ASTM D4052-11 | 0.85-0.89 |
| Viscosity (cP, 30 °C) | 1.39 | IK-40 Rheometer Brookfield | 2.3-6.0 $mm^2/s$ |
| Acid Number (mg KOH / gr) | 2.53 | SNI 7182:2012 part 9.13 | 0.5 max |
| Saponification (mg KOH / g) | 285.00 | SNI 7182:2012 part 9.15 | - |
| Iod number (g-I2 / 100 gr) | 2.53 | SNI 7182:2012 part 9.16 | 115 g-$I_2$/100 gr max |
| Total Glycerol (%) | 0.33 | SNI 7182:2012 part 9.14 | 0,24 max |
| Yield (%) | 74.69 | Calculation | - |

The saponation number is a measure for seeing the reaction between KOH and methyl ester. It is also a measure of the average relative molecular mass (chain length) of all the fatty acids present in biodiesel / methyl esters. The soaping number is usually influenced by the molecular weight of the material. The saponification number of the methyl ester resulting from the transesterification process tends to increase with increasing transesterification time, but the increase is only slightly. This means that the transesterification time has very little effect on the formation of soap from the methyl ester product.

The iodine number represents the degree of saturation. The greater the iodine number, the higher the degree of unsaturation. High iodine numbers are an unfavorable property for fuels. The iodine number of methyl ester obtained meets the quality requirements of biodiesel (a maximum of 115 g-I2 / 100 gr) because it ranges from 1.03-5.79 g-I2 / 100 gr.

Furthermore, laboratory testing of the methyl ester content was carried out to see the level of purity of the purified methyl ester, where the laboratory test results mean that the methyl ester content of the sample was 97.46%.Furthermore, the methyl ester composition of the purified samples was tested using GC-MS. The test results can be seen in Table 4 below.

**Table 4.** Composition of Methyl Ester from VCO Residue

| Composition of Methyl Ester | Amount (%) |
|---|---|
| Caprylic Acid (C8) | 5.01 |
| Capric Acid (C10) | 6.57 |
| Lauric Acid (C12) | 34.49 |
| Myristic Acid (C14) | 20.97 |
| Palmitic Acid (C16-0) | 13.67 |
| Stearate (C18-0) | 5.73 |
| Linoleic Acid (C18-2) | 3.16 |
| Oleic Acid (C18-1) | 10.41 |

From Table 4, it can be seen that the results of the transesterification of the fatty acid methyl ester composition test using GC-MS produce 8 (eight) compounds with the main compound being lauric acid methyl ester. This is in line with the results of research which explains that the composition of the transesterified fatty acid methyl ester from VCO produces 8 (eight) compounds with the main compound islauric acid [15].

### 3. 3 Diethanolamide Surfactant

The yield of diethanolamide surfactant obtained from the amidation process is by reacting methyl ester (fixed amount) and dietanolamine with several ratios (1: 1,25; 1: 1,5; 1: 1,75 and 1: 2) using a NaOH catalyst 0.25, 0.5, 0.75% can be seen in Table 5.

From Table 5, it can be seen that the yield of DEA surfactant increases with increasing concentration of NaOH catalyst. The average yields of DEA surfactants using 0.25%, 0.5% and 0.75% NaOH catalyst were 93.6%, 94.52% and 94.60%. NaOH catalyst is a homogeneous catalyst, which has the same phase as the reactants. The function of this catalyst is to break the bonds in the methyl ester to make the reaction easier. If the concentration is too low, the possibility of the reaction progressing also slower.

Treatment ratio of methyl ester and dietanolamine were 1: 1,25; 1: 1,5; 1: 1,75 and 1: 2, and yielded yields: 93.51%, 94.70%, 94.18% and 94.58%. This means that the highest yield was obtained in the ratio of methyl ester and dietanolamine 1: 1.5.

The physical properties of the surfactants can be seen in Table 6. It can be seen that the DEA surfactants produced by comparison treatment of methyl ester and diethanolamine produce DEA surfactants with the same consistency, namely semi-solid viscous liquid with clear yellow color. Likewise for the treatment of using 0.25, 0.5 and 0.75 NaOH catalysts produced DEA surfactants with the same consistency, namely semi-solid viscous liquid with a clear yellow color.

**Table 5.** Yield of dietanolamide surfactant

| Ratio of Methyl Ester and Diethanolamine (A) | Concentration of NaOH (B) | Yield (%) |
|---|---|---|
| (1) 1:1.25 | (1) 0.25 % | 92,80 |
| (2) 1:1.5 | | 95,20 |
| (3) 1:1.75 | | 93.07 |
| (4) 1:2 | | 93.33 |
| Average B (1) | | 93.60 |
| (1) 1:1.25 | (2) 0.5 % | 94.40 |
| (2) 1:5 | | 93.70 |
| (3) 1:1.75 | | 94.80 |
| (4) 1:2 | | 95.20 |
| Average B (2) | | 94,52 |
| (1) 1:1.25 | (3) 0.75 % | 93.33 |
| (2) 1:1.5 | | 95.20 |
| (3) 1:1.75 | | 94.67 |
| (4) 1:2 | | 95.20 |
| Average B (3) | | 94.60 |
| Average A (1) | | 93.51 |
| Average A (2) | | 94.7 |
| Average A (3) | | 94.18 |
| Average A (4) | | 94.58 |

**Table 6.** Observation Results of the Consistency and Appearance of DEA Surfactants on Several Comparisons of ME and Diethanolamine

| Treatment | Consistency | Color / appearance |
|---|---|---|
| Comparison of ME and Diethanolamine | | |
| 1:1.25 | Semi solid / viscous liquid | Clear yellow |
| 1:5 | Semi solid / viscous liquid | Clear yellow |
| 1:1.75 | Semi solid / viscous liquid | Clear yellow |
| 1:2 | Semi solid / viscous liquid | Clear yellow |
| Use of a NaOH catalyst (%) | | |
| 0.25 | Semi solid / viscous liquid | Clear yellow |
| 0.5 | Semi solid / viscous liquid | Clear yellow |
| 0.75 | Semi solid / viscous liquid | Clear yellow |
| 1.0 | Solid | A little light brown |

The use of 1.0% NaOH catalyst produces DEA surfactant with a solid consistency and a slightly light brown color. This means that the use of 1.0% NaOH catalyst is not recommended because it produces DEA surfactants with a consistency and color that does not match the appearance standards (consistency and color) of DEA surfactants on the market.

The results of chemical testing of DEA surfactants and the average test results with catalyst concentration treatment can be seen in Table 7. It can be seen that the average viscosity of DEA surfactants decreases with the increase of NaOH catalyst from 0.25% -0.75%. namely 411.38 cP, 400.92 cP and 367.34, respectively. The average viscosity of DEA surfactants decreased with increasing amounts of dietanolamine, namely as follows. At a ratio of 1: 1.25 produces a viscosity of 440.46 cP, 1: 1.5 produces a viscosity of 423.27 cP, 1: 1.75 produces a viscosity of 358.91 cP and 1: 2 viscosity of 354.86 cP.

The degree of acidity (pH) is one of the characteristics of a surfactant. Each type of surfactant has a different pH, for example, the pH of the surfactant dietanolamide ranges between 9 and 10. In using surfactants, pH needs to be considered because it will affect the activity of the surfactants even though there are some types of surfactants that are not affected by changes in pH. All surfactant products resulting from the oxidation reaction in all treatments have a pH of 10.

The soaping number is expressed as the number (milligrams) of KOH needed to lather one gram of fat or oil. This figure represents the crude molecular weight of fats and oils. Oil which is composed of short carbon chain fatty acids means that it has a relatively small molecular weight and will have a large number of saplings. Conversely, if the oil has a large molecular weight, the saponification number is relatively small [16]. The DEA surfactants produced in this study have saponification numbers between 35.22-52.95 mg KOH / g. The higher the amount of dietanolamine used in comparison with the methyl ester results in a decreasing number of saponification levels.

Total glycerol content indicates the total amount of glycerol both in the bound state (mono-, di- and triglycerides) and in the free state in the DEA surfactant product. The total amount of glycerol present in the DEA surfactant product ranges from 0.08% -0.29%. This is because the methyl ester raw material from the transesterification process used in the amidation process contains glycerol 0.33% (see Table 7).

**Table 7.** The results of the viscosity test, pH, saponification number and glycerol content of the surfactant dietanolamide

| Treatment | | | | | |
|---|---|---|---|---|---|
| ME and Diethanolamine ratio (A) | Concentration of NaOH catalyst (B) | Viscosity (cP) | pH | Saponification numbers (mg KOH/g) | Glycerol levels (%) |
| (1) 1:1.25 | (1) 0.25 % | 374.66 | 10 | 51.79 | 0.15 |
| (2) 1:1.5 | | 392.92 | 10 | 50.36 | 0.12 |
| (3) 1:75 | | 455.24 | 10 | 43.37 | 0.10 |
| (4) 1:2 | | 422.71 | 10 | 41.76 | 0.10 |
| Average B (1) | | 411.38 | 10 | 46.82 | 0.12 |
| (1) 1:1.25 | (2) 0.5 % | 436.83 | 10 | 46.82 | 0.08 |
| (2) 1:5 | | 422.91 | 10 | 41.16 | 0.12 |
| (3) 1:75 | | 354.55 | 10 | 37.25 | 0.29 |
| (4) 1:2 | | 389.40 | 10 | 35.22 | 0.19 |
| Average B (2) | | 400,92 | 10 | 40,11 | 0,17 |
| (1) 1:1.25 | (3) 0.75 % | 509.89 | 10 | 52.95 | 0.20 |
| (2) 1:1.5 | | 440.06 | 10 | 38.89 | 0.18 |
| (3) 1:1.75 | | 266.94 | 10 | 36.63 | 0.22 |
| (4) 1:2 | | 252.47 | 10 | 35.80 | 0.18 |
| Average B (3) | | 367.34 | 10 | 41.68 | 0.20 |
| Average A (1) | | 440.46 | 10 | 50.52 | 0.14 |
| Average A (2) | | 423.27 | 10 | 43.47 | 0.14 |
| Average A (3) | | 358.91 | 10 | 39.08 | 0.20 |
| Average A (4) | | 354.86 | 10 | 37.59 | 0.16 |

## 4. Conclusion

The phenolic compounds extracted VCO produced DEA surfactant and glycerol by-products.The optimal esterification time in making methyl ester (purified) with the highest yield is esterification time of 1.5 hours using 1.25% NaOH catalyst, with an average yield of 57.92%, and 0.5 hours of esterification time using KOH catalyst 1% with an average yield of 74.69%.The appropriate ratio of reactants to produce surfactant DEA is the ratio of methyl ester and dietanolamine 1: 1.5 with yield about 94.58%, viscosity about 423.27 cP, pH about 10, Saponification number 43.47 mg KOH / g and total glycerol content. 0.14%.

## References

[1] Paria S and Khilar K C 2004 A review on experimental studies of surfactant adsorption at the hydrophilic solid–water interface *Advances in colloid and interface science* **110** (3) 75-95

[2] Swadesi B, Marhaendrajana T, Siregar H P and Mucharam L 2015 The Effect of Surfactant Characteristics on IFT to Improve Oil Recovery in Tempino Light Oil Field Indonesia *Journal of Engineering & Technological Sciences* **47** (3) 250-265

[3] Dayrit F M 2015 The properties of lauric acid and their significance in coconut oil *Journal of the American Oil Chemists' Society* **92** (1) 1-15

[4] Marina A M, Man Y C and Amin I 2009 Virgin coconut oil: emerging functional food oil *Trends in Food Science & Technology* **20** (10) 481-487
[5] Tangwatcharin P and Khopaibool P 2012 Activity of virgin coconut oil, lauric acid or monolaurin in combination with lactic acid against Staphylococcus aureus *Southeast Asian Journal of Tropical Medicine & Public Health* **43** (4) 969-985
[6] Adewuyi A 2019 Synthesis and surface-active property of diethanolamide and epoxidised diethanolamide surfactant from the seed oil of Baphia nitida *Arabian Journal of Chemistry* **12** (7) 1545-1551
[7] Khanmohammadi M, Kojidi M H, Garmarudi A B, Ashuri A and Soleymani M 2009 Quantitative monitoring of the amidation reaction between coconut oil and diethanolamine by attenuated total reflectance Fourier transform infrared spectrometry *Journal of surfactants and detergents* **12** (1) 37-41
[8] Utami M, Wijaya K, Syoufian A and Wicaksono S H 2020 Catalytic Synthesis of Diethanolamide Surfactant from Used Cooking Oil *EKSAKTA: Journal of Sciences and Data Analysis* **20** (2) 133-138
[9] Badan Standardisasi Nasional 2012 *Biodisel SNI 7182:2012* Jakarta: BSN
[10] Badan Standardisasi Nasional 2008 *SNI 7381:2008 Minyak Kelapa Murni (VCO)* Jakarta: BSN
[11] Tiwari A K, Kumar A and Raheman H 2007 Biodiesel production from jatropha oil (Jatropha curcas) with high free fatty acids: an optimized process *Biomass and bioenergy* **31** (8) 569-575
[12] Hu S, Luo X, Wan C and Li Y 2012 Characterization of crude glycerol from biodiesel plants *Journal of agricultural and food chemistry* **60** (23) 5915-5921
[13] Selim M Y 2009 Reducing the viscosity of Jojoba Methyl Ester diesel fuel and effects on diesel engine performance and roughness *Energy Conversion and Management* 50 (7) 1781-1788
[14] Benjapornkulaphong S, Ngamcharussrivichai C and Bunyakiat K 2009 Al2O3-supported alkali and alkali earth metal oxides for transesterification of palm kernel oil and coconut oil *Chemical Engineering Journal* **145** (3) 468-474
[15] Marina A M, Che Man Y B, Nazimah S A H and Amin I 2009 Chemical properties of virgin coconut oil *Journal of the American Oil Chemists' Society* **86** (4) 301-307
[16] Wahyuningsih T D, Kurniawan Y S, Amalia S, Wardhani T A and Muriningsih C E S 2019 Diethanolamide derivatives as a potential enhanced oil recovery agent from indonesian castor oil and used frying oil: Isolation, synthesis, and evaluation as nonionic biosurfactants *Rasayan Journal Chemistry* **12** (2) 741-748

IOP Conf. Series: Materials Science and Engineering 1115 (2021) 012074 doi:10.1088/1757-899X/1115/1/012074

# Molecular Docking Studies of (2E)-1(1-Methyl-1H-Indol-3-yl)-3-Phenyl-Prop-2-en-one) as Antimalarial and Its Synthesis using Dimethylsulfate

**K Mustikasari*, J E Harap and T B Susilo**[1]

Department of Chemistry, Faculty of Science and Mathematics, University of Lambung Mangkurat, South Borneo, Indonesia 70714

*Corresponding author: kmustikasari@ulm.ac.id

**Abstract.** Malaria is a disease with high prevalence in tropical and subtropical areas. The case of parasite resistance to proguanil, an antimalarial drug, has prompted the discovery of new drug candidates through molecular docking studies. In this study, we used the derivative of chalcone (2E)-1-(1-methyl-1H-indol-3-yl)-3-phenyl-prop-2-en-1-one as a test ligand and some comparison ligands to see its interaction with *Plasmodium falciparum dihydrofolate reductase – thymidylate synthase* (*Pf*DHFR-TS) enzyme. Furthermore, the compound (2E)-1-(1-methyl-1H-indol-3-yl)-3-phenyl prop-2-en-1-one was synthesized by reacting (2E)-1-(1H-indol-3-yl)-3-phenyl prop-2-en-1-one and dimethyl sulfate through an *N*-alkylation reaction. The (2E)-1-(1-methyl-1H-indol-3-yl)-3-phenyl-prop-2-en-1-one has the higher $\Delta G_{bind}$ (-7.53 kcal/mole) than native ligan WR99210(-8.83 kcal/mole), but has the lower than proguanil (-6.75 kcal/mole) and (*E*)-3-(3,4-dimethoxy phenyl)-1-(2-hydroxy-4-methoxy-5-(prenyl)phenyl)-prop-2-en-1-one (-7.29 kcal/mole). However, there was only pi interaction of amino acid residue of the *Pf*DHFR-TS ligand complexes ($PHE^{58}$, $ILE^{112}$, $LEU^{119}$, $ALA^{16}$, $ILE^{14}$), in contrast to the WR99210 and proguanil complexes. This is similar to the interaction of the (*E*)-3-(3,4-dimethoxyphenyl)-1-(2-hydroxy-4-methoxy-5-(prenyl)phenyl)-prop-2-en-1-one-*Pf*DHFR-TS complex which is active as an antimalarial *in vitro*, so that the test ligand still has a chance to be used as an antimalarial candidate. The compound (2E)-1-(1-methyl-1H-indol-3-yl)-3-phenyl prop-2-en-1-one was successfully synthesized with 41.55% yield.

## 1. Introduction

Malaria is a contagious disease that causes the death of thousands of millions of people every year. The number of cases of malaria infection worldwide is estimated at 219 million, and as many as 435,000 cases have died [1]. There were 62.8% of malaria cases in Southeast Asia which were caused by the parasite *Plasmodium falciparum* and cases of resistance of these parasites to the anti-folate class of malaria drugs, namely proguanil, were found [2-5]. Therefore, it is necessary to develop a drug design to overcome this resistance problem.

Compound (2E)-1-(1-methyl-1H-indol-3-yl)-3-phenyl-prop-2-en-1-one **(1)** is a derivative of chalcone. Previous research has shown that chalcone and its derivatives active as antimalarials *in vitro*[6-11]. Prediction of the anti-malarial activity of chalcone compounds through the molecular docking method [12-15] has been carried out for the compound (E) -3- (3,4-dimethoxy phenyl) -1- (2-

hydroxy-4-methoxy-5-(prenyl) phenyl)-prop-2-en-1-one **(2)** against the enzyme *Plasmodium falciparum dihydrofolate reductase – thymidylate synthase* (*Pf*DHFR-TS), an enzyme that plays a role in the multiplication of Plasmodium cells. The study succeeded in proving that the chalcone **(2)** compound had a similar interaction with the natural inhibitor of the *Pf*DHFR-TS enzyme, namely the compound1,6-dihydro-6,6-dimethyl-1-(3-(2,4,5-trichlorophenoxy)propoxy]-1,3,5-triazine-2,4 diamine (WR99210) **(3)**[10].

In this paper we report an *in silico* study of compound (2E) -1- (1-methyl-1H-indol-3-yl) -3-phenyl-prop-2-en-1-one **(1)** as an antimalarial to enzymes *Pf*DHFR-TS and its synthesis through the *N*-alkylation reaction using dimethylsulfate. This compound was synthesized by Martel-frachet *et al.* [16] through the Claisen-Schmidt condensation reaction but has never been tested as an antimalarial.

## 2. Methodology

### 2.1 Molecular Docking Study

The VMD 1.9.2 was used for the identification process of the *Pf*DHFR-TS protein complex. The natural ligand WR99210 obtained from protein bank data (https://www.rcsb.org/structure/1J3I). The A.D.T 1.5.6 and PyMOL 2.3 software were used for the structure optimization process. The ligands were drawn using Marvin Sketch® 16.9.12. Furthermore, AutoDock Tools®4.2 and AutoDock Vina®1.1.2 were used for optimization ligand-protein [17]. Also, the visualization of docking results was used the Discovery Studio 19.0 Client (DSV 19.0).

### 2.2. Synthesis (2E)-1-(1H-indol-3-yl)-3-phenyl-prop-2-en-1-one

The mixture of 3-acetylindole (3.14 mmol) and benzaldehyde (3.14 mmol) in ethanol (10 ml) was added with sodium hydroxide (10%). It was stirred for 48 hours in room temperature and was added cold aquadest (40 ml), respectively. Also, the mixture was filtered and obtained a yellow solid with a yield of 44.68%. The compound was characterized by $^{1}$H and $^{13}$CNMR spectrometers.

(2E)-1-(1H-indol-3-yl)-3-phenyl-prop-2-en-1-one **(4)**$^{1}$ H NMR ($CDCl_3$): 7.65 (d, $J$ = 16 Hz, 1H) 8.36 (d,$J$ = 7.5 Hz, 1H), 7.21-7.27 (m, 2H), 7.40-7.46 (m, 3H), 7.52 (d, 1H), 7.85-7.88 (m, 3H), 8.76 (s, 1H); $^{13}$C NMR ($CDCl_3$): 117.7, 139.5, 112.3, 121.6, 121.8, 121.9, 123.1, 124.6, 125.9, 128.3, 128.8, 129.8, 135.2, 137.0, 183.6.

### 2.3. Synthesis (2E)-1-(1-methyl-1H-indol-3-yl)-3-phenyl-prop-2-en-1-one (1)

The synthesis process of (2E) -1-(1H-indol-3-yl) -3-phenyl-prop-2-en-1-one **(4)** was carried out by the indole *N*-alkylation procedure. (2E)-1-(1H-indol-3-yl)-3-phenyl-prop-2-en-1-one **(4)** (0.202 mmol) dissolved in DMSO solvent (10 ml) then added with NaOH (0.606 mmol). After stirring for 1 hour at room temperature, dimethyl sulfate (0.606 mmol) was added. Stirring is carried out for up to 3 hours. The reaction was stopped by adding cold distilled water (40 ml). The precipitate obtained was filtered and dried, until a yellow solid was obtained with a yield of 41.55%. The compound was characterized by $^{1}$H, $^{13}$C NMR spectrometer, and mass.

(2E)-1-(1-methyl-1H-indol-3-yl)-3-phenyl-prop-2-en-1-one **(1)** $^{1}$ H NMR ($CDCl_3$): 7.78 (d, $J$ = 16 Hz, 1H), 8.34 (d, $J$ = 8 Hz, 1H), 7.25-7.27 (m, 7H), 7.82-7.85 (m, 2H), 8.75 (s, 1H),7.25-7.27 (m, 7H), 7.82-7.85 (m, 2H), 8.75 (s, 1H), 3.91 (s, 3H); $^{13}$C NMR ($CDCl_3$):33.4, 116.6, 139.6, 110.8, 121.6, 121.9, 121.9, 123.2, 124.5, 125.9, 128.3, 128.9, 129.9, 135.2, 137.4, 183.1. MS (EI) m/z (%)204 (30), 189 (5), 161 (100), 133 (30), 119 (50).

IOP Conf. Series: Materials Science and Engineering 1115 (2021) 012074 doi:10.1088/1757-899X/1115/1/012074

## 3. Results and Discussion

### 3.1. Prediction of(2E)-1-(1-methyl-1H-indol-3-yl)-3-phenyl-prop-2-en-1-one(1) Compound Activity as Antimalarial with molecular docking

The antimalarial activity of a compound can be predicted by using molecular docking simulations. Before the target compound or ligands were tethered to the *Pf*DHFR-TS enzyme, the optimization of the natural ligand WR99210 in the active site was performed. The redocking results showed that the RMSD value was smaller than two (0.78), so the docking process was considered valid and could continue. Interaction of (2E) -1- (1-methyl-1H-indol-3-yl) -3-phenyl-prop-2-en-1-one **(1)** with the *Pf*DHFR-TS enzyme as a test ligand compared with (E) -3- (3,4-dimethoxy phenyl) -1- (2-hydroxy-4-methoxy-5- (prenyl) phenyl) -prop-2-en-1-one **(2)**, proguanil (malaria drug), and WR99210 interactions. (E) -3- (3,4-dimethoxy phenyl) -1- (2-hydroxy-4-methoxy-5- (prenyl) phenyl) -prop-2-en-1-one **(2)** and WR99210 active as antimalarial *in vitro*. The results of these interactions are shown in Table 1.

**Table 1**. Result of Study Molecular Docking with enzyme *Pf*DHFR-TS

| **Compound** | **Parameters** | **Type of interaction** | |
|---|---|---|---|
| | **Binding energy (kcal mole$^{-1}$)** | **Hydrogen bonding** | **Π bonding and *van der waals*** |
| (2E)-1-(1-methyl-1H-indol-3-yl)-3-phenyl-prop-2-en-1-one**(1)**(Test Ligand) | -7.53 | - | *$PHE^{58}$, $ILE^{112}$, $LEU^{119}$, $ALA^{16}$*, $ILE^{14}$ |
| (*E*)-3-(3,4-dimethoxy phenyl)-1-(2-hydroxy-4-methoxy-5-(prenyl)phenyl)-prop-2-en-1-one**(2)** | -7.29 | - | *$PHE^{58}$, $ILE^{112}$, $PRO^{113}$, $MET^{55}$, $ALA^{16}$* |
| Proguanil (malaria drug) | -6.75 | *$ILE^{164}$, $ILE^{14}$* | $TYR^{57}$, *$ILE^{112}$, $MET^{55}$, $PRO^{113}$* |
| WR99210 (Natural Ligand) | -8.83 | $ILE^{164}$, $ILE^{14}$, $TYR^{170}$, $ASP^{54}$, $CYS^{15}$ | $ALA^{16}$, $PHE^{58}$, $LEU^{119}$, $PRO^{113}$, $PHE^{116}$, $ILE^{112}$, $MET^{55}$ |

The 2D visualization of docking results can be seen in Figure 1.

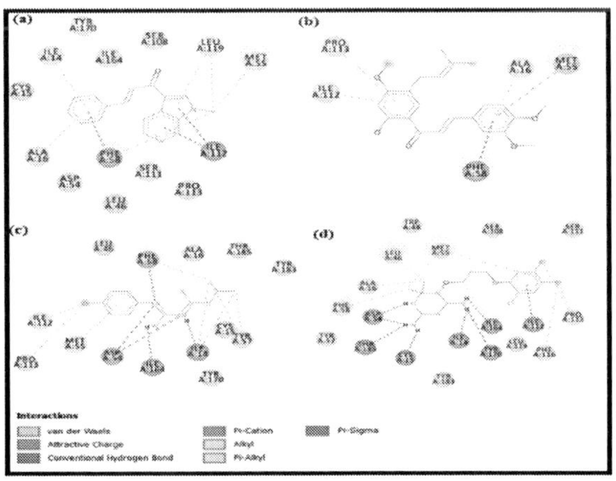


**Figure 1**. 2D visualization of docking result chalcone **(1)** **(a)**, chalcone **(2)** **(b)**, proguanil **(c)**, and WR99210 **(d)**

Table 1 show that the complex compound (2E) -1- (1-methyl-1H-indol-3-yl) -3-phenyl-prop-2-en-1-one **(1)** has a binding energy of -7.53 kcal / mole, this value is greater than the binding energy of the natural ligand complex (-8.83 kcal / mole). However, the energy value that binds the complex compound (2E) -1- (1-methyl-1H-indol-3-yl) -3-phenyl-prop-2-en-1-one **(1)** is smaller than the binding energy of (E) -3- (3,4-dimethoxy phenyl) -1- (2-hydroxy-4-methoxy-5- (prenyl) phenyl) - prop-2-en-1-one **(2)** (- 7.29 kcal / mole) and proguanil (-6.75 kcal / mole). The value of binding energy indicates the stronger bond or interaction, which is between the ligand and the enzyme. However, there is not hydrogen bond interaction between the test ligands (2E) -1- (1-methyl-1H-indol-3-yl) -3-phenyl-prop-2-en - 1-one **(1)** with the *Pf*DHFR-TS enzyme. The interactions were occured in pi and *van der Waals* bond, that were interacting with residues of acid amino $PHE^{58}$, $ILE^{112}$, $LEU^{119}$, $ALA^{164}$ similar to WR99210. This type of interaction without hydrogen bonds is similar to that of the complex (E) -3- (3,4-dimethoxy phenyl) -1- (2-hydroxy-4-methoxy-5- (prenyl) phenyl) -prop-2 -en-1-one **(2)**. The compound (E) -3- (3,4-dimethoxy phenyl) -1- (2-hydroxy-4-methoxy-5- (prenyl) phenyl) -prop-2-en-1-one **(2)** was shown to be active as antimalarial *in vitro* with an $IC_{50}$ of $1.08 \pm 0.405$ μM[10].

### 3.2. Synthesis (2E)-1-(1H-indol-3-yl )-3-phenyl prop-2-en-1-one(4)

Synthesis of the compound (2E) -1- (1H-indol-3-yl) -3-phenyl prop-2-en-1-one **(4)** was carried out as an initial step in the synthesis of the compound (2E) -1- (1-methyl- 1H-indol-3-yl) -3-phenyl-prop-2-en-1-one **(1)** (Figure 2). This synthesis was used the Claisen-Schmidth condensation reaction by reacting acetyl indole with benzaldehyde in alkaline conditions to produce the yellow solid compound (2E) -1- (1H-indol-3-yl) -3-phenyl prop-2-en-1-one **(4)**. The $^{1}H$ and $^{13}C$ NMR spectra in Tables 2 and 3 show that this compound was successfully synthesized.

**Table 2.** $^{1}H$ NMR Spectra of (2E) -1- (1H-indol-3-yl) -3-phenyl prop-2-en-1-one **(4)**

| Proton | $^{1}$H-NMR (ppm) 500 MHz<br>(2E)-1-(1H-indol-3-yl )-3-phenyl prop-2-en-1-one (4) Product of synthesis | (2E)-1-(1H-indol-3-yl )-3-phenyl prop-2-en-1-one(4) [18] |
|---|---|---|
| -C<u>H</u>=CH- | 7.65 (d, *J* = 16 Hz, 1H) 8.36 (d, *J* = 7.5 Hz, 1H) | 7.64 (d, *J* = 15.8 Hz, 1H) 8.35 (d, *J* = 7.6 Hz, 1H) |
| Ar<u>H</u> | 7.21-7.27 (m, 2H)<br>7.40-7.46 (m, 3H)<br>7.52 (d, 1H)<br>7.85-7.88 (m, 3H)<br>8.76 (s, 1H) | 7.21-7.27 (m, 2H)<br>7.40-7.44 (m, 3H)<br>7.48 (d,1H)<br>7.83-7.86 (m, 3H)<br>8.74 (s, 1H) |

**Table 3.** $^{13}C$ NMR Spectra of (2E) -1- (1H-indol-3-yl) -3-phenyl prop-2-en-1-one **(4)**

| Carbon | $^{13}$C-NMR (ppm) 500 MHz<br>(2E)-1-(1H-indol-3-yl )-3-phenyl prop-2-en-1-one (2) Product of synthesis | (2E)-1-(1H-indol-3-yl )-3-phenyl prop-2-en-1-one(2) [18] |
|---|---|---|
| -<u>C</u>H=CH- | 117.7<br>139.5 | 117.7<br>139.5 |
| Ar<u>C</u> | 112.3<br>121.6<br>121.8<br>121.9 | 112.1<br>121.6<br>121.7<br>121.8 |

| | | |
|---|---|---|
| | 123.1 | 123.1 |
| | 124.6 | 124.6 |
| | 125.9 | 125.9 |
| | 128.3 | 128.3 |
| | 128.8 | 128.8 |
| | 129.8 | 129.7 |
| | 135.2 | 135.2 |
| | 137.0 | 136.8 |
| -C=O | 183.6 | 183.5 |

### 3.3. Synthesis (2E)-1-(1-methyl-1H-indol-3-yl)-3-phenyl-prop-2-en-1-one(1)

Synthesis of the compound (2E)-1-(1-methyl-1H-indol-3-yl)-3-phenyl-prop-2-en-1-one **(1)** was carried out by reacting the compound (2E)-1-(1H -indol-3-yl)-3-phenyl prop-2-en-1-one **(4)** results from the previous reaction. This reaction uses dimethylsulfate in alkaline conditions through an *N*-alkylation reaction. Scheme of synthesis as shown in Figure 2.

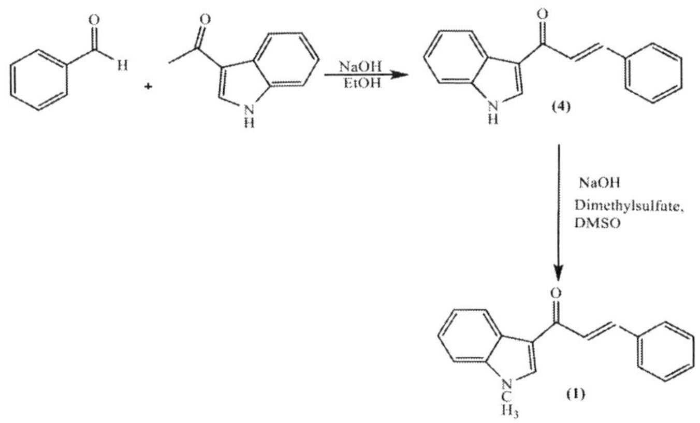


**Figure 2**. Sceme of synthesis (2E)-1-(1-methyl-1H-indol-3-yl)-3-phenyl-prop-2-en-1-one **(1)**

The successful synthesis of (2E) -1- (1-methyl-1H-indol-3-yl) -3-phenyl-prop-2-en-1-one **(1)** is shown in the $^{1}$H NMR and $^{13}$C NMR spectra which is presented in Table 4 and 5. The resulting product was a yellow solid with a yield of 41.55%.

**Table 4**. $^{1}$H NMR Spectra of (2E) -1- (1-methyl-1H-indol-3-yl) -3-phenyl prop-2-en-1-one **(1)**

| **Proton** | **$^{1}$H-NMR (ppm)** | |
|---|---|---|
| | **(2E)-1-(1-methyl-1H-indol-3-yl)-3-phenyl-prop-2-en-1-one(1) Product of synthesis($CDCl_3$, 500 MHz)** | **(2E)-1-(1-methyl-1H-indol-3-yl)-3-phenyl-prop-2-en-1-one (1) [16] ($CDCl_3$, 400 MHz)** |
| -CH=CH- | 7.78 (d, *J* = 16 Hz 1H)<br>8.34 (d, *J* = 8 Hz 1H) | 7.66 (d, *J* = 16 Hz 1H)<br>8.36 (d, *J* = 7,5 Hz 1H) |
| ArH | 7.25-7.27 (m, 7H)<br>7.82-7.85 (m, 2H)<br>8.75 (s, 1H) | 7.27-7.42 (m, 7H)<br>7.86-7.88 (s, 1H)<br>8.51 (m, 1H) |
| -$CH_3$ | 3.91 (s, 3H) | 3.87 (s, 3H) |

**Table 5.** $^{13}$C NMR Spectra of (2E) -1- (1-methyl-1H-indol-3-yl) -3-phenyl prop-2-en-1-one **(1)**

| Carbon | $^{13}$C-NMR (ppm) (2E)-1-(1-methyl-1H-indol-3-yl)-3-phenyl-prop-2-en-1-one(1) Product of Synthesis (500 MHz) | (2E)-1-(1-methyl-1H-indol-3-yl)-3-phenyl-prop-2-en-1-one (1) [16] (100 MHz) |
|---|---|---|
| -C̲H=CH- | 116.6 | 111.0 |
| | 139.6 | 111.4 |
| ArC̲ | 110.8 | 119.0 |
| | 121.6 | 120.5 |
| | 121.9 | 121.9 |
| | 121.9 | 122.1 |
| | 123.2 | 126.7 |
| | 124.5 | 127.1 |
| | 125.9 | 128.0 |
| | 128.3 | 128.5 |
| | 128.9 | 128.8 |
| | 129.9 | 135.0 |
| | 135.2 | 135.3 |
| | 137.4 | 136.6 |
| -C̲=O | 183.1 | 190.2 |
| -C̲H$_3$ | 33.4 | 42.4 |

## 4. Conclusion

Molecular docking study of the compound (2E)-1- (1-methyl-1H-indol-3-yl) -3-phenyl-prop-2-en-1-one **(1)** with the *Pf*DHFR-TS enzyme produces binding energy of - 7.53 kcal/mole. $\Delta G_{bind}$ is larger than the original ligand WR99210 (-8.83 kcal / mole), but has less than proguanil (-6.75 kcal / mole) and (E) -3- (3,4-dimethoxy phenyl) -1- (2- hydroxy-4-methoxy-5- (prenyl) phenyl) -prop-2-en-1-one**(2)** (-7.29 kcal / mole). However, there was only pi interaction at the binding site of the remaining *Pf*-DHFR-TS ligand complexes (PHE$^{58}$, ILE$^{112}$, LEU$^{119}$, ALA$^{16}$, ILE$^{14}$), in contrast to the WR99210 and proguanil complexes. This is similar to the interaction on the residue (E) -3- (3,4-dimethoxy phenyl) -1- (2-hydroxy-4-methoxy-5- (prenyl) phenyl) -prop-2-en -1-one**(2)** - *Pf*DHFR-TS complex is active as an *in vitro* antimalarial so that the test ligand still has a chance to be used as an antimalarial candidate. The compound (2E) -1- (1-methyl-1H-indol-3-yl) -3-phenyl prop-2-en-1-one **(1)** was successfully synthesized with a yield of 41.55%.

## Acknowledgments

The authors would like to acknowledge the Lambung Mangkurat University through the Compulsory Research Grants for Lecturers for funding this research.

## References

[1] World Health Organization. 2018. http://www.who.int/malaria/publications/world-malaria-report-2018/report/en/. (Accessed 20 Mei 2019)
[2] McKeage K and Scott L J 2003 Atovaquone/proguanil *Drugs* **63** (6) 597-623
[3] Bushman M, Antia R, Udhayakumar V and de Roode J C 2018 Within-host competition can delay evolution of drug resistance in malaria *PLoS biology* **16** (8) e2005712

[4] Menard D and Dondorp A 2017 Antimalarial drug resistance: a threat to malaria elimination *Cold Spring Harbor perspectives in medicine* **7** (7) a025619
[5] Calic P P, Mansouri M, Scammells P J and McGowan S 2020 Driving antimalarial design through understanding of target mechanism *Biochemical Society Transactions* **48** (5) 2067-2078
[6] Mishra N, Arora P, Kumar B, Mishra L C, Bhattacharya A, Awasthi S K and Bhasin V K 2008 Synthesis of novel substituted 1, 3-diaryl propenone derivatives and their antimalarial activity in vitro *European journal of medicinal chemistry* **43** (7) 1530-1535
[7] Sharma N, Thomas S, Golden E B, Hofman F M, Chen T C, Petasis N A and Louie S G 2012 Inhibition of autophagy and induction of breast cancer cell death by mefloquine, an antimalarial agent *Cancer letters* **326** (2) 143-154
[8] Sharma N, Mohanakrishnan D, Shard A, Sharma A, Sinha A K and Sahal D 2012 Stilbene–chalcone hybrids: design, synthesis, and evaluation as a new class of antimalarial scaffolds that trigger cell death through stage specific apoptosis *Journal of medicinal chemistry* **55** (1) 297-311
[9] Syahri J, Yuanita E, Nurohmah B A, Armunanto R and Purwono B 2017 Chalcone analogue as potent anti-malarial compounds against Plasmodium falciparum: Synthesis, biological evaluation, and docking simulation study *Asian Pacific Journal of Tropical Biomedicine* **7** (8) 675-679
[10] Yadav N, Dixit S K, Bhattacharya A, Mishra L C, Sharma M, Awasthi S K and Bhasin V K 2012 Antimalarial activity of newly synthesized chalcone derivatives in vitro *Chemical biology & drug design* **80** (2) 340-347
[11] Pereira G A, Souza G C, Santos L S, Barata L E, Meneses C C, Krettli A U and Alves C N 2017 Synthesis, antimalarial activity in vitro, and docking studies of novel neolignan derivatives *Chemical biology & drug design* **90** (3) 464-472
[12] Dohutia C, Chetia D, Gogoi K and Sarma K 2017 Design, in silico and in vitro evaluation of curcumin analogues against Plasmodium falciparum *Experimental parasitology* **175** 51-58
[13] Thillainayagam M, Anbarasu A and Ramaiah S 2016 Comparative molecular field analysis and molecular docking studies on novel aryl chalcone derivatives against an important drug target cysteine protease in Plasmodium falciparum *Journal of theoretical biology 403* 110-128
[14] Hameed A, Masood S, Hameed A, Ahmed E, Sharif A and Abdullah M I 2019 Anti-malarial, cytotoxicity and molecular docking studies of quinolinyl chalcones as potential anti-malarial agent *Journal of computer-aided molecular design* **33** (7) 677-688
[15] Shi J H, Pan D Q, Wang X X, Liu T T, Jiang M and Wang Q 2016 Characterizing the binding interaction between antimalarial artemether (AMT) and bovine serum albumin (BSA): Spectroscopic and molecular docking methods *Journal of Photochemistry and Photobiology B: Biology* **162** 14-23
[16] Morris G M, Huey R and Lindstrom W 2009 MF Sanner RK Belew DS Goodsell and AJ Olson *J. Computational Chemistry* **16** 2785-91
[17] Engel M, Fröhner W, Stroba A and Biondi R M 2014 *U.S. Patent No. 8,912,186*. Washington, DC: U.S. Patent and Trademark Office
[18] Ugwu D I, Okoro U C, Ukoha P O, Okafor S, Ibezim A and Kumar N M 2017 Synthesis, characterization, molecular docking and in vitro antimalarial properties of new carboxamides bearing sulphonamide *European journal of medicinal chemistry* **135** 349-369

# The Effectiveness of the Absorption of CO, $CH_4$, $CO_2$, $H_2S$ and Pb Gasess by a Mixture of Sansiviera and Coffee Powder

**M V Oviantari*, I M Gunamantha and I A P A Deviana Pratiwi**

Department of Chemistry, Universitas Pendidikan Ganesha, Indonesia

*Corresponding author: oviantari@gmail.com

**Abstract.** This study aims to analyze the effectiveness of the absorption of CO, $CH_4$, $CO_2$, $H_2S$ and Pb Gasses by a mixture of Sansiviera and Coffe Powder. Gas measurements were carried out using a Multitec 250 gas detector, while Pb levels were analyzed using AAS. The results showed that the mixture of Sansevieria and coffee powder effectively absorbed $CH_4$, $CO_2$, $H_2S$ and Pb gases with the highest effectiveness, respectively 87.5%, 71.06%, 55.5% and 59% with the composition of sansiviera: coffee in the sample respectively 50:50; 50: 0; 30:20; 40:10. Meanwhile, the mixture of sansiviera and coffee was slightly effective in absorbing CO gas with the highest effectiveness of 15.65% at the composition of 50: 0.

## 1. Introduction

Motorized vehicles are one of the main assets used by Indonesian people to carry out their daily activities, such as doing business, schooling, and so on today [1]-[3]. Therefore, the number of motorized vehicles in Indonesia is growing rapidly. Based on data from the Central Statistics Agency (BPS) in 2017 with an increase in the number of vehicles by 10-15% each year, the predicted number of vehicles in Indonesia in 2020 is 133,533,603.5; -139,603,312.75 million.

Motorized vehicles can not only ease human activity in moving places, but can also have negative impacts, such as congestion, pollution in the form of incomplete combustion in the form of vehicle fumes containing Pb gas, suspended particulate matter (SPM), nitrogen oxides (NOx), sulfur oxides ($SO_2$), hydrocarbons (HC), carbon monoxide (CO), and photochemical oxides (Ox) [4]-[7]. The presence of this pollution in the air will cause health problems, especially disturbances in the physiological functions of the body's organs, such as the lungs, blood vessels, eye and skin irritation. Generally, dust particles in the air will cause chronic respiratory disease, bronchial asthma, pulmonary emphiesma and even lung cancer. High levels of Pb in the air will also interfere with the formation of red blood cells. Disruption of enzyme function is a symptom of premature poisoning which is shown in the formation of red blood cells which eventually causes other health problems, such as anemia and kidney damage. Meanwhile, the high CO gas in the air will cause carboxyhemoglobin (COHb) to form in the blood, if the affinity of CO is greater than oxygen to Hb it will interfere with the Hb function to carry oxygen throughout the body. This will cause the available oxygen to decrease, and if you do not immediately get fresh air, shortness of breath will occur, which can cause death. For NOx, SOx and H2S pollutants in the environment that exceed the allowable threshold, will cause irritation and inflammation of the respiratory organs. In addition, all these pollutants can also damage the environment because they are one of the greenhouse gases that can cause global warming [9]-[11].

Many things have been done to reduce these negative impacts, including an increase in the number of quality mass vehicles for the community, an increase in the number of broadleaf trees in high-traffic areas. The use of masks will also greatly help people to maintain health in an environment with lots of air pollution. In this research, we want to develop materials that absorb some of these pollutants in order to maintain environmental quality so that public health can improve.

## 2. Method

### 2.1 Tools and Materials

The tools used in this study included a knife, oven, analytical balance, gas detector (Multitec 250 brand), beaker glass (Herma brand). While the samples in this study were Sansevieria and dried coffee.

### 2.1 Subjects and Objects

In this study, the subject was a mixture of the sansiviera and coffee, while the objects in the study were $CH_4$, $CO_2$, $H_2S$, CO, and Pb gases which were able to be absorbed by the mixture.

### 2.2 Preparation

The sansiviera is cut into small pieces using a knife, then oven at 105 ° C for 21 hours, then blend until smooth using a blender. Meanwhile, 2 grams of coffee is put into 10mL of water, soaked for 5 minutes, then filtered, the dregs are taken, after which the dregs are set aside to dry at room temperature.

### 2.3 Data collection

Samples in the form of a mixture of sansiviera and coffee powder were placed on the test media with different variations to measure the gas levels of $CH_4$, $CO_2$, $H_2S$ and CO. The variations in the mass of the sansiviera and coffee powder are presented in Table 1. The gas used is gas produced from motorized vehicles. The test media used for the previous measurement was filled sansiviera and coffee powder and then the gas was measured by taking the vehicle gas and then put it in the test medium and then tested using a gas detector with the brand Multitec 520. The gas was measured at the start. increments and every hour after the initial measurement. Meanwhile, the measurement of Pb levels is carried out using AAS which will be carried out at the Analytical Laboratory at Udayana University, Denpasar.

**Table 1.** Variation mixture of sansiviera and coffee

| Variation | Sansiviera (gr) | Coffee (gr) |
|---|---|---|
| I | 0 | 50 |
| II | 10 | 40 |
| III | 20 | 30 |
| IV | 30 | 20 |
| V | 40 | 10 |
| VI | 50 | 0 |
| VII | 50 | 50 |
| VIII | 0 | 50 |

## 2.4 Data Analysis

The data generated in the study were gas levels of $CH_4$, $CO_2$, $H_2S$, CO, and Pb before and after application. The data was then searched for the effectiveness of the mixture of coffee and sansiviera on the absorption of $CH_4$, $CO_2$, $H_2S$, CO, and Pb gas is sought with the formula as follows:

$$\text{Absorption} = t0 - t1$$

The effectiveness of absorption is sought by the following formula

$$Effektiveness = \frac{t0-t1}{t0} \text{x}100\%$$

Note: t0 is the absorption at the first contact time
t1 is the absorption at the end contact time.

## 3 Result and Discusion

The absorption levels of CO, $CH_4$, $CO_2$, $H_2S$ and Pb by organic absorbent mixtures of the coffee and sansiviera are presented respectively in Table 2, Table 3, Table 4, Table 5 and Table 6.

**Table 2.** Absorption of CO Levels in Mixed Variations of Sansiviera and Coffee

| No | Variation Coffee (gr) : Sansiviera (gr) | CO Levels | | Absorption CO Level (ppm) | %E | x/m (ppm/g) |
|---|---|---|---|---|---|---|
| | | $t_0$ (ppm) | $t_1$ (ppm) | | | |
| 1. | 50 : 0 | 417 | 417 | 0 | 0 | 0 |
| 2. | 40 : 10 | 436 | 405 | 31 | 7.11 | 0.62 |
| 3. | 30 : 20 | 441 | 400 | 41 | 9.29 | 0.82 |
| 4. | 20 : 30 | 436 | 381 | 55 | 12.61 | 1,.1 |
| 5. | 10 : 40 | 445 | 382 | 63 | 14.15 | 1.26 |
| 6. | 0 : 50 | 447 | 377 | 70 | 15.65 | 1.4 |
| 7. | 50 : 50 | 447 | 379 | 68 | 15.21 | 1.36 |

**Table 3.** Absorption of $CH_4$ Levels in Mixed Variations of Sansiviera and Coffee

| No | Variation Coffee (gr) : Sansiviera (gr) | $CH_4$ Levels | | Absorption $CH_4$ Level (ppm) | %E | x/m (ppm/g) |
|---|---|---|---|---|---|---|
| | | $t_0$ (ppm) | $t_1$ (ppm) | | | |
| 1. | 50 : 0 | 0.40 | 0.10 | 0.3 | 75 | 0.006 |
| 2. | 40 : 10 | 0.30 | 0.14 | 0.16 | 53.33 | 0.0032 |
| 3. | 30 : 20 | 0.25 | 0 | 0.25 | 100 | 0.005 |
| 4. | 20 : 30 | 0.30 | 0.14 | 0.16 | 53.33 | 0.0032 |
| 5. | 10 : 40 | 0.10 | 0.00 | 0.1 | 100 | 0.002 |
| 6. | 0 : 50 | 0.35 | 0.13 | 0.22 | 62.8 | 0.004 |
| 7. | 50 : 50 | 0.40 | 0.05 | 0,.35 | 87.5 | 0.007 |

If the concentrations of CO, $CH_4$ and $H_2S$ gases are compared in Tables 2, 3 and 5, it turns out that the gas from motorbike combustion contains much higher CO gas than other gases with an average level of 438.42ppm, then $H_2S$ gas (20.28ppm) and $CH_4$ gas (0.3ppm).

Table 2 shows that the greater the amount of sansivera, the greater the effectiveness of absorption of CO levels. This means that sansivera is able to absorb CO gas with the highest effectiveness (15.65%), namely in the VI variation with the amount of coffee: sansivera = 0: 50. Table 2 also shows that coffee is unable to absorb CO gas, which can be seen in variation I, namely the composition of coffee: sansivera = 50: 0, the effectiveness is 0%.

**Table 4.** Absorption of $CO_2$ Levels in Mixed Variations of Sansiviera and Coffee

| No | Variation Coffee (gr) : Sansiviera (gr) | $CO_2$ Levels (%) $t_0$ | $t_1$ | Absorption $CO_2$ (%) | %E | x/m (ppm/g) |
|---|---|---|---|---|---|---|
| 1. | 50 : 0 | 2.25 | 1.80 | 1.07 | 47.55 | 0.0214 |
| 2. | 40 : 10 | 3.5 | 3.13 | 0.35 | 10 | 0.007 |
| 3. | 30 : 20 | 1.70 | 1.55 | 0.15 | 8.82 | 0.003 |
| 4. | 20 : 30 | 1.65 | 1.63 | 0.02 | 1.2 | 0.0004 |
| 5. | 10 : 40 | 1.60 | 1.54 | 0.06 | 3.75 | 0.0012 |
| 6. | 0 : 50 | 1.65 | 1.55 | 0.1 | 71.06 | 0.002 |
| 7. | 50 : 50 | 2.26 | 2.00 | 0.26 | 11.50 | 0.0052 |

**Table 5.** Absorption of $H_2S$ Levels in Mixed Variations of Sansiviera and Coffee

| No | Variation Coffee (gr) : Sansiviera (gr) | $H_2S$ Levels $t_0$ (ppm) | $t_1$ (ppm) | Absorption $H_2S$ (ppm) | %E | x/m (ppm/g) |
|---|---|---|---|---|---|---|
| 1. | 50 : 0 | | | 0 | 0 | 0 |
| 2. | 40 : 10 | 36 | 20 | 16 | 44,44 | 0,32 |
| 3. | 30 : 20 | 18 | 8 | 10 | 55,55 | 0,2 |
| 4. | 20 : 30 | 23 | 15 | 8 | 34,78 | 0,16 |
| 5. | 10 : 40 | 17 | 14 | 3 | 17,64 | 0,06 |
| 6. | 0 : 50 | 24 | 19 | 5 | 20,83 | 0,1 |
| 7. | 50 : 50 | 24 | 14 | 10 | 41,66 | 0,2 |

**Tabel 6.** Absorption of Pb Levels in Mixed Variations

| Mixed Code | Variation Coffe (gr) : Sansiviera (gr) | $C_o$ (ppm) | $C_e$ (ppm) | Adsorption Pb Levels | % E | x/m |
|---|---|---|---|---|---|---|
| 1 | 50 : 0 | 12.5 | 39.74 | 27.24 | 68% | 0.55 |
| 2 | 0 : 50 | 38.97 | 55.57 | 16.6 | 29% | 0.33 |
| 3 | 40 : 10 | 17.97 | 53.90 | 35.93 | 66% | 0.72 |
| 4 | 30 : 20 | 23.44 | 59.71 | 36.27 | 60% | 0.73 |
| 5 | 20 : 30 | 28.91 | 65.32 | 36.41 | 55% | 0.74 |
| 6 | 10 : 40 | 34.38 | 84.65 | 50.27 | 59% | 1.05 |

In Table 3, it can be seen that coffee and sansivera are able to absorb $CH_4$ gas. This can be seen in the effectiveness given to a mixture of variations 1 and 6 with the composition of coffee: sansivera, respectively 50: 0 and 0:50, namely 75% and 62.8%. The highest effectiveness is in mixtures with variations 3 and 5 with the composition of coffee: sansivera 30:20 and 10:40, respectively, that is 100%. Sansivera absorbs CO2 gas better than coffee, which can be seen in Table 4 in a mixture of variations 1 and 6. The effectiveness of the mixture with variation 6 also provides the highest effectiveness in absorbing CO2 gas. The mixture of variation 6 which only contains sansivera gives an effectiveness of absorbing 71.06 CO2, while the mixture with variation 1 which only contains coffee only gives an effectiveness of 47.55%. In Table 5, it can be seen that the mixture with variation 3, namely the composition of coffee: sansivera = 30:20, provides the highest effectiveness in absorbing H2S gas. While in Table 6 shows that coffee is very good at absorbing Pb gas, this can be seen in the mixture of variation 1 which only contains coffee which provides the highest effectiveness in absorbing Pb gas, which is 68%. The Table 7 shows the data that the composition of the coffee and sansiviera blends that provide the highest effectiveness.

**Tabel 7.** Absorption of CO Levels in Mixed Variations

| Gasses | Variation Coffe (gr) : Sansiviera (gr) | Gasses Levels | | Absorption Levels | %E | x/m |
|---|---|---|---|---|---|---|
| | | $t_0$ | $t_1$ | | | |
| CO | 0 : 50 | 447ppm | 377ppm | 70ppm | 15,65 | 1,4 (ppm/g) |
| $CH_4$ | 30 : 20 | 0,25ppm | 0,00ppm | 0,25ppm | 100 | 0,005 (ppm/g) |
| $CH_4$ | 10 : 40 | 0,10 ppm | 0,00 ppm | 0,1ppm | 100 | 0,002 (ppm/g) |
| $CO_2$ | 50 : 50 | 2,26% | 2,00% | 0,26% | 11,50 | 0,0052 (%/g) |
| $H_2S$ | 30 : 20 | 18 ppm | 8 ppm | 10 ppm | 55,55 | 0,2 (ppm/g) |

Note :
t0: rate of first contact time
t1: rate of final contact time

Based om Table 7, it can be seen that the organic fragrances mixed from the coffee and sansiviera are not only able to absorb CO and $CO_2$ gases and neutralize air but also reduce $CH_4$, $H_2S$ and Pb levels in motor vehicle fumes. The ability of the sansiviera to absorb pollutants is because in one strand it contains the active ingredient, pregnan glycoside, which is able to reduce pollutants into organic acids, sugars, and several amino acid compounds (Prasetyo, 2013). The use of coffee grounds as a mixture of the tongue-in-law for this organic fragrance is because coffee itself is able to neutralize air (Muspa, 2017). So that this organic fragrance will be very useful if used because it has many benefits.

## 4. Conclusion

Based on the description of the discussion above, it can be concluded that the mixture of Sansevieria and coffee powder effectively absorbed $CH_4$, $CO_2$, $H_2S$ and Pb gases with the highest effectiveness, respectively 87.5%, 71.06%, 55.5% and 59% with the composition of sansiviera: coffee in the sample respectively 50:50; 50: 0; 30:20; 40:10. Meanwhile, the mixture of sansiviera and coffee was slightly effective in absorbing CO gas with the highest effectiveness of 15.65% at the composition of 50: 0.

## References

[1] Xie D F, Gao Z Y, Zhao X M and Li K P 2009 Characteristics of mixed traffic flow with non-motorized vehicles and motorized vehicles at an unsignalized intersection *Physica A: Statistical Mechanics and its Applications* **388** (10) 2041-2050
[2] Tiwari G 1999 Towards a sustainable urban transport system: Planning for non-motorized vehicles in cities *Transport and Communications Bulletin for Asia and the Pacific* **68** 49-66
[3] Mehta K, Mehran B and Hellinga B 2015 Evaluation of the passing behavior of motorized vehicles when overtaking bicycles on urban arterial roadways *Transportation Research Record* **2520** (1) 8-17
[4] Resosudarmo B P and Napitupulu L 2004 Health and economic impact of air pollution in Jakarta *Economic Record* **80** S65-S75
[5] Listyarini S and Warlina L 2017 The Development Model of the Policy Alternatives in Controlling Air Pollution in the Jakarta Province, Indonesia *International Journal of Environmental Science and Development* **8** (7) 496
[6] Kusumaningtyas S D A, Aldrian E, Wati T, Atmoko D and Sunaryo S 2018 The recent state of ambient air quality in Jakarta *Aerosol and Air Quality Research* **18** (9) 2343-2354
[7] Kuncoro A H, Mellyanawaty M, Sambas A, Maulana D S, Subiyanto and Mamat M 2020 Air Quality Monitoring System in the City of Tasikmalaya based on the Internet of Things (IoT) *Jour of Adv Research in Dynamical & Control Systems* **12** (2) 2473-2479
[8] Herndon J M 2018 Air pollution, not greenhouse gases: The principal cause of global warming *J Geog Environ Earth Sci Intn* **17** (2) 1-8
[9] Jacobson M Z 2017 Roadmaps to transition countries to 100% clean, renewable energy for all purposes to curtail global warming, Air Pollution, and Energy Risk *Earth's Future* **5** (10) 948-952
[10] Jacobson M Z 2009 Review of solutions to global warming, air pollution, and energy security *Energy & Environmental Science* **2** (2) 148-173
[11] Herndon J M 2018 Air pollution, not greenhouse gases: The principal cause of global warming *J Geog Environ Earth Sci Intn* **17** (2) 1-8
[12] Gupta R, Somanathan E and Dey S 2017 Global warming and local air pollution have reduced wheat yields in India *Climatic Change* **140** (3-4) 593-604

IOP Conf. Series: Materials Science and Engineering 1115 (2021) 012076 doi:10.1088/1757-899X/1115/1/012076

# Absorption of Remazol Red dye from textile waste using activated carbon from coconut shell

**N L P A Saraswati*** **and I D K Sastrawidana**

Department of Chemistry, Faculty of Mathematics and Natural Sciences, Universitas Pendidikan Ganesha, Udayana street 11 Singaraja, Bali

Corresponding author: putu.ananda@undiksha.ac.id

**Abstract.** The use of dyes in textile industry has great potential in polluting the environment, particularly due to the remaining dyes in wastewater are very difficult to decompose naturally. This pollution can be reduced by conducting special treatment on textile waste to decrease the dyes level. One of the most popular is the adsorption method. The purpose of this research was to study the adsorption of Remazol Red dye in textile waste by activated carbon synthesized from coconut shell. Activated carbon was given two different treatments: without activation and with activation at 900 °C. The results showed that the activated carbon synthesized had water content of 22.52% (without activation) and 9.49% (with activation). The adsorption of these activated carbons against Remazol Red dye took place optimally at the contact time of 30 minutes by following the Langmuir adsorption isotherm model. The adsorption process occurs homogeneously with the formation of monolayer with a $q_m$ value (maximum monolayer capacity) of 226.2 mg/g (without activation) and 250 mg/g (with activation).

## 1. Introduction

The textile industry is currently known to provide the largest contribution to water pollution, far greater than the ink, paper, leather and printing industries [1]-[4]. This is due to the use of various synthetic dyes that will inevitably result in non-biodegradable waste which pollutes the aquatic ecosystems and create negative impact on human health [5]-[7]. One of the most widely used dyes in textile industry is Remazol Red, a water soluble compound that belongs to anthracene derivative group. Recently, various efforts, techniques, and methods have been developed to reduce the levels of this dye in textile industrial waste. Among them, adsorption is preferred due to its simple procedure with relatively low cost and abundant adsorbents availability. The trend of research that based on environmentally friendly materials gives major influence in the selection of the adsorbent. Activated carbon is one the powerful candidate mainly due to the abundant carbon sources in nature, both from plants and animals [8]-[10]. Bali as a coastal area which is rich in coconut plantation products has great potential in developing activated carbon-based absorbent from coconut shell.

Adsorbents derived from natural materials generally have poor adsorption properties, thus it is necessary to carry out an activation process to obtain better surface characteristics, such as large surface area, number of pores, and pore size. Apart from these surface characteristics, the adsorption effectiveness is also influenced by contact time, solution concentration, and temperature. In general, the adsorption behavior of a system can be described well using an adsorption isotherm model. Choosing the right model will make it easier to determine the optimum conditions that can be achieved by the studied materials.

In this research, activated carbons derived from coconut shell were synthesized and a series of analyzes were carried out to determine the suitable model for the adsorption of Remazol Red solution. The charcoals obtained from coconut shell were given two different treatments: without activation and with activation at 900 °C. Remazol red solutions were made by dissolving certain amount of Remazol Red powder with distilled water. The study of isotherm adsorption model was carried out by mixing the charcoal with Remazol Red solution in various concentration and different contact time to obtain the optimum condition. The concentration of Remazol Red solutions after the adsorption was measured using UV-Vis spectrophotometry technique.

## 2. Experimental

### 2.1. Material and devices

The materials used in this study were coconut shell, Remazol Red dye, iodine, potassium iodide, sodium thiosulfate pentahydrate, sodium carbonate, hydrochloric acid, sodium hydroxide, distilled water, and filter paper. The equipment used is the laboratory glassware, analytical balance, ovens, and desiccators. Characterization was carried out using UV-Vis spectrophotometer.

### 2.2. Preparation of activated carbon from coconut shell

The coconut shells are washed with water, cut into small pieces, and then burned in a container. The combustion product was crushed with a ball mill. The result was sieved to obtain fine activated carbon. In this study, the effect of activation temperature was analysed on the adsorption capacity of activated carbon. The activation was carried out at a temperature of 900 °C.

### 2.3. Preparation or Remazol Red solutions

500 mL of 500 ppm Remazol Red main solution was prepared by dissolving 0.25 grams of Remazol Red powder with distilled water to a volume of 500 mL in a measuring flask. The preparation of various concentrations of Remazol Red solution is done by diluting a certain amount of the stock solution with distilled water.

### 2.4. Characterization of activated carbon

The mass of activated carbon before and after heating is weighed to calculate the water content in it. The water content of activated carbon was characterized by heating it in an oven at 105 °C for 2 hours. Activated carbon was also characterized for its adsorption ability by testing the iodine number using a solution of $I_2$ with KI in an acidic atmosphere. Iodometric titrations were carried out with $Na_2S_2O_3$ standard solution

### 2.5. Study of optimum contact time

A total of 25 mL of 200 ppm Remazol Red solution was put into a 100 mL Erlenmeyer flask. To the solution, 25 mg of activated carbon was added and contacted at room temperature with time variation of 10; 20; 30; 50; 75, and 90 minutes. The remaining Remazol Red concentrations were then analyzed using a UV-Vis spectrophotometer.

### 2.6. Study of adsorption isotherm model

Remazol Red solution with a concentration of 50 ppm, 100 ppm, 150 ppm, 200 ppm, 250 ppm, and 300 ppm were prepared. A total of 25 mL of the prepared solution was put into a 100 mL Erlenmeyer flask, and 25 mg of activated carbon was added. The contact process was carried out for 30 minutes. The remaining Remazol Red concentrations were then analyzed using a UV-Vis spectrophotometer.

## 3. Results and Discussion

### 3.1. Characteristics of synthesized charcoal

The water content of coconut shell charcoal without and with activation is shown in Table 1. Based on the analysis, inactivated charcoal has a water content of 22.52%. After being activated, charcoal has lower water content, namely 9.49%. Activation at 900 °C reduces the water content to more than double the original water content of the charcoal.

**Table 1.** Water content in charcoal obtained

| Sample | Initial weigh (g) | Final weigh (g) | Water content (%) |
|---|---|---|---|
| Without activation | 1.9437 | 1.5059 | 22.52 |
| With activation at 900 °C | 3.2506 | 2.9420 | 9.49 |

In this research, the measurement of iodine number was also carried out. This aims to analyze the surface characteristics of the charcoals. A large iodine number indicates the good surface porous to support adsorption process. The national standard for charcoal as activated carbon in adsorption is having iodine number greater than 750 mg/g. The determination of iodine number was carried out using the Freundlich model, by plotting log $q_e$ (charcoal adsorption capacity at equilibrium) with log $C_e$ (concentration of iodine solution at equilibrium). The graph for determining the iodine number of activated carbon produced from coconut shells without and with activation is shown in Figure 1.

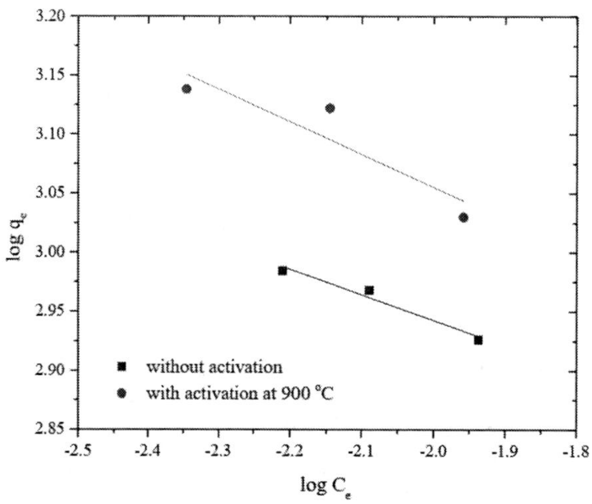


**Figure 1.** Determination of iodine number

Based on the Figure 1, the mathematical equation for activated carbon without activation is y = -0.214x + 2.513 with linearity (R2) 0.966. Mathematical equation for activated carbon with activation y = -0.277x + 2.501 with linearity (R2) 0.845. Referring to the international procedure established by ASTM, the definition of the iodine number is the value of the adsorption capacity ($q_e$) when the iodine ($C_e$) solution concentrates 0.02 N, so the iodine numbers obtained for charcoal without activation is 752.6 mg/g and for charcoal with activation at 900 °C is 936.7 mg/g.

### 3.2. Optimum contact time

The adsorption process does not occur all at once, but slowly with the absorption rate depending on the interaction that occurs between the adsorbate and the adsorbent. The relation between contact time and charcoal absorption without and with activation of Remazol Red dye is shown in Figure 2.

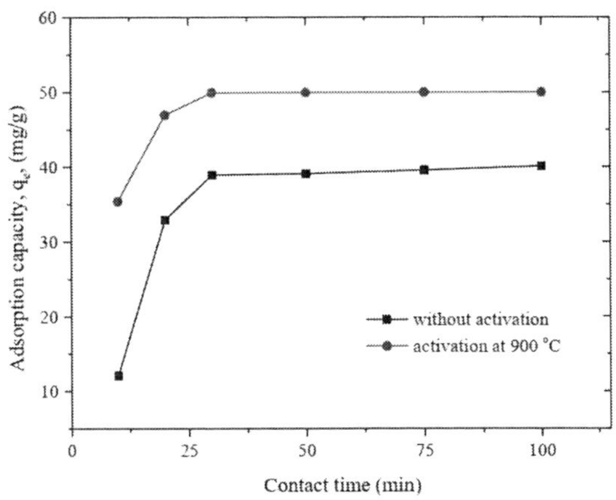


**Figure 2.** Relation of contact time with charcoal absorption of Remazol Red dye

Based on Figure 2, it can be seen that the adsorption capacity increases with increasing contact time until it reaches equilibrium or maximum absorption. Both types of charcoal produced in this study have the optimum contact time for Remazol Red dye adsorption in 30 minutes. After 30 minutes, the adsorption process was still on-going but the absorption capacity did not increase significantly.

### 3.3. Reamazol Red adsorption isotherm model with activated carbon

In this study, Freundlich and Langmuir adsorption isotherm model was used to determine the most suitable model for the adsorption of Remazol Red dye from coconut shell-based activated carbon. The Freundlich adsorption isotherm model was obtained by distributing log $q_e$ to log $C_e$ according to the equation: $\log q_e = \log K_F + 1/n \log C_e$. The graphs obtained using Freundlich adsorption isotherm model for both charcoal are given in Figure 3.

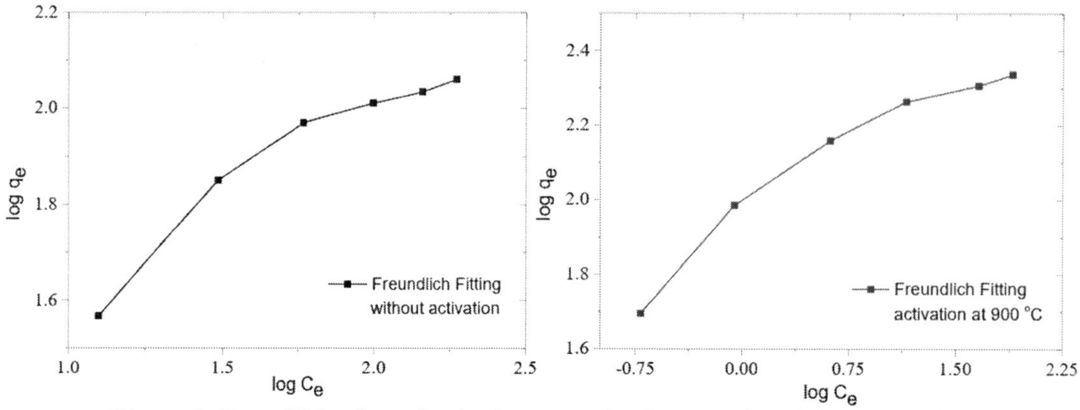


**Figure 3.** Freundlich adsorption isotherm graph of Remazol Red dye adsorption

The Langmuir adsorption isotherm model is obtained by plotting $\frac{C_e}{q_e}$ to $C_e$ according to: $\frac{C_e}{q_e} = \frac{1}{q_m} C_e + \frac{1}{q_m K_L}$. The curves obtained using the Langmuir adsorption isotherm model are given in Figure 4.

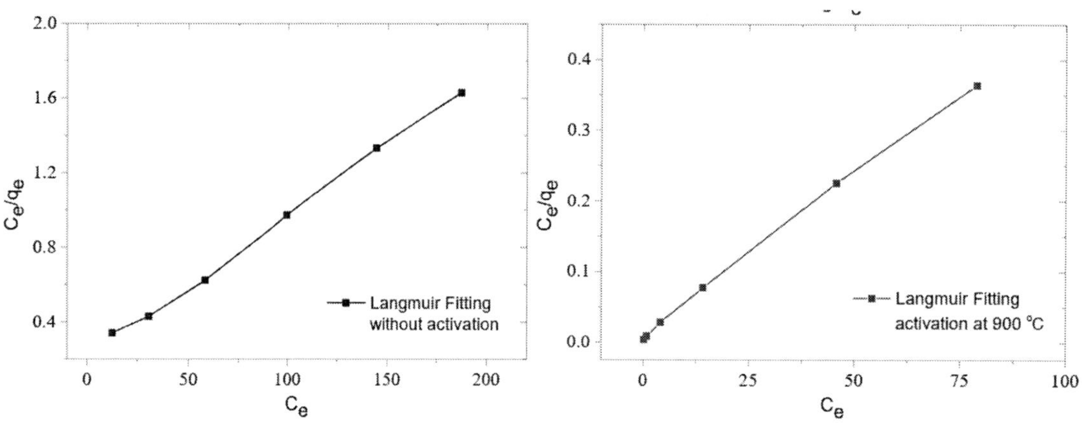


**Figure 4.** Langmuir adsorption isotherm graph of Remazol Red dye adsorption

Linearity indicates an isotherm model that is more suitable for the system being analyzed. Based on the two curves above, the linearity ($R^2$) of the Langmuir adsorption isotherm model is greater than that of the Freundlich adsorption isotherm model. Thus, the adsorption process of Remazol Red by the two types of activated carbon produced in this study followed the Langmuir adsorption isotherm model. The adsorption process occurs homogeneously with the formation of a monolayer (single layer) on the surface of the activated carbon without any interaction between adsorbate molecules. Adsorbent from charcoal with activation at 900 °C showed greater Langmuir adsorption parameters with $q_m$ dan $K_L$ value 218.34 mg/g and 0.56, respectively, compared to adsorbent without activation with $q_m$ dan $K_L$ 131.75 mg/g and 0.04, respectively.The adsorption parameters of the Freundlich and Langmuir models are given in Table 2.

**Table 2.** Adsorption parameters with Freundlich and Langmuir adsorption isotherm models

| Sampel | Freundlich model | | | Langmuir model | | |
|---|---|---|---|---|---|---|
| | n | $K_F$ | $R^2$ | $K_L$ | $q_m$ | $R^2$ |
| Without activation | 2.5 | 34.0 | 0.952 | 0.07 | 226.2 | 0.965 |
| With activation at 900 °C | 4.1 | 82.6 | 0.921 | 0.41 | 250.0 | 0.990 |

When the result of these charcoals compared, it can be concluded that sample with activation showed higher Langmuir parameters and thus the adsorption capacity compare to the sample without activation. We strongly suggest this result was due to the formation of better charcoal surface morphology after activation at high temperature, which is also supported by the water content data. Charcoal with activation at 900 °C contain less water and high pores to support the adsorption process.

## 4. Conclusion

Based on the results of the activities carried out, it can be concluded that the activated carbon produced from coconut shells has characteristics in the form of a moisture content of 22.52% (without activation) and 9.49% (with activation), and an iodine number of 752.6 (without activation) and 936.7 (with activation). The adsorption isotherm model that is suitable for the adsorption of Remazol Red dye with activated carbon produced is the Langmuir model with KL values of 0.07 (without activation) and 0.41 (with activation). The optimum contact time for activated carbon with Remazol Red dye is 30 minutes.

### References

[1] Franco M A 2017 Circular economy at the micro level: A dynamic view of incumbents' struggles and challenges in the textile industry *Journal of Cleaner Production* **168** 833-845
[2] Ciardelli G and Ranieri N 2001 The treatment and reuse of wastewater in the textile industry by means of ozonation and electroflocculation *Water research* **35** (2) 567-572
[3] Desore A and Narula S A 2018 An overview on corporate response towards sustainability issues in textile industry *Environment, Development and Sustainability* **20** (4) 1439-1459
[4] Ahmed S S, Akter T and Ma Y 2018 Green supply chain management (GSCM) performance implemented by the textile industry of Gazipur district, Dhaka *Logistics* **2** (4) 21
[5] Schwarzenbach R P, Egli T, Hofstetter T B, Von Gunten U and Wehrli B 2010 Global water pollution and human health *Annual review of environment and resources* **35** 109-136
[6] Horrigan L, Lawrence R S and Walker P 2002 How sustainable agriculture can address the environmental and human health harms of industrial agriculture *Environmental health perspectives* **110** (5) 445-456
[7] Alberti M, Booth D, Hill K, Coburn B, Avolio C, Coe S and Spirandelli D 2007 The impact of urban patterns on aquatic ecosystems: An empirical analysis in Puget lowland sub-basins. *Landscape and urban planning* **80** (4) 345-361
[8] Baseri J R, Palanisamy P N and Sivakumar P 2012 Preparation and characterization of activated carbon from Thevetia peruviana for the removal of dyes from textile waste water *Advances in Applied Science Research* **3** (1) 377-383
[9] Jiang J and Sun K 2017 Review on preparation technology of activated carbon and its application *Chemistry and Industry of Forest Products* **37** (1) 1-13
[10] Al-Ghouti M A and Da'ana D A 2020 Guidelines for the use and interpretation of adsorption isotherm models: A review *Journal of hazardous materials* **393** 122383

# Synthesis of Chitosan-Alginate Beads and Its Application for Alkyl Benzene Sulfonates (ABS) Adsorption

**N P S Ayuni*, I N Sukarta and P A Wulandari**

Department of Chemistry, Faculty of Mathematics and Natural Sciences, Universitas Pendidikan Ganesha, Indonesia.

*Corresponding author: sri.ayuni@undiksha.ac.id

**Abstract.** ABS is the largest group of anionic surfactants. Anionic surfactants (AS), especially ABS are used extensively due to their impacts on ecosystems and are usually disposed of after their usein the environment. Therefore, they represent one of the main causes of water pollution. Methods have been proposed to remove surfactants such as adsorption with beads. This study aims to analyze the results of the synthesis and characterization of chitosan-alginate beads and determine the efficiency adsorption of ABS. The synthesized beads were characterized by swelling tests. To find out the maximum efficiency of ABS, ABS adsorption with chitosan-alginate beads was carried out by varying pH (5, 6, 7, 8, and 9), time (15, 30, 45, 60, 75, and 90 minutes) and ABS concentration (10, 20, 30, 40, and 50 mg/L). The results of the adsorption solution were measured by a UV-Vis spectrophotometer.Theresults of theswelling test onbeadsfor 24 hoursreachedanaverage of 148.18%.The maximum efficiency adsorption of chitosan alginate beads was determined to be 46.46% with an initial ABS concentration of 30 mg/L at pH 6 in 60 minutes.

## 1. Introduction

Surfactants are a diverse group of chemicals with cleaning properties and consist of two heads with a different polarity or solubility in water: a polar head group, which is well solvated in water, and a non-polar hydrocarbon tail, which is not easy to dissolve in water. surfactants combine hydrophobic and hydrophilic properties in one molecule. Synthetic surfactants are economically important chemicals. They are widely used in household cleaning detergents, personal care products, textiles, paints, polymers, pesticide formulations, pharmaceuticals, mining, oil recovery, and pulp and paper industries. The world production of synthetic surfactants amounts to 7.2 million tons annually [1,2].

Surfactants are classified by their ionic activity in water into four types: anionic, cationic, non-anionic, and amphoteric [3]. ABS is the largest group of anionic surfactants. Anionic surfactants (AS), especially ABS are used extensively due to their impacts on ecosystems and are usually disposed of after their usein the environment [4,5]. Although a wide range of physical and chemical processes such as chemical precipitation, ion exchange, membrane filtration, solvent extraction, etc. are available for the removal of surfactans; most of these methods are not practicable. Therefore, the development of low-cost methods to remove surfactanshas received the attention of scientists worldwide and biosorptionhas been recognized as a cost-effective method of removal of surfactans contaminants in water using low cost bioadsorbents such as activated carbons, burned brick particles, clay, some aquatic weeds, agricultural and biological wastes, chitosan, and zeolites, etc. Chitosan is the second abundant polycationic biopolymer after cellulose derived from chitin. Chitosan is insoluble in air, but it is a repetitive structure of the amine group that allows it to be dissolved in dilute acids. Under acidic conditions, amine groups can easily be protonated and become positively charged. In this experiment, chitosan was free with alginate. Combining chitosan with alginate which is formed into chitosan-alginate beads is an effective way to achieve ABS levels in laundry waste. Chitosan-alginate in the form of beads has a larger pore size, better diffusion properties, and is hydrodynamic thereby increasing the adsorption ability.The present study aims is to obtain a characterization of chitosan alginate beads and determine the efficiency adsorption of ABS

## 2. Materials and Method

The materials used were distilled water, chloroform, methylene blue, chitosan, alginate, acetic acid 5% solution, sodium hydroxide solution, ABS, sulfuric acid 6 N, sodium dihydrogen phosphate monohydrate, and isopropyl alcohol. The instruments used in this research are UV-Vis spectrophotometer, analytical balance, magnetic stirrer, and glassware.

### 2.1 Synthesis Chitosan-Alginate Beads

The chitosan solution is made by dissolving 1 gram of chitosan in anacetic acid solution 5% of 90 mL in a 500 mL beaker [6]. Then the alginate solution is made by dissolving 0.1 gram in 10 mL of distilled water [7]. The chitosan solution and alginate solution are then mixed and stirred until homogeneous. This solution is then stirred with a stirrer for 24 hours. The mixture of chitosan alginate solution is then printed using a dropper by dropping in NaOH 2 M solution and soaking for 3 hours. The chitosan alginate beads formed were filtered. After that, the beads are washed with distilled water until the beads are neutral. Chitosan alginate beads are then dried using an oven at 50 °C for 2 hours [8].

### 2.2. Determination of pH, Contact Time, and Optimum ABS Solution Concentration

The 0.1 grams of chitosan-alginate beads were added to 15 mL of 30 mg/L ABS solution with varying pH (5, 6, 7, 8, and 9), then stirred for 60 minutes at 150 rpm [9]. The filtrate obtained was extracted by the MBAS procedure. Then the optimum contact time was determined with variations (15, 30, 45, 60, 75, and 90 minutes) with the adsorbent mass of 0.1 gram in 15 mL of 30 mg/L ABS solution at the optimum pH. The filtrate obtained was extracted by the MBAS procedure (5 mL of the filtrate was taken and put into a 50 mL volumetric flask, then added 10 mL of 100 mg/L methylene blue solution and distilled water until the limit mark. The mixture of this solution was extracted with 10 mL of chloroform. The organic phase (the resulting chloroform) was measured for its absorbance value using a UV-Vis spectrophotometer at a wavelength of 652 nm). pH and contact time that produce maximum adsorption efficiency are used to determine the optimum concentration of ABS solution (10, 20, 30, 40, and 50 mg/L) [8].

### 2.3 Data Analysis

The results of the synthesis of chitosan-alginate beads were then characterized. The results of characterization include the water adsorption test (swelling test) (Equation1). The data will be analyzed descriptively. ABS adsorption efficiency data using a UV-Vis spectrophotometer will be analyzed quantitatively with the results obtained are absorbance data and ABS concentration. From these data it can be calculated the ABS adsorbed concentration and the efficiency (%E) of the variation in contact time and ABS concentration. The maximum adsorption efficiency of ABS with variations in pH, variations in contact time, and ABS concentration was calculated using Equation 2. The values obtained from these data will be presented in graphical form and analyzed descriptively.

$$\% \, Swelling \, Test = \left(\frac{W_{wet}-W_{dry}}{W_{dry}}\right) \times 100\% \quad (1)$$

$$\%E = \frac{C_o - C_s}{C_o} \times 100\% \quad (2)$$

$C_o$ was the initial concentration (mg/L), $C_s$ was the residual concentration (mg/L)

## 3. Results and Discussion

Alginate beads with a mass ratio of chitosan and sodium alginate of 0.1: 0.01 (g /g) and a volume ratio of acetic acid and distilled water 9:1 (v/v). Beads were synthesized from chitosan stirring in acetic acid solution 5% and alginate in distilled water. The results of chitosan alginate beads show in Figure 1.

**Figure 1.** Chitosan-Alginate Beads

In an acidic atmosphere, chitosan forms polycationic, while alginate forms polyanionic. Polycationics and polyanionic will react to form polyelectrolyte through ionic reactions [10]. The chitosan material modified with alginate aims to increase the active group. When chitosan and alginate are combined into one, there will be a reaction between the carboxyl group of the alginate and the amine group from chitosan to form an amide group so that it can increase the adsorption ability [11]. To determine the interaction between the carboxyl group and the amine group of these two materials

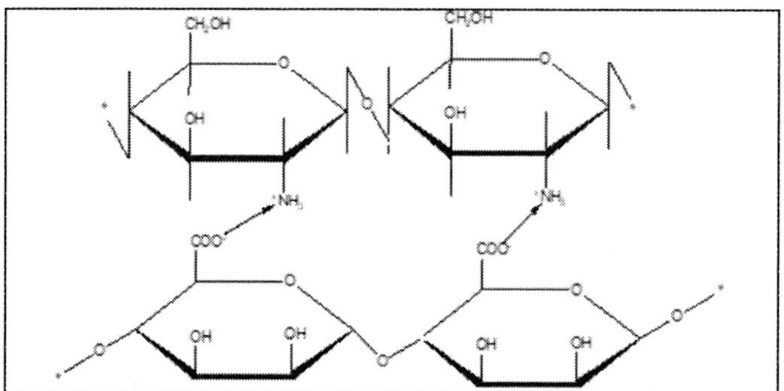

**Figure 2.** Interaction the carboxyl group of the alginate and the amine group from chitosan

The swelling test was carried out to determine the capacity of the beads to absorb water (hydrophilic nature) so that they were able to expanded due to the presence of pores between polymer bonds [8]. Water absorption was carried out using the gravimetric method, namely by calculating the difference between the dry weight and the wet weight of the beads. From the research results, the swelling test on beads for 24 hours reached an average of 148.18% (Table 1). These results indicate that the chitosan-alginate beads can expand in water so that it can be used as an adsorbent [8]. When a biopolymer comes into contact with a liquid, there will be expansion due to thermodynamics. When the beads expand, the mobility of the polymer chain increases so that the penetration of the solvent is easier. The development of chitosan-alginate beads is due to the ionic interaction between the amino group ($NH_3^+$) of chitosan and the carboxyl group ($COO^-$) of the alginate. Chitosan-alginate beads have a large degree of swelling which indicates that they are good at absorbing water. The amount of water that can be absorbed in the beads is due to the pores of the beads. The swelling test of the beads can show the hydrophilicity of the beads. The more chitosan composition will increase the hydrophilicity of the beads so that the number of diffused ABS molecules will increase.

**Table 1.** Swelling Test of Chitosan-Alginate Beads

| Replication | $W_{wet}$ (g) | $W_{dry}$ (g) | % |
|---|---|---|---|
| 1 | 0.1012 | 0.2518 | 148.81 |
| 2 | 0.1004 | 0.2483 | 147.31 |
| 3 | 0.1035 | 0.2571 | 148.41 |
| | **Average** | | **148.18** |

Figure 3 shows that the adsorption of ABS by chitosan-alginate beads shows an optimum pH efficiency at pH 6 with an efficiency of 48.25%. The decreasing of the acidity of the food solution will increase the protonation by neutralizing the negative charge from the adsorbent surface so that it will facilitate the diffusion process in the adsorbent area. When the beads adsorption process which has the main $NH_2$ group will undergo a protonation process to form $NH_3^+$ so that there will be adsorption of solution cations by the beads through the exchange of $H^+$ ions which occur electrostatically. In this study, the pH 6 beads reached the optimum condition because the $NH_2$ group is in the form of $NH_3^+$ so that it acts as an electron-pair donor that produces complex compounds. pH above 6, there is a decrease in adsorption efficiency because pH above 6 contains a lot of OH- ions, which results in a rejecting reaction with the adsorbent, and precipitation will occur so that the absorption capacity is reduced.

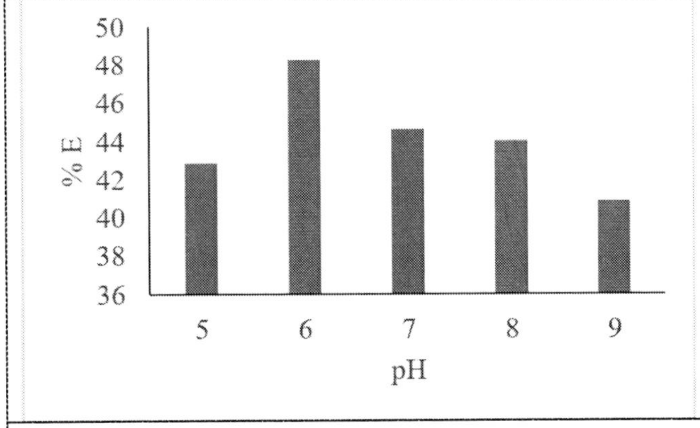


**Figure 3.** The Adsorption of ABS by Chitosan-Alginate Beads with pH Variation

Figure 4 shows that the chitosan-alginate beads can adsorb optimally in 60 minutes with an efficiency value of 68.16%. At the contact time of 15, 30, and 45 minutes there is still a space that has not been filled by the ABS contained in the beads because the active groups in the beads have not interacted optimally. In the 60th minute the space in the adsorbent is filled optimally, this is because the longer the contact time, the more likely the adsorbed substance will be absorbed. Meanwhile, at the contact time of 75 and 90 minutes, there was a decrease in adsorption ability, with increasing contact time, the absorption of the beads would decrease so that the beads were unable to bind the ABS molecules, and the ABS molecules as adsorbate would be released from the beads.

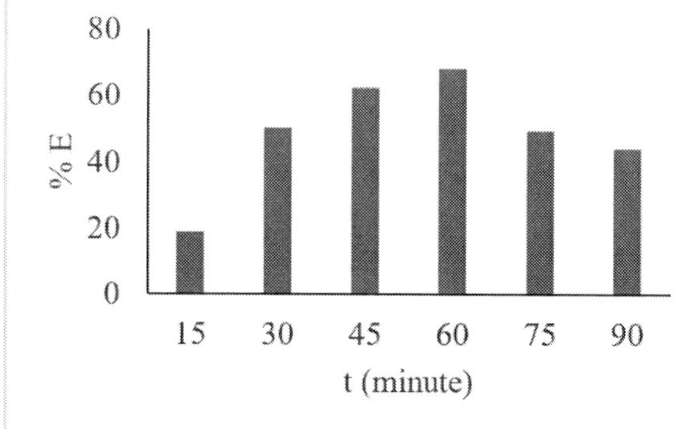


**Figure 4.** The Adsorption of ABS by Chitosan Alginate Beads with Time Contact Variation

Figure 5 shows that the optimum concentration efficiency is at a concentration of 30 mg/L with an adsorption efficiency value of 46.46%, in this state the empty spaces on the adsorbent have been filled optimally by ABS because the active groups of the beads bind to ABS. Concentrations of 10 and 20 mg/L have lower adsorption efficiency values because of the large number of empty spaces in the adsorbent that have not been filled by ABS. At concentrations of 40 and 50 mg/L, there was a decrease in the value of adsorption efficiency because the concentration was too high which caused the amount of ABS in the solution to be not proportional to the number of adsorbents so that the absorption decreased. The addition of ABS concentration will decrease the adsorbent absorption because in this condition the adsorbent is saturated. The concentration of ABS is closely related to the number of active sites of the adsorbent, if the number of active sites on the adsorbent surface is large enough compared to the amount of ABS, the absorption efficiency will be high.

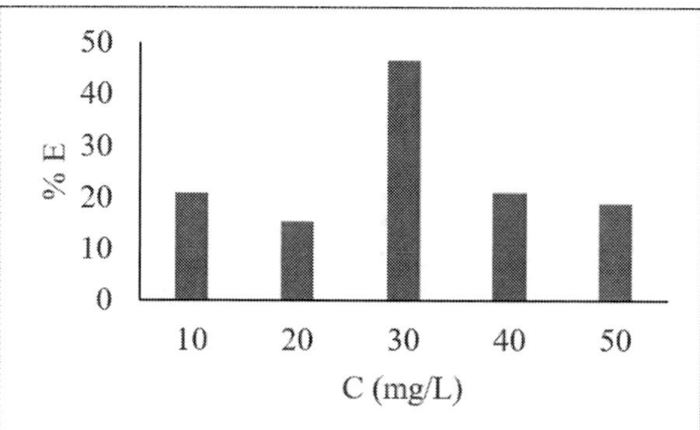


**Figure 5.** The Adsorption of ABS by Chitosan-Alginate Beads with Concentration Variation

## 4. Conclusions

The results of thes welling test onbeadsfor 24 hoursreachedanaverage of 148.18%. The maximum efficiency adsorption of chitosan alginate beads was determined to be 46.46% with an initial ABS concentration of 30 mg/L at pH 6 in 60 minutes.

## Acknowledgments

This work was supported by Research and Community Service Institution Universitas Pendidikan Ganesha under Penelitian Terapan Scheme 2020 (contract number: 793/UN48.16/LT/2020).

## References

[1] Ying G G 2006 Fate, behavior and effects of surfactants and their degradation products in the environment *Environment international* **32** (3) 417-431
[2] Mungray A K and Kumar P 2009 Fate of linear alkylbenzene sulfonates in the environment: a review *International Biodeterioration & Biodegradation* **63** (8) 981-987
[3] Ghoochani M, Shekoohiyan S, Mahvi A H, Haibati B and Norouzi M 2011 Determination of detergent in Tehran ground and surface water *American-Eurasian J Agric Environ Sci* **10** (3) 464-469
[4] Verge C, Moreno A, Bravo J and Berna J L 2001 Influence of water hardness on the bioavailability and toxicity of linear alkylbenzene sulphonate (LAS) *Chemosphere* **44** (8) 1749-1757
[5] Brandt K K, Hesselso M, Roslev P, Henriksen K and So J 2001Toxic effects of linear alkylbenzene sulfonate on metabolic activity, growth rate, and microcolony formation ofNitrosomonas and nitrosospira strains *Applied and environmental microbiology* **67** (6) 2489-2498
[6] Nitsae M, Madjid A, Hakim L and Sabarudin A 2016 Preparation of chitosan beads using tripolyphosphate and ethylene glycol diglycidyl ether as crosslinker for Cr (VI) adsorption *Chemistry and Chemical Technology* **10** 105-114
[7] Fajarwati F I, Kurniawan M A, Fatima M N and Fikrina R 2018 Penghilangan Zat Warna menggunakan Kompleks Polielektrolit Kitosan-Alginat *JPSCR: Journal of Pharmaceutical Science and Clinical Research* **3** (1) 36-42
[8] Kurniyati R, Sumarni W and Latifah, L. (2015). Pengaruh chitosan beads dan chitosan beads ikat silang asam sitrat sebagai penurun kadar fosfat dan ABS (Alkyl Benzene Sulfonate) pada limbah laundry *Indonesian Journal of Chemical Science* **4** (1) 36-41
[9] Gheorghita Puscaselu R, Lobiuc A, Dimian M and Covasa M 2020 Alginate: From food industry to biomedical applications and management of metabolic disorders *Polymers* **12** (10) 2417
[10] Komariah A, Sriatun S and Pardoyo P 2017 Adsorpsi Alkil Benzena Sulfonat Menggunakan Zeolit Termodifikasi Cetyltrimethylammonium. *Jurnal Kimia Sains dan Aplikasi*, *20*(1), 13-18.
[11] Kang Y G, Vu H C, Le T T and Chang Y S 2018 Activation of persulfate by a novel Fe (II)-immobilized chitosan/alginate composite for bisphenol A degradation *Chemical Engineering Journal* **353** 736-745

IOP Conf. Series: Materials Science and Engineering 1115 (2021) 012078 doi:10.1088/1757-899X/1115/1/012078

# Phytochemical Screening and Antioxidant Activity of *Hippobroma longiflora* Extracts

**N W Martiningsih[1,*], I W Mudianta[1] and I A P Suryanti[2]**

[1]Department of Chemistry, Faculty of Mathematics and Natural Sciences, Universitas Pendidikan Ganesha, Indonesia
[2]Department of Biology and Marine Fisheries, Faculty of Mathematics and Natural Sciences, Universitas Pendidikan Ganesha, Indonesia

*Corresponding author: wayan.martiningsih@undiksha.ac.id

**Abstract.** This study aims to identify the secondary metabolite compounds and to test the antioxidant activity of ethanol extract, n-hexane fraction, and chloroform fraction of *Hippobroma longiflora* leaves. Extraction was carried out by the maceration method using ethanol. The resulting crude ethanol extract was then partitioned with n-hexane and chloroform. Each extract and fraction was then tested by phytochemical screening. Antioxidant activity testing was carried out using the 2,2-diphenyl-1-picrylhydrazyl (DPPH) method. The phytochemical screening test showed that the ethanol extract contained flavonoids, saponins, triterpenoids, and alkaloids; the n-hexane fraction contained steroids, and alkaloids; and the chloroform fraction contained flavonoids, steroids, and alkaloids. Analysis of the antioxidant activity revealed the following Inhibitory Concentration ($IC_{50}$ values): 9.57 ppm of ethanol extract, 99.59 ppm of n-hexane fraction, 48.54 ppm of chloroform fraction, and 4.30 ppm of ascorbic acid. Based on these results, the antioxidant activity of the ethanol extract was more potential than the n-hexane and chloroform fractions, but smaller than ascorbic acid.

## 1. Introduction

The rapid development of technology and industry has changed the behavior and lifestyle of the community. Unhealthy lifestyles such as consuming fast food, lack of time to exercise, and smoking habits as well as environmental factors such as vehicle fumes and solar radiation can cause an increase in the number of free radicals in the body. This condition will trigger oxidative stress [1]. The magnitude of the negative effect of free radicals on human health causes the body to require an intake that contains an antioxidant compound that will be able to capture and neutralize free radicals. It happens because antioxidants can provide electron pairs to free radicals so that they can stop further reactions that cause oxidative stress and cell damage to be avoided [2]–[4].

There are various kinds of food ingredients that have natural antioxidant properties such as apples, tomatoes, and green tea. Besides, there are also synthetic antioxidants that are produced through several chemical reactions, such as Tert-Butyl Hydroxy Quinone (THBQ), Butyl HydroxyAnisol (BHA), and Butyl Hydroxy Toluene (BHT). However, these synthetic antioxidants have side effects on the body and harmful to human health so that their use must be monitored [5].

*Hippobroma longiflora* is a plant that has potential as a natural antioxidant and its use has not been widely studied for reducing free radical activity in the body. The classification of *H. longiflora* plants is as follows:

| | |
|---|---|
| Kingdom | : Plantae |
| Subkingdom | : Tracheobionta |
| Superdivision | : Spermatophyta |
| Division | : Magnoliophyta |
| Class | : Magnoliopsida |
| Subclass | : Asteridae |
| Order | : Asterales |
| Family | : Campanulaceae |
| Genus | : Hippobroma |
| Species | : *Hippobroma longiflora* (L.) |
| Synonym | : |
| | *Isotoma longiflora* (L.) |
| | *Isotoma longiflora var. runcinata* (Hassk) |
| | *Isotoma runcinata* Hassk |
| | *Laurantia longiflora* (L.) |
| | *Laurantia longiflora var. runcinata* (Hassk) |
| | *Lobelia longiflora* L |
| | *Rapuntium longiflora* (L.) |
| | *Solenopsis longiflora* (L.) |

The leaves of *H. longiflora* contain alkaloids, phenolics, flavonoids, triterpenoids, steroids, coumarins, saponins, and tannins [6]-[7]. The activity of alkaloids and flavonoids has been reported to have many pharmacological effects including anti-inflammatory, antioxidant, anticancer, antidiabetic, antibacterial, antimalarial, antitumor, antimicrobial, antifungal, anti-insecticide, and antiseptic [8]-[9]. Flavonoids are a group of polyphenolic compounds that act as antioxidants, namely to capture free radicals [10]. The presence of polar and non-polar compounds in *H. longiflora* is the reason to research the activity fractionation of leaves of *H. longiflora* by using three types of solvents, namely ethanol, n-hexane, and chloroform. This study aim to identify the secondary metabolite compounds and test the antioxidant activity of ethanol extract, n-hexane fraction, and chloroform fraction of *H. longiflora* leaves.

## 2. Materials and methods

### 2.1. Materials

The chemicals used for the extraction are ethanol, n-hexane, and chloroform. Dragendorff reagent, Liebermann-Burchard reagent, Mayer reagent, Bouchardat reagent, Wagner reagent, sulphuric acid, sodium hydroxide, magnesium ribbon, hydrochloric acid, and ferric chloride for the phytochemical screening test. Methanol, DPPH, and ascorbic acid for antioxidant activity test. The tools which used in this study included a rotary vacuum evaporator, UV-VIS spectrophotometer, and other glass tools.

### 2.2. Plant sample

Fresh leaves of *H. longiflora* were collected from Singaraja Region, Bali, Indonesia. The botanical identification and authentication were confirmed at the Indonesian Institute of Sciences (LIPI) UPT Bali Botanical Garden Eka Karya Bali.

### 2.3. Preparation of samples

*H. longiflora* leaves are taken from the plant and then washed in running water. Leaves that have been washed are cut into small pieces, then air-dried without being exposed to sunlight at room temperature for 7 days. The dried leaves are then ground to a powder.

### 2.4. Extraction procedure

The crude powder that has been weighed (100 gm) is macerated with ethanol. Every 24 hours the extract is filtered and replaced with a solvent. This extraction was repeated until the last extract contained no metabolites. The obtained ethanol filtrate is evaporated to produce a crude ethanol extract. Furthermore, the crude extract (10 gm) is dissolved in 100 mL of water. This water extract was partitioned with n-hexane, then separated and the n-hexane layer was evaporated, to obtain the n-hexane fraction (2,7 gm). The residue (water layer) is partitioned with chloroform and separated. The chloroform layer was evaporated to obtain the chloroform fraction (1,1 gm).

### 2.5. Preliminary phytochemicals screening

Each extract and fraction are tested by phytochemical screening. The test is performed according to various standard methods such as flavonoids, alkaloids, tannins, saponins, and steroids/triterpenoids [11]-[12]. The results are represented as + or – which indicate the presence or absence of classes of compounds in the extract and fractions.

### 2.6. Antioxidant activity

Antioxidant activity testing used the 2,2-diphenyl-1-picrylhydrazyl (DPPH) method with the ascorbic acid as a positive control. In this research, the DPPH method was described by [13], with minor modification. Five concentration series were created for each extract and fractions. Each concentration was taken 2 milliliter and added as much as 1 milliliter of DPPH (0.1 mM). The blank (DPPH in methanol) and positive controls were prepared in the same way without any extract. Then, the mixture was left to stand for 30 minutes in a dark place and the absorbance was measured at $\lambda$ 517 nm using a UV-Vis spectrophotometer. The repeated of this treatment is three times. Antioxidant activity can be expressed in units of percent inhibition. This value is obtained by the following equation.

$$\%\ \text{inhibition}=\frac{A_{control} - A_{sample}}{A_{control}} \times 100$$

After the inhibition percentage of each extract and fractions has been obtained, it is followed by a linear regression calculation (x, y) to obtain the $IC_{50}$ (inhibition concentration) value. x is the concentration (μg/mL), while y is the percentage of antioxidant activity (%). The $IC_{50}$ value is obtained by the equation $y = ax + b$.

## 3. Results and Discussion

The phytochemical screening test showed the ethanol extract which contained flavonoids, saponins, triterpenoids, and alkaloids; the n-hexane fraction contained steroids and alkaloids; and the chloroform fraction contained flavonoids, steroids, and alkaloids. These phytochemical compounds are known to support bioactivity, thus responsible for the antioxidant activities. Phytochemical screening results showed in Table 1.

**Table 1.**Phytochemical screening of *H. longiflora* extract and its fractions.

| Phytochemical constituents | Reagent | Result | | |
|---|---|---|---|---|
| | | Ethanol extract | n-Hexane fraction | Chloroform fraction |
| Flavonoids | Mg + HCl | + | - | + |
| Tannins | $FeCl_3$ 1% | - | - | - |
| Saponins | Distilled water + HCl 1% | + | - | - |
| Steroid/Triterpenoids | $CH_3COOH + H_2SO_4$ | + | + | + |
| Alkaloids | $H_2SO_4$+Meyer | + | - | + |
| | $H_2SO_4$+Wagner | + | - | + |
| | $H_2SO_4$+Dragendroff | + | + | + |
| | $H_2SO_4$+Bounchardat | + | + | + |

Remarks: + = presence; - = absence

Antioxidant activity for each extract and fractions was evaluated using the DPPH method. This method was chosen because it is a simple, fast, and easy method for screening the radical scavenging activity of several compounds. In addition, this method has proven to be accurate and practical [14]. DPPH is a free radical that is stable at room temperature and is purple. If DPPH is reacted with free radical-reducing compounds (for example flavonoids), the intensity of the purple color will decrease and if the amount of free radical-reducing compounds that react is large, the DPPH can change color to yellow. This color change can be measured its absorbance with a UV-Vis spectrophotometer. DPPH provides strong absorption at a wavelength of 517 nm with a dark violet color. The free radical scavenging causes the electrons to pair then causes a decoloration proportional to the number of electrons taken [15].

**Table 2.**Antioxidant activity of *H. longiflora* extract and its fractions.

| Sample | Concentration (μg/mL) | % Inhibition | The regression equation | $IC_{50}$(μg/mL) |
|---|---|---|---|---|
| Crude ethanol extract | 10 | 52.36 | $y = 3.1229x + 20.101$ <br> $R^2 = 0.9919$ | 9.57 |
| | 8 | 44.43 | | |
| | 6 | 37.80 | | |
| | 4 | 32.48 | | |
| | 2 | 27.11 | | |
| n-Hexane fraction | 100 | 50.35 | $y = 0.2516x + 24.942$ <br> $R^2 = 0.999$ | 99.59 |
| | 80 | 44.66 | | |
| | 60 | 40.07 | | |
| | 40 | 35.16 | | |
| | 20 | 29.94 | | |
| Chloroform fraction | 50 | 50.47 | $y = 0.5434x + 23.624$ <br> $R^2 = 0.9926$ | 48.54 |
| | 40 | 45.88 | | |
| | 30 | 40.37 | | |
| | 20 | 33.33 | | |
| | 10 | 29.57 | | |

| | | | | |
|---|---|---|---|---|
| Ascorbic acid | 5 | 53.33 | $y = 4.8417x + 29.184$ $R^2 = 0.9987$ | 4.30 |
| | 4 | 48.75 | | |
| | 3 | 43.35 | | |
| | 2 | 39.22 | | |
| | 1 | 33.89 | | |

In samples containing antioxidant compounds, the higher concentration means the more compounds will donate electrons or hydrogen atoms to the DPPH free radicals. This reaction also causes color fading in the DPPH. DPPH, which was originally dark purple, when reacted with large amounts of antioxidant compounds, will turn yellow. The DPPH color change is also related to the energy possessed by DPPH free radicals. In the form of radicals, DPPH tends to be unstable (reactive) and has large energy because it always reacts looking for its electron pair, but after getting an electron pair, DPPH becomes more stable (low energy).

The results of the analysis of the antioxidant activity for each *H. longiflora* extract and fraction and compared with ascorbic acid as a positive control are presented in Table 2. Based on this table, it can be seen that there is an increase in the percentage of inhibition along with the increase in the concentration value. This shows that there is a relationship between an increase in the concentration of the test sample and an increase in free radical reduction. This relationship is given by the linear regression equation of the sample. Based on ref [16], a compound is stated as a very strong free anti-radical if the $IC_{50}$ value is <10 μg/mL, strong if the $IC_{50}$ value is between 10-50 μg/mL, moderate if the $IC_{50}$ value ranges from 50-100 μg/mL, weak when the $IC_{50}$ value ranges from 100-250 μg/mL. Moreover, it will be inactive when the $IC_{50}$ is above 250 μg/mL. Based on the results of the antioxidant activity test, it was found that the ethanol extract of *H.longiflora* leaves had very strong antioxidant activity, n-hexane fraction had moderate antioxidant activity, and the chloroform fraction had strong antioxidant activity. The results showed that the antioxidant activity of the ethanol extract was more potential than the n-hexane and chloroform fractions, but smaller than ascorbic acid. The flavonoids and alkaloids compounds contain in ethanol extract might be responsible for the activity. This type of compound has been reported to have antioxidant activity [17].

## 4. Conclusion

To sum up, ethanol extract, n-hexane fraction, and chloroform fraction of *H. longiflora* have the potential and can be developed as a source of natural antioxidants. Further research is needed for the isolation and identification of the compounds responsible for antioxidants.

## Acknowledgments

This work was supported by Research and Community Service Institution Universitas Pendidikan Ganesha under *Penelitian Dasar* scheme 2020 (contract number: 579/UN48.16/LT/2020).

## References

[1] Parwata I M O A, Manuaba I B P, Sutirta-Yasa I W P and Wita I W 2016 Gaharu Leaf Extract Water Reduce MDA and 8-OHdG Levels and Increase Activities SOD and Catalase in Wistar Rats Provided Maximum Physical Activity *Bali Med J* **5** (3) 434-438

[2] Tavanai E and Mohammadkhani G 2017 Role of antioxidants in prevention of age-related hearing loss: a review of literature *European Archives of Oto-Rhino-Laryngology* **274** (4) 1821-1834

[3] Kikuzaki H, Hisamoto M, Hirose K, Akiyama K and Taniguchi H 2002 Antioxidant properties of ferulic acid and its related compounds *Journal of agricultural and food chemistry* **50** (7) 2161-2168

[4] Asif M 2015 Chemistry and antioxidant activity of plants containing some phenolic compounds *Chemistry international* **1** (1) 35-52
[5] Lourenço S C, Moldão-Martins and Alves V D 2019 Antioxidants of natural plant origins: From sources to food industry applications *Molecules* 24 4132
[6] Anjelina S H 2020 Antibacterial Activity of Ethanolic Extract of Kitolod (Hippobroma longiflora) Leaf Against Staphylococcus aureus and Salmonella typhi *Asian Journal of Pharmaceutical Research and Development* **8** (1) 52-54
[7] Hapsari A, Asti D, Hidayati R, Kumalla N and Suhendi A 2016 The Potency of Kitolod (Isotoma longiflora (L) Presl.) Herb Extract as a Cure for Cervical Cancer an in Vitro Study of Hela Cells *Int. Conf. Sci. Technol. Humanit* 109–114
[8] Kumar S and Pandey A K 2013 Chemistry and biological activities of flavonoids: an overview *The scientific world journal* **2013** 162750
[9] Pietta P G 2000 Flavonoids as antioxidants *Journal of natural products* **63** (7) 1035-1042
[10] Kumar T S, Sampath M, Sivachandran S V, Shanmugam S and Rajasekaran P 2009 Optimal process for the extraction and identification of flavonoids from the leaves of Polyalthia longifolia using L16 Orthogonal design of experiment *International Journal of Biological and Chemical Sciences* **3** (4) 736-745
[11] Alabri T H A, Al Musalami A H S, , Hossain M A, Weli A M and Al-Riyami Q 2014 Comparative study of phytochemical screening, antioxidant and antimicrobial capacities of fresh and dry leaves crude plant extracts of Datura metel L *Journal of King Saud University-Science* **26** (3) 237-243
[12] Mithraja M J, Marimuthu J, Mahesh M, Paul Z M and Jeeva S 2012 Inter–specific variation studies on the phyto–constituents of Christella and Adiantum using phytochemical methods *Asian Pacific Journal of Tropical Biomedicine* **2** (1) S40-S45
[13] Rasika J 2013 Antimicrobial Activity of Moringa oleifera and Its Synergism with Cleome Viscosa *Int. J. Life Sciences* **1** (3) 182-189
[14] Mugesh G, Panda A, Singh H B, Punekar N S and Butcher R J 2001 Glutathione peroxidase-like antioxidant activity of diaryl diselenides: a mechanistic study *Journal of the American Chemical Society* **123** (5) 839-850
[15] Molyneux P 2004 The use of the stable free radical diphenylpicrylhydrazyl (DPPH) for estimating antioxidant activity *Songklanakarin J. sci. technol* **26** (2) 211-219
[16] Phongpaichit S, Nikom J, Rungjindamai N, Sakayaroj J, Hutadilok-Towatana N, Rukachaisirikul V and Kirtikara K 2007 Biological activities of extracts from endophytic fungi isolated from Garcinia plants *FEMS Immunology & Medical Microbiology* **51** (3) 517-525
[17] Quezada N, Asencio M, Del Valle J M, Aguilera J M and Gómez B 2004 Antioxidant activity of crude extract, alkaloid fraction, and flavonoid fraction from Boldo (Peumus boldus Molina) leaves *Journal of food science* **69** (5) C371-C376

# Activity recognition (AR) to detect dementia using wi-fi based wireless sensor network with receive signal srength indicator (RSSI) method

**H Rubiani[1], E Samsoleh[2] and S Fitri[3]**

[1,2]Department of Electical Engineering, Universitas Muhammadiyah Tasikmalaya, Tamansari Gobras street, West Java, Indonesia.
[2]Department of Technology and Information Education, Universitas Muhammadiyah Tasikmalaya, Tamansari Gobras street, West Java, Indonesia.

Corresponding author: hanirubiani@gmail.com

**Abstract.**
The existence of an increase in human life is the main goal of the growing development of information technology. There are several technologies in the field of Wireless Sensor Network (WSN) which are the main features, one of which is activity recognition. Using AR allows understanding human behavior when carrying out an activity. We have entered a period of population aging, namely an increase in life expectancy which increases with increasing old age. The elderly is susceptible to a disease called dementia. One of the symptoms of dementia is repetition of simple activities such as wandering. These symptoms can be detected by the AR system by classifying the current activity patterns of the elderly and comparing them with previously stored patterns to show early symptoms such as wandering. In efficient AR location research and Wi-Fi based detection methods using the Receive Signal Strength Indicator (RSSI) method and location detection algorithms with K-NN modeling with k = 1. With cross validation showing k = 1 yields an accuracy of 75.73% and distance The minimum average error is 0.298 meters which is better than k = 2 to k = 10.To see the pattern of object movement, in this case, the 85 year old elderly applies the Random Way Point (RWP) model as a model to infer or conclude dementia with calculating the value of the $A_{total}$ based on the location of the object and the time it was observed for 24 hours, and the result is that the amount of 3.139 exceeds the limit of normal conditions of 1.81.

## 1. Introduction

Currently there is an increase in life expectancy followed by an increase in the number of elderly people, or it is said that they have entered a period of aging population. The elderly is susceptible to dementia. Repetition of simple actions such as wandering is a symptom of dementia. Such symptoms can be detected by an Activity Recognition (AR) system. AR in daily life for the elderly must be done indirectly and in a comfortable way without disturbing the daily activities of the elderly. Technology that has been widely used for position tracking is Global Positioning System (GPS). However, GPS has a disadvantage, namely low accuracy when used indoors. Therefore, indoor position tracking technology was developed with the concept of the Indoor Positioning System. In this study, a Wi-Fi based Indoor Positioning System will be used using the signal strength of several access points recorded in the room. The results of recording the Received Signal Strength Indication (RSSI) signal to digitize the behavior that can be tracked from an individual in this study are the elderly in their daily activities. Then explain the modeling method using the k- nearest neighbor for elderly observation data and the Random Way Point modeling method (RWP) to detect dementia symptoms.

## 2. Activity recognition

Human activity recognition (HAR) has been widely used in previous research and has value both in theory and practice in solving problems related to time series. These studies include in smart home [1] [2], human computer interaction [3] [4] and health monitoring [5] [6];

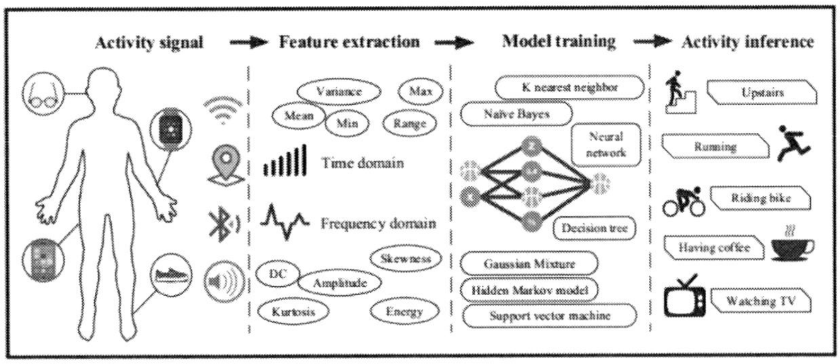


**Figure 1.** Sensor-based AR illustration using pattern recognition [7]

### 2.1. Low level activity recognition

Running, lying down and walking are some examples of low-level activity. A number of studies for low-level WIFI-based activities have been carried out, mainly utilizing received signal strength (RSS) [8] [9]. The RSS-based HAR uses a variety of on-site-recorded RSS to identify people's activities. Low-level activity recognition is based on data collected by sensors. If the subject is moving, we can analyze the sensor data packet to determine the duration of the movement. But there are many high-level activities that cannot be determined based on packet data analysis.

### 2.2. High level activity recognition

Eating, watching television, studying, sleeping, etc. Is an example of a high level activity. High-level activity is predicted using time and place. For example, if someone is in the bedroom from 11 at night until 6:00 in the morning and in observing there is no movement whatsoever, it can be assumed that the person is sleeping [10].

## 3. Human activity recognition based on wi-fi

Since 2000 [11] for the first time proposed Radar based on received signal strength as an indoor localization system. Then [12] proposed the USRP as a hardware device to record the RSS value of Wi-fi signal and use the WiFi RSS value to recognize four activities including standing, lying down, walking, and crawling. The result of accuracy in recognizing these four activities reaches more than 80% [13] proposed WiGes that uses Wi-Fi signal strength for activity recognition. Based on existing tools, WiGest performs gesture recognition by analyzing the rising and falling edges of the RSS signal. Accuracy can be as high as 87.5% for one access point, and can increase up to 96% if there are three access points.

## 4. Receive signal strength indicator (RSSI)

There are several methods of tracking locations, namely triangulation, lateration, angulation, scene analysis and proximity. Angulation places objects computationally relative to a reference point of several angles. Lateration is divided into Time of Arival (TOA), Time Difference of Arival (TDOA) and Receive Signal Strength Indicator (RSSI). The TOA and TDOA schemes have several disadvantages, for the environment in the building it is difficult to find the Line of Sight (LOS) channel between the transmitter and receiver (see Table 1).

**Table 1.** Wi-fi signal strength parameters.

| Signal strength level (signal bar) | Category | Value of signal strength (dBm) |
|---|---|---|
| 5 | Excellent | > -60 |
| 4 | Good | -60 to -70 |
| 3 | Fair | -71 to -80 |
| 2 | Bad | -81 to -90 |
| 1 | Very bad | < -90 |

## 5. Location tracking algorithm

### 5.1. Nearest neighbour

Classification is the process of finding a model or function that describes or distinguishes a concept or data class, with the aim of being able to estimate the class of an object whose label is unknown. The model itself can be an "if-then" rule, a decision tree, a mathematical formula or a Neural Network. One of the classification methods is Nearest Neighbor (NN). The NN method is a method for classifying objects based on the learning data that is closest to the object. The working principle of NN is to find the closest distance between the data to be evaluated and the closest neighbors in the training data. Near or far neighbors are usually calculated based on the euclidean distance. The euclidian distance formula used in the NN method to make decisions by calculating the shortest distance between the input and each data where the determination of the weight or label is carried out randomly or randomly, namely:

$$d = \sqrt{\sum_{i=1}^{p} (rss_q - RSS_q)^2} \quad (1)$$

*5.1.1.* Nearest neighbor

The special case in which classification is predicted based on the closest learning data (in other words, k = 1) is called the Nearest Neighbor method [14] [15].

*5.1.2.* K - nearest neighbor

The steps in the k- NN method are as follow:

- determine the value of k.
- calculate the distance between the new data andeach data in the classification or label section.
- classify the new data into the majority label data.
- classification is carried out on the k closet label data. The value of k is greater than 1.

## 6. Mobility model

There are several mobility models divided into 4 parts, namely Random Models, Models with Temporal Dependency, Models with Spatial Dependency and Models with Geographic Restriction.

### 6.1. Random way point (RWP)

Human mobility can be explained by the RWP model. RWP is a mobility model that has a movement pattern that tends to gather in one simulation area so that it can cause a density wave to occur. In the RWP model, the nodes are randomly distributed in the network. Each node spreads independently of the other nodes. The procedure used by RWP in starting the movement is that each node chooses a destination randomly. Apart from that, each node also chooses a movement speed randomly from the [Speedmin, ..., Speedmax] interval. After the node arrives at its destination, each node will pause according to the predetermined pause time before moving back to using the same procedure [16]. The movement pattern of each node that uses the RWP mobility model tends to converge on one simulation area. Which can cause a density wave or a buildup of nodes in one simulation area (see Figure 2).

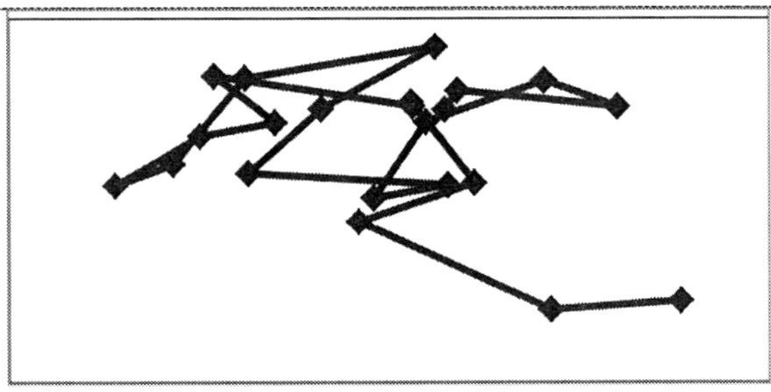

**Figure 2.** Random way point (RWP) model

## 7. Methodology and design system

Note that as a general principle, for large tables font sizes can be reduced to make the table fit on a page or fit to the width of the text.The research methodology to be carried out includes the following steps: Stages of the Activity Recognition (AR) system and determining the location of objects, in this case the elderly, including preparation of the research location by making a plan of the research location area, then the stages of the RSSI method which are divided into 2 stages, namely the training stage and the positioning stage, the positioning stage includes the testing process with method modeling. The positions are predictable and the minimum average error distance can be calculated. The training data that has been previously collected becomes a reference pattern for determining the location of the elderly. The elderly observed were 85 years old, and the location of the observation which was the research location was in a house. Observation of the elderly's location movement was carried out for 24 hours using Random way Point (RWP) modeling to infer dementia symptoms.

### 7.1. Preparation of research site

Preparation of the research location is the first step in this research. There are several things that need to be prepared before starting Activity Recognition (AR) and determining the location, including making a location plan that will be used as a location for data collection from the elderly, in this case in a house.

*7.1.1.* Making a site plan

At this stage, it is done by collecting information about the plan of the research location, after getting a picture of the plan that will be used as the research location, then making a location plan using the Visio software with the position and distance between rooms adjusted to the actual conditions. A house with a size of (24 x 14) meters which is used as the research location *(see Figure 3)*.

**Figure 3.** Plan of house for research location

## 7.2. Determination of location by receive signal strength indicator (RSSI)

At this stage it is divided into 3 steps, namely the training stage, then from the results of the training data measurement visualization is made which aims to see the radio signal strength map and the last is the positioning stage.

*7.2.1. Data collection*

In this process, RSSI measurements were carried out from several access points installed in the research location house. RSSI measurements are determined at coordinate points located in the middle of each (2x2) m area, there are 83 coordinate points shown in Figure 4. RSSI measurement of Wi-Fi signal using a laptop that has NetSurveyor software installed as a Wi-Fi signal strength recorder. RSSI recording is carried out for approximately 2 minutes at each coordinate point.

*7.2.2. RSSI Visualization.*

RSSI visualization is used to provide an overview of the received signal strength map. From the visualization, it can be seen that the distribution of signal strength read at the research location, the strongest signal strength is usually close to the source. Visualization is done using Rapidminer software. The results of RSSI measurements collected for approximately 2 minutes contained 50 data at each coordinate point, so the total data collected was 50 x 83 coordinate points = 4150 (see Figures 6-9).

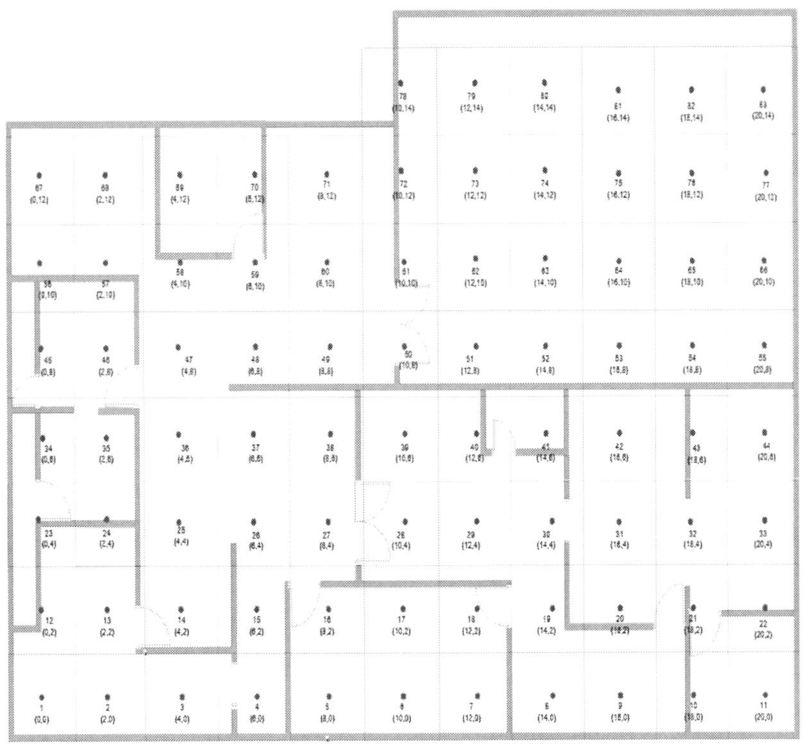

**Figure 4.** Location of coordinat points

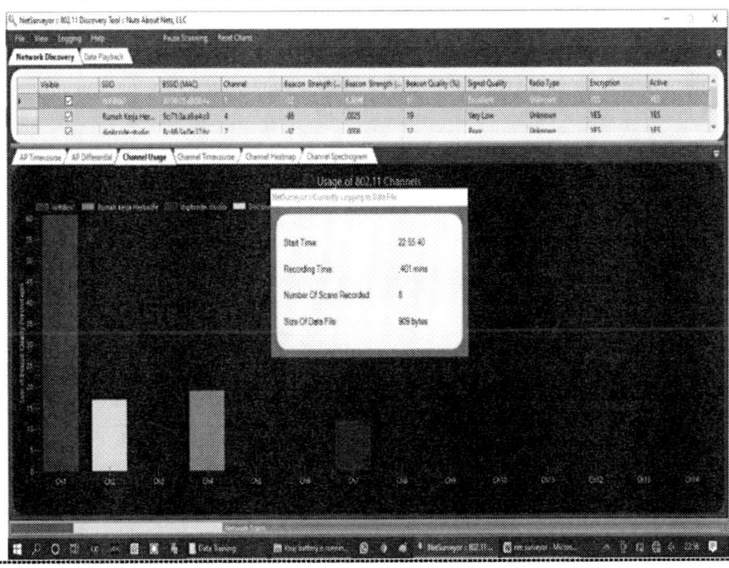

**Figure 5.** Netsurveyor display

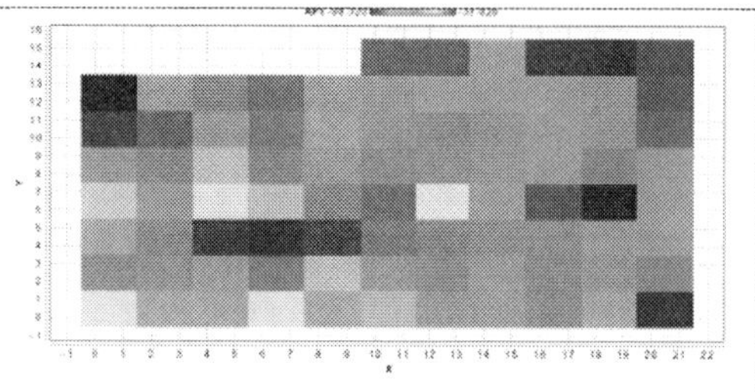

**Figure 6.** Visualization of access point 1.

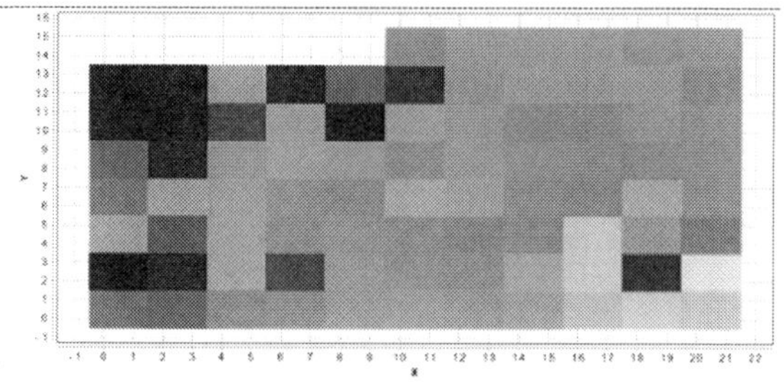

**Figure 7.** Visualization of access point 2.

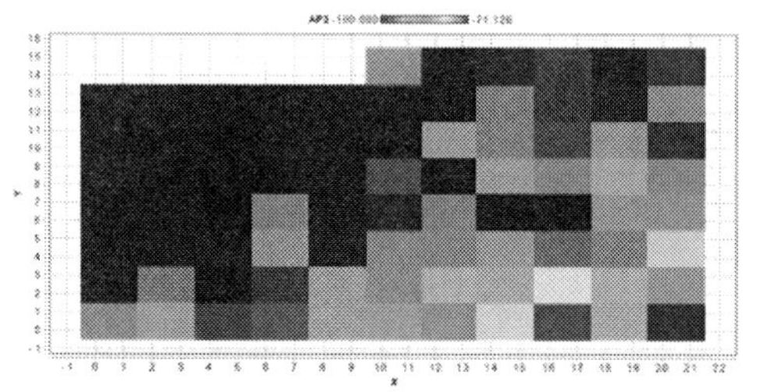

**Figure 8.** Visualization of access point 3.

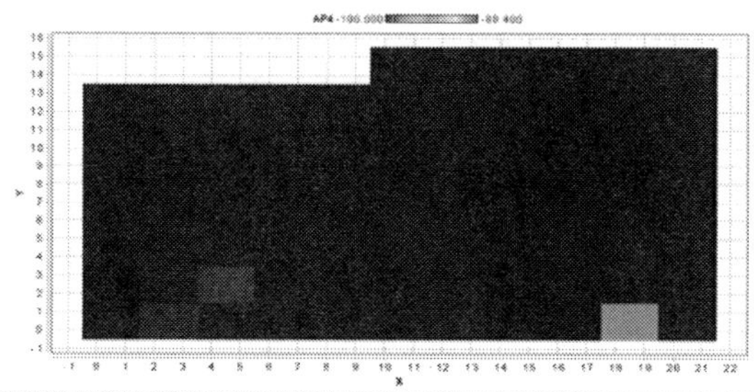

**Figure 9.** Visualization of access point 4.

### 7.3. Cross validation data training

To measure the accuracy of the classification algorithm, methods that can be used include cross validation. To develop applications (development) based on the model created, Rapid Miner is used. Cross validation is a standard test performed to predict the error rate. The training data is divided randomly into several parts with the same ratio then the error rate is calculated part by part, then calculate the average of all error rates to get the overall error rate.

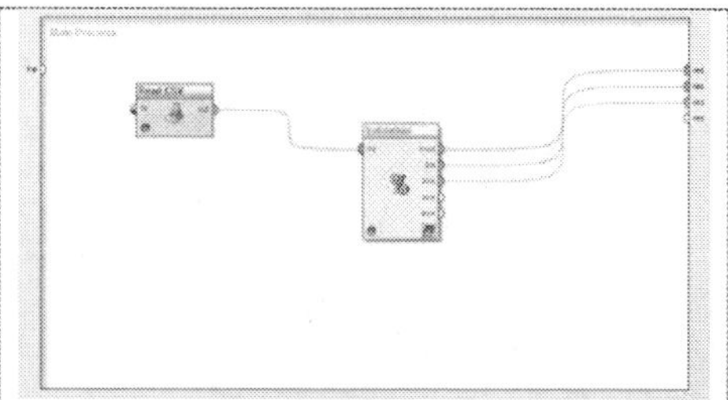

**Figure 10.** Main process cross vlidation.

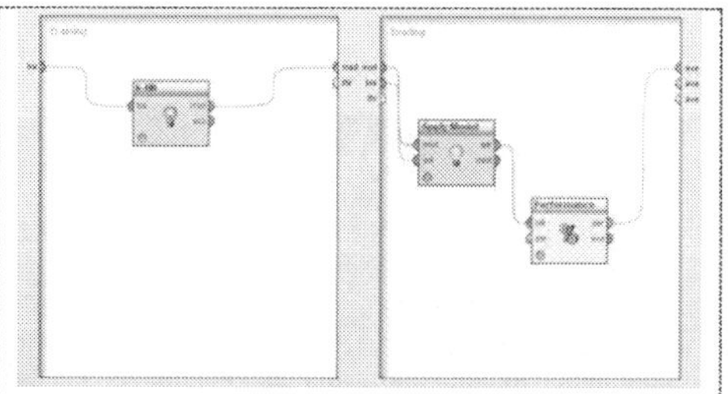

**Figure 11.** Training and testing cross validation.

### 7.4. Positioning

At this stage, there are several steps that must be taken, including data measurement testing, modeling methods used and position prediction.

*7.4.1.* Measurenment data testing.

RSSI measurement data testing in the context of testing. The walking speed in the testing data measurement is adjusted to the speed of people walking in general

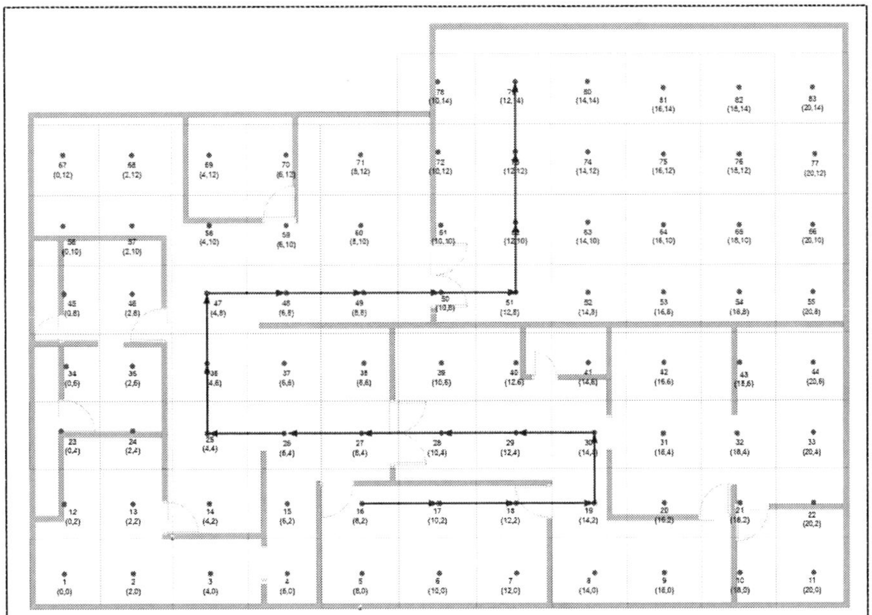

**Figure 12.** The test location

*7.4.2. Algorithm modeling.*

Modeling method using Rapid Miner is to look for position predictions so that the minimum average error distance can be calculated. The data entered is in the form of training data which contains attributes and classifications. The attribute in this study is the sensor node, where the value of the attribute is the signal strength value that is read at each reference point and the classification shows the coordinates (x, y) of the reference points. The testing data contains only attributes. All data formats use the * csv () format in Excel software.

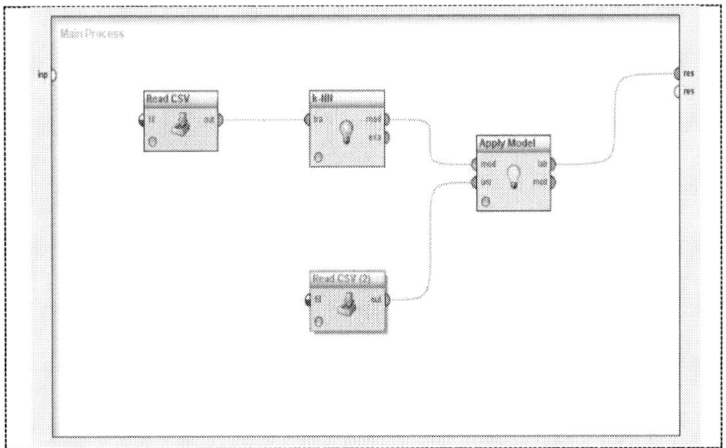


**Figure 13.** Modeling method using k–nearest neighbour

IOP Conf. Series: Materials Science and Engineering 1115 (2021) 012079 doi:10.1088/1757-899X/1115/1/012079

### 7.5. Steps of activity recognition

At this stage, the elderly who will be used as experimental samples are monitored for their daily activities within a certain period of time. Location tracking data will be collected and become a pattern that will be used as a reference in classifying the activity patterns of the elderly by comparing them with previously collected patterns.

## 8. Result and discussion

### 8.1. The result of accuracy k-nearest neighbour

To find out how accurate the training data which contains the Wi-Fi signal strength pattern at the research location has been previously collected using the K-NN algorithm, using the Rapid Miner software

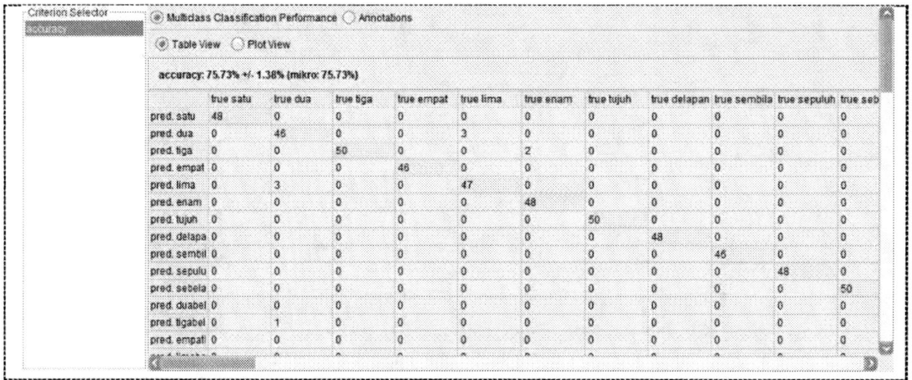


| | true satu | true dua | true tiga | true empat | true lima | true enam | true tujuh | true delapan | true sembila | true sepuluh | true seb |
|---|---|---|---|---|---|---|---|---|---|---|---|
| pred. satu | 48 | 0 | 0 | 0 | 0 | 0 | 0 | 0 | 0 | 0 | 0 |
| pred. dua | 0 | 46 | 0 | 0 | 3 | 0 | 0 | 0 | 0 | 0 | 0 |
| pred. tiga | 0 | 0 | 50 | 0 | 0 | 2 | 0 | 0 | 0 | 0 | 0 |
| pred. empat | 0 | 0 | 0 | 46 | 0 | 0 | 0 | 0 | 0 | 0 | 0 |
| pred. lima | 0 | 3 | 0 | 0 | 47 | 0 | 0 | 0 | 0 | 0 | 0 |
| pred. enam | 0 | 0 | 0 | 0 | 0 | 48 | 0 | 0 | 0 | 0 | 0 |
| pred. tujuh | 0 | 0 | 0 | 0 | 0 | 0 | 50 | 0 | 0 | 0 | 0 |
| pred. delapa | 0 | 0 | 0 | 0 | 0 | 0 | 0 | 48 | 0 | 0 | 0 |
| pred. sembil | 0 | 0 | 0 | 0 | 0 | 0 | 0 | 0 | 46 | 0 | 0 |
| pred. sepulu | 0 | 0 | 0 | 0 | 0 | 0 | 0 | 0 | 0 | 48 | 0 |
| pred. sebela | 0 | 0 | 0 | 0 | 0 | 0 | 0 | 0 | 0 | 0 | 50 |
| pred. duabel | 0 | 0 | 0 | 0 | 0 | 0 | 0 | 0 | 0 | 0 | 0 |
| pred. tigabel | 0 | 1 | 0 | 0 | 0 | 0 | 0 | 0 | 0 | 0 | 0 |
| pred. empatl | 0 | 0 | 0 | 0 | 0 | 0 | 0 | 0 | 0 | 0 | 0 |

**Figure 14.** k nearest neighbour result for k=1

The results obtained for K = 1 are shown in Figure 14 which shows the figure of 75.73% for k = 1. Then try using k = 2 to k = 10. And the results are shown in table 2. From the table, it can be seen that for k = 1, it shows an accuracy value of 75.73%, better than k=2 to k = 10.

**Table 2.** k-nearest neighbour results for k=1 to k=10.

| k | acuracy |
|---|---|
| 1 | 75.73 % +- 1.38 % (mikro: 75.73%) |
| 2 | 61.57 % +/- 1.30% (mikro: 61.57%) |
| 3 | 51.01% +/- 1.49% (mikro: 51.01%) |
| 4 | 50.46% +/- 1.14% (mikro: 50.46%) |
| 5 | 54.84% +/- 1.75% (mikro: 54.84%) |
| 6 | 53.81% +/- 1.67% (mikro: 53.81%) |
| 7 | 53.98% +/- 2.01% (mikro: 53.98%) |
| 8 | 53.81% +/- 1.72% (mikro: 53.81%) |
| 9 | 53.71% +/- 1.88% (mikro: 53.71%) |
| 10 | 52.82% +/- 1.27% (mikro: 52.82%) |

IOP Conf. Series: Materials Science and Engineering 1115 (2021) 012079 doi:10.1088/1757-899X/1115/1/012079

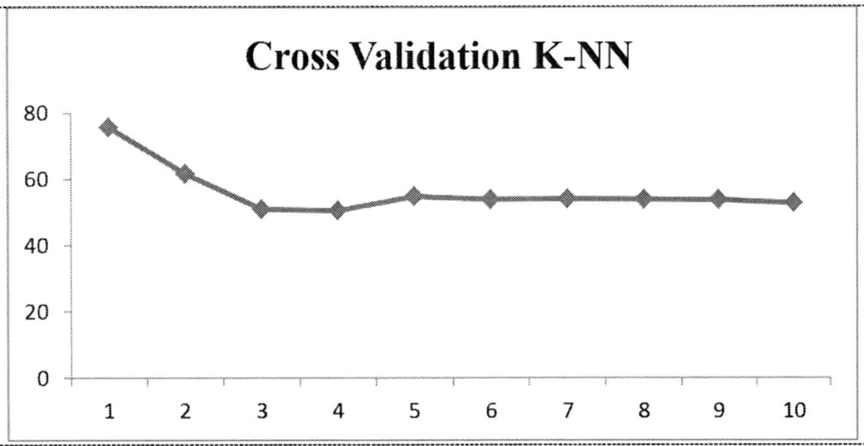


**Figure 15.** k nearest neighbour result for k=1 to k=10

## 8.2. Minimun average error distance result

Determination of the position of the object, namely the reference point represented by physical coordinates in the form (x, y). The minimum average error distance is obtained from the comparison between the actual RSS measurements, namely the testing data with the previously stored measurements. Position prediction is obtained from the positioning stage using a modeled algorithm, namely the KNN. From the results of the prediction of the position of each testing data that becomes the target that is missed in the testing data measurement, the minimum average error distance value can be calculated.

$$Minimum\ average error = \frac{\sqrt{\sum_{i=1}^{N} (x-x_i)^2 + (y-y_i)^2}}{N} \quad (2)$$

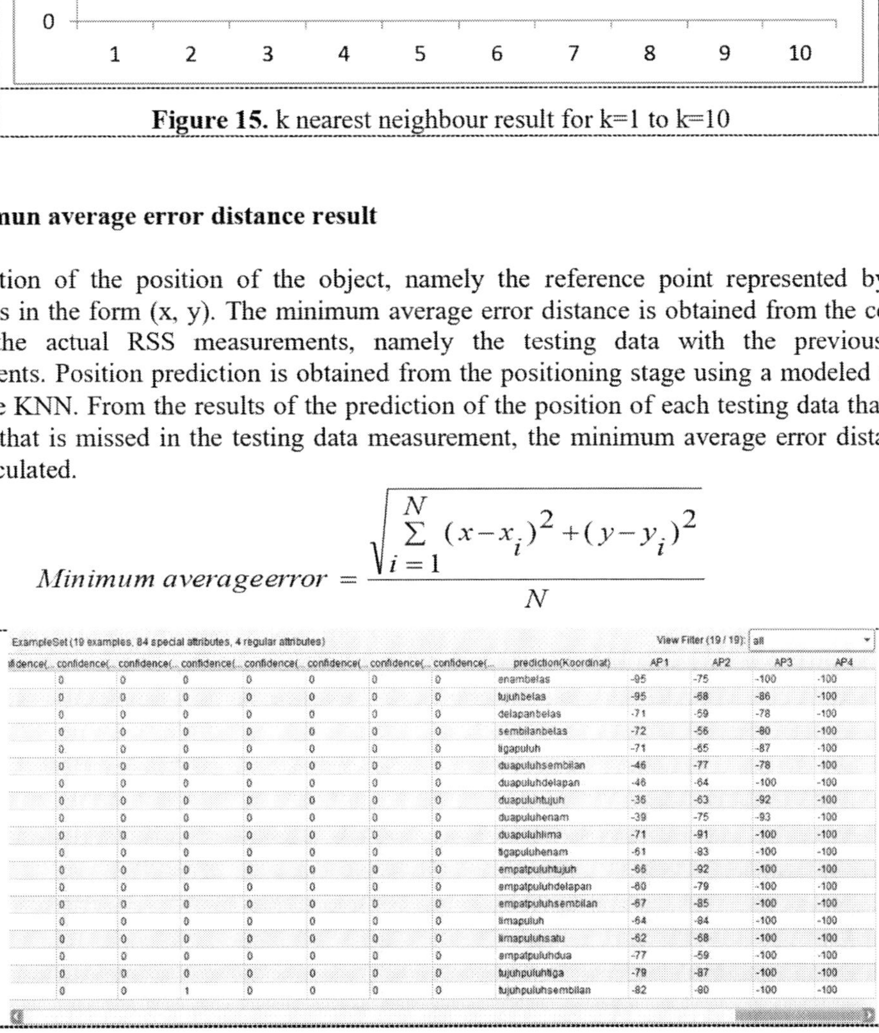

ExampleSet (19 examples, 84 special attributes, 4 regular attributes) View Filter (19 / 19): all

| ...dence(... | confidence(... | confidence(... | confidence(... | confidence(... | confidence(... | confidence(... | confidence(... | prediction(Koordinat) | AP1 | AP2 | AP3 | AP4 |
|---|---|---|---|---|---|---|---|---|---|---|---|---|
| | 0 | 0 | 0 | 0 | 0 | 0 | 0 | enambelas | -95 | -75 | -100 | -100 |
| | 0 | 0 | 0 | 0 | 0 | 0 | 0 | tujuhbelas | -95 | -68 | -86 | -100 |
| | 0 | 0 | 0 | 0 | 0 | 0 | 0 | delapanbelas | -71 | -59 | -78 | -100 |
| | 0 | 0 | 0 | 0 | 0 | 0 | 0 | sembilanbelas | -72 | -56 | -80 | -100 |
| | 0 | 0 | 0 | 0 | 0 | 0 | 0 | tigapuluh | -71 | -65 | -87 | -100 |
| | 0 | 0 | 0 | 0 | 0 | 0 | 0 | duapuluhsembilan | -46 | -77 | -78 | -100 |
| | 0 | 0 | 0 | 0 | 0 | 0 | 0 | duapuluhdelapan | -48 | -64 | -100 | -100 |
| | 0 | 0 | 0 | 0 | 0 | 0 | 0 | duapuluhtujuh | -36 | -63 | -92 | -100 |
| | 0 | 0 | 0 | 0 | 0 | 0 | 0 | duapuluhenam | -39 | -75 | -93 | -100 |
| | 0 | 0 | 0 | 0 | 0 | 0 | 0 | duapuluhlima | -71 | -91 | -100 | -100 |
| | 0 | 0 | 0 | 0 | 0 | 0 | 0 | tigapuluhenam | -61 | -83 | -100 | -100 |
| | 0 | 0 | 0 | 0 | 0 | 0 | 0 | empatpuluhtujuh | -66 | -92 | -100 | -100 |
| | 0 | 0 | 0 | 0 | 0 | 0 | 0 | empatpuluhdelapan | -80 | -79 | -100 | -100 |
| | 0 | 0 | 0 | 0 | 0 | 0 | 0 | empatpuluhsembilan | -67 | -85 | -100 | -100 |
| | 0 | 0 | 0 | 0 | 0 | 0 | 0 | limapuluh | -64 | -84 | -100 | -100 |
| | 0 | 0 | 0 | 0 | 0 | 0 | 0 | limapuluhsatu | -62 | -68 | -100 | -100 |
| | 0 | 0 | 0 | 0 | 0 | 0 | 0 | empatpuluhdua | -77 | -59 | -100 | -100 |
| | 0 | 0 | 0 | 0 | 0 | 0 | 0 | tujuhpuluhtiga | -79 | -87 | -100 | -100 |
| | 0 | 0 | 1 | 0 | 0 | 0 | 0 | tujuhpuluhsembilan | -82 | -80 | -100 | -100 |

**Figure 16.** Location prediction

**Table 3.** Minimum average distance results for k=1 to k=10.

| k | Minimum average distance (meters) |
|---|---|
| 1 | 0.298 |
| 2 | 0.298 |
| 3 | 0.298 |
| 4 | 3.184 |
| 5 | 4.578 |
| 6 | 4.683 |
| 7 | 4.157 |
| 8 | 3.922 |
| 9 | 3.637 |
| 10 | 3.637 |

From the results k = 1 to k = 10, it shows that the minimum error distance is at k = 1, k = 2 and k = 3 with the same results, which is 0.298 meters. For this reason, the location prediction used is determined, namely k = 1

### 8.3. Object tracking location

The object in this study is an elderly 85 years old who is observed movement using sensors based on Wi-Fi signal strength. Then compared with the previously collected pattern, namely training data. And the results of the mapping of the tracking patterns of objects or the elderly during a period of 24 hours using the k-nearest neighbor algorithm with k = 1 are shown in Figure 17.

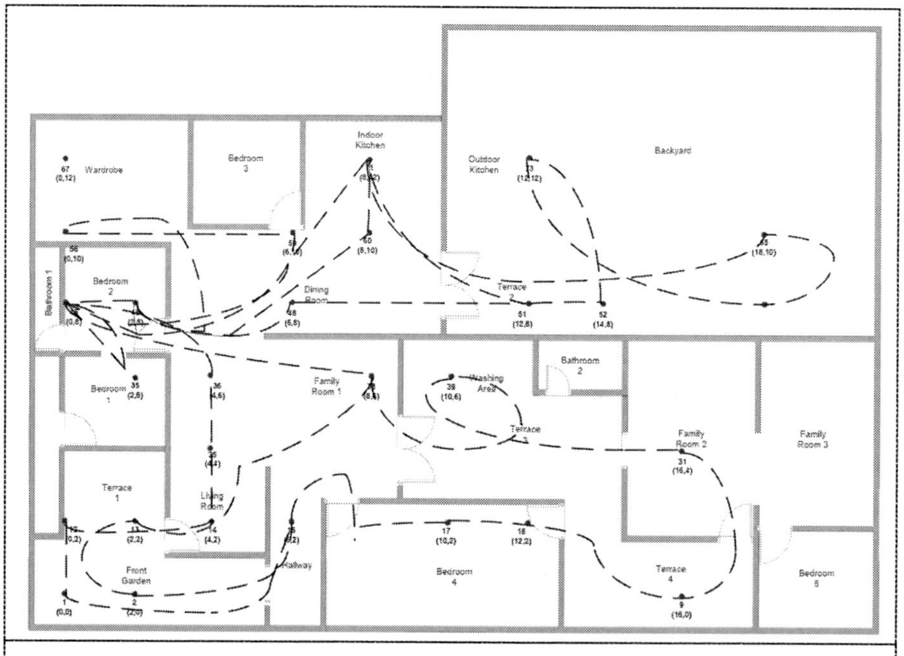


**Figure 17.** Object tracking location

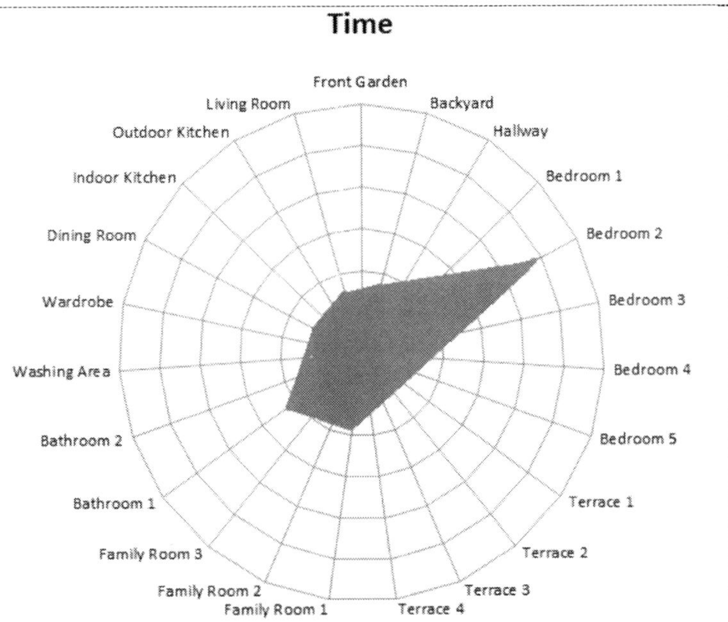


**Figure 18.** Time graph

### 8.4. Mobility model using random way point (RWP)

To create an appropriate detection model, in this study data collection is carried out in terms of geographic location information, namely the coordinate points that have been determined at the time of measuring training data and the amount of time spent at the location, the time domain is closely related to object mobility, especially when detecting movement activity.

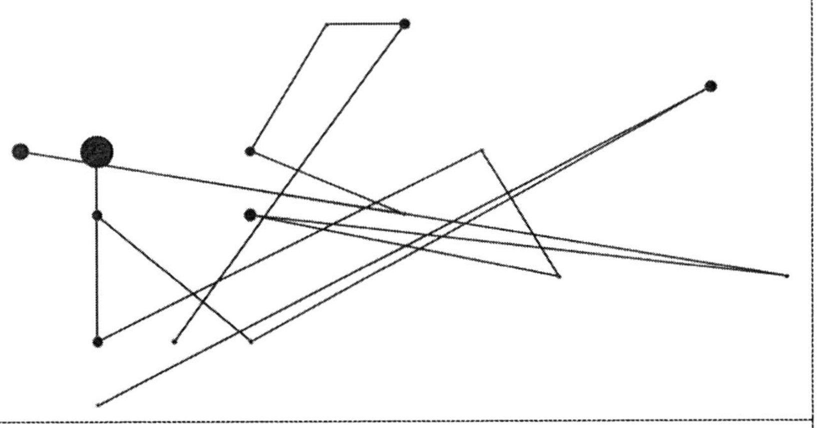

**Figure 19.** Mobility model using random way point (RWP)

Certain points of an RWP model indicate the location or area of movement of the object. The size of the point is determined by the length of time living in the area during the observation period, namely the total time of 24 hours. Based on research that has been done previously, namely [5], that for normal human movement has a value between $0.53 \leq A_{total} \leq 1.81$. Beyond this value there is a different pattern of movement with dementia patients who go to more locations than in normal cases. In interpreting there are different characteristics of human mobility, namely human mobility in general has a specific or certain purpose of visit and spends a certain amount of time in a certain time. In cases of dementia, memory loss can cause frequent wandering or wandering even though it seems to be looking for places with clear objectives such as bedrooms, bathrooms and dining rooms.

**Figure 20.** Mobility of object on site using RWP model

The calculation of the totals uses equations (3), (4) and (5) where n shows the number of rooms and R as the level of the room.

$$A_{t\_exceed} = \frac{\pi}{n} \sum (r_i^2 - \frac{R_i^2}{4}) \quad , r > \frac{R}{2} \tag{3}$$

$$A_{t\_daily} = \pi * (\frac{R}{2})^2 \tag{4}$$

$$A_{total} = \beta * \frac{A_{t\_exceed}}{A_{t\_daily}} \tag{5}$$

The ratio of daily changes over normal time ($A_{t\_exceed}$) and $A_{total}$ calculated heuristically with observed values. The table shows that over time the suspected symptoms of dementia are indicated by random walking by exceeding 1.81, which is 3.139.

**Table 4.** $A_{total}$ results.

| | | | | |
|---|---|---|---|---|
| $A_{t\ exceed}$ | 0.139 | ... | 0.468 | 5.022 |
| $A_{t\_exceed}/A_{t\ daily}$ | 0.006 | ... | 0.020 | 0.209 |
| $A_{total}$ | 0.087 | ... | 0.293 | 3.139 |

## 9. Conclusion

The results of tracking objects within 24 hours show that the object or the elderly is in several rooms within a certain period of time. And the longest time spent in the room in Bedroom 2 and several other spaces in this case the object experiences such a thing as wandering. In the case of the object being observed, an elderly person aged 85 years from modelling using the random way point shows an Atotal value of 3.139 which exceeds the normal case, which is more than 1.81. In this case indicates an indication of dementia symptoms

### Acknowledgments

This work was supported by Department of Electrical Engineering, Faculty of Engineering Universitas Muhammadiyah Tasikmalaya and Lembaga Penelitian dan Pengabdian kepada Masyarakat (LPPM) Universitas Muhammadiyah Tasikmalaya.

## References

[1] Ahmed S H, Kim D 2016 Named data networking-based smart home. *ICT Express* **2** 130–4
[2] Kumar S 2014 Ubiquitous Smart Home System Using Android Application. *IJCNC* **6** 33–43
[3] Rautaray S S, Agrawal A 2015 Vision based hand gesture recognition for human computer interaction: a survey. *Artificial Intelligence Review* **43** 1–54
[4] Yeo H-S, Lee B-G, Lim H 2013 Hand tracking and gesture recognition system for human-computer interaction using low-cost hardware. *Multimedia Tools and Applications* **74** 2687–715.
[5] Kim K-J, Hassan M, Na S-H, Huh E 2010 Dementia Wandering Detection and Activity Recognition Algorithm Using Tri-Axial Accelerometer Sensors *Proceedings of the 4th International Conference on Ubiquitous Information Technologies & Applications 2009* 1-5
[6] Kalantarian H, Sideris C, Mortazavi B, Alshurafa N, Sarrafzadeh M 2017 Dynamic Computation Offloading for Low-Power Wearable Health Monitoring Systems. *IEEE Trans Biomed Eng* **64** (3) 621-628
[7] Jindong Wang, Yiqiang Chen, Shuji Hao, Xiaohui Peng, Lisha Hu 2019 Deep Learning for Sensor-based Activity Recognition: A Survey. *Pattern Recognition Letters* **119** 3–11
[8] Barsocchi P 2013 Position Recognition to Support Bedsores Prevention. *IEEE Journal of Biomedical and Health Informatics* **17** 53–9
[9] Guraliuc AR, Barsocchi P, Potortì F, Nepa P 2011 Limb movements classification using wearable wireless transceivers. *IEEE Transactions on Information Technology in Biomedicine* **15** 474–80
[10] Dimitrios Lymberopoulos, Athanasios Bamis, Andreas Savvides 2008 Extracting spatiotemporal human activity patterns in assisted living using a home sensor network. *Proceedings of the 1st international conference on PErvasive Technologies Related to Assistive Environments* Article **29** 1–8.
[11] Bahl P, Padmanabhan VN 2000 RADAR: An in-building RF-based user location and tracking system. *Proceedings IEEE INFOCOM 2000. Conference on computer communications. Nineteenth annual joint conference of the IEEE computer and communications societies (Cat. No. 00CH37064)* **2** 775–84
[12] Sigg S, Shi S, Buesching F, Ji Y, Wolf L 2013 Leveraging RF-channel fluctuation for activity recognition: Active and passive systems, continuous and RSSI-based signal features. *Proceedings of International Conference on Advances in Mobile Computing & Multimedia* 43–52.
[13] Abdelnasser H, Youssef M, Harras KA 2015 Wigest: A ubiquitous wifi-based gesture recognition system. *2015 IEEE conference on computer communications (INFOCOM), IEEE* 1472–80.

[14] Shadiq M A 2009 *Keoptimalan Naïve Bayes Dalam Klasifikasi.* Bandung: Jurnal Universitas Pendidikan Indonesia.

[15] Dasarathy B V 1991 *Nearest neighbor (NN) norms: NN pattern classification techniques*. IEEE Computer Society Tutorial.

[16] Laqtib S, El Yassini K, Houmer M, El Ouadghiri MD, Hasnaoui ML 2016 Impact of mobility models on optimized link state routing protocol in MANET. *International Conference on Wireless Networks and Mobile Communications (WINCOM), IEEE* 104–9

# Overproduction and Purification of Organomercuric Lyase (MerB) from Mercury-resistant Bacteria *Pseudomonas aeruginosa* Isolate 4B2

**Fatimawali[1,] *, B J Kepel[2] and T E Tallei[3]**

[1]Department of Pharmacy, Faculty of Mathematic and Natural Sciences, Sam Ratulangi University, Indonesia
[2]Department of Chemistry, Faculty of Medicine, Sam Ratulangi University, Indonesia
[3]Department of Biology, Faculty of Mathematic and Natural Sciences, Sam Ratulangi University, Indonesia

*Corresponding author: fatimawali@unsrat.ac.id

**Abstract.** Mercury is a very toxic element even though there is very little concentration in the body. Although all chemical forms of mercury are toxic, public health attention is focused on organic mercury which is the most toxic form of mercury. Organic mercury can, however, be detoxified by organomercuric lyase (MerB) protein derived from mercury resistant bacteria. This study aims to overproduce of MerB protein by transforming merB gene into E. coli BL-21(DE3). Nucleotide sequence of *mer*B gene of mercury resistant bacteria *Pseudomonas aeruginosa* isolates 4B2, optimized by using gene program designers (www.dna20/com) then commercially synthesized and cloned in pET16b expression plasmid vector. Plasmid pET16b_merB (syntetic gene) was transformed into E. coli BL21(DE3) to produce MerB protein recombinant, induced with isopropyl-β-D-thiogalactopyranoside (IPTG) and purified by imidazol. Overproduction and purification of MerB protein was successfully performed in E. coli BL21 mediated by plasmid pET16b, resulting MerB protein with a molecular weight of 25.6 kDa, with the optimum at 37°C incubation temperature, incubation time of 3 hours and 0.1 mM IPTG induction. MerB protein obtained can be used in further research on the enzymatic detoxification of organic mercury.

## 1. Introduction

Natural resources are one of the backbone of state income and also direct income to the community. This has made the natural environment widely exploited, for example to obtain gold as a source of income. Besides being able to bring direct benefits to the community in the form of community gold mining, this exploitation of gold can sometimes be detrimental due to uncontrolled activities. People exploit the environment only to get the gold they want but sacrifice environmental health aspects, for example the use of mercury in the gold amalgamation process. The results of mercury waste from community gold mining processing are processed in a simple manner which only ends up in the disposal of mercury waste into the environment which is a source of livelihood and will have an impact on human and animal health [1]. An important mechanism of microbial resistance to mercury is by reducing mercury compounds to volatile mercury metal [2].

Several studies have reported that bacteria that live in environments contaminated with mercury, become resistant to mercury. Zulaika et al. (2019) found several genera of Bacillus, Staphylococcus and Acetobacter bacteria that were resistant to 25mg / L $HgCl_2$ [3]. Fatimawali et al. (2019) regarding the use of bacteria in the mercury detoxification process [4]. The ability of mercury-resistant bacteria to reduce mercury has been reported in ref [5], that *Bacillus sp.* can reduce $HgCl_2$ to 74.5%. Zhang et al. (2012), reported the ability of *Pseudomonas putida* SP-1 MRB to reduce 89% of toxic $Hg^{2+}$ to less toxic $Hg^{o}$ [6]. Anna Kowalczyk et al. (2016) found *Pseudomonas syringae* which can reduce almost all mercury metal in liquid medium after 120 hours incubation [7]. Even the detoxification ability of organic mercury has been reported in ref [4], that *Pseudomonas sp.* AA isolate can reduce the concentration of phenyl mercury in the media up to 74.99%.

The ability of bacteria to detoxify mercury because bacteria contain mercury detoxifying genes merA and merB. The *mer*A gene can reduce inorganic mercury to mercury metal called the mercurireductase gene, while the merB gene can break the C-Hg bond from organic mercury called organomercury lyase [8]. The use of mercury reductase and organomercury lyase to overcome the problem of inorganic and organic mercury pollution in the environment has a very high potential. *Klebsiella pneumoniae* A1.1.1. isolate which contains the *mer*A gene and encodes the MerA protein and can be cloned and overproduced to produce recombinant MerA protein and can detoxify inorganic mercury [9]. The purpose of this study was to obtain the MerB organomerkuric lyase enzyme which can be used in the mercury detoxification process.

## 2. Materials and Methods

Transformation was carried out on two *E. coli* (*E. coli* Top10 and *E. coli* BL21 (DE3) with ampicillin concentrations of 100µg/mL and 25µg/mL. The amount of plasmid inserted was 100 ng (stock concentration 40ng/µL) plasmid pJExpress_SOD as a competent control, duplo done. To ensure that the transformed gene is the merB gene inserted, sequencing is carried out.

MerB protein overproduction of the 10 mL scale was carried out to see the expression of the MerB protein. Performed with IPTG induction conditions of 0.1 mM; 0.5 mM and 1 mM at 150 rpm for 3 hours of incubation. Compared with E. coli BL21 (DE3) without plasmid. MerB protein overproduction at a scale of 200 mL, carried out at 37°C for 3 hours of incubation at 150 rpm with 0.1 mM IPTG induction using a 1 L Erlenmeyer flask. The cell suspension was sonicated with an amplitude of 60%, for 15 seconds on and 15 seconds off, for 20 minutes, then centrifuged at 10,000 rpm at 4°C for 10 minutes to produce 4 mL crude merB protein.

Purification optimization was carried out with 4 mL crude extract. Purification optimization stage with column equilibrium using 5 mM imidazole. MerB protein can be purified by Nickel column chromatography. Crude protein with a certain amount that has been calculated against the capacity of the column, is put into the column and washed with a buffer containing imidazole with 5 mM imidazole concentration. MerB protein is eluted with 250 mM imidazole or eluted with imidazole in a stepwise concentration of 50 - 500 mM. The elution results were collected as much as 1.5 mL and were analyzed by SDS-PAGE. The fraction containing the most MerB protein was collected for further analysis.

## 3. Result and Discussion

### 3.1 Transform of plasmid pET16b_merB to *E. coli* BL21 and *E. coli* Top10

The merB gene used in this study was made using synthetic DNA techniques after a codon optimization process for the *E. coli* expression system was carried out. The advantage of this technique is that it can increase the efficiency of DNA design, while codon optimization is carried out to increase the efficiency of protein expression in the host cell used[10].Transformation of plasmid into *E. coli* BL21(DE3) and *E. coli* Top10 before overproduction was conducted, the presence of plasmid containing *mer*B gene inside *E. coli* BL21 (DE3) was confirmed. The existence of colonies that grew on the positive control transformation and no colonies that grew on the negative control can also indicate the successful transformation and manufacture of competent cells, including in this case indicates the absence of contamination[11].

Migration and restriction analysis were performed to confirm the presence of the plasmid pET16b-merB in Transforman. The migration analysis shows a suitable result where the migration distance of pET16b-merB (6300 pb) in both Top10 and BL21 has a longer migration distance than Rtase (7300pb) as a comparison. The restriction analysis showed that the *merB* gene fragments in *E. coli* BL21 (DE3) matched the bands at 5700 bp and 630 bp sizes, although in *E. coli* Top10 it showed that the plasmid was not completely cut so that the 5700s still saw two bands. The BamHI enzyme cleaves the plasmid at the GGATCC site at the tip 5', while the HindIII enzyme cuts at the AAGCTT site at the 3' end. Both restriction enzymes this results in a sticky end cutting pattern. This helps the ligation process between the inserted DNA and the expression vector to be more efficient because the tapered end that is free from the restriction enzyme cutting can pair with the complement of the other taper through hydrogen bonds [12]. The migration and transformation results was shown in Figure 1.

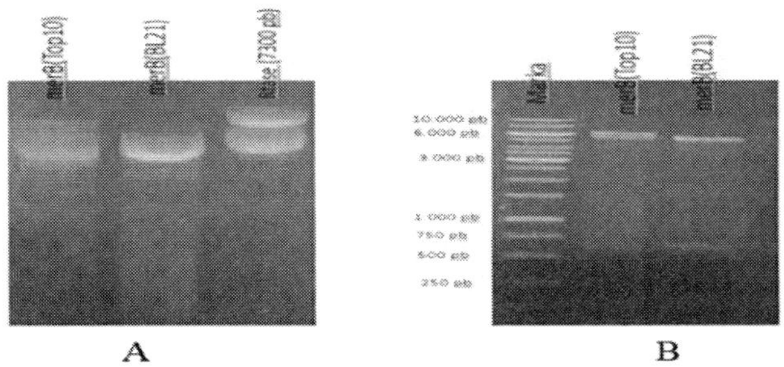


**Figure 1** (A) Results of migration analysis and (B) Restriction analysis

### 3.2 Sequencing of the transformed *merB* gene

Organomercuric liase gene (*mer*B) used in this study was cloned previously into pET-16b plasmid. The length of the plasmid was 5711 bp as seen in Figure 2, harboring *mer*B gene with the size of 630 bp which was synthesized based on the size of *mer*B gene of *Pseudomonas aeruginosa* 4B2 isolate obtained from Tatelu North Sulawesi [13], as seen in Figure 3.

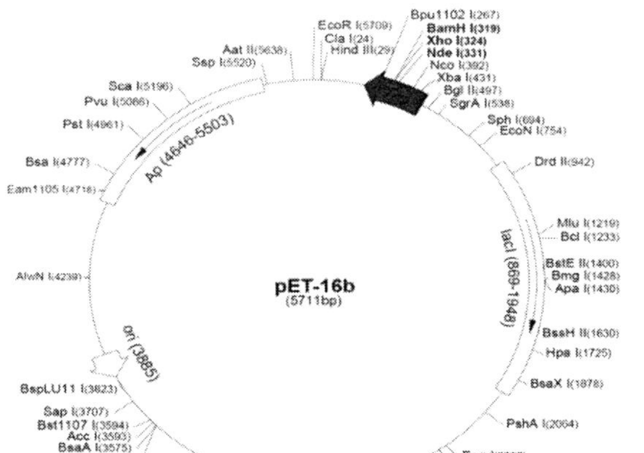


**Figure 2** pET-16b Plasmid with a size of 5711 bp carries the *mer*B gene

ATG AAA CTG GCC CCG TAT ATT CTG GAA CTG TTA ACA TCT GTT AAT CGT ACC
AAT GGA ACT GCC GAT CTG CTG GTG CCG CTG CTG CGC GAA TTG GCC AAA
GGC CGT CCG GTC AGC CGC ACC ACC CTT GCC GGT ATT CTT GAT TGG CCA
GCA GAA CGC GTA GCG GCT GTT TTA GAA CAG GCA ACG AGT ACG GAA TAT
GAC AAG GAT GGC AAT ATT ATC GGG TAT GGA TTG ACT TTA CGC GAA ACA
AGC TAT GTT TTT GAA ATT GAT GAT CGC AGA CTG TAT GCT TGG TGT GCG TTG
GAT ACT CTG ATT TTT CCT GCA TTG ATC GGA AGA ACC GCG AGA GTT AGC
TCT CAC TGC GCT GCG ACC GGC GCT CCA GTC AGT CTT ACT GTT AGT CCT TCT
GGC ATC CAA GCC GTG GAA CCT GCT GGG ATG GCG GTC TCT TTA GTA CTG
CCA CAG GAA GCG GCG GAT GTT CGT CAG TCA TTT TGC TGC CAT GTG CAC
TTC TTT GCT TCA GTG CCT ACG GCC GAA GAT TGG GCA TCG AAA CAT CAG
GGA CTT GAA GGC TTG GCT ATC GTG TCT GTC CAT GAA GCC TTC GGC CTT
GGC CAA GAA TTT AAT CGG CAT TTA TTG CAG ACA ATG TCA AGC CGC ACC
CCG TGA

**Figure 3** *Mer*B gene sequence of *Pseudomonas aeruginosa* 4B2 isolate obtained From Tatelu North Sulawesi (optimation result)

The sequencing results obtained from the nucleotide sequence genes that were signed online via the website (https://www.ebi.ac.uk/Tools/services/rest/clustalo/). The result of *mer*B sequencing nucleotide sequence alignment is translated to amino acid deduction online via the website (https://web.expasy.org/translate/) as shown in Figure 4.

Based on the results of the alignment, it can be concluded that the nucleotide sequence of the merB gene sequences with synthetic is suitable and matches 100%, it can be seen in the results of the reduced amino acid alignment which can be seen in Figure 4. The first 21 amino acids in the sequencing results are amino acids derived from vectors. plasmid pET-16b according to the design / synthesis.

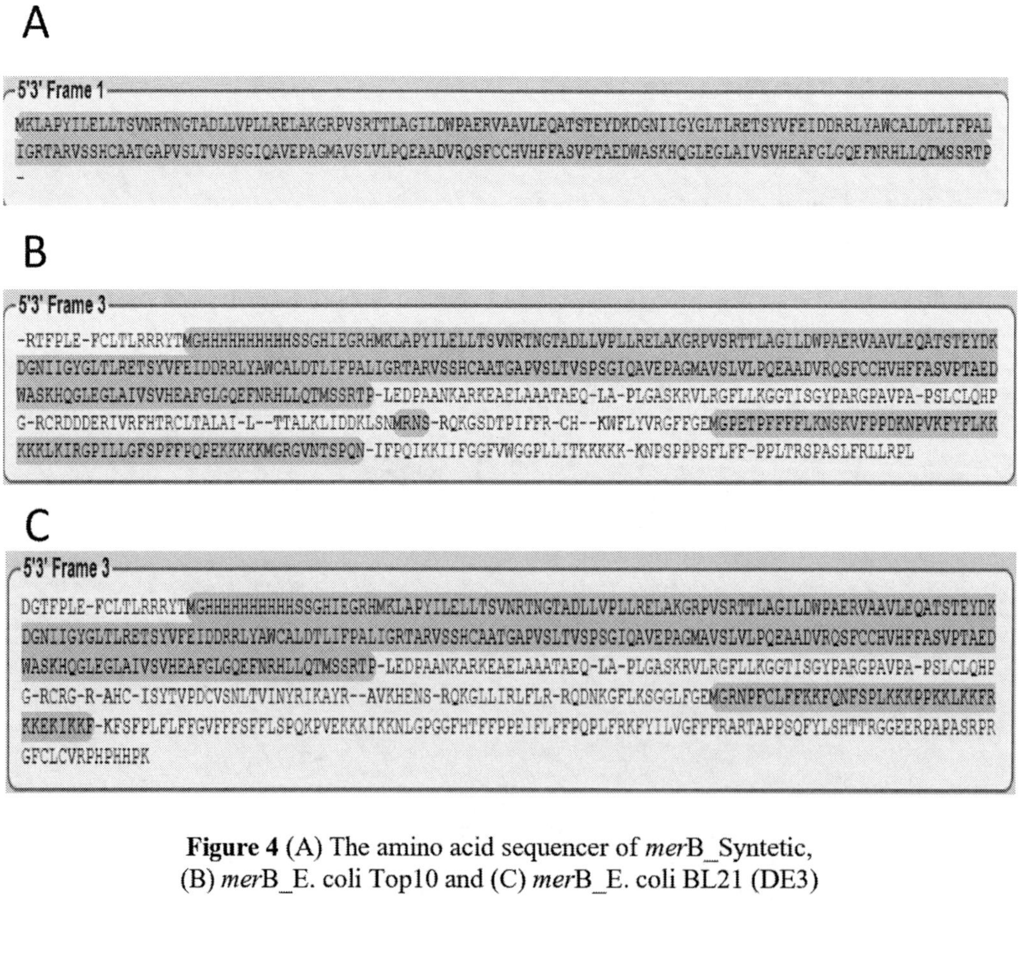


**Figure 4** (A) The amino acid sequencer of *mer*B_Syntetic, (B) *mer*B_E. coli Top10 and (C) *mer*B_E. coli BL21 (DE3)

## 3. 3 Overproduction and Purification of MerB Protein

Overproduction of 200 mL scale was carried out at optimum conditions, at 37°C for 3 hours of incubation at 150 rpm with IPTG induction of 0.1 mM, to see the condition of the protein in a dissolved or undissolved state. Based on the SDS-PAGE results, it is known that the MerB protein is in a dissolved state which can be seen from the SDS band of MerB protein (25.6 kDa) in the supernatant and not present in the debris.

Purification of affinity-tagged MerB protein expressed during overproduction was purified on Ni-NTA beads and eluted with washing buffers containing 25, 50, 100, 200 and 500 mM Imidazol, respectively. Approximately 25.6 kDa MerB protein was detected after resolving on 10% SDS PAGE (Figure 5). The merB gene encodes for organomercury liase (MerB), an enzyme that plays a role in breaking the Hg-C bonds in organic mercury compounds into inorganic mercury compounds and *mer*A gene encodes for mercury reductase (MerA), an enzyme that plays a role in the reduction of highly toxic ionic $Hg^{2+}$ into nontoxic volatile metallic $Hg^{\circ}$ species [9]. The use of these two protein enzymes can detoxify the mercury that pollutes the environment.

Optimization of purification was carried out with as much as 4 mL of crude extract which was equivalent to the amount of MerB protein ± 26.4 mg, therefore crude extract can still be added because the column binding capacity is up to 40 mg protein. In the optimization stage of purification, the column equilibrium was consecutively using 5 mM imidazole. wash using lew buffer 4 times 1mL each, 15mM imidazole twice each with 1.5mL. Imidazole 25 mM 2 times each as much as 1.5 ml. Imidazole 50 mM 2 times each as much as 1.5 ml. Imidazole 100 mM 2 times 1.5 mL each). Imidazole 200 mM 3 times each amounting to 1.5 mL. Imidazole 500mM 4 times 1.5mL each.

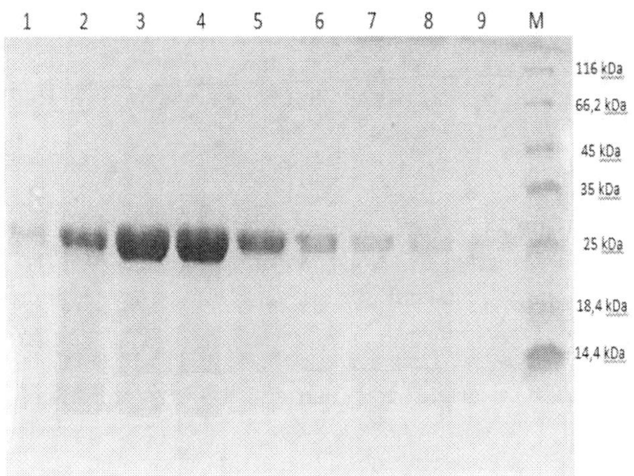


**Figure 5.** MerB Protein Purification Results under Optimum Conditions

Based on Figure 5, it is known that the column equilibrium concentration with 5 mM imidazole, wash imidazole concentration to attract impurities with 50 mM, MerB protein elution concentration with imidazole 100 mM and column striping concentration with imidazole 500 mM.

## 4. Conclusion

Overproduction and purification of MerB protein was successfully performed in E. coli BL21 mediated by plasmid pET16b, resulting MerB protein with a molecular weight of 25.6 kDa, with the optimum at 37°C incubation temperature, incubation time of 3 hours and 0.1

mM IPTG induction. MerB protein obtained can be used in further research on the enzymatic detoxification of organic mercury.

## Acknowledgment

The authors owe a profound debt of thanks to Directorate General of Higher Education, Ministry of Education and Culture for funding this research through PDUPT Grant under Higher Education Decentralization scheme for fiscal year 2018-2020.

## References

[1] Figueiredo N L, Canário J, Duarte A, Serralheiro M L and Carvalho C 2014 Isolation and characterization of mercury-resistant bacteria from sediments of Tagus Estuary (Portugal): Implications for environmental and human health risk assessment *Journal of Toxicology and Environmental Health, Part A* **77** (1-3)155-168

[2] Naguib M M, El-Gendy A O and Khairalla A S 2018 Microbial diversity of mer operon genes and their potential rules in mercury bioremediation and resistance *The Open Biotechnology Journal* **12** (1) 56-77

[3] Zulaika E, Andry P U M, Fitria A N and Prasetyo E N 2019 The Potential of Bacillus cereus S1 as an Environmentally Friendly Bioaccumulator of Gold Nanoparticle Waste *IOP Conference Series: Materials Science and Engineering* **546** (6) 062036

[4] Fatimawali K B and Tallei T E 2019 Potential of organic mercury-resistant bacteria isolated from mercury contaminated sites for organic mercury remediation *Pak J Biol Sci* **22** (1) 45-50

[5] Zulaika E and Sembiring L 2015 Mercury reductase activity of an indigenous mercury resistant bacterial isolate (Bacillus sp. S1) from kalimas-surabaya as a potential reducing agent for mercurial ion ($Hg^{2+}$) *KnE Life Sciences* 446-450

[6] Zhang X, Xu D, Zhu C, Lundaa T and Scherr K E 2012 Isolation and identification of biosurfactant producing and crude oil degrading Pseudomonas aeruginosa strains *Chemical Engineering Journal* **209** 138-146

[7] Kowalczyk A, Wilińska M, Chyc M, Bojko M and Latowski D 2016 Isolation, screening and identification of mercury resistant bacteria from mercury contaminated soil *E3S Web of Conferences* **10** 0042

[8] Rojas L A, Yáñez C, González M, Lobos S, Smalla K and Seeger M 2011 Characterization of the metabolically modified heavy metal-resistant Cupriavidus metallidurans strain MSR33 generated for mercury bioremediation *PloS one* **6** (3) e17555

[9] Kepel, B., & Tallei, T. E. 2015 Overproduction of Mercuric Reductase from Mercury-Resistant Bacteria Klebsiella Pneumoniae Isolate A1.1.1 *Indonesian Journal of Pharmacy* **26** (3) 141-146

[10] Villalobos A, Ness J E, Gustafsson C, Minshull J and Govindarajan S 2006 Gene Designer: a synthetic biology tool for constructing artificial DNA segments *BMC bioinformatics* **7** (1) 1-8

[11] Tu Z, He G, Li K X, Chen M J, Chang J, Chen L and Wu X 2005 An improved system for competent cell preparation and high efficiency plasmid transformation using different Escherichia coli strains *Electronic Journal of Biotechnology* **8** (1) 113-120

[12] Brown T 2010 *Gene cloning & DNA analysis an introduction* Oxford: Wiley- Blackwell

[13] Kepel B J 2012 Isolasi dan Karakterisasi Gen merB pada Bakteri Pseudomonas sp. Sebagai Gen Resistensi Merkuri Organik *Jurnal Kedokteran YARSI* **20** (2) 69-80

IOP Conf. Series: Materials Science and Engineering 1115 (2021) 012081 doi:10.1088/1757-899X/1115/1/012081

# Antioxidant activity and total fenol content white saffron (*Curcuma mangga Val*)

**S Maryam* and Martiningsih**

Department of Chemistry, Faculty of Mathematic and Natural Science, Universitas Pendidikan Ganesha, Bali, Indonesia

*Corresponding author: siti.maryam@undiksha.aci.id

**Abstract.** White saffron (*Curcuma mangga Val*) is one of the rhizomes that is widely used as traditional medicine or health drinks, because of the presence of antioxidant compounds in the form of curcuminoids. Curcuminoids are compounds that have fenol groups, have the ability to reduce free radicals and can function to repair damaged cells. This study aims to determine the antioxidant activity and total fenol content of white saffron (*Curcuma mangga Val*) so as to determine the potential of white saffron (*Curcuma mangga Val*) as an antioxidant. The sample in this study was white saffron (*Curcuma mangga Val*) powder. Analysis of antioxidant activity using the DPPH method and analysis of total fenol content using the Folin Ciucalceu test. The results of the mean analysis of the samples obtained, the antioxidant activity ($IC_{50}$) of white saffron (*Curcuma mangga Val*) was 60.61 ppm and the total fenol level was 87.73 mg/g. From the magnitude of the $IC_{50}$ value, it can be said that white saffron (*Curcuma mangga Val*) is a strong antioxidant category, so it can be suggested that white saffron (*Curcuma mangga Val*) can be used as a functional drink and can be a supplement for degenerative diseases.

## 1. Introduction

The use of traditional medicines in Indonesia has been used as an alternative medicine for a long time and there is a belief that traditional medicines are safer because they come from plants or plants [1]. Plants that have the potential as ingredients for traditional medicines contain phytochemical components that play a role in both prevention and treatment. Traditional medicine is formulated traditionally and its use and utilization is obtained based on previous experience, one of which is white saffron (*Curcuma mangga Val*).

White saffron (*Curcuma mangga Val*) is one of the rhizomes that is currently used as a functional drink, because it is healthy due to its antioxidant content such as curcuminoids, flavonoids and total fenols [2]. Curcuma consists of several species including *Curcuma xanthorriza*, *Curcuma domestica*, *Curcuma mangga*, *Curcuma zedoaria*, *Curcuma heyneana* and *Curcuma aeruginoza* [3]-[4]. These species of Curcuma have similar chemical components of curcuminoids, flavonoids and essential oils that have potential as antioxidants [5].

The presence of fenolic groups in Curcuma compounds causes strong activity in biological systems [4]. White saffron (*Curcuma mangga Val*) contains secondary metabolites such as curcuminoids consisting of curcumin and its derivatives which include desmetoxicurcumin and bisdesmetoxycurcumin [6] as well as terpenoids, alkaloids, and saponins [7]-[8]. Chemical compounds that are classified as antioxidants and are widely available in saffron, among others, come from the polyfenols, flavonoids, vitamin C, vitamin E, β-carotene, and catechins [9]. Some plants have proven potential as antioxidants because they contain various substances such as carotene, flavonoids and fenolic components, as well as vitamins C and E [10]-[12].

Reactive oxygen species (ROS) and free radicals can cause severe damage to normal cells of the body. This damage can occur to DNA, proteins, and other macromolecules. This damage is the beginning of various diseases, especially heart disease and cancer. There are numerous studies proving that because this disease is mediated by oxidative stress and disrupts the balance between pro-oxidant and antioxidant factors, antioxidants can play an important role in preventing or slowing the progression of this condition [13]. One way to overcome this health problem is to consume foods that are high in antioxidants.

The components in each plant are strongly influenced by the place where each plant grows, this is because the content in plants depends on the absorption of nutrients in the soil and metabolic processes in plants [14]. Fenolic compounds such as flavonoids can be found in almost all types of plants. Flavonoids in plants act as protection against stresses from the environment [15].

The presence of antioxidant components in the form of fenols and flavonoids will have an impact on the antioxidant components in plants. Thus in this study, the antioxidant levels and antioxidant activity of white saffron (*Curcuma mangga Val*) will be found in the Buleleng area, which is a tropical area with hot weather. Antioxidant activity will have an impact on the strength and ability of an ingredient to act as an antioxidant.

## 2. Materials and Method

There are several parts to this method, including the materials and tools and the sample preparation process (sample preparation).

### 2.1. Materials

Samples in the form of white saffron (*Curcuma mangga Val*) powder can be found in plantations in the Buleleng area of Bali. Chemical materials for analysis are ethanol, aquades, *2,2-diphenyl-1-picrylhydrazyl* (DPPH), $Na_2CO_3$, Folin-ciocalteu reagent, gallic acid, ethanol, methanol, acetone, acetic acid, distilled water, acid buffer, $FeCl_3.6H_2O$, $NaCO_3$, 10% $NaNO_2$, 10% $AlCl_3.6H_2O$ and 10% NaOH.

### 2.2. Equipment

The equipment used is a knife, washbasin, cutting board and tampah. Analytical equipment is oven, centrifuge, magnetic stirrer, vortex, spectrophotometer, analytical balance, glass appliance, and vacuum rotary evaporator.

### 2.3. White SaffronSample Preparation *(Curcuma mangga Val)*

#### *2.3.1. Making white saffron powder*

The first step in making white saffron (Curcuma mangga Val) rhizome extract is to make white saffron powder.There are several steps in making white saffron powder (Curcuma mangga Val), including:

1. Sort the white saffron from the dirt and wash it with running water
2. Peel the white saffron rhizome
3. Thinly sliced pure white saffron
4. Dry the white saffron slices, by aerating them to dry
5. Blend the dried slices of white saffron in a blender until a powder forms

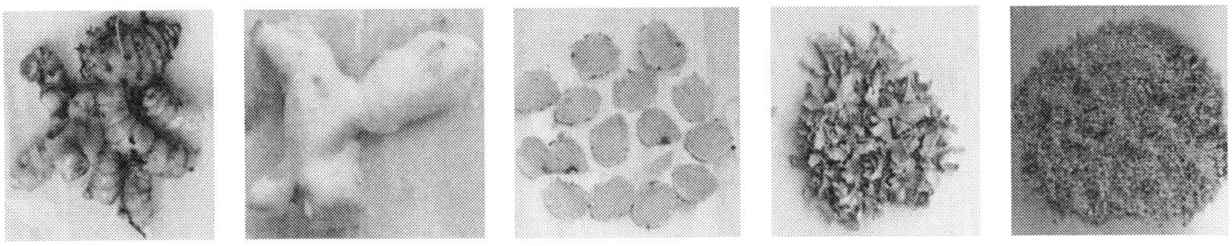

**Figure 1.** The process of making white saffron powder (*Curcuma mangga Val*). From left to right: white saffron, peeled white saffron, white saffron slices, dried white saffron, and white saffron powder.

### *2.3.2. Maceration of white saffron powder (Curcuma mangga Val)*

a. White saffron powder (*Curcuma mangga Val*) was weighed 250 grams and put into a 1 L beaker, then added with 90% ethanol with a ratio of 1: 3 (w/v). Then macerated with ethanol solvent for 3 x 24 hours and every 24 hours the ethanol solvent was replaced.
b. The maceration results were then filtered using Whatman filter paper no. 1 so that the resulting filtrate.
c. The combined filtrate is then dried with a vacuum rotary evaporator at a temperature of 40° C to obtain a thick extract.
d. The viscous extract obtained was then tested for phytochemicals (total phenol) using the Folin Ciocalteu method, using gallic acid as the standard, and the antioxidant activity test using the DPPH method with the Xu and Chang method [16]. The analysis was carried out 3 times each.

### *2.3.3. Determination of total phenol levels*

Total phenol levels was determined by the Folin Ciocalteu method using gallic acid as the standard. 50 μl of sample, added with 250 μl of Folin ciocalteu solution, then let stand for 1 minute and added 750 μl of 20% $NaCO_3$, then pulverized, and added distilled water to a volume of 5 ml. After 5 minutes of incubation at room temperature, the absorbance was measured at λ 760 nm. Gallic acid was used as a standard and a calibration curve was made with gallic acid 31.875 to 510 mg/L with r = 0.99. The calculation result of total phenol is mg Gallic Acid Equivalent (EAG) per gram of dry extract. The analysis was carried out on 3 samples each with 3 replications.

### *2.3.4. Antioxidant activity test with the DPPH method*

Samples in the form of white saffron extract (*Curcuma mangga Val*) with various concentrations (150, 200, 250, 300, 350 ppm) were taken as much as 2 ml and put into a test tube wrapped in aluminum foil then added 1 ml of 0.004% DPPH solution. Furthermore, the solution was shaken until homogeneous and left for 30 minutes at room temperature without light. Followed by measuring the absorbance at a wavelength of 517 nm using a UV-Vis spectrophotometer. The present value of inhibition represented by the $IC_{50}$ value is calculated by the following formula:

$$\%\,inhibition = \frac{control\ absorbance - sample\ absorbance}{control\ absorbance}\ x\ 100$$

The result of the percentage of inhibition is included in the linear equation with the equation Y = aX + b. Where Y = percentage of inhibition, a = gradient, X = concentration (μg/ml), b = constant. To calculate the value of $IC_{50}$, the equation becomes: 50 = aX + b, X = 50 - w/a. The price of X is $IC_{50}$ in units of μg / ml (ppm).

### 2.3.5. *Data analysis*

The study was conducted with 3 repetitions in order to obtain the mean value of one sample. In this study, three samples were used. The data obtained from the absorbance value were tested for antioxidant activity and the total fenol content was also calculated.

## 3. Result and Discussion

### 3.1. Antioxidant activity of white saffron *(Curcuma mangga Val)*

The results of the analysis of antioxidant activity using the DPPH method of white saffron (*Curcuma mangga Val*) can be seen in Table 1.

**Table 1.** Antioxidant activity ($IC_{50}$) of white saffron (*Curcuma mangga Val*)

| No | White Saffron Sample | Antioxidant Activity ($IC_{50}$) (ppm) |
|---|---|---|
| 1 | Sample 1.1 | 60.60 |
| 2 | Sample 1.2 | 60.88 |
| 3 | Sample 1.3 | 60.87 |
| 4 | Sample 2.1 | 60.17 |
| 5 | Sample 2.2 | 61.08 |
| 6 | Sample 2.3 | 60.76 |
| 7 | Sample 3.1 | 59.98 |
| 8 | Sample 3.2 | 60.66 |
| 9 | Sample 3.3 | 60.78 |
| | Mean | 60.61 |

The ability of a material to function as an antioxidant is determined by the presence of groups that act to scavenge free radicals. White saffron (*Curcuma mangga Val*) is a medicinal plant that contains bioactive compounds, which can be seen from its antioxidant activity. In this study, Table 1 shows that white saffron (*Curcuma mangga Val*) has an $IC_{50}$ value of 60.61 ppm.The mean value of white saffron antioxidant activity from each sample (samples 1, 2 and 3) is shown in the Figure 2.

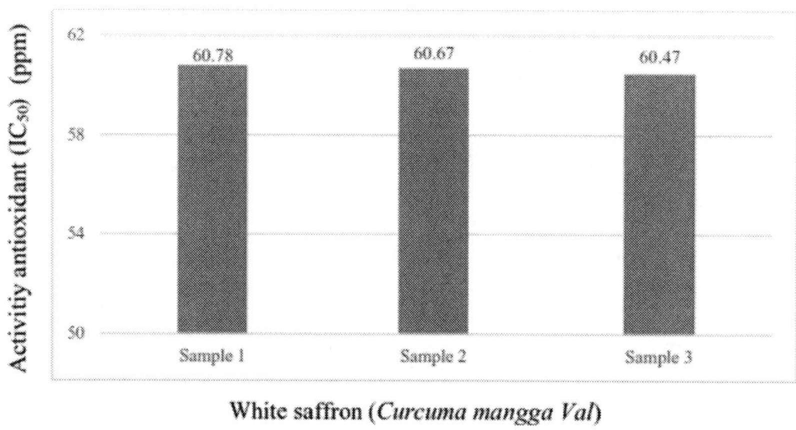


**Figure 2.** Histogram of antioxidant activity ($IC_{50}$) of white saffron (*Curcuma mangga Val*)

In another study it was also stated that white saffron (*Curcuma mangga Val*) has potential as an antioxidant through NO and $H_2O_2$ capture activity [17]. White saffron (*Curcuma mangga Val*) has curcuminoid compounds, such as curcumin [11]. Curcumin is capable of scavenging free radicals such as superoxide anions and hydroxyl radicals, which are initiators of lipid peroxidation. The resulting antioxidant activity of 60.61 ppm means that white saffron (*Curcuma mangga Val*) has strong antioxidants. This situation is in accordance with what was said by Molyneux, the $IC_{50}$ value <50 ppm is a very strong antioxidant, $IC_{50}$ = 50-100 ppm is strong, 100-150 ppm moderate, 150-200 ppm is weak and $IC_{50}$> 200 ppm is categorized as very weak [18].

### 3.2. Total fenol content of white saffron *(Curcuma mangga Val)*

Based on the fenol content test in white saffron (*Curcuma mangga Val*) using the ciucalceu folin test, the total fenol content in white saffron can be seen in Table 2.

**Table 2.**Total fenol content of white saffron (*Curcuma mangga Val*)

| No | White Saffron Sample | Phenol (mg/gr) |
|---|---|---|
| 1 | Sample 1.1 | 87.27 |
| 2 | Sample 1.2 | 88.01 |
| 3 | Sample 1.3 | 87.37 |
| 4 | Sample 2.1 | 88.09 |
| 5 | Sample 2.2 | 87.44 |
| 6 | Sample 2.3 | 87.39 |
| 7 | Sample 3.1 | 88.07 |
| 8 | Sample 3.2 | 87.93 |
| 9 | Sample 3.3 | 87.98 |
| | Mean | 87.73 |

Fenolic compounds are secondary metabolites found in an organism to prevent damage or reduce damage to the membrane components that occur in living organisms [11]. In this study, from three samples that were repeated three times for each sample, the total fenol content in white saffron (*Curcuma mangga Val*) was 87.73 mg/g. From the total fenol content in white saffron, it can be said that white saffron contains good fenol compounds. Figure 3 is a histogram of the fenol content of each sample.

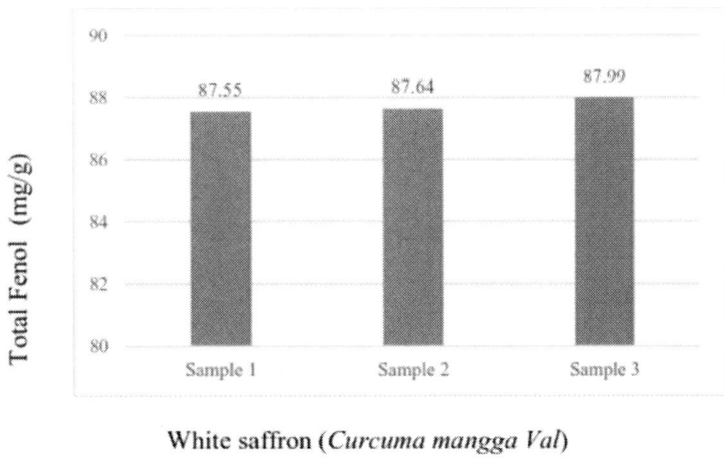


**Figure 3.** Histogram of total phenol levels of white saffron (*Curcuma mangga Val*)

The hydroxyl group of fenols is able to scavenge free radicals, able to reduce the radical properties of reactive oxygen compounds such as superoxides, peroxide radicals, hydroxyl radicals and feroxynitrite. Fenolic compounds are known to protect plants from herbivores and the main function of most fenolic compounds is to protect plants from damage due to excessive light by acting as antioxidants, and their levels vary according to the conditions or growth of the plant. Seeing the potential that exists in white saffron (*Curcuma mangga Val*), it is advisable to use white saffron (*Curcuma mangga Val*) as a functional drink.

## 4. Conclusion

The results of research that has been done, it shows that white saffron (*Curcuma mangga Val*) has a fairly high antioxidant activity, namely 60.61 ppm and a total phenol of 87.73 mg / g, so white saffron is very well used for the prevention of diseases caused by free radicals.

## References

[1] Siipi H 2013 Is natural food healthy? *Journal of agricultural and environmental ethics* **26** (4) 797-812
[2] Jayaprakasha G K, Rao L J and Sakariah K K 2006 Antioxidant activities of curcumin, demethoxycurcumin and bisdemethoxycurcumin *Food chemistry* **98** (4) 720-724
[3] Zou X, Dai Z, Ding C, Zhang L, Zhou Y and Yang R 2011 Relationships among six medicinal species of Curcuma assessed by RAPD markers *Journal of Medicinal Plants Research* **5** (8) 1349-1354
[4] Ahsan H, Parveen N, Khan N U and Hadi S M 1999 Pro-oxidant, anti-oxidant and cleavage activities on DNA of curcumin and its derivatives demethoxycurcumin and bisdemethoxycurcumin *Chemico-biological interactions* **121** (2) 161-175
[5] Kusuma I W, Kuspradini H, Arung E T, Aryani F, Min Y H, Kim J S and Kim Y U 2011 Biological activity and phytochemical analysis of three Indonesian medicinal plants, Murraya koenigii, Syzygium polyanthum and Zingiber purpurea *Journal of Acupuncture and Meridian Studies* **4** (1) 75-79
[6] Rajkumari S and Sanatombi K 2017 Nutritional value, phytochemical composition, and biological activities of edible Curcuma species: A review *International journal of food properties* **20** (3) S2668-S2687
[7] Alara O R, Abdurahman N H, Ukaegbu C I, Azhari N H and Kabbashi N A 2018 Metabolic profiling of flavonoids, saponins, alkaloids, and terpenoids in the extract from Vernonia cinerea leaf using LC-Q-TOF-MS *Journal of Liquid Chromatography & Related Technologies* **41** (11) 722-731
[8] Mir M A, Parihar K, Tabasum U and Kumari E 2016 Estimation of alkaloid, saponin and flavonoid, content in various extracts of Crocus sativa *Journal of Medicinal Plants Studies* **4** (5) 171-174
[9] Pisoschi A M and Pop A 2015 The role of antioxidants in the chemistry of oxidative stress: A review *European journal of medicinal chemistry* **97** 55-74
[10] Maryam S 2019 The flavonoid levels in subtituted noodles of tempe flour and carrot extract. *Journal of Physics: Conference Series* 1317 (1) 012038
[11] Balasundram N, Sundram K and Samman S 2006 Phenolic compounds in plants and agri-industrial by-products: Antioxidant activity, occurrence, and potential uses *Food chemistry* **99** (1) 191-203
[12] Krishnaiah D, Sarbatly R and Nithyanandam R 2011 A review of the antioxidant potential of medicinal plant species *Food and bioproducts processing* **89** (3) 217-233
[13] Mandal V, Mohan Y and Hemalatha S 2007 Microwave assisted extraction—an innovative and

promising extraction tool for medicinal plant research *Pharmacognosy reviews* **1** (1) 7-18

[14] Shahidi F and Zhong Y 2015 Measurement of antioxidant activity *Journal of functional foods* **18** 757-781

[15] Antolovich M, Prenzler P D, Patsalides E, McDonald S and Robards K 2002 Methods for testing antioxidant activity *Analyst* **127** (1) 183-198

[16] Sethi S, Joshi A, Arora B, Bhowmik A, Sharma R R and Kumar P 2020 Significance of FRAP, DPPH, and CUPRAC assays for antioxidant activity determination in apple fruit extracts *European Food Research and Technology* **246** (3) 591-598

[17] Pujimulyani D, Yulianto W A, Setyowati A, Arumwardana S, Kusuma H S W, Sholihah I A and Maruf A 2020 Hypoglycemic Activity of Curcuma mangga Val. Extract via Modulation of GLUT4 and PPAR-γ mRNA Expression in 3T3-L1 Adipocytes *Journal of Experimental Pharmacology* **12** 363

[18] Molyneux P 2004 The use of the stable free radical diphenylpicrylhydrazyl (DPPH) for estimating antioxidant activity *Songklanakarin J. sci. technol* **26** (2) 211-219

# Synthesis of $Ag_3PO_4$/HAp from Red Snapper Bone (*Lutjanus spp.*) For Photodegradation of Methylene Blue

**V P J Lairah, A D Wuntu and H F Aritonang**

Department of Chemistry, Faculty of Natural Science and Mathematics, Sam Ratulangi University, Manado, Indonesia

Corresponding author: vanialairah08@gmail.com

**Abstract.** Fish bone is one part of the fish body that is often not used in the fish processing industry, so it needs to be used as HAp. For the composite material to be used in the study of the photodegradation kinetics model of methylene blue, it must go through synthesis. This research went through a calcination process to obtain fish bone powder at a temperature of 600°C. The purpose of this study was to determine the kinetics model of the photodegradation of methylene blue by $Ag_3PO_4$/HAp composites from 600°C calcined fish bones with a ratio of Ag;HAp = 5:1 and 2.5:1. The synthesis was carried out by reacting HAp with acid and $AgNO_3$ and then characterized by the *X-Ray Diffraction* (XRD) method. The results showed that the composite formed was $Ag_3PO_4$/HAp and at the peak of the 2θ samples diffraction patterns with standard diffraction patterns were very similar. The determination of this kinetics model resulted in the highest $R^2$ value in the second order kinetics model with a mole ratio of Ag:HAp = 2.5:1, namely 0.001 so that the photodegradation of methylene blue took place following the second order kinetics model.

## 1. Introduction

North Sulawesi is one of the places that has marine biological resources with various types of fish contained in it. Fish bones are more often not used in the fish processing industry and fish bones contain the highest calcium content among fish body parts [1]. Fish bones contain 60-70% minerals such as HAp and 30% organic components in the form of collagen protein [2].

In Indonesia, there are several types of snapper, including (*Red snapper, Lutjanus sanguine*) and dark greenish red snapper, known as giant (*Seaperch* or *seabass, Lates calcarifer*). The red snapper comes from the *Lutjanidae* tribe, while the snapper is from the Centropomidae family [3]. Red snapper (*Lutjanus spp.*) Is one type of marine fish that has important economic value and has the potential to be cultivated. The red snapper, *Lutjanus argentimaculatus* or "*red snapper*", spreads across Indo-Pacific waters, Line islands in North Africa to Australian waters and Ryukyu Islands, Japan [4].

Dyestuff waste produced from the textile industry is generally non-biodegradable organic compounds, which can cause environmental pollution, especially the aquatic environment. The types of dyes used in the textile industry today are very diverse, and usually do not consist of one type of dye, therefore handling textile waste becomes very complicated and requires several steps until the waste is completely safe to be released into the aquatic environment [5].

Methylene blue can reduce the intensity of light entering the waters and slow down the photosynthesis of aquatic biota, besides that, methylene blue can cause serious health problems if it enters the human body. Therefore, a waste treatment technology is needed that is able to accelerate the decomposition of dye waste [6].

As an alternative, a photodegradation method was developed using a photocatalyst material and light radiation whose energy is equal to or greater than the band gap energy of the photocatalyst. With this photodegradation method, the dye will be broken down into simpler components that are safer for the environment [7].

In this research, the synthesis of $Ag_3PO_4$/HAp composites from calcined red snapper bone treated with acid and $AgNO_3$ with different Ag:HAp ratio, namely 5:1 and 2,5:1 was carried out for later use in the study of photodegradation kinetics using methylene blue dye. This study aims to determine the kinetics model of the photodegradation of methylene blue by calcined $Ag_3PO_4$/HAp composites with a ratio of Ag:HAp = 5:1 and 2.5:1.

## 2. Methodology

### 2.1. Sample Preparations

Preparation of fish bones begins with boiling the fish bones. Boiling is carried out at a temperature of 100 60C for 60 minutes. After that, the remaining meat is cleaned by brushing it. The fish bones are then air-dried for about 1 day to reduce the water content during the steaming process. The dried fish bones are then in the oven at 100°C for 24 hours. After the fish bones were dry, they were calcined at 600°C for 5 hours. Fish bones crushed and sieved with a 100 mesh sieve.

### 2.2. Composite Synthesis $Ag_3PO_4$ /Hap

Synthesis of $Ag_3PO_4$/HAp composites using the modified [8] method. Fish bones that have been processed into 600°C flour weigh 2.8519 g, mixed into 100 mL $HNO_3$ 0,01 M solution and stirred for 2 hours then filtered with filter paper and rinsed with distilled water until the pH returns to neutral, after In an oven with a temperature of 105°C for approximately 2 hours, after drying the material is put in 100 mL of 0,15 M $AgNO_3$ solution. This mixture is a mixture with a mole ratio of Ag:HAp = 2.5:1.The mixture is then stirred using a magnetic stirrer for 24 hours. Furthermore, the mixture is filtered with filter paper, then in the oven for 1 night with a temperature of 100°C. The same procedure was carried out using 100 mL of 0.28 M $AgNO_3$ solution to produce a material with an Ag:HAp = 5:1 ratio. The material was further characterized by XRD and used for the photodegradation of methylene blue.

### 2.3. Photodegradation Kinetics of Methylene Blue

The photodegradation kinetics experiment of methylene blue was carried out following the procedure of [9]. The composites were weighed 0,03 g each and put in 6 glass bottles, then added 15 mL of 6 ppm of methylene blue. The mixture was then placed in a photodegradation box and exposed to a commercial 100 W incandescent lamp (Phillips E27 220-240V A55 CL) for a time range of 10 minutes to 240 minutes. After the specified time, the mixture was centrifuged and the remaining methylene blue concentration in the solution was determined using a UV-Vis spectrophotometer (Shimadzu UV-1800) at a wavelength of 664 nm. Existing data were then analyzed using kinetics models of order-1 and order-2 [10].

The first order kinetics model is expressed in the following equation: $-\ln\frac{C_t}{C_0} = k\,t$.

The second order kinetics model is expressed in the following equation: $\frac{1}{C_t} = \frac{1}{C_0} + k\,t$.

## 3. Results and Discussion

### 3.1. Preparation and Synthesis

Calcination of red snapper bones aims to remove organic compounds contained in the bones by heating at a temperature of 600°C. Uncalcinated bones generally contain approximately 30% organic material and 70% inorganic material in the form of hydroxyapatite (HAp). Organic material will burn out at heating 550 C [11] and leave inorganic material in the bones.

The fish bones that have been calcined at 600˚C are crushed and sieved with a 100 mesh sieve to obtain fish bone powder which is then reacted with $HNO_3$ and $AgNO_3$ so that a composite $Ag_3PO_4$/HAp at 600˚C with a mole ratio of Ag:HAp 5:1 (A1-1) and 2.5:1 (A1-2) shown in Figure 1.

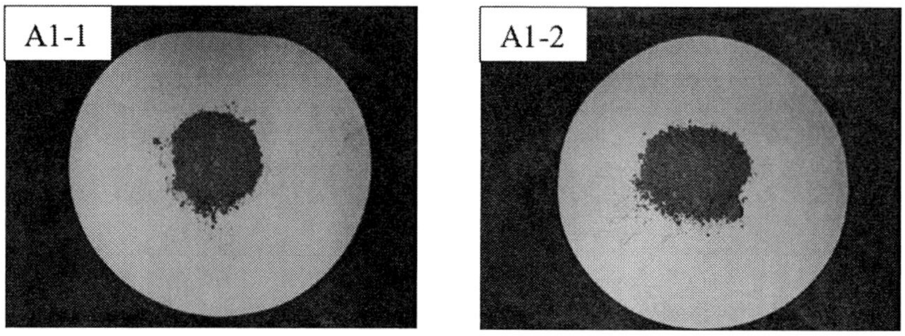


**Figure 1.** $Ag_3PO_4$/HAp composite at 600˚C at a mole ratio of Ag:HAp = 5:1 (A1-1) and 2.5:1 (A1-2)

### 3.2. Analysis of X-Ray Diffraction (XRD)

The *X-Ray Diffraction* (XRD) method provides information about the structure and degree of crystallinity of a solid produced in the form of a diffraction pattern (diffractogram). XRD testing was conducted to determine two types of data, namely qualitative and quantitative. XRD analysis aims to determine the presence of a peak at 2θ position of the HAp sample and the dispersion of $Ag_3PO_4$ on the red snapper bone, the dispersion of the sample can be determined by comparing the sample with HAp and $Ag_3PO_4$ standards. On the diffractogram, the sample and standard diffraction patterns are clearly visible in the presence of peaks in the 2θ region, as shown in Figure 2.

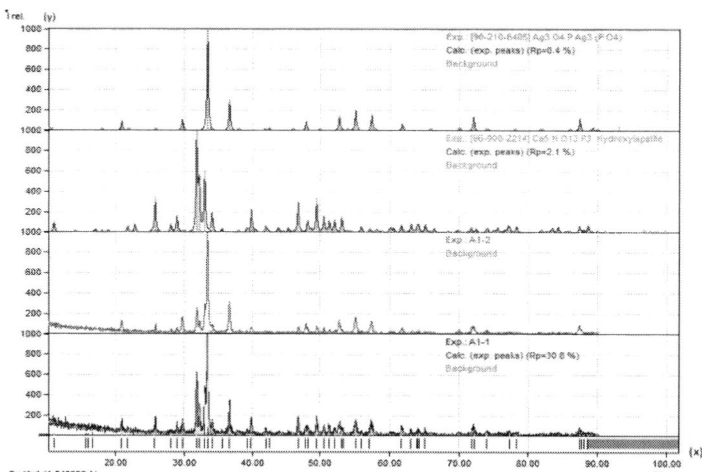


**Figure 2.** Composite $Ag_3PO_4$/HAp diffractogram at 600°C at the mole ratio of Ag:HAp (5:1 and 2.5:1) with HAp and $Ag_3PO_4$ standards

The diffraction patterns of the samples were compared with the standard diffraction patterns from the *Crystal Open Database* (COD). Qualitatively Figure 2 shows the presence of HAp and $Ag_3PO_4$ in the sample by looking at the 2θ peaks very similar to the standard. In addition to the peaks of HAp and $Ag_3PO_4$, the sizes of HAp and $Ag_3PO_4$ particles can be calculated using *Full Width at Half* data. X-ray diffraction can be used to determine the crystallite size with a certain phase [12]. The determination refers to the main peaks of the diffractogram pattern through the Debye Scherrer equation approach which is formulated in Equation (1).

$$D = \frac{K\ \lambda}{\beta\ \mathrm{Cos}\ \theta} \tag{1}$$

Based on the results of the XRD data, the particle size was obtained by calculating FWHM (*Full width Half Maximum*). The combined particle sizes of HAp and $Ag_3PO_4$ found in the samples are presented in Table 1.

**Table 1.** The size of the HAp and $Ag_3PO_4$ particles in the sample

| Ratio | $D_{(nm)}$ | |
|---|---|---|
| | **HAp** | **$Ag_3PO_4$** |
| 5 : 1 | 426.7283738 | 380.526588 |
| 2.5 : 1 | 405.139647 | 320.9982077 |

### 3.3. Photodegradation Kinetics of Methylene Blue

Photodegradation kinetics were used to determine the rate of photodegradation of methylene blue which was influenced by time. The rate test was carried out by estimating the reaction order by plotting –ln Ct/Co vs t for the 1st order reaction kinetics model and 1/Ct vs t for the 2nd order reaction kinetics model (Figures 3, 4, 5 and 6).

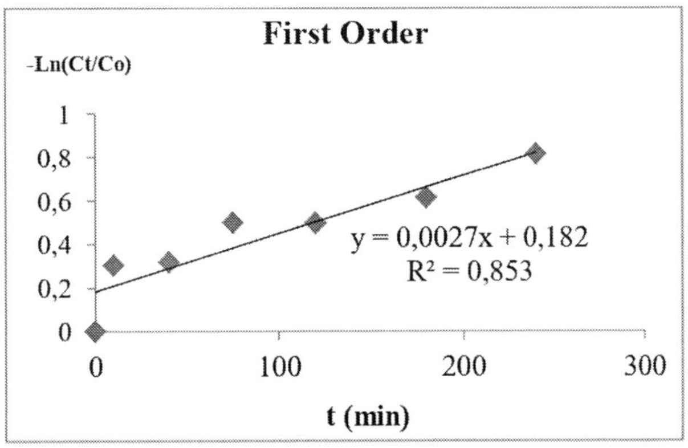


**Figure 3.** Kinetics model of 1st order photodegradation of methylene blue by $Ag_3PO_4$/HAp composite from 600°C calcined fish bone with Ag:HAp mole ratio (5:1)

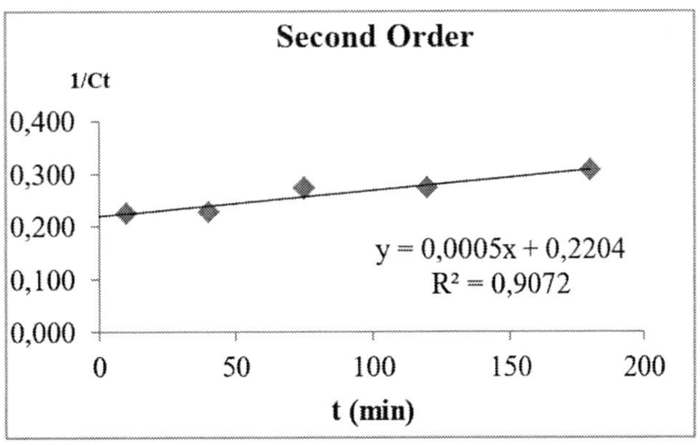


**Figure 4.** Kinetics model of 2nd order photodegradation of methylene blue by $Ag_3PO_4$/HAp composite from 600°C calcined fish bone with Ag:HAp mole ratio (5:1)

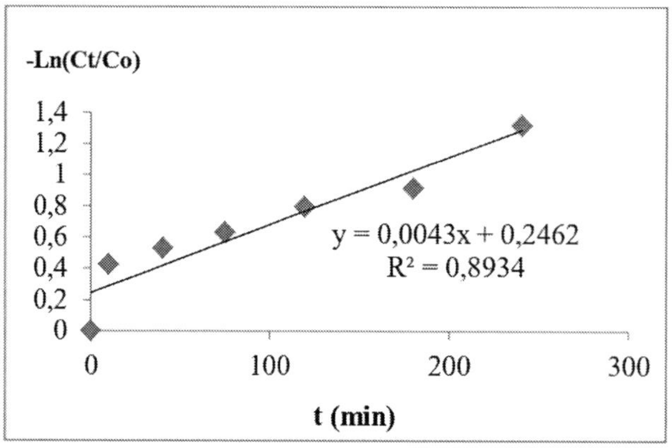


**Figure 5.** Kinetics model of 1st order photodegradation of methylene blue by $Ag_3PO_4$/HAp composite from 600°C calcined fish bone with Ag:HAp mole ratio (2.5:1)

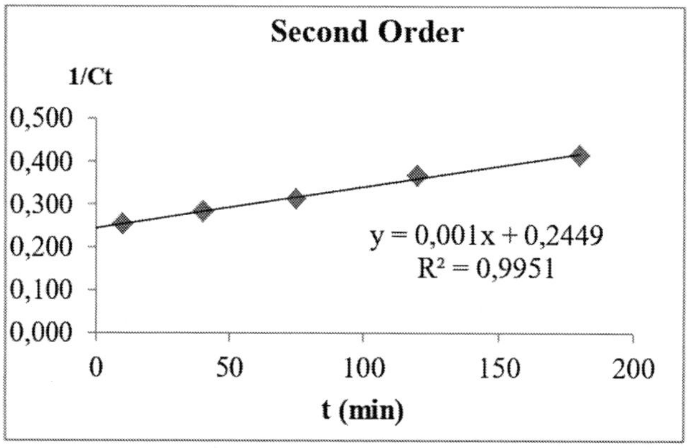


**Figure 6.** Kinetics model of second order photodegradation of methylene blue by $Ag_3PO_4$/HAp composite from 600°C calcined fishbone with Ag:HAp mole ratio (2.5:1)

The calculation results show that the $Ag_3PO_4$/HAp composite material with a mole ratio of Ag:HAp 5:1 and 2.5:1 degrades methylene blue following the second order kinetics model. The determination of this kinetics model is based on the linearity of the curve according to the equation of the 1st and 2nd order kinetics models where the curve in the 2nd order kinetics model produces the greatest R2 (*coefficient of determination*) Table 2.

**Table 2.** Reaction rate constants for the photodegradation of methylene blue by $Ag_3PO_4$/HAp composites

| Composite | Orde 1 | | Orde 2 | |
|---|---|---|---|---|
| | $R^2$ | K | $R^2$ | K |
| 5 : 1 | 0.853 | 0.0027 | 0.9072 | 0.0005 |
| 2.5 : 1 | 0.8934 | 0.0043 | 0.9951 | 0.001 |

In general, the reaction order for a certain substance is not the same as the coefficient in the stoichiometric equation of the reaction [12] and a second-order kinetics model of the reaction rate which is directly proportional to the square of the concentration of the reactants. The values of the photodegradation reaction rate constants and the kinetics model determination coefficients for the two $Ag_3PO_4$/HAp composites with a temperature of 600°C in the Ag:HAp mole ratio (2.5:1 and 5:1) are shown in Table 2.

Table 2 shows that the fastest photodegradation of methylene blue occurred in composites synthesized from calcined fish bones at a temperature of 600°C ikan at an Ag:HAp ratio of 2.5: 1. This characteristic seems to be related to the particle size produced by the different Ag:HAp ratios. The ratio of Ag:HAp = 2.5:1 resulted in smaller particles Table 1 which were estimated using the Debye Scherrer equation. The smaller particle size provides a larger contact surface area between this $Ag_3PO_4$/HAp composite and methylene blue. This allows more methylene blue to interact with the catalyst surface and undergo photodegradation on the photocatalyst surface per unit time so that the photodegradation process takes place faster.

## 4. Conclusion

$Ag_3PO_4$/HAp composites which act as photocatalysts can be synthesized by treating nitric acid and $AgNO_3$ on calcined red snapper bone 600°C tulang. Treatment Ag:HAp ratio = 2.5:1 resulted in composites with smaller particle sizes and Photodegradation of methylene blue by composite $Ag_3PO_4$/HAp took place following the second order reaction kinetics model and the photodegradation of methylene blue was faster in composites that were synthesized using a ratio of Ag/HAp = 2.5: 1.

## References

[1] Jayathilakan K, Sultana K, Radhakrishna K and Bawa A S 2012 Utilization of byproducts and waste materials from meat, poultry and fish processing industries: a review *Journal of food science and technology* **49** (3) 278-293
[2] Mery D, Lillo I, Loebel H, Riffo V, Soto A, Cipriano A and Aguilera J M 2011 Automated fish bone detection using X-ray imaging *Journal of Food Engineering* **105** (3) 485-492
[3] Carter C H R I S, Glencross B R E T T, Katersky R S and Bermudes M 2010 The snooks (family: centropomidae) *Finfish Aquaculture Diversification. CABI, Oxfordshire* 323-336
[4] Andriyono S, Alam J, Kwak D H and Kim H W 2018 Complete mitochondrial genome of brownstripe red snapper, Lutjanus vitta (Perciformes: Lutjanidae) *Mitochondrial DNA Part B* **3** (2) 1129-1130
[5] Rai H S, Bhattacharyya M S, Singh J, Bansal T K, Vats P and Banerjee U C 2005 Removal of dyes from the effluent of textile and dyestuff manufacturing industry: a review of emerging techniques with reference to biological treatment *Critical reviews in environmental science and technology* **35** (3) 219-238
[6] Hanafiah M A K M, Ngah W S W, Zolkafly S H, Teong L C and Majid Z A A 2012 Acid Blue 25 adsorption on base treated Shorea dasyphylla sawdust: Kinetic, isotherm, thermodynamic and spectroscopic analysis *Journal of Environmental Sciences* **24** (2) 261-268
[7] Corrent S, Cosa G, Scaiano J C, Galletero M S, Alvaro M and Garcia H 2001 Intrazeolite photochemistry. 26. photophysical properties of nanosized TiO2 clusters included in zeolites Y, β, and Mordenite. *Chemistry of materials* **13** (3) 715-722
[8] Piccirillo C, Pinto R A, Tobaldi D M, Pullar R C, Labrincha J A, Pintado M M E and Castro P M L 2015 Light induced antibacterial activity and photocatalytic properties of Ag/Ag3PO4-based material of marine origin *Journal of Photochemistry and Photobiology A: Chemistry* **296** 40-47
[9] Taringan A K, Wuntu A D and Aritonang H F 2017 Kinetika Fotodegradasi Remazol Yellow Menggunakan Fotokatalis ZnO dan ZnO-Ag *Jurnal MIPA* **6** (2) 68-71
[10] Wu C H and Chern J M 2006 Kinetics of photocatalytic decomposition of methylene blue *Industrial & engineering chemistry research* **45** (19) 6450-6457
[11] Hafiludin H 2011 Karakteristik Proksimat dan Kandungan Senyawa Kimia Daging Putih dan Daging Merah Ikan Tongkol (Euthynnus affinis) *Jurnal Kelautan: Indonesian Journal of Marine Science and Technology* **4** (1) 1-10
[12] Zak A K, Majid W A, Abrishami M E and Yousefi R 2011 X-ray analysis of ZnO nanoparticles by Williamson–Hall and size–strain plot methods *Solid State Sciences* **13** (1) 251-256

# The Clique Numbers and Chromatic Numbers of The Coprime Graph of a Dihedral Group

**A G Syarifudin, I G A W Wardhana*, N W Switrayni, and Q Aini**

Department of Mathematics, Faculty of Mathematics and Natural Science University of Mataram, Indonesia.

*Corresponding author: adhitya.wardhana@unram.ac.id

**Abstract.** The graph has many properties and characterizations. One interesting topic to discuss is the clique numbers and chromatic numbers. This research will determine the clique numbers and chromatic numbers of the coprime graph of the dihedral group. One of the main results is if $n = 2^k$ then the chromatics numbers of the coprime graph of the dihedral group are 2, and if $n$ is odd composite numbers, then the clique numbers and chromatics numbers are $(m + 2)$.

## 1. Introduction

Many mathematicians study the representation graph from an algebraic structure in recent years. Mostly studied the representation graph from an algebraic structure called group. Another study is the representation graph from an algebraic structure called ring and module [1]. The coprime graph introduced in ref [1], he gives the definitions of a graph as a representation of a group. In ref [2], Dorbini classified is a complete $t$-partite graph or planar graph and automorphism group of $\Gamma_G$. In addition, Mansoori (2016) gives a dual version of the coprime graph called non-coprime [3].

In 2020, two different studies about the coprime graph were given in ref [4]. Syarifudin et al. (2020) studied some characterization of a dihedral group's coprime graph. He gives a sufficient condition whenever the coprime graph forms a complete bipartite graph, a complete tripartite graph, and $m$-partite graph. He also studies other properties such as radius, diameter, and girth [4], [6]. On the other hand. Juliana et al. (2020) give some characterization of the coprime graph of an integer modulo group [5]. Masriani et al. (2020) give a dual version of Juliana's studied on a non-coprime graph of an integer modulo group [7].

In another study about the representation graph of a group, Chelvama et al. give the clique numbers and chromatics numbers of the dihedral group's commuting graph [6]. Abdollahi et al. previously give some properties of the representation graph of a group called non-commuting graph [9] and non-cyclic graph [10].

This paper aims to find the clique numbers and chromatic numbers of the dihedral group's coprime graph.

IOP Conf. Series: Materials Science and Engineering 1115 (2021) 012083 doi:10.1088/1757-899X/1115/1/012083

## 2. Methodology

The methodology of this research is by making a pattern from some example of the dihedral group for some $\boldsymbol{n}$, and the conjecture will be made by the pattern. And by deductive, we will prove the conjecture. If the conjecture is correct, then we state it as a Theorem. If wrong, then we make another conjecture.

## 3. Result and Discussion

This section will discuss the coprime graph of a dihedral group and some characterization like clique numbers and chromatic numbers. Before that, several definitions that support this paper will be presented.

A dihedral group is a group generated by two elements, more details in the following definition.

**Definition** 1 ([3]) Group $G$, named dihedral group with order $2n$, $n \geq 3$ and $n \in \mathbb{N}$, is a group generated by $a, b \in G$ with properties

$$G = \langle a, b | a^n = e, b^2 = e, bab^{-1} = a^{-1} \rangle$$

A dihedral group with order $2n$ denoted by $D_{2n}$.
The following definition of the order of each element of a group, as follows.

**Definition 2** ([3]) If $(G,*)$ any group. Let $a$ any elements of $G$. The least positive integer $m$ with $a^m = e$ ($e$ identity in $G$) then $m$ named order of $a$, and denoted by $|a| = m$.

The following definitions are a point in constructing the coprime graph of a dihedral group.

**Definition 3** ([1]) Let $G$ finite group, the coprime graph of $G$ denoted by $\Gamma_G$ is a graph with vertices are elements of $G$ and two distinct vertices $x$ and $y$ are adjacent if and only if $(|x|, |y|) = 1$.

The following definition describes the clique number of a graph, explained in the next Theorem.

**Definition 4** The subset of the vertices of $\Gamma_G$ is called a clique if the induced subgraph of a subset is a complete graph. The maximum size of a clique in a graph $\Gamma_G$ is called the clique number of $\Gamma_G$ and denoted by $\omega(\Gamma_G)$.

This last definition discusses the chromatic number of a graph that is related to graph coloring.

**Definition 5** Let $k > 0$ be an integer. A $k$-vertex of $\Gamma_G$ coloring of a graph $\Gamma_G$ is an assignment of $k$ colors to the vertices of $\Gamma_G$ such that no two adjacent vertices have the same color. The chromatic number $\chi(\Gamma_G)$ of a graph is the minimum $k$ for which $\Gamma_G$ has a vertex coloring.

### 2.1 Coprime Graph of $D_{2n}$

The first Theorem about the form of the coprime graph of $D_{2n}$, with $n$ odd prime number explained in the next Theorem.

***Theorem 1*** ([3]) Let If $n$ is an odd prime number, then the coprime graph of $D_{2n}$is complete tripartite.
**Proof**. Let $n$ is an odd prime number, Define three partition by $V_1 = \{e\}$, $V_2 = \{a, a^2, \dots, a^{p-1}\}$ and $V_3 = \{ b, ab, a^2b, \dots, a^{p-1}b\}$. Clearly $|e| = 1$, as $n$ is a prime number. Since the order of $a$ is $n$ and the order of $b$ is 2, then we have $|a| = |a^2| = \cdots |a^{p-1}| = p$, and $|b| = |ab| = |a^2b| = \cdots = |a^{p-1}b| =$

2. Hence for each$x, y \in V_i$ then $(|x|,|y|) = p \neq 1$ or $(|x|,|y|) = 2 \neq 1$, $i \in \{2,3\}$. Then for any $u \in V_i$ and $v \in V_j$ where $i \neq j$ we have $(|u|,|v|) = 1$, thus $u$ and $v$ are adjacent, then the coprime graph of this dihedral group is complete tripartite. ∎

The next Theorem explains the coprime graph of $D_{2n}$ is complete bipartite when $n = 2^k$, for some $k \in N$.

***Theorem 2*** ([3]) Let $n = 2^k$, for some $k \in N$ then the coprime graph of $D_{2n}$ is a complete bipartite.

**Proof**. Let $D_{2n} = \{e, a, a^2, \dots, a^{n-1}, b, ab, \dots, a^{2^k-1}b\}$ is a dihedral group. Define two partitions by $V_1 = \{e\}$, and $V_2 = \{a, a^2, \dots, a^{2^k-1}, b, ab, a^2b, \dots, a^{2^k-1}b\}$. Clearly $|e| = 1$, as Let $n = 2^k$ then we have $|a| = |a^2| = \cdots = |a^{2^k-1}| = 2^s$and $|b| = |ab| = |a^2b| = \cdots = \left|a^{2^k-1}b\right| = 2$. Hence for each $x, y \in V_2$ then $(|x|,|y|) = 2^s \neq 1$ for some $s \in \mathbb{N}$, thus $x$ and $y$ are not adjacent. Since $|e| = 1$ and for any $v \in V_2$ we have $(|e|,|v|) = 1$. We have $e$, and $v$ are adjacent, then the coprime graph of this dihedral group is complete bipartite. ∎

This Theorem explains about the coprime graph of $D_{2n}$ is complete tripartite for some $n = p^k$, explained in the next Theorem.

***Theorem 3*** ([3]) Let $n = p^k$ for some $k \in N$ and $p$ is a prime number, $p \neq 2$, then the coprime graph of $D_{2n}$ is complete tripartite.

**Proof.** Let $D_{2n} = \{e, a, a^2, \dots, a^{p^k-1}, b, ab, a^2b, \dots, a^{p^k-1}b\}$ is a dihedral group. Define three partitions by $V_1 = \{e\}, V_2 = \{a, a^2, a^3 \dots, a^{p^k-1}\}$ and $V_3 = \{b, ab, a^2b, \dots, a^{p^k-1}b\}$. Clearly $|e|$=1, since $n = p^k$, $p \neq 2$ then we have $|a| = |a^2| = \cdots |a^{p^k-1}| = p^s$ and $|b| = |ab| = |a^2b| = \cdots = |a^{p-1}b| = 2$. Hence for each $x, y \in V_i$ then $(|x|,|y|) = p^s \neq 1$ for some $s \in \mathbb{N}$ or $(|x|,|y|) = 2 \neq 1$, $i \in \{2,3\}$. Then for any $u \in V_i$ and $v \in V_j$ where $i \neq j$ then $(|u|,|v|) = 1$. Hence $u$ and $v$ are adjacent, then the coprime graph of the dihedral group is a complete tripartite graph. ∎

The last Theorem in this section about the form of the coprime group of $D_{2n}$ where $n$ that the following Theorem gives more generalize.

***Theorem 4*** ([7]) Let $n = p_1^{k_1} p_2^{k_2} p_3^{k_3} \dots p_m^{k_m}$ where $1 \leq i \leq m$, $p_i$ are distinct prime number, and $p_i \neq 2$ then the coprime graph of $D_{2n}$ is $(m+2)$-partite.

**Proof**. Let $D_{2n}$ a dihedral group with $n = p_1^{k_1} p_2^{k_2} p_3^{k_3} \dots p_m^{k_m}$ where $1 \leq i \leq m$, $p_i$ are distinct prime number, $p_i \neq 2$. We define some set, the first set is a set of elements with order 1, the second set is a set of elements with order 2, or even, the third set is a set of an element with order $p_1$and odd, and the $(m+2)^{th}$ set is a set of elements with order $p_m$ and odd and $p_j$ not divide $p_m$ where $1 \leq j \leq m - 1$, clearly these sets are a partition of $D_{2n}$. Let $x, y \in V_i$, thus $p_i | ord(x)$ and $p_i | ord(y)$, so $(ord(x), ord(y)) \neq 1$, then $x$ and $y$ are not adjacent. So, the coprime graph of $D_{2n}$ is $m + 2$-partite. ∎

## 2.2 The Clique Number and Chromatic Number

The first Theorem in this section about the clique number of the coprime graph of $D_{2n}$.

***Theorem 5*** Let $D_{2n}$ group dihedral. If $\Gamma_{D_{2n}}$ is $k$-partite graph then $\omega(\Gamma_{D_{2n}}) = k$

**Proof**. Let $\Gamma_{D_{2n}}$ is $k$-partite graph, it means that $\Gamma_{D_{2n}}$ has $k$ set of partitions. Let $V_1, V_2, \dots, V_k$ where $V_i$ and $V_j$ adjacent when $i \neq j$ and $1 \leq i, j \leq k$. Because $V_i$ and $V_j$ are adjacent, then we have any $a \in V_i$

and $b \in V_j$ are adjacent so that it is formed complete subgraph $C_k$ in the coprime graph of the dihedral group then $\omega(\Gamma_{D_{2n}}) = k$.

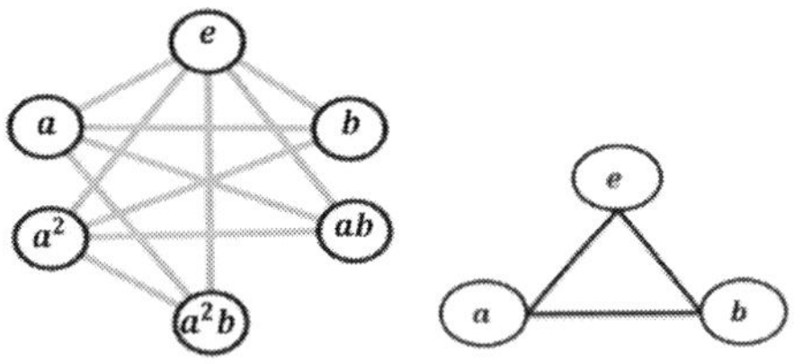

**Figure 1**. Example of $\omega(\Gamma_{D_{2.3}})$

An example of Theorem 5 can be seen in Figure 1. The next Theorem about the chromatic number of a graph, as follows.

***Theorem 6*** Let $D_{2n}$ group dihedral. If $\Gamma_{D_{2n}}$ is $k$-partite graph then $\chi(\Gamma_{D_{2n}}) = k$
**Proof.** Let $\Gamma_{D_{2n}}$ is $k$-partite graph, it means that $\Gamma_{D_{2n}}$ has $k$ set of partitions. Let $V_1, V_2, \ldots, V_k$ where $V_i$ and $V_j$ adjacent when $i \neq j$ and $1 \leq i, j \leq k$. These $V_i$ and $V_j$ are adjacent, so the color of $V_i$ and $V_j$ must different. Hence for each $a, b \in V_i$, because $a$ and $b$ are not adjacent then $a$ and $b$ have the same color, furthermore each element in $V_i$ have the same color, hence $\chi(\Gamma_{D_{2n}}) = k$.

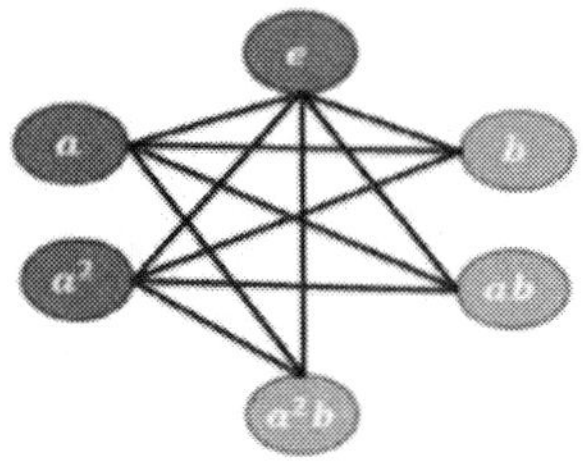

**Figure 2** Example of $\chi(\Gamma_{D_{2.3}})$

An example of Theorem 6 can be seen in Figure 2. Some of the corollaries of the preceding Theorem are as follows.

**Corollary 1** Let $D_{2n}$ group dihedral. If $n = p_1^{k_1} p_2^{k_2} p_3^{k_3} \ldots p_m^{k_m}$ where $1 \leq i \leq m$, $p_i$ are distinct prime number, and $p_i \neq 2$, then $\omega(\Gamma_{D_{2n}}) = \chi(\Gamma_{D_{2n}}) = m + 2$.

**Corollary 2** Let $D_{2n}$ group dihedral. If $n = 2^k$, for a $k \in \mathbb{N}$, then $\chi(\Gamma_{D_{2n}}) = 2$.

## 4. Conclusion

The obtained result shows that the clique number and the chromatic number of the coprime graph of a dihedral group where $n$ composite number specifically $n = p_1^{k_1} p_2^{k_2} p_3^{k_3} \ldots p_m^{k_m}$ is $m + 2$ and the clique number & chromatic number of the coprime graph of a dihedral group where $n = 2^k$ is 2.

## Acknowledgments

We would like to thanks KPBI Matematika Murni of Mataram University and Gamatika Research Club for their advice and support.

## References

[1] Ma X L, Wei H Q and Yang L Y 2014 The Coprime Graph of a Group *International Journal of Group Theory* **3** (3) 13–23

[2] Dorbidi H M 2016 A Note on The Coprime Graph of a Group *International Journal of Group Theory* **5** (4) 17–22

[3] Mansoori F, Erfanian A and Toulue B 2016 Non-coprime Graph of a Finite Group *AIP Conference Proceeding* **1750** 050017

[4] Syarifudin A G, Wardhana I G A W, Switrayni N W and Aini Q 2020 Some Properties of Coprime Graph of Dihedral Group When is Prime Power *Journal of Fundamental Mathematics and Applications* **3** (1) 34-38

[5] Juliana R, Wardhana I G A W and Irwansyah, 2020, Coprime graph of integers modulo n group and its subgroups, *Journal of Fundamental Mathematics and Applications* **3**1 (1) 34-38

[6] Syarifudin A G, Nurhabibah, Malik D P and Wardhana I G A W, 2021, Some Characterizations Of Coprime Graph Of Dihedral Group D 2n, *Journal of Physics: Conference Series* **1722** (1) 012051

[7] Masriani, Juliana R, Syarifudin A G, Wardhana I G A W, Irwansyah, Switrayni N W 2020, Some Result Of Non-Coprime Graph Of Integers Modulo N Group For N A Prime Power, *Journal of Fundamental Mathematics and Applications* **3**1 (2) 107-111

[8] Syarifudin A G and Wardhana I G A W, 2019, Subgrup Non Trivial Dari Grup Dihedral *Eigen Mathematics Journal* **2** (2) 73-76

[9] Abdollahi A, Akbari S and Maimani H R 2006 Non-commuting graph of a group *J. Algebra*, 298 468-492

[10] Abdollahi A and Hassanabadi A M 2007 Non-cyclic graph of a group *Comm. Algebra*, 35 2057-2081

# Some Characteristics of Prime Cyclic Ideal On Gaussian Integer Ring Modulo

**W U Misuki, I G A W Wardhana* and N W Switrayni**

Department of Mathematics, Faculty of Mathematics and Science, Universitas Mataram

*Corresponding author: adhitya.wardhana@unram.ac.id

**Abstract.** Gaussian modulo integer is a complex number $a + ib$ where $a, b \in \mathbb{Z}_n$. Some of the characteristics of the prime ideal on Gaussian integer are a trivial ideal $\{0\}$ is a prime ideal, and $I$ ideal prime if and only if $I$ almost prime ideal. These characteristics do not necessarily apply to modulo Gaussian integers. In this paper, we give some characteristics of the prime ideal on modulo Gaussian integer.

## 1. Introduction

Prime numbers always have been an exciting topic among mathematicians until today. The abstraction of prime numbers first introduced by Dedekind in 1871, called prime ideal [1]. Recently, some mathematicians have brought the prime ideal generalization study around the world in different algebraic structures.

Bhatwadekar and Sharma (2009) introduced a generalization of a prime ideal in ring theory, called almost prime ideal [2]. Darani (2011) give more detailed properties around almost prime ideal [3] and new insight about almost prime ideal [4]. Maulana et al. (2019) study prime ideal and almost prime ideal properties on Gaussian integer [5]. Yuwaningsih (2020) give more general properties of almost prime submodule [6]. Additional, Khashan and Bani-Ata (2021) gives a prime ideal version in module theory in other algebraic structures, called almost prime submodule [7].

This study is based on Maulana's research in 2019, aiming to find the prime cyclic prime ideal properties on a particular ring called Gaussian integer modulo.

## 2. Methodology

The methodology of this research is by making a pattern from some example of the dihedral group for some $n$, and the conjecture will be made by the pattern. And by deductive, we will prove the conjecture. If the conjecture is correct, then we state it as a Theorem. If wrong, then we make another conjecture.

## 3. Results and Discussion

**Definition 1**. [9] A non-empty set $R$ with two binary operations, addition $(+)$ and multiplication $(\times)$ defined on $R$ is called a ring if it satisfies the following axioms.

i. $(R, +)$ is a commutative group.
ii. The $\times$ operation is associative, that is $(a \times b) \times c = a \times (b \times c)$ for all $a, b, c \in R$.
iii. The distributive law applies to $R$, that is: for all $a, b, c \in R$ then $a \times (b + c) = (a \times b) + (a \times c)$ and $(a + b) \times c = (a \times c) + (b \times c)$.

Ring $R$ is said to be commutative if the operation $\times$ is commutative, i.e., $a \times b = b \times a$, for every $a, b \in R$. Ring $R$ is said to have unity to the multiplication operation if there is an element $1 \in R$ so that $a \times 1 = 1 \times a = a$, for every $a \in R$.

In a ring with unity, all the elements do not have to have an inverse to multiplication. Elements that have an inverse to multiplication are called units.

The ideal is an essential sub-structure of a ring, the ideal concept of a ring is analogous to the normal subgroup concept of a group.

**Definition 2.** [9] Let $R$ is a ring, a non-empty subset $I$ of $R$ called ideal if

i. The set $I$ is a subgroup of $R$, that is $a, b \in I$ applies $a - b \in I$.
ii. The set $I$ closed to the multiplication by each element in ring $R$, that is, if $a \in I$, $r \in R$ then applies $ar \in I$ and $ra \in I$.

**Theorem 1.** [9] If $R$ a ring with unity and $I$ is an ideal of $R$ containing a unit, then $I = R$.

The abstraction of prime numbers in the ring was introduced by Dedekind in 1871, called the ideal prime. The following theorem gives it.

**Definition 3.** [9] An ideal $I \neq R$ in the commutative ring $R$ is a prime ideal if $ab \in I$ implies $a \in I$ or $b \in I$ for $a, b \in R$.

Furthermore, in 2009, the prime ideal was generalized to the ideal almost prime by Bhatwadekar. The definition is as follows.

**Definition 4.** [2] An ideal $I \neq R$ in the commutative ring $R$ is an almost prime ideal if $ab \in I - I^2$ implies $a \in I$ or $b \in I$ for $a, b \in R$.

Complex numbers are a generalization of real numbers, as well as Gaussian integers, which are the generalization of integers.

**Definition 5.** [9] Gaussian integer is complex number $a + ib$ where $a, b \in \mathbb{Z}$.

Maulana found the characterizations of Gaussian prime numbers in the following theorem.

**Teorema 2.** [5] Let $\alpha = a + ib \in \mathbb{Z}[i]$, $\alpha$ satisfies one of the following properties

i. For $a \neq 0$, $b = 0$, $a$ prime number in $\mathbb{Z}$ and $|a| \equiv 3 \ (mod\ 4)$.
ii. For $a = 0$, $b \neq 0$, $b$ prime number in $\mathbb{Z}$ and $|b| \equiv 3 \ (mod\ 4)$.
iii. For $a$, $b \neq 0$, $a^2 + b^2$ is a prime number in $\mathbb{Z}$.

If and only if $\alpha$ is Gaussian prime number.

**Definition 6.** [1] Modulo Gaussian integer is a complex number $a + ib$ where $a, b \in \mathbb{Z}_n$.

Every non-zero ring $R$ has at least two ideals, the improper ideal $R$, and a trivial ideal $\{0\}$. The other ideal in $R$, we named it proper non-trivial ideal. By definition of prime ideal, it is clear the improper ideal R is not prime ideal, and then a trivial ideal $\{0\}$ is not necessarily prime ideal on modulo Gaussian integer [10]-[11].

A trivial ideal $\{0\}$ is prime ideal when modulo $n$ is Gaussian prime numbers, as shown in the following theorem.

**Theorem 3.** Let $I = \langle\bar{a}\rangle$ ideal of $\mathbb{Z}_n[i]$. $(a, n) = 1$ if and only if $I = \mathbb{Z}_n[i]$.

**Proof.** ($\Rightarrow$) Because $(a, n) = 1$ then there are $p, q \in \mathbb{Z}[i]$ so that $pa + qn = 1$ or $pa = 1 - qn$. By dividing the two sides with $n$, obtained $\overline{pa} = \bar{1} - \overline{qn}$ or $\overline{pa} = \bar{1} - \bar{0}$. Thus $\bar{1} = \overline{pa}$, so that $\bar{1} = \bar{p}(\bar{a}) \in \langle\bar{a}\rangle$. Because $\bar{1}$ is unit in $\mathbb{Z}_n[i]$ then by theorem 1 obtained $I = \mathbb{Z}_n[i]$.
($\Leftarrow$) Let $I = \mathbb{Z}_n[i]$ and suppose $(a, n) = p \neq 1$. Thus $p|a$ and $p|n$ then $\langle\bar{a}\rangle \subset \langle\bar{p}\rangle$. Karena $p|n$, there $q \in \mathbb{Z}[i]$ so that $n = pq$. From that, $\langle\bar{p}\rangle \neq \mathbb{Z}_n[i]$. Therefore $I \neq \mathbb{Z}_n[i]$. ∎

Some ideal on modulo Gaussian integer is equivalent, following the theorem is given.

**Theorem 4.** Let $I = \langle\bar{a}\rangle$ ideal of $\mathbb{Z}_n[i]$. If $(a, n) = p$ then $\langle\bar{a}\rangle = \langle\bar{p}\rangle$.

**Proof.** Let $(a, n) = p$ then $p|a$ and $p|n$. For $\bar{x} \in \langle\bar{a}\rangle$ obtained $\bar{x} = \overline{ka}$ or $\bar{x} - \overline{ka} = \bar{0}$ for $k \in \mathbb{Z}_n[i]$. By multiplying the two sides with multiples of $n$ obtained $x - ka = ln$ or $x = ka + ln$ so that $x = krp + lsp = (kr + ls)p$ for $r, s \in \mathbb{Z}_n[i]$. Therefore $\bar{x} = (\overline{kr + ls})\bar{p} \in \langle\bar{p}\rangle$. Otherwise for $\bar{x} \in \langle\bar{p}\rangle$, then $\bar{x} = \overline{kp}$ or $\bar{x} - \overline{kp} = \bar{0}$. Furthermore, the two sides multiplying with multiples of $n$ so that $x - kp = ln$ or $x = ln + kp$. Because $(a, n) = p$ obtained $ra + sn = p$ for $r, s \in \mathbb{Z}[i]$ obtained $x = ln + k(ra + sn) = ln + kra + ksn$. Therefore $\bar{x} = \overline{kra}$ or $\bar{x} = \overline{kr}(\bar{a}) \in \langle\bar{a}\rangle$. So $\langle\bar{a}\rangle = \langle\bar{p}\rangle$. ∎

Ideal $I = \langle\bar{a}\rangle$ on modulo Gaussian integer is prime ideal when great common divisor $n$ and $a$ is Gaussian prime number, this is given by the following theorem.

**Theorem 5.** Let $I = \langle\bar{a}\rangle$ non-zero ideal in $\mathbb{Z}_n[i]$. If $(a, n) = p$ is Gaussian prime number then $I = \langle\bar{a}\rangle$ is a prime ideal.

**Proof.** It is clear $\langle\bar{a}\rangle \neq \mathbb{Z}_n[i]$. Case 1. Suppose $a$ is Gaussian prime number then $a = p$ so that $a|n$ or $n = aq$. Thus for any $\overline{xy} \in \langle\bar{a}\rangle$ then $\overline{xy} = \overline{ka}$. By multiplying the two sides with multiples of $n$ obtained $xy - ka = ln$ or $xy = ka + ln$. Because $n = aq$ then $xy = ka + laq = (k + lq)(a)$. Because $a$ is a Gaussian prime number, it must $a|x$ or $a|y$, obtained $\bar{x} \in \langle\bar{a}\rangle$ or $\bar{y} \in \langle\bar{a}\rangle$. Therefore $I$ is a prime ideal. Case 2. Suppose $a$ is not a Gaussian prime number, and because $(a, n) = p$ then by theorem 4 obtained $\langle\bar{a}\rangle = \langle\bar{p}\rangle$. By case 1, $\langle\bar{p}\rangle$ is a prime ideal. Therefore $\langle\bar{a}\rangle$ also prime. ∎

In the Gaussian ring integer, $I$ is a prime ideal if and only if $I$ is an almost prime ideal. On modulo Gaussian integer, this property does not apply to the opposite, for example, in $\mathbb{Z}_{12}[i]$, $I = \langle\bar{4}\rangle$ is almost prime ideal, but $I$ is not a prime ideal. The following is the theorem given.

**Theorem 6.** Let $I = \langle\bar{a}\rangle$ non-zero ideal on $\mathbb{Z}_n[i]$. If $I$ is a prime ideal, then $I$ also almost prime ideal.

**Proof.** Let any $\overline{xy} \in I - I^2$. Because $I - I^2 \subset I$ and $I$ is prime ideal then it is clear $\bar{x} \in I$ or $\bar{y} \in I$. So $I$ is an almost prime ideal. ∎

## 4. Conclusion

In rings Gaussian Integer Modulo, we have some characterizations of the cyclic prime ideal of $\mathbb{Z}_n[i]$. If given $I = \langle\bar{a}\rangle$, a non-zero ideal in $\mathbb{Z}_n[i]$ and If $(a, n) = p$, where $p$ is a Gaussian prime number, then $I = \langle\bar{a}\rangle$ is a prime ideal. And if $I$ is a prime ideal, then we found that $I$ is almost prime ideal, but it is not the opposite.

## References

[1] Wardhana I G A W, Astuti P, Muchtadi-Alamsyah I 2016, On Almost Prime Submodule of Finitely Generated Module Over Principal Ideal Domain *JP Journal of Algebra, Number Theory and Applications* **38** (2) 121–128

[2] Bhatwadekar S M and Sharma P K 2009 Unique Factorization, and Birth of Almost Prime *Communication In Algebra* **33** (1) 43-49

[3] Darani A Y 2011 Almost Primal Ideals in Commutative Rings *Chiang Mai J. Sci.* **38** (2) 161-165

[4] Rostami E and Nekooei R 2016 On a question about almost prime ideals *Turkish Journal of Mathematics* **40** (3) 620-623

[5] Maulana F, Wardhana I G A W and Switrayni N W 2019 Evivalensi Ideal Hampir Prima dan Ideal Prima pada Bilangan Bulat Gauss *Eigen Mathematics Journal* **2** (1) 1-5Yuwaningsih D A 2020 Some Properties of Left Weakly Jointly Prime (R, S)-Submodules *Journal of the Indonesian Mathematical Society* **26** (2) 234-241

[6] Maulana F, Wardhana I G A W and Switrayni N W 2019 Evivalensi Ideal Hampir Prima dan Ideal Prima pada Bilangan Bulat Gauss *Eigen Mathematics Journal* **2** (1) 1-5

[7] Yuwaningsih D A 2020 Some Properties of Left Weakly Jointly Prime (R, S)-Submodules *Journal of the Indonesian Mathematical Society* **26** (2) 234-241

[8] Khashan H A and Bani-Ata A B 2021 J-ideals of commutative rings *International Electronic Journal of Algebra* **29** (29) 148-164

[9] Fraleigh J B 2014 *A First Course in Abstract Algebra* New York: Pearson Education Limited

[10] Waterhouse W C 2002 Quadratic Polynomials And Unique Factorization *Am. Math. Monthly* **109** (1) 70–71

[11] Anderson D D and Bataineh M 2008 Generalizations of Prime Ideals *Comm Algebra* **36** 686-696

IOP Conf. Series: Materials Science and Engineering 1115 (2021) 012085 doi:10.1088/1757-899X/1115/1/012085

# Impact of Computer Simulation Assisted Virtual Experiment Module in Learning Hydrogen Atom in Senior High School

**I B P Mardana**[*]

Department of Physics Education, Ganesha University of Education, Indonsesia

*Corresponding author: putu.mardana@undiksha.ac.id

**Abstract.** This study aims to (1) develop virtual-experimental software based on augmented reality as an interactive learning medium, which can be used as a teaching tool to improve the students' mastery of physics concepts, (2) test the comparative advantage of Hydrogen Atom learning with a virtual approach experiment based on augmented reality with ordinary learning models on senior high school students' learning outcomes. The research was categorized as quasi-experimental research which used a non-equivalent pre-test post-test control group. The research instrument used was a learning outcome test. The research data were analyzed by using covariance analysis (ANACOVA). The results of this study showed that the learning outcomes of students who studied with the AR-based virtual in the experimental learning model group were higher than students who studied in the conventional learning model group.

## 1. Introduction

Physics as one of the elements in science that has a very vital and strategic role in the development of technology in the future [1]. Until now, physics learning at the senior school level is generally still seen as difficult subject and becomes a subject that students are not interested in. It is due to the lack of attractiveness of the physics learning package in the classroom, which leads to low learning outcomes achieved by students. One of the factors that suspected to be the main problem of the low learning outcomes achieved by students is misconceptions.

Misconceptions are prior knowledge that is formed through informal learning in the process of understanding everyday experiences and interpreting a concept that is incompatible with the scientific concepts. This prior knowledge is generally resistant in learning [2]. Based on ref [3], learning concepts don't pay attention to students' initial concepts will make students' misconceptions more complex and tend to persist, that might influence students' learning outcomes. Various physics learning approaches have been developed as a strategy of changing students 'concepts of physics (strategy of conceptual change) to direct students' conceptions.

Various physics learning approaches have been developed as a strategy of changing students 'physics concepts (strategy of conceptual change) to lead students' conceptions from everyday knowledge to scientific knowledge. That is a learning approach which is based on an inquiry approach through experimental activities. Computers can be used functionally to support physics learning activities with a virtual-experimental approach [4]-[6]. Virtual-experiment that based on augmented reality can be utilized to visualize abstract physical reality with animation modeling [7]. In addition, learning with the use of computer simulations will provide many advantages in the learning process such as fun, so that students are naturally motivated to engage and participate actively in the learning [8]-[10]. From the background above, a virtual-experimental based augmented reality was developed and tested its' comparative advantage of Hydrogen Atom learning

## 2. Methodology

This study was categorized as quasi-experimental research which used a non-equivalent pre-test post-test control group. The population in this study were all students of class II State of Senior High School 3 in Singaraja, as many as 136 students in the 2018/2019 academic year. The research sample was taken by simple random sampling of 54 students. In which there 28 students in the experimental groups (class II4), and in the control group (class II3) as many as 26 people. This research procedure comprises the following stages; (1) identifying students' initial learning outcomes (pre-test), (2) developing virtual experimental software based on augmented reality, (3) designing learning tools, (4) implementing learning designs and (5) evaluating outcome learning (post-test) (see Table 1).

**Table 1.** Research Paradigm

| Group | Pretest | Treatment | Posttest |
|---|---|---|---|
| Experiment | Q1 | X1 | Q2 |
| Control | Q3 | - | Q4 |

The research instrument used was a learning outcome test, which was validated through expert judgment and content validity. Learning outcome tests were arranged based on student success indicators that were adjusted to the learning material, at the cognitive levels C1, C2, C3, and C4. The test is constructed in the form of an essay test which consists of 10 items. Each item has a rating range of 0-4. The value of content validity by using Gregory's analysis is obtained VI = 0.87. Furthermore, the results of the learning outcomes instrument trial are as shown in Table 2. The initial learning outcome data were obtained before treatment was given to both classes using the pre-test, while the physics learning outcomes data were collected from the result of post-test after the treatment was given to both students in the control class and experimental class.

**Table 2.** The Trial Results of the Learning Outcomes Test

| No. | Item Internal Consistency (r > 0.30) | | Difference Index (IDB > 0.20) | | Item Difficulty Index (IKB = 0.30-0.70) | |
|---|---|---|---|---|---|---|
| | r | criteria | IDB | criteria | IKB | criteria |
| 1 | 0.340 | consistent | 0.25 | low | 0.50 | medium |
| 2 | 0.356 | consistent | 0.20 | low | 0.70 | easy |
| 3 | 0.565 | consistent | 0.75 | high | 0.50 | medium |
| 4 | 0.356 | consistent | 0.29 | low | 0.56 | medium |
| 5 | 0.352 | consistent | 0.25 | low | 0.44 | medium |
| 6 | 0.400 | consistent | 0.20 | low | 0.60 | medium |
| 7 | 0.556 | consistent | 0.3 | low | 0.60 | medium |
| 8 | 0.658 | consistent | 0.40 | medium | 0.20 | difficult |
| 9 | 0.690 | consistent | 0.54 | medium | 0.52 | medium |
| 10 | 0.665 | consistent | 0.50 | mediam | 0.50 | medium |

The research data were analyzed by using covariance analysis (ANACOVA). The significance of differences in the physics learning outcomes between groups of AR-based virtual experimental learning models and conventional learning models were analyzed by using the LSD method.

## 3. Results and Discussion

The results of the development of an experimental virtual module based on augmented reality (AR) as an interactive learning media, are shown in Figure 1. The validation of content experts, media experts, and design experts of experimental virtual module are good qualified. This experimental virtual module was implemented in the experimental class (II4) in Hydrogen Atom learning, while in the control class (II3), the learning process is carried out conventionally.

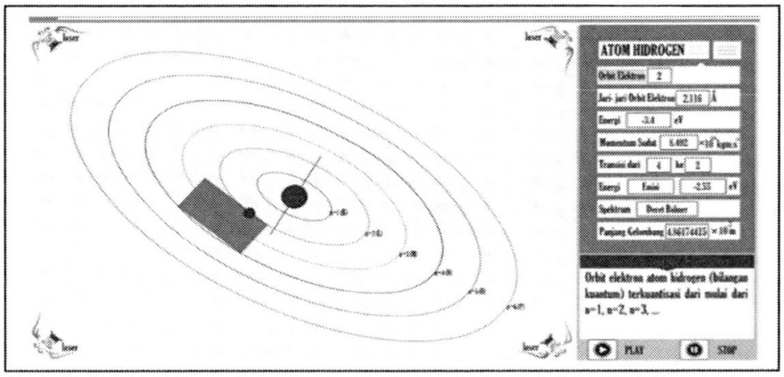


**Figure 1.** Virtual Augmented Reality-based Experiment Module

The results of the data on students' preliminary learning outcomes (pre-test) and learning outcomes after the implementation of learning (post-test) in the conventional learning model group and the AR-based virtual experimental learning model group are shown in Table 3. The average score (M) of students' initial learning outcomes (pretest) in the conventional learning model group is 16.15 which is in very low qualifications with a standard deviation of 3.69. Meanwhile, the average value of students' initial learning outcomes (pretest) in the group learning model with virtual experiment AR- based is 16.43 which is in very low qualifications with a standard deviation of 6.25. Those results indicate that the average score of students' early learning outcomes (pretest) in the virtual experiment learning model group with AR is higher than the conventional learning model group. Furthermore, the average score (M) of students' initial learning outcomes (posttest) in the conventional learning model group is 55.58 which is in sufficient qualifications with the standard deviation of 8.52. Meanwhile, the average score of students' learning outcomes after treatment (posttest) with the AR-based virtual in the experimental learning model group is 60.00 that is in sufficient qualification with the standard deviation of 8.33.

**Table 3.** The Average Value and Standard Deviation of Students' Learning Outcomes

| Class | Learning Model | Pre-test | | | Post-test | | |
|---|---|---|---|---|---|---|---|
| | | M | SD | qualification | M | SD | qualification |
| II4 | Conventional | 16.15 | 3.69 | very low | 55.58 | 8.52 | sufficient |
| II3 | Virtual Eksperimen | 16.43 | 6.25 | very low | 60.00 | 8.33 | sufficient |

The normality test of the data distribution was tested on all data of physics' learning outcomes. It was tested to both the conventional learning model group and the AR-based experimental virtual learning model group, as shown in Table 4.

**Table 4.** Tests of Normality

| Test | Model | Kolmogorov-Smirnov | | | Shapiro-Wilk | | |
|---|---|---|---|---|---|---|---|
| | | Statistic | df | Sig. | Statistic | df | Sig. |
| *Pretest* | virtual experiment AR-based | 0.129 | 28 | 0.200* | 0.981 | 28 | 0.869 |
| | conventional | 0.145 | 26 | 0.170 | 0.967 | 26 | 0.541 |
| *Posttest* | virtual experiment AR-based | 0.094 | 28 | 0.200* | 0.966 | 28 | 0.468 |
| | conventional | 0.112 | 26 | 0.200* | 0.962 | 26 | 0.431 |

The variance homogeneity test between groups aims to measure whether the data group of students' learning outcomes has the same variant between the experimental group and the control group. The homogeneity test of variance between groups were tested by using Leneve's test on the equality of error variance (see Table 5).

**Table 5.** Test of Homogeneity of Variance

| | | Levene Statistic | df1 | df2 | Sig. |
|---|---|---|---|---|---|
| Pre-test learning outcomes | Based on Mean | 3,806 | 1 | 52 | 0.056 |
| | Based on Median | 3.246 | 1 | 52 | 0.077 |
| | Based on Median and with | 3.246 | 1 | 42 | 0.079 |
| | Based on trimmed mean | 3.642 | 1 | 52 | 0.062 |
| Post-test learning outcomes | Based on Mean | 0.339 | 1 | 52 | 0.563 |
| | Based on Median | 0.210 | 1 | 52 | 0.649 |
| | Based on Median and with | 0.210 | 1 | 48 | 0.649 |
| | Based on trimmed mean | 0.264 | 1 | 52 | 0.609 |

The linearity test was conducted to show the relation between the covariate variable (students' initial learning outcomes) and the dependent variable (students' learning outcomes). The results of the linearity test for the conventional learning model group and the AR-based virtual experimental learning model is shown in Table 6. The effect of initial physics learning outcomes on the dependent variable in this study, namely students' physics learning outcomes, shows the statistical value of F * = 16.664 ($p<0.005$). The results indicate that there is a significant effect between the covariates on students' physics learning outcomes.

The effect of the independent variable on the dependent variable, namely students' physics learning outcomes, obtained a statistical value of F * = 4.359 with a significance of 0.042, and the calculated R-Squared value of 0.297.

Ho: [$\mu$1 Y1] = [$\mu$2 Y2]: there is no difference in learning outcomes between students who are taught using the AR-based experimental virtual learning model and students who are taught using conventional learning models. (rejected)

Ha: [$\mu$1 Y1] $\neq$ [$\mu$2 Y2]: there are differences in learning outcomes between students who are taught using the AR-based experimental virtual learning model and students who are taught using conventional learning models. (accepted).

IOP Conf. Series: Materials Science and Engineering 1115 (2021) 012085 doi:10.1088/1757-899X/1115/1/012085

**Table 6.** Test of Linierity

| Model | | | | Sum of Squares | df | Mean Square | F | Sig. |
|---|---|---|---|---|---|---|---|---|
| Conventional | Posttest * Pretest | Between Groups | (Combined) | 71.751 | 5 | 14.350 | 2.824 | 0.044 |
| | | | Linearity | 27.708 | 1 | 27.708 | 5.453 | 0.030 |
| | | | Deviation from Linearity | 44.043 | 4 | 11.011 | 2.167 | 0.110 |
| | | Within Groups | | 101.633 | 20 | 5.082 | | |
| | | Total | | 173.385 | 25 | | | |
| virtual experiment AR- based | Posttest * Pretest | Between Groups | (Combined) | 188.355 | 8 | 23.544 | 4.007 | 0.006 |
| | | | Linearity | 121.102 | 1 | 121.102 | 20.609 | 0.000 |
| | | | Deviation from Linearity | 67.252 | 7 | 9.607 | 1.635 | 0.186 |
| | | Within Groups | | 111.645 | 19 | 5.876 | | |
| | | Total | | 300.000 | 27 | | | |

The analysis of the significance of the difference in the mean value of students' learning outcomes through LSD was obtained at $\alpha = 0.05$, $\Delta\mu = \mu\ (I) - \mu\ (J) = 1.681$. This means that the learning outcomes achieved by students who learned with the use of AR-based virtual experimental learning model are significantly higher than students who were taught by using conventional learning models. The results of this analysis indicate that the learning outcomes achieved by students who were in the AR-based virtual experimental learning model class are better than students in the conventional learning model class. The test of hypothesis can be seen in Table 7.

**Table 7.** Test of Hypothesis

| Source | Type III Sum of Squares | df | Mean | F | Sig. |
|---|---|---|---|---|---|
| Corrected Model | 187.656 | 2 | 93.828 | 10.749 | 0.000 |
| Intercept | 1527.792 | 1 | 1527.792 | 175.033 | 0.000 |
| Pretest | 145.456 | 1 | 145.456 | 16.664 | 0.000 |
| Model | 38.047 | 1 | 38.047 | 4.359 | 0.042 |
| Error | 445.159 | 51 | 8.729 | | |
| Total | 29568.000 | 54 | | | |
| Corrected-item | 632.815 | | | | |

The results of this study prove that there are differences in the students' learning outcomes between students who were taught by using the AR-based experimental virtual learning model and students with conventional learning models. The AR-based experimental virtual learning model is superior to conventional learning models in terms of achieving students' learning outcomes. The results of the descriptive analysis show that the initial learning outcomes of students who learned with AR-based virtual experimental learning models and conventional learning models have different mean values. Students who participated in the AR-based virtual experimental learning model group have an average score of 16.43 with a standard deviation of 6.25. Meanwhile, students who were in the conventional learning model group have an average value of 16.15 with a standard deviation of 3.69.

In general, the pretest mean score of students who were taught by using the AR-based virtual experimental learning model was higher than students taught by using conventional learning model. However, the pretest average score obtained by students in both groups is still consider as very low.

The low students' learning outcomes can be caused by many factors, both factors from within and outside the individual. There was an increase in the average value of the students' posttest results after the treatment was carried out. Descriptively, the group of students who learned in the AR-based virtual experimental learning model class had higher learning outcomes than students who were in the conventional learning model class. This is based on the average level of the data presented in Table 3, where the average learning outcomes of students who were in the AR-based virtual experimental learning model class is 60.00 which is categorized as sufficient, while in the conventional learning model class the average studens't learning outcomes is 55.58 which is categorized as sufficient. Although the students' learning outcomes in these two groups were in the same category, the average value of students' learning outcomes who took part in the AR-based virtual experimental learning model class was still higher than the group of students who were in the conventional learning model class.

Based on the result of covariance analysis in this study, it appears that there is a significant difference between the AR-based experimental virtual learning model group and the conventional learning model group on students' learning outcomes. This difference can be seen from the statistical value F * = 4.359 with a significance number of 0.042 which is lower than the significance level of 0.05 ($p < 0.05$), and the calculated R-Squared value of 0.297. This statistical value means that there are differences in learning outcomes between students who are taught by using the AR-based virtual experimental learning model and students who are taught by using conventional learning models in learning physics. Therefore, it can be said that students' learning outcomes are significantly influenced by the learning model used by the teacher. These results indicate that the learning outcomes achieved by students who were in the AR-based virtual experimental learning model class are better than students who took learning in the conventional learning model class. The results of the follow-up LSD test show that the learning outcomes of students who studied in the AR-based virtual experimental learning model group were higher than students who studied in the conventional learning model group (LSD = 1.614; $\Delta\mu$ = 1.681). Accordingly, the effect of AR-based virtual experimental learning model on the students' learning outcomes is statistically superior when compared to the conventional learning models. Based on the results of descriptive analysis and one-way covariance analysis, it can be concluded that the AR-based virtual experimental learning model has a better effect than the conventional learning model.

## 4. Conclusion

There are differences in students' learning outcomes between students who were taught by using the AR-based virtual experimental learning model and students who were taught by using the cooperative learning model. This difference can be seen from the statistical value F * = 4.359 with a significance number of 0.042 which is lower than the significance level of 0.05 ($p < 0.05$), and the calculated R-Squared value of 0.297. The group of students who studied using the AR-based experimental virtual learning model showed relatively higher learning outcomes than the group of students who studied using the conventional learning model. The results of the follow-up LSD test show that the students' learning outcomes in the AR-based virtual experimental learning model group were higher than students who studied in the conventional learning model group (LSD = 1.614; $\Delta\mu$ = 1.681).

In the learning process by using the AR-based virtual experimental learning model, the teacher should be able to use learning media such as simulations or animations that can make students are motivated to engage in the learning and interested in participating in the learning process. The use of experimental virtual learning model based on AR should be carried out continuously so that students are able to improve their both skills in developing their reasoning and in solving problems. This will help students to enhance their learning outcomes.

## References

[1] Akhlis I and Wahyuni S 2019 The development of web-based multimedia physic learning tool to enhance student's character *Journal of Physics: Conference Series* **1321** (3) 032083
[2] Pine K, Messer D and St John K 2001 Children's misconceptions in primary science: A survey of teachers' views *Research in Science & Technological Education* **19** (1) 79-96
[3] Leonard M J, Kalinowski S T and Andrews T C 2014 Misconceptions yesterday, today, and tomorrow *CBE—Life Sciences Education* **13** (2) 179-186
[4] Zacharia Z C and Olympiou G 2011 Physical versus virtual manipulative experimentation in physics learning *Learning and instruction* **21** (3) 317-331
[5] Siswanto S, Gumilar S, Yusiran Y and Trisnowati E 2018 Scientific approach-integrated virtual simulation: a physics learning design to enhance student's science process skills (sps) *Unnes Science Education Journal* **7** (1) 7-12
[6] Gordienko Y, Bekenov L, Baskova O, Gatsenko O, Zasimchuk E and Stirenko S 2015 IMP science gateway: from the portal to the hub of virtual experimental labs in e-science and multiscale courses in e-learning *Concurrency and Computation: Practice and Experience* **27** (16) 4451-4464
[7] Wang X 2012 Augmented Reality: A new way of augmented learning *eLearn* **2012** (10) 238017
[8] Di Serio Á, Ibáñez M B and Kloos C D 2013 Impact of an augmented reality system on students' motivation for a visual art course *Computers & Education* **68** 586-596
[9] Lee Y F, Guo Y and Ho H J 2008 Explore effective use of computer simulations for physics education *Journal of Computers in Mathematics and Science Teaching* **27** (4) 443-466
[10] Jimoyiannis A and Komis V 2001 Computer simulations in physics teaching and learning: a case study on students' understanding of trajectory motion *Computers & education* **36** (2) 183-204

IOP Conf. Series: Materials Science and Engineering 1115 (2021) 012086 doi:10.1088/1757-899X/1115/1/012086

# Citrus Farmers Institutional Support on Technology Adoption of Integrated Management of Healthy Citrus Orchard in Garut Regency

**R Indrasti*[1], J B M Rawung[2], N Roufiq[3] and N R Sudolar[4]**

[1]Indonesian Center for Agricultural Technology Assessment and Development, Indonesia
[2]North Sulawesi Assessment Institute for Agricultural Technology, Indonesia
[3]Center for Horticultural Research and Development, Indonesia
[4]Jakarta Assessment Institute for Agricultural Technology, Indonesia

*Corresponding author: ritaindrasti@yahoo.com

**Abstract.** The study was aimed to identify factors that influence institutional success, to analyze the potential of farmers institutional, and to identify the dissemination level of Integrated Management of Healthy Citrus Orchard (IMHCO). The data were obtained by interviewing 25 citrus farmers in Garut Regency which randomly selected as respondents in 2017. The data taken were primary and secondary data. The data were complemented with the information on the institutional role of farmers and the role of IMHCO. The data were analyzed descriptively and non-parametrically. The results showed that the farmers institutional were considered very important but their human resources (farmers, farmer group and Extension) have not focused on this sustainability. Institutional aspects that required to be considered and implemented were input, production and marketing. The role of IMHCO was very relevant to increase citrus production, but the support of human resources was unreliable. Therefore, it is required a strategy to improve the adoption level of IMHCO technology in Garut Regency.

## 1. Introduction

Physics Citrus is one of the fruit crops which very popular and become an important commodity. Citrus can support the economy for both domestic and export markets. Indonesia has a diverse citrus type, especially tangerine and siam. Many types of tangerines are known along with their regional names, such as Keprok Tawangmangu (Central Java), Keprok Garut (West Java), Keprok Berasitepu (North Sumatra), Keprok Batu 55 (East Java), Keprok Tejakula (Bali), Keprok Selayar (South Sulawesi), Keprok Sioumpu (Southeast Sulawesi), and Keprok Madura (Madura). However, on later development, it was dominated by types of mandarin citrus [1]-[2]. Garut is the main producer of citrus in West Java. Garut Regency once have been experienced its peak as a producer of citrus. Citrus is a priority for the Garut society, and provide a significant economic contribution to the area. The large demand of citrus from the domestic market is a potential market for national citrus production. The citrus production does not depend on the season and almost available any time. The increasing number of population and improved lifestyle were influenced the demand for various fruits, including citrus. High demand of citrus encouraged the farmer to cultivate the citrus from one generation to another generation.

Nevertheless, it cannot be denied that the CVPD (Citrus Vein Phloem Degenaration) or so called Huanglongbin (HLB) disease attack was caused damage to citrus plants in Garut Regency. In ref [3], the CVPD attack affect on nutrient deficiencies and citrus quality. The Local Government recommended the farmers to plant the citrus in non-endemic locations, otherwise provide disease preventive strategies. Bassanezi and Gottwald (2008) reveal that it is important to implement the regional control management for the disease immediately after detected [4]. Regional disease management highly affects the probability and efficacy of slowing the epidemic.

In Indonesia, one of the ways to prevent CVPD attack is an integrated management of healthy citrus orchard (IMHCO). The technology of IMHCO comprises five component of technology, namely: (1) using virus-free citrus seeds, (2) pest and disease control, (3) sanitation, (4) Optimal plant maintenance, and (5) Consolidation of orchard management [5]. The use of citrus seeds labeled free of disease, especially CVPD, can minimize damage on citrus plants and ultimately the farmers also saved from the suffer of huge losses. Symptoms of CVPD infection similar to symptoms of nutrient deficiency due to metabolic disorders in plant tissue. However, IMHCO should be supported by skilled human resources and solid institutional. If it is not supported by reliable human resources, the farmers will continue to suffer losses.

Institutions play an important role to support a farm to succeed. The existence of these institutions can get solutions in dealing with problems. The institution is a tendency of interaction in socio-economic matters relating to two or more actors of social interaction which includes applicable rules to be agreed upon [6]. In real terms, the conditions of citrus farming in Indonesia are very diverse and the farms in each location have greatly varied characteristics. Institutional plays an important role in the existence of farmer groups (Poktan) and farmer group associations (Gapoktan). Besides Poktan institutional, there are several factors that can affect the citrus farming institutional, namely input institutional, the existence of citrus seed breeder groups and integrated pest control groups (IPM). Within the institutional aspect, the one important things is the need for partnership between citrus farmers and several companies. The partnership pattern in citrus farming may refer to the nucleus-plasma pattern, the cooperation pattern in providing capital, through the Multi-Business Cooperative and Village Credit Institution. The functions of agricultural institutions are very diverse, including as a mobilizer, collector, distributor of production facilities, and a motivator of interest and attitude [7].

## 2. Methodology

The study was conducted in Garut Regency in 2017. Primary and secondary data were collected. Respondents of 25 people were chosen randomly. Primary data was collected in a Focus Group Discussion (FGD) using the Rapid Appraisal of Agricultural Knowledge Systems (RAAKS) methodology [8]. Information was carried out by in-depth interviews with competent community leaders. Secondary data includes data obtained from literature, reports, publications and relevant institution, as well as through limited discussions with relevant authorized officials. The data were analyzed non-parametrically.

## 3. Results and Discussion

### 3.1 Farmer Respondents Characteristics

Garut Regency is a citrus production center area which spread over several districts. Spread in the sub districts of Samarang, Pasirwangi, Cisurupan, Wanaraja, Karapawitan, Tarogong, Cigedug, Cilawu, Banyuresmi, Bayongbong and Leles. Within 2016-2019, the citrus production in Garut region have been fluctuating, as displayed on Figure 1.

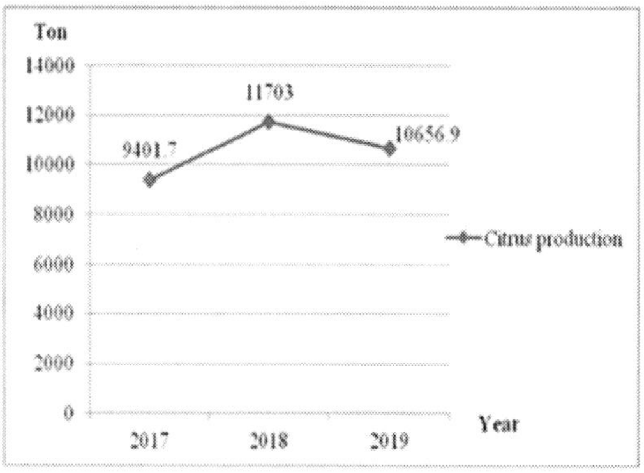


**Figure 1.** The production of citrus in Garut region during 2016-2019

The respondent profile shows that the average of respondents' age was 42.6 years old. They were sufficient experienced in dealing with their work. Half of respondents have been experienced on citrus farming more than 10 years. Their educational background was varies greatly from elementary to high school, but on average they only had junior high school education. Several reasons related to such conditions namely: (1) they cannot afford school fees, (2) there are limited interesting on school, (3) they want to help their parents earn a living quickly, and (4) they are continue their parents' work as farmers. The citrus cultivation pattern of the Garut farmers were on average. The data regarding the characteristics of citrus farming in Garut presented in Table 1.

**Table 1.** Characteristics of farmer respondents

| **Characteristics** | Percentage (%) |
|---|---|
| **Farm size (ha)** | |
| < 1 | 58.33 |
| 1 | 16.67 |
| > 1 | 25.00 |
| **Main source income** | |
| Citrus cultivation | 8.33 |
| Crops cultivation | 66.67 |
| Entrepreneur | 16.67 |
| Government Employee | 8.33 |
| **The distance from house to the farm (km)** | |
| < 1 | 8.33 |
| 1 | 58.33 |
| > 1 | 33.33 |
| **Know the IMHCO** | |
| Yes | 33.33 |
| No | 66.67 |
| **Problem** | |
| Pests attack and diseases | 100.00 |
| Labelled seeds | 33.33 |
| Drought | 33.33 |
| **The main determinant on farming decision** | |
| Government officials | 33.33 |
| Self initiative | 33.33 |
| Fellow farmer groups | 33.33 |

| | |
|---|---|
| **Experiences on citrus farming (years)** | |
| < 10 | 25.00 |
| 10 | 25.00 |
| > 10 | 50.00 |

**Table 1. Processed data, 2017.**

The farm size and land management vary widely, but they remain consistent in farming citrus. Generally the respondents earn primary income from crops cultivation, and only less than 10 percent dependent their income on citrus farming. Regarding the farm size, more than half of the respondents cultivate the citrus on small scale, with the farm size less than one hectare. Regarding the distance from the house to the farm was averaged less than one kilometer, only a few farmer have more than one kilometer away. There are no problems faced regarding the accessibility to the farm. Likewise the ease of access to agricultural input such as labeled seeds and fertilizers. Nevertheless, the pest attack, diseases and environmental stress were the essential problems faced by the citrus farmer. Pests attack of mites and trips commonly faced in the citrus field. Whereas the diseases often found were sooty mildew and aphids. In some area, the farmer faced problem to get water and experienced the drought.

Regarding the technology of IMHCO to improve the citrus productivity, more than half of the respondents did not know about the IMHCO. It is indicated that the dissemination of IMHCO technology need to be accelerated. The similar condition has been reported in South Sulawesi [9]. Citrus farmers required to understand and implement the IMHCO on their citrus farming practices. The respondents argued that the socialization of IMHCO was insufficient. That was impacted on the unequal implementation of IMHCO among farmers. One program to improve farmer's skill and knowledge is farmer field school (FFS) training program. Victor et al. (2017) reported that FFS training program was efficient to increase farmer's knowledge level significantly [10]. The government recommendation, the self-initiative, or the suggestions from fellow group members were equally influenced the farmer on making decision. It was indicated that the role of human resources may have great effect to the institutional sustainability.

## 3.2 Institutional Performance of the Farmer Groups (Poktan) and Farmer Groups Association (Gapoktan)

The farmers institutional that exist in five sub-districts have the similar performance. Generally, all of the respondents have desire on the regular group meetings, an increase in farmer income, and higher quality of citrus yields. Recording was the one important thing in farmer group. Nevertheless, almost all farmer groups still have weaknesses in terms of recording, both production and financial records. The performance of farmer group institutional displayed on Table 2.

**Table 2**. Farmer Group Institutional Performance

| No | Activities | Performance (%) | | Significance Level[1] |
|---|---|---|---|---|
| | | Agreed | Disagreed | |
| 1. | Intensive of farmer group meetings | 40 | 60 | ** |
| 2. | Farmer group benefits the member | 68 | 32 | ** |
| 3. | Numerous problems in farmer group | 28 | 72 | ** |
| 4. | Communication ease in farmer group | 68 | 32 | ** |

[1]Test with binomial non-parametric method. * = significantly different, ** = highly significantly different

The results showed that only 40% respondents stated that the farmer group meetings was active, while 60% respondents did not perform the active farmer group meeting because there was no problem in the group. They can solve the problem privately without involving other groups. The statement that the existence of farmer groups was considered benefits the farmers stated by 68% respondents. Whereas other respondents considered that it was useless because there were self-centered of each farmer groups. There were still many citrus farmer groups that carried out their own management. Only 28% respondents stated that there was no serious problem, because the respondents were already independent and had broad insights. While the most of the respondents faced numerous problems on their farming business. It is the government's role to assist the farmer to overcome their problems. The extension agent or assistant capability on group coaching also affected on the institutional sustainability.

Regarding communication in farmer group, only 32% respondents faced the difficulties on communication in the farmer group. The respondent's background seemed effect on the response of the farmer on the information. The likelihood of adoption on agricultural technology increase until 20% when the educational background of the recipient at least a high school education. Victor et al. (2017) revealed that the age, farm size, and educational background may influence on the farmer response on the agricultural technologies [10]. The young farmer and the large land holders were tended to be more receptive to adopt the innovation.

Farmers' marketing institutional in Garut Regency seemed to be very weak. The farmers did not have a strong bargaining position, and the price was controlled by collectors and outside traders. On financial institutional, there were some difficulties faced by the farmer to access existing capital institutions.

## 4. Conclusion

The farmer institutional plays an important role toward the sustainability of citrus farming in Garut. Nonetheless, human resources capacity is the main key to achieve the institutional success. Besides human resources capacity, there were several factors which determined effects on institutional success, then need to be considered seriously, namely input, production and marketing. The number of input, the plants management to improve the production, and the wide market for the product has been concerned as the essential factors on citrus farming sustainability. The farmers institutional seemed potential to organize the farmers to obtain the target. Lastly, the intensive assistance on IMHCO technology was required to support the citrus sustainability.

## References

[1] Koehler-Santos P, Dornelles A L C and Freitas L B D 2003 Characterization of mandarin citrus germplasm from Southern Brazil by morphological and molecular analyses *Pesquisa Agropecuária Brasileira* **38** (7) 797-806

[2] Moya J L, Gómez-Cadenas A, Primo-Millo E and Talon M 2003 Chloride absorption in salt-sensitive Carrizo citrange and salt-tolerant Cleopatra mandarin citrus rootstocks is linked to water use *Journal of Experimental Botany* **54** (383) 825-833

[3] Yuniti I G A D, Wirawan I G P, Wijaya I N and Sritamin M 2017 CVPDr DNA fragment affect differences in resistant to citrus vein phloem degeneration (CVPD) disease, nutrient deficiencies and quality of fruits *International Journal of Biosciences and Biotechnology* **5** (1) 60-70

[4] Bassanezi R B and Gottwald T R 2009 Epidemiology of HLB and potential pathways for introduction. *Taller internacional de plagas cuarentenarias de los citricos Villahermosa, Tabasco* 27-31

[5] Suharyanto S, Arya N N and Sutami P 2017 Knowledge, Attitude and Behaviour of Farmer on Integrated Crop Management for Healthy Citrus Orchard (ICMHCO) in Gianyar District *KnE Life Sciences* 670-679

[6] Ahnquist J, Wamala S P and Lindstrom M 2012 Social determinants of health–a question of social or economic capital? Interaction effects of socioeconomic factors on health outcomes

*Social science & medicine* **74** (6) 930-939
[7] Rada N E, Buccola S T and Fuglie K O 2011 Government policy and agricultural productivity in Indonesia *American journal of agricultural economics* **93** (3) 867-884
[8] Salomon M and Seegers S 1996 Rapid appraisal of agricultural knowledge systems (Raaks) and its use in irrigation management research *Training Report* **18** 1-10
[9] Zamzami L and Dwiastuti M E 2013 Technology adoption of integrated management of healthy citrus orchard (IMHCO) at citrus production centers in South Sulawesi *Proceedings International Conference on Tropical Horticulture* 364-369
[10] Victor A N, Luqman M, Shiwei X, Wen Y and Majeed M Z 2017 Farmer field school's training on knowledge level of citrus growers regarding improved production practices *Ciência Rural* **47** (10) 1-12

# Penalized Logistic Regression Model to Predict a Results of RT-PCR by Using Blood Laboratory Test

**A Dwinata[1,2], K A Notodiputro[2] and B Sartono[2,] ***

[1]Mathematics Education Study Program, Raja Ali Haji Maritime University, Indonesia
[2]Department of Statistics, IPB University, Indonesia

*Corresponding author: bagusco@apps.ipb.ac.id

**Abstract.** Statistical modelling to determine the effect of several predictor variables on the binary response variable is known as multiple logistic regression model. The addition of a penalty function to the model is done to improve prediction accuracy. Penalized logistic regression shrinks the regression coefficient to zero. This penalized logistic regression model will be used to predict a result of RT-PCR by using the features of blood laboratory tests. This research uses LASSO and elastic net penalties function. This study aims to determine the prediction performance of the RT-PCR test using logistic regression with LASSO and elastic net penalties. The data from the RT-PCR test were used as the binary response variable. Patient age quantile and 27 features of laboratory blood test were used as predictor variables. The results of this research showed that prediction performance of a RT-PCR test using LASSO logistic regression was better than elastic net logistic regression. The LASSO logistic regression model had a good performance for predicting the RT-PCR test with 88% accuracy and 93% AUC. Based on the result of LASSO logistic regression model, the features of laboratory blood tests that affect a RT-PCR test were leukocytes, basophils, RDW and C-reactive protein.

## 1. Introduction

Statistical modelling to determine the effect of several predictor variables on the binary response variable is known as multiple logistic regression model. Multiple logistic regression coefficients are estimated using the maximum likelihood method [1]. The more predictor variables used in modelling will increase the chances of multicollinearity between the predictor variables. Multicollinearity tends to increase the variance of regression coefficients [1]. The addition of predictor variables that are not related to the response variable in modelling will cause overfitting problem [2]. So, prediction error will increase if the predictor variables in modelling are not related to the response variables. The addition of a penalty function to a model is done to improve a prediction accuracy [2]. Adding a penalty to the logistic regression model is done by adding the penalty function to the log likelihood function. Penalized logistic regression shrinks the regression coefficient to zero. There are three types of penalties, namely ridge regression, Least Absolute Shrinkage and Selection Operators (LASSO) and Elastic Net [3].

Ridge regression use $L_2$ penalty that is $\lambda \sum_{j=1}^{p} \beta_j^2$, LASSO use $L_1$ penalty that is $\lambda \sum_{j=1}^{p} |\beta_j|$ and elastic net combines $L_1$ and $L_2$ penalties that is $\lambda \sum_{j=1}^{p} (1-\alpha)\beta_j^2 + \alpha|\beta_j|$. In general, ridge regression involves all predictor variables in the model, but the regression coefficient value is smaller than a model without penalty [2]. LASSO reduces overfitting and also selects predictor variables [4]. Elastic net uses the ridge regression penalty to handle the problem of high correlation and the LASSO penalty to do variables selection [3]. The shrinkage parameter ($\lambda$) directly regulates the bias and variance and regulates the shrinkage strength of the regression coefficient. The value of $\lambda$ can be calculated using cross validation [1]. The parameter $\alpha$ on the elastic net adjusts for the penalty, when $\alpha = 0$ a ridge regression penalty is obtained and when $\alpha$ approaches 1, a LASSO penalty is obtained. Elastic net penalty which is a linear combination of the ridge regression penalty and LASSO is used as a penalty function in the R programming system with the glmnet package [5]. The solution of the elastic net log likelihood function is obtained using descent coordinates [3]. The addition of LASSO and elastic net penalties can shrink the regression coefficients and select variables in modeling. The correlation between the response variable and certain predictor variables can also depend on the values of the other predictor variables. When important covariates are associated with correlated variables, the simulation results show that the predicted performance of LASSO and elastic net increases [6]. The logistic regression model with LASSO and elastic net penalty have been performed on gene expression data of patients that suffer colorectal cancer and hepatocellular carcinoma [7]. The results of this study have a high accuracy.

In this study, penalized logistic regression models will be used to predict a result of RT-PCR by using the features of blood laboratory tests. The RT-PCR stands for Reverse Transcription Polymerase Chain Reaction. The RT-PCR is the most reliable test tool for coronavirus disease (COVID-19). The COVID-19 is spreading rapidly from China to countries in the world. The transmission of COVID-19 has exponential growth. The true level of spread of COVID-19 is not yet known due to various limitations in conducting the COVID-19 test. The RT-PCR test takes a long time approximately three until four days to find out the results. While, the status of a patient infected with COVID-19 or not is quickly required for prevention of transmission and clinical treatment measures. Apart from the RT-PCR test, a rapid diagnostic test can provide results in a short time. However, rapid diagnostic tests are very susceptible to accuracy problems so that the tests performed are not efficient. Research to overcome the limitations of the COVID-19 test kit is still ongoing. Models using routine laboratory test results offer opportunities for early and rapid identification of patients infected with COVID-19 [8]. The blood laboratory test results can be known within two to three hours. The blood laboratory test results from a patient include many features. This research interested in selecting the features of blood test laboratory results using logistic regression with LASSO and elastic net penalties. So, the logistic regression model with LASSO and elastic net penalties can be used to obtain a simpler model. This research aims at the model performance of the RT-PCR test result predictions by using the blood laboratory test.

## 2. Data and Research Method

### 2.1. Data

The data in this study were secondary data published by Einstein Data4u on March 28, 2020 which can be accessed via https://www.kaggle.com/einsteindata4u/covid19. This data was the RT-PCR test results and laboratory test results from the patient at the Israelita Albert Einstein Hospital, in São Paulo, Brazil. Many variables in the data contain missing values. Before this data was used, the data cleaning process was carried out. Firstly, we deleted data on patients who did not perform laboratory tests. The data cleaning process was continued by deleting variables containing more than 80% missing values. At the final stage of the data cleaning process, variables with zero diversity were not used. After performing the data cleaning process, we obtained the data on 75 patients who performed RT-PCR tests and laboratory tests. This study used 28 predictor variables in modelling, patient age quantile and 27 variables of blood laboratory test results. The features of blood laboratory test results were the haematocrit, haemoglobin, platelets, mean platelet volume, red blood cells, lymphocytes, mean corpuscular haemoglobin concentration (MCHC), leukocytes, basophils, mean corpuscular haemoglobin (MCH), eosinophils, mean corpuscular volume. (MCV), monocytes, red blood cell distribution width (RDW), neutrophils, urea, C-reactive protein, creatinine, potassium, sodium, $pCO_2$ (venous blood gas analysis), Hb saturation (venous blood gas analysis), base excess ( venous blood gas analysis), $pO_2$ (venous blood gas analysis), total $CO_2$ (venous blood gas analysis), pH (venous blood gas analysis), $HCO_3$ (venous blood gas analysis).

### 2.2. Research Method

The results of the RT-PCR test were currently very reliable for examining a person who was infected by COVID-19. The data from the RT-PCR test were used as the binary response variable. A positive RT-PCR test result was categorized as one and a negative RT-PCR test result was categorized as zero. The predictor variables were the patient age quantile and 27 features of blood laboratory test results from patients who performed a RT-PCR test. This study used a logistic regression model with LASSO and Elastic Net penalties to select the variables that affect the RT-PCR test results. Penalized logistic regression model with LASSO and elastic net penalties used the R programming system with the glmnet package. LASSO logistic regression modelling with $\alpha = 1$ was carried out in the following steps:

a. the data was divided based on the leave one out cross validation procedure, 80% as modeling data and 20% as validation data
b. we determined optimal $\lambda$ (shrinkage parameter) with 10-fold cross validation
c. we estimated the coefficient $\beta$ using modelling data
d. we perform analysis on the predictor variables of modelling results

Furthermore, we conducted elastic net logistic regression model with $0 < \alpha < 1$, which was carried out in the following steps:

a. the data was divided based on the leave one out cross validation procedure, 80% as modeling data and 20% as validation data
b. we determined the optimal $\alpha$ and $\lambda$ with 10-fold cross validation
c. we estimated the coefficient $\beta$ using modelling data
d. we perform analysis on the predictor variables of modelling results

This study also investigated the selection of the best model to predict the RT-PCR test results. Selection of the best model was based on accuracy and AUC performances. The steps for selecting the best model were as follows:

a. we performed logistic regression modelling using LASSO and Elastic Net penalties with 30 repetitions on modelling data. A sample taken as modelling data and validation data were arranged as one observation. So, there were 30 pairs of modelling and validation data
b. we measured model performance with accuracy and AUC values using validation data for 30 repetitions
c. we performed the mean difference test for accuracy and AUC values with paired sample t-test
d. we determined the best model to predict the results of a RT-PCR test

## 3. Results and Discussion

### 3.1. Results

#### *3.1.1. LASSO Logistic Regression Model*

The data of the patient was randomly partitioned. They were 61 patient data (80%) as modelling data and 14 patient data (20%) as validation data. LASSO logistic regression required $\lambda$ (shrinkage parameter) to shrink the regression coefficient. The optimum $\lambda$ was obtained by 10-fold cross-validation.

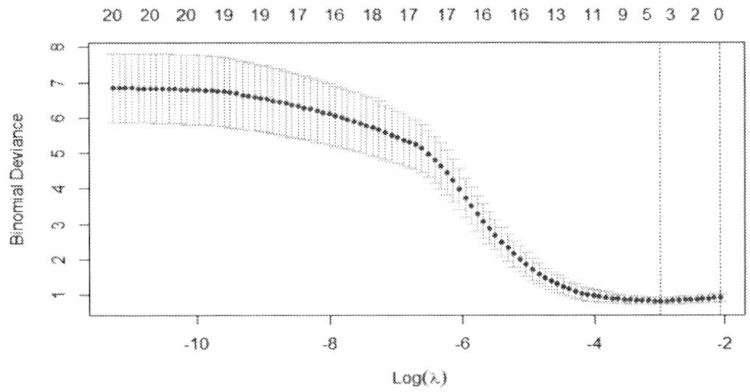


**Figure 1.** Cross validation plot for optimizing the LASSO shrinkage parameter

Figure 1 shows the binomial deviance value for each log λ. The vertical line showed the optimum log λ. The first vertical line represented the value of $\lambda_{-min}$ which was 0.05 and the second vertical line represented the value of $\lambda_{1se}$ which was 0.13. The optimum λ selected in this study when the binomial deviance was minimum. So, $\lambda_{-min}$ was used as the shrinkage parameter in the LASSO logistic regression model. The value of the upper horizontal line from Figure 1 shows the number of LASSO coefficients with nonzero values for each log λ. By using the $\lambda_{-min}$ value, there are four LASSO coefficients with nonzero values. The predictor variables with non-zero LASSO coefficient were shown in Table 1.

**Table 1.** LASSO logistic regression coefficient

| Variables | Coefficient |
|---|---|
| Leukocytes | -0.9116 |
| Basophils | -0.4482 |
| Red blood cell distribution width (RDW) | -0.0015 |
| C-reactive Protein | 0.3936 |

The plot of the LASSO logistic regression coefficient spread for each log λ can be seen in Figure 2. All values of the LASSO logistic regression coefficient go to zero as the increases of the shrinkage parameter.

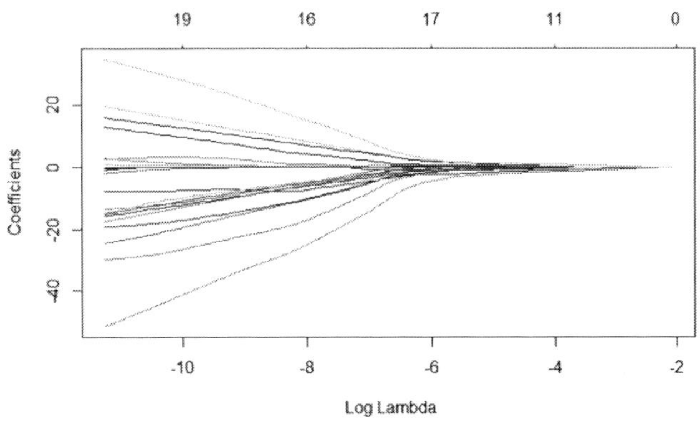


**Figure 2.** Plot of LASSO logistic regression coefficients for each shrinkage parameter

### 3.1.2. *Elastic Net Logistic Regression Model*

The data of the patient was randomly partitioned. They were 61 patient data (80%) as modelling data and 14 patient data (20%) as validation data. The optimum $\alpha$ in the modelling data was obtained by 10-fold cross-validation based on the minimum CV. Based on modelling data, the elastic net logistic regression analysis used $\alpha = 0.3$. Elastic net logistic regression required $\lambda$ (shrinkage parameter) to shrink the regression coefficient.

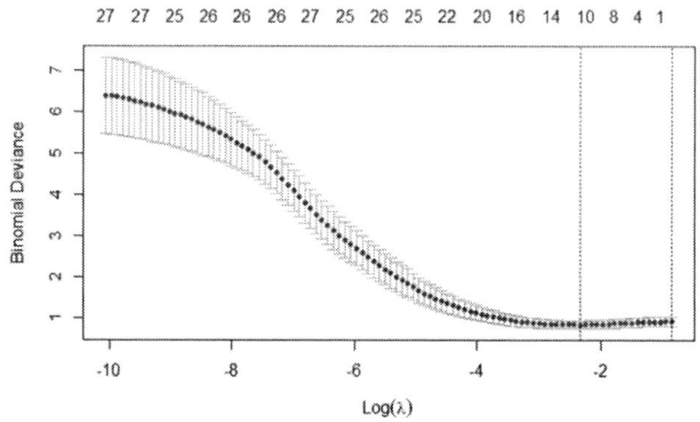


**Figure 3.** Cross validation plot for optimizing the elastic net shrinkage parameter

Figure 3 shows the binomial deviance value for each log $\lambda$. The vertical line showed the optimum log $\lambda$. The first vertical line represented the value of $\lambda_{-min}$ which was 0.096 and the second vertical line represented the value of $\lambda_{1se}$ which was 0.43. In this study, the minimum binomial deviance value was obtained when log $\lambda_{-min}$. So, the value of $\lambda_{-min}$ was used as a parameter shrinkage of elastic net logistic regression. The upper horizontal line in Figure 3 showed the number of elastic net coefficients with nonzero values for each log $\lambda$. By using the log $\lambda_{-min}$, there are 11 elastic net logistic regression coefficients which have non-zero values. The predictor variables with non-zero elastic net coefficient were shown in Table 2.

**Table 2.** Elastic net logistic regression coefficients

| Variables | Coefficients |
|---|---|
| Leukocytes | -0,5462 |
| Basophils | -0,3504 |
| Platelets | -0,1969 |
| Red blood cell distribution width (RDW) | -0,1693 |
| Eosinophils | -0,0892 |
| Potassium | -0,0674 |
| Urea | -0,0550 |
| Mean corpuscular volume (MCV) | -0,0093 |
| pH (venous blood gas analysis) | 0,0168 |
| Monocytes | 0,1028 |
| C-reactive Protein | 0,3241 |

The plot of the elastic net logistic regression coefficient spread for each log $\lambda$ can be seen in Figure 4. All values of the elastic net logistic regression coefficient go to zero as the increases of the shrinkage parameter.

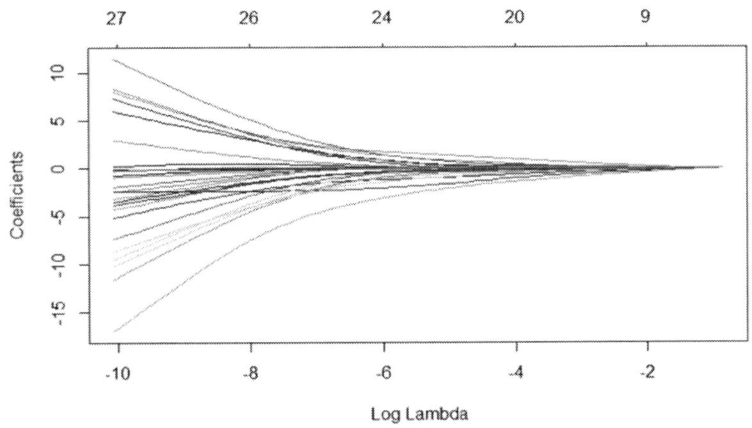


**Figure 4.** Plot of elastic net logistic regression coefficients for each shrinkage parameter

*3.1.3. Selection of the best model*

The samples of modelling data and validation data for 30 repetitions was carried out with the set.seed setting during the sampling process. So, there are 30 sets of modelling data and validation data that would be used to predict the results of a RT-PCR test. The sampling process was randomly carried out. They were 61 patient data or 80% as modelling data and 14 patient data or 20% as validation data. LASSO logistic regression model used α = 1, while the elastic net logistic regression used the optimum $0 < \alpha < 1$. The optimum α was obtained based on the modeling data taken for each repetition.

Furthermore, LASSO and elastic net logistic regression modelling was carried out for each repetition. Modelling performance of LASSO and elastic net logistic regression were carried out using the accuracy and AUC values based on the validation data. The comparison of accuracy and AUC values of 30 repetitions for each model was shown in Figure 5 and Figure 6.

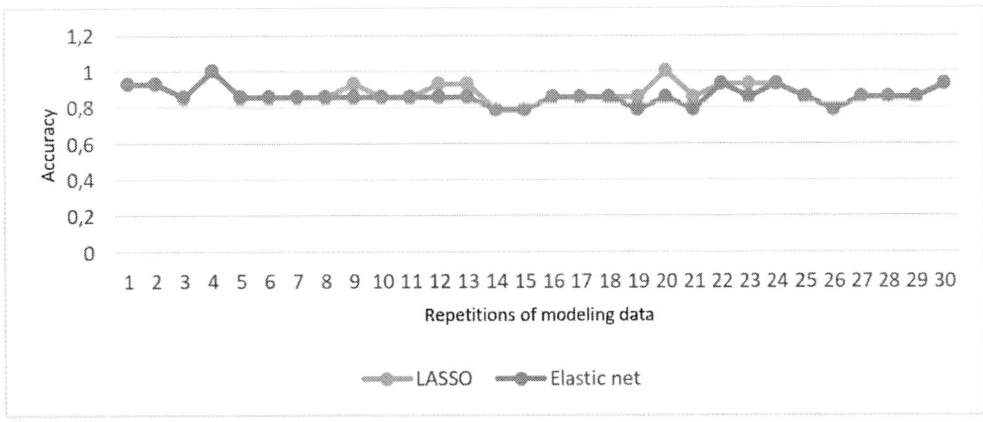


**Figure 5.** Accuracy value of LASSO and elastic net logistic regression model

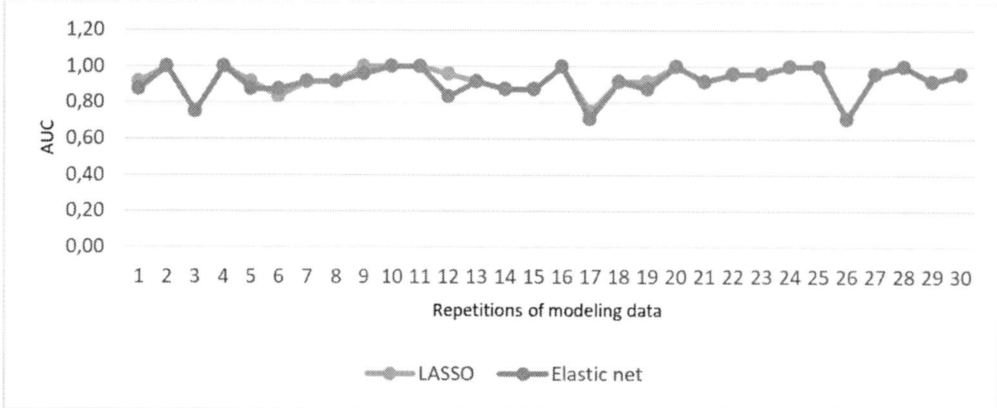


**Figure 6.** AUC value of LASSO and elastic net logistic regression model

The accuracy average for logistic regression with LASSO and elastic net penalties were 0.88 and 0.86, respectively. The mean AUC values for the LASSO and elastic net logistic regression were 0.93 and 0.92, respectively. The criteria for selecting the best model was the model with the greatest accuracy and AUC value. Based on the average of the two goodness criteria of the model, the LASSO logistic regression model was better than the elastic net logistic regression model.

To state that the LASSO logistic regression is better than the Elastic Net logistic regression statistically, the mean difference test was carried out using paired sample t-test. The results of the paired sample t-test for the two models used could be seen in Table 3. The hypothesis was used to see the differences significantly in accuracy/AUC of the two models. The hypothesis was as follows:
$H_0$: the accuracy/AUC average of the LASSO logistic regression was less than or equal to the accuracy /AUC average of the Elastic Net logistic regression.

$H_1$: the accuracy/AUC average of the LASSO logistic regression was more than the accuracy/AUC average of the Elastic Net logistic regression.

**Table 3.** The results of the paired sample t-test of accuracy and AUC

| | t-stat | p-value |
|---|---|---|
| **Accuracy** | 2.8045 | 0.0045 |
| **AUC** | 1.8810 | 0.0350 |

The t-test result in Table 3 showed the p-value for accuracy and AUC less than 0.05 at significance level 5%. It could be concluded that the accuracy/AUC average of LASSO logistic regression was more than the accuracy/AUC average of elastic net logistics regression. So, the LASSO logistic regression model had a better performance for predicting the results of the RT-PCR test.

### 3.2. Discussion

The logistic regression analysis with LASSO and elastic net penalties could select 28 predictor variables that were used to predict the results of the RT-PCR test. Figure 2 and Figure 4 showed that the LASSO logistic regression places more exact regression coefficients at zero than the elastic net logistic regression. The predictor variables selected by LASSO penalty in Table 1 showed that four predictor variables that affect the results of RT-PCR. They are leukocytes, basophils, RDW and C-reactive protein. C-reactive protein had a positive coefficient. It means that the higher the C-reactive Protein level, the more likely a person is to get positive RT-PCR test result. While leukocytes, basophils and RDW had a negative coefficient. It can be interpreted that the less a number of leukocytes, basophils and RDW caused the tendency to get positive RT-PCR test results increases.

The results of the Elastic Net logistic regression analysis have the same four predictor variables as the LASSO penalty, the other seven predictor variables that influence the prediction of RT-PCR results are platelets, eosinophil, potassium, urea, MCV, pH (venous blood gas analysis) and monocytes. C-reactive protein, monocytes and pH (venous blood gas) had a positive coefficient. It means the higher the pH (venous blood gas analysis), the number of monocytes and the C-reactive protein level caused the more likely a person is to get a positive RT-test results. Conversely leukocyte, basophil, platelets, RDW, eosinophils, potassium, urea and MCV had a negative coefficient. It means that the smaller the number of leukocytes, basophil, platelets, RDW, eosinophils, potassium, urea and MCV caused the tendency to obtain a positive RT-PCR test results increases.

In the early stages of a patient infected with COVID-19, previous research has explained that the patient's C-reactive protein levels have a positive correlation with lung lesions and can reflect the severity of the disease [9]. The higher the level of C-reactive protein the patient has, the more lung lesions will be so that the patient's disease gets worse. Patients with COVID-19 pneumonia have a statistically different number of leukocytes and platelets from non-COVID-19 pneumonia patients [10]. The results of this study also showed that decreasing the number of leukocytes and platelets can increase the probability of a person getting a positive RT-PCR test. Venous blood gas analysis provides useful information regarding lung function and acid-base balance [11]. The higher the pH (venous blood gas analysis), the more likely it is to get a positive RT-PCR test. So, if there is an increase in pH it can be an indicator of impaired lung function. MCV is a measure of the average size of red blood cells, if too low, it can indicate that the patient has some type of anaemia. The lower the MCV value, the more likely it is to get a positive RT-PCR test result. Urea and potassium are mineral substances that are essential for organs such as the heart or kidneys to function normally. The lower the urea and potassium levels in a person's body, there will be interference with heart or kidney function.

The results of this research showed that prediction performance of a RT-PCR test using LASSO logistic regression was better than elastic net logistic regression. The LASSO logistic regression model had a good performance for predicting the RT-PCR test with 88% accuracy and 93% AUC. Based on the result of LASSO logistic regression model, the features of laboratory blood tests that affect a RT-PCR test were leukocytes, basophils, RDW and C-reactive protein. The odds ratio of leukocytes was 0.4. It meant that the odds of a positive RT-PCR test result given an increase of a unit leukocytes ($10^3/\mu l$) 0.4 times was lower than before the increase. The odds ratio of basophils was 0.64. It meant that the odds of a positive RT-PCR test result given an increase in 1% basophils 0.64 times was lower than before the increase. The odds ratio of RDW was 0.99. It meant that the odds of a positive RT-PCR test result given an increase of 1% RDW 0.99 times was lower than before the increase. The odds ratio of C-reactive protein was 1.48. It meant that the odds of a positive RT-PCR test result given an increase in one milligram of C-reactive protein per liter of blood (mg/L) 1.48 times was higher than before the increase.

## 4. Conclusion

Selection of variables to predict the results of the RT-PCR test using LASSO logistic regression resulted four predictor variables with non-zero regression coefficients, namely Leukocytes, Basophils, Red blood cell distribution width (RDW) and C-reactive protein. The selection of variables to predict the results of the RT-PCR test using elastic net logistic regression resulted eleven predictor variables with non-zero regression coefficients, namely platelets, eosinophils, mean corpuscular volume (MCV), monocytes, urea, potassium, pH (venous blood gas analysis) and four other predictor variables were equal to the predictor variables of the LASSO logistic regression model outcome.

The results of the model goodness measurement analysis showed that the LASSO logistic regression model is better than the elastic net logistic regression model for predicting the RT-PCR test results. The laboratory results of blood tests as predictor variables have high accuracy and are fast to predict the results of the RT-PCR test.

## References

[1] James G, Witten D, Hastie T and Tibishirani R 2013 *An Introduction to Statistical Learning with Applications in R* New York: Springer
[2] Hastie T, Tibshirani R and Friedman J 2008 *Springer Series in Statistics* New York: Springer
[3] Zou H and Hastie T 2005 Regularization and variable selection via the elastic net *J. R. Stat. Soc. Ser. B Stat. Methodol.* **67** 301–320
[4] Tibshirani R 2011 Regression shrinkage and selection via the lasso: A retrospective *J. R. Stat. Soc. Ser. B Stat. Methodol.* **73** 273–282
[5] Friedman J, Hastie T and Tibshirani R 2010 Regularization paths for generalized linear models via coordinate descent *J. Stat. Softw.* **33** 1–22
[6] Sirimongkolkasem T and Drikvandi R 2019 On Regularisation Methods for Analysis of High Dimensional Data *Ann. Data Sci.* **6** 737–763
[7] Susanti A 2019 Perbandingan Metode Regresi Logistik Terpenalti pada Data Ekspresi Gen [thesis]. Institut Pertanian Bogor
[8] Yang H S, Hou Y, Vasovic L V, Steel P, Chadburn A, Racine-Brzostek S E, Velu P, Cushing M M, Loda M, Kaushal R, Zhao Z and Wang F 2020 Routine laboratory blood tests predict SARS-CoV-2 infection using machine learning *Clin. Chem.* **66** (11) 1396-1404
[9] Wang L 2020 C-reactive protein levels in the early stage of COVID-19 *Med. Mal. Infect.* **50** 332–334
[10] Cheng Z, Lu Y, Cao Q, Qin L, Pan Z, Yan F and Yang W 2020 Clinical features and chest CT manifestations of coronavirus disease 2019 (COVID-19) in a single-center study in Shanghai, China *Am. J. Roentgenol.* **215** 121–126
[11] Rieser T M 2013 Arterial and Venous Blood Gas Analyses *Top. Companion Anim. Med.* **28** 86–90

# Comparison of Spatial Weights in Modeling Inflation and Money Supply Using the Spatial Vector Autoregressive Model with Calendar Variations

**E Sumarminingsih[1,]*, Setiawan[2], A Suharsono[2] and B N Ruchjana[3]**

[1] Department of Statistics, University of Brawijaya, Indonesia
[2] Department of Statistics, Institut Teknologi Sepuluh Nopember, Indonesia
[3] Department of Mathematics, University of Padjadjaran, Indonesia.

*Corresponding author: eni_stat@ub.ac.id

**Abstract.** In modeling inflation and the money supply simultaneously using SpVAR with calendar variations, it is necessary to have the appropriate spatial weight. There are several spatial weights that can be used, including uniform weight, inverse distance weight and weight based on normalization of cross correlation. This study aims to compare the three spatial weights for modeling inflation and the money supply simultaneously using the SpVAR model with calendar variations. The results of this study indicate that the weight based on normalization of cross correlation is the best spatial weight for modeling inflation and the money supply simultaneously using the SpVAR model with calendar variations because it produces the model with the smallest Mean Square Error (MSE)..

## 1. Introduction

In economic modeling such as inflation and money supply modeling, space-time models such as Generalized Space-Time Autoregression (GSTAR), Vector Autoregressive (VAR) and Spatial Vector Autoregressive (SpVAR) have been used. This is done because of the space-time relationship in these economic variables. Modeling inflation and the money supply using the space-time model have been done by some researchers include [1], [2], [3], [4], [5], [6], [7] and [8].

Economic data is usually monthly data. This monthly data can be influenced by calendar variations. One of the calendar variation is Holiday variation which refers to fluctuations in economic activity due to the effect of certain holidays or religious holidays that are determined based on the lunar calendar so that the holidays occur at different dates and months each year. Research conducted by [3], [9], [10], [11] shows that inflation and money supply in Indonesia are influenced by the effects of calendar variations.

In space time modeling, spatial weights play an important role because spatial weights represent spatial dependencies between locations. In addition, spatial weights can also determine the results of parameter estimation in spatial or space-time models. Therefore determining the appropriate spatial weight is a necessity. Some spatial weights that are often used include uniform weight, binary weight, inverse of distance weight based on semi-variogram or covariogram of variable between locations, weight based on the normalization of cross-correlation between locations at the appropriate time lag [12]. In this study spatial weights were compared, namely uniform weights, inverse distance weights and Weight based on the normalization of cross-correlation between locations at the appropriate time lag in inflation and the money supply modeling using SpVAR with calendar variations.

## 2. Methodology

### 2.1. Data

The data used in this study are monthly inflation and money supply data for 12 years (January 2003-December 2014) from the cities of Surabaya, Malang, Kediri and Jember. The inflation data was obtained from the Statistics Indonesia's website (www.bps.go.id), while the money supply data was obtained from Bank Indonesia.

### 2.2. Spatial Vector Autoregressive Model with Calendar Variations

Spatial Vector Autoregressive Model with temporal order $p$ and spatial order 1 [SpVAR ($p$, 1)] is the development of the Vector Autoregressive model with order $p$ [VAR ($p$)]. The SpVAR model ($p$, 1) adds spatial elements to the VAR ($p$) model. This results in an increase in the index of the variables for the SpVAR model. The SpVAR model with calendar variations is an extension of the SpVAR model that accommodates the effects of calendar variations on time series data. The model was formed by adding components of exogenous variables to accommodate the effects of calendar variation and expressed as [3]

$$\mathbf{z}_t = \mathbf{\Gamma}\mathbf{x}_t + \mathbf{B}_1\mathbf{z}_{t-1} + \cdots + \mathbf{B}_p\mathbf{z}_{t-p} + \boldsymbol{\varepsilon}_t \tag{1}$$

where $\mathbf{\Gamma} = diag\left(\boldsymbol{\gamma}_{11},\ldots,\boldsymbol{\gamma}_{N1},\ldots,\boldsymbol{\gamma}_{1K},\ldots,\boldsymbol{\gamma}_{NK}\right)$, $\boldsymbol{\gamma}_{nk} = \left[\gamma_{nk1} \quad \gamma_{nk2} \quad \cdots \quad \gamma_{nkm}\right]$

$\mathbf{x}_t = \left[\mathbf{x}_{11t} \quad \mathbf{x}_{21t} \quad \cdots \quad \mathbf{x}_{N1t} \quad \cdots \quad \mathbf{x}_{1Kt} \quad \mathbf{x}_{2Kt} \quad \cdots \quad \mathbf{x}_{NKt}\right]'$, $\mathbf{x}_{nkt} = \left[x_{nk1t} \quad x_{nk2t} \quad \cdots \quad x_{nkmt}\right]'$ is an $m$ by 1 vector and contains a dummy variable for the calendar variation effect for an equation $z_{nkt}$.

where $\mathbf{z}_t = \left[z_{11t}\ z_{12t} \cdots z_{1Kt} \cdots z_{N1t}\ z_{N2t} \cdots z_{NKt}\right]'$,

$$\mathbf{B}_i = \begin{bmatrix} b_{11}^{(i)} & b_{12}^{(i)} & \cdots & b_{1(NK)}^{(i)} \\ b_{21}^{(i)} & b_{22}^{(i)} & \cdots & b_{2(NK)}^{(i)} \\ \vdots & \vdots & \ddots & \vdots \\ b_{(NK)1}^{(i)} & b_{(NK)2}^{(i)} & \cdots & b_{(NK)(NK)}^{(i)} \end{bmatrix}, \text{ and } \boldsymbol{\varepsilon}_t = \left[\varepsilon_{11t}\ \varepsilon_{12t} \cdots \varepsilon_{1Kt} \cdots \varepsilon_{N1t}\ \varepsilon_{N2t} \cdots \varepsilon_{NKt}\right]'.$$

The more locations and /or variables, the more parameters are also estimated. [13] Di Giacinto, 2010) then do restrictions on the coefficient matrix, so that the parameters that are estimated are not too many. The restriction is to form a matrix

$$\mathbf{B}_i = \begin{bmatrix} B_{11}^{(i)} & B_{12}^{(i)} & \cdots & B_{1K}^{(i)} \\ B_{21}^{(i)} & B_{22}^{(i)} & \cdots & B_{2K}^{(i)} \\ \vdots & \vdots & \ddots & \vdots \\ B_{K1}^{(i)} & B_{K2}^{(i)} & \cdots & B_{KK}^{(i)} \end{bmatrix} \tag{2}$$

where

$$\mathbf{B}_{kr}^{(i)} = \mathbf{\Phi}_{kr}^{(i0)} + \mathbf{\Phi}_{kr}^{(i1)}\mathbf{W}_{kr}, \mathbf{\Phi}_{kr}^{(il)} = diag\left[\phi_{1kr}^{(il)},\ldots,\phi_{Nkr}^{(il)}\right]; k,r = 1,\ldots,K; i = 1,\ldots,p \tag{3}$$

where $\mathbf{W}_{kr}$ ($k$ and $r$ represent the $k$-th and the $r$-th variables) is an $N$ by $N$ spatial matrix weight, with elements that are already known and are assumed to be non-negative and positive if the locations $i$ and $j$ are neighbors. Spatial weights are assumed to remain constant over time, while the autoregressive coefficient is assumed to vary between locations.

### 2.3. Spatial Weight

Spatial weights have a significant role in SpVAR modeling, because spatial weights show connectivity between locations. Generally spatial weights are expressed in the form of spatial weight matrices. Some spatial weights that can be used in spatial econometric analysis include uniform weight, inverse distance weight and weight based on normalization of cross correlation [12]. The spatial weight formulas are:

1. Uniform weight

$$w_{kr}(i,j)=\frac{1}{n_i},$$

where $n_i$ is the number of locations adjacent to location $i$.

2. Inverse distance weight

$$w_{kr}(i,j)=\frac{1}{d_{ij}},$$

where $d_{ij}$ is the distance between location $i$ and location $j$.

3. Weight based on normalization of cross correlation between locations at the appropriate time lag

$$w_{kr}(i,j)=\frac{corr_{kr,ij}(l)}{\sum_{h\neq i} corr_{kr,ih}(l)} \tag{4}$$

where $i \neq j$ and fulfill $\sum_{j\neq i}\left|w_{kr}(i,j)\right|=1$ where $l$ is time lag and

$$corr_{rk,ij}(l)=\frac{\sum_{t=k+1}^{T}\left[Z_{irt}-\bar{Z}_{ir}\right]\left[Z_{jk(t-l)}-\bar{Z}_{jk}\right]}{\sqrt{\left(\sum_{t=1}^{T}\left[Z_{irt}-\bar{Z}_{ir}\right]\right)^2\left(\sum_{t=1}^{T}\left[Z_{jk(t-l)}-\bar{Z}_{jk}\right]\right)^2}}, \tag{5}$$

## 3. Research Procedure

This research was carried out with the following procedure

1. Stationarity test is carried out using panel unit root test from [14]
2. Determination of calendar variation dummy variables. There are 8 dummy variables used in this study

$$x_{bi,t}=\begin{cases}1, & \text{if } t \text{ is Ramadhan month when Eid al Fitr in week-}i\\ 0, & \text{for other month}\end{cases}$$

$$x_{i,t}=\begin{cases}1, & \text{if } t \text{ is Eid al Fitr month when Eid al Fitr in week-}i\\ 0, & \text{for other month}\end{cases}$$

For $i=1,2,3,4$

Furthermore, the model was formed

$$\mathbf{z}_t=\mathbf{\Gamma x}_t+\mathbf{e}_t \tag{6}$$

Where $\boldsymbol{x}$ is a vector of calendar variation variable. Estimating the parameter model (6) using the generalized least square (GLS) method.

3. Determination of spatial weights. There are three spatial weights used in this study, namely uniform weight, inverse distance weight and weight based on normalization of cross correlation
4. Order selection. To determine the order, an SpVAR ($p$, 1) model is formed from the residual of the model (6).Next, the following hypotheses are tested sequentially

$$H_0^1 : \mathbf{\Phi}^{(M)} = \mathbf{0} \text{ vs } H_1^1 : \mathbf{\Phi}^{(M)} \neq \mathbf{0},$$

where M is a positive integer $\mathbf{\Phi}^{(M)} = \left[\mathbf{\Phi}_1^{(M)} \quad \mathbf{\Phi}_2^{(M)} \quad \cdots \quad \mathbf{\Phi}_K^{(M)}\right]'$, $\mathbf{\Phi}_k^{(M)} = \left[\phi_{1k}^{(M)} \quad \phi_{2k}^{(M)} \quad \cdots \quad \phi_{Nk}^{(M)}\right]'$,

and $\mathbf{\Phi}_{nk}^{(M)} = \left[\phi_{nk1}^{(M0)} \quad \phi_{nk1}^{(M1)} \quad \phi_{nk2}^{(M0)} \quad \phi_{nk2}^{(M1)} \quad \cdots \quad \phi_{nkK}^{(M0)} \quad \phi_{nkK}^{(M1)}\right]'$. if the null hypothesis is not rejected, then a hypothesis test is conducted for the order $M$-1. Every null hypothesis is tested conditional on the previous null hypothesis being true. This procedure stops if one null hypothesis is rejected. If $H_0^i$ rejected, then $\hat{p} = M - i + 1$ is chosen as the order.

5. Estimating the parameters is done by the Full Information Maximum Likelihood (FIML) method.
6. Test the significance of the parameters.
7. Selection of the best model. The smallest MSE was chosen as the best model selection criteria.

## 4. Result and Discussion

Stationary test result for inflation is stationary because the $p$-value is less than 5%. Furthermore, the stationarity test result for money supply is the $p$-value more than 5% so that the money supply is not stationary. Differencing is done on money supply so that it is stationary. Stationary test results on money supply that have been differenced has $p$-value less than 5% for stasionarity test, so the differenced money supply is stationary.

The next step is estimating parameters of model (6). After the model is obtained, the residual of the model (6) is calculated to be used in order determination. In the procedure of determining the order, determined a maximum order of three. The test statistic value for third order is 36.9273 while the value of $\chi^2_{(0.10,32)} = 42.5847$, so the null hypothesis fails to be rejected. Furthermore, testing the hypothesis for second order and obtained the test statistic value is 73.78977, so the null hypothesis is rejected. Therefore, the order used is two.

There are three spatial weights used in this study, namely uniform weights, distance inverse weights and weights based on cross correlations. Because it consists of 4 locations, the uniform spatial weight is

$$\mathbf{W}_{11} = \mathbf{W}_{12} = \mathbf{W}_{21} = \mathbf{W}_{22} = \begin{bmatrix} 0 & 0.33 & 0.33 & 0.33 \\ 0.33 & 0 & 0.33 & 0.33 \\ 0.33 & 0.33 & 0 & 0.33 \\ 0.33 & 0.33 & 0.33 & 0 \end{bmatrix}$$

The inverse distance weight is

$$\mathbf{W}_{11} = \mathbf{W}_{12} = \mathbf{W}_{21} = \mathbf{W}_{22} = \begin{bmatrix} 0 & 0.46 & 0.33 & 0.21 \\ 0.46 & 0 & 0.38 & 0.20 \\ 0.38 & 0.47 & 0 & 0.15 \\ 0.38 & 0.38 & 0.24 & 0 \end{bmatrix}.$$

Based on the correlation between inflation and money supply in four cities at lag one and the formula in equations (4) and (5), the weight based on normalization of cross correlation between locations at time lag 1 is

$$\mathbf{W}_{11}(1) = \begin{bmatrix} 0 & 0 & 0 & 1 \\ 0.33 & 0 & 0.30 & 0.37 \\ 0 & 0 & 0 & 0 \\ 0 & 0 & 0 & 0 \end{bmatrix}, \mathbf{W}_{12}(1) = \begin{bmatrix} 0 & 0 & 0 & 0 \\ 0 & 0 & 0 & 0 \\ 0 & 0 & 0 & 0 \\ 0 & 1 & 0 & 0 \end{bmatrix},$$

$$\mathbf{W}_{21}(1)=\begin{bmatrix}0&0&0&0\\0&0&0&0\\0&0&0&0\\0&0&0&0\end{bmatrix},\mathbf{W}_{22}(1)=\begin{bmatrix}0&0.22&0.34&0.44\\0.26&0&0.32&0.42\\0.35&0&0&0.65\\0.35&0.21&0.44&0\end{bmatrix}.$$

Spatial weights based on correlations for lag two is

$$\mathbf{W}_{11}(2)=\mathbf{W}_{12}(2)=\mathbf{W}_{21}(2)=\begin{bmatrix}0&0&0&0\\0&0&0&0\\0&0&0&0\\0&0&0&0\end{bmatrix},$$

Because the correlation between inflation and inflation and between inflation and the money supply at lag two is not significant. Whereas the spatial weight for the money supply and money supply at lag 2 is

$$\mathbf{W}_{22}(2)=\begin{bmatrix}0&0&0.44&0.56\\0.34&0&0.31&0.35\\0.46&0&0&0.54\\0.49&0&0.51&0\end{bmatrix}.$$

The results of parameter estimation for exogenous variables for the three weights are almost the same, especially for significant parameters. While the results of parameter estimation for endogenous variables also show many similar parameter estimators. Although there are some results that show different results, the difference is not large.

Furthermore MSE from the SpVAR Model with calendar variations of each weight is presented in Table 1.

Table 1. MSE of SpVAR(2,1) Model with Calendar Variations with some Spatial Weight

| Equation | Mean Square Error | | |
|---|---|---|---|
| | Uniform Weight | Weight Based on Cross Correlation | Invers Distance Weight |
| Inflation in Surabaya | 0.467466 | 0.452595 | 0.467253 |
| Inflation in Malang | 0.528772 | 0.508202 | 0.531206 |
| Inflation in Kediri | 0.925423 | 0.924263 | 0.926935 |
| Inflation in Jember | 0.64156 | 0.633298 | 0.644928 |
| Money Supply in Surabaya | 0.732718 | 0.737825 | 0.734882 |
| Money Supply in Malang | 0.056566 | 0.057929 | 0.056964 |
| Money Supply in Kediri | 0.123183 | 0.124271 | 0.123245 |
| Money Supply in Jember | 0.026587 | 0.027222 | 0.026522 |
| **Total MSE** | **3.502275** | **3.465605** | **3.511935** |

Based on Table 1, the SpVAR (2,1) model with calendar variation with correlation weights has four equations that has the smallest MSE. While the SpVAR(2,1) model with calendar variation with uniform weights has one equation that has the smallest MSE and the SpVAR (2.1) model with calendar variation with inverse distance weight has three equations with the smallest MSE. The smallest total MSE is owned by the SpVAR (2.1) model with calendar variation with correlation weights, so this model is chosen as the best model. However, it should be noted that the MSE values of the three models are almost the same, the difference is smaller than 1 percent. The SpVAR(2.1) model with calendar variations with correlation weights can be written as

$$\begin{aligned}\hat{z}_{11t} &= 3.6929^{*}x_{b1t} + 0.8688^{*}x_{b2t} - 0.0261x_{b3t} - 0.0332x_{b4t} \\ &-0.4313x_{1t} + 0.1824x_{2t} + 0.6721x_{3t} + 0.3225x_{4t} \\ &-0.0612z_{11t-1} + 0.0190z_{12t-1} - 0.1306^{*}z_{11t-2} - 0.0027z_{12t-2} + 0.1578^{*}z_{41t-1}\end{aligned} \quad (7)$$

$$\begin{aligned}\hat{z}_{21t} &= 3.8343^{*}x_{b1t} + 0.8620^{*}x_{b2t} - 0.1489x_{b3t} - 0.2271x_{b4t} \\ &-0.3906x_{1t} - 0.0254x_{2t} + 0.2459x_{3t} + 0.0896x_{4t} \\ &-0.1622^{*}z_{21t-1} + 0.0358z_{22t-1} - 0.1421^{*}z_{21t-2} + 0.0076z_{22t-2} \\ &+0.2938^{*}\left(0.33z_{11t-1} + 0.3z_{31t-1} + 0.37z_{41t-1}\right),\end{aligned} \quad (8)$$

$$\begin{aligned}\hat{z}_{31t} &= 5.4487^{*}x_{b1t} + 0.6678x_{b2t} - 0.0652x_{b3t} + 0.1533x_{b4t} \\ &-0.4588x_{1t} + 0.4804x_{2t} + 0.9013x_{3t} + 0.6198x_{4t} \\ &+0.0388z_{31t-1} + 0.0692z_{32t-1} - 0.1693^{*}z_{31t-2} + 0.0551z_{32t-2},\end{aligned} \quad (9)$$

$$\begin{aligned}\hat{z}_{41t} &= 4.0208^{*}x_{b1t} + 0.5444x_{b2t} - 0.2446x_{b3t} - 0.2079x_{b4t} \\ &-0.8825x_{1t} + 0.3345x_{2t} + 0.4212x_{3t} + 0.4062x_{4t} \\ &+0.1704^{*}z_{41t-1} + 0.0704z_{42t-1} - 0.1636^{*}z_{41t-2} + 0.1060z_{42t-2} + 0.0330z_{22t-1},\end{aligned} \quad (10)$$

$$\begin{aligned}\hat{z}_{12t} &= 2.0688^{*}x_{b1t} + 1.3611^{*}x_{b2t} + 0.2624x_{b3t} + 0.2033x_{b4t} \\ &-0.8966x_{1t} + 0.5818x_{2t} + 1.8395^{*}x_{3t} + 1.9287^{*}x_{4t} \\ &-0.1215^{*}z_{11t-1} - 0.6208^{*}z_{12t-1} + 0.0398z_{11t-2} - 0.4488^{*}z_{12t-2} \\ &-0.0592^{*}\left(0.22z_{22t-1} + 0.34z_{32t-1} + 0.44z_{42t-1}\right) + 0.4810\left(0.44z_{32t-2} + 0.56z_{42t-2}\right),\end{aligned} \quad (11)$$

$$\begin{aligned}\hat{z}_{22t} &= 0.6353^{*}x_{b1t} + 0.4095^{*}x_{b2t} + 0.2894^{*}x_{b3t} + 0.0706x_{b4t} \\ &-0.3066^{*}x_{1t} + 0.0706x_{2t} + 0.7295^{*}x_{3t} + 0.5459^{*}x_{4t} \\ &-0.0273^{*}z_{21t-1} - 0.6610^{*}z_{22t-1} + 0.0061z_{21t-2} - 0.3111^{*}z_{22t-2} \\ &+0.0438\left(0.26z_{12t-1} + 0.32z_{32t-1} + 0.42z_{42t-1}\right) \\ &+0.0108\left(0.34z_{12t-2} + 0.31z_{32t-2} + 0.35z_{42t-2}\right),\end{aligned} \quad (12)$$

$$\begin{aligned}\hat{z}_{32t} &= 0.9280^{*}x_{b1t} + 0.5516^{*}x_{b2t} + 0.2045x_{b3t} + 0.1972x_{b4t} \\ &-0.4283^{*}x_{1t} + 0.0287x_{2t} + 0.8160^{*}x_{3t} + 0.7430^{*}x_{4t} \\ &-0.0336^{*}z_{31t-1} - 0.5046^{*}z_{32t-1} - 0.0182z_{31t-2} - 0.2896^{*}z_{32t-2} \\ &-0.434^{*}\left(0.35z_{12t-1} + 0.65z_{42t-1}\right) + 0.0015\left(0.46z_{12t-2} + 0.54z_{42t-2}\right),\end{aligned} \quad (13)$$

$$\begin{aligned}\hat{z}_{42t} &= 0.4287^{*}x_{b1t} + 0.2844^{*}x_{b2t} + 0.1516x_{b3t} + 0,1485x_{b4t} \\ &-0.3352^{*}x_{1t} - 0.0532x_{2t} + 0.2394^{*}x_{3t} + 0.3056^{*}x_{4t} \\ &-0.0029z_{41t-1} - 0.5603^{*}z_{42t-1} - 0.0032z_{41t-2} - 0.2333^{*}z_{42t-2} \\ &-0.0155\left(0.35z_{12t-1} + 0.21z_{22t-1} + 0.44z_{32t-1}\right) - 0.0167\left(0.49z_{12t-2} + 0.51z_{32t-2}\right).\end{aligned} \quad (14)$$

where $z_{nkt}$ is the $k$-th variable at $n$-th city at time $t$. In this research, $k$=1 is inflation and $k$=2 is money supply. While $n$=1 is Surabaya, $n$=2 is Malang, $n$=3 is Kediri and $n$=4 is Jember.

Equation (7) – (14) show that significant calendar variation variables from the eight equations vary. But there is one significant exogenous variable in all equations, namely $x_{b1t}$ with a positive coefficient. This shows that if Eid al-Fitr occurs in the first week, inflation and money supply in Surabaya, Malang, Kediri and Jember in the one month before Eid-al Fitr increase. The coefficient is quite large especially in the inflation equation, this shows that in a month before Eid-al fitr, inflation rises quite high if Eid occurs in the first week. The highest increase in inflation occurred in Surabaya. In addition, there are two calendar variation variables that affect money supply in Surabaya, Malang, Kediri and Jember, namely $x_{3t}$ and $x_{4t}$ with positive coefficients as well. This shows that if Eid al-Fitr occurs in the third or fourth week, then the money supply in the month of Eid al-Fitr increases. The highest coefficient for $x_{3t}$ and $x_{4t}$ occurs for the money supply in Surabaya. This means that the highest increase in the money supply at the time of Eid al-Fitr month occurred in Surabaya.

Equation (7) – (10) also shows that inflation in a city is related to the inflation of the city itself in the past. In addition, inflation in a city this month is also related to inflation of another city one or two months before, as happened in Surabaya and Malang. Based on Equation (11) – (14), it can also be seen that the current money supply in a city is related to the money supply in the city itself one and two months before. In addition, the money supply in a city is also related to the money supply in other cities as happened in Surabaya and Kediri. Meanwhile the money supply in Surabaya, Malang and Kediri was influenced by inflation in the city itself one month earlier.

## 4. Conclusion

In this study, it is concluded that the spatial weight does not significantly affect the estimation of the SpVAR model parameters with calendar variations for modeling inflation and money supply in Surabaya, Malang, Kediri and Jember. This study also shows that the money supply is influenced by inflation, but not vice versa. In addition, it can be shown the existence of space-time relationship on inflation and money supply and the influence of Eid al-Fitr on inflation and money supply.

### References

[1] Sumarminingsih E, Setiawan, Suharsono A, and Ruchjana B N 2019 Modeling Inflation and Money Supply using Spatial Vector Autoregressive Model with Calendar Variation: Restricted vs Non-restricted Coefficient *IOP Conference Series: Material Science and Engineering* **546** 052075.

[2] Prillantika J R, Apriliani E and Wahyuningsih N 2018 Comparison between GSTAR and GSTAR-Kalman Filter models on inflation rate forecasting in East Java *Journal of Physics: Conference Series* **974** 012039

[3] Sumarminingsih E, Matoha S, Suharsono A, and Ruchjana B N 2018 Spatial Vector Autoregressive Model with Calendar Variation for East Java Inflation and Money Supply *Appl. Math. Inf. Sci.* **12** 6 pp. 1157–1163.

[4] Bonar H, Ruchjana B N and Darmawan G 2017 Development of generalized space time autoregressive integrated with ARCH error (GSTARI–ARCH) model based on consumer price index phenomenon at several cities in North Sumatera province *AIP Conference Proceedings* **1827** 020009

[5] Sumarminingsih E, Setiawan, Suharsono A, and Ruchjana B N 2016 *International Conference on Theoretical and Applied Statistics*.

[6] Suhartono, Wahyuningrum S R, Setiawan and Akbar M S 2016 GSTARX-GLS model for spatio-temporal data forecasting *Malaysian J. Math. Sci.* **10** (S) pp. 91–103

[7] Apriliadara M, Suhartono and Prastyo D D 2016 VARI-X model for currency inflow and outflow forecasting with Eid Fitr effect in Indonesia *AIP Conference Proceedings* **1746**.

[8] Sumarminingsih E 2015 Spatio–Temporal Analysis of Inflation Rate in East Java, Indonesia: Vector Autoregressive Approach *Int. J. Appl. Math. Stat* **53** 16–23.

[9] Ahmad I S, Setiawan, Suhartono and Masun N H 2015 Forecasting of monthly inflow and outflow currency using time series regression and ARIMAX: The Idul Fitri effect *AIP Conference Proceedings* **1691** 050002

[10] Setiawan, Suhartono, Ahmad I S and Rahmawati N I 2015 Configuring calendar variation based on time series regression method for forecasting of monthly currency inflow and outflow in Central Java *AIP Conference Proceedings* **1691** 050024
[11] Wulansari R E, Suryanto E, Ferawati K, Andalita I and Suhartono 2014 Penerapan Time Series Regression with Calendar Variation Effect pada Data Netflow Uang Kartal Bank Indonesia Sebagai Solusi Kontrol Likuiditas Perbankan di Indonesia *Stat. Forum Teor. dan Apl. Stat.*, **14** (2) 59-68
[12] Suhartono and Subanar 2006 The Optimal Determination Of Space Weight in Gstar Model by Using Cross-Correlation Inference *Quantitative Methods* **2** (2) 45–53.
[13] Di Giacinto V 2010 On vector autoregressive modeling in space and time *J. Geogr. Syst.* **12** (2) 125–154
[14] Levin A, Lin C F and Chu C S J 2002 Unit root tests in panel data: asymptotic and finite-sample properties *J. Econom.* **108** (1) 1–24

IOP Conf. Series: Materials Science and Engineering 1115 (2021) 012089 doi:10.1088/1757-899X/1115/1/012089

# A study on group lasso for grouped variable selection in regression model

**E Sunandi[1,2], K A Notodoputro[2], and B Sartono[2*]**

[1]Statistics Program, Department of Mathematics, Universitas Bengkulu, Indonesia
[2]Department of Statistics, Institut Pertanian Bogor (IPB), Indonesia

*Corresponding author: bagusco@apps.ipb.ac.id

**Abstract.** Estimation of regression parameters using the Least Squares (LS) method could not be performed when the number of explanatory variables exceeds the number of observations. An approach that can solve the problem is the LASSO (Least Absolute Shrinkage and Selection Operator) method. This method produces a stable model but with slight bias as the trade-off. Yuan and Lin [6] introduced the Group LASSO method which can be used when there are grouped structure in the variables. This current paper provided a study of the performance of the Group LASSO method through a simulation with several different scenarios. Furthermore, the Group LASSO method was applied to the Human Development Index (HDI) data of Bengkulu Province in 2019. The simulation yieled that the Group LASSO analysis was better than LASSO in term of its Mean Squared Error of Prediction (MSEP), False Negative Rate (FNR) and R-Squared. In the application of the approach to the HDI data, our result was in line with the simulation results that the analysis of Group LASSO was better than LASSO with MSEP Group LASSO of 0.25 and R-Squared of 98%.

## 1. Introduction

Regression analysis is the most common method to examine the relationship between some explanatory variables (X) and a response variable (Y) [1]. The analysts frequently use least squares (LS) method which works to find model estimate by minimizing the sum of the squared errors. If X is the design matrix from the dataset, LS can be implemented when the X'X matrix is nonsingular [2]. For the case when the number of variables exceed the number of observations, it is impossible to have non-singular X'X so that LS could not work any longer. One way to solve this problem is the penalized regression method which estimates the parameters by minimizing of the sum of squared errors that are penalized [3]. The well-known penalized regression methods is the LASSO (Least Absolute Shrinkage and Selection Operator) method which was popularized by Tibshirani [4].

LASSO works by adding constraints to the LS method. The LASSO shrinks the solution of LS to zero even right to zero especially for the coefficients of less important variables. Therefore, LASSO could work as selection procedure to result simpler and easier modelto interpret. However, this approach leads some bias in the estimation. Previous research has applied the LASSO method to data that has many explanatory variables [5]. As a result, the regression coefficients generated from the LASSO method are abler to select explanatory variables than the common regression method. Apart from that, LASSO can also solve multicollinearity problems in regression analysis.

The original version of LASSO method has a limitation in the precision when it is applied in data with grouped variable. Yuan and Lin [6] introduced a new method called Group LASSO to handle this situation. This study provided a study of the performance of the Group LASSO method through a simulation with three different scenarios. We also implemented the Group LASSO method to the Human Development Index (HDI) data of Bengkulu Province in 2019 which includes several groups of variables as explanatory variables. Those variables are Expenditure per Capita Per Month (ECM), Expected Years of Schooling (EYS), Gross Regional Domestic Product (GRDP), and Life Expectancy Rate (LER) in Bengkulu Province [7]. Goodness of fit measures used in this study are Mean Squared Error of Prediction (MSEP), R-squared ($R^2$), False Discovery Rate (FDR), and False Negative Rate (FNR).

## 2. Method and Data

### 2.1 Method

*2.1.1 Standard Normal Variate (SNV) Method*

Standard Normal Variate (SNV) is the most common method for normalize the data with the following formula [8]

$$SNV_k = \frac{x_k - \overline{x_k}}{S_k}; \quad k = 1, \dots, p = \sum_{j=1}^{J} p_j, \tag{1}$$

where $SN_k$ is the k-th explanatory variable from the standardization, $x_k$ is the k-th explanatory variable before standardization, $\overline{x_k}$ is the mean of the k-th explanatory variables before standardization, and $S_k$ is the deviation of the explanatory variable before standardization. Normalization is needed to standardize the explanatory variables because the coefficient estimator obtained cannot be compared with the results if the original variable is not standardized. It is known as non-equivariant [9].

*2.1.2 Group LASSO*

In many problems, regression is used to find important explanatory factors in predicting response variables, where each explanatory factor can be represented by a group of variables/factors. Several studies on the Group LASSO have been conducted before [10-13]. Consider common regression model with J groups

$$Y = \sum_{j=1}^{J} X_j \beta_j + \varepsilon, \tag{2}$$

where Y is an $n \times 1$ vector, $\varepsilon \sim N(0, \sigma^2 I)$, $X_j$ is an $n \times p_j$ matrix corresponding to the j-th group and $\beta_j$ is a coefficient vector of size $p_j$, $j = 1, \dots, J$. Denoting $X = (X_1, X_2, \dots, X_J)$ and $\beta = (\beta'_1, \dots, \beta'_J)$ equestion (2) can be written as $Y = X\beta + \varepsilon$ .

LASSO method is a special case of equestion (2) when $p_1 = \cdots = p_J = 1$. According to Tibshirani [4], the LASSO estimator is defined as

$$\hat{\beta}^{lasso} = arg \min_{\beta} \left\{ \frac{1}{2N} \|\boldsymbol{y} - \boldsymbol{X}\beta\|^2 + \lambda \|\beta\|_1 \right\}, \tag{3}$$

where $\lambda$ is a tuning parameter and $\|.\|_1$ stands for the vector $l_1$-norm [4]. When X is orthonormal, the Group LASSO estimator with tuning parameter $\lambda$ can be given as [6]

$$\beta_j = \left(1 - \frac{\lambda\sqrt{p_j}}{\|X_j'Y\|}\right)_+ X_j'Y; j = 1, \ldots, J. \tag{4}$$

As $\lambda$ descends from $\infty$ ke 0, Group LASSO follows a piecewise linear solution path with changepoints at $\lambda = \frac{\|X_j'Y\|}{\sqrt{p_j}}; j = 1, \ldots, J.$

*2.1.3 Leave-One-Out Cross-Validation (LOOCV)*

Cross validation is used to determine the optimal $\lambda$ value so that the best model is obtained. LOOCV is a method that deals with cross validation [14]. The k-fold CV is a form of LOOCV. In the k-fold CV method, observations are randomly divided into k folds of relatively the same size. the first fold is set as the validation data while the other as the training data. The value of the Mean Squared Error $(MSE_1)$ is calculated based on the predetermined validation data. This step is repeated k times for each fold of observations that are different as validation data. Therefore, k estimated MSE values were obtained, $MSE_1, MSE_2, \ldots, MSE_k$. The formula of k-fold CV, i.e.

$$CV_k = \frac{1}{k}\sum_{l=1}^{k} MSE_l. \tag{5}$$

In this study data divided into $t$ training data sets and $v$ validation data sets where $n = t + v$ [9]. The training data set is used to determine the estimated β value. The validation data set is used to test the goodness of the model from the results of the predicted β value in the training data set. The cross-validation process used validation data.

*2.1.4 Goodness of Fit Measures*

The Goodness of fit measures that are used in this research are Mean Squared Error of Prediction (MSEP), R-squared ($R^2$), and several measures that are measured from the confusion matrix. The main measure used to see the goodness of the resulting model is MSEP, according to references from Yuan and Lin [6], who used the error value as the main measure when comparing Group LASSO and LASSO models. The MSEP value is calculated based on validation data. Meanwhile, the R-Squared value is used to see how far the explanatory variables in the resulting model can explain the response variable. The R-Squared value is calculated based on training data. Through this measure, it can be concluded that the best model between the LASSO and Group LASSO analyzes, i.e. with a smaller MSEP value. The MSEP formula is as follows [13]

$$MSEP = \sum_{i=1}^{v} \frac{(y_i - \hat{y}_i)^2}{v}, \tag{6}$$

where $v$ is the number of validation data, $y_i$ is the response variable and $\hat{y}_i$ is the predicted value of the response variable. Furthermore, $R^2$ is a percentage value that shows how much the explanatory variables in the model can explain the response variable. Its value is calculated based on training data. The formula of $R^2$ is [15]

$$R^2 = \frac{\sum_{i=1}^{t}(y_i - \hat{y}_i)^2}{\sum_{i=1}^{t}(y_i - \bar{y}_i)^2}. \tag{7}$$

**Table 1**. Confusion Matrix

| | | Actual data | |
|---|---|---|---|
| | | Positive | Negative |
| **Predicted data** | Positive | True Positive (TP) | False Positive (FP) |
| | Negative | False Negative (FN) | True Negative (TN) |
| | | TP+FN | FP+TN |

Meanwhile, the regression parameters ($\beta$) are set with the classification of non-zero beta and zero beta. In this study, it was also investigated whether non-zero beta would be estimated by non-zero beta or conversely. The accuracy of the regression parameter estimator is measured by using confusion matrix. Confusion matrix can be interpreted as a tool that has a function to analyze whether the model is good at recognizing the estimator values from different classification classes (see Table 1) [16]. In this study, the focus is the beta classification (non-zero beta and zero beta). In a confusion matrix, there are four terms as a representation of the results of the classification process. They are True Positive (TP), True Negative (TN), False Positive (FP) and False Negative (FN). The True Negative (TN) value is the number of zero beta detected correctly. The False Positive (FP) is zero beta but is detected as non-zero beta. True Positive (TP) is the number of non-zero beta detected correctly. False Negative (FN) is number of non-zero beta, but detected as zero beta [17].

There are two values that are measured using the confusion matrix, i.e the False Discovery Rate (FDR) and the False Negative Rate (FNR). The FDR value describes the probability of the zero-beta error that is suspected as non-zero beta. The FDR value can be obtained by Equation 8. The FNR value shows the probability of non-zero beta error which is estimated to be zero beta. The FNR value is obtained by Equation 9. The smaller the FDR and FNR values describe the method used the better in the parameter estimation [18].

$$FDR = \frac{FP}{FP + TP}, \tag{8}$$

$$FNR = \frac{FN}{TP + FN}. \tag{9}$$

## 2.2 Data

The data used in this study are simulation data and empiric data. Simulation provided the performance of the LASSO and Group LASSO methods. The simulation scenarios can be seen in Table 2.
The formula used to generate Y data is as follows.

- Simulation 1: three continuous groups are represented by a third-order polynomial and 3 categorical groups is trichotomized as 0, 1 or 2.

$$Y = 0.5 + 0.940X_1 + 0.422X_1^2 + 0.822X_1^3 + 0825X_2 + 0.786X_2^2 + 0.361X_2^3 + 0.956X_3 + 0.473X_3^2 + 0.453X_3^3 + 0.359[Z_1 = 1] + 0.532[Z_1 = 2] + 0.002[Z_2 = 1] + 0.533[Z_2 = 2] + 0.824[Z_3 = 1] + 0.577[Z_3 = 2] + \text{epsilon}.$$

- Simulation 2: three continuous groups are represented by a third-order polynomial and 3 categorical groups is trichotomized as 0, 1 or 2; 3 continuous groups are represented by a third-order polynomial with zero beta.

$$Y = 0.5 + 0.764X_1 + 0.889X_1^2 + 0.634X_1^3 + 0.922X_2 + 0.472X_2^2 + 0.251X_2^3 + 0.311X_3 + 0.419X_3^2 + 0.420X_3^3 + 0.746[Z_1 = 1] + 0.200[Z_1 = 2] + 0.856[Z_2 = 1] + 0.030[Z_2 = 2]0.901[Z_3 = 1] + 0.893[Z_3 = 2] + \text{epsilon}.$$

- Simulation 3: three continuous groups are represented by a third-order polynomial and 3 categorical groups is trichotomized as 0, 1 or 2; 3 continuous groups are represented by a third-order polynomial with zero beta.

$$Y = 0.5 + 0.700X_1 + 0.462X_1^2 + 0.499X_1^3 + 0.490X_2 + 0.461X_2^2 + 0.155X_2^3 + 0.056X_3 + 0.819X_3^2 + 0.329X_3^3 + 0.179[Z_1 = 1] + 0.290[Z_1 = 2] + 0.839[Z_2 = 1] + 0.389[Z_2 = 2] + 0.541[Z_3 = 1] + 0.074[Z_3 = 2] + \text{epsilon}.$$

**Table 2.** The Simulation Scenarios

| **Simulation 1 ($\frac{p}{n} < 1$)** | |
|---|---|
| Number of observations (n) | 100 |
| Number of variables (p) | 18 |
| Number of groups (J) | 6 |
| Number of simulation samples (S) | 1000 |
| **Simulation 2 ($\frac{p}{n} < 1$)** | |
| Number of observations (n) | 50 |
| Number of variables (p) | 36 |
| Number of groups (J) | 12 |
| Number of simulation samples (S) | 1000 |
| **Simulation 3 ($\frac{p}{n} > 1$)** | |
| Number of observations (n) | 30 |
| Number of variables (p) | 36 |
| Number of groups (J) | 12 |
| Number of simulation samples (S) | 1000 |

Meanwhile, empiric data was obtained from the Statistics of Bengkulu Province in the book "Bengkulu Province in Figures 2020". As the unit of observation are districts/cities in Bengkulu Province. As for the variables used in this study, as a response variable, namely the Human Development Index (HDI) of Bengkulu Province in 2019. The Human Development Index (HDI) is a summary measure of average achievement in key dimensions of human development: a long and healthy life, being knowledgeable, and have a decent standard of living. Based on the HDI compiler indicators, several groups of variables used are Expenditure Per Capita Per Month (ECM), Expected Years of Schooling (EYS), Gross Regional Domestic Product (GRDP), and Life Expectancy Rate (LER) in Bengkulu Province [7]. All groups are continuous variables. The definitions of the variables of each group can be seen in Table 3.

**Tabel 3**. Details of the Variables in the Group

| Y | Group | Definition |
|---|---|---|
| HDI | 1. Expenditure Per Capita Per Month (ECM) | Costs incurred (in rupiah) for the consumption of all household members for a month are divided by the number of household members. |
| | 2. Expected Years of Schooling (EYS) | Length of schooling (in years) that the child is expected to experience at a certain age in the future. |
| | 3. Gross Regional Domestic Product (GRDP) | Ability of an area to create the output (value added) at a certain time (in billions). |
| | 4. Life Expectancy Rate (LER) | The average life year that will still be lived by someone who has reached the age of x, in a certain year, in a situation of mortality that prevails in his community. |

Based on a preliminary analysis that there were non-linear effects of both HDI and each explanatory variable. it is known that all groups of variables are represented by a third-order polynomial (alpha 5.5%). This can be seen based on the p-value of each effect in Table 4 [19]. So that for modeling with the Group LASSO in the application of HDI data, polynomial effects up to order 3 of each group are used as explanatory variables.

**Table 4**. The Significance of the polynomial effects Explanatory Variables

| Group | p-value | | |
|---|---|---|---|
| | Linear | Quadratic | Cubic |
| ECM | 0.001 | 0.001 | 0.001 |
| EYS | 0.000 | 0.000 | 0.000 |
| GRDP | 0.000 | 0.000 | 0.000 |
| LER | 0.054 | 0.001 | 0.001 |

*2.2.1 Data Analysis Procedures*

This research method is carried out in two stages. The first is to examine the performance of the LASSO and Group LASSO methods through simulation. The second stage is the application of the LASSO and Group LASSO methods to the 2019 Bengkulu Province HDI data. The simulation algorithm are as follows.

1. Determined the number of groups J.
2. Determined $p = \sum_{j=1}^{J} p_j$ (number of variables), n (number of observations), S (number of simulation samples).
3. Generated dummy variables with size $nxa$ for each of 3 categories.
4. Generated X $nxb$ $(a + b = p)$ matrix from Normal Multivariate distribution (0, sigma). set the correlation between X is high $(rho = 0.9)$. Furthermore, the matrix X is standardized becomes X*.
5. Generated $\beta \sim uniform(0,1)$.
6. Generated $epsilon \sim N(0,0.3)$.
7. Set $\beta_0 = 0.5$.
8. Calculated the generated Y $nx1$ vector with the formula $Y = 0.51_n + X^*\beta + epsilon$.
9. Analyzed Grup LASSO and LASSO:
   a) selected 80% training data and 20% testing data,
   b) determined the best lambda, and
   c) estimated the coefficient of each explanatory variable.
10. Determined the goodness of fit measures.
11. Interpretated results.

Analogous to the simulation flow, the real data analysis procedure uses the Group LASSO and LASSO method as follows.

1. Data exploration.
2. Data standardization.
3. Analyzed Grup LASSO and LASSO:
   a) selected 80% training data and 20% testing data,
   b) determined the best lambda, and
   c) estimated the coefficient of each explanatory variable.
4. Determined the the goodness of fit measures.
5. Interpretated results.

# 3. Results and Discussion

## 3.1 Simulation Results

The simulation was designed to assess the performance of the Group LASSO method with LASSO as a comparison method. In this study, a simulation was carried out with three scenarios, each generating 1000 times of simulation data. The detailed simulation results are summarized in Table 5. In simulation 1, all beta generated are non-zero beta at condition $p < n$. Based on the simulation results can be seen that FDR and FNR values in the LASSO and Group LASSO analysis are the same, i.e 0%. Group LASSO and LASSO estimated all beta values that are not equal to zero. If seen from the R-Squared value, both analyzes have the same ability to explain the diversity of the response variables very well ($R^2 = 98\%$). When viewed from the MSEP value, the Group LASSO analysis has a smaller value (MSEP = 1.76) than compared to LASSO.

In simulation 2 at the condition of $p < n$, the beta parameter was set as non-zero beta and zero beta, with a ratio of 50:50. The results of simulation 2 were different from simulation 1. For the Group LASSO analysis, the FDR and FNR values were similar, i.e 37.50%. It means, the Group LASSO was quite capable of classifying the beta estimator according to the true beta conditions. Meanwhile, LASSO was very good estimating the non-zero beta correctly. This was indicated by the FDR value of 0%. The FNR value was still relatively large (42.86%). This indicates that LASSO was weak in estimating zero beta correctly. From the MSEP value can be seen that the Group LASSO analysis has a much smaller MSEP value (MSEP = 18.49%) than compared to LASSO (MSEP = 28.75%) with the same R-squared value ($R^2 = 96\ \%$).

Simulation 3 is designed to be analogous to simulation 2, but the condition is $p > n$. In LASSO analysis, the probability of non-zero beta error estimated correctly is small (FDR = 6.25%). However, the probability of non-zero beta estimated incorrectly is quite large (53.13%). This indicates that LASSO has limitations in estimating zero beta. Furthermore, it can be seen from the R-Squared value, the Group LASSO analysis has a slightly larger value ($R^2 = 85\%$) than LASSO ($R^2 = 83\%$). Also, similar to simulations 1 and 2, the Group LASSO analysis has a much smaller MSEP value, even half of the LASSO MSEP value. Overall in 3 simulations, analysis Group LASSO has better performance than LASSO.

**Table 5**. The Goodness of Fit Measures

| | | Simulation 1 | Simulation 2 | Simulation 3 |
|---|---|---|---|---|
| Group LASSO | FDR | 0% | 37.50% | 37.50% |
| | FNR | 0% | 37.50% | 41.18% |
| | MSEP | 1.76 | 18.49 | 84.46 |
| | R-Squared | 0.98 | 0.96 | 0.85 |
| LASSO | FDR | 0% | 0% | 6.25% |
| | FNR | 0% | 42.86% | 53.13% |
| | MSEP | 2.26 | 28.75 | 160.17 |
| | R-Squared | 0.98 | 0.96 | 0.83 |

## 3.2 Empirical Application to HDI Data

The application of the regression model using the LASSO and Group LASSO methods was carried out at the 2019 HDI of Bengkulu Province. This application was adjusted to the conditions in simulation 3, i.e $p > n$. There are 10 the city/district as units of observation ($n = 10$) with 12 explanatory variables ($p = 12$) which are divided into 4 groups. Next, Table 6 shows a summary statistical description of the variables used. In the Table 6, it can be seen that the HDI scores in Bengkulu Province are in the range of 66.69-80.35.

**Table 6**. Statistics Description of the variables

| | HDI | ECM | EYS | GRDP | LER |
|---|---|---|---|---|---|
| Mean | 69.29 | 10138.00 | 13.36 | 7208.70 | 67.38 |
| Median | 67.90 | 9987.50 | 13.00 | 4883.16 | 67.79 |
| Standard Deviation | 4.10 | 1627.26 | 1.00 | 5947.66 | 1.80 |
| Minimum | 66.69 | 8209.00 | 12.56 | 3346.49 | 63.12 |
| Maximum | 80.35 | 14030.00 | 16.01 | 23200.95 | 70.04 |

Then the LASSO and Group LASSO analyzes were carried out. From the results of the LASSO analysis, several variables affect the HDI score in Bengkulu Province. They are Expenditure Per Capita Per Month (ECM), Expected Years of Schooling (EYS) and GRDP. This can be seen from the coefficient of variables. The variable that has a non-zero coefficient is the variable that affects the HDI Score in Bengkulu Province. All variables that explain the HDI score in Bengkulu Province have positive coefficients. This shows that the higher value of the explanatory variable reflects higher HDI score in Bengkulu Province.

**Table 7**. Parameter Estimator in the LASSO and Group LASSO

| | | LASSO | GROUP LASSO |
|---|---|---|---|
| | Intersept | 69.29 | 69.29 |
| ECM | Linear | . | 0.56 |
| | Quadratic | . | 0.53 |
| | Cubic | 0.21 | 0.51 |
| EYS | Linear | . | 0.73 |
| | Quadratic | . | 0.72 |
| | Cubic | 1.17 | 0.72 |
| GRDP | Linear | 1.66 | . |
| | Quadratic | . | . |
| | Cubic | . | . |
| LER | Linear | . | 0.16 |
| | Quadratic | . | 0.16 |
| | Cubic | . | 0.16 |

The coefficient value of each explanatory variable in the LASSO and Group LASSO analysis can be seen in Table 7. In line with the results of the LASSO analysis, the Group LASSO results showed that the group of variables that can be indicator the HDI score in Bengkulu Province is all variables in the Expenditure Per Capita Per Month (ECM), Expected Years of Schooling (EYS), and Life Expectancy Rate (LER). These variables are polynomial effects up to order 3 of each group.

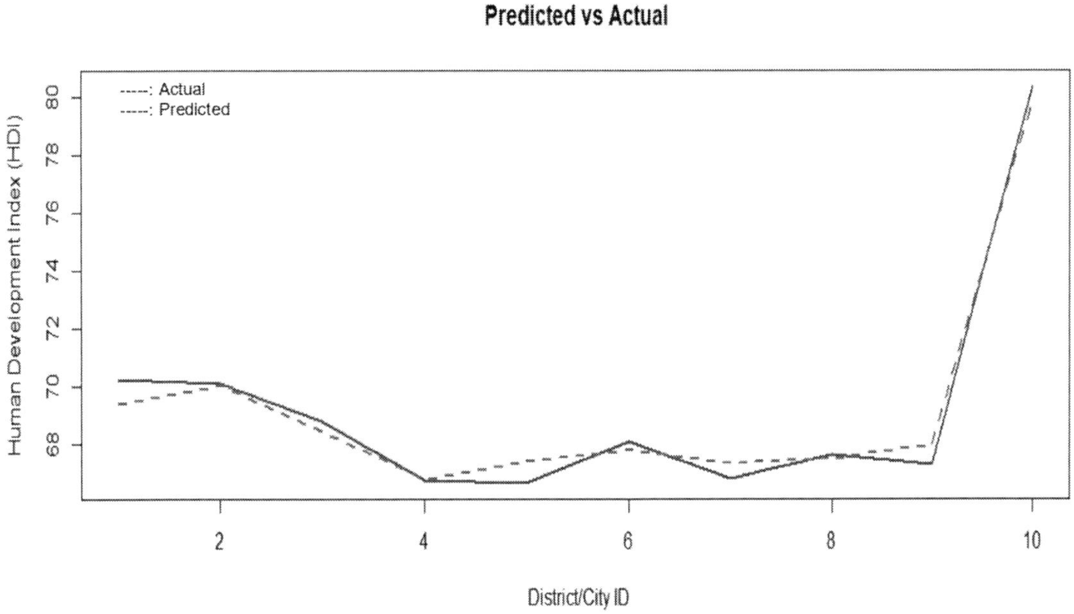


**Figure 1.** Distribution of Actual and Predicted HDI for Districts/Cities in Bengkulu Province

Chen et al. [11] conducted a recognized regression analysis through the Group LASSO and LASSO methods in identifying factors associated with tuberculosis cases in West Java. The results showed that the Group LASSO method is better than the LASSO [10]. This is indicated by the MSE value of the Group LASSO which is smaller than the LASSO. Furthermore, Yunus et al. [10] conducted a study on the characteristics of the Group LASSO in handling high correlated data. The simulation results showed the Group LASSO was better than LASSO and LS technique when group of is present included in a group [11].

In this further discussion, we see the significant differences between LASSO and Group LASSO. LASSO model uses far fewer variables which is not used in Group LASSO model. The measure of the goodness of these two models is presented in Table 8. Based on the results of the 10-fold cross validation, the lambda ($\lambda$) was 0.178. The model obtained by the Group LASSO analysis had an MSEP value of 0.25. This value was smaller than the MSEP LASSO value (1.40). Meanwhile, the Group LASSO R-squared value was 98%. This means that through Group LASSO analysis the variables used are able to explain HDI diversity very well. This is in line with the predicted vs actual comparison graph in Figure 1. In the figure, it can be seen that the predicted data pattern is almost the same as the actual data.

**Table 8**. Goodness of fit Measures of Group LASSO and LASSO Analysis

| Method | MSEP | $R^2$ |
|---|---|---|
| Group LASSO | 0.25 | 0.98 |
| LASSO | 1.40 | 0.91 |

It can be seen from Table 8 that the Group LASSO model is better than LASSO with a lower MSEP value and a higher R-Squared. Referring to this value, it can be said that the Group LASSO is better than LASSO.

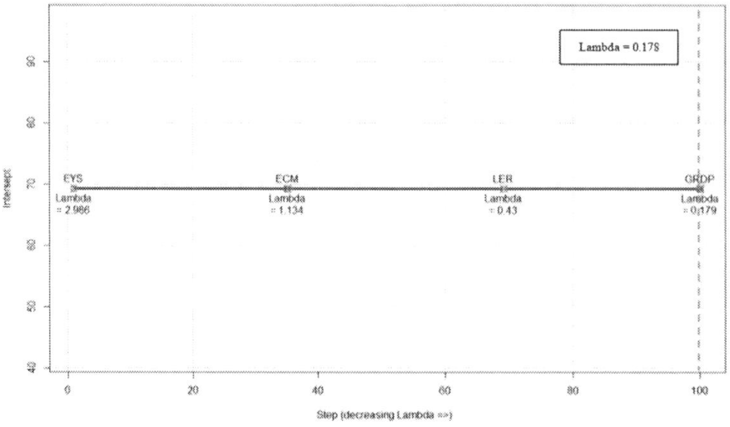


**Figure 2.** All possible lambda ($\lambda$) in the Group LASSO analysis

The Group LASSO analysis can also determine which group of variables has the most influence and does not affect the HDI score in Bengkulu Province. This can be seen in the graph presented in Figure 2. The red line in the graph above, shows all possible lambda. The horizontal axis (x) represents the step, that is, the index for each possible lambda, displayed from 0 to 100, where as the step index increases, the lambda value decreases further. The vertical axis is the intercept value in the model, during the x-step [8]. The greater the lambda value, the more explanatory variables are not included in the model. Based on Figure 2, if the lambda value is $1.134 < \lambda \leq 2.986$, then only the EYS variable group is included in the model. If the lambda value is $0.430 < \lambda \leq 1.134$, then the EYS and ECM variable groups will be included in the model. If the lambda value is $0.179 < \lambda \leq 0.430$, then the EYS, ECM, and LER variable groups will be entered in the model. Finally, if the lambda value is $\lambda \geq 0.179$, then all of the grouped variables will be included in the model. Based on the results of the 10-fold cross validation, the best lambda value that minimizes the MSEP average value, and is used is 0.178. This resulted in the GRDP variable group not being included in the Group LASSO model. These results also indicate that the group order of the most important variables on HDI score in Bengkulu Province is Expenditure Per Capita Per Month (ECM), Expected Years of Schooling (EYS), and Life Expectancy Rate (LER).

## 4. Conclusion

From the simulation results in 3 scenarios (when $p < n$ and $p > n$) showed that the Group LASSO was better than LASSO. This can be seen from the value of MSEP and FNR Group LASSO which were smaller than LASSO and the value of R-Squared Group LASSO which was greater than LASSO. In the application of HDI data, it is in line with the simulation results that the analysis of Group LASSO was better than LASSO with MSEP Group LASSO of 0.25 and R-Squared of 98%. Based on the best lambda value (0.178), the order of the most important to insignificant group of variables, which was an indicator of the HDI score in Bengkulu Province was Expenditure Per Capita Per Month (ECM), Expected Years of Schooling (EYS), and Life Expectancy Rate (LER).

## References

[1] Draper N R and Smith H 1998 Applied Regression Analysis New York: John Wiley Sons
[2] Setiawan and Sutikno 2010 Bayes Wavelet Regression Approach to Solve Problems in Multivariable Calibration Modeling *IPTEK: The Journal for Technology and Science* **21** (2) 1-8

[3] Hastie T, Tibshirani R and Wainwright M 2015 Statistical Learning with Sparsity: The Lasso and Generalizations London: CRC Press
[4] Tibshirani R 1996 Regression Shrinkage and Selection Via the Lasso *Journal of the Royal Statistical Society. Series B (Methodological)* **58** (1) 267-288
[5] Rajaratnam B, Roberts S, Sparks D, Yu H 2019 Influence Diagnostics for High-Dimensional Lasso Regression Journal of Computational and Graphical Statistics **28** (4) 877-890
[6] Yuan M and Lin Y 2006 Model selection and estimation in regression with grouped variables *J. R. Stat. Soc. Ser. B* **68** (1) 49-67
[7] Kuswardani D A 2020 *Bengkulu Province in Figures 2020* Bengkulu: Badan Pusat Statistik (BPS)
[8] Grisanti E, Totska M, Huber S, Calderon C K, Hohmann M, Lingenfelser D and Otto M 2018 Dynamic localized SNV, Peak SNV, and partial peak SNV: Novel standardization methods for preprocessing of spectroscopic data used in predictive modeling *Journal of Spectroscopy* **2018** 5037572
[9] James G, Witten D, Hastie T and Tibshirani R 2013 *An Introduction to Statiscal Learning with Applications in R* Berlin: Springer
[10] Chen S, Notodiputro K A, Rahardiantoro S 2020 Penerapan Analisis Lasso dan Group Lasso dalam Mengidentifikasi Faktor-Faktor yang Berhubungan dengan Tuberkulosis di Jawa Barat Indonesia *Indonesian Journal of Statistics and Its Applications* **4** (1) 39-54
[11] Yunus M, Saefudin A, Soleh A M 2017 Characteristics of group LASSO in handling high correlated data *Applied Mathematical Sciences* **11** (20) 953-961
[12] Rahmat H S, Wijayanto H, Afendi F M, Sartono B, Anisa R and Septianingsih D A 2018 Variable-Group Selection on Estimated Metabolites of Curcuma aeruginosa Related to Antioxidant Activity by Using Group Lasso Regression *International Conference on Mathematics and Mathematics Education 2018, Advances in Social Science, Education and Humanities Research* **285** 128-130
[13] Rachmawati R N, Pusponegoro N H, Muslim A, Notodiputro K A, Sartono B 2017 Group LASSO for Rainfall Data Modeling in Indramayu District, West Java, Indonesia *Procedia Computer Science* **116** 190-197
[14] Riau B P R, Afendi F M, Anisa R 2019 Selection of Compound Group to Identify the Authenticity One of Jamu Product Using the Group Lasso for Logistic Regression *Journal of Physics: Conference Series* **1341** 092020
[15] Darlington R B and Hayes A F 2017 *Regression Analysis and Linear Models: Concepts, Application and Implementation* New York: The Guilford Press
[16] Han J, Kamber M, Pei J 2012 *Data Mining: Concepts and Techniques* Netherland: Elsevier Inc.
[17] Narkhede S 2018 Understanding Confusion Matrix *Towards Data Science* **180** (1) 1-12
[18] Trevethan R 2017 Sensitivity, Specificity, and Predictive Values: Foundations, Pliabilities, and Pitfalls in Research and Practice Front Public Health **5** (307) 1-7
[19] Widhiarso W 2010 *Catatan pada Uji Linieritas Hubungan Fakultas Psikologi Universitas Gadjah Mada (UGM)* Yogyakarta: UGM

# Propagation and Characteristic of Tsunami Wave in Lembeh Strait, North Sulawesi, Due to the Earthquake in the Molucca Sea

**G H Tamuntuan*[1], B B Mambu[1], G Pasau[1], A T Mandagi[2], D P Pandara[1], and Ferdy[1]**

[1]Department of Physics, Faculty of Mathematics and Natural Sciences, Sam Ratulangi University, Jl. Kampus Unsrat, Manado – Indonesia
[2]Department of Civil Engineering, Faculty of Engineering, Sam Ratulangi University, Indonesia

*Corresponding author: gtamuntuan@gmail.com

**Abstract.** Modeling the propagation of a tsunami triggered by an earthquake with a hypocenter in the double subduction of the Maluku Sea using the TUNAMI-N2 program has been carried out. The objectives of this research are to find out the model of tsunami propagation towards Bitung City, how the effect of Lembeh Island on tsunami waves, and the characteristics of the waves that occur in the Lembeh Strait. The modeling uses two earthquake scenarios with different magnitudes which are 7.5 *Mw* (scenario A) and 8.0 *Mw* (scenario B). The results show that the wave propagation model towards Bitung City in the two scenarios is relatively the same, but has a different arrival time and wave height depending on the earthquake energy of the tsunami generator. The arrival time of the tsunami waves in Lembeh strait area, the strait between Bitung City and Lembeh Island, was relatively slower indicating that Lembeh Island could serve as a tsunami barrier for Bitung City. The characteristics of the tsunami waves in Lembeh Strait are also different from the areas outside the strait. This is due to the effects of diffraction and wave superposition that occur in the strait area. The characteristics of the waves that occur inside and outside the Lembeh Strait are important things to know in order to determine a more optimal tsunami disaster mitigation strategy in Bitung City.

## 1. Introduction

North Sulawesi is one of the disaster-prone areas in Indonesia. The Molucca Sea, which is located in the eastern part of this region, is a tectonically complex place and has high earthquake intensity [1]. Seismic activity in the Molucca Sea occurs due to the double subduction of the Molucca Sea plate, where to the west of this plate is sub ducted under the Sangihe plate and to the east under the Halmahera plate [2]. Until 2000, this region was the host of 12.6% of destructive earthquakes and about 30% of tsunami cases that have occurred in Indonesia [1]. This makes the area around the Molucca Sea as a zone that is very potential to be affected by the tsunami disaster, so that various mitigation efforts need to be done.

The city of Bitung, the largest port city in North Sulawesi, is located on the east coast of the northern peninsula of Sulawesi Island, facing the Molucca Sea. The uniqueness of this city is the presence of Lembeh Island between the city and the Molucca Sea. The interesting thing is that Bitung City and Lembeh Island are only bordered by the relatively narrow Lembeh Strait. Although the shape, size and distance of the island from the coast in relation to the characteristics of tsunami waves have been studied [3, 4], the study of the effects of the existence of islands which are relatively close to the coast and the characteristics of the interference of tsunami waves that occur in relatively narrow is very rare.

One way of studying the propagation pattern and characteristics of a tsunami wave is through modeling. TUNAMI-N2 is a model and software developed by the Disaster Control Research Center, Tohoku University, Japan [5]. This model has been widely used for the tsunami simulation process in various places in the world [6, 7, 8]. In this study, a tsunami propagation model triggered by an earthquake in the Molucca Sea was carried out using TUNAMI-N2. Further analysis has been carried out on a number of tide gauges or imaginary points in Bitung City to determine the characteristics of the tsunami waves that occur in the open area (facing directly to the Molucca Sea) and along the Lembeh Strait, an area obstructed by Lembeh Island.

## 2. Material and Methods

Tsunami wave modeling and propagation simulations are carried out using the TUNAMI-N2 software. The data input is based on two earthquake scenarios, each with a magnitude of 7.5 *Mw* and 8.0 *Mw*. The epicenter of the earthquake is assumed to be in the Molucca Sea and opposite the City of Bitung, which is at coordinates 1.239 ° N and 126,213 ° East. The earthquake is assumed to occur at a depth of 10 km and form a fault with a strike equal to 208 °, dip equal to 53 °, slip angle equal to 106 °. The fault parameter data used in the study were calculated based on earthquake magnitude and supported by marine bathymetry data around Sulawesi Island obtained from GEBCO (General Bathymetric Chart of the Ocean) and topographic data of Bitung City from SRTM (Shuttle Radar Topograhy Mission). Subsequent modeling was carried out on four different layers as shown in Figure 1.

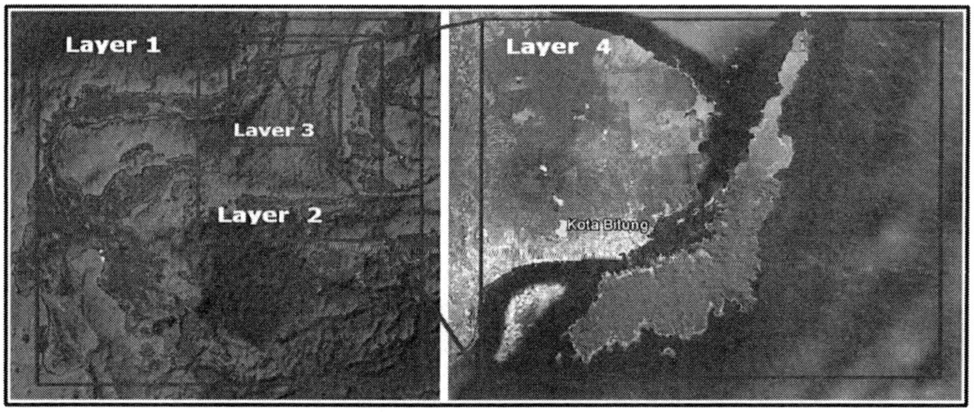


**Figure 1.** Modeling layers for analysis of tsunami propagation in Bitung City. Layer four shows Bitung City with Lembeh Island in front.

To obtain the height and arrival time of the tsunami waves in the Bitung area, ten tsunami wave observation points or tide gauges have been added as illustrated in Figure 2. These ten points cover several villages on the coast of Bitung City which are inhabited by many residents, namely Girian Bawah (TG1), Wangurer Barat (TG2), Wangurer Timur (TG2), Paceda (TG3), Madidir Unet (TG3), Madidir Ure (TG4) ), Madidir Weru (TG5), Pakadoodan (TG6), Bitung Barat One (TG6), Kadoodan (TG6), Bitung Timur I (TG7), Bitung Timur (TG8), Winenet Dua (TG9), Aertembaga I (TG10) and

Aertembaga II (TG10). TG1 to TG6 are outside the Lembeh Strait, while TG7 to TG10 are inside the strait.

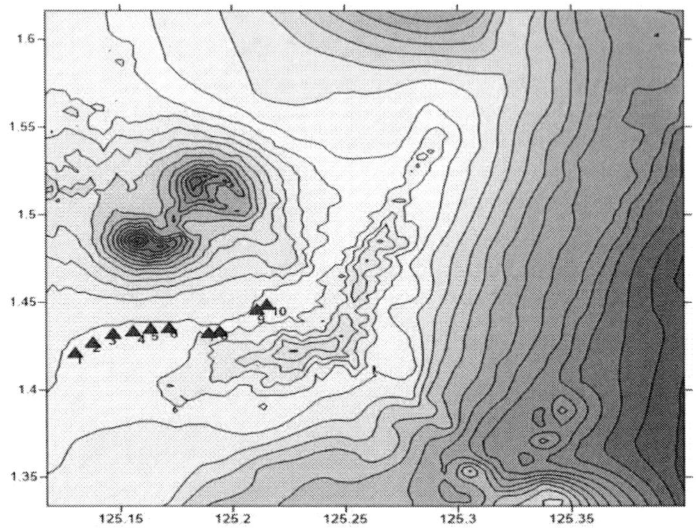


**Figure 2.** Tide Gauge at 10 points in Bitung City.

## 3. Results and Discussion

The simulation process begins by calculating the fault parameters using the empirical equation of Wells and Coppersmith (1994) [9]. Although there are quite a number of regression equations between magnitude and fault parameters [10], the Wells and Coppersmith equations are more commonly used in previous studies [11]. The fault parameters based on the two earthquake magnitude scenarios developed in this research are presented in Table 1. The simulation results with TUNAMI-N2 show that the tsunami propagation pattern is relatively the same. The difference is that the greater the magnitude of the earthquake that triggers the tsunami, the greater the speed of propagation of the waves. Figure 3 illustrates the wave propagation pattern based on Scenario B at four layers at different times. In the example presented, it can be seen that until the 17th minute (1,020 s) the part of Bitung City which is in the inner area of the Lembeh Strait has not been relatively affected by the tsunami waves, while the part of the city which is outside the strait has been attacked by the tsunami tidal waves. This indicates that the existence of Lembeh Island is very effective in reducing the arrival time of tidal waves, especially in the city of Bitung which is in the Lembeh Strait.

**Table 1.** Earthquake scenario and fault parameters

| Scenario / Magnitude ($Mw$) | Fault Parameter* | | |
|---|---|---|---|
| | Length (km) | Width (km) | Slip (m) |
| A / 7.5 | 73.28 | 29.17 | 7.35 |
| B / 8.0 | 151.36 | 46.77 | 14.50 |

*** calculated based on the Wells and Coppersmith equation (1994)**

In Scenario A, the maximum recorded tsunami wave height ($H_{max}$) detected on the entire tide gauge varies between 0.78 m (tide gauge 10) to 3.56 m (tide gauge 6), while the wave arrival time varies between 900 s on tide gauge 1 and the longest. which is 1,810 s on the tide gauge 9. In TG1 - TG6, the peak of the first wave arrived at the 1,200 to 1,290 second, while in the TG7 - TG10 it arrived at the 1,460 to 2,050 second.

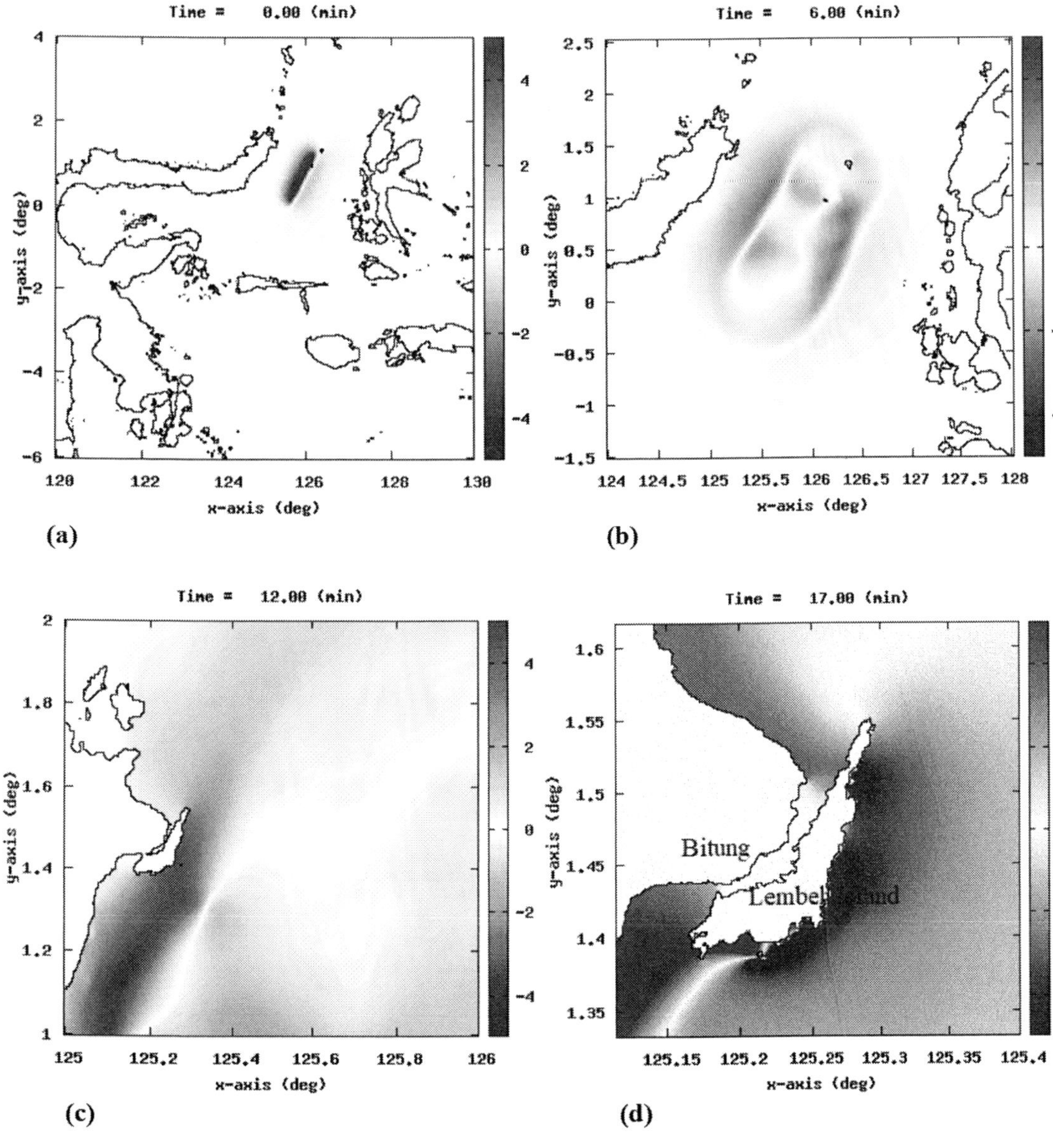


**Figure 3**. (a) Tsunami onset at 0 minutes, (b) Tsunami propagation at 6 minutes, (c) Tsunami propagation at 12th minute and (d) Tsunami reaches Bitung City at 17th minute

In Scenario B, the $H_{max}$ varies between 3.66m (on tide gauge 9) to 14.98m (on tide gauge 6). The arrival time of the wave on tide gauge 1 occurs at 710 seconds and on tide gauge 10 at 1.590 seconds. The variation of the arrival time of the first wave peaks at TG1 - TG6 is 1180 s to 1,260 s. The arrival time of the first wave peaks at TG7 and TG8 is around 1,470 s, while in TG9 and TG10 it is around 1,920 s. The difference in earthquake energy in the two scenarios causes $H_{max}$ in Scenario B to be 4.2

IOP Conf. Series: Materials Science and Engineering 1115 (2021) 012090 doi:10.1088/1757-899X/1115/1/012090

times higher than in Scenario A. In general, both scenarios show three tsunami waves occurring over a period of 80 minutes. However, in contrast to scenario A, the waves recorded in TG1 - TG8 in scenario B are relatively in the same phase, except that in TG9 and TG10 they have a phase difference of about 180 ° to the waves in TG1 - TG8.

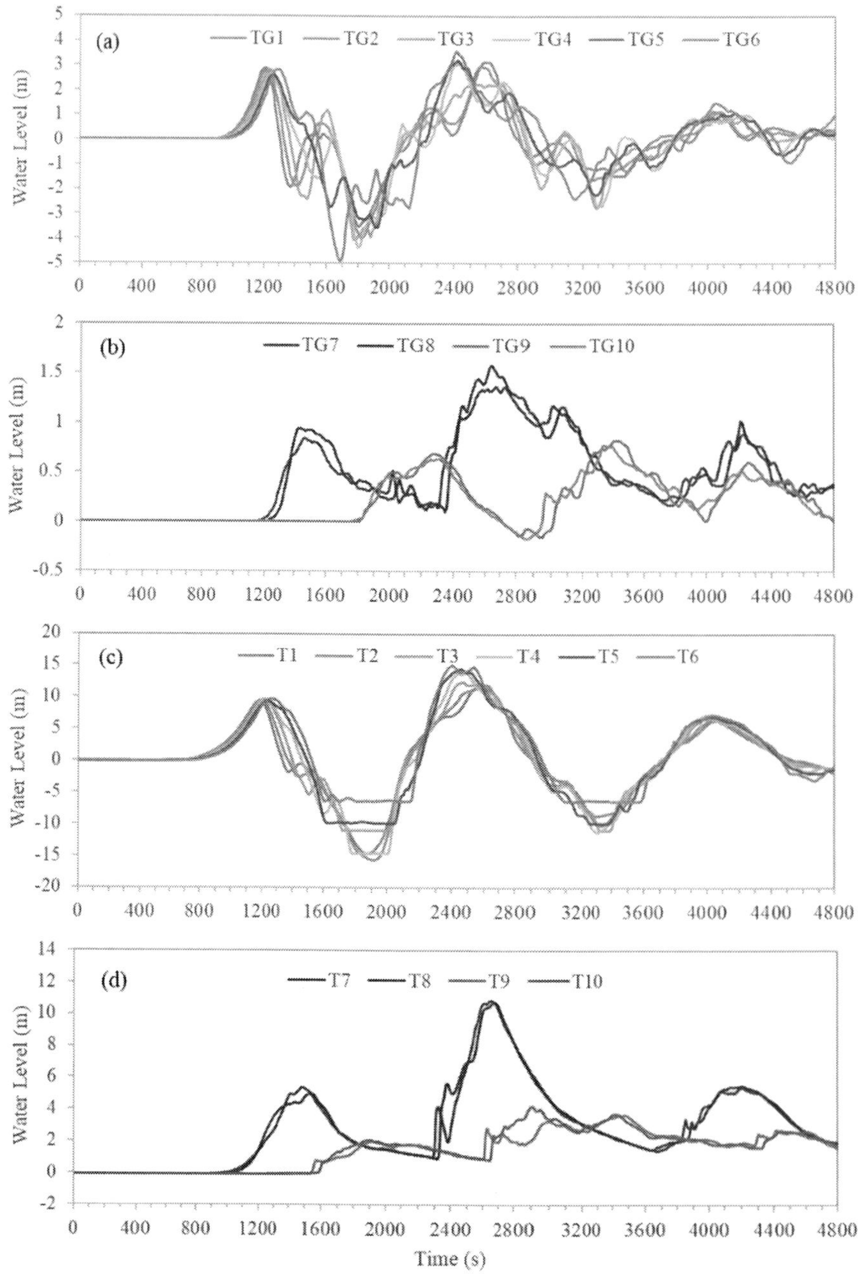


**Figure 4**. Water level fluctuations recorded in the tide gauge (a) TG1 - TG6 in Scenario A, (b) TG7 - TG10 in Scenario A, (c) TG1 - TG6 in Scenario B, and (d) TG7 - TG10 in Scenario B.

Waves passing through the island will experience diffraction. The case in this study shows that the direction of the tsunami wave propagation is relatively perpendicular to Lembeh Island so that it experiences diffraction on both sides of the island. Figure 3d shows the diffraction wave moving in from the northeast to the southwest and vice versa. The water level fluctuation pattern on an imaginary 10 tide gauge is shown in Figure 4. The water level fluctuation due to the tsunami waves at the 6 tide gauge outside the Lembeh Strait, namely TG1 - TG6, shows the cycle of the tide and low tide levels which oscillate against the still water level. In Scenario A, the arrival times of the tsunami waves peaks at the tide gauges are relatively different (Figure 4a). The stronger the earthquake that triggered the tsunami, the more uniform the water level fluctuation pattern due to the tsunami on the tide gauge 1 - 6 and the smaller the difference in the arrival time of the wave crests (Figure 4c).

Figures 4b and 4d are the patterns of water level fluctuation due to the tsunami waves on the tide gauge 7 - 10 which is located relative to the Lembeh Strait. Unlike the 1 - 6 tide gauge, the sea level on the 7 - 10 tide gauge fluctuates above the SWL. This occurs because the masses of water entering from both sides of the strait are relatively confined within a certain period of time and experience interference. The wave crests are caused by constructive interference, while the troughs are caused by destructive interference. The water level above the SWL over a longer period of time indicates that the part of Bitung City which is in the middle of the Lembeh Strait will experience longer inundation.

The existence of offshore islands that are relatively far from the coast in several studies can amplify the height of the tsunami waves due to the interference effect that occurs behind the island [12]. Due to the Lembeh Strait, it is relatively narrow (the distance between the coast of Bitung City is relatively close to Lembeh Island) so that the tsunami waves that form in the strait have unique characteristics. The characteristics of the tsunami waves that occur in this section, in particular for tide gauges 9 and 10, are different compared to other tide gauges. On tide gauges 9 and 10, the tsunami wavelength is relatively smaller and has a tsunami wave $H_{max}$ of only about 20 - 40% of the $H_{max}$ recorded on other tide gauges. The uniqueness of the characteristics of the tsunami waves in the central part of the Lembeh Strait is closely related to the energy reduction and diffraction effects of the tsunami waves due to the existence of Lembeh Island.

## 4. Conclusions

TUNAMI-N2 modeling based on earthquake data and position as discussed in this study indicate that the arrival time of tsunami waves in Bitung City is around 900 s for a 7.5 *Mw* earthquake and 710 s for an 8.0 *Mw* earthquake. In the worst case scenario, the height of the tsunami waves in Bitung City especially on the coast outside the Lembeh Strait can reach 14.98 m. There are differences in the characteristics of the tsunami waves on the coast outside and inside the Lembeh Strait. The arrival time of the tsunami waves on the coast that is in the Lembeh Strait takes longer than the coast of Bitung City which is outside the Lembeh Strait, while the maximum wave height that occurs is only around 20 - 40%. This indicates that Lembeh Island can slow down the arrival time and reduce the tsunami wave height. However, the fluctuation of the water level in the interior of the strait is relatively above the SWL so that the impact on standing water in the zone has a much longer duration. The difference in characteristics between waves that occur inside and outside the Lembeh Strait is an important thing that needs to be studied further to determine a more optimal tsunami disaster mitigation strategy in Bitung City.

## Acknowledgments

This research was fully funded by Ministry of Research and Technology through Penelitian Strategis Nasional Institusi Research Grant (2018) to GP & GT.

IOP Conf. Series: Materials Science and Engineering 1115 (2021) 012090 doi:10.1088/1757-899X/1115/1/012090

## References

[1] Hamzah L, Puspito N T and Imamura F 2000 Tsunami catalog and zones in Indonesia *Journal of Natural Disaster Science* **22** (1) 25-43

[2] Widiwijayanti C, Tiberi C, Deplus C, Diament M, Mikhailov V and Louat R 2004 Geodynamic evolution of the northern Molucca Sea area (Eastern Indonesia) constrained by 3-D gravity field inversion *Tectonophysics* **386** (3-4) 203-222

[3] Fu D, Liu B, Wang G, Zheng J and Zeng Y 2020 An analytical study of long wave propagation around two circular islands *Ocean Engineering* **211** 107617

[4] Stefanakis T S, Contal E, Vayatis N, Dias F and Synolakis C E 2014 Can small islands protect nearby coasts from tsunamis? An active experimental design approach *Proceedings of the Royal Society A: Mathematical, Physical and Engineering Sciences* **470** (2172) 20140575

[5] Imamura F, Yalciner A C and Ozyurt G 2006 *Tsunami Modelling Manual (TUNAMI model)* Japan: Tohoku University

[6] Pasau G, Tamuntuan G and Tanauma A 2019 Numerical modelling for tsunami wave propagation (case study: Manado bays) *IOP Conference Series: Materials Science and Engineering* **567** (1) 012005

[7] Ha D M, Tkalich P, Soon C E and Megawati K 2009 Tsunami propagation scenarios in the South China Sea *Journal of Asian Earth Sciences* **36** (1) 67-73

[8] Tinti S, Armigliato A, Manucci A, Pagnoni G, Zaniboni F, Yalciner A C and Altinok Y 2006 The generating mechanisms of the August 17, 1999 Izmit bay (Turkey) tsunami: regional (tectonic) and local (mass instabilities) causes *Marine Geology* **225** (1-4) 311-330

[9] Wells D L and Coppersmith K J 1994 New empirical relationships among magnitude, rupture length, rupture width, rupture area, and surface displacement *Bulletin of the seismological Society of America* **84** (4) 974-1002

[10] Stirling M, Goded T, Berryman K and Litchfield N 2013 Seismic hazard of the Canterbury region, New Zealand: New earthquake source model and methodol- ogy *Bull. New Zeal. Natl. Soc. Earthquake Eng.* **41** 51–65

[11] Asim K M, Javed F, Hainsl S and Iqbal T 2019 Fault Parameters-Based Earthquake Magnitude Estimation Using Artificial Neural Networks *Seismological Research Letters* **90** (4) 1544-1551

[12] Kornei K 2017 Offshore islands might not shield coastlines from tsunami waves *Eos* **98** 1-11

# AUTHOR INDEX